T0328216

CHROMATIN

CHROMATIN
Structure, Function, and History

RANDALL H. MORSE
Laboratory of Molecular Genetics, Wadsworth Center, NY State Department of Health, Albany, NY, United States;
Department of Biomedical Sciences, UAlbany School of Public Health, Albany, NY, United States

Academic Press is an imprint of Elsevier
125 London Wall, London EC2Y 5AS, United Kingdom
525 B Street, Suite 1650, San Diego, CA 92101, United States
50 Hampshire Street, 5th Floor, Cambridge, MA 02139, United States

Notices

Knowledge and best practice in this field are constantly changing. As new research and experience broaden our understanding, changes in research methods, professional practices, or medical treatment may become necessary.

Practitioners and researchers must always rely on their own experience and knowledge in evaluating and using any information, methods, compounds, or experiments described herein. In using such information or methods they should be mindful of their own safety and the safety of others, including parties for whom they have a professional responsibility.

To the fullest extent of the law, neither the Publisher nor the authors, contributors, or editors, assume any liability for any injury and/or damage to persons or property as a matter of products liability, negligence or otherwise, or from any use or operation of any methods, products, instructions, or ideas contained in the material herein.

ISBN 978-0-12-814809-9

For information on all Academic Press publications
visit our website at https://www.elsevier.com/books-and-journals

Publisher: Andre Wolff
Acquisitions Editor: Michelle Fisher
Editorial Project Manager: Franchezca Cabural
Production Project Manager: Jayadivya Saiprasad
Cover Designer: Miles Hitchen
Cover Image Credit: Sam Bowerman

Typeset by STRAIVE, India

Contents

Preface

It is a happy coincidence that the publication of this book occurs 50 years after the formulation of the particle model of the nucleosome, a coincidence that makes the "*History*" part of the title especially apropos. Progress in chromatin research over this span has been remarkable. In 1988, Ken van Holde wrote in the Preface to his textbook on *Chromatin*, "we have reached a point where the picture of chromatin *structure* . . . has clarified to the point where a coherent presentation is possible... [but] the same cannot be said of chromatin *function*." The decade following was a period of explosive growth, marked by major discoveries relating to histone variants, chromatin remodeling and modification, and correspondingly chromatin function. Much of this growth was documented in the three editions of Alan Wolffe's *Chromatin Structure and Function* published before his untimely death in 2001. Advances in technology beginning in the late 1990s opened an era of genome-wide interrogation that led to new insights into chromatin structure and function, and we are now witnessing the fruits of the revolution in cryoelectron microscopy in allowing high-resolution determination of large complexes involved in histone modification, chromatin remodeling, and transcription.

While new and unexpected discoveries continue to be made (e.g., the concept of biomolecular condensates in nuclear function), the field of chromatin research can now fairly be described as mature. The somewhat audacious intent of this book is to summarize the progress that has led to this point. By providing some historical background and experimental detail, I hope to have avoided devolving to a dry recitation of the current "facts" as we know them. Instead, I have aimed to provide the reader with an appreciation of the way in which the knowledge available at a given time combined with available technology to govern, as well as limit, the advances made. Furthermore, progress has not been linear, and in the belief that there is wisdom to be gained from learning of missteps and mistakes as well as successes, examples of the former have not been excluded from the discussion.

The book is organized into eight chapters, beginning with the early history of the field and moving from an emphasis on structure in Chapters 1–5 to function in the last three chapters, although this is far from a clean division. Readers primarily interested in a general and historical perspective may wish to focus on the early sections of each chapter, while those interested in more detail can find it in later sections and the list of references. Clearly, coverage is incomplete; among topics receiving what may be perceived as short shrift are DNA replication and repair, nonhistone chromatin components such as the high-mobility group proteins, linker histones, and chromatin structure and function in plants. Similarly, my choices of cited literature are doubtless imperfect, and I apologize in advance for omissions or misattributions. I also expect that I have failed in places to be consistent in my assumed knowledge of the reader; I won't say this was unavoidable, but I was unable to avoid it.

I have benefited greatly from discussions with colleagues and would like to thank

Trevor Archer, Giacomo Cavalli, Steve Henikoff, Matt Hirschey, Lis Knoll, Craig Peterson, Mitch Smith, and Ken Zaret for helpful input. I am especially grateful to David Clark, Sharon Dent, Steve Hanes, Jeff Hansen, Jeff Hayes, Philipp Korber, Paul Talbert, Joe Wade, and Fred Winston for their detailed comments on portions of the manuscript. In spite of their help, I will claim any errors remaining as my own. I thank Karolin Luger and Sam Bowerman for their extremely generous provision of images used in figures for Chapters 1 and 2, as well as the cover image; Ryan Treen, Janice Pata, and Julio Abril-Garrido for help with Figures 5.11, 6.1, and 8.1, respectively; and Lis Knoll for help with numerous figures. I would also like to extend my appreciation to the folks at Elsevier for their support and for their patience in the face of multiple missed deadlines. Finally, thanks to Sue, David, and Sarah for their encouragement throughout the long course of this project.

PART 1

Chromatin structure

CHAPTER

1

Early history

"Chromatin" was first defined by the German biologist Walther Flemming as a cellular structure made visible under the microscope by its affinity for aniline dyes. Chromatin is now most broadly defined as the complex of DNA and associated proteins found in eukaryotic (and archaebacterial) cells. In most types of eukaryotic cells, the basic unit of chromatin is the nucleosome. (An important exception is spermatozoa, in which DNA is packaged by small basic proteins called protamines.) Fig. 1.1 depicts three views of the *nucleosome core particle*, a well-defined complex in which 147 bp of DNA is wrapped around a protein core of two copies each of the core histones H2A, H2B, H3, and H4. Nucleosomes are found in all eukaryotes except the unicellular dinoflagellates (Rizzo and Nooden, 1972; Talbert and Henikoff, 2012). Fish, yeast, pea plants, and humans all have nucleosomes that are nearly identical in their construction. For this reason, the term "chromatin" is often understood principally to refer to DNA packaged into nucleosomes, with other, nonhistone proteins viewed as ancillary.

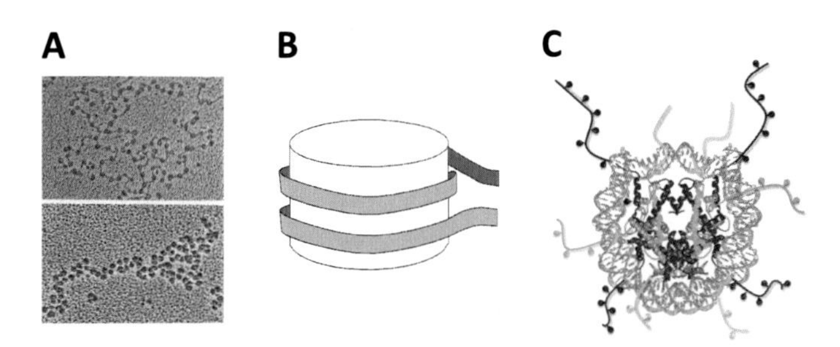

FIG. 1.1 The nucleosome. (A) Electron micrographs of decondensed (top) and condensed (bottom) chromatin. (B) Cartoon of a nucleosome, showing DNA wrapping around the disk-shaped histone octamer. (C) High-resolution structure of the nucleosome core particle. Histones are color-coded: *orange*, H2A; *red*, H2B; *blue*, H3; and *green*, H4. *Credit: (A) Rockefeller Press. (C) Courtesy of Karolin Luger.*

In cells, nucleosome core particles are separated by variable lengths of DNA referred to as *linker DNA*, which is often associated with linker histones, such as histone H1 and H5 (but not present in the figure) (Fig. 1.2). (Coinage of the term "linker" has been attributed to Noll

Chromatin. **https://doi.org/10.1016/B978-0-12-814809-9.00007-X**

and Kornberg (Noll and Kornberg, 1977; Van Holde, 1988, p. 29); however, although that paper refers to the DNA that links nucleosomes, the term "linker" does not appear. It may first have appeared in print in Kornberg's (1977) review article, where the linker is explicitly defined as the DNA connecting nucleosomes.) The structure including core histones, linker DNA, and a single linker histone is properly called the *nucleosome*, although *nucleosome* and *nucleosome core particle* are sometimes used interchangeably in the literature. Here, we shall use the term *nucleosome core particle* in its original sense of the complex resulting from digestion by micrococcal nuclease and containing 146–147 bp of DNA and two copies each of the four core histones H2A, H2B, H3, and H4.

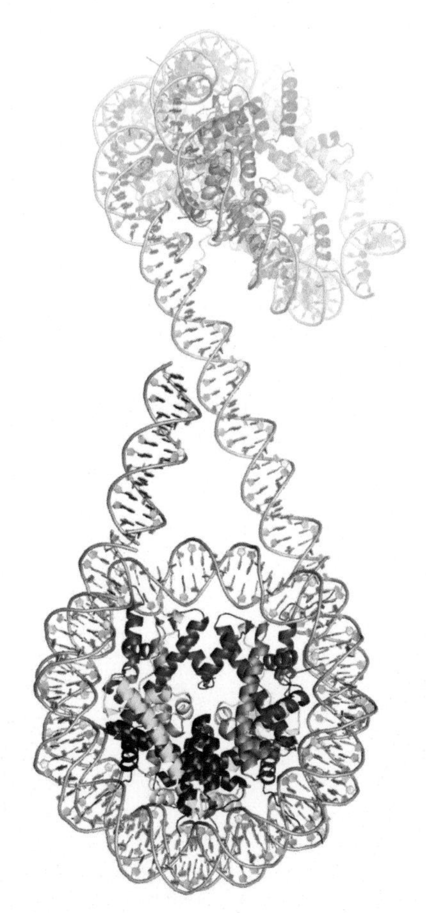

FIG. 1.2 Two nucleosomes plus the linker DNA connecting them. *Credit: Courtesy of Karolin Luger.*

Unlike eukaryotes, bacteria do not have nuclei, and they do not have nucleosomes. Archaea, which exist somewhere between these two kingdoms, lack nuclei but have histone-like proteins, and their DNA is organized into chromatin whose basic unit is a structure easily recognizable as a close relative of the eukaryotic nucleosome (Mattiroli et al., 2017).

Why did nucleosomes arise as such a ubiquitous structure in evolution? The facile answer is that they help to organize the DNA to allow it to fit within the dimensions of the cell. A "typical" bacterium, say a cylindrical *E. coli* cell that is 1 μm wide by 2 μm in length, contains about 4 million bp of DNA. This amount of DNA, if stretched taut, would be about 1.4 mm in length, and so must be compacted over a 1000-fold to fit within the dimensions of the cell (actually, a bit less, since the *E. coli* genome is circular). In contrast, the human diploid genome consists of about 6.4 billion base pairs of DNA, which when stretched out would be about 2 m in length. Since human cell nuclei are approximately 10 μm in diameter, compaction close to five orders of magnitude is required. It therefore seems reasonable to postulate that nucleosomes could help achieve this stuffing of our genetic material into its cellular suitcase.

This idea may be right, though (a) it is difficult to prove since experiments involving evolution are challenging at best and (b) the nucleosome itself compacts DNA only about fivefold. Moreover, comparing the compaction required for human and bacterial genomes is perhaps disingenuous, leaving out as it does the genome sizes of simpler organisms such as yeasts, which have genomes in the tens of megabases, only a fewfold larger than bacterial complements. It could be that this size difference is sufficient that an advantage was gained by incorporating the genome into chromatin and that this played an eventual role in the advent of metazoan organisms with their much larger genomes. Indeed, metazoan chromosomes are organized into higher-order structures that are compacted sufficiently to fit within cell nuclei, and nucleosomes assist in this folding, as described in Chapter 3.

If solving compaction were the only function of chromatin, it would amount to little more than packing material, and indeed this viewpoint has held sway at various times since the components of chromatin were first identified (Kadonaga, 2019). Since the early 1970s, however, when the nucleosome was identified as the fundamental unit in chromatin, researchers in the field have recognized that the close apposition of DNA to histones in chromatin was likely to have important ramifications for DNA function. In the ensuing decades, an enormous body of work has substantiated this view, revealing that nucleosomes and related proteins in the cell have not only evolved to allow DNA to maintain its normal suite of functions but have been co-opted in numerous subtle and sophisticated ways to facilitate and regulate DNA repair, replication, and transcription (Kornberg and Lorch, 2020; Paranjape et al., 1994; Struhl, 1999).

1869–1950: Identification of nucleic acid and the "dark ages" of chromatin research

I cannot claim to be able to read German; therefore, the history related here relies to a considerable extent on secondary sources. In particular, I have leaned heavily on the accounts in Ken van Holde's opus, *Chromatin* (see Sidebar: Ken van Holde); the account of the discovery of DNA by Alfred Mirsky; and the excellent synopses of Friedrich Miescher and his contemporaries by Ralf Dahm (Dahm, 2005, 2010; Mirsky, 1968; Van Holde, 1988). For an engrossing and extended recounting of the early history of genetics, readers are referred to the opening chapters of *The Gene*, by Siddhartha Mukherjee (Mukherjee, 2016).

Sidebar: Ken van Holde

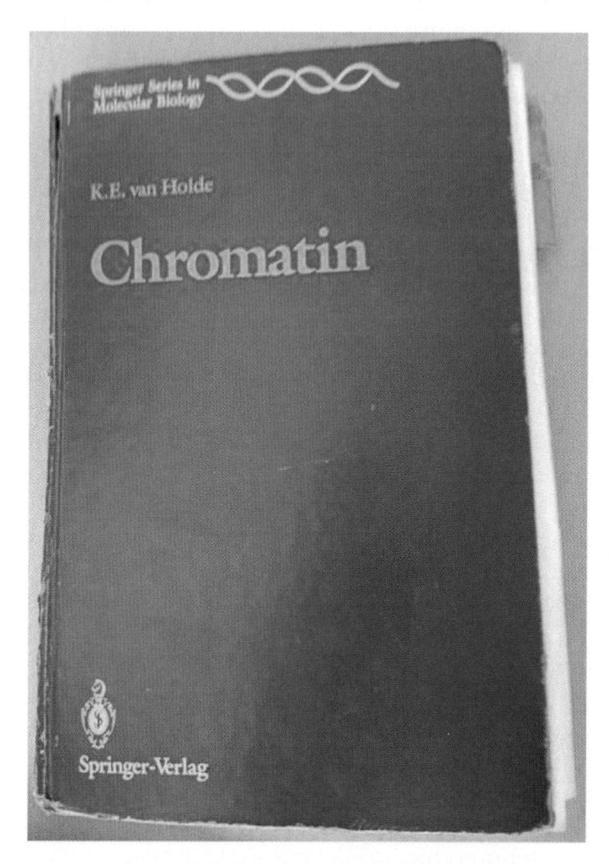

Karolin Luger was given this copy by Jon Widom's lab manager after his untimely passing.

It will be impossible to include sketches of—indeed, even to mention explicitly—all of the many scientists who have contributed to our knowledge of chromatin biology, and no slight is intended to those not included in this regard. However, it seems particularly apt that Ken van Holde, a tall, remarkably unassuming, chain-smoking chromatin enthusiast, be singled out for mention. Ken was an outstanding biophysical chemist whose contributions to solving the structure of the nucleosome are evident from the main text; he also (and not less importantly) played a major role in the development of sedimentation equilibrium centrifugation to study the size and shape of biomolecules, a method that has been front and center in the study of chromatin folding (Chapter 3) (van Holde, 2008). But Ken is especially known in the chromatin community for his 1989 book, simply entitled "Chromatin." This monumental and comprehensive treatise reviews, analyzes, and critiques research on chromatin encompassing close to 2000 primary literature reports and remains an invaluable resource for anyone interested in the detailed early history of the field. Ken was a faculty member at Oregon State University for over 40 years, served as an Editor at the Journal of Biological Chemistry, and was elected to the National Academy of Sciences in 1989. He passed away in November 2019 at the age of 91.

Two distinct lines of inquiry mark the beginnings of the modern era of chromatin biology. These began in the late 1800s and did not coalesce into a fully realized synthesis until 100 years later when the elucidation of the nucleosome as the fundamental unit of chromatin finally emerged. Fittingly, one of these lines was structural, the other functional.

To understand structure, one must first know the components present in the assemblage under study. In the case of chromatin, this began with investigations by Friedrich Miescher in the 1860s on the chemical composition of components found in the nuclei of eukaryotic cells. The study of molecules found in living cells, which went by the name of physiological chemistry, was a fledgling science at this time. The term "protein" first appeared in the scientific literature in 1838, encompassing compounds such as silk, albumin, and glutenin that were found in abundance in specific cell types, and that were therefore amenable to analysis with the methods then available (Vickery, 1950), while Mendeleev's first attempt at describing the elements in a periodic table debuted in 1869 (Gordin, 2019). It was in this historical context that Miescher, whose father and uncle were both physicians and professors of anatomy and physiology at the University of Basel, joined the lab of Felix Hoppe-Seyler with the aim of using chemistry to investigate "the last remaining questions concerning the development of tissues," as his uncle, Wilhelm His, had written (Dahm, 2005). Miescher chose well; Hoppe-Seyler's lab was the first to be strictly dedicated to studies in what we now term biochemistry (Mirsky, 1968).

Having prepared himself for his work with Hoppe-Seyler by first spending a semester in the laboratory of Alfred Strecker learning techniques in organic chemistry, Miescher set out to obtain a "simple and independent cell type" for his experiments (Dahm, 2005), a reductionist approach readily appreciated 150 years later. For this purpose, he chose to use lymphocytes, which he isolated from pus on fresh surgical bandages from the surgery clinic nearby Hoppe-Seyler's Tübingen lab. Miescher first used a painstaking process that included washing the bandages with dilute sodium sulfate, filtering through a sheet, and allowing the cells to settle at $1 \times g$ (centrifuges not yet being in use) to isolate undegraded, intact leukocytes, and then set out to isolate nuclei from these cells. Nuclei were known from microscopic investigations, but little was known about their composition or function, and no one had purified nuclei from cells. Working by himself—he was Hoppe-Seyler's only student at this time—Miescher developed his own protocol by trial and error. He eventually succeeded in isolating clean nuclei that did not stain with iodine, used at the time to detect cytoplasm, and then subjected these to extraction to characterize them chemically.

Miescher's initial studies on leukocytes focused on proteins and lipids, identifying them as the principal cytoplasmic components and describing some of their properties. Critically, however, he discovered that extraction of his purified nuclei using a mildly alkaline solution, followed by acidification using acetic or hydrochloric acid, yielded an interesting precipitate. The material in the precipitate could be redissolved in a mildly basic solution, but not in water, acidic, or salt solution, and so appeared not to conform to what was known of the properties of proteins or lipids. Furthermore, the substance was not digested by pepsin, and contained about 2.5% phosphorus and no sulfur, again distinguishing it from protein preparations. Miescher's mysterious substance was in fact the first instance of isolated chromatin.

Miescher himself appreciated, at least as much as he could in the context of the times, the significance of his discovery. He gave the preparation a name, "nuclein" and wrote that it appeared to be "an entity *sui generis* not comparable to any hitherto known group" (Dahm, 2005). He had discovered something new! He went on to identify nuclein in other cell types (following a template for doing science that we continue today—discovery followed by determination of generality) and speculated that it might be central to cell function and possibly involved in pathological processes. Miescher reported his discovery in a manuscript that he submitted to Hoppe-Seyler, who repeated Miescher's experiments before accepting his paper (Dahm, 2005; Miescher, 1871). The work was published together with a separate paper by Hoppe-Seyler confirming Miescher's results, and another by a student of Hoppe-Seyler's that showed that nuclein was also found in the nucleated erythrocytes of geese and snakes (Hoppe-Seyler, 1871; Plosz, 1871).

Miescher performed further studies on nuclein at Basel University, where he was hired as Chair of Physiology following his habilitation. He chose to use salmon sperm for refining his isolation of nuclein, based on the large proportion of these cells occupied by the nucleus, and the convenient proximity of the university to the Rhine River and its flourishing population of migrating salmon. Eventually, he was able to obtain nearly pure preparations of nuclein with a measured phosphorus component nearly identical to what we now know to be the value for pure DNA. Miescher further showed that his substance was a "multibasic acid" that is bound in sperm to a basic molecule that he termed "protamin." (Miescher appears to have used the term "nuclein" to refer both to what we would now call chromatin and to DNA, as well as to the specialized form of chromatin found in sperm, in which protamines rather than histones are the protein components.) A few years later, Albrecht Kossel, another student of Hoppe-Seyler, prepared nuclein from geese erythrocytes. Kossel separated the basic, protein component from the acidic component and found that this preparation clearly differed in its physical and chemical properties from Miescher's *protamin*; he named these proteins *histon* (Kossel, 1884). Kossel would be awarded the Nobel Prize in Medicine in 1910. (Miescher died of tuberculosis in 1895 at age 51, 6 years before the first Nobels were awarded.) The properties of chromatin were further clarified by work around the turn of the century, in which basic solubility properties of nucleohistone, as it came to be called (Lilienfeld, 1894), and its electrostatic ("salt-like" at the time) nature were described in detail. In accord with the latter property, histones and DNA were shown to be, respectively, positively and negatively charged (Huiskamp, 1901).

Cytological work taking place in this same era provided the first clues linking structure with function (Mirsky, 1968; Van Holde, 1988). Walther Flemming applied new staining techniques using aniline dyes to identify remarkable structures in cell nuclei that changed in morphology as cells divided; he named this stained material *chromatin* and also coined the term *mitosis* for the process of cell division (Fig. 1.3). Contemporaneous studies by Oscar Hertwig and Hermann Fol, again relying on microscopy and judicious use of stains, revealed that fertilization in the sea urchin and the starfish was accompanied by fusion of sperm and egg nuclei, and Edouard van Beneden followed with observations of chromosome replication during fertilization in the threadworm *Ascaris*. These inquiries into cell division and fertilization suggested that cell nuclei and the chromosomes they contained were good candidates for providing continuity of form and function, and by inference the elements of heredity, from mother to daughter cell and intergenerationally.

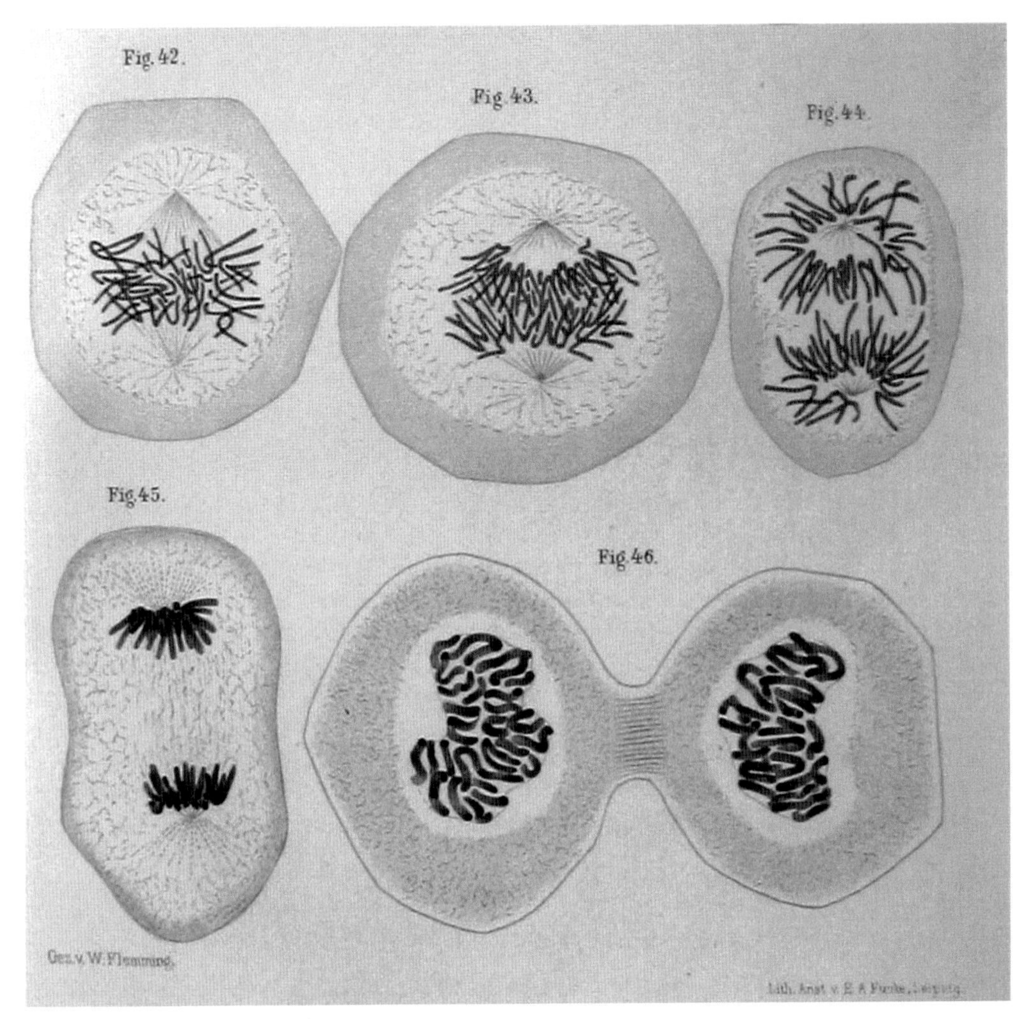

FIG. 1.3 Drawings of a cell undergoing mitosis, revealing the chromosomes as distinct entities that replicate along with the cells. *Credit: From Walther Flemming's* Zellsubstanz, Kern und Zeitheilung, *1882.*

"The circle was closed," as van Holde puts it, by the investigations of the botanist Eduard Zacharias, which according to Miescher's uncle Wilhelm His "were the first to combine the histological concept of chromatin with the chemical substance nuclein" (Dahm, 2005; His, 1897). Zacharias followed Miescher's protocol for isolating nuclein to show that when cells were treated in a way that left nuclein intact, such as digestion with pepsin-hydrochloric acid (obtained as an extract from pig stomach), the resulting material could still be stained with dyes that were used to visualize chromosomes, while removal of nuclein by extraction with dilute alkali resulted in the loss of stainable material (Mirsky, 1968; Zacharias, 1881). These studies, then, linked the visible, stainable entities that had only recently been shown to be likely carriers of hereditary information with Miescher's chemically defined but poorly understood, nonproteinaceous component found in nuclei of all species that had been examined for its presence.

It is easy for us now to appreciate that the foregoing investigations represented the first lifting of the curtain on DNA as the material basis for the inheritance of traits. Nor was this possibility missed by scientists at the time. Indeed, Miescher wrote that "If one ... wants to assume that a single substance ... is the specific cause of fertilization, then one should undoubtedly first and foremost consider nuclein" (Dahm, 2005). Even more tellingly, four seminal publications from Oscar Hertwig, Albrecht von Kölliker, Eduart Strasburger, and August Weismann in 1884 and 1885 recognized the significance of the inheritance of chromosomes both in cell division and fertilization, with Hertwig writing "I believe that I have at least made it highly probable that nuclein is the substance that is responsible not only for fertilization but also for the transmission of hereditary characteristics" (Dahm, 2005; Hertwig, 1885).

In spite of this early burst of enthusiasm, the idea of nuclein as the repository of hereditary information fell away over the decades that followed, while efforts to understand its chemical structure yielded answers only after many years of work. Kossel demonstrated in the 1880s that the nonproteinaceous component of nuclein was composed of four bases and sugar molecules, a critical step in the long path to the elucidation of the structure of DNA (Dahm, 2005), but decades would pass before the chemical structures of these building blocks were entirely solved, and still more time before DNA would be recognized as the repository of genetic information. In the meantime, the scientific community gradually came to agree that a relatively simple molecule of only 4 bases could not compete with proteins, with their 20 constituent amino acids, in its potential for conferring the complexity and variability of living organisms. Miescher himself came to doubt that sufficient diversity could exist in a single substance to account for the vast diversity among different species, or even to account for individual differences within a species (Dahm, 2005), and as late as the 1947 Cold Spring Harbor Symposium, Stedman and Stedman (quoted in Van Holde (1988)) wrote:

> There is probably general agreement that the essential component of the chromosome must belong to the proteins, for no other known class of naturally occurring chemical compounds would be capable of possessing the properties or existing in the variety necessary to account for their genetic function.

It did not help that ideas about the structure of nucleic acid in the first decades of the twentieth century were dominated by the "tetranucleotide hypothesis." This idea, that the nucleic acid component of nuclein consisted of a simple linear or circular molecule of the four bases, stemmed from work showing that the four bases were found in approximately equal abundance in preparations of nucleic acid, leading to the simplifying assumption that they were in fact exactly equal in abundance. The concept of long-chain polymers—macromolecules, as we call them now—was not part of the scientific lexicon in the early 1900s when chemists were elucidating the structures of the chemical components of nucleic acids, and so the idea of nucleic acid as a simple tetranucleotide held wide currency into the 1930s. Eventually, the adoption of methods for preparing undegraded nucleic acids and for their analysis by filtration and ultracentrifugation revealed a substance of very high molecular weight (estimated to be on the order of 1×10^6 Da) (Astbury and Bell, 1938; Signer et al., 1938), and it was further demonstrated that this large molecule could be depolymerized by a nuclease derived from pancreas (Schmidt and Levene, 1938). In the latter work, the authors wrote: "Complete depolymerization of the acid to a single tetranucleotide has not yet been accomplished by chemical means." It is difficult to overstate the weight of first impressions and entrenched ideas.

(On the other hand, van Holde points out that had chemists appreciated the true complexity of nuclein molecules, they likely would have "turned from them in horror" (Van Holde, 1988). As it was, the view of DNA as a simple tetranucleotide emboldened chemists to tackle the basic structure of nucleic acids in the early part of the twentieth century, allowing them to define the structures of the bases and backbone linkages. This information was later critical to the work of Watson and Crick in their "discovery" of the two-stranded antiparallel form of DNA found in cells.)

Further damaging the notion of nucleic acid as a carrier of hereditary information, cytological investigations and later on studies of the UV absorption of nuclei, indicated that its abundance in cells appeared to change during the cell cycle (Caspersson, 1936; Strasberger, 1909). How could such an inconstant component play the role of providing a heritable, continuous foundation for the transmission of inherited characteristics? Such was the thinking for the first decades of the twentieth century.

Thus, in spite of the prescient insights that followed Miescher and colleagues' work on nuclein, it was not until the demonstration (following work 16 years earlier by Griffith (1928)) in 1944 by Avery, MacLeod, and McCarty that DNA, not protein or RNA, contained the transforming material governing bacterial morphology and pathogenicity, that the tide began irreversibly to turn (Avery et al., 1944). (Amazingly, no Nobel Prize was awarded for this work.) Chargaff's demonstration (Chargaff, 1950) that the four bases were not present in exactly equal abundance, and in fact varied considerably in DNA from calf thymus, yeast, and tubercle bacillus, was perhaps the nail in the coffin with regard to the tetranucleotide hypothesis. Shortly following Chargaff's seminal observations, Watson and Crick's stunning publication of a model for DNA structure (Watson and Crick, 1953) can fairly be said to mark the culmination of the paradigm shift, begun by the work of Griffith and then Avery, MacLeod, and McCarthy, from the idea of protein as the carrier of hereditary information to the recognition of DNA as fulfilling that role.

Given the more than half century that elapsed between the first chemical characterization of nucleic acid and the elucidation of DNA structure, along with the back seat that nucleic acid took to protein in attempts at identifying the molecular basis of heredity, it is not surprising that studies of chromatin and the histones languished for the first part of the twentieth century. An indicator of the murkiness of chromatin studies in this period can be found in the biographical memoir of Edgar Stedman from the Royal Society, which states that Stedman, as a new entrant into studies of the cell nucleus and its contents in the 1940s, "was perplexed by the uncritical way the terms 'nucleus', 'chromatin', and 'chromosomes' were often used synonymously" (Cruft, 1976). Characterized as "the Dark Ages" of chromatin research by Don and Ada Olins (Olins and Olins, 2003), this era was marked by a handful of major advances, such as the discovery of the polytene chromosomes of *Drosophila* and the correlation between bands within those chromosomes and specific genes (defined as units of heredity), as well as the belated recognition that nucleic acids comprised vast polymers, as already described. But virtually no progress was made in understanding the chemical nature of the histone proteins, and the idea that chromatin could confer regulatory layers upon DNA was far in the future. (Van Holde provides a thorough and interesting discussion of some of the confusion and false starts of this period (Van Holde, 1988).) Even in 1988, van Holde wrote that the view that histones were most likely to function as "non-specific agents in a general mechanism for chromatin inactivation ... is widely held to the present day." Much has changed since those words were written.

Prelude to the nucleosome: 1950s and 1960s

The *histon* (histones) first isolated by Albrecht Kossel from geese erythrocytes in 1884 were found to differ in their elemental composition and solubility properties from the *protamin* discovered in salmon sperm heads by Miescher (Doenecke and Karlson, 1984; Kossel, 1884; Miescher, 1874). The discovery of the basic amino acids arginine and lysine in the 1880s prompted Kossel to reinvestigate the histones and protamines and enabled him to demonstrate that lysine, arginine, and a newly identified amino acid, which he designated histidine, were all constituents of these fundamental nuclear substances (Doenecke and Karlson, 1984). By 1910, Kossel was able to assert in his Nobel Prize lecture that substances similar to the histones and protamines "occur in a salt-like interaction with nucleic acids in the tissues of a wide variety of lower and higher animals" (quoted in Doenecke and Karlson (1984)).

Following this early work, apart from determinations that they were present in a variety of organisms and tissue types, almost nothing more was learned about the histones until the 1950s. Fundamental questions, such as whether and how much histones differed between species or in different tissue types, and what their relationship was to the nucleic acid component of nuclei, remained unanswered during these "Dark Ages" of chromatin research. Even the question of whether protamines constituted an entity distinct from the histones, or instead represented an extreme evolutionary variant, remained open until the 1940s (Stedman and Stedman, 1951). This changed with the publication in 1951 of a seminal paper by the husband and wife team of Edgar and Ellen Stedman (Stedman and Stedman, 1951). The paper described several major advances in understanding the nature of the basic histones and protamines:

- Purification from disparate species and tissue types—avian erythrocytes, ox thymus, liver, and spleen, chicken thymus, salmon liver, and wheat germ—revealed that histones were present in all cell types examined, and thus likely ubiquitous in eukaryotes.
- In contrast to earlier studies, histones were in all cases prepared after isolating nuclei from the cells in question, thereby demonstrating that the proteins were uniformly nuclear.
- Fractionation of histone preparations using protocols involving differential solubility, followed by analysis of amide nitrogen and arginine by chemical methods, tyrosine by a colorimetric test, and amino acid composition using paper chromatography, showed that the histones were not homogeneous, but rather could be separated into two groups described as a main and subsidiary histone fraction. Chromatographic analysis revealed the main group to be rich in arginine and the subsidiary fraction in lysine residues. Further, it was suggested that these groups themselves comprised more than one species of histone, setting the stage for further work that culminated in the identification and characterization of the core and linker histones that we know today.

This is a fascinating paper to read from today's historical perspective. It is clear that it represents a massive amount of work; some of the frustrations encountered are elaborated in the extensive and discursive descriptions of methods, so different from the cryptic and abbreviated experimental protocols typically found in today's manuscripts:

"*... difficulties have sometimes, although by no means invariably, been encountered {with} fowl erythrocyte nuclei. In such cases treatment of the nuclei with water has caused the latter to coalesce to a sticky mass which is impenetrable to the acid added subsequently. When this has occurred, it* *has been necessary to dry the nuclear material with alcohol, which eventually causes it to set to a hard, brittle mass, and to resume the extraction of the latter with acid after grinding it to a powder. For this reason ... a low yield of histone almost certainly results."* And so forth. Close reading of the manuscript also provides insight into some of the contentious issues of the day; refutation of claims by Mirsky and Ris (Mirsky and Ris, 1947) that thread-like structures they had isolated from cell nuclei could be described as "chromosomes" is especially pointed.

Advances in analytical approaches, along with improvements in biochemical methods of fractionation, led to the identification of additional histone fractions and eventually individual histones beginning in the late 1950s. The introduction of gel electrophoresis and the use of ion-exchange chromatography were particularly important in this regard (Luck et al., 1958). Obtaining pure specimens remained a hurdle, however, until the establishment of polyacrylamide gel electrophoresis (PAGE) as an analytical tool in the 1960s. Marking the maturation of the field, the First World Conference on Histone Biology and Chemistry, organized by James Bonner from Caltech and supported by funds from the Rockefeller Foundation and the National Science Foundation, was held in 1963 at Rancho Santa Fé in California. This marked the beginning of serious advances in understanding the histone proteins, as Bonner has stated that "{t}he take-home message from this conference was that there was nobody in the world that was ... making any sensible chemical contribution to the study of histone chemistry and/or enzymology" (http://oralhistories.library.caltech.edu/15/1/OH_Bonner_J.pdf). Steady strides were made in the 1960s in the characterization of the histones, led by E.W. Johns and co-workers (Johns, 1964; Phillips and Johns, 1965). A major innovation was the use of low pH PAGE, and the addition of urea to gels, allowing clear separation of all four core histones and linker histones on a single gel (Fig. 1.4) (Johns, 1967; Panyim and Chalkley, 1969a,b).

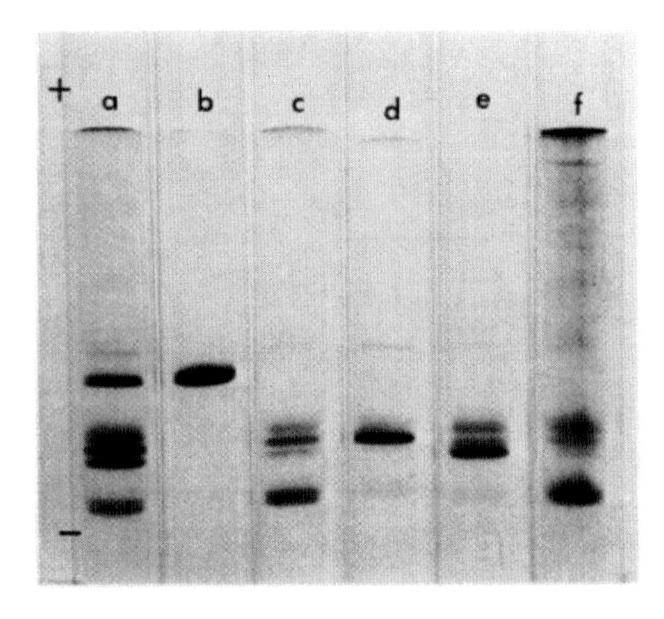

FIG. 1.4 Separation of histones by electrophoresis using a polyacrylamide gel containing 2.5M urea and 0.9N acetic acid. (A) Whole calf thymus histone; (B) linker histone H1; (C) histone H4 (mostly); (D) histone H2B; (E) histone H2A; (F) histone H3. *Credit: Panyim, S., Chalkley, R., 1969a. The heterogeneity of histones. I. A quantitative analysis of calf histones in very long polyacrylamide gels. Biochemistry 8, 3972–3979, Panyim, S., Chalkley, R., 1969b. High resolution acrylamide gel electrophoresis of histones. Arch. Biochem. Biophys. 130, 337–346.*

The ability to obtain clean preparations allowed the first complete sequencing of histones, reported in 1969, with the startling result that histone H4 from calf thymus and pea plants differed by only two amino acid residues (DeLange et al., 1969a,b; Ogawa et al., 1969). On the one hand, this rather incredible conservation of sequence implied that the histones were under stringent selective pressure across millions of years of evolution, and must therefore have previously unsuspected functional properties beyond simply sticking to DNA. Or, as stated in a Nature News and Views summary, "This is the first evidence that histones constitute anything other than glue" (Anonymous, 1969). On the other hand, however, the relative invariance of histone sequences among species and different cell types proved strongly discouraging to proponents of the histones as specific regulators of gene expression. How could histones guide the differential expression of genes if they were always the same? For the next 20 years, the transcription field largely held to the view that histones were unimportant to the specific regulation of genes.

With regard to nucleoprotein, now more generally referred to as chromatin, researchers in the 1960s began applying X-ray diffraction methods, encouraged by the success of this approach to understanding the structure of DNA. These studies required chromatin preparations of high quality. Advances made in the late 1950s, particularly the use of low salt extraction methods and the introduction of chelating agents such as citrate and arsenate to sequester divalent cations and thereby inactivate troublesome nucleases present in the isolates, provided the necessary means to achieve such preparations (discussed in detail in Van Holde (1988)). Application of X-ray diffraction methods to these improved preparations yielded a few reflections that were lost upon removal of the histones by salt extraction and therefore were specific to chromatin (Pardon and Wilkins, 1972; Pardon et al., 1967; Richards and Pardon, 1970). From these data, a model of chromatin as a superhelix of 100 Å diameter and 120 Å pitch was proposed and came to be known as the Pardon-Wilkins model (Pardon and Wilkins, 1972). The disposition of the histones relative to the DNA remained completely undetermined in this model, and the model itself turned out to be entirely incorrect.

A model for the nucleosome: A new era of chromatin biology begins

Ironically, the nucleases that had been an impediment to quality chromatin preparations for so long provided major clues to solving the subunit structure of chromatin. Two early studies by Williamson (Williamson, 1970) and Clark and Felsenfeld (Clark and Felsenfeld, 1971) heralded a rush of results to follow in the next couple of years. Williamson isolated nucleic acid from cytoplasmic extracts prepared from cells of embryonic mouse livers and showed that the DNA in these preparations, when subjected to gel electrophoresis, migrated as a "ladder" of fragments that were "multiples of a unit molecular weight of 135,000," or about 200 bp (Fig. 1.5). (We now know that these cytoplasmic DNA fragments derive from cells that have undergone apoptosis (Henikoff and Church, 2018).) Furthermore, the DNA ladder found in cytoplasmic extracts could also be observed in purified nuclear DNA, though the degree of fragmentation was lower. Present-day researchers are accustomed to using DNA ladders as size markers in gel electrophoresis experiments, but the regular patterns of bands observed by Williamson were novel and remarkable (Fig. 1.5). These findings provided evidence for a subunit structure of chromatin containing ~200 bp of DNA in a nuclease-resistant complex with protein, and Williamson suggested that the periodic fragmentation he observed might "result from DNase activity," with "specificity … imposed by some other component of the chromatin, such as protein." Nonetheless, his paper did not have an immediate major impact on the chromatin field, perhaps because his findings of cytoplasmic DNA fragments were also tied up with questions regarding whether DNA or mRNA was the primary means by which DNA sequence information was converted into proteins. Clark and Felsenfeld, meanwhile, employing the deliberate application of nucleases, reported that about 50% of DNA in chromatin was accessible to digestion while the remainder was resistant, implying that chromatin had a nonuniform structure, contrary to the structurally homogeneous Pardon-Wilkins model. They did not, however, make the conceptual jump to propose a particulate nature to chromatin. Biophysical studies from various labs, particularly linear dichroism measurements that contradicted predictions from the Pardon-Wilkins model, were also yielding results incompatible with a uniform coiled model of chromatin, further troubling the waters (Van Holde, 1988).

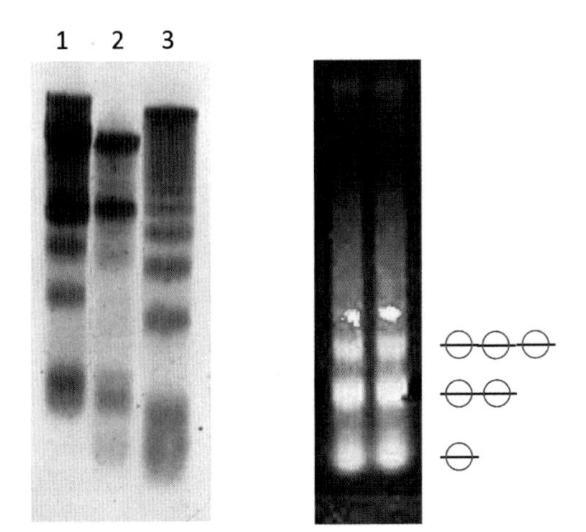

FIG. 1.5 (Left) DNA ladders produced by endogenous nuclease activity (Williamson, 1970). Lane 1, total cytoplasmic nucleic acid; lane 2, same treated with DNase; lane 3, same treated with RNase. (Right) Yeast chromatin digested with MNase; the bands corresponding to DNA associated with one, two, and three nucleosomes are clearly visible and are schematized at the right. *Credit: Williamson, R., 1970. Properties of rapidly labelled deoxyribonucleic acid fragments isolated from the cytoplasm of primary cultures of embryonic mouse liver cells. J. Mol. Biol. 51, 157–168.*

In 1973, Hewish and Burgoyne published a breakthrough result that was the first among several, obtained independently by multiple laboratories, that together irrevocably demolished the model of a uniform supercoiled structure for chromatin and replaced it with the current model of the nucleosome (Hewish and Burgoyne, 1973). Dean Hewish was a

PhD student at Flinders University of South Australia in the Burgoyne laboratory, which had been utilizing isolated rat liver nuclei in an effort to establish an in vitro system to study DNA replication. The lab had observed that certain nuclei preparations suffered degradation of nuclear DNA in the presence of magnesium and calcium ions, due to the presence of an endogenous Ca-Mg-dependent nuclease; they also found that the resulting DNA fragments were not as small as might have been expected had degradation been complete. Having seen the Williamson paper, Hewish and Burgoyne wondered whether the degradation in their system was related to the fragments that Williamson had reported (Hewish, 1982). Analysis by gel electrophoresis of DNA from rat liver nuclei that were allowed to incubate for various lengths of time in the presence of calcium and magnesium ions yielded dramatic results: a ladder of regularly spaced fragments was observed, with a greater abundance of the smaller fragments seen with increasing times of incubation. Although they did not report the size of these fragments, as Williamson had done, the fragments clearly appeared to be multiples of a unit size. Moreover, treatment of naked DNA with a crude preparation of the endogenous nuclease did not yield such a ladder, and Hewish and Burgoyne consequently inferred the existence of a "simple, basic, repeating substructure" in chromatin.

The next critical evidence for the nucleosome model came from electron microscopic observations. In the winter of 1972–73, Don and Ada Olins had returned to Oak Ridge National Laboratories in Tennessee from a sabbatical spent at Kings College in London. During their sabbatical, they had worked on isolating nuclei from chicken erythrocytes as a source of chromatin to examine with the electron microscope. Standard methods for preparing chromatin gave variable results, often yielding spreads that resembled "a bad day at a macaroni factory" (Olins and Olins, 2003). Fortunately for the Olinses, new methods developed by Oscar Miller at Oak Ridge, involving the use of detergent to visualize RNA polymerase molecules engaged with DNA in nucleoli—the famous "Miller spreads"—provided a much better means of preparing chromatin for visualization in the EM (Miller Jr. and Bakken, 1972). These preparations eventually provided the now-famous images of nucleosomes as "beads on a string" (Fig. 1.6), which were examined by biophysical methods and termed "ν (nu) bodies" (Olins and Olins, 1974), a name which did not quite stick. Chris Woodcock at the University of Massachusetts at Amherst independently made similar observations, which he reported at a meeting of the American Society for Cell Biology in 1973 that the Olinses attended, but he did not manage to publish his findings until 1976 (Woodcock et al., 1976a,b). (Van Holde cites the rather cranky, and likely forever anonymous, reviewer who believed that Woodcock's results "would necessitate rewriting our basic textbooks" and "{d}efinitely should not be published anywhere!" (Van Holde, 1988).) Independent evidence for the existence of compact particles comprising chromatin came from van Holde and co-workers, who measured circular and electric dichroism and hydrodynamic properties of particles obtained by nuclease digestion of chromatin (Rill and Van Holde, 1973; Sahasrabuddhe and Van Holde, 1974). Their results indicated that a length of DNA of about 110 bp, which would have an extended length of about 375 Å, was compacted to a diameter of 80 Å, and that the histone proteins were responsible for this compaction (Sahasrabuddhe and Van Holde, 1974).

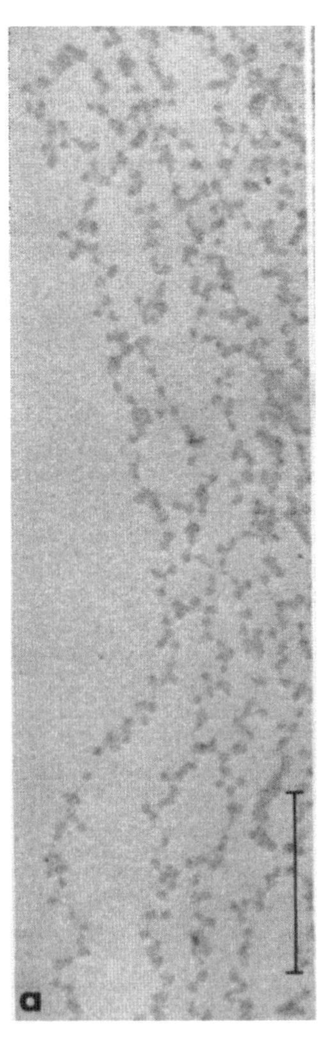

FIG. 1.6 Electron microscope image of rat thymus chromatin, showing the famous "beads on a string" structure. The scale bar is 0.2 μm. *Credit: Olins, A.L., Olins, D.E., 1974. Spheroid chromatin units (v bodies). Science 183, 330–332.*

The nuclease digestion results together with the Olinses' observations provided strong evidence for a repetitive subunit structure as the basis for chromatin. The nature of this subunit was unclear, and for at least some was clouded by preconceptions that such a simple structure must be insufficient for the complex gene regulation characteristic of metazoans (Kornberg and Klug, 1981). The final breakthrough to a clear model of the nucleosome leaned heavily on biochemical experiments performed at the MRC in Cambridge by Roger Kornberg and Jean Thomas, and a clear exposition of the model was presented in a seminal paper by

Kornberg in 1974 (Kornberg, 1974; Kornberg and Thomas, 1974). A key advance was the ability to reconstitute and therefore analyze nucleosomes using purified histones and DNA (see Sidebar: Reconstituting nucleosomes); combining such analyses with parallel studies on particles made by treatment of native chromatin with nuclease, especially the commercially available micrococcal nuclease, provided a convincing demonstration of the particle nature and composition of the nucleosome. Preparation and separation of the highly basic histones had proved difficult, but gentler methods allowed the purification of separate fractions including H3 and H4 on the one hand and H2A and H2B on the other, which we now know to comprise distinct substructures within the nucleosome (van der Westhuyzen and von Holt, 1971). Thomas and Kornberg were able to reconstitute chromatin from histones prepared by this gentler method together with DNA. The resulting reconstituted material yielded X-ray diffraction patterns consistent with that seen using native chromatin preparations (Kornberg and Thomas, 1974). Cross-linking studies performed with histones isolated in this way led to the discovery that H2A and H2B existed as a stable dimer, whereas H3 and H4 formed a tetramer in solution, while cross-linking of isolated chromatin fragments yielded multiple products up to the size of the histone octamer, but few species of larger size. These results, together with nuclease digestion experiments, electron microscopy, and biochemical and biophysical analysis of chromatin isolated from cells or reconstituted in vitro, provided the basis for a model of the nucleosome, comprising about 200 bp of DNA wrapped around an octamer of two copies each of H2A, H2B, H3, and H4, as the unit structure of chromatin (Kornberg, 1974) (see Sidebar: Credit where credit is due). A mere 3 years would pass before the first published report of an X-ray crystal structure of the nucleosome core particle, but it would be nearly 25 years before this structure was determined at high resolution (Finch et al., 1977; Luger et al., 1997), as described in the next chapter.

Sidebar: Reconstituting nucleosomes

The ease with which nucleosomes can be reconstituted from purified histones and DNA is remarkable; indeed, nucleosomes can aptly be described as "self-assembling." Dozens of protocols for reconstituting chromatin have been published over the years. All of them rely on controlling the association of the positively charged histones with negatively charged DNA to avoid problems with aggregation and precipitation; methods include using tRNA or polyglutamine as a "charge buffer," employment of nuclear extracts found to harbor inherent nucleosome assembly capability, and the method used by Thomas and Kornberg (1974), in which purified histones are mixed with DNA in a buffered solution containing 2 M NaCl, which is then dialyzed to a final salt concentration of 150 mM. Purification of DNA is now routine, no longer requiring the procurement of pus-soaked bandages, and histones are similarly readily obtained in quantity and free from contaminants (though, as discussed in Chapters 2 and 3, are preferably expressed as recombinant proteins from *E. coli* for some purposes). Here is one (undetailed) recipe for preparing histones from chicken erythrocytes, beginning with chicken blood, modified from original work from the van Holde and Stein and Bina labs (Tatchell and Van Holde, 1977; Stein and Bina, 1984):

- Prepare histones:
  - Wash cells several times in phosphate-buffered saline solution, spinning down at low speed in a centrifuge after each wash

 - Take up washed cells in 0.7M NaCl/50mM NaP_i, pH4.7; homogenize with Dounce
 - Add hydroxyapatite, wash, and spin down; repeat 4–5×
 - Take up pellet in 2.5M NaCl/50mM NaPi pH7.0
 - Spin down, save histone-containing supernatant
- Reconstitute nucleosomes
 - Mix histones and DNA at 0.8 weight/weight ratio in 2M NaCl/10mM Tris pH8
 - Dialyze 2–4h at 4°C against 0.8M NaCl/10mM Tris pH8; and against 0.17M NaCl/10mM Tris pH8

Sidebar: Credit where credit is due

Fame is a preoccupation of our time, and scientists more often than not are motivated in their work not only by deep-seated curiosity but also by a hunger for some measure of recognition among their fellow scientists. In enterprises involving groups of co-workers, "{C}redit depends on perception, and is a collective social phenomenon" (M. Buchanan in *Nature* (2018) 563:624, reviewing Barbasi, *The Formula: The Universal Laws of Success*). With regard to the nucleosome model, the work most often cited is Roger Kornberg's article in Science, which provided a succinct summary of relevant data supporting the new and specific particle model for the subunit structure of chromatin proposed in the article (Kornberg, 1974). This paper did indeed mark a paradigm shift in research on chromatin, but of course relied extensively on contributions from many sources, as outlined in the main text. Critical results from Don and Ada Olins, Markus Noll, and Hewish and Burgoyne are cited; additionally, although Williamson's, 1970 paper is not cited, it evidently played an important part in fitting the pieces together, as the 200bp DNA ladder reported in that work was later referenced by Kornberg as the final key to the formulation of the nucleosome model (Van Holde, 1988, pp. 23–24). Meanwhile, independent investigations from Chris Woodcock, Ken van Holde, E.-M. Bradbury, and Pierre Chambon and colleagues (the latter group coining the term "nucleosome"), among others, converged on this same model. Many of the relevant papers have been recognized as classic contributions, and some of the major contributors were chosen to chair the Gordon Research Conference on Nuclear Proteins in the early years of this meeting, the major venue for sharing research on chromatin in the years before it gained its current popularity.

References

Anonymous, 1969. Nature 223, 892.

Astbury, W.T., Bell, F.O., 1938. X-ray study of thymonucleic acid. Nature 141, 747–748.

Avery, O.T., Macleod, C.M., McCarty, M., 1944. Studies on the chemical nature of the substance inducing transformation of pneumococcal types: induction of transformation by a desoxyribonucleic acid fraction isolated from pneumococcus type III. J. Exp. Med. 79, 137–158.

Caspersson, T., 1936. Uber den chemischen Aufbau der Strukturen des Zellkernes. Skand. Arch. Physiol. 73 (Suppl 8), 1–151.

Chargaff, E., 1950. Chemical specificity of nucleic acids and mechanism of their enzymatic degradation. Experientia 6, 201–209.
Clark, R.J., Felsenfeld, G., 1971. Structure of chromatin. Nat. New Biol. 229, 101–106.
Cruft, H.J., 1976. Edgar Stedman. Biograph. Memoirs Fellows R. Soc. 22, 529–553.
Dahm, R., 2005. Friedrich Miescher and the discovery of DNA. Dev. Biol. 278, 274–288.
Dahm, R., 2010. From discovering to understanding. Friedrich Miescher's attempts to uncover the function of DNA. EMBO Rep. 11, 153–160.
DeLange, R.J., Fambrough, D.M., Smith, E.L., Bonner, J., 1969a. Calf and pea histone IV. 3. Complete amino acid sequence of pea seedling histone IV; comparison with the homologous calf thymus histone. J. Biol. Chem. 244, 5669–5679.
DeLange, R.J., Fambrough, D.M., Smith, E.L., Bonner, J., 1969b. Calf and pea histone IV. II. The complete amino acid sequence of calf thymus histone IV; presence of epsilon-N-acetyllysine. J. Biol. Chem. 244, 319–334.
Doenecke, D., Karlson, P., 1984. Albrecht Kossel and the discovery of histones. Trends Biochem. Sci. 9, 404–405.
Finch, J.T., Lutter, L.C., Rhodes, D., Brown, R.S., Rushton, B., Levitt, M., Klug, A., 1977. Structure of nucleosome core particles of chromatin. Nature 269, 29–36.
Gordin, M.D., 2019. Ordering the elements. Science 363, 471–473.
Griffith, F., 1928. The significance of pneumococcal types. J. Hyg. (Lond.) 27, 113–159.
Henikoff, S., Church, G.M., 2018. Simultaneous discovery of cell-free DNA and the nucleosome ladder. Genetics 209, 27–29.
Hertwig, O., 1885. Das Problem der Befruchtung und der Isotropie des Eies. Eine Theorie der Vererbung. Jenaische Z. Naturwiss. 18, 276–318.
Hewish, D.R., 1982. Citation classic. Curr. Contents 5, 20.
Hewish, D.R., Burgoyne, L.A., 1973. Chromatin sub-structure. The digestion of chromatin DNA at regularly spaced sites by a nuclear deoxyribonuclease. Biochem. Biophys. Res. Commun. 52, 504–510.
His, W., Miescher, F., 1897. In: His, W.E.A. (Ed.), Die Histochemischen und Physiologischen Arbeiten von Friedrich Miescher. F.C.W. Vogel, Leipzig, pp. 5–32 (F. Miescher).
Hoppe-Seyler, F., 1871. Ueber die chemische Zusammensetzungdes Eiters. Med. Chem. Unters. 4, 486–501.
Huiskamp, W., 1901. Ueber die Elektrolyse der Salze des Nucleohistons und Histons. Hoppe Seylers Z. Physiol. Chem. 34, 32–54.
Johns, E.W., 1964. Studies on histones. 7. Preparative methods for histone fractions from calf thymus. Biochem. J. 92, 55–59.
Johns, E.W., 1967. The electrophoresis of histones in polyacrylamide gel and their quantitative determination. Biochem. J. 104, 78–82.
Kadonaga, J.T., 2019. The transformation of the DNA template in RNA polymerase II transcription: a historical perspective. Nat. Struct. Mol. Biol. 26, 766–770.
Kornberg, R.D., 1974. Chromatin structure: a repeating unit of histones and DNA. Science 184, 868–871.
Kornberg, R.D., 1977. Structure of chromatin. Annu. Rev. Biochem. 46, 931–954.
Kornberg, R.D., Klug, A., 1981. The nucleosome. Sci. Am. 244, 52–64.
Kornberg, R.D., Lorch, Y., 2020. Primary role of the nucleosome. Mol. Cell 79, 371–375.
Kornberg, R.D., Thomas, J.O., 1974. Chromatin structure: oligomers of the histones. Science 184, 865–868.
Kossel, A., 1884. Ueber einen peptoartigen bestandheil des zellkerns. Z. Physiol. Chem. 8, 511–515.
Lilienfeld, L., 1894. Zur Chemie der Leucocyten. Z. Physiol. Chem. 18, 473–486.
Luck, J.M., Rasmussen, P.S., Satake, K., Tsvetikov, A.N., 1958. Further studies on the fractionation of calf thymus histone. J. Biol. Chem. 233, 1407–1414.
Luger, K., Mader, A.W., Richmond, R.K., Sargent, D.F., Richmond, T.J., 1997. Crystal structure of the nucleosome core particle at 2.8 A resolution [see comments]. Nature 389, 251–260.
Mattiroli, F., Bhattacharyya, S., Dyer, P.N., White, A.E., Sandman, K., Burkhart, B.W., Byrne, K.R., Lee, T., Ahn, N.G., Santangelo, T.J., et al., 2017. Structure of histone-based chromatin in Archaea. Science 357, 609–612.
Miescher, F., 1871. Ueber die chemische Zusammensetzung der Eiterzellen. Med. Chem. Unters. 4, 411–460.
Miescher, F., 1874. Die Spermatozoen einiger Wirbeltiere. Ein Beitrag zur Histochemie. Verh. Naturforsch. Ges. Basel 6, 138–208.
Miller Jr., O.L., Bakken, A.H., 1972. Morphological studies of transcription. Acta Endocrinol. Suppl. (Copenh.) 168, 155–177.
Mirsky, A.E., 1968. The discovery of DNA. Sci. Am. 218, 78–88.

Mirsky, A.E., Ris, H., 1947. Isolated chromosomes. J. Gen. Physiol. 31, 1–6.
Mukherjee, S., 2016. The Gene: An Intimate History. Scribner.
Noll, M., Kornberg, R.D., 1977. Action of micrococcal nuclease on chromatin and the location of histone H1. J. Mol. Biol. 109, 393–404.
Ogawa, Y., Quagliarotti, G., Jordan, J., Taylor, C.W., Starbuck, W.C., Busch, H., 1969. Structural analysis of the glycine-rich, arginine-rich histone. 3. Sequence of the amino-terminal half of the molecule containing the modified lysine residues and the total sequence. J. Biol. Chem. 244, 4387–4392.
Olins, A.L., Olins, D.E., 1974. Spheroid chromatin units (v bodies). Science 183, 330–332.
Olins, D.E., Olins, A.L., 2003. Chromatin history: our view from the bridge. Nat. Rev. Mol. Cell Biol. 4, 809–814.
Panyim, S., Chalkley, R., 1969a. The heterogeneity of histones. I. A quantitative analysis of calf histones in very long polyacrylamide gels. Biochemistry 8, 3972–3979.
Panyim, S., Chalkley, R., 1969b. High resolution acrylamide gel electrophoresis of histones. Arch. Biochem. Biophys. 130, 337–346.
Paranjape, S.M., Kamakaka, R.T., Kadonaga, J.T., 1994. Role of chromatin structure in the regulation of transcription by RNA polymerase II. Annu. Rev. Biochem. 63, 265–297.
Pardon, J.F., Wilkins, M.H., 1972. A super-coil model for nucleohistone. J. Mol. Biol. 68, 115–124.
Pardon, J.F., Wilkins, M.H., Richards, B.M., 1967. Super-helical model for nucleohistone. Nature 215, 508–509.
Phillips, D.M., Johns, E.W., 1965. A fractionation of the histones of group F2a from calf Thymus. Biochem. J. 94, 127–130.
Plosz, P., 1871. Ueber das chemische Verhalten der Kerne der Vogel-und Schlangenblutkorperchen. Med. Chem. Unters. 4, 461–462.
Richards, B.M., Pardon, J.F., 1970. The molecular structure of nucleohistone (DNH). Exp. Cell Res. 62, 184–196.
Rill, R., Van Holde, K.E., 1973. Properties of nuclease-resistant fragments of calf thymus chromatin. J. Biol. Chem. 248, 1080–1083.
Rizzo, P.J., Nooden, L.D., 1972. Chromosomal proteins in the dinoflagellate alga Gyrodinium cohnii. Science 176, 796–797.
Sahasrabuddhe, C.G., Van Holde, K.E., 1974. The effect of trypsin on nuclease-resistant chromatin fragments. J. Biol. Chem. 249, 152–156.
Schmidt, G., Levene, P.A., 1938. The effect of nucleophosphatase on "native" and depolymerized Thymonucleic acid. Science 88, 172–173.
Signer, R., Caspersson, T., Hammarsten, E., 1938. Molecular shape and size of thymonucleic acid. Nature 141, 122.
Stedman, E., Stedman, E., 1951. The basic proteins of cell nuclei. Philos. Trans. R. Soc. Lond. 235, 565–595.
Stein, A., Bina, M., 1984. A model chromatin assembly system. Factors affecting nucleosome spacing. J. Mol. Biol. 178 (2), 341–363.
Strasberger, E., 1909. The minute structure of cells in relation to heredity. In: Seward, A.C. (Ed.), Darwin and Modern Science. Cambridge Press, Cambridge, pp. 102–111.
Struhl, K., 1999. Fundamentally different logic of gene regulation in eukaryotes and prokaryotes. Cell 98, 1–4.
Talbert, P.B., Henikoff, S., 2012. Chromatin: packaging without nucleosomes. Curr. Biol. 22, R1040–R1043.
Tatchell, K., Van Holde, K.E., 1977. Reconstitution of chromatin core particles. Biochemistry 16 (24), 5295–5303.
van der Westhuyzen, D.R., von Holt, C., 1971. A new procedure for the isolation and fractionation of histones. FEBS Lett. 14, 333–337.
Van Holde, K.E., 1988. Chromatin. Springer-Verlag.
van Holde, K.E., 2008. Learning how to be a scientist. J. Biol. Chem. 283 (8), 4461–4463.
Vickery, H.B., 1950. The origin of the word protein. Yale J. Biol. Med. 22, 387–393.
Watson, J.D., Crick, F.H., 1953. Molecular structure of nucleic acids; a structure for deoxyribose nucleic acid. Nature 171, 737–738.
Williamson, R., 1970. Properties of rapidly labelled deoxyribonucleic acid fragments isolated from the cytoplasm of primary cultures of embryonic mouse liver cells. J. Mol. Biol. 51, 157–168.
Woodcock, C.L., Safer, J.P., Stanchfield, J.E., 1976a. Structural repeating units in chromatin. I. Evidence for their general occurrence. Exp. Cell Res. 97, 101–110.
Woodcock, C.L., Sweetman, H.E., Frado, L.L., 1976b. Structural repeating units in chromatin. II. Their isolation and partial characterization. Exp. Cell Res. 97, 111–119.
Zacharias, E., 1881. Ueber die chemische Beschaffenheit des Zellkerns. Bot. Zeitung 39, 169–176.

CHAPTER

2

The nucleosome unveiled

Abbreviations

EM electron microscopy
MNase micrococcal nuclease
NCP nucleosome core particle
NMR nuclear magnetic resonance
SHL superhelix location
SV40 Simian Virus 40
Taf TBP-associated factor
TBP TATA-binding protein

From model to structure: The nucleosome core particle

The model of the nucleosome as a unit structure that emerged in the early 1970s, together with the ability to reconstitute such structures in vitro, set off a race to obtain a high-resolution crystal structure of the nucleosome (or more properly, the nucleosome core particle, as discussed below). This proved to be a difficult task. Crystallization of nucleosomes was reported in 1977 (Finch et al., 1977), and the structure was resolved to 7 Å in 1984 (Richmond et al., 1984), but it was another 13 years before a structure providing detailed resolution of both protein and DNA in the nucleosome was reported (Luger et al., 1997). The first efforts along these lines used nucleosome core particles prepared by micrococcal nuclease (MNase) digestion of isolated chromatin. Progress in obtaining a high-resolution structure from these preparations was hindered by the heterogeneous nature of core particle preparations. Crystallization depended on obtaining relatively homogeneous lengths of DNA associated with the purified nucleosomes and on removal of linker histone. This was challenging in the era preceding the use of polymerase chain reaction (PCR) technology; it was first accomplished by using a three-step protocol to prepare nucleosome core particles. First, chromatin was lightly digested with MNase to prepare long fragments; then linker histone was removed by extraction with 0.45 M NaCl; and finally, the resulting chromatin fragments were redigested with MNase to produce core particles that

Chromatin. **https://doi.org/10.1016/B978-0-12-814809-9.00001-9**

were sufficiently homogeneous to allow crystal formation (Finch et al., 1977). Nucleosome core particles prepared using this protocol and subsequent refinements allowed the preparation of crystals that diffracted to 7 Å (Richmond et al., 1984), but the eventual high-resolution structure required major modifications. Instead of isolating nucleosomes from cells, which produced particles that were inherently heterogeneous for DNA sequence, a single DNA sequence along with recombinant histones expressed in bacteria, and hence lacking posttranslational modifications, was eventually used to reconstitute nucleosomes that diffracted at high resolution (Luger et al., 1997). In the interim between first crystallization and determination of the nucleosome core particle structure at high resolution, biophysical, biochemical, and enzymatic methods provided additional structural information. Some of this work was aimed at elucidating basic structural properties—size, shape, and topology—of the particle that was now established as the basic unit of chromatin. Beyond this, there was great interest in gaining information on the structure of the protein moiety of the nucleosome, the path of DNA, and the nature of the interactions between DNA and histones. We now consider these aspects in turn.

Size and shape

Electron microscopy (EM) experiments and measurements of hydrodynamic properties provided a rough picture of the nucleosome as containing 100–200 bp of DNA with the shape of a flattened disc (Langmore and Wooley, 1975; Sahasrabuddhe and Van Holde, 1974). X-ray diffraction of partially proteolyzed nucleosomes prepared from rat liver chromatin after removal of the linker histone H1, although only providing a resolution of about 20 Å, confirmed and refined this picture, yielding a model of the nucleosome core particle as a flat particle, quaintly termed a "platysome," with dimensions of $110 \times 110 \times 57$ Å (Finch et al., 1977). X-ray diffraction and EM results led to the suggestion of an approximate mirror symmetry, "as might arise from the arrangement of eight similar histone molecules about a true dyad" (Finch et al., 1977). Such pseudodyadic symmetry was borne out by later, higher resolution determination (Richmond et al., 1984). This symmetry can best be visualized by imagining an observer (very small) as sitting on the central base of the DNA that wraps around the histones (Fig. 2.1). Such an observer, looking "downward" at the histone core, would see a near perfect twofold rotational symmetry, and if the observer traveled in either direction along the DNA helix, s/he would be treated to identical views of the histones as they passed by. The DNA itself, of course, would not be perfectly symmetric, unless its sequence were expressly chosen to be a palindrome. (Such a palindromic sequence, derived from human α-satellite DNA, was in fact used to obtain the high-resolution structure of the nucleosome, but was found to be slightly offset from a perfectly symmetric configuration.) Because of this deviation from perfect twofold symmetry, the term "pseudodyadic" aptly describes the symmetry of the nucleosome.

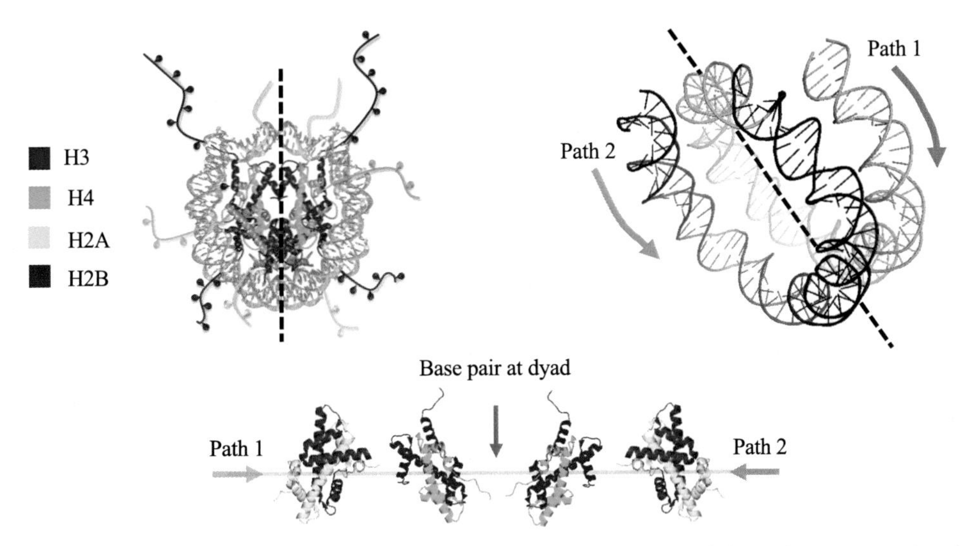

FIG. 2.1 Dyad symmetry of the nucleosome. Upper left shows the dyad axis of the nucleosome; rotating the nucleosome 180 degrees around this axis leaves the structure unchanged. Upper right shows that following the path of the DNA from either end is equivalent, while the bottom illustrates that equivalent "views" of the protein core are seen upon traveling from either end or from the center in either direction. *Credit: Courtesy of Karolin Luger and Elisabeth Knoll.*

The path of the DNA

The initial model of the nucleosome as a particle comprising eight histone proteins and about 200 bp of DNA left open the question of the disposition of the DNA: was it coated by the histone proteins, or somehow wound around the outside (see Sidebar: Ball of confusion)? Kornberg suggested that "DNA in a repeating unit would follow some path on the tetramer" (referring to the $(H3\text{-}H4)_2$ tetramer) in his initial model of the nucleosome (Kornberg, 1974), but did not speculate further regarding what that path might be (Thomas and Kornberg, 1975; Van Holde, 1988). Two nearly concurrent models explicitly placed the DNA on the outside of the histones (Baldwin et al., 1975; Van Holde et al., 1974). DNase I digestion experiments provided concrete, if inferential, evidence for this hypothesis (Noll, 1974). Limited digestion of chromatin with this enzyme, which cuts one strand on the DNA double helix, yielding "nicks" in the DNA, resulted in a set of fragments that appeared as a ladder when visualized by denaturing gel electrophoresis (Fig. 2.2). This ladder of fragments was similar in appearance to that seen upon MNase digestion of chromatin but very different in size: the spacing between bands, rather than being about 200 bp, was 10 bp, very close to the helical repeat length of DNA (Noll, 1974). This was interpreted as reflecting periodic accessibility to the enzyme of the phosphodiester linkage on the individual DNA strands, as would be expected if the helix were laying on a surface. Indeed, cleavage of DNA adhered to a crystalline surface by DNase I or by chemical cleavage using hydroxyl radicals also occurs according to the helical periodicity of DNA (Liu and Wang, 1978; Rhodes and Klug, 1980; Tullius and Dombroski, 1985). (Hydroxy radical cleavage was developed as an alternative to nucleolytic enzymes as a tool for studying DNA-protein interactions by Tom Tullius at Johns Hopkins University (Tullius and Dombroski, 1985). These radicals are easily generated in the Fenton reaction between Fe(II)·EDTA and hydrogen peroxide, and offer the advantages of being less hindered in their approach to DNA by steric interactions distant from the double helix, and of being much less sensitive to DNA sequence in their cleavage specificity than nucleases.) Subsequent crystallization of DNase I in complex with a nicked DNA octanucleotide revealed the DNA to be curved away from the enzyme, consistent with the ability of DNase I to nick DNA that is sharply curved on the surface of the nucleosome (Suck et al., 1988).

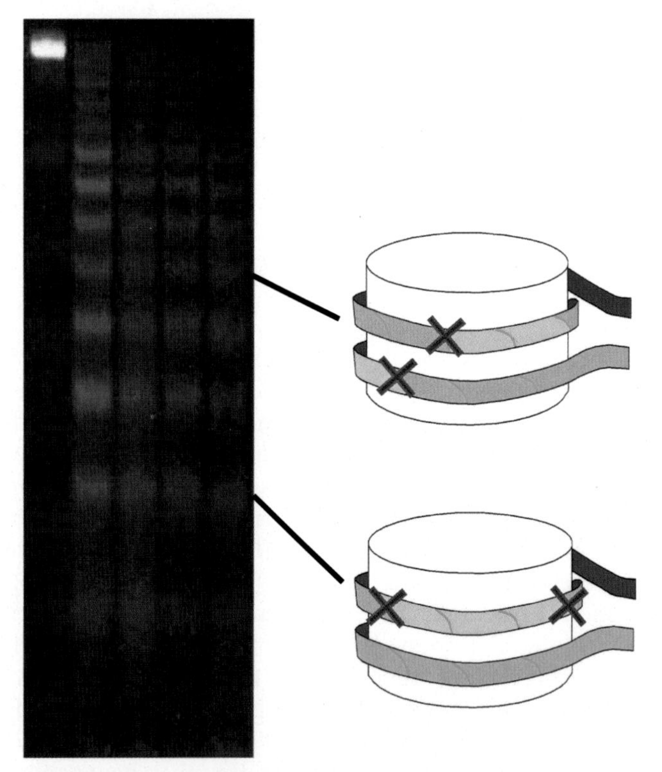

FIG. 2.2 Single-stranded DNA fragments resulting from DNase digestion of nucleosome core particles prepared from chicken erythrocyte nuclei. The cartoon on the right depicts how the individual bands result from cleavage of DNA on the nucleosome surface; the colored-in fragments result from cleavage at the exposed minor groove sites marked by *red* "×"s and together with other fragments created by cleavage at sites separated by the same number of DNA helical turns, generate the indicated bands. *Adapted from Tatchell, K., Van Holde, K.E., 1977. Reconstitution of chromatin core particles. Biochemistry 16, 5295–5303.*

Sidebar: Ball of confusion

Research on chromatin from ~1969 to 1974 produced a flood of results that culminated in a structural model for the nucleosome, as recounted in Chapter 1. As always, new data leads to new ideas, and there was no shortage of ideas about the structure of the nucleosome. The period following the paradigm shift leading to the particle model of the nucleosome was an ideal time for speculation about the structure: there was enough data to justify specific proposals, but not so much as to place severe constraints on imagination. Leading scientists of the time stuck their necks out with specific models and deserve credit for doing so; such models in turn provided grist for new experiments, to support or refute them. Francis Crick, who got quite a few things right in his career, proposed in 1971 a model for chromosomes in which a small fraction of eukaryotic genomes existed as double-stranded, "fibrous" DNA while the rest was "globular" DNA that included unwound, single-stranded regions and exerted regulatory control over gene expression (as based in part on the influential work of Britten and Davidson) (Britten and Davidson, 1969; Crick, 1971). Histones were proposed as the most likely candidate for unwinding regulatory DNA. When EM and nuclease experiments over the following years revealed the particle nature of the nucleosome, Crick and Klug (a Nobel laureate and future laureate), confronted with the energetic difficulty in bending 200 bp of DNA smoothly to a radius compatible with EM results, suggested that large kinks in nucleosomal DNA might solve this problem; the model was refined by Sobell and colleagues (Crick and Klug, 1975; Sobell et al., 1976). However, subsequent detailed energetic calculations supported the plausibility of a smooth deformation of DNA in the nucleosome (Levitt, 1978; Sussman and Trifonov, 1978). Another idea, focused on accounting for the pattern of DNA fragments resulting from limited nuclease digestion, posited a single site on nucleosomal DNA that was sensitive to nuclease (Cantor, 1976). Other models included histones being organized into an extensive, linear overlapping array (Chalkley and Hunter, 1975); a helical polymer based upon a repeated structure in which 9.6 turns of DNA were wrapped around a central core composed of one copy of each of the core histones (Hyde and Walker, 1975); and a tetrameric, heterotypic complex comprised of one copy of each of the four core histones (Weintraub et al., 1975).

The emergence of the particle model for the nucleosome altered the landscape for models of the chromosome, and researchers proposed several compact models for the nucleosome (Baldwin et al., 1975; Kornberg, 1974; Olins and Olins, 1974; Oudet et al., 1975; Van Holde et al., 1974). The Kornberg model posited a particle in which most of the 200 bp of DNA in the repeating unit would follow a path on the $(H3\text{-}H4)_2$ tetramer, with the remainder of the DNA connecting the tetramers following a path defined by possibly oligomeric H2A-H2B dimers (Kornberg, 1974). A nearly contemporaneous model from van Holde and colleagues explicitly proposed DNA wrapping around a hydrophobic protein core complex built from the C-terminal halves of the four core histones, comprising "about eight protein molecules," with the histone N-termini wrapped around the outside of the DNA helix (see figure) (Van Holde et al., 1974). Baldwin et al. used neutron diffraction to show that the DNA was wrapped around a protein core in the nucleosome and to obtain corresponding dimensions of the DNA and protein moieties (Baldwin et al., 1975). Their model (see figure) stipulated, as did that of van Holde, that the protein core was formed by the apolar segments of the histones, while the basic, N-terminal portions of the histones complexed with the DNA on the outside. In the end, the debate was fully settled by the high-resolution structures of the histone octamer and nucleosome core particle.

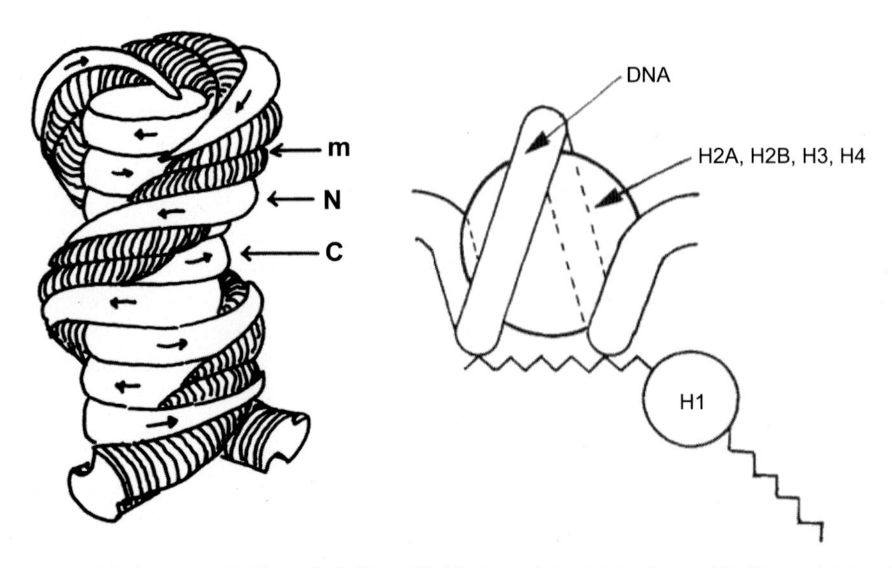

Nucleosome models from van Holde et al. (left) and Baldwin et al. (right). In the van Holde model, the DNA is shaded, "N" and "C" indicate the N- and C-terminal regions of the histones, and the minor groove of the DNA is indicated by "m."

More definitive evidence for a structure in which DNA was wrapped around a histone core derived from neutron scattering experiments. In these experiments, the ratio of H_2O to D_2O is varied so that the scattering from the aqueous solvent, which differs greatly between H_2O and D_2O, matches the scattering of DNA or protein. This approach—analogous to viewing an image in red and green with either red or green illumination—gives maximal contrast to the unmatched component, thus allowing the radius of the DNA and protein components to be separately determined (Hjelm et al., 1977; Pardon et al., 1975). These experiments indicated that the histone-containing part of the nucleosome had a smaller radius of gyration than the nucleic acid component, and led to a model of the nucleosome as a flattened disk of about 110 Å diameter and 50 Å height, with DNA wrapping in a superhelix around the histone core (Fig. 1.1) (Van Holde, 1988). This model was consistent with early images obtained by transmission EM (Langmore and Wooley, 1975), and was eventually corroborated and refined by the 7 Å structure published in 1984 (Richmond et al., 1984).

Another important issue to be resolved in these early studies of nucleosome structure was the length of DNA associated with the histones. As discussed earlier, this length was first established as being about 200 bp, based on the repeat length measured following nuclease digestion of chromatin. The disposition of this 200 bp relative to the histones was assessed by experiments using controlled digestion of chromatin by MNase. Monitoring the products of digestion during a time-course of MNase treatment revealed that the unit particle, containing about 200 bp of DNA, was digested further to yield a relatively stable product containing about 145 bp of DNA and an octamer of the core histones H2A, H2B, H3, and H4 (an entity termed the "nucleosome core particle"), with the linker histones (which at this time were referred to as the "lysine-rich histones") being lost during digestion (Bakayev et al., 1975;

Noll and Kornberg, 1977; Sollner-Webb and Felsenfeld, 1975; Van Holde et al., 1974, 1975; Whitlock Jr. and Simpson, 1976). The level of interest in gaining a precise understanding of the length of DNA associated with the nucleosome core can be gleaned from the number of publications reporting measurements of this parameter: van Holde cites 11 papers published from 1978 to 1980 reporting core DNA lengths of 145–146 bp, with yeast DNA being a slight outlier at 148 bp (Van Holde, 1988). But these publications served another purpose as well: the same length of DNA, within two or three base pairs, was found to be associated with nucleosomes from such diverse sources as HeLa cells, chicken erythrocytes, yeast, and pea plants. Even more impressively, core particles reconstituted from purified core histones and synthetic poly(dA·dT) or poly(dG·dC) oligonucleotides were digested to the same length, 146 ± 1 bp, as naturally derived chromatin (Simpson and Kunzler, 1979). Such concordance in the length of DNA associated with core particles from diverse organisms was not entirely surprising, as the histones were known at this time to be highly conserved. Nonetheless, these findings firmly established the nucleosome core particle as a highly conserved structure across eukaryotes, and additionally justified extrapolating findings derived from reconstituted core particles to those found in nature as well as for comparing results across species.

The experiments showing loss of the lysine-rich histones concomitantly with the digestion of nucleosomes to a stable length of ~145 bp implied that the nucleosome comprised a "core particle," identical to the monomer product resulting from extensive MNase digestion, together with "linker DNA," to which the lysine-rich, "linker histones" were bound. Further analysis was done by purifying chromatin fractions digested with micrococcal nuclease over sucrose gradients; this allowed the separation of fractions containing mononucleosomes, dinucleosomes, and so forth. Analysis by denaturing gel electrophoresis of the protein and DNA content in the fractions corresponding to monomer particles revealed that digestion of the monomer particle showed a pause at about 160 bp, followed by a stronger pause at about 140 bp, with linker histone being present only in the larger particles (Varshavsky et al., 1976). This larger particle, eventually shown to contain 168 bp of DNA, was isolated from chicken erythrocytes and termed the "chromatosome" (Simpson, 1978). Careful biochemical experiments revealed the chromatosome to contain two copies of each of the four core histones H2A, H2B, H3, and H4, but only one copy of the linker histone H1 or H5, and ~168 bp of DNA (Bates and Thomas, 1981; Hayashi et al., 1978; Simpson, 1978). These experiments also indicated that the additional DNA was distributed evenly, about 10 bp on each side of the nucleosome core DNA.

The location of linker histones in the core particle was first deduced by consideration of the enzymatic digestion experiments, which, together with biophysical data, indicated interactions with the ends of the DNA in the core particle. Thermal denaturation and circular dichroism measurements suggested that the ends of the DNA in the core particle dissociated at around 40°C and melted into single-stranded chains at about 60°C, while the inner portion of the nucleosomal DNA did not dissociate and melt until about 80°C (Van Holde, 1988). The presence of linker histone altered this melting, affecting the transition corresponding to the outermost DNA in the core particle, thus providing further evidence for its location close to the DNA entry and exit points (Simpson, 1978).

Additional detail on the path of DNA in the nucleosome was provided by the 7 Å structure published by Tim Richmond, Aaron Klug, and co-workers from the MRC in 1984 (Richmond et al., 1984). This structure allowed tentative assignments of the location of the histones, and established definitively that the DNA was wound in 1.8 turns around the outside of the histones in a left-handed superhelix comprising about 80 bp/turn. Such a tight winding immediately implies distortion of the DNA double helix; a smooth bending of the DNA to this dimension would compress the edge-to-edge separation of adjacent base pairs from 3.4 Å, on average, to 2.9 Å on the inner, histone-proximal side of the double helix, while the outside separation would increase to 3.9 Å. Indeed, the X-ray diffraction data showed the minor groove of the DNA helix varying from 7 to 13 Å in width while the major groove varied from 11 to 20 Å (Richmond et al., 1984).

The 7 Å structure also revealed that the path of the DNA was not completely smooth. Rather, notable bends in the DNA helix were seen on both sides of the dyad center of symmetry, removed by one and four helical turns. The nomenclature adopted to describe the winding of the DNA along the histone core designated the origin, termed superhelix location zero, or SHL0, at the location where the major groove faces inward at the dyad; the next points of inward-facing major grooves are designated SHL+1 and SHL-1, while locations of inward-facing minor grooves that directly contact histone residues are at SHL-6.5, -5.5, … 5.5, and 6.5. Using this nomenclature, bends were observed at SHL$\pm$1 and SHL$\pm$4 (Richmond et al., 1984). An altered structure at these positions was consistent with data showing reduced cleavage at these sites, relative to other sites, by DNase I (Lutter, 1978; Noll, 1977; Simpson and Whitlock, 1976; Tatchell and Van Holde, 1977). In these experiments, the relative sensitivities of individual cleavage sites were mapped by end-labeling core particle DNA with radioactive phosphate (^{32}P) before DNase treatment, in what were essentially the first footprinting experiments (Simpson and Whitlock, 1976). Gel electrophoresis of the fragments resulting from light digestion, so that core particles were cleaved on average only about once, followed by autoradiography, allowed the distance of cleavage sites from the DNA end to be precisely mapped. The results showed that cleavage occurred with maxima approximately every 10 bp, and that sites 30, 60, 80, and 110 bp from the 5′ end were relatively resistant to cutting. These sites correspond to SHL$\pm$1 and SHL$\pm$4, indicating that the altered structure observed in the crystal at these sites likely provided a poorer substrate for DNase I than the less distorted structure found elsewhere on the nucleosome surface. In later work, DNA kinks at SHL$\pm$1.5 were reported based on enhanced cleavage by singlet oxygen and binding of the intercalator methylene blue; however, the base opening at these kinks was deduced to be 90 degrees, a more extreme deformation than that observed in the crystal structure (Hogan et al., 1987). Hydroxyl radical footprinting provided additional evidence for a distortion of the DNA helix at 1–2 helical repeats on either side of the dyad, both for a nucleosome having a defined DNA sequence and for mixed sequence nucleosomes (Hayes et al., 1991, 1990). In addition, the hydroxy radical footprinting experiments indicated that the central 3 helical turns of the nucleosomal DNA, encompassing 1.5 superhelical turns on either side of the dyad, exhibited a helical period of ~10.7 bp/turn, while the outer regions had a periodicity of ~10.0 bp/turn (Hayes et al., 1991, 1990). Similar results were obtained by determination of the frequency of thymine dimer formation by photofootprinting experiments, in which a central region with a periodicity of ~10.5 bp/turn was found to be flanked by regions of ~10.0 bp/turn (Gale and Smerdon, 1988). The shift in helical periodicity determined in these experiments implied that there must be a region of DNA distortion at the junction of the two regions. These DNA distortions were quickly recognized as having the potential to govern the propensity of specific sequences to adopt precise positions relative to the histone octamer (see Chapter 4).

Another aspect of the DNA path in the nucleosome that deserves consideration is its topology. The winding of DNA around the histones in the nucleosome immediately suggested that there must be a specific DNA topology associated with nucleosomes, and researchers recognized that this could be investigated using nucleosomes on closed circular DNA (ccDNA) molecules. A closed circular piece of DNA that can be laid flat, or nearly so, with no windings around itself is in its lowest energy topological state and is referred to as "relaxed" (Fig. 2.3). If the double helix is cut, twisted 360 degrees, and resealed, this introduces a supercoil, a process requiring energy. More formally, the number of supercoils can be considered as the linking number or the net number of crossings-over of a circular molecule of DNA upon itself when projected on a plane. Treatment of supercoiled DNA with the enzyme topoisomerase I (or an "untwisting extract" containing this enzyme), which nicks and religates DNA, results in a Boltzmann distribution of topoisomers centered on the most relaxed state—zero supercoils for naked DNA.

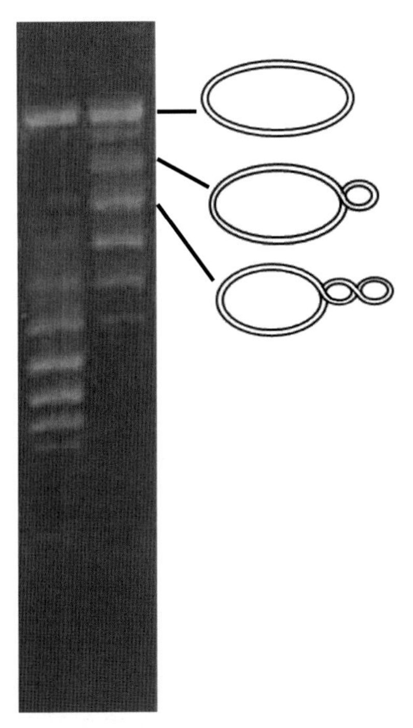

FIG. 2.3 Topoisomers resolved by agarose gel electrophoresis. Plasmid pBR322 was treated under different conditions with a nicking-closing extract to generate different distributions of topoisomers. The relaxed plasmid was electrophoresed under conditions that allow individual topoisomers—molecules differing by integer numbers of supercoils—to be resolved, as illustrated on the right. For more detail see Morse (1991).

Researchers interested in the nucleosome realized that if closed circular DNA packaged into nucleosomes were relaxed, and the histones then removed, any supercoils remaining must have been constrained by nucleosomes. This strategy was applied using closed circular Simian Virus 40 (SV40) minichromosomes isolated from virally infected African green monkey kidney cells, in which the SV40 DNA is packaged into nucleosomes. Isolated SV40 minichromosomes were relaxed using extracts containing topoisomerase I, and the results were compared with the number of nucleosomes determined by EM (Keller, 1975; Shure and Vinograd, 1976). Supercoils were counted by electrophoresing the DNA on gels that, remarkably, allow the separation of topoisomers differing from one another by single supercoils, thus permitting the exact average linking number to be determined (Fig. 2.3); the comparison indicated that each nucleosome constrained from 1 to 1.25 negative supercoils of DNA. In a more refined strategy, Chambon and colleagues again used isolated SV40 minichromosomes and also prepared reconstituted minichromosomes using purified SV40 DNA and purified histones (Germond et al., 1975). The nucleosome distributions were again determined by counting particles on the SV40 DNA in the electron microscope and compared to the supercoil distributions obtained after treatment of these native or reconstituted minichromosomes using untwisting extract, followed by deproteinization. These experiments revealed that each nucleosome constrained one negative, or left-handed, supercoil of DNA. Later experiments using minichromosomes isolated from yeast cells, and nucleosome arrays reconstituted in vitro onto closed circular DNA, also yielded a value of very close to 1.0 negative supercoil per nucleosome (Pederson et al., 1986; Simpson et al., 1985).

Germond et al. recognized that a single superhelical turn of about 200 bp of DNA, as might be predicted from the above experiments revealing one negative supercoil per nucleosome, was inconsistent with the measured dimensions of the nucleosome, and proposed that a structure in which DNA was folded or kinked could better fit their result (Germond et al., 1975). However, no such kinked or folded structure was observed in the early, low-resolution nucleosome structures, and it was quickly realized that the ~1.7 turns of DNA that wrapped around the core histone octamer was discordant with the value of −1 supercoil induced per nucleosome (Finch et al., 1977). This problem was formalized as the "linking number paradox" and a variety of solutions were proposed (Prunell, 1998). Careful consideration of topology led to the idea that the paradox could be partly resolved by accounting for the curved path of DNA in space, or "surface twist," and by an overtwisting of nucleosomal DNA (Finch et al., 1977; Klug and Lutter, 1981; White et al., 1988). These ideas were confirmed by measurements of DNA helical repeat on the nucleosome surface, which revealed a mean value of about 10.2 bp/turn compared to 10.5 bp/turn in solution, by calculations of surface twist (reviewed in Segura et al. (2018)), and by direct observation of nucleosome structure at 2.8 Å resolution (Luger et al., 1997). However, these corrections still leave a discrepancy of ~0.3 in the linking number to be accounted for. Recent work suggests that this may be resolved by considering effects of heterogeneous linker lengths, or by adjustment of linking number constraint per nucleosome using newer experimental approaches (Nikitina et al., 2017; Segura et al., 2018).

How important are these DNA deformations in the nucleosome? With the elucidation of the structure of the nucleosome, many researchers realized that the interactions of DNA-binding proteins, including proteins involved in transcription, with DNA incorporated into these particles, could be affected by steric occlusion but also by altered DNA structure (Morse and Simpson, 1988). It is easy to understand how nucleosomal DNA is a poor substrate for being recognized by sequence-specific binding proteins: critical sequence elements could be facing inwards toward the histones, and thus rendered "invisible" to binding proteins; they could be occluded by the neighboring strand of DNA; or they could fit poorly onto the appropriate protein surface because of their deformation by the sharp bending that occurs in the nucleosome. Conversely, the altered conformation of nucleosomal DNA can create an improved binding site. An example of this is provided by viral integrases, including that of HIV, for which the wide major groove of DNA on the outside of the nucleosome creates a preferred binding site that is particularly enhanced at the distorted SHL ± 3.5 locations (Maskell et al., 2015; Muller et al., 1993; Pruss et al., 1994). We will return to interactions of sequence-specific binding to nucleosomal DNA later when we consider in depth the interaction of transcription factors with chromatin.

Histones in the nucleosome

We shall discuss the histone proteins at greater length later; for now, it is sufficient to mention their most conspicuous characteristics: they are small, they are basic, and they are highly conserved (Table 2.1). The "canonical" core histones are synthesized during S phase, concomitant with the doubling of the genome, and comprise the majority of the histones found in chromatin. Variant histones exist that are minor, albeit important constituents of chromatin, and which serve disparate functions; these will be discussed in Chapter 5. The canonical core histones sequenced to date vary in length from just over 100 to about 140 amino acids in most organisms, although variants that are longer than this are also known. For comparison, the mean protein size in the budding yeast *Saccharomyces cerevisiae* is close to 500 aa, and the median is about 400 aa (neglecting open reading frames whose protein products would be <50 aa; most of these remain uncharacterized) (Zhang, 2000). The basic residues Lys and Arg comprise 20%–25% of the core histone protein mass (Table 2.1) and are important both in defining sites of interaction with DNA and as sites of posttranslational modification, as will be discussed later. The histones are among the most highly conserved of all proteins (Fig. 2.4). As mentioned in Chapter 1, the first reports of histone sequence revealed the startling fact that histone H4 from pea plant and calf thymus differed in only 2 of 102 amino acids (DeLange et al., 1969a,b; Ogawa et al., 1969). As additional histone sequences were obtained, some by direct sequencing and some by deduced translation of cloned histone genes (or better, to rule out sequencing of pseudogenes, of reverse-transcribed mRNA), it became clear that histones did show evolutionary divergence. (Sequences of histones from a large number of species can now be viewed and compared online at https://www.ncbi.nlm.nih.gov/research/HistoneDB2.0/ (Draizen et al., 2016).) Nonetheless, strong conservation was the rule.

TABLE 2.1 Properties of the histones.

Histone		MW (kD)	# of aa residues	Lys	Arg
H1	Calf thymus	22.5	224	66	3
	Yeast	27.9	258	58	4
H2A	Calf thymus	14	129	14	12
	Yeast	14	132	11	10
H2B	Calf thymus	13.8	125	20	8
	Yeast	14.3	131	19	6
H3	Calf thymus	15.3	135	13	18
	Yeast	15.4	136	16	17
H4	Calf thymus	11.2	102	11	14
	Yeast	11.4	103	11	14

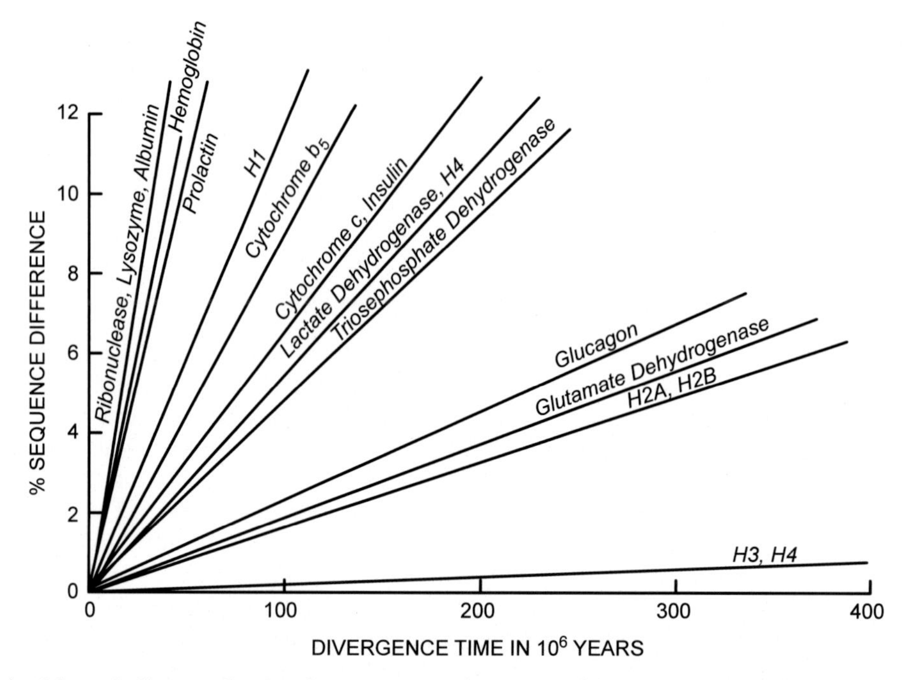

FIG. 2.4 Schematic diagram showing the approximate divergence of the core histones relative to that of several other proteins. From Van Holde (1988), where it was reprinted from Isenberg (1978). Although the data is far from current, the figure provides a roughly accurate representation of the relative conservation of the histones. *From Busch, H. (ed.), 1978. The Cell Nucleus, vol. 4, pp. 135–154. Copyright 1978 Academic Press.*

The earliest crystal structures of the histone octamer and nucleosome core particle lacked sufficient resolution to follow the paths of the histone peptide chains or to visualize the amino acid side chains, or even to ascertain how the histones were arranged (Finch et al., 1977). Structural information on the histones was sought by the use of various biophysical techniques, including nuclear magnetic resonance (NMR), circular dichroism, infrared spectroscopy, and Raman spectroscopy. NMR investigations of large macromolecules can provide information on the presence of highly structured and therefore relatively immobile domains vis-à-vis more freely mobile regions, as found for random coil conformations (or what are now generally referred to as intrinsically disordered domains). The spectroscopic methods yield estimates of the proportion of proteins having α-helical or β-sheet conformations. Numerous studies of these types were conducted in the early days of research on chromatin structure, with the underlying assumption that information on the structure of individual histones or histone subcomplexes would translate into understanding relevant to the structure of the nucleosome. However, results of such studies were confounded by a notable dependence of structural features and aggregation on salt concentration, as well as the fact that structural features varied depending on whether the histones were present alone, in complex with other histones, or in a nucleosomal context (discussed at length in Van Holde (1988)). Because of these and other uncertainties, investigations on individual histone conformation, although producing controversies that now seem as esoteric as ancient philosophical arguments, yielded modest insight into nucleosome structure. Instead, information on the disposition of the histones in the nucleosome derived mainly from three sources: biochemical studies on histone-histone interactions, cross-linking studies that delineated histone-histone and histone-DNA interactions in the nucleosome, and investigations of the effect of limited proteolysis on nucleosome structure.

As discussed earlier, evidence for the existence of a stable H2A-H2B complex, and especially of an $(H3\text{-}H4)_2$ tetramer, was critical in the development of the original nucleosome model, as it pointed to a stable structure of defined composition, as opposed to a nonstoichiometric coating of DNA with histones (D'Anna Jr. and Isenberg, 1974; Kelley, 1973; Kornberg, 1974; Kornberg and Thomas, 1974; Roark et al., 1974). Two studies performed before the formulation of the particle model of the nucleosome provided some of the first evidence for the tripartite structure of the nucleosome that eventually emerged. These studies reported that histones could be progressively removed from chromatin (that is, separated from the DNA) by increasing salt concentrations, with H2A and H2B being lost first at NaCl concentrations above 0.8 M, followed by H3 and H4 at about 1.2 M NaCl (Ilyin et al., 1971; Ohlenbusch et al., 1967). At about the same time, the gentle extraction method developed by van der Westhuyzen et al. revealed that when histones were extracted and then eluted over a size exclusion column at 2 M NaCl, histones H3 and H4 eluted as a single peak followed by a second peak that included H2A and H2B, with the four histones being present at approximately equimolar ratio (van der Westhuyzen and von Holt, 1971). This isolation method was exploited by Eickbush and Moudrianakis, who showed that the histone octamer, although stable at 2 M NaCl, could be dissociated into two H2A-H2B dimers and an $(H3\text{-}H4)_2$ tetramer by various treatments, including decreased NaCl concentration or increased temperature (Eickbush and Moudrianakis, 1978). These findings indicated that the H2A-H2B dimer and $(H3\text{-}H4)_2$ tetramer were held together by mostly hydrophobic interactions distinct from the weaker interactions responsible for octamer formation.

Cross-linking studies, in which chemical agents or UV light were used to induce protein-protein cross links, corroborated the solution studies (Kornberg, 1977). Long cross-linkers such as dimethyl suberimidate, which can span 10–20 Å, are able to cross-link the entire octamer; this was the cross-linking reagent used by Kornberg and Thomas (Kornberg and Thomas, 1974). However, while this was important in establishing the particle nature of the nucleosome, it provides little information on the internal arrangement of the histones (Kornberg and Thomas, 1974). Shorter cross-linking agents, such as tetranitromethane, revealed associations between H2A and H2B, H3 with H3, and H2B with H4, which corresponded to the strongest solution interactions reported by Isenberg and colleagues (D'Anna Jr. and Isenberg, 1974; Martinson and McCarthy, 1975; Van Holde, 1988). These various results were not as straightforward initially as they are in hindsight. Early studies were sometimes confounded by nonspecific aggregation, and long polymers of cross-linked histones were frequently reported upon treatment of chromatin with cross-linking agents. Furthermore, the use of varying solution conditions and analytical methods produced disparate findings of associations and conformational changes that were difficult to integrate into an encompassing model. Interactions of H4 were found not only with H3, but also H2A and H2B, and the observations of cross-linked oligomers of H2A and H2B led to the initial suggestion that H2A-H2B polymers might connect globular $(H3\text{-}H4)_2$ tetramers along the chromatin fiber (D'Anna Jr. and Isenberg, 1974; Kornberg, 1974, 1977; Thomas and Kornberg, 1975).

Despite the difficulty of reconciling the torrent of biochemical and cross-linking data relevant to histone interactions, the notion of the $(H3\text{-}H4)_2$ tetramer as central to nucleosome structure was firmly established by the late 1970s, supported especially by the preferential removal of H2A-H2B from chromatin at 0.8–1.2M NaCl, by the sedimentation experiments of Roark and colleagues and the cross-linking studies of Thomas and Kornberg, and by the careful biochemical work of Eickbush and Moudrianakis (Eickbush and Moudrianakis, 1978; Ilyin et al., 1971; Kornberg and Thomas, 1974; Ohlenbusch et al., 1967; Roark et al., 1974). This notion was buttressed by experiments in which chromatin reconstituted from different admixtures of histones led to the finding that the $(H3\text{-}H4)_2$ tetramer was sufficient to reconstitute a particle with DNA having properties similar to a core particle. Specifically, these particles protected from 70 to 120bp from MNase digestion, depending on conditions of reconstitution, resembled core particles under the electron microscope, and constrained DNA supercoiling similarly to native nucleosomes (Bina-Stein, 1978; Camerini-Otero and Felsenfeld, 1977; Camerini-Otero et al., 1976; Oudet et al., 1978). Although the possibility of particles containing one or two $(H3\text{-}H4)_2$ tetramers may have led to some uncertainties as to the nature of the particles obtained (see discussion in Van Holde (1988, pp. 254–255)), the absence of any such structures obtained by reconstitution using only H2A and H2B constituted strong evidence for the tetramer being central to the nucleosome core particle, with H2A and H2B stabilizing the ends of the DNA. Early low-resolution structural determination of the histone octamer, in conjunction with cross-linking studies that mapped contacts between histones and DNA in purified core particles, supported the concept of the $(H3\text{-}H4)_2$ tetramer defining the central turn of the DNA helix in the nucleosome, with H2A-H2B dimers assembling onto each face of the tetramer (Klug et al., 1980; Mirzabekov et al., 1978). This arrangement was eventually confirmed by the first high resolution determination of the structure of the histone octamer, which also yielded the first detailed information on the structure of the histones in the nucleosome, as discussed below (Arents et al., 1991).

Another important aspect of histone configuration in the nucleosome that emerged from early studies was the existence of flexible, lysine-rich N-termini that are dispensable for the essential structure of the core particle. The high concentration of positively charged residues, particularly lysine residues, in the amino terminal regions of all four histones and the carboxy terminal regions of H2A and H3 gave rise to the idea that these parts of the histones, particularly the N-terminal regions, might be especially important in securing DNA to the histone octamer via major groove interactions (Van Holde et al., 1974). In contrast to the highly basic amino terminal regions, the remaining parts of the core histones contained nearly equal numbers of acidic and basic amino acid residues and also harbored the majority of hydrophobic residues. This suggested that these regions, which conformed more closely to typical globular proteins in their composition, were likely to comprise the structured core of the nucleosome. However, the idea that the histone "tails" secured DNA via interactions with the major groove was refuted first by experiments showing that both the major and minor grooves of nucleosomal DNA were accessible to chemical reagents, and later by demonstration that DNA from phage T4, in which the major groove is occupied by glucose moieties from glucosylated hydroxymethylcytosine, could be reconstituted into nucleosomes that retained normal physical properties (McGhee and Felsenfeld, 1979, 1982).

Arguing against the idea that the histone N-terminal regions were important to nucleosome structure, but supporting the distinction between the highly basic termini and the more apolar central regions, several studies revealed that physical characteristics of the core particle did not depend on the histone termini. The proteolytic enzyme trypsin cleaves peptide chains at sites following lysine or arginine residues (unless they are followed by proline), and isolated, unfolded histones in low ionic strength solutions are correspondingly digested into small fragments upon extensive trypsinization. In contrast, treatment of chromatin or core particles with trypsin results in a relatively rapid reduction in size of the histones, reflecting loss of the amino-terminal regions of all four histones and short C-terminal regions of H2A and H2B, after which longer exposure to trypsin results in little further change (Fig. 2.5) (Bohm et al., 1981, 1980; Bohm and Crane-Robinson, 1984; Weintraub and Van Lente, 1974). Similar protection of the "globular" regions of the histones was seen after digestion with chymotrypsin (Sollner-Webb et al., 1976), which cleaves the polypeptide chain at sites following aromatic residues; thus, the limited digestion observed was inferred to result from steric protection in the nucleosome rather than some enzymatic peculiarity. These findings indicated the existence of two distinct domains of histones in the nucleosome: an accessible, flexible (see below) region corresponding to the N-termini of all four core histones as well as short C-terminal regions of H2A and H2B, and a protected globular region that was more highly structured and therefore less mobile. Further supporting this notion, the major sites known at this time for acetylation and phosphorylation (from protein sequencing studies) were located in these same N-terminal regions, which were therefore presumed to be more accessible than residues located in the central globular regions (Bohm and Crane-Robinson, 1984).

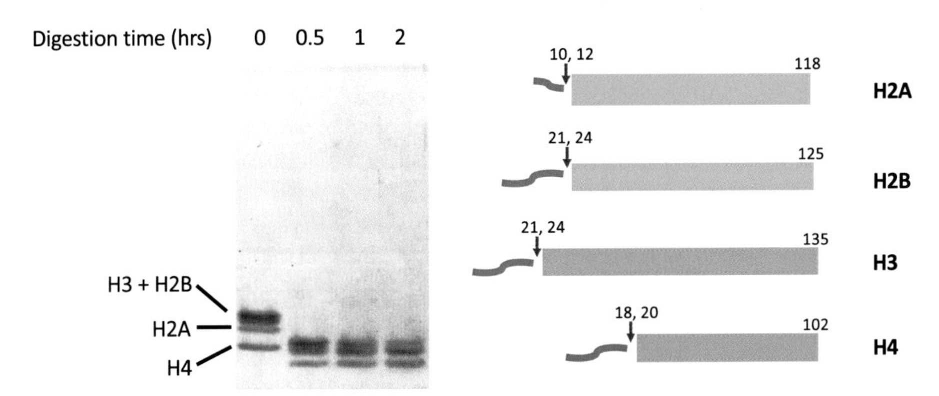

FIG. 2.5 (Left) Trypsinization of nucleosomes yields relatively stable histone fragments. In this experiment, reconstituted minichromosomes were digested for the indicated times with trypsin before stopping the digestion with an inhibitor. (Right) Prominent cleavage sites of nucleosomal histones, based on amino acid analysis and Edman degradation of peptide fragments following trypsinization. The number of amino acids in the full-length peptides is shown at the right for each histone (Bohm and Crane-Robinson, 1984). *(Left) From Morse, R.H., Cantor, C.R., 1986. Effect of trypsinization and histone H5 addition on DNA twist and topology in reconstituted minichromosomes. Nucleic Acids Res. 14, 3293–3310.*

Remarkably, despite the loss of approximately 30% of the histone mass and 40%–50% of the positive charge of the histones after limit trypsinization, the physical properties of the nucleosome do not show major changes (Bohm and Crane-Robinson, 1984). Trypsinized core particles exhibited sedimentation coefficients similar to intact core particles, particularly after accounting for the loss of histone mass, and the stability of intact core particles to increased salt concentrations was unchanged after trypsinization (Ausio et al., 1989; Lilley and Tatchell, 1977; Whitlock Jr. and Simpson, 1977; Whitlock Jr. and Stein, 1978). Enzymatic assays also indicated that nucleosome structure suffered only minor perturbation upon trypsinization: resistance of nucleosomal DNA to MNase exhibited little change, digestion by DNase I still resulted in a 10bp ladder of fragments, and accessibility to hydroxy radical was essentially indistinguishable between intact and trypsinized core particles (Hayes et al., 1991; Lilley and Tatchell, 1977; Morse and Cantor, 1986; Whitlock Jr. and Simpson, 1977). Topology of nucleosomal DNA was also independent of the histone tails: the same supercoiling was observed for closed circular plasmid DNA reconstituted into nucleosomes before and after removal of the histone tails by trypsinization (Morse and Cantor, 1986). Furthermore, histones could be dissociated from trypsinized nucleosomes and reconstituted to yield complexes with sedimentation velocity, DNase I digestion susceptibility, and cross-linking patterns consistent with those of canonical core particles, demonstrating that histones lacking amino terminal regions retained the features needed to assemble with DNA into a complex resembling the nucleosome core particle (Whitlock Jr. and Stein, 1978). Thus, the histone "tails" were apparently dispensable for packaging DNA into chromatin.

All this is not to say that the trypsin-sensitive tails have no role whatsoever in nucleosome structure; some changes in relative sensitivity of specific sites in nucleosomal DNA to DNase I digestion were observed following trypsinization, and the kinetics of digestion by MNase or DNase I were altered as well (Lilley and Tatchell, 1977). The thermal stability of core particles was also substantially altered upon loss of the histone tails, consistent with the histone termini interacting with specific regions of nucleosomal DNA (Ausio et al., 1989). Indeed, the high-resolution determination of the structure of the nucleosome core particle revealed that 14 histone arginine residues interact with the minor groove of DNA where it faces inwards toward the octamer; 4 of these residues derive from the histone tails, and about 1/3 of the tail domains defined in trypsinization experiments exhibited defined structure in the nucleosome (Luger et al., 1997).

Nuclear magnetic resonance (NMR) spectroscopy provided independent evidence for distinct structural roles for the histone N-termini and the globular regions (Cary et al., 1978; Lilley et al., 1977; van Emmerik and van Ingen, 2019). Protein and ^{13}C NMR spectra of small molecules display sharp peaks, or resonances, for individual atoms; these resonances are characterized by their chemical shifts—variable dependence on the magnetic field strength—that depend on the local environment in the molecule in which the atoms giving rise to them reside. In larger molecules, the slow tumbling of the molecule in solution causes a broadening of the resonance lines. Accordingly, comparison of line widths corresponding to ^{1}H or ^{13}C resonances from particular amino acids, measured in native and unfolded protein configurations, can yield information regarding mobility of those residues. Comparison of the ^{1}H and ^{13}C NMR spectra of histones that were in a completely unfolded, random coil configuration (obtained by thermal denaturation) with the spectra of intact chromatin indicated the presence of histone domains that possessed considerable mobility arising from random-coil peptide; these were interpreted as being likely due to N-terminal "tails" that extended from highly structured globular regions (Lilley et al., 1977). It was these NMR experiments that first gave rise to the concept of histone "tails." In another study, selective broadening of peaks due to amino acid residues enriched in or exclusive to the globular region of histones (leucine, isoleucine, valine, arginine) was observed in native core particles when compared to chemically denatured histones, while the remaining sharp peaks, due to amino acids that remained mobile in the core particle, were enriched in lysines and alanines, which are concentrated in the N-termini. These findings led to the inference that N-terminal tails of the histones extended from globular C-terminal regions possessing a defined tertiary structure (Cary et al., 1978). Although the H4 amino terminal region appeared to have decreased mobility in the core particle, this region did not appear similarly restricted in an $(H3\text{-}H4)_2$ tetramer, based on the relatively unaffected peaks due to glycine $\alpha\text{-}CH_2$ groups, 59% of which are in the basic N-terminal third of H3 and H4 (Moss et al., 1976). From this latter result the authors concluded that the N-terminal regions of H3 and H4, and possibly short regions at the C-termini, were likely not involved in formation of the $(H3\text{-}H4)_2$ tetramer. A direct link between the flexible histone termini inferred from NMR studies and the regions accessible to protease was eventually shown by Walker (Walker, 1984). At salt concentrations close to physiological (0.2M NaCl), proton NMR spectra of core chromatin yielded only very broad peaks, indicating that the entire histone polypeptide chains were relatively immobile. Raising the salt concentration to 0.6M, while not causing dissociation of chromatin, resulted in the appearance of several relatively narrow spectral lines. When this experiment was repeated with chromatin that was first subjected to a limit trypsin digest and dialysis to remove the proteolyzed tails, no change in the spectra was observed upon raising the salt concentration from 0.2 to 0.6M, clearly demonstrating that the trypsinized tails were responsible for the peaks observed with intact chromatin, and therefore that these same tails corresponded to the flexible portions of the histones detected by NMR in earlier work.

The model these findings led to, of a structured protein core of the nucleosome comprising the apolar, protease-resistant regions of the histones, with flexible amino terminal (and for H2A and H3, carboxy terminal) tails extending from this structured region, was eventually confirmed by the high-resolution structure of the nucleosome (Luger et al., 1997).

The nucleosome unveiled: High-resolution structure of the core particle

The first detailed information on histone structure in the nucleosome came from X-ray crystallography of the histone octamer in the absence of DNA. Evidence for a stable complex of histones, variously construed as an octamer or heterotypic tetramer, had been obtained from cross-linking studies and from isolation of a non-cross-linked histone complex from chromatin in 2M NaCl (Campbell and Cotter, 1976; Chung et al., 1978; Pardon et al., 1977; Thomas and Butler, 1978; Thomas and Kornberg, 1975; Weintraub et al., 1975). As discussed earlier, Eickmann and Moudrianakis demonstrated that histone complexes extracted from calf thymus chromatin with 2M NaCl at pH7.5 comprised octamers of two copies each of H2A, H2B, H3, and H4 (Eickbush and Moudrianakis, 1978). Subsequently, Moudrianakis and colleagues were able to crystallize octamers from concentrated ammonium sulfate solution, yielding a structure resolved to 3.3Å (Burlingame et al., 1984, 1985). The structure revealed a tripartite organization—a central $(H3\text{-}H4)_2$ tetramer flanked by H2A-H2B dimers—that was consistent with previous biochemical and biophysical data. However, although the histone octamer structure was to provide compelling insight into the structure of the nucleosome, its arrival was marked by controversy due to an analytical gaffe combined with some very bad luck (Wang et al., 1994). The initial report indicated a "size and shape … radically different from those determined in earlier studies" (Burlingame et al., 1985). This was no exaggeration, as the octamer was reported to be a prolate ellipsoid (like an American football or a rugby ball) with a length of 110 Å and width of about 70 Å, as opposed to the wedge-shaped disk (often compared to a hockey puck) with thickness of 55 Å and maximum diameter of 110 Å reported in earlier and contemporaneous low- and medium-resolution diffraction studies of the nucleosome (Burlingame et al., 1985; Finch et al., 1977; Richmond et al., 1984; Uberbacher and Bunick, 1985b). Since the DNA in the latter structure comprised nearly half of the total volume, the two structures appeared incompatible, leading to a fractious exchange in the literature that resolved nothing (Klug et al., 1985; Moudrianakis et al., 1985a,b; Uberbacher and Bunick, 1985a). Klug and colleagues noted that the heavy atom derivative employed in the structural determination occupied a "special position" (with coordinates very close to (1/3, 1/3, 0)) in the crystal, which could lead to inaccurate definition of the boundaries of the protein molecules in the density map (Klug et al., 1985). They were right; it eventually emerged that a redetermination of the heavy atom position, in which it was adjusted to a position 2.27 Å from the initial assignment, sufficed to produce a structure entirely in accord with nucleosome core structures reported previously (Arents et al., 1991; Wang et al., 1994).

Despite this rocky beginning, the revised histone octamer structure, solved to a resolution of 3.1 Å, proved to be a landmark paper, providing unprecedented detail on the structure and interactions of histones in the octamer and allowing a refined modeling of the path of DNA in the nucleosome (Arents et al., 1991; Arents and Moudrianakis, 1993). The four core histones were all shown to be elongated in structure, rather than globular, with extensive contacts between histones H2A and H2B, and between H3 and H4. All 4 histones possess a common structural element, termed the "histone fold," comprising a long central α-helix (~27 amino acids) flanked on either side by a loop segment and a shorter helix (~10 amino acids) (Fig. 2.6). The long central helices of the histones paired as heterodimers—H2A and H2B, and H3 and H4—lay across one another at an angle of about 30 degrees, providing a dimerization interface termed a "handshake" motif (Arents and Moudrianakis, 1993). The presence of the histone fold motif in all four core histones suggested a common evolutionary origin. This notion was supported by subsequent studies that identified this motif in other proteins (Baxevanis et al., 1995; Burley et al., 1997), and was definitively borne out by work showing a highly similar structure adopted by dimers of histone B from the archaeon *Methanothermus fervidus* (Fig. 2.6) (Decanniere et al., 2000; Mattiroli et al., 2017) (see also Chapter 5). This ancient motif serves as a widely used dimerization interface in both DNA-binding and nonbinding complexes, and is closely tied to both chromatin and processes occurring in chromatin. Prominent examples occur in the TFIID and the SAGA complexes. In TFIID, several of the TBP-associated factors (Tafs) contain histone-fold domains and interact to form heterodimers (Patel et al., 2018). The SAGA complex, which like TFIID delivers TBP to facilitate transcriptional activation but also modifies nucleosomal histones to regulate transcription, has at the center of its structure a deformed octamer in which Tafs, which are present in both TFIID and SAGA, as well as the SAGA-specific subunit Spt3, contribute histone fold and handshake-like motifs (Papai et al., 2020; Wang et al., 2020).

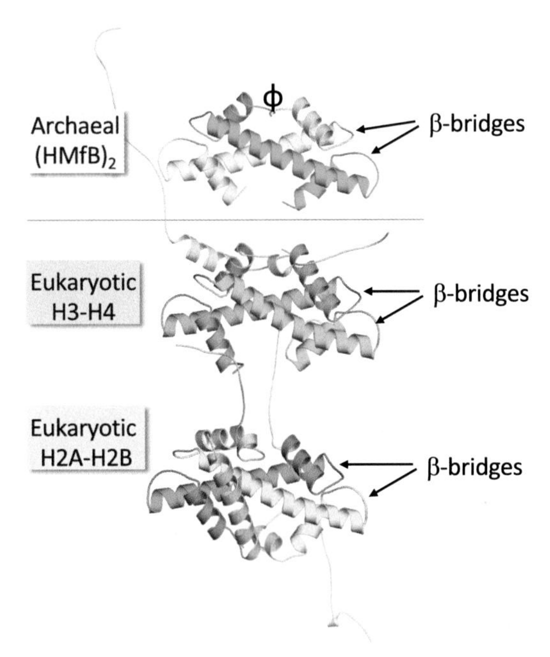

FIG. 2.6 The histone fold. See text for details. *Credit: Courtesy of Karolin Luger.*

Beyond the structures and interactions of the histones, the octamer structure also revealed a remarkable topography of 12 repeated "paired" motifs well suited to organizing DNA (Arents and Moudrianakis, 1993). One motif, termed a parallel β bridge, occurs at each end of the paired long α-helices of the histones, where the short loops connecting the long α-helix with the shorter flanking α-helices are in close proximity (Fig. 2.6). These paired loops each contain two invariant positively charged amino acids (either Lys-Arg or Lys-Lys), and the bridges are arranged on the octamer surface with a spacing consistent with their making close contact with the negatively charged phosphates of the DNA superhelix. Altogether then, the β-bridges had the potential to make 8 contacts with the 14 helical turns of DNA in the nucleosome. The second paired motif occurs once in each histone dimer pair, where the N-terminal ends of the first (most N-terminal) helix of the histones in each pair are in close proximity. These "paired ends of helix I" motifs have positive dipole moments associated with them, and occupy positions that would allow them to contact DNA at an additional 4 locations; thus, between the β-bridges and paired ends of helices, electrostatic contacts could be made with the central 12 of the 14 turns of DNA in the nucleosome core. Huiskamp, who in 1901 showed that the binding of nucleic acid and histone components of nuclein "must probably be expressed as a salt-like one" (Huiskamp, 1901), would surely have marveled at the structural detail underlying this association.

The long-awaited high-resolution structure of the nucleosome core particle, a testament to persistence, was at last realized in 1997 (Fig. 2.7) (Luger et al., 1997). This milestone required the use of a defined DNA sequence and recombinant histones (to avoid heterogeneity due to differential modification state), as well as advances in crystallographic techniques (Rhodes, 1997). The resulting structure confirmed features of the octamer structure, including the histone fold and "handshake" arrangement in the H2A-H2B and H3-H4 dimers (Fig. 2.6), and revealed in detail how the octamer organizes DNA into the nucleosome. Of the 146bp of DNA in the nucleosome, 121bp are organized by the histone fold domains; the remaining DNA at the entry and exit points of the nucleosome is bound by an N-terminal α-helix in H3 that is not part of the H3 histone fold region. Contacts are made between histones and DNA every 10bp, at the sites where the minor groove is at its most inward-facing. These contacts involve hydrogen bonds to phosphates in the DNA chain made by main-chain amide nitrogen atoms near the end of histone α-helices, as well as with basic and hydroxyl side chain groups; nonpolar contacts with deoxyribose groups; electrostatic interactions between phosphate groups and the positively charged dipole at the N-termini of histone α-helices (the "paired end of helix I" motif); and interaction of all 14 inward-facing minor grooves with arginine residues of the histones. Ten of these latter contacts occur through arginines in histone fold domains, with the remaining four occurring through arginines in the histone tails, consistent with the lower mobility of arginine residues compared to lysine residues observed in NMR experiments (Walker, 1984). Recalling that the central base pair of DNA, located at the dyad, is defined as superhelix location zero (SHL0), the central portion of DNA, from SHL+3 to SHL−3, is organized by the $(H3\text{-}H4)_2$ tetramer, while SHL +4 to +6 and −4 to −6 are bound by H2A-H2B dimers. The low impact of DNA sequence on these interactions is compatible with the requirement for packaging the majority of the genome into nucleosomes. The structure also confirmed the DNA distortions observed in the 7Å structure and revealed an average DNA periodicity of 10.2bp/turn, reduced from the 10.54bp/turn observed in solution. This altered periodicity was in remarkable agreement with studies on the helical periodicity obtained using hydroxyl radical and photofootprinting cited earlier (Gale and Smerdon, 1988; Hayes et al., 1991).

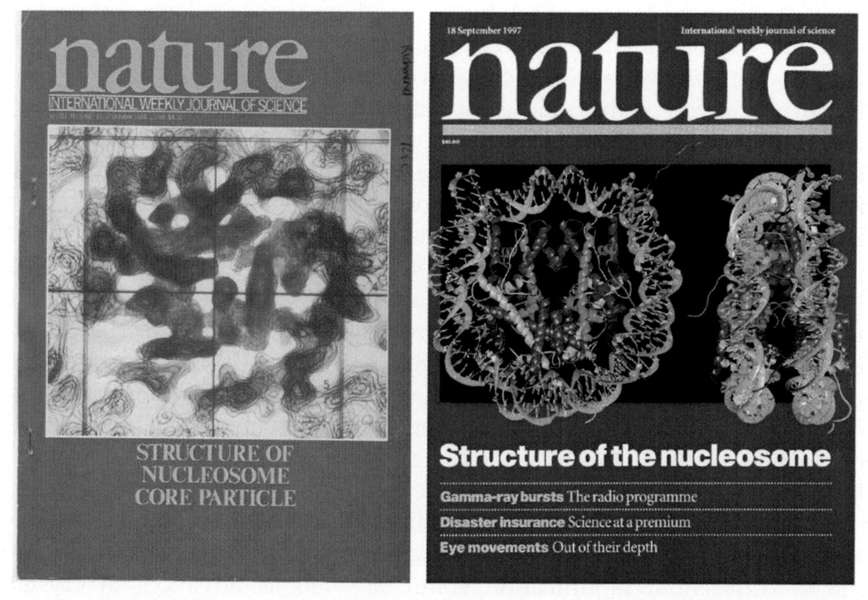

FIG. 2.7 Covers of the journal Nature depicting the structure of the nucleosome core particle solved to 7Å (left, in 1984) and to 2.8Å (right, in 1997).

The high-resolution structure of the nucleosome also provided new insight into the disposition of the histone tails. In addition to the contacts made by arginine residues in the tails with the minor groove of DNA, the structure showed that the N-terminal regions of H2A and H3 extended outward from the octamer between the two gyres of DNA. The H3 tail passes through a channel formed by the minor grooves of the outermost region of DNA and the central (SHL −0.5 to −1) turn, while the H2A tail extends from a region offset from the H3 tails by 20 bp. Electron density for most N-terminal regions of the histones was generally weak, consistent with the random coil nature of these regions inferred from biophysical and proteolytic digestion studies discussed earlier.

One intriguing observation was an interaction between a segment of the H4 amino terminal region with an "acidic patch" formed by H2A and H2B on an adjacent particle. This region of H4 encompasses amino acid residues 16–24 and makes multiple salt bridges and hydrogen bonds via its basic side chains (three Lys and one Arg residue) with negatively charged side chains in the acidic patch, six belonging to H2A and two to H2B. Although this internucleosomal interaction could have been passed off as a crystallization artifact, early studies (discussed in more detail in Chapter 3) had demonstrated that the histone tails were important for higher order folding of chromatin (Allan et al., 1982). This suggested that this internucleosomal contact could contribute to higher order folding of chromatin, and that acetylation of the H4 amino terminal region, which alters its charge, could affect such interactions and thereby modify higher order chromatin structure. This prediction proved to be accurate, as will be discussed in Chapter 3.

The high-resolution structure of the nucleosome was far from a lucky fluke, but rather was the result, as should be clear, of years of hard work and innovation. It would be overstating the case to say that obtaining such structures is now routine, but the groundbreaking work of Tim Richmond, Karolin Luger, and colleagues certainly set the stage for obtaining structures of variant nucleosomes alone and in complex with interacting proteins. Since the 1997 nucleosome structure was published, over 150 nucleosome core particle (NCP) structures have been deposited in the Protein Data Bank, with over 30 of these comprising NCPs in complex with partner proteins or peptides (Kale et al., 2019). The structures include reconstituted nucleosome core particles incorporating variant histones or linker histones, NCPs in complex with specifically interacting nonhistone proteins, nucleosomes in the process of being transcribed by RNA polymerase II, and nucleosome dimers and tetramers. All these structures have of course provided important insights into chromatin structure and function.

References

Allan, J., Harborne, N., Rau, D.C., Gould, H., 1982. Participation of core histone "tails" in the stabilization of the chromatin solenoid. J. Cell Biol. 93, 285–297.

Arents, G., Moudrianakis, E.N., 1993. Topography of the histone octamer surface: repeating structural motifs utilized in the docking of nucleosomal DNA. Proc. Natl. Acad. Sci. U. S. A. 90, 10489–10493.

Arents, G., Burlingame, R.W., Wang, B.C., Love, W.E., Moudrianakis, E.N., 1991. The nucleosomal core histone octamer at 3.1 A resolution: a tripartite protein assembly and a left-handed superhelix. Proc. Natl. Acad. Sci. U. S. A. 88, 10148–10152.

Ausio, J., Dong, F., van Holde, K.E., 1989. Use of selectively trypsinized nucleosome core particles to analyze the role of the histone "tails" in the stabilization of the nucleosome. J. Mol. Biol. 206, 451–463.

Bakayev, V.V., Melnickov, A.A., Osicka, V.D., Varshausky, A.J., 1975. Studies on chromatin. II. Isolation and characterization of chromatin subunits. Nucleic Acids Res. 2, 1401–1419.

Baldwin, J.P., Boseley, P.G., Bradbury, E.M., Ibel, K., 1975. The subunit structure of the eukaryotic chromosome. Nature 253, 245–249.

Bates, D.L., Thomas, J.O., 1981. Histones H1 and H5: one or two molecules per nucleosome? Nucleic Acids Res. 9, 5883–5894.

Baxevanis, A.D., Arents, G., Moudrianakis, E.N., Landsman, D., 1995. A variety of DNA-binding and multimeric proteins contain the histone fold motif. Nucleic Acids Res. 23, 2685–2691.

Bina-Stein, M., 1978. Folding of 140-base pair length DNA by a core of arginine-rich histones. J. Biol. Chem. 253, 5213–5219.

Bohm, L., Crane-Robinson, C., 1984. Proteases as structural probes for chromatin: the domain structure of histones. Biosci. Rep. 4, 365–386.

Bohm, L., Crane-Robinson, C., Sautiere, P., 1980. Proteolytic digestion studies of chromatin core-histone structure. Identification of a limit peptide of histone H2A. Eur. J. Biochem. 106, 525–530.

Bohm, L., Briand, G., Sautiere, P., Crane-Robinson, C., 1981. Proteolytic digestion studies of chromatin core-histone structure. Identification of the limit peptides of histones H3 and H4. Eur. J. Biochem. 119, 67–74.

Britten, R.J., Davidson, E.H., 1969. Gene regulation for higher cells: a theory. Science 165, 349–357.

Burley, S.K., Xie, X., Clark, K.L., Shu, F., 1997. Histone-like transcription factors in eukaryotes. Curr. Opin. Struct. Biol. 7, 94–102.

Burlingame, R.W., Love, W.E., Moudrianakis, E.N., 1984. Crystals of the octameric histone core of the nucleosome. Science 223, 413–414.

Burlingame, R.W., Love, W.E., Wang, B.C., Hamlin, R., Nguyen, H.X., Moudrianakis, E.N., 1985. Crystallographic structure of the octameric histone core of the nucleosome at a resolution of 3.3 A. Science 228, 546–553.

Camerini-Otero, R.D., Felsenfeld, G., 1977. Supercoiling energy and nucleosome formation: the role of the arginine-rich histone kernel. Nucleic Acids Res. 4, 1159–1181.

Camerini-Otero, R.D., Sollner-Webb, B., Felsenfeld, G., 1976. The organization of histones and DNA in chromatin: evidence for an arginine-rich histone kernel. Cell 8, 333–347.

Campbell, A.M., Cotter, R.I., 1976. The molecular weight of nucleosome protein by laser light scattering. FEBS Lett. 70, 209–211.

Cantor, C.R., 1976. A possible explanation for the nuclease limit digestion pattern of chromatin. Proc. Natl. Acad. Sci. U. S. A. 73, 3391–3393.

Cary, P.D., Moss, T., Bradbury, E.M., 1978. High-resolution proton-magnetic-resonance studies of chromatin core particles. Eur. J. Biochem. 89, 475–482.

Chalkley, R., Hunter, C., 1975. Histone-histone propinquity by aldehyde fixation of chromatin. Proc. Natl. Acad. Sci. U. S. A. 72, 1304–1308.

Chung, S.Y., Hill, W.E., Doty, P., 1978. Characterization of the histone core complex. Proc. Natl. Acad. Sci. U. S. A. 75, 1680–1684.

Crick, F., 1971. General model for the chromosomes of higher organisms. Nature 234, 25–27.

Crick, F.H., Klug, A., 1975. Kinky helix. Nature 255, 530–533.

D'Anna Jr., J.A., Isenberg, I., 1974. A histone cross-complexing pattern. Biochemistry 13, 4992–4997.

Decanniere, K., Babu, A.M., Sandman, K., Reeve, J.N., Heinemann, U., 2000. Crystal structures of recombinant histones HMfA and HMfB from the hyperthermophilic archaeon *Methanothermus fervidus*. J. Mol. Biol. 303, 35–47.

DeLange, R.J., Fambrough, D.M., Smith, E.L., Bonner, J., 1969a. Calf and pea histone IV. 3. Complete amino acid sequence of pea seedling histone IV; comparison with the homologous calf thymus histone. J. Biol. Chem. 244, 5669–5679.

DeLange, R.J., Fambrough, D.M., Smith, E.L., Bonner, J., 1969b. Calf and pea histone IV. II. The complete amino acid sequence of calf thymus histone IV; presence of epsilon-N-acetyllysine. J. Biol. Chem. 244, 319–334.

Draizen, E.J., Shaytan, A.K., Marino-Ramirez, L., Talbert, P.B., Landsman, D., Panchenko, A.R., 2016. HistoneDB 2.0: a histone database with variants—an integrated resource to explore histones and their variants. Database (Oxford) 2016, baw014.

Eickbush, T.H., Moudrianakis, E.N., 1978. The histone core complex: an octamer assembled by two sets of protein-protein interactions. Biochemistry 17, 4955–4964.

Finch, J.T., Lutter, L.C., Rhodes, D., Brown, R.S., Rushton, B., Levitt, M., Klug, A., 1977. Structure of nucleosome core particles of chromatin. Nature 269, 29–36.

Gale, J.M., Smerdon, M.J., 1988. Photofootprint of nucleosome core DNA in intact chromatin having different structural states. J. Mol. Biol. 204, 949–958.

Germond, J.E., Hirt, B., Oudet, P., Gross-Bellark, M., Chambon, P., 1975. Folding of the DNA double helix in chromatin-like structures from simian virus 40. Proc. Natl. Acad. Sci. U. S. A. 72, 1843–1847.

Hayashi, K., Hofstaetter, T., Yakuwa, N., 1978. Asymmetry of chromatin subunits probed with histone H1 in an H1-DNA complex. Biochemistry 17, 1880–1883.

Hayes, J.J., Tullius, T.D., Wolffe, A.P., 1990. The structure of DNA in a nucleosome. Proc. Natl. Acad. Sci. U. S. A. 87, 7405–7409.

Hayes, J.J., Clark, D.J., Wolffe, A.P., 1991. Histone contributions to the structure of DNA in the nucleosome. Proc. Natl. Acad. Sci. U. S. A. 88, 6829–6833.

Hjelm, R.P., Kneale, G.G., Sauau, P., Baldwin, J.P., Bradbury, E.M., Ibel, K., 1977. Small angle neutron scattering studies of chromatin subunits in solution. Cell 10, 139–151.

Hogan, M.E., Rooney, T.F., Austin, R.H., 1987. Evidence for kinks in DNA folding in the nucleosome. Nature 328, 554–557.

Huiskamp, W., 1901. Ueber die Elektrolyse der Salze des Nucleohistons und Histons. Hoppe Seylers Z. Physiol. Chem. 34, 32–54.

Hyde, J.E., Walker, I.O., 1975. A model for chromatin sub-structure incorporating symmetry considerations of histone oligomers. Nucleic Acids Res. 2, 405–421.

Ilyin, Y.V., Varshavsky, A.Y., Mickelsaar, U.N., Georgiev, G.P., 1971. Studies on deoxyribonucleoprotein structure. Redistribution of proteins in mixtures of deoxyribonucleoproteins, DNA and RNA. Eur. J. Biochem. 22, 235–245.

Isenberg, I., 1978. Protein-protein interactions of histones. In: Busch, H. (Ed.), The Cell Nucleus. Academic Press, pp. 135–154.

Kale, S., Goncearenco, A., Markov, Y., Landsman, D., Panchenko, A.R., 2019. Molecular recognition of nucleosomes by binding partners. Curr. Opin. Struct. Biol. 56, 164–170.

Keller, W., 1975. Determination of the number of superhelical turns in simian virus 40 DNA by gel electrophoresis. Proc. Natl. Acad. Sci. U. S. A. 72, 4876–4880.

Kelley, R.I., 1973. Isolation of a histone IIb1-IIb2 complex. Biochem. Biophys. Res. Commun. 54, 1588–1594.

Klug, A., Lutter, L.C., 1981. The helical periodicity of DNA on the nucleosome. Nucleic Acids Res. 9, 4267–4283.

Klug, A., Rhodes, D., Smith, J., Finch, J.T., Thomas, J.O., 1980. A low resolution structure for the histone core of the nucleosome. Nature 287, 509–516.

Klug, A., Finch, J.T., Richmond, T.J., 1985. Crystallographic structure of the octamer histone core of the nucleosome. Science 229, 1109–1110.

Kornberg, R.D., 1974. Chromatin structure: a repeating unit of histones and DNA. Science 184, 868–871.

Kornberg, R.D., 1977. Structure of chromatin. Annu. Rev. Biochem. 46, 931–954.

Kornberg, R.D., Thomas, J.O., 1974. Chromatin structure; oligomers of the histones. Science 184, 865–868.

Langmore, J.P., Wooley, J.C., 1975. Chromatin architecture: investigation of a subunit of chromatin by dark field electron microscopy. Proc. Natl. Acad. Sci. U. S. A. 72, 2691–2695.

Levitt, M., 1978. How many base-pairs per turn does DNA have in solution and in chromatin? Some theoretical calculations. Proc. Natl. Acad. Sci. U. S. A. 75, 640–644.

Lilley, D.M., Tatchell, K., 1977. Chromatin core particle unfolding induced by tryptic cleavage of histones. Nucleic Acids Res. 4, 2039–2055.

Lilley, D.M., Pardon, J.F., Richards, B.M., 1977. Structural investigations of chromatin core protein by nuclear magnetic resonance. Biochemistry 16, 2853–2860.

Liu, L.F., Wang, J.C., 1978. DNA-DNA gyrase complex: the wrapping of the DNA duplex outside the enzyme. Cell 15, 979–984.

Luger, K., Mader, A.W., Richmond, R.K., Sargent, D.F., Richmond, T.J., 1997. Crystal structure of the nucleosome core particle at 2.8 A resolution [see comments]. Nature 389, 251–260.

Lutter, L.C., 1978. Kinetic analysis of deoxyribonuclease I cleavages in the nucleosome core: evidence for a DNA superhelix. J. Mol. Biol. 124, 391–420.

Martinson, H.G., McCarthy, B.J., 1975. Histone-histone associations within chromatin. Cross-linking studies using tetranitromethane. Biochemistry 14, 1073–1078.

Maskell, D.P., Renault, L., Serrao, E., Lesbats, P., Matadeen, R., Hare, S., Lindemann, D., Engelman, A.N., Costa, A., Cherepanov, P., 2015. Structural basis for retroviral integration into nucleosomes. Nature 523, 366–369.
Mattiroli, F., Bhattacharyya, S., Dyer, P.N., White, A.E., Sandman, K., Burkhart, B.W., Byrne, K.R., Lee, T., Ahn, N.G., Santangelo, T.J., et al., 2017. Structure of histone-based chromatin in Archaea. Science 357, 609–612.
McGhee, J.D., Felsenfeld, G., 1979. Reaction of nucleosome DNA with dimethyl sulfate. Proc. Natl. Acad. Sci. U. S. A. 76, 2133–2137.
McGhee, J.D., Felsenfeld, G., 1982. Reconstitution of nucleosome core particles containing glucosylated DNA. J. Mol. Biol. 158, 685–698.
Mirzabekov, A.D., Shick, V.V., Belyavsky, A.V., Bavykin, S.G., 1978. Primary organization of nucleosome core particle of chromatin: sequence of histone arrangement along DNA. Proc. Natl. Acad. Sci. U. S. A. 75, 4184–4188.
Morse, R.H., 1991. Topoisomer heterogeneity of plasmid chromatin in living cells. J. Mol. Biol. 222, 133–137.
Morse, R.H., Cantor, C.R., 1986. Effect of trypsinization and histone H5 addition on DNA twist and topology in reconstituted minichromosomes. Nucleic Acids Res. 14, 3293–3310.
Morse, R.H., Simpson, R.T., 1988. DNA in the nucleosome. Cell 54, 285–287.
Moss, T., Cary, P.D., Crane-Robinson, C., Bradbury, E.M., 1976. Physical studies on the H3/H4 histone tetramer. Biochemistry 15, 2261–2267.
Moudrianakis, E.N., Love, W.E., Burlingame, R.W., 1985a. Crystallographic structure of the octamer histone core of the nucleosome. Science 229, 1113.
Moudrianakis, E.N., Love, W.E., Wang, B.C., Xuong, N.G., Burlingame, R.W., 1985b. Crystallographic structure of the octamer histone core of the nucleosome. Science 229, 1110–1112.
Muller, H.P., Pryciak, P.M., Varmus, H.E., 1993. Retroviral integration machinery as a probe for DNA structure and associated proteins. Cold Spring Harb. Symp. Quant. Biol. 58, 533–541.
Nikitina, T., Norouzi, D., Grigoryev, S.A., Zhurkin, V.B., 2017. DNA topology in chromatin is defined by nucleosome spacing. Sci. Adv. 3, e1700957.
Noll, M., 1974. Internal structure of the chromatin subunit. Nucleic Acids Res. 1, 1573–1578.
Noll, M., 1977. DNA folding in the nucleosome. J. Mol. Biol. 116, 49–71.
Noll, M., Kornberg, R.D., 1977. Action of micrococcal nuclease on chromatin and the location of histone H1. J. Mol. Biol. 109, 393–404.
Ogawa, Y., Quagliarotti, G., Jordan, J., Taylor, C.W., Starbuck, W.C., Busch, H., 1969. Structural analysis of the glycine-rich, arginine-rich histone. 3. Sequence of the amino-terminal half of the molecule containing the modified lysine residues and the total sequence. J. Biol. Chem. 244, 4387–4392.
Ohlenbusch, H.H., Olivera, B.M., Tuan, D., Davidson, N., 1967. Selective dissociation of histones from calf thymus nucleoprotein. J. Mol. Biol. 25, 299–315.
Olins, A.L., Olins, D.E., 1974. Spheroid chromatin units (v bodies). Science 183, 330–332.
Oudet, P., Gross-Bellard, M., Chambon, P., 1975. Electron microscopic and biochemical evidence that chromatin structure is a repeating unit. Cell 4, 281–300.
Oudet, P., Germond, J.E., Sures, M., Gallwitz, D., Bellard, M., Chambon, P., 1978. Nucleosome structure I: all four histones, H2A, H2B, H3, and H4, are required to form a nucleosome, but an H3-H4 subnucleosomal particle is formed with H3-H4 alone. Cold Spring Harb. Symp. Quant. Biol. 42 (Pt 1), 287–300.
Papai, G., Frechard, A., Kolesnikova, O., Crucifix, C., Schultz, P., Ben-Shem, A., 2020. Structure of SAGA and mechanism of TBP deposition on gene promoters. Nature 577, 711–716.
Pardon, J.F., Worcester, D.L., Wooley, J.C., Tatchell, K., Van Holde, K.E., Richards, B.M., 1975. Low-angle neutron scattering from chromatin subunit particles. Nucleic Acids Res. 2, 2163–2176.
Pardon, J.F., Worcester, D.L., Wooley, J.C., Cotter, R.I., Lilley, D.M., Richards, R.M., 1977. The structure of the chromatin core particle in solution. Nucleic Acids Res. 4, 3199–3214.
Patel, A.B., Louder, R.K., Greber, B.J., Grunberg, S., Luo, J., Fang, J., Liu, Y., Ranish, J., Hahn, S., Nogales, E., 2018. Structure of human TFIID and mechanism of TBP loading onto promoter DNA. Science 362, eaau8872.
Pederson, D.S., Venkatesan, M., Thoma, F., Simpson, R.T., 1986. Isolation of an episomal yeast gene and replication origin as chromatin. Proc. Natl. Acad. Sci. U. S. A. 83, 7206–7210.
Prunell, A., 1998. A topological approach to nucleosome structure and dynamics: the linking number paradox and other issues. Biophys. J. 74, 2531–2544.
Pruss, D., Bushman, F.D., Wolffe, A.P., 1994. Human immunodeficiency virus integrase directs integration to sites of severe DNA distortion within the nucleosome core. Proc. Natl. Acad. Sci. U. S. A. 91, 5913–5917.

Rhodes, D., 1997. Chromatin structure. The nucleosome core all wrapped up. Nature 389, 231–233.
Rhodes, D., Klug, A., 1980. Helical periodicity of DNA determined by enzyme digestion. Nature 286, 573–578.
Richmond, T.J., Finch, J.T., Rushton, B., Rhodes, D., Klug, A., 1984. Structure of the nucleosome core particle at 7 A resolution. Nature 311, 532–537.
Roark, D.E., Geoghegan, T.E., Keller, G.H., 1974. A two-subunit histone complex from calf thymus. Biochem. Biophys. Res. Commun. 59, 542–547.
Sahasrabuddhe, C.G., Van Holde, K.E., 1974. The effect of trypsin on nuclease-resistant chromatin fragments. J. Biol. Chem. 249, 152–156.
Segura, J., Joshi, R.S., Diaz-Ingelmo, O., Valdes, A., Dyson, S., Martinez-Garcia, B., Roca, J., 2018. Intracellular nucleosomes constrain a DNA linking number difference of −1.26 that reconciles the Lk paradox. Nat. Commun. 9, 3989.
Shure, M., Vinograd, J., 1976. The number of superhelical turns in native virion SV40 DNA and minicol DNA determined by the band counting method. Cell 8, 215–226.
Simpson, R.T., 1978. Structure of the chromatosome, a chromatin particle containing 160 base pairs of DNA and all the histones. Biochemistry 17, 5524–5531.
Simpson, R.T., Kunzler, P., 1979. Chromatin and core particles formed from the inner histones and synthetic polydeoxyribonucleotides of defined sequence. Nucleic Acids Res. 6, 1387–1415.
Simpson, R.T., Whitlock, J.P., 1976. Mapping DNAase l-susceptible sites in nucleosomes labeled at the 5′ ends. Cell 9, 347–353.
Simpson, R.T., Thoma, F., Brubaker, J.M., 1985. Chromatin reconstituted from tandemly repeated cloned DNA fragments and core histones: a model system for study of higher order structure. Cell 42, 799–808.
Sobell, H.M., Tsai, C.C., Gilbert, S.G., Jain, S.C., Sakore, T.D., 1976. Organization of DNA in chromatin. Proc. Natl. Acad. Sci. U. S. A. 73, 3068–3072.
Sollner-Webb, B., Felsenfeld, G., 1975. A comparison of the digestion of nuclei and chromatin by staphylococcal nuclease. Biochemistry 14, 2915–2920.
Sollner-Webb, B., Camerini-Otero, R.D., Felsenfeld, G., 1976. Chromatin structure as probed by nucleases and proteases: evidence for the central role of histones H3 and H4. Cell 9, 179–193.
Suck, D., Lahm, A., Oefner, C., 1988. Structure refined to 2A of a nicked DNA octanucleotide complex with DNase I. Nature 332, 464–468.
Sussman, J.L., Trifonov, E.N., 1978. Possibility of nonkinked packing of DNA in chromatin. Proc. Natl. Acad. Sci. U. S. A. 75, 103–107.
Tatchell, K., Van Holde, K.E., 1977. Reconstitution of chromatin core particles. Biochemistry 16, 5295–5303.
Thomas, J.O., Butler, P.J., 1978. The nucleosome core protein. Cold Spring Harb. Symp. Quant. Biol. 42 (Pt 1), 119–125.
Thomas, J.O., Kornberg, R.D., 1975. An octamer of histones in chromatin and free in solution. Proc. Natl. Acad. Sci. U. S. A. 72, 2626–2630.
Tullius, T.D., Dombroski, B.A., 1985. Iron(II) EDTA used to measure the helical twist along any DNA molecule. Science 230, 679–681.
Uberbacher, E.C., Bunick, G.J., 1985a. Crystallographic structure of the octamer histone core of the nucleosome. Science 229, 1112–1113.
Uberbacher, E.C., Bunick, G.J., 1985b. X-ray structure of the nucleosome core particle. J. Biomol. Struct. Dyn. 2, 1033–1055.
van der Westhuyzen, D.R., von Holt, C., 1971. A new procedure for the isolation and fractionation of histones. FEBS Lett. 14, 333–337.
van Emmerik, C.L., van Ingen, H., 2019. Unspinning chromatin: revealing the dynamic nucleosome landscape by NMR. Prog. Nucl. Magn. Reson. Spectrosc. 110, 1–19.
Van Holde, K.E., 1988. Chromatin. Springer-Verlag.
Van Holde, K.E., Sahasrabuddhe, C.G., Shaw, B.R., 1974. A model for particulate structure in chromatin. Nucleic Acids Res. 1, 1579–1586.
Van Holde, K.E., Shaw, B.R., Lohr, D., Herman, T.M., Kovacic, R.T., 1975. Proc. Tenth FEBS Meeting. vol. 38 American Elsevier, North Holland.
Varshavsky, A.J., Bakayev, V.V., Georgiev, G.P., 1976. Heterogeneity of chromatin subunits in vitro and location of histone H1. Nucleic Acids Res. 3, 477–492.
Walker, I.O., 1984. Differential dissociation of histone tails from core chromatin. Biochemistry 23, 5622–5628.

Wang, B.C., Rose, J., Arents, G., Moudrianakis, E.N., 1994. The octameric histone core of the nucleosome. Structural issues resolved. J. Mol. Biol. 236, 179–188.
Wang, H., Dienemann, C., Stutzer, A., Urlaub, H., Cheung, A.C.M., Cramer, P., 2020. Structure of the transcription coactivator SAGA. Nature 577, 717–720.
Weintraub, H., Van Lente, F., 1974. Dissection of chromosome structure with trypsin and nucleases. Proc. Natl. Acad. Sci. U. S. A. 71, 4249–4253.
Weintraub, H., Palter, K., Van Lente, F., 1975. Histones H2a, H2b, H3, and H4 form a tetrameric complex in solutions of high salt. Cell 6, 85–110.
White, J.H., Cozzarelli, N.R., Bauer, W.R., 1988. Helical repeat and linking number of surface-wrapped DNA. Science 241, 323–327.
Whitlock Jr., J.P., Simpson, R.T., 1976. Removal of histone H1 exposes a fifty base pair DNA segment between nucleosomes. Biochemistry 15, 3307–3314.
Whitlock Jr., J.P., Simpson, R.T., 1977. Localization of the sites along nucleosome DNA which interact with NH2-terminal histone regions. J. Biol. Chem. 252, 6516–6520.
Whitlock Jr., J.P., Stein, A., 1978. Folding of DNA by histones which lack their NH2-terminal regions. J. Biol. Chem. 253, 3857–3861.
Zhang, J., 2000. Protein-length distributions for the three domains of life. Trends Genet. 16, 107–109.

CHAPTER

3

Chromatin compaction

Abbreviations

3C	chromosome conformation capture
CTCF	CCCTC binding factor
EM	electron microscopy
ES cell	embryonic stem cell
FISH	fluorescent in situ hybridization
HP1	heterochromatin protein 1
PCR	polymerase chain reaction
PEV	position effect variegation
SAXS	small angle X-ray scattering
TAD	topologically associated domain
TEACl	triethanolamine chloride
TPE	telomere position effect

Chromosomes were first identified in the late 1800s, but before the elucidation of DNA as the central component of these structures, their nature and composition remained a mystery. Even after the double-helical nature of DNA was determined, an enormous gap remained between models of chromosome structure and data that would support or refute those models. The emergence of the nucleosome as the primary building block of chromatin allowed questions concerning the structure and function of chromosomes to be formulated with a specificity and level of detail that was not previously possible. The nucleosome paradigm provided a new foundation for models of chromosome structure: any model envisioned for the high-level folding of DNA into chromosomes must begin not with a naked polymer of nucleotides, but with a chain of nucleosomes.

For the first decades "postnucleosome," chromatin folding was studied primarily using direct visualization by electron microscopy (EM), and by physical methods including scattering of X-rays, neutrons, and light; circular dichroism; electric field dichroism; and hydrodynamic methods. Today, the principal methods used to study "higher-order" chromatin structure include EM (especially cryo-EM), X-ray crystallography, hydrodynamic approaches, NMR, and molecular modeling. Advances in obtaining high-resolution structures using cryo-EM have rendered it an especially important tool in such studies, while imaging studies have benefited greatly from the introduction of novel fluorophores and improved detection technology. Some of the older methods, such as dichroism studies, have largely fallen out of favor. Other approaches, used less widely, include atomic force microscopy and single molecule studies using techniques such as optical tweezers (Cui and Bustamante,

Chromatin. **https://doi.org/10.1016/B978-0-12-814809-9.00005-6**

2000; Leuba et al., 1994). Most recently, high throughput techniques combined with chromosome conformation capture (3C) methods have yielded major new insights into the spatial arrangement of chromosomes in intact nuclei.

(A note regarding the term "higher-order chromatin structure": as noted by Grigoryev and Woodcock, this term is sometimes tacitly interpreted to mean that folded chromatin is highly ordered (Grigoryev and Woodcock, 2012). However, as we shall see, chromatin folding in vivo is quite heterogeneous, and the most condensed structures, such as metaphase chromosomes, are poorly understood. Thus, "higher-order structure" (or "folding") should be understood in terms of its intended meaning, in which "order" is used in the sense as in "order of magnitude." Even this usage could be viewed as questionable, insofar as it implies a regular hierarchy of folded structures, as described in textbook views of chromatin structure (Fig. 3.1); structural studies performed over the past decade, as discussed later, argue against such a regular hierarchy in most cell types (Fussner et al., 2011; Hansen, 2012; van Holde and Zlatanova, 1995)).

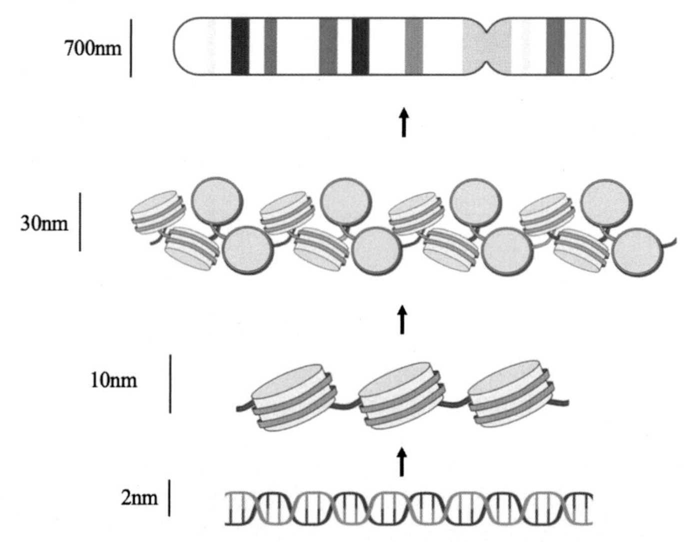

FIG. 3.1 "Textbook" version of chromatin folding, in which compaction of DNA leading to the highly condensed metaphase chromosome is envisioned as occurring in distinct, hierarchical stages. *Credit: Courtesy of Elisabeth Knoll.*

Early investigations of chromatin folding were mostly performed using fragments of chromatin isolated from lysed cell nuclei. These methods could only provide low-resolution information on chromatin structure, and the use of isolated chromatin fragments was open to criticism regarding artifacts due to methods of preparation, particularly for studies using transmission electron microscopy (see the incisive and fairly merciless critique by van Holde and Zlatanova (1995)). Increased sophistication in the application of these methods and the generation and controlled condensation of chromatin have allowed substantial progress to be made in the understanding of mechanisms by which chromatin can fold into higher-order structures. However, as we shall see, how chromatin *can* fold and how it actually *does* fold in cells are not necessarily the same. In this chapter, we outline the progress that has been made in understanding chromatin folding since the emergence of the nucleosome model. This is a very large topic, and the review provided here is necessarily incomplete. Rather, it is intended to acquaint the reader with an outline of work in this area, and an appreciation of the strides that have been made in the decades since the nucleosome model was first proposed.

Early observations

The nucleosome reduces the effective length of linear DNA only about fivefold, while compressing 1–3 m of DNA into a metazoan cell nucleus requires compaction of over four orders of magnitude. It was thus recognized early on that multiple layers of additional folding would be required to generate the compact structure of the metaphase chromosome (Ris and Kubai, 1970). Early investigations focused on the compaction of extended nucleosome arrays, isolated as chromatin fragments from lysed nuclei. Studies using isolated chromatin fragments showed that linear chains of nucleosomes in low salt solution could be visualized by electron microscopy, as in the original studies revealing the "particles on a string" configuration (Olins and Olins, 1974; Woodcock, 1973). These chains underwent compaction in the presence of monovalent or divalent cations (Mg^{+2} or Na^{+}) to form fibers of diameter 25–35 nm (Finch and Klug, 1976; Renz et al., 1977). (Early literature generally referred to the "300 Å fiber"; this eventually gave way to common usage of the "30 nm fiber.") These structures, formed and examined in vitro, were of great interest as they resembled structures found in whole-mount preparations of chromosomes and sectioned nuclei (Davies, 1968; Ris and Kubai, 1970); fibers 25–30 nm in diameter had also been observed in chromosome spreads from newt erythrocytes and from cultured human cells as early as 1966 (Gall, 1966). Moreover, the 30 nm fiber formed in vitro at physiologically relevant salt concentrations (about 150 mM monovalent cation; usually Na^{+}), further supporting its likely relevance to naturally occurring chromosomes.

An influential study by Fritz Thoma and colleagues began with the isolation of chromatin fragments from rat liver chromatin following light MNase digestion, yielding fragments comprising 20–100 nucleosomes, which were then fixed in glutaraldehyde in buffers of various ionic strengths (Thoma and Koller, 1977, 1981; Thoma et al., 1979). The structure and inferred folding of these fragments, and the effects of cations and linker histone H1, were then systematically examined by electron microscopy. What resulted were compelling images of highly regular structures of chromatin in varying degrees of compaction (Fig. 3.2). Perhaps the most remarkable image was of chromatin containing histone H1 at very low ionic strength (1 mM triethanolamine chloride (TEACl), 0.2 mM EDTA), where a zigzag fiber of nucleosomes was observed (Fig. 3.2A). At slightly higher ionic strength (5 mM TEACl), a zigzag arrangement of nucleosomes was still observed, but the nucleosomes were more closely packed and linker DNA was mostly obscured (Fig. 3.2B). These structures contrasted with those seen in chromatin from which H1 was removed by use of an ion exchange resin. At 5 mM TEACl, the DNA in H1-depleted chromatin appeared to enter and exit on opposite sides of the nucleosome particles, whereas in the zigzag fibers of H1-containing chromatin the DNA entry and exit points occur close together (Fig. 3.2A). At low ionic strength (1 mM TEACl), the particle structures characteristic of nucleosomes were lost from H1-depleted chromatin, supporting the idea that linker histone stabilized nucleosome structure, as had been proposed from biochemical studies (Simpson, 1978). Finally, at ionic strengths from 20 to 100 mM (approaching physiologically relevant ionic strength), H1-containing chromatin condensed into irregular

25–35 nm fibers (Fig. 3.2D and F). These structures were not observed for H1-depleted chromatin, which instead aggregated to form irregular clumps.

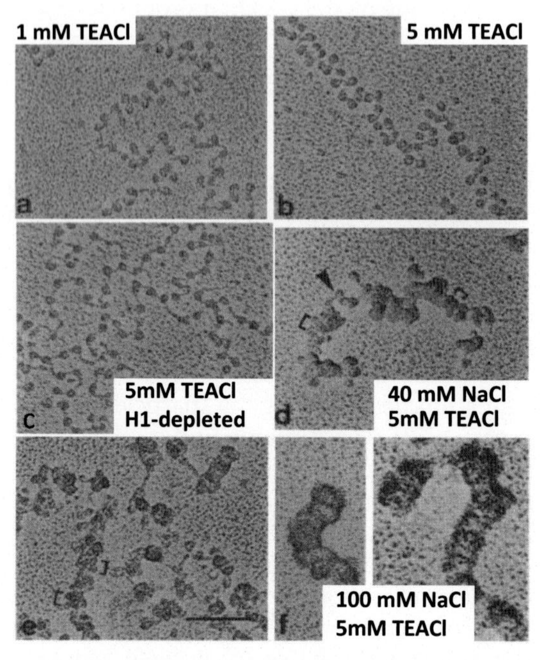

FIG. 3.2 EM collage of chromatin fibers. Rat liver chromatin fibers were fixed at (A) 1 mM TEACl, (B and C) 5 mM TEACl, (D and E) 5 mM TEACl plus 40 mM NaCl, or 5 mM TEACl plus 100 mM NaCl, with the inclusion of 0.2 mM EDTA at pH 7.0 for all samples. Linker histone was depleted prior to fixation of the sample in panel (C). Panel (E) is the same as (D), but freeze-dried after absorption, washed in water, dehydrated in ethanol, and air-dried. The scale bar in (E) is 100 nm. *Modified from Thoma, F., Koller, T., Klug, A., 1979. Involvement of histone H1 in the organization of the nucleosome and of the salt-dependent superstructures of chromatin. J. Cell Biol. 83, 403–427.*

The apparent regularity of the structures observed by Thoma and colleagues in the EM may have exerted an outsized influence on the field for the next several decades, as enormous efforts were made to ascertain the structure of the 30 nm fiber which, it now appears, may not be a major component of metaphase chromosomes (Hansen, 2012). Nonetheless, these investigations yielded important insights into the principles governing the folding of nucleosome arrays, showing that the chromatin fiber underwent two major transitions, the first from a zigzag fiber with extended linkers to a flat ribbon with a similar arrangement of nucleosomes, and the second from the flat ribbon to a condensed, somewhat irregular 30 nm fiber. Linker histone stabilized the nucleosome at low ionic strength by binding near the entry and exit points of DNA in the nucleosome and appeared to be essential to the formation of a well-organized 30 nm fiber at higher ionic strength.

Subsequent work built on these results and benefited from the introduction of regular nucleosome arrays reconstituted in vitro and the increasing sophistication of sedimentation analysis and cryo-electron microscopy as analytical tools. Although early studies of chromatin folding yielded significant information on the structures and mechanisms involved, they were limited by the heterogeneous nature of preparations of chromatin fragments. Fragments isolated from lysed nuclei varied in size and linker lengths and were heterogeneous with regard to DNA sequence, histone modifications, and the presence of linker histones and other chromatin-associated proteins. Chromatin could also be reconstituted in vitro by mixing purified histones with plasmid or random sequence DNA (before the use of PCR to obtain defined sequence fragments in large quantities) at 2M NaCl followed by stepwise dialysis to 100–200mM NaCl (see Sidebar in Chapter 1: Reconstituting nucleosomes). However, such reconstituted chromatin generally did not exhibit the fairly regular spacing between nucleosomes characteristic of nuclear chromatin; rather, clustered, close-packed nucleosomes tended to form, yielding templates not well-suited for serving as proxy for native chromatin (Fulmer and Fasman, 1979; Spadafora et al., 1978; Steinmetz et al., 1978).

This limitation was overcome to a degree by the construction of a tandem array of a sequence that had been shown to yield a strongly positioned nucleosome—that is, a nucleosome having a single, well-defined arrangement of histones with respect to the DNA sequence—when reconstituted with histones isolated from chicken erythrocytes (Simpson et al., 1985). (Nucleosome positioning is discussed in depth in Chapter 4.) Twelve to eighteen copies of the positioning sequence were carried on a plasmid; when excised by restriction endonuclease cleavage, the array could be reconstituted with histones to give a highly homogeneous chromatin fragment, as determined by micrococcal nuclease digestion, electron microscopy, and topological analysis (Fig. 3.3) (Simpson et al., 1985). Arrays were constructed having repeat lengths from 172 to 207bp, allowing systematic investigation of the impact of parameters such as salt concentration, presence of linker histones, histone tails, and repeat length on folding and oligomerization of short chromatin fragments. Later studies extended this approach by adoption of a different nucleosome-positioning sequence, replacing the 5S RNA gene sequence used by Simpson and colleagues with the Widom 601 sequence (see Chapter 4), and by use of recombinant histones, further improving homogeneity of the reconstituted arrays (Carruthers et al., 1999; Lowary and Widom, 1998).

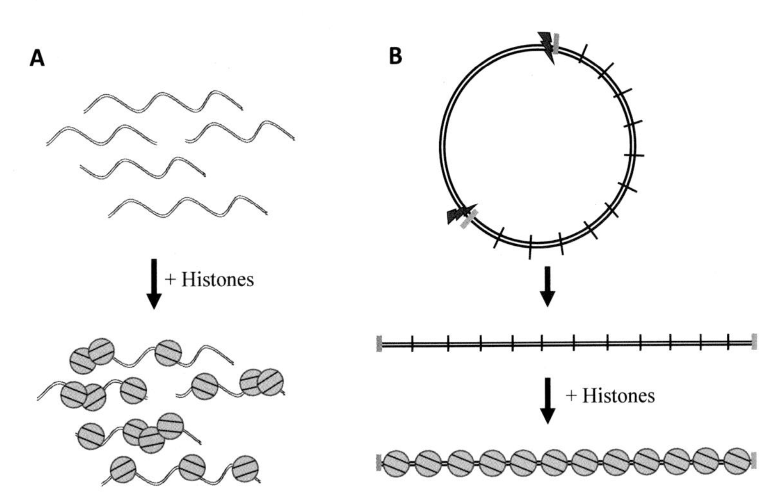

FIG. 3.3 Reconstitution of nucleosome arrays. (A) Reconstitution of nucleosomes by salt dialysis onto linear DNA fragments yields irregularly spaced clusters of nucleosomes. (B) A tandem array of 12 copies of a nucleosome positioning sequence is excised from a plasmid by restriction enzyme digestion and purified; reconstitution of the fragment with histones yields an array of regularly spaced nucleosomes. *Credit: Courtesy of Elisabeth Knoll.*

In one of the first investigations using these reconstituted arrays, Jeff Hansen and colleagues showed that following reconstitution of the 12-mer arrays with histones, about half of the nucleosomes occupied a single position in the repeat, while the remainder were generally offset by multiples of 10bp (Hansen et al., 1989). This "rotational positioning" implied that adjacent nucleosomes in the array were oriented with the same relative surface orientation but displaced by one or more helical turns of DNA, as opposed to being differentially rotated relative to one another. Further studies showed that reconstituted arrays recapitulated properties of isolated chromatin fragments with regard to folding in NaCl solutions, and appeared identical when visualized by cryo-EM, supporting their relevance for studies of chromatin folding (Carruthers et al., 1998; Carruthers and Hansen, 2000).

Folding and oligomerization of 12-mer nucleosome arrays were investigated principally using the analytical ultracentrifuge (or the "Model E" as it was fondly referred to). In sedimentation velocity experiments, measurement of the movement of solutes—in this case, chromatin fragments—at very high centrifugal fields (up to 300,000×g) is analyzed using hydrodynamic theory to yield information on size, shape, and interactions of macromolecules. Additional information can be gained from sedimentation equilibrium measurements, in which lower centrifugal fields are used such that the macromolecules under investigation establish an equilibrium gradient in which movement due to sedimentation is balanced by flux due to diffusion. Early analyses established that the dependence of sedimentation on ionic strength, known to affect chromatin folding from EM experiments, was different for templates having more than six nucleosomes than for shorter chromatin fragments; this was interpreted as indicating that six or more nucleosomes were required to form a stable higher-order structure (Pearson et al., 1983). Consistent with these findings, reconstituted arrays containing 12 nucleosomes were shown to fold reversibly into higher-order structures in a salt-dependent fashion (Hansen et al., 1989).

Although linker histones are important in chromatin folding, as discussed below, nucleosome arrays reconstituted without linker histones associate and fold via the same pathways, but with different salt dependence (Carruthers and Hansen, 2000; Carruthers et al., 1999). Hydrodynamic experiments utilizing nucleosome arrays, with or without linker histones, revealed complexes varying in their sedimentation coefficients ($s_{20,w}$) in a salt-dependent fashion (Fig. 3.4). In 10mM Tris buffer, arrays sedimented at 29S, as predicted for chromatin fibers present in a "beads-on-a-string" configuration. This value increased to near 40S in >150mM NaCl or ~1mM $MgCl_2$ (Hansen et al., 1989; Schwarz and Hansen, 1994); additional analyses revealed that the 29S and 40S species exist in a reversible equilibrium, with the 40S species representing a moderately compacted form that is inherently unstable (Hansen, 2002). In the presence of 1–2mM $MgCl_2$, a species sedimenting at ~55S is observed, corresponding to the 30nm fiber (Fig. 3.4) (Hansen, 2002).

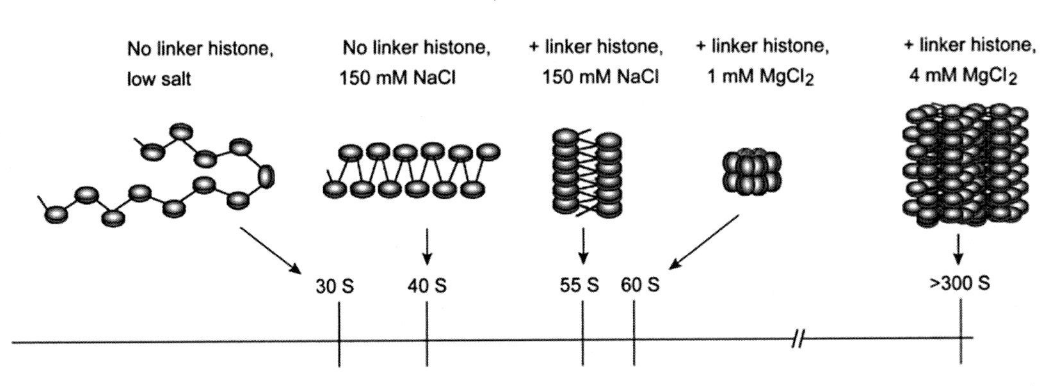

FIG. 3.4 Correspondence between states of chromatin folding in vitro and sedimentation coefficients obtained for 12-mer arrays of nucleosomes measured at indicated cation concentrations and with or without linker histone, as indicated. Note that transitions between structures with sedimentation of 60S or lower are reversible, and all of these structures can self-associate at high $MgCl_2$ to form structures with sedimentation value >300S. *From Grigoryev, S.A., Woodcock, C.L., 2012. Chromatin organization—the 30 nm fiber. Exp. Cell Res. 318, 1448–1455.*

An additional transition of reconstituted arrays occurs at higher $MgCl_2$ concentrations and corresponds to the association of arrays to form large, oligomeric structures sedimenting at values >300S (Hansen, 2002; Schwarz et al., 1996). These structures do not correspond to precipitated chromatin, as the transition can be reversed by returning them to a buffer lacking Mg^{2+} (Schwarz et al., 1996). Oligomerization can occur even with arrays that cannot fold into 30nm fibers because they lack uninterrupted runs of more than six or seven nucleosomes, demonstrating that self-association of nucleosome arrays does not require prior folding into intermediate levels such as the 30nm fiber (Lowary and Widom, 1998). Evidence for such self-association was found in the earliest studies of higher-order chromatin structure, which reported association, as visualized by EM, of isolated nucleosomes dialyzed against deionized water (Finch and Klug, 1976). The association was reversed upon the addition of 25mM ammonium sulfate, and was interpreted as showing that reversible internucleosomal aggregation "was not dependent on the continuity of the DNA along a nucleofilament" (Finch and Klug, 1976). Array oligomerization was monitored for many years by a simple assay in which arrays were pelleted in the microfuge and the fraction remaining in solution determined, but little was known about the structure of the large complexes that precipitated in this assay. Large chromatin oligomers have more recently been characterized using several physical and enzymatic approaches (Gibson et al., 2019; Maeshima et al., 2016; Strickfaden et al., 2020). The oligomers, also referred to as condensates based on these newer findings, were found to exhibit a globular shape and to vary greatly in size, spanning four orders of magnitude in *S* value. Small angle X-ray scattering did not detect 30nm fibers within the oligomers, consistent with the ability of interrupted arrays to self-associate as mentioned above, and their mechanical properties suggested they behave as a gel- or solid-like state. These results, together with evidence for *trans*-interactions among chromatin fibers, have potential relevance for the organization of interphase chromatin in vivo (Hansen, 2002; Hansen et al., 2018).

Factors influencing chromatin folding

Chromatin compaction in vitro has been studied primarily in regard to three parameters: cation concentration, linker histones, and histone tails. In addition, chromatin condensation can be effected by nonhistone proteins or protein complexes, often in combination with histone modifications, as in mammalian heterochromatin, yeast telomeric chromatin, or the Barr bodies comprising inactive X-chromosomes in female mammals.

Cations

As discussed previously, early studies revealed condensation of chromatin containing linker histone H1 in the presence of low concentrations of Mg^{+2} (~0.5–2 mM) or higher concentrations of monovalent cations (Finch and Klug, 1976; Renz et al., 1977). Free Mg^{+2} concentration in human cells is estimated at 0.5–1.2 mM (de Baaij et al., 2015), and the compaction observed in solutions containing monovalent cations also occurred at concentrations close to physiological, making plausible the notion that folding transitions observed in in vitro experiments could be relevant in vivo. A theoretical analysis in which Manning's polyelectrolyte theory (a framework in which simplifying assumptions are used to quantitatively model the interaction between a polyion such as DNA and counterions from a surrounding medium (Manning, 1969)) was used to model the folding of chromatin fibers indicated that folding is primarily governed by electrostatic interactions (Clark and Kimura, 1990). At low ionic strength, segments of internucleosomal (linker) DNA, whose charge is not shielded by the histones, repel one another electrostatically, thus limiting condensation. Monovalent cations screen this repulsive charge, thus facilitating chromatin condensation, while divalent cations are effective at much lower concentrations in part due to specific binding to DNA. Similarly, the binding of linker histone (discussed further in the following) reduces the effective negative charge of linker DNA, thereby allowing chromatin folding at lower salt concentrations than occurs for linker histone-depleted chromatin. A more recent theoretical treatment argues that conditions in the nucleus situate chromatin close to a "tipping point" between highly condensed and more extended configurations (Korolev et al., 2009). However, as we shall see, understanding the structural nature of condensed chromatin remains a challenge.

Linker histones

Early EM observations indicated that the highly ordered structures observed in chromatin fibers in 1–2 mM $MgCl_2$ or above about 80 mM NaCl depended on the presence of linker histone (Finch and Klug, 1976; Thoma et al., 1979). Condensation was observed in H1-depleted chromatin, but rather than forming ordered 30 nm fibers, H1-depleted chromatin folded into irregular clumps when fixed in the presence of 1–2 mM $MgCl_2$ (Finch and Klug, 1976; Oudet et al., 1975; Thoma et al., 1979). Later studies confirmed and extended these results, showing that H1-containing chromatin folded into 30 nm fibers at 0.5–2 mM $MgCl_2$, while interfiber interactions, such as might facilitate oligomerization to form higher-order structures, were observed at 3–5 mM $MgCl_2$ even with H1-depleted chromatin (Hansen, 2002; Ohno et al., 2018; Schwarz and Hansen, 1994). However, compaction and oligomerization of H1-depleted chromatin (using reconstituted 12-mer arrays of 208 bp nucleosome positioning sequence), as reflected by sedimentation at 55S in the analytical ultracentrifuge, occurred only for a fraction of the molecules under study (Schwarz and Hansen, 1994). In contrast, the same 12-mer array was converted into a homogeneous, maximally folded species when bound with one molecule of the chicken erythrocyte linker histone H5 per nucleosome (Carruthers et al., 1998). Furthermore, although H1-depleted chromatin can condense into irregular structures in the presence of divalent cations such as $MgCl_2$, reconstituted arrays lacking linker histones do not oligomerize in monovalent salt solutions (Hansen et al., 1989; Schwarz and Hansen, 1994). Thus, histone H1 (or other linker histones) appears not to be strictly necessary for the compaction of chromatin into higher-order structures but rather affects the equilibrium by which those structures are established and maintained, such that linker histones are required for the stable formation of regular higher-order structures (Hansen, 2002). Supporting the relevance of these results to chromatin folding in vivo, knockout of the three most highly expressed H1 subtypes in mouse splenocytes resulted in partial decompaction of chromatin, most prominently in regions having the highest linker histone densities (Willcockson et al., 2021).

What is the mechanism by which linker histones facilitate higher-order chromatin folding? Not surprisingly, they do so mainly by shielding the charge of the DNA and thus, similarly to the effect of monovalent and divalent cations, counteracting the repulsive forces that prevent the close approach of helical segments. As described in Chapter 2, linker histones protect about 20 bp of DNA beyond the 146 bp protected in nucleosome core particles from digestion by MNase, and additionally increase the length of DNA, relative to core particles, that is stabilized to thermal denaturation (Simpson, 1978). In addition, DNA near the dyad axis was less susceptible to digestion by DNase in the chromatosome than in nucleosome core particles. These and other observations led investigators to conclude that linker histones associate with DNA as it enters and exits the nucleosome core particle. This idea, based on enzymatic, biochemical, and biophysical data, received further support from EM observations (Thoma et al., 1979). First, chromatin fragments from rat liver that were depleted of H1 by treatment with an ion exchange resin lost their nucleosomal appearance at low salt (1 mM TEACl), whereas H1-containing fragments exhibited characteristic nucleosome structures. This suggested that H1 stabilized nucleosomes at low ionic strength. Second, H1-containing chromatin revealed a zigzag-shaped structure at 5 mM TEACl (as mentioned earlier), in which DNA entry and exit points on the nucleosome appeared proximate, whereas in H1-depleted chromatin, DNA can be seen entering and exiting on opposite sides of core particles (Fig. 3.2). This suggested that linker histones bound to the DNA end in the core particle and that the structure thus formed rendered the ordered 30 nm fiber more stable under conditions of moderate salt concentration. A subsequent study utilized electron cryomicroscopy, in which samples prepared in a given ionic environment are rapidly frozen and imaged in the frozen hydrated state, to identify a stem motif characteristic of H1-containing chromatin; this motif was then used as a basis for modeling the 30 nm fiber as a zigzag arrangement of nucleosomes (see below) (Bednar et al., 1998).

Given the importance of linker histones in chromatin folding and their ubiquity in metazoan chromatin, understanding how they interact with the nucleosome core particle has been a major goal since the nucleosome model was first formulated. The linker histones most often referred to in the literature are histones H1 and H5. Histone H1 derives from chromatin prepared from rat liver, calf thymus, and other sources frequently utilized in early work aimed at identifying the constituents of chromatin, while histone H5 is a closely related variant found in chicken erythrocytes, which also provide a reliable source of histones for reproducible chromatin preparations. H1 comprises a large family of proteins having a common evolutionary origin that show much more variability across species than do the core histones (Fig. 2.4) and is expressed as multiple variants in many organisms (Cole, 1984; Hergeth and Schneider, 2015). Linker histones were first identified as "lysine-rich histones" that were clearly distinct from what are now known as the core histones in their chemical make-up (Table 2.1) and could be stripped from chromatin at around 0.6M NaCl, lower than salt concentrations at which H2A and H2B were lost (Ohlenbusch et al., 1967; Tatchell and Van Holde, 1977).

Two main lines of evidence, as mentioned earlier in Chapter 2, gave strong indications that H1 binds to DNA extraneous to the core particle proper. First was the development of nondenaturing and two-dimensional gel electrophoresis to analyze the protein constituents of heterogeneous preparations of digested chromatin (Todd and Garrard, 1977; Varshavsky et al., 1976). This approach revealed that in a mix of mononucleosomes containing variable lengths of associated DNA, H1 was present only in the particles with longer DNA, with the difference in DNA length between nucleosomes lacking and containing H1 being 20–30bp. The second line of evidence was provided by experiments showing that controlled digestion of nucleosome particles with MNase was accompanied by loss of H1 as the DNA was trimmed (Noll and Kornberg, 1977; Whitlock Jr. and Simpson, 1976), along with the isolation and characterization of the H1-containing "chromatosome," the term coined to define the particles resulting from limited MNase digestion that contained, in addition to the normal complement of core histones, one molecule of linker histone and 168 ± 8bp of DNA (Hao et al., 2021; Simpson, 1978). Together with the EM data discussed previously, these observations pointed to linker histones associating with the DNA entering and exiting the core particle and thereby facilitating the folding of chromatin into higher-order structures.

Early studies revealed linker histones H1 and H5 both to consist of a central, structured region of ~75 amino acid residues that is resistant to digestion by trypsin, flanked by a shorter, basic amino-terminal region (20–35 residues) and a long C-terminal region (~100 residues) with a very high density (close to 50%) of lysine and arginine residues (hence the initial designation as "lysine-rich" histone) (Aviles et al., 1978; Hartman et al., 1977). Hydrodynamic and circular dichroism studies showed the central, trypsin-resistant domain of H5 to adopt a compact, globular configuration and to contain all of the α-helical structures present in the intact H5 molecule (Aviles et al., 1978). Nucleosomes reconstituted with the central, globular domain of histone H1 exhibited the same protection of linker DNA as did chromatosomes, indicating that this moiety bound to the nucleosome independently of the N- and C-terminal arms, in spite of the abundance of lysine and arginine residues in the latter regions (Allan et al., 1980). X-ray crystal structure determination of the globular region of H5 revealed a bundle of three alpha helices followed by a β-sheet formed by a β-hairpin at the C-terminus and a three-residue stretch between helices I and II in the β-strand conformation (Ramakrishnan et al., 1993). This structure was termed a "winged-helix" fold and is related to both the DNA-binding domain of the prokaryotic catabolite gene activator protein (CAP) and the eukaryotic transcriptional activator HNF3 (Clark et al., 1993; Clore et al., 1987).

These structural and biochemical studies revealed much about linker histones but also underscored what was not understood: linker histones resembled protein domains that bound specific DNA sequences but did not themselves bind specific sequences; they possessed a long C-terminal region that was loaded with lysine and arginine residues, yet was dispensable for binding to the nucleosome; they were asymmetric, but bound to the symmetrical nucleosome. The high lysine and arginine content of linker histones suggested that they should bind DNA, and this was supported by the protection of an extra 20bp beyond the core particle mentioned previously. This binding should be nonspecific, like that of the core histones, and therefore electrostatic in nature (again consistent with the abundance of lysine and arginine residues); but it should also be constrained to a specific location with respect to the core particle, implying some interaction with the core histones. One clue to the nature of linker histone binding to the nucleosome is derived from the demonstration that the globular region of H1 exhibited a quantitative and qualitative preference for binding to superhelical rather than relaxed DNA; this suggested that linker histones might recognize special DNA configurations, such as the proximate strands entering and exiting the nucleosome (Singer and Singer, 1976). Later work showed directly that linker histones could bind to two DNA duplexes, via two sites in the globular domain (Goytisolo et al., 1996; Zhou et al., 1998). One site corresponds to the region that is structurally similar to the DNA-binding domain of the CAP protein, as mentioned previously, while the second site comprises a cluster of four positively charged residues 25Å away, on the opposite face of the domain (Goytisolo et al., 1996).

The protection by linker histones of an additional two helical turns of DNA against MNase digestion, and the excess protection conferred against DNase I cleavage at the dyad axis, suggested an "on-dyad" disposition of H1/H5, in which one turn of the helix at the entry and exit points of the nucleosome was associated with linker histone (Simpson, 1978; Staynov and Crane-Robinson, 1988). However, two independent lines of inquiry from the early 1990s led to distinct, alternative models in which linker histones bound asymmetrically, or "off-dyad" to core particles. (Speaking rigorously, linker histone binding must perforce be asymmetric, as only one copy of this symmetry-lacking histone is associated with a nucleosome.) In the first, incorporation of H1 or H5 into nucleosomes reconstituted using the 5S rDNA sequence from *Xenopus borealis*, which had been shown to be a strongly nucleosome-positioning sequence, was reported to lead to asymmetric protection of an additional 20bp at one end of the nucleosomal DNA sequence against MNase digestion while leaving the accessibility of DNA near the nucleosome dyad to DNase I or hydroxyl radical unaltered (Hayes and Wolffe, 1993). Subsequent experiments using site-directed cleavage to monitor DNA accessibility and cross-linking to determine DNA-histone contacts (but no direct structural determination) led to proposal of a model in which linker histone bound on the inside of one gyre, or turn, of DNA on the nucleosome by inserting a portion of the winged helix domain between the DNA and core histones at a site about 65bp from the nucleosome dyad (Hayes et al., 1996;

Pruss et al., 1996). This model generated considerable controversy, and turned out to be incorrect, due to the reconstituted chromatosomes having translational settings other than the single dominant position deduced by the authors (An et al., 1998b; Panetta et al., 1998; Thomas, 1999; Travers, 1999).

Although this first off-dyad model of linker histone binding fell victim to a technical artifact, independent evidence has pointed to a less radical model of asymmetric linker histone binding to the nucleosome. Travers and colleagues cloned and sequenced several hundred DNA sequences purified from nucleosome core particles or chromatosomes to glean the rules for DNA sequence-directed positioning of nucleosomes (see Chapter 4) (Satchwell and Travers, 1989; Travers and Muyldermans, 1996). (At the time, these were seriously "high throughput" experiments!) Computational analysis of the resulting chromatosome sequences revealed asymmetrical features: first, a short sequence preferentially found at the dyad of core particles was offset from the central position in chromatosome sequences, being positioned ~93bp from one end and ~76bp from the other (Satchwell and Travers, 1989); second, a short DNA sequence of the form NGGR (where N is any base, G is guanine, and R is A or G) was enriched at only one terminus of chromatosome sequences (Travers and Muyldermans, 1996). Additional experimental support was derived from studies of positioned chromatosomes in which the 20bp of extra protection against MNase digestion of the incorporated DNA occurred at only one terminus and not the other (An et al., 1998a; Wong et al., 1998). Off-dyad localization of linker histones was further supported by a subsequent study in which residues in the mouse linker variant $H1^0$ that are involved in nucleosomal contacts were identified using a combination of mutagenesis and fluorescence recovery after photobleaching (FRAP) (Brown et al., 2006). The resulting information was then used to model H1 association with the nucleosome; the modeling indicated that one binding site interacted with DNA approximately one helical turn away from the nucleosome dyad, while the second binding site interacted with one DNA terminus approximately 15bp from the end of the nucleosomal DNA (Brown et al., 2006). More recent work has suggested that symmetric or asymmetric binding modes may be dictated by linker histone subtype: NMR-based structural investigations of the *Drosophila* linker histone H1 lacking its N-terminal domain in complex with a nucleosome core particle indicated an asymmetric binding mode in which binding of H1 was off the dyad axis and principally associated with only one linker DNA segment (Zhou et al., 2013). The same group reported on-dyad binding of the globular domain of H5 to a nucleosome formed from the same 167bp sequence as used in the earlier study, based both on a 3.5Å crystal structure and NMR data; in contrast to *Drosophila* H1, H5 was found to interact with both linker DNA segments (Fig. 3.5) (Zhou et al., 2015). This more symmetrical mode of association was also reported for both *Xenopus* linker histone H1.0 and human H1.5 by a cryo-EM study, and also in solution studies based on cross-linking and hydroxyl radical footprinting data, using a longer (197bp) DNA segment that would have had ample length to form an asymmetric structure (Bednar et al., 2017). Mutagenesis studies then indicated that the substitution of five surface-exposed residues in gH5 (the globular domain of H5) with the corresponding residues from *Drosophila* resulted in the mutant gH5 adopting an off-dyad binding mode (Zhou et al., 2016).

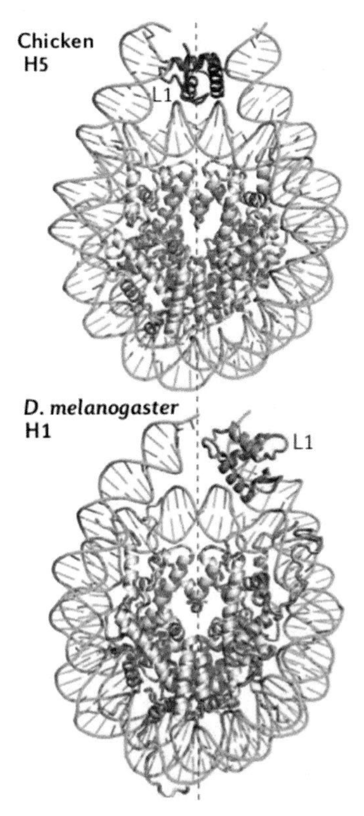

FIG. 3.5 X-ray crystal structure of the nucleosome core particle bound to the globular domain of linker histone H5 (top), showing on-dyad association of linker histone, in contrast to the off-dyad binding in the NMR structural model of the nucleosome associated with linker histone H1 from *D. melanogaster* (bottom). The dyad axis is indicated by the *dashed line. From Fyodorov et al., 2018. Nat. Rev. Mol. Cell Biol. 19, 192–206.*

While the studies cited above suggest the possibility of asymmetric linker histone binding for at least some linker histone subtypes, the preponderance of evidence indicates that symmetric binding is generally favored (Hao et al., 2021). Early work pointed to symmetric localization of linker histones, based principally on protection against nuclease digestion of DNA conferred by incorporation of linker histones into nucleosomes (Allan et al., 1980; Staynov and Crane-Robinson, 1988). Later studies used cryo-EM, hydroxyl radical footprinting, and modeling to examine mononucleosomes, dinucleosomes, and trinucleosomes that included intact H1 or truncated, recombinant versions, and concluded that the globular region of H1 contacted the DNA minor groove at the nucleosome center and made contacts with 10bp entering and exiting the nucleosome, and a cryo-EM structural determination of chromatosomes containing three linker histone isoforms revealed on-dyad binding in all three cases (Bednar et al., 2017; Meyer et al., 2011; Syed et al., 2010; Zhou et al., 2021). On-dyad binding of linker histone was also found to dominate in vivo in a high-resolution cryo-EM study of mononucleosomes isolated from both metaphase and interphase chromatin from *Xenopus* egg extract (Arimura et al., 2021). It is possible that specific experimental conditions may have favored asymmetric binding in some studies (Hao et al., 2021). For example, the reports of asymmetric binding of specific linker histones were based on NMR determinations that employed linker histones lacking some or all of the N-terminal and C-terminal domains, thus missing potential interactions of the C-terminal domain that have been shown to contribute to linker DNA organization and to favor symmetric binding by restraining the dynamics of the central, globular domain of H1 (Meyer et al., 2011; Syed et al., 2010; Wu et al., 2021). Another cryo-EM study reporting asymmetric linker histone binding may have been subject to bias resulting from the use of glutaraldehyde to cross-link chromatin fibers before imaging (Song et al., 2014; Zhou et al., 2018). The question of whether linker histones bind preferentially or exclusively on-dyad (symmetrically) or not is not insignificant, as the structures available for higher-order folding of chromatin differ for symmetrically and asymmetrically bound linker histones (see below) (Bednar et al., 2017; Zhou et al., 2016). Resolution of this issue is certain to benefit from the gains in imaging technology made over the past decade.

Histone tails

In addition to monovalent or divalent cations and linker histones, the histone "tails," particularly that of H4, are important for condensation of chromatin. A seminal study by James Allan, Hannah Gould, and co-workers at King's College in London used controlled trypsin digestion to remove the unstructured N-termini from the core histones to examine their role in the formation of the 30 nm chromatin fiber (Allan et al., 1982). Chromatin fragments were prepared from chicken erythrocytes, stripped of linker histones H1 and H5, and reconstituted with H1 or H5 either with or without removal of the core histone tails by treatment with trypsin. Despite the dispensability of the core histone tails for the integrity of the nucleosome core particle, as described in Chapter 2, the tails proved essential for the compaction of chromatin into the 30 nm fiber, as indicated by sedimentation, electric dichroism, and EM (Allan et al., 1982). Compact structures were largely restored (at appropriate salt concentrations) when stripped and trypsinized chromatin was reconstituted with H5 in the presence of a 94-residue polypeptide from the carboxy-terminal end of calf thymus H1. This fragment is distinct from the globular core of H1, as described previously, and contains 40 lysine residues; the authors inferred that this basic, unstructured polypeptide facilitated chromatin condensation by electrostatic shielding of DNA charge and that the basic, unstructured core histone tails likely functioned in the same way.

The importance of the histone tails in chromatin condensation was confirmed and extended in subsequent studies utilizing the nucleosome arrays described earlier. Reconstitution of nucleosome arrays having twelve 208 bp repeat units of the *Lytechinus variegatus* 5S rRNA gene with histones lacking the flexible N-termini was achieved first by treating intact nucleosomes with trypsin, purifying the trypsinized histones, and using these to reconstitute arrays (Ausio et al., 1989; Garcia-Ramirez et al., 1992). Comparison of reconstituted arrays having or lacking histone N-termini by sedimentation velocity analysis and EM yielded the following results:

- In low salt conditions (5 mM triethanolamine or 10 mM Tris) and in the absence of linker histone, partial unwrapping of DNA occurs in trypsinized nucleosomes (Garcia-Ramirez et al., 1992); the addition of Mg^{+2} ions restores proper oligonucleosome structure under these conditions (Fletcher and Hansen, 1995).

- Mg^{+2}-dependent folding of oligonucleosomes into higher-order structures (sedimenting at ~55S and corresponding to the 30nm fiber; see Fig. 3.4) requires the histone tails, whether the arrays lack or contain linker histone (Carruthers and Hansen, 2000; Fletcher and Hansen, 1995).
- Tailless arrays fail to oligomerize (and so do not yield structures sedimenting at >300S; Fig. 3.4) (Schwarz et al., 1996).
- Oligomerization, but not the formation of 30nm fiber structures, is observed in arrays in which either the H2A-H2B or the H3-H4 tails are absent (Moore and Ausio, 1997; Tse and Hansen, 1997).

The ability of Mg^{+2} ions to substitute for the histone tails in the formation of normal oligonucleosome structure at low salt suggested that the tails contribute to this structure via an electrostatic mechanism, whereas the inability of Mg^{+2} ions to facilitate oligomerization or formation of the 30nm fiber from tailless arrays indicated that these transitions depend at least in part on a nonelectrostatic mechanism (Hansen, 2002). (However, Richmond and colleagues observed near-maximal compaction of fibers at intermediate and high $MgCl_2$ concentrations even from reconstituted arrays of the Widom 601 sequence lacking all of the histone tails (Dorigo et al., 2003).) These inferences are consistent with the original suggestion of Allan et al. that the histone tails facilitate chromatin folding by shielding DNA charge, while also serving to bridge adjacent nucleosomes in the chromatin fiber (Allan et al., 1982).

The use of reconstituted arrays to investigate higher-order chromatin structure was later improved upon by using recombinant histones, thereby eliminating possible heterogeneity due to posttranslational modifications and allowing the assembly of arrays containing mutant histones, including histones lacking part or all of the flexible N-termini (Dorigo et al., 2003; Fan et al., 2004). Arrays constructed using recombinant *Xenopus laevis* core histones and a 12-mer (having a 177bp repeat unit) of the Widom 601 sequence condense into structures sedimenting at ~53S in the presence of ~1mM $MgCl_2$, and this compaction is affected very little in the absence of the H2A, H2B, or H3 tails. In contrast, arrays in which the H4 tail is absent show maximum sedimentation of ~44S. Experiments using partial deletions of the H4 amino terminus showed residues 14–19 to be critical in the formation of a compact chromatin fiber (Dorigo et al., 2003). Additional evidence for the importance of the H4 amino terminus in chromatin folding derived from a study reporting Heterochromatin Protein 1 (HP1)-dependent chromatin folding (Fan et al., 2004). Nucleosome arrays lacking linker histones are able to condense into higher-order fibers independently of divalent cations in the presence of HP1, which is localized to and important for the formation of highly condensed heterochromatin in mouse cells (Fan et al., 2004; Nielsen et al., 2001). Using a similar approach to that of Dorigo et al., in which recombinant wild-type and mutant histones were used to reconstitute 12-mer arrays, Tremethick, Luger and colleagues showed that HP1-dependent chromatin condensation also depends on the H4 tail, but not on that of H3 (Fan et al., 2004).

Arrays reconstituted using recombinant histones also leant themselves well to examining the effects of the histone tails on interarray oligomerization, or self-association. One method for monitoring oligomerization is by determining the Mg^{+2} concentration at which 50% of the arrays precipitate from solution. Using this method, Richmond and colleagues showed that oligomerization occurred normally for arrays lacking H2A, H2B, or H3 tails, but occurred at higher Mg^{+2} (3.4 mM vs 1.5 mM) for arrays lacking the H4 tail (Dorigo et al., 2003). Gordon et al. subsequently reported an exhaustive study on the oligomerization of arrays comprising all 16 combinations of tail mutants and wild-type histones (Gordon et al., 2005). Among arrays lacking a single histone tail, they also found the strongest effect when the H4 amino terminus was absent. However, by using arrays lacking combinations of histone tails, they also showed that all four histone tails contributed to array oligomerization in an additive and independent (rather than synergistic) fashion. This finding was in contrast to the effects of the histone tails on the formation of the 30 nm fiber, where deletion of the H4 tail inhibited folding, but additive effects were not observed upon deletion of other histone tails (Dorigo et al., 2003). Clearly, the histone tails function differently in compaction of chromatin to form the 30 nm fiber than they do during self-association. These findings also constituted additional evidence that self-association, or condensation, of chromatin fibers occurs independently of higher-order folding.

The requirement for the histone H4 amino terminus, and specifically the critical importance of residues 14–19, for compaction of chromatin raised the possibility that interactions between the H4 tail and the "acidic patch" formed by negatively charged residues in H2A and H2B, observed in the high-resolution X-ray crystal structure of the nucleosome (see Chapter 2) (Luger et al., 1997), could contribute to formation of the 30 nm fiber. This interaction occurs internucleosomally, rather than within a single nucleosome, consistent with interactions that could contribute to chromatin folding, and involved amino acid residues 16–24 in the H4 tail (Luger et al., 1997). Additional evidence for this hypothesis was provided by experiments showing that substitution of Cys residues at H4-V21 and H2A-E64 in 12-mer arrays allowed cross-links to be formed in the presence of glutathione; cross-linking depended on compaction, which could be induced by the inclusion of divalent cation or linker histone H1 (Dorigo et al., 2004). Furthermore, a naturally occurring variant of histone H2A, called H2A.Z (see Chapter 5: Histones and Histone Variants), possesses an expanded acidic patch and when incorporated into nucleosome arrays, enables folding into a more compacted structure than observed with H2A; mutations that reduce the size of the acidic patch inhibit this compaction (Fan et al., 2004). Another histone H2A variant found in human cells, H2A.Bbd, has three of the residues comprising the acidic patch altered to nonacidic residues (Chadwick and Willard, 2001); arrays that include this variant are resistant to condensation (Zhou et al., 2007). Nucleosomes containing H2A.Bbd differ in some important respects from canonical nucleosomes (see Chapter 5), potentially complicating interpretation of these results. However, mutagenizing residues in H2A.Bbd to restore the acidic patch to that found in canonical H2A allowed folding of arrays containing the mutated H2A.Bbd to an extent approaching that seen with wild-type arrays, providing convincing additional evidence as to the requirement of the acidic patch for the formation of the 30 nm fiber. We shall see later that the functional properties of H2A.Z and H2A.Bbd are consonant with their effects on the formation of higher-order chromatin structures (Chapter 5).

The interaction between the positively charged H4 tail and the H2A-H2B acidic patch is not a simple electrostatic attraction between two formless blobs. Rather, as has been emphasized by Luger, Hansen, and colleagues, the acidic patch is a highly structured, or "contoured," surface, and the interaction with the H4 tail is complex (Chodaparambil et al., 2007;

Kalashnikova et al., 2013). Lysine 16 (K16) of H4 has a particularly important role in this interaction, forming multiple salt bridges with aspartic acid (D) and glutamic acid (E) residues (Luger et al., 1997). Acetylation of H4K16 is predicted in modeling studies to disrupt these salt bridges (Yang and Arya, 2011), and has been reported to disrupt chromatin compaction (Allahverdi et al., 2011; Robinson et al., 2008; Shogren-Knaak et al., 2006). Shogren-Knaak, Peterson, and colleagues addressed this issue by using an innovative approach to construct arrays incorporating a single histone modification, acetylated lysine 16 of histone H4 (H4K16ac) (Shogren-Knaak et al., 2006). To achieve this, a chemical ligation strategy was employed in which recombinant histone H4 lacking its N-terminal 22 residues and having R23 substituted with Cys (H4(23-102)R23C) was reductively ligated to H4(1–22) having a C-terminal thioester and acetylated H4K16. A control, unacetylated H4 was constructed in the same way, as well as a mutant lacking its N-terminal 19 residues (H4Δ1–19). These variants of histone H4 were incorporated into 12-mer nucleosome arrays and the folding of the resulting arrays was investigated using sedimentation velocity analysis. The results showed that while the control arrays (having R23C) and arrays incorporating wild-type H4 condensed into structures sedimenting at about 54S in the presence of 1 mM $MgCl_2$, those having H4K16ac displayed the same defect as arrays containing H4Δ1–19 (Shogren-Knaak et al., 2006). The H4K16ac modification also interfered with array oligomerization, as seen for arrays lacking the H4 tail (Dorigo et al., 2003; Shogren-Knaak et al., 2006).

Structural analyses of nucleosome core particles in complex with several proteins or peptides derived from proteins that interact closely with nucleosomes have revealed precise interactions with the acidic patch formed by H2A and H2B (Kalashnikova et al., 2013). These include the nonhistone high mobility group (HMGN) proteins, which compete with linker histone for nucleosomal binding and promote chromatin decompaction (Postnikov and Bustin, 2010); Sir3, which is required for formation of condensed chromatin structures in yeast that repress genes at telomeres and silent mating type loci (see later section on Yeast Heterochromatin) (Rusche et al., 2003); the latency-associated nuclear antigen (LANA) of Kaposi's sarcoma herpes virus, which tethers viral episomes to host chromosomes to allow their replication by the host cell replication apparatus (Ballestas et al., 1999); and Regulator of Chromosome Condensation 1 (RCC1, also known as RanGEF), which binds nucleosomes and creates a Ran concentration gradient that is important for various chromosomal processes in mitosis (Clarke and Zhang, 2008). Although these proteins lack homology in their amino acid sequences, they display overlapping complementarity with the acidic patch, such that the interacting regions must bind in a mutually exclusive manner not only with respect to one another but also with respect to the H4 tail. A recent study has considerably expanded the catalog of nucleosome-binding proteins whose interactions depend on the integrity of the acidic patch (Skrajna et al., 2020). In this study, a small "library" of seven nucleosomes, one having wild-type histones and the remainder mutated in groups of seven amino acids on various exposed surfaces of the octamer disk or lacking the amino termini was used to pull down proteins from lysates obtained from murine embryonic stem cells. Mutation of the acidic patch was the most disruptive toward interactions as determined by mass spectrometry after pulldown; over half of the interactions with a variety of proteins involved in nuclear processes including transcription, cell cycle control, and epigenetic regulation were reduced by mutation of the acidic patch. The importance of the acidic patch is further underscored by findings that the only lethal point mutations in H2A or H2B uncovered in yeast libraries in which all individual core histone residues were mutated to alanine correspond to a cluster in the acidic patch (Matsubara et al., 2007; Nakanishi et al., 2008).

Consistent with the varying roles of these interacting proteins, Hansen and colleagues have hypothesized that proteins interacting with nucleosomes via the acidic patch compete with the H4 tail for binding and are thus able to remodel the higher-order structure of chromatin and alter its functionality in a manner that depends on the properties of the nonhistone-interacting domains of the interacting protein (Kalashnikova et al., 2013). The exact nature of the altered higher-order folding properties instigated by these and other nucleosome-binding proteins is a current area of investigation.

The structure of compact chromatin

If determining the structure of the nucleosome was challenging, elucidating the higher-order structure of chromatin has been even more so. We will consider in the next section the question of chromatin structure in living cells; here, we review models of higher-order chromatin structure deduced from in vitro experiments. We have seen that at the time that the nucleosome model was formulated, isolated fragments of chromatin were observed to fold at physiologically relevant salt concentrations into structures that resembled those seen in whole-mount preparations of chromosomes and sectioned nuclei (Davies, 1968; Gall, 1966; Ris and Kubai, 1970). The resemblance made it reasonable to presume that the structures observed in vitro, which comprised fibers of about 30 nm diameter and appeared in some studies to have a highly regular structure (Fig. 3.2) (Thoma and Koller, 1977, 1981; Thoma et al., 1979), were likely to be representative of chromatin structures in living cells. This presumption, and the idea that higher-order structures were important not just for constraining chromosomal DNA into the small volume of the nucleus, but also were likely to function in gene repression, motivated the quest to understand at the molecular level the condensed structures generated from fragments of chromatin isolated from cell nuclei and from reconstitution of nucleosome arrays.

Two widely divergent structural models were proposed for folded chromatin in the mid-1970s, in spite of the fact, as noted by van Holde, that the electron micrographs used to support the models appeared essentially identical "to the untutored eye" (Van Holde, 1988). One model, first proposed by Klug and colleagues at the MRC (Finch and Klug, 1976), was a highly regular "solenoidal" structure, formed by a helix of nucleosomes. (Interestingly, Taylor et al. had suggested in their classic 1957 paper on chromosome duplication that "large metaphase chromosomes ...perhaps are twice coiled, a helix within a helix" (Taylor et al., 1957).) This model was based on the appearance of occasional diagonal striations in the EM, suggesting a helical structure, which was posited to have a pitch of about 11 nm based on the appearance of an 11 nm reflection in early X-ray diffraction studies. The model appeared to find support in the highly regular structures observed by Thoma et al. (Thoma et al., 1979). The second model posited that nucleosomes aggregated into clumps, or "superbeads," the linear aggregation of which formed the fibers observed in the EM (Kiryanov et al., 1976; Renz et al., 1977). Variations on these and additional themes were proposed by other labs in a flurry of model-building; van Holde illustrates no fewer than five models for the 30 nm fiber (Van Holde, 1988).

Although aspects of the superbead model may apply to chromatin in the cell nucleus—in particular, the idea that the level of compaction can vary heterogeneously along the chromatin fiber (see next section)—several studies argued that the observed clumps of nucleosomes were artifacts of specimen preparation, as reviewed by van Holde and Butler (Butler, 1983; Van Holde, 1988). In addition, early biophysical investigations and, as we have seen, EM studies, supported a solenoidal, or coiled, model for chromatin condensation. Electric dichroism measurements, in which the absorbance of polarized UV light by chromatin fibers that have been oriented in an electric field is measured, were consistent with chromatin from a variety of sources adopting a 30 nm fiber conformation in which nucleosomes were oriented with their "faces" (the surfaces of the nucleosome disk) at an angle from 20 to 35 degrees relative to the fiber axis—that is, approximately parallel to the fiber (but see the summary of the review by van Holde and Zlatanova below) (Felsenfeld and McGhee, 1986). Neutron scattering experiments provided information on the dimensions of the fiber (as they did for the nucleosome itself) (Suau et al., 1979); together with the electric dichroism experiments, the data provided evidence for a 30 nm fiber in which nucleosomes are radially positioned, as in Fig. 3.6.

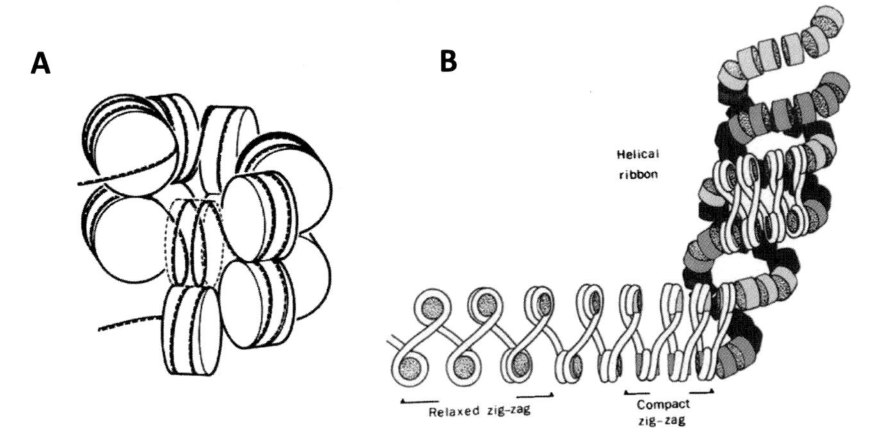

FIG. 3.6 (A) One-start and (B) two-start models of folding of chromatin to give 30 nm fibers. *From Butler, 1984. EMBO J. 3, 2599 and modified from Woodcock, C.L., Frado, L.L., Rattner, J.B., 1984. The higher-order structure of chromatin: evidence for a helical ribbon arrangement. J. Cell Biol. 99, 42–52. (Rockefeller Press).*

For most of the 40+ years since the first model of the 30 nm fiber, the debate on its structure has largely, if not entirely, focused on two competing solenoidal models. The first, a simple, "one-start" helix of nucleosomes, was proposed by Finch and Klug (Fig. 3.6) (Finch and Klug, 1976). In this model, consecutive nucleosomes in the fiber are simply stacked one on top of the next, but at an angle relative to the helical axis, in the coiled higher-order structure. The alternative model, variations of which were formulated independently by several groups, is a "two-start" helix in which two stacks of nucleosomes, separated by the linker DNA between consecutive nucleosomes, form a ribbon-like arrangement that condenses longitudinally to generate the 30 nm fiber (Staynov, 1983; Williams et al., 1986; Woodcock et al., 1984; Worcel et al., 1981). In the extended configuration observed at low salt concentration, the linker DNA follows a zigzag path between two chains of nucleosomes, with each nucleosome in the linear chain interacting most closely not with its immediate upstream and downstream neighbors, but with the second nucleosome on either side (Fig. 3.6). Put differently, if the nucleosomes in the linear chain are numbered consecutively, one stack of nucleosomes in the folded configuration is formed from the even-numbered nucleosomes and the other from the odd-numbered nucleosomes.

Numerous predictions were made of the properties of one-start and two-start nucleosome arrays, and experimental tests of those predictions were carried out, without resolving the "true" structure of the 30 nm fiber (Fussner et al., 2011; Ohno et al., 2018; Tremethick, 2007). Two particularly influential lines of inquiry carried out in the early 2000s in the labs of Tim Richmond and Daniela Rhodes yielded distinctly different conclusions. The Richmond lab investigated chromatin templates reconstituted using recombinantly expressed histones and arrays of nucleosome positioning sequences, as discussed earlier. Using arrays based on a relatively short (167 bp) repeat length, the same cross-linking studies that indicated proximity of the H4 tail and H2A-H2B acidic patch, coupled with EM observations, also provided evidence for a two-start helix of nucleosomes (Dorigo et al., 2004). Evidence for a two-start helix was observed for arrays of lengths from 10 to 48 nucleosomes, whether or not histone H1 was incorporated into the array. In a subsequent study, the same lab obtained crystals of a tetranucleosome folded into a compact structure in the presence of high Mg^{+2} concentration (without linker histone), the structure of which was solved by X-ray diffraction to a resolution of 9 Å (Schalch et al., 2005). Because of the low resolution, the structure was solved by molecular replacement using the high-resolution structure of the nucleosome core particle. Most critically, the electron density for the linker DNA was of sufficient quality to allow reliable assignment of the linker path, which was incompatible with a one-start helix, and generally consistent with a two-start helix. However, although the tetramer structure shows that a two-start helix can be formed at least from short chromatin fragments (notably, too short to form a one-start solenoidal structure) and under particular conditions, caution is due in extending the results more generally. First, the tetranucleosome crystals were obtained "under conditions providing maximum array compaction," namely, by equilibrating tetranucleosome arrays in 25 mM KCl and 90 mM $MgCl_2$ with 12.5 mM KCl and 45 mM $MgCl_2$, and then transferring to a solution for cryo-cooling that included 12.5 mM KCl and 40 mM $MgCl_2$ (Schalch et al., 2005). Such high Mg^{+2} concentrations may have induced structural changes uncharacteristic of "standard" 30 nm fibers, particularly when employed in conjunction with such minimal arrays. Second, modeling chromatin fibers based on the structural results indicated a fiber diameter of 24–25 nm, rather than the 30 nm expected, when the most direct modeling approach of simply stacking tetranucleosomes was employed. In addition, the expected interaction between the H4 tail and the H2A-H2B acidic patch was not compatible with this model. The latter problem could be overcome by making reasonable adjustments to the model, but additional concerns have been raised regarding the possibility of artifacts due to crystal packing and, in the case of the 12-mer array studied by cross-linking and EM (Dorigo et al., 2003), due to cross-linking used to stabilize the condensed chromatin configuration prior to EM visualization. Bai and colleagues have pointed out that the cross-link used for this purpose, between H2A E64 and H4 V21, each mutated to Cys to allow a disulfide cross-link to be made under oxidizing conditions, is incompatible with the crystal structure obtained for the tetranucleosome (Zhou et al., 2018). They also showed that the cross-linked 12-mer array had a much larger sedimentation coefficient than the noncross-linked form, indicating that the cross-link likely captured a transiently formed, more compact conformation of the chromatin fiber that is not representative of the population as a whole. Indeed, Zhou et al. conclude that even reconstituted arrays having well-positioned nucleosomes exist as an ensemble of structures having variable degrees of compaction (Zhou et al., 2018).

Are these concerns fatal with regard to the evidence favoring a two-start helix? Perhaps not. Although Bai and colleagues make the case for structural heterogeneity of reconstituted nucleosome arrays, they also posit ladder-like stacking of nucleosomes in these arrays that could readily fold into a two-start helix. In addition, a subsequent cryo-EM study corroborated and refined the cross-linking and structural studies from the Richmond lab (Song et al., 2014). In this study, arrays of 12 and 24 nucleosomes with repeat lengths of 177 and 187 bp were reconstituted together with histone H1, and the structure was solved at 11 Å resolution. The results revealed a left-handed, two-start helix, consistent with that reported for the tetranucleosome, albeit with some differences from the original model. Specifically, both the fiber diameter and the separation between the two stacks of nucleosomes increased with increasing linker length, consistent with other work (see below) (Robinson et al., 2006). The helix was found to be built on a tetranucleosomal unit; contacts within each unit are made between H2A/H2B four-helix bundles from neighboring nucleosomes, while the H4 tail and the H2A-H2B acidic patch mediate contacts between neighboring tetranucleosome units.

In contrast to these findings, the Rhodes lab reported evidence for a one-start, solenoidal helix (Robinson et al., 2006). Their studies used long repeats, varying from about 55 to 72 nucleosomes, and having repeat lengths varying in 10 bp increments from 177 to 237 bp. These arrays were reconstituted with histone H5 and their compaction was measured by EM. The results indicated that the arrays folded into fibers in 1.6 mM $MgCl_2$ having diameters from 32.1 to 33.6 nm, with the value jumping about 10 nm for repeat lengths of 217 bp and larger. These dimensions and the density of nucleosomes along the fiber were found to be incompatible with a two-start helix; rather, they were most consistent with a one-start solenoid in which a high nucleosome density is attained by interdigitating solenoidal turns, as proposed by Daban and Bermúdez (Daban and Bermudez, 1998). Rhodes and colleagues also reported that arrays having short repeat lengths of 167 bp, as are found in yeast and neuronal cells but which are considerably shorter than the more typical repeat lengths of ~180–210 bp found in most species and cell types, do not bind linker histone at stoichiometric ratios and in the absence of linker histone give rise to narrow, highly regular 21 nm fibers, similar to the 24–25 nm diameter reported for the tetranucleosome studied by Richmond and colleagues (Routh et al., 2008).

In sum, more than 40 years after the first solenoidal model of chromatin compaction was proposed, the structure of the 30 nm fiber remains controversial and unresolved. Principles of higher levels of chromatin folding and organization, as in metaphase chromosomes, are largely unknown, although, as we shall see, newer methods have begun to shed light on this issue. But the problem may be even worse than this sounds. The increasingly well-defined systems used to examine chromatin folding, from isolated chromatin fragments, to reconstituted arrays assembled using regularly repeated DNA sequences and recombinant histones, to small fragments amenable to crystallization, in a sense represent reductionism gone wild. Although researchers have long recognized the inability of these systems to capture the inherent heterogeneity of chromatin as it exists in the nucleus, with its irregular linker lengths, variable histone modifications and presence of histone variants, and motley associated nonhistone proteins, critical regard of the application of the in vitro findings to living systems has often been absent. What relation do the structures that have been observed outside the cell have to those actually found in living cell nuclei?

Compact chromatin in living cells

As we have seen, following the first observations and proposal of a model for condensation of the chromatin fiber into higher-order structures, prodigious efforts were made to understand the nature of chromatin folding, in particular of the 30nm fiber. These investigations employed various methods for examining chromatin fragments isolated from nuclei and were extended to regular nucleosome arrays reconstituted from purified components. A major question not directly addressed by such studies is whether the structures observed in vitro were the same as those existing in situ. In light of recent work aimed at determining the nature of higher-order chromatin structures as they exist in vivo, which is discussed in the following, it is worth first revisiting a critical review of this topic published by Ken van Holde and Jordanka Zlatanova in 1995, nearly 20 years after the first model of the 30nm fiber was proposed (van Holde and Zlatanova, 1995). The main points elaborated in the review are as follows:

- Few experiments had examined chromatin fibers in situ, and those few often used cells having low transcriptional activity, such as chicken erythrocytes and starfish sperm, which may not be representative in terms of their chromatin structure.
- Two decades of experiments using electron microscopy had failed to produce convincing evidence for a regular helical structure present in anything larger than short segments of chromatin. The authors write: "Paper after paper presents multiple (presumably selected) images of fibers in which tiny bits are pointed out by arrows as indicating whatever model the authors champion." The problem with this, of course, is that most of the chromatin strands in such studies do not show strong visual indication of conforming to specific models of condensed chromatin. A careful EM study indicated nonuniform diameters for chicken erythrocyte chromatin (Gerchman and Ramakrishnan, 1987) (although another study using cryo-EM found 30nm fibers to be present (Woodcock, 1994)), while studies reporting evidence for helical structures found these only in limited regions of chromatin (Bartolome et al., 1994; Williams et al., 1986).
- Data from low angle X-ray and neutron scattering experiments were more consistent with a generally irregular fiber than with a pervasive, regular helical structure.
- Linear electric dichroism and flow dichroism measurements interpreted as evidence for helical structures having nucleosomes at a fixed angle relative to the helical axis were equally consistent with randomly oriented nucleosomes. Although such studies did provide evidence for the orientation of nucleosomes relative to the fiber, they did not imply the existence of a regular structure throughout.

- Considerations of the mechanism by which an open fiber (at low salt concentration) having variable linker lengths might fold into a more condensed structure suggested an "accordion-like," longitudinal contraction that would not likely produce any kind of regular helical structure.

These concerns did not appear to disturb researchers engaged in intensive studies aimed at determining specific structures for the 30nm fiber. However, technological advances permitted new approaches to examining chromatin structure, and several studies using these newer methods as well as longer-established approaches contradict the notion of a regular 30nm chromatin fiber in vivo. Cryo-EM has been particularly important in this regard. Classical transmission EM studies typically involved the lysis of nuclei in low salt buffer, followed by chemical treatment with a cross-linking agent (most commonly glutaraldehyde) to fix structures in place; the fixed chromatin was then dehydrated with alcohol and embedded into the plastic. Each of these treatments carries the danger of introducing artifacts (see Van Holde (1988, p. 391)). Cryo-EM circumvents many of the associated problems of sample preparation by using rapid cooling to vitrify aqueous samples; that is, to freeze them into a glass-like solid rather than allowing microcrystals associated with slow freezing to form, which can disrupt the microscopic cellular structure. These samples are then imaged in the EM, after being cut into thin sections, with no chemical fixation or staining. The application of cryo-EM to human mitotic chromosomes, used in conjunction with correction for a potential artifact that can be introduced during the processing of the images, did not support the existence of 30nm fibers in situ (Eltsov et al., 2008). However, earlier work using small angle X-ray scattering (SAXS) of HeLa mitotic chromosomes had reported a reflection at 30nm suggested to be caused by side-by-side packing of 30nm fibers (Langmore and Paulson, 1983), and this result was reproduced by the authors of the cryo-EM study, creating an apparent conflict between the two sets of results (Nishino et al., 2012). The conflict was resolved when the authors showed that chromatin fibers prepared as in the earlier study, and which gave rise to SAXS peaks as had been reported, were associated with ribosomal aggregates that were responsible for the 30nm reflection (Nishino et al., 2012). As a positive control, cryo-EM was able to detect 30nm fibers in nuclei from chicken erythrocytes and starfish sperm (which, as mentioned above, exhibit exceptionally low transcriptional activity) (Nishino et al., 2012; Woodcock, 1994), showing that the 30nm fiber was not inherently difficult to detect by this method.

Another method used to examine condensed chromatin in a more "natural" state is electron spectroscopic imaging (ESI). In this technique, the requirement for heavy metal contrast agents, which can affect the apparent size of the structure under examination, is avoided by using electrons having energies that allow interactions with specific elements (most relevantly, phosphorus and nitrogen) to be visualized (Bazett-Jones, 1992). The use of ESI to visualize phosphate atoms enables high contrast images of the DNA in chromatin structures to be obtained, and revealed the presence of 30nm fibers in starfish sperm nuclei, but not in mouse somatic cells, with observed fibers in the latter being no wider than 10nm (Bazett-Jones, 1992; Fussner et al., 2012).

Two other studies, both employing groundbreaking advances in imaging techniques, also support a model in which chromosome compaction in nuclei does not involve the formation of a 30nm regular fiber. One study employed stochastic optical reconstruction microscopy

(STORM), also known as super-resolution microscopy, to examine chromatin fibers in nuclei of human fibroblasts and induced pluripotent stem cells (Ricci et al., 2015). The results indicated that rather than being arrayed in 30nm fibers, nucleosomes condensed into groups of varying size, termed "clutches" (perhaps analogous to superbeads) that were interspersed with nucleosome-depleted regions. The size of the clutches varied according to cell type, and larger clutches were more associated with histone H1 and heterochromatin, while smaller clutches were enriched in associated RNA polymerase II. Another study employed chromatin electron microscopy tomography, or ChromEMT, which leveraged advances in labeling and image acquisition and analysis to enable high-resolution determination of the structure of chromatin fibers (Ou et al., 2017). Once again, no evidence was found for a 30nm fiber configuration; instead, heterogeneous arrays of nucleosomes with diameters varying from 5 to 24nm were observed in both interphase and mitotic nuclei from human small airway epithelial cells, with no apparent higher-order structure being present. An important experiment in this study addressed the apparent inconsistency with SAXS and cryo-EM studies that reported 30–40nm fibers (Langmore and Paulson, 1983; Langmore and Schutt, 1980; Scheffer et al., 2011). When chicken erythrocyte nuclei were lysed under low salt conditions and treated with 2mM $MgCl_2$ to induce chromatin compaction, fibers having a diameter from 30 to 40nm were observed by ChromEMT, as in the earlier studies (Ou et al., 2017). This result raises the prospect that the relatively regular, helical fibers of ~30nm observed in earlier EM studies could be preparative artifacts, possibly due to the extensive dilution of chromatin fibers favoring intrafiber over interfiber contacts. Most recently, the path of DNA in chromatin fibers from partially unfolded mitotic chromosomes was analyzed by cryo-electron tomography (Beel et al., 2021). Segmentation of tomograms was performed manually, which, although limiting throughput, allowed increased accuracy; the resulting analysis revealed linker trajectories to be irregular, concordant with 30nm structures being absent or nearly so from mitotic chromosomes.

Eltsov et al. proposed a model for chromatin condensation based on the principle that compaction requires nucleosome-nucleosome interactions (Eltsov et al., 2008). Under dilute conditions, as when chromatin is isolated from nuclei, such interactions will readily occur within a single fiber, but not between fibers, thereby favoring the formation of the 30nm fiber configuration. However, within the confines of the nucleus, interfiber interactions are much more likely; these interactions will interrupt intrafiber interactions, and thus interfere with the formation of the 30nm fiber. The authors argue that their observations are most consistent with the forces of interfiber and intrafiber nucleosomal associations being similar, a condition known as a "melt" in polymer physics; they therefore refer to this model as a "polymer melt," in which no global secondary DNA folding exists, and postulate that the high density of chromatin in the nucleus is attained by interdigitation of adjacent fibers occurring through internucleosome interactions (Eltsov et al., 2008). Such interdigitation has been observed directly in a structural study in which nucleosome fiber crystals were obtained beginning with dinucleosome templates having single-stranded, "sticky" ends allowing ligation into long chains (Adhireksan et al., 2020). These chains formed open, zigzag conformations that interdigitated with one another into compact structures. The zigzag conformations appear to result from energetic constraints on the bending of linker DNA, which have long been considered to be a critical factor in dictating the structure of folded chromatin. The nature of the interactions facilitating interdigitation with two distinct nucleosome dimer templates depended on linker length, in spite of both templates yielding chains that formed zigzag conformations: in one case, the nucleosome cores in one fiber contact the linker DNA in its partner, while in the other, differing by 2 and 4 bp in the length of the two linker DNA segments present in the dinucleosome unit, nucleosome cores are juxtaposed while linkers are present in an isolated zone between the two fibers. The consequences of such variable fiber-fiber interactions that depend on linker length for chromatin compaction in vivo, in which linker lengths are heterogeneous, remain to be explored.

The polymer melt model, which is supported by cryo-EM and SAXS studies of HeLa mitotic chromosomes (Chicano et al., 2019) implies that chromatin is dynamic, with nucleosome arrays constantly moving and rearranging locally. Notably, fluorescence recovery after photobleaching (FRAP) experiments indicate that histone H1 exhibits mobility with a half-time of a few minutes in mitotic chromosomes, and in chromosomes of both mouse and human cells, which is rapid compared to the cell cycle (Chen et al., 2005; Horowitz et al., 1994). Such dynamic behavior is difficult to reconcile with a model of a static 30 nm fiber in which linker histone is confined to the interior of the helix, as prescribed in most such models, but easily accommodated in the polymer melt model. The polymer melt model is also consistent with the nuclear organization of chromatin being fractal; that is, having similar characteristics regardless of the magnification at which it is viewed (Bancaud et al., 2012).

This revision in ideas regarding chromatin condensation in the cell nucleus necessitates a reevaluation of ideas based on the supposition of the 30 nm fiber as the lowest level of chromatin compaction in the cell. For example, the effects of altering the properties of the acidic patch of H2A-H2B on chromatin folding are consistent with the functional effects of histone variants with altered acid patch properties, as discussed earlier; thus, the role of the interaction between the acidic patch and the H4 tail needs to be reinterpreted in terms of condensation of chromatin in a state more disordered than the 30 nm fiber, as in the polymer melt model. Inasmuch as the polymer melt model, as well as liquid-liquid phase separation that is discussed further on, depend on nucleosome-nucleosome interactions as does the condensation yielding the 30 nm fiber, accommodating accumulated knowledge regarding the role of the acidic patch, histone tails, histone acetylation, and linker histones into this new framework does not require major conceptual revision. Furthermore, the 30 nm fiber does appear to be present in at least some cell types having low transcriptional activity (Fussner et al., 2012; Woodcock, 1994); future studies are likely to address the generality of these observations. Models for chromatin condensation continue to evolve, and the next few years appear likely to yield new insights into this longstanding issue.

Chromosome architecture

Chromosomes were first reported by Walther Flemming in 1882 as stained bodies visible during the mitotic phase of cell division (Fig. 1.3); the stained structures disappeared as cells "regressed" back to interphase (Paweletz, 2001). This transition reflected the condensation of chromatin during mitosis, which ranges from ~4- to 50-fold (in metaphase relative to interphase chromatin) in mammalian cells (Belmont, 2006), and its decondensation following anaphase. The conspicuous condensation and evident structure of the mitotic chromosome predisposed researchers generally to consider chromatin folding in that context; indeed, textbooks typically depict the mitotic chromosome as the uppermost level of chromosome condensation (Fig. 3.1). Nonetheless, considerable compaction is also required to confine DNA into the dimensions of the nucleus even during interphase, and it is during interphase that chromatin structure is most relevant to transcriptional regulation and hence to cell function and identity. In spite of the apparently random appearance of the interphase nucleus, it was recognized early on that interphase chromosomes might well be subject to some type of organization, and that understanding the principles underlying any such organization was a worthy goal. Comings, for example, asserted in a 1968 review that the obvious order imposed upon chromosomes during mitosis or meiosis and subsequent cell division argued that the nucleus was "a well-ordered place," and suggested that such order could be maintained by chromatin being attached to the nuclear membrane at specific sites dictated by where DNA synthesis was initiated within each replicon (Comings, 1968). Benyajati and Worcel, in a groundbreaking analysis of chromosome structure in *Drosophila* cells (discussed below), stated that it "seems probable that interphase chromosomes may have a defined structure and that nuclei are not just bags full of chromatin" (Benyajati and Worcel, 1976). This interest in chromosome structure and nuclear organization has continued in the decades since. In this section, we briefly discuss efforts to understand the organization of both mitotic and interphase chromosomes in the cell nucleus.

Chromosome territories

The idea that individual chromosomes occupied distinct territories within the eukaryotic nucleus, rather than intermixing, was invoked in the earliest years of cytological studies (Boveri, 1909; Rabl, 1885) and reviewed in Cremer and Cremer (2010) and Cremer et al. (1993). The failure to observe distinct chromosome territories, or CTs, by EM in the late 1960s temporarily raised the profile of alternative models in which chromosome strands were intermixed in the nucleus, but new evidence in favor of chromosome territories emerged in the 1970s. Indirect evidence for distinct chromosomal domains came from studies in which DNA in living Chinese hamster cells was damaged by a highly focused UV laser beam, followed by the incorporation of radiolabeled thymidine by the cell's damage repair apparatus. These experiments showed that damage was generally localized to a few chromosomes, suggesting at most limited intermixing (Cremer and Cremer, 2010). More direct and convincing evidence came with the advent of fluorescent in situ hybridization, or FISH, in which prespecified loci or genomic regions are visualized by hybridization of fluorescently labeled RNA or DNA probes following fixation of cells. Most compellingly, "chromosome painting" experiments, in which two or more composite probe pools are labeled with different fluorochromes, permitted visualization of individual chromosomes in fixed nuclei. Generation of these probe pools was initially made possible by the use of fluorescent-activated cell sorting to isolate individual chromosomes, which could then be used to construct corresponding DNA libraries. Methods were developed to deplete or to suppress signal from repetitive sequences present in the libraries, which would by nature not be chromosome-specific. More recently, the ability to synthesize large, customized pools of oligonucleotides free of repetitive sequences has allowed construction of probe libraries with which select genomic regions—for example, all of the exons on a particular chromosome—can be visualized (Boyle et al., 2011).

Using these methods in conjunction with confocal microscopy and sophisticated 3D image reconstruction revealed individual chromosomes to occupy distinct regions of the cell nucleus (Fig. 3.7) (Cremer and Cremer, 2001; Ried et al., 1998). For example, human chromosomes 18 and 19 were shown to be preferentially located near the nuclear periphery and more internally, respectively (Croft et al., 1999). Similar preferential radial positioning was later observed for the complete complement of human chromosomes (Boyle et al., 2001).

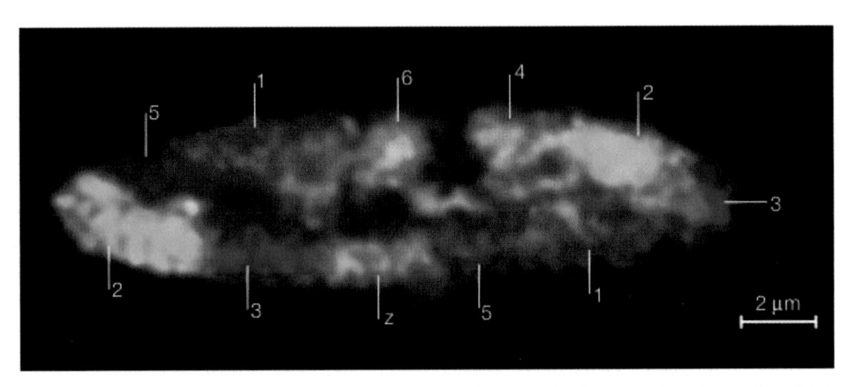

FIG. 3.7 Chromosome painting of chromosomes in the chicken. Mid-plane section through a chicken fibroblast nucleus reveals distinct territories for individual chromosomes. Homologs of chromosomes 1–3 and 5 occupy separate locations; only one homolog for each of chromosomes 4 and 6 is visible in this section. *From Cremer, T., Cremer, C., 2001. Chromosome territories, nuclear architecture and gene regulation in mammalian cells. Nat. Rev. Genet. 2, 292–301.*

Although visualization experiments indicate that chromosomes occupy relatively constrained volumes in the cell nucleus rather than being spread throughout with concomitant intermingling, the term "chromosome territory" is misleading insofar as it suggests that chromosomes are situated uniquely in the nuclear volume. There is no unique three-dimensional map of chromosomes in the nucleus of a given cell type; chromosomes that are close neighbors in a mother cell may not be so proximate in the daughter cells following cell division, and vice versa (Walter et al., 2003). As this dynamic repositioning of chromosomes implies, chromosome territories reflect averages over cell populations; a particular chromosome may generally be more associated than another with the nuclear periphery, for example, but its exact position will vary from cell to cell (Dixon et al., 2016).

The radial positioning of chromosomes is related, to a degree, to gene density. As mentioned earlier, chromosome painting established that human chromosome 18, which is relatively gene poor, is localized toward the nuclear periphery in lymphoblastoid cells, while the gene-rich chromosome 19 occupies a more central position in the nucleus (Croft et al., 1999); extension of this work to include the full complement of human chromosomes confirmed the generality of this observation (Boyle et al., 2001). Similarly, gene-rich and gene-poor regions of the same chromosome exhibit preferential localization closer to the nuclear axis or periphery, respectively (Boyle et al., 2011; Kupper et al., 2007). Peripheral localization of gene-poor domains has been corroborated in DamID experiments (Guelen et al., 2008). In these experiments, the *E. coli* DNA adenine methyltransferase was tethered to the nuclear periphery by its expression as a fusion with lamin B1, resulting in adenine methylation of GATC sequences near the nuclear lamina. Enrichment of methylated sequences using a PCR protocol, followed by sequencing, identified about 1300 lamin-associated domains, or LADs, that were shown to be relatively gene-poor. That such localization could influence gene expression was suggested by experiments in which genes were artificially relocalized to the nuclear periphery by use of a gene fusion between an inner nuclear membrane protein and the *E. coli* Lac repressor, in conjunction with a reporter gene having nearby LacO sites (Finlan et al., 2008; Kumaran and Spector, 2008; Reddy et al., 2008). Artificially tethering reporter genes to the nuclear periphery in this way resulted in repression in some instances. However, repression was variable and not always observed; similarly, when genes are examined across the nucleus, localization and gene activity are not well correlated (Meaburn and Misteli, 2008). Altogether, these findings indicate that although a functional link exists between gene activity and nuclear localization, the strength of the link depends on other, unknown factors to a variable extent.

Some preferential positioning of chromosomal regions is also seen within individual chromosome territories. While each chromosome appears to occupy, in general, a distinct territory and not intermingle much with other chromosomes, looping out of transcriptionally active regions from the core chromosome territory has been observed, most conspicuously in experiments in which custom FISH probes that cover all the exons of a given chromosome are used together with chromosome painting probes that allow the core CT to be visualized (Boyle et al., 2011). This outward migration from core CTs is also supported by experiments based on chromosome conformation capture, or 3C, approaches, which are discussed later on. These experiments show that interchromosomal interactions, which require proximity between sequences present on different chromosomes, occur predominantly between sequences observed to loop out from core CTs in FISH experiments (Kalhor et al., 2011; Mahy et al., 2002). Looping out of transcriptionally active regions of chromatin from core CTs may allow active genes to gain access to transcriptional machinery that is largely excluded from core CTs; this may assist their inclusion in liquid condensates that are enriched for Pol II and other components of the transcriptional apparatus, and that can form dynamic nuclear compartments (discussed in the section on liquid-liquid condensates).

Loops

Following the solution of the structure of the DNA double helix, how this structure was arranged in the chromosomes of living cells became a major question. Taylor and colleagues addressed this problem by labeling DNA in English broad bean seedlings, chosen for their morphologically distinct, large chromosomes, with tritiated thymidine and following the labeled DNA through multiple cell divisions (Taylor et al., 1957). (The use of tritiated thymidine was an important technical advance, as it had been shown not to be incorporated into RNA, and in contrast to other available radioisotopes, its low energy emission allowed sufficient resolution to distinguish individual chromosomes by autoradiography.)

They observed that the initially uniformly incorporated label was maintained in a single chromosome through succeeding chromosome duplications and nuclear divisions, save for an occasional chromatid exchange, indicating that the semiconservative mechanism of DNA replication proposed by Watson and Crick, and later demonstrated by Meselson and Stahl, was correct (Meselson and Stahl, 1958; Taylor et al., 1957; Watson and Crick, 1953). However, Taylor et al. considered the idea of the chromosome as consisting of a single, continuous double helix of DNA to be "inconceivable," stating that "chromosomes are much more likely to be composed of multistranded units." Meselson and Stahl, using density weight centrifugation following incorporation of ^{15}N into *E. coli* chromosomes, subsequently provided their own demonstration of semiconservative DNA replication while also showing that the *E. coli* chromosome was a single continuous molecule of DNA (Meselson and Stahl, 1958). Kleinschmidt et al. published a study in *Zeitsschrift für Naturforschung* in 1961 that included beautiful micrographs of DNA spilling out of lysed *M. lysodeikticus* as apparently unbroken, massively long molecules, leading them to conclude that "the total amount of DNA most probably [exists] as a single threadlike unit" (Kleinschmidt et al., 1961). Nonetheless, the debate over the continuity of DNA in large eukaryotic chromosomes continued through the 1960s.

Joe Gall, initially intrigued by the remarkable size and morphology of meiotic chromosomes from immature oocytes of the newt, termed "lampbrush chromosomes," used microscopy to observe the kinetics of chromosome breakage by DNase in the chromosome loops that extend laterally from the main chromosome axis (Gall, 1963). From a careful analysis of the results, he concluded that the data were most simply explained if the lampbrush chromatid consisted of a single very long DNA double helix. Ten years later, a study from Bruno Zimm's lab reported viscoelastic measurements that provided evidence that *Drosophila* chromosomes each contained a long, continuous molecule of DNA (Kavenoff and Zimm, 1973). This work avoided potential artifacts of isolation and measurement in prior studies and essentially settled the debate. Interestingly, the biophysical approach of the Zimm lab helped to inspire the invention of pulsed-field gel electrophoresis by David Schwartz, which allowed the separation of whole-chromosome-sized DNA on agarose gels (Schwartz and Cantor, 1984).

Given that chromosomal DNA consisted of continuous double helices extending up to millions of base pairs in length, how were these enormous molecules organized? Several studies conducted in the mid-1970s provided evidence for the existence of topologically independent chromosome domains on the order of tens of kilobases. Benyajati and Worcel monitored changes in the sedimentation coefficient of intact interphase chromosomes from *Drosophila* cells, prepared by gentle lysis of nuclei in 0.9 M NaCl and 0.4% Nonidet P40, during a time course of DNase I treatment (Benyajati and Worcel, 1976). The use of the nonionic detergent, Nonidet P40, was critical, as it had been shown to allow the removal of the nuclear envelope and most of the phospholipids while preserving the general ultrastructural characteristics of chromatin observed in the cell and leaving the normal complement of histones, nonhistone proteins, and RNA intact (Hancock, 1974). (For these same reasons, Nonidet P40 is commonly used in experiments in which chromatin in cell nuclei is digested with MNase or other externally added cleavage reagents too large to passively enter the cell nucleus.) The results of following the changes in sedimentation coefficient during the course of DNase I treatment showed that the values gradually decreased until they reached a plateau, and the authors' analysis suggested a model in which chromosomal domains are constrained into supercoiled loops averaging ~85,000 bp in size. This interpretation was based on the notion that large supercoiled loops were initially present in the isolated chromatin, despite the presence of active topoisomerase in the cells utilized that might be expected to relax these loops. These loops would be relaxed upon being nicked by DNase I, with resulting change in sedimentation properties. The presence of supercoils was rationalized as being due to the high salt concentration inhibiting topoisomerase while also inducing, along with the temperature shift from 26°C to 4°C, a change in DNA topology that resulted in the generation of supercoils (Depew and Wang, 1975). In retrospect, it seems likely that at least some loss of histones occurred at the high salt concentration that was used, especially of H2A-H2B, with accompanying perturbation of nucleosome structure and alteration in supercoiling (Cook et al., 1976; Ohlenbusch et al., 1967). Nonetheless, the study provided some of the first strong evidence for chromosomal loop domains in interphase nuclei. Contemporaneous studies from Peter Cook and colleagues used γ-irradiation to nick DNA in nuclear structures prepared similarly to those investigated by Benyajati and Worcel, and monitored the change in sedimentation rate and the binding of ethidium bromide, which differs between supercoiled and relaxed DNA (Cook and Brazell, 1975, 1976). They concluded that DNA in chromosomes existed as topologically isolated domains in a wide variety of cell types, including HeLa cells, chick embryo fibroblasts, *Drosophila* cells, and the African clawed frog *Xenopus laevis* (Cook and Brazell, 1976). Their initial estimates of the size of these domains were much higher than those deduced by Benyajati and Worcel but were later refined to similar values (e.g., 220 kb in HeLa cells) (Cook and Brazell, 1978).

Independent evidence for the existence of anchored loops of DNA in interphase chromatin was derived from EM experiments (Lebkowski and Laemmli, 1982a,b; McCready et al., 1979). Nuclear preparations made as described earlier, or by extraction with digitonin along with either 2 M NaCl or a mixture of heparin and dextran sulfate to remove the histones, allowed visualization of a "halo" of DNA loops surrounding a dense, fibrous structure termed by Laemmli the nuclear scaffold (Fig. 3.8A). Examination of metaphase chromosomes from HeLa cells similarly revealed DNA fibers that appeared to be anchored to create independent domains, but in this case to a "core ribbon" or "scaffold" that was best resolved after the removal of histones (Fig. 3.8B) (Mace Jr. et al., 1977; Marsden and Laemmli, 1979; Paulson and Laemmli, 1977; Stubblefield and Wray, 1971). The dimensions of the loops varied according to the method of preparation: at moderate ionic strength or in the presence of 1 mM Mg^{+2}, fibers of 20–30 nm were observed; chelation of divalent cations caused swelling of chromosomes to occur, and the thick fibers relaxed to 10 nm fibers that were "probably a linear array of nucleosomes" (Marsden and Laemmli, 1979). Conversely, at high ionic strength, fibers of ~50 nm thickness were observed, which were ascribed to twisting of the 30 nm fiber upon itself (Mace Jr. et al., 1977; Marsden and Laemmli, 1979). The size of the loops was similar between interphase and metaphase nuclei preparations and in good accord with estimates derived from digestion experiments (Benyajati and Worcel, 1976; Cook and Brazell, 1975; Igo-Kemenes et al., 1977).

The scaffold underlying the metaphase chromosome was extraordinarily stable at 2 M NaCl, enabling its isolation following treatment of chromosome preparations with MNase

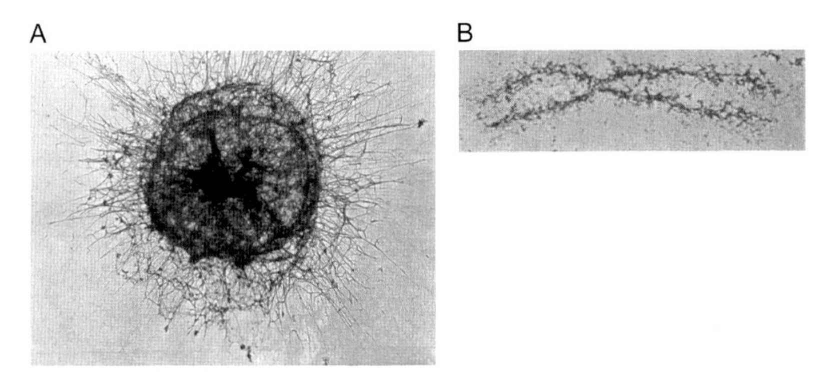

FIG. 3.8 (A) Electron micrograph of a HeLa nucleoid spread. The more densely staining "nuclear cage" is about 15 mM in diameter. (B) Electron micrograph of a chromosomal scaffold prepared from metaphase HeLa chromosomes by MNase digestion followed by dilution into 2 M NaCl. *(A) From Jackson, et al., 1984. J. Cell Sci. Suppl. 1, 59–79. (B) From Adolph, K.W., Cheng, S.M., Paulson, J.R., Laemmli, U.K., 1977. Isolation of a protein scaffold from mitotic HeLa cell chromosomes. Proc. Natl. Acad. Sci. U. S. A. 74, 4937–4941.*

and removal of the histones (Adolph et al., 1977). Biochemical analysis identified topoisomerase II and a subunit of the condensin complex as major components of the metaphase chromosome scaffold (Earnshaw et al., 1985; Saitoh et al., 1994). This identification was later substantiated by visualization of condensin (by immunofluorescence) and of a topoisomerase II fusion with enhanced green fluorescent protein, which showed axial localization of both in native chromosomes (Maeshima and Laemmli, 2003; Ono et al., 2003; Tavormina et al., 2002). Functional evidence for the involvement of topoisomerase II in chromosome condensation was obtained from in vivo and in vitro experiments. The use of a cold-sensitive mutant of topoisomerase II in fission yeast showed that inactivation of the enzyme prevented normal chromosome condensation, which was restored upon return to the permissive temperature (Uemura et al., 1987). Chromosome condensation could be recapitulated using cell extracts, and condensation in this system also required topoisomerase II (Adachi et al., 1991; Hirano and Mitchison, 1991). However, once chromosomes were assembled in vitro, their structure was not perturbed by the extraction of topoisomerase II. Thus, topoisomerase II was required for establishment, but not for maintenance, of condensed chromosome structure, and it was not a requisite structural component of the scaffold (Hirano and Mitchison, 1993).

The question of the nature, and indeed the existence, of the interphase nuclear scaffold, also sometimes referred to as the nuclear matrix or skeleton, is more problematic. The nuclear matrix was first reported as a proteinaceous structure, approximating the shape and some morphological features of the nucleus, that remained after removal of chromatin by treatment with 2 M NaCl, and gained considerable attention when it was reported to be associated with newly replicated DNA (Berezney and Coffey, 1974, 1975). Substantial effort was spent in analyzing this structure (and variants thereof, prepared by modifications of the isolation procedure) and identifying its component parts, but it also was the subject of skepticism from the beginning. Principal objections were the lack of evidence for the nuclear matrix existing in living cells and the possibility that it represented an experimental artifact formed by aggregation or spurious association of proteins during the very nonphysiological isolation procedure; additional problematic issues are raised in reviews by Hancock and by Razin and colleagues (Hancock, 2000; Razin et al., 2014). A possibility raised in these reviews is that the functions attributed to the nuclear matrix, namely, in providing a platform for organizing nuclear processes including transcription and replication, could instead be supported by other mechanisms, including molecular crowding. The recent concept of liquid condensates, discussed later, may be relevant to this idea.

Characterization of the components of the nuclear matrix has reproducibly identified the major nuclear lamina proteins, lamins A, B, and C; other proteins have been variously identified as matrix components, depending on the method of preparation (Razin et al., 2014). Intriguingly, the metaphase scaffold component, topoisomerase II, has been reported also to be associated with the nuclear matrix (Berrios et al., 1985). Another link between the putative interphase nuclear scaffold and the metaphase chromosome scaffold involves the condensin complex. As mentioned earlier, subunits of the condensin complex were identified as metaphase chromosome scaffold components; in particular, ScII (Scaffold component II) was discovered to be a homolog of a member of the "structural maintenance of chromosome," or SMC, family in yeast, that is involved in chromosome condensation (Maeshima and Eltsov, 2008; Saitoh et al., 1994). Two condensin complexes may play distinct roles in chromosome condensation in metaphase and interphase, as their cellular localization differs and the effect of knockdown is specific, with depletion of condensin II but not of condensin I delaying initiation of prophase chromosome condensation (Hirota et al., 2004).

In sum, although the nature and even the existence of a stable nuclear matrix are uncertain, considerable evidence indicates that in both interphase and metaphase nuclei, chromosomes are organized in part as topologically independent loops of DNA. Additional evidence for this model derives from more recent experiments using high throughput sequencing methods, as discussed in the next section.

Topologically associating domains or TADs

We have seen how the visualization of chromosomes or chromosomal regions using microscopy-based methods has established the existence of chromosome territories. Higher-resolution information on the spatial relationship between genomic regions has been obtained by using FISH to measure the distance between two loci labeled with different fluorescent probes. This is a fairly laborious approach in which data gathered from observations of individual cells is collated and processed to obtain distance measurements between specified loci, and resolution is limited to separations of >100 kb (Gilbert et al., 2004). An alternative and complementary approach is Chromosome Conformation Capture, or 3C (Dekker et al., 2002). This method is based on measuring the relative frequency with which fragments of DNA that are distant from one another in the linear chromosomal sequence, or that reside on different chromosomes entirely, can be ligated together following fixation and cleavage by restriction enzymes (Fig. 3.9). Since ligation depends directly on proximity, this approach provides a measure of how frequently genomic regions interact, and hence the spatial organization of the genome.

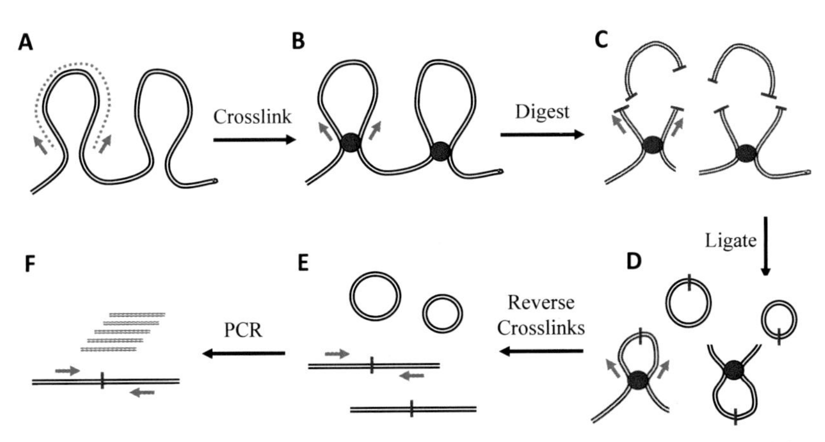

FIG. 3.9 The chromatin conformation capture (3C) protocol. Looped-out regions of chromatin (A) are captured by cross-linking (B); *blue arrows* depict regions corresponding to primers used for eventual analysis. Following the removal of noncross-linked proteins from DNA, the cross-linked chromatin is digested with a restriction enzyme (C), followed by ligation, which results in close apposition of chromosomal sequences that were originally far from each other (D). Cross-links are then reversed and the ligated DNA is purified to allow analysis by PCR (E and F); the indicated primers, originally far apart in the chromosomal sequence (B), yield a short product indicating that these sequences were spatially proximate when cross-linked. *Credit: Courtesy of Elisabeth Knoll.*

By nature, 3C does not provide direct measurements of the distance between chromosomal loci but rather assesses the relative frequency of close contact between two loci. Extensions of the original method have expanded its use to allow the mapping of interactions of megabase regions of chromosomes with one another, and even of interactions genome-wide (see Sidebar: 3C: Precursors and extensions). These methods initially were mostly applied to elucidating the physical basis of functional interactions, for example, between enhancers and their target promoters at distances from a few thousand to on the order of a million base pairs (Dekker and Misteli, 2015). More recently, they have provided insights into the organizational principles of the genome. Initial studies showed that contacts between even relatively distant regions within a given chromosome are much more frequent than interchromosomal contacts, consistent with the existence of chromosome territories (Lieberman-Aiden et al., 2009). The same studies also found more interchromosomal contacts among gene-rich chromosomes than between gene-rich and gene-poor chromosomes, consistent with the preferential radial positioning of chromosomes according to gene density discussed earlier. Furthermore, transcriptionally active and inactive chromosomal regions appear segregated in space: active regions, referred to as the A compartment and corresponding to open chromatin, preferentially associate with other active regions while inactive regions, referred to as the B compartment and corresponding to closed chromatin, are generally localized to distinct subnuclear regions. The interactions among active regions may represent proximity between domains looped out from compact, inactive core chromosome territories, as discussed earlier, while less actively transcribed genes may be sequestered in the more condensed, core regions of their respective chromosome territories. Interactions among active genes are not specific, but vary between cells, and appear to represent a sort of functional compartmentalization.

Sidebar: 3C: Precursors and extensions

Two reports that presaged the 3C approach were published in 1988 and 1993. Mukherjee et al. showed in an in vitro experiment that the cyclization of two linear fragments of DNA by T4 DNA ligase was enhanced when the bacterial replication initiator protein of the plasmid R6K brought the two segments together by binding sequences important for R6K replication (Mukherjee et al., 1988). The approach was later adapted to show estrogen-stimulated enhancement of interaction between the rat *prolactin* promoter and enhancer, which are separated by about 1500 bp (Cullen et al., 1993). Although the latter experiment showed that the assay could be performed on a minichromosomal template present in isolated nuclei, and introduced the use of PCR amplification as part of the detection protocol, it was another 9 years before Job Dekker and colleagues in Nancy Kleckner's lab showed the feasibility of ligation-mediated amplification as a means of identifying interactions between chromosomal DNA sequences separated by thousands of base pairs (Dekker et al., 2002).

With high throughput methods becoming available at about the same time as 3C was introduced to the wider scientific community, adaptations in the original protocol that allowed more extensive mapping of chromosomal proximity soon followed. The major variants include.

- 4C: Whereas the original 3C method examines the relative frequency of interactions between preidentified DNA sequences, in 4C all interactions with a given region are mapped by first circularizing the initial ligation products, which are then amplified using primers pointing outward from the "bait" sequence (Simonis et al., 2006; Zhao et al., 2006). The resulting fragments are then quantified using microarray technology or by direct sequencing and mapped to their chromosomal locations, yielding information on all of the sequences interacting with a selected bait, or target, sequence. (In the original publications, this extension was given different names—in one case, chromosome conformation capture-on-chip, in the other, circular chromosome conformation capture. Happily, both can be appropriately abbreviated as 4C.)
- 5C (Chromosome Conformation Capture Carbon Copy): This method extends 3C by allowing the detection of multiple interactions between pairs of defined regions (Dostie et al., 2006). In this approach, multiplexing permits high throughput analysis of thousands of potential junctions between interacting regions to be assessed simultaneously, making it useful for interrogating the nature of interactions between, for example, promoter and enhancer regions spanning thousands of base pairs.
- Hi-C: In this "all-to-all" approach, genome-wide interactions are captured by sequencing the entire collection of cross-linked and ligated fragments obtained using the 3C technique (Fig. 3.10) (Lieberman-Aiden et al., 2009). Sensitivity is gained by selective enrichment of ligated products; this is achieved by first filling in the overhangs, or "sticky ends," that are generated by restriction enzyme digestion (e.g., with *Hin*dIII) with nucleotides, one of which is biotinylated. The blunt-end fragments thus generated are then ligated and fragments containing biotinylated junctions are purified on streptavidin beads before library construction and sequencing. Although this approach allows unbiased determination of all genome-wide interchromosomal and intrachromosomal interactions, resolution for the large genomes of metazoan cells is typically limited by sequencing depth to ~0.1–1 Mb.
- ChIA-PET (Chromatin Interaction Analysis with Paired-End Tags): This method modifies the Hi-C technique by the inclusion of an immunoprecipitation step, using an antibody against a protein of interest, before ligation and subsequent 3C processing (Fullwood et al., 2009). This allows greater read depth for the regions of interest, resulting in a focused interaction map of interactions

involving the selected protein at a higher resolution than Hi-C allows. In the prototype publication, genome-wide interactions involving the estrogen receptor α were determined in the human breast cancer cell line MCF-7, revealing extensive interactions among ER-α bound promoters through long-range chromatin looping (Fullwood et al., 2009).

The existence of topologically associating domains, or TADs, was first revealed by experiments using 5C to investigate the organization of the inactive mammalian X chromosome (Sexton et al., 2012), and by Hi-C studies mapping intrachromosomal and interchromosomal contacts genome-wide in *Drosophila* embryonic nuclei and human and mouse embryonic stem cells (Dixon et al., 2012; Nora et al., 2012). Data from Hi-C experiments are typically represented in the form of heat maps, in which the axes correspond to chromosomal positions (Fig. 3.10).

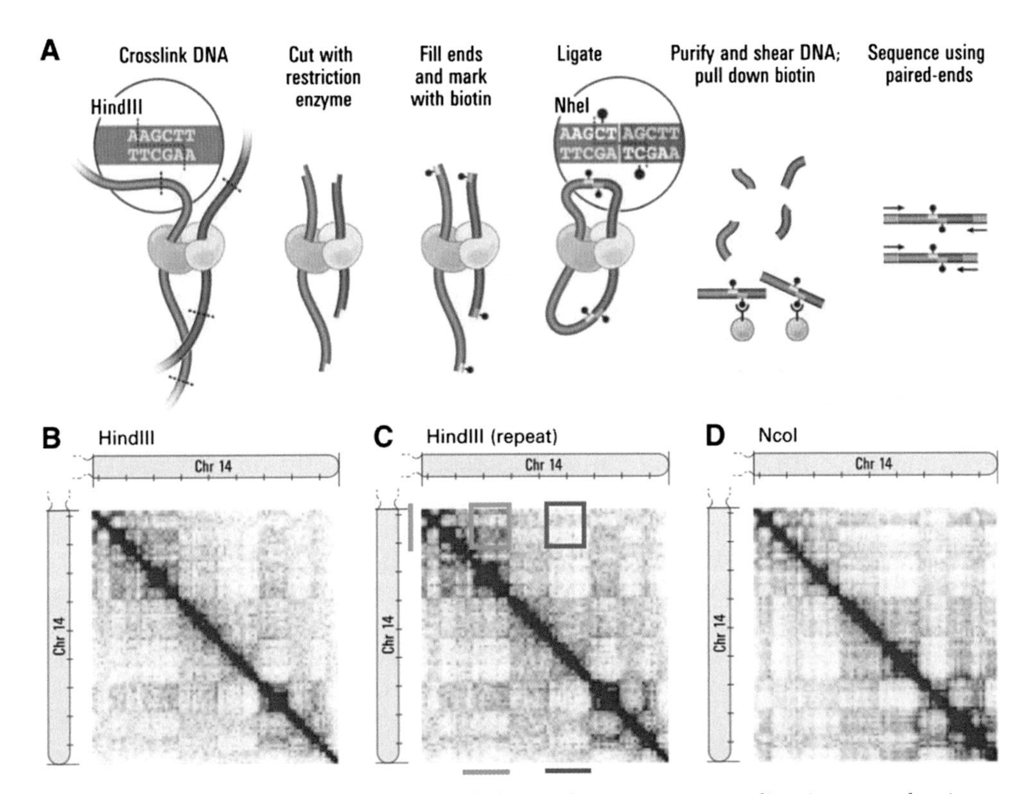

FIG. 3.10 (A) Overview of Hi-C. Following cross-linking and restriction enzyme digestion, gapped regions generated by restriction enzyme digestion are filled in with the inclusion of biotinylated nucleotides. Ligation is performed under dilute conditions to minimize ligation of noncross-linked fragments, and DNA is further purified using streptavidin beads before sequencing. (B–D) Data is shown for intrachromosomal interactions of Chr 14 from replicate experiments using *Hind*III (B–C) or alternatively using *Nco*I (D) for the initial restriction digestion. Pixels represent interactions between 1 Mb regions. As an example, stronger interactions are seen between the "upper" end of the chromosome *(green bar)* with the nearby region (*green bar* at the bottom; the interaction is indicated in the region of the *green box*) than with the more distant region indicated by the *blue bar* and *box*. *From Lieberman-Aiden, E., van Berkum, N.L., Williams, L., Imakaev, M., Ragoczy, T., Telling, A., Amit, I., Lajoie, B.R., Sabo, P.J., Dorschner, M.O., et al., 2009. Comprehensive mapping of long-range interactions reveals folding principles of the human genome. Science 326, 289–293.*

Experimental data is divided into regions of appropriate size, typically 0.1–1 Mb and the heat map is colored according to the frequency of ligation products observed between regions; thus, interactions between two regions *i* and *j* would correspond to the intersection of row *i* and column *j* in the heat map. Interactions are, of course, commutative—if *i* interacts with *j*, then *j* interacts with *i*—so the heat maps are symmetrical across the diagonal, while the interactions that map closest to the diagonal are short-range and therefore the most frequently observed. Interactions between chromosomally distant regions are far from the diagonal and are very rare. The most informative region is close to the diagonal, where a patchwork is often observed that reflects the heterogeneity in long-range (>50 kb in mammalian cells) interactions. The general pattern of stronger interactions separated by domains with few interactions is observed for a variety of metazoan cell types and is interpreted as representing regions that are generally in close proximity separated by chromosomal segments that show little long-range interaction, and therefore are relatively isolated from more distant chromosomal regions.

TADs identified using 3C-based methods reflect interactions averaged over many cells. Their existence in the mammalian X-chromosome and the dosage-compensated *C. elegans* X-chromosome was corroborated by FISH experiments that showed chromosomal segments within TADs to be closer together, for a given linear chromosomal separation, than segments from different TADs (Crane et al., 2015; Nora et al., 2012). A later study used advances in imaging techniques to determine the spatial organization of Chr21 in human IMR90 cells, revealing a strong correlation between the spatial distance thus measured with contact frequency measured by Hi-C (Wang et al., 2016). Similarly, imaging of *Drosophila* chromosomes showed that TADs containing repressed genes formed discrete compartments, separated by regions of less condensed chromatin (Szabo et al., 2018). The development of methods to perform high-resolution imaging using multiple fluorescent probes in high throughput fashion has allowed direct comparison of the measurement of chromosomal interactions and proximity in single cells with population averages (Bintu et al., 2018; Cattoni et al., 2017; Finn et al., 2019). These studies show that in both insect and mammalian cells, interactions between distant chromosomal regions that reflect TADs occur at low frequency in individual cells. In contrast, the larger compartments comprising active and inactive genes are conserved on a cell-by-cell basis (Stevens et al., 2017). Thus, current data suggests that TADs can best be thought of as small chromosomal domains that vary over time, and therefore at any given moment vary from cell to cell. These domains then reflect the infrequent, but nonetheless nonrandom, interactions among particular chromosomal regions whose preferential association is observable when averaged over a large population of cells. These variable interactions occur in the context of larger, stable chromosomal compartments, and are likely to vary because of both intrinsic and extrinsic stochastic influences (Finn and Misteli, 2019).

The "boundary regions" between TADs were found to be enriched for binding of the zinc finger protein, CTCF (CCCTC binding factor), a conserved metazoan sequence-specific DNA-binding protein that can functionally separate, or "insulate," gene promoters when situated between them (Dixon et al., 2012; Nora et al., 2012; Phillips and Corces, 2009; Rao et al., 2014). Considerable evidence supports a model in which chromatin loops in metazoan cells are formed by extrusion from sites bound by the cohesin complex, with CTCF-bound sites forming blocks that act as borders for loops (Dekker and Mirny, 2016). Deletion of a specific boundary element in the X-chromosome resulted in new contacts being observed between the neighboring TADs, and deleting or altering the configuration of CTCF sites can disrupt TAD organization (de Wit et al., 2015; Hanssen et al., 2017; Nora et al., 2012). However, CTCF binding sites are also frequently present within TADs, so although they may contribute to TAD segregation, they are not sufficient to demarcate TAD boundaries. Thus, TADs may comprise domains containing small sets of CTCF-anchored loops, with borders established by some combination of CTCF-bound sites, transcriptionally active sites, and unknown factors (van Steensel and Furlong, 2019). Furthermore, TADs or TAD-like organizations has been observed in cells lacking CTCF. The nematode *C. elegans* lacks CTCF entirely, and although its chromosomes are generally disorganized compared to those of mammalian or *Drosophila* cells, the X chromosome in XX hermaphrodites exhibits TADs that are controlled by the same dosage compensation complex that balances its transcription to the level seen in XO males (Crane et al., 2015). This dosage compensation complex is closely related to the Condensin I complex, which like cohesin is intimately involved in chromosome condensation and sister chromatid cohesion in mitosis and meiosis. Evidently, protein complexes that organize chromosomes during mitosis and meiosis have co-evolved or been adapted to participate in the spatial organization of chromosomes during interphase as well. The budding yeast *S. cerevisiae*, which lacks CTCF and whose 12 Mb genome contains a much lower fraction of intergenic DNA than is found in metazoan organisms, also exhibits self-associating domains when mapped using Hi-C, but these are much smaller than those observed in metazoans, being 2–10 kb in size (Hsieh et al., 2015).

TADs themselves proved to be related to previously established genomic compartments. TADs correspond closely to replication sites, regions of ~1 Mb first visualized as clusters of replicons following BrdU pulse labeling of early S phase nuclei and fluorescence imaging (Jackson and Pombo, 1998; Ma et al., 1998; Pope et al., 2014). Domains of active or inactive genes comprise multiple TADs, being considerably larger, while the low-interacting boundaries between TADs also demarcate lamin-associated domains (LADs) in some instances (Dixon et al., 2012).

Remarkably, TAD boundaries were found to be strongly conserved between human fibroblasts and ES cells, and between mouse ES cells and cortex, in spite of differential gene expression between the cell types (Dixon et al., 2012). However, regions within TADs display cell-type-specific interactions, consistent with differential regulation leading to distinct enhancer-promoter interactions in different cell types (Dowen et al., 2014; Nora et al., 2012; Phillips-Cremins et al., 2013; Rao et al., 2014). Boundaries also showed evolutionary conservation between humans and mice, with 54% of human boundaries also being boundaries in mice, and 76% of mouse boundaries being boundaries in humans (compared to 21% and 29% at random) (Dixon et al., 2012).

In addition to the evidence for chromosome organization deriving from imaging and biochemical approaches, functional studies have supported the existence of spatially defined chromosomal domains:

- Chromosomal translocations, in which broken chromosomes fuse to generate novel, hybrid chromosomes, occur preferentially between chromosomes that are physically proximate in the cell type in which they occur (discussed in Bickmore (2013); Dekker and Misteli (2015); see also Roukos et al. (2013)).

- Three naturally occurring, heritable mutations that cause limb malformation in human families were attributed to alterations in TAD boundary sequences that result in ectopic expression of genes in the developing limb bud (Lupianez et al., 2015).
- Interactions between activator-bound enhancers and gene promoters tend to be constrained to "insulated neighborhoods" that are bounded by CTCF sites; perturbation of neighborhood boundaries has frequently been found to be accompanied by altered gene expression in the flanking regions, as well as altered physical interactions (reviewed in Hnisz et al. (2016)).

Although these findings support, from a functional standpoint, the existence of TADs, a large-scale study on the effect on gene expression of extensive chromosomal rearrangements in *Drosophila melanogaster* "balancer" chromosomes found that disruption of TADs affected fewer than 20% of nearby genes (Ghavi-Helm et al., 2019). While not disputing the existence of TADs, these results underscore that chromosome organization is heterogeneous across a population of cells, and that relative chromosomal location is not an absolute arbiter of gene expression.

Liquid-liquid condensates

A recent development regarding chromatin compaction in the cell nucleus is the identification of membraneless nuclear compartments that function as phase-separated liquid-liquid condensates, also sometimes referred to as biomolecular condensates. These condensates behave much like oil droplets in water, although they are much more complex in composition and are dynamic in nature: they tend toward spherical shape (thereby maximizing volume to the surface area, and hence achieving the highest possible ratio of molecular interactions within the droplet relative to heterotypic interactions), and they can fuse with like droplets or divide. Some long-known cellular compartments, such as the nucleolus, conform to the liquid-liquid condensate model (Brangwynne et al., 2011).

Liquid-liquid condensates have been proposed to play a critical role in transcriptional activation by providing compartments that facilitate the recruitment of gene targets to locations having high local concentrations of components of the transcription apparatus (Hnisz et al., 2017). These phase-separated droplets can be visualized in vitro by the use of fluorescently tagged proteins, which condense to reveal spherical droplets that are separate from the bulk aqueous phase. In vivo evidence for liquid-liquid condensates derives from observing punctate foci in the nuclei of cells expressing fluorescently labeled proteins. Some of the proteins most central to transcriptional activation, such as transcriptional activators, subunits of the Mediator complex, and Pol II, are seen in such experiments to segregate into phase-separated droplets in vitro and to be present in puncta in living cells (Boija et al., 2018; Cho et al., 2018; Sabari et al., 2018).

More directly relevant to chromatin folding, human heterochromatin protein 1 alpha (HP1α) can form phase-separated droplets when bound to DNA in vitro (Larson et al., 2017; Strom et al., 2017). Chromatin fragments prepared as short reconstituted nucleosome arrays, as described earlier, also were reported to exhibit properties of liquid-liquid condensates in vitro (Gibson et al., 2019). Phase separation was observed by using fluorescently tagged histones and requires the presence of the histone tails (although the importance of individual tails has not yet been reported), is promoted by linker histone H1, and is antagonized by histone acetylation. Internal mixing was measured by monitoring fluorescence recovery after photobleaching within droplets and was found to occur relatively slowly, on a time scale of 10–20 min. Injecting fluorescently labeled nucleosome arrays into nuclei of HeLa cells that had been pretreated with trichostatin A to induce hyperacetylation resulted in the formation of fluorescent puncta. These puncta were not observed if the HeLa cells were not pretreated with trichostatin A; the authors interpreted these results as indicating that the unmodified nucleosome arrays were dispersed in existing chromatin architecture when the nuclei also contained unacetylated histones but segregated into liquid condensates when the endogenous chromatin was hyperacetylated and therefore less able to form such condensates to absorb the injected material. A subsequent study examined the mechanical properties and mobility of proteins within condensed chromatin in vitro and in vivo and found that chromatin fragments formed liquid condensates only under very specific conditions, and more generally behaved as a solid; specifically, the mobility of DNA relative to neighboring segments over a range of 10–1000 nm, tracked using a fluorescently labeled nucleotide incorporated into the genomic DNA, took place only over a time scale from minutes to hours (Strickfaden et al., 2020). Solid-like chromatin could then act as a scaffold around which nonhistone proteins could coalesce into liquid condensates. Although it is not yet clear how the more heterogeneous chromatin in living cells behaves with regard to phase separation, these studies provide a basis for studying the compartmentalization of chromatin and other components in vitro and in vivo.

The principles underlying the formation, function, and presumed specificity of liquid-liquid condensates in the cell nucleus are an active area of investigation in this rapidly evolving area and will not be discussed here. Likewise, the relationship of these membraneless compartments to chromosome territories, nuclear compartments, and TADs, is currently unclear, although it seems plausible that condensates involved in transcription may be related to TADs enriched for active genes, and to regions that are looped out from chromosome territories and active in transcription. Furthermore, the notion that heterochromatin may be compartmentalized by the formation of phase-separated condensates raises many questions regarding the mechanisms of its formation and function, including the relation of putative heterochromatin condensates to nuclear localization and their segregation from transcriptionally active euchromatin. It should also be noted that questions have been raised regarding the criteria used as evidence for liquid-liquid condensates and regarding the rigor of some studies reporting liquid-liquid phase separation of biological compartments in vivo (Erdel et al., 2020; McSwiggen et al., 2019; Peng and Weber, 2019). The role of liquid-liquid condensates in nuclear function is an active and rapidly evolving area of research and is sure to remain so for some time.

Heterochromatin

Although this chapter has concentrated on structural aspects of chromatin condensation, it would not be complete without some discussion of heterochromatin beyond its brief mention in the preceding section. Heterochromatin is now generally viewed as a compact state characterized by specific histone modifications and associated nonhistone chromatin proteins, and which is generally refractory to transcription. Initially, however, heterochromatin was identified cytologically and named by Emil Heitz at the University of Hamburg, who reported his findings in 1928 (Heitz, 1928; Passarge, 1979). Having developed a novel method for obtaining chromosome preparations from moss, Heitz reported that five out of nine chromosomes from *Pellia epiphylla*, a species of liverwort, contained regions that remained condensed throughout interphase while the remainder of the chromosomal cohort was lost to observation at late telophase. He proposed that the latter regions be referred to as euchromatin, and showed that heterochromatin and euchromatin were similarly present in a large variety of mosses and flowering plants. Remarkably, both of these appellations have remained in use ever since; just as remarkably, Heitz received little recognition for his work during his lifetime and did not obtain an academic position commensurate with his achievements until 1955, when he joined the faculty at the Max Planck Institute in Tübingen. This in spite of major discoveries including

- Reporting that sex chromosomes, discovered in 1902, were enriched in heterochromatin relative to autosomes
- Distinguishing between constitutive and facultative heterochromatin
- Determining that the nucleolus (the only structural component of the nucleus that could be resolved by light microscopy at the time) was not part of the chromosomes, but rather served as an organizing center associated with specific chromosomal regions
- Discovering giant chromosomes in larval salivary glands of insects
- Connecting his cytological observations of *Drosophila* chromosomes with genetic studies of Thomas Hunt Morgan and colleagues to identify heterochromatic regions as being gene-poor relative to euchromatin, and suggesting that heterochromatin conferred a functional as well as structural role.

Heitz's career was in large part a casualty of wartime politics and the rise of the Nazi party in Germany. Heitz's faculty appointment at Hamburg ended in 1937 when the Nazi-ruled administration, deciding his 4 years served in the German military in WWI did not outweigh the crime of having Jewish ancestry, discontinued his salary and withdrew his appointment. His life and career had already been impacted by the status of his hometown on the border between France and Germany, with its shifting status following WWI. Heitz did gain some measure of recognition in his final decade; nonetheless, his relative obscurity was bemoaned in print again 40 years following a summary of his major achievements (Berger, 2019; Passarge, 1979).

In the years following Heitz's seminal publication, heterochromatin was found to be nearly ubiquitous in plants and animals, most often being observed near the centromere ("pericentric heterochromatin"), at chromosome ends, and proximate to the nucleolus. Heterochromatin comprised a large fraction of the genomes of many species; in *Drosophila*, for example, about one-third of chromosomal DNA is packaged as heterochromatin. A distinction was made early on between constitutive heterochromatin and facultative heterochromatin, the latter first being defined as regions that became heterochromatic for only one of the two chromosomal homologs during development. Heitz considered heterochromatin to be a substance, but findings that heterochromatin could not be detected cytologically during early development in *Drosophila* controverted this idea; instead, it was suggested that the condensed appearance of heterochromatin was due to an accessory factor or factors (Brown, 1966; Cooper, 1959).

The first insights into the molecular nature of heterochromatin emerged in the 1960s. Recognition of the complementary nature of double-stranded DNA prompted experiments showing that the two strands, once separated, could specifically reassociate. This was first observed for simple polymers of polyadenylic acid and polyuridylic acid (also demonstrating the existence of double-stranded RNA) and later shown for complex, mixed-sequence DNA (Doty et al., 1960; Marmur and Lane, 1960; Rich and Davies, 1956; Warner, 1957). Quantitative analysis of such reassociation of DNA fragments from protozoa, plants, and metazoan organisms revealed that in all cases, a fraction of the DNA annealed at the rate expected for single copy sequences from a genome of the appropriate size, while another fraction annealed much more rapidly—typically by two or more orders of magnitude (Britten and Kohne, 1968). For example, about 40% of calf thymus DNA was shown to consist of repetitive sequences present in about 100,000 copies in the genome. These repetitive sequences could be physically separated by centrifugation on cesium chloride density gradients, allowing their purification and use as radiolabeled probes for in situ hybridization. Hybridization of these "satellite" sequences, so named because of the "satellite" peak they formed in CsCl gradients, to metaphase chromosomes resulted in labeling of pericentric heterochromatin, while hybridization to interphase chromosomes labeled dense chromatin at the nuclear periphery (Jones, 1970; Pardue and Gall, 1970; Rae and Franke, 1972). Thus, heterochromatin was shown to be heavily enriched in repetitive DNA. Moreover, repetitive DNA sequences failed to hybridize to RNA, supporting the idea that heterochromatin was transcriptionally inert (Flamm et al., 1969).

Another major advance in elucidating the nature of heterochromatin was the identification and cloning of the gene encoding heterochromatin protein 1, or HP1, from *Drosophila melanogaster* (James and Elgin, 1986). This was achieved by a workmanlike approach that yielded gold: proteins eluted from nuclei of *Drosophila* embryos with increasing concentrations of potassium thiocyanate were purified by electrophoresis, and antibodies were prepared against a subset of proteins released at 2M KSCN. Staining of polytene chromosomes by one such antibody revealed its target protein to be present in known regions of heterochromatin, including the chromocenter, a region of condensed chromatin observed in interphase cells by Heitz and comprising heterochromatin from multiple chromosomes (Heitz, 1935; James et al., 1989).

The cloning of HP1 allowed a direct test of the putative link between the structure and function of heterochromatin. Heterochromatin was posited to be repressive toward gene expression and also inhibited the recombination of chromosomal loci occurring during meiotic crossing over (Barton, 1951; Brown, 1966). A tractable approach to examine perturbation of the repressive effect of heterochromatin (for example, by deficiency in HP1 expression) was provided by the phenomenon of position effect variegation, or PEV (Elgin and Reuter, 2013; Lewis, 1950). (For additional historical detail on PEV, see Spradling and Karpen (1990).) PEV was first reported by Herman Muller, who identified mutant flies after X-ray irradiation in which the characteristic red eye color was altered to a variegated phenotype in which patches of both white (mutant) and red facets were present (Fig. 3.11) (Muller, 1930). These variegated phenotypes were due to mutations that were linked to chromosomal rearrangements. The breakpoints found in these mutant flies occurred near the borders of heterochromatic regions, and genes showing variegated expression were located close to the breakpoints (Muller, 1930; Schultz, 1936). "Position effect variegation" thus refers to the variegation of phenotype (eye color here, but also observed for other morphological characteristics) caused by an altered chromosomal position of the affected gene.

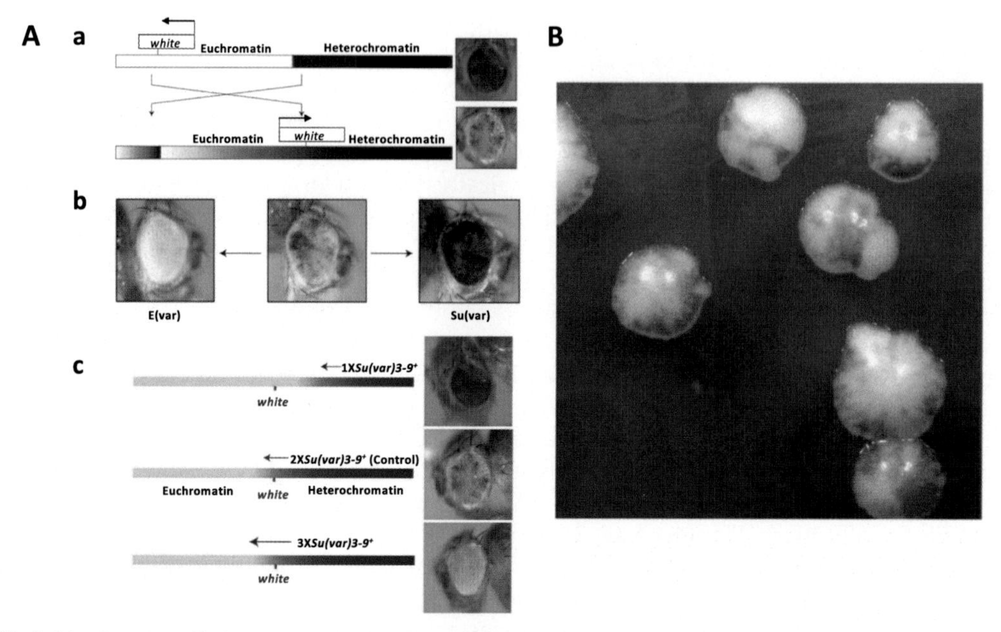

FIG. 3.11 Position effect variegation in (A) *Drosophila* and (B) yeast. (A-a) Chromosomal rearrangement moving the *white* gene from its normal euchromatic location to a site adjacent to heterochromatin, resulting in a phenotypic switch from a *red* eye to one having variegated pigment. (b) *Su(var)* mutations suppress heterochromatin-mediated silencing, restoring the wild-type phenotype; *E(var)* mutations enhance PEV, increasing the mutant phenotype. (c) Copy number of some *Su(var)* mutant genes affects the extent of heterochromatin spread to the rearranged *white* gene, with a corresponding effect on phenotype. (B) Metastable silencing of a telomeric copy of the *ADE2* gene in yeast. Yeast colonies were grown from single cells in media having low adenine content; silencing of the gene causes the buildup of an intermediate that yields *red* pigmentation. The sectoring seen within colonies results from switching between the silent and expressed states. *(A) From Elgin, S.C., Reuter, G., 2013. Position-effect variegation, heterochromatin formation, and gene silencing in Drosophila. Cold Spring Harb. Perspect. Biol. 5, a017780. (B) From Gartenberg, M.R., Smith, J.S., 2016. The nuts and bolts of transcriptionally silent chromatin in Saccharomyces cerevisiae. Genetics 203, 1563–1599.*

The rearrangements giving rise to PEV were termed "eversporting displacements" by Muller, who suggested that the responsible *white* gene might undergo frequent genetic changes during development. Muller was wrong on this point; the repressed phenotype (loss of *white* expression, resulting in white eyes—*Drosophila* genes being named after their mutant phenotypes) resulted from the spreading of heterochromatin into the translocated region containing the *white* gene, while the variegation arose from this spreading being variable on a cell to cell basis. The use of a *lacZ* reporter inducible by heat shock as a PEV reporter revealed that silencing mediated by heterochromatin occurred early in embryogenesis; variegated expression was then postulated to occur by derepression in a subset of cells during development (Lu et al., 1996). Clonal inheritance of the silenced or derepressed state would then result in the observed mosaic pattern of red and white patches. Thus, the status—silent or derepressed—of a gene affected by PEV was viewed as being metastable: while derepression of the silenced state could occur at a low frequency, the expression state was generally heritable. The relationship between heterochromatin spreading and gene repression was shown directly by an experiment using a strain in which the *ecs* gene was translocated into the vicinity of centromeric heterochromatin (Zhimulev et al., 1986). The *ecs* gene is important in the response to the hormone ecdysone, one effect of which is to induce chromosome puffs visible in the polytene chromosomes of the salivary gland. Examination of the translocated *ecs* locus in cells in which puffs did not develop in response to ecdysone revealed the presence of heterochromatin blocks that impaired access to the region; such blocks were not seen in cells exhibiting chromosome puffs after ecdysone treatment. Analogous effects on gene expression caused by juxtaposition to heterochromatin have been observed in a variety of species, including yeast and mammals. In flies, increased dosage of heterochromatin in mutants exhibiting PEV, as caused for example by an extra copy of the heterochromatic Y chromosome, suppressed the extent of the observed variegation, leading to the idea that some heterochromatin component required for gene repression could be titrated out by extra heterochromatin (Gowen and Gay, 1934; Schultz, 1936). (As a side note, Schultz refers to heterochromatin as "inert chromosome regions," but allows in a footnote that Heitz's nomenclature of heterochromatin and euchromatin is probably better.)

A natural approach for geneticists seeking to understand PEV, and thus presumably gain new insights into heterochromatin, was to seek mutants that suppressed or enhanced the effect (Fig. 3.11). Such mutants were first identified by Thomas Hunt Morgan and colleagues in 1937, and additional suppressor of variegation, or *Su(var)*, and enhancer, or *E(var)*, mutants were reported in a subsequent study (Morgan et al., 1937, 1941). The obvious question, then, following the identification of HP1 as a constituent of heterochromatin was whether mutations thereof would affect PEV. Two independent screens had identified *Su(var)* mutations that were mapped to the same chromosomal locus as that containing the gene encoding HP1 (Sinclair et al., 1983; Wustmann et al., 1989); these mutations were shown to correspond directly to mutations in HP1 (Eissenberg et al., 1990, 1992). Thus, mutations in a gene whose product was localized to heterochromatin were found to suppress the reduced expression of genes juxtaposed to heterochromatin by chromosomal rearrangements, thereby providing a direct link between heterochromatin structure and function.

Abundant additional evidence supported this linkage. The identification of numerous independent *Su(var)* mutants led investigators to propose that multiple genes contributed in *trans* to heterochromatin spreading (Henikoff, 1979; Sinclair et al., 1983; Spofford, 1967). Reduction of histone gene copy through chromosomal deletions suppressed PEV, implicating a chromatin-mediated mechanism, and cloning of *Su(var)3–7* revealed it to be a zinc finger protein likely to have DNA-binding activity (Moore et al., 1983; Reuter et al., 1990). Tartof and colleagues identified loci that enhanced or suppressed PEV when duplicated or deficient and made the general observation that modifiers of PEV fell into two classes: one enhanced PEV when duplicated and suppressed variegation when mutated or deficient, and the other class, of which only two members had been identified, behaved oppositely. Consideration of these findings and of the idea that PEV resulted from the spreading of heterochromatin onto genes juxtaposed to such regions by translocation led them to propose a model in which *Su (var)* mutants encoded structural components of heterochromatin (Locke et al., 1988).

According to this model, mutation or reduced expression of any such component would impair the spreading of heterochromatin by mass action and consequently suppress PEV, while overexpression of such components would enhance the effect (Tartof et al., 1989). This model captures some essential features of PEV; however, it does not account for differential effects seen for different reporters and at distinct heterochromatic loci. It must therefore be the case that other gene-specific features also influence the spreading of heterochromatin.

Altogether, genetic screens have identified several hundred PEV mutants, whose effects are due to mutations in ~150 genes (Elgin and Reuter, 2013; Schotta et al., 2003). The small minority that has been cloned mostly comprises chromosomal proteins or modifiers thereof. The role that these proteins play in the establishment and maintenance of heterochromatin is incompletely understood, as is the molecular architecture of heterochromatin. Moreover, several major domains of heterochromatin are found in *Drosophila*: at regions flanking centromeres (pericentric heterochromatin), at telomeres, the entire Y chromosome, and the small (4.3 Mb) fourth chromosome. These different heterochromatic regions exhibit varied responses to PEV modifiers, indicating complexity in heterochromatin properties despite observed commonalities, such as visual morphology. For example, HP1 deficiencies do not confer PEV on reporter genes located in telomeric heterochromatin (Fanti et al., 1998), and a test of 70 modifiers of PEV revealed substantial variability in their effects on the expression of PEV reporters according to the heterochromatic location of the reporter gene (Donaldson et al., 2002). Telomeric heterochromatin, in fact, behaves so distinctly that its effect on expression is referred to as telomeric position effect, or TPE, to distinguish it from classical PEV.

Nonetheless, some critical aspects have been uncovered that allow a simple model for heterochromatinizaton to be formulated (Fig. 3.12). In this model, constitutive heterochromatin is established and maintained through the combined action of a histone-modifying protein and a "reader" protein, HP1a (renamed from the original HP1 designation after the identification of additional HP1 paralogs in *Drosophila* (Smothers and Henikoff, 2001)), that binds to the modified histone. The discovery of this modifying protein, SU(VAR)3–9, as a methyltransferase that methylates lysine 9 of histone H3 (H3K9), is intimately tied into the flood of research on histone modifications that occurred beginning in 1996, and detailed discussion is deferred to Chapter 7. (Although one might assume that SU(VAR)3–9 is named after its target, H3K9, this is not the case. *Su(var)* mutants are named after the chromosome to which they localize; thus, *Su(var)3–9* is the ninth *Su(var)* mutant identified on the third chromosome. Weird, right?) The essential features of the model are that histone H3 undergoes dimethylation and trimethylation at Lys9 by SU(VAR)3–9, resulting in the binding of HP1a, which interacts specifically with H3K9me2/3-containing chromatin. HP1a, in turn, binds SU(VAR)3–9, as shown by co-IP and yeast two-hybrid analysis (Schotta et al., 2002); thus, recruitment of this modifying enzyme to existing heterochromatin by HP1a facilitates the spread of H3K9me2/3 and heterochromatinization of the adjacent chromatin. The zinc finger protein SU(VAR)3–7 has also been shown to associate with heterochromatin and to interact with HP1a (Cleard et al., 1997). Association of SU(VAR)3–7, SU(VAR)3–9, HP1a, and nucleosomes containing H3K9me2 and H3K9me3 at visually defined regions of heterochromatin has been confirmed both by immunostaining and by ChIP-seq (Cleard et al., 1997; Elgin and Reuter, 2013; James et al., 1989; Schotta et al., 2002). The higher resolution afforded by ChIP-seq, in comparison to cytogenetic analysis, reveals that the boundaries between heterochromatin and euchromatin, marked by the border between highly enriched and much lower binding of HP1a, SU(VAR)3–9, and the H3K9me2/3 modifications, varies by as much as several hundred kb in different cell types (Elgin and Reuter, 2013). This indicates that the heterochromatin border is not absolutely fixed by any specific sequence, perhaps being dictated instead, at least in part, by the expression level of heterochromatin proteins in the specific cell type, as suggested by the mass action model of Tartof and colleagues. In sum, although the precise architecture has yet to be elucidated, the core structure of nontelomeric, constitutive heterochromatin in *Drosophila* appears to be formed by these three proteins and their interactions with nucleosomes bearing the H3K9me2/3 modifications.

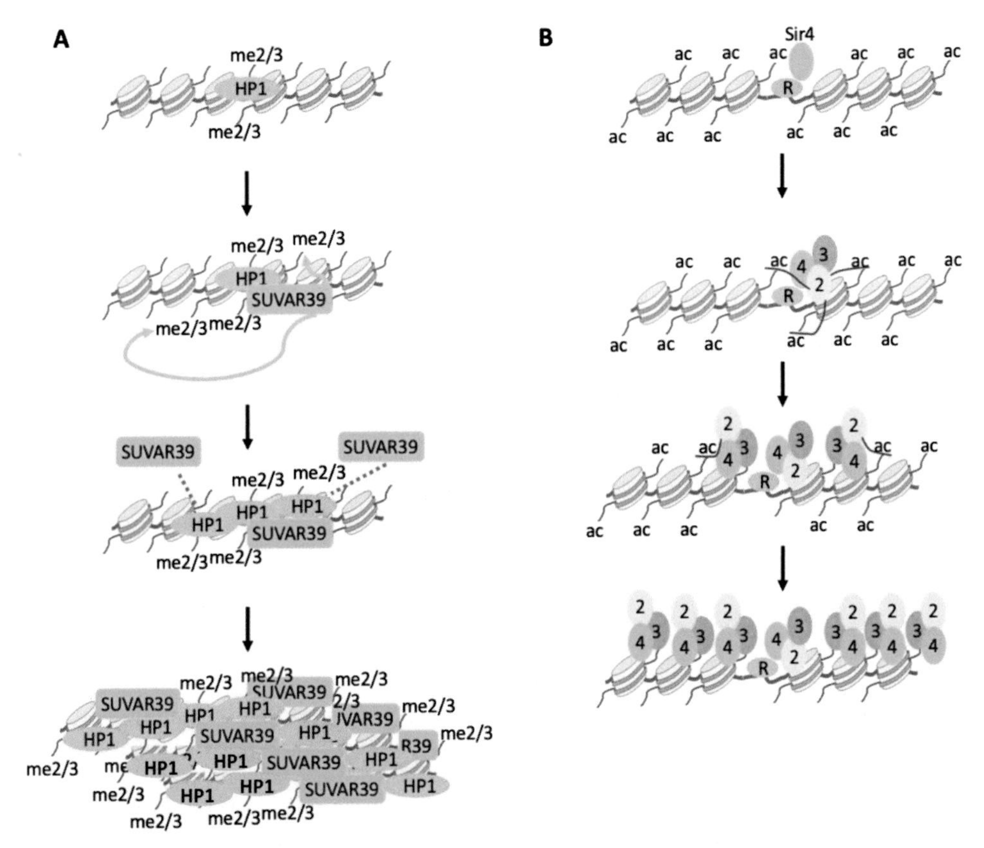

FIG. 3.12 Heterochromatin spreading. (A) In *Drosophila*, HP1a is recruited to nucleosomes dimethylated or trimethylated at H3K9. HP1 then recruits the histone methyltransferase SUVAR39, which methylates H39 in nearby nucleosomes, resulting in additional recruitment of SUVAR39 and the spreading of heterochromatin. (B) In budding yeast, Sir4 is recruited by Rap1 at telomeres or Sir1 at silent mating type loci; Sir4 recruits Sir2 and Sir3, and Sir2 deacetylates H4K16 at nearby nucleosomes. Sir3 binds at deacetylated nucleosomes, recruiting Sir2/4 with the resulting expansion of heterochromatin.

The reciprocal positive feedback between modification of H3K9 by SU(VAR)3–9 and binding of HP1a depicted in Fig. 3.12 provides a mechanistic basis for maintenance of heterochromatin, but does not explain its initial formation. A key characteristic of heterochromatin is the presence of repetitive DNA sequences; such sequences, frequently comprising remnants of transposable elements (TEs), appear capable of serving as nucleation sites for the formation of heterochromatin, which can then spread to the surrounding chromatin by the mechanism illustrated in Fig. 3.12. Experiments in which a *white* gene reporter fused to a transposable element termed *1360* was inserted into various genomic sites supported this idea, revealing that the inclusion of *1360* or another TE, *Invader4*, allowed TE-dependent gene silencing at many such insertions (Sentmanat and Elgin, 2012). ChIP analysis at one such insertion site revealed the presence of HP1a that depended on the presence of *1360*. Experiments first conducted in fission yeast (*S. pombe*), later extended to other organisms including insects, worms, and plants, have revealed a complex mechanism for the formation of repeat-induced formation of heterochromatin. In this mechanism, transcription of repeated sequences causes recruitment of molecularachinery by the nascent RNA that, in a feedback cycle, inhibits expression of those same sequences through the formation of heterochromatin; this phenomenon is known as RNA interference, or RNAi (Martienssen and Moazed, 2015; Volpe et al., 2002). RNAi appears to have evolved to suppress TE activity and to protect genome integrity, particularly in the face of DNA damage (Allshire and Madhani, 2018); indirect evidence also supports the idea that a version of this mechanism contributes to the formation of heterochromatin at repetitive sequences in *Drosophila* (Elgin and Reuter, 2013). Heterochromatin formation in plants and mammals is further complicated by reciprocal interactions between methylation of DNA and of histone H3 Lys9 (Hashimshony et al., 2003; Jackson et al., 2002; Lehnertz et al., 2003).

The principal constituent protein components of heterochromatin, HP1a and SU(VAR)3–9, are widely conserved across eukaryotes, consistent with the near ubiquity of heterochromatin. Some exceptions do occur: a clear homolog of HP1a has not been found in *C. elegans*, and budding yeast (*S. cerevisiae*) lack homologs of HP1 and SU(VAR)3–9 and do not exhibit H3K9 methylation, although these features are found in heterochromatin in the distantly related fission yeast, *S. pombe*. HP1 and SU(VAR)3–9, where tested, play roles in heterochromatin propagation similar to that found in *Drosophila*.

Heterochromatin in budding yeast

Saccharomyces cerevisiae, though lacking cytologically obvious heterochromatin as well as the characteristic molecular building blocks thereof, exhibit their own position effect, as manifested by variable silencing of genes dictated by special chromatin structures in two genomic locations. One of these is proximate to telomeres, and therefore the silencing is referred to as the telomere position effect, while the other occurs in the region of the silent mating type loci (Brand et al., 1985; Gottschling et al., 1990; Schnell and Rine, 1986). As with PEV in *Drosophila*, silencing is not absolute and is metastable (especially at telomeric chromatin): genetically identical cells may express or not express a reporter gene located in one of these silencing loci, and may occasionally, but infrequently, switch states (Fig. 3.11). Researchers studying PEV in yeast and *Drosophila* experienced a productive synergy: a molecular model for heterochromatin in budding yeast preceded that for flies, but leaned on ideas regarding the spreading of heterochromatin that were based on studies in *Drosophila*.

S. cerevisiae can exist as **a** or **α** haploids, which can mate to produce **a/α** diploids; in haploid cells, silenced copies of the **a** and **α** genes are retained at the silent mating type cassettes,

HMLα and *HMRa*, while the active copy of the **a** or **α** gene is carried at the mating type, or *MAT*, locus. These three loci all reside on the relatively small (~320 kb) Chromosome III, and "mating type switching" occurs when the HO endonuclease introduces a cut at the *MAT* locus, facilitating recombination that results in interconversion between *MATa* and *MATα*. Demonstration that a sequence from *HMR* could silence *LEU2* and *TRP1* alleles when placed at distances up to 2600 bp away established *HMR* as a general "silencer" rather than a gene-specific repressor (Brand et al., 1985). Although the silent mating type loci are only ~1.5–3 kb in length, much smaller than the cytologically visible domains of heterochromatin in *Drosophila* and other eukaryotes, the similarity to *Drosophila* heterochromatin in functional and physical characteristics, including replication late in S phase, the ability to confer gene silencing, and positioning near the gene envelope, led the field to categorize them as heterochromatic (Franke, 1974; Thompson et al., 1993).

The popularity of budding yeast as a model organism, as with *Drosophila*, is in large part based on its genetic tractability. Correspondingly, the characterization of heterochromatin in yeast began with genetic screens that identified components critical for position-dependent silencing. The "Silent-information regulator" genes *SIR1–4* (also initially named *MAR* [Mating-type regulator], or *CMT* [Change of mating type genes]) were identified as mutants that altered mating behavior in a manner consistent with loss of silencing at the *HM* loci (Haber and George, 1979; Klar et al., 1979; Rine and Herskowitz, 1987; Rine et al., 1979). (The adopted nomenclature was that of Jasper Rine, who identified all four *SIR* genes in his PhD thesis work.) Sequence elements critical for silencing at the *HM* loci were also identified and found to include binding sites for the general regulatory factor Rap1, and silencing, as measured by mating assays and expression of *HM*-silenced genes, was impaired in yeast harboring mutants of Rap1 (Abraham et al., 1984; Brand et al., 1987; Kurtz and Shore, 1991; Laurenson and Rine, 1992; Shore and Nasmyth, 1987). (Dissection of sequences required for *HM* silencing in fact revealed a complicated picture, with specific sequence elements contributing redundantly to silencing. Thus, the effect of *rap1* and other mutants on *HM*-dependent silencing was tested in genetic backgrounds in which other, distinct elements were also disabled.)

Rap1 was also known to bind to the $C_{1-3}A$ repeats (very different from its binding site at the *HM* loci and other sites in the genome (Pina et al., 2003)) found at yeast telomeres (Berman et al., 1986; Buchman et al., 1988; Longtine et al., 1989), which were found also to confer position-dependent silencing. Telomere position effect in yeast was first observed as repression of each of four reporter genes inserted within 5 kb or less from a chromosome end (Gottschling et al., 1990). Telomeres were known to be packaged as heterochromatin in other organisms, likely as a means to protect chromosome ends from aberrant fusion (Brown, 1966; Traverse and Pardue, 1989); however, the discovery of TPE in yeast was more serendipitous than this parallel might lead one to expect. Rather, the study was initiated in the course of an investigation of telomere properties when efforts to insert a marker gene adjacent to a $C_{1-3}A$ tract at a chromosome end led to the realization that the telomeric location of the marker gene caused its transcriptional properties to be altered.

The recognition of TPE together with the knowledge that Rap1 binding occurred at both *HM* loci and at telomeres prompted Gottschling and colleagues to test the effect of mutants known to affect silencing at *HM* loci on telomeric silencing (Aparicio et al., 1991). Telomeric silencing of *URA3* and *ADE2* reporters was relieved by *sir2*, *sir3*, and *sir4* mutations, but not by a *sir1* mutation that suppressed silencing at *HM* (Aparicio et al., 1991). A subsequent report showed that *rap1* mutants that affected *HM*-dependent silencing also abrogated the telomere position effect (Kyrion et al., 1993). These and other studies indicated a common basic mechanism of silencing at the *HM* and telomeric loci that was reinforced by additional means, including a contribution of Sir1, at the *HM* loci.

Specific cleavage by the endogenous yeast HO endonuclease at the *MAT*, but not *HMLα* or *HMRa* loci, suggested that silencing at the *HM* loci could reflect a physical state of the chromatin at these regions (Strathern et al., 1982). This physical correlate to the position effect exerted on gene expression at both the *HM* loci and at telomeres was supported by experiments monitoring the accessibility of GATC sequences to the *E. coli dam* methyltransferase in yeast (Gottschling, 1992; Singh and Klar, 1992). Somewhat remarkably, expression of *dam* methyltransferase, in spite of its methylating adenine residues at all accessible GATC sites in the yeast genome, is tolerated by yeast. Methylation of select sites was assayed by the use of restriction enzymes sensitive to GATC methylation: *Dpn*I cuts only GATC sites methylated on both strands, while *Mbo*I cleaves only unmethylated GATC sites. Using these enzymes in conjunction with Southern blotting, strong inhibition to methylation was observed at both telomeric and silent mating type loci, which in both cases was relieved in *sir* mutants. Furthermore, *rap1* mutants that disrupted position effect increased accessibility to *dam* methyltransferase at both telomeric and silent mating type loci, thus forging a direct connection between physical and functional characteristics of these putative heterochromatin domains (Kyrion et al., 1993). These results were corroborated by another study showing that restriction endonuclease cleavage in isolated yeast nuclei was inhibited at the *HML* and *HMR* loci, but not at the *MAT* locus, in a *SIR*-dependent manner (Loo and Rine, 1994).

In *Drosophila*, histones were implicated in PEV and heterochromatin structure via mutations that affected gene dosage; in yeast, point mutants and partial deletions of the tails of histones H3 and H4 were found to affect gene silencing mediated by position effects, thus implicating chromatin in the special structures responsible (Aparicio et al., 1991; Johnson et al., 1990; Megee et al., 1990; Park and Szostak, 1990). (Studies using histone mutants are discussed at greater length in Chapter 5.) Point mutants of *sir3* were identified as suppressors of histone H4 point mutations that impaired position effects, suggesting a direct interaction between Sir3 and the histone H4 amino terminus (Johnson et al., 1990). This interaction was confirmed biochemically by showing that glutathione-*S*-transferase (GST) fusions to histone H3 or H4 amino termini, immobilized on glutathione Separose beads, retained Sir3 and Sir4 expressed in vitro (Hecht et al., 1995). Retention of Sir3 and Sir4 was not seen with amino termini of H2A or H2B, which are not required for silencing, nor with histone H3 or H4 tail mutants that abrogate silencing. Moreover, immunoprecipitation of histones from yeast cellular extracts using Sir3 antibodies is disrupted in yeast having mutations in histone H4 that suppress silencing (Hecht et al., 1996).

These various findings indicated that Rap1, the Sir proteins, and the histone H3 and H4 amino termini establish a specialized chromatin structure at telomeres and the *HM* loci that reduce access to the incorporated DNA and thereby repress transcription of genes present in these domains. A flurry of studies revealed aspects of the interactions among these components that led to a succinct model of yeast heterochromatin structure:

- Rap1 interacts with Sir3 and Sir4 via its C-terminal domain, and a *rap1* mutant lacking the C-terminal domain fails to silence (Moretti et al., 1994).
- A fusion protein in which the Rap1 C-terminal domain is attached to the Gal4 DNA-binding domain (DBD) can silence a reporter gene placed near a Gal4 binding site, and this silencing depends on Sir2–4 (Buck and Shore, 1995).
- Sir3 and Sir4 interact in vitro and in vivo (Hecht et al., 1996; Moretti et al., 1994; Strahl-Bolsinger et al., 1997).
- Recruitment of Sir3 or Sir4, using the DBD of Gal4 or LexA, to a telomeric site can silence an adjacent reporter gene in a *rap1* C-terminal domain mutant that is defective for position effect (Lustig et al., 1996; Marcand et al., 1996).
- Overexpression of Sir3, but not Sir2 or Sir4, extends TPE inward to sites as far as 20 kb from the chromosome ends (Renauld et al., 1993).
- Sir2, Sir3, and Sir4 cross-link to chromatin at *HML*, *HMR*, and telomeres, and spread inward from telomeres when Sir3 is overexpressed. Sir3 cross-linking to *HML*, *HMR*, and telomeric chromatin is greatly decreased by silencing-defective mutations in Rap1 or histone H4 (Hecht et al., 1996; Strahl-Bolsinger et al., 1997).
- Telomeric sequences co-localize with Rap1, Sir3, and Sir4 at foci often seen near the nuclear periphery in fixed diploid yeast cells; these foci are lost in cells lacking Sir3 or Sir4, and in mutants of the H3 or H4 amino terminus (Gotta et al., 1996; Hecht et al., 1995; Palladino et al., 1993).
- Sir4 recruitment by Rap1 to telomere ends (observed by ChIP) and by Sir1 at the silent mating type loci can occur in the absence of Sir2 or Sir3, but spreading requires Sir2 and Sir3 as well (Hoppe et al., 2002; Luo et al., 2002; Rusche et al., 2002).

These various findings suggested a model for telomeric heterochromatin in yeast, greatly informed by Tartof's model for heterochromatin spreading in *Drosophila*, in which Rap1, bound to telomeric chromatin by sequence-specific interactions, recruits the Sir proteins via direct interactions (Grunstein, 1997). Cooperative interactions among Sir3, Sir4, and the histone H3 and H4 amino termini then result in recruitment of the Sir2/3/4 complex and spreading from its nucleation site; Sir3, but not Sir2 or Sir4, is limiting for the expansion of telomeric heterochromatin from the Rap1-binding $C_{1-3}A$ repeat sites. A major surprise that yielded a key insight into the spreading mechanism was the discovery that Sir2 is a histone deacetylase that targets Lys16 of histone H4, in the region that is important for interaction with Sir3 (Gartenberg and Smith, 2016). This finding was consistent with the observed hypoacetylation of histones at both the silent mating type and telomeric loci; hypoacetylation had also been observed at the inactive X chromosome in female mammalian cells (Braunstein et al., 1996). This discovery, coupled with findings that the initial association of the Sir complex required Sir4 while its spread required the deacetylase activity of Sir2, led to a refinement of the spreading model (Fig. 3.12) (Hoppe et al., 2002; Luo et al., 2002; Rusche et al., 2002). This model postulates that recruitment of Sir4 by Rap1 at telomeres or by Sir1 at the silent mating type loci (Triolo and Sternglanz, 1996), together with interactions with the histone H3 and H4 amino termini, facilitates the association of Sir2 and Sir3; deacetylation of adjacent nucleosomes by Sir2 at Lys16 of histone H4 then creates a preferred binding site for additional Sir3, leading to Sir2 and Sir4 recruitment, and so forth, until spreading is stopped by a barrier, such as a nucleosome gap, or is hampered by limiting amounts of

Sir3. This spreading model has obvious conceptual similarities to that of the spreading of heterochromatin in *Drosophila*.

Numerous subsequent studies have elucidated molecular details supporting this model (Gartenberg and Smith, 2016). In addition to the N-terminal region of histone H4, Lys79 of histone H3 was also shown to be critical for silencing (Ng et al., 2002; van Leeuwen et al., 2002). Biochemical and structural studies indicated a basis for these findings by showing that Sir3 interacted with these same regions of H3 and H4 (Armache et al., 2011; Onishi et al., 2007). Another important finding was that recombinant Sir2, Sir3, and Sir4 form a complex in a 1:1:1 ratio (Cubizolles et al., 2006; Liou et al., 2005). In spite of this remarkable progress and a model that captures myriad experimental observations, a defined structure for yeast heterochromatin is yet to be determined. Efforts to reconstruct a heterochromatin complex in vitro have yielded insights but thus far proved refractory to high-resolution structural analysis. An early study demonstrated by gel shift analyses that recombinant Sir3 could bind to naked DNA, mononucleosomes, and a 12-mer nucleosome array, in the latter two instances yielding supershifted species at approximately 1:1 ratios of Sir3:nucleosome (Georgel et al., 2001). (This study and the others cited here did not employ Rap1, Sir1, or otherwise take steps to specifically recruit the Sir complex to nucleosome array templates; evidently, the high molar ratio, compared to the relative abundance of Sir proteins to genomic DNA in vivo, is sufficient to facilitate assembly.) Sir3 interaction with nucleosome arrays was reduced by the removal of the H3 and H4 amino termini; however, the Sir3-array complex did not impede access of restriction endonuclease to linker regions. Analysis of nucleosome arrays complexed with Sir3 by sedimentation velocity and atomic force microscopy indicated the formation of a Sir3 fiber somewhat narrower and less compact than the 30 nm fiber, suggesting a structure in which Sir3, at a 2:1 ratio relative to nucleosomes, coats the 10 nm fiber (Swygert et al., 2014). A follow-up study using Sir2/3/4 together with nucleosome arrays revealed a nearly uniform, compact fiber, the integrity of which depended on both H4K16 and Sir3-Sir4 interactions (Swygert et al., 2018). In another report, assemblage of complexes using Sir2, Sir3, and Sir4 together with chromatin templates resulted in the formation of filaments of about 20 nm diameter visible by EM (Onishi et al., 2007). These assemblages were refractory to digestion by restriction enzymes and inhibited transcription by both RNA polymerase II and III (Johnson et al., 2009). Inhibition to both enzymatic cleavage and transcription was relieved by mutation of H4 Lys16 to Ala, despite having little effect on the association of a preassembled Sir2/3/4 complex with the chromatin template. Some details may be coming clear—two independent investigations conclude that Sir3 acts as a bridge between nucleosomes, with Sir3 monomers binding on opposite faces of each nucleosome (Behrouzi et al., 2016; Swygert et al., 2018)—but challenges clearly remain. Moreover, additional complexities regarding the in vivo situation remain unresolved: the silencing domains associated with TPE vary considerably in extent among yeast telomeres and are not continuous (Fourel et al., 1999; Pryde and Louis, 1999). In vitro studies obviously do not account for effects caused by the congregation of telomeres at the nuclear periphery, and whether and how telomere clustering affects access to the transcription machinery—for example, perhaps by sequestering telomeric chromatin apart from biomolecular condensates where transcription occurs—remains to be addressed. It may in fact be incorrect to view heterochromatin, whether that of yeast or metazoan organisms, as a static structure; consideration of the dynamic nature of Sir protein association may be necessary to formulate a model that accounts for both spreading and discontinuity of telomeric heterochromatin (Cheng and Gartenberg, 2000).

References

Abraham, J., Nasmyth, K.A., Strathern, J.N., Klar, A.J., Hicks, J.B., 1984. Regulation of mating-type information in yeast. Negative control requiring sequences both 5′ and 3′ to the regulated region. J. Mol. Biol. 176, 307–331.

Adachi, Y., Luke, M., Laemmli, U.K., 1991. Chromosome assembly in vitro: topoisomerase II is required for condensation. Cell 64, 137–148.

Adhireksan, Z., Sharma, D., Lee, P.L., Davey, C.A., 2020. Near-atomic resolution structures of interdigitated nucleosome fibres. Nat. Commun. 11, 4747.

Adolph, K.W., Cheng, S.M., Paulson, J.R., Laemmli, U.K., 1977. Isolation of a protein scaffold from mitotic HeLa cell chromosomes. Proc. Natl. Acad. Sci. U. S. A. 74, 4937–4941.

Allahverdi, A., Yang, R., Korolev, N., Fan, Y., Davey, C.A., Liu, C.F., Nordenskiold, L., 2011. The effects of histone H4 tail acetylations on cation-induced chromatin folding and self-association. Nucleic Acids Res. 39, 1680–1691.

Allan, J., Hartman, P.G., Crane-Robinson, C., Aviles, F.X., 1980. The structure of histone H1 and its location in chromatin. Nature 288, 675–679.

Allan, J., Harborne, N., Rau, D.C., Gould, H., 1982. Participation of core histone "tails" in the stabilization of the chromatin solenoid. J. Cell Biol. 93, 285–297.

Allshire, R.C., Madhani, H.D., 2018. Ten principles of heterochromatin formation and function. Nat. Rev. Mol. Cell Biol. 19, 229–244.

An, W., Leuba, S.H., van Holde, K., Zlatanova, J., 1998a. Linker histone protects linker DNA on only one side of the core particle and in a sequence-dependent manner. Proc. Natl. Acad. Sci. U. S. A. 95, 3396–3401.

An, W., van Holde, K., Zlatanova, J., 1998b. Linker histone protection of chromatosomes reconstituted on 5S rDNA from *Xenopus borealis*: a reinvestigation. Nucleic Acids Res. 26, 4042–4046.

Aparicio, O.M., Billington, B.L., Gottschling, D.E., 1991. Modifiers of position effect are shared between telomeric and silent mating-type loci in *S. cerevisiae*. Cell 66, 1279–1287.

Arimura, Y., Shih, R.M., Froom, R., Funabiki, H., 2021. Structural features of nucleosomes in interphase and metaphase chromosomes. Mol. Cell 81, 4377–4397.e4312.

Armache, K.J., Garlick, J.D., Canzio, D., Narlikar, G.J., Kingston, R.E., 2011. Structural basis of silencing: Sir3 BAH domain in complex with a nucleosome at 3.0 A resolution. Science 334, 977–982.

Ausio, J., Dong, F., van Holde, K.E., 1989. Use of selectively trypsinized nucleosome core particles to analyze the role of the histone "tails" in the stabilization of the nucleosome. J. Mol. Biol. 206, 451–463.

Aviles, F.J., Chapman, G.E., Kneale, G.G., Crane-Robinson, C., Bradbury, E.M., 1978. The conformation of histone H5. Isolation and characterisation of the globular segment. Eur. J. Biochem. 88, 363–371.

Ballestas, M.E., Chatis, P.A., Kaye, K.M., 1999. Efficient persistence of extrachromosomal KSHV DNA mediated by latency-associated nuclear antigen. Science 284, 641–644.

Bancaud, A., Lavelle, C., Huet, S., Ellenberg, J., 2012. A fractal model for nuclear organization: current evidence and biological implications. Nucleic Acids Res. 40, 8783–8792.

Bartolome, S., Bermudez, A., Daban, J.R., 1994. Internal structure of the 30 nm chromatin fiber. J. Cell Sci. 107 (Pt 11), 2983–2992.

Barton, D.W., 1951. Localized chiasmata in the differentiated chromosomes of the tomato. Genetics 36, 374–381.

Bazett-Jones, D.P., 1992. Electron spectroscopic imaging of chromatin and other nucleoprotein complexes. Electron Microsc. Rev. 74, 37–61.

Bednar, J., Horowitz, R.A., Grigoryev, S.A., Carruthers, L.M., Hansen, J.C., Koster, A.J., Woodcock, C.L., 1998. Nucleosomes, linker DNA, and linker histone form a unique structural motif that directs the higher-order folding and compaction of chromatin. Proc. Natl. Acad. Sci. U. S. A. 95, 14173–14178.

Bednar, J., Garcia-Saez, I., Boopathi, R., Cutter, A.R., Papai, G., Reymer, A., Syed, S.H., Lone, I.N., Tonchev, O., Crucifix, C., et al., 2017. Structure and dynamics of a 197 bp nucleosome in complex with linker histone H1. Mol. Cell 66, 384–397.e388.

Beel, A.J., Azubel, M., Mattei, P.J., Kornberg, R.D., 2021. Structure of mitotic chromosomes. Mol. Cell 81, 4369–4376.e4363.

Behrouzi, R., Lu, C., Currie, M.A., Jih, G., Iglesias, N., Moazed, D., 2016. Heterochromatin assembly by interrupted Sir3 bridges across neighboring nucleosomes. elife 5, e17556.

Belmont, A.S., 2006. Mitotic chromosome structure and condensation. Curr. Opin. Cell Biol. 18, 632–638.

Benyajati, C., Worcel, A., 1976. Isolation, characterization, and structure of the folded interphase genome of Drosophila melanogaster. Cell 9, 393–407.

Berezney, R., Coffey, D.S., 1974. Identification of a nuclear protein matrix. Biochem. Biophys. Res. Commun. 60, 1410–1417.

Berezney, R., Coffey, D.S., 1975. Nuclear protein matrix: association with newly synthesized DNA. Science 189, 291–293.

Berger, F., 2019. Emil Heitz, a true epigenetics pioneer. Nat. Rev. Mol. Cell Biol. 20, 572.

Berman, J., Tachibana, C.Y., Tye, B.K., 1986. Identification of a telomere-binding activity from yeast. Proc. Natl. Acad. Sci. U. S. A. 83, 3713–3717.

Berrios, M., Osheroff, N., Fisher, P.A., 1985. In situ localization of DNA topoisomerase II, a major polypeptide component of the Drosophila nuclear matrix fraction. Proc. Natl. Acad. Sci. U. S. A. 82, 4142–4146.

Bickmore, W.A., 2013. The spatial organization of the human genome. Annu. Rev. Genomics Hum. Genet. 14, 67–84.

Bintu, B., Mateo, L.J., Su, J.H., Sinnott-Armstrong, N.A., Parker, M., Kinrot, S., Yamaya, K., Boettiger, A.N., Zhuang, X., 2018. Super-resolution chromatin tracing reveals domains and cooperative interactions in single cells. Science 362, eaau1783.

Boija, A., Klein, I.A., Sabari, B.R., Dall'Agnese, A., Coffey, E.L., Zamudio, A.V., Li, C.H., Shrinivas, K., Manteiga, J.C., Hannett, N.M., et al., 2018. Transcription factors activate genes through the phase-separation capacity of their activation domains. Cell 175, 1842–1855.e1816.

Boveri, T., 1909. Die blastomerenkerne von Ascaris meglocephala und die theorie der chromosomenindividualität. Arch. Zellforsch. 3, 181–268.

Boyle, S., Gilchrist, S., Bridger, J.M., Mahy, N.L., Ellis, J.A., Bickmore, W.A., 2001. The spatial organization of human chromosomes within the nuclei of normal and emerin-mutant cells. Hum. Mol. Genet. 10, 211–219.

Boyle, S., Rodesch, M.J., Halvensleben, H.A., Jeddeloh, J.A., Bickmore, W.A., 2011. Fluorescence in situ hybridization with high-complexity repeat-free oligonucleotide probes generated by massively parallel synthesis. Chromosom. Res. 19, 901–909.

Brand, A.H., Breeden, L., Abraham, J., Sternglanz, R., Nasmyth, K., 1985. Characterization of a "silencer" in yeast: a DNA sequence with properties opposite to those of a transcriptional enhancer. Cell 41, 41–48.

Brand, H.A., Micklem, G., Nasmyth, K., 1987. A yeast silencer contains sequences that can promote autonomous plasmid replication and transcriptional action. Cell 51, 709–719.

Brangwynne, C.P., Mitchison, T.J., Hyman, A.A., 2011. Active liquid-like behavior of nucleoli determines their size and shape in *Xenopus laevis* oocytes. Proc. Natl. Acad. Sci. U. S. A. 108, 4334–4339.

Braunstein, M., Sobel, R.E., Allis, C.D., Turner, B.M., Broach, J.R., 1996. Efficient transcriptional silencing in *Saccharomyces cerevisiae* requires a heterochromatin histone acetylation pattern. Mol. Cell. Biol. 16, 4349–4356.

Britten, R.J., Kohne, D.E., 1968. Repeated sequences in DNA. Hundreds of thousands of copies of DNA sequences have been incorporated into the genomes of higher organisms. Science 161, 529–540.

Brown, S.W., 1966. Heterochromatin. Science 151, 417–425.

Brown, D.T., Izard, T., Misteli, T., 2006. Mapping the interaction surface of linker histone H1(0) with the nucleosome of native chromatin in vivo. Nat. Struct. Mol. Biol. 13, 250–255.

Buchman, A.R., Kimmerly, W.J., Rine, J., Kornberg, R.D., 1988. Two DNA-binding factors recognize specific sequences at silencers, upstream activating sequences, autonomously replicating sequences, and telomeres in *Saccharomyces cerevisiae*. Mol. Cell. Biol. 8, 210–225.

Buck, S.W., Shore, D., 1995. Action of a RAP1 carboxy-terminal silencing domain reveals an underlying competition between HMR and telomeres in yeast. Genes Dev. 9, 370–384.

Butler, P.J., 1983. The folding of chromatin. CRC Crit. Rev. Biochem. 15, 57–91.

Carruthers, L.M., Hansen, J.C., 2000. The core histone N termini function independently of linker histones during chromatin condensation. J. Biol. Chem. 275, 37285–37290.

Carruthers, L.M., Bednar, J., Woodcock, C.L., Hansen, J.C., 1998. Linker histones stabilize the intrinsic salt-dependent folding of nucleosomal arrays: mechanistic ramifications for higher-order chromatin folding. Biochemistry 37, 14776–14787.

Carruthers, L.M., Tse, C., Walker 3rd, K.P., Hansen, J.C., 1999. Assembly of defined nucleosomal and chromatin arrays from pure components. Methods Enzymol. 304, 19–35.

Cattoni, D.I., Cardozo Gizzi, A.M., Georgieva, M., Di Stefano, M., Valeri, A., Chamousset, D., Houbron, C., Dejardin, S., Fiche, J.B., Gonzalez, I., et al., 2017. Single-cell absolute contact probability detection reveals chromosomes are organized by multiple low-frequency yet specific interactions. Nat. Commun. 8, 1753.

Chadwick, B.P., Willard, H.F., 2001. A novel chromatin protein, distantly related to histone H2A, is largely excluded from the inactive X chromosome. J. Cell Biol. 152, 375–384.

Chen, D., Dundr, M., Wang, C., Leung, A., Lamond, A., Misteli, T., Huang, S., 2005. Condensed mitotic chromatin is accessible to transcription factors and chromatin structural proteins. J. Cell Biol. 168, 41–54.

Cheng, T.H., Gartenberg, M.R., 2000. Yeast heterochromatin is a dynamic structure that requires silencers continuously. Genes Dev. 14, 452–463.

Chicano, A., Crosas, E., Oton, J., Melero, R., Engel, B.D., Daban, J.R., 2019. Frozen-hydrated chromatin from metaphase chromosomes has an interdigitated multilayer structure. EMBO J. 38, e99769.

Cho, W.K., Spille, J.H., Hecht, M., Lee, C., Li, C., Grube, V., Cisse, I.I., 2018. Mediator and RNA polymerase II clusters associate in transcription-dependent condensates. Science 361, 412–415.

Chodaparambil, J.V., Barbera, A.J., Lu, X., Kaye, K.M., Hansen, J.C., Luger, K., 2007. A charged and contoured surface on the nucleosome regulates chromatin compaction. Nat. Struct. Mol. Biol. 14, 1105–1107.

Clark, D.J., Kimura, T., 1990. Electrostatic mechanism of chromatin folding. J. Mol. Biol. 211, 883–896.

Clark, K.L., Halay, E.D., Lai, E., Burley, S.K., 1993. Co-crystal structure of the HNF-3/fork head DNA-recognition motif resembles histone H5. Nature 364, 412–420.

Clarke, P.R., Zhang, C., 2008. Spatial and temporal coordination of mitosis by ran GTPase. Nat. Rev. Mol. Cell Biol. 9, 464–477.

Cleard, F., Delattre, M., Spierer, P., 1997. SU(VAR)3-7, a Drosophila heterochromatin-associated protein and companion of HP1 in the genomic silencing of position-effect variegation. EMBO J. 16, 5280–5288.

Clore, G.M., Gronenborn, A.M., Nilges, M., Sukumaran, D.K., Zarbock, J., 1987. The polypeptide fold of the globular domain of histone H5 in solution. A study using nuclear magnetic resonance, distance geometry and restrained molecular dynamics. EMBO J. 6, 1833–1842.

Cole, R.D., 1984. A minireview of microheterogeneity in H1 histone and its possible significance. Anal. Biochem. 136, 24–30.

Comings, D.E., 1968. The rationale for an ordered arrangement of chromatin in the interphase nucleus. Am. J. Hum. Genet. 20, 440–460.

Cook, P.R., Brazell, I.A., 1975. Supercoils in human DNA. J. Cell Sci. 19, 261–279.

Cook, P.R., Brazell, I.A., 1976. Conformational constraints in nuclear DNA. J. Cell Sci. 22, 287–302.

Cook, P.R., Brazell, I.A., 1978. Spectrofluorometric measurement of the binding of ethidium to superhelical DNA from cell nuclei. Eur. J. Biochem. 84, 465–477.

Cook, P.R., Brazell, I.A., Jost, E., 1976. Characterization of nuclear structures containing superhelical DNA. J. Cell Sci. 22, 303–324.

Cooper, K.W., 1959. Cytogenetic analysis of major heterochromatic elements (especially Xh and Y) in Drosophila melanogaster, and the theory of "heterochromatin". Chromosoma 10, 535–588.

Crane, E., Bian, Q., McCord, R.P., Lajoie, B.R., Wheeler, B.S., Ralston, E.J., Uzawa, S., Dekker, J., Meyer, B.J., 2015. Condensin-driven remodelling of X chromosome topology during dosage compensation. Nature 523, 240–244.

Cremer, T., Cremer, C., 2001. Chromosome territories, nuclear architecture and gene regulation in mammalian cells. Nat. Rev. Genet. 2, 292–301.

Cremer, T., Cremer, M., 2010. Chromosome territories. Cold Spring Harb. Perspect. Biol. 2, a003889.

Cremer, T., Kurz, A., Zirbel, R., Dietzel, S., Rinke, B., Schrock, E., Speicher, M.R., Mathieu, U., Jauch, A., Emmerich, P., et al., 1993. Role of chromosome territories in the functional compartmentalization of the cell nucleus. Cold Spring Harb. Symp. Quant. Biol. 58, 777–792.

Croft, J.A., Bridger, J.M., Boyle, S., Perry, P., Teague, P., Bickmore, W.A., 1999. Differences in the localization and morphology of chromosomes in the human nucleus. J. Cell Biol. 145, 1119–1131.

Cubizolles, F., Martino, F., Perrod, S., Gasser, S.M., 2006. A homotrimer-heterotrimer switch in Sir2 structure differentiates rDNA and telomeric silencing. Mol. Cell 21, 825–836.

Cui, Y., Bustamante, C., 2000. Pulling a single chromatin fiber reveals the forces that maintain its higher-order structure. Proc. Natl. Acad. Sci. U. S. A. 97, 127–132.

Cullen, K.E., Kladde, M.P., Seyfred, M.A., 1993. Interaction between transcription regulatory regions of prolactin chromatin. Science 261, 203–206.

Daban, J.R., Bermudez, A., 1998. Interdigitated solenoid model for compact chromatin fibers. Biochemistry 37, 4299–4304.

Davies, H.G., 1968. Electron-microscope observations on the organization of heterochromatin in certain cells. J. Cell Sci. 3, 129–150.

de Baaij, J.H., Hoenderop, J.G., Bindels, R.J., 2015. Magnesium in man: implications for health and disease. Physiol. Rev. 95, 1–46.

de Wit, E., Vos, E.S., Holwerda, S.J., Valdes-Quezada, C., Verstegen, M.J., Teunissen, H., Splinter, E., Wijchers, P.J., Krijger, P.H., de Laat, W., 2015. CTCF binding polarity determines chromatin looping. Mol. Cell 60, 676–684.

Dekker, J., Mirny, L., 2016. The 3D genome as moderator of chromosomal communication. Cell 164, 1110–1121.

Dekker, J., Misteli, T., 2015. Long-range chromatin interactions. Cold Spring Harb. Perspect. Biol. 7, a019356.

Dekker, J., Rippe, K., Dekker, M., Kleckner, N., 2002. Capturing chromosome conformation. Science 295, 1306–1311.

Depew, D.E., Wang, J.C., 1975. Conformational fluctuations of DNA helix. Proc. Natl. Acad. Sci. U. S. A. 72, 4275–4279.

Dixon, J.R., Selvaraj, S., Yue, F., Kim, A., Li, Y., Shen, Y., Hu, M., Liu, J.S., Ren, B., 2012. Topological domains in mammalian genomes identified by analysis of chromatin interactions. Nature 485, 376–380.

Dixon, J.R., Gorkin, D.U., Ren, B., 2016. Chromatin domains: the unit of chromosome organization. Mol. Cell 62, 668–680.

Donaldson, K.M., Lui, A., Karpen, G.H., 2002. Modifiers of terminal deficiency-associated position effect variegation in Drosophila. Genetics 160, 995–1009.

Dorigo, B., Schalch, T., Bystricky, K., Richmond, T.J., 2003. Chromatin fiber folding: requirement for the histone H4 N-terminal tail. J. Mol. Biol. 327, 85–96.

Dorigo, B., Schalch, T., Kulangara, A., Duda, S., Schroeder, R.R., Richmond, T.J., 2004. Nucleosome arrays reveal the two-start organization of the chromatin fiber. Science 306, 1571–1573.

Dostie, J., Richmond, T.A., Arnaout, R.A., Selzer, R.R., Lee, W.L., Honan, T.A., Rubio, E.D., Krumm, A., Lamb, J., Nusbaum, C., et al., 2006. Chromosome conformation capture carbon copy (5C): a massively parallel solution for mapping interactions between genomic elements. Genome Res. 16, 1299–1309.

Doty, P., Marmur, J., Eigner, J., Schildkraut, C., 1960. Strand separation and specific recombination in deoxyribonucleic acids: physical chemical studies. Proc. Natl. Acad. Sci. U. S. A. 46, 461–476.

Dowen, J.M., Fan, Z.P., Hnisz, D., Ren, G., Abraham, B.J., Zhang, L.N., Weintraub, A.S., Schujiers, J., Lee, T.I., Zhao, K., et al., 2014. Control of cell identity genes occurs in insulated neighborhoods in mammalian chromosomes. Cell 159, 374–387.

Earnshaw, W.C., Halligan, B., Cooke, C.A., Heck, M.M., Liu, L.F., 1985. Topoisomerase II is a structural component of mitotic chromosome scaffolds. J. Cell Biol. 100, 1706–1715.

Eissenberg, J.C., James, T.C., Foster-Hartnett, D.M., Hartnett, T., Ngan, V., Elgin, S.C., 1990. Mutation in a heterochromatin-specific chromosomal protein is associated with suppression of position-effect variegation in *Drosophila melanogaster*. Proc. Natl. Acad. Sci. U. S. A. 87, 9923–9927.

Eissenberg, J.C., Morris, G.D., Reuter, G., Hartnett, T., 1992. The heterochromatin-associated protein HP-1 is an essential protein in Drosophila with dosage-dependent effects on position-effect variegation. Genetics 131, 345–352.

Elgin, S.C., Reuter, G., 2013. Position-effect variegation, heterochromatin formation, and gene silencing in Drosophila. Cold Spring Harb. Perspect. Biol. 5, a017780.

Eltsov, M., Maclellan, K.M., Maeshima, K., Frangakis, A.S., Dubochet, J., 2008. Analysis of cryo-electron microscopy images does not support the existence of 30-nm chromatin fibers in mitotic chromosomes in situ. Proc. Natl. Acad. Sci. U. S. A. 105, 19732–19737.

Erdel, F., Rademacher, A., Vlijm, R., Tunnermann, J., Frank, L., Weinmann, R., Schweigert, E., Yserentant, K., Hummert, J., Bauer, C., et al., 2020. Mouse heterochromatin adopts digital compaction states without showing hallmarks of HP1-driven liquid-liquid phase separation. Mol. Cell 78, 236–249.e237.

Fan, J.Y., Rangasamy, D., Luger, K., Tremethick, D.J., 2004. H2A.Z alters the nucleosome surface to promote HP1alpha-mediated chromatin fiber folding. Mol. Cell 16, 655–661.

Fanti, L., Giovinazzo, G., Berloco, M., Pimpinelli, S., 1998. The heterochromatin protein 1 prevents telomere fusions in Drosophila. Mol. Cell 2, 527–538.

Felsenfeld, G., McGhee, J.D., 1986. Structure of the 30 nm chromatin fiber. Cell 44, 375–377.

Finch, J.T., Klug, A., 1976. Solenoidal model for superstructure in chromatin. Proc. Natl. Acad. Sci. U. S. A. 73, 1897–1901.

Finlan, L.E., Sproul, D., Thomson, I., Boyle, S., Kerr, E., Perry, P., Ylstra, B., Chubb, J.R., Bickmore, W.A., 2008. Recruitment to the nuclear periphery can alter expression of genes in human cells. PLoS Genet. 4, e1000039.

Finn, E.H., Misteli, T., 2019. Molecular basis and biological function of variability in spatial genome organization. Science 365, eaaw9498.

Finn, E.H., Pegoraro, G., Brandao, H.B., Valton, A.L., Oomen, M.E., Dekker, J., Mirny, L., Misteli, T., 2019. Extensive heterogeneity and intrinsic variation in spatial genome organization. Cell 176, 1502–1515.e1510.

Flamm, W.G., Walker, P.M., McCallum, M., 1969. Some properties of the single strands isolated from the DNA of the nuclear satellite of the mouse (*Mus musculus*). J. Mol. Biol. 40, 423–443.

Fletcher, T.M., Hansen, J.C., 1995. Core histone tail domains mediate oligonucleosome folding and nucleosomal DNA organization through distinct molecular mechanisms. J. Biol. Chem. 270, 25359–25362.

Fourel, G., Revardel, E., Koering, C.E., Gilson, E., 1999. Cohabitation of insulators and silencing elements in yeast subtelomeric regions. EMBO J. 18, 2522–2537.

Franke, W.W., 1974. Structure, biochemistry, and functions of the nuclear envelope. Int Rev Cytol (Suppl 4), 71–236.

Fullwood, M.J., Liu, M.H., Pan, Y.F., Liu, J., Xu, H., Mohamed, Y.B., Orlov, Y.L., Velkov, S., Ho, A., Mei, P.H., et al., 2009. An oestrogen-receptor-alpha-bound human chromatin interactome. Nature 462, 58–64.

Fulmer, A.W., Fasman, G.D., 1979. Analysis of chromatin reconstitution. Biochemistry 18, 659–668.

Fussner, E., Ching, R.W., Bazett-Jones, D.P., 2011. Living without 30nm chromatin fibers. Trends Biochem. Sci. 36, 1–6.

Fussner, E., Strauss, M., Djuric, U., Li, R., Ahmed, K., Hart, M., Ellis, J., Bazett-Jones, D.P., 2012. Open and closed domains in the mouse genome are configured as 10-nm chromatin fibres. EMBO Rep. 13, 992–996.

Gall, J.G., 1963. Kinetics of deoxyribonuclease action on chromosomes. Nature 198, 36–38.

Gall, J.G., 1966. Chromosome fibers studied by a spreading technique. Chromosoma 20, 221–233.

Garcia-Ramirez, M., Dong, F., Ausio, J., 1992. Role of the histone "tails" in the folding of oligonucleosomes depleted of histone H1. J. Biol. Chem. 267, 19587–19595.

Gartenberg, M.R., Smith, J.S., 2016. The nuts and bolts of transcriptionally silent chromatin in *Saccharomyces cerevisiae*. Genetics 203, 1563–1599.

Georgel, P.T., Palacios DeBeer, M.A., Pietz, G., Fox, C.A., Hansen, J.C., 2001. Sir3-dependent assembly of supramolecular chromatin structures in vitro. Proc. Natl. Acad. Sci. U. S. A. 98, 8584–8589.

Gerchman, S.E., Ramakrishnan, V., 1987. Chromatin higher-order structure studied by neutron scattering and scanning transmission electron microscopy. Proc. Natl. Acad. Sci. U. S. A. 84, 7802–7806.

Ghavi-Helm, Y., Jankowski, A., Meiers, S., Viales, R.R., Korbel, J.O., Furlong, E.E.M., 2019. Highly rearranged chromosomes reveal uncoupling between genome topology and gene expression. Nat. Genet. 51, 1272–1282.

Gibson, B.A., Doolittle, L.K., Schneider, M.W.G., Jensen, L.E., Gamarra, N., Henry, L., Gerlich, D.W., Redding, S., Rosen, M.K., 2019. Organization of chromatin by intrinsic and regulated phase separation. Cell 179, 470–484.e421.

Gilbert, N., Boyle, S., Fiegler, H., Woodfine, K., Carter, N.P., Bickmore, W.A., 2004. Chromatin architecture of the human genome: gene-rich domains are enriched in open chromatin fibers. Cell 118, 555–566.

Gordon, F., Luger, K., Hansen, J.C., 2005. The core histone N-terminal tail domains function independently and additively during salt-dependent oligomerization of nucleosomal arrays. J. Biol. Chem. 280, 33701–33706.

Gotta, M., Laroche, T., Formenton, A., Maillet, L., Scherthan, H., Gasser, S.M., 1996. The clustering of telomeres and colocalization with Rap1, Sir3, and Sir4 proteins in wild-type *Saccharomyces cerevisiae*. J. Cell Biol. 134, 1349–1363.

Gottschling, D.E., 1992. Telomere-proximal DNA in *Saccharomyces cerevisiae* is refractory to methyltransferase activity in vivo. Proc. Natl. Acad. Sci. U. S. A. 89, 4062–4065.

Gottschling, D.E., Aparicio, O.M., Billington, B.L., Zakian, V.A., 1990. Position effect at *S. cerevisiae* telomeres: reversible repression of pol II transcription. Cell 63, 751–762.

Gowen, J.W., Gay, E.H., 1934. Chromosome constitution and behavior in Eversporting and mottling in Drosophila Melanogaster. Genetics 19, 189–208.

Goytisolo, F.A., Gerchman, S.E., Yu, X., Rees, C., Graziano, V., Ramakrishnan, V., Thomas, J.O., 1996. Identification of two DNA-binding sites on the globular domain of histone H5. EMBO J. 15, 3421–3429.

Grigoryev, S.A., Woodcock, C.L., 2012. Chromatin organization—the 30 nm fiber. Exp. Cell Res. 318, 1448–1455.

Grunstein, M., 1997. Molecular model for telomeric heterochromatin in yeast. Curr. Opin. Cell Biol. 9, 383–387.

Guelen, L., Pagie, L., Brasset, E., Meuleman, W., Faza, M.B., Talhout, W., Eussen, B.H., de Klein, A., Wessels, L., de Laat, W., et al., 2008. Domain organization of human chromosomes revealed by mapping of nuclear lamina interactions. Nature 453, 948–951.

Haber, J.E., George, J.P., 1979. A mutation that permits the expression of normally silent copies of mating-type information in *Saccharomyces cerevisiae*. Genetics 93, 13–35.

Hancock, R., 1974. Interphase chromosomal deoxyribonucleoprotein isolated as a discrete structure from cultured cells. J. Mol. Biol. 86, 649–663.

Hancock, R., 2000. A new look at the nuclear matrix. Chromosoma 109, 219–225.

Hansen, J.C., 2002. Conformational dynamics of the chromatin fiber in solution: determinants, mechanisms, and functions. Annu. Rev. Biophys. Biomol. Struct. 31, 361–392.

Hansen, J.C., 2012. Human mitotic chromosome structure: what happened to the 30-nm fibre? EMBO J. 31, 1621–1623.

Hansen, J.C., Ausio, J., Stanik, V.H., van Holde, K.E., 1989. Homogeneous reconstituted oligonucleosomes, evidence for salt-dependent folding in the absence of histone H1. Biochemistry 28, 9129–9136.

Hansen, J.C., Connolly, M., McDonald, C.J., Pan, A., Pryamkova, A., Ray, K., Seidel, E., Tamura, S., Rogge, R., Maeshima, K., 2018. The 10-nm chromatin fiber and its relationship to interphase chromosome organization. Biochem. Soc. Trans. 46, 67–76.

Hanssen, L.L.P., Kassouf, M.T., Oudelaar, A.M., Biggs, D., Preece, C., Downes, D.J., Gosden, M., Sharpe, J.A., Sloane-Stanley, J.A., Hughes, J.R., et al., 2017. Tissue-specific CTCF-cohesin-mediated chromatin architecture delimits enhancer interactions and function in vivo. Nat. Cell Biol. 19, 952–961.

Hao, F., Kale, S., Dimitrov, S., Hayes, J.J., 2021. Unraveling linker histone interactions in nucleosomes. Curr. Opin. Struct. Biol. 71, 87–93.

Hartman, P.G., Chapman, G.E., Moss, T., Bradbury, E.M., 1977. Studies on the role and mode of operation of the very-lysine-rich histone H1 in eukaryote chromatin. The three structural regions of the histone H1 molecule. Eur. J. Biochem. 77, 45–51.

Hashimshony, T., Zhang, J., Keshet, I., Bustin, M., Cedar, H., 2003. The role of DNA methylation in setting up chromatin structure during development. Nat. Genet. 34, 187–192.

Hayes, J.J., Wolffe, A.P., 1993. Preferential and asymmetric interaction of linker histones with 5S DNA in the nucleosome. Proc. Natl. Acad. Sci. U. S. A. 90, 6415–6419.

Hayes, J.J., Kaplan, R., Ura, K., Pruss, D., Wolffe, A., 1996. A putative DNA binding surface in the globular domain of a linker histone is not essential for specific binding to the nucleosome. J. Biol. Chem. 271, 25817–25822.

Hecht, A., Laroche, T., Strahl-Bolsinger, S., Gasser, S.M., Grunstein, M., 1995. Histone H3 and H4 N-termini interact with SIR3 and SIR4 proteins: a molecular model for the formation of heterochromatin in yeast. Cell 80, 583–592.

Hecht, A., Strahl-Bolsinger, S., Grunstein, M., 1996. Spreading of transcriptional repressor SIR3 from telomeric heterochromatin. Nature 383, 92–96.

Heitz, E., 1928. Das Heterochromatin der Moose. I. Jahrb. Wiss. Bot. 69, 762–818.

Heitz, E., 1935. Chromosomenstruktur und Gene. Z. Indukt. Vererb. 70, 402–447.

Henikoff, S., 1979. Position effects and variegation enhancers in an autosomal region of *DROSOPHILA MELANOGASTER*. Genetics 93, 105–115.

Hergeth, S.P., Schneider, R., 2015. The H1 linker histones: multifunctional proteins beyond the nucleosomal core particle. EMBO Rep. 16, 1439–1453.

Hirano, T., Mitchison, T.J., 1991. Cell cycle control of higher-order chromatin assembly around naked DNA in vitro. J. Cell Biol. 115, 1479–1489.

Hirano, T., Mitchison, T.J., 1993. Topoisomerase II does not play a scaffolding role in the organization of mitotic chromosomes assembled in Xenopus egg extracts. J. Cell Biol. 120, 601–612.

Hirota, T., Gerlich, D., Koch, B., Ellenberg, J., Peters, J.M., 2004. Distinct functions of condensin I and II in mitotic chromosome assembly. J. Cell Sci. 117, 6435–6445.

Hnisz, D., Day, D.S., Young, R.A., 2016. Insulated neighborhoods: structural and functional units of mammalian gene control. Cell 167, 1188–1200.

Hnisz, D., Shrinivas, K., Young, R.A., Chakraborty, A.K., Sharp, P.A., 2017. A phase separation model for transcriptional control. Cell 169, 13–23.

Hoppe, G.J., Tanny, J.C., Rudner, A.D., Gerber, S.A., Danaie, S., Gygi, S.P., Moazed, D., 2002. Steps in assembly of silent chromatin in yeast: Sir3-independent binding of a Sir2/Sir4 complex to silencers and role for Sir2-dependent deacetylation. Mol. Cell. Biol. 22, 4167–4180.

Horowitz, R.A., Agard, D.A., Sedat, J.W., Woodcock, C.L., 1994. The three-dimensional architecture of chromatin in situ: electron tomography reveals fibers composed of a continuously variable zig-Zag nucleosomal ribbon. J. Cell Biol. 125, 1–10.

Hsieh, T.H., Weiner, A., Lajoie, B., Dekker, J., Friedman, N., Rando, O.J., 2015. Mapping nucleosome resolution chromosome folding in yeast by Micro-C. Cell 162, 108–119.

Igo-Kemenes, T., Greil, W., Zachau, H.G., 1977. Preparation of soluble chromatin and specific chromatin fractions with restriction nucleases. Nucleic Acids Res. 4, 3387–3400.

Jackson, D.A., Pombo, A., 1998. Replicon clusters are stable units of chromosome structure: evidence that nuclear organization contributes to the efficient activation and propagation of S phase in human cells. J. Cell Biol. 140, 1285–1295.

Jackson, J.P., Lindroth, A.M., Cao, X., Jacobsen, S.E., 2002. Control of CpNpG DNA methylation by the KRYPTONITE histone H3 methyltransferase. Nature 416, 556–560.

James, T.C., Elgin, S.C., 1986. Identification of a nonhistone chromosomal protein associated with heterochromatin in Drosophila melanogaster and its gene. Mol. Cell. Biol. 6, 3862–3872.

James, T.C., Eissenberg, J.C., Craig, C., Dietrich, V., Hobson, A., Elgin, S.C., 1989. Distribution patterns of HP1, a heterochromatin-associated nonhistone chromosomal protein of Drosophila. Eur. J. Cell Biol. 50, 170–180.

Johnson, L.M., Kayne, P.S., Kahn, E.S., Grunstein, M., 1990. Genetic evidence for an interaction between SIR3 and histone H4 in the repression of the silent mating loci in *Saccharomyces cerevisiae*. Proc. Natl. Acad. Sci. U. S. A. 87, 6286–6290.

Johnson, A., Li, G., Sikorski, T.W., Buratowski, S., Woodcock, C.L., Moazed, D., 2009. Reconstitution of heterochromatin-dependent transcriptional gene silencing. Mol. Cell 35, 769–781.

Jones, K.W., 1970. Chromosomal and nuclear location of mouse satellite DNA in individual cells. Nature 225, 912–915.

Kalashnikova, A.A., Porter-Goff, M.E., Muthurajan, U.M., Luger, K., Hansen, J.C., 2013. The role of the nucleosome acidic patch in modulating higher order chromatin structure. J. R. Soc. Interface 10, 20121022.

Kalhor, R., Tjong, H., Jayathilaka, N., Alber, F., Chen, L., 2011. Genome architectures revealed by tethered chromosome conformation capture and population-based modeling. Nat. Biotechnol. 30, 90–98.

Kavenoff, R., Zimm, B.H., 1973. Chromosome-sized DNA molecules from Drosophila. Chromosoma 41, 1–27.

Kiryanov, G.I., Manamshjan, T.A., Polyakov, V.Y., Fais, D., Chentsov, J.S., 1976. Levels of granular organization of chromatin fibres. FEBS Lett. 67, 323–327.

Klar, A.J., Fogel, S., Macleod, K., 1979. MAR1-a regulator of the HMa and HMalpha loci in *SACCHAROMYCES CEREVISIAE*. Genetics 93, 37–50.

Kleinschmidt, A., Lang, D., Zahn, R.K., 1961. Über die intrazelluläre formation von Bakterien-DNS. Z. Naturforsch. 16b, 730–739.

Korolev, N., Berezhnoy, N.V., Eom, K.D., Tam, J.P., Nordenskiold, L., 2009. A universal description for the experimental behavior of salt-(in)dependent oligocation-induced DNA condensation. Nucleic Acids Res. 37, 7137–7150.

Kumaran, R.I., Spector, D.L., 2008. A genetic locus targeted to the nuclear periphery in living cells maintains its transcriptional competence. J. Cell Biol. 180, 51–65.

Kupper, K., Kolbl, A., Biener, D., Dittrich, S., von Hase, J., Thormeyer, T., Fiegler, H., Carter, N.P., Speicher, M.R., Cremer, T., et al., 2007. Radial chromatin positioning is shaped by local gene density, not by gene expression. Chromosoma 116, 285–306.

Kurtz, S., Shore, D., 1991. RAP1 protein activates and silences transcription of mating-type genes in yeast. Genes Dev. 5, 616–628.

Kyrion, G., Liu, K., Liu, C., Lustig, A.J., 1993. RAP1 and telomere structure regulate telomere position effects in *Saccharomyces cerevisiae*. Genes Dev. 7, 1146–1159.

Langmore, J.P., Paulson, J.R., 1983. Low angle X-ray diffraction studies of chromatin structure in vivo and in isolated nuclei and metaphase chromosomes. J. Cell Biol. 96, 1120–1131.

Langmore, J.P., Schutt, C., 1980. The higher order structure of chicken erythrocyte chromosomes in vivo. Nature 288, 620–622.

Larson, A.G., Elnatan, D., Keenen, M.M., Trnka, M.J., Johnston, J.B., Burlingame, A.L., Agard, D.A., Redding, S., Narlikar, G.J., 2017. Liquid droplet formation by HP1alpha suggests a role for phase separation in heterochromatin. Nature 547, 236–240.

Laurenson, P., Rine, J., 1992. Silencers, silencing, and heritable transcriptional states. Microbiol. Rev. 56, 543–560.

Lebkowski, J.S., Laemmli, U.K., 1982a. Evidence for two levels of DNA folding in histone-depleted HeLa interphase nuclei. J. Mol. Biol. 156, 309–324.

Lebkowski, J.S., Laemmli, U.K., 1982b. Non-histone proteins and long-range organization of HeLa interphase DNA. J. Mol. Biol. 156, 325–344.

Lehnertz, B., Ueda, Y., Derijck, A.A., Braunschweig, U., Perez-Burgos, L., Kubicek, S., Chen, T., Li, E., Jenuwein, T., Peters, A.H., 2003. Suv39h-mediated histone H3 lysine 9 methylation directs DNA methylation to major satellite repeats at pericentric heterochromatin. Curr. Biol. 13, 1192–1200.

Leuba, S.H., Yang, G., Robert, C., Samori, B., van Holde, K., Zlatanova, J., Bustamante, C., 1994. Three-dimensional structure of extended chromatin fibers as revealed by tapping-mode scanning force microscopy. Proc. Natl. Acad. Sci. U. S. A. 91, 11621–11625.

Lewis, E.B., 1950. The phenomenon of position effect. Adv. Genet. 3, 73–115.

Lieberman-Aiden, E., van Berkum, N.L., Williams, L., Imakaev, M., Ragoczy, T., Telling, A., Amit, I., Lajoie, B.R., Sabo, P.J., Dorschner, M.O., et al., 2009. Comprehensive mapping of long-range interactions reveals folding principles of the human genome. Science 326, 289–293.

Liou, G.G., Tanny, J.C., Kruger, R.G., Walz, T., Moazed, D., 2005. Assembly of the SIR complex and its regulation by O-acetyl-ADP-ribose, a product of NAD-dependent histone deacetylation. Cell 121, 515–527.

Locke, J., Kotarski, M.A., Tartof, K.D., 1988. Dosage-dependent modifiers of position effect variegation in Drosophila and a mass action model that explains their effect. Genetics 120, 181–198.

Longtine, M.S., Wilson, N.M., Petracek, M.E., Berman, J., 1989. A yeast telomere binding activity binds to two related telomere sequence motifs and is indistinguishable from RAP1. Curr. Genet. 16, 225–239.

Loo, S., Rine, J., 1994. Silencers and domains of generalized repression. Science 264, 1768–1771.

Lowary, P.T., Widom, J., 1998. New DNA sequence rules for high affinity binding to histone octamer and sequence-directed nucleosome positioning. J. Mol. Biol. 276, 19–42.

Lu, B.Y., Bishop, C.P., Eissenberg, J.C., 1996. Developmental timing and tissue specificity of heterochromatin-mediated silencing. EMBO J. 15, 1323–1332.

Luger, K., Mader, A.W., Richmond, R.K., Sargent, D.F., Richmond, T.J., 1997. Crystal structure of the nucleosome core particle at 2.8 A resolution [see comments]. Nature 389, 251–260.

Luo, K., Vega-Palas, M.A., Grunstein, M., 2002. Rap1-Sir4 binding independent of other sir, yKu, or histone interactions initiates the assembly of telomeric heterochromatin in yeast. Genes Dev. 16, 1528–1539.

Lupianez, D.G., Kraft, K., Heinrich, V., Krawitz, P., Brancati, F., Klopocki, E., Horn, D., Kayserili, H., Opitz, J.M., Laxova, R., et al., 2015. Disruptions of topological chromatin domains cause pathogenic rewiring of gene-enhancer interactions. Cell 161, 1012–1025.

Lustig, A.J., Liu, C., Zhang, C., Hanish, J.P., 1996. Tethered Sir3p nucleates silencing at telomeres and internal loci in *Saccharomyces cerevisiae*. Mol. Cell. Biol. 16, 2483–2495.

Ma, H., Samarabandu, J., Devdhar, R.S., Acharya, R., Cheng, P.C., Meng, C., Berezney, R., 1998. Spatial and temporal dynamics of DNA replication sites in mammalian cells. J. Cell Biol. 143, 1415–1425.

Mace Jr., M.L., Daskal, Y., Busch, H., Wray, V.P., Wray, W., 1977. Isolated metaphase chromosomes: scanning electron microscopic appearance of salt-extracted chromosomes. Cytobios 19, 27–40.

Maeshima, K., Eltsov, M., 2008. Packaging the genome: the structure of mitotic chromosomes. J. Biochem. 143, 145–153.

Maeshima, K., Laemmli, U.K., 2003. A two-step scaffolding model for mitotic chromosome assembly. Dev. Cell 4, 467–480.

Maeshima, K., Rogge, R., Tamura, S., Joti, Y., Hikima, T., Szerlong, H., Krause, C., Herman, J., Seidel, E., DeLuca, J., et al., 2016. Nucleosomal arrays self-assemble into supramolecular globular structures lacking 30-nm fibers. EMBO J. 35, 1115–1132.

Mahy, N.L., Perry, P.E., Bickmore, W.A., 2002. Gene density and transcription influence the localization of chromatin outside of chromosome territories detectable by FISH. J. Cell Biol. 159, 753–763.

Manning, G.S., 1969. Limiting laws and counterion condensation in polyelectrolyte solutions. I. Colligative properties. J. Chem. Phys. 51, 924–933.

Marcand, S., Buck, S.W., Moretti, P., Gilson, E., Shore, D., 1996. Silencing of genes at nontelomeric sites in yeast is controlled by sequestration of silencing factors at telomeres by rap 1 protein. Genes Dev. 10, 1297–1309.

Marmur, J., Lane, D., 1960. Strand separation and specific recombination in deoxyribonucleic acids: biological studies. Proc. Natl. Acad. Sci. U. S. A. 46, 453–461.

Marsden, M.P., Laemmli, U.K., 1979. Metaphase chromosome structure: evidence for a radial loop model. Cell 17, 849–858.

Martienssen, R., Moazed, D., 2015. RNAi and heterochromatin assembly. Cold Spring Harb. Perspect. Biol. 7, a019323.

Matsubara, K., Sano, N., Umehara, T., Horikoshi, M., 2007. Global analysis of functional surfaces of core histones with comprehensive point mutants. Genes Cells 12, 13–33.

McCready, S.J., Akrigg, A., Cook, P.R., 1979. Electron-microscopy of intact nuclear DNA from human cells. J. Cell Sci. 39, 53–62.

McSwiggen, D.T., Mir, M., Darzacq, X., Tjian, R., 2019. Evaluating phase separation in live cells: diagnosis, caveats, and functional consequences. Genes Dev. 33, 1619–1634.

Meaburn, K.J., Misteli, T., 2008. Locus-specific and activity-independent gene repositioning during early tumorigenesis. J. Cell Biol. 180, 39–50.

Megee, P.C., Morgan, B.A., Mittman, B.A., Smith, M.M., 1990. Genetic analysis of histone H4: essential role of lysines subject to reversible acetylation. Science 247, 841–845.

Meselson, M., Stahl, F.W., 1958. The replication of DNA in *Escherichia coli*. Proc. Natl. Acad. Sci. U. S. A. 44, 671–682.

Meyer, S., Becker, N.B., Syed, S.H., Goutte-Gattat, D., Shukla, M.S., Hayes, J.J., Angelov, D., Bednar, J., Dimitrov, S., Everaers, R., 2011. From crystal and NMR structures, footprints and cryo-electron-micrographs to large and soft structures: nanoscale modeling of the nucleosomal stem. Nucleic Acids Res. 39, 9139–9154.

Moore, S.C., Ausio, J., 1997. Major role of the histones H3-H4 in the folding of the chromatin fiber. Biochem. Biophys. Res. Commun. 230, 136–139.

Moore, G.D., Sinclair, D.A., Grigliatti, T.A., 1983. Histone gene multiplicity and position effect variegation in *DROSOPHILA MELANOGASTER*. Genetics 105, 327–344.

Moretti, P., Freeman, K., Coodly, L., Shore, D., 1994. Evidence that a complex of SIR proteins interacts with the silencer and telomere-binding protein RAP1. Genes Dev. 8, 2257–2269.

Morgan, T.H., Bridges, C.B., Schultz, J., 1937. Constitution of the germinal material in relation to heredity. Yearb. Carnegie Inst. 36, 298–305.

Morgan, T.H., Schultz, J., Curry, V., 1941. Investigations on the constitution of the germinal material in relation to heredity. Yearb. Carnegie Inst. 40, 282–287.

Mukherjee, S., Erickson, H., Bastia, D., 1988. Enhancer-origin interaction in plasmid R6K involves a DNA loop mediated by initiator protein. Cell 52, 375–383.

Muller, J., 1930. Types of visible variations induced by X-rays in *Drosophila*. J. Genet. 22, 299–334.

Nakanishi, S., Sanderson, B.W., Delventhal, K.M., Bradford, W.D., Staehling-Hampton, K., Shilatifard, A., 2008. A comprehensive library of histone mutants identifies nucleosomal residues required for H3K4 methylation. Nat. Struct. Mol. Biol. 15, 881–888.

Ng, H.H., Feng, Q., Wang, H., Erdjument-Bromage, H., Tempst, P., Zhang, Y., Struhl, K., 2002. Lysine methylation within the globular domain of histone H3 by Dot1 is important for telomeric silencing and sir protein association. Genes Dev. 16, 1518–1527.

Nielsen, A.L., Oulad-Abdelghani, M., Ortiz, J.A., Remboutsika, E., Chambon, P., Losson, R., 2001. Heterochromatin formation in mammalian cells: interaction between histones and HP1 proteins. Mol. Cell 7, 729–739.

Nishino, Y., Eltsov, M., Joti, Y., Ito, K., Takata, H., Takahashi, Y., Hihara, S., Frangakis, A.S., Imamoto, N., Ishikawa, T., et al., 2012. Human mitotic chromosomes consist predominantly of irregularly folded nucleosome fibres without a 30-nm chromatin structure. EMBO J. 31, 1644–1653.

Noll, M., Kornberg, R.D., 1977. Action of micrococcal nuclease on chromatin and the location of histone H1. J. Mol. Biol. 109, 393–404.

Nora, E.P., Lajoie, B.R., Schulz, E.G., Giorgetti, L., Okamoto, I., Servant, N., Piolot, T., van Berkum, N.L., Meisig, J., Sedat, J., et al., 2012. Spatial partitioning of the regulatory landscape of the X-inactivation centre. Nature 485, 381–385.

Ohlenbusch, H.H., Olivera, B.M., Tuan, D., Davidson, N., 1967. Selective dissociation of histones from calf thymus nucleoprotein. J. Mol. Biol. 25, 299–315.

Ohno, M., Priest, D.G., Taniguchi, Y., 2018. Nucleosome-level 3D organization of the genome. Biochem. Soc. Trans. 46, 491–501.

Olins, A.L., Olins, D.E., 1974. Spheroid chromatin units (v bodies). Science 183, 330–332.

Onishi, M., Liou, G.G., Buchberger, J.R., Walz, T., Moazed, D., 2007. Role of the conserved Sir3-BAH domain in nucleosome binding and silent chromatin assembly. Mol. Cell 28, 1015–1028.

Ono, T., Losada, A., Hirano, M., Myers, M.P., Neuwald, A.F., Hirano, T., 2003. Differential contributions of condensin I and condensin II to mitotic chromosome architecture in vertebrate cells. Cell 115, 109–121.

Ou, H.D., Phan, S., Deerinck, T.J., Thor, A., Ellisman, M.H., O'Shea, C.C., 2017. ChromEMT: visualizing 3D chromatin structure and compaction in interphase and mitotic cells. Science 357, eaag0025.

Oudet, P., Gross-Bellard, M., Chambon, P., 1975. Electron microscopic and biochemical evidence that chromatin structure is a repeating unit. Cell 4, 281–300.

Palladino, F., Laroche, T., Gilson, E., Axelrod, A., Pillus, L., Gasser, S.M., 1993. SIR3 and SIR4 proteins are required for the positioning and integrity of yeast telomeres. Cell 75, 543–555.

Panetta, G., Buttinelli, M., Flaus, A., Richmond, T.J., Rhodes, D., 1998. Differential nucleosome positioning on Xenopus oocyte and somatic 5 S RNA genes determines both TFIIIA and H1 binding: a mechanism for selective H1 repression. J. Mol. Biol. 282, 683–697.

Pardue, M.L., Gall, J.G., 1970. Chromosomal localization of mouse satellite DNA. Science 168, 1356–1358.

Park, E.C., Szostak, J.W., 1990. Point mutations in the yeast histone H4 gene prevent silencing of the silent mating type locus HML. Mol. Cell. Biol. 10, 4932–4934.

Passarge, E., 1979. Emil Heitz and the concept of heterochromatin: longitudinal chromosome differentiation was recognized fifty years ago. Am. J. Hum. Genet. 31, 106–115.
Paulson, J.R., Laemmli, U.K., 1977. The structure of histone-depleted metaphase chromosomes. Cell 12, 817–828.
Paweletz, N., 2001. Walther Flemming: pioneer of mitosis research. Nat. Rev. Mol. Cell Biol. 2, 72–75.
Pearson, E.C., Butler, P.J., Thomas, J.O., 1983. Higher-order structure of nucleosome oligomers from short-repeat chromatin. EMBO J. 2, 1367–1372.
Peng, A., Weber, S.C., 2019. Evidence for and against liquid-liquid phase separation in the nucleus. Noncoding RNA 5.
Phillips, J.E., Corces, V.G., 2009. CTCF: master weaver of the genome. Cell 137, 1194–1211.
Phillips-Cremins, J.E., Sauria, M.E., Sanyal, A., Gerasimova, T.I., Lajoie, B.R., Bell, J.S., Ong, C.T., Hookway, T.A., Guo, C., Sun, Y., et al., 2013. Architectural protein subclasses shape 3D organization of genomes during lineage commitment. Cell 153, 1281–1295.
Pina, B., Fernandez-Larrea, J., Garcia-Reyero, N., Idrissi, F.Z., 2003. The different (Sur)faces of Rap1p. Mol. Gen. Genomics. 268, 791–798.
Pope, B.D., Ryba, T., Dileep, V., Yue, F., Wu, W., Denas, O., Vera, D.L., Wang, Y., Hansen, R.S., Canfield, T.K., et al., 2014. Topologically associating domains are stable units of replication-timing regulation. Nature 515, 402–405.
Postnikov, Y., Bustin, M., 2010. Regulation of chromatin structure and function by HMGN proteins. Biochim. Biophys. Acta 1799, 62–68.
Pruss, D., Bartholomew, B., Persinger, J., Hayes, J., Arents, G., Moudrianakis, E.N., Wolffe, A.P., 1996. An asymmetric model for the nucleosome: a binding site for linker histones inside the DNA gyres [see comments]. Science 274, 614–617.
Pryde, F.E., Louis, E.J., 1999. Limitations of silencing at native yeast telomeres. EMBO J. 18, 2538–2550.
Rabl, C., 1885. Über Zellteilung. In: Morphologisches Jahrbuch. 10, pp. 214–330.
Rae, M.M., Franke, W.W., 1972. The interphase distribution of satellite DNA-containing heterochromatin in mouse nuclei. Chromosoma 39, 443–456.
Ramakrishnan, V., Finch, J.T., Graziano, V., Lee, P.L., Sweet, R.M., 1993. Crystal structure of globular domain of histone H5 and its implications for nucleosome binding. Nature 362, 219–223.
Rao, S.S., Huntley, M.H., Durand, N.C., Stamenova, E.K., Bochkov, I.D., Robinson, J.T., Sanborn, A.L., Machol, I., Omer, A.D., Lander, E.S., et al., 2014. A 3D map of the human genome at kilobase resolution reveals principles of chromatin looping. Cell 159, 1665–1680.
Razin, S.V., Iarovaia, O.V., Vassetzky, Y.S., 2014. A requiem to the nuclear matrix: from a controversial concept to 3D organization of the nucleus. Chromosoma 123, 217–224.
Reddy, K.L., Zullo, J.M., Bertolino, E., Singh, H., 2008. Transcriptional repression mediated by repositioning of genes to the nuclear lamina. Nature 452, 243–247.
Renauld, H., Aparicio, O.M., Zierath, P.D., Billington, B.L., Chhablani, S.K., Gottschling, D.E., 1993. Silent domains are assembled continuously from the telomere and are defined by promoter distance and strength, and by SIR3 dosage. Genes Dev. 7, 1133–1145.
Renz, M., Nehls, P., Hozier, J., 1977. Involvement of histone H1 in the organization of the chromosome fiber. Proc. Natl. Acad. Sci. U. S. A. 74, 1879–1883.
Reuter, G., Giarre, M., Farah, J., Gausz, J., Spierer, A., Spierer, P., 1990. Dependence of position-effect variegation in Drosophila on dose of a gene encoding an unusual zinc-finger protein. Nature 344, 219–223.
Ricci, M.A., Manzo, C., Garcia-Parajo, M.F., Lakadamyali, M., Cosma, M.P., 2015. Chromatin fibers are formed by heterogeneous groups of nucleosomes in vivo. Cell 160, 1145–1158.
Rich, A., Davies, D., 1956. A new two stranded helical structure: polyadenylic acid and polyuridylic acid. J. Am. Chem. Soc. 78, 3548–3549.
Ried, T., Schrock, E., Ning, Y., Wienberg, J., 1998. Chromosome painting: a useful art. Hum. Mol. Genet. 7, 1619–1626.
Rine, J., Herskowitz, I., 1987. Four genes responsible for a position effect on expression from HML and HMR in *Saccharomyces cerevisiae*. Genetics 116, 9–22.
Rine, J., Strathern, J.N., Hicks, J.B., Herskowitz, I., 1979. A suppressor of mating-type locus mutations in *Saccharomyces cerevisiae*: evidence for and identification of cryptic mating-type loci. Genetics 93, 877–901.
Ris, H., Kubai, D.F., 1970. Chromosome structure. Annu. Rev. Genet. 4, 263–294.
Robinson, P.J., Fairall, L., Huynh, V.A., Rhodes, D., 2006. EM measurements define the dimensions of the "30-nm" chromatin fiber: evidence for a compact, interdigitated structure. Proc. Natl. Acad. Sci. U. S. A. 103, 6506–6511.

Robinson, P.J., An, W., Routh, A., Martino, F., Chapman, L., Roeder, R.G., Rhodes, D., 2008. 30 nm chromatin fibre decompaction requires both H4-K16 acetylation and linker histone eviction. J. Mol. Biol. 381, 816–825.

Roukos, V., Voss, T.C., Schmidt, C.K., Lee, S., Wangsa, D., Misteli, T., 2013. Spatial dynamics of chromosome translocations in living cells. Science 341, 660–664.

Routh, A., Sandin, S., Rhodes, D., 2008. Nucleosome repeat length and linker histone stoichiometry determine chromatin fiber structure. Proc. Natl. Acad. Sci. U. S. A. 105, 8872–8877.

Rusche, L.N., Kirchmaier, A.L., Rine, J., 2002. Ordered nucleation and spreading of silenced chromatin in *Saccharomyces cerevisiae*. Mol. Biol. Cell 13, 2207–2222.

Rusche, L.N., Kirchmaier, A.L., Rine, J., 2003. The establishment, inheritance, and function of silenced chromatin in *Saccharomyces cerevisiae*. Annu. Rev. Biochem. 72, 481–516.

Sabari, B.R., Dall'Agnese, A., Boija, A., Klein, I.A., Coffey, E.L., Shrinivas, K., Abraham, B.J., Hannett, N.M., Zamudio, A.V., Manteiga, J.C., et al., 2018. Coactivator condensation at super-enhancers links phase separation and gene control. Science 361, eaar3958.

Saitoh, N., Goldberg, I.G., Wood, E.R., Earnshaw, W.C., 1994. ScII: an abundant chromosome scaffold protein is a member of a family of putative ATPases with an unusual predicted tertiary structure. J. Cell Biol. 127, 303–318.

Satchwell, S.C., Travers, A.A., 1989. Asymmetry and polarity of nucleosomes in chicken erythrocyte chromatin. EMBO J. 8, 229–238.

Schalch, T., Duda, S., Sargent, D.F., Richmond, T.J., 2005. X-ray structure of a tetranucleosome and its implications for the chromatin fibre. Nature 436, 138–141.

Scheffer, M.P., Eltsov, M., Frangakis, A.S., 2011. Evidence for short-range helical order in the 30-nm chromatin fibers of erythrocyte nuclei. Proc. Natl. Acad. Sci. U. S. A. 108, 16992–16997.

Schnell, R., Rine, J., 1986. A position effect on the expression of a tRNA gene mediated by the SIR genes in *Saccharomyces cerevisiae*. Mol. Cell. Biol. 6, 494–501.

Schotta, G., Ebert, A., Krauss, V., Fischer, A., Hoffmann, J., Rea, S., Jenuwein, T., Dorn, R., Reuter, G., 2002. Central role of Drosophila SU(VAR)3-9 in histone H3-K9 methylation and heterochromatic gene silencing. EMBO J. 21, 1121–1131.

Schotta, G., Ebert, A., Dorn, R., Reuter, G., 2003. Position-effect variegation and the genetic dissection of chromatin regulation in Drosophila. Semin. Cell Dev. Biol. 14, 67–75.

Schultz, J., 1936. Variegation in Drosophila and the inert chromosome regions. Proc. Natl. Acad. Sci. U. S. A. 22, 27–33.

Schwartz, D.C., Cantor, C.R., 1984. Separation of yeast chromosome-sized DNAs by pulsed field gradient gel electrophoresis. Cell 37, 67–75.

Schwarz, P.M., Hansen, J.C., 1994. Formation and stability of higher order chromatin structures. Contributions of the histone octamer. J. Biol. Chem. 269, 16284–16289.

Schwarz, P.M., Felthauser, A., Fletcher, T.M., Hansen, J.C., 1996. Reversible oligonucleosome self-association: dependence on divalent cations and core histone tail domains. Biochemistry 35, 4009–4015.

Sentmanat, M.F., Elgin, S.C., 2012. Ectopic assembly of heterochromatin in *Drosophila melanogaster* triggered by transposable elements. Proc. Natl. Acad. Sci. U. S. A. 109, 14104–14109.

Sexton, T., Yaffe, E., Kenigsberg, E., Bantignies, F., Leblanc, B., Hoichman, M., Parrinello, H., Tanay, A., Cavalli, G., 2012. Three-dimensional folding and functional organization principles of the Drosophila genome. Cell 148, 458–472.

Shogren-Knaak, M., Ishii, H., Sun, J.M., Pazin, M.J., Davie, J.R., Peterson, C.L., 2006. Histone H4-K16 acetylation controls chromatin structure and protein interactions. Science 311, 844–847.

Shore, D., Nasmyth, K., 1987. Purification and cloning of a DNA binding protein from yeast that binds to both silencer and activator elements. Cell 51, 721–732.

Simonis, M., Klous, P., Splinter, E., Moshkin, Y., Willemsen, R., de Wit, E., van Steensel, B., de Laat, W., 2006. Nuclear organization of active and inactive chromatin domains uncovered by chromosome conformation capture-on-chip (4C). Nat. Genet. 38, 1348–1354.

Simpson, R.T., 1978. Structure of the chromatosome, a chromatin particle containing 160 base pairs of DNA and all the histones. Biochemistry 17, 5524–5531.

Simpson, R.T., Thoma, F., Brubaker, J.M., 1985. Chromatin reconstituted from tandemly repeated cloned DNA fragments and core histones: a model system for study of higher order structure. Cell 42, 799–808.

Sinclair, D.A.R., Mottus, R.C., Grigliatti, T.A., 1983. Genes which suppress position-effect variegation in *Drosophila melanogaster* are clustered. Mol. Gen. Genet. 191, 326–333.

Singer, D.S., Singer, M.F., 1976. Studies on the interaction of H1 histone with superhelical DNA: characterization of the recognition and binding regions of H1 histones. Nucleic Acids Res. 3, 2531–2547.

Singh, J., Klar, A.J., 1992. Active genes in budding yeast display enhanced in vivo accessibility to foreign DNA methylases: a novel in vivo probe for chromatin structure of yeast. Genes Dev. 6, 186–196.

Skrajna, A., Goldfarb, D., Kedziora, K.M., Cousins, E.M., Grant, G.D., Spangler, C.J., Barbour, E.H., Yan, X., Hathaway, N.A., Brown, N.G., et al., 2020. Comprehensive nucleosome interactome screen establishes fundamental principles of nucleosome binding. Nucleic Acids Res. 48, 9415–9432.

Smothers, J.F., Henikoff, S., 2001. The hinge and chromo shadow domain impart distinct targeting of HP1-like proteins. Mol. Cell. Biol. 21, 2555–2569.

Song, F., Chen, P., Sun, D., Wang, M., Dong, L., Liang, D., Xu, R.M., Zhu, P., Li, G., 2014. Cryo-EM study of the chromatin fiber reveals a double helix twisted by tetranucleosomal units. Science 344, 376–380.

Spadafora, C., Oudet, P., Chambon, P., 1978. The same amount of DNA is organized in in vitro-assembled nucleosomes irrespective of the origin of the histones. Nucleic Acids Res. 5, 3479–3489.

Spofford, J.B., 1967. Single-locus modification of position-effect variegation in *Drosophila melanogaster*. I. White variegation. Genetics 57, 751–766.

Spradling, A.C., Karpen, G.H., 1990. Sixty years of mystery. Genetics 126, 779–784.

Staynov, D.Z., 1983. Possible nucleosome arrangements in the higher-order structure of chromatin. Int. J. Biol. Macromol. 5, 3–9.

Staynov, D.Z., Crane-Robinson, C., 1988. Footprinting of linker histones H5 and H1 on the nucleosome. EMBO J. 7, 3685–3691.

Steinmetz, M., Streeck, R.E., Zachau, H.G., 1978. Closely spaced nucleosome cores in reconstituted histone.DNA complexes and histone-H1-depleted chromatin. Eur. J. Biochem. 83, 615–628.

Stevens, T.J., Lando, D., Basu, S., Atkinson, L.P., Cao, Y., Lee, S.F., Leeb, M., Wohlfahrt, K.J., Boucher, W., O'Shaughnessy-Kirwan, A., et al., 2017. 3D structures of individual mammalian genomes studied by single-cell hi-C. Nature 544, 59–64.

Strahl-Bolsinger, S., Hecht, A., Luo, K., Grunstein, M., 1997. SIR2 and SIR4 interactions differ in core and extended telomeric heterochromatin in yeast. Genes Dev. 11, 83–93.

Strathern, J.N., Klar, A.J., Hicks, J.B., Abraham, J.A., Ivy, J.M., Nasmyth, K.A., McGill, C., 1982. Homothallic switching of yeast mating type cassettes is initiated by a double-stranded cut in the MAT locus. Cell 31, 183–192.

Strickfaden, H., Tolsma, T.O., Sharma, A., Underhill, D.A., Hansen, J.C., Hendzel, M.J., 2020. Condensed chromatin behaves like a solid on the mesoscale in vitro and in living cells. Cell 183, 1772–1784.e1713.

Strom, A.R., Emelyanov, A.V., Mir, M., Fyodorov, D.V., Darzacq, X., Karpen, G.H., 2017. Phase separation drives heterochromatin domain formation. Nature 547, 241–245.

Stubblefield, E., Wray, W., 1971. Architecture of the Chinese hamster metaphase chromosome. Chromosoma 32, 262–294.

Suau, P., Bradbury, E.M., Baldwin, J.P., 1979. Higher-order structures of chromatin in solution. Eur. J. Biochem. 97, 593–602.

Swygert, S.G., Manning, B.J., Senapati, S., Kaur, P., Lindsay, S., Demeler, B., Peterson, C.L., 2014. Solution-state conformation and stoichiometry of yeast Sir3 heterochromatin fibres. Nat. Commun. 5, 4751.

Swygert, S.G., Senapati, S., Bolukbasi, M.F., Wolfe, S.A., Lindsay, S., Peterson, C.L., 2018. SIR proteins create compact heterochromatin fibers. Proc. Natl. Acad. Sci. U. S. A. 115, 12447–12452.

Syed, S.H., Goutte-Gattat, D., Becker, N., Meyer, S., Shukla, M.S., Hayes, J.J., Everaers, R., Angelov, D., Bednar, J., Dimitrov, S., 2010. Single-base resolution mapping of H1-nucleosome interactions and 3D organization of the nucleosome. Proc. Natl. Acad. Sci. U. S. A. 107, 9620–9625.

Szabo, Q., Jost, D., Chang, J.M., Cattoni, D.I., Papadopoulos, G.L., Bonev, B., Sexton, T., Gurgo, J., Jacquier, C., Nollmann, M., et al., 2018. TADs are 3D structural units of higher-order chromosome organization in Drosophila. Sci. Adv. 4, eaar8082.

Tartof, K.D., Bishop, C., Jones, M., Hobbs, C.A., Locke, J., 1989. Towards an understanding of position effect variegation. Dev. Genet. 10, 162–176.

Tatchell, K., Van Holde, K.E., 1977. Reconstitution of chromatin core particles. Biochemistry 16, 5295–5303.

Tavormina, P.A., Come, M.G., Hudson, J.R., Mo, Y.Y., Beck, W.T., Gorbsky, G.J., 2002. Rapid exchange of mammalian topoisomerase II alpha at kinetochores and chromosome arms in mitosis. J. Cell Biol. 158, 23–29.

Taylor, J.H., Woods, P.S., Hughes, W.L., 1957. The organization and duplication of chromosomes as revealed by autoradiographic studies using tritium-labeled Thymidinee. Proc. Natl. Acad. Sci. U. S. A. 43, 122–128.

Thoma, F., Koller, T., 1977. Influence of histone H1 on chromatin structure. Cell 12, 101–107.

Thoma, F., Koller, T., 1981. Unravelled nucleosomes, nucleosome beads and higher order structures of chromatin: influence of non-histone components and histone H1. J. Mol. Biol. 149, 709–733.
Thoma, F., Koller, T., Klug, A., 1979. Involvement of histone H1 in the organization of the nucleosome and of the salt-dependent superstructures of chromatin. J. Cell Biol. 83, 403–427.
Thomas, J.O., 1999. Histone H1: location and role. Curr. Opin. Cell Biol. 11, 312–317.
Thompson, J.S., Hecht, A., Grunstein, M., 1993. Histones and the regulation of heterochromatin in yeast. Cold Spring Harb. Symp. Quant. Biol. 58, 247–256.
Todd, R.D., Garrard, W.T., 1977. Two-dimensional electrophoretic analysis of polynucleosomes. J. Biol. Chem. 252, 4729–4738.
Travers, A., 1999. The location of the linker histone on the nucleosome. Trends Biochem. Sci. 24, 4–7.
Travers, A.A., Muyldermans, S.V., 1996. A DNA sequence for positioning chromatosomes. J. Mol. Biol. 257, 486–491.
Traverse, K.L., Pardue, M.L., 1989. Studies of he-T DNA sequences in the pericentric regions of Drosophila chromosomes. Chromosoma 97, 261–271.
Tremethick, D.J., 2007. Higher-order structures of chromatin: the elusive 30 nm fiber. Cell 128, 651–654.
Triolo, T., Sternglanz, R., 1996. Role of interactions between the origin recognition complex and SIR1 in transcriptional silencing. Nature 381, 251–253.
Tse, C., Hansen, J.C., 1997. Hybrid trypsinized nucleosomal arrays: identification of multiple functional roles of the H2A/H2B and H3/H4 N-termini in chromatin fiber compaction. Biochemistry 36, 11381–11388.
Uemura, T., Ohkura, H., Adachi, Y., Morino, K., Shiozaki, K., Yanagida, M., 1987. DNA topoisomerase II is required for condensation and separation of mitotic chromosomes in S. pombe. Cell 50, 917–925.
Van Holde, K.E., 1988. Chromatin. Springer-Verlag.
van Holde, K., Zlatanova, J., 1995. Chromatin higher order structure: chasing a mirage? J. Biol. Chem. 270, 8373–8376.
van Leeuwen, F., Gafken, P.R., Gottschling, D.E., 2002. Dot1p modulates silencing in yeast by methylation of the nucleosome core. Cell 109, 745–756.
van Steensel, B., Furlong, E.E.M., 2019. The role of transcription in shaping the spatial organization of the genome. Nat. Rev. Mol. Cell Biol. 20, 327–337.
Varshavsky, A.J., Bakayev, V.V., Georgiev, G.P., 1976. Heterogeneity of chromatin subunits in vitro and location of histone H1. Nucleic Acids Res. 3, 477–492.
Volpe, T.A., Kidner, C., Hall, I.M., Teng, G., Grewal, S.I., Martienssen, R.A., 2002. Regulation of heterochromatic silencing and histone H3 lysine-9 methylation by RNAi. Science 297, 1833–1837.
Walter, J., Schermelleh, L., Cremer, M., Tashiro, S., Cremer, T., 2003. Chromosome order in HeLa cells changes during mitosis and early G1, but is stably maintained during subsequent interphase stages. J. Cell Biol. 160, 685–697.
Wang, S., Su, J.H., Beliveau, B.J., Bintu, B., Moffitt, J.R., Wu, C.T., Zhuang, X., 2016. Spatial organization of chromatin domains and compartments in single chromosomes. Science 353, 598–602.
Warner, R.C., 1957. Studies on polynucleotides synthesized by polynucleotide phosphorylase. III. Interaction and ultraviolet absorption. J. Biol. Chem. 229, 711–724.
Watson, J.D., Crick, F.H., 1953. Genetical implications of the structure of deoxyribonucleic acid. Nature 171, 964–967.
Whitlock Jr., J.P., Simpson, R.T., 1976. Removal of histone H1 exposes a fifty base pair DNA segment between nucleosomes. Biochemistry 15, 3307–3314.
Willcockson, M.A., Healton, S.E., Weiss, C.N., Bartholdy, B.A., Botbol, Y., Mishra, L.N., Sidhwani, D.S., Wilson, T.J., Pinto, H.B., Maron, M.I., et al., 2021. H1 histones control the epigenetic landscape by local chromatin compaction. Nature 589, 293–298.
Williams, S.P., Athey, B.D., Muglia, L.J., Schappe, R.S., Gough, A.H., Langmore, J.P., 1986. Chromatin fibers are left-handed double helices with diameter and mass per unit length that depend on linker length. Biophys. J. 49, 233–248.
Wong, J., Patterton, D., Imhof, A., Guschin, D., Shi, Y.B., Wolffe, A.P., 1998. Distinct requirements for chromatin assembly in transcriptional repression by thyroid hormone receptor and histone deacetylase. EMBO J. 17, 520–534.
Woodcock, C.L.F., 1973. Ultrastructure of inactive chromatin. J. Cell Biol. 59, 368a.
Woodcock, C.L., 1994. Chromatin fibers observed in situ in frozen hydrated sections. Native fiber diameter is not correlated with nucleosome repeat length. J. Cell Biol. 125, 11–19.
Woodcock, C.L., Frado, L.L., Rattner, J.B., 1984. The higher-order structure of chromatin: evidence for a helical ribbon arrangement. J. Cell Biol. 99, 42–52.

Worcel, A., Strogatz, S., Riley, D., 1981. Structure of chromatin and the linking number of DNA. Proc. Natl. Acad. Sci. U. S. A. 78, 1461–1465.

Wu, H., Dalal, Y., Papoian, G.A., 2021. Binding dynamics of disordered linker histone H1 with a Nucleosomal particle. J. Mol. Biol. 433, 166881.

Wustmann, G., Szidonya, J., Taubert, H., Reuter, G., 1989. The genetics of position-effect variegation modifying loci in *Drosophila melanogaster*. Mol. Gen. Genet. 217, 520–527.

Yang, D., Arya, G., 2011. Structure and binding of the H4 histone tail and the effects of lysine 16 acetylation. Phys. Chem. Chem. Phys. 13, 2911–2921.

Zhao, Z., Tavoosidana, G., Sjolinder, M., Gondor, A., Mariano, P., Wang, S., Kanduri, C., Lezcano, M., Sandhu, K.S., Singh, U., et al., 2006. Circular chromosome conformation capture (4C) uncovers extensive networks of epigenetically regulated intra- and interchromosomal interactions. Nat. Genet. 38, 1341–1347.

Zhimulev, I.F., Belyaeva, E.S., Fomina, O.V., Protopopov, M.O., Bolshakov, V.N., 1986. Cytogenetic and molecular aspects of position effect variegation in *Drosophila melanogaster*. 1. Morphology and genetic activity of the 2AB region in chromosome rearrangement *T(1;2)dorvar7*. Chromosoma 94, 429–504.

Zhou, Y.B., Gerchman, S.E., Ramakrishnan, V., Travers, A., Muyldermans, S., 1998. Position and orientation of the globular domain of linker histone H5 on the nucleosome. Nature 395, 402–405.

Zhou, J., Fan, J.Y., Rangasamy, D., Tremethick, D.J., 2007. The nucleosome surface regulates chromatin compaction and couples it with transcriptional repression. Nat. Struct. Mol. Biol. 14, 1070–1076.

Zhou, B.R., Feng, H., Kato, H., Dai, L., Yang, Y., Zhou, Y., Bai, Y., 2013. Structural insights into the histone H1-nucleosome complex. Proc. Natl. Acad. Sci. U. S. A. 110, 19390–19395.

Zhou, B.R., Jiang, J., Feng, H., Ghirlando, R., Xiao, T.S., Bai, Y., 2015. Structural mechanisms of nucleosome recognition by linker histones. Mol. Cell 59, 628–638.

Zhou, B.R., Feng, H., Ghirlando, R., Li, S., Schwieters, C.D., Bai, Y., 2016. A small number of residues can determine if linker histones are bound on or off dyad in the Chromatosome. J. Mol. Biol. 428, 3948–3959.

Zhou, B.R., Jiang, J., Ghirlando, R., Norouzi, D., Sathish Yadav, K.N., Feng, H., Wang, R., Zhang, P., Zhurkin, V., Bai, Y., 2018. Revisit of reconstituted 30-nm nucleosome arrays reveals an Ensemble of Dynamic Structures. J. Mol. Biol. 430, 3093–3110.

Zhou, B.R., Feng, H., Kale, S., Fox, T., Khant, H., de Val, N., Ghirlando, R., Panchenko, A.R., Bai, Y., 2021. Distinct structures and dynamics of Chromatosomes with different human linker histone isoforms. Mol. Cell 81, 166–182.e166.

CHAPTER

4

Nucleosomes in context: Positioning, occupancy, and spacing

Abbreviations

ATAC Access for Transposase-Accessible Chromatin
ChIP chromatin immunoprecipitation
CTCF CCCTC binding factor
FAIRE Formaldehyde-Assisted Isolation of Regulatory Sequences
GAF GAGA factor
GRF general regulatory factor
MMTV mouse mammary tumor virus
MNase micrococcal nuclease
MPE methidiumpropyl-EDTA
MTase methyltransferase
NDR nucleosome-depleted region
PCR polymerase chain reaction
PIC preinitiation complex
RSC Remodels the Structure of Chromatin
SV40 Simian Virus 40
TBP TATA-binding protein
TF transcription factor
TSS transcription start site

Overview

The formulation of the particle model of the nucleosome in 1974 immediately raised the question of how the histone octamer is disposed relative to specific DNA sequences. Earlier models in which histones coated the DNA more or less uniformly did not readily support a regulatory role for chromatin; now it could be asked whether nucleosomes played specific roles in regulating DNA access, thereby governing function. Roger Kornberg, in his seminal 1974 paper, suggested that the "full significance of the repeating unit of histones and DNA may lie in the relation of the units to base sequences in the DNA," and outlined in a footnote how this might be explored by using the small genome of a virus such as SV40 (Kornberg, 1974). It did not require a deep consideration of nucleosome structure to appreciate that the close apposition of DNA to histones posed a potential obstacle to cellular machineries involved in transcription, replication, recombination, and repair, whereas linker DNA would

Chromatin. **https://doi.org/10.1016/B978-0-12-814809-9.00003-2**

be expected to be more accessible to those machineries, as it is to nucleases. How this problem is reconciled in living cells is considered in depth later on, but its recognition provided powerful motivation for determining the existence and extent of nucleosome positioning, and the rules by which such positioning is governed. Furthermore, the very question of whether chromatin structure was important to function, or rather played a passive role as packaging material, hinged on whether specific DNA sequences were randomly incorporated into nucleosomes or were specifically disposed. If the former, it would be difficult to argue against a general mechanism that allowed nucleosomes to be "transparent" to the cellular machineries, while the latter would conversely suggest that chromatin might indeed be an active, functional component in eukaryotes, and would raise new questions regarding which DNA sequences were and were not incorporated into nucleosomes and how this was determined.

Reviews on the topic of nucleosome positioning over the years provide a window into how conceptual and experimental approaches have evolved since the particle model of the nucleosome was proposed (Baldi et al., 2020; Chereji and Clark, 2018; Clark, 2010; Eissenberg et al., 1985; Hughes and Rando, 2014; Igo-Kemenes et al., 1982; Jansen and Verstrepen, 2011; Jiang and Pugh, 2009b; Kornberg, 1981; Lieleg et al., 2015; Segal and Widom, 2009b; Simpson, 1991; Struhl and Segal, 2013; Thoma, 1992; Van Holde, 1988; Wolffe, 1998). A major advance took place with the advent of high-throughput approaches, first using microarrays and later next-generation sequencing, to investigate nucleosome positioning across entire genomes (Albert et al., 2007; Bernstein et al., 2004; Kaplan et al., 2009; Lee et al., 2004, 2007; Mavrich et al., 2008b; Valouev et al., 2008; Whitehouse et al., 2007; Yuan et al., 2005). In this chapter, I discuss methods for investigating nucleosome positioning, from the earliest efforts to current approaches, and how these methods have been employed to understand determinants of positioning. As we are interested in this book not only in the current state of our understanding of chromatin structure and function, but in the sometimes meandering path that has brought us to that state, our discussion includes some investigative excursions that proved to be less productive than first hoped—for example, the idea of using sequence data alone to predict nucleosome positioning. Concurrent studies investigated the effects of nucleosome positioning on chromatin function, most prominently on transcriptional regulation; additional discussion of this topic is found in Chapter 8.

Rotational vs translational nucleosome positioning

The relationship of DNA sequence to the histones in the nucleosome is best captured by the term "nucleosome positioning." Early papers often referred to "nucleosome phasing" when discussing nonrandom arrangements of nucleosomes with respect to DNA sequence, but the term was frequently used in a vague and inconsistent way (see van Holde for a fuller discussion of this issue) (Van Holde, 1988). "Nucleosome phasing" is now used most often to refer to arrangements in which a group of nucleosomes exhibits a regular arrangement relative to a specific alignment point, such as a transcription start site (Baldi et al., 2020). Nucleosome positioning can be considered in the context of the two extremes of perfect positioning and random positioning (Fig. 4.1). In the case of perfect positioning, a specific DNA sequence adopts

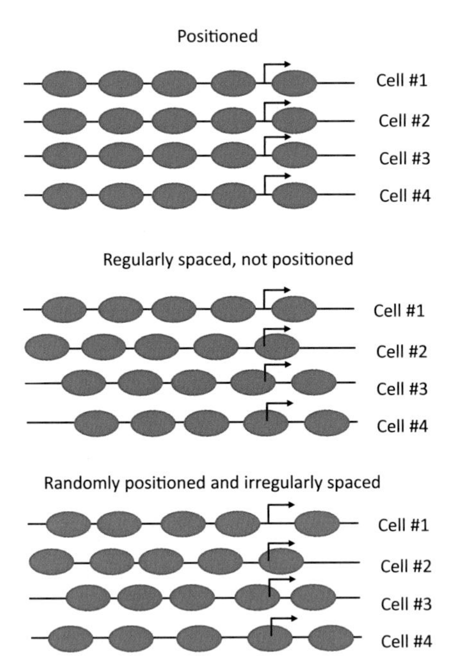

FIG. 4.1 Nucleosome positioning. Chromatin segments from a group of cells, or from in vitro reconstituted chromatin, may show strong positioning (top), regular spacing without sequence-specific positioning (middle) or be randomly positioned and irregularly spaced (bottom).

the same location relative to the histones 100% of the time, and nucleosomal and linker sequences will occupy two completely nonoverlapping sets. In contrast, if nucleosomes in a given region of chromatin are randomly positioned, every base will be found in nucleosomal and linker DNA in the same proportion. In between these two extremes exist various intermediate possibilities.

Two types of positioning may be distinguished. While translational positioning refers to the location of DNA sequence relative to the histone octamer, essentially as defined above, rotational positioning refers to the helical orientation of DNA relative to the histone octamer (Fig. 4.2). Two nucleosomes in which the DNA is shifted 10bp relative to the histone octamer have altered translational position (by 10bp), but equivalent rotational position, as the same sequences will have minor grooves facing outward (and thus accessible to solvent or DNase I) or inward relative to the histones. Because of the way DNA sequence impacts nucleosome positioning, as we discuss below, distinct translational positions that maintain rotational position need not differ greatly in relative free energy, and are correspondingly often observed both in vitro and in vivo (Cole et al., 2012). In contrast, nucleosomes having translational positions offset by only a few bp, which would greatly alter rotational position, are rarely observed.

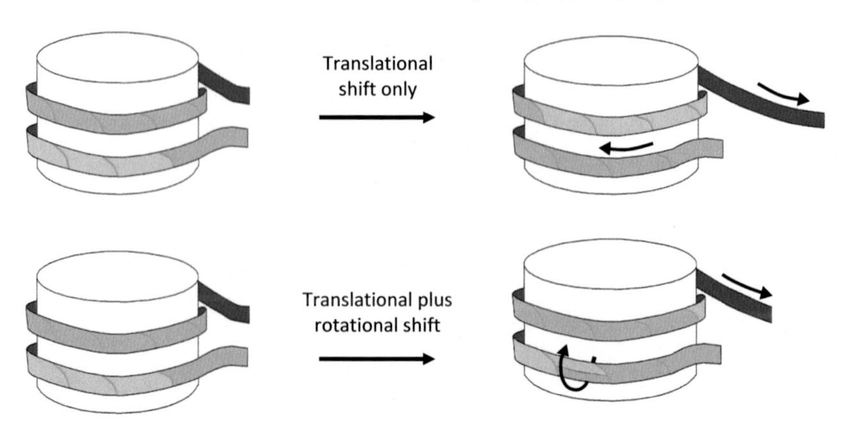

FIG. 4.2 Translational vs rotational nucleosome positioning. The two nucleosomes depicted at the top, in which the DNA is shifted by an integral multiple of the helical turn of DNA, differ in translational but not rotational positioning, while the two nucleosomes at bottom, in which the DNA has been shifted by about one half of a helical turn (about 5bp), differ in both.

Nucleosome spacing

Distinct from nucleosome positioning is nucleosome spacing. While nucleosome positioning is always defined relative to the DNA sequence, spacing is defined relative to the neighboring nucleosome(s), which need not be precisely positioned relative to the DNA sequence. Nonetheless, spacing and positioning can be intimately connected: precise positioning of a single nucleosome will dictate the positioning of surrounding nucleosomes, depending on how regular their spacing is. Even in the absence of any extrinsic mechanism governing spacing, a single strongly positioned nucleosome is expected to be flanked by regular arrays of nucleosomes occupying nonrandom locations, based only on statistical considerations (Fig. 4.3; Kornberg and Stryer, 1988). Conversely, even if positioning is absent, arrays of

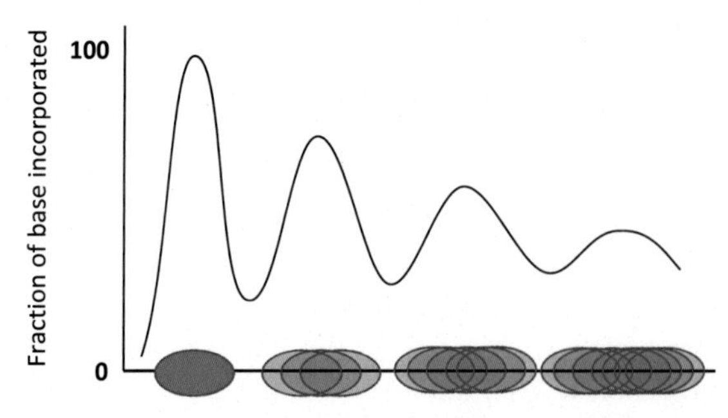

FIG. 4.3 Statistical positioning of nucleosomes. A strongly positioned nucleosome (on the left) causes surrounding nucleosomes to adopt a limited number of positions, thereby leading to a "phased array" of nucleosomes.

nucleosomes may exist in a given region of chromatin if spacing is sufficiently regular and the histone octamer sufficiently mobile with respect to the incorporated DNA (Fig. 4.1). Such randomly positioned, regularly spaced arrays of nucleosomes have been reported in nontranscribed regions of chromatin in *Drosophila* and human cells (Baldi et al., 2018b; Lai et al., 2018).

While the first studies that showed ladders of DNA fragments resulting from nucleolytic digestion of chromatin indicated a repeating unit of about 200 bp (Chapter 2), dozens of studies published from 1977 to 1980 revealed variable average spacing in MNase-digested chromatin from different sources (Van Holde, 1988). Chromatin from various protists showed average repeat lengths varying from 154 bp in *Aspergillus nidulans* and 165 bp in *Saccharomyces cerevisiae* to 225 bp in *Euglena gracilis*, corresponding to linker lengths of 10 to 80 bp, while repeat lengths in rat cortical neurons averaged 174 bp and those in rat liver averaged about 200 bp (Compton et al., 1976; Ermini and Kuenzle, 1978; Magnaval et al., 1980; Morris, 1976; Thomas and Furber, 1976). The latter observations immediately led to the inference that positioning must vary across tissues, as it would be impossible to maintain the same positions while varying the spacing (Kornberg, 1981). This conclusion was further supported by findings that nucleosome spacing changed during developmental programs such as sea urchin embryogenesis and avian erythropoiesis (Arceci and Gross, 1980; Chambers et al., 1983; Spadafora et al., 1976; Weintraub, 1978).

Early efforts to identify the determinants of nucleosome spacing were largely met with frustration. Reconstitution of nucleosomes in vitro using various protocols resulted in clusters of close-packed nucleosomes separated by regions of unincorporated DNA, and addition of linker histone, a reasonable candidate for dictating spacing, did not resolve this problem (Annunziato and Seale, 1983; Fulmer and Fasman, 1979; Spadafora et al., 1978; Steinmetz et al., 1978). Although in vitro assembly at physiological ionic strength of nucleosome arrays with 200 bp spacing, with spacing dependent on the presence of linker histone H5, was achieved using poly(dA-dT) templates, this could not be extended to mixed sequence templates (Kunzler and Stein, 1983; Stein and Bina, 1984). However, cell lysates obtained from mature *Xenopus* eggs or from *Drosophila* embryos were capable of assembling DNA and histones into properly spaced nucleosome arrays, suggesting a route to identification of factors that determined nucleosome spacing in vivo (Becker and Wu, 1992; Glikin et al., 1984; Kamakaka et al., 1993; Laskey et al., 1977; Nelson et al., 1979). One such factor, nucleoplasmin, was identified as potentially playing such a role and indeed could facilitate nucleosome assembly from purified histones and DNA at physiological ionic strength (see Chapter 5; Earnshaw et al., 1980; Laskey et al., 1978). However, nucleoplasmin failed to promote correct nucleosome spacing when used as a purified component in a nucleosome assembly reaction. Eventually, a distinct line of research led to the discovery of ATP-utilizing chromatin remodeling complexes; these have proved to be critical in establishing nucleosome spacing in vivo (Kadonaga, 2019; Prajapati et al., 2020; Struhl and Segal, 2013). This is discussed in more detail later in this chapter, in the context of chromatin remodelers and genome-wide nucleosome positioning.

Methods for determining nucleosome positioning

Determination of the average nucleosome *spacing* in a given chromatin sample is relatively straightforward: chromatin is digested with MNase, resulting in a ladder of DNA fragments having a repeat length that depends on the average length of DNA separating nucleosome core particles (see Fig. 1.5). Close-packed nucleosomes that include linker histones will generate fragments about 165, 330, 495 bp, and so forth, while chromatin having average linker lengths of 80 bp yields fragments averaging 225, 450, 675, ... bp. While some subtleties, such as the extent of MNase digestion and the possibility of nucleosome movement may come into play, execution and interpretation of such experiments are nonetheless fairly uncomplicated. In contrast, accurate determination of nucleosome *positioning* is more challenging.

Methods for determination of nucleosome positioning rely on either of two common, conceptually similar approaches (Fig. 4.4). In the first, chromatin is treated with a chemical or

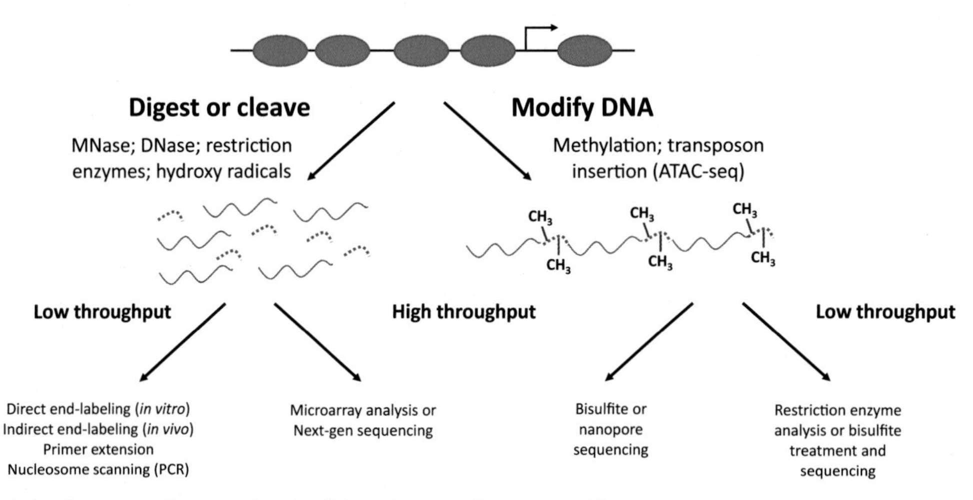

FIG. 4.4 Summary of approaches for determining nucleosome positioning.

enzymatic reagent that differs in its ability to cleave or digest nucleosomal versus linker DNA; the DNA remaining after such treatment is then analyzed to determine whether clear boundaries exist between digested (depleted) sequences and undigested (protected) sequences, and if so, where such boundaries lie. This approach is exemplified by experiments in which chromatin is treated with MNase to remove linker DNA, followed by end-label analysis or primer extension to determine nucleosome location at the single-gene level or by the use of microarrays or high-throughput sequencing to determine nucleosome locations across an entire genome. In the second approach, rather than cleaving or digesting DNA, the reagent leaves a mark of some sort that differentiates nucleosomal and linker DNA, for example by methylating or inserting a transposon into nonnucleosomal DNA.

Various reagents have been employed in studies of nucleosome positioning. Micrococcal nuclease (MNase; also sometimes referred to as staphylococcal nuclease, particularly in the early literature and still in some vendor's catalogs) was and remains the most popular of these reagents. Limited digestion of chromatin by MNase was first shown by Markus Noll to produce a ladder of DNA fragments that are multiples of a unit length (Noll, 1974), and as we have seen, it was soon recognized that this pattern reflected the ability of MNase to preferentially digest linker DNA while leaving the histone-associated core DNA intact. This property, together with the commercial availability of MNase, ensured its early adoption for studies of chromatin structure and is the basis for its continued popularity today.

One might have expected that exonucleases would be ideal for detecting nucleosome borders, although it would first require endonucleolytic cleavage between nucleosomes. Carl Wu used exonuclease III to identify sites bound by regulatory proteins (Wu, 1985), and this method works well for mapping locations of proteins binding to specific DNA sequences; for example, the Hager lab adapted its use to map sites of regulatory protein binding in genes activated by the steroid hormone receptor, while more recently, exonuclease treatment has been exploited to sharpen signals in ChIP-seq experiments (Archer et al., 1992; Cordingley et al., 1987; Rhee and Pugh, 2011; Wu, 1985). However, although sequence-specific DNA-binding proteins severely impede exonuclease trimming of DNA, the histone-DNA contacts in the nucleosome do so less efficiently. Instead, exonuclease treatment of nucleosomes generates a series of fragments corresponding to an initial pause at the outer edge of the nucleosome, with additional pause sites at 10bp intervals (Prunell and Kornberg, 1978; Riley and Weintraub, 1978). This behavior reflects the histone-DNA interactions that occur with each turn of the DNA helix throughout the nucleosome core particle (Chapter 2). Transient disruption of the outermost of these electrostatic interactions—sometimes referred to as "breathing" of the nucleosomal DNA—allows exonuclease digestion of the outermost DNA, which can proceed for another 10bp until it encounters the next obstacle, which can again be relieved by transient disruption, and so forth. This mechanism of access to nucleosomal DNA was demonstrated convincingly by experiments showing gradually decreasing accessibility of nucleosomal DNA to restriction enzymes from the border toward the "interior" (Polach and Widom, 1995). The difference in the abilities of MNase and exonuclease to digest nucleosomal DNA is analogous to the difference between the strong impediment that nucleosomes present to binding by many sequence-specific binding proteins and the inability to prevent passage by RNA polymerase by invading from one end (Felsenfeld, 1992). The upshot is that while exonuclease can be useful in determining the borders of histone-DNA contacts in a nucleosome at high resolution (Cole et al., 2016; Rhee et al., 2014), MNase remains the reagent of choice for monitoring nucleosome positioning.

MNase cleaves DNA by acting as a single strand endonuclease; the double-stranded breaks observed in MNase ladders arise from cleavages on the complementary strands that are in close proximity (Modak and Beard, 1980). MNase also possesses exonucleolytic activity, digesting exposed DNA ends until blocked by a histone octamer or other obstruction, but exhibits much less exonucleolytic trimming of nucleosomal DNA than does exonuclease III (Modak and Beard, 1980). Although MNase will digest essentially any DNA linker, it nonetheless possesses considerable bias in its cleavage reaction, such that AT-rich sequences are digested more rapidly than GC-rich sequences (Dingwall et al., 1981; Horz and Altenburger, 1981; Keene and Elgin, 1981; Von Hippel and Felsenfeld, 1964). This sequence selectivity necessitates the use of naked DNA controls in experiments in which MNase is used to probe chromatin structure. Indeed, early experiments reported a regular pattern of cleavage of chromatin by MNase at repeated satellite DNA sequences that was interpreted as reflecting strong nucleosome positioning; however, this interpretation was called into question when it was found that the same pattern of cleavage was observed using naked DNA as substrate

(Fittler and Zachau, 1979; Musich et al., 1977). Other cleavage reagents have been tested in attempts to circumvent confounding issues of MNase cleavage bias. Perhaps the most successful of these has been methidiumpropyl-EDTA (MPE), which in the presence of Fe(II), O_2, and reducing agent cleaves DNA with low sequence specificity and displays selectivity for nonnucleosomal DNA similar to that shown by MNase (Cartwright et al., 1983). This reagent, first used to determine nucleosome positioning at individual gene loci, has more recently been adopted for genome-wide analysis of nucleosome positioning (Ishii et al., 2015). Nonetheless, MNase remains the first choice of many investigators for studying nucleosome positioning.

In addition to the preference for particular DNA sequences exhibited by MNase and other DNA cleavage reagents, another concern in interpreting results of experiments aimed at determining nucleosome positioning is the possibility of nucleosome movement, or "sliding," during sample preparation or analysis (Simpson, 1991). Underscoring this concern, Morton Bradbury's lab showed that histone octamers assembled onto the sea urchin 5S rDNA positioning sequence (discussed below) adopted multiple positions separated by 10bp increments, thereby having altered translational positioning but identical rotational setting, and that histone octamers displayed mobility relative to DNA sequence even at low salt concentration (Meersseman et al., 1992; Pennings et al., 1991). Approaches to contend with this issue in early investigations included recognizing that chromatin samples should be isolated and analyzed as rapidly as possibly, without freezing or other long-term storage steps, and that conditions that were favorable for nucleosome mobility, such as elevated salt concentration or temperature, should be avoided. Researchers also favored experiments in which nucleosome positioning was analyzed using shorter digestion times that yielded chromatin fragments that were kilobases in length, rather than mononucleosome preparations that were more likely to be subject to nucleosome movement during a limit digest. Formaldehyde cross-linking was introduced as a way to prevent nucleosome migration during chromatin preparation (Fragoso et al., 1995; Jackson, 1978; Solomon and Varshavsky, 1985), and cross-linking is widely used in genome-wide assessments of nucleosome positioning (discussed below). Nonetheless, one indication that nucleosome mobility was not so likely to be a major problem in interpreting nucleosome positioning in nuclear chromatin was that consistent results were generally observed over a range of digestion times and incubation temperatures, with or without cross-linking.

Another problem arises in approaches that rely on isolation of mononucleosomal DNA to map nucleosome positions, such as the use of microarrays or high-throughput sequencing to determine genome-wide nucleosome positioning. Because the exonuclease activity of MNase shows some sequence dependence, trimming of the DNA to the 147bp most closely associated with the histone octamer is not uniform (Horz and Altenburger, 1981); furthermore, the outermost DNA in the nucleosome may also show sequence-dependent differential digestion. Both of these issues create uncertainty in identifying the outside border of nucleosomes. Furthermore, nucleosomes are not uniform in their susceptibility to MNase digestion, so that when a large population is assayed, such as all of the nucleosomes in a genome, DNA sequences present in mononucleosomes will differ at early and late digestion times (Fig. 4.5; Henikoff et al., 2011). In one careful analysis, Chereji, Clark, and colleagues showed that mononucleosomal DNA from early

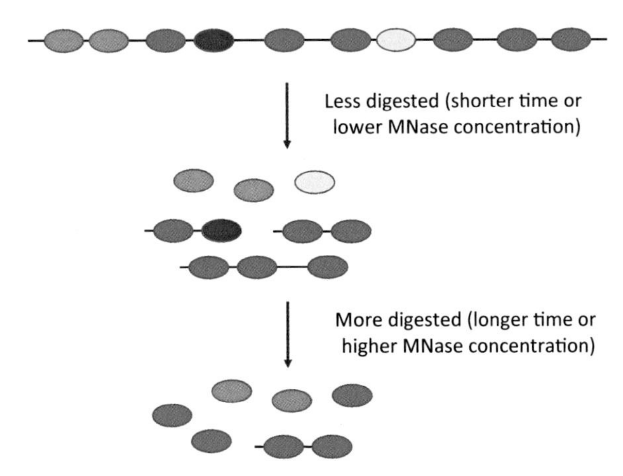

FIG. 4.5 MNase digestion of chromatin first releases mononucleosomes flanked by A/T rich sequences (*yellow* and *orange* nucleosomes). Upon prolonged digestion, A/T rich nucleosomes are digested and lost sooner than G/C rich nucleosomes, whether initially flanked by A/T rich linkers *(yellow)* or not *(red).*

stages of digestion was enriched in A/T nucleotides, while longer digestion resulted in enrichment in G/C-rich sequences (Chereji et al., 2017). We will touch on some of these issues again when genome-wide approaches to determining nucleosome positioning are considered.

A clever approach taken to circumvent the idiosyncracies of MNase and the possibility of nucleosome movement during chromatin preparation was to probe chromatin in its natural state in vivo by taking advantage of the inhibition of methyltransferase (MTase) activity on DNA that is incorporated into nucleosomes (Fig. 4.4). (In a related approach, the restriction enzyme *Hin*fI was inducibly expressed in yeast cells to probe chromatin structure, but this method was not widely adopted (Iyer and Struhl, 1995; Mai et al., 2000).) Probing this "natural state" still necessitates consideration of the possibility of nucleosome movement, especially given the presence in the cellular milieu of chromatin remodelers that actively move histones with respect to the DNA, but it does eliminate artifacts due to nucleosome mobility during chromatin isolation and digestion or modification. In early experiments, methylation of the purified DNA was detected by combining cleavage by methylation-sensitive restriction enzymes with primer extension analysis, but this gave way to the more efficient method of using bisulfite to convert deoxycytidine but not ^{5me}C to deoxyuridine, followed by sequencing. Comparison of the sequence obtained following bisulfite treatment to the known sequence reveals methylated C residues as still being detected as cytosine, while unmethylated cytosines are read as uridine. The first studies employing this method induced expression in yeast of the *E. coli dam* MTase, which methylates adenine embedded in GATC sequences, and showed that telomere-proximal DNA was refractory to methylation while promoters of active genes were susceptible (Gottschling, 1992; Singh and Klar, 1992). A follow-up report established that positioned nucleosomes inhibited *dam* methylation in budding yeast cells and corroborated results of previous analyses of nucleosome positioning (Kladde and Simpson, 1994). A shortcoming of the method is the relative paucity of GATC sequences, which occur only about every 256 bp on average in random sequence DNA; the use of SssI methyltransferase, which methylates CpG residues, allowed much higher-resolution determination of chromatin accessibility (Kladde et al., 1996). This higher resolution approach also allows footprinting of nonnucleosomal proteins, such as transcription factors (TFs), on DNA; they can be distinguished from nucleosomes by the size of the protected region. Recently, the use of MTases has been coopted in experiments aimed at determining cell-to-cell variability in the arrangement of nucleosomes; we will return to this approach when we discuss genome-wide mapping of nucleosomes.

Although in vivo expression of MTases allowed interrogation of chromatin structure without the potential artifacts accompanying chromatin isolation, the approach is technically more challenging than methods using MNase as probe. Methyltransferase must be expressed inducibly and tightly repressed until needed, as uncontrolled methylation of DNA is typically deleterious to cellular health. Moreover, experiments involving varied approaches to monitoring nucleosome positioning generally yielded consistent results, largely relieving early concerns over potential experimental artifacts; thus, most investigators continued utilizing MNase as the primary reagent to assess chromatin structure. A recent innovation that is conceptually related to in vivo expression of methyltransferase (Fig. 4.4) is ATAC-seq (Access for Transposase-Accessible Chromatin using sequencing). This approach monitors accessibility in chromatin to a hyperactive transposase derived from the Tn5 transposase on a genome-wide level (Buenrostro et al., 2013). However, like its cousins DNase-seq and FAIRE-seq (Formaldehyde-Assisted Isolation of Regulatory Sequences), ATAC-seq is useful for identifying functional elements residing in open chromatin, but it is not suitable for determining nucleosome positioning (Boyle et al., 2008; Simon et al., 2012).

In addition to MNase and MPE, two other cleavage reagents, DNase I and hydroxyl radical, have enjoyed wide application for probing rotational nucleosome positioning. DNase I has sometimes been used to assess translational positioning of nucleosomes, as it does exhibit preferential cleavage of linker DNA relative to nucleosomal DNA, but it is of limited utility in this regard for the same reason that it is useful in monitoring rotational positioning: its ability to cut DNA on the surface of the nucleosome. Both DNase I and hydroxyl radical, generated by the Fenton reaction in the presence of ascorbic acid, hydrogen peroxide, and Fe(II), preferentially cleave phosphate linkages of DNA at the locations where the minor groove is facing outward from the surface of the nucleosome. This same preference is observed for cleavage by DNase I of DNA adhered to a flat surface of crystalline calcium phosphate (Rhodes and Klug, 1980) and is consistent with the crystal structure of DNase I in complex with DNA, which shows the enzyme bending the double helix away from the cleavage site in the minor groove (Suck et al., 1988). DNase I cleavage of nucleosomal minor groove sites that do not face directly outward falls off greatly in comparison, likely due to steric occlusion by the adjacent DNA helix and the histones (Klug and Lutter, 1981). The much smaller hydroxyl radical, in contrast to DNase I, is able to cleave sites in nucleosomal DNA regardless of their location on the surface, but shows preference for the most accessible, outward-facing targets (Hayes et al., 1990). Careful comparison of cleavage patterns of naked to nucleosomal DNA by both of these reagents has contributed greatly to our understanding of alterations in DNA structure, such as changes in helical periodicity and irregularities in its path, accompanying its incorporation into nucleosomes (Hayes et al., 1990, 1991; Lutter, 1978). More pertinent to the present discussion, the high-resolution information on nucleosome positioning obtained by use of DNase I and hydroxyl radical cleavage was critical to some of the first studies that documented nucleosome positioning.

Local analysis of nucleosome positioning

The concept behind the use of these various cleavage reagents in determining nucleosome positioning is simple: DNA sequences of >140 bp that are protected against digestion in chromatin are likely to be nucleosomal. (The same idea applies, of course, to modification by reagents such as MTases.) To determine which sequences are digested and which are protected, cleavage sites and regions of protected DNA must be mapped onto the DNA sequence. Different methods for doing so were used in early experiments, depending on whether the studies involved nucleosomes assembled in vitro or chromatin isolated from nuclei, in which the sequence being analyzed must be distinguished from the bulk of nuclear chromatin.

In vitro analyses of nucleosome positioning have mostly relied on the use of "end-labeled" DNA segments incorporating a radioactive label at the 5′ end of one or both DNA strands. This approach was introduced by Simpson and Whitlock at the National Institutes of Health (see Sidebar: Chromatin research at the NIH), who used T4 polynucleotide kinase in conjunction with γ-[^{32}P]-ATP to label the 5′-ends of DNA in intact isolated nucleosomes from HeLa cells (Simpson and Whitlock, 1976). (MNase cleavage leaves a 5′-hydroxyl group on the cleaved DNA, which can be phosphorylated by T4 polynucleotide kinase.) These end-labeled nucleosomes were then treated with DNase I and the sites of cleavage mapped relative to the radiolabel by separating the resulting DNA fragments using polyacrylamide gel electrophoresis (PAGE) followed by autoradiography. Inclusion of radiolabeled size standards allowed determination of the sizes of the end-labeled fragments and thus the location of cleavage sites on the nucleosome, albeit on a very large collection of DNA molecules differing in sequence. This method was subsequently modified to map cleavage sites on chromatin reconstituted in vitro by end-labeling a purified DNA fragment to be used in the reconstitution; cleavage with a restriction enzyme having a site near one end of the DNA allowed purification of a suitably sized fragment labeled only at one end (Fig. 4.6). Following reconstitution into

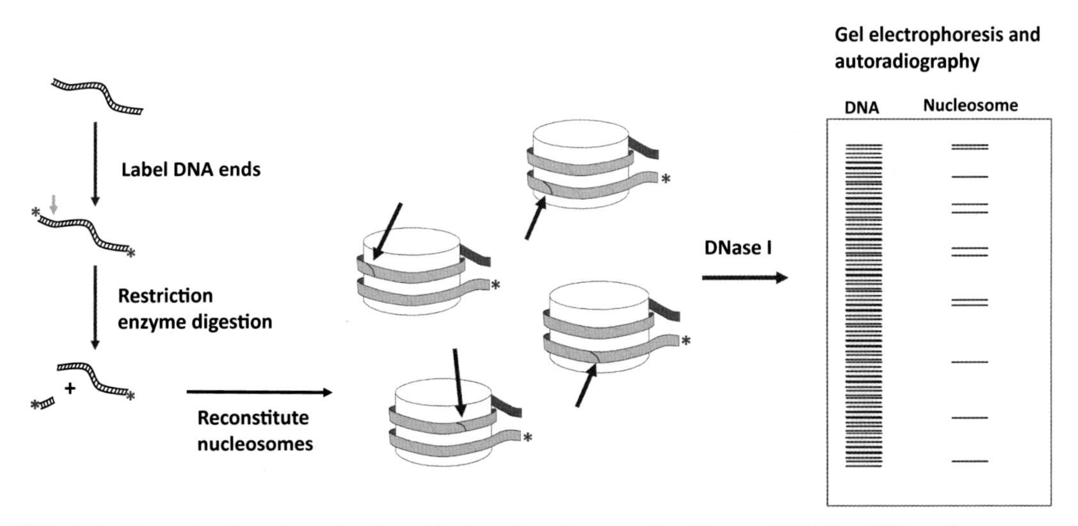

FIG. 4.6 Determination of rotational positioning of nucleosomes by direct end-labeling. DNase cleavages are indicated by *arrows*.

nucleosomes, nuclease cleavage sites could then be mapped relative to the end-label by PAGE and autoradiography. A similar approach was later used in DNase footprinting experiments aimed at determining sites of interaction of DNA-binding proteins such as the bacterial *lac* repressor with defined sequence DNA (Galas and Schmitz, 1978).

Sidebar: Chromatin research at the NIH

In a retrospective article, Jim Kadonaga wrote that "chromatin was not a popular subject in the late 1980s and early 1990s," and that, with regard to studying the role of chromatin in transcription, he was given helpful advice such as "Doesn't 'chromatin' mean 'artifact'?" (Kadonaga, 2019). Kadonaga's view may have been colored by his focus on transcription, as for a time researchers in that field did seem, on the whole, hostile to incursions into their area by chromatin researchers. At the NIH, however, it did not feel that way at all (as must surely have been the case also at the MRC in Cambridge, among other "chromatin oases"); on the contrary, seminal work from several labs made it a hotbed of chromatin studies, and the progeny of these labs have gone on to their own successes in the field (some of those are indicated in parentheses following the parent lab). Gary Felsenfeld's lab (Richard Axel, Bev Emerson, David Clark, Vasily Studitsky, Joan Boyes) performed early biophysical studies and later identified insulator elements, culminating in the identification of CTCF. Carl Wu (Peter Becker, Toshi Tsukiyama, Snow Shen), after coinventing the indirect end-label assay and discovering DNase hypersensitive sites at promoters of active genes while still at Harvard, conducted pioneering studies on chromatin remodeling complexes. Gordon Hager's lab (Trevor Archer, Emery Bresnick) did seminal work on the role of chromatin in gene regulation, using the steroid-inducible MMTV promoter as an initial model and expanding into genome-wide studies and pioneer factors in more recent years. Michael Bustin (David Landsman) pioneered studies on the high mobility group (HMG) proteins, nonhistone chromatin proteins that contribute to chromatin structure and function. Bill Bonner's lab performed seminal studies on histone variants, especially H2A.X, and Alan Wolffe (Genevieve Almouzni, Jeff Hayes, Axel Imhof, Hitoshi Kurumizaka, Kiyoe Ura, Paul Wade) led studies on nucleosome structure using hydroxyl radicals as probe and on the relationship between chromatin structure and function, in addition to writing numerous reviews and an influential textbook that helped propel chromatin biology into the mainstream. Tragically, Alan was struck by a bus and killed while out running at a meeting in Brazil in 2001 at the age of 41. And Bob Simpson (Sharon Dent, Ann Dean, Mike Kladde, Julie Cooper, Fritz Thoma, Hiro Shimizu, and myself) contributed pioneering biochemical and biophysical studies on nucleosome structure and nucleosome positioning, and investigated the role of positioned nucleosomes in transcriptional regulation. Bob moved to Penn State where he continued work in chromatin biology until his death in 2004 at age 65 in a home accident.

Although this footprinting approach was useful for in vitro studies, the impossibility of introducing a radiolabel into a specific site in nuclear DNA necessitated a different approach for investigating nucleosome positioning in vivo. Two labs independently arrived at the same solution to this problem; the method they invented is termed *indirect end-label* analysis (Fig. 4.7; Nedospasov and Georgiev, 1980; Wu, 1980). This approach can be used to analyze sites cleaved by nuclease or chemical probes in isolated chromatin or intact nuclei. Nedospasov and Georgiev employed it to study SV40 minichromosome structure, while Carl

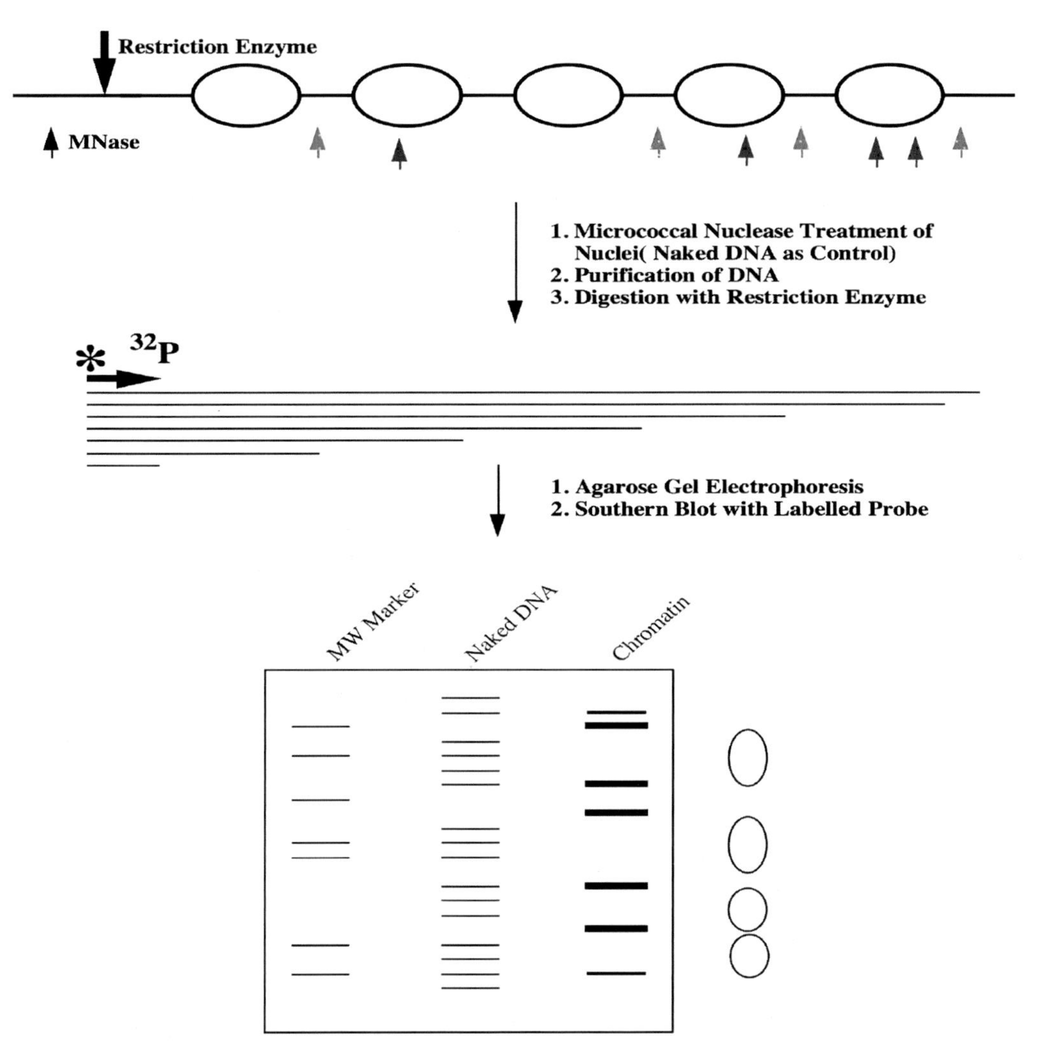

FIG. 4.7 Mapping nucleosome position by indirect end-labeling. *Green arrows* represent sites cleaved in both naked DNA and chromatin, while *red arrows* are sites cleaved in naked DNA but protected in chromatin. See text for details.

Wu used the method to identify DNase hypersensitive sites at the 5′ ends of heat shock genes in *Drosophila* chromatin. As schematized in Fig. 4.7, chromatin is treated with MNase or another appropriate reagent at a level that produces DNA fragments that range in size from a few hundred to several thousand base pairs. To identify locations of the cleavage sites relative to a specific genomic location, the digested chromatin is first treated with protease and extracted to remove the histones and the purified DNA digested with a restriction enzyme. The resulting DNA fragments are fractionated by electrophoresis on a nondenaturing agarose gel, the fragments transferred to a membrane by Southern blotting, and the resulting blot hybridized with a radiolabeled probe abutting the restriction site. This allows visualization of DNA fragments hybridizing to the probe that are cut at one end by the restriction enzyme and at the other end by MNase. Comparison of the cleavage pattern observed for naked DNA and chromatin may reveal sites that are cut in naked DNA but protected in chromatin; protected regions on the order of 150–200bp are indicative of nucleosome occupancy (Fig. 4.8).

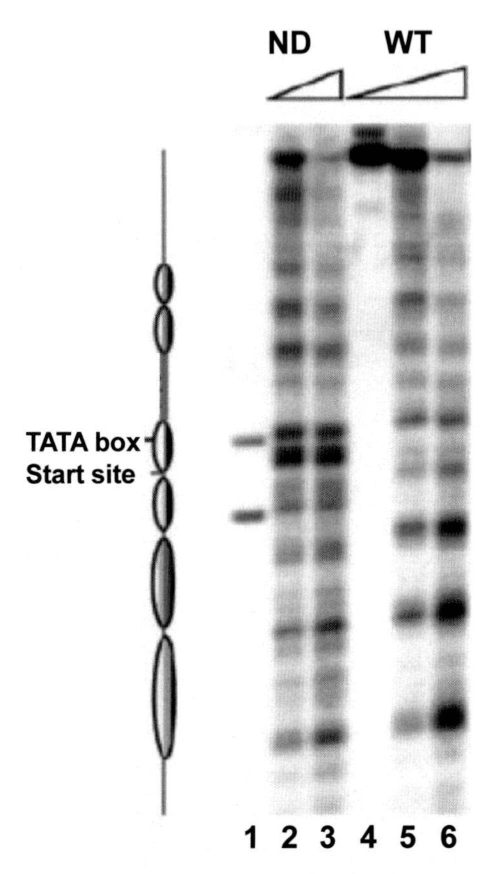

FIG. 4.8 Indirect end-label mapping of chromatin in the yeast *RNR3* gene. Lane 1 shows markers from restriction enzyme digests; lanes 2–3 show naked DNA digested with increasing amounts of MNase from left to right; lanes 4–6 show undigested yeast chromatin (lane 4) and chromatin digested with increasing amounts of MNase from left to right. Ovals represent positioned nucleosomes. *From Li, B. and Reese, J.C., 2001. Ssn6-Tup1 regulates RNR3 by positioning nucleosomes and affecting the chromatin structure at the upstream repression sequence. J. Biol. Chem. 276, 33788.*

Conversely, cleavage patterns for DNA and chromatin that are identical, or nearly so, suggest either randomly positioned nucleosomes or the absence of nucleosomes; these possibilities can be distinguished by determining whether mononucleosomal DNA contains the sequences of interest, using hybridization techniques or high-throughput sequencing (see below). Intermediate results, typified by partial protection in which a cleavage site found in naked DNA is still cut in chromatin, but less efficiently relative to other cleavage sites, are also sometimes observed. Such partial protection can indicate that multiple nucleosome positions exist in the population being assayed, but sufficient preference for some positions is seen that cleavage is not random (i.e., is not identical to the pattern observed for naked DNA). Similarly, protection or partial protection may be observed for regions of DNA that do not conform to integral multiples of nucleosome length (e.g., 230bp). This can arise from a population in which nucleosomes adopt a limited number of overlapping preferred positions. Such mixed positioning is sometimes referred to as being due to "fuzzy" nucleosomes; the use of high-throughput techniques to monitor nucleosome positioning genome-wide has allowed a more precise definition of fuzzy nucleosomes, as discussed later.

Use of nondenaturing agarose gel electrophoresis in conjunction with indirect end-label analysis permits determination of nucleosome positioning with resolution limited to about ±20bp. To obtain higher-resolution information, Simpson and colleagues mapped single-stranded MNase and DNase I cleavage sites in yeast chromatin and naked DNA controls by primer extension of a radiolabeled oligonucleotide probe (Shimizu et al., 1991). The method was adapted from a protocol used to obtain DNase footprints of sequence-specific binding proteins at high resolution and took advantage of the use of *Taq* polymerase to achieve multicycle linear amplification (Axelrod and Majors, 1989; Gralla, 1985). Use of this method to analyze MNase cleavage requires much lower levels of digestion than for indirect end-labeling, as single-stranded cuts of nucleosomal DNA occur at much lower levels than required for the double-stranded cuts mapped on nondenaturing agarose gels. Comparison of the cleavages induced in chromatin and DNA on large polyacrylamide gels (sequencing gels) allowed resolution of protected regions, and by inference of nucleosome positions, to the base pair level.

An alternative method to determine nucleosome positioning at higher resolution utilized primer extension to identify the borders of individual nucleosomes in reconstituted and native chromatin (Buttinelli et al., 1993; Fragoso et al., 1995). Following MNase digestion, purified DNA was fractionated on denaturing gels and single-stranded DNA of nucleosome length was purified, thereby eliminating internally nicked mononucleosomal DNA; Fragoso et al. further added a formaldehyde cross-linking step to prevent nucleosome migration. These and related approaches revealed that in some cases nucleosomes were positioned with base pair precision, while in others multiple translational frames with a common rotational phase were observed for nucleosomes that appeared well-positioned in indirect end-labeling experiments (Buttinelli et al., 1993; Costanzo et al., 1995; Fragoso et al., 1995; Yenidunya et al., 1994).

A related approach, termed nucleosome-scanning analysis, was devised by Sekinger et al. (2005). In this approach, mononucleosomal DNA is analyzed by qPCR using overlapping primer pairs to quantify the relative protection of DNA in chromatin against MNase digestion (Fig. 4.9). As with the indirect end-labeling and primer extension methods, it is critical to compare amplification of isolated, MNase-protected fragments against total DNA to control for differences in efficiencies of DNA recovery and amplification; the method and necessary controls have been thoroughly discussed (Infante et al., 2012). Although this approach is technically somewhat easier than the primer extension methods discussed above, it does not allow precise determination of nucleosome borders. It is also not notably cheaper or easier than using high-throughput sequencing to map nucleosomes genome-wide, and has largely been supplanted by the latter approach, which we discuss in the next section.

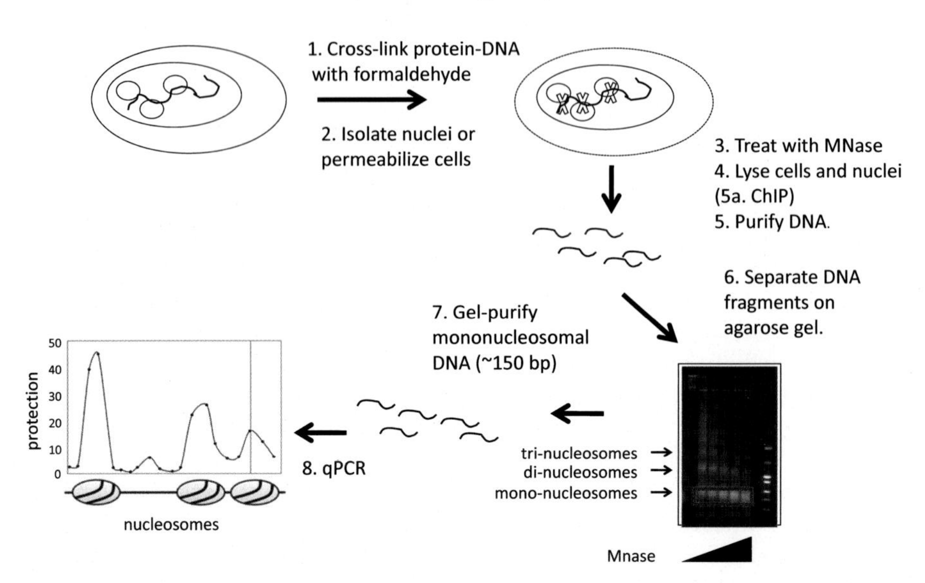

FIG. 4.9 Nucleosome scanning assay. Following purification of mononucleosomal DNA (Step 7), qPCR results from individual amplicons (Step 8) are plotted along the chromosomal coordinate to indicate relative nucleosome occupancy; nucleosome midpoints are assigned at the corresponding peaks *(vertical line). From Infante, J.J., Law, G.L., Young, E.T., 2012. Analysis of nucleosome positioning using a nucleosome-scanning assay. Methods Mol. Biol. 833, 63.*

Global analysis of nucleosome positioning

Use of enzymatic and chemical probes together with direct and indirect end-label analyses, and variants of these approaches, provided the foundation for our understanding of nucleosome positioning and functional roles of chromatin, as discussed below and later in Chapter 8. However, these approaches suffered a major limitation in that they only permitted interrogation of small regions of a genome—for example, a single gene promoter—in a given experiment. Labor-intensive application of these standard techniques permitted mapping of larger genomic regions, but still fell far short of mapping entire chromosomes or genomes (Ercan and Simpson, 2004).

This picture changed dramatically with the advent of microarray technology, followed by high-throughput sequencing approaches. DNA microarrays, in which single-stranded oligonucleotides were spotted onto glass slides robotically, were first introduced by Pat Brown, Joe DeRisi, and coworkers at Stanford University (DeRisi et al., 1997). (Pat Brown went on to found Impossible Foods, Inc., one of the first companies devoted to producing plant-based meat substitutes.) These arrays were initially designed for the determination of mRNA transcript abundance on a genome-wide scale, with the aim of overcoming the limitations of assessing transcript abundance on a gene-by-gene basis, and were first applied to the budding yeast *Saccharomyces cerevisiae* using the following steps:

- Isolated mRNA was reverse transcribed to make cDNA, which was fragmented and labeled with a fluorescent dye. Typically, a control and an experimental sample were compared by using distinct fluorescent labels, such as Cy3 and Cy5.

- The labeled cDNA was applied to the microarray under conditions that allowed hybridization.
- After washing the microarray to remove nonhybridized material, the microarray was scanned to measure the fluorescence intensity of the spots on the array. The variation in intensity between control and experimental samples, or the absolute intensities when appropriately controlled for differences in hybridization efficiency, yielded information on changes in mRNA abundance between samples or on relative abundance within a sample.

The first microarrays were spotted using in-house apparatus with a few thousand oligonucleotides; they are now sold by various manufacturers and can contain several million oligonucleotides, allowing the construction of "tiling" arrays that provide near base pair resolution (David et al., 2006). Nonetheless, their use has mostly, if not entirely, been superseded by high-throughput sequencing approaches, discussed below.

Following numerous publications in which microarrays were used to gauge mRNA abundance genome-wide in various organisms under different conditions, their use was expanded to monitor genome-wide binding of TFs in "ChIP-chip" experiments (Lieb et al., 2001). The first applications of microarrays to chromatin were performed using low-resolution arrays (Fig. 4.10; Bernstein et al., 2004; Lee et al., 2004). In these experiments, mononucleosomal

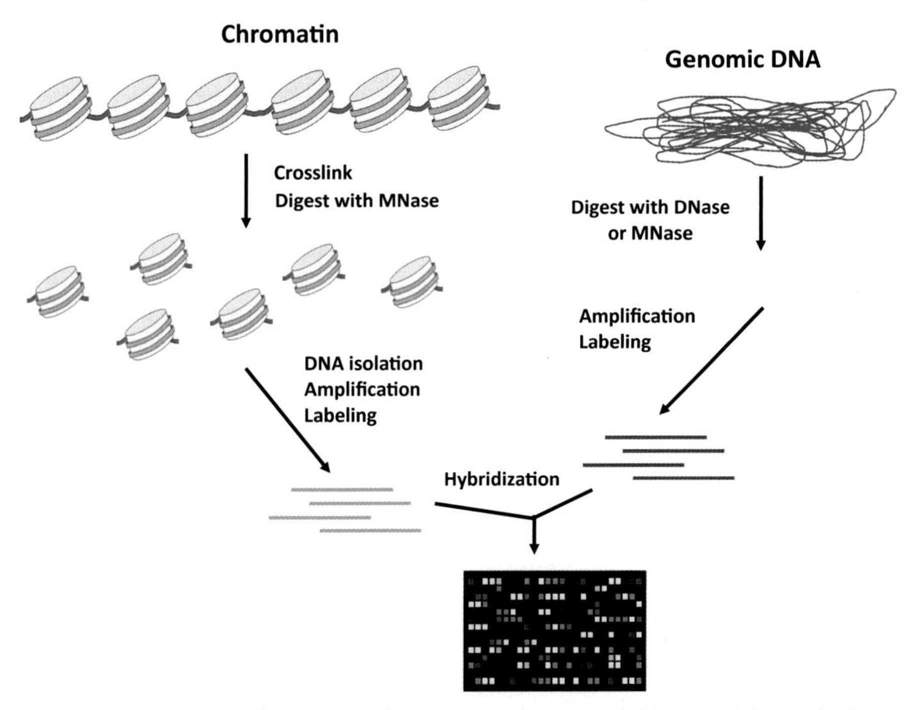

FIG. 4.10 Using microarrays to determine nucleosome occupancy on a global scale. A "two-color" experiment is depicted, in which nucleosomal and naked DNA samples are labeled with different dyes and hybridized to the same microarray. Alternatively, nucleosomal and naked DNA are separately hybridized and the signals compared computationally.

DNA was prepared by MNase digestion of yeast chromatin and the signal compared with naked DNA that had been digested to a similar extent; the relative abundance of nucleosomal DNA was then measured across the yeast genome using microarrays in which the reporting oligonucleotides covered the genome rather sparsely. Consequently, although it was determined that promoters of transcriptionally active genes were relatively depleted of nucleosomes, consistent with a large body of work on nuclease sensitivity (Elgin, 1988; Gross and Garrard, 1988), nucleosome positions could not be ascertained. This finding underscores an important aspect of high-throughput approaches for mapping nucleosomes: what is typically reported is not nucleosome position, but *nucleosome occupancy*. This is usually considered as the relative abundance of nucleosomal DNA, measured as fluorescence intensity on microarrays or as the number of reads in next-generation sequencing experiments, at all sites in the genome; by definition it varies from zero (no measurable fluorescence or no reads corresponding to nucleosomal DNA) to 100%, corresponding to the maximum intensity or read number. *Absolute* nucleosome occupancy refers to the fractional occupancy out of the entire population of cells monitored in which a given sequence is occupied by a nucleosome; this is not generally assessed in high-throughput determinations, although recent work describes an avenue toward this end (Chereji et al., 2019; Oberbeckmann et al., 2019). Mapping nucleosome occupancy (relative or absolute) does, however, provide information on average positioning, as indicated by occupancy peaks flanked by minima that are separated by about 150 bp.

Following the early experiments performed using low-density microarrays, higher-resolution arrays were developed first for the genome of budding yeast. "Home-made" arrays were printed that covered about 5% of the *S. cerevisiae* genome at 20 bp resolution (Yuan et al., 2005); soon after, Affymetrix introduced microarrays that contained 6.5 million probes 25 nucleotides in length that provided resolution across the yeast genome (excluding repetitive sequences such as transposable elements and telomeres) to about 4 bp (David et al., 2006; Lee et al., 2007).

The much larger size of metazoan genomes compared to yeast precluded the production of microarrays having similarly (<10 bp) high resolution and complete genome coverage, although partial coverage (for a single chromosome or promoter regions across the genome, for example) could be achieved (Ozsolak et al., 2007). However, the concurrent advances in high-throughput sequencing led to the adoption of sequencing approaches as the preferred method for genome-wide mapping of nucleosomes in both yeast and metazoans (Albert et al., 2007; Mavrich et al., 2008b; Schones et al., 2008; Valouev et al., 2008). The method for constructing nucleosome occupancy maps across a genome by high-throughput sequencing is conceptually similar to that used with microarrays: mononucleosomal DNA is typically prepared from MNase-digested chromatin, followed by library construction and sequencing; mapping of the resulting reads onto the genome (and comparison with results using naked DNA) yields a spectrum in which peaks and valleys indicate regions of high and low nucleosome occupancy. This approach is thus usually referred to as "MNase-seq." (A nice illustration of concordance between the nucleosome-scanning assay and MNase-seq at several yeast loci is provided in Small et al. (2014); see Fig. 4.15.) An immunoprecipitation step using antihistone antibodies or epitope-tagged histones may be used to ensure that MNase-protected DNA is nucleosomal and not protected by association with nonhistone protein complexes (Chereji et al., 2017). Alternatively, the location of nucleosomes containing specific histone modifications or histone variants can be ascertained by inclusion of an immunoprecipitation step, following MNase treatment, using appropriate antibodies (e.g., Albert et al., 2007). Chromatin can be treated with the cross-linking agent formaldehyde to inhibit nucleosome movement during digestion and isolation, and mononucleosomal DNA is typically size-selected by gel electrophoretic fractionation (Rando, 2010).

Caveats discussed earlier regarding sequence preference of MNase apply equally to MNase-seq (Chung et al., 2010; Rando, 2010), and several studies emphasize the importance of using mononucleosomal DNA obtained from treatment of chromatin over a range of MNase concentrations or digestion times (Chereji et al., 2017; DeGennaro et al., 2013; Givens et al., 2012; Mieczkowski et al., 2016; Rando, 2010; Rizzo et al., 2012; Weiner et al., 2010). An alternative approach to MNase-seq that circumvents some of the potential artifacts caused by MNase sequence preferences was introduced by the Widom lab (Fig. 4.11; Brogaard et al., 2012). In the first application of this chemical mapping approach, which derived from earlier

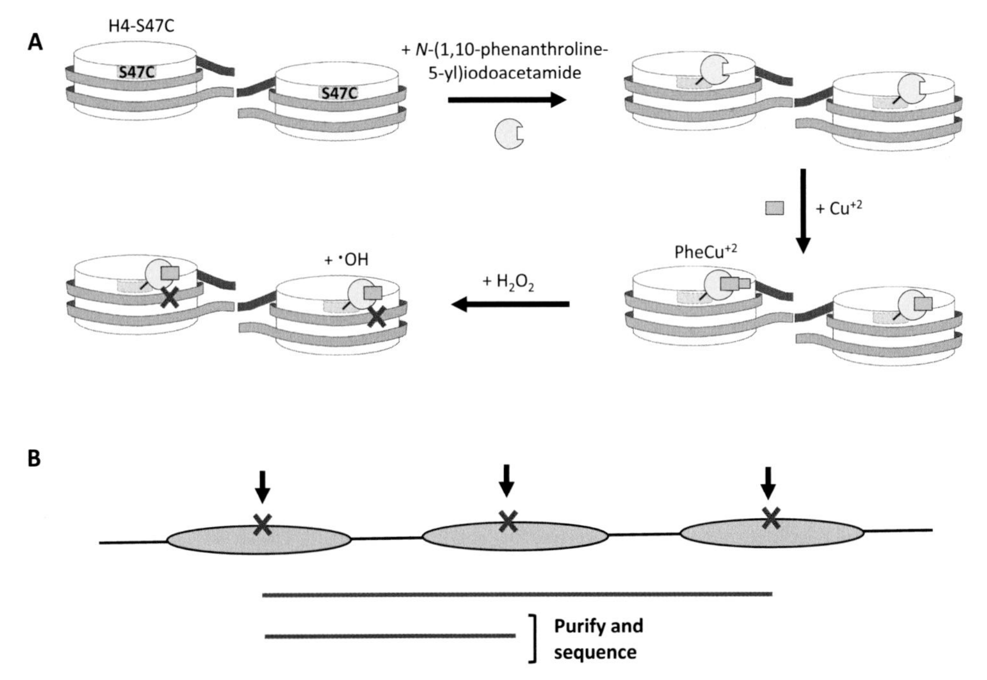

FIG. 4.11 Chemical mapping of nucleosomes. (A) Substitution of H4 Ser47 by a cysteine residue allows coupling of a phenanthroline derivative, which chelates added cupric ions. The chelated copper catalyzes DNA cleavage near its vicinity when hydrogen peroxide is added, producing cleavages near the nucleosome dyad. (B) Purification and sequencing of the smallest resulting fragment provides high-resolution information on the positions of the centers of adjacent nucleosomes.

work aimed at increasing the accuracy of mapping nucleosome positioning in vitro (Flaus et al., 1996), a histone H4 mutation, H4S47C, was introduced into budding yeast (*Saccharomyces cerevisiae*); no other Cys residues are present in yeast histones. This unique cysteine residue allowed coupling of a sulfhydryl-reactive phenanthroline derivative to histone H4 in permeabilized yeast cells. Added copper ions (as copper sulfate, for example) are chelated by the H4-coupled phenanthroline, and addition of hydrogen peroxide then produces short-lived hydroxyl radicals that cleave the DNA close to the H4S47C site, which is situated close to the nucleosome dyad. Fractionation of DNA purified from digested genomic chromatin allows purification of a DNA fragment arising from cleavages at the dyad of adjacent nucleosomes. Sequencing of this material allows mapping of nucleosomes at high resolution and has been applied to budding yeast and also to *S. pombe* and murine embryonic stem cells

(Brogaard et al., 2012; Moyle-Heyrman et al., 2013; Voong et al., 2016). More recently, the Henikoff lab modified this approach by using a different site for the mutation to cysteine, namely H3Q85C; cleavage at both histone H3 sites in a nucleosome releases a 51 bp fragment that allows mapping of nucleosomes at base pair precision, and avoids potential complications caused by cleavage of linker regions by free copper ions or via cysteine residues belonging to DNA-associated, nonhistone proteins (Chereji et al., 2018).

These various methods for mapping nucleosome positioning on a global scale have proven to be generally robust (Jiang and Pugh, 2009a) and have yielded considerable insights into genome organization and the role of chromatin in gene regulation. We discuss some of the findings based on these techniques a little farther on.

First studies of nucleosome positioning

Early investigations of nucleosome positioning in vivo tended to focus on repetitive DNA sequences, such as the ribosomal RNA gene clusters and satellite DNA (so named because the repetitive DNA comprising this class was first observed as a separately migrating, "satellite" band in centrifugal separations of nuclear DNA; later, the definition was expanded to include tandemly repeated sequences even when not separable by centrifugation techniques (Pech et al., 1979)). The presence of these elements as multiple copies in a genome rendered them more amenable to detection and analysis, and therefore more tractable for determination of nucleosome positioning. Several such early studies found evidence for nonrandom positioning of nucleosomes, but not for a single dominant positioned array (Eissenberg et al., 1985). The idea that some sequences could be more favorable than others for nucleosome formation, as suggested by these findings of in vivo nonrandom positioning, was supported by experiments showing that *Hpa*II restriction fragments from the plasmid pBR322 that had similar size but different sequence varied considerably in the efficiency with which they were incorporated into nucleosomes in vitro (Linxweiler and Hörz, 1984). Early studies on single-copy genes did not support strongly positioned nucleosomes, and were in some cases interpreted as indicating a lack of nucleosomes over transcribed regions (Eissenberg et al., 1985). These initial reports, however, suffered from a lack of naked DNA controls, which together with the bias by MNase for cutting at A/T rich sequences, made results at best difficult to interpret (Fittler and Zachau, 1979; McGhee and Felsenfeld, 1983).

A major step forward in the investigation of nucleosome positioning was the use of unique sequence DNA to reconstitute nucleosome core particles for analysis in vitro. The first such study reported used restriction fragments harboring sequences from the *E. coli lac* control region, isolated (prior to the advent of PCR technology) from a *lac*-containing plasmid (Chao et al., 1979). Two fragments, of size 203 and 144 bp, were radioactively end-labeled and reconstituted by salt dialysis into nucleosomes with histones isolated from calf thymus. Both fragments readily reconstituted nucleosomes despite their bacterial origin, consistent with other work showing that no specific eukaryotic sequences were required for nucleosome formation (Axel et al., 1974). The reconstituted nucleosomes were purified from sucrose gradients, digested with DNase I or exonuclease III, and the cleavage pattern analyzed by gel electrophoresis and autoradiography. The results revealed decidedly nonrandom cleavage patterns, particularly for DNase I, and were interpreted as indicating two or three predominant positions. However, the inherent sequence specificity of nucleases was not appreciated at this time, and nonnucleosomal controls were consequently lacking.

The first definitive evidence for specific positioning of a nucleosome on a unique DNA sequence was provided by Simpson and Stafford (1983). Stafford's lab had cloned the 5S rRNA gene from the sea urchin *Lytechinus variegatus* and provided a 260 bp segment including the gene to Simpson, who used the end-labeled fragment to reconstitute nucleosomes (without linker histone), which were purified by sucrose gradient fractionation prior to DNase I cleavage and analysis. Secondary restriction enzyme cleavage of the end-labeled fragment prior to reconstitution allowed mapping of cleavage sites from each end of the fragment (Fig. 4.6). The results were dramatic: the nucleosomal fragment exhibited a cleavage pattern with periodicity of 10 bp in which cleavage sites present in naked DNA were strongly protected (Fig. 4.12). The pattern of protection extended from 20 to 165 bp inside the fragment, from the "left" end (orientation was assigned in the direction of 5S rRNA transcription), indicating a translationally positioned nucleosome occupying this region. The strongly periodic

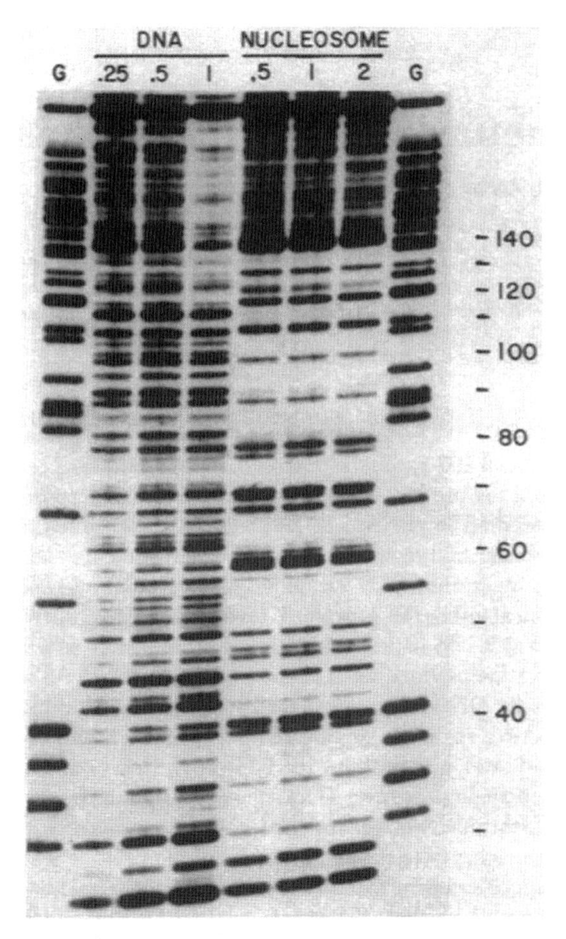

FIG. 4.12 Nucleosome positioning on the 5S rRNA sequence from *L. variegatus*. Reconstituted nucleosome core particles and naked DNA were digested with DNase at indicated concentrations (in units/mL); lanes labeled G contain samples cleaved at guanyl residues with dimethyl sulfate. Sizes of the fragments are indicated on the right. *From Simpson, R.T., Stafford, D.W., 1983. Structural features of a phased nucleosome core particle. Proc. Natl. Acad. Sci. USA 80, 51–55.*

cleavage pattern also pointed to precise rotational positioning. These results showed that in spite of the ability of DNA from nearly any source to associate with histones to form nucleosomal particles (Axel et al., 1974), DNA sequence could nonetheless dictate the relative disposition of histones to DNA. Corroborating this result, Gary Felsenfeld's lab showed in a follow-up study to the Chao investigations of nucleosome positioning on *lac*-containing DNA sequences, with inclusion of appropriate naked DNA controls, that prokaryotic DNA could also dictate preferential nucleosome positioning, although this appeared less homogeneous than that observed with the 5S rRNA fragment (Ramsay et al., 1984).

These findings laid the groundwork for questions that have occupied the field for the decades following: How prevalent is positioning in vivo? What are the functional consequences of nucleosome positioning in living cells? What are the mechanisms that dictate nucleosome positioning?

Nucleosome positioning in vivo

Investigations of nucleosome positioning in vivo proceeded in parallel with those being done by reconstitution in vitro. Some of the early studies on nucleosome positioning at multicopy sequences, such as satellite DNA and 5S rDNA repeats, have already been mentioned. The introduction of the indirect end-labeling approach allowed researchers to begin to map nucleosome positions on single-copy genes in various organisms. Many early studies employed the model organism budding yeast (*Saccharomyces cerevisiae*), as its smaller genome (~100-fold smaller than human) allowed better signal to noise on southern blots used in indirect end-label experiments than did chromatin derived from metazoan organisms. Thus, some of the first examples of nucleosome positioning were obtained for the *PHO5* and *HSP82* loci in *S. cerevisiae* (Almer and Horz, 1986; Bergman and Kramer, 1983; Szent-Gyorgyi et al., 1987). The relatively small genome (~180 Mb) of *Drosophila melanogaster*, although still ~10-fold larger than that of yeast, also leant itself to such studies, with analyses of the *hsp70* and *hsp26* loci being early examples (Thomas and Elgin, 1988; Udvardy and Schedl, 1984). Nucleosome positioning was observed in mammalian cells as well, at the mouse mammary tumor virus promoter and at the mouse β-major globin locus; in the former case the technical challenge of achieving sufficient signal to noise ratio in indirect end-label experiments was mitigated by the use of multicopy episomes, as discussed below (Benezra et al., 1986; Richard-Foy and Hager, 1987). These loci were chosen in part because of their inducible, or developmentally regulated, expression, as there was intense interest in learning more about the relationship between chromatin structure and gene activation. Evidence for well-positioned nucleosomes in the region upstream of the coding region (i.e., the promoter region) was found for all of these examples under conditions in which the gene was inactive, and increased accessibility to various cleavage reagents, including DNase I, MNase, MPE-Fe(II), and restriction enzymes, was observed at the promoters when the genes were active or, for the β-globin gene, in a cell line in which the gene is inducible but not in one in which it is permanently repressed (Almer et al., 1986; Benezra et al., 1986; Bergman and Kramer, 1983; Richard-Foy and Hager, 1987; Thomas and Elgin, 1988). In contrast, coding regions appeared to lack positioned nucleosomes in these examples, although more recent genome-wide investigations have shown otherwise. Conflicting results were reported as to whether transcribed regions generally contained randomly positioned nucleosomes or were nucleosome-free; we return to this issue shortly.

Researchers were keenly interested in investigating mechanisms of nucleosome positioning in vivo, which required the ability to edit gene sequences in the cell. In the mid-1980s, when these first studies were taking place, methods for editing DNA in vivo were just being pioneered in yeast and were unavailable for metazoan cells. An alternative, first developed by Fritz Thoma and Bob Simpson for yeast and by Gordon Hager's lab, also at the National Institutes of Health, for mouse mammary tumor virus (MMTV) in mammalian cells, was the use of episomal plasmids (Ostrowski et al., 1983; Thoma et al., 1984). For these studies, specific sequences from yeast or MMTV were inserted into plasmids that could be cloned in *E. coli* and their sequences manipulated using standard techniques for recombinant DNA. When introduced into yeast cells or mouse cell lines, respectively, these plasmids were packaged into nucleosomes and stably replicated via origins of replication, from yeast and bovine papilloma virus respectively, encoded on the plasmids. Importantly, nucleosome positions determined on such stable episomes have generally been found to be the same as seen at endogenous chromosomal loci (e.g., Bresnick et al., 1992; Fascher et al., 1993). Thus, nucleosome positioning and the effects of specific sequence manipulation thereon could be readily investigated in these tractable systems.

The yeast minichromosome that was used in this way originated with a 1453bp DNA fragment containing the *TRP1* gene. The yeast *TRP1* gene was cloned by complementation of *E. coli* mutants that lacked *N*-(5′-phosphoribosyl) anthranilate isomerase (the product of the *TRP1* gene); when transformed back into yeast *trp1*- mutants, colonies capable of growing on medium lacking tryptophan were obtained at about 100-fold higher yield than normally observed for linear DNA fragments that were integrated into the *S. cerevisiae* genome by homologous recombination (Struhl et al., 1979). The reason that eventually emerged for this high transformation rate was that a replication origin, called an autonomously replicating sequence (ARS) in yeast, sits adjacent to the *TRP1* gene in *S. cerevisiae*; this, together with ligase activity present in the yeast cell nucleus, allowed the fragment to be converted into a closed circular molecule and replicated. This small circular episome, dubbed TRP1ARS1, was shown to be packaged into chromatin and maintained at about 100 copies per cell (Zakian and Scott, 1982), making it an excellent vehicle to study and manipulate nucleosome positioning in vivo.

Initial characterization of the TRP1ARS1 minichromosome by mapping MNase cleavage sites using indirect end-labeling revealed two nucleosomal regions separated by two MNase-sensitive regions of about 180bp each (Fig. 4.13; Thoma et al., 1984). One of the MNase-sensitive regions resided 5′ of the *TRP1* gene, while the other encompassed the *ARS1* region 3′ of *TRP1*. The *TRP1* coding region, which is transcribed at very low levels, included four nucleosome-sized regions that showed partial protection against MNase digestion, while three regions of ~160bp located downstream of the 3′ end of the *TRP1* gene exhibited MNase cleavages in naked DNA that were strongly protected in chromatin (Fig. 4.13). Electron microscopy and topological assays showed that TRP1ARS1 was occupied by seven nucleosomes; hence, the region downstream of the *TRP1* gene was interpreted as being incorporated into three strongly positioned nucleosomes, with the *TRP1* coding region interpreted to include four "unstable" nucleosomes. (The region downstream of *TRP1* was initially called the region of unknown function, or UNF, and eventually was found to include the upstream portion of the *GAL3* promoter, which is repressed when yeast are grown using glucose as carbon source, as they are for most experiments.) More than 20years later, these assignments were corroborated

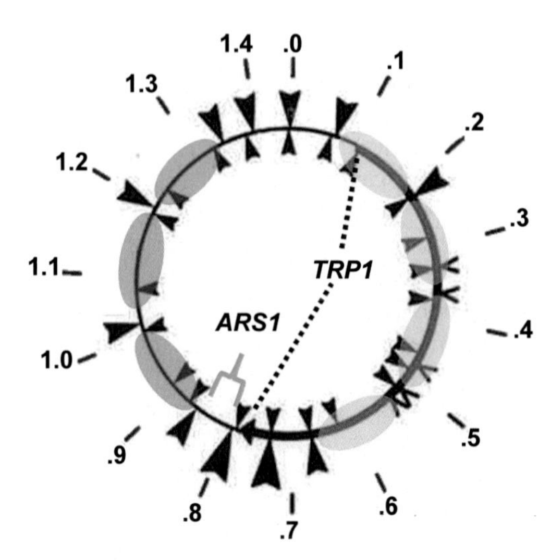

FIG. 4.13 The TRP1ARS1 yeast minichromosome. *Arrowheads* indicate sites cleaved by MNase, with larger *arrowheads* indicating stronger cleavages; cuts in chromatin are shown on the outside, and sites cut in naked DNA are shown inside the *circle*. Distances in kilobase pairs are indicated; nucleosomes are represented by *orange* ("unstable") and *blue* (strongly positioned) ellipses, and the *TRP1* gene (*thick arrow* extending from 0.1 to ~0.8 map units) and *ARS1* region are also indicated. *Modified from Thoma, F., Bergman, L.W., Simpson, R.T., 1984. Nuclease digestion of circular TRP1ARS1 chromatin reveals positioned nucleosomes separated by nuclease-sensitive regions. J. Mol. Biol. 177, 715.*

in the first genome-wide determinations of nucleosome occupancy in yeast, which showed at the corresponding chromosomal locus one well-positioned and four "fuzzy" nucleosome occupying the *TRP1* coding sequence and three well-positioned nucleosomes farther downstream (Kaplan et al., 2009; Lee et al., 2007). Strong nucleosome positioning was also observed in a study in which the *GAL1-GAL10* intergenic region was inserted into a yeast minichromosome; in this case, altering the DNA sequences in the minichromosome showed that nucleosomes adopted specific positions relative to the upstream activation sequence, which was postulated to act as a boundary element (Fedor et al., 1988). Studies on the MMTV promoter, carried on a stable episome in NIH3T3-derived and other tissue culture cells, similarly revealed patterns of protection in chromatin against digestion by MNase and MPE-Fe(II) consistent with strongly positioned nucleosomes (Richard-Foy and Hager, 1987). These early demonstrations of positioned nucleosomes provided the foundation for subsequent investigations of in vivo mechanisms of nucleosome positioning and the effect of *trans*-acting factors on chromatin structure that will be discussed later on.

Continuing investigations provided numerous examples of nucleosomes being positioned at specific genomic sites (Table 4.1). Mapping of nucleosomes using primer extension methods revealed instances in which nucleosomes were positioned with base pair precision in vivo in yeast and mammalian systems (McPherson et al., 1993; Shimizu et al., 1991), while instances were also reported in which nucleosomes showed heterogeneity in their translational position (An et al., 1998; Buttinelli et al., 1993; Fragoso et al., 1995). However, it required

TABLE 4.1 Early examples of positioned nucleosomes.

Organism	Locus	References
Budding yeast (*S. cerevisiae*)	*PHO5*	Bergman and Kramer (1983) and Almer and Horz (1986)
	HSP82	Szent-Gyorgyi et al. (1987)
	STE6, BAR1	Shimizu et al. (1991)
	STE2, STE3	Ganter et al. (1993)
	ADH2	Verdone et al. (1996)
	CHA1	Moreira and Holmberg (1998)
Fission yeast (*S. pombe)*	*ade6*	Bernardi et al. (1991)
Drosophila	*hsp26*	Thomas and Elgin (1988)
	hsp70	Udvardy and Schedl (1984)
Mouse	β-Major globin	Benezra et al. (1986)
	MMTV LTR	Richard-Foy and Hager (1987)
	Serum albumin enhancer	McPherson et al. (1993)
Human	pS2 (MCF7 cells)	Sewack and Hansen (1997)
Tobacco	β-Phaseolin	Li et al. (1998)

the advent of genome-wide approaches for determining nucleosome occupancy in vivo to obtain a full and coherent picture of nucleosome positioning in living cells—a picture that continues to evolve. As mentioned earlier, analysis of nucleosomal DNA from budding yeast using low-resolution DNA microarrays revealed that promoter regions were less densely occupied by nucleosomes than were coding regions (which, contrary to some early reports, were found to be incorporated into nucleosomes) or 3′ regions of genes (Bernstein et al., 2004; Lee et al., 2004). The first results based on high-resolution microarrays supported these findings while providing much more detail (Lee et al., 2007; Whitehouse et al., 2007; Yuan et al., 2005). These studies revealed that a large fraction of nucleosomes in *S. cerevisiae* were well-positioned. Furthermore, they reported that nucleosomes were positioned relative to the transcription start sites (TSSs) of mRNA-encoding genes in a stereotypical pattern, first observed over about 5% of the yeast genome and later confirmed genome-wide, in which most promoters displayed a nucleosome-depleted region (NDR; also sometimes referred to as NFR, or nucleosome-free region; see Sidebar: NDR or NFR?) flanked by arrays of strongly positioned nucleosomes extending in both directions (Fig. 4.14). The downstream array of nucleosomes exhibited more marked positioning than the upstream array, suggesting that it was set by a barrier of some sort defined by the TSS (Lee et al., 2007; Mavrich et al., 2008a). Understanding the nature of this barrier has been a central preoccupation of researchers in this area; we will return to this issue when we consider mechanisms of nucleosome positioning.

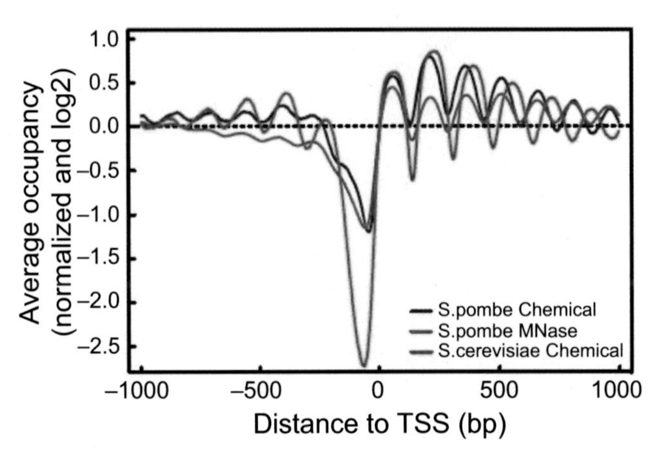

FIG. 4.14 Nucleosome arrays surround transcription start sites in yeast. Nucleosomes were mapped in budding yeast (*S. cerevisiae*) and fission yeast (*S. pombe*), using chemical mapping or MNase-seq, as indicated, and occupancy plotted relative to transcription start sites averaged over all genes. Occupancies are calculated relative to genome averages, so both positive and negative values result. Different spacing of nucleosomes is observed downstream of the TSS in *S. cerevisiae* and *S. pombe*. *From Moyle-Heyrman, G., Zaichuk, T., Xi, L., Zhang, Q., Uhlenbeck, O.C., Holmgren, R., Widom, J., Wang, J.P., 2013. Chemical map of Schizosaccharomyces pombe reveals species-specific features in nucleosome positioning. Proc. Natl. Acad. Sci. USA 110, 20158.*

Sidebar: NDR or NFR?

Constitutively nonnucleosomal, or nucleosome-free, regions were identified in early studies that focused on specific loci. These regions, identified by their sensitivity to nucleases and hence originally defined as DNase hypersensitive sites, often corresponded to linker DNA flanked by strongly positioned nucleosomes, as in the yeast *PHO5* promoter and the *Drosophila hsp26* promoter (Almer and Horz, 1986; Thomas and Elgin, 1988). Some nuclease-sensitive regions, however, such as those in the TRP1ARS1 minichromosome and at the *AKY2* and *GCY1* promoters, were long enough to accommodate a nucleosome and yet apparently were free, or nearly free, of histones (Angermayr and Bandlow, 2003; Angermayr et al., 2002; Thoma et al., 1984). When extended (>150 bp) regions of low nucleosome occupancy were discovered to be commonplace at yeast promoters, their designation as nucleosome-free regions, or NFRs, became widely adopted (Yuan et al., 2005). Ollie Rando pointed out in a review that "some researchers prefer that the term NFR be qualified, because microarray studies do not provide absolute amounts and thus can only reveal relative depletion," leading to the introduction of the less stringent designation of such regions as nucleosome-depleted (NDRs) (Lee et al., 2007; Rando, 2007). Whether such regions are ever truly free of nucleosomes or bound histones has been a matter of some debate, and bears on the nature of "fragile nucleosomes," which are discussed later in this chapter (Barnes and Korber, 2021). Here I employ the cautious approach of referring to such regions as NDRs, as any region that is entirely free of nucleosomes is by definition depleted, while proof of complete absence is much more challenging.

Genome-wide mapping of nucleosome occupancy by high-throughput sequencing corroborated results of microarray studies (Field et al., 2008; Mavrich et al., 2008a; Shivaswamy et al., 2008). A careful analysis of datasets from six independent genome-wide determinations of nucleosome occupancy in yeast, two conducted using high-resolution microarrays and four using sequencing approaches, revealed a reassuring concordance across the data (Jiang and Pugh, 2009a). Using all six datasets, a reference set of 59,915 nucleosomes across the yeast genome was constructed, using a cutoff of at least 5% occupancy (meaning that the number of reads to "call" a nucleosome had to be at least 5% of the maximum number observed for a positioned nucleosome). This compares to the number of nucleosomes counted in the six cited studies ranging from 48,126 to 70,871. More than 95% of called nucleosomes had >50% occupancy relative to the maximum observed, while some regions were found to be nucleosome-free, inasmuch as no reads were found within the regions, compared to an average of 160 tags in nucleosomal regions. Whether these putative nucleosome-free regions, which principally occur at promoter regions, in fact lack histone binding or are occupied by noncanonical nucleosomes especially susceptible to MNase digestion, has been an ongoing topic of discussion (Brahma and Henikoff, 2019, 2020; Chereji et al., 2017; Kubik et al., 2017).

In addition to revealing widespread nucleosome positioning in vivo, genome-wide mapping of nucleosome positioning revealed the widespread presence of "fuzzy," or delocalized, nucleosomes (Yuan et al., 2005). "Unstable" positioned nucleosomes on the TRP1ARS1 minichromosome were noted previously, and nucleosomes exhibiting multiple preferred, but still nonrandom, positions were inferred at various genomic loci by indirect end-label analysis that showed partial protection of MNase cleavage sites compared to naked DNA, and in primer extension experiments that revealed multiple closely spaced positions for nucleosomes in vivo and in vitro (An et al., 1998; Buttinelli et al., 1993; Costanzo et al., 1995; Dong et al., 1990; Fragoso et al., 1995; Pennings et al., 1991; Thoma et al., 1984). Fuzzy, or delocalized, nucleosomes were first defined as showing protection over a range of about 160 bp or more, as compared to the ~147 bp of protection expected of a highly positioned nucleosome (Yuan et al., 2005). This does not mean that individual nucleosomes protected 160 bp or more against digestion, but rather that a region of >160 bp showed at least partial protection against MNase digestion; the implication is that individual nucleosomes are present at multiple positions over the region. Nucleosome mapping by MNase-seq allowed a more precise definition and quantitative measure of fuzziness, based on the standard deviations of identified sequence tag locations around the nucleosome midpoint (Mavrich et al., 2008a). Nucleosomes more distal to the TSS displayed increasing fuzziness, and this was corroborated by showing that the standard deviations of nucleosome positions across all six datasets displayed this same trend (Jiang and Pugh, 2009a). Nucleosomes occupying translational frames offset by integral multiples of the DNA helical repeat (i.e., preserving rotational positioning) were also detected in yeast using the chemical cleavage strategy based on H4S47C mutant yeast discussed above (Brogaard et al., 2012). Heterogeneity in translational positioning was also found in studies examining nucleosome positioning in single cells, as described in the next section.

Extensive nucleosome positioning was also observed in the fission yeast *S. pombe*, with some important differences relative to budding yeast (*S. pombe* and *S. cerevisiae* are very distantly related, being separated by >300 million years of evolution (Sipiczki, 2000)). As in

S. cerevisiae, low nucleosome occupancy was evident upstream of transcription start sites, followed by a slowly decaying pattern of positioned nucleosomes similar to that seen in *S. cerevisiae*. However, the array of nucleosomes upstream of the TSS was diminished in *S. pombe*, whether monitored by MNase-seq or by chemical mapping (Fig. 4.14; Givens et al., 2012; Lantermann et al., 2010; Moyle-Heyrman et al., 2013). A small flood of papers has reported genome-wide nucleosome occupancy in metazoan organisms from *C. elegans* and *Drosophila melanogaster* to *Arabidopsis* to human cells of various tissue types (Teif, 2016). Early reports are summarized in Table 4.2; numerous additional studies have been

TABLE 4.2 Early genome-wide determinations of nucleosome occupancy.

Organism	Method	Condition	Number of nucleosomes mapped	References
S. cerevisiae	"Home-made" microarray; low resolution	Standard growth conditions	n/a	Bernstein et al. (2004)
	"Home-made" microarray; low resolution	Standard growth conditions	n/a	Lee et al. (2004)
	"Home-made" microarray; 20bp resolution covering ~5% of genome	Standard growth conditions	n/a	Yuan et al. (2005)
	Tiling arrays; ~5bp resolution	Standard growth conditions	n/a	Lee et al. (2007)
	Tiling arrays; ~5bp resolution	Wild type and *isw2*Δ	n/a	Whitehouse et al. (2007)
	MNase-seq after H2A.Z ChIP	Standard growth conditions	322,000	Albert et al. (2007)
	MNase-seq	Standard growth conditions	380,000	Field et al. (2008)
	MNase-seq	±Heat shock	515,000 (NHS) and 1.04M (HS)	Shivaswamy et al. (2008)
	MNase-seq after H3 or H4 ChIP	Standard growth conditions	1.2M	Mavrich et al. (2008a)
	MNase-seq	Wild type and *rpb1-1 ts* mutant	1.5–4.2M	Weiner et al. (2010)
	Chemical mapping	Standard growth conditions	105M	Brogaard et al. (2012)
S. pombe	Tiling array; ~20bp resolution	Standard growth conditions	n/a	Lantermann et al. (2010)
	MNase-seq	Standard growth conditions	Not reported	Givens et al. (2012)
	Chemical mapping	Standard growth conditions	95M	Moyle-Heyrman et al. (2013)

TABLE 4.2 Early genome-wide determinations of nucleosome occupancy—cont'd

Organism	Method	Condition	Number of nucleosomes mapped	References
Multiple yeast species	MNase-seq	*S. cerevisiae, S. paradoxus,* and hybrid diploids	Not reported	Tirosh et al. (2010a)
	MNase-seq	12 Ascomycete species	Not reported	Tsankov et al. (2010)
	MNase-seq	4 Ascomycete species	10–40 M per species	Tsui et al. (2011)
C. elegans	MNase-seq	Mixed-stage wild type	284,000	Johnson et al. (2006)
	MNase-seq	Mixed-stage wild type	44 M	Valouev et al. (2008)
D. melanogaster	MNase-seq after H2A.Z ChIP	Whole embryos	653,000	Mavrich et al. (2008b)
	Tiling arrays; 36 bp spacing	Whole embryos	n/a	Mavrich et al. (2008b)
Oryzias latipes (medaka)	MNase-seq	Blastula	37 M	Sasaki et al. (2009)
Homo sapiens	Tiling arrays	Various cell lines	n/a	Ozsolak et al. (2007)
	MNase-seq; ChIP against H2A.Z, modified histones	20 Histone modifications and H2A.Z in CD4+ T cells	2.1 to 13.6 M	Barski et al. (2007)
	MNase-seq	Resting and activated CD4+ T cells	155 and 142 M	Schones et al. (2008)
	MNase-seq	CD4+ and CD4+ T cells and granulocytes	342–584 M	Valouev et al. (2011)

performed since, including some that reveal genome-wide patterns of nucleosome positioning that do not conform to the stereotypical pattern dictated by transcriptional start sites (Maree et al., 2017; Wedel et al., 2017). Although a considerable number of online tools exist for analyzing genome-wide datasets on nucleosome occupancy and viewing results of such analyses (Teif, 2016), these are still rapidly evolving, and they appear and disappear from the web with disconcerting frequency. Furthermore, nucleosome occupancy data for large genomes such as that of humans are generally low in coverage, as on the order of a billion reads is needed to cover the human genome to similar density as that obtained for high quality nucleosome maps of yeast. Nonetheless, some conclusions can be drawn from published work on genome-wide nucleosome occupancy in the cells of plants and animals (Barski et al., 2007; Gaffney et al., 2012; Mavrich et al., 2008b; Schones et al., 2008; Teif et al., 2012; Valouev et al., 2008, 2011; Zhang et al., 2015). First, in contrast to the small genome (~12 Mb) of budding yeast, the larger genomes (>100 Mb) of metazoan organisms are mostly not packaged into well-positioned nucleosomes, although regular spacing of nucleosomes is generally observed (Baldi et al., 2018a). Second, functional elements, and particularly promoters of transcriptionally active genes, are depleted of nucleosomes, consistent with many early studies correlating such elements with DNase hypersensitive sites (Gross and Garrard, 1988). An important difference between NDRs in yeast and mammalian cells is that whereas the great majority of genes in yeast, even those exhibiting the lowest level of transcription, possess NDRs immediately upstream of the transcription start site, in mammalian cells this is observed for genes that are expressed or that are associated with a preinitiation complex (PIC), but not for inactive genes lacking a PIC (Ozsolak et al., 2007; Valouev et al., 2011). Third, the NDRs associated with functional elements are often flanked by well-positioned nucleosomes, from which are propagated arrays of nucleosomes whose positioning diminishes (fuzziness increases) over a range of ~5–10 nucleosomes. These observations in combination explain the greater proportion of the yeast genome being incorporated into positioned nucleosomes: yeast have a compact genome, with short intergenic regions and genes that are not interrupted by long introns, so that NDRs at gene promoters together with the arrays they generate add together to a substantial fraction of the genome. In contrast, arrays generated by promoter elements, or paused RNA polymerase molecules, in metazoans decay over a relatively short length compared to the characteristically long regions transcribed by Pol II that give rise to mRNA transcripts. This effect is further exacerbated by the larger internucleosomal spacing characteristic of metazoan chromatin.

Although positioned nucleosomes may be relatively rare in large metazoan genomes, this does not imply they are not functionally important. In fact, the preferred association of positioned nucleosomes with functional elements suggests just the opposite, although one must be wary of mistaking effect for cause. Besides the stereotypical pattern of strongly positioned nucleosomes and NDRs surrounded by arrays of nucleosomes that is frequently observed at the proximal promoters of active genes transcribed by Pol II, locally nucleosome-free regions surrounded by arrays of positioned nucleosomes have also been found at replication origins, TF binding sites, and CTCF binding sites, noted in Chapter 3 to be implicated in formation of topologically associating domains (TADs) (Eaton et al., 2010; Fu et al., 2008; Gaffney et al., 2012; Wang et al., 2012).

Heterogeneity in nucleosome positioning: Single cell approaches

Properties measured in large groups of cells necessarily reflect bulk population averages, but researchers have long been interested in how the molecular features underlying phenotypes are distributed across the cells in a population. Recent technical advances, particularly high-throughput sequencing and the ability to sequence single molecules of DNA that are kilobases in length, have opened new avenues to this problem. With regard to nucleosome positioning and occupancy, two distinct questions may be posed: (1) In what fraction of a population of cells is a given sequence or site incorporated into a nucleosome?; and (2) How are nucleosome positions and occupancies across a given region related to each other in a single cell? Early efforts toward addressing these questions used a similar strategy of monitoring accessibility to methyltransferases in chromatin, as described earlier. One study monitored methylation by M.HhaI, which methylates cytosines embedded in GCGC sequences, in the yeast *PHO5* promoter (with GCGC sites engineered into each promoter nucleosome sequence), while the other footprinted chromatin in mammalian promoters using the SssI MTase, which methylates the more prevalent CpG dinucleotide sequence (Fatemi et al., 2005; Jessen et al., 2006). Both studies found heterogeneity in nucleosome occupancy; the use of SssI MTase allowed sufficiently high resolution to also reveal variability in nucleosome position. A subsequent investigation employed a variation of this strategy to map nucleosome occupancies over promoter regions encompassing several nucleosomes in individual yeast cells (Small et al., 2014). In this study, permeable spheroplasts were prepared from yeast cells and treated with the methyltransferase M.CviPI, which methylates cytosines occurring in GpC dimers. Single cells were then isolated by serial dilution (more tractable with yeast than with mammalian cells), the DNA extracted and amplified, and the pooled PCR products subjected to bisulfite sequencing to identify methylated GpC residues. Patterns of protection against methylation were consistent with the presence of positioned nucleosomes at the *PHO5*, *CHA1*, and *CYS3* promoters and revealed heterogeneity in positioning and occupancy that nonetheless were consistent with bulk (population-based) determinations of nucleosome positioning (Fig. 4.15). Furthermore, nucleosomes were found to have

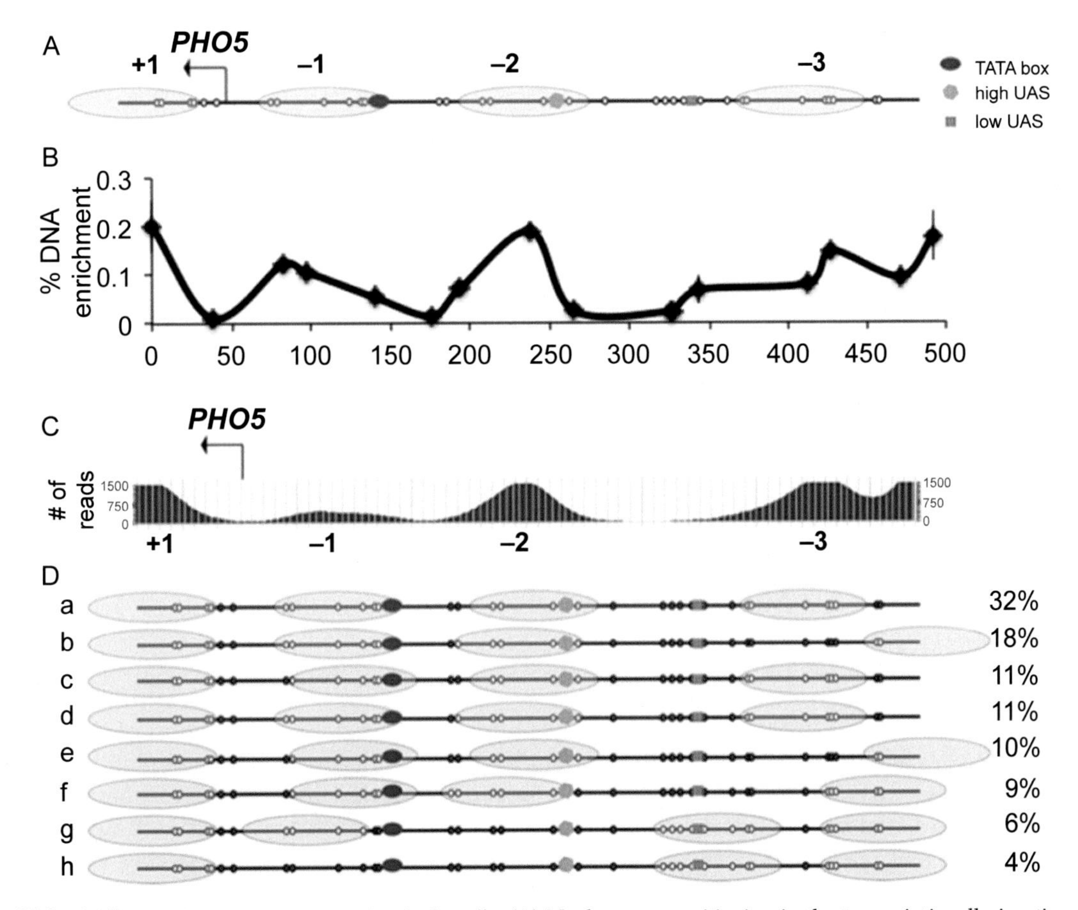

FIG. 4.15 Nucleosome occupancy in single cells. (A) Nucleosome positioning in the transcriptionally inactive *PHO5* promoter as determined from earlier work. (B) Nucleosome scanning determination of nucleosome occupancy in the *PHO5* promoter. (C) MNase-seq determination of nucleosome occupancy in the *PHO5* promoter. (D) Protection against GpC methylation in 806 individual cells and inferred nucleosome positions. *From Small, E.C., Xi, L., Wang, J.P., Widom, J., Licht, J.D., 2014. Single-cell nucleosome mapping reveals the molecular basis of gene expression heterogeneity. Proc. Natl. Acad. Sci. USA 111, E2462.*

minor shifts in translational positioning, as seen in studies cited earlier, although rotational positioning could not be assessed by this method. It should also be noted that caution must be exercised in interpreting methylase protection as being due to nucleosomes, as protection may also be conferred by the PIC or other large protein complexes.

These early efforts toward determination of chromatin structure in single cells were relatively low throughput, as they relied on sequencing of amplicons corresponding to specific targets, after bisulfite treatment, rather than interrogating an entire genome. Advances in sequencing technology allowed introduction of Nucleosome Occupancy and Methylome sequencing, or NOMe-seq (Kelly et al., 2012). In this approach, nuclei are treated with M. CviPI, thereby methylating cytosines at accessible GpC sites, after which purified genomic DNA fragments are treated with bisulfite and then subjected to high-throughput sequencing. This allowed simultaneous determination of endogenous methylation of CpG sites and of GpC methylation by M.CviPI as a probe for nucleosome occupancy, thereby allowing these two parameters to be correlated on a single-molecule basis. Consistent with earlier work, Kelly et al. reported phased nucleosome arrays surrounding CTCF binding sites, and NDRs upstream of the TSS of active genes, with more highly active genes exhibiting increased nucleosome depletion.

Further progress has been enabled by methods permitting sequencing of single molecules several kilobases in length without amplification, such as nanopore and Pacbio sequencing. Nanopore sequencing, which can directly detect both m^5C and N^6 methylated adenine (Rand et al., 2017; Simpson et al., 2017), has recently been used to determine accessibility of yeast chromatin to M.CviPI and mammalian chromatin to N^6 methylation of adenine over regions of several kilobases, revealing heterogeneity in both nucleosome positioning and occupancy and allowing relationships between chromatin status of genomic elements separated by thousands of base pairs to be ascertained (Shipony et al., 2020; Stergachis et al., 2020; Wang et al., 2019). Although technical hurdles remain, such methods are likely to be expanded by use of additional DNA modifying enzymes and coupling with additional assays of the physical and functional state of the genome, and doubtless will provide new insights into heterogeneity of chromatin structure and stochastic contributions to phenotype.

Mechanisms of nucleosome positioning

At the beginning of this chapter, it was noted that investigations into nucleosome positioning were motivated by the idea that the location of particular DNA sequences with respect to the histones seemed likely to affect the accessibility and therefore the function of those sequences. We have seen that some DNA sequences do indeed adopt specific orientations with respect to the histone octamer, both rotationally and translationally, when incorporated into nucleosomes, and that this is true both in vitro and in vivo. We have also seen that stereotypical patterns of nucleosome arrays have been observed in vivo, using approaches allowing genome-wide determination of nucleosome occupancy. The locations of nucleosomes in these arrays are generally determined with respect to functional elements, and not necessarily by DNA sequence. We now examine the mechanisms that determine the positions of individual nucleosomes, and of nucleosome arrays that are observed in living cells. These mechanisms involve not only signals that favor individual nucleosomes residing in particular locations, but also DNA-encoded elements and extrinsic factors that exclude nucleosomes and that determine the spacing between nucleosomes, which governs the regularity of nucleosome arrays.

Potential determinants of nucleosome positioning were first considered on a mostly theoretical basis. Alternative DNA structures, such as Z-DNA and cruciform structures, were considered as possibly excluding nucleosomes and thereby establishing a boundary from which a phased array of nucleosomes might originate (Eissenberg et al., 1985). In support of special DNA sequences or structures providing order to nucleosome arrays by excluding nucleosomes, reconstitution experiments indicated that an 80 bp run of poly(dA).(dT) was resistant to incorporation into nucleosomes, although a 20 bp stretch was not refractory to nucleosome formation (Kunkel and Martinson, 1981; Rhodes, 1979; Simpson and Kunzler, 1979). Another idea was that sequence-specific DNA-binding proteins might direct nearby placement of histones (Eissenberg et al., 1985; Igo-Kemenes et al., 1982); in favor of this notion, a protein was identified that bound preferentially to α-satellite DNA and was proposed as possibly dictating the preferential nucleosome positioning reported for these sequences (Strauss and Varshavsky, 1984). Evidence for nonrandom nucleosome positioning was also reported at yeast centromeres and adjacent to telomeric complexes present in macronuclear DNA of *Oxytricha*, suggesting that these special chromosomal structures could dictate ordering of nearby nucleosomes as well (Bloom and Carbon, 1982; Gottschling and Cech, 1984). The idea that a boundary element could establish an array of well-positioned nucleosomes was predicted on theoretical grounds, and evidence for such arrays was obtained at the *GAL1-10* and *HSP82* loci using indirect end-labeling methodology long before such arrays were identified in genome-wide studies (Fedor et al., 1988; Kornberg and Stryer, 1988; Szent-Gyorgyi et al., 1987). Folding of chromatin into higher-order structures was reported to affect nucleosome positioning, in experiments using the TRP1ARS1 minichromosome system described earlier (Thoma and Zatchej, 1988), while the discovery of chromatin remodelers that are able to move or evict nucleosomes prompted investigation of their contributions to nucleosome positioning in vivo.

Investigations over the years following these early reports supported both *cis* (direct effects of DNA sequence on histone-DNA interactions) and *trans* (e.g., DNA-binding proteins) factors as contributing to nucleosome positioning. Ingenious experiments using methods undreamt of in the 1980s have dissected the extent of *cis* and *trans* contributions to nucleosome positioning in vivo, and genome-wide approaches have been used to study the role in both nucleosome positioning and spacing of chromatin remodelers, whose existence was unknown even if vaguely suspected in this early epoch. We now discuss these various contributing elements in turn.

Role of DNA sequence

The first demonstrations of nucleosome positioning were soon followed by experiments aimed at identifying DNA sequences that governed nucleosome placement (FitzGerald and Simpson, 1985; Ramsay, 1986). Sequence changes were introduced into the segments used in the original reports (5S rDNA-containing and *lac*-containing DNA segments) and the effect on positioning of nucleosomes reconstituted with the mutated sequences then ascertained using exonuclease III or DNase I in conjunction with end-labeled DNA. Neither study found any definitive positioning determinant, but both reports found that altering the DNA sequence over the ~40bp centered on the nucleosome dyad resulted in a loss of positioning. Notably, this region coincides with that found to exhibit structural discontinuities—i.e., DNA kinks or bends—within the nucleosome, as discussed in Chapter 2, suggesting that DNA sequences amenable to such deformation could contribute to nucleosome positioning. Related experiments were performed in yeast using the TRP1ARS1 minichromosome to examine determinants of nucleosome positioning in vivo (Thoma, 1986; Thoma and Simpson, 1985). Insertion of a 75bp DNA segment, too small to assemble into a nucleosome on its own, between two of the three strongly positioned nucleosomes in TRP1ARS1 resulted in the original nucleosomal sequences remaining protected from MNase digestion, while the inserted sequence was accessible to nuclease. Insertion of either a 155 or 300bp sequence also resulted in retention of the original three positioned nucleosomes; however, both sequences showed protection most consistent with a single nucleosome, rather than the larger sequence accommodating two nucleosomes as might have been expected. The 300bp sequence included the same 5S rDNA sequence shown to strongly position a nucleosome in vitro, and the same position was observed in vivo, leaving nonnucleosomal regions on either side accessible to nuclease (Thoma and Simpson, 1985). These results provided a strong indication that DNA sequence could dictate nucleosome positioning. Further work, however, revealed a more complex picture, as placement of a 560bp segment of TRP1ARS1 into a context in which the segment was flanked by two DNA sequences known to be hypersensitive to DNase I in vivo resulted in the segment being incorporated into four closely packed nucleosomes, with the pattern of protection against MNase digestion, and thus the positioning of nucleosomes, differing from that seen in TRP1ARS1 (Thoma, 1986). Taken together, these experiments indicated that DNA sequence played a role in nucleosome positioning both in vitro and in vivo, but that the sequence determinants were likely to be complex; that additional determinants, such as factors in the cell causing DNase hypersensitivity, also influenced nucleosome positioning; and that a complicated hierarchy of contributing factors was likely to govern the location of nucleosomes in living cells.

DNA bending and nucleosome positioning

The lack of a specific positioning element was consistent with an idea first proposed by Trifonov that was based on consideration of the energetics required for the tight, anisotropic bending of DNA required to wrap 80bp of DNA in a single turn around the histone octamer. Trifonov speculated that periodic bending of DNA, determined by regular occurrence of specific dinucleotide pairs that favored such bending, could favor particular rotational orientation of DNA relative to the histone octamer surface (Trifonov, 1980). Indeed, correlation analysis revealed a 10.5bp periodicity for specific dinucleotides in eukaryotic DNA derived from chromatin (Trifonov and Sussman, 1980). Around this same time, X-ray crystallography of short segments of DNA—most famously, perhaps, the "Dickerson-Drew dodecamer"—revealed that the structure of the DNA helix varied in its geometry in a sequence-dependent fashion, such that specific dinucleotides were characterized by differences in tilt, roll, and minor groove width (Dickerson and Drew, 1981; Drew et al., 1981). Horace Drew, who had participated in this work as a graduate student at Caltech, then moved to Andrew Travers's lab at the MRC in Cambridge and helped lead pioneering studies that, in parallel with work from other labs, lent additional support to the idea of DNA structural features contributing to the placement of DNA in nucleosomes:

- A study using a 169 bp DNA segment revealed a similar configuration whether the segment was ligated into a covalently closed, and thus considerably bent, circular molecule or incorporated into a nucleosome. In either case, measurement of DNase I cleavages indicated that short runs of (A,T) had minor grooves facing inward and short runs of (G,C) had minor grooves facing outward (G-C base pairs are characterized by a wide minor groove and A-T base pairs by a narrow minor groove) (Drew and Travers, 1985).
- Intrinsically curved DNA fragments favored nucleosome formation (Hsieh and Griffith, 1988; Pennings et al., 1989; Wolffe and Drew, 1989).
- Sequence analysis of 177 DNA segments isolated and cloned from chicken erythrocyte nucleosome core particles showed rotational preference for AAA and AA sequences to have inward-facing minor grooves, and GGC and AGC to have outward-facing minor grooves (Satchwell et al., 1986). A subsequent study analyzed 199 nucleosomal sequences isolated from yeast, as well as nucleosomal sequences from mouse and from in vitro reconstitution of nucleosomes using chemically synthesized "random" DNA and mouse genomic DNA (Segal et al., 2006). Examination of dinucleotide frequencies yielded results similar to those of Satchwell et al.: AA/TT/AT dinucleotides were found most frequently at 10 bp intervals, with minor grooves facing inward, while GC dinucleotides were enriched with opposite phase. Many large-scale analyses of nucleosome positioning have since corroborated these results, with AA/AT/TA/TT dinucleotides disposed toward having inward-facing minor grooves and CG/GC dinucleotides having outward-facing minor grooves (e.g., Brogaard et al., 2012; Moyle-Heyrman et al., 2013).

Shrader and Crothers put the notion of DNA bendability contributing to nucleosome positioning on a quantitative footing by performing a competitive reconstitution experiment to compare the abilities of defined DNA sequences to assemble into nucleosomes (Shrader and Crothers, 1989). Radiolabeled DNA fragments about 200 bp in length were assembled into nucleosomes by mixing in a 1 M NaCl solution with a large excess of chromatin from chicken erythrocytes that had been stripped of linker histones, followed by successive dilutions to lower the salt concentration to 0.1 M. Reconstitution was performed in the presence of a fourfold excess (relative to the chicken erythrocyte chromatin) of unlabeled naked chicken erythrocyte DNA; the relative degree to which the labeled DNA was incorporated into nucleosomes was then determined by using native (i.e., without SDS) gel electrophoresis to separate the nucleosomal fragments from those that were not incorporated into nucleosomes. Using this protocol, the relative free energies of incorporation for different fragments could then be calculated. (Although questions were raised as to whether this approach could accurately measure relative free energies of assembly of DNA into nucleosomes at physiological salt concentration (Drew, 1991), various evidence suggests that the equilibrium measured at high salt by this approach closely correlates with relative free energies at physiological salt concentration (Shrader and Crothers, 1989; Thastrom et al., 1999). Another study, however, indicated that relative affinities of specific sequences for binding to the histone octamer depends on both temperature and the concentration of histone octamer during assembly (Wu and Travers, 2005). An alternative approach involves the use of the histone chaperone Nap1, which allows competitive reconstitution to be performed at physiological salt concentration (Andrews and Luger, 2011). For additional discussion of the limitations of competitive reconstitution experiments for determining relative energetics of nucleosome formation with variable DNA sequences, see (Barnes and Korber (2021).) The results of these experiments showed that DNA segments containing the 5S rDNA segment from the sea urchin *L. variegatus* (used by Simpson in the first clear demonstration of nucleosome positioning in vitro) or from the frog *Xenopus laevis*, also shown to act as a nucleosome-positioning sequence (Rhodes, 1985), had more favorable free energies for nucleosome incorporation than bulk DNA by about 1.4 kcal/mol. The sequences most favorable for nucleosome formation contained five repeats of a 20 bp sequence with tandemly repeated $(A/T)_3N_2(G/C)_3N_2$ decamers (where N is any base); these "TG" and "GT" sequences had free energy of reconstitution of 2.8 kcal/mol lower than bulk mononucleosomal DNA, which equates to about 100-fold preference. Later experiments by Widom and colleagues showed that specific DNA sequences could vary in their propensity to be incorporated into nucleosomes by 1000-fold or more (Thastrom et al., 1999), while other reports indicated considerably lower energetic differences (Andrews et al., 2010; Lorch et al., 2014). Analysis of the cleavage pattern of reconstituted nucleosomes by Shrader and Crothers using hydroxy radical showed that, as predicted, (G/C) trimers were oriented with their wide minor grooves facing away from the histones and (A/T) trimers conversely oriented. Subsequent work showed that preferred incorporation of 5S rDNA sequences was observed at high salt (1 M NaCl), conditions under which only subnucleosomal particles containing H3-H4 tetramers and 120 bp of DNA, but lacking H2A-H2B, are formed, and at temperatures up to 75°C, indicating that the sequence-dependent anisotropic bending of DNA favoring nucleosome assembly could conceivably govern assembly of positioned nucleosomes under a wide range of conditions (Bashkin et al., 1993; Hansen et al., 1991).

Genome-wide studies have provided an abundance of data to test the contributions of DNA sequence to nucleosome positioning. Here it is important to distinguish *positioning* from *occupancy*; identifying DNA sequence contributions to the former, especially rotational positioning, requires high-resolution determination of positioning, as displacement by even a single base pair substantially alters rotational positioning. Two studies that fulfilled this criterion identified periodic distribution of dinucleotides in nucleosomes consistent with earlier studies (Brogaard et al., 2012; Gaffney et al., 2012). The first of these studies, mentioned earlier, examined nucleosome positioning in yeast using chemical mapping in H4S47C mutant yeast, while the second employed paired-end sequencing to obtain sequences of 147 bp fragments resulting from MNase digestion of chromatin from human lymphoblastoid cell lines. In addition, a distribution of DNase I nicks was observed with helical periodicity extending beyond the putative core region of the nucleosome; the simplest explanation for this is the existence of multiple translational positions that retain the same rotational position.

Altogether, these and other reports established that DNA sequence, and in particular differences in the energy required to deform a specific sequence to allow its bending around the histone octamer, could contribute to nucleosome positioning. Sequence effects that reflect DNA bendability, however, would be expected to affect rotational positioning and exert much less, if any, effect on translational positioning. This idea is supported by evidence, cited earlier, that nucleosomes both in vivo and in vitro have been found to adopt overlapping translational frames having the same rotational setting. In addition, it should be emphasized that although enrichment was observed for di- and tri-nucleotides having compressed or expanded minor grooves facing inward or outward on the nucleosome surface, this was only found by examination of sequences of large numbers of positioned nucleosomes; the tendency for such enrichment in individual nucleosomes is exceedingly modest. Thus, although reconstitution experiments with defined sequences established that sequence features *could* determine nucleosome positioning, and particularly rotational positioning, the extent to which such features actually contribute to nucleosome positioning in vivo is not addressed by such experiments. As we shall see, determinants of nucleosome positioning must be considerably more complex than can be accounted for by DNA bendability alone.

Poly(dA)-poly(dT) tracts and nucleosome exclusion

Early experiments, mentioned above, indicated that sufficiently long stretches of poly(dA)-poly(dT) or poly(dG)-poly(dC) were refractory to incorporation into nucleosomes in vitro (Kunkel and Martinson, 1981; Prunell, 1982; Rhodes, 1979; Simpson and Kunzler, 1979). Interest in the role of such homopolymeric tracts, especially of poly(dA)-poly(dT), in influencing chromatin structure and function in vivo was heightened by findings that poly(dA)-poly (dT) promoter elements facilitated transcriptional activity in yeast, and the suggestion that they might do so by excluding nucleosomes from promoter regions (Struhl, 1985). However, later studies appeared to controvert, or at least temper, these findings, as it was shown that these same poly(dA)-poly(dT)-containing promoter elements could be incorporated into nucleosomes in vitro (Losa et al., 1990), as could A-tracts as long as 69 bp (Fox, 1992). Even longer, completely homopolymeric sequences could be incorporated into nucleosomes by altering the reconstitution conditions to employ a low-salt exchange protocol (Puhl et al., 1991). In addition, a high-resolution X-ray crystal structure determination of a nucleosome incorporating a 16 bp poly(dA)-poly(dT) element revealed little effect on nucleosome structure (Bao et al., 2006).

Although it is evident from these various studies that homopolymeric tracts *can* be incorporated into nucleosomes, the more salient issue is whether such sequences are quantitatively less favorable for nucleosome formation than mixed-sequence DNA. In vitro reconstitution experiments and studies in yeast indicate that homopolymeric tracts tend to be excluded from nucleosomes when competing with mixed sequence DNA for incorporation. One example of poly(dA)-poly(dT) sequences being relatively unfavorable for nucleosome formation is provided by an investigation of the Alu family of short interspersed elements (SINEs) present in human cells (Englander and Howard, 1996). This study reported that whereas several evolutionarily distinct Alu elements contained in sequences of greater than nucleosomal length were incorporated into well-positioned nucleosomes when reconstituted using salt dialysis, a $T_{14}A_{11}$ sequence adjacent to an element resulting from a relatively recent transposition event caused the Alu sequence to be excluded (Englander and Howard, 1996). Instead, reconstitution resulted in a nucleosome incorporating sequences from a different region of the fragment used. Removal of the $T_{14}A_{11}$ sequence restored positioning relative to the Alu element. Thus, competition for nucleosome formation occurred between different locations on the same fragment, and the presence or absence of the $T_{14}A_{11}$ sequence affected the outcome. More recently, the effect of A-tracts on in vitro nucleosome formation was examined quantitatively (Lorch et al., 2014). Poly(dA)-poly(dT) tracts of 7, 10, and 14 bp were introduced into the 5S rDNA nucleosome positioning sequence at different locations and the affinity for nucleosome formation of the resulting sequences compared to the parent using the same competition assay devised by Shrader and Crothers. The affinity of the altered sequences differed at most by 2.5-fold from the parent 5S rDNA sequence and had almost no effect except when placed about two helical turns from the dyad. A conceptually similar test was performed in yeast using TRP1ARS1-based minichromosomes; insertion of sequences containing relatively long tracts of A or T, such as $A_{15}TATA_{16}$ or A_{34} disrupted an array of positioned nucleosomes, while shorter tracts such as A_5TATA_4 were incorporated into positioned nucleosomes (Shimizu et al., 2000).

As was the case for di- and tri-nucleotide orientational preference in nucleosomes, genome-wide experiments have provided more comprehensive data on the role of poly (dA)-poly(dT) tracts on nucleosome occupancy. The earliest such experiments, performed at low resolution, identified poly(dA)-poly(dT) tracts as being enriched in nucleosome-depleted promoters in budding yeast (Bernstein et al., 2004), and this was confirmed in subsequent high-resolution measurements of nucleosome occupancy (Yuan et al., 2005). Examination of additional structural and sequence parameters, using a larger (complete genomic) dataset of nucleosome positions in *S. cerevisiae*, revealed strong correlation of GC content, free energy of bending, major groove size and other DNA structural features with the stereotypical pattern of low nucleosome occupancy at promoters; all of these properties are closely related to poly(dA)-poly(dT) content (Lee et al., 2007).

A compelling question arising from these various findings was, to what extent could DNA sequence and structure be used to predict nucleosome occupancy and positioning in vivo? Efforts toward constructing computational models to predict nucleosome occupancy were greatly stimulated by the availability of high-throughput datasets. Two early computational models were based on distinct sequence datasets from ~200 individual mononucleosome DNA sequences (Ioshikhes et al., 2006; Segal et al., 2006). Both studies identified dinucleotide probability distributions as being correlated with positioning, and signatures corresponding to positioned nucleosomes were found to be enriched in the yeast genome. The predictive accuracy achieved by the models was low: one model reproduced the trough of nucleosome occupancy present in promoter regions (Fig. 4.14) (although this may have largely been due to exclusion of nucleosomes by poly(dA)-poly(dT), rather than the propensity for AA and TT dinucleotides to rotationally position nucleosomes) and predicted positioned nucleosomes within 30bp for ~77% of those observed experimentally at proximal promoter regions (Ioshikhes et al., 2006), while the other reported a modest improvement over chance (45% vs 32%) of predicting the approximate position (within 35bp) of yeast nucleosomes as determined at high resolution in vivo (Segal et al., 2006). A more stringent analysis of Pearson correlation coefficient between occupancy predicted by these models and that observed in vivo in yeast or *C. elegans*, or by in vitro reconstitution on total yeast genomic DNA (see below) indicated that performance was not significantly better than random ($r < 0.1$) (Stein et al., 2010; Tillo and Hughes, 2009). In a subsequent study, high-throughput sequencing was used to determine nucleosome occupancy in yeast and a new model based on this data was developed that included dinucleotide periodicity as well as preference for 5-mer sequences to be included or excluded from nucleosomes (Field et al., 2008). The new model performed much better at identifying nucleosome centers mapped in yeast, human, and fly, despite the greater GC content of the latter two organisms, indicating that inherent sequence preferences for nucleosome formation were important for nucleosome placement and occupancy in vivo. However, this conclusion was questioned based on methodological concerns (Stein et al., 2010); moreover, the model included parameters based on all possible 5-mers as well as on dinucleotide frequencies, resulting in over 2000 parameters and a corresponding difficulty in assessing what, really, is important in determining nucleosome occupancy. A follow-up analysis showed that most of the features governing predictable nucleosome occupancy could be winnowed to 14 parameters, with the most important being %GC content and short A-tracts (Tillo and Hughes, 2009). Notably, dinucleotide frequency was not a significant contributor to nucleosome occupancy prediction, although it is important to recognize that this is not inconsistent with it being important for rotational positioning.

Other models also supported poly(dA)-poly(dT) sequences acting as nucleosome-excluding elements and thereby governing nucleosome occupancy and positioning. One such study identified sequences that favored or disfavored nucleosome formation, based on data from Yuan et al. (2005), and used *k*-mers from these sequences to train a support vector machine to predict the nucleosome formation potential of a given DNA sequence (Peckham et al., 2007). The authors found that C/G content was most predictive for nucleosome formation, while A/T content and A-tracts were most important in inhibiting nucleosome formation. Despite the exclusion of dinucleotide periodicity, the model performed well in predicting nucleosome occupancy, especially when applied to fragments with strongest nucleosome forming or inhibiting signals. This suggested that DNA sequence information was most important in governing occupancy for a subset of nucleosomes. In support of this idea, a computational model based principally on distinguishing nucleosome-binding from nucleosome-disfavoring DNA sequences (especially poly(dA)-poly(dT) tracts) performed well at predicting nucleosome enrichment or depletion, but did not do well at predicting exact nucleosome positions (Yuan and Liu, 2008). Yet another model based on sequence-dependent flexibility and intrinsic curvature succeeded at recapitulating stereotypical nucleosome occupancy patterns observed in yeast and *Drosophila* (Miele et al., 2008).

The preceding discussion points to poly(dA)-poly(dT) tracts playing an important role in nucleosome occupancy through their ability to disfavor nucleosome formation. However, most of the supporting evidence for this derives from studies using or modeled on data from *Saccharomyces cerevisiae*, due to its experimental tractability and small genome size. What about other genomes? Enrichment of GC content in nucleosomes, and depletion of AT content, have also been observed in genome-wide mapping of nucleosomes from *C. elegans* and correspondingly, nucleosome occupancy was predicted similarly well in yeast and worms by models based mostly or entirely on nucleosome-favoring and disfavoring sequences (Tillo and Hughes, 2009). A-tracts are, however, less prevalent in NDRs from human and flies (Mavrich et al., 2008b; Radman-Livaja and Rando, 2010; Valouev et al., 2008, 2011), although Valouev et al. used data from in vivo nucleosome mapping and from nucleosomes reconstituted onto human genomic DNA to identify a subset of strongly positioned nucleosomes characterized by high GC content at the nucleosome dyad and AT content at sites flanking the nucleosome, and termed these "container sites" (Valouev et al., 2011). Even among closely related (<50 Myr) yeast species, the extent to which A-tracts contribute to NDRs was found to vary (Tsankov et al., 2010), while in the distantly related (>300 Myr) *S. pombe*, A-tracts were slightly *depleted* relative to non-NDR regions and conversely were enriched in nucleosomal sequences (Lantermann et al., 2010; Moyle-Heyrman et al., 2013).

Taken together, these various results indicate that poly(dA)-poly(dT) tracts are capable of acting as nucleosome exclusion elements, but contribute to nucleosome occupancy to a limited and variable extent in vivo, with most of the supporting evidence deriving from studies using budding yeast. The organismal variability of (polydA)-(polydT) tracts as nucleosome excluding elements suggests that some yeast-specific factor may contribute to their enrichment in NDRs in *S. cerevisiae*; we will return to this idea in the discussion of chromatin remodelers later in this chapter. But what about the structural considerations that led to the idea of these sequences being refractory to nucleosome formation in the first place? Initial studies that investigated incorporation of homopolymeric sequences—both poly(dA)-poly (dT) and poly(dG)-poly(dC) tracts—into nucleosomes were motivated by structural variability observed in the first X-ray crystal structures of DNA oligonucleotides, as mentioned earlier, together with observations that a region near the origin of replication of the SV40 minichromosome possessed several G- or C-tracts and was accessible to restriction endonucleases, suggesting the absence of nucleosomes (Reddy et al., 1978; Simpson and Kunzler, 1979; Varshavsky et al., 1978). These observations led to the idea that structural features of specific DNA sequences could govern their affinity for forming nucleosomes. Many of the subsequent papers addressing sequence dependence of nucleosome formation refer to A-tracts as being "stiff" or resistant to bending, but this is not borne out by detailed studies of the mechanical properties and dynamics of sequences containing AA dinucleotides (discussed in Segal and Widom (2009a)). However, A-tracts do possess unusual structural features (for example, a shorter helical repeat and the narrow minor groove that has been mentioned earlier), and these are length-dependent, consistent with the increasing nucleosome depletion observed in yeast with increasing A-tract length (Field et al., 2008; Raveh-Sadka et al., 2012; Segal and Widom, 2009a). A recent study employed a high-throughput cyclization assay to examine the contribution of DNA bendability to nucleosome occupancy apart from sequence considerations (Basu et al., 2021). Sequences most able to accommodate the mechanical deformation associated with their cyclization when included in a 50bp segment were enriched near nucleosome dyads in yeast, while those having low bendability were more characteristic of linker DNA. This segregation was not, however, due solely to intrinsic histone-DNA interactions, but also reflected inhibition of the ability of the chromatin remodeler INO80 to move nucleosomes to incorporate sequences most resistant to mechanical deformation. It seems plausible that although structural features of poly(dA)-poly(dT) tracts increase the energetic cost of the distortion needed for incorporation into nucleosomes, this cost is not so high as to prevent their incorporation entirely. Furthermore, different organisms have apparently made, during the course of evolution, different adjustments to override or accentuate the tendency of these tracts to be incorporated into nucleosomes, and have employed other mechanisms to set patterns of nucleosome positioning that have been revealed in genome-wide determinations.

Cis vs *trans* determinants of nucleosome positioning and occupancy

We have seen that specific DNA sequences are capable of adopting preferred rotational, and to some extent translational, positioning when incorporated into nucleosomes, and that such sequences can influence nucleosome positioning in vivo. In addition, certain sequence signatures, particularly A-tracts that function as nucleosome disfavoring elements in some contexts and dinucleotides with wide minor groove facing outward or narrow minor groove facing inward, are evident in large scale datasets of in vivo nucleosome positions. However, we have also seen that sequences that are incorporated into positioned nucleosomes in vitro may or may not be similarly incorporated in vivo, depending on context (Thoma, 1986; Thoma and Simpson, 1985). Several ensuing studies provided additional evidence against DNA sequence playing a dominant role in nucleosome positioning in vivo:

- Placement of the *S. cerevisiae URA3* gene into the fission yeast *Schizosaccharomyces pombe*, or of the *S. pombe ade6* gene into *S. cerevisiae*, resulted in nucleosome positions that differed from their native context (Bernardi et al., 1992).
- A 100bp sequence containing repeated strong "TG" nucleosome positioning sequences identified by Shrader and Crothers was cloned into the TRP1ARS1 plasmid at locations corresponding to the center or edge of a positioned nucleosome, or in a nuclease sensitive region, and introduced into *S. cerevisiae* (Tanaka et al., 1992). Mapping nucleosomes on the modified TRP1ARS1 minichromosomes by the indirect end-label method showed that in no case did the TG sequence position nucleosomes as it did in vitro.
- Widom and colleagues identified a number of putative nucleosome positioning sequences using a competitive nucleosome assembly protocol to test naturally occurring nucleosomal DNA sequences and random sequences of nucleosomal length (Thastrom et al., 1999). One of the highest affinity sequences, the "601" sequence, gained popularity for its use in assembling nucleosome arrays when cloned in multiple tandem copies (Chapter 3). However, when the 601 sequence was introduced into yeast either within an open reading frame or an intergenic region, nucleosomes were not positioned over the inserted element (Gaykalova et al., 2011; Perales et al., 2011).
- Nucleosome-scanning revealed distinct positioning over the *S. cerevisiae HIS3* coding sequence, compared to its native state, whether inserted into the genome of *S. pombe* or reconstituted into nucleosomes in vitro (Sekinger et al., 2005). In contrast, the low nucleosome occupancy observed at the *HIS3* promoter in vivo is reproduced after in vitro reconstitution and in *S. pombe* (Sekinger et al., 2005) and is accurately captured by a computational model that emphasizes the role of nucleosome-disfavoring sequences (Peckham et al., 2007).

These examples demonstrate that factors present in vivo can override nucleosome positioning information inherent in DNA sequences. However, being anecdotal in nature, they cannot provide general principles of how nucleosome positions and occupancies are governed in vivo. To obtain a more comprehensive picture of the extent to which *cis*-acting elements—that is, DNA sequence elements that directly affect histone-DNA sequence preferences—as compared to *trans*-acting factors present in living cells contribute to nucleosome positioning and occupancy, broader approaches are required. One such approach, as discussed in the preceding section, is to test how well computational models based on sequence features alone can reproduce in vitro or in vivo data on nucleosome positioning and occupancy; as we have seen, they perform moderately well in predicting occupancy in yeast and *C. elegans*, but poorly in predicting translational positions. An alternative approach to address this question was taken by asking how closely nucleosome reconstitution onto genomic DNA could recapitulate positioning and occupancy observed in vivo. This approach was taken independently by two groups, who used somewhat different methods and obtained results that at first appeared diametrically opposed, with one concluding that "intrinsic DNA sequence preferences of nucleosomes have a central role in determining the organization of nucleosomes *in vivo*" (Kaplan et al., 2009) and the other reporting in their title that "Intrinsic histone-DNA interactions are not the major determinant of nucleosome positions *in vivo*" (Zhang et al., 2009). Both reports utilized salt dialysis (see Chapter 1 Sidebar: Reconstituting nucleosomes) to reconstitute chromatin using purified yeast genomic DNA, sequenced isolated mononucleosomal DNA following MNase digestion, and mapped these fragments onto the genome. Both studies observed moderate correlation between in vitro and in vivo maps, with Pearson correlation coefficients of 0.74 and 0.54, respectively. Differences in experimental methods likely explained some of the differences between the two studies; in addition, questions were raised regarding the statistical methods employed by Kaplan et al. (2009), and additional confusion was caused by a lack of clarity in distinguishing nucleosome occupancy from nucleosome positioning (Kaplan et al., 2010; Pugh, 2010; Segal, 2008; Stein et al., 2010; Zhang et al., 2010). The net outcome was that in vitro reconstitution recapitulated, albeit weakly, low nucleosome occupancy at promoters that contain poly(dA)-poly(dT) tracts and also captured strong rotational positioning, but translational positioning correlated poorly with that found in vivo.

At this point it is clear that a host of experiments, high throughput and otherwise, indicate that factors operating in vivo are important in determining nucleosome occupancy and positioning in living cells. What are these factors, and how can their influence be quantified? The latter question was approached by combining high-throughput methods with genetic analysis. We have mentioned experiments in which nucleosome positioning was compared between in vitro reconstituted templates and in vivo contexts, or between the same DNA sequences in heterologous hosts. Tirosh and Barkai scaled up this approach to a genome-wide level, employing the closely related (~15 Myr) yeast species *S. cerevisiae* and *S. paradoxus* (Tirosh et al., 2010a). The genome sequence of *S. cerevisiae* had been reported in 1996 (Goffeau et al., 1996), while draft sequences of several close relatives in the Hemiascomycota family were reported in 2003, allowing for the first time the possibility of performing comparative genomics on a genome-wide scale (Cliften et al., 2003; Kellis et al., 2003). The genomes of *S. paradoxus* and *S. cerevisiae* proved to be ~85% identical, allowing ready identification of orthologous genes but also permitting sequence tags from these species to be assigned to the correct host. Furthermore, the two species could be mated to form interspecific hybrids containing the complete genomes of each parent. Mapping of nucleosomal sequences in the two parents allowed differences in nucleosome positioning and occupancy to be identified; differences that were maintained in the hybrid (that is, the alleles corresponding to the two parents differed in nucleosome occupancy or position) could be attributed to *cis* effects, whereas differences that were lost in the hybrid, where the two genomes are subject to the same environment, must be due to *trans* effects (Fig. 4.16). About 10% of

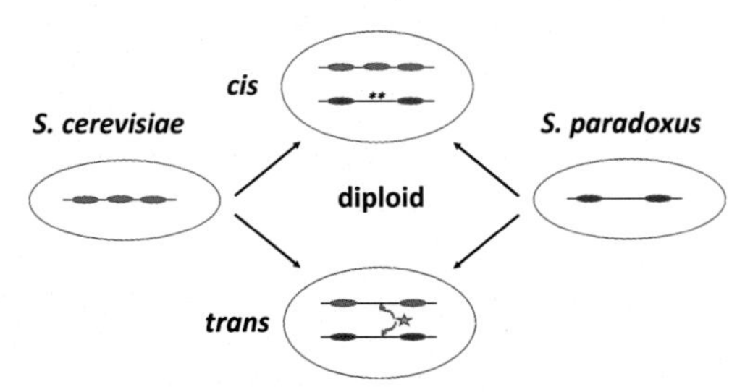

FIG. 4.16 Determination of *cis* vs *trans* effects on nucleosome occupancy. Schematized here are homologous regions of *S. cerevisiae* and *S. paradoxus* genomes, with the central region being occupied by a nucleosome in *S. cerevisiae* but not in *S. paradoxus*. If the difference is due to a *cis* effect, it will be preserved in the diploid cross; if it is due, for example, to a *trans*-acting factor that removes the nucleosome in *S. paradoxus* but is absent or nonfunctional in *S. cerevisiae*, the region will be unoccupied in the diploid, which will express the *trans*-acting factor from the corresponding *S. paradoxus* gene.

mapped nucleosomes exhibited reliable inter-species differences, with ~2000 showing altered occupancy, ~300 showing altered position, and ~170 nucleosomes being lost/gained between species. Of these diverged nucleosomes, about 70% were classified as being due to *cis* effects, with the remaining 30% due to *trans* effects. Differences in nucleosome occupancy, but not positioning, caused by *cis* effects were largely attributed to changes in the presence of A/T tracts, and could be predicted based on the nucleosome favoring or disfavoring properties of 5-mers (Field et al., 2008). In contrast, shifts in nucleosome position were not predicted using the same approach. These results indicate that changes in nucleosome occupancy and position between recently diverged species were mainly due to changes in nucleosome-disfavoring AT-rich sequences. A subsequent investigation extended these findings by mapping nucleosomes across artificial chromosomes introduced into *S. cerevisiae* that contained large segments (averaging ~140 kb) containing sequences from three much more distantly related yeast species (Hughes et al., 2012). Again, nucleosome depletion was preserved in heterologous sequences in a manner dependent on the presence of poly(dA)-poly(dT), while other features of nucleosome positioning were not preserved in the heterologous environment.

A related study mapped nucleosomes in the micronucleus and macronucleus of the ciliated protozoan *Tetrahymena thermophila* (Xiong et al., 2016). *Tetrahymena* cells harbor two copies of their genome, one contained in a transcriptionally active macronucleus and the other in the transcriptionally silent germ-line micronucleus. Unlike the case for the yeast hybrid study, macronuclear and micronuclear genomes cannot, of course, be distinguished by sequence; however, they can be physically isolated and subjected to MNase-seq. Xiong et al. discovered abundant well-positioned nucleosomes in macronuclei downstream of transcriptional start sites, reminiscent of the pattern observed in yeast and *C. elegans*; micronuclei, in contrast, were characterized by delocalized nucleosomes. Nonetheless, nucleosome occupancy was well-correlated between micro- and macronuclei, and *cis*-determinants were found to contribute strongly to nucleosome occupancy based on computational modeling. How then to reconcile these apparently contradictory findings? The authors suggest that the nucleosome distribution in the transcriptionally inactive micronuclei may be a delocalized version of that found in macronuclei, dictated to first approximation by DNA sequence features; the more strongly positioned arrays in micronuclei would result from *trans* effects, principally barrier elements that set origins for the arrays, and remodelers recruited during transcription (see next section).

In principle, *cis* and *trans* contributions to nucleosome positioning and occupancy could be gauged in metazoan cells by comparing nucleosome maps in different cell types or on large DNA segments from heterologous species (McManus et al., 1994). Such an experiment would require a very large number of reads (Gaffney et al., 2012; Teif, 2016), and to my knowledge has not been reported. The expectation, based on the relatively low degree of nucleosome positioning observed in metazoan genomes (other than *C. elegans*) and the considerable extent of *trans* determination of nucleosome positioning and occupancy reported in other experiments, is that DNA sequence is likely to play a relatively minor role in nucleosome positioning and occupancy in metazoan cells.

The major *cis* contributors to nucleosome positioning and occupancy in vivo, as we have seen, are nucleosome-disfavoring sequences that affect nucleosome occupancy and periodically distributed dinucleotides that affect rotational positioning. We next discuss the two major *trans* contributors, DNA-binding proteins and chromatin remodelers.

Factors that contribute to nucleosome organization in *trans*

Nucleosome-positioning proteins

A remarkable early example of a sequence-specific DNA-binding protein impacting nucleosome position derived from studies using the TRP1ARS1 minichromosome system (Roth et al., 1990). This study took advantage of yeast mating-type specific genes to investigate the possible role of chromatin structure in gene regulation. Budding yeast exist in the haploid state as either **a** or **α** cells, which can mate to form diploids. A handful of genes are regulated according to mating type, and genes expressed in **a** cells and not in **α** cells, such as *STE6*, are repressed in **α** cells by a heterodimer of the Mcm1 and α2 proteins that binds to sites in the corresponding promoters (Keleher et al., 1988). The binding site for α2 occurs ~200 bp upstream of the promoter at its repressed target genes, suggesting that repression could occur by a chromatin-mediated mechanism (as opposed to direct occlusion of TBP or Pol II binding, for example). Sharon Roth (now Dent), as a postdoctoral fellow in Bob Simpson's lab at the NIH, sought to test this idea by examining the chromatin structure near α2 binding sites in **a** and **α** cells. Remarkably, mapping nucleosomes on TRP1ARS1 derivatives harboring α2/Mcm1 binding sites from the *STE6* gene by indirect end-labeling revealed an array of strongly positioned nucleosomes in **α** but not **a** cells (Fig. 4.17). Subsequent work demonstrated that

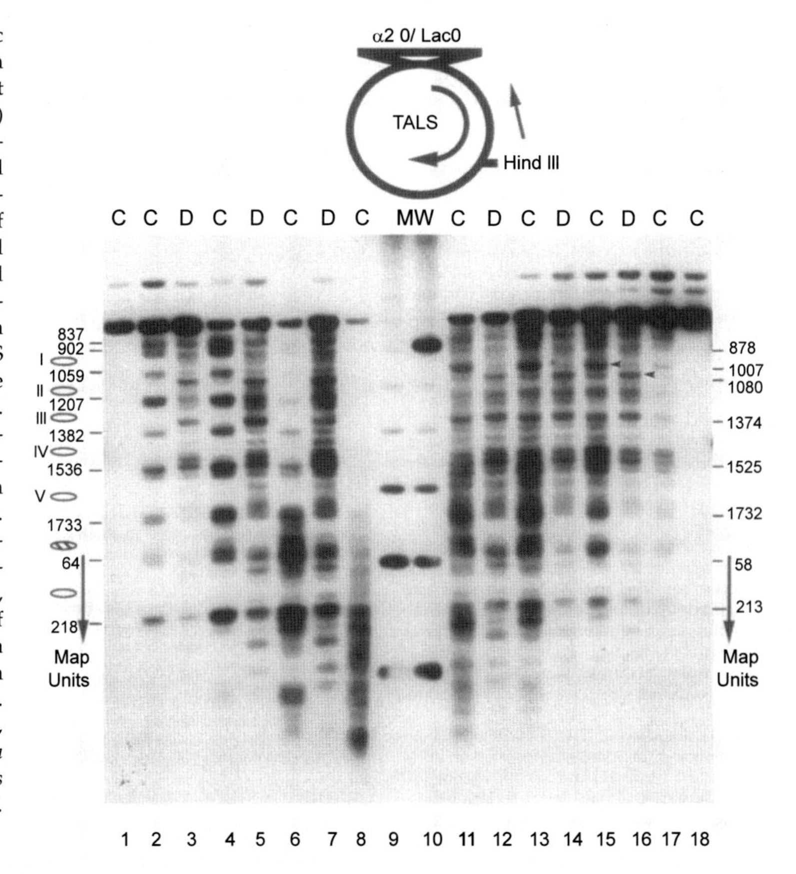

FIG. 4.17 Cell-type specific nucleosome positioning in yeast. Chromatin from yeast **α** cells (left) or **a** cells (right) harboring the TALS minichromosome, and of naked DNA, was digested with increasing concentrations of MNase (toward center) and mapped from the indicated Hind III site by indirect end-labeling. The *curved arrow* in the schematic of the TALS minichromosome indicates the *TRP1* gene (see also Fig. 4.13). Map units on TALS are indicated, and the deduced positions of TALS nucleosomes in **α** cells are shown at the left. The *arrowheads* (lanes 15–16) indicate partial protection in chromatin against MNase digestion, suggesting partial positioning of nucleosomes in this region in **a** cells. The central lanes contain molecular weight standards. *From Roth, S.Y., Dean, A., Simpson, R.T., 1990. Yeast alpha 2 repressor positions nucleosomes in TRP1/ARS1 chromatin. Mol. Cell Biol. 10, 2247–2260.*

positioning conferred by α2/Mcm1 occurred at native promoters repressed by α2, was precise at the base pair level, and required a corepressor complex (Tup1-Cyc8, also known as Tup1-Ssn6) and the amino terminus of histone H4 (Cooper et al., 1994; Ganter et al., 1993; Roth et al., 1992; Shimizu et al., 1991).

Surprisingly, no other example has emerged in which such precise positioning is conferred by a sequence-specific DNA-binding protein. Homologous corepressor complexes in *Drosophila* and in plants interact with nucleosomes and induce compaction of chromatin, but do not appear to confer positioning (Ma et al., 2017; Sekiya and Zaret, 2007). For this reason, perhaps, interesting questions regarding the mechanism of nucleosome positioning by α2/Mcm1 remain unanswered; for example, the structure of the complex together with the Tup1-Cyc8 corepressor complex and a nucleosome might now be approached by cryo-EM, but this has not been reported. Nonetheless, the connection between α2/Mcm1 and nucleosome positioning is firmly established and has provided important insights into transcriptional repression involving chromatin.

Nucleosome-excluding proteins

The idea that sequence-specific DNA-binding proteins could function directly to exclude nucleosomes is clearly connected to the question of how TFs gain access to DNA in chromatin; we will consider this question in more detail in Chapter 8. Here we will focus on the role of DNA-binding proteins in governing nucleosome occupancy without going into detail on the functional ramifications. The creation of nucleosome depleted regions, or NDRs, by such proteins can lead to positioning of nearby nucleosomes; this has best been established in the case of barrier elements at transcription start sites in yeast from which nucleosome arrays are propagated. The mechanisms underlying nucleosome positioning in such arrays, along with chromatin remodelers that collaborate in their propagation, are discussed in the next section.

Prior to observation of widespread NDRs in yeast and at transcriptionally active promoters in metazoans, promoter accessibility had been inferred first from DNase hypersensitivity (Elgin, 1988; Gross and Garrard, 1988) and corroborated by indirect end-labeling analyses of chromatin structure showing extended regions lacking nucleosomes. Promoters identified as containing extended NDRs (over 100 bp in length) included the budding yeast *GAL1-10*, *ADH2*, and *CHA1* promoters, the AreA promoter from *Aspergillus nidulans*, the *Drosophila HSP26* promoter, and the human interferon-β promoter (Agalioti et al., 2000; Lohr and Lopez, 1995; Moreira and Holmberg, 1998; Muro-Pastor et al., 1999; Thomas and Elgin, 1988; Verdone et al., 1996). In most such examples, the NDR included binding sites for the principal activators of the genes in question, making those activators obvious candidates for effecting nucleosome exclusion. However, deletion of the primary activators of the yeast *GAL1-10* and *CHA1* genes did not alter the NDR seen under nonactivating conditions, suggesting that activators might not generally be responsible for creating NDRs (Lohr and Hopper, 1985; Moreira and Holmberg, 1998). (We shall see later, in Chapter 8, that activators do generally play a critical role in the establishment of chromatin structures that are generated specifically in the transcriptionally active state.)

One means of testing whether TFs—especially transcriptional activators, which by the mid-1980s were recognized as often having primacy in gene activation—were capable of altering chromatin structure on their own was to determine whether activators could bind to nucleosomal templates reconstituted in vitro, and whether such binding was compatible with nucleosome structure. These experiments typically involved reconstituting a well-positioned nucleosome using sequences such as the 5S rRNA gene sequence discussed earlier, with a binding site or sites for a particular activator embedded at specific nucleosome locations, and then challenging the nucleosome with the activator. Standard footprinting techniques and native gel electrophoresis were then used to monitor factor binding and nucleosome integrity. Experiments of this kind showed that some TFs were essentially completely inhibited from binding nucleosomal sites (heat shock factor, NF1, c-Myc homodimers) (Archer et al., 1992; Taylor et al., 1991; Wechsler et al., 1994), while others (TFIIIA, Gal4, Sp1, USF, and Myc-Max heterodimers) formed ternary complexes with nucleosomes rather than displacing the histones and generally bound with 10- to 1000-fold reduced affinity (Chen et al., 1994; Hayes and Wolffe, 1992; Taylor et al., 1991; Wechsler et al., 1994). Displacement of histones from ternary complexes of nucleosomes and Gal4, USF, and Sp1 did occur in the presence of nonspecific competitor DNA or nucleoplasmin, which had been identified as a histone chaperone in Xenopus extracts by virtue of its ability to prevent histone aggregation during nucleosome assembly (Earnshaw et al., 1980; Laskey et al., 1978). This suggested that at least some TFs might be capable of binding to nucleosomal sites with concomitant eviction of histones in vivo (Chen et al., 1994; Owen-Hughes and Workman, 1996; Workman and Kingston, 1992).

In vitro reconstitution experiments are limited in their ability to test mechanisms occurring in the complex milieu of cell nuclei. To test whether TFs could occupy nucleosomal binding sites and possibly outcompete histones for binding to DNA in vivo, the TRP1ARS1 minichromosome again provided a tractable system. To set up a direct competition between nucleosome positioning signals and TFs, a tRNA gene was inserted into TRP1ARS1 derivatives such that the tRNA transcription start site was predicted to be incorporated into a positioned nucleosome, using either a "native" TRP1ARS1 nucleosome or a nucleosome positioned by the α2/Mcm1 complex as described above. Northern analysis showed that the constructions allowed tRNA expression, while indirect end-label analysis following MNase digestion revealed disruption of nucleosome positioning (Morse et al., 1992). As a control, a 2bp mutation was introduced that inactivated the tRNA gene; this restored the predicted positioned nucleosome. These results showed that TFs could outcompete histones for binding to DNA in vivo, but did not determine which TFs were responsible or the mechanism by which histone displacement occurred. A similar strategy was used to demonstrate that the transcriptional activator Gal4 could disrupt nucleosome positioning via its binding site, whereas Gcn4, when expressed at normal endogenous levels, could not (Morse, 1993; Yu and Morse, 1999). In a related study, deletion of a strong binding site for heat shock factor (HSF) from the yeast *HSP82* promoter that is normally present in an NDR resulted in a nucleosome occupying the NDR (Gross et al., 1993). This nucleosome incorporated two low-affinity binding sites for HSF, and overexpression of HSF restored the NDR in the mutated promoter. A contrasting example was provided by another yeast transcriptional activator, Pho4, which could occupy a site in the linker DNA between two nucleosomes in the *PHO5* promoter, but whose binding was inhibited at a nucleosomal site (Fascher et al., 1993; Venter et al., 1994). Taken together, these various experiments conducted using budding yeast showed that when expressed at their normal levels, some TFs are capable of disrupting nucleosome positioning (tRNA transcription factors, Gal4, HSF) while others (Gcn4, Pho4) were not; that in at least one case, a TF (HSF) was indeed responsible for excluding nucleosomes (at the *HSP82* promoter); but that in some cases, NDR formation did not require the associated primary activators (*CHA1* and *GAL1-10* promoters).

In a more complex example, multiple TFs were found to be responsible for an extended NDR of ~300 bp at the *CLN2* promoter (Bai et al., 2011). This promoter lacks any long poly (dA)-poly(dT) tracts, and in vitro reconstitution did not indicate it to be refractory to nucleosome formation. Rather, binding sites for multiple TFs present in the promoter were found to contribute redundantly to NDR formation. This conclusion was arrived at by placing various segments of the *CLN2* promoter into a heterologous locus in yeast cells and performing nucleosome scanning assays. Various *CLN2* segments were sufficient to allow NDR formation, and only mutation of multiple TF binding sites permitted nucleosome occupancy of the corresponding segments. In particular, Reb1, Mcm1, and Rsc3 were sufficient to induce NDR formation, while the TFs SBF (a heterodimer of Swi4 and Swi6) and Rme1 that bind to the *CLN2* promoter and contribute to its expression were not. Rsc3 is a subunit of the chromatin remodeling complex RSC (*R*emodels the *S*tructure of *C*hromatin) and is discussed further in the next section, while Reb1 and Mcm1, together with Abf1 and Rap1, belong to the family of General Regulatory Factors, or GRFs, found in yeast (Fourel et al., 2002). This family comprises a set of essential proteins that participate in disparate genetic functions, including transcriptional activation and silencing, establishment of telomere structure, replication, recombination, and repair (Fourel et al., 2002; Morse, 2000; Shore, 1994). Early studies identified binding sites for GRFs at many individual promoters, including many constitutively active "housekeeping" genes such as those encoding ribosomal proteins. GRFs were found to be important for full transcriptional activity, and in some cases were shown to synergize with other DNA-binding TFs for transcriptional activation, in studies that focused on individual genes (Chasman et al., 1990; de Winde and Grivell, 1992; Della Seta et al., 1990; Devlin et al., 1991). These findings were later extended by genome-wide investigations (Badis et al., 2008; Bussemaker et al., 2001; Lieb et al., 2001; Miyake et al., 2004; Pilpel et al., 2001; Yarragudi et al., 2007).

Abundant evidence has been found for GRFs playing an important role in the establishment of NDRs in yeast. Some of the first evidence for this derived from a study employing the yeast minichromosome system that identified a short sequence in the *GAL1-10* promoter responsible for the NDR and flanking nucleosome arrays (Fedor et al., 1988). Subsequent work identified Reb1 (referred to as GRF2 in early work) as binding to this sequence, suggesting that it could be responsible for creating the NDR (Chasman et al., 1990). An Abf1 binding site near the *ARS1* replication origin was found to act as a barrier element, preventing nucleosome incursion into a nearby region of *ARS1* (Venditti et al., 1994), while another study suggested that Rap1 could function to open chromatin at the *HIS4* promoter, as mutation of the Rap1 binding site caused a decrease in MNase sensitivity (Devlin et al., 1991). Rap1 and Abf1 were shown directly to prevent the formation of a positioned nucleosome at their binding sites in experiments analogous to those testing the ability of Gal4 and Gcn4 to outcompete the histones for binding, using the TRP1ARS1 minichromosome system (Yarragudi et al., 2004; Yu and Morse, 1999). In both cases, strong binding sites allowed MNase cleavages in the vicinity of the binding sites nearly identical to those seen using a naked DNA template, while 2 bp mutations that abrogated Rap1 or Abf1 binding allowed nucleosome positioning, as ascertained by protection against MNase digestion, over the inserted sequences.

Widespread promoter binding of Reb1, Rap1, and Abf1 was demonstrated in early ChIP-chip (i.e., microarray) experiments (Harbison et al., 2004; Lieb et al., 2001), and the binding motif for Rap1 was found to be enriched in NDRs in one of the first low-resolution microarray determinations of nucleosome occupancy (Bernstein et al., 2004). The latter observation, although consistent with Rap1 contributing to NDRs, showed only correlation and not causation. Direct roles for Reb1, Rap1, and Abf1 in NDR formation on a genome-wide scale was established by using temperature sensitive (*ts*) mutants that abrogated binding at the restrictive temperature of 37°C or degron mutants that allowed controlled degradation of Reb1 and Abf1 (Badis et al., 2008; Ganapathi et al., 2011; Hartley and Madhani, 2009). Increased nucleosome occupancy was observed upon inactivation or depletion of each factor at the factor's respective binding sites, demonstrated by use of high-resolution microarrays, thereby demonstrating that these three GRFs act to exclude nucleosomes from their binding sites.

Two recent investigations performed high-throughput assays of the nucleosome-excluding capability of a large number of yeast TFs (Levo et al., 2017; Yan et al., 2018). Levo et al. integrated into the yeast genome variants of two sequences, one (derived from the *HIS3* promoter) possessing an NDR (referred to as the "open" configuration) and the second deriving from the portion of the *GAL1-10* promoter that is normally part of the nucleosome array that flanks the NDR (referred to as the "closed" configuration) (Fig. 4.18). Nucleosome occupancy was monitored by determining the extent of methylation of cytosine residues within CpG sequences by the SssI methylase; the results showed that the *HIS3* sequence

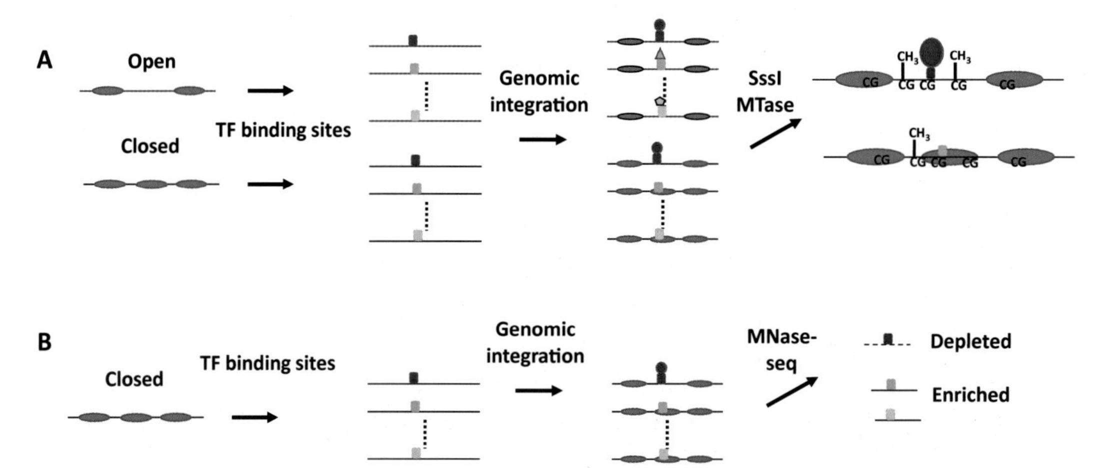

FIG. 4.18 High-throughput analysis of TF nucleosome competition. (A) Levo et al. (2017) introduced TF binding sites into two sequences shown to have open and closed chromatin configurations, and integrated the resulting ~1600 constructs upstream of a YFP reporter gene. Nucleosome occupancy was assessed by measuring accessibility to the SssI methylase. In the example shown, the TF represented by a *red oval* is able to displace a nucleosome, while the *green triangle and gold pentagon* are not. Note that the TF itself may be able to occlude a CpG site from methylation. (B) Yan et al. (2018) used a different sequence that forms "closed" chromatin in vivo as a template for construction of ~17,000 variants containing various combinations of TF binding sites and introduced into yeast *en masse*. MNase digestion of the pooled library (and of naked genomic DNA as control) followed by amplicon sequencing resulted in sequences depleted of nucleosomes by TF binding being depleted and those occupied by nucleosomes being relatively enriched.

retained its NDR and the *GAL1-10* sequence was incorporated into nucleosomes—the former was methylated over a region of ~150bp, while the latter showed protection against methylation. Binding sites for 12 different TFs or TF complexes were then introduced into these sequences within the NDR or the corresponding nucleosomal region and accessibility to SssI methylation again determined. Remarkably, introduction of binding sites for Rap1, Abf1, and Reb1 resulted in large decreases in nucleosome occupancy, as did Gal4 binding sites under activating conditions (i.e., in galactose medium). In contrast, several non-GRF TFs, including Gcn4, were unable to perturb nucleosome occupancy.

Yan et al. used a modified yeast *HO* promoter as their "background" sequence; similar to the *GAL1-10* intergenic promoter region, the *HO* promoter in its repressed state is characterized by an array of positioned nucleosomes, and modifications were introduced so that this normally cell-cycle activated promoter remained permanently repressed and nucleosomal. Placing binding sites for 104 TFs into a nucleosomal site in this promoter (with mutated binding sites as negative controls), followed by its integration into the yeast genome, allowed these TFs to be tested for nucleosome exclusion capability. A large number (>16,000) of such constructs were pooled prior to integration, and the resulting transformed yeast treated with MNase (after spheroplasting and permeabilizing nuclei). Mononucleosomal DNA was then purified and the region containing the introduced TF bindng sites amplified by PCR, followed by high-throughput sequencing. The relative abundance of the sequences containing various TF sites (normalized to their abundance in the genomic DNA from the pooled transformants) then provided a measure of nucleosome integrity: sequences containing sites for TFs that did not perturb the nucleosome would be abundant, and sequences containing sites for TFs that displaced nucleosomes would be depleted. TFs were categorized into three categories based on the results: the first group comprised TFs that caused nucleosome displacement via a single binding site, the second group could displace nucleosomes but only via multiple binding sites, while the third group could not displace a nucleosome. Consistent with the previous studies that we have discussed, the first category comprised the classical GRFs Abf1, Rap1, and Reb1, and two other TFs (Cbf1, Mcm1) that similarly are abundant, essential, and associated with many promoters, and the origin recognition protein Orc1. The second group consisted of a disparate group of canonical TFs, while the majority of TFs tested—about two-thirds—were unable to effect nucleosome exclusion. Yan et al. also concluded, based on experiments in which TFs were overexpressed or weaker binding sites employed, that the essential characteristics required for nucleosome exclusion were that a factor be present in high concentration and bind to its site with high affinity, consistent with earlier suggestions based on much more limited data (Yu and Morse, 1999).

In metazoans, the relationship between nucleosome occupancy and trans-acting factors has mainly been investigated in the context of open chromatin structure induced upon transcriptional activation, especially during development. Aside from previously mentioned results that showed a correlation between NDRs and the presence of a preinitiation complex (Ozsolak et al., 2007), little is known regarding nucleosome exclusion from promoter proximal regions of constitutively active genes. With respect to inducible genes, early work indicated a special role for the *Drosophila* GAGA factor (GAF), so named for its affinity for $(GA)_n$ sequences, in facilitating transcriptional activation in the context of chromatin of genes induced by heat shock or during development (Biggin and Tjian, 1988; Gilmour et al., 1989; Granok et al., 1995; Lu et al., 1993). Subsequent studies indicated that GAF can interact with the chromatin remodeling complexes NURF and ISWI, discussed in Chapter 6, and disrupt chromatin structure of in vitro assembled templates (Okada and Hirose, 1998; Tsukiyama and Wu, 1995). RNAi-mediated knockdown of GAF in *Drosophila* S2 cells resulted in increased nucleosome occupancy proximal to a subset of GAF-binding sites; thus, GAF appears to exclude nucleosomes near its binding sites via its ability to recruit chromatin remodeling complexes to GAGA elements (Fuda et al., 2015).

In mammalian cells, TFs dubbed "pioneer factors" have been identified as binding and increasing accessibility to local regions of chromatin (Klemm et al., 2019; Mayran and Drouin, 2018; Zaret, 2020; Zaret and Carroll, 2011). However, the effects of pioneer factors in vivo have generally been assessed by monitoring DNase I sensitivity, accessibility to other TFs, or other assays that do not yield direct information on nucleosome occupancy or position. We therefore defer discussion of metazoan pioneer factors until later, when we shall consider them in the context of TF access in Chapter 8. For now, we simply recognize that pioneer factors are strong candidates for altering nucleosome occupancy of transcriptionally induced gene promoters, through mechanisms that are still under active study.

Remodelers and nucleosome positioning

Chromatin remodeling complexes will be discussed in depth in Chapter 6. Here, I briefly summarize their contribution to nucleosome positioning in vivo, and in the next section, I discuss their role in establishing the stereotypical nucleosome arrays that are found at active eukaryotic gene promoters and at other specific genomic locations (Fig. 4.14).

Chromatin remodeling enzymes comprise a family of multisubunit complexes that utilize energy provided by ATP to alter histone-DNA interactions. The prototype chromatin remodeler is the SWI/SNF complex from yeast, which was discovered through its role as a global regulator of transcriptional activation and characterized using genetic and biochemical approaches (Winston and Carlson, 1992). Subsequently, additional remodeling complexes were discovered in yeast and higher eukaryotes, in some cases by identifying proteins homologous to SWI/SNF complex members and in other cases by purifying complexes from extracts by using activity assays. In addition to the SWI/SNF complex, six remodelers are known in yeast: ISW1, ISW2, CHD1, INO80, RSC, and FUN80. Purification of various of these complexes allowed their action on chromatin templates to be assessed in vitro, which revealed that some complexes altered histone-DNA contacts in a fashion that increased accessibility to nucleosomal DNA and could lead to histone eviction, while others could slide the histones along the DNA, thus altering nucleosome positioning.

Early in vivo studies concentrated on the role of chromatin remodeling complexes in altering chromatin structure during transcriptional activation. In contrast to remodelers functioning during activation, Isw2 in yeast was revealed to have a widespread role in transcriptional repression as determined by microarray analysis that showed an increase in transcript levels for many genes in *isw2*△ yeast (Fazzio et al., 2001). Indirect end-label analysis showed that DNase I sensitivity and nucleosome positioning were altered at specific derepressed loci upon deletion of Isw2; at the *POT1* locus, nucleosome positioning was shifted in *isw2*△ yeast from a less to a more energetically favorable position, as judged by competitive nucleosome reconstitution assays (Whitehouse and Tsukiyama, 2006). Additional analysis utilizing mutant sequences indicated that Isw2 promoted incorporation of normally unfavorable dA-dT-rich sequences into nucleosomes; thus, Isw2 overrides nucleosome sequence preferences to incorporate promoter elements into nucleosomes in vivo. Subsequently, analysis of nucleosome positioning using high-resolution microarrays showed that loss of Isw2 led to nucleosome repositioning at about 7% of all promoters; the shift resulted in the +1 nucleosome shifting slightly downstream in these promoters (~15 bp on average), resulting in an expanded NDR (Whitehouse et al., 2007). In addition, ~250 genes showed a shift in nucleosome position at the 3′ end, with the nucleosome again shifting away from the intergenic region, in this case moving into the gene body by a small amount.

The pioneering study of Whitehouse et al., together with other genome-wide reports of nucleosome occupancy discussed earlier, reframed the question of what governs nucleosome positioning in vivo. Rather than focusing on how individual well-positioned nucleosomes were established in cells, investigators turned their efforts to understanding how the stereotypical patterns observed in these early studies were determined.

Barrier elements, nucleosome spacing, and chromatin remodelers

As discussed earlier, genome-wide investigations of multiple organisms, including budding yeast (*S. cerevisiae*) and fission yeast (*S. pombe*), *Drosophila*, *C. elegans*, plants, and *H. sapiens* (Table 4.2), revealed the presence of nucleosome-depleted regions (NDRs), defined by their MNase sensitivity, upstream of the transcription start sites of active genes (in yeast, at inactive genes as well). These NDRs are flanked by well-positioned nucleosomes from which are propagated arrays of 5–10 nucleosomes whose positioning decays with increasing distance from the TSS-proximal nucleosome (Fig. 4.14). These observations, coupled to Kornberg and Stryer's argument that nucleosome arrays would be expected on a purely statistical basis to propagate from well-positioned nucleosomes (Kornberg and Stryer, 1988), led to the idea that the observed nucleosome arrays are likely initiated from a "barrier element" that defines the position of the TSS-proximal, or "+1" nucleosome (Mavrich et al., 2008a). Positions of flanking nucleosomes would then be determined stochastically, possibly in combination with some cellular mechanism for setting the spacing between nucleosomes in the array. This in turn prompted the questions as to what exactly determines the barrier dictating the position of the +1 nucleosome (and presumably the −1 nucleosome, present upstream of the NDR); and what sets the spacing between nucleosomes in the downstream array.

Consistent with the idea of nucleosome arrays being established by a barrier element, arrays similar to those seen at promoters of Pol II-transcribed genes are also seen at other functional elements such as replication origins, CTCF binding sites, and tRNA genes (Albert et al., 2007; Chereji et al., 2017; Eaton et al., 2010; Fu et al., 2008). In each of these examples, the arrays flank a nucleosome-depleted region, suggesting that any protein or complex capable of establishing an NDR could suffice to generate a nucleosome array. However, while the nature of the barrier element appears obvious in these instances—the origin replication complex (ORC), CTCF, and the TFIIIB-TFIIIC complex that recruits RNA polymerase III to tRNA genes, respectively—it is less clear what is responsible for the NDRs present at the promoters of mRNA-encoding genes.

One possibility that occurred to researchers was that the preinitiation complex, comprising the general transcription factors TFIIA, -B, -D (including TBP), -E, -H, Pol II, and the Mediator complex, or some component of it could prevent nucleosome occupancy at the TSS and act as a barrier to establish the NDR. Precise positioning of the +1 nucleosome and a well-defined array can be visualized in plots of genome-wide averages of nucleosome occupancy whether centered on the TSS or the closest (+1) nucleosome, and a high-resolution ChIP-seq study (see

Sidebar: ChIP-seq variations) revealed a tight coupling between the locations of the transcription complex and the +1 nucleosome (Rhee and Pugh, 2012). However, the NDR is present even at the least transcribed genes in yeast, suggesting that factors other than the transcription machinery must be involved in its genesis, at least in *S. cerevisiae* (Gkikopoulos et al., 2011; Mavrich et al., 2008a). A direct test of the role of transcription in the formation of the NDR and positioning of the flanking nucleosomes was performed by Rando, Friedman, and colleagues, who measured nucleosome occupancy by MNase-seq after inactivation of Pol II in *rpb1-1 ts* yeast (Weiner et al., 2010; see Sidebar: Deletions and depletions). This study showed that Pol II inactivation caused a moderate narrowing of the NDR due to incursion of the −1 nucleosome, and a slight downstream shift of the +1 nucleosome (into the coding region) of ~10 bp on average, while two later reports found little or no change in nucleosome organization following depletion of Pol II or TBP from the nucleus (Kubik et al., 2015; Tramantano et al., 2016; Weiner et al., 2010). Thus, the transcription machinery appears not to make a major contribution to NDR formation or organization of the downstream array in yeast. Furthermore, reconstitution of chromatin with yeast genomic DNA and purified remodeling complexes, described below, recapitulates NDR formation (albeit less pronounced) in the absence of the transcription machinery (Krietenstein et al., 2016).

Sidebar: ChIP-seq variations

The discovery in the 1980s of proteins that enhanced transcription in eukaryotes by binding to specific DNA sequences in promoters raised a knotty problem: how to determine the collection of genes under control of a particular activator? Genetic experiments could help to identify the set of genes expressed under specific conditions, such as the *GAL* regulon in yeast (Johnston, 1987), but experiments to determine de novo the cohort activated by a specific DNA-binding activator were technically challenging in this era. Differentially expressed genes could be identified using subtractive cDNA techniques, and promoter sequences, once cloned, could be tested biochemically for their ability to bind and be footprinted by the activator in question. The advent of genome-wide approaches, first microarrays and then high-throughput sequencing, expedited this process greatly by allowing ready identification of differentially expressed transcripts on a genome-wide level, and by providing an avenue to determine all the binding sites, using ChIP-seq, for a particular DNA-binding protein.

Although ChIP-seq has been an enormous advance, it has shortcomings. Averaging signals across the genome can allow determination of binding of a particular factor, or of nucleosomes in the case of MNase-seq, at 10–20 bp resolution relative to specific landmarks, such as a putative recognition sequence or transcription start sites. However, resolution at specific sites is limited, proteins that interact with DNA transiently can be difficult to detect, and DNA-associating proteins may vary in their properties in ways that affect their detection by ChIP. To overcome these issues, variations of ChIP-seq have been introduced. ChIP-exo, and the related method ChIP-nexus, use exonuclease to digest immunoprecipitated DNA up to the point where covalent crosslinking of the target protein to DNA stops the exonuclease. Determination by next generation sequencing of the DNA-ends thus generated permits identification of target binding sites at high resolution and with low background (He et al., 2015; Rhee and Pugh, 2011). ChIP-exo relies on the same cross-linking and immunoprecipitation protocols as standard ChIP; two methods that employ alternative means of target detection are Chromatin Endogenous Cleavage followed by high-throughput sequencing, or ChEC-seq (Schmid et al., 2004; Zentner et al., 2015); and Cleavage Under Targets & Release Using Nuclease, or CUT&RUN (Skene and Henikoff, 2017). In ChEC-seq, the protein of interest is fused to MNase, which is activated by addition of calcium ions to nuclei or permeabilized cells. The stringent dependence of MNase on Ca^{+2} permits expression of the fusion protein without damaging the cell, and the method, in contrast to ChIP, does not require cross-linking, chromatin solubilization, or antibodies. For reasons that are not well understood, ChEC-seq and ChIP-seq appear in some cases to be complementary, with each method allowing detection of specific interactions to which the other is blind (Albert et al., 2019; Grunberg et al., 2016). One limitation of ChEC-seq is that it does not permit targeting of specifically modified proteins, such as phosphorylated or acetylated histones or other proteins. This limitation is overcome in CUT&RUN, in which an antibody, which can recognize an unmodified or modified protein, is coupled to MNase, similarly to ChEC-seq. In CUT&RUN, which was modified from an earlier protocol termed Chromatin ImmunoCleavage (ChIC) (Schmid et al., 2004), unfixed nuclei are coupled to magnetic beads, incubated first with a specific antibody and then with a protein A-MNase fusion, which binds to the antibody. MNase is then activated by addition of Ca^{+2} (Skene and Henikoff, 2017). After a short (seconds to minutes) incubation on ice, MNase is inactivated by chelation and short fragments of protein-antibody bound DNA are released and recovered from the supernatant following centrifugation. Commercially sold kits are now available for these various methods, and they are being adapted for use with very low cell number, and even in single cells.

Sidebar: Deletions and depletions

Many of the experiments described in this chapter involve the removal of a protein of interest from the cell or cell nucleus. The most obvious approach to accomplish this is by use of mutants in which the gene encoding a given protein is deleted or encodes a loss-of-function mutant. Gene deletions are readily created in *S. cerevisiae*, due to its efficient cellular machinery for homologous recombination. Yeast can be transformed with a linear DNA fragment having a selectable marker gene flanked by $\geq$40bp of sequence on either side of the gene to be deleted, yielding transformants with an efficiency of $\sim 10^{-5}$–10^{-6} per μg of DNA. The selectable marker can encode either an antibiotic marker such as KanMX or a gene that compensates for auxotrophy, such as *URA3* in a *ura3* yeast strain with selection on medium lacking uracil. Homologous recombination in mammalian cells is much less efficient, and strategies employing expressed *flp* recombinase have been used to create transgenic mouse gene knockouts, while gene knockdowns using siRNA are also popular, in spite of sometimes yielding only partial depletion. The exploitation of the CRISPR-Cas system, following approaches using engineered zinc finger nucleases and TALENS, is the most recent innovation applied to creation of gene knockouts and mutations in higher eukaryotes.

Deletion mutants suffer two major limitations: they cause chronic, rather than acute, loss of the targeted protein, with possible compensatory effects occurring as a consequence; and they cannot be used for essential genes. Temperature sensitive, or *ts*, mutants, have been a popular means to overcome these limitations since Lee Hartwell's isolation of ~400 yeast *ts* mutants (Hartwell, 1967). *Ts* mutants have their own disadvantages; they can be leaky, or slow to respond to temperature shifts, and the exposure to elevated temperature can induce a stress response with substantial physiological effects, as for example repression of genes encoding ribosomal proteins. Newer methods to circumvent these deficiencies include the use of inducible degron tags, which allow targeted degradation of target proteins, typically in a time span from under an hour to several hours, and "anchor away" technology, which leverages the coupling of moieties from the human FK506 and mTOR proteins in the presence of rapamycin to evict a target protein from the nucleus to the cytoplasm in a matter of minutes; "engineered" approaches to rapid depletion of proteins have also been devised for use in metazoan cells (Haruki et al., 2008; Jaeger and Winter, 2021; Nabet et al., 2018; Nishimura et al., 2009). The use of these methods in budding yeast requires strains engineered for that purpose, with corresponding exigencies. Nonetheless, all of these methods, including permanent gene deletions, have provided much insight into molecular processes occurring in vivo.

In contrast to the situation in yeast, the transcription machinery does appear to contribute to nucleosome depletion near promoters in higher eukaryotes. Metazoans differ from *S. cerevisiae* in that pausing of Pol II, where polymerase accumulates at or shortly downstream of the TSS and productive transcription is limited or absent, is observed at a substantial fraction of genes (Core and Adelman, 2019). Pausing is typically documented by ChIP-seq experiments that show sharp peaks of Pol II just downstream of the TSS, and requires Negative Elongation Factor (NELF), which is present in metazoans but absent in yeast. Correspondingly, depletion of NELF by RNAi alleviates Pol II pausing, and the Adelman lab found that NELF depletion also increased nucleosome occupancy at several genes occupied by paused polymerase in *Drosophila* S2 cells (Gilchrist et al., 2008). In a subsequent genome-wide study, RNAi knockdown of Pol II resulted in increased nucleosome occupancy at genes having paused polymerase, while little change was observed for genes at which pausing was not observed (Gilchrist et al., 2010). Low nucleosome density was also observed between the TSSs of closely separated (<300bp) divergently transcribed genes in human and mouse cells by MNase-seq, and high accessibility in these regions was confirmed by DNase hypersensitivity and FAIRE-seq, in which fragmented DNA sequences that are cross-linked to proteins are separated from unbound fragments by differential partitioning between aqueous and phenol phases (Chen et al., 2016; Core et al., 2014; Scruggs et al., 2015). However, the effect on the NDRs at these sites of Pol II depletion was not examined, and whether the mechanism by which these NDRs are formed is the same as that at paused polymerase sites is not known.

Two main avenues have been taken toward understanding how nucleosome arrays are established at Pol II-transcribed promoters in yeast, the most tractable system for approaching this problem. One approach, exemplified by the study examining the effect of deletion of Isw2 cited earlier (Whitehouse et al., 2007), is to remove candidate factors from the yeast nucleus and determine the effect on genome-wide nucleosome occupancy. The other approach is to attempt to reconstitute this pattern on yeast genomic DNA in vitro. Results from such studies indicate that in yeast, a complex interplay between chromatin remodeling complexes and the general regulatory factors dictates genome-wide patterns of nucleosome occupancy.

We have already noted that inactivation or depletion of yeast GRFs such as Abf1, Reb1, and Rap1 causes increased nucleosome occupancy at their respective binding sites, which occur nearly exclusively in NDRs. However, NDRs are not completely eliminated and are unaffected at many promoters in these experiments, indicating that other factors may contribute to nucleosome depletion (de Jonge et al., 2020; Kubik et al., 2018; Paul et al., 2015). Removal or depletion of chromatin remodeling proteins similarly has shown overlapping or redundant contributions to nucleosome occupancy patterns. RSC is unique among the remodelers found in yeast in being essential and in its association with a specific binding motif, and inactivation of RSC using a temperature sensitive mutant resulted in increased nucleosome occupancy at several hundred promoters (Badis et al., 2008). Subsequent studies showed that depletion of RSC caused a narrowing of the NDR at many genes and loss ("filling in") at others; globally, nucleosome arrays shifted slightly upstream relative to TSSs while the spacing between nucleosomes in the arrays was unaltered (Ganguli et al., 2014; Kubik et al., 2015). NDRs most altered by depletion of RSC were associated with RSC, based on ChIP-seq experiments, and were enriched for G/C rich sequences and A-tracts, consistent with the CGCG motif identified as the core RSC binding site and the stimulation of RSC activity by A-tracts (Badis et al., 2008; Kubik et al., 2018; Lorch et al., 2014). Furthermore, A-tracts exhibit a remarkable asymmetry in yeast that corresponds to the preferred directionality of nucleosome movement conferred by RSC (Fig. 4.19; de Boer and Hughes, 2014; Wu and Li, 2010). This asymmetry favors shifting

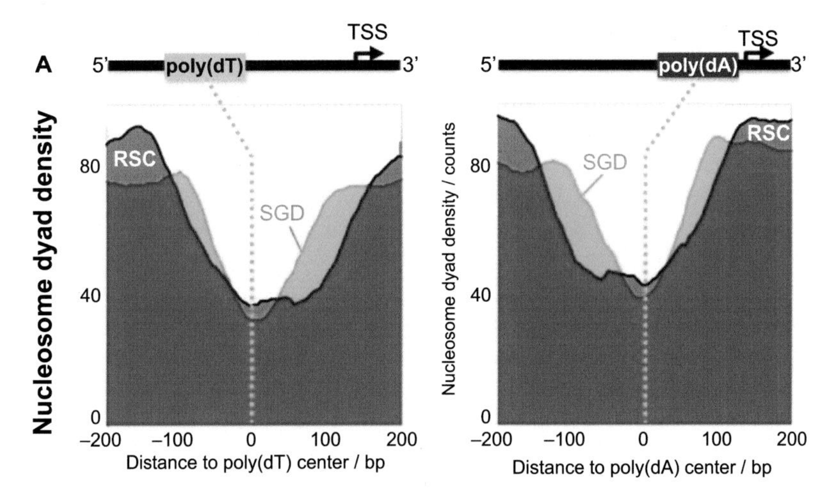

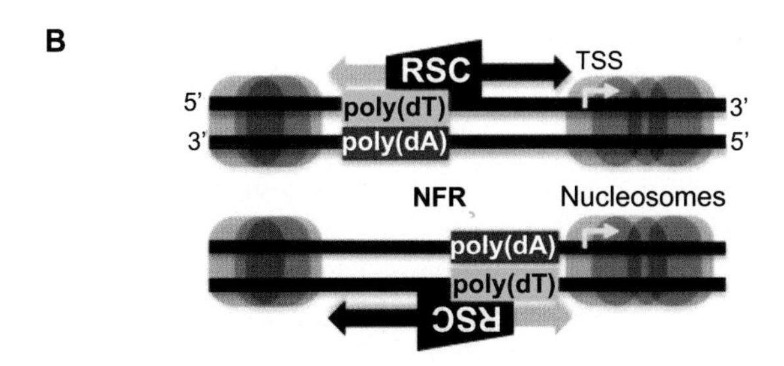

FIG. 4.19 Directional effect of RSC at NDRs. (A) Density of nucleosome centers (dyads) relative to the centers of poly(dT) or poly(dA) sequences, defined as runs ≥6bp on the sense strand and occurring uniquely within NDRs, in yeast genomic DNA reconstituted into chromatin by salt gradient dialysis (SGD) with *(dark green)* or without *(light green)* RSC. Note the greater encroachment of *light green* on the 3′ side of the poly(dT) (or 5′ of poly(dA)) sequences. (B) Schematic illustrating asymmetric action of RSC depending on the orientation of poly(dT) sequences; the stronger effect on moving nucleosomes away from the NDR is depicted as *dark black arrows. From Krietenstein, N., Wal, M., Watanabe, S., Park, B., Peterson, C.L., Pugh, B.F., Korber, P., 2016. Genomic nucleosome organization reconstituted with pure proteins. Cell 167, 709–721.e712.*

of nucleosomes away from the NDR by RSC (Krietenstein et al., 2016; Kubik et al., 2018). Comparison of nucleosome occupancy after depletion of RSC, GRFs (Abf1, Rap1, and Reb1), or both (double depletions of RSC and individual GRFs) revealed promoters depending on RSC, on GRFs, on both and on neither (Kubik et al., 2018). Recalling the case of the *CLN2* promoter, where a wide NDR depended on Reb1, Mcm1, and RSC, it seems likely that NDRs that persist even in doubly depleted yeast are governed by other or additional, redundantly acting factors.

It is worth emphasizing the linkage between RSC and asymmetric poly(dA)-poly(dT) tracts and NDR formation in yeast (for a more extensive discussion, see Barnes and Korber (2021)). We have seen that a large effort over many years explored the possibility that poly(dA)-poly (dT) tracts might act as nucleosome exclusion elements by virtue of their resistance to structural deformation. Some experiments, both in vitro and in vivo, provided confirmatory evidence for this hypothesis, while others raised doubts. In particular, the variability among species with regard to enrichment of such tracts in NDRs (with *S. pombe* being especially contrarian, having NDRs that are depleted of A-tracts) was difficult to reconcile with the idea of any such sequence effect being universal, while the asymmetry of A-tracts in budding yeast promoters also seemed incongruous with this notion (de Boer and Hughes, 2014). The discovery of RSC-dependent directional movement of nucleosomes stimulated by asymmetric A/T-tracts in yeast now appears to have reconciled these incongruities, and while structural contributions of poly(dA)-poly(dT) tracts do indeed appear to play a role in preferential sequence incorporation into nucleosomes, it appears to be, in general, a relatively minor one.

With regard to the remaining yeast remodelers, ISW2 contributes to the position of nucleosomes at 5′ and 3′ ends of a subset of genes, as described above (Whitehouse et al., 2007), while deletion of ISW1 resulted in nucleosomes near the middle of gene coding regions shifting upstream (toward the 5′ NDR) in about half of all genes (Tirosh et al., 2010b). Variable effects on nucleosome organization have been reported upon deletion of individual remodelers in several studies, but *isw1*, *ino80*, and *chd1* mutants consistently showed the largest effects, especially on the downstream nucleosome array (Gkikopoulos et al., 2011; Kubik et al., 2019; Ocampo et al., 2016; van Bakel et al., 2013; Yen et al., 2012). Additive or synergistic effects were seen between some pairs of remodelers, specifically ISW1-CHD1 and ISW2-INO80, indicating that these remodelers function redundantly at some promoters (Gkikopoulos et al., 2011; Kubik et al., 2019; Ocampo et al., 2016). Redundancy was especially marked between ISW1 and CHD1, as *isw1 chd1* mutant yeast exhibited globally disordered nucleosome positioning (Gkikopoulos et al., 2011; Ocampo et al., 2016).

The in vivo studies discussed above indicate that NDR formation is primarily dictated by GRFs and the RSC remodeling complex, while the downstream array is set by the barrier formed at the NDR with spacing and regularity governed by multiple chromatin remodelers. If this is correct, an ambitious but potentially conclusive experiment would be to reconstitute nucleosome arrays in vitro using yeast genomic DNA and appropriate purified remodelers and GRFs. Earlier we discussed experiments in which nucleosomes were reconstituted onto yeast genomic DNA by salt dialysis and analyzed using MNase-seq (Kaplan et al., 2009; Zhang et al., 2009). These reconstitutions did not reproduce the stereotypical nucleosome array pattern observed in vivo, but did result in nucleosome depletion at promoter regions, albeit in a much less pronounced fashion than observed in vivo, whether performed at a 1:1 (near physiological) or 0.4 histone:DNA weight ratio (Kaplan et al., 2009; Zhang et al., 2009, 2011). NDR occupancy was highest at binding sites for Abf1, Reb1, and to lesser extent

Rap1, supporting their role in NDR formation. Nonetheless, these experiments and others recounted earlier that focused on single gene loci indicated that intrinsic histone-DNA preferences could not account for the nucleosome array pattern seen in yeast.

To investigate this issue further, the Korber and Pugh labs conducted reconstitution experiments employing an ATP-dependent assembly system derived from cell-free yeast extracts (Zhang et al., 2011). Early studies indicated that chromatin reconstituted using salt gradient dialysis resulted in close-packed nucleosomes, rather than reproducing physiological spacing (Spadafora et al., 1978; Steinmetz et al., 1978). Extracts from *Xenopus* and *Drosophila* cells were prepared that could facilitate assembly of regularly spaced nucleosomes—nucleosome arrays—in an ATP-dependent fashion (Becker and Wu, 1992; Glikin et al., 1984; Kamakaka et al., 1993; Laskey et al., 1977). Korber, working with the late Wolfram Hörz, developed a similar system based on yeast cell extracts in work aimed at reconstituting nucleosome positioning at the well-studied yeast *PHO5* promoter (Korber and Hörz, 2004). Reconstitution of chromatin by salt dialysis using yeast genomic DNA resulted in modest depletion of nucleosomes from promoter regions, consistent with earlier work, while incubation of the reconstituted chromatin with the yeast whole cell extract and ATP led to a more depleted NDR and conspicuous arrays of nucleosomes having the same spacing observed in vivo but with reduced peak amplitudes (Zhang et al., 2011). If ATP was omitted, arrays were not seen, consistent with the idea that ATP-dependent remodelers with the ability to dictate nucleosome spacing were critical to establishing regular arrays that emanated from NDRs. These conclusions were reinforced and extended in a subsequent study that used purified components to reconstitute yeast genomic chromatin (Krietenstein et al., 2012, 2016). In good agreement with the in vivo experiments discussed above, RSC was most effective in generating an NDR, but did not generate well-positioned −1 and +1 nucleosomes, while including INO80 produced well-positioned +1 and −1 nucleosomes. Curiously, addition of INO80 to reconstitutions that employed an *isw1△ isw2△ chd1△* extract resulted in a detectable downstream nucleosome array, although inclusion of INO80 together with ISW1, ISW2, and RSC did not in the absence of extract; thus, some other component of the extract also must contribute to array formation. The study also showed that inclusion of Abf1 together with RSC led to an enhanced NDR at Abf1-binding promoters, while inclusion of ISW1 or ISW2 both enhanced the NDR and generated a weak downstream nucleosome array. These results, while not entirely concordant with in vivo studies using depletion mutants, generally accord with the idea that multiple remodelers, sometimes in collaboration with GRFs like Abf1, cooperate to generate the stereotypical nucleosome arrays observed in vivo.

Unsurprisingly, chromatin remodeling complexes, some of which are related to those found in yeast, are also present in higher eukaryotes. The role of these complexes in governing nucleosome arrays has begun to be investigated only recently. As in yeast, chromatin remodeling complexes have been reported to be important in determining nucleosome occupancy and the regularity of nucleosome arrays in higher eukaryotes. In one of the first such studies, knockdowns by RNA interference of four chromatin remodeling complexes in *Drosophila* S2 cells revealed that the NURD, (P)BAP, INO80, and ISWI complexes have largely non-overlapping targets, in contrast to the redundancy observed for yeast chromatin remodelers (Moshkin et al., 2012). Sequence features of ISWI targets were relatively favorable for nucleosome formation, while the opposite was true for the targets of the other three remodelers. Nucleosome occupancy following knockdown of the remodelers was monitored using microarrays; rather than performing MNase digestion, the authors performed ChIP against H3 and H2B after sonication of chromatin. Although this approach did not permit analysis of nucleosome positioning, relative occupancy could still be ascertained, and regions of higher and lower occupancy were corroborated using FAIRE-seq. The results showed that knockdown of ISWI resulted in greater nucleosome occupancy at ISWI binding sites, with corresponding decreased accessibility; knockdowns of the other three remodelers showed opposite effect. Thus, ISWI moves nucleosomes away from sites that show preferred occupancy, while NURD, (P)BAP, and INO80 facilitate inclusion of relatively disfavored sequences into nucleosomes.

With regard to nucleosome arrays in higher eukaryotes, remodeling complexes that are able to "slide" the histone octamer relative to DNA in vitro, thereby altering nucleosome positions, have been reported as contributing to spacing and regularity in *Drosophila* and mammalian cells (Baldi et al., 2018a; Barisic et al., 2019; Fyodorov et al., 2004; Scacchetti et al., 2018; Wiechens et al., 2016). Knockdown of SNF2H, which is a component of five distinct remodeling complexes referred to as the ISWI complexes in mammalian cells, resulted in decreased nucleosome positioning around CTCF sites in HeLa cells, while CRISPR-generated SNF2H knockout murine embryonic stem cells exhibited altered nucleosome spacing and recapitulated the dampening of nucleosome arrays surrounding CTCF sites, while leaving arrays around binding sites for the TF REST unaffected (Barisic et al., 2019; Wiechens et al., 2016). Conversely, deletion of BRG1, a component of the mammalian SWI/SNF complex, reduced nucleosome positioning surrounding REST sites but not around CTCF sites. Correspondingly, CTCF association but not REST association was reduced in SNF2H knockout cells, while the converse was true in BRG1 knockout cells. These results indicate specific dependence on chromatin remodelers for binding of these factors and imply that their role in nucleosome array formation is not direct but is dependent on the factors whose binding they facilitate.

In contrast to the roles of SNF2H and BRG1 in contributing to nucleosome array formation by facilitating binding of proteins that establish barriers that set the array origins, Peter Becker's lab has reported two examples in *Drosophila* embryos in which distinct factors contribute to barrier formation and the propagation of regularly spaced nucleosomes from the barriers formed (Baldi et al., 2018a). Although regularly spaced arrays of nucleosomes have been reported at some individual gene loci (Moreira and Holmberg, 1998; Szent-Gyorgyi et al., 1987), more typically they are uncovered only by averaging nucleosome occupancy data over many gene loci. This requires identification of candidate barrier elements, such as transcription start sites or CTCF binding sites, on which to center nucleosome occupancy data. Recognizing this limitation, Baldi et al. used spectral density estimation, which uses Fourier transformation applied to nucleosome occupancy data to identify regions in which a single "frequency" of nucleosome periodicity dominates, to uncover about 10,000 regular nucleosome arrays with median length of four nucleosomes. About half the identified arrays corresponded to those occurring at TSSs; a search for over-represented sequence motifs in the other half identified Suppressor of Hairy Wing, or su(Hw), which like CTCF functions as an insulator and participates in the organization of topologically associated domains (TADs), and a sequence, ATACG, that is a binding site for a previously uncharacterized zinc finger protein that the authors dubbed Phaser. Reconstitution of genomic DNA into chromatin using a preblastoderm embryo extract recapitulated nucleosome arrays surrounding su(Hw) and Phaser binding sites; mutation of the ATACG site resulted in loss of surrounding nucleosome arrays. The authors then sought to determine whether either the ACF or CHRAC complexes, which were known to promote the formation of regularly spaced nucleosomes in reconstituted chromatin, contributed to array regularity. Embryos lacking functional *Acf*, which is critical to both ACF and CHRAC activity, exhibited dampened arrays surrounding both su(Hw) and Phaser sites, but did not eliminate the NDR at the sites themselves or alter arrays surrounding TSSs. Most remarkably, reconstitution of chromatin using extracts prepared from *Acf-* embryos resulted in the formation of positioned nucleosomes on either side of su(Hw) and Phaser binding sites, but essentially eliminated the surrounding arrays. Although the effect of *Acf* loss on binding of su(Hw) and Phaser was not directly tested, the latter finding strongly suggests that su(Hw) and Phaser establish barriers and strongly positioned flanking nucleosomes, while ACF/CHRAC are required for regular spacing of surrounding nucleosomes that results in regular array propagation. ACF/CHRAC are not, however, required for array propagation at active genes, where presumably other remodeling complexes may play this role. Utilization of statistical methods to identify regular nucleosome arrays in vivo in unbiased fashion is likely to be important in future studies, as is the strategic combination of in vivo and reconstitution approaches to dissect the contributors to barrier formation and array propagation.

NDRs and "fragile" nucleosomes

Analysis of nucleosome arrays has also raised questions regarding the nature of NDRs. Earlier in this chapter we discussed the importance of assessing nucleosome occupancy and positioning in experiments employing MNase over a range of digestion extents. Experiments in which genome-wide nucleosome occupancy was assessed after limited and more extensive MNase digestion has revealed that NDRs show protection against limited MNase digestion in *S. cerevisiae*, *S. pombe*, *Drosophila* S2 cells, maize (*Zea mays*), and mouse liver and embryonic stem cells; more extensive MNase digestion resulted in the absence or near absence of fragments mapping to NDRs (Chereji et al., 2016; Ishii et al., 2015; Iwafuchi-Doi et al., 2016; Kubik et al., 2015; Moyle-Heyrman et al., 2013; Vera et al., 2014; Voong et al., 2016). The Henikoff lab first showed that subnucleosomal fragments (i.e., < 120 bp) within NDRs are protected against low levels of MNase digestion, and that in some cases protection could be ascribed to binding of TFs (Henikoff et al., 2011). Further investigation showed that protected regions (at low levels of MNase digestion) of nucleosomal as well as of subnucleosomal length that were lost upon more extensive digestion were bound by histones, indicating the presence of "fragile" nucleosomes (Brahma and Henikoff, 2019; Ishii et al., 2015; Kubik et al., 2015; Voong et al., 2016). Studies using conventional ChIP-seq methods similarly indicated fragile nucleosomes, although some of these studies were compromised by the lack of replicate experiments (Floer et al., 2010; Kubik et al., 2015, 2017; Xi et al., 2011), while evidence for fragile nucleosomes in *S. pombe* and mouse ESCs was obtained by using chemical cleavage via H3S47C-coupled hydroxyl radical generation, as described earlier (Fig. 4.20; Ishii et al., 2015; Moyle-Heyrman et al., 2013; Voong et al., 2016). In mouse ESCs, the sensitivity of fragile nucleosomes appears to be due in part to sequence-dependent preferential digestion of nucleosomal DNA by MNase. Only modest depletion of poly(dA)-poly(dT) tracts in nucleosomal compared to genomic DNA was seen by chemical cleavage ($\sim$10% reduction for A-tracts of 20 bp), whereas a fourfold reduction was seen based on MNase-seq; likewise, poly(dG)-poly(dC) tracts also conferred sensitivity of nucleosomal DNA to MNase but were very little reduced compared to genomic DNA when assessed by chemical cleavage (Voong et al., 2016). These findings suggest that NDRs are, in many cases, actually occupied by "fragile" nucleosomes whose MNase sensitivity is in part determined by the sequence of the incorporated DNA.

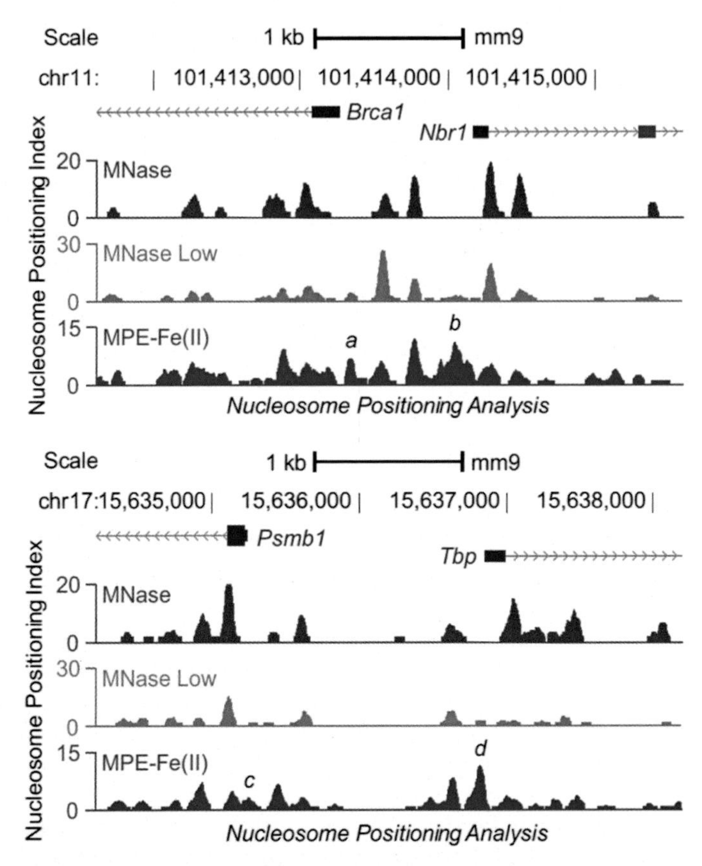

FIG. 4.20 "Fragile" nucleosomes are susceptible to MNase digestion but revealed by chemical mapping. Browser screenshots are shown for nucleosome occupancy at two loci after digestion with MNase under standard conditions, digestion with low concentration of MNase, or digestion with MPE-Fe(II). Note the peaks labeled a–d in the MPE-Fe(II) lanes that are also visible, albeit as low peaks, in the low MNase lanes; these correspond to putative fragile nucleosomes. *From Ishii, H., Kadonaga, J.T., Ren, B., 2015. MPE-seq, a new method for the genome-wide analysis of chromatin structure. Proc. Natl. Acad. Sci. USA 112, E3457.*

Counter to this interpretation, however, several studies reported being unable to detect histones in these regions in budding yeast by ChIP-exo, MNase-ChIP-seq, or H3- or H4-directed chemical cleavage, leading those authors to suggest that nonhistone protein complexes were responsible for the observed MNase resistance (Chereji et al., 2017, 2018; Rhee et al., 2014). Recent work from the Henikoff lab is aimed at resolving this dilemma (Brahma and Henikoff, 2019). First, using CUT&RUN (see Sidebar: ChIP-seq variations), particles were excised by MNase coupled to RSC; when the associated DNA was separated into nucleosomal (121–200 bp) and subnucleosomal (<120 bp) fractions and mapped to the yeast genome following sequencing, the results showed that full-length, nucleosomal DNA associated with RSC was enriched at the −1 and +1 nucleosomes, while subnucleosomal DNA was mainly present at NDRs. Furthermore, combining CUT&RUN with ChIP-seq, particles released by MNase coupled to RSC were found to contain histones and were enriched for the histone variant H2A.Z and for specific histone modifications that are known to be associated with promoter regions (Brahma and Henikoff, 2019). Based on these results and previous work on interactions of RSC with nucleosomes, fragile nucleosomes (at least in yeast) were proposed to be partially unwrapped nucleosomes engaged by RSC, possibly lacking one H2A-H2B dimer (Brahma and Henikoff, 2019, 2020). RSC association could mask histone epitopes or tags used in ChIP experiments, thereby explaining why histone association was not detected in some experiments (Brahma and Henikoff, 2020; Kubik et al., 2017; Lorch et al., 2014). Arguing against this interpretation, however, is the low fraction of RSC-associated nucleosomes recovered in these experiments, and the lack of enrichment for RSC at promoters reported by others (Rawal et al., 2018). Should we still then be referring to these MNase sensitive promoters as nucleosome-depleted regions? In fact, these regions do still appear to be relatively depleted of histones, and certainly of canonical nucleosomes; the term would still seem to apply, although not necessarily in the way originally understood.

The existence of "fragile" nucleosomes implies that not all nucleosomes are equal. Two features that contribute to distinctive properties of individual nucleosomes are histone modifications and the presence of histone variants. We next turn to a discussion of the latter.

References

Agalioti, T., Lomvardas, S., Parekh, B., Yie, J., Maniatis, T., Thanos, D., 2000. Ordered recruitment of chromatin modifying and general transcription factors to the IFN-beta promoter. Cell 103, 667–678 (in process citation).

Albert, I., Mavrich, T.N., Tomsho, L.P., Qi, J., Zanton, S.J., Schuster, S.C., Pugh, B.F., 2007. Translational and rotational settings of H2A.Z nucleosomes across the *Saccharomyces cerevisiae* genome. Nature 446, 572–576.

Albert, B., Tomassetti, S., Gloor, Y., Dilg, D., Mattarocci, S., Kubik, S., Hafner, L., Shore, D., 2019. Sfp1 regulates transcriptional networks driving cell growth and division through multiple promoter-binding modes. Genes Dev. 33, 288–293.

Almer, A., Horz, W., 1986. Nuclease hypersensitive regions with adjacent positioned nucleosomes mark the gene boundaries of the PHO5/PHO3 locus in yeast. EMBO J. 5, 2681–2687.

Almer, A., Rudolph, H., Hinnen, A., Horz, W., 1986. Removal of positioned nucleosomes from the yeast PHO5 promoter upon PHO5 induction releases additional upstream activating DNA elements. EMBO J. 5, 2689–2696.

An, W., van Holde, K., Zlatanova, J., 1998. Linker histone protection of chromatosomes reconstituted on 5S rDNA from *Xenopus borealis*: a reinvestigation. Nucleic Acids Res. 26, 4042–4046.

Andrews, A.J., Luger, K., 2011. A coupled equilibrium approach to study nucleosome thermodynamics. Methods Enzymol. 488, 265–285.

Andrews, A.J., Chen, X., Zevin, A., Stargell, L.A., Luger, K., 2010. The histone chaperone Nap1 promotes nucleosome assembly by eliminating nonnucleosomal histone DNA interactions. Mol. Cell 37, 834–842.

Angermayr, M., Bandlow, W., 2003. Permanent nucleosome exclusion from the Gal4p-inducible yeast GCY1 promoter. J. Biol. Chem. 278, 11026–11031.

Angermayr, M., Oechsner, U., Gregor, K., Schroth, G.P., Bandlow, W., 2002. Transcription initiation in vivo without classical transactivators: DNA kinks flanking the core promoter of the housekeeping yeast adenylate kinase gene, AKY2, position nucleosomes and constitutively activate transcription. Nucleic Acids Res. 30, 4199–4207.

Annunziato, A.T., Seale, R.L., 1983. Chromatin replication, reconstitution and assembly. Mol. Cell. Biochem. 55, 99–112.

Arceci, R.J., Gross, P.R., 1980. Histone variants and chromatin structure during sea urchin development. Dev. Biol. 80, 186–209.
Archer, T.K., Lefebvre, P., Wolford, R.G., Hager, G.L., 1992. Transcription factor loading on the MMTV promoter: a bimodal mechanism for promoter activation. Science 255, 1573–1576.
Axel, R., Melchior Jr., W., Sollner-Webb, B., Felsenfeld, G., 1974. Specific sites of interaction between histones and DNA in chromatin. Proc. Natl. Acad. Sci. USA 71, 4101–4105.
Axelrod, J.D., Majors, J., 1989. An improved method for photofootprinting yeast genes in vivo using Taq polymerase. Nucleic Acids Res. 17, 171–183.
Badis, G., Chan, E.T., van Bakel, H., Pena-Castillo, L., Tillo, D., Tsui, K., Carlson, C.D., Gossett, A.J., Hasinoff, M.J., Warren, C.L., et al., 2008. A library of yeast transcription factor motifs reveals a widespread function for Rsc3 in targeting nucleosome exclusion at promoters. Mol. Cell 32, 878–887.
Bai, L., Ondracka, A., Cross, F.R., 2011. Multiple sequence-specific factors generate the nucleosome-depleted region on CLN2 promoter. Mol. Cell 42, 465–476.
Baldi, S., Jain, D.S., Harpprecht, L., Zabel, A., Scheibe, M., Butter, F., Straub, T., Becker, P.B., 2018a. Genome-wide rules of nucleosome phasing in Drosophila. Mol. Cell 72, 661–672.e664.
Baldi, S., Krebs, S., Blum, H., Becker, P.B., 2018b. Genome-wide measurement of local nucleosome array regularity and spacing by nanopore sequencing. Nat. Struct. Mol. Biol. 25, 894–901.
Baldi, S., Korber, P., Becker, P.B., 2020. Beads on a string-nucleosome array arrangements and folding of the chromatin fiber. Nat. Struct. Mol. Biol. 27, 109–118.
Bao, Y.H., White, C.L., Luger, K., 2006. Nucleosome core particles containing a poly(dA center dot dT) sequence element exhibit a locally distorted DNA structure. J. Mol. Biol. 361, 617–624.
Barisic, D., Stadler, M.B., Iurlaro, M., Schubeler, D., 2019. Mammalian ISWI and SWI/SNF selectively mediate binding of distinct transcription factors. Nature 569, 136–140.
Barnes, T., Korber, P., 2021. The active mechanism of nucleosome depletion by poly(dA:dT) tracts in vivo. Int. J. Mol. Sci. 22, 8233.
Barski, A., Cuddapah, S., Cui, K., Roh, T.Y., Schones, D.E., Wang, Z., Wei, G., Chepelev, I., Zhao, K., 2007. High-resolution profiling of histone methylations in the human genome. Cell 129, 823–837.
Bashkin, J., Hayes, J.J., Tullius, T.D., Wolffe, A.P., 1993. Structure of DNA in a nucleosome core at high salt concentration and at high temperature. Biochemistry 32, 1895–1898.
Basu, A., Bobrovnikov, D.G., Qureshi, Z., Kayikcioglu, T., Ngo, T.T.M., Ranjan, A., Eustermann, S., Cieza, B., Morgan, M.T., Hejna, M., et al., 2021. Measuring DNA mechanics on the genome scale. Nature 589, 462–467.
Becker, P.B., Wu, C., 1992. Cell-free system for assembly of transcriptionally repressed chromatin from Drosophila embryos. Mol. Cell. Biol. 12, 2241–2249.
Benezra, R., Cantor, C.R., Axel, R., 1986. Nucleosomes are phased along the mouse beta-major globin gene in erythroid and nonerythroid cells. Cell 44, 697–704.
Bergman, L.W., Kramer, R.A., 1983. Modulation of chromatin structure associated with derepression of the acid phosphatase gene of *Saccharomyces cerevisiae*. J. Biol. Chem. 258, 7223–7227.
Bernardi, F., Koller, T., Thoma, F., 1991. The ade6 gene of the fission yeast Schizosaccharomyces pombe has the same chromatin structure in the chromosome and in plasmids. Yeast 7, 54.
Bernardi, F., Zatchej, M., Thoma, F., 1992. Species specific protein–DNA interactions may determine the chromatin units of genes in *S. cerevisiae* and in *S. pombe*. EMBO J. 11, 1177–1185.
Bernstein, B.E., Liu, C.L., Humphrey, E.L., Perlstein, E.O., Schreiber, S.L., 2004. Global nucleosome occupancy in yeast. Genome Biol. 5, R62.
Biggin, M.D., Tjian, R., 1988. Transcription factors that activate the Ultrabithorax promoter in developmentally staged extracts. Cell 53, 699–711.
Bloom, K.S., Carbon, J., 1982. Yeast centromere DNA is in a unique and highly ordered structure in chromosomes and small circular minichromosomes. Cell 29, 305–317.
Boyle, A.P., Davis, S., Shulha, H.P., Meltzer, P., Margulies, E.H., Weng, Z., Furey, T.S., Crawford, G.E., 2008. High-resolution mapping and characterization of open chromatin across the genome. Cell 132, 311–322.
Brahma, S., Henikoff, S., 2019. RSC-associated subnucleosomes define MNase-sensitive promoters in yeast. Mol. Cell 73, 238–249.e233.
Brahma, S., Henikoff, S., 2020. Epigenome regulation by dynamic nucleosome unwrapping. Trends Biochem. Sci. 45, 13–26.
Bresnick, E.H., Rories, C., Hager, G.L., 1992. Evidence that nucleosomes on the mouse mammary tumor virus promoter adopt specific translational positions. Nucleic Acids Res. 20, 865–870.

Brogaard, K., Xi, L., Wang, J.P., Widom, J., 2012. A map of nucleosome positions in yeast at base-pair resolution. Nature 486, 496–501.

Buenrostro, J.D., Giresi, P.G., Zaba, L.C., Chang, H.Y., Greenleaf, W.J., 2013. Transposition of native chromatin for fast and sensitive epigenomic profiling of open chromatin, DNA-binding proteins and nucleosome position. Nat. Methods 10, 1213–1218.

Bussemaker, H.J., Li, H., Siggia, E.D., 2001. Regulatory element detection using correlation with expression. Nat. Genet. 27, 167–171.

Buttinelli, M., Di Mauro, E., Negri, R., 1993. Multiple nucleosome positioning with unique rotational setting for the *Saccharomyces cerevisiae* 5S rRNA gene in vitro and in vivo. Proc. Natl. Acad. Sci. USA 90, 9315–9319.

Cartwright, I.L., Hertzberg, R.P., Dervan, P.B., Elgin, S.C., 1983. Cleavage of chromatin with methidiumpropyl-EDTA.iron(II). Proc. Natl. Acad. Sci. USA 80, 3213–3217.

Chambers, S.A., Vaughn, J.P., Shaw, B.R., 1983. Shortest nucleosomal repeat lengths during sea urchin development are found in two-cell embryos. Biochemistry 22, 5626–5631.

Chao, M.V., Gralla, J., Martinson, H.G., 1979. DNA sequence directs placement of histone cores on restriction fragments during nucleosome formation. Biochemistry 18, 1068–1074.

Chasman, D.I., Lue, N.F., Buchman, A.R., LaPointe, J.W., Lorch, Y., Kornberg, R.D., 1990. A yeast protein that influences the chromatin structure of UASG and functions as a powerful auxiliary gene activator. Genes Dev. 4, 503–514.

Chen, H., Li, B., Workman, J.L., 1994. A histone-binding protein, nucleoplasmin, stimulates transcription factor binding to nucleosomes and factor-induced nucleosome disassembly. EMBO J. 13, 380–390.

Chen, Y., Pai, A.A., Herudek, J., Lubas, M., Meola, N., Jarvelin, A.I., Andersson, R., Pelechano, V., Steinmetz, L.M., Jensen, T.H., et al., 2016. Principles for RNA metabolism and alternative transcription initiation within closely spaced promoters. Nat. Genet. 48, 984–994.

Chereji, R.V., Clark, D.J., 2018. Major determinants of nucleosome positioning. Biophys. J. 114, 2279–2289.

Chereji, R.V., Kan, T.W., Grudniewska, M.K., Romashchenko, A.V., Berezikov, E., Zhimulev, I.F., Guryev, V., Morozov, A.V., Moshkin, Y.M., 2016. Genome-wide profiling of nucleosome sensitivity and chromatin accessibility in *Drosophila melanogaster*. Nucleic Acids Res. 44, 1036–1051.

Chereji, R.V., Ocampo, J., Clark, D.J., 2017. MNase-sensitive complexes in yeast: nucleosomes and non-histone barriers. Mol. Cell 65 (565–577), e563.

Chereji, R.V., Ramachandran, S., Bryson, T.D., Henikoff, S., 2018. Precise genome-wide mapping of single nucleosomes and linkers in vivo. Genome Biol. 19, 19.

Chereji, R.V., Bryson, T.D., Henikoff, S., 2019. Quantitative MNase-seq accurately maps nucleosome occupancy levels. Genome Biol. 20, 198.

Chung, H.R., Dunkel, I., Heise, F., Linke, C., Krobitsch, S., Ehrenhofer-Murray, A.E., Sperling, S.R., Vingron, M., 2010. The effect of micrococcal nuclease digestion on nucleosome positioning data. PLoS One 5, e15754.

Clark, D.J., 2010. Nucleosome positioning, nucleosome spacing and the nucleosome code. J. Biomol. Struct. Dyn. 27, 781–793.

Cliften, P., Sudarsanam, P., Desikan, A., Fulton, L., Fulton, B., Majors, J., Waterston, R., Cohen, B.A., Johnston, M., 2003. Finding functional features in *Saccharomyces* genomes by phylogenetic footprinting. Science 301, 71–76.

Cole, H.A., Nagarajavel, V., Clark, D.J., 2012. Perfect and imperfect nucleosome positioning in yeast. Biochim. Biophys. Acta 1819, 639–643.

Cole, H.A., Cui, F., Ocampo, J., Burke, T.L., Nikitina, T., Nagarajavel, V., Kotomura, N., Zhurkin, V.B., Clark, D.J., 2016. Novel nucleosomal particles containing core histones and linker DNA but no histone H1. Nucleic Acids Res. 44, 573–581.

Compton, J.L., Bellard, M., Chambon, P., 1976. Biochemical evidence of variability in the DNA repeat length in the chromatin of higher eukaryotes. Proc. Natl. Acad. Sci. USA 73, 4382–4386.

Cooper, J.P., Roth, S.Y., Simpson, R.T., 1994. The global transcriptional regulators, SSN6 and TUP1, play distinct roles in the establishment of a repressive chromatin structure. Genes Dev. 8, 1400–1410.

Cordingley, M.G., Riegel, A.T., Hager, G.L., 1987. Steroid-dependent interaction of transcription factors with the inducible promoter of mouse mammary tumor virus in vivo. Cell 48, 261–270.

Core, L., Adelman, K., 2019. Promoter-proximal pausing of RNA polymerase II: a nexus of gene regulation. Genes Dev. 33, 960–982.

Core, L.J., Martins, A.L., Danko, C.G., Waters, C.T., Siepel, A., Lis, J.T., 2014. Analysis of nascent RNA identifies a unified architecture of initiation regions at mammalian promoters and enhancers. Nat. Genet. 46, 1311–1320.

Costanzo, G., Di Mauro, E., Negri, R., Pereira, G., Hollenberg, C., 1995. Multiple overlapping positions of nucleosomes with single in vivo rotational setting in the Hansenula polymorpha RNA polymerase II MOX promoter. J. Biol. Chem. 270, 11091–11097.

David, L., Huber, W., Granovskaia, M., Toedling, J., Palm, C.J., Bofkin, L., Jones, T., Davis, R.W., Steinmetz, L.M., 2006. A high-resolution map of transcription in the yeast genome. Proc. Natl. Acad. Sci. USA 103, 5320–5325.

de Boer, C.G., Hughes, T.R., 2014. Poly-dA:dT tracts form an in vivo nucleosomal turnstile. PLoS One 9, e110479.

de Jonge, W.J., Brok, M., Lijnzaad, P., Kemmeren, P., Holstege, F.C., 2020. Genome-wide off-rates reveal how DNA binding dynamics shape transcription factor function. Mol. Syst. Biol. 16, e9885.

de Winde, J.H., Grivell, L.A., 1992. Global regulation of mitochondrial biogenesis in *Saccharomyces cerevisiae*: ABF1 and CPF1 play opposite roles in regulating expression of the QCR8 gene, which encodes subunit VIII of the mitochondrial ubiquinol-cytochrome c oxidoreductase. Mol. Cell. Biol. 12, 2872–2883.

DeGennaro, C.M., Alver, B.H., Marguerat, S., Stepanova, E., Davis, C.P., Bahler, J., Park, P.J., Winston, F., 2013. Spt6 regulates intragenic and antisense transcription, nucleosome positioning, and histone modifications genome-wide in fission yeast. Mol. Cell. Biol. 33, 4779–4792.

Della Seta, F., Ciafre, S.A., Marck, C., Santoro, B., Presutti, C., Sentenac, A., Bozzoni, I., 1990. The ABF1 factor is the transcriptional activator of the L2 ribosomal protein genes in *Saccharomyces cerevisiae*. Mol. Cell. Biol. 10, 2437–2441.

DeRisi, J.L., Iyer, V.R., Brown, P.O., 1997. Exploring the metabolic and genetic control of gene expression on a genomic scale. Science 278, 680–686.

Devlin, C., Tice-Baldwin, K., Shore, D., Arndt, K.T., 1991. RAP1 is required for BAS1/BAS2- and GCN4-dependent transcription of the yeast HIS4 gene. Mol. Cell. Biol. 11, 3642–3651.

Dickerson, R.E., Drew, H.R., 1981. Structure of a B-DNA dodecamer. II. Influence of base sequence on helix structure. J. Mol. Biol. 149, 761–786.

Dingwall, C., Lomonossoff, G.P., Laskey, R.A., 1981. High sequence specificity of micrococcal nuclease. Nucleic Acids Res. 9, 2659–2673.

Dong, F., Hansen, J.C., van Holde, K.E., 1990. DNA and protein determinants of nucleosome positioning on sea urchin 5S rRNA gene sequences in vitro. Proc. Natl. Acad. Sci. USA 87, 5724–5728.

Drew, H.R., 1991. Can one measure the free energy of binding of the histone octamer to different DNA sequences by salt-dependent reconstitution? J. Mol. Biol. 219, 391–392.

Drew, H.R., Travers, A.A., 1985. DNA bending and its relation to nucleosome positioning. J. Mol. Biol. 186, 773–790.

Drew, H.R., Wing, R.M., Takano, T., Broka, C., Tanaka, S., Itakura, K., Dickerson, R.E., 1981. Structure of a B-DNA dodecamer: conformation and dynamics. Proc. Natl. Acad. Sci. USA 78, 2179–2183.

Earnshaw, W.C., Honda, B.M., Laskey, R.A., Thomas, J.O., 1980. Assembly of nucleosomes: the reaction involving *X. laevis* nucleoplasmin. Cell 21, 373–383.

Eaton, M.L., Galani, K., Kang, S., Bell, S.P., MacAlpine, D.M., 2010. Conserved nucleosome positioning defines replication origins. Genes Dev. 24, 748–753.

Eissenberg, J.C., Cartwright, I.L., Thomas, G.H., Elgin, S.C., 1985. Selected topics in chromatin structure. Annu. Rev. Genet. 19, 485–536.

Elgin, S.C., 1988. The formation and function of DNase I hypersensitive sites in the process of gene activation. J. Biol. Chem. 263, 19259–19262.

Englander, E.W., Howard, B.H., 1996. A naturally occurring T14A11 tract blocks nucleosome formation over the human neurofibromatosis type 1 (NF1)-Alu element. J. Biol. Chem. 271, 5819–5823.

Ercan, S., Simpson, R.T., 2004. Global chromatin structure of 45,000 base pairs of chromosome III in a- and alpha-cell yeast and during mating-type switching. Mol. Cell. Biol. 24, 10026–10035.

Ermini, M., Kuenzle, C.C., 1978. The chromatin repeat length of cortical neurons shortens during early posnatal development. FEBS Lett. 90, 167–172.

Fascher, K.D., Schmitz, J., Horz, W., 1993. Structural and functional requirements for the chromatin transition at the PHO5 promoter in *Saccharomyces cerevisiae* upon PHO5 activation. J. Mol. Biol. 231, 658–667.

Fatemi, M., Pao, M.M., Jeong, S., Gal-Yam, E.N., Egger, G., Weisenberger, D.J., Jones, P.A., 2005. Footprinting of mammalian promoters: use of a CpG DNA methyltransferase revealing nucleosome positions at a single molecule level. Nucleic Acids Res. 33, e176.

Fazzio, T.G., Kooperberg, C., Goldmark, J.P., Neal, C., Basom, R., Delrow, J., Tsukiyama, T., 2001. Widespread collaboration of Isw2 and Sin3-Rpd3 chromatin remodeling complexes in transcriptional repression. Mol. Cell. Biol. 21, 6450–6460.

Fedor, M.J., Lue, N.F., Kornberg, R.D., 1988. Statistical positioning of nucleosomes by specific protein-binding to an upstream activating sequence in yeast. J. Mol. Biol. 204, 109–127.

Felsenfeld, G., 1992. Chromatin as an essential part of the transcriptional mechanism. Nature 355, 219–224.

Field, Y., Kaplan, N., Fondufe-Mittendorf, Y., Moore, I.K., Sharon, E., Lubling, Y., Widom, J., Segal, E., 2008. Distinct modes of regulation by chromatin encoded through nucleosome positioning signals. PLoS Comput. Biol. 4, e1000216.

Fittler, F., Zachau, H.G., 1979. Subunit structure of alpha-satellite DNA containing chromatin from African green monkey cells. Nucleic Acids Res. 7, 1–13.

FitzGerald, P.C., Simpson, R.T., 1985. Effects of sequence alterations in a DNA segment containing the 5 S RNA gene from Lytechinus variegatus on positioning of a nucleosome core particle in vitro. J. Biol. Chem. 260, 15318–15324.

Flaus, A., Luger, K., Tan, S., Richmond, T.J., 1996. Mapping nucleosome position at single base-pair resolution by using site-directed hydroxyl radicals. Proc. Natl. Acad. Sci. USA 93, 1370–1375.

Floer, M., Wang, X., Prabhu, V., Berrozpe, G., Narayan, S., Spagna, D., Alvarez, D., Kendall, J., Krasnitz, A., Stepansky, A., et al., 2010. A RSC/nucleosome complex determines chromatin architecture and facilitates activator binding. Cell 141, 407–418.

Fourel, G., Miyake, T., Defossez, P.A., Li, R., Gilson, E., 2002. General regulatory factors (GRFs) as genome partitioners. J. Biol. Chem. 277, 41736–41743.

Fox, K.R., 1992. Wrapping of genomic polydA.polydT tracts around nucleosome core particles. Nucleic Acids Res. 20, 1235–1242.

Fragoso, G., John, S., Roberts, M.S., Hager, G.L., 1995. Nucleosome positioning on the MMTV LTR results from the frequency-biased occupancy of multiple frames. Genes Dev. 9, 1933–1947.

Fu, Y., Sinha, M., Peterson, C.L., Weng, Z., 2008. The insulator binding protein CTCF positions 20 nucleosomes around its binding sites across the human genome. PLoS Genet. 4, e1000138.

Fuda, N.J., Guertin, M.J., Sharma, S., Danko, C.G., Martins, A.L., Siepel, A., Lis, J.T., 2015. GAGA factor maintains nucleosome-free regions and has a role in RNA polymerase II recruitment to promoters. PLoS Genet. 11, e1005108.

Fulmer, A.W., Fasman, G.D., 1979. Analysis of chromatin reconstitution. Biochemistry 18, 659–668.

Fyodorov, D.V., Blower, M.D., Karpen, G.H., Kadonaga, J.T., 2004. Acf1 confers unique activities to ACF/CHRAC and promotes the formation rather than disruption of chromatin in vivo. Genes Dev. 18, 170–183.

Gaffney, D.J., McVicker, G., Pai, A.A., Fondufe-Mittendorf, Y.N., Lewellen, N., Michelini, K., Widom, J., Gilad, Y., Pritchard, J.K., 2012. Controls of nucleosome positioning in the human genome. PLoS Genet. 8, e1003036.

Galas, D.J., Schmitz, A., 1978. DNAse footprinting: a simple method for the detection of protein-DNA binding specificity. Nucleic Acids Res. 5, 3157–3170.

Ganapathi, M., Palumbo, M.J., Ansari, S.A., He, Q., Tsui, K., Nislow, C., Morse, R.H., 2011. Extensive role of the general regulatory factors, Abf1 and Rap1, in determining genome-wide chromatin structure in budding yeast. Nucleic Acids Res. 39, 2032–2044.

Ganguli, D., Chereji, R.V., Iben, J.R., Cole, H.A., Clark, D.J., 2014. RSC-dependent constructive and destructive interference between opposing arrays of phased nucleosomes in yeast. Genome Res. 24, 1637–1649.

Ganter, B., Tan, S., Richmond, T.J., 1993. Genomic footprinting of the promoter regions of STE2 and STE3 genes in the yeast *Saccharomyces cerevisiae*. J. Mol. Biol. 234, 975–987.

Gaykalova, D.A., Nagarajavel, V., Bondarenko, V.A., Bartholomew, B., Clark, D.J., Studitsky, V.M., 2011. A polar barrier to transcription can be circumvented by remodeler-induced nucleosome translocation. Nucleic Acids Res. 39, 3520–3528.

Gilchrist, D.A., Nechaev, S., Lee, C., Ghosh, S.K., Collins, J.B., Li, L., Gilmour, D.S., Adelman, K., 2008. NELF-mediated stalling of Pol II can enhance gene expression by blocking promoter-proximal nucleosome assembly. Genes Dev. 22, 1921–1933.

Gilchrist, D.A., Dos Santos, G., Fargo, D.C., Xie, B., Gao, Y., Li, L., Adelman, K., 2010. Pausing of RNA polymerase II disrupts DNA-specified nucleosome organization to enable precise gene regulation. Cell 143, 540–551.

Gilmour, D.S., Thomas, G.H., Elgin, S.C., 1989. Drosophila nuclear proteins bind to regions of alternating C and T residues in gene promoters. Science 245, 1487–1490.

Givens, R.M., Lai, W.K., Rizzo, J.M., Bard, J.E., Mieczkowski, P.A., Leatherwood, J., Huberman, J.A., Buck, M.J., 2012. Chromatin architectures at fission yeast transcriptional promoters and replication origins. Nucleic Acids Res. 40, 7176–7189.

Gkikopoulos, T., Schofield, P., Singh, V., Pinskaya, M., Mellor, J., Smolle, M., Workman, J.L., Barton, G.J., Owen-Hughes, T., 2011. A role for Snf2-related nucleosome-spacing enzymes in genome-wide nucleosome organization. Science 333, 1758–1760.

Glikin, G.C., Ruberti, I., Worcel, A., 1984. Chromatin assembly in Xenopus oocytes: in vitro studies. Cell 37, 33–41.

Goffeau, A., Barrell, B.G., Bussey, H., Davis, R.W., Dujon, B., Feldmann, H., Galibert, F., Hoheisel, J.D., Jacq, C., Johnston, M., et al., 1996. Life with 6000 genes. Science 274, 546 (563–547).

Gottschling, D.E., 1992. Telomere-proximal DNA in *Saccharomyces cerevisiae* is refractory to methyltransferase activity in vivo. Proc. Natl. Acad. Sci. USA 89, 4062–4065.

Gottschling, D.E., Cech, T.R., 1984. Chromatin structure of the molecular ends of Oxytricha macronuclear DNA: phased nucleosomes and a telomeric complex. Cell 38, 501–510.

Gralla, J.D., 1985. Rapid "footprinting" on supercoiled DNA. Proc. Natl. Acad. Sci. USA 82, 3078–3081.

Granok, H., Leibovitch, B.A., Shaffer, C.D., Elgin, S.C., 1995. Chromatin. Ga-ga over GAGA factor. Curr. Biol. 5, 238–241.

Gross, D.S., Garrard, W.T., 1988. Nuclease hypersensitive site in chromatin. Annu. Rev. Biochem. 57, 159–197.

Gross, D.S., Adams, C.C., Lee, S., Stentz, B., 1993. A critical role for heat shock transcription factor in establishing a nucleosome-free region over the TATA-initiation site of the yeast HSP82 heat shock gene. EMBO J. 12, 3931–3945.

Grunberg, S., Henikoff, S., Hahn, S., Zentner, G.E., 2016. Mediator binding to UASs is broadly uncoupled from transcription and cooperative with TFIID recruitment to promoters. EMBO J. 35, 2435–2446.

Hansen, J.C., van Holde, K.E., Lohr, D., 1991. The mechanism of nucleosome assembly onto oligomers of the sea urchin 5 S DNA positioning sequence. J. Biol. Chem. 266, 4276–4282.

Harbison, C.T., Gordon, D.B., Lee, T.I., Rinaldi, N.J., Macisaac, K.D., Danford, T.W., Hannett, N.M., Tagne, J.B., Reynolds, D.B., Yoo, J., et al., 2004. Transcriptional regulatory code of a eukaryotic genome. Nature 431, 99–104.

Hartley, P.D., Madhani, H.D., 2009. Mechanisms that specify promoter nucleosome location and identity. Cell 137, 445–458.

Hartwell, L.H., 1967. Macromolecule synthesis in temperature-sensitive mutants of yeast. J. Bacteriol. 93, 1662–1670.

Haruki, H., Nishikawa, J., Laemmli, U.K., 2008. The anchor-away technique: rapid, conditional establishment of yeast mutant phenotypes. Mol. Cell 31, 925–932.

Hayes, J.J., Wolffe, A.P., 1992. Histones H2A/H2B inhibit the interaction of transcription factor IIIA with the *Xenopus borealis* somatic 5S RNA gene in a nucleosome. Proc. Natl. Acad. Sci. USA 89, 1229–1233.

Hayes, J.J., Tullius, T.D., Wolffe, A.P., 1990. The structure of DNA in a nucleosome. Proc. Natl. Acad. Sci. USA 87, 7405–7409.

Hayes, J.J., Clark, D.J., Wolffe, A.P., 1991. Histone contributions to the structure of DNA in the nucleosome. Proc. Natl. Acad. Sci. USA 88, 6829–6833.

He, Q., Johnston, J., Zeitlinger, J., 2015. ChIP-nexus enables improved detection of in vivo transcription factor binding footprints. Nat. Biotechnol. 33, 395–401.

Henikoff, J.G., Belsky, J.A., Krassovsky, K., MacAlpine, D.M., Henikoff, S., 2011. Epigenome characterization at single base-pair resolution. Proc. Natl. Acad. Sci. USA 108, 18318–18323.

Horz, W., Altenburger, W., 1981. Sequence specific cleavage of DNA by micrococcal nuclease. Nucleic Acids Res. 9, 2643–2658.

Hsieh, C.H., Griffith, J.D., 1988. The terminus of SV40 DNA replication and transcription contains a sharp sequence-directed curve. Cell 52, 535–544.

Hughes, A.L., Rando, O.J., 2014. Mechanisms underlying nucleosome positioning in vivo. Annu. Rev. Biophys. 43, 41–63.

Hughes, A.L., Jin, Y., Rando, O.J., Struhl, K., 2012. A functional evolutionary approach to identify determinants of nucleosome positioning: a unifying model for establishing the genome-wide pattern. Mol. Cell 48, 5–15.

Igo-Kemenes, T., Horz, W., Zachau, H.G., 1982. Chromatin. Annu. Rev. Biochem. 51, 89–121.

Infante, J.J., Law, G.L., Young, E.T., 2012. Analysis of nucleosome positioning using a nucleosome-scanning assay. Methods Mol. Biol. 833, 63–87.

Ioshikhes, I.P., Albert, I., Zanton, S.J., Pugh, B.F., 2006. Nucleosome positions predicted through comparative genomics. Nat. Genet. 38, 1210–1215.

Ishii, H., Kadonaga, J.T., Ren, B., 2015. MPE-seq, a new method for the genome-wide analysis of chromatin structure. Proc. Natl. Acad. Sci. USA 112, E3457–E3465.

Iwafuchi-Doi, M., Donahue, G., Kakumanu, A., Watts, J.A., Mahony, S., Pugh, B.F., Lee, D., Kaestner, K.H., Zaret, K.S., 2016. The pioneer transcription factor FoxA maintains an accessible nucleosome configuration at enhancers for tissue-specific gene activation. Mol. Cell 62, 79–91.

Iyer, V., Struhl, K., 1995. Poly(dA:dT), a ubiquitous promoter element that stimulates transcription via its intrinsic DNA structure. EMBO J. 14, 2570–2579.

Jackson, V., 1978. Studies on histone organization in the nucleosome using formaldehyde as a reversible cross-linking agent. Cell 15, 945–954.
Jaeger, M.G., Winter, G.E., 2021. Fast-acting chemical tools to delineate causality in transcriptional control. Mol. Cell 81, 1617–1630.
Jansen, A., Verstrepen, K.J., 2011. Nucleosome positioning in *Saccharomyces cerevisiae*. Microbiol. Mol. Biol. Rev. 75, 301–320.
Jessen, W.J., Hoose, S.A., Kilgore, J.A., Kladde, M.P., 2006. Active PHO5 chromatin encompasses variable numbers of nucleosomes at individual promoters. Nat. Struct. Mol. Biol. 13, 256–263.
Jiang, C., Pugh, B.F., 2009a. A compiled and systematic reference map of nucleosome positions across the *Saccharomyces cerevisiae* genome. Genome Biol. 10, R109.
Jiang, C., Pugh, B.F., 2009b. Nucleosome positioning and gene regulation: advances through genomics. Nat. Rev. Genet. 10, 161–172.
Johnson, S.M., Tan, F.J., McCullough, H.L., Riordan, D.P., Fire, A.Z., 2006. Flexibility and constraint in the nucleosome core landscape of Caenorhabditis elegans chromatin. Genome Res. 16, 1505–1516.
Johnston, M., 1987. A model fungal gene regulatory mechanism: the GAL genes of *Saccharomyces cerevisiae*. Microbiol. Rev. 51, 458–476.
Kadonaga, J.T., 2019. The transformation of the DNA template in RNA polymerase II transcription: a historical perspective. Nat. Struct. Mol. Biol. 26, 766–770.
Kamakaka, R.T., Bulger, M., Kadonaga, J.T., 1993. Potentiation of RNA polymerase II transcription by Gal4-VP16 during but not after DNA replication and chromatin assembly. Genes Dev. 7, 1779–1795.
Kaplan, N., Moore, I.K., Fondufe-Mittendorf, Y., Gossett, A.J., Tillo, D., Field, Y., LeProust, E.M., Hughes, T.R., Lieb, J.-D., Widom, J., et al., 2009. The DNA-encoded nucleosome organization of a eukaryotic genome. Nature 458, 362–366.
Kaplan, N., Moore, I., Fondufe-Mittendorf, Y., Gossett, A.J., Tillo, D., Field, Y., Hughes, T.R., Lieb, J.D., Widom, J., Segal, E., 2010. Nucleosome sequence preferences influence in vivo nucleosome organization. Nat. Struct. Mol. Biol. 17, 918–920.
Keene, M.A., Elgin, S.C., 1981. Micrococcal nuclease as a probe of DNA sequence organization and chromatin structure. Cell 27, 57–64.
Keleher, C.A., Goutte, C., Johnson, A.D., 1988. The yeast cell-type-specific repressor alpha 2 acts cooperatively with a non-cell-type-specific protein. Cell 53, 927–936.
Kellis, M., Patterson, N., Endrizzi, M., Birren, B., Lander, E.S., 2003. Sequencing and comparison of yeast species to identify genes and regulatory elements. Nature 423, 241–254.
Kelly, T.K., Liu, Y., Lay, F.D., Liang, G., Berman, B.P., Jones, P.A., 2012. Genome-wide mapping of nucleosome positioning and DNA methylation within individual DNA molecules. Genome Res. 22, 2497–2506.
Kladde, M.P., Simpson, R.T., 1994. Positioned nucleosomes inhibit Dam methylation in vivo. Proc. Natl. Acad. Sci. USA 91, 1361–1365.
Kladde, M.P., Xu, M., Simpson, R.T., 1996. Direct study of DNA-protein interactions in repressed and active chromatin in living cells. EMBO J. 15, 6290–6300.
Klemm, S.L., Shipony, Z., Greenleaf, W.J., 2019. Chromatin accessibility and the regulatory epigenome. Nat. Rev. Genet. 20, 207–220.
Klug, A., Lutter, L.C., 1981. The helical periodicity of DNA on the nucleosome. Nucleic Acids Res. 9, 4267–4283.
Korber, P., Hörz, W., 2004. In vitro assembly of the characteristic chromatin organization at the yeast PHO5 promoter by a replication-independent extract system. J. Biol. Chem. 279, 35113–35120.
Kornberg, R.D., 1974. Chromatin structure: a repeating unit of histones and DNA. Science 184, 868–871.
Kornberg, R., 1981. The location of nucleosomes in chromatin: specific or statistical. Nature 292, 579–580.
Kornberg, R.D., Stryer, L., 1988. Statistical distributions of nucleosomes: nonrandom locations by a stochastic mechanism. Nucleic Acids Res. 16, 6677–6690.
Krietenstein, N., Wippo, C.J., Lieleg, C., Korber, P., 2012. Genome-wide in vitro reconstitution of yeast chromatin with in vivo-like nucleosome positioning. Methods Enzymol. 513, 205–232.
Krietenstein, N., Wal, M., Watanabe, S., Park, B., Peterson, C.L., Pugh, B.F., Korber, P., 2016. Genomic nucleosome organization reconstituted with pure proteins. Cell 167, 709–721.e712.
Kubik, S., Bruzzone, M.J., Jacquet, P., Falcone, J.L., Rougemont, J., Shore, D., 2015. Nucleosome stability distinguishes two different promoter types at all protein-coding genes in yeast. Mol. Cell 60, 422–434.

Kubik, S., Bruzzone, M.J., Albert, B., Shore, D., 2017. A reply to "MNase-sensitive complexes in yeast: nucleosomes and non-histone barriers," by Chereji et al. Mol. Cell 65, 578–580.

Kubik, S., O'Duibhir, E., de Jonge, W.J., Mattarocci, S., Albert, B., Falcone, J.L., Bruzzone, M.J., Holstege, F.C.P., Shore, D., 2018. Sequence-directed action of RSC remodeler and general regulatory factors modulates+1 nucleosome position to facilitate transcription. Mol. Cell 71, 89.

Kubik, S., Bruzzone, M.J., Challal, D., Dreos, R., Mattarocci, S., Bucher, P., Libri, D., Shore, D., 2019. Opposing chromatin remodelers control transcription initiation frequency and start site selection. Nat. Struct. Mol. Biol. 26, 744–754.

Kunkel, G.R., Martinson, H.G., 1981. Nucleosomes will not form on double-stranded RNa or over poly(dA).poly(dT) tracts in recombinant DNA. Nucleic Acids Res. 9, 6869–6888.

Kunzler, P., Stein, A., 1983. Histone H5 can increase the internucleosome spacing in dinucleosomes to nativelike values. Biochemistry 22, 1783–1789.

Lai, B., Gao, W., Cui, K., Xie, W., Tang, Q., Jin, W., Hu, G., Ni, B., Zhao, K., 2018. Principles of nucleosome organization revealed by single-cell micrococcal nuclease sequencing. Nature 562, 281–285.

Lantermann, A.B., Straub, T., Stralfors, A., Yuan, G.C., Ekwall, K., Korber, P., 2010. Schizosaccharomyces pombe genome-wide nucleosome mapping reveals positioning mechanisms distinct from those of *Saccharomyces cerevisiae*. Nat. Struct. Mol. Biol. 17, 251–257.

Laskey, R.A., Mills, A.D., Morris, N.R., 1977. Assembly of SV40 chromatin in a cell-free system from Xenopus eggs. Cell 10, 237–243.

Laskey, R.A., Honda, B.M., Mills, A.D., Finch, J.T., 1978. Nucleosomes are assembled by an acidic protein which binds histones and transfers them to DNA. Nature 275, 416–420.

Lee, C.K., Shibata, Y., Rao, B., Strahl, B.D., Lieb, J.D., 2004. Evidence for nucleosome depletion at active regulatory regions genome-wide. Nat. Genet. 36, 900–905.

Lee, W., Tillo, D., Bray, N., Morse, R.H., Davis, R.W., Hughes, T.R., Nislow, C., 2007. A high-resolution atlas of nucleosome occupancy in yeast. Nat. Genet. 39, 1235–1244.

Levo, M., Avnit-Sagi, T., Lotan-Pompan, M., Kalma, Y., Weinberger, A., Yakhini, Z., Segal, E., 2017. Systematic investigation of transcription factor activity in the context of chromatin using massively parallel binding and expression assays. Mol. Cell 65, 604–617.e606.

Li, G., Chandler, S.P., Wolffe, A.P., Hall, T.C., 1998. Architectural specificity in chromatin structure at the TATA box in vivo: nucleosome displacement upon beta-phaseolin gene activation. Proc. Natl. Acad. Sci. USA 95, 4772–4777.

Lieb, J.D., Liu, X., Botstein, D., Brown, P.O., 2001. Promoter-specific binding of Rap1 revealed by genome-wide maps of protein-DNA association. Nat. Genet. 28, 327–334.

Lieleg, C., Krietenstein, N., Walker, M., Korber, P., 2015. Nucleosome positioning in yeasts: methods, maps, and mechanisms. Chromosoma 124, 131–151.

Linxweiler, W., Hörz, W., 1984. Reconstitution of mononucleosomes: characterization of distinct particles that differ in the position of the histone core. Nucleic Acids Res. 12, 9395–9413.

Lohr, D., Hopper, J.E., 1985. The relationship of regulatory proteins and DNase I hypersensitive sites in the yeast GAL1-10 genes. Nucleic Acids Res. 13, 8409–8423.

Lohr, D., Lopez, J., 1995. GAL4/GAL80-dependent nucleosome disruption/deposition on the upstream regions of the yeast GAL1-10 and GAL80 genes. J. Biol. Chem. 270, 27671–27678.

Lorch, Y., Maier-Davis, B., Kornberg, R.D., 2014. Role of DNA sequence in chromatin remodeling and the formation of nucleosome-free regions. Genes Dev. 28, 2492–2497.

Losa, R., Omari, S., Thoma, F., 1990. Poly(dA).poly(dT) rich sequences are not sufficient to exclude nucleosome formation in a constitutive yeast promoter. Nucleic Acids Res. 18, 3495–3502.

Lu, Q., Wallrath, L.L., Granok, H., Elgin, S.C., 1993. (CT)n (GA)n repeats and heat shock elements have distinct roles in chromatin structure and transcriptional activation of the Drosophila hsp26 gene. Mol. Cell. Biol. 13, 2802–2814.

Lutter, L.C., 1978. Kinetic analysis of deoxyribonuclease I cleavages in the nucleosome core: evidence for a DNA superhelix. J. Mol. Biol. 124, 391–420.

Ma, H., Duan, J., Ke, J., He, Y., Gu, X., Xu, T.H., Yu, H., Wang, Y., Brunzelle, J.S., Jiang, Y., et al., 2017. A D53 repression motif induces oligomerization of TOPLESS corepressors and promotes assembly of a corepressor-nucleosome complex. Sci. Adv. 3, e1601217.

Magnaval, R., Valencia, R., Paoletti, J., 1980. Subunit organization of Euglena chromatin. Biochem. Biophys. Res. Commun. 92, 1415–1421.

Mai, X., Chou, S., Struhl, K., 2000. Preferential accessibility of the yeast his3 promoter is determined by a general property of the DNA sequence, not by specific elements. Mol. Cell. Biol. 20, 6668–6676.

Maree, J.P., Povelones, M.L., Clark, D.J., Rudenko, G., Patterton, H.G., 2017. Well-positioned nucleosomes punctuate polycistronic pol II transcription units and flank silent VSG gene arrays in Trypanosoma brucei. Epigenetics Chromatin 10, 14.

Mavrich, T.N., Ioshikhes, I.P., Venters, B.J., Jiang, C., Tomsho, L.P., Qi, J., Schuster, S.C., Albert, I., Pugh, B.F., 2008a. A barrier nucleosome model for statistical positioning of nucleosomes throughout the yeast genome. Genome Res. 18, 1073–1083.

Mavrich, T.N., Jiang, C., Ioshikhes, I.P., Li, X., Venters, B.J., Zanton, S.J., Tomsho, L.P., Qi, J., Glaser, R.L., Schuster, S.-C., et al., 2008b. Nucleosome organization in the Drosophila genome. Nature 453, 358–362.

Mayran, A., Drouin, J., 2018. Pioneer transcription factors shape the epigenetic landscape. J. Biol. Chem. 293, 13795–13804.

McGhee, J.D., Felsenfeld, G., 1983. Another potential artifact in the study of nucleosome phasing by chromatin digestion with micrococcal nuclease. Cell 32, 1205–1215.

McManus, J., Perry, P., Sumner, A.T., Wright, D.M., Thomson, E.J., Allshire, R.C., Hastie, N.D., Bickmore, W.A., 1994. Unusual chromosome structure of fission yeast DNA in mouse cells. J. Cell Sci. 107 (Pt 3), 469–486.

McPherson, C.E., Shim, E.Y., Friedman, D.S., Zaret, K.S., 1993. An active tissue-specific enhancer and bound transcription factors existing in a precisely positioned nucleosomal array. Cell 75, 387–398.

Meersseman, G., Pennings, S., Bradbury, E.M., 1992. Mobile nucleosomes—a general behavior. EMBO J. 11, 2951–2959.

Mieczkowski, J., Cook, A., Bowman, S.K., Mueller, B., Alver, B.H., Kundu, S., Deaton, A.M., Urban, J.A., Larschan, E., Park, P.J., et al., 2016. MNase titration reveals differences between nucleosome occupancy and chromatin accessibility. Nat. Commun. 7, 11485.

Miele, V., Vaillant, C., d'Aubenton-Carafa, Y., Thermes, C., Grange, T., 2008. DNA physical properties determine nucleosome occupancy from yeast to fly. Nucleic Acids Res. 36, 3746–3756.

Miyake, T., Reese, J., Loch, C.M., Auble, D.T., Li, R., 2004. Genome-wide analysis of ARS (autonomously replicating sequence) binding factor 1 (Abf1p)-mediated transcriptional regulation in *Saccharomyces cerevisiae*. J. Biol. Chem. 279, 34865–34872.

Modak, S.P., Beard, P., 1980. Analysis of DNA double- and single-strand breaks by two dimensional electrophoresis: action of micrococcal nuclease on chromatin and DNA, and degradation in vivo of lens fiber chromatin. Nucleic Acids Res. 8, 2665–2678.

Moreira, J.M., Holmberg, S., 1998. Nucleosome structure of the yeast CHA1 promoter: analysis of activation-dependent chromatin remodeling of an RNA-polymerase-II-transcribed gene in TBP and RNA pol II mutants defective in vivo in response to acidic activators. EMBO J. 17, 6028–6038.

Morris, N.R., 1976. Nucleosome structure in Aspergillus nidulans. Cell 8, 357–363.

Morse, R.H., 1993. Nucleosome disruption by transcription factor binding in yeast. Science 262, 1563–1566.

Morse, R.H., 2000. RAP, RAP, open up! New wrinkles for RAP1 in yeast. Trends Genet. 16, 51–53.

Morse, R.H., Roth, S.Y., Simpson, R.T., 1992. A transcriptionally active tRNA gene interferes with nucleosome positioning in vivo. Mol. Cell. Biol. 12, 4015–4025.

Moshkin, Y.M., Chalkley, G.E., Kan, T.W., Reddy, B.A., Ozgur, Z., van Ijcken, W.F., Dekkers, D.H., Demmers, J.A., Travers, A.A., Verrijzer, C.P., 2012. Remodelers organize cellular chromatin by counteracting intrinsic histone-DNA sequence preferences in a class-specific manner. Mol. Cell. Biol. 32, 675–688.

Moyle-Heyrman, G., Zaichuk, T., Xi, L., Zhang, Q., Uhlenbeck, O.C., Holmgren, R., Widom, J., Wang, J.P., 2013. Chemical map of Schizosaccharomyces pombe reveals species-specific features in nucleosome positioning. Proc. Natl. Acad. Sci. USA 110, 20158–20163.

Muro-Pastor, M.I., Gonzalez, R., Strauss, J., Narendja, F., Scazzocchio, C., 1999. The GATA factor AreA is essential for chromatin remodelling in a eukaryotic bidirectional promoter. EMBO J. 18, 1584–1597.

Musich, P.R., Maio, J.J., Brown, F.L., 1977. Subunit structure of chromatin and the organization of eukaryotic highly repetitive DNA: indications of a phase relation between restriction sites and chromatin subunits in African green monkey and calf nuclei. J. Mol. Biol. 117, 657–677.

Nabet, B., Roberts, J.M., Buckley, D.L., Paulk, J., Dastjerdi, S., Yang, A., Leggett, A.L., Erb, M.A., Lawlor, M.A., Souza, A., et al., 2018. The dTAG system for immediate and target-specific protein degradation. Nat. Chem. Biol. 14, 431–441.

Nedospasov, S.A., Georgiev, G.P., 1980. Non-random cleavage of SV40 DNA in the compact minichromosome and free in solution by micrococcal nuclease. Biochem. Biophys. Res. Commun. 92, 532–539.

Nelson, T., Hsieh, T.S., Brutlag, D., 1979. Extracts of Drosophila embryos mediate chromatin assembly in vitro. Proc. Natl. Acad. Sci. USA 76, 5510–5514.

Nishimura, K., Fukagawa, T., Takisawa, H., Kakimoto, T., Kanemaki, M., 2009. An auxin-based degron system for the rapid depletion of proteins in nonplant cells. Nat. Methods 6, 917–922.

Noll, M., 1974. Subunit structure of chromatin. Nature 251, 249–251.

Oberbeckmann, E., Wolff, M., Krietenstein, N., Heron, M., Ellins, J.L., Schmid, A., Krebs, S., Blum, H., Gerland, U., Korber, P., 2019. Absolute nucleosome occupancy map for the *Saccharomyces cerevisiae* genome. Genome Res. 29, 1996–2009.

Ocampo, J., Chereji, R.V., Eriksson, P.R., Clark, D.J., 2016. The ISW1 and CHD1 ATP-dependent chromatin remodelers compete to set nucleosome spacing in vivo. Nucleic Acids Res. 44, 4625–4635.

Okada, M., Hirose, S., 1998. Chromatin remodeling mediated by Drosophila GAGA factor and ISWI activates fushi tarazu gene transcription in vitro. Mol. Cell. Biol. 18, 2455–2461.

Ostrowski, M.C., Richard-Foy, H., Wolford, R.G., Berard, D.S., Hager, G.L., 1983. Glucocorticoid regulation of transcription at an amplified, episomal promoter. Mol. Cell. Biol. 3, 2045–2057.

Owen-Hughes, T., Workman, J.L., 1996. Remodeling the chromatin structure of a nucleosome array by transcription factor-targeted trans-displacement of histones. EMBO J. 15, 4702–4712.

Ozsolak, F., Song, J.S., Liu, X.S., Fisher, D.E., 2007. High-throughput mapping of the chromatin structure of human promoters. Nat. Biotechnol. 25, 244–248.

Paul, E., Tirosh, I., Lai, W., Buck, M.J., Palumbo, M.J., Morse, R.H., 2015. Chromatin mediation of a transcriptional memory effect in yeast. G3 (Bethesda) 5, 829–838.

Pech, M., Igo-Kemenes, T., Zachau, H.G., 1979. Nucleotide sequence of a highly repetitive component of rat DNA. Nucleic Acids Res. 7, 417–432.

Peckham, H.E., Thurman, R.E., Fu, Y., Stamatoyannopoulos, J.A., Noble, W.S., Struhl, K., Weng, Z., 2007. Nucleosome positioning signals in genomic DNA. Genome Res. 17, 1170–1177.

Pennings, S., Muyldermans, S., Meersseman, G., Wyns, L., 1989. Formation, stability and core histone positioning of nucleosomes reassembled on bent and other nucleosome-derived DNA. J. Mol. Biol. 207, 183–192.

Pennings, S., Meersseman, G., Bradbury, E.M., 1991. Mobility of positioned nucleosomes on 5 S rDNA. J. Mol. Biol. 220, 101–110.

Perales, R., Zhang, L., Bentley, D., 2011. Histone occupancy in vivo at the 601 nucleosome binding element is determined by transcriptional history. Mol. Cell. Biol. 31, 3485–3496.

Pilpel, Y., Sudarsanam, P., Church, G.M., 2001. Identifying regulatory networks by combinatorial analysis of promoter elements. Nat. Genet. 29, 153–159.

Polach, K.J., Widom, J., 1995. Mechanism of protein access to specific DNA sequences in chromatin: a dynamic equilibrium model for gene regulation. J. Mol. Biol. 254, 130–149.

Prajapati, H.K., Ocampo, J., Clark, D.J., 2020. Interplay among ATP-dependent chromatin remodelers determines chromatin organisation in yeast. Biology (Basel) 9, 190.

Prunell, A., 1982. Nucleosome reconstitution on plasmid-inserted poly(dA).poly(dT). EMBO J. 1, 173–179.

Prunell, A., Kornberg, R.D., 1978. Relation of nucleosomes to DNA sequences. Cold Spring Harb. Symp. Quant. Biol. 42 (Pt 1), 103–108.

Pugh, B.F., 2010. A preoccupied position on nucleosomes. Nat. Struct. Mol. Biol. 17, 923.

Puhl, H.L., Gudibande, S.R., Behe, M.J., 1991. Poly[d(A.T)] and other synthetic polydeoxynucleotides containing oligoadenosine tracts form nucleosomes easily. J. Mol. Biol. 222, 1149–1160.

Radman-Livaja, M., Rando, O.J., 2010. Nucleosome positioning: how is it established, and why does it matter? Dev. Biol. 339, 258–266.

Ramsay, N., 1986. Deletion analysis of a DNA sequence that positions itself precisely on the nucleosome core. J. Mol. Biol. 189, 179–188.

Ramsay, N., Felsenfeld, G., Rushton, B.M., McGhee, J.D., 1984. A 145-base pair DNA sequence that positions itself precisely and asymmetrically on the nucleosome core. EMBO J. 3, 2605–2611.

Rand, A.C., Jain, M., Eizenga, J.M., Musselman-Brown, A., Olsen, H.E., Akeson, M., Paten, B., 2017. Mapping DNA methylation with high-throughput nanopore sequencing. Nat. Methods 14, 411–413.

Rando, O.J., 2007. Chromatin structure in the genomics era. Trends Genet. 23, 67–73.

Rando, O.J., 2010. Genome-wide mapping of nucleosomes in yeast. Methods Enzymol. 470, 105–118.

Raveh-Sadka, T., Levo, M., Shabi, U., Shany, B., Keren, L., Lotan-Pompan, M., Zeevi, D., Sharon, E., Weinberger, A., Segal, E., 2012. Manipulating nucleosome disfavoring sequences allows fine-tune regulation of gene expression in yeast. Nat. Genet. 44, 743–750.

Rawal, Y., Chereji, R.V., Qiu, H., Ananthakrishnan, S., Govind, C.K., Clark, D.J., Hinnebusch, A.G., 2018. SWI/SNF and RSC cooperate to reposition and evict promoter nucleosomes at highly expressed genes in yeast. Genes Dev. 32, 695–710.

Reddy, V.B., Thimmappaya, B., Dhar, R., Subramanian, K.N., Zain, B.S., Pan, J., Ghosh, P.K., Celma, M.L., Weissman, S.M., 1978. The genome of simian virus 40. Science 200, 494–502.

Rhee, H.S., Pugh, B.F., 2011. Comprehensive genome-wide protein-DNA interactions detected at single-nucleotide resolution. Cell 147, 1408–1419.

Rhee, H.S., Pugh, B.F., 2012. Genome-wide structure and organization of eukaryotic pre-initiation complexes. Nature 483, 295–301.

Rhee, H.S., Bataille, A.R., Zhang, L., Pugh, B.F., 2014. Subnucleosomal structures and nucleosome asymmetry across a genome. Cell 159, 1377–1388.

Rhodes, D., 1979. Nucleosome cores reconstituted from poly (dA-dT) and the octamer of histones. Nucleic Acids Res. 6, 1805–1816.

Rhodes, D., 1985. Structural analysis of a triple complex between the histone octamer, a Xenopus gene for 5S RNA and transcription factor IIIA. EMBO J. 4, 3473–3482.

Rhodes, D., Klug, A., 1980. Helical periodicity of DNA determined by enzyme digestion. Nature 286, 573–578.

Richard-Foy, H., Hager, G.L., 1987. Sequence-specific positioning of nucleosomes over the steroid-inducible MMTV promoter. EMBO J. 6, 2321–2328.

Riley, D., Weintraub, H., 1978. Nucleosomal DNA is digested to repeats of 10 bases by exonuclease III. Cell 13, 281–293.

Rizzo, J.M., Bard, J.E., Buck, M.J., 2012. Standardized collection of MNase-seq experiments enables unbiased dataset comparisons. BMC Mol. Biol. 13, 15.

Roth, S.Y., Dean, A., Simpson, R.T., 1990. Yeast alpha 2 repressor positions nucleosomes in TRP1/ARS1 chromatin. Mol. Cell. Biol. 10, 2247–2260.

Roth, S.Y., Shimizu, M., Johnson, L., Grunstein, M., Simpson, R.T., 1992. Stable nucleosome positioning and complete repression by the yeast alpha 2 repressor are disrupted by amino-terminal mutations in histone H4. Genes Dev. 6, 411–425.

Sasaki, S., Mello, C.C., Shimada, A., Nakatani, Y., Hashimoto, S., Ogawa, M., Matsushima, K., Gu, S.G., Kasahara, M., Ahsan, B., et al., 2009. Chromatin-associated periodicity in genetic variation downstream of transcriptional start sites. Science 323, 401–404.

Satchwell, S.C., Drew, H.R., Travers, A.A., 1986. Sequence periodicities in chicken nucleosome core DNA. J. Mol. Biol. 191, 659–675.

Scacchetti, A., Brueckner, L., Jain, D., Schauer, T., Zhang, X., Schnorrer, F., van Steensel, B., Straub, T., Becker, P.B., 2018. CHRAC/ACF contribute to the repressive ground state of chromatin. Life Sci. Alliance 1, e201800024.

Schmid, M., Durussel, T., Laemmli, U.K., 2004. ChIC and ChEC; genomic mapping of chromatin proteins. Mol. Cell 16, 147–157.

Schones, D.E., Cui, K., Cuddapah, S., Roh, T.Y., Barski, A., Wang, Z., Wei, G., Zhao, K., 2008. Dynamic regulation of nucleosome positioning in the human genome. Cell 132, 887–898.

Scruggs, B.S., Gilchrist, D.A., Nechaev, S., Muse, G.W., Burkholder, A., Fargo, D.C., Adelman, K., 2015. Bidirectional transcription arises from two distinct hubs of transcription factor binding and active chromatin. Mol. Cell 58, 1101–1112.

Segal, M.R., 2008. Re-cracking the nucleosome positioning code. Stat. Appl. Genet. Mol. Biol. 7, 14.

Segal, E., Widom, J., 2009a. Poly(dA:dT) tracts: major determinants of nucleosome organization. Curr. Opin. Struct. Biol. 19, 65–71.

Segal, E., Widom, J., 2009b. What controls nucleosome positions? Trends Genet. 25, 335–343.

Segal, E., Fondufe-Mittendorf, Y., Chen, L., Thastrom, A., Field, Y., Moore, I.K., Wang, J.P., Widom, J., 2006. A genomic code for nucleosome positioning. Nature 442, 772–778.

Sekinger, E.A., Moqtaderi, Z., Struhl, K., 2005. Intrinsic histone-DNA interactions and low nucleosome density are important for preferential accessibility of promoter regions in yeast. Mol. Cell 18, 735–748.

Sekiya, T., Zaret, K.S., 2007. Repression by Groucho/TLE/Grg proteins: genomic site recruitment generates compacted chromatin in vitro and impairs activator binding in vivo. Mol. Cell 28, 291–303.

Shimizu, M., Roth, S.Y., Szent-Gyorgyi, C., Simpson, R.T., 1991. Nucleosomes are positioned with base pair precision adjacent to the alpha 2 operator in *Saccharomyces cerevisiae*. EMBO J. 10, 3033–3041.

Sewack, G.F., Hansen, U., 1997. Nucleosome positioning and transcription-associated chromatin alterations on the human estrogen-responsive pS2 promoter. J. Biol. Chem. 272, 31118–31129.
Shimizu, M., Mori, T., Sakurai, T., Shindo, H., 2000. Destabilization of nucleosomes by an unusual DNA conformation adopted by poly(dA) small middle dotpoly(dT) tracts in vivo. EMBO J. 19, 3358–3365.
Shipony, Z., Marinov, G.K., Swaffer, M.P., Sinnott-Armstrong, N.A., Skotheim, J.M., Kundaje, A., Greenleaf, W.J., 2020. Long-range single-molecule mapping of chromatin accessibility in eukaryotes. Nat. Methods 17, 319–327.
Shivaswamy, S., Bhinge, A., Zhao, Y., Jones, S., Hirst, M., Iyer, V.R., 2008. Dynamic remodeling of individual nucleosomes across a eukaryotic genome in response to transcriptional perturbation. PLoS Biol. 6, e65.
Shore, D., 1994. RAP1: a protean regulator in yeast. Trends Genet. 10, 408–412 (review. 44 refs).
Shrader, T.E., Crothers, D.M., 1989. Artificial nucleosome positioning sequences. Proc. Natl. Acad. Sci. USA 86, 7418–7422.
Simon, J.M., Giresi, P.G., Davis, I.J., Lieb, J.D., 2012. Using formaldehyde-assisted isolation of regulatory elements (FAIRE) to isolate active regulatory DNA. Nat. Protoc. 7, 256–267.
Simpson, R.T., 1991. Nucleosome positioning: occurrence, mechanisms, and functional consequences. Prog. Nucleic Acid Res. Mol. Biol. 40, 143–184.
Simpson, R.T., Kunzler, P., 1979. Chromatin and core particles formed from the inner histones and synthetic polydeoxyribonucleotides of defined sequence. Nucleic Acids Res. 6, 1387–1415.
Simpson, R.T., Stafford, D.W., 1983. Structural features of a phased nucleosome core particle. Proc. Natl. Acad. Sci. USA 80, 51–55.
Simpson, R.T., Whitlock, J.P., 1976. Mapping DNAase l-susceptible sites in nucleosomes labeled at the 5′ ends. Cell 9, 347–353.
Simpson, J.T., Workman, R.E., Zuzarte, P.C., David, M., Dursi, L.J., Timp, W., 2017. Detecting DNA cytosine methylation using nanopore sequencing. Nat. Methods 14, 407–410.
Singh, J., Klar, A.J., 1992. Active genes in budding yeast display enhanced in vivo accessibility to foreign DNA methylases: a novel in vivo probe for chromatin structure of yeast. Genes Dev. 6, 186–196.
Sipiczki, M., 2000. Where does fission yeast sit on the tree of life? Genome Biol. 1, REVIEWS1011.
Skene, P.J., Henikoff, S., 2017. An efficient targeted nuclease strategy for high-resolution mapping of DNA binding sites. elife 6, e21856.
Small, E.C., Xi, L., Wang, J.P., Widom, J., Licht, J.D., 2014. Single-cell nucleosome mapping reveals the molecular basis of gene expression heterogeneity. Proc. Natl. Acad. Sci. USA 111, E2462–E2471.
Solomon, M.J., Varshavsky, A., 1985. Formaldehyde-mediated DNA-protein crosslinking: a probe for in vivo chromatin structures. Proc. Natl. Acad. Sci. USA 82, 6470–6474.
Spadafora, C., Bellard, M., Compton, J.L., Chambon, P., 1976. The DNA repeat lengths in chromatins from sea urchin sperm and gastrule cells are markedly different. FEBS Lett. 69, 281–285.
Spadafora, C., Oudet, P., Chambon, P., 1978. The same amount of DNA is organized in in vitro-assembled nucleosomes irrespective of the origin of the histones. Nucleic Acids Res. 5, 3479–3489.
Stein, A., Bina, M., 1984. A model chromatin assembly system. Factors affecting nucleosome spacing. J. Mol. Biol. 178, 341–363.
Stein, A., Takasuka, T.E., Collings, C.K., 2010. Are nucleosome positions in vivo primarily determined by histone-DNA sequence preferences? Nucleic Acids Res. 38, 709–719.
Steinmetz, M., Streeck, R.E., Zachau, H.G., 1978. Closely spaced nucleosome cores in reconstituted histone.DNA complexes and histone-H1-depleted chromatin. Eur. J. Biochem. 83, 615–628.
Stergachis, A.B., Debo, B.M., Haugen, E., Churchman, L.S., Stamatoyannopoulos, J.A., 2020. Single-molecule regulatory architectures captured by chromatin fiber sequencing. Science 368, 1449–1454.
Strauss, F., Varshavsky, A., 1984. A protein binds to a satellite DNA repeat at three specific sites that would be brought into mutual proximity by DNA folding in the nucleosome. Cell 37, 889–901.
Struhl, K., 1985. Naturally occurring poly(dA-dT) sequences are upstream promoter elements for constitutive transcription in yeast. Proc. Natl. Acad. Sci. 82, 8419.
Struhl, K., Segal, E., 2013. Determinants of nucleosome positioning. Nat. Struct. Mol. Biol. 20, 267–273.
Struhl, K., Stinchcomb, D.T., Scherer, S., Davis, R.W., 1979. High-frequency transformation of yeast: autonomous replication of hybrid DNA molecules. Proc. Natl. Acad. Sci. USA 76, 1035–1039.
Suck, D., Lahm, A., Oefner, C., 1988. Structure refined to 2A of a nicked DNA octanucleotide complex with DNase I. Nature 332, 464–468.

Szent-Gyorgyi, C., Finkelstein, D.B., Garrard, W.T., 1987. Sharp boundaries demarcate the chromatin structure of a yeast heat-shock gene. J. Mol. Biol. 193, 71–80.

Tanaka, S., Zatchej, M., Thoma, F., 1992. Artificial nucleosome positioning sequences tested in yeast minichromosomes: a strong rotational setting is not sufficient to position nucleosomes in vivo. EMBO J. 11, 1187–1193.

Taylor, I.C., Workman, J.L., Schuetz, T.J., Kingston, R.E., 1991. Facilitated binding of GAL4 and heat shock factor to nucleosomal templates: differential function of DNA-binding domains. Genes Dev. 5, 1285–1298.

Teif, V.B., 2016. Nucleosome positioning: resources and tools online. Brief. Bioinform. 17, 745–757.

Teif, V.B., Vainshtein, Y., Caudron-Herger, M., Mallm, J.P., Marth, C., Hofer, T., Rippe, K., 2012. Genome-wide nucleosome positioning during embryonic stem cell development. Nat. Struct. Mol. Biol. 19, 1185–1192.

Thastrom, A., Lowary, P.T., Widlund, H.R., Cao, H., Kubista, M., Widom, J., 1999. Sequence motifs and free energies of selected natural and non-natural nucleosome positioning DNA sequences. J. Mol. Biol. 288, 213–229.

Thoma, F., 1986. Protein-DNA interactions and nuclease-sensitive regions determine nucleosome positions on yeast plasmid chromatin. J. Mol. Biol. 190, 177–190.

Thoma, F., 1992. Nucleosome positioning. Biochim. Biophys. Acta 1130, 1–19.

Thoma, F., Simpson, R.T., 1985. Local protein-DNA interactions may determine nucleosome positions on yeast plasmids. Nature 315, 250–252.

Thoma, F., Zatchej, M., 1988. Chromatin folding modulates nucleosome positioning in yeast minichromosomes. Cell 55, 945–953.

Thoma, F., Bergman, L.W., Simpson, R.T., 1984. Nuclease digestion of circular TRP1ARS1 chromatin reveals positioned nucleosomes separated by nuclease-sensitive regions. J. Mol. Biol. 177, 715–733.

Thomas, G.H., Elgin, S.C., 1988. Protein/DNA architecture of the DNase I hypersensitive region of the Drosophila hsp26 promoter. EMBO J. 7, 2191–2201.

Thomas, J.O., Furber, V., 1976. Yeast chromatin structure. FEBS Lett. 66, 274–280.

Tillo, D., Hughes, T.R., 2009. G+C content dominates intrinsic nucleosome occupancy. BMC Bioinform. 10, 442.

Tirosh, I., Sigal, N., Barkai, N., 2010a. Divergence of nucleosome positioning between two closely related yeast species: genetic basis and functional consequences. Mol. Syst. Biol. 6, 365.

Tirosh, I., Sigal, N., Barkai, N., 2010b. Widespread remodeling of mid-coding sequence nucleosomes by Isw1. Genome Biol. 11, R49.

Tramantano, M., Sun, L., Au, C., Labuz, D., Liu, Z., Chou, M., Shen, C., Luk, E., 2016. Constitutive turnover of histone H2A.Z at yeast promoters requires the preinitiation complex. elife 5, e14243.

Trifonov, E.N., 1980. Sequence-dependent deformational anisotropy of chromatin DNA. Nucleic Acids Res. 8, 4041–4053.

Trifonov, E.N., Sussman, J.L., 1980. The pitch of chromatin DNA is reflected in its nucleotide sequence. Proc. Natl. Acad. Sci. USA 77, 3816–3820.

Tsankov, A.M., Thompson, D.A., Socha, A., Regev, A., Rando, O.J., 2010. The role of nucleosome positioning in the evolution of gene regulation. PLoS Biol. 8, e1000414.

Tsui, K., Dubuis, S., Gebbia, M., Morse, R.H., Barkai, N., Tirosh, I., Nislow, C., 2011. Evolution of nucleosome occupancy: conservation of global properties and divergence of gene-specific patterns. Mol. Cell Biol. 31, 4348–4355.

Tsukiyama, T., Wu, C., 1995. Purification and properties of an ATP-dependent nucleosome remodeling factor. Cell 83, 1011–1020.

Udvardy, A., Schedl, P., 1984. Chromatin organization of the 87A7 heat shock locus of *Drosophila melanogaster*. J. Mol. Biol. 172, 385–403.

Valouev, A., Ichikawa, J., Tonthat, T., Stuart, J., Ranade, S., Peckham, H., Zeng, K., Malek, J.A., Costa, G., McKernan, K., et al., 2008. A high-resolution, nucleosome position map of *C. elegans* reveals a lack of universal sequence-dictated positioning. Genome Res. 18, 1051–1063.

Valouev, A., Johnson, S.M., Boyd, S.D., Smith, C.L., Fire, A.Z., Sidow, A., 2011. Determinants of nucleosome organization in primary human cells. Nature 474, 516–520.

van Bakel, H., Tsui, K., Gebbia, M., Mnaimneh, S., Hughes, T.R., Nislow, C., 2013. A compendium of nucleosome and transcript profiles reveals determinants of chromatin architecture and transcription. PLoS Genet. 9, e1003479.

Van Holde, K.E., 1988. Chromatin. Springer-Verlag.

Varshavsky, A.J., Sundin, O.H., Bohn, M.J., 1978. SV40 viral minichromosome: preferential exposure of the origin of replication as probed by restriction endonucleases. Nucleic Acids Res. 5, 3469–3477.

Venditti, P., Costanzo, G., Negri, R., Camilloni, G., 1994. ABFI contributes to the chromatin organization of *Saccharomyces cerevisiae* ARS1 B-domain. Biochim. Biophys. Acta 1219, 677–689.
Venter, U., Svaren, J., Schmitz, J., Schmid, A., Horz, W., 1994. A nucleosome precludes binding of the transcription factor Pho4 in vivo to a critical target site in the PHO5 promoter. EMBO J. 13, 4848–4855.
Vera, D.L., Madzima, T.F., Labonne, J.D., Alam, M.P., Hoffman, G.G., Girimurugan, S.B., Zhang, J., McGinnis, K.M., Dennis, J.H., Bass, H.W., 2014. Differential nuclease sensitivity profiling of chromatin reveals biochemical footprints coupled to gene expression and functional DNA elements in maize. Plant Cell 26, 3883–3893.
Verdone, L., Camilloni, G., Di Mauro, E., Caserta, M., 1996. Chromatin remodeling during *Saccharomyces cerevisiae* ADH2 gene activation. Mol. Cell. Biol. 16, 1978–1988.
Von Hippel, P.H., Felsenfeld, G., 1964. Micrococcal nuclease as a probe of DNA conformation. Biochemistry 3, 27–39.
Voong, L.N., Xi, L., Sebeson, A.C., Xiong, B., Wang, J.P., Wang, X., 2016. Insights into nucleosome organization in mouse embryonic stem cells through chemical mapping. Cell 167, 1555–1570.e1515.
Wang, J., Zhuang, J., Iyer, S., Lin, X., Whitfield, T.W., Greven, M.C., Pierce, B.G., Dong, X., Kundaje, A., Cheng, Y., et al., 2012. Sequence features and chromatin structure around the genomic regions bound by 119 human transcription factors. Genome Res. 22, 1798–1812.
Wang, Y., Wang, A., Liu, Z., Thurman, A.L., Powers, L.S., Zou, M., Zhao, Y., Hefel, A., Li, Y., Zabner, J., et al., 2019. Single-molecule long-read sequencing reveals the chromatin basis of gene expression. Genome Res. 29, 1329–1342.
Wechsler, D.S., Papoulas, O., Dang, C.V., Kingston, R.E., 1994. Differential binding of c-Myc and Max to nucleosomal DNA. Mol. Cell. Biol. 14, 4097–4107.
Wedel, C., Forstner, K.U., Derr, R., Siegel, T.N., 2017. GT-rich promoters can drive RNA pol II transcription and deposition of H2A.Z in African trypanosomes. EMBO J. 36, 2581–2594.
Weiner, A., Hughes, A., Yassour, M., Rando, O.J., Friedman, N., 2010. High-resolution nucleosome mapping reveals transcription-dependent promoter packaging. Genome Res. 20, 90–100.
Weintraub, H., 1978. The nucleosome repeat length increases during erythropoiesis in the chick. Nucleic Acids Res. 5, 1179–1188.
Whitehouse, I., Tsukiyama, T., 2006. Antagonistic forces that position nucleosomes in vivo. Nat. Struct. Mol. Biol. 13, 633–640.
Whitehouse, I., Rando, O.J., Delrow, J., Tsukiyama, T., 2007. Chromatin remodelling at promoters suppresses antisense transcription. Nature 450, 1031–1035.
Wiechens, N., Singh, V., Gkikopoulos, T., Schofield, P., Rocha, S., Owen-Hughes, T., 2016. The chromatin remodelling enzymes SNF2H and SNF2L position nucleosomes adjacent to CTCF and other transcription factors. PLoS Genet. 12, e1005940.
Winston, F., Carlson, M., 1992. Yeast SNF/SWI transcriptional activators and SPT/SIN chromatin connection. Trends Genet. 8, 387–391.
Wolffe, A.P., 1998. Chromatin: Structure & Function, third ed. Academic Press, San Diego.
Wolffe, A.P., Drew, H.R., 1989. Initiation of transcription on nucleosomal templates. Proc. Natl. Acad. Sci. USA 86, 9817–9821.
Workman, J.L., Kingston, R.E., 1992. Nucleosome core displacement in vitro via a metastable transcription factor-nucleosome complex. Science 258, 1780–1784.
Wu, C., 1980. The 5′ ends of Drosophila heat shock genes in chromatin are hypersensitive to DNase I. Nature 286, 854–860.
Wu, C., 1985. An exonuclease protection assay reveals heat-shock element and TATA box DNA-binding proteins in crude nuclear extracts. Nature 317, 84–87.
Wu, R., Li, H., 2010. Positioned and G/C-capped poly(dA:dT) tracts associate with the centers of nucleosome-free regions in yeast promoters. Genome Res. 20, 473–484.
Wu, C., Travers, A., 2005. Relative affinities of DNA sequences for the histone octamer depend strongly upon both the temperature and octamer concentration. Biochemistry 44, 14329–14334.
Xi, Y.X., Yao, J.H., Chen, R., Li, W., He, X.W., 2011. Nucleosome fragility reveals novel functional states of chromatin and poises genes for activation. Genome Res. 21, 718–724.
Xiong, J., Gao, S., Dui, W., Yang, W., Chen, X., Taverna, S.D., Pearlman, R.E., Ashlock, W., Miao, W., Liu, Y., 2016. Dissecting relative contributions of cis- and trans-determinants to nucleosome distribution by comparing Tetrahymena macronuclear and micronuclear chromatin. Nucleic Acids Res. 44, 10091–10105.
Yan, C., Chen, H., Bai, L., 2018. Systematic study of nucleosome-displacing factors in budding yeast. Mol. Cell 71, 294–305.e294.

Yarragudi, A., Miyake, T., Li, R., Morse, R.H., 2004. Comparison of ABF1 and RAP1 in chromatin opening and transactivator potentiation in the budding yeast *Saccharomyces cerevisiae*. Mol. Cell. Biol. 24, 9152–9164.

Yarragudi, A., Parfrey, L.W., Morse, R.H., 2007. Genome-wide analysis of transcriptional dependence and probable target sites for Abf1 and Rap1 in *Saccharomyces cerevisiae*. Nucleic Acids Res. 35, 193–202.

Yen, K., Vinayachandran, V., Batta, K., Koerber, R.T., Pugh, B.F., 2012. Genome-wide nucleosome specificity and directionality of chromatin remodelers. Cell 149, 1461–1473.

Yenidunya, A., Davey, C., Clark, D., Felsenfeld, G., Allan, J., 1994. Nucleosome positioning on chicken and human globin gene promoters in vitro. Novel mapping techniques. J. Mol. Biol. 237, 401–414.

Yu, L., Morse, R.H., 1999. Chromatin opening and transactivator potentiation by RAP1 in *Saccharomyces cerevisiae*. Mol. Cell. Biol. 19, 5279–5288.

Yuan, G.C., Liu, J.S., 2008. Genomic sequence is highly predictive of local nucleosome depletion. PLoS Comput. Biol. 4, e13.

Yuan, G.C., Liu, Y.J., Dion, M.F., Slack, M.D., Wu, L.F., Altschuler, S.J., Rando, O.J., 2005. Genome-scale identification of nucleosome positions in *S. cerevisiae*. Science 309, 626–630.

Zakian, V.A., Scott, J.F., 1982. Construction, replication, and chromatin structure of TRP1 RI circle, a multiple-copy synthetic plasmid derived from *Saccharomyces cerevisiae* chromosomal DNA. Mol. Cell. Biol. 2, 221–232.

Zaret, K.S., 2020. Pioneer transcription factors initiating gene network changes. Annu. Rev. Genet. 54, 367–385.

Zaret, K.S., Carroll, J.S., 2011. Pioneer transcription factors: establishing competence for gene expression. Genes Dev. 25, 2227–2241.

Zentner, G.E., Kasinathan, S., Xin, B., Rohs, R., Henikoff, S., 2015. ChEC-seq kinetics discriminates transcription factor binding sites by DNA sequence and shape in vivo. Nat. Commun. 6, 8733.

Zhang, Y., Moqtaderi, Z., Rattner, B.P., Euskirchen, G., Snyder, M., Kadonaga, J.T., Liu, X.S., Struhl, K., 2009. Intrinsic histone-DNA interactions are not the major determinant of nucleosome positions in vivo. Nat. Struct. Mol. Biol. 16, 847–852.

Zhang, Y., Moqtaderi, Z., Rattner, B.P., Euskirchen, G., Snyder, M., Kadonaga, J.T., Liu, X.S., Struhl, K., 2010. Reply to "evidence against a genomic code for nucleosome positioning". Nat. Struct. Mol. Biol. 17, 920–923.

Zhang, Z., Wippo, C.J., Wal, M., Ward, E., Korber, P., Pugh, B.F., 2011. A packing mechanism for nucleosome organization reconstituted across a eukaryotic genome. Science 332, 977–980.

Zhang, T., Zhang, W., Jiang, J., 2015. Genome-wide nucleosome occupancy and positioning and their impact on gene expression and evolution in plants. Plant Physiol. 168, 1406–1416.

CHAPTER

5

Histones and variants

Abbreviations

CAF-1	chromatin assembly factor 1
ChIP	chromatin immunoprecipitation
EM	electron microscopy
ESC	embryonic stem cell
FACT	facilitates chromatin transcription
GFP	Green Fluorescent Protein
HJURP	Holliday junction recognizing protein
MNase	micrococcal nuclease
NAP-1	nucleosome assembly protein 1
NCP	nucleosome core particle
NDR	nucleosome-depleted region
PARP	poly-ADP-ribose polymerase
PCNA	proliferating cell nuclear antigen
RC	replication-coupled
RI	replication-independent
RSC	remodels the structure of chromatin
SV40	Simian Virus 40

The first edition of Alan Wolffe's book on *Chromatin Structure and Function*, published in 1992, devotes just four pages to the topic of histone variants, most of them dedicated to a discussion of variants of H2A and H2B and linker histones expressed during development in sea urchin and the African clawed frog *Xenopus*. That same year Sally Elgin's lab reported that a specialized histone variant, H2AvD, is essential in *Drosophila*, and John Pehrson identified the very strange macroH2A (Pehrson and Fried, 1992; van Daal and Elgin, 1992). Clues from studies of the ciliated protozoan *Tetrahymena* and fractionation of murine chromatin had only recently suggested a possible functional role in transcription for the histone variants *hv1* and H2A.Z, respectively, while sequencing data revealed both these variants to be related to H2AvD. The discovery of a histone H3 variant, CENP-A, uniquely associated with the centromere in both yeast and mammals, revealed a direct connection between a histone variant and a specific chromosomal function. These initial revelations regarding histone variants were the first glimpses of a molecular *terra incognita*, whetting researchers' appetites for further exploration. There is still much that is unknown regarding variants of the canonical core histones, but the work of the past 30 years, including the application of novel techniques and genome-wide approaches, has provided powerful insights into their structures and

Chromatin. **https://doi.org/10.1016/B978-0-12-814809-9.00004-4**

functions. This chapter provides a survey and history of these interesting contributors to variations on the theme of the nucleosome.

General considerations

An excellent resource for the exploration of the range of variant histones is the Histone Database (https://www.ncbi.nlm.nih.gov/research/HistoneDB2.0/), which includes data on canonical and variant histones from a wide variety of organisms (Draizen et al., 2016).

Variations in the histones need to be considered in the context of three important pieces of background information.

First, as mentioned in Chapter 2, the core histones (H2A, H2B, H3, and H4) are highly conserved in evolution (Fig. 5.1; Delange, 1978). This conservation expedited identification of histones across species once cloned histone genes from sea urchin and *Drosophila* became available, as discussed below. The conservation of the core histones has also allowed the phylogenies of variants to be distinguished from those of the canonical versions; for example, the variant histone H2A.Z shows greater conservation across species than it does with the canonical H2A within species. Linker histones, in contrast, vary much more widely in sequence, in some cases (budding yeast in particular) diverging so much from the canonical structure and sequence that their identification required years of investigation, finally facilitated by the availability of whole genome sequences (Landsman, 1996).

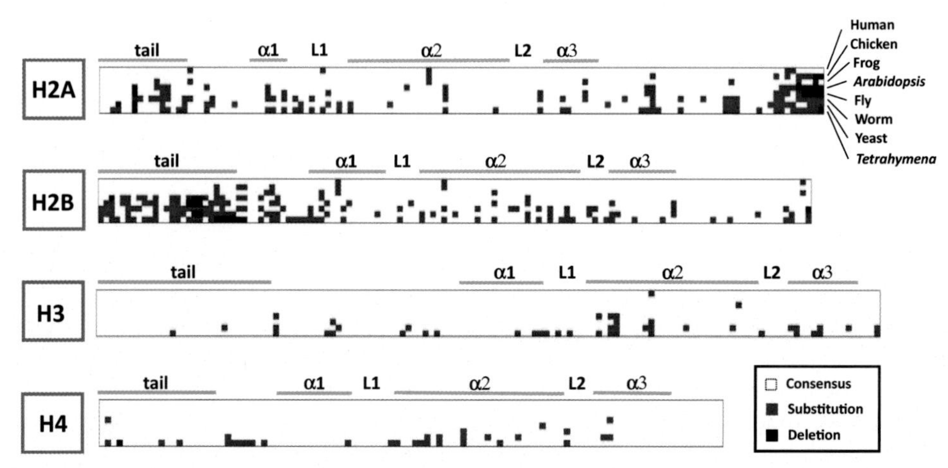

FIG. 5.1 Conservation of canonical core histone sequences. Histone sequences were aligned for canonical core histone sequences from *H. sapiens*, *Gallus gallus*, *Xenopus laevis*, *Arabidopsis thaliana*, *Drosophila melanogaster*, *Caenorhabditis elegans*, *Saccharomyces cerevisiae*, and *Tetrahymena thermophila* using the HistoneDB 2.0 database (Draizen et al., 2016). Each square or "pixel" represents an amino acid; white conforms to the consensus sequence, *red* indicates a substitution, and *black* indicates a deletion. Unstructured N-terminal tails, α-helices, and loops of the histone fold motif are indicated. Inserted amino acids (H2A: *Tetrahymena*, yeast, *Arabidopsis*, worm; H2B: yeast, *Arabidopsis*; H4, *Tetrahymena*) are not shown.

Second, the canonical histone genes are present in most organisms as multicopy families: while budding yeast have only two copies of the genes encoding each of the four core histones, dozens to hundreds of histone gene copies are present in mice, frogs, insects, chickens, sea urchins, and plants (Kedes, 1979; Probst et al., 2020). This was first established by using hybridization kinetics to gauge copy number, and eventually by whole genome sequencing. Specialized variants, in contrast, are often present in single copies; this has made genetic experiments addressing function more tractable for such variants than for canonical histones, which have mainly been subjected to genetic manipulations in yeast. However, with the advent of CRISPR/Cas technology, which can target common sequences in multicopy gene families or be used to delete large chromosomal regions, researchers are beginning to explore the effects of altered core histone sequences in metazoan organisms (Copur et al., 2018; Zhang et al., 2019).

Third, the bulk of histone synthesis occurs during S-phase, coincident with the replication of the genome and the requirement for a new complement of histones for its packaging. (The need for a sudden, large burst of histone synthesis at this juncture seems likely to have favored the high copy number of the histone genes in metazoans.) The coupling of DNA and histone synthesis was initially demonstrated by staining DNA in rat fibroblasts and liver cells with Feulgen dye followed by staining of associated basic proteins (predominantly histones) with fast green dye and observing that increased staining of histones was seen in cells showing greater DNA content (ploidy) (Bloch and Godman, 1955). In subsequent studies, histone synthesis was shown to occur in the S-phase by monitoring the incorporation of radiolabeled amino acid and thymidine (to follow protein and DNA synthesis) in the macronuclei of the ciliated protozoan *Euplotes eurystomus* and in HeLa cells (Prescott, 1966; Robbins and Borun, 1967). The *Euplotes* study benefited from the large size of these cells, which allowed visual identification of dividing cells, while synchronization of HeLa cells was achieved by now-classic mitotic shake-off methods. Histone mRNA synthesis was later convincingly shown to occur late in the G1 phase in yeast, by using temperature-sensitive cell cycle mutants (Hereford et al., 1982).

While S-phase specific expression is characteristic of canonical core histones across species, metazoan core histone genes are unique in lacking introns and in being the only eukaryotic mRNA species that are not polyadenylated, instead terminating in a conserved stem-loop structure that regulates their processing (Marzluff et al., 2008). The lack of introns in these "replication-coupled," or RC, histone genes is explained by the requirement for a specific secondary structure for their processing; insertion of an intron has been demonstrated to interfere with the proper formation of the 3′ end of the mRNA (Pandey et al., 1990). In contrast, functionally specialized histones, including some developmentally expressed variants, are generally synthesized throughout the cell cycle, as first reported by Wu and Bonner (1981). (Some specialized variants, such as H3.3 in sea urchins, are expressed both in S-phase and throughout the cell cycle (Marzluff et al., 2006).) The genes encoding replication-independent, or RI, histones often contain introns, and the mRNAs that they produce are polyadenylated, thereby providing an additional criterion distinguishing these "replacement" variants from the bulk of the core histones produced in metazoan cells. The distinction between "replication-coupled" and "replication-independent" histones (referring, of course, to replication of the cell's DNA complement) has long been recognized (Zweidler, 1984), and immediately suggests the possibility of distinct means for their incorporation into chromatin.

Histone deposition and chromatin maturation following DNA replication

Consideration of the mechanisms by which histone variants that are synthesized throughout the cell cycle are incorporated into chromatin necessitates a more general discussion of how chromatin is assembled in living cells. Histones can be incorporated into chromatin following DNA synthesis in S-phase (or during DNA repair), which accounts for the majority of histone deposition; a smaller fraction of nucleosomes present at any given time in a cell contains histones deposited in a replication-independent manner, either following histone eviction during processes such as transcription or DNA repair or via direct exchange, as discussed later on. Histones that are incorporated in a replication-dependent fashion derive from two sources: new histones that are synthesized in a burst at the beginning of S-phase and histones that are present in the parental chromatin. Evidence for the recycling of parental histones was first derived from studies in which histone synthesis was blocked by inhibiting translation in vivo by treatment with cycloheximide (Seale, 1976; Seale and Simpson, 1975; Weintraub, 1976). Replication continued after such treatment, albeit at a reduced rate, as shown by the incorporation of radiolabeled thymidine, while histone synthesis was blocked. Under these conditions, newly synthesized DNA, monitored by incorporation of radiolabel, displayed patterns of protection consistent with the presence of nucleosomes when subjected to digestion by MNase, while also exhibiting increased susceptibility to digestion. These findings indicated that histones from parental nucleosomes could be transferred to newly replicated DNA, but that newly synthesized histones were needed to allow complete assembly into chromatin. More compelling evidence for the recycling of parental histones, as it did not involve perturbing replication, came from studies showing that histones labeled isotopically or with fluorescent probes persisted in cells through multiple divisions (Leffak, 1984; Prior et al., 1980; Wu and Bonner, 1981). This conclusion was confirmed in numerous later studies, both in vivo and by in vitro replication-coupled nucleosome assembly assays using nuclear extracts (see below; reviewed in Annunziato (2012)) and provided the underpinning for studies on the mechanism of nucleosome assembly in living cells.

A burning question in early studies on nucleosome assembly following DNA replication was whether nucleosomes were assembled in a dispersive fashion, in which nucleosomes containing parental and newly synthesized histones assembled on both newly synthesized DNA strands, or whether the histones segregated such that one daughter DNA molecule was packaged with parental histones and the other with newly synthesized histones. Even before the discovery of the vast array of specific histone modifications and the enzymes that carry them out, the potential for heritable regulatory information to reside in chromatin was recognized, hence the importance attached to determining patterns of histone inheritance. Initial studies were controversial and contradictory (reviewed in Annunziato (2005, 2012)).

MNase digestion of chromatin assembled following replication in the presence of cycloheximide yielded short ladders characteristic of clustered nucleosomes and were interpreted as indicating conservative nucleosome assembly (Seale, 1976; Weintraub, 1976). Clustering of newly assembled nucleosomes was also supported by electron microscopy (EM) (Cremisi et al., 1978). Other reports, however, mostly using radiolabel to monitor newly synthesized histones and DNA, supported dispersive segregation of nucleosomes following replication (e.g., Fowler et al., 1982; Jackson et al., 1975; Russev and Hancock, 1982). Dispersive segregation of nucleosomes in HeLa cells was recently demonstrated on a genome-wide basis (Alabert et al., 2015). In this work, synchronized cells were released into S-phase in the presence of lysine and arginine labeled with heavy isotope (^{13}C-labeled lysine and ^{13}C, ^{15}N-labeled arginine), and biotin-labeled dUTP was used to label newly synthesized DNA. The biotin label allowed the isolation of newly replicated chromatin with streptavidin beads; mass spectrometry revealed that old and new histones were present in nearly equimolar abundance, indicating dispersive segregation of nucleosomes. The most parsimonious explanation for these (and other) various results at present is that parental nucleosomes are segregated more or less randomly to the daughter strands of replicated DNA, but that this occurs in a clustered fashion; the likelihood of a given parental nucleosome segregating to one or the other daughter strand is greater if the neighboring nucleosome was deposited on that same strand (Annunziato, 2012). Whether this means that cooperative assembly is driven by the assembly of nucleosomes from newly synthesized histones, or by the re-assembly of parental nucleosomes, is currently unclear.

A related question is whether histones from individual nucleosomes segregate and mix during replication of chromatin, or whether octamers are transferred intact from parent chromosome to daughter. This issue was tackled in early studies using various labeling protocols, with sometimes conflicting results (Annunziato, 2005; van Holde, 1988). The preponderance of evidence supported nucleosome disruption, specifically the segregation of H2A-H2B from $(H3\text{-}H4)_2$ tetramers in parental nucleosomes, during replication of chromatin. Support for this model included evidence for the presence of hybrid nucleosomes containing parental H2A-H2B dimers and newly synthesized H3 and H4, and vice versa, and for the sequential assembly first of $(H3\text{-}H4)_2$ tetramers followed by deposition of H2A-H2B dimers (discussed further below) (Annunziato et al., 1982; Jackson, 1987a,b; Jackson and Chalkley, 1981; Jackson et al., 1981; Russev and Hancock, 1981; Sogo et al., 1986). These and other studies, however, were generally not free of caveats, such as perturbation of DNA replication or histone synthesis, incomplete fractionation of newly synthesized and parental chromatin, or inferences based on indirect evidence. Nonetheless, this model was also consistent with the lower affinity binding of H2A-H2B dimers compared to $(H3\text{-}H4)_2$ tetramers in the nucleosome (Eickbush and Moudrianakis, 1978; Ohlenbusch et al., 1967), and with studies on chromatin maturation described below. Definitive evidence for nucleosomes containing parentally derived H2A-H2B and newly synthesized $(H3\text{-}H4)_2$ tetramers came from experiments in which expression of Flag-tagged histone H3 was induced and then shut off, allowing immunopurification of nucleosomes containing "old" H3 (Fig. 5.2; Xu et al., 2010).

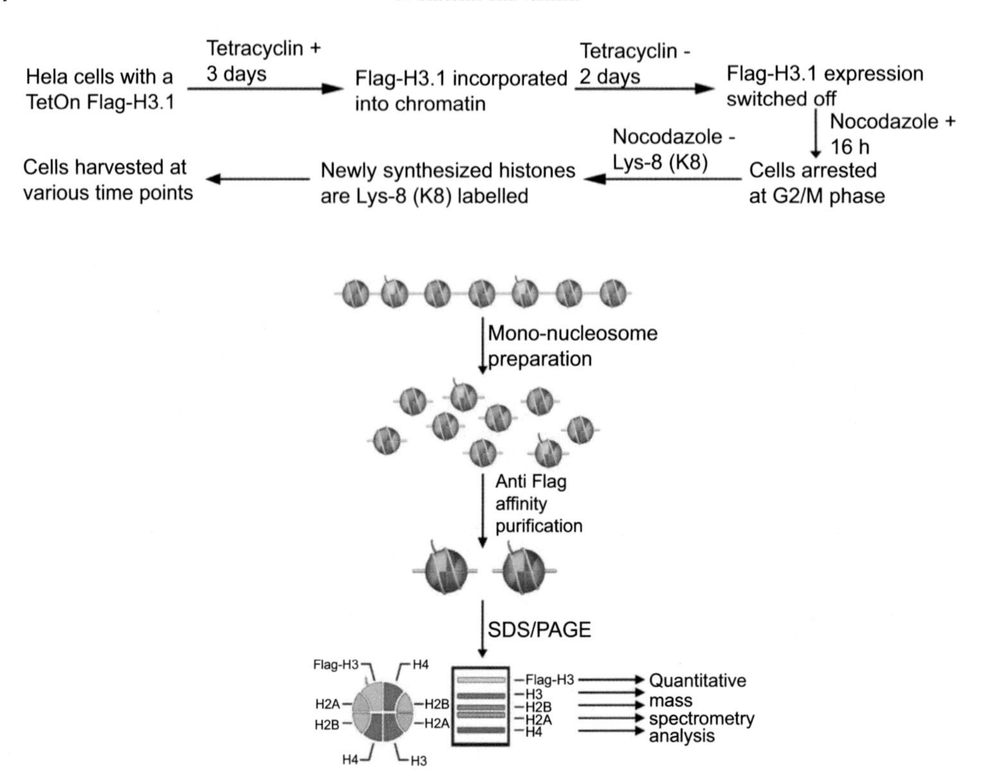

FIG. 5.2 Scheme for determining the distribution of parental and newly synthesized histones in nucleosomes containing parental, or "old," histone H3. *From Xu, M., Long, C., Chen, X., Huang, C., Chen, S., Zhu, B., 2010. Partitioning of histone H3-H4 tetramers during DNA replication-dependent chromatin assembly. Science 328, 94.*

Following the shutoff of Flag-H3 expression, cells were arrested at G2/M and released into a medium containing lysine labeled with ^{13}C and ^{15}N to mark newly synthesized histones. Mass spectrometry of isolated nucleosomes containing "old" histone H3 showed that while very little new H3 or H4 was associated with these nucleosomes, about 50% of the H2A and H2B associated with old histone H3 was newly synthesized. Thus new and old H3 and presumably H4 are not mixed in $(H3\text{-}H4)_2$ tetramers, while parental H2A-H2B dimers appear to be deposited essentially randomly to new and old $(H3\text{-}H4)_2$ tetramers in replicating chromatin. These results have important implications for the inheritance of histone modifications.

Some of the first studies on histone deposition focused on chromatin assembly during S-phase and gave rise to the concept of chromatin maturation. EM of replicating chromatin fibers from *Drosophila* embryos revealed particles closely resembling those first reported by Woodcock and Don and Ada Olins (Chapter 2) occurring at both unreplicated and newly replicated DNA close to replication forks, suggesting that nucleosomes did not dissociate in advance of replication and were rapidly reassembled onto newly replicated chromatin (McKnight and Miller, 1977). A follow-up study used immunomicroscopy to show that the particles observed included histones H3 and H2B (McKnight et al., 1978). However, a caveat of these studies is that nuclei in the syncytial (i.e., lacking cell membranes; essentially this is a bag of nuclei) *Drosophila* blastoderm embryo divide extremely rapidly, with an S-phase of about 3 min (Rabinowitz, 1941), and the embryos contain a large supply of maternally derived histones (Cermelli et al., 2006); thus, nucleosome assembly in these nuclei could represent a special case differing from more slowly dividing cells. Presumptive nucleosomes were also observed by EM close to replication forks of SV40 minichromosomes; although viral replication is rapid, it occurs without the benefit of a large pool of histones and would be less likely to involve specialized assembly factors that might operate in rapidly dividing embryonic cells, thus supporting the likely generality of rapid assembly of nucleosomes following DNA synthesis (Seidman et al., 1978). A subsequent study monitored nucleosome deposition in replicating SV40 chromatin by using [^{3}H]-thymidine in combination with exonuclease treatment (recall from Chapter 4 that nucleosomes block exonuclease cleavage of DNA) to determine the average distance of nucleosomes from the replication fork and reported an average distance of 125 to 300 bp from either 3′ or 5′ ends of nascent DNA chains to the first nucleosome (Herman et al., 1981). These results were supported by later experiments in which psoralen cross-linking, which occurs efficiently on naked DNA but is inhibited by nucleosomes, was used in combination with EM to again investigate replicating SV40 chromatin (Sogo et al., 1986).

A maturation process of newly assembled chromatin was revealed in studies using nucleases to probe chromatin structure. Seale employed a pulse-chase protocol in which replicating chromatin was briefly labeled with [^{3}H]-thymidine, and the susceptibility of the labeled chromatin to DNase I was compared to that of bulk chromatin (Seale, 1975). These experiments showed that newly synthesized DNA in HeLa cells was more sensitive than bulk chromatin to DNase I digestion, but recovered to exhibit similar resistance as bulk chromatin in ~15 min. Similar results were obtained using MNase: newly synthesized DNA was digested to mononucleosomes more rapidly than bulk chromatin, but within ~10 min of synthesis was restored to a state having MNase sensitivity of bulk chromatin (Hildebrand and Walters, 1976). Protease sensitivity and lability in the presence of 0.45 M NaCl also indicated a maturation phase of newly replicated chromatin consistent with the nuclease digestion data (Seale, 1981).

Although the notion of a maturation process following the assembly of chromatin after DNA replication was supported by work from several laboratories (reviewed in Annunziato (2012)), there was a problem. The rapid rate of DNA synthesis (~3 kb/min) (Annunziato and Seale, 1983a) would mean that on the order of 100 nucleosomes observed in the EM on newly replicated DNA should be altered in some way, but nucleosomes near the replication fork appeared normal. A potential solution to this conundrum came from a study that examined the staged assembly of nucleosomes in cultured *Drosophila* cells (Worcel et al., 1978). When chromatin from *Drosophila* tissue culture cells was lightly digested with MNase and fractionated by high-speed centrifugation through sucrose gradients, newly synthesized chromatin, labeled with a pulse of [^{3}H]-thymidine, was found to sediment more slowly than bulk chromatin. Using a pulse-chase protocol, it was then found that newly synthesized histones co-sedimented with newly replicated DNA, and that nascent H3-H4, H2A-H2B, and H1 were associated with different fractions in the gradient. Digestion products sedimenting at 10-15S contained H3 and H4, but lacked H2A-H2B, while more rapidly sedimenting species contained all four core histones. The authors concluded that histones H3 and H4 were deposited first on nascent chromatin, followed after 2–10 min by H2A-H2B, and finally, after 10–20 min, by linker histone H1. These findings were in good accord with considerable work indicating that an $(H3\text{-}H4)_2$ tetramer formed the core of the nucleosome; in particular, reconstituted tetramer particles resembled nucleosomes when analyzed by nuclease or protease digestion or low-angle X-ray diffraction, and in their topological properties (Bina-Stein and Simpson, 1977; Boseley et al., 1976; Camerini-Otero and Felsenfeld, 1977;

Camerini-Otero et al., 1976; Sollner-Webb et al., 1976). Furthermore, analysis of intermediates formed during nucleosome reconstitution by salt dialysis revealed the initial assembly of DNA with H3 and H4 to form a nucleosome-like structure, followed at a lower salt concentration by incorporation of H2A and H2B (Wilhelm et al., 1978). Nucleosome tetramers are nearly indistinguishable from intact nucleosomes in the EM (Bina-Stein, 1978; Wilhelm et al., 1978), and so might be taken for canonical nucleosomes in the micrographs of replicating chromatin. Taken together, these findings indicate a scenario in which $(H3\text{-}H4)_2$ tetramers are rapidly assembled onto newly synthesized DNA, followed by the incorporation of H2A and H2B and linker histone to complete the assembly of mature chromatin. This scenario was supported by later studies of nucleosome assembly facilitated by histone chaperones, as described in the next section.

The experiments described above still left unresolved the apparent presence of H2B in the particles near the replication fork observed by immunomicroscopy (McKnight et al., 1978). One possible explanation is that this apparent paradox represents an anomaly of the rapidly dividing nuclei of the syncytial embryo, as suggested above. Another possibility, proffered recently by the Kadonaga group, is based on in vitro experiments indicating the formation of a structure dubbed the prenucleosome during the initial stages of reconstitution (Fei et al., 2015; Khuong et al., 2015; Torigoe et al., 2011). These structures were identified as stable histone-DNA assemblages that formed rapidly upon incubation of a complex of histones and a histone chaperone (see next section) with a DNA template, and which did not confer supercoiling on the DNA but could not be competed away by the addition of a second DNA template. Upon addition of an ATP-dependent motor protein, such as the *Drosophila* chromatin remodeler ATP-utilizing chromatin assembly and remodeling factor (ACF), these prenucleosomes are converted into canonical nucleosomes in a much slower (~10–15 min) reaction. While the idea that the structures observed by EM near replication forks are in vivo counterparts of these in vitro generated prenucleosomes is appealing, it contradicts the interpretation of several groups that nucleosomes near the fork lack H2A and H2B (which itself contradicts the immunomicroscopy results of McKnight and colleagues) (e.g. Seale, 1981; Senshu and Yamada, 1980); in addition, the lack of supercoiling observed in prenucleosomes is difficult to reconcile with the condensed structures observed in the EM.

An alternative explanation for the maturation of newly assembled chromatin comes from studies examining posttranslational modification of newly synthesized histones. Histone H4 in particular was discovered to be rapidly diacetylated in the cytoplasm following its synthesis and then deacetylated in the nucleus over a time span of ~20–30 min (Jackson et al., 1976; Ruiz-Carrillo et al., 1975). Inhibition of the deacetylation reaction by treatment of HeLa cells with sodium butyrate did not prevent assembly of newly synthesized DNA into chromatin, but caused prolonged sensitivity of nascent chromatin to DNase I; the maturation process was stalled and could be restored by removal of butyrate (Annunziato and Seale, 1983b). Curiously, however, when assayed by MNase digestion, maturation of nascent chromatin proceeded normally. It was proposed that histone acetylation might affect higher-order chromatin structure, thereby producing differential susceptibility to DNase I and MNase, and indeed we saw in Chapter 3 that histone acetylation impacts chromatin condensation. A follow-up study reported that butyrate treatment induced depletion of linker histone H1 from newly replicated chromatin, supporting the idea that deacetylation of histone H4 was required to allow condensation of chromatin to a state exhibiting the properties of mature chromatin (Perry and Annunziato, 1989). Although this is in many regards a satisfying explanation for observations of chromatin maturation, it leaves unresolved the kinetics of H2A-H2B deposition versus that of $(H3\text{-}H4)_2$. A full understanding of the process of the maturation of newly assembled chromatin, and the relation of the "prenucleosome" to the structures observed close to active replication forks, will require a better characterization of those structures in living cells.

Histone chaperones

That complex three-dimensional structures can be effectively encoded in linear arrays of amino acids is a remarkable fact of biology. Similarly, sufficient information resides in the sequences of the four core histones to allow their assembly into nucleosomes, under the right conditions, simply by mixing them with DNA (Chapter 1, Reconstituting chromatin sidebar). The principal requirement for this process to occur efficiently in vitro is the neutralization of the positively charged histones, to prevent their nonspecific aggregation with negatively charged DNA. This can be accomplished by initiating the process at high salt, typically 2M NaCl, or by the use of other counterions such as tRNA or polyglutamic acid (Nelson et al., 1981). However, chromatin assembled in this fashion does not resemble that found in cells since the spacing of nucleosomes is random or grossly uneven, with clusters of nucleosomes being interrupted by relatively long stretches of naked DNA (Spadafora et al., 1978). Moreover, these methods of reconstitution are decidedly nonphysiological, and although examples of nucleosome assembly at physiological salt concentrations have been reported, these differed from physiological conditions in other regards and in some cases led to a mixture of soluble nucleosomes and insoluble precipitates (Ruiz-Carrillo et al., 1979; Stein et al., 1979). Compelled both by the desire to reconstitute native-like chromatin in the test tube and to understand the process by which chromatin was assembled in living cells, researchers turned to investigating nuclear extracts as nucleosome assembly systems.

One of the first approaches to utilizing extracts to facilitate chromatin assembly took advantage of the eggs and oocytes of the African clawed frog *Xenopus* (for an extensive discussion of this topic, see Wolffe (1998)). The popularity of the eggs and oocytes of *Xenopus laevis* as an experimentally manipulable model for the study of embryonic development can in part be attributed to research performed in the 1950s by John Gurdon, who was then a graduate student at Oxford University. Researchers at the time were keen on knowing whether the nuclear contents, and specifically the DNA, from differentiated tissues retained all the information needed to allow the complete development of a fertilized egg into an adult organism. Briggs and King addressed this question by injecting enucleated eggs from the American leopard frog, *Rana pipiens*, with cell nuclei from later stages of development, and reported that blastula nuclei thus transplanted allowed normal development (production of swimming tadpoles), but nuclei from gastrula or later stages had reduced capability of doing so (Briggs and King, 1952; King and Briggs, 1955). Gurdon, however, performed nuclear transplantation experiments using eggs from *Xenopus laevis*, which develops more quickly than *Rana* and can regenerate lost limbs, and reported that nuclei even from adult intestinal epithelial cells could support complete development, albeit at a very low success rate (see Sidebar: Persistence) (Gurdon, 1962b).

Sidebar: Persistence

John Gurdon was well prepared to overcome adversity by the time he undertook his nuclear transplantation experiments as a graduate student in Michael Fischberg's laboratory in the Department of Zoology at Oxford University (Maayan and Cohmer, 2012). His schoolmaster at Eton College labeled the 15-year-old Gurdon the worst pupil he had ever taught after his first semester of science instruction, stating that it would be a waste of time for him to continue such pursuits. Gurdon turned to the study of classics but was required also to study zoology as a condition of admittance to Christ Church College at Oxford University, and achieved first-class honors. His first choice of professor to work with during his doctoral studies rejected him, and he wound up in Fischberg's lab. His first efforts to perform nuclear transplantation using *Xenopus* eggs failed because the micropipette needed to manipulate the nuclease could not penetrate the tough gel coating of the egg. In the end, it required 726 nuclear transfer experiments for the young student to produce 10 feeding tadpoles (Gurdon, 1962b). Given his response to these challenges, it should come as no surprise that Gurdon did not back down from his refutation of Briggs's and King's assertion that development could not be faithfully recapitulated using nuclei from adult tissues; one paper was uncompromisingly titled "Adult Frogs Derived from the Nuclei of Single Somatic Cells" (Gurdon, 1962a). Fifty years later, having been honored with Knighthood and having an institute named after him at Cambridge University, Gurdon was awarded, along with Shinya Yamanaka, the Nobel Prize in Medicine or Physiology.

These nuclear transplantation experiments, which were the first to be done outside of protozoa and were precursors to modern stem cell biology, underscore the advantages of *Xenopus* eggs and oocytes. These cells are very large, 1.2–1.4 mm in diameter (thus their special advantage for nuclear transplantation), and can readily be obtained in large quantities (>5000 eggs or oocytes from a single frog). *Xenopus* particularly excels in this regard, as its reproductive capabilities do not vary with season, in contrast to *Rana*, and it can be induced to lay eggs by injection of the dorsal lymph sac with human chorionic gonadotropin, while *Rana* requires injection of homogenized pituitary glands (not from humans!) for this purpose (Jones and Smith, 2008). These oocytes (and eggs) are storehouses of the components needed for early development of the embryo, as transcription does not begin until the mid-blastula transition, after about 12 synchronous cell divisions (4096 cells) (Newport and Kirschner, 1982). The provisions thus stored include ribosomes, tRNA molecules, RNA and DNA polymerases, and core histones in excess of those present in somatic cells by four to five orders of magnitude, making these cells outstanding vehicles for studying transcription, translation, and the cell cycle in addition to basic investigations of embryonic development (Almouzni and Wolffe, 1993; Brown, 2004).

Of particular interest with regard to chromatin assembly, core histones were found to be present at ~15,000-fold excess in *Xenopus* eggs relative to somatic cells (Adamson and Woodland, 1974). Laskey and colleagues went about testing extracts from unfertilized eggs of *Xenopus* for their ability to assemble nucleosomes onto exogenous DNA. Extracts were prepared by collecting and homogenizing eggs, centrifuging at low speed to remove lipids and a yolk pellet, and finally performing high-speed centrifugation and collection of the clear upper supernatant (Laskey et al., 1977). This extract proved capable of packaging SV40 closed circular DNA molecules into nucleosomes, as assayed by the generation of a ladder of fragments after treatment with MNase, and the incorporation of negative superhelical turns (see Chapter 2, section on "The path of the DNA"). Subsequent fractionation of extracts allowed the identification of an abundant, acidic protein later dubbed nucleoplasmin as a critical factor in this in vitro, extract-mediated nucleosome assembly reaction (Laskey and Earnshaw, 1980; Laskey et al., 1978); nucleoplasmin was consequently the first protein to be termed a histone chaperone (Laskey et al., 1978). Histone chaperones have been defined more rigorously since the discovery of nucleoplasmin as "factors that associate with histones and stimulate a reaction involving histone transfer without being part of the final product" (De Koning et al., 2007). Histone chaperones have also sometimes been considered more narrowly as "being defined by their ability to bind histones and deposit them on DNA," or more broadly as "proteins that specifically bind and protect positively charged histones from making spurious ionic interactions with proteins and DNA without the use of ATP hydrolysis" (Grover et al., 2018). The first definition (from De Koning et al. (2007)) seems most useful with respect to chromatin assembly, as it includes factors that escort histones from cytoplasm to nucleus, or that participate in pathways in which histones are transferred between complexes before being assembled with DNA into nucleosomes, while emphasizing pathways that eventually lead to such assembly as opposed to proteins or complexes that function only as storage depots for histones.

Although nucleoplasmin was initially believed to be sufficient to catalyze nucleosome assembly in the context of *Xenopus* extracts, subsequent work caused this picture to be revised. Fractionation of *Xenopus* oocyte extracts revealed that H3 and H4 were sequestered in a complex eventually found to contain proteins N1 and N2, which had first been identified in pioneering studies of nuclear import (Bonner, 1975; De Robertis et al., 1978; Kleinschmidt et al., 1985; Kleinschmidt and Franke, 1982). Despite its highly acidic character, being composed of about 30% aspartic acid and glutamic acid residues, nucleoplasmin, as shown by immunoprecipitation experiments, associated only with H2A and H2B (Dilworth et al., 1987; Kleinschmidt et al., 1985). Depletion of either complex from extracts abrogated nucleosome assembly, and incorporation of H2A and H2B required prior deposition of H3 and H4 (Dilworth et al., 1987; Kleinschmidt et al., 1990), consistent with the ability of H2A and H2B, but not H3 and H4, to exchange from chromatin fibers in vitro at approximately physiological conditions (Louters and Chalkley, 1984). Altogether, then, these experiments revealed an assembly pathway occurring at a physiological salt concentration that followed the same order as had been determined using reconstitution with purified histones and high salt concentrations. This further validated the notion of the $(H3\text{-}H4)_2$ tetramer as comprising the core of the nucleosome, with its assembly with DNA being a prerequisite for the deposition of H2A-H2B.

Although the assembly of chromatin using extracts represented a major advance, in some ways this process was still very nonphysiological. First, assembly reactions were conducted using massive amounts of DNA—over a thousand-fold excess relative to that found in the egg or oocytes. In contrast, in the developing embryo, the stored histones are incorporated into chromatin over the course of multiple cell divisions, with assembly being coupled to DNA synthesis. Second, assembly in vitro required on the order of an hour to proceed to completion, too slow to accommodate the in vivo assembly in which the first 12 cell divisions of the *Xenopus* embryo take place over the course of about seven hours. Third, the spacing between nucleosomes assembled using *Xenopus* extracts did not reflect the physiological spacing of ~190 bp. Fourth, the histones sequestered by nucleoplasmin and N1/N2 contained a higher proportion of the replication-independent histones than was observed in somatic cells (Dilworth et al., 1987). These discrepancies raised the possibility that nucleoplasmin and N1/N2 might function principally as storage proteins in vivo and served only to prevent aggregation, similarly to other large, acidic molecules, during the assembly reaction. Faced with these issues and seeking to understand the natural mechanisms of chromatin assembly, investigators turned to systems that would allow the coupling of DNA replication to nucleosome assembly.

The closed circular DNA of the SV40 virus, which we saw in Chapters 2 and 4 was used in early studies of nucleosome topology and positioning, proved useful in characterizing extracts capable of coupling nucleosome assembly to DNA replication. SV40 was used to test and successfully establish extracts capable of replicating DNA in vitro; replication of SV40 using cytosolic extracts required only the addition of SV40 T antigen (Li and Kelly, 1984). Bruce Stillman, who would later succeed Jim Watson as director of Cold Spring Harbor Laboratory, showed that adding nuclear extract from human embryonic kidney 293 cells to the cytosolic extract resulted in replication-coupled assembly of nucleosomes, as demonstrated by induction of negative supercoils, a characteristic ladder of DNA fragments resulting from MNase digestion, and electron microscopic images of beads-on-a-string particles (Stillman, 1986). Using an approach similar to that employed to characterize the packaging of newly synthesized DNA into chromatin described earlier, Stillman showed that newly replicated SV40 molecules were preferentially packaged into chromatin over nonreplicated SV40; this work also demonstrated that nucleosome assembly did not depend on the template being packaged into chromatin before replication. The supercoiling assay was then used in conjunction with biochemical fractionation to identify a small protein complex that was responsible for the replication-coupled assembly of nucleosomes (Smith and Stillman, 1989). Assembly occurred in a staged fashion, with H3-H4 being deposited before H2A-H2B. The complex was designated as chromatin assembly factor 1, or CAF-1, and later shown to comprise subunits termed p48, p60, and p150 (Kaufman et al., 1995). Antibodies generated against CAF-1 subunits allowed the visualization of CAF-1 association with replication foci during S-phase in HeLa cells, cementing its role in replication-coupled chromatin assembly (Krude, 1995).

Concurrently with the work using human cells, a cell-free system from *Xenopus* eggs was reported that supported DNA replication and assembly of replicated DNA into chromatin (Almouzni and Mechali, 1988). Like the human cell-derived extract, the *Xenopus* extract preferentially packaged replicated DNA into nucleosomes over nonreplicated DNA, and did so in a staged fashion, with H3-H4 being incorporated before H2A-H2B in a separable, slower step (Almouzni et al., 1990). In a separate line of research aimed at constructing chromatin templates for use in transcription studies, extracts from eggs and embryos of *Drosophila* were being employed in attempts to assemble chromatin in which nucleosomes were not close-packed but adopted physiological spacing, as described in Chapter 4 (Becker and Wu, 1992; Kamakaka et al., 1993). These extracts proved highly useful in constructing templates for investigations of the influence of chromatin in transcription but, like the *Xenopus* egg extracts that relied on nucleoplasmin as a histone chaperone, they did not couple assembly with DNA replication. Nonetheless, biochemical fractionation of these extracts allowed the identification of several histone chaperones, including nucleoplasmin, nucleosome assembly protein 1 (NAP-1) (see below), CAF-1, and remodeling activities necessary for achieving physiological nucleosome spacing, as described in Chapter 4 (Ito et al., 1996a,b; Tyler et al., 1996). Moreover, replication-coupled nucleosome assembly could be achieved by depleting the extract of CAF-1 and supplementing it with partially purified CAF-1 in more limited amounts (Kamakaka et al., 1996). Homologs of CAF-1 subunits were also found in budding yeast; thus CAF-1 was firmly established as a participant in a highly conserved mechanism for the assembly of newly replicated DNA into chromatin (Annunziato, 2012; Kaufman et al., 1997). Somewhat surprisingly, although deletion of CAF-1 subunits in yeast causes defects in heterochromatin-associated transcriptional silencing and increased sensitivity to UV-induced DNA damage, such deletion was found not to be lethal, indicating the existence of alternative pathways of nucleosome assembly in vivo (Enomoto et al., 1997; Kaufman et al., 1997). One such pathway turns out to be provided by the H3-H4 chaperone Asf1, discussed below. Reassuringly, CAF-1 is essential in human and *Drosophila* cells (although it is not essential in *Arabidopsis*), with its depletion resulting in defects in S-phase chromatin assembly (Hoek and Stillman, 2003; Kaya et al., 2001; Song et al., 2007; Ye et al., 2003). Given the extreme evolutionary conservation of the core histones, it is hardly surprising to find that the mechanism of their deposition following DNA replication would also be conserved.

CAF-1 associates with H3-H4 via its p48 subunit, while the p150 subunit binds proliferating cell nuclear antigen (PCNA), the processivity clamp for DNA polymerase, and thus couples deposition of H3-H4 to DNA replication (Shibahara and Stillman, 1999; Verreault et al., 1996). Subsequent investigations led by Geneviève Almouzni, using cell extracts from a *Drosophila* cell-free system, showed that nucleosome assembly following repair of DNA damage via the NER (nucleotide excision repair) pathway also depends on CAF-1 and its interaction with PCNA (Gaillard et al., 1996; Moggs et al., 2000). CAF-1, however, is not the only chaperone that associates with H3-H4 in cells. Jessica Tyler, Jim Kadonaga, and colleagues sought to identify novel factors involved in replication-coupled nucleosome assembly from the *Drosophila* embryo extracts mentioned previously (Tyler et al., 1999). These investigators found that when excess naked DNA was added to the human 293 cell extracts used by Stillman and colleagues, the extracts failed to assemble circular SV40 DNA into chromatin even when supplemented with previously identified assembly factors (CAF-1, NAP-1, or nucleoplasmin) or additional histones; replication-coupled assembly could be restored, however, by addition of a crude extract from *Drosophila* embryos. Purification of the assembly activity from the extract revealed it to be a complex of a single protein, dASF1, with histones H3 and H4, which was named Replication-Coupled Assembly Factor, or RCAF. Microsequencing revealed that H3 and H4 present in RCAF were modified by acetylation at sites that are acetylated in newly synthesized histones (see Sidebar: Modifications of newly synthesized histones). Specifically, Lys5 and Lys12 of histone H4 and Lys14 of H3 were acetylated; as these residues are deacetylated rapidly following their deposition onto newly replicated DNA (Annunziato, 2012), this observation suggested that the newly identified complex could function to escort H3 and H4 from the cytoplasm into the nucleus and facilitate their incorporation into chromatin. The third protein in the complex was sequenced and found to have homology to the yeast protein Asf1 (Antisilencing function 1), which had been identified in genetic screens by its ability to derepress transcriptional silencing at the silent mating type loci and telomeres when overexpressed (Le et al., 1997; Singer et al., 1998), and therefore this protein was named dASF1. (Asf1 from human cells was also known as CIA, for CCG1-interacting protein A (Umehara et al., 2002).) In vitro assembly reactions performed using the dASF1-H3-H4 complex, CAF-1, or both revealed that the two complexes function synergistically, at least in the context of the extract-mediated assembly reaction. In vivo, yeast cells lacking Asf1 are viable but exhibit defects in cell cycle progression, while *cac1* cells (lacking CAF-1 activity) did not show a growth defect; *asf1 cac1* double mutants had a doubling time greater than *asf1* cells, suggesting functional redundancy between CAF-1 and Asf1 in yeast (Tyler et al., 1999). The nonessential nature of Asf1 in budding yeast appears to be the exception, as Asf1 was found to be essential in *S. pombe*, *Drosophila*, and vertebrate cells, with depletion leading to defects in S-phase progression (Groth et al., 2005; Sanematsu et al., 2006; Schulz and Tyler, 2006; Umehara et al., 2002). Further supporting a direct role in replication-coupled nucleosome assembly, Asf1 was observed by immunofluorescence at sites of active replication in *Drosophila* cells (Schulz and Tyler, 2006).

Sidebar: Modifications of newly synthesized histones

Readers of this book are likely well aware of the scope and impact of histone modifications on chromatin biology (Chapter 7, Histone modifications). Acetylation of Lys5 and Lys12 of histone H4 are likely the first such modifications to be directly tied to a specific cellular process, specifically that of incorporation of newly synthesized histones into chromatin (Annunziato, 2012). First observed as a reversible modification occurring during spermatogenesis in trout (harking back to Miescher's early studies on nuclein) (Louie et al., 1974), acetylation of newly synthesized histone H4 was detected in dividing duck erythroblasts and in rat hepatoma cells by very brief (1 min) pulse labeling with tritiated leucine followed by analysis using gel electrophoresis (Jackson et al., 1976; Ruiz-Carrillo et al., 1975). These studies demonstrated first that H4 was "modified"; that is, the prominent radiolabeled histone band was the second band above the parent, unmodified H4 seen by Coomassie staining, and that the modification was specifically acetylation. Subsequent work revealed that the diacetylation of newly synthesized H4 was widely conserved, ranging from yeast and *Tetrahymena* to frogs and sea urchins, among others (Annunziato, 2012). In contrast, acetylation of newly synthesized H3 varies among species, and little modification is seen in mammalian cells (Annunziato, 2012). Microsequencing newly synthesized H4 pulse-labeled with tritiated lysine revealed the modification to be specific for Lys5 and Lys12 in *Drosophila* and HeLa cells, later confirmed for a variety of additional organisms (first observed at the cognate positions in *Tetrahymena*, Lys4, and Lys11) (Chicoine et al., 1986; Sobel et al., 1995). Despite the near universality of this modification, it does not appear essential to the process of either nuclear import of the histones or for their assembly into chromatin, as shown by the viability of yeast expressing only tail-less histone H4 and the incorporation of H4 lacking these residues into chromatin when expressed in *Xenopus* embryos, as described later in this chapter (Freeman et al., 1996; Kayne et al., 1988). However, mutation of these residues does affect growth, lengthening S-phase in yeast (Megee et al., 1990), and tail-less H4 expressed in *Physarum* competes poorly against wild-type H4 for incorporation into chromatin in *Physarum* (Ejlassi-Lassallette et al., 2011). Although these and other experiments support the importance of this conserved modification for incorporation of newly synthesized histones into chromatin, it is something of an irony that the precise molecular mechanism underlying this function remains elusive more than 40 years after these modifications were first reported.

A surprising aspect of nucleosome assembly that emerged from studies on CAF-1 and Asf1 was the finding that histones H3 and H4 were chaperoned as a heterodimer. Classic studies of the histone octamer, described in Chapter 2, indicated that the H3-H4 complex was most stable in tetramer form (Eickbush and Moudrianakis, 1978), and a multitude of biochemical studies supported an $(H3\text{-}H4)_2$ tetramer as the core component of the nucleosome, as we have discussed. This dogma was first challenged by experiments in which H3 and H4, along with associated proteins, were isolated from HeLa cell nuclear extracts in predeposition complexes along with chaperone proteins including Asf1 and components of the CAF-1 complex (Tagami et al., 2004). Isolation was performed by expressing H3 containing C-terminal Flag and HA epitope tags and using sequential immunoaffinity purification to purify the tagged histone and associated proteins. Although the cells expressed only a low proportion of epitope-tagged H3 relative to untagged, canonical H3, immunoprecipitation of the tagged H3 did not recover untagged H3, implying that tetramers were absent from the predeposition complexes. Soon after this was reported, Asf1 was reported to exist as a 1:1:1 complex with

H3 and H4 (English et al., 2005). These findings engendered some controversy, but structural studies helped quell skepticism, showing that Asf1 binding occluded the C-terminal region of H3 that is critical for forming the four-helix bundle in the $(H3\text{-}H4)_2$ tetramer (English et al., 2006; Natsume et al., 2007).

A major obstacle to a detailed and precise elucidation of the pathway from histone synthesis to deposition into newly replicated chromatin is the rapidity of the process; pulse-chase experiments in mammalian tissue culture cells indicated that histones are deposited into chromatin less than a minute after their synthesis (Bonner et al., 1988). Nonetheless, a large body of structural, biochemical, and cell biological research elucidating interactions among Asf1, CAF-1, replication proteins and histones H3-H4 has allowed the formulation of a model for a complex set of pathways for the deposition of newly synthesized and parental H3 and H4 following DNA replication or damage (reviewed in Grover et al. (2018), Gurard-Levin et al. (2014), Pardal et al. (2019), Sauer et al. (2018)). A simplified version of such models is depicted in Fig. 5.3; some variations on this basic model occur in a species-specific manner. Folding, posttranslational modification, and heterodimerization of H3 and H4 are first assisted by heat shock proteins and NASP (nuclear autoantigenic sperm protein) family members, which are related to the *Xenopus* N1/N2 proteins described earlier (Campos et al., 2010; O'Rand et al., 1992). Folded and processed H3-H4 dimers are then likely transferred from the NASP-containing complex

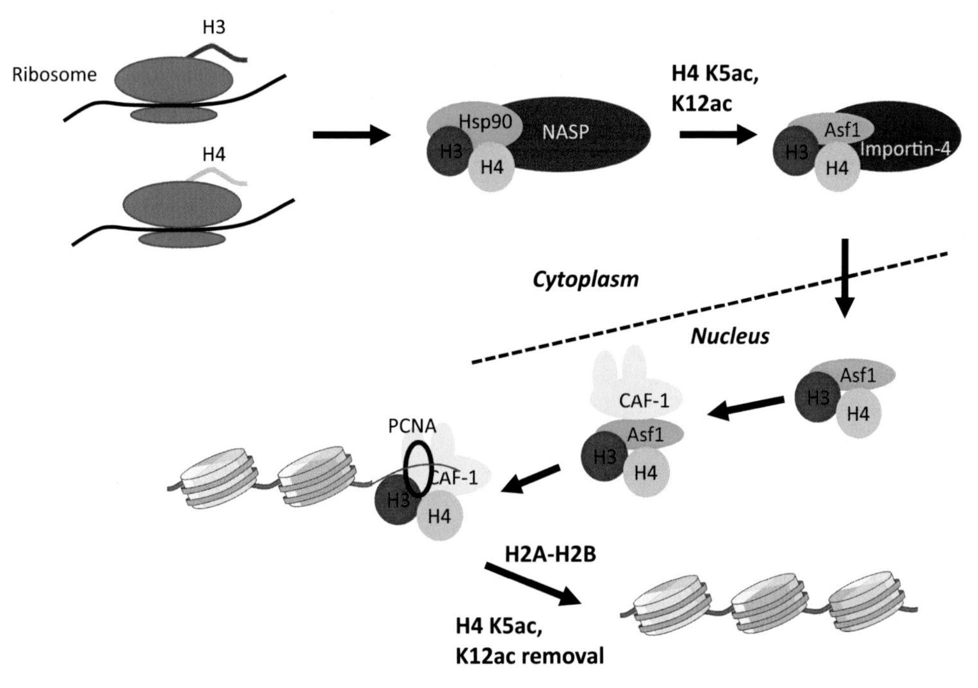

FIG. 5.3 Simplified schematic of replication-coupled histone deposition. Newly synthesized H3 and H4 are folded and assembled into a heterodimer with assistance from Hsp90 and NASP and are acetylated before handoff to Asf1 and Importin-4, the latter of which facilitates nuclear import. In the nucleus, the heterodimer is again handed off to CAF-1, which interacts with PCNA, resulting in $(H3\text{-}H4)_2$ deposition. This is followed rapidly by the removal of the H4K5ac and K12ac marks and incorporation of H2A-H2B, which may derive from newly synthesized histones or from a parental nucleosome, to complete the assembly of a nucleosome on the newly replicated DNA. Some intermediate steps and details have been omitted for clarity.

to Asf1, based on the observation of a quaternary complex in vitro containing NASP, Asf1, and H3-H4, and the identification of a complex present in cytosolic extracts from HeLa cells containing Asf1, the H3-H4 dimer, and the nuclear import protein, or karyopherin, importin-4 (Campos et al., 2010). The mechanism by which H3-H4 dimers are transferred from the NASP-containing complex to Asf1 before nuclear import remains to be elucidated.

Following nuclear import, histones are transferred from the Asf1-H3-H4 complex to CAF-1; the association of CAF-1 with PCNA, as already mentioned, then facilitates the assembly of nucleosomes at the replication fork. This "handoff" mechanism was first proposed based on several pieces of circumstantial evidence (Mello et al., 2002):

- Interaction between CAF-1 and Asf1 was shown by pulldown assays using HeLa cell nuclear extracts; in vitro assays demonstrated a direct interaction between Asf1 and the middle subunit (p60 in humans and p105 in *Drosophila*; also called Cac2) of CAF-1 and showed that binding of Asf1 and CAF-1 was enhanced by H3-H4 association with Asf1 (Mello et al., 2002; Tyler et al., 2001). (A wrinkle in this finding is that Asf1 exists as two paralogs in most vertebrates; although both proteins were associated with CAF-1 in pulldown assays using HeLa cell extracts, later work demonstrated a preferential interaction between one paralog, Asf1b, and the p60 subunit of CAF-1, while Asf1a was found to interact preferentially with the H3.3 chaperone HIRA (see "H3.3" section) (Abascal et al., 2013; Daganzo et al., 2003; Tagami et al., 2004).)
- In nucleosome assembly reactions coupled either to DNA replication or repair carried out using extracts from human cells, Asf1 synergized with CAF-1 but assembly depended on CAF-1 (Mello et al., 2002; Sharp et al., 2001).
- CAF-1 and PCNA were recruited to damaged DNA in pulldown assays, but Asf1 was not (Mello et al., 2002).

Additional support for this model derived from structural and biochemical studies that revealed that the p60 subunit of CAF-1 binds on the opposite face of Asf1 from H3-H4, indicating the possibility of a tripartite complex that could be involved in the handoff of H3-H4 (although how an H3-H4 dimer would then be transferred from the far side of Asf1 to CAF-1 remains unclear) (English et al., 2006; Liu et al., 2012b; Malay et al., 2008; Tang et al., 2006). In addition, epitope-tagged histone H3 mutants defective in their interactions with Asf1, when expressed in HeLa cells, were incorporated into chromatin inefficiently in comparison to wild-type tagged H3 (Galvani et al., 2008). Although numerous references in the literature refer to transfer of H3 and H4 from Asf1 to CAF-1 during nucleosome assembly as an established fact, it was not until 2014 that perhaps the most compelling evidence for this was presented when Groth and colleagues showed that blocking the interaction between Asf1 and CAF-1 by overexpression of a peptide binding Asf1 at the interaction interface caused human cells to arrest in S-phase with concomitant inhibition of DNA replication (Mejlvang et al., 2014). Expression of a mutant version of the interfering peptide that did not bind to Asf1 had no such effect. The mechanism by which the H3-H4 dimer is transferred from Asf1 to CAF-1 remains to be determined; Asf1 also apparently has some capability to facilitate nucleosome assembly independently of CAF-1 in budding yeast, given the mutant phenotypes mentioned earlier.

Once newly synthesized H3-H4 dimers have been transferred to CAF-1, CAF-1 is directed to replication forks via its interaction with PCNA, as mentioned earlier. Experiments using extracts to facilitate nucleosome assembly showed that PCNA marks replicated DNA and stimulates CAF-1-dependent nucleosome assembly, and PCNA and the large subunit of CAF-1 were found to colocalize at sites of replication in cells (Shibahara and Stillman, 1999). Additional evidence for the importance of the CAF-1 interaction with PCNA in replication-coupled nucleosome assembly came from a screen identifying *cac1* mutants (encoding the p150 subunit of CAF-1) defective for telomeric silencing in budding yeast (Krawitz et al., 2002). This silencing depends on CAF-1, and *cac1* mutants thus identified included some defective for binding to PCNA; when these mutants were tested in nuclear assembly reactions in vitro, they exhibited a substantially reduced preference for replicated DNA relative to wild-type CAF-1. A similar dependence on PCNA for CAF-1-mediated nuclear assembly was reported at sites of DNA damage; in an in vitro system in which nucleosome assembly was coupled to DNA repair, depletion of PCNA resulted in disruption of chromatin assembly (Moggs et al., 2000).

Histone deposition is also likely aided by CAF-1 interactions with DNA. Cac1 possesses a winged-helix domain and a distinct coiled-coil domain that bind DNA in a nonsequence-specific fashion, and in mutant studies conducted with budding yeast, the winged helix domain was found to synergize with the PCNA-binding surface in heterochromatic silencing and DNA damage response (Sauer et al., 2017; Zhang et al., 2016). Furthermore, mutation of the corresponding winged helix domain resulted in reduced recruitment of p150 to replication forks in murine cells (Zhang et al., 2016). Thus CAF-1 interactions both with PCNA and with DNA contribute to its recruitment to replication forks and sites of DNA damage and contribute to the deposition of $(H3\text{-}H4)_2$ tetramers at those sites. Recent work has provided new insight into the mechanism of deposition: in the absence of H3-H4 binding, an intramolecular interaction occludes the winged helix domain of Cac1 and thereby inhibits its binding to DNA; binding of H3-H4 releases this interaction and allows the formation of a transient CAF-1-histone-DNA intermediate containing two CAF-1 complexes (Mattiroli et al., 2017b). This intermediate leads to the formation of an $(H3\text{-}H4)_2$ tetramer from the two dimers in the complexes and assembly of the tetramer with DNA.

Deposition of newly synthesized histones in S-phase only accounts for about half of the nucleosomes formed on nascent DNA; the rest of the complement derives from parental histones that are transiently removed from the replicating DNA and then redeposited. While the mechanism by which this occurs has not been definitively established, experiments have established constraints and provided clues into the process. Groth, Almouzni, and colleagues, searching for proteins that might partner with Asf1 in histone chaperoning or nucleosome assembly processes, identified the replicative helicase components Mcm2, -4, -6, and -7 in a complex with Asf1 and H3-H4 from HeLa cells expressing epitope-tagged Asf1 in pulldown experiments (Groth et al., 2007). (MCM, or Minichromosome maintenance, proteins, were first identified in yeast in genetic screens for proteins involved in faithful propagation of centromere-containing minichromosomes to daughter cells (Chong et al., 1996; Tye, 1994). Subsequent studies showed that the MCM complex was highly conserved across eukaryotes and intimately involved in DNA replication.) Expression of an epitope-tagged mutant version of Asf1 that is unable to bind H3-H4 abrogated pulldown of MCM proteins, indicating that histones H3-H4 were likely to act as a bridge between Asf1 and the MCM complex. Biochemical experiments had shown that Mcm2 binds to histone H3 through a conserved N-terminal region termed the histone binding domain, or HBD (Ishimi et al., 1998), and X-ray crystal structure results showed that the HBD of Mcm2, although capable of binding a $(H3\text{-}H4)_2$ tetramer, interacts only with a single dimer of H3-H4 (Huang et al., 2015; Richet et al., 2015). Mcm2 contacts surfaces of H3-H4 that interact with DNA and with H2A-H2B, consistent with a model in which binding would occur only after nucleosome disruption, possibly by direct action of the replicative helicase. A mimic of an Mcm2-H3-H4-Asf1 complex was constructed by connecting the Mcm2 HBD and Asf1 with a flexible linker, and the structure of the resulting complex revealed a 1:1:1:1 complex, consistent with disruption of the $(H3\text{-}H4)_2$ tetramer upon Asf1 binding as discussed earlier (Huang et al., 2015). Taken together, the biochemical and structural findings point to a model in which parental nucleosomes are disrupted during replication, allowing Mcm2 to bind $(H3\text{-}H4)_2$ tetramers and either transfer them directly back to the replicated DNA or hand them off, with dissociation of the tetramer into dimers, to Asf1 for deposition onto the nascent DNA chain. Consistent with this model, histones H3 and H4 bearing modifications associated with parental histones were found associated with Asf1 in human cells, as were H3 and H4 having predeposition marks (see Sidebar: Modifications of newly synthesized histones) in separate complexes (Jasencakova et al., 2010). Furthermore, parental histones were shown to assemble close to their initial position both in vitro using *Xenopus* extracts that utilize the MCM complex and in yeast (Madamba et al., 2017; Radman-Livaja et al., 2011). However, even this relatively complex model is over-simplified. Evidence for chaperone function residing in additional components of the replication machinery, specifically RPA (Replication protein A) and the POLE3 and POLE4 subunits of DNA Polymerase ε, has been reported (Bellelli et al., 2018; Liu et al., 2017), and numerous mechanistic questions remain unanswered: What is the advantage of (presumably) dissociating $(H3\text{-}H4)_2$ tetramers into dimers only to reassociate them into tetramers upon re-assembly? How is Asf1 recruited to the replication fork to help with the re-assembly of parental histones? Are there separate Asf1-dependent and -independent pathways of tetramer re-assembly involving Mcm2, and what dictates their use? How does Asf1 release bound histone dimers to re-form the tetramer core of the nucleosome? This is a decidedly active area of research, and it would not be surprising if these and other outstanding questions are being answered even as this is being written.

Absent from the discussion thus far is the question of how H2A-H2B dimers are deposited into nascent chromatin, and conversely how they are dissociated from parental nucleosomes during replication, an obligatory prerequisite to removing H3-H4. We have discussed early evidence for a delay in the deposition of H2A-H2B relative to $(H3\text{-}H4)_2$ tetramer formation following DNA replication, and more recent findings that newly synthesized and parentally derived H2A-H2B dimers are associated with new and old $(H3\text{-}H4)_2$ tetramers in approximately equal proportion. Additional early evidence for the independent deposition of H2A-H2B derived from experiments in which replication was inhibited by treatment with hydroxyurea (discussed in Annunziato (2012)). This treatment allows continued synthesis of histones, albeit at reduced rates, and incorporation of H2A-H2B into chromatin via histone exchange was observed under these conditions, while little deposition of newly synthesized H3-H4 was seen. These various results establish that H2A-H2B disassembly and deposition are separable from that of the central $(H3\text{-}H4)_2$ tetramer.

Two major players in the removal and deposition of H2A-H2B dimers are Nap1 (also known as NAP-1) and the FACT, or Facilitates Chromatin Transcription, complex. Nap1 was one of the earliest histone chaperones to be discovered and was identified by an experimental route similar to that which led to the discovery of CAF-1. Extracts from human and murine cells were shown to assemble nucleosome-like particles when supplemented with exogenous histones. Fractionation of these extracts, using the ability to confer supercoiling on closed circular DNA as an assay, led to the identification of a 59 kDa protein as the active factor for assembly (Ishimi et al., 1983; Senshu and Yamada, 1980). Further characterization revealed that the factor, initially named AP-1 (for Assembly Protein-1), could form a complex with all four core histones or with just H2A-H2B and that the complex containing all four core histones could assemble nucleosomes at physiological ionic strength; this property was exploited in numerous in vitro studies on chromatin. To avoid confusion with the transcription factor AP-1 (a heterodimer of c-Jun and c-Fos), the factor was renamed NAP-1 (Ishimi and Kikuchi, 1991). An antibody against human NAP-1 was used to identify reactive proteins in, yeast, frog, and *Drosophila* cells, and the yeast protein was shown to have chaperone activity equivalent to the human version. In a parallel line of experimentation, the Kadonaga group isolated and cloned the *Drosophila* homolog of NAP-1 from extracts used to assemble chromatin having properly spaced nucleosomes and showed that it bound H2A-H2B in vivo and shuttled from cytoplasm to nucleus in S-phase (Bulger et al., 1995; Ito et al., 1996a). H2A was also found in complex with NAP-1 in cytosolic extracts from HeLa cells (Chang et al., 1997).

To date, the structure of Nap1 bound to H2A-H2B has been obtained only at a relatively low (6.7 Å) resolution (Aguilar-Gurrieri et al., 2016), although the structure of yeast Nap1 (also used in the co-crystal structure) was solved at 3.0 Å (Park and Luger, 2006). The co-crystal structure showed that a Nap1 homodimer binds a single dimer of H2A-H2B and contacts the same histone fold surface that binds DNA, likely thereby preventing adventitious and nonproductive interaction of H2A-H2B with DNA (Andrews et al., 2010). MNase digestion experiments together with genome-wide analysis of nucleosome occupancy in *nap1*△ yeast revealed a modest reduction in the size of fragments resulting from MNase digestion, and reduced precision of nucleosome positioning (increased "fuzziness"; see Chapter 4) relative to wild-type cells (Aguilar-Gurrieri et al., 2016). These findings were interpreted as indicating an increased prevalence of "hexasomes," or nucleosomes lacking one copy of H2A-H2B, leading the authors to propose that Nap1 deposits the second copy of H2A-H2B during nucleosome assembly.

FACT, the second major player in the deposition of H2A-H2B, was first purified from HeLa cell extracts in a search for factors that allowed transcription of chromatinized templates in vitro using purified transcription factors (discussed at more length in Chapter 8) (Orphanides et al., 1998). FACT comprises a heterodimer of Structure-specific binding protein 1 (SSRP1; in yeast, the homologous subunit is Pob3) and Spt16 (Orphanides et al., 1999), and is highly conserved among eukaryotes. Spt16, also known as Cdc68, was discovered by genetic screens in yeast as playing roles in transcriptional and cell cycle regulation (Malone et al., 1991; Prendergast et al., 1990), while Pob3 (Pol I binding 3) was first characterized as forming a stable heterodimer with Spt16/Cdc68 that associated with the catalytic subunit of DNA polymerase α, which is responsible for initiating polymerization of Okazaki fragments during lagging strand synthesis (Brewster et al., 1998; Wittmeyer and Formosa, 1997). This was the first biochemical clue, occurring before the discovery of FACT in HeLa cells, that the complex might play a role in replication. The discovery that this same complex facilitated transcription of chromatin led to tests of its ability to bind histones, with results showing that FACT possesses histone chaperone activity and binds both nucleosomes and free histones, with a preference for binding H2A-H2B (Belotserkovskaya et al., 2003; Orphanides et al., 1999). Additional evidence for FACT functioning in nucleosome dynamics during DNA replication was provided by findings that FACT is associated with the large complex, termed the replisome, that assembles at the replication fork during DNA synthesis (Gambus et al.,

2006; Tan et al., 2006). Screening for components of the replisome that can bind histones in extracts from yeast, and thus potentially function as histone chaperones during DNA replication, revealed that FACT and Mcm2 cooperatively bound histones released into the extracts by digestion of chromatin (Foltman et al., 2013). Moreover, the bound histones lacked modifications characteristic of newly synthesized histones, indicating that FACT together with Mcm2 could function to disassemble and reassemble nucleosomes at the replication fork (bearing in mind that FACT, although preferably binding H2A-H2B, is also able to associate with $(H3\text{-}H4)_2$ tetramers (Stuwe et al., 2008)). A detailed structural and biochemical analysis suggests a model in which the acidic region of Spt16 first interacts with the H2B amino-terminal region, after which displacement of H2A from DNA allows further interaction and removal of H2A-H2B from the $(H3\text{-}H4)_2$ tetramer during nucleosome disassembly (Tsunaka et al., 2016). However, whether and how FACT and Nap1 interact with each other and/or with Asf1 in the processes of nucleosome disassembly and re-assembly, and the molecular details of parental histone transfer, remain to be elucidated. Moreover, unlike Asf1 and Nap1, FACT is essential for viability in yeast, mice, and *Arabidopsis* (Cao et al., 2003; Formosa, 2008; Lejeune et al., 2007; Lolas et al., 2010; Malone et al., 1991; Wittmeyer and Formosa, 1997). Whether this reflects a more critical role in histone chaperone function during replication or is due to its multiple roles in the transcription and replication of chromatin is currently unknown.

Histone variation: Early work

In Chapter 1, the 1951 publication by Stedman and Stedman was cited as establishing the histones as ubiquitous nuclear proteins that could be separated into two major fractions, corresponding to what we now know to be the core histones (the "main" histones) and the linker histones (the "subsidiary" fraction) (Stedman and Stedman, 1951). However, the paper had a larger impact on the field than that might suggest: based on analyses showing heterogeneity in histones isolated from different tissues and organisms, the Stedmans proposed that the cohorts of histones found in different cell types could exert regulatory influence on gene expression. This proposal was embraced by the field with some enthusiasm; a 1971 review states that it was "customary to speak of the great multiplicity of histones and even to think in terms of one type of histone for each gene" (DeLange and Smith, 1971). This idea of a panoply of histones regulating gene expression "stimulated a search for species and tissue specificity" among the histones (Bustin and Cole, 1968). For quite some time, this search was not very fruitful, hindered by diverse approaches for isolation of histones that made comparison of results from different laboratories difficult, by the tendency of histones to aggregate, and by the paucity of analytical methods for their characterization. Although histones could be analyzed by polyacrylamide gel electrophoresis, the similar sizes and charges of H2A, H2B, and H3 made them difficult to separate from each other, let alone from minor sequence variants. Experiments before 1969 therefore relied principally on analysis of amino acid composition and comparison of chromatographic patterns obtained following tryptic digestion. Blunt as these methods were, they did allow the identification of multiple subtypes of linker histone ("very lysine-rich" histone) first in calf thymus, and subsequently in rabbit thymus, liver, and mammary gland, providing some of the first rigorous hints into histone diversity (Bustin and Cole, 1968; Kinkade and Cole, 1966).

As described in Chapter 1, the utilization of gentler techniques for the preparation of chromatin and the isolation of histones, together with innovations in electrophoretic methods, allowed the reliable identification of the individual core histones while avoiding ambiguities associated with oxidation or degradation. These advances also enabled investigators to obtain histones in sufficient quantity and purity to allow the direct determination of amino acid sequence, although this remained a laborious process involving enzymatic hydrolysis and sequencing of tryptic and chymotryptic fragments by Edman degradation (Delange, 1978; Hsiang and Cole, 1978). These same approaches also opened the way to the identification of histones differing in primary structure.

Sea urchins: Variant histones expressed during development

A principal early model for investigation of histone variants occurring during development was, perhaps surprisingly, the sea urchin. This would certainly not be the first choice of investigators today, as these species are essentially intractable with regard to genetic manipulation. However, sea urchin embryos were the subject of seminal investigations in embryology in the late 19th and early 20th centuries (Gilbert and Barresi, 2016), and the well-defined stages occurring during embryogenesis together with the accessibility of externally developing embryos made them an attractive choice for studying histone composition. Sea urchins were subjects of early investigations of both the histones and their genes: the histone genes of the sea urchin *Psammenichus milaris*, which are present in ~2000 copies per cell, were among the last genes to be isolated by physicochemical rather than molecular biological means (by banding after CsCl gradient centrifugation) (Birnstiel et al., 1974; Birnstiel, 2002), and cleavage stage histone genes expressed early in development (see below) were the first protein-coding genes to be cloned and sequenced (Cohn et al., 1976; Schaffner et al., 1978; Sures et al., 1978).

Early investigations reported slight changes in mobility or relative abundance of core and linker histones, monitored using conventional gel electrophoresis, during embryogenesis of various species of sea urchins, suggesting that the histone complement changed during development (Easton and Chalkley, 1972; Hill et al., 1971; Seale and Aronson, 1973). This finding was put on firmer footing by a report from the laboratory of Leonard Cohen at Fox Chase Cancer Center (Cohen et al., 1975). These investigators exposed embryos of *Strongylocentrus purpuratus* (the spiny purple sea urchins familiar to anyone who has explored tide pools along the West Coast of the United States) to tritiated leucine for periods of 4–5 h at different stages during the first 40 h of development, allowed the embryos to grow to blastula stage or later, purified histones and analyzed them on Triton-acid-urea gels followed by autoradiography. (For those of us accustomed to working with microcurie amounts of radioisotope, the levels of tritiated leucine used, varying from 1.5 to 4 curies/mL, are impressive and would certainly raise the eyebrows of radiation safety officers today.) This allowed histones synthesized at specific stages of development to be visualized by autoradiography and compared to the total complement of histones observed by staining. The results revealed switches in the forms of H1, H2A, and H2B synthesized, while no such variation was observed for histones H3 or H4. This was a remarkable finding, indicating as it did that *"there must be multiple forms of nucleosomes"* (Cohen et al., 1975) (their emphasis). Subsequent investigations from several labs using different species of urchin revealed that "cleavage stage" variants of H1, H2A, and H2B, expressed in the oocyte and during the first few divisions of the embryo, predominated until about the 8–16 cell stage (with some species variability), when new "α" variants are synthesized, followed by the "late stage" variants after about another three cell divisions. When linker histone is included, some species produced as many as 24 variant histones (van Holde, 1988, pp. 100ff). In contrast to linker histone, H2A, and H2B, no evidence was found for variants of H3 or H4.

Cloning and sequencing sea urchin histone genes eventually confirmed these findings, while adding additional detail (Mandl et al., 1997; Marzluff et al., 2006). H4 is indeed invariant; early and late stage H3 are also identical while, intriguingly, cleavage stage H3 differs in just three amino acids that are otherwise not observed in the animal kingdom (two are found in plants and ciliated protozoa while one appears to be unique) (Mandl et al., 1997). Greater divergence is observed among the cleavage, early and late-stage linker histone, and H2A and H2B (Fig. 5.4).

H2A and H2B diverge especially in the N-terminal regions, consistent with the biochemical results and with the generally greater evolutionary divergence of these "outer" histones of the nucleosome. As with cleavage stage H3, H2A and H2B include amino acid substitutions that are unique, or nearly so, to these histones. In addition, the cleavage stage histones, including CS H4, which is identical in sequence to early and late stage H4, differ from early, late, and sperm (see below) histones in being replication-independent in their synthesis and in their mRNAs being poly-adenylated and containing introns (Marzluff et al., 2006).

What physiological significance does this developmental switching among histone variants have? Van Holde wrote that the physiological function "remains obscure," and speculated that it could reflect an evolutionary selection based on producing sufficient quantities of histones with the variations in sequence observed simply being tolerated (van Holde, 1988, p. 103). This was a reasonable speculation at the time, but there now appears to be more to it than this. In addition to the histone variants produced during embryogenesis, sea urchins also express specialized variants of H1 and H2B in the male germline (Poccia and Green, 1992). Unlike the protamine-bound DNA found in the sperm of many organisms, sea urchin sperm chromatin is packaged into nucleosomes that retain the canonical beads-on-a-string structure found in somatic cells and that protect 146bp of DNA from MNase digestion (see Sidebar: Protamines). However, the sperm-specific histones SpH1 and SpH2B are larger than the histones found in somatic cells of the sea urchin, and the repeat length in sperm cells is ~250bp, the largest observed (Arceci and Gross, 1980; Keichline and Wassarman, 1979; van Holde, 1988). This sperm chromatin is more highly condensed than somatic chromatin and is neither transcribed nor (of course) replicated. Upon fertilization, both SpH1 and SpH2B of the sperm chromatin, now localized to the male pronucleus, are phosphorylated on their extended N-termini, and SpH1 is replaced by the cleavage stage variant of H1 (Green and Poccia, 1985). The exchange of linker histone is followed by decondensation of the chromatin, a decrease in repeat length to ~200bp, and transcriptional activation of the paternal genome before the first cell division is completed (Green and Poccia, 1985; Maxson and Egzie, 1980; Poccia et al., 1984). The obvious conclusion is that the sperm and cleavage stage variant histones have evolved to facilitate chromatin remodeling that is critical for the transition from silent to active following fertilization. In this regard, it is noteworthy that the cleavage stage H1 is homologous to H1M, also known as B4 and most recently designated as H1.8, a variant linker histone from *Xenopus laevis*, which similarly serves as a replacement histone during the development of the frog (Mandl et al., 1997; Smith et al., 1988).

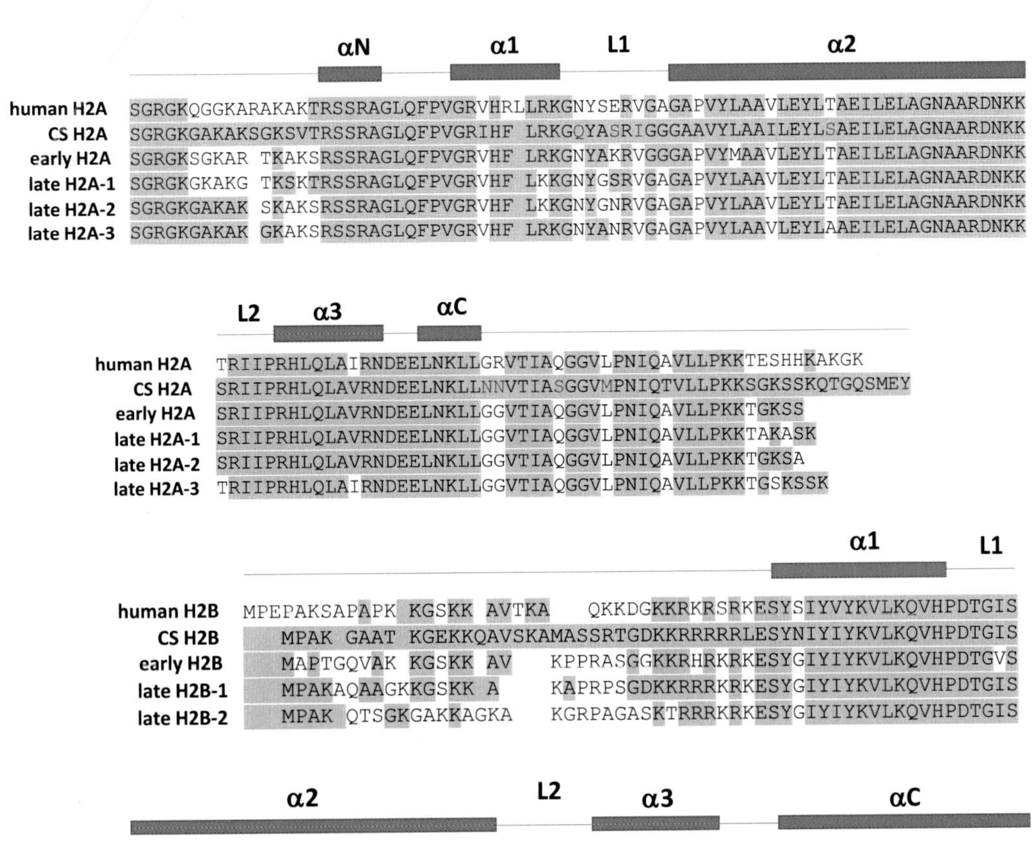

FIG. 5.4 Developmental variants of sea urchin (*P. miliaris*) histones H2A and H2B. Unaltered residues relative to cleavage stage (CS) H2A and H2B are highlighted in *gray*; amino acids indicated in *red* are rarely or never observed in other species at the corresponding positions. The consensus sequences for human canonical H2A and H2B are shown for reference. *Blue boxes* indicate the α-helices of the H2A and H2B histone folds and N- and C-terminal α-helices; the L1 and L2 loops that contact DNA in the nucleosome are also indicated (Luger et al., 1997). *Data from Mandl, B., Brandt, W.F., Superti-Furga, G., Graninger, P.G., Birnstiel, M.L., Busslinger, M., 1997. The five cleavage-stage (CS) histones of the sea urchin are encoded by a maternally expressed family of replacement histone genes: functional equivalence of the CS H1 and frog H1M (B4) proteins. Mol. Cell. Biol. 17, 1189–1200, and the HistoneDB 2.0 database (Draizen, E.J., Shaytan, A.K., Marino-Ramirez, L., Talbert, P.B., Landsman, D., Panchenko, A.R., 2016. HistoneDB 2.0: a histone database with variants—an integrated resource to explore histones and their variants. Database (Oxford) 2016).*

Additional lines of evidence support the idea that H1M (B4) facilitates the transition from inert sperm chromatin to active chromatin in the zygote. Like CS H1 in the sea urchin, H1M (B4) participates in the remodeling of sperm chromatin, as observed using frog egg extracts (Dimitrov et al., 1994; Ohsumi and Katagiri, 1991). In vitro reconstitution experiments show that H1M (B4) is incorporated into chromatin in the same fashion as classical linker histone, but with lower affinity, and the resulting chromatin structure is less efficient at repressing gene transcription than that formed using classical linker histone (Nightingale et al., 1996; Ura et al., 1996). Altogether, then, it appears likely that sperm and cleavage stage histones have evolved to allow the facile remodeling of repressed sperm chromatin to highly transcriptionally active chromatin in the fertilized embryo. However, apart from this broad conclusion, the importance of specific residues that are peculiar to these variant histones remains unknown.

Sidebar: Protamines

The packaging of DNA into chromatin in sperm cells presents a special case. DNA in sperm is condensed ~40x relative to in somatic cells; this condensation facilitates sperm motility by creating a hydrodynamically efficient sperm head and also protects against radiation damage the precious single genome copy intended to produce a new organism (Balhorn, 2007; Nili et al., 2009; Rathke et al., 2010). Although some metazoans such as sea urchins use sperm-specific histones to effect this compaction, most use a special class of proteins called protamines. Friedrich Miescher took advantage of the availability of pure populations of sperm cells from salmon to study the protein constituents of their nuclei and coined the term "protamin" to refer to the "organic base" associated with the nuclein in these cells (Miescher, 1874). Miescher did not differentiate the "protamin" from the basic constituent found in the nuclein he had isolated from leukocytes; elucidating this distinction fell to Albrecht Kossel, who pioneered early studies of histones and related proteins (Kossel, 1896a, b). The protamines are members of a large and heterogeneous family of sperm nuclei basic proteins whose job it is to package chromatin in the highly condensed nuclei of sperm cells (Lewis et al., 2003). In mammals, protamines are expressed during spermiogenesis, the haploid phase of spermatogenesis, along with the transition nuclear proteins, or TNPs; the latter proteins function as intermediaries during the remodeling of nucleosomal chromatin to protamine-packaged chromatin found in mature sperm. The protamines are characterized by short runs of arginine residues (composition >30%, and as high as 70%, Arg), and enrichment for Lys, His, Ser, Thr, and Gly residues; phosphorylation of Ser and Thr residues is important in the initial deposition of protamines during spermiogenesis (Papoutsopoulou et al., 1999; Wu et al., 2000). In addition, protamines of eutherian mammals contain Cys residues that contribute to chromatin condensation via intermolecular cross-links. The structure of chromatin conferred by protamines is very different from that found in somatic cells: in contrast to histones, protamines wrap around the DNA helix, probably binding in the major groove, with stabilization being conferred by contacts between Arg residues and DNA phosphate groups (Balhorn, 2007; Hud et al., 1994). The mechanism by which such structures (protamines being highly heterogeneous) lead to the extreme condensation found in sperm continues to be investigated (e.g., Ukogu et al., 2020).

Not all the DNA in mammalian sperm is packaged by protamines; in humans, ~85% of sperm DNA is associated with protamines and ~15% with histones (Gatewood et al., 1990). The latter cohort includes key developmental genes, microRNA-encoding genes, and imprinted genes (Brykczynska et al., 2010; Hammoud et al., 2009). Technical advances, especially high throughput sequencing and the ability to perform gene knockdown experiments have allowed researchers to delve more deeply into the mechanisms of chromatin remodeling during spermiogenesis, the location of canonical nucleosomes in sperm chromatin, and the role of remodeling in retaining or erasing epigenetic information encoded in chromatin (Kota and Feil, 2010).

Other species

The cloning of sea urchin histone genes, coupled with their strong evolutionary conservation, led directly to the identification and characterization of histone genes in other organisms (e.g., *Drosophila melanogaster* and several marine invertebrates (Kedes, 1979; Lifton et al., 1978)) through cross-hybridization experiments in which the cloned histone genes were used as probes. The cloning of histone genes in budding yeast (*Saccharomyces cerevisiae*) nicely illustrates these early efforts (Smith, 1984). In this case, initial attempts to use radiolabeled sea urchin DNA encoding H2A and H2B to identify clones in a plasmid library by colony hybridization were not fruitful. A second approach was therefore taken: a set of 100 recombinant clones identified as encoding low molecular weight poly(A) + mRNA (yeast histone mRNAs, unlike those found in metazoans, are polyadenylated) was used to isolate the corresponding mRNA species using R-loop hybridization. The RNA thus hybridized was recovered and translated in vitro, and the proteins produced were compared with native yeast histones by two-dimensional gel electrophoresis, eventually resulting in the identification of the genes encoding H2A and H2B (Hereford et al., 1979). Cloning of the yeast H3-H4 genes (which, like the H2A-H2B genes, are divergently transcribed) followed a different path. Phage lambda libraries containing yeast genomic DNA were hybridized using clones of sea urchin and *Drosophila* histone genes as probes. Numerous clones from the library were identified, most of which comprised false positives resulting from hybridization to sequences outside the coding region of the histone gene probes. The simple (yet elegant) solution was to identify clones that hybridized to both sea urchin and *Drosophila* probes, which culminated in the successful identification of the two yeast H3-H4 gene cassettes (Smith and Murray, 1983).

In a relatively short span of time, the newly developed tools of molecular genetics allowed the determination of histone gene sequences, and by inference those of the proteins, from a large variety of eukaryotes, from *Tetrahymena* to frogs to chickens to humans (Stein et al., 1984). Sequences of core histones, sometimes by direct protein sequencing and oftentimes inferred from the gene sequences, affirmed the high degree of conservation first seen in the remarkable similarity of H4 between pea plants and calf thymus (Fig. 5.1; DeLange and Smith, 1971). Histones H3 and H4 exhibited stronger conservation than H2A and H2B, reflecting in part their more central role in nucleosome structure: recall from Chapter 2 that an $(H3\text{-}H4)_2$ tetramer is sufficient to wrap 120bp of DNA in a structure similar to an intact nucleosome. A curious finding that remained a puzzle for some time was the extreme conservation seen for the amino termini of histones H3 and H4 (Fig. 5.1). Biophysical and biochemical experiments discussed in Chapter 2 demonstrated that the amino termini of all four core histones, and the C-termini of H2A and H2B, are dispensable for the nucleosomal packaging of DNA. Removal of the tails by trypsinization did not alter the path of the DNA or the protection against enzymatic digestion conferred by the nucleosome; why, then, were they so highly conserved? Further compounding this mystery, the first experiments examining the in vivo function of the histone tails showed that yeast could survive the loss of the amino termini of individual histones, although effects on growth and mating were observed, as discussed in greater depth later on (Kayne et al., 1988; Morgan et al., 1991; Wallis et al., 1983). We now know that the conservation of the histone tails, particularly those of H3 and H4, reflects their critical functions as platforms for posttranslational modifications throughout the eukaryotic kingdom.

Although the histones are highly conserved in their sequences and nucleosome structure is correspondingly nearly independent of the source of the constituent histones (Harp et al., 2000; Luger et al., 1997), biophysical studies have revealed some species-specific differences in nucleosome properties. Nucleosomes from budding yeast were suspected early on to differ from those from metazoans in some respects, based on both sequence divergence (Fig. 5.1) and the much higher fraction of transcribed genes and lower fraction of condensed chromatin in yeast. Researchers also noted, both in publications and informally, the greater difficulty in reconstituting nucleosomes using histones from yeast compared to those from chicken erythrocytes or calf thymus; based on reconstitution experiments using combinations of yeast and calf histones, histone H3 from yeast was proposed as responsible for the apparently lower stability of yeast nucleosomes (Lee et al., 1982). The idea that yeast nucleosomes were less constrained than those from metazoans was supported by studies showing lower stability of yeast nucleosomes towards salt-dependent or thermal unfolding than chicken or bovine nucleosomes (Lee et al., 1982; Pineiro et al., 1991). In addition, the thermal untwisting seen in naked DNA, which gives rise to temperature-dependent changes in the supercoiling of closed circular DNA molecules, was observed to be suppressed in nucleosomal DNA containing metazoan histones but much less so in yeast chromatin (Ambrose et al., 1987; Keller et al., 1978; Morse and Cantor, 1985; Morse et al., 1987; Saavedra and Huberman, 1986). A high-resolution X-ray crystal structure of nucleosome core particles assembled using recombinant, and thus unmodified, yeast histones revealed that protein-protein and protein-DNA interactions were nearly unaltered between yeast and *Xenopus* nucleosome core particles (NCPs) (White et al., 2001). However, differences were also noted: yeast NCP crystals grew more slowly and were fragile relative to those from *Xenopus*, the paths of the H2A C-terminal tail and the H4 N-terminal tail were completely different, and stabilizing interactions between H2A-H2B dimers present in *Xenopus* NCPs were lost in yeast. Reconstitution of hybrid nucleosomes using yeast and human histones revealed a more relaxed conformation in nucleosomes containing yeast H2A and H2B, while the yeast H3 sequence, specifically yeast-specific residues in the α3 helix close to the C-terminus, was shown to be the principal contributor to the low efficiency of octamer reconstitution using yeast histones (Leung et al., 2016).

Genetic experiments support functional correlates to the altered biophysical properties of yeast nucleosomes. Efforts to engineer yeast expressing "humanized" histones, using strategies discussed in the "Histone mutants" section (see Fig. 5.7), recovered such mutants at very low frequency, suggesting that their survival required the accumulation of suppressor mutations (Truong and Boeke, 2017). The most critical differences with regard to yeast viability were in the H3 α3 helix and the H2A C-terminal and N-terminal regions (McBurney et al., 2016; Truong and Boeke, 2017). Amino acid residues specific to the yeast α3 helix of histone H3 and at the H2A-H2B dimer-dimer interface were noted as being conserved in several other protozoan species, suggesting a common evolutionary constraint on nucleosome stability or interactions with histone chaperones (Leung et al., 2016; White et al., 2001).

With regard to variation in histone subtype during development, it turns out that sea urchins present an unusual case: although variant core histones were identified in multiple species by biochemical methods and their primary sequences determined, the developmental switching observed in sea urchins is atypical. Some of the invertebrate cousins of sea urchins, including the mud snail *Ilyanassa* and the surf clam *Spisula*, do exhibit developmental switching among core histone subtypes (Gabrielli and Baglioni, 1975; Mackay and Newrock, 1982).

However, most eukaryotes were found not to exhibit such switching. Zweidler and co-workers identified variants of H2A, H2B, and H3 differing in only a few amino acids in the chicken and in mammalian tissues—for example, murine H2B.1 and H2B.2 differed only in having glycine or serine at residue 75, whereas three H2A variants differing at eight residues were found—but these variants were found throughout development and in multiple tissue types (Franklin and Zweidler, 1977; Urban and Zweidler, 1983; Zweidler, 1984).

We have seen that before the advent of reliable methods for histone isolation and characterization in the 1960s, the Stedmans' 1951 paper precipitated widespread enthusiasm for the notion that a panoply of histones might be central to the variety of gene regulation observed across tissues and throughout development of multicellular organisms. This notion was severely undercut by the revelation that only four highly conserved histones comprised the core structure of the nucleosome, which was itself (nearly) ubiquitous in eukaryotes. The early experiments of direct sequencing of histones, and then cloning and sequencing histone genes, identified subtle (and sometimes not so subtle) variations in the sequence of the histone proteins as well as differential (replication-coupled and replication-independent) regulation of histone synthesis, re-opening the idea of the potential for gene regulation involving histones. One of the first strong indications for such functional specialization derived from studies of the ciliated protozoan *Tetrahymena thermophilus*, largely from Marty Gorovsky's lab at the University of Rochester. As mentioned in Chapter 4 in the context of genome-wide nucleosome positioning, *Tetrahymena* cells possess a transcriptionally active macronucleus and a transcriptionally inactive, replicative micronucleus. Characterization of histones isolated from micro- and macronuclei revealed that both contained the same two variants of H2A, but that a novel histone related to H2A and designated as hv1 was specific to the transcriptionally active macronucleus (Allis et al., 1980, 1986). Furthermore, hv1 was first detected by antibody staining in newly formed macronuclei following conjugation not until 8 h after their formation, about the same time at which new RNA synthesis begins, while a later study showed positive staining for hv1 in micronuclei only during the early stages of conjugation, just before their becoming transcriptionally active (Stargell et al., 1993; Wenkert and Allis, 1984). Based on these findings, it was suggested that this histone variant might play a role in the formation of transcriptionally active chromatin. The discovery of evolutionarily conserved, nonallelic variant histones also pointed to the histones playing "more than a passive role in chromatin function" (van Holde, 1988). Van Holde further observed:

> In a certain sense, the Stedman hypothesis ... seems to be vindicated. There are many kinds of histones, and they do exhibit some cell-type specificity. It is probable that these differences play a functional role, but we do not, as yet, understand that role at a molecular level.

These early studies thus were precursors to a tidal wave of work on histone variants that goes on to this day. Much of this work has focused on a handful of variants that are found through most of the eukaryotic kingdom, that have evolutionary lineages distinct from the major, replication-coupled, canonical core histones, and that, in some cases, have conserved and specialized functional roles. We will discuss these in some detail, but will first consider the evolutionary history of the histones and the little that is known regarding nonallelic variants of the replication-coupled major core histones.

Evolution of the histones

The histones, being common to all eukaryotes save the dinoflagellates (see Sidebar: How the dinoflagellates (may have) lost their histones), have ancient origins. Our current understanding of those origins can fairly be said to have had its birth in the 1970s when two genuine paradigm shifts led to a major revision in our conception of the "tree of life" (Fig. 5.5). The first

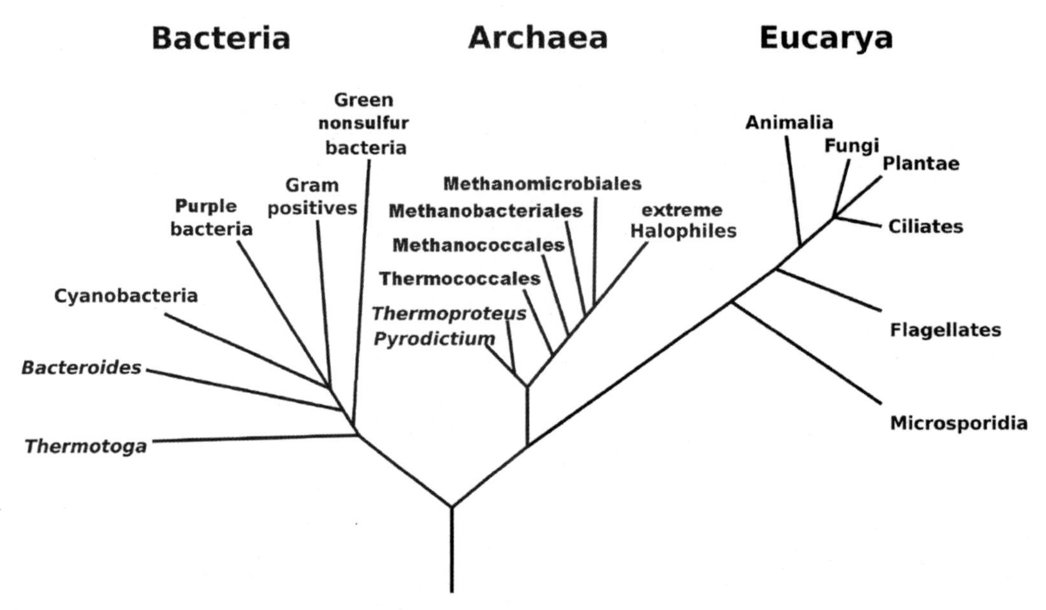

FIG. 5.5 From Malucioni. *Own work, CC BY-SA 3.0, https://commons.wikimedia.org/w/index.php?curid=24740337.*

of these shifts occurred with the discovery of archaea and the proposal that they constituted a third kingdom distinct from both bacteria and eukaryotes (Woese and Fox, 1977). Before this discovery, textbooks and working scientists divided life into prokaryotic and eukaryotic organisms, principally defined by the absence or presence of nuclei and membrane-bound organelles (Sapp, 2005; Stanier and Van Niel, 1962). (Following the assertion by Stanier and Van Niel that the distinction between prokaryotes and eukaryotes was foretold decades earlier by Chatton (1925, 1938), Chatton has frequently been cited as the first to promote, or at least glimpse, this distinction. Sapp argues convincingly that the "singular prescience" ascribed to Chatton by Stanier and Van Niel is misplaced and has only been exaggerated with time.) The second paradigm shift was due to Lynn Margulis (and first published under her married name), who formulated the endosymbiosis hypothesis to explain the origin of eukaryotic organelles such as chloroplasts and mitochondria as arising from symbiotic mergers between bacteria (Sagan, 1967). Together these breakthrough insights precipitated novel ideas regarding how eukaryotes acquired their distinctive means of packaging DNA and the requisite machinery for handling transactions involving DNA thus packaged.

Sidebar: How the dinoflagellates (may have) lost their histones

The absence of nucleosomes in dinoflagellates, a highly divergent eukaryote, has long been noted (Rizzo and Nooden, 1972). Neither nuclease digestion nor EM experiments revealed evidence of nucleosomal structure in these unicellular algae, and efforts to characterize their nuclear proteins did not uncover anything resembling histone proteins van Holde, 1988, pp.178ff). Dinoflagellate chromosomes have a liquid crystalline structure, quite unlike other eukaryotes that package their genomes in nucleosomes (Rill et al., 1989). However, transcriptome and partial-genome sequencing revealed the expression of histone genes in several dinoflagellate species (Marinov and Lynch, 2015). These histones are expressed at relatively low levels and diverge considerably from those found elsewhere in the eukaryotic kingdom, although the histone fold domains are readily identified (Gornik et al., 2012; Marinov and Lynch, 2015). Investigation of an early branching dinoflagellate, *Hematodinium* sp., and comparison with a close relative (*Perkinsus marinus*) that packages its genome into conventional nucleosomes, revealed the presence of a novel family of DNA packaging proteins unique to the dinoflagellates (Gornik et al., 2012). These proteins are not found in eukaryotes outside of the dinoflagellates, but are present in a group of algal viruses of the Phycodnaviridae family and hence were termed dinoflagellate/viral nucleoproteins, or DVNPs. These findings suggest a plausible scenario in which the DVNPs were introduced into the dinoflagellate lineage by horizontal gene transfer as a consequence of viral infection, with accompanying loss of canonical histone gene expression. Lending support to this idea, the expression of DVNPs in yeast leads to their competing with histones for DNA binding with concomitant impairment of growth; this toxicity can be relieved by histone depletion, and DVNP expression in yeast leads to a reduction in histone expression (Irwin et al., 2018). Thus the introduction of DVNPs into the dinoflagellate common ancestor may have driven histone depletion and thus given rise to the entire lineage.

Alternative means of packaging DNA are also found among archaeal species: whereas histone fold domain proteins appear predominant in most archaeal species, including recently discovered, deeply branching superphyla, genes encoding these proteins are mostly absent from Crenarcheota, which express DNA bridging Alba proteins at high levels (Henneman et al., 2018). Structural data indicate that archaeal Alba proteins coat DNA to form filamentous structures (Laursen et al., 2021); the evolutionary contingencies that determine the extent to which DNA in archaeal species is packaged by archaeal histones, Alba proteins, or even HU proteins are as yet unknown.

Bacteria, archaea, and eukaryotes all compact their DNA with the help of DNA-binding proteins. Bacteria condense their genomes into a nucleoprotein complex termed the *nucleoid* with help from nucleoid-associated proteins; prominent among these are the HU and H-NS proteins (Dillon and Dorman, 2010). These and other nucleoid-associated proteins bear no discernible resemblance to histones, either structurally or in their amino acid sequence although they are, as one would expect for DNA-binding proteins, rich in arginine and lysine residues (Laine et al., 1980). Although particle-like structures have been observed for partially unfolded nucleoids by EM, treatment with nuclease does not yield a ladder-like pattern of cleavage as seen for eukaryotic chromatin (Griffith, 1976). Archaea, in contrast, use both HU-related proteins and HMfA and HMfB proteins to compact their DNA; the latter two proteins are descended from the same ancestral proteins that gave rise to eukaryotic histones (Bhattacharyya et al., 2018; Sandman and Reeve, 2006; Talbert and Henikoff, 2010; Talbert et al., 2019).

The first archaeal packaging protein to be sequenced was Hta from the extremophile *Thermoplasma acidophilum* (DeLange et al., 1981). Comparison of the amino acid sequence of this 89-residue protein with the *E. coli* HU proteins revealed notable homology, while a weak homology of the amino-terminal region of Hta with histone H2A was also recognized. Thus Hta appeared likely to be evolutionarily intermediate between the HU proteins and eukaryotic histones. A more dramatic discovery (reported in a paper communicated by Carl Woese) was that of two nearly identical proteins from the hyperthermophile *Methanothermus fervidus* that bound to double-stranded DNA and increased its resistance to thermal denaturation (Sandman et al., 1990). These proteins, designated HMfA (Histone *M. fervidus*) and HMfB (initially HMf-1 and -2), compact and alter the path of bound DNA as seen by EM, and show much greater homology to eukaryotic histones than to HU-1 or Hta. Subsequent studies identified and obtained sequences of archaeal histones from numerous (>100) additional species (Henneman et al., 2018; Nishida and Oshima, 2017; Sandman and Reeve, 2006).

Structural determination provided a definitive measure of the similarity of putative archaeal histones and eukaryotic core histones. As discussed in Chapter 2, the high-resolution structure of the histone octamer revealed the existence of the histone fold, comprising three α–helices separated by two short β-loops (Arents et al., 1991) (Fig. 2.6). NMR studies of the HMfB homodimer indicated a histone fold structure similar to that observed for eukaryotic histones (Starich et al., 1996), a conclusion confirmed by the X-ray crystal structure (Decanniere et al., 2000). Despite relatively low amino acid homology between archaeal and eukaryotic histones, the archaeal $(HMfA)_2$ dimer structure could be superimposed on the H3-H4-DNA structure from the nucleosome without leading to steric clashes between HmfA and DNA (Decanniere et al., 2000). Moreover, the resulting structure suggested that two Arg, one Lys, and one Thr residue were likely to contact and bind to the DNA backbone (Decanniere et al., 2000). These residues are highly conserved among archaeal histones and have structural homologs in all four eukaryotic core histones, and their role in DNA binding was confirmed by mutagenesis experiments (Soares et al., 2000). These structural similarities between archaeal and eukaryotic histone-DNA interactions were eventually validated by the X-ray crystal structure of an $(HMfB)_2$-DNA complex, which revealed conserved amino acid side chain interactions with DNA (Fig. 5.6; Mattiroli et al., 2017a). Outside the histone fold domain, however, the similarity between archaeal and eukaryotic histones ends; the amino- and carboxy-terminal regions present in nucleosomal histones and critical in chromatin regulation by posttranslational modifications are entirely lacking in almost all archaeal histones (Henneman et al., 2018).

As discussed in Chapter 2, in the nucleosome, heterodimers are formed between H2A and H2B, and between H3 and H4, by interactions between the central long α-helices that form a "handshake" motif. The $(H3\text{-}H4)_2$ tetramer at the center of the nucleosome is stabilized by interactions among two of the α-helices from each of the H3 monomers; the interacting surface is referred to as a four-helix bundle (4HB). The H2A-H2B dimers are similarly bound in the nucleosome by 4HB interactions between H2B and H4 (Luger et al., 1997). In contrast, both homodimer and heterodimer formation have been observed for archaeal histones; reflecting the high conservation of these proteins, heterodimers can form even between some archaeal histones from separate species (Li et al., 2000; Sandman et al., 1994). In some archaeal species, such as *Haloferax volcanii* (first isolated in 1975 from the Dead Sea), two histone-fold domains are linked to form a single protein in which the two domains interact via 4HBs to form a

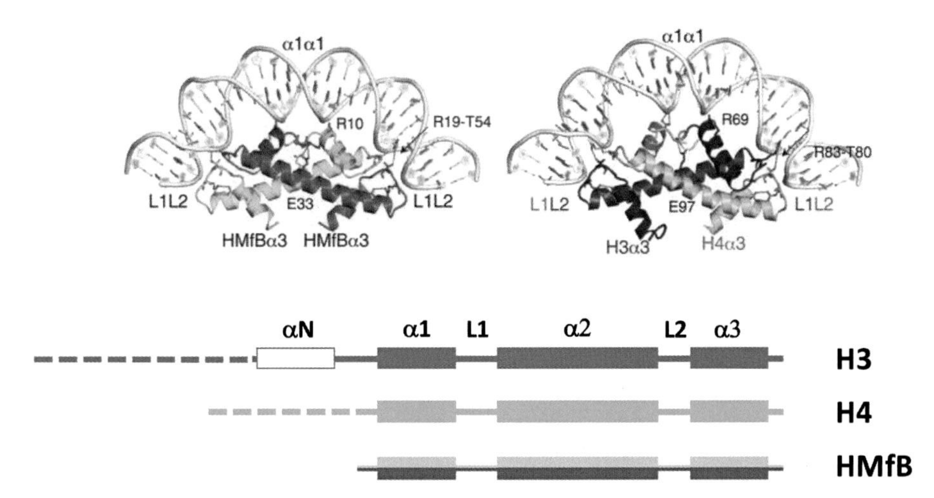

FIG. 5.6 *Top*: Histone fold domains of $(HMfB)_2$ and H3-H4 dimers in complex with 30bp of DNA. The two identical HMfB monomers are colored differently for clarity. Conserved interactions with amino acids from the L1-L2 loops and the α1 helices are indicated: R, arginine; T, threonine; E, glutamic acid. *Bottom*: schematic depiction of histones H3 and H4 and HMfB; the *open box* in histone H3 is an α-helix that is not part of the histone fold, and the *dashed lines* represent the unstructured amino-terminal histone tails. *From Mattiroli, F., Bhattacharyya, S., Dyer, P.N., White, A.E., Sandman, K., Burkhart, B.W., Byrne, K.R., Lee, T., Ahn, N.G., Santangelo, T.J., et al., 2017a. Structure of histone-based chromatin in Archaea. Science 357, 609.*

dimer-like structure; this has been proposed as an evolutionary intermediate that was a precursor to the diversification of eukaryotic histones with their accompanying specific pairwise interactions (Malik and Henikoff, 2003).

Dimerization of archaeal histones involves interactions between the central α-helices similar to that seen in eukaryotic histones, and archaeal homodimers—for example, $(HMfA)_2$—can further interact via 4HBs to form tetramers that can compact about 60bp of DNA (Pereira et al., 1997). In some cases, hydrophobic patches on the outside of such tetramers foster association between tetramers, leading to the formation of long chains of DNA-bound tetramers referred to as "hypernucleosomes" (Henneman et al., 2018; Mattiroli et al., 2017a; Sandman and Reeve, 2006). In addition, to supporting structural similarities between archaeal and eukaryotic histone-DNA interactions, the X-ray crystal structure also revealed that the superhelical path of the DNA is similar in the two structures (Mattiroli et al., 2017a). However, unlike eukaryotic histones, the basic structure formed by archaeal histones is a *trimer* of HMfB with 90bp of DNA. This property was confirmed by nuclease protection and solution studies both with HMfB and with histone A from *Thermococcus kodakerensis* (HtkA) (Mattiroli et al., 2017a). This difference, which permits the formation of an extended polymer of DNA with archaeal histones, is due to the symmetry of archaeal histone complexes, wherein each of the monomers can participate in 4HB interactions with monomers from the flanking dimers, whereas the asymmetric composition of histone dimers in eukaryotic nucleosomes precludes such interactions. Thus the ancestral histone fold domain protein that gave rise to archaeal and eukaryotic histones diverged in this important property; ironically, the quasicontinuous hypernucleosome of archaeal species bears a resemblance to the continuous protein-DNA superhelix once envisioned for eukaryotic chromatin.

Archaeal chromatin shares additional properties with that of eukaryotes. Not surprisingly, given that DNA bending in association with archaeal histones is similar to that in the nucleosome, $(HMfB)_2$ dimers favor "positioned" complexes with DNA governed by structural features similar to those seen in positioned nucleosomes (Bailey et al., 2000). These similarities extend to genome-wide occupancy patterns (Ammar et al., 2012). MNase-seq carried out with *Hfx. Volcanii*, modified to map DNA fragments of 50–60 bp protected from digestion, revealed a pattern of NDRs and positioned regions of protection ("nucleosomes" in the publication) flanking transcription start sites. This could suggest that archaeal and eukaryotic chromatin evolved first for regulatory advantage; alternatively, archaeal chromatin may have originated as a means of preventing denaturation of DNA under extreme conditions of temperature or salt concentration, with TSS exposure co-evolving as a strong selective advantage. How NDRs are established in archaeal species is mysterious at present. We saw in Chapter 4 that chromatin remodeling complexes are important in establishing NDRs in eukaryotic cells, but so far as is known, these complexes are not present in archaea.

As is typical with questions of evolution, uncertainties abound. The ancestral species that gave rise to archaea and eukaryotes appeared between 1.5 and 3 billion years ago (Koumandou et al., 2013). Whether the precursor of this species arose as a distinct lineage before or after endosymbiotic events leading to the origination of chloroplasts and mitochondria, and the involvement of ancient bacterial, archaeal, or even viral species in those events, is currently under debate (Talbert et al., 2019); the origin of the histone fold domain is correspondingly murky. The idea that the histone fold domain originated after the divergence of the archaeal and eukaryal branches from bacteria, with the innovation of flexible amino and carboxy termini subject to regulatory posttranslational modifications arising only after archaeal-eukaryal divergence, is complicated by the discovery of lysine-rich amino termini among archaeal histones (Henneman et al., 2018). The putative histone-fold domain proteins have been identified using bioinformatic approaches in bacterial species, leading to the suggestion that histones predate the divergence of archaea from bacteria (Alva and Lupas, 2019). Structural determination of Aq_328 from the hyperthermophilic bacterium *Aquifex aeolicus* shows it to be a histone-fold protein, supporting this idea (Qiu et al., 2006). However, Aq_328 and its close relatives appear not to be DNA-binding proteins, and it has been argued that Aq_328 may have been acquired from an archaeal species by horizontal transfer (Qiu et al., 2006; Sandman and Reeve, 2006). Furthermore, the putative bacterial histone fold proteins are more similar to archaeal than to eukaryotic histone fold domain proteins, perhaps bringing us back to the HU-Hta-histone axis. Very recently, histones have been identified as essential components of chromatin in the bacterial species *Bdellovibrio bacteriavorus* and *Leptospira interrogans* (Hocher et al., 2023). However, these histones bind to DNA end-on, in complete contrast to archaeal and eukaryotic histones, and encase DNA in linear structures. This structural divergence is certainly remarkable and remains to be further explored.

Histones have been discovered even in viruses: the Marseilleviridiae family of giant viruses, first discovered infecting *Acanthamoeba* in a Paris water tower and possessing a genome of 368 Mb, encodes proteins comprising linked histone fold domains (akin to those of *Hfx. volcanii*) that fold into obligate "heterodimers" (Boyer et al., 2009). One of these proteins resembles an H2A-H2B doublet and the other an H3-H4 doublet, and each also possesses an unstructured tail (Erives, 2017). Assembling these viral histones together with the Widom nucleosome positioning sequence produced structures virtually identical to eukaryotic nucleosomes (Liu et al., 2021; Valencia-Sanchez et al., 2021). Whether these viral histones were acquired from an archaeon or a proto-eukaryote, and whether they might have been the primary source of eukaryotic histones, is uncertain (Alva and Lupas, 2019; Erives, 2017; Talbert et al., 2022). The large and rapidly expanding library of genome sequences from bacterial, archaeal, and protozoan species is likely to add grist to these various debates for years to come.

Canonical histones: Nonallelic variants

As a postdoctoral fellow in Charles Cantor's lab at Columbia University in the early 1980s, I undertook a project to investigate the effect of expressing sea urchin histones in cultured monkey kidney cells. Although the project never got off the ground (and taught me that the care and feeding of sea urchins was a nontrivial enterprise), its conceptual basis was not unreasonable. Variation in the sequences of the core histones had been observed between and within species, developmental switching among sea urchin histones was known, and the idea that histone variants might have regulatory roles in transcription was intriguing. Here we consider nonallelic variants of the canonical core histones and describe experiments performed over the course of several decades that have examined, with increasing sophistication, the effects of altering histone sequence on chromatin structure and function.

We have noted that histone genes exist in multiple copies, sometimes numbering in the hundreds, in different cell types. These genes do not all encode identical copies of the histones. Early work established that the core histones frequently included minor variants, including histones that were incorporated into chromatin in both replication-coupled and replication-independent fashion. Subtle sequence differences among variants expressed within a given species were discovered; for example, the mouse H2A.2 variant was found to differ from H2A.1 only in the substitution of Ser for Thr at residue 16, Met for Leu at residue 51, and Lys for Arg at residue 99, while H2B variants were identified differing only in the substitution of a serine for Gly75 (Franklin and Zweidler, 1977; Zweidler, 1984). An H3 variant was identified from calf thymus differing from the major H3 species only in the substitution of a serine for Cys96; this residue was identified as Ser in shark, chicken, and carp, while pea plant embryos expressed variants having Ser96 and Ala96 (Brandt et al., 1974; Patthy and Smith, 1973). No H4 variants were found across multiple species (Zweidler, 1984).

The full scope of nonallelic variations in the core histones has emerged from whole-genome sequencing (Draizen et al., 2016). Analysis of the completed genome sequences of humans and mice supports the lack of nonallelic variants of histone H4: the 26 copies of human and mouse H4 encode the identical protein (Marzluff et al., 2002). Similarly, both mouse and human genomes contain genes encoding only two replication-dependent H3 variants, H3.1 and H3.2, differing by a single amino acid; these are the same variants identified in early studies by direct sequencing of tryptic fragments (Franklin and Zweidler, 1977). The minor variant, H3.2, containing Ser96 rather than Cys96, is expressed at levels approximately proportional to its lower gene copy number in both humans and mice (Marzluff et al., 2002). In contrast, 10–12 variants are found among the genes encoding H2A and H2B. Similarly, the sea urchin H2A genes expressed late in development encode seven different primary sequences with alterations at 23 amino acid residues, while those for H2B encode ten distinct sequences with variable substitutions at 27 residues (Marzluff et al., 2006). Among the human and mouse variants, some residues vary only in a single allele, while others are altered in multiple alleles, and some of the variant residues found in human and mouse H2A also vary in sea urchin H2A. Whether these variable residues have functional significance or are simply sites that are under less selective pressure is unclear. Greater variability is observed in the DNA sequences than of the corresponding encoded amino acid sequences of the multiple histone gene alleles; differential regulation of alleles, the human versions of which have recently undergone revisions in nomenclature, is poorly understood (Flaus et al., 2021; Seal et al., 2022).

Experiments to test the possible functional specificity of nonallelic histone subtypes were first conducted in budding yeast. Only two copies of each of the four core histones are present in yeast, and while the copies of H3 and H4 encode identical histones, the two H2B alleles differ at four amino residues in the N-terminal region while the H2A alleles differ only in having Thr-Ala or Ala-Thr at residues 124–125 (Choe et al., 1982; Smith and Andresson, 1983; Wallis et al., 1980). This raised the immediate question of whether these small sequence changes represented differences in functional roles of the H2A- and H2B-encoding alleles. Michael Grunstein's lab at UCLA first tested whether either of the H2B subtypes had indispensable roles in growth, mating, sporulation, or germination by constructing yeast strains in which a frameshift mutation was engineered into one or the other gene encoding H2B (Rykowski et al., 1981). The only effect observed was that *h2b1⁻* yeast grew more slowly after germination. In contrast, sporulation of diploid yeast heterozygous for functional copies of *HTB1* and *HTB2* never yielded *htb1⁻ htb2⁻* spores, demonstrating that H2B is essential for viability. Similarly, the individual alleles encoding H2A are dispensable; furthermore, all four combinations of *hta* and *htb* mutants were viable, showing that H2A and H2B subtypes could associate interchangeably in nucleosomes (Kolodrubetz et al., 1982). Yeast *hta1⁻* mutants grew more slowly than wild-type yeast or *hta2⁻* mutants; it was later found that increased transcription of the divergent *HTA1-HTB1* compensated for reduced histone levels in *hta2-htb2* mutants, while such dosage compensation was not observed in *hta1-htb1* mutants (Norris and Osley, 1987).

These first, pioneering experiments on the effects of histone gene mutations were precursors to numerous investigations that yielded tremendous insight into the functional roles of histones in chromatin. In the next section, we describe some of these subsequent studies, illustrating the increasing sophistication and comprehensive scope of genetic approaches to histone function.

Histone mutants

In the late 1980s and early 1990s, a series of elegant and now classic investigations from Michael Grunstein's lab at UCLA in which histones were genetically altered in vivo, together with related studies from Mitch Smith at the University of Virginia, Fred Winston at Harvard, and others, contributed to major advances in understanding the role of chromatin in gene regulation. These foundational studies set the precedent for subsequent work in which engineered histone mutants have been used to study chromatin structure and function.

We have already discussed investigations demonstrating that each of the alleles encoding H2A and H2B in yeast is dispensable. A major surprise that followed this work was that yeast engineered to lack the flexible amino-terminal region of any of the core histones remained viable, although effects on growth, cell cycle progression, and mating were observed in some cases (Kayne et al., 1988; Mann and Grunstein, 1992; Morgan et al., 1991; Schuster et al., 1986; Wallis et al., 1983). Viability was assessed by several approaches. Two classical approaches from yeast genetics are depicted in Fig. 5.7. (The approaches actually used were somewhat

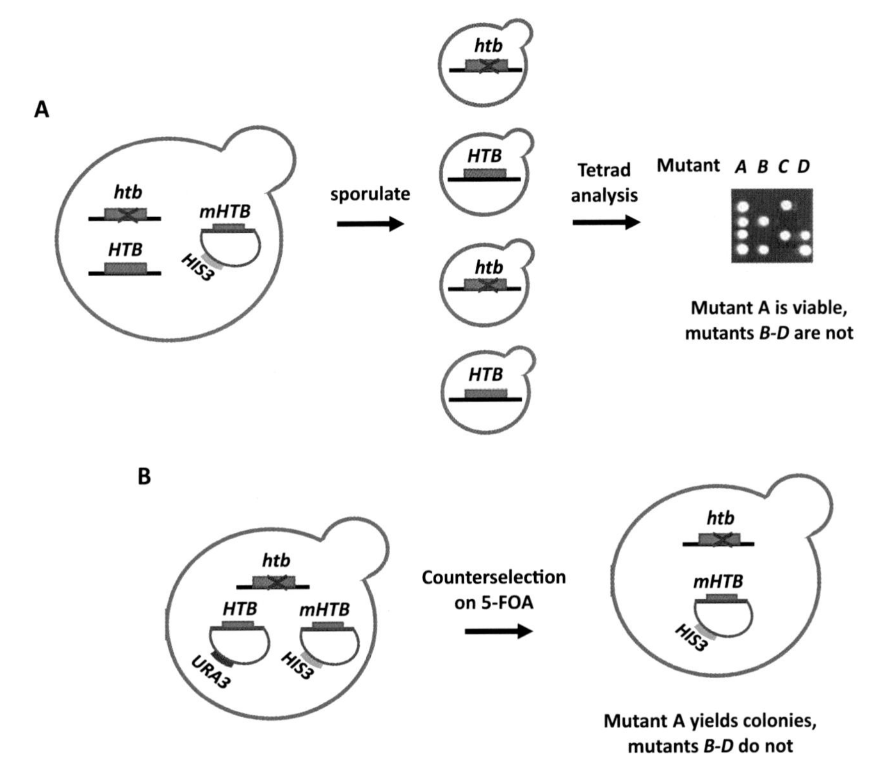

FIG. 5.7 Simplified version of two protocols to assess the viability of histone mutants. (A) In the upper scheme, a diploid yeast strain is engineered to have one chromosomal copy of gene *HTB*, encoding histone H2B, knocked out, and to carry a plasmid expressing a mutant version of *HTB*. Sporulation yields two colonies with intact chromosomal *HTB* and two deletion mutants; if the mutant version of *HTB* is propagated to all four progeny, all four spores will grow if the mutant is viable and only two will grow if it is not. (B) In the lower scheme, a haploid strain is constructed with the chromosomal *HTB* knocked out and "covered" by a plasmid-borne copy with a counter-selectable marker gene; a second plasmid carries the mutant version of *HTB*. Counter-selection (by treatment with a drug that is converted to a toxic intermediate by the *URA3*-encoded enzyme) yields surviving colonies only if the mutant is viable. In reality these screens are complicated by the presence of two copies of each histone gene and the use of plasmids that segregate similarly to chromosomes upon sporulation.

more complicated than shown.) In the first, diploid yeast were constructed to have one wild-type allele for a given histone gene and one mutant copy (expressing, for example, an N-terminally deleted histone.) Sporulation of these diploids yields four haploid cells; all four will be viable if the mutant supports life, but only two spores will produce colonies if the mutant is inviable. In the second approach, yeast harboring plasmids expressing a wild-type histone gene and a mutant copy, and having the chromosomal copies removed, are tested for the ability to survive the loss of the plasmid harboring the wild-type histone gene; this is generally referred to as a "plasmid shuffle" experiment. A novel approach employed by the Grunstein lab was a "glucose shift" experiment, in which two plasmid-encoded copies of a histone gene were present in yeast lacking its chromosomal copies. The wild-type histone gene was placed under galactose control (using a promoter activated in the presence of galactose), and the mutant histone gene was under the control of its promoter. In galactose medium, the wild-type histone is expressed and the yeast can grow normally; if the mutant histone gene cannot support viability, the yeast will not be able to grow in glucose medium, in which the expression of the wild-type histone is repressed (Kayne et al., 1988).

The results of these various experiments showed that not only was the tail of each individual core histone, defined as the region that was lost upon trypsinization of nucleosomes as discussed in Chapter 2 (see Fig. 2.5), not required for near-normal growth, but that tail-less histones could be incorporated into nucleosomes in vivo. This was a remarkable finding, particularly with regard to histone H4; as stated in Kayne et al. (1988), the 28 amino-terminal amino acids in H4 that proved to be dispensable in yeast have "remained almost invariant in more than 2 billion years of evolution." In contrast, simultaneous deletion of the amino-terminal regions of H2A and H2B or H3 and H4 was lethal, suggesting that the tails functioned redundantly in processes essential for viability (Morgan et al., 1991; Schuster et al., 1986). Larger deletions that included what we now know to include portions of the histone fold, not yet known at the time, were lethal; for example, while the H4△(4-28) mutant was viable, H4△(4-34) was not (Fig. 5.8). Mutants in which the three C-terminal amino acids of histone H4 were deleted proved to be viable, but deletions of six or more amino acids were lethal, as were deletions of even small portions of the central, structured regions of the histones (Kayne et al., 1988; Mann and Grunstein, 1992; Schuster et al., 1986).

Closely following these first genetic experiments on histone function in vivo were studies employing point mutants. Highly conserved lysine residues in the H4 tail (Lys 5, 8, 12, and 16) were known to be subject to reversible acetylation (Davie et al., 1981; Nelson, 1982), prompting investigators to examine the effects of mutating these residues (Durrin et al., 1991; Megee et al., 1990; Park and Szostak, 1990). Because acetylation of lysine residues results in neutralization of their positive charge, lysines were mutated to both glutamine and arginine, thereby either mimicking or preventing charge neutralization. Mutation of any individual lysine residue had little effect on growth or mating efficiency, while mutation of all four lysines resulted in severe defects in both (Durrin et al., 1991; Megee et al., 1990; Park and Szostak, 1990). (Different outcomes of mutating four Lys to Arg residues, in one case resulting in lethality and another not, likely resulted from strain-specific responses to this mutation (Dai et al., 2008).) Induced transcription from the *GAL1*, *PHO5*, or *CUP1* promoters, ascertained by using *lacZ* fusions and monitoring β-galactosidase activity, was similarly relatively unperturbed by single lysine mutations but was reduced fourfold to fivefold by mutation of three lysine residues to arginine, and about 50-fold for the *GAL1* and *PHO5* promoters by mutating all four residues to Arg (Durrin et al., 1991). Mutation of all four Lys to Arg elicited more severe phenotypes than mutation to Gln, while the latter mutant was more similar to deletion mutants that encompassed all four Lys residues; this suggested that the positively charged lysine residues exerted a repressive effect on growth and transcription that

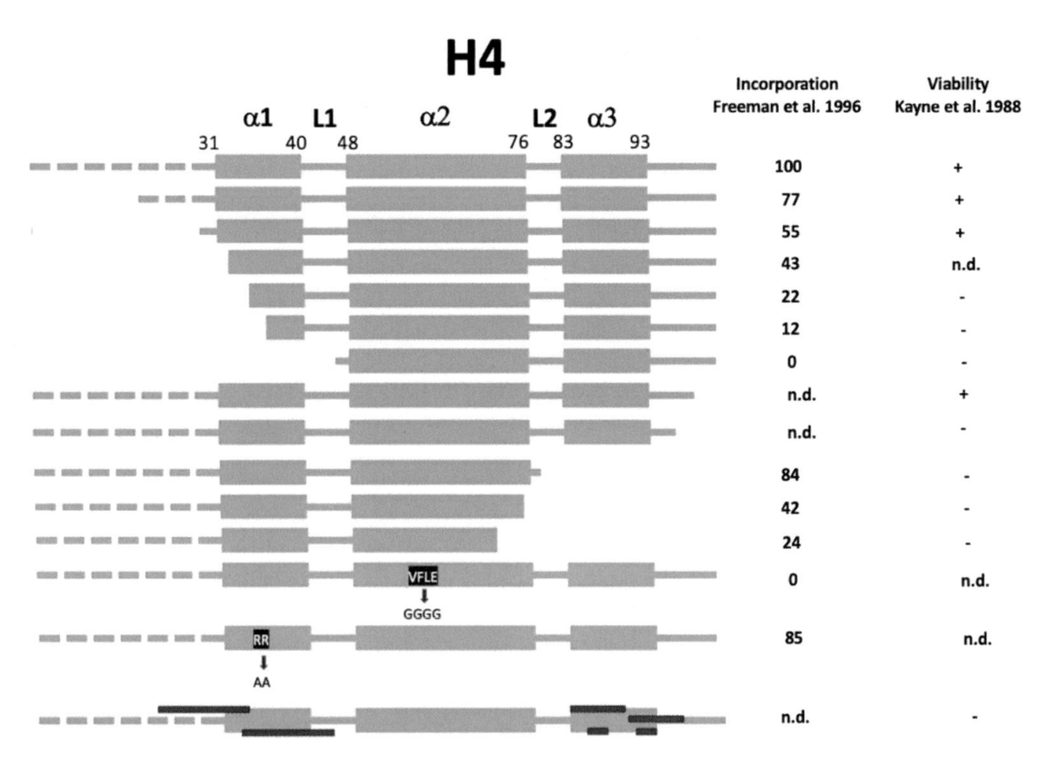

FIG. 5.8 The effect of mutations in histone H4 on incorporation into nucleosomes and on the viability of yeast. Shown are the N-terminal deletions, C-terminal deletions, and substitution mutants tested for incorporation into nucleosomes in *Xenopus* embryos; the ability of these mutants to support the viability of yeast is inferred from similar mutants examined by Kayne et al. (1988). In addition, deletions causing loss of viability are indicated by *red rectangles* in the bottommost depiction.

could be alleviated by charge neutralization or removal. These experiments provided the first evidence for specific functional roles for the modifiable lysine residues in the histone amino termini. A later, exhaustive study corroborated the redundant function of lysine residues in the histone H4 amino terminus by constructing all 15 combinations of Lys to Arg mutations at these sites and assessing, using microarrays, the effect on genome-wide transcription (Dion et al., 2005). Although transcription was only assessed during growth in rich medium, so that effects on inducible transcription were not examined, effects of single and double Lys to Arg mutations were relatively mild with the exception of those that included the K16R mutant. Concordant with the earlier reports, mutation of three or four Lys residues to Arg exhibited much stronger effects, with transcription of up to ~1200 genes being affected.

Genetic experiments in which histone mutations led to inviability could not distinguish whether the corresponding mutants failed to be incorporated into nucleosomes or were needed for a critical function in chromatin. The question of which regions of histones H3 and H4 are required for incorporation into nucleosomes was addressed in a clever approach that exploited the efficient assembly of chromatin in *Xenopus* embryos during the phase of rapid cell division following fertilization (Freeman et al., 1996). *Xenopus* embryos were injected with mRNA encoding full-length or mutant versions of histones H3 or H4 bearing an epitope tag, along with tritiated lysine and arginine to allow radiolabeling of newly synthesized histones. After 3 h, embryos were treated with MNase, and nucleosomes were purified by sucrose gradient centrifugation. Electrophoresis of histones from these purified nucleosomes allowed the determination of the ratio of endogenous, full-length histones to the mutant versions in the assembled chromatin, and thus of the efficiency of incorporation of mutant histones. The results showed that deletions of the H4 amino terminus extending to 28 amino acids still allowed relatively efficient incorporation into nucleosomes, in accord with the yeast genetic experiments (Fig. 5.8). Deletions that encroached into the first α-helix of the histone fold (known at this point from the structure of the histone octamer) showed progressively reduced incorporation, with no incorporation observed for the △1-45 mutant in which the first α-helix is entirely absent. Other mutations that disrupted the histone fold domain also reduced incorporation into chromatin. Interestingly, a C-terminal mutant lacking residues required for viability in yeast was incorporated with near wild-type efficiency, suggesting a critical function for this unstructured region of H4 in yeast chromatin. Similar experiments were conducted to examine regions of H3 required for incorporation into chromatin. Somewhat surprisingly, deletion of the N-terminal 63 amino acids, which includes the entire short N-terminal α-helix (which is not, however, part of the histone fold; Fig. 5.6), reduced incorporation only about threefold compared to intact H3. Mutants that interrupted the integrity of the first or second α-helix of the histone fold also reduced but did not eliminate incorporation, while deletion of the C-terminal α-helix (α3) resulted in complete loss of incorporation. Altogether, these various experiments utilizing engineered histone mutants, along with the structural determination of the histone octamer and nucleosome core particle in the same decade, greatly advanced the understanding of the structural and functional contributions of the histones to nucleosome assembly and chromatin function.

Since these pioneering genetic investigations of histone function, numerous investigators have used mutant histones to examine their role in transcription, replication, repair, and recombination of DNA in yeast. A logical extension of this approach was the development of means to simultaneously interrogate large numbers of histone mutants. Two nearly contemporaneous publications addressed this challenge by constructing libraries of yeast mutants in which single amino acid residues in the core histones were replaced with alanines and introduced by plasmid shuffle into yeast lacking chromosomal copies of the histone genes (Fig. 5.7; Matsubara et al., 2007; Nakanishi et al., 2008). A third study examined only histones H3 and H4, but additionally included substitutions of lysines with glutamine and arginine as well as a series of N-terminal deletions of H3 and H4, and integrated the mutants into the *HHT2-HHF2* chromosomal locus rather than relying on plasmid-borne copies of the genes (Dai et al., 2008). (Wild-type alanine residues, apart from those in the tails of H3 and H4 which were mutated to serine in the third study, were left unperturbed in all three studies.) It is important to recognize that at many sites, alanine substitution is a relatively nonperturbing mutation; for example, alanine substitutions in the α-helices of the histone fold are not expected to disrupt the structure. (It would be an interesting exercise to test whether this assertion was supported by predicted structures using AlphaFold (Jumper et al., 2021).) Nonetheless, the first two studies identified a total of 19 residues whose mutation to alanine was lethal; the third study found a larger number of lethal mutations, possibly reflecting a more stringent response to having the mutated genes chromosomally integrated. All three studies found essential H3 and H4 residues clustered near the dyad or the DNA entry-exit point. The five essential residues identified in H2A and H2B were located on the nucleosome surface in the region of the acidic patch that mediates interaction with the H4 amino terminus in the nucleosome crystal structure (Luger et al., 1997). As we discussed in Chapter 3, this interaction is important for chromatin compaction, and the acidic patch also is an interaction surface for other proteins (Kalashnikova et al., 2013); evidently at least some such interaction is critical for viability in yeast. Consistent with the nonessential nature of the histone amino termini, all the essential residues identified are located in the structured core of the nucleosome; furthermore, the majority (16/19) have not been found to be subject to posttranslational modification (Huang et al., 2014). No lysine residues, which are common targets for acetylation and methylation (see Chapter 7: Histone modifications), were found to be essential. Thus at the most basic level of sustaining life (for yeast), the critical function of essential amino acids in the histones appears to be structural, rather than regulatory.

As is true for many mutations, some histone mutations that allow the growth of yeast in rich medium are deleterious under other conditions, such as exposure to methylmethanesulfonate, which evokes the DNA damage response, or other perturbations. In addition, many of the viable mutants exhibit defects in fitness; in a competitive growth experiment, ~40% of the mutants showed reduced growth (Dai et al., 2008). These histone mutant libraries have proved useful in studies of the role of chromatin in DNA repair and transcription, and histone modification pathways, among other investigations.

Although the potential utility of genetic manipulation of histones in higher eukaryotes was recognized early on, the difficulty of simultaneously mutating most or all the multiple copies of a given histone gene, as opposed to the two copies in yeast, was an insurmountable obstacle for many years. A breakthrough occurred in 2010 with a report of a heterozygous mutant of *Drosophila melanogaster* in which one copy of the *HisC* cluster, which encodes all copies of the genes for H2A, H2B, H3, H4, and the linker histone H1, was deleted (Gunesdogan et al., 2010). Homozygotes produced by crossing these mutants die early in embryogenesis after the maternal supply of histones is exhausted, but they can be rescued by a transgene cassette encoding multiple copies of the histone gene unit. This novel tool has been exploited in investigations of the role of histone modification and histone level on transcription, cell differentiation, and formation of heterochromatin in *Drosophila* (Hodl and Basler, 2012; McKay et al., 2015; Pengelly et al., 2013; Yung et al., 2015). Such experiments nonetheless remain technically challenging, as multiple plasmids or large, bacterial artificial chromosomes (BACS), together with complex crossing procedures, are required for complementation of the histone genes. In a modified approach, CRISPR/Cas genome editing has recently been used in conjunction with engineered in vivo recombination to construct *Drosophila* mutants harboring specific histone mutations or altered gene copy numbers (Zhang et al., 2019). Using this approach, 19 of 40 tested Ala substitutions resulted in inviability, with lethality variously occurring at the embryonic, larval, and pupal stages for specific mutants. Evidently the demands of cellular differentiation, similarly to environmental perturbations in protozoans such as yeast, result in stricter dependence on chromatin structure and function relative to that seen for yeast growing in rich medium. Similar approaches to those recently developed in flies are surely forthcoming in mammalian systems but will require overcoming the additional barrier of the histones being encoded in multiple gene clusters (Marzluff et al., 2002).

Specialized variants

To this point, our discussion has centered on the canonical core histones and allelic variants and mutants thereof. We now examine specialized histone variants (Fig. 5.9 and Table 5.1).

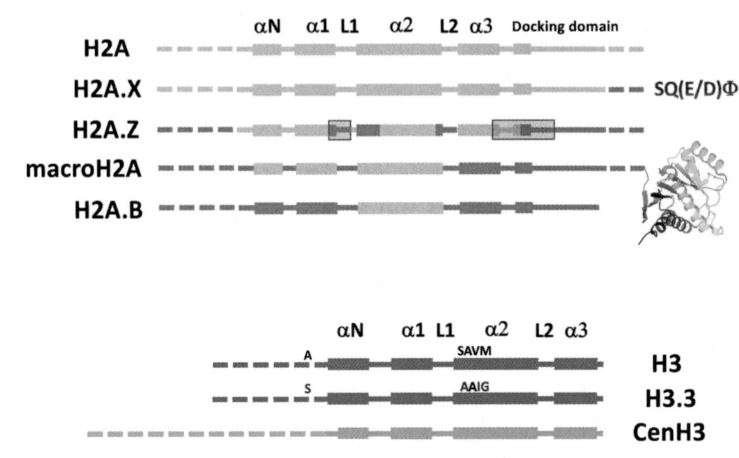

FIG. 5.9 Specialized histone variants. The canonical versions of H2A and H3 are shown with variants below. Altered shades indicate regions of substantial divergence (>30%) relative to the canonical versions, while the *green* region in CenH3 is unique to the variant. *Shaded boxes* indicate regions of H2A.Z in which amino acid substitutions occur that affect H2A dimerization (L1 loop) or nucleosome stability (docking domain). The unique macrodomain C-terminal portion is not shown to scale, as it comprises two-thirds of macroH2A (PDB ID: 1YD9).

TABLE 5.1 Histone variants.

Core histone	Variant	Similarity to canonical	Function	Range	Fraction of total histone	Genomic localization	Discovery
H2A	H2A.X	~80%	DNA ds break repair	Universal	10%–15% (mammals)	Widespread	West and Bonner (1980)
	H2A.Z	~60%	Transcription	Universal	5%–10% (vertebrates)	Promoters, enhancers, heterochromatin	West and Bonner (1980)
	MacroH2A	65% (histone fold)	Unknown	Metazoa	3% in rat liver	Heterochromatin esp Xi	Pehrson and Fried (1992)
	H2A.Bbd	<50% (48% in humans)	Unknown	Mammals	v. low except testis, brain	Promoters	Chadwick and Willard (2001)
H3	H3.3	97%	Replacement	Universal	10%–25% (100% in yeast)	Promoters, enhancers, gene bodies, heterochromatin	Franklin and Zweidler (1977)
	CenH3	50%–60%	Centromere identity	Universal	v. low	Centromeres	Palmer et al. (1987)

These variants are expressed throughout the cell cycle, are present in only one or two gene copies, and have molecular phylogenies that distinguish them from the canonical core histones. In metazoan cells, they are also distinguished from the canonical core histones by their genes often containing introns and by their corresponding mRNAs being polyadenylated. While H2A and H3 have evolved to yield multiple specialized variants, H2B and H4 exhibit almost no functional diversification. (Testis-specific variants of H2B, as well as of H2A and H3, are present in metazoans, while variants of H4 have been reported only in lower eukaryotes such as trypanosomes and tunicates (Moosmann et al., 2011; Siegel et al., 2009). A recent report identifies a novel variant of histone H4, H4G, specific to hominid cells (Long et al., 2019). However, the gene encoding H4G lies within one of the clusters of canonical core histone genes, and the protein, which differs in amino acid sequence at several sites important for interaction with H2A in the nucleosome, appears incapable of being incorporated into a nucleosome in vitro (Long et al., 2019). H4G thus appears not to qualify as a specialized histone variant in the sense of contributing to variant chromatin.) One possible factor contributing to this dichotomy is based on structural considerations (Chakravarthy et al., 2004; Henikoff and Smith, 2015; Talbert and Henikoff, 2010). Both H2A and H3 make homodimeric contacts as well as contacting other histones in the histone octamer, while H2B and H4 only contact other histones. Mutation of residues could occur at the homodimerization interface, thereby favoring homodimer formation over heterodimerization between canonical and variant histone, thus allowing the "clean" evolution of variant nucleosomes. However, this argument is weakened by observations of hybrid nucleosomes containing both canonical and variant histones in vitro and in vivo (Luk et al., 2010). Luger and colleagues also consider that alterations in H3 and H2A would be more likely than changes in H4 or H2B to affect higher-order chromatin structure or interactions with linker histones, modulation of which might offer evolutionary advantages. An alternative vantage might be to consider whether variants of H2B and H4 would more likely be subject to negative selection; however, given that only a handful of variants have actually arisen over millions of years, we may also be wise to recognize the limitations of brains that survive only for a few decades to consider such questions.

As we have seen, variations in primary amino acid sequence for core and linker histones within a given species were observed from the very earliest studies (Bustin and Cole, 1968; Cohen et al., 1975; Franklin and Zweidler, 1977), and the distinction between replication-coupled and replication-independent histone synthesis was reported as early as 1981 (Wu and Bonner, 1981). However, while some of the earliest reported histone variants turned out to be specialized in function, this was not recognized immediately (West and Bonner, 1980). The *Tetrahymena* variant of H2A, hv1, was reported to be specific to the transcriptionally active macronucleus in 1980 (Allis et al., 1980), and its Drosophila homolog shown to be essential in 1992 (van Daal and Elgin, 1992), while the centromeric histone H3 variant, CenH3, was discovered in 1987 and demonstrated to be a *bona fide* histone variant 4 years later (Palmer et al., 1987, 1991). Nonetheless, the first reviews to bring these findings together in a discussion of histone variants, motivated also by the ability to compare differences in amino acid sequence in the context of the structure of the histone octamer (Arents et al., 1991), were not published until 1995 and 1996 (Wolffe, 1995; Wolffe and Pruss, 1996). Numerous excellent reviews on histone variants have been published in the years since, including some focusing on structure (Bönisch and Hake, 2012; Kurumizaka et al., 2021; Shaytan et al., 2015), evolutionary aspects (Malik and Henikoff, 2003; Talbert and Henikoff, 2010; Talbert et al., 2019), nucleosome dynamics (Talbert and Henikoff, 2017), and the role of variants in human disease and cancer (Buschbeck and Hake, 2017; Martire and Banaszynski, 2020). In this section, we discuss in turn the specialized histone variants depicted in Figs. 5.9 and 5.10 and Table 5.1 and also examine some less widespread histone variants.

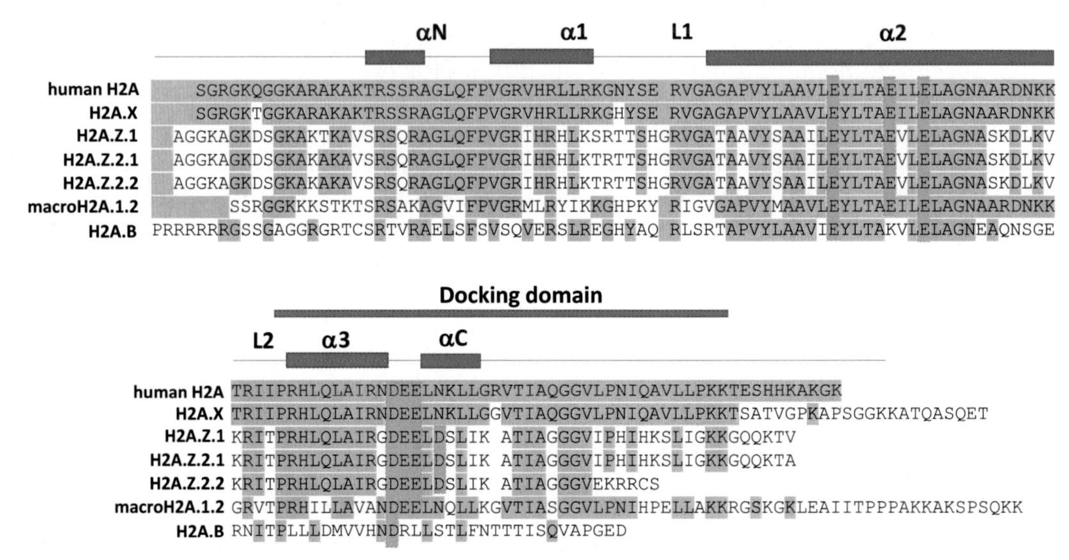

FIG. 5.10 Human histone H2A variants. The amino acid sequence of the canonical version of H2A is shown with variants below. Residues shaded in *gray* are identical to those found in canonical H2A; residues highlighted in *light red* comprise the acidic patch (note the extended patch represented by the N to D alteration in the C-terminal a-helix of H2A.Z variants). Only the histone fold portion of macroH2A.1.2 is shown.

H2A.Z

H2A.Z was first identified and characterized, along with H2A.X, in tissue culture cells from mouse, human, and chicken by two-dimensional electrophoresis and analysis of tryptic fragments (West and Bonner, 1980). Both these variants were shown to be present in core particles obtained from MNase digestion of chromatin, establishing them as *bona fide* histones (Hatch et al., 1983; West and Bonner, 1980). The similar behavior of H2A.Z from mouse and sea urchin as assayed by two-dimensional electrophoresis, along with their similar tryptic peptide patterns, both of which distinguished the variant from the major H2A species, led Bonner and colleagues to hypothesize that H2A.Z could represent an evolutionarily distinct lineage from canonical H2A (Wu et al., 1982). Molecular phylogeny based on the determination of complete amino acid sequences from multiple species eventually confirmed this notion, showing that the H2A.Z family diverged from canonical H2A very early in evolution (Malik and Henikoff, 2003; Thatcher and Gorovsky, 1994; van Daal et al., 1990).

Two observations lent special interest to H2A.Z early on. First, acid extraction of proteins from the transcriptionally active macronucleus and inactive micronucleus from *Tetrahymena thermophila* yielded a novel minor histone variant, designated hv1, exclusive to the macronucleus and thus associated with transcriptionally active chromatin (Allis et al., 1980). Initial characterization suggested that hv1 was related to H2A, and subsequent partial sequencing confirmed this, while also revealing its close similarity to the chicken variant H2A.F and calf H2A.Z (Allis et al., 1986). (The chick H2A.F variant was first isolated by screening a cDNA library from chicken embryonic RNA for clones that hybridized poorly with histone gene probes (Harvey et al., 1983); subsequent cloning and sequence comparison with sea urchin and mammalian H2A.Z showed H2A.F to be most closely related to these variants, with the sea urchin variant consequently being named H2A.F/Z (Ernst et al., 1987; Hatch and Bonner, 1988).) In parallel with this observation, the murine homolog M1, later identified as an H2A.Z, was found to be present in a chromatin fraction highly enriched in transcribed sequences and absent in the nontranscribing fraction, suggesting a role for this variant in transcriptional regulation that was not unique to *Tetrahymena* (Gabrielli et al., 1981). The second observation came from Sally Elgin's lab. Using a cDNA clone of the *hv1* gene, the *Drosophila* homolog was identified and shown to exist in a single copy; sequence analysis revealed that the *Drosophila* protein, named H2AvD, was also homologous to chicken H2A.F, sea urchin H2A.F/Z, and mammalian H2A.Z (van Daal et al., 1988). The monoallelic nature of the *Drosophila* gene allowed the effect of a deletion mutant to be ascertained, resulting in the startling finding that H2AvD was essential in the fruit fly (van Daal and Elgin, 1992). This was the first demonstration of a histone variant being essential in any organism, and given the minor fraction of H2A that it comprised (Table 5.1), pointed to a functional role more specific than simple packaging of bulk chromatin. H2A.Z (hv1) was subsequently shown also to be essential in *Tetrahymena* (Liu et al., 1996), while its deletion in the mouse caused early embryonic lethality (Faast et al., 2001).

The obvious potential utility of budding yeast for studying the function of histone variants led various investigators to attempt to identify H2A.Z in *Saccharomyces cerevisiae*, but it was not until the (nearly) complete genome sequence was available that *HTZ1* (initially named *HTA3*), encoding yeast H2A.Z, was identified using a simple bioinformatic approach (Jackson et al., 1996). (A similar path led to the identification of the yeast version of linker histone H1 (Landsman, 1996).) Independently of its sequence-based identification, *HTZ1* was also found in two separate yeast genetic screens (Dhillon and Kamakaka, 2000; Santisteban et al., 2000). Dhillon and Kamakaka identified Htz1 in budding yeast by virtue of its ability to restore transcriptional silencing in a defective mutant when overexpressed; later work showed that H2A.Z functions as a barrier between heterochromatin and euchromatin in yeast (Meneghini et al., 2003). Santisteban et al. identified H2A.Z (Htz1) as a suppressor of a specific histone H4 mutant. Tyr72 and Tyr88 in histone H4 were known from the histone octamer structure to contact one H2A-H2B dimer, and Tyr98 the other (Arents et al., 1991; Luger et al., 1997); targeted mutagenesis of these residues led to lethality, temperature sensitivity, or other phenotypes, depending on the nature of the mutation (as later confirmed by high throughput genetic screening) (Dai et al., 2008; Matsubara et al., 2007; Nakanishi et al., 2008; Santisteban et al., 1997). A Y98H mutant that exhibited a *ts* phenotype was used to identify suppressors from a cDNA library that would allow growth at the restrictive temperature; in addition to finding that wild-type H4 suppressed the phenotype, the *HTA3* gene identified by Jackson et al. was found also to act as a suppressor. Overexpression of the canonical H2A gene *HTA1* did not suppress the Y98H *ts* phenotype, and other evidence also indicated regulatory functions and localization of H2A.Z that distinguished it from the canonical H2A proteins.

In contrast to the essential nature of H2A.Z in *Tetrahymena*, *Drosophila*, and mouse, deletion of H2A.Z is tolerated in both budding and fission yeast (Carr et al., 1994; Santisteban et al., 2000). Deletion in *S. pombe* resulted in slow growth and defects in chromosome segregation during cell division, while *S. cerevisiae htz1*△ mutant cells also displayed slow growth, as well as temperature sensitivity and transcriptional regulatory defects. In addition to these phenotypes, other reports found cells lacking H2A.Z to be defective in cell cycle control, DNA damage repair, and DNA replication, and implicated H2A.Z in centromere structure and function (Greaves et al., 2007; Henikoff and Smith, 2015; Zlatanova and Thakar, 2008).

The variety of cellular functions apparently impacted by H2A.Z heightened interest in this variant, and investigators sought explanations for its functional contributions by examining its impact on nucleosome structure and chromatin compaction, and by determining its chromosomal localization and mechanism of deposition. To identify indispensable features of H2A.Z in *Drosophila*, David Tremethick and colleagues parsed the H2A.Z sequence into seven regions exhibiting divergence from the canonical H2A sequence and tested the ability of H2A.Z mutants in which each region in turn was replaced by the corresponding sequence from H2A to complement deletion of *His2AvD* in developing embryos (Clarkson et al., 1999). The only region that completely failed to complement when mutated encompassed the extended C-terminal α-helix in the center of the H2A docking domain (Figs. 5.9 and 5.10). This domain creates a large interaction surface that stabilizes the structure of the nucleosome core particle by facilitating interaction of the H3 N-terminal extended helix with the outermost turn of nucleosomal DNA and by interacting with the C-terminal region of histone H4 (Luger et al., 1997). Replacement of a more C-terminal region also located in the docking domain and harboring seven amino acid changes between H2A and H2A.Z was also extremely deleterious, allowing eclosure (i.e., emergence of pupa) in <10% of examined mutants. Two other regions, the amino-terminal portion that contains modifiable lysine residues and the beginning of the α2 helix of the histone fold, exhibited negative effects when mutated to their H2A counterparts but to a lesser extent than the docking domain mutants.

Given the requirement for H2A.Z for viability, the ability of flies to tolerate the substitution of structurally important regions of H2A.Z with the corresponding regions of H2A despite numerous changes in the amino acid sequence was something of a surprise. However, this tolerance found support in the high-resolution X-ray crystal structure of a nucleosome containing recombinant murine H2A.Z and recombinant H2B, H3, and H4 from *Xenopus laevis* (Suto et al., 2000). This structure showed little difference in the path of the DNA or protein-protein interactions within nucleosomes containing H2A and H2A.Z, despite the high degree of amino acid divergence. The most significant structural differences were found in the docking domain, whose interaction with H3 was compromised by the loss of three hydrogen bonds, and between H2A.Z-H2B dimers. The latter difference was due to alterations in the L1 loop that, while not affecting protein-DNA interactions or H2A(.Z)-H2B interactions, was predicted to produce a steric clash between heterotypic dimers (H2A-H2B with H2A.Z-H2B) (Fig. 5.11). This prediction, however, was refuted by experiments in which

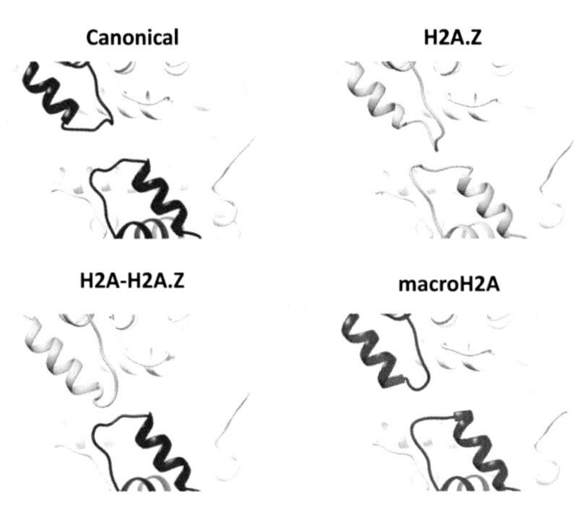

FIG. 5.11 Close-up views of the L1 loop region of H2A subunits in nucleosome core particles containing bacterially expressed histones derived from humans. Regions are shown from nucleosomes containing canonical H2A (*blue;* PDB ID: 5Y0C), H2A.Z.1 (*gold;* PDB ID: 3WA9), heterotypic H2A-H2A.Z.1 (PDB ID: 5B31), and macroH2A.1.2 (*magenta;* PDB ID: 1U35).

nucleosomes were reconstituted from mixtures of $(H3\text{-}H4)_2$ tetramers, His-tagged H2A-H2B dimers, and untagged H2A.Z-H2B dimers (Chakravarthy et al., 2004). When the reconstituted nucleosomes were purified using the His-tag, approximately equal amounts of H2A.Z and H2A were observed by gel electrophoresis, indicating the formation of hybrid nucleosomes rather than the segregation of H2A and H2A.Z. Hybrid nucleosomes were also observed in vivo in yeast, flies, and human cells (Luk et al., 2010; Viens et al., 2006; Weber et al., 2010), and structural studies showed that the H2A.Z L1 loop structure was altered in the context of a heterotypic H2A-H2A.Z-containing nucleosome, avoiding the predicted steric clash (Fig. 5.11; Horikoshi et al., 2016).

Could the "subtle destabilization" caused by the divergence of the docking domain of H2A.Z be sufficient to account for its functional roles in vivo? Initial efforts to test the effect of H2A.Z on nucleosome stability yielded variable results, with stabilization and destabilization both being reported (reviewed in Bönisch and Hake (2012)). Stability was monitored by various methods in the different studies, including analytical ultracentrifugation and gel filtration in varying salt concentrations, fluorescence resonance energy transfer, and immunoblot analysis of yeast chromatin following salt washes (Abbott et al., 2001; Park et al., 2004; Thakar et al., 2009; Thambirajah et al., 2006; Zhang et al., 2005). Moreover, experiments differed in their use of histones from different species (e.g., yeast vs human vs *Xenopus*) and the use of reconstituted chromatin with various DNA templates or chromatin isolated from cells. It therefore seems fair to agree that the incorporation of H2A.Z might modulate nucleosome stability depending on DNA sequence, posttranslational modifications, and nucleosome composition (Bönisch and Hake, 2012).

The effect of H2A.Z incorporation on higher-order folding of chromatin was investigated using the nucleosome array system that was introduced in Chapter 3 (Fig. 3.3) (Fan et al., 2002). Nucleosomes were reconstituted into 12-mer arrays with equal efficiency using either canonical histones or with H2A.Z in place of H2A. Analytical centrifugation showed that the two arrays behaved indistinguishably under low salt conditions in which the arrays adopt an unfolded configuration, consistent with the structure of the H2A.Z-containing nucleosome being essentially identical to that of a canonical nucleosome. However, in the presence of divalent cations (Mg^{+2} or Zn^{+2}), the H2A.Z-containing nucleosome array exhibited enhanced intramolecular folding, as shown by increased amounts of material sedimenting at 55S with increasing divalent cation concentration. As discussed in Chapter 3, another transition observed in this system at high Mg^{+2} concentrations is array oligomerization to yield soluble material that pellets rapidly upon centrifugation; this transition was substantially reduced by the incorporation of H2A.Z. (A contemporaneous publication reported impaired folding of arrays containing H2A.Z; this contrasting result was most likely attributable to differences in experimental methodology, most critically the use of monovalent cations rather than divalent cations to induce folding (Abbott et al., 2001; Fan et al., 2002).)

Another feature noted in the crystal structure of the H2A.Z-containing nucleosome was an extended acidic patch (Suto et al., 2000). While not directly affecting the structure of the core particle, the acidic patch formed by one amino acid from H2B and six amino acids in H2A, three of which lie in the docking domain, was observed to contact the H4 amino-terminal region in the nucleosome core particle crystal lattice; indeed, the contact was reported to be necessary for crystallization (Luger et al., 1997). This internucleosomal contact was suggested to potentially facilitate chromatin compaction, an idea supported by the requirement for the histone H4 amino terminus for higher-order folding of nucleosomal arrays (Chapter 3; Dorigo et al., 2003; Gordon et al., 2005). It was therefore proposed that the enhanced compaction observed for nucleosome arrays containing H2A.Z was facilitated by the extended acidic patch. To test this idea, nucleosome arrays were constructed containing H2A.Z in which the acidic patch was mutated to the H2A sequence; these arrays were found to fold identically to arrays containing canonical H2A, both with regard to intrafiber compaction and interfiber oligomerization (Fan et al., 2004). Subsequent experiments showed that nucleosome arrays containing the H2A.Bbd variant, in which three of the six amino acids comprising the acidic patch are altered to nonacidic residues (Fig. 5.10), are deficient in intrafiber compaction and exhibit enhanced oligomerization; mutagenesis experiments in which acidic patches were extended or reduced yielded consistent results (Zhou et al., 2007). The complementary effects seen upon fiber compaction and interfiber association when the acidic patch was altered led to a proposed model in which intrafiber interactions between the H4 amino-terminal region and the acidic patch are in competition with interfiber interactions, also involving the H4 tail (Fan et al., 2004). Chemical cross-linking experiments later confirmed the interaction between the acidic patch and the H4 tail in both intra- and interfiber interactions; moreover, interarray interactions, monitored by measuring cross-links between two different arrays, increased under conditions favoring array oligomerization at the expense of intrafiber cross-links, in accord with the competition model (Kan et al., 2009; Sinha and Shogren-Knaak, 2010). The importance of the acidic patch to in vivo function was underscored by experiments showing that overexpression of H2A.Z slowed the development of the frog *Xenopus laevis* following gastrulation, but altering the acidic patch suppressed this effect (Ridgway et al., 2004). It is no stretch to imagine that H2A.Z is similarly essential for human development, and as of this writing (6/3/2021), the gnomAD database indicates no heterozygous loss of function mutants and only four missense mutants in ~140,000 human exomes (heterozygous mutations observed were A6S, V18I, T42S, and S43N) (Karczewski et al., 2020).

In Chapter 3, we saw that the relevance of in vitro studies of chromatin folding, particularly those relying on regular nucleosome arrays, has been called into question, as mounting evidence has shown that structures similar to the 30 nm fibers observed in such experiments are rare or absent in vivo (Maeshima et al., 2010). How, then, should findings regarding the influence of H2A.Z and other histone variants on chromatin folding that rely on this artificial in vitro system be interpreted? As pointed out by Bönisch and Hake, the effects of extension or contraction of the acidic patch on internucleosomal interactions with the H4 tail are likely to apply even in the absence of a regular 30 nm fiber (Bönisch and Hake, 2012). In vivo, interfiber contacts are likely to dominate due to the very high concentration of nucleosomes (Maeshima et al., 2010), and these contacts should be stabilized by an increased acidic patch and decreased by a contracted acidic patch; thus H2A.Z should facilitate such contacts and favor compaction of chromatin in vivo.

As mentioned above, the importance of interaction between the H4 tail and the acidic patch and the relevance of that interaction to H2A.Z function are also supported by in vivo experiments such as those examining *Xenopus* development. Consistent with a role in chromatin compaction, H2A.Z is enriched in centromeric heterochromatin in *Drosophila* and pericentric heterochromatin during mammalian development and is a structural component of the centromere in the latter (Greaves et al., 2007; Leach et al., 2000; Rangasamy et al., 2003; Swaminathan et al., 2005). Furthermore, H2A.Z is important for heterochromatin function: mutations in H2A.Z disrupt transcriptional silencing mediated by heterochromatin in *Drosophila*, and H2A.Z knockdown by RNAi in mammalian cells perturbs localization of heterochromatin protein 1-α (HP1-α) to regions of heterochromatin in chromosome arms in human Cos-7 cells (Rangasamy et al., 2004; Swaminathan et al., 2005). H2A.Z plays a different role related to heterochromatin in budding yeast; the compact yeast genome lacks the large heterochromatin regions found in metazoan organisms but possesses two regions that are structurally and functionally closely related to those regions, namely, the silent mating type loci and subtelomeric regions (Chapter 3). H2A.Z is enriched near telomeres in yeast, and loss of H2A.Z causes heterochromatic gene repression to spread into nearby regions, indicating that H2A.Z is needed to establish boundaries between heterochromatin and euchromatin in yeast (Meneghini et al., 2003). The mechanisms by which heterochromatin is established and acts in gene repression are complex, and H2A.Z constitutes just one piece among others that include linker histones, histone modifications, and nonhistone proteins such as HP1-α, in its formation and function (Ryan and Tremethick, 2018).

Studies on the genomic localization of H2A.Z revealed surprises that are still being sorted out. Early studies on this variant showed nonrandom localization on *Drosophila* polytene chromosomes by immunostaining (Leach et al., 2000; van Daal and Elgin, 1992). Chromatin immunoprecipitation (ChIP) experiments were performed to determine H2A.Z localization in yeast, first at individual loci, then using low-resolution microarrays (ChIP-chip) and eventually by high-resolution microarray and by ChIP-seq (Albert et al., 2007; Guillemette et al., 2005; Li et al., 2005; Millar et al., 2006; Raisner et al., 2005; Santisteban et al., 2000; Zhang et al., 2005). The most startling observation came from a high-resolution microarray analysis of H2A.Z occupancy over yeast chromosome III, using the same tiling microarrays that were used to analyze nucleosome positioning over this part of the yeast genome (Chapter 4; Raisner et al., 2005). This analysis revealed that the majority of genes on Chr III had H2A.Z-containing nucleosomes on either side of the characteristic nucleosome-depleted region (NDR) (Fig. 5.12); aside from genes near silenced telomeric regions, within the silent mating type cassette, or assigned as "dubious," only seven genes lacked H2A.Z at their TSS regions, and some of these displayed low but nonzero H2A.Z occupancy in later ChIP-seq

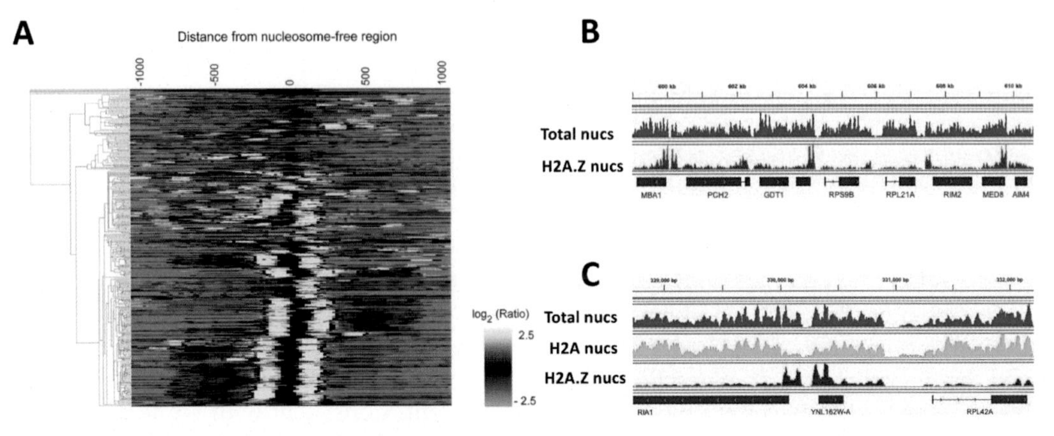

FIG. 5.12 H2A.Z localization in yeast. (A) Heat map of H2A.Z localization on yeast Chr III. Each row represents a gene promoter, centered on the NDR, with H2A.Z occupancy shown for 1 kb in each direction. (B) Browser scan of total nucleosome occupancy and H2A.Z occupancy over a region of yeast Chr II. (C) Expanded view at the *RIA1* promoter. *(A) From Raisner, R.M., Hartley, P.D., Meneghini, M.D., Bao, M.Z., Liu, C.L., Schreiber, S.L., Rando, O.J., Madhani, H.D., 2005. Histone variant H2A.Z marks the 5′ ends of both active and inactive genes in euchromatin. Cell 123, 233. (B) Data from Tramantano, M., Sun, L., Au, C., Labuz, D., Liu, Z., Chou, M., Shen, C., Luk, E., 2016. Constitutive turnover of histone H2A.Z at yeast promoters requires the preinitiation complex. elife 5.*

experiments (Tramantano et al., 2016). This pattern was confirmed across the entire yeast genome by ChIP-seq, with lower occupancy also observed over coding sequences and flanking tRNA genes (Albert et al., 2007). Some quantitative differences were seen in the data, with relatively lower occupancy of H2A.Z at the −1 nucleosome and higher occupancy over coding regions seen in the ChIP-seq results than with high-resolution microarrays; more recent results using ChIP-seq support relatively strong occupancy at −1 and +1 nucleosomes relative to other locations, such as transcribed regions, as originally reported (see Fig. 4.14 and Fig. 5.12) (e.g., Bagchi et al., 2020; Tramantano et al., 2016). Recent CUT&RUN experiments (see ChIP-seq variations sidebar, Chapter 4) indicate H2A.Z to be confined exclusively to the nucleosomes flanking NDRs; the authors suggested that H2A.Z occupancy reported further downstream in coding regions could be attributable to background signal (Brahma and Henikoff, 2019). However, ChIP-exo experiments indicate that H2A.Z deposition can occur over coding regions under at least some conditions, as a mutant (*swc2*△; see below) that reduces H2A.Z at the +1 nucleosome in yeast exhibits increased H2A.Z occupancy over coding regions (Yen et al., 2013).

Enrichment of H2A.Z at the +1 nucleosome was observed in *Drosophila* embryos, root tissue from *Arabidopsis thaliana*, and *C. elegans*, but in contrast to *S. cerevisiae*, these organisms did not exhibit H2A.Z enrichment at the first upstream, or −1, nucleosome (Mavrich et al., 2008; Whittle et al., 2008; Zilberman et al., 2008), whereas the pattern observed in human T cells was more like that seen in yeast, with H2A.Z-enriched nucleosomes being present on either side of the TSS (Barski et al., 2007). A possible explanation for the enrichment of H2A.Z at the −1 nucleosome in yeast derives from an analysis showing that H2A.Z enrichment upstream of the nucleosome-depleted region is associated with divergent transcription, either of canonical Pol II-transcribed genes or with antisense transcription (Bagchi et al., 2020). Thus, it is possible that H2A.Z-enriched −1 nucleosomes in yeast may be more accurately considered as +1 nucleosomes for antisense transcription.

The highly specific localization of H2A.Z nucleosomes near transcription start sites in yeast immediately raised the possibility that transcription played a causative role in their occupancy pattern. However, multiple studies found that H2A.Z occupancy in yeast was negatively correlated or uncorrelated with transcription rates (Bagchi et al., 2020; Guillemette et al., 2005; Raisner et al., 2005; Zhang et al., 2005). Raisner et al. went further and tested whether H2A.Z could be deposited at promoters of repressed genes by inducing an epitope-tagged allele of *HTZ1* and examining occupancy at two constitutively repressed promoters; both showed H2A.Z enrichment after induction of the tagged protein, demonstrating that transcription is not required for H2A.Z deposition at promoters. Disparate results have been observed in other systems: H2A.Z occupancy at promoters correlated with transcription in *Drosophila* (Mavrich et al., 2008), and correlation was also found in *C. elegans* embryos except at very highly transcribed genes (Whittle et al., 2008). Low H2A.Z occupancy was also found at highly transcribed genes in yeast; this anticorrelation was attributed both in yeast and *C. elegans* to nucleosome eviction occurring at highly transcribed genes (H2A.Z occupancy was not normalized to nucleosome occupancy in these experiments). A positive correlation between transcription and H2A.Z occupancy at promoters was reported in human T cells and neural precursors, while in murine embryonic stem cells, H2A.Z was enriched in promoters of genes not expressed in ES cells but required for differentiation (Barski et al., 2007; Creyghton et al., 2008; Schones et al., 2008). The relationship between H2A.Z and transcription is evidently complex and likely reflects both positive and negative influences related to transcription including nucleosome depletion at active gene promoters, the influence of the NDR on H2A.Z deposition (see below), and stimulation of histone turnover at promoters by the PIC (Tramantano et al., 2016).

A discovery critical to addressing the mechanism underlying the localization of H2A.Z is that deposition of H2A.Z is an active process that requires the activity of the ATP-utilizing SWR1 complex, referred to variously as SWR1, SWR-Com, SWR-C, or SWR.C (Kobor et al., 2004; Krogan et al., 2003; Mizuguchi et al., 2004). The SWR1 complex was identified independently by three labs using three different approaches: Nevan Krogan (while still a graduate student at the University of Toronto) and colleagues identified *HTZ1* and the genes encoding three subunits of what turned out to be the SWR1 complex in a genetic screen for deletion mutants that showed synthetic growth defects in combination with three different genes involved in transcriptional elongation through chromatin. The other two reports used biochemical purification followed by mass spectrometry to identify proteins associating with H2A.Z or Swr1, which at the time was known as an uncharacterized protein related by sequence to members of the Swi2/Snf2 chromatin remodeling ATPase family. All three studies found that similar phenotypes, including changes in genome-wide transcription, resulted from deleting either SWR1 subunit-encoding genes or *HTZ1*, indicating that SWR1 and H2A.Z functioned in the same pathway. Additionally, all three studies showed that defects in SWR1 reduced H2A.Z occupancy at highly occupied sites in vivo; this was later shown to be true on a genome-wide basis (Li et al., 2005; Zhang et al., 2005). Consistent with a dedicated role in H2A.Z deposition, genome-wide localization of SWR1 components was found to strongly overlap that of H2A.Z (Venters and Pugh, 2009; Zhang et al., 2005). Biochemical experiments demonstrated that SWR1 catalyzed the replacement of H2A-H2B dimers in nucleosomes with H2A.Z-H2B dimers in an ATP-dependent process (Fig. 5.13; Mizuguchi et al., 2004). Together, these experiments established SWR1 as the responsible entity for mediating

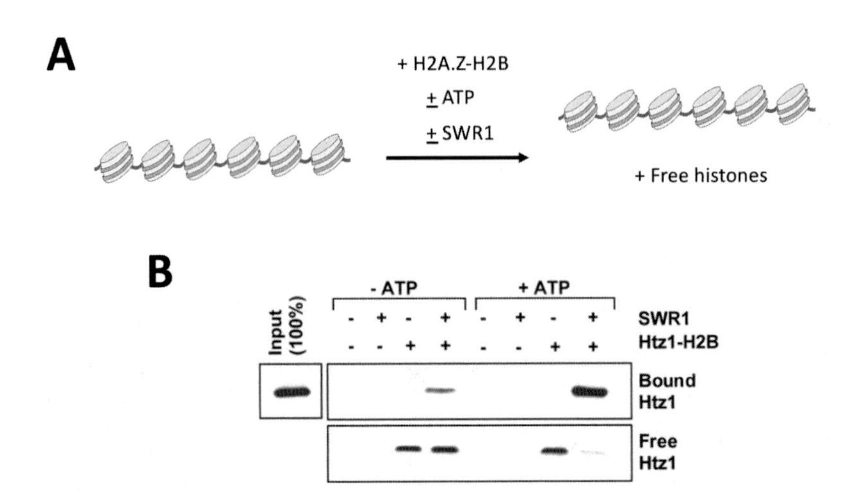

FIG. 5.13 SWR1 catalyzes the ATP-dependent incorporation of H2A.Z-H2B into chromatin. (A) Assay for incorporation of H2A.Z-H2B. A short nucleosome array is mixed with free H2A.Z-H2B dimers (or H2A-H2B dimers as a control) in the presence or absence of ATP and SWR1. Free and bound histones are separated and analyzed by gel electrophoresis and Western blotting (using Flag-tagged H2A.Z). (B) Results showing that incorporation requires SWR1 and is enhanced by ATP. Control experiments showed less efficient incorporation of H2A-H2B, similar to that seen with H2A.Z-H2B in the absence of ATP. *From Mizuguchi, G., Shen, X., Landry, J., Wu, W.H., Sen, S., Wu, C., 2004. ATP-driven exchange of histone H2AZ variant catalyzed by SWR1 chromatin remodeling complex. Science 303, 343.*

H2A.Z incorporation by exchanging H2A.Z-H2B dimers for H2A-H2B dimers in assembled nucleosomes. Subsequently, the *Drosophila* homolog Domino and human homologs *SRCAP* and *Ep400* also were shown to catalyze incorporation of H2A.Z into chromatin, while ANP32E, a member of the SWR1-related p400/TIP60 complex, was found to act as an H2A.Z chaperone functioning to remove H2A.Z-H2B from enhancer and insulator sites in mammalian cells (Obri et al., 2014; Ruhl et al., 2006; Scacchetti et al., 2020).

Despite a high-resolution cryo-EM structure of SWR1 bound to a nucleosome, the mechanism by which SWR1 catalyzes the replacement of H2A-H2B dimers in a nucleosome with H2A.Z-H2B dimers remains poorly understood (Willhoft et al., 2018). However, structural and biochemical experiments have provided considerable insight into the basis for the specificity of this reaction. Two subunits of the SWR1 complex, Swc2 and the catalytic subunit Swr1, directly bind H2A.Z, with preference for H2A.Z-H2B dimers over H2A-H2B dimers; specificity depends on the αC helix of H2A.Z (Figs. 5.9 and 5.10; Hong et al., 2014; Wu et al., 2005). In addition, the exchange reaction is facilitated by an H2A.Z-H2B specific chaperone, Chz1, and Swr1 binds to the opposite face of an H2A.Z-H2B dimer from Chz1, with the two "chaperones" able to form a tetrameric complex with H2A.Z-H2B (Hong et al., 2014; Luk et al., 2007). These results suggest a possible "hand-off" mechanism in which H2A.Z-H2B is escorted to nucleosome-bound SWR1 by Chz1, via the Swr1 subunit, followed by ATP-dependent eviction of H2A-H2B and replacement by H2A.Z-H2B.

The dependence of H2A.Z incorporation on SWR1 changed the question of how H2A.Z was localized to one of how SWR1 was localized. Acetylation of histone tails was found to positively influence H2A.Z deposition in vivo: deletion of either of two histone acetyltransferases, Eaf1 and Elp3, or both together, caused a modest (1.5- to 3-fold) decrease in H2A.Z occupancy at a number of loci (Raisner et al., 2005), as did deletion of histone acetyltransferases Gcn5 or Sas3 or mutation of specific lysine residues in the histone tails, with the latter reported to vary widely in their gene-specific effects (Zhang et al., 2005). Proteins containing a sequence motif designated as Bromodomain have an affinity for acetylated histones; one such protein expressed in yeast, Bdf1, is a subunit of SWR1. Deletion of Bdf1 does not notably affect cell growth or H2A.Z deposition; however, when Bdf1 is deleted in combination with reduced expression of its paralog Bdf2 (deletion of which similarly exhibits little effect, while the double mutant was reported to be inviable), a decrease in H2A.Z occupancy resulted (Raisner et al., 2005). Targeting of SWR1 thus appears to contribute to promoter localization of H2A.Z, as acetylation of histone tails in yeast is enriched at these regions; however, this does not explain enrichment at other regions such as inactive gene promoters or subtelomeric regions. A more accurate determination of the effect of loss of Bdf1 on H2A.Z deposition might be obtained by using rapid depletion methods (see Deletions and depletions sidebar, Chapter 4), but this has not been reported at present.

H2A.Z is greatly enriched at the ~5000 "+1" nucleosomes in the yeast genome over the other ~55,000 nucleosomes; it seems unlikely that the affinity afforded by acetylated histone tails, which are not as highly localized as H2A.Z, could be solely responsible for that enrichment. Initial studies sought sequence determinants for H2A.Z deposition by mutagenizing the *SNT1* yeast promoter to identify sequences required for H2A.Z enrichment, as monitored by ChIP (Raisner et al., 2005). These experiments indicated that a short sequence encompassing a Reb1 binding site was critical for H2A.Z deposition; insertions of a 22 bp sequence containing a Reb1 binding site into the inactive *PRM1* coding sequence resulted in the establishment of an NDR flanked by two H2A.Z-containing nucleosomes, as shown by ChIP and nucleosome scanning assays (see Chapter 4). A follow-up study reported loss of NDR formation at this inserted Reb1 site upon inactivation of the RSC complex with concomitant loss of H2A.Z enrichment, while two native promoters that exhibited partial "filling" of nucleosomes in the NDR upon RSC inactivation showed a corresponding partial decrease in H2A.Z occupancy (Hartley and Madhani, 2009). Reb1 is a member of the class of general regulatory factors that bind to a large number of yeast promoters and are capable of establishing NDRs, as discussed in Chapter 4. The results showing that Reb1 binding in cooperation with RSC can create an NDR flanked by H2A.Z-containing nucleosomes, together with experiments showing that GRFs cooperate with the RSC complex to establish NDRs at many yeast genes, suggests that this mechanism may pertain on a genome-wide basis. In support of this model, depleting or inactivating Reb1 or the additional GRFs, Abf1, and Rap1, leads to increased nucleosome occupancy at NDRs (see Chapter 4), but the effect of their depletion on H2A.Z occupancy has not yet been assessed.

Experiments conducted in vitro and in vivo indicate that the presence of nonnucleosomal DNA at NDRs is a major factor in localized deposition of H2A.Z. Comparison of SWR1 binding to dinucleosomes with a short or long linker (20 vs 140 bp) showed that SWR1 preferentially binds the latter, with about a 10-fold increase in affinity over most of the range of concentrations employed (Ranjan et al., 2013). This compares with an ~threefold enhanced affinity of SWR1 for nucleosomes with acetylated tails in the same assay. Nucleosomes with long linkers showed greater affinity than naked DNA alone, showing synergy between naked DNA and nucleosomes for SWR1 binding. The affinity of SWR1 for nucleosomal templates with long linkers was shown to depend strongly on the Swc2 subunit, and Swc2 was found associated with NDRs in ChIP-exo assays (Ranjan et al., 2013; Yen et al., 2013). Consistent with these results, Swr1 and H2A.Z deposition was reduced and localization was nearly abrogated in *swc2*△ yeast. Furthermore, H2A.Z occupancy was reduced at nucleosomes flanking NDRs in vivo as linker lengths decreased from 200 to 50 bp. This preference for long nonnucleosomal regions for H2A.Z deposition may also explain its presence flanking tRNA genes, which are also depleted of nucleosomes in yeast (Albert et al., 2007). Later work has reported specific sequence effects as well as environmental effects (temperature dependence) on H2A.Z deposition (Sun et al., 2020).

Consistent with the idea that H2A.Z deposition is governed by SWR1 affinity for NDRs flanked by nucleosomes having acetylated histone tails, perhaps modulated by sequence-specific effects, interactions between H2A.Z and transcription factors have also been reported in other systems (Giaimo et al., 2019). In *C. elegans*, the pioneer factor PHA-4, which excludes nucleosomes from its binding site and decompacts chromatin in vivo (Hsu et al., 2015), is required for recruitment of H2A.Z to promoters of genes expressed during early foregut development at the onset of their transcription (Updike and Mango, 2006). However, the relationship between NDRs and H2A.Z deposition in higher eukaryotes remains largely unexplored.

The localization of H2A.Z, its requirement for viability in metazoans and for normal growth in yeast, and its differential characteristics with respect to chromatin compaction compared to canonical H2A, all underscore its functional importance across species. This functional importance appears not to be restricted to a single process, as has already been alluded to. In addition to its roles in transcription and heterochromatin regulation, as mentioned earlier, H2A.Z has been implicated in the response to DNA damage (Kalocsay et al., 2009; Xu et al., 2012), meiotic recombination (Yamada et al., 2018), and nuclear assembly after mitosis (Moreno-Andres et al., 2020), among other processes (reviewed in Giaimo et al., 2019; Henikoff and Smith, 2015; Scacchetti and Becker, 2021). Although in some cases indirect effects may be responsible for phenotypes caused by loss of H2A.Z, for example, a cell cycle delay caused by misregulation of cyclin gene expression (Dhillon et al., 2006), evidence suggests that H2A.Z may exert effects directly by interacting with cellular components to recruit them to specific sites in chromatin (Adam et al., 2001; Yamada et al., 2018).

With respect to transcription, dozens of studies have implicated H2A.Z in various aspects of regulation (Giaimo et al., 2019). Although H2A.Z occupancy does not uniformly correlate with transcription rates (mostly assessed by monitoring transcript levels), as noted above, knockdown of H2A.V (the more recent designation for the H2A.Z ortholog) in *Drosophila*, in which H2A.V occupancy does correlate with expression levels, alters transcription of thousands of genes (Scacchetti et al., 2020). In contrast, loss of H2A.Z or SWR1 in yeast causes relatively modest decreases in transcript levels for only ~100–200 genes (Kobor et al., 2004; Krogan et al., 2003; Meneghini et al., 2003; Mizuguchi et al., 2004). However, this result likely underestimates the true impact of H2A.Z in transcription in yeast. First, H2A.Z functions redundantly with the SWI/SNF remodeling complex in the transcriptional activation of some genes (e.g., *PHO5* and *GAL1*) (Santisteban et al., 2000). Second, the large-scale analysis of the effects of loss of H2A.Z or SWR1 on transcript levels does not allow assessment of induced genes, such as *PHO5* in low phosphate conditions or the *GAL* genes in galactose medium, some of which are affected by loss of H2A.Z (Adam et al., 2001; Santisteban et al., 2000). Third, transcript levels can be buffered by decreased mRNA decay rates that compensate for decreased transcription rates (Timmers and Tora, 2018). More accurate determination of the effect of H2A.Z on transcription could be made by monitoring Pol II occupancy or nascent mRNA transcription following rapid depletion of SWR1, coupled with assessment of H2A.Z occupancy (see Deletions and Depletions, Chapter 4).

In vertebrate cells, H2A.Z is encoded by two genes, *H2A.Z.1* (*H2AFZ*) and *H2A.Z.2* (*H2AFV*), which produce H2A.Z variants (first identified by mass spectrometry (Coon et al., 2005)) differing from each other at three amino acids. The first of these, *H2A.Z.1*, is the gene that was shown to be essential for embryonic development in the mouse (Faast et al., 2001). Despite the modest difference in sequence, knockout experiments have revealed differential function in chicken DT40 cells and rat neurons (Dunn et al., 2017; Matsuda et al., 2010), and knockdown of the specific isoforms indicates overlapping but distinct effects on gene expression (Lamaa et al., 2020). Adding to this complexity, the *H2A.Z.2* gene in primates produces two splice variants designated H2A.Z.2.1 and H2A.Z.2.2 (Bonisch et al., 2012; Wratting et al., 2012). H2A.Z.2.2 differs from H2A.Z.2.1 in having its C-terminus truncated by 14 amino acids and altered in its terminal six amino acids, and this alteration results in considerable destabilization of nucleosomes containing this variant. This has led to speculation that H2A.Z.2.2 might confer structural properties to nucleosomes that alter chromatin function in specialized cell types in which its expression is high (Bönisch and Hake, 2012). Given that a considerable fraction of H2A.Z.2.2 exists as a soluble pool not incorporated into chromatin in cells where it is expressed, a possible (and not mutually exclusive) alternative is that it may function in some way outside of chromatin, as with the H4G variant mentioned earlier.

While the H2A.Z.2.2 splice variant clearly destabilizes nucleosomes in which it is incorporated, H2A.Z may also affect nucleosome stability under particular circumstances and thereby facilitate transcription initiation. While H2A.Z-containing nucleosomes appear nearly indistinguishable in their structure and properties from those containing canonical H2A, incorporation of H2A.Z together with the H3 variant H3.3 (see "H3.3" section) appears to destabilize nucleosomes under some conditions. H2A.Z-containing nucleosomes isolated from vertebrate chromatin are more sensitive to salt-dependent disruption than those containing canonical H2A, and those also containing H3.3 even more so (Jin and Felsenfeld, 2007). This sensitivity may depend on specific posttranslational modifications or additional factors associated with these nucleosomes, as nucleosomes reconstituted from recombinant histones showed only subtle effects on stability when H2A.Z was incorporated together with either canonical H3 or H3.3 (Horikoshi et al., 2016; Thakar et al., 2009). Nonetheless, the effect observed for nucleosomes isolated from vertebrate chromatin was corroborated in experiments showing that nucleosomes exhibiting sensitivity to MNase digestion—"fragile" nucleosomes (see Chapter 4)—were enriched for H2A.Z and H3.3, and were found at transcriptional start sites and enhancers (Jin et al., 2009). The discrepancy between in vitro and in vivo results remains to be resolved.

In metazoan organisms, H2A.Z may also affect transcriptional regulation at postinitiation steps, but evidence for its role in this is inconsistent. In *Drosophila* and mammalian cells, Pol II pauses 25–50 bp downstream of its initiation site at a large fraction of genes at the expense of productive elongation as discussed in Chapter 4 (Core and Adelman, 2019). Incorporation of H2A.Z (H2avD/H2A.V) into nucleosomes was shown to lower the energy barrier for the passage of Pol II in *Drosophila*, and H2A.Z occupancy was found to anticorrelate with Pol II pausing (Weber et al., 2014). In contrast, mammalian H2A.Z-containing nucleosomes impeded Pol II passage more than those containing canonical H2A in a single molecule study, and H2A.Z occupancy in mammalian cells positively correlated with pausing (Chen et al., 2019; Day et al., 2016). However, this positive correlation was hypothesized to arise from a feedback loop in which Pol II pausing stimulated H2A.Z deposition, and knockdown of H2A.Z by RNAi resulted in increased pausing of Pol II, consistent with H2A.Z-containing nucleosomes being less inhibitory to Pol II passage (Day et al., 2016). A more recent study used a tagged degron approach to deplete H2A.Z in 8h in murine embryonic stem cells and reported alleviation of pausing as a result (Mylonas et al., 2021). Determining whether these varying results reflect real differences between the mammalian and insect systems, indirect effects due to the slower action of RNAi-mediated depletion compared to the degron approach, or other technical issues, will require further study.

H2A.X

H2A.X, like H2A.Z, was first identified by its distinct electrophoretic behavior and shown to be synthesized throughout the cell cycle and incorporated into nucleosomes (West and Bonner, 1980; Wu and Bonner, 1981). H2A.X exists as a minor fraction of total H2A protein, slightly more abundant than H2A.Z (Table 5.1), and is conserved across species, with some notable anomalies: in yeast (both *S. cerevisiae* and *S. pombe*), the coding sequence for the C-terminal sequence that characterizes H2A.X (Fig. 5.9; see below) is appended to each of the two genes encoding canonical H2A (but not that for H2A.Z), while in *Drosophila*, H2AvD is a hybrid that possesses core sequence placing it in the H2A.Z family together with the characteristic H2A.X C-terminal sequence (Redon et al., 2002).

The first complete amino acid sequence for an H2A.X species was inferred from sequencing the human cDNA (Mannironi et al., 1989). The sequence revealed a 96% homology with the first 120 amino acids of human H2A.1, while the C-terminal region diverged completely, instead aligning with C-terminal sequences from yeast, *Aspergillus*, *Tetrahymena*, and *Drosophila* H2AvD. These sequences showed the presence of a terminal conserved motif comprising Ser-Gln followed by an acidic and aliphatic residue (SQ(E/D)Φ; Figs. 5.9 and 5.10). Subsequent sequencing of H2A.X in other species confirmed the conservation of this motif and also revealed an increasingly extended C-terminus over the evolutionary "ladder": protists < plants < insects < mammals (Redon et al., 2002). This same study revealed another unusual feature of H2A.X: while the gene encoding this variant in human cells, *H2AFX*, lacks introns and has a stem-loop structure characteristic of replication-dependent histones, H2A.X is also expressed outside of S-phase with transcription reading through the stem-loop to a downstream poly(A). This dual RI and RC expression pattern may reflect the need to incorporate H2A.X throughout the genome at sufficient levels for function (see below), while also allowing its incorporation even in terminally differentiated cell lineages (Bonner et al., 2008).

The conservation of the H2A.X C-terminus was especially striking in light of the variability of this region among canonical H2A histones (Fig. 5.1) and led to the speculation that the conservation of the carboxy-terminal sequence of H2A.X could reflect its participation in an essential function, with the corollary that it should be conserved across all eukaryotic species (Mannironi et al., 1989). This prediction was largely borne out, although H2A.X appears to be absent in nematodes and bdelloid rotifers, indicating that it is not uniformly essential (https://www.ncbi.nlm.nih.gov/research/HistoneDB2.0/index.fcgi/browse/).

What essential function might H2A.X perform? At a chromatin meeting in 1996, Bill Bonner shared that he had some exciting findings on H2A.X but could not divulge them until his paper was accepted. The news was indeed big: In a landmark 1998 paper, it was shown that H2A.X was phosphorylated at Ser139 (in the SQ(E/D)Φ motif) following exposure of cells to ionizing radiation (Rogakou et al., 1998). Phosphorylation was detectable within 1 min of exposure and was maximal at 10 min, and quantitation of the phosphorylated species indicated that sufficient γ-H2AX, as the phosphorylated species was named, was generated to encompass on the order of 2 Mb of DNA per double-stranded break. These findings, which were based on biochemical analyses, were corroborated by experiments using antibodies generated against phosphorylated H2A.X to demonstrate localized foci of γ-H2AX following DNA damage in mammals, frogs, flies, and yeast (Fig. 5.14; Redon et al., 2002; Rogakou et al., 1999).

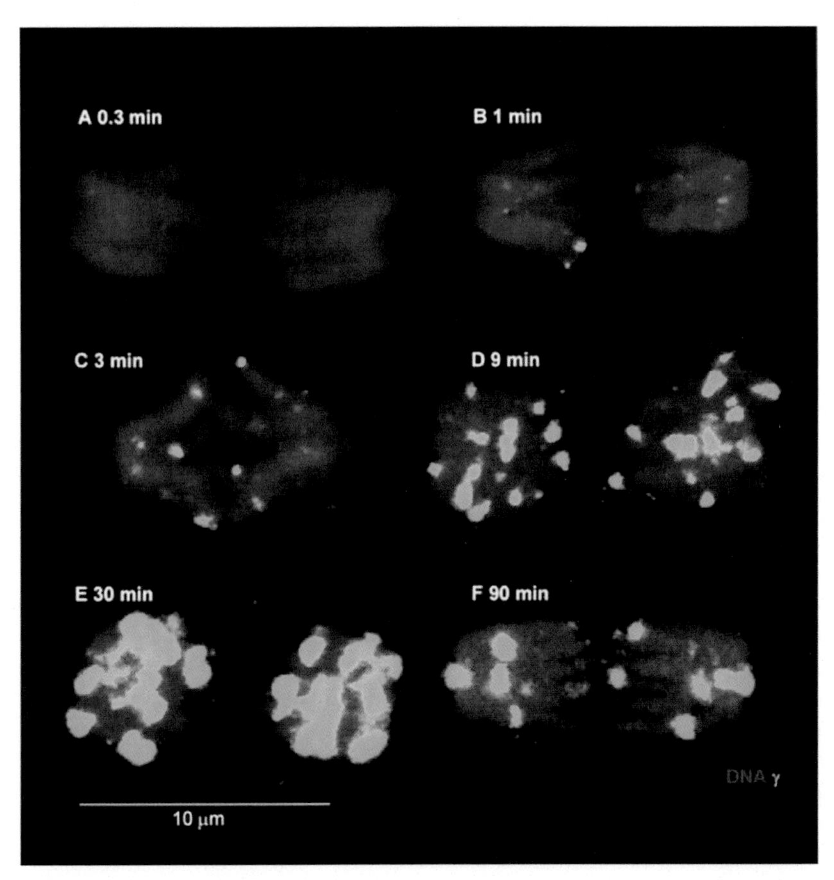

FIG. 5.14 γ-H2AX foci developing on *M. muntjak* mitotic chromosomes (maximum projections). *M. muntjak* cell cultures were exposed to 0.6 Gy on ice, covered with growth media at 37°C, and allowed to recover for 0.3 (A), 1 (B), 3 (C), 9 (D), 30 (E), and 90 min (F) before fixation. The *green channel* is amplified to the same extent in all samples to visualize the nascent foci at 1 and 3 min; however, this results in some overexposed foci in the 9-, 30-, and 90-min samples. *From Rogakou, E.P., Boon, C., Redon, C., Bonner, W.M., 1999. Megabase chromatin domains involved in DNA double-strand breaks in vivo. J. Cell Biol. 146, 905.*

Induction of γ-H2AX was observed following induction of double-stranded DNA breaks by various means, and damage-signaling (ATM or ataxia telangiectasia mutated) and DNA repair factors (BRCA1, Rad50, Rad51) form foci that co-localize with γ-H2AX foci following radiation damage (Paull et al., 2000). Loss of ATM nearly abolished γ-H2AX formation, as shown in murine *atm*$^{-/-}$ fibroblasts (Burma et al., 2001). Although these findings clearly implicated H2AX phosphorylation in the DNA damage response pathway, γ-H2AX is not absolutely essential for DNA repair; in yeast, mutation of Ser129 (in the SQ(D/E)Φ motif) of H2A does not abolish repair, but it does increase sensitivity to several DNA-damaging agents (Downs et al., 2000). This is likely due to the existence of alternative DNA repair pathways. Similarly, knockout of H2A.X in mice did not eliminate irradiation-induced cell cycle checkpoints, but H2AX$^{-/-}$ mice were radiation sensitive and exhibited pleiotropic phenotypes indicative of chromosome instability and deficient repair of DNA damage (Celeste et al., 2002).

Mechanistically, γ-H2AX-containing nucleosomes in mammalian cells act as a platform for the recruitment and retention of factors involved in DNA repair, beginning with MDC1 (Mediator of Damaged Checkpoint 1) (Stucki et al., 2005). MDC1-bound chromatin serves to recruit DNA repair proteins, including ATM, recruitment of which leads to phosphorylation of H2A.X in surrounding chromatin, producing the observed large domains of γ-H2AX characteristic of double-strand breaks in DNA (Kloeber and Lou, 2021; Lou et al., 2006). H2A.X, then, has evolved to function as a sensor that recruits the DNA repair machinery only when modified to respond to DNA damage. This is just one example of a histone modification that acts to allow nucleosomes to recruit proteins to chromatin under specific conditions (Chapter 7, Histone modifications).

H2A.B and the short H2A variants

Close to the time that the first draft of the human genome was completed, more and more raw sequence data was becoming available in public databases. Chadwick and Willard searched the GenBank human expressed sequence tag (EST) database for sequences related to that encoding H2A, doubtless intrigued by the connection between the macroH2A variant (see below) and X chromosome inactivation, a topic Willard had been engaged in for some time. Their search uncovered a distantly related, truncated version of H2A (Chadwick and Willard, 2001). The mRNA had a poly(A) tail and lacked the stem-loop structure characteristic of canonical, replication-coupled histones. The inferred protein sequence indicated a truncated version of H2A with 48% homology to the canonical version, with the highest homology in the first two α-helices of the histone fold region and considerable divergence everywhere else (Figs. 5.9 and 5.10). Overexpression of epitope-tagged versions of this H2A variant in HEK293 cells was used to demonstrate its incorporation into nucleosomes, verifying its identity as a bona fide histone; this was later confirmed for endogenous H2A.B in mature human sperm (Ishibashi et al., 2010). Immunofluorescence experiments using the overexpressed H2A variant in primary fibroblasts revealed uniform distribution in cell nuclei with the exception of an excluded region in the Barr body, corresponding to the inactive X chromosome, of female-derived cells. Based on this finding, the variant was named H2A.Bbd, for Barr body-deficient; however, this appears to be a misnomer, as subsequent work showed H2A.

Bbd to be present on the inactive X chromosome following meiotic sex chromosome inactivation in the mouse testis, the tissue where it is most highly expressed (Soboleva et al., 2011). H2A. Bbd is now generally referred to as H2A.B (Talbert et al., 2012), and is restricted to eutherian mammals where it plays a prominent role in the male germline (see below) (Jiang et al., 2020).

H2A.B differs from canonical H2A in several major features (Fig. 5.10). First, in addition to the substantial divergence in the primary acid sequence, it is truncated relative to H2A, missing the C-terminal 19 amino acids including nine residues that comprise the C-terminal portion of the docking domain. Second, three of the six residues comprising the acidic patch in H2A are altered to nonacidic residues in H2A.B, leading to the murine homolog initially being referred to as "H2A.Lap1," for "Lacks acidic patch" (Soboleva et al., 2011). (The mouse ortholog is now generally called H2A.B.3.) Third, although H2A.B retains the characteristic, highly basic N-terminal tail, none of the known modifiable residues are conserved, with the four lysine residues found in the canonical H2A tail all absent in H2A.B.

The consequences of these changes in sequence are profound. H2A is unique among the core histones in possessing an extended flexible C-terminal region (although H3 and H2B C-termini are subject to proteolytic digestion, these unstructured regions are considerably shorter than that of H2A). This C-terminal tail encompasses the docking domain that contacts the N-terminal α-helix of H3 and organizes DNA as it enters and exits the nucleosome core; removal of 15 aa from this tail causes destabilization of the histone octamer under high salt conditions and moderate decrease of nucleosome stability in vitro and in vivo (Eickbush et al., 1988; Vogler et al., 2010). Correspondingly, nucleosomes reconstituted using H2A.B instead of H2A protect only 118 ± 2 bp of DNA against MNase digestion, compared to the 146 bp in a canonical nucleosome (Bao et al., 2004), while cryo-EM and atomic force microscopy indicated that about 130 bp of DNA is organized into H2A.B-containing nucleosomes (Doyen et al., 2006). The latter experiments also showed a drastic change in the angle between entering and exiting DNA, to about 180° in the H2A.B-containing nucleosome as compared to the V-shaped structure in canonical nucleosomes, as also supported by FRET experiments showing an increased distance between the DNA ends (Bao et al., 2004). This would be expected to result in a decreased ability of linker histone to bind to these variant nucleosomes, as the linker DNA would no longer be well positioned to interact with linker histone at the dyad, and this was confirmed experimentally (Shukla et al., 2011). Domain-swapping experiments in which chimeric versions of H2A were incorporated into nucleosomes showed that these altered properties of H2A.B-containing nucleosomes were caused by the extensive alterations in the entire docking domain, and not just the shortened C-terminal region.

With regard to the diminished acidic patch of H2A.B, we have already discussed how this change was used in experiments that demonstrated the importance of interactions between the acidic patch and the H4 tail in chromatin compaction (Chapter 3; see also "H2A.Z" section) (Fan et al., 2004; Zhou et al., 2007). Thus H2A.B incorporation not only creates a less compact, less stable nucleosome but also inhibits the folding of such nucleosomes into higher-order structures.

Finally the alteration of modifiable residues in the H2A.B amino terminus is reflected in the current lack of evidence for posttranslational modification of H2A.B (Jiang et al., 2020). Recent evidence indicates an entirely novel function for the H2A.B amino-terminal tail: immunoprecipitation of H2A.B.3 from mouse testis chromatin uncovered interactions with

RNA-processing factors, including splicing factors, and further experiments revealed the N-terminal tail to be an RNA-binding module (Soboleva et al., 2017). Moreover, splicing events are altered in H2A.B.3 knockout mice, and also by exogenous expression of H2A.B in HeLa cells (Anuar et al., 2019; Tolstorukov et al., 2012). Based on these results, it was suggested that H2A.B may have been evolutionarily coopted to facilitate regulated mRNA splicing in the testis and brain (Jiang et al., 2020).

Genome-wide localization of H2A.B.3 in mouse testis revealed it to be incorporated into nucleosomes at the TSS of active genes concurrent with gene activation, consistent with a role in facilitating access to the transcription machinery by creating less stable nucleosomes; this was also found to be true in the brain (Soboleva et al., 2011, 2017). To date, however, no chaperones have been identified that could be responsible for this specificity, and the underlying mechanism remains to be solved.

Three additional short H2A histones, H2A.L, H2A.P, and H2A.Q, are testis-specific variants that evolved, together with H2A.B, from the common ancestor of all eutherian mammals (Molaro et al., 2018). These short H2A histones have evolved rapidly compared to canonical H2A, and their presence correspondingly varies across eutherian species; for example, H2A.P but not H2A.Q is found in humans, with the converse being true in mice. Murine H2A.L.1 and H2A.L.2 are expressed during spermatogenesis and evidently facilitate transitions in chromatin structure that culminate in the packaging of DNA primarily with protamines in mature sperm (Govin et al., 2007; Jiang et al., 2020). Maximal expression of H2A.B.3 precedes that of H2A.L.2, while the latter pairs specifically with the H2B testis-specific variant TH2B and is essential for the recruitment of transition proteins that enable eventual packaging of DNA with protamines in mature spermatozoa (Barral et al., 2017; Soboleva et al., 2011). Correspondingly, H2A.L.2 male knockout mice are infertile, as they do not produce viable mature sperm (Barral et al., 2017). As with H2A.B, H2A.L.2-containing nucleosomes, assembled in vitro from purified histones, protect only ~120bp of DNA from digestion by MNase, consistent with the importance of the docking domain in stabilizing histone-DNA interactions at the nucleosomal entry and exit points (Barral et al., 2017). Much less is known regarding H2A.P and H2A.Q, although it is intriguing that the amino-terminal region of H2A.L, like that of H2A.B, is an RNA-binding module (Hoghoughi et al., 2020). In contrast, the N-terminus of H2A.P lacks the enrichment for arginine residues seen in H2A.B and H2A.L that is characteristic of RNA-binding proteins, and both H2A.P and H2A.Q diverge considerably in their N-termini from H2A.B and H2A.L, which possess similar N-terminal regions (Molaro et al., 2018). The extent to which these alterations reflect functional divergence remains to be seen. It is also interesting to note that the H2A.Z primate-specific splice variant H2A.Z.2.2 mentioned earlier possesses a C-terminal region truncated by 14 amino acids, with the remaining C-terminal residues also divergent from the H2A consensus, thereby destabilizing nucleosomes in which it is incorporated similarly to H2A.B and H2A.L (Bonisch et al., 2012). However, H2A.Z.2.2 has an enlarged acidic patch, while that of H2A.B is reduced, so that they have opposite effects on chromatin compaction (Fig. 5.10). H2A.Z.2.2 is enriched in brain tissue, while H2A.B is expressed in the brain as well as the testis (Bonisch et al., 2012; Soboleva et al., 2011). Whether there are functional links between these two short H2A variants is currently unknown.

MacroH2A

Discovery of the macroH2A variant, together with that of CenH3 (see below), startled the world of chromatin researchers. These two histone variants (or "deviants"! (Wolffe, 1995)) varied so markedly from their canonical counterparts that they raised the novel prospect of histones adapting to fulfill roles far beyond that of packaging DNA. MacroH2A was first identified as a 42 kD protein associated with mononucleosomes from rat liver that were purified by sedimentation in 0.5 M NaCl to remove linker histones and most nonhistone proteins (Pehrson and Fried, 1992). Two proteins were observed to migrate much more slowly than the canonical core histones on polyacrylamide gels and could be resolved from each other using gels with a low degree of cross-linking. Peptide sequencing revealed that the proteins were partially similar to H2A and partly novel; cloning and sequencing of the encoding genes identified these proteins, now named macroH2A.1 and macroH2A.2, as proteins having their N-terminal third homologous to H2A, with the C-terminal two-thirds (residues 123–367) being completely unrelated in sequence. MacroH2A.1 was later found to be expressed as two isoforms produced by alternative splicing and differing only in a short segment of the macro domain (Rasmussen et al., 1999).

Pehrson and Fried noted in their original report a region outside of the histone-fold domain of macroH2A (aa 180–208) that had features of a leucine zipper, a helical structure that forms a dimerization interface in some transcription factors (Landschulz et al., 1988; Pehrson and Fried, 1992). Based in part on this finding, they hypothesized that macroH2A might position nucleosomes by interacting with sequence-specific DNA-binding proteins, or perhaps by binding directly to DNA via the C-terminal region. Subsequent work, however, pointed to a very different role for macroH2A: immunofluorescence experiments using an antibody developed against macroH2A revealed a conspicuous enrichment at one of the two X chromosomes in female murine and human cells (Costanzi and Pehrson, 1998). Dosage compensation of sex chromosomes—the process by which equivalent gene expression is achieved for sex chromosomes that are unequally distributed in male and female animals—is accomplished in mammals by inactivation of one of the two X chromosomes in females (Lee, 2011). Examination of macroH2A in murine and human cells having X-chromosome abnormalities showed that the X chromosome enriched for macroH2A was inactive. ChIP-seq experiments confirmed the enrichment of macroH2A on the inactive X chromosome, while also showing depletion at active genes, including genes on the inactive X chromosome that escape inactivation (Changolkar et al., 2010). However, the function of macroH2A with regard to X-inactivation appears to be complex. The preferred deposition of macroH2A occurs late in the process of X-inactivation, indicating that it is not required for the initiation of X-inactivation (Rasmussen et al., 2001). Moreover, mice lacking both macroH2A.1 and macroH2A.2 did not exhibit obvious defects in X-inactivation, and other studies showed reactivation of the inactive X chromosome in only a small fraction of cells upon loss of macroH2A (Hernandez-Munoz et al., 2005; Pasque et al., 2011; Pehrson et al., 2014). Evolutionary considerations also make it clear that X-inactivation cannot be the principal function of this variant, as it is present in nonmammalian species, including sea urchin, tick, bird, fish, and reptile species (but not in plants or protozoa), and is in fact essential for proper development in zebrafish (Buschbeck et al., 2009; Talbert et al., 2012).

A broader repressive role for macroH2A is suggested by several studies. The association of macroH2A with heterochromatin domains has been observed in various reports (reviewed in Pehrson et al. (2014), Douet et al. (2017), Rivera-Casas et al. (2016)), and ChIP-seq experiments mentioned above revealed macroH2A enrichment at genes in the liver, particularly a group involved in lipid metabolism, whose transcript levels increased upon macroH2A depletion (Changolkar et al., 2010). MacroH2A also is found at repressed genes that are activated during differentiation in human male pluripotent cells and is inhibitory to the reprogramming of nuclei following nuclear transfer in frogs, underscoring its involvement in developmental regulation (Buschbeck et al., 2009; Pasque et al., 2011). In this regard, macroH2A may function in association with the polycomb group repressive complex, which is also implicated in X-chromosome inactivation (Buschbeck et al., 2009; Pasque et al., 2011). The exact nature of this linkage remains obscure. In HepG2 cells, simultaneous knockdown of macroH2A.1 and macroH2A.2 leads to major disruption of heterochromatin structure, but only mild de-repression of gene expression (Douet et al., 2017). These and other results suggest that macroH2A functions together with other players in the maintenance of repressive heterochromatin structures in a variety of contexts. Mechanistically, macroH2A may exert a repressive effect by generating more stable nucleosomes (see below), by hindering access of transcription factors to DNA near the nucleosome dyad axis, and by resisting chromatin remodeling complexes (Angelov et al., 2003), although conflicting results on the effect of macroH2A on remodeling have been reported (Chang et al., 2008).

Structural studies revealed that little change is conferred to the nucleosome core particle upon incorporation of macroH2A despite its substantial difference in primary amino acid sequence (~35% deviation in the histone fold region; Figs. 5.9 and 5.10; Chakravarthy et al., 2005), and, in contrast to H2A.Z and H2A.B, the acidic patch is completely conserved. The only notable structural difference between canonical and macroH2A-containing nucleosomes with regard to the core, or histone-fold, region, is in the L1 loop (Fig. 5.11). This difference was shown to be responsible for the increased stability of histone octamers containing macroH2A, a property possibly consistent with a repressive role in vivo (Chakravarthy and Luger, 2006). The L1 loop, it will be recalled, is important in establishing contacts between the histone octamer and the DNA as it enters and exits the nucleosome; consistent with its impact on macroH2A-containing nucleosomes, DNase digestion revealed altered protection near the dyad of a positioned nucleosome containing macroH2A (Angelov et al., 2003). Surprisingly, reconstitution assays indicated a preference for macroH2A to be incorporated into hybrid nucleosomes having one copy of canonical H2A over homotypic nucleosomes, in contrast to H2A.Z and H2A.Bbd, which form stochastic mixtures of homotypic and hybrid nucleosomes with canonical H2A in reconstitution experiments (Chakravarthy et al., 2004; Chakravarthy and Luger, 2006). This preferential behavior is dictated by the L1 loop of macroH2A; however, the reconstitution experiments used recombinantly expressed macroH2A lacking its C-terminal domains—that is, "macroless" macroH2A—leaving open the possibility that interactions of the macro-domain with other proteins in living cells might supersede this preference, whose significance at present is unclear with regard to function.

Another curious feature of macroH2A is that its localization to the inactive X-chromosome in mammals is controlled by sequences in the histone fold domain, including the L1 loop, rather than the macro-domain, with multiple regions conferring X_i localization when incorporated into GFP-tagged H2A (Nusinow et al., 2007b). These regions are buried upon incorporation into the nucleosome but are exposed along one surface in macroH2A-H2B dimers, suggesting that preferential incorporation could be accomplished by interaction with chaperones or remodeling factors. However, macroH2A-specific deposition factors are not known at present.

What of the other two-thirds of macroH2A, the macro-domain? While macroH2A likely originated with the divergence of metazoa (Rivera-Casas et al., 2016; Talbert et al., 2012), macro-domain proteins can be identified in bacteria, archaea, eukarya, and viruses (Han et al., 2011; Pehrson and Fuji, 1998). Structural studies and molecular phylogeny reveal the conserved macro domain to comprise residues 161–367, with aa 123–166 constituting a lysine-rich linker that connects the histone domain to the macro domain (Chakravarthy et al., 2005; Pehrson and Fuji, 1998). The X-ray crystal structure of a segment including most of the macro domain (aa 180–367) revealed an α/β structure, in which a seven-strand β-sheet is protected on one side by α-helices (Chakravarthy et al., 2005). This structure also showed that the original prediction of a leucine zipper, based on sequence analysis, was incorrect, as the relevant segment (aa 180–208) instead constitutes the first two strands of the β-sheet.

Macro-domain proteins engage in a wide variety of functions, which does not help to narrow down the role of this domain in macroH2A. Some macro-domain proteins can bind to NAD metabolites; macroH2A.1.1 is included in this group, while macroH2A.2 and the splice variant macroH2A.1.2 are precluded from binding these metabolites due to altered metabolite binding pockets (Kozlowski et al., 2018; Kustatscher et al., 2005). One such NAD metabolite is poly(ADP-ribose), which can be attached to proteins, including H2A and macroH2A, by poly-ADP-ribose polymerases (PARPs). Some confusion exists in the literature regarding the effect of macroH2A variants on PARP activity: an early investigation reported that macroH2A1.2, but not macroH2A.1, could interact with PARP-1 via its nonhistone domain, while the nonhistone domains from all three macroH2A variants were reported to be capable of inhibiting PARP activity in an in vitro assay (Nusinow et al., 2007a). These findings were controverted in other studies, which reported that only the nonhistone domain of macroH2A.1.1 was capable of binding NAD metabolites (Kustatscher et al., 2005; Timinszky et al., 2009). A recent report from the Ladurner group supported this specificity, showing that only the macroH2A.1.1 nonhistone domain retained PARP-1 in an assay using column-bound macroH2A nonhistone domains and nuclear extracts; binding to PARP-1 present in the extracts was found to depend on ADP-ribose binding and resulted in inhibition of PARP activity (Kozlowski et al., 2018; Timinszky et al., 2009). In contrast, all three macroH2A variants can contribute to heterochromatin condensation through their linker regions (Timinszky et al., 2009). A direct connection between macroH2A and metabolic regulation was reported in a study showing that macroH2A1.1 expression is induced during myogenic differentiation and, by sequestering PARP-1, serves to reduce NAD+ consumption and thus preserve NAD+ pools necessary for respiration (Posavec Marjanovic et al., 2017). Cellular metabolism can affect histone modifications and chromatin structure in myriad ways; more surprisingly, this example shows that the converse can also be true.

CenH3

CenH3 shares with macroH2A the distinction of early identification as an extreme histone variant, as it has only 50%–60% homology to canonical H3, with an N-terminus and extended C-terminal region not found in any other histone variant (Fig. 5.9 and Table 5.1). CenH3 was not found in a search for variant histones; rather, its discovery resulted from a search whose goal was to isolate components that function at the centromere during cell division. Centromeres are chromosomal regions that serve as assembly sites for the kinetochore (in yeast, the spindle pole body), providing attachment sites for microtubules that partition chromosomes to the daughter cells. Early studies sought to identify the components responsible for centromere function, following the same logic that motivated Friedrich Miescher to isolate nuclein—we must know the parts involved to understand the mechanism. Centromeres exist in only one copy per chromosome, presenting a challenge to their isolation; however, researchers discovered that certain autoimmune disease patients produced antibodies that recognized centromeres when used to immunostain nuclei. These antibodies were used to identify three centromere components, dubbed CENP-A, -B, and -C, with molecular weights of 17, 80, and 140 kDa (Earnshaw and Rothfield, 1985). The smallest of these proteins, CENP-A, could be purified in a nucleosome core particle preparation and exhibited biochemical properties similar to those of the core histones; most notably, when acid-extracted HeLa cell nuclear proteins were eluted at increasing salt concentrations from a cation exchange column, CENP-A was eluted at a NaCl concentration of ~1.45 M together with histones H3 and H4, while H2A and H2B and other nuclear proteins eluted at much lower [NaCl] (Palmer and Margolis, 1985; Palmer et al., 1987). In a clever approach to CENP-A purification, Palmer et al. took advantage of discovering that while the core histones were supplanted by protamines in bull spermatozoa, CENP-A was quantitatively retained; partial sequencing of the purified protein revealed regions similar to histone H3, while other segments were completely novel (Palmer et al., 1991). Cloning and sequencing of the human CENP-A gene confirmed its identity as a histone H3 variant, and expression of epitope-tagged CENP-A derivatives in mammalian cells demonstrated that the histone-fold portion was responsible for its targeting to the centromere (Sullivan et al., 1994).

The cloning of CENP-A, now more generally known as CenH3, did not immediately lead to its cognate being identified in other organisms by classical cloning avenues (e.g., Southern blotting at reduced stringency), as this variant turns out to have evolved far more rapidly than canonical H3 or other histone variants (Talbert and Henikoff, 2010). Thus the *S. cerevisiae* homolog of CENP-A, Cse4, was identified via a mutation causing chromosome missegregation (Stoler et al., 1995), while CenH3-encoding genes were identified in *C. elegans* (HCP-3) and *Drosophila* (CID) by searching for H3-like proteins among expressed sequence tags in GenBank (Buchwitz et al., 1999; Henikoff et al., 2000), and the *S. pombe* CenH3 (Mis6) was found by use of degenerate PCR applied to a genomic DNA library (Takahashi et al., 2000). Henikoff and colleagues have speculated that competition among chromosomal sequences to act as centromeres has resulted in an evolutionary "arms race" in which CenH3 must evolve rapidly to maintain its function in centromere identity (Henikoff and Smith, 2015).

As with other histone variants, the major questions posed by the identification of CenH3 concerned its impact on nucleosome structure, the mechanism by which it achieved its specific function in chromatin, and what directed its chromosomal localization. Evidence has been reported for CenH3 occupying a tetrasome comprising one copy each of H2A, H2B, CenH3, and H4 with DNA wrapped in a right-handed sense around the histone core (Henikoff and Furuyama, 2012); countering this result, CenH3 can be incorporated into nucleosomes in vitro that are highly similar to canonical nucleosomes (Tachiwana et al., 2011a). In support of CenH3-containing nucleosomes having an unusual structure, DNA near the entry and exit points shows enhanced susceptibility to digestion by nucleases, and appears disordered in the crystal structure, with only 121 bp being resolved (Conde e Silva et al., 2007; Tachiwana et al., 2011a). This loosening of the entering and exiting DNA, which is also reflected by changes in topology, was attributed to CenH3 possessing an αN helix shortened from three helical turns in canonical H3 to two in CenH3, with the preceding loop region being completely disordered in the crystal structure. It will be recalled that this region of H3, in conjunction with interactions with H2A, dictates histone-DNA contacts at the nucleosomal entry and exit points. This region also includes Arg49 in H3, which is altered to Lys in cenH3; Arg49 is one of the 14 Arg residues that intercalates into the minor groove of DNA and thereby stabilizes histone-DNA interactions in the nucleosome (Luger et al., 1997). The effect of this region on CenH3-containing nucleosomes was confirmed by cryo-EM of cenH3-containing nucleosomes in which the CenH3 αN helix was replaced by that of canonical H3, resulting in stabilization of the DNA-histone contacts at the entry and exit of the nucleosome (Roulland et al., 2016). In addition, CenH3-containing nucleosomes were shown not to bind linker histone H1, but binding was restored in nucleosomes containing chimeric cenH3 in which the αN helix was replaced by that from canonical H3, both in vitro and in vivo. Replacement of native CENP-A with the chimeric CenH3 in HeLa cells resulted in mitotic and cytokinetic defects that are similar to those seen upon depletion of CENP-A, despite a lack of effect on association of the centromeric platform protein CENP-C, providing strong evidence for the importance of the structural alteration in cenH3-containing nucleosomes conferred by the αN helix on centromere function. The phenotypic effect was attributed to the inability of several centromere-associated proteins to associate with the more compact, linker histone-bound chromatin presumably formed in the presence of the canonical H3 αN helix.

Although centromeres are specified by a specific DNA sequence in budding yeast, this is not the case in higher eukaryotes, where CenH3 and the complex of proteins associated with it specify its location (Black and Cleveland, 2011). This implies that the mechanism of CenH3 deposition must underlie the maintenance of centromeres across cell divisions. *SCM3* was shown to be critical to this process in yeast following its identification as a high-copy suppressor of a *ts* mutant of *CSE4*, the gene encoding CenH3 (Stoler et al., 2007). High-copy expression of *SCM3* suppressed three independent *ts* mutants of *CSE4*, all having point mutations in the histone-fold domain, and Scm3 was shown to interact with Cse4 and to be localized at centromeres, while depletion of Scm3 resulted in cell cycle arrest and mislocalization of Cse4. Independently, Scm3 was identified as interacting with Cse4 in biochemical experiments and implicated as a chaperone that incorporates Cse4-H4 dimers into nucleosomes at yeast centromeres (Camahort et al., 2007; Mizuguchi et al., 2007). Functional counterparts of Scm3 were subsequently found in other organisms; consistent with the rapid evolution of CenH3, these counterparts exhibit little or no evolutionary relatedness to Scm3. In mammalian cells, the Holliday junction recognizing protein, HJURP, acts as a CenH3-H4 chaperone to incorporate CenH3 into centromeric nucleosomes. HJURP was discovered as a CenH3-

interacting protein in two independent screens aimed at finding proteins that interacted with nonnucleosomal, epitope-tagged CENP-A (Dunleavy et al., 2009; Foltz et al., 2009). As in the studies with Scm3, HJURP specifically recognized CenH3-H4 and not canonical H3-H4 dimers, and depletion of HJURP (by RNAi) resulted in CENP-A mislocalization and mitotic defects. Although other centromere-associated proteins are also necessary for the incorporation of CenH3 in both yeast and human cells, these proteins do not interact directly with CenH3 (Fujita et al., 2007; Hayashi et al., 2004). Rather, these proteins act indirectly to facilitate CenH3 incorporation; for example, in human cells, HJURP associates with the centromere via interaction with Mis18BP, a part of the Mis18 complex that was first discovered in *S. pombe* (Hayashi et al., 2004), and thereby facilitates the incorporation of the CENP-A-H4 dimer (Barnhart et al., 2011).

Comparison of HJURP and Scm3 sequences did not directly reveal homology, but an in-depth analysis of fungal Scm3 sequences and mammalian HJURP sequences showed that they were indeed distantly related (Sanchez-Pulido et al., 2009). However, although the analysis also found HJURP homologs in frogs and birds, no relative of HJURP was found in insects or nematodes. Indeed, in *Drosophila* CAL1, first identified as being required for mitotic spindle assembly (Goshima et al., 2007), has apparently evolved to replace HJURP/Scm3 as a CenH3 chaperone (Chen et al., 2014), while in *C. elegans*, CenH3 itself may have adapted to the apparent absence of HJURP/Scm3 (de Groot et al., 2021). *C. elegans* CENP-A possesses an extended N-terminal tail that is needed for centromere chromatin assembly; this tail interacts with KNL-2, the worm homolog of Mis18BP, which in vertebrates recruits CENP-A by interacting with HJURP (Barnhart et al., 2011). CenH3 clearly occupies a unique niche as a histone variant that has evolved rapidly to maintain a function essential to eukaryotic life.

H3.3

Histone H3.3 was discovered as a mammalian variant histone by Franklin and Zweidler based on the altered chromatographic behavior, relative to H3.1 and H3.2, of a single proteolytic fragment following digestion of the purified histones with thermolysin (Franklin and Zweidler, 1977). Amino acid analysis led to the identification of substitutions of Ile and Gly in H3.3 for Val and Met in H3.1 and H3.2 at positions 89 and 90; sequencing eventually revealed four substitutions relative to the canonical H3 sequence (Fig. 5.9; Franklin and Zweidler, 1977; Zweidler, 1984). The altered sequence was sufficient to allow separation by two-dimensional electrophoresis and was important in determining that some histones, including H3.1 and H3.2, were expressed only in S-phase while others, such as H3.3, were expressed throughout the cell cycle (Wu and Bonner, 1981). However, it was not for another 20 years that the import of H3.3 as a replacement histone was fully appreciated. In the interim, replication-independent (RI) H3.3 variants were found across the eukaryotic kingdom, with identical sequence variants present in birds, mammals, and *Drosophila*, and recognizable H3.3 variants, characterized by similar substitutions relative to canonical H3, in protists and plants. As with other RI variant histones, vertebrate genes encoding H3.3 contain introns and express mRNAs that are polyadenylated. One of the four amino acid changes in mammalian H3.3 is in the flexible N-terminus (Ala31Ser), while the other three are at the base of the central α-helical domain, with residues 87–90 comprising Ser-Ala-Val-Met in H3.1 and H3.2 replaced by Ala-Ala-Ile-Gly in H3.3 (Fig. 5.9). Variations on this theme are found in distantly related organisms (Fig. 5.15; Probst et al., 2020).

	L1 α2
Human H3.1	. . . DFKTDLRFQSSAVMALQEACEAYLVGLFEDTNSCAIHA . . .
Human H3.3	. . . DFKTDLRFQSAAIGALQEACEAYLVGLFEDTNSCAIHA . . .
Xenopus H3.1	. . . DFKTDLRFQSSAVMALQEASEAYLVGLFEDTNSCAIHA . . .
Xenopus H3.3	. . . DFKTDLRFQSHAVLALQEASEAYLVGLFEDTNSCAIHA . . .
Drosophila H3.1	. . . DFKTDLRFQSSAVMALQEASEAYLVGLFEDTNSCAIHA . . .
Drosophila H3.3	. . . DFKTDLRFQSAAIGALQEASEAYLVGLFEDTNSCAIHA . . .
Aradopsis H3.1	. . . DFKTDLRFQSSAVAALQEAAEAYLVGLFEDTNSCAIHA . . .
Aradopsis H3.3	. . . DFKTDLRFQSHAVLALQEAAEAYLVGLFEDTNSCAIHA . . .
C. elegans H3.1	. . . DFKTDLRFQSSAVMALQEAAEAYLVGLFEDTNLCAIHA . . .
C. elegans H3.3	. . . DFKTDLRFQSAAIGALQEASEAYLVGLFEDTNLCAIHA . . .
Neurospora H3.1/H3.3	. . . DFKSDLRFQSSAIGALQESVESYLVSLFEDTNLCAIHA . . .
Plasmodium H3.1	. . . DFKSDLRFQSQAVLALQEAAEAYLVGLFEDTNSCAIHA . . .
Plasmodium H3.3	. . . EYKTDLRFQSQAILALQEAAEAYLVGLFEDTNSCAIHA . . .
Tetrahymena H3.1	. . . DFKADLRFQSSAVLALQEAAEAYLVGLFEDTNSCAIHA . . .
Tetrahymena H3.3	. . . EMKSDIRFQSQAILALQEAAEAYLVGLFEDTNSCAIHA . . .

FIG. 5.15 Comparison of sequences of the RC histone H3.1 with the RI histone H3.3 across species. Amino acids altered in H3.3 relative to H3.1 for a given species are highlighted in white; note the prevalence of altered residues at aa 87–90 (SAVM in human H3.1). Neurospora expresses a single histone H3 (apart from a CenH3) which, similarly to the *Ascomycetes*, has substitutions relative to the SAVM sequence characteristic of human H3.1 and typical of H3.3 sequences; residues differing from human H3.1 at positions 89–90 are highlighted.

Phylogenetic analysis indicated that H3.3 evolved independently as an RI variant at least four times, and it was suggested that its importance lay in its replication independence and not in any alteration to the structure of H3.3-containing nucleosomes (Talbert and Henikoff, 2010; Thatcher et al., 1994; Waterborg, 2012). This hypothesis was supported by experiments with *Tetrahymena*: knockout of the RI histone hv2, the *Tetrahymena* H3.3 equivalent, was not lethal and resulted in no obvious growth defects (Yu and Gorovsky, 1997), while the major, RC (replication coupled) H3 could also be replaced by high-level expression of hv2 (Cui et al., 2006). Ascomycetes, including *S. cerevisiae*, have lost the canonical H3 and retain only H3.3 as both RC and RI H3 (Ahmad and Henikoff, 2002; Henikoff and Smith, 2015). Consistent with the *Tetrahymena* results, yeast cells expressing H3 in which the α2 region is replaced by that from human H3.1 exhibit normal growth in a rich medium (counter to a prediction made 14 years earlier) (McBurney et al., 2016; Smith, 2002). Structural studies confirmed that H3.3-containing nucleosomes are nearly identical to canonical nucleosomes (Tachiwana et al., 2011b).

Insight into the role of the H3.3-specific amino acids near the base of the α2 helix emerged from contemporaneous studies of histone deposition in *Drosophila* and *Xenopus* (Ahmad and Henikoff, 2002; Ray-Gallet et al., 2002). In the *Drosophila* study, RC H3 and RI H3.3 were tagged with Green Fluorescent Protein (GFP) and acutely expressed (over a two-hour period) from a heat shock inducible promoter. Fluorescent imaging confirmed that canonical H3 was incorporated only in a replication-dependent fashion, with deposition completely abrogated when replication was blocked. H3.3, in contrast, was incorporated into chromatin at low levels both during and outside of S-phase. Expressing GFP-tagged H3 mutants revealed that single amino changes in the α2 helix towards those present in H3.3 allowed RI deposition; thus all three H3.3-specific residues in the α2 helix must be altered to prevent RI deposition. This specificity was eventually shown to result, not surprisingly, from interactions between the H3.3 α2 helix and its specific chaperone, the HIRA complex. The *HIR1* and *HIR2* genes were first identified in yeast as being required for proper cell cycle-regulated expression of the divergent H2A-H2B loci, by repressing their transcription outside of S-phase (Osley and Lycan, 1987). Deletion of the *HIR* genes in yeast together with *Cac2*, encoding the p60 subunit of CAF-1, caused synergistic defects in silencing at telomeric and silent mating type loci, hinting at a role for *HIR1* and *HIR2* in heterochromatin function (Kaufman et al., 1998). Almouzni and colleagues cloned the *Xenopus* homolog of the mammalian HIRA protein, which had previously been identified as containing features similar to Hir1 and Hir2, and showed that it facilitated nucleosome assembly using purified components in an in vitro assay (Lamour et al., 1995; Ray-Gallet et al., 2002). Using the well-established *Xenopus* egg extract assembly system, they then showed that immunodepletion of HIRA from extracts severely compromised nucleosome assembly. This was somewhat surprising in light of the presence of other assembly factors, including CAF-1, Asf1, N1/N2, and nucleoplasmin in the extract. Indeed, these factors were still capable of facilitating nucleosome assembly that was coupled to DNA synthesis (examined in the context of DNA repair), while HIRA was specifically required for nucleosome assembly that was uncoupled to DNA replication or damage repair. Based on these results, HIRA was proposed to participate in a chromatin assembly pathway distinct from the RC pathways involving CAF-1 or Asf1. This proposal was supported by experiments in which the affinity purification scheme described earlier, which identified Asf1 and CAF-1 association with predeposition H3 and H4, revealed that epitope-tagged H3.3 expressed in HeLa cells was associated with HIRA (Tagami et al., 2004).

Yeast Hir1 was found to interact with Asf1, and the human homolog HIRA was shown to exist in a complex with two other components, CABIN1 (calcineurin binding protein 1) and UBN1 (ubinuclein 1); as with HIRA, support for CABIN1 and UBN1 being integral to the HIRA complex also derived from their homology to their counterparts in yeast (Balaji et al., 2009; Banumathy et al., 2009; Rai et al., 2011; Sharp et al., 2001; Sutton et al., 2001; Tagami et al., 2004; Tang et al., 2006). Biochemical and structural experiments revealed that HIRA itself does not directly bind H3.3, but rather facilitates interaction between H3.3 and UBN1, while it is UBN1 that specifies binding via interactions with H3.3-specific amino acids, particularly G90 (Ricketts et al., 2015).

The HIRA complex is not the only chaperone that handles H3.3. An early indication of an alternative pathway for H3.3 deposition was the finding that the deposition of Flag-tagged H3.3 was apparently unaffected by a loss of function *hira* mutant in *Drosophila* embryos (Bonnefoy et al., 2007). ChIP-seq examination of H3.3 deposition in *hira*$^{-/-}$ murine embryonic stem cells revealed extensive deposition independent of HIRA (Goldberg et al., 2010). Immunoaffinity purification followed by mass spectrometry identified the ATRX and DAXX proteins as associating with H3.3 in both wild-type and *hira*$^{-/-}$ ESCs. Knockout of *Atrx* resulted in loss of H3.3 deposition at telomeres, demonstrating the existence of two independent pathways, one dependent on HIRA and the other on ATRX, responsible for deposition of H3.3 at distinct genomic locations. Other reports showed that ATRX and DAXX associate in a complex with H3.3 and facilitate H3.3 deposition at telomeres and pericentric heterochromatin and that DAXX interacts directly with the short amino acid sequence at the base of the α2 helix (Drane et al., 2010; Lewis et al., 2010; Wong et al., 2010). Structural comparison of DAXX and UBN1 in complex with H3.3-H4 dimers revealed similar interactions allowing discrimination of H3.3 from H3.1, despite dissimilar amino acid sequences between the two proteins (Elsasser et al., 2012; Liu et al., 2012a; Ricketts et al., 2015).

Consistent with its role as a replacement histone, the relative abundance of H3.3 can vary widely in different cell types: in terminally differentiated neurons, H3.3 replaces H3.1 and H3.2 with a half-life of 142 days, eventually reaching levels of ~90% of all H3, whereas in dividing cells H3.3 accounts for about 25% of total H3 (McKittrick et al., 2004; Pina and Suau, 1987). As with H2A.Z, the localization of H3.3 was found to be nonuniform. Enrichment of GFP-tagged H3.3 was observed at active rDNA loci and chromosomal "puffs" corresponding to induced heat shock loci in *Drosophila* (Ahmad and Henikoff, 2002; Schwartz and Ahmad, 2005). Deposition of H3.3 was inferred to require ongoing transcription, as H3.3 incorporation was observed continually at rDNA loci but ceased at HSP70 genes upon cessation of heat shock. Moreover, H3.3 was found to replace canonical H3 marked by the repressive Lys9 methylation modification at rDNA loci upon transcriptional activation, suggesting that replacement of canonical H3 by H3.3 might establish a chromatin state, possibly heritable, favorable for transcription. The introduction of microarray technology, described in Chapter 4, prompted experiments that confirmed genome-wide H3.3 enrichment at transcribed genes while also demonstrating enrichment at transcription factor binding sites in *Drosophila* S2 tissue culture cells (Mito et al., 2005, 2007). ChIP-seq experiments further extended these findings to HeLa cells and murine ESCs (Goldberg et al., 2010; Jin et al., 2009). In addition, enrichment was observed at transposons in S2 cells and at transcription factor binding sites and telomeric and pericentric heterochromatin in ESCs (Drane et al., 2010; Goldberg et al., 2010; Mito et al., 2005; Wong et al., 2010). Deposition at telomeres depended on ATRX, as mentioned above, while H3.3 enrichment at gene bodies was abrogated in ESCs lacking HIRA. More recently, distinct HIRA-dependent pathways have been identified for H3.3 deposition, one for deposition of newly synthesized H3.3 that requires UBN1, and the other for recycling of H3.3 from existing nucleosomes that are displaced by transcription that does not require UBN1 (Torne et al., 2020).

What determines the sites of H3.3 deposition? Genomic sites of enrichment correspond to regions exhibiting high rates of histone turnover. Turnover of histone H3 in yeast was first measured by adoption of a method used by Wolfram Hörz and colleagues to monitor nucleosome reassembly following transcriptional shutoff at the *PHO5* promoter (Schermer et al., 2005). Expression of Flag-tagged H3 was induced in cells constitutively expressing Myc-H3 in G1-arrested cells; ChIP, using antibodies to both the Flag and Myc epitopes, followed by microarray analysis allowed determination of the Flag/Myc ratio, with a higher ratio corresponding to higher turnover (Dion et al., 2007; Rufiange et al., 2007). These experiments revealed the highest turnover at promoters, with replacement over coding regions being lower and correlating with polymerase density. Henikoff and colleagues invented a different approach, termed CATCH-IT (for covalent attachment of tags to capture histones and identify turnover), to assess histone turnover in *Drosophila* cells (Deal et al., 2010). In this method, cells were treated with a methionine surrogate, azidohomoalanine, that was incorporated into newly synthesized histones (and other proteins). Purified nucleosomes containing the surrogate were chemically coupled to biotin and affinity-purified with streptavidin; the associated nucleosomal DNA was then analyzed on microarrays, and the highest turnover was observed at regulatory elements and gene bodies.

The observation of high histone turnover at regulatory elements that also exhibited enrichment for H3.3 suggested that H3.3 deposition might operate through a simple gap-filling mechanism. This idea was supported by results showing that HIRA directly binds DNA and is recruited to sites of induced nucleosome depletion (Ray-Gallet et al., 2011; Schneiderman et al., 2012). In sum, RI deposition of H3.3 at regulatory regions and transcribed genes is fostered by its association with the HIRA chaperone complex via contacts between UBN1 and the H3.3-specific sequence at the base of the α2 helix and by recruitment of HIRA to accessible DNA regions created by eviction of nucleosomes. H3.3 can also be recycled from "old" nucleosomes and deposited in a process involving HIRA and Asf1 but not requiring UBN1, while its deposition at heterochromatic sites occurs by a distinct mechanism in which a DAXX-ATRX complex functions as a chaperone.

With regard to function, despite the close structural resemblance of H3.3-containing nucleosomes to those containing canonical H3, H3.3 incorporation has been reported to affect nucleosome stability in a manner with potential functional consequences. H3.3-containing nucleosomes isolated from vertebrate cells are sensitive to disruption at elevated salt concentrations, and this sensitivity is enhanced when H3.3 is partnered with H2A.Z, as mentioned in our discussion of H2A.Z (Jin and Felsenfeld, 2007). Such dual variant nucleosomes are also more sensitive to MNase digestion and are greatly enriched at proximal promoters of active genes (Jin et al., 2009). These findings suggest a model in which these variants evolved in part to favor the formation of labile nucleosomes at sites where they could more easily allow access of TFs to regulatory elements. This model, however, has important caveats: first, as mentioned earlier, H2A.Z-H3.3 doubly variant nucleosomes reconstituted using recombinant, and therefore unmodified, histones do not exhibit increased instability or significant structural alterations (Horikoshi et al., 2016; Thakar et al., 2009), and second, knockout of H3.3 in murine ESCs did not reduce accessibility of regulatory elements that are enriched for H3.3 in wild-type ESCs (Martire et al., 2019).

Functional effects of the substitution of serine or threonine for alanine at position 31 of H3.3 have also been reported (Martire and Banaszynski, 2020). These effects may be especially important in the context of heterochromatin; for example, H3.3S31p (H3.3 phosphorylated at Ser31) was localized near centromeric regions in HeLa cells (Hake et al., 2005). In *Arabidopsis*, H3.3T31 inhibits the methyltransferase ATXR5 from methylating H3K27 (Jacob et al., 2014). H3K27 methylation is a particularly important modification during the development of metazoan organisms, as we shall discuss later; suppression of this modification by H3.3 incorporation in *Arabidopsis* may protect H3.3-containing genes from being incorporated into repressive heterochromatin. More recently, defects in developing embryos of *Xenopus* caused by depletion of H3.3 were found to be rescued by expression of H3.3 mutants if and only if Ser31 was not mutated or was replaced by the phospho-mimic Asp (Sitbon et al., 2020).

In metazoans, H3.3 is especially important for development and reproduction (Elsaesser et al., 2010). Early studies in *Drosophila* showed that the H3.3 chaperone HIRA was essential for the postfertilization remodeling of sperm nuclei, which entailed the replacement of sperm chromosomal proteins such as protamines by maternally provided H3.3 (Bonnefoy et al., 2007; Loppin et al., 2005). Flies lacking both H3.3 genes (*HIS3.3A* and *HIS3.3B*) are viable but sterile, revealing a specialized role in development (Hodl and Basler, 2009; Sakai et al., 2009). Early embryonic lethality and a variety of developmental defects, including germline defects, have been reported upon loss or mutation of H3.3-encoding genes in mice, *Xenopus*, and *Arabidopsis* (reviewed in Delaney et al., 2018). Surprisingly, an H3.3 null mutant in *C. elegans* is viable and fertile and displays no obvious developmental defects or delays, with the only effect observed being an altered response to thermal stress (Delaney et al., 2018). A possible explanation for this apparent anomaly is that canonical H3 expression can compensate for the absence of H3.3; supporting this possibility, in *Drosophila*, transcriptional defects caused by loss of H3.3 are rescued by expression of an additional H3.2 transgene (Sakai et al., 2009). However, H3.3-specific residues are essential for male fertility in flies, implying that differences in the incorporation or function of H3.3 must account for the differing requirements in *C. elegans*. Moreover, *hira*- worms exhibit morphological defects not seen in H3.3 null mutants, implying some function of HIRA beyond its role as a H3.3 chaperone; this is certainly the case in yeast, where Hir1 was first identified as a transcriptional repressor (Osley and Lycan, 1987). Understanding the molecular mechanisms connecting H3.3 to its physiological functions remains a highly active area of research (Buschbeck and Hake, 2017; Martire and Banaszynski, 2020).

A truly remarkable finding regarding H3.3 was reported in 2012. In one study, sequencing of 48 samples from patients with pediatric glioblastoma multiforme, an aggressive brain cancer, revealed somatic mutations in H3.3, DAXX, or ATRX in 44% (21/48) of the samples analyzed (Schwartzentruber et al., 2012). Mutations in *H3F3A* encoding H3.3 were confined to H3K27M (nine samples), H3G34R (five), and H3G34V (one), sometimes occurring in conjunction with ATRX/DAXX mutations and sometimes not. The second study reported that 39 of 50 pediatric diffuse intrinsic pontine gliomas contained a K27M mutation in H3.3 (Wu et al., 2012). Numerous studies have explored the molecular mechanism by which these mutations could give rise to this devastating cancer; we will return to this topic in Chapter 8: Chromatin and transcription.

References

Abascal, F., Corpet, A., Gurard-Levin, Z.A., Juan, D., Ochsenbein, F., Rico, D., Valencia, A., Almouzni, G., 2013. Subfunctionalization via adaptive evolution influenced by genomic context: the case of histone chaperones ASF1a and ASF1b. Mol. Biol. Evol. 30, 1853–1866.

Abbott, D.W., Ivanova, V.S., Wang, X., Bonner, W.M., Ausio, J., 2001. Characterization of the stability and folding of H2A.Z chromatin particles: implications for transcriptional activation. J. Biol. Chem. 276, 41945–41949.

Adam, M., Robert, F., Larochelle, M., Gaudreau, L., 2001. H2A.Z is required for global chromatin integrity and for recruitment of RNA polymerase II under specific conditions. Mol. Cell. Biol. 21, 6270–6279.

Adamson, E.D., Woodland, H.R., 1974. Histone synthesis in early amphibian development: histone and DNA syntheses are not co-ordinated. J. Mol. Biol. 88, 263–285.

Aguilar-Gurrieri, C., Larabi, A., Vinayachandran, V., Patel, N.A., Yen, K., Reja, R., Ebong, I.O., Schoehn, G., Robinson, C.V., Pugh, B.F., et al., 2016. Structural evidence for Nap1-dependent H2A-H2B deposition and nucleosome assembly. EMBO J. 35, 1465–1482.

Ahmad, K., Henikoff, S., 2002. The histone variant H3.3 marks active chromatin by replication-independent nucleosome assembly. Mol. Cell 9, 1191–1200.

Alabert, C., Barth, T.K., Reveron-Gomez, N., Sidoli, S., Schmidt, A., Jensen, O.N., Imhof, A., Groth, A., 2015. Two distinct modes for propagation of histone PTMs across the cell cycle. Genes Dev. 29, 585–590.

Albert, I., Mavrich, T.N., Tomsho, L.P., Qi, J., Zanton, S.J., Schuster, S.C., Pugh, B.F., 2007. Translational and rotational settings of H2A.Z nucleosomes across the Saccharomyces cerevisiae genome. Nature 446, 572–576.

Allis, C.D., Glover, C.V., Bowen, J.K., Gorovsky, M.A., 1980. Histone variants specific to the transcriptionally active, amitotically dividing macronucleus of the unicellular eucaryote, Tetrahymena thermophila. Cell 20, 609–617.

Allis, C.D., Richman, R., Gorovsky, M.A., Ziegler, Y.S., Touchstone, B., Bradley, W.A., Cook, R.G., 1986. hv1 is an evolutionarily conserved H2A variant that is preferentially associated with active genes. J. Biol. Chem. 261, 1941–1948.

Almouzni, G., Mechali, M., 1988. Assembly of spaced chromatin promoted by DNA synthesis in extracts from Xenopus eggs. EMBO J. 7, 665–672.

Almouzni, G., Wolffe, A.P., 1993. Nuclear assembly, structure, and function: the use of Xenopus in vitro systems. Exp. Cell Res. 205, 1–15.

Almouzni, G., Clark, D.J., Mechali, M., Wolffe, A.P., 1990. Chromatin assembly on replicating DNA in vitro. Nucleic Acids Res. 18, 5767–5774.

Alva, V., Lupas, A.N., 2019. Histones predate the split between bacteria and archaea. Bioinformatics 35, 2349–2353.

Ambrose, C., McLaughlin, R., Bina, M., 1987. The flexibility and topology of simian virus 40 DNA in minichromosomes. Nucleic Acids Res. 15, 3703–3721.

Ammar, R., Torti, D., Tsui, K., Gebbia, M., Durbic, T., Bader, G.D., Giaever, G., Nislow, C., 2012. Chromatin is an ancient innovation conserved between Archaea and Eukarya. elife 1, e00078.

Andrews, A.J., Chen, X., Zevin, A., Stargell, L.A., Luger, K., 2010. The histone chaperone Nap1 promotes nucleosome assembly by eliminating nonnucleosomal histone DNA interactions. Mol. Cell 37, 834–842.

Angelov, D., Molla, A., Perche, P.Y., Hans, F., Cote, J., Khochbin, S., Bouvet, P., Dimitrov, S., 2003. The histone variant macroH2A interferes with transcription factor binding and SWI/SNF nucleosome remodeling. Mol. Cell 11, 1033–1041.

Annunziato, A.T., 2005. Split decision: what happens to nucleosomes during DNA replication? J. Biol. Chem. 280, 12065–12068.

Annunziato, A.T., 2012. Assembling chromatin: the long and winding road. Biochim. Biophys. Acta 1819, 196–210.

Annunziato, A.T., Seale, R.L., 1983a. Chromatin replication, reconstitution and assembly. Mol. Cell. Biochem. 55, 99–112.

Annunziato, A.T., Seale, R.L., 1983b. Histone deacetylation is required for the maturation of newly replicated chromatin. J. Biol. Chem. 258, 12675–12684.

Annunziato, A.T., Schindler, R.K., Riggs, M.G., Seale, R.L., 1982. Association of newly synthesized histones with replicating and nonreplicating regions of chromatin. J. Biol. Chem. 257, 8507–8515.

Anuar, N.D., Kurscheid, S., Field, M., Zhang, L., Rebar, E., Gregory, P., Buchou, T., Bowles, J., Koopman, P., Tremethick, D.J., et al., 2019. Gene editing of the multi-copy H2A.B gene and its importance for fertility. Genome Biol. 20, 23.

Arceci, R.J., Gross, P.R., 1980. Histone variants and chromatin structure during sea urchin development. Dev. Biol. 80, 186–209.

Arents, G., Burlingame, R.W., Wang, B.C., Love, W.E., Moudrianakis, E.N., 1991. The nucleosomal core histone octamer at 3.1 A resolution: a tripartite protein assembly and a left-handed superhelix. Proc. Natl. Acad. Sci. USA 88, 10148–10152.

Bagchi, D.N., Battenhouse, A.M., Park, D., Iyer, V.R., 2020. The histone variant H2A.Z in yeast is almost exclusively incorporated into the +1 nucleosome in the direction of transcription. Nucleic Acids Res. 48, 157–170.

Bailey, K.A., Pereira, S.L., Widom, J., Reeve, J.N., 2000. Archaeal histone selection of nucleosome positioning sequences and the procaryotic origin of histone-dependent genome evolution. J. Mol. Biol. 303, 25–34.

Balaji, S., Iyer, L.M., Aravind, L., 2009. HPC2 and ubinuclein define a novel family of histone chaperones conserved throughout eukaryotes. Mol. BioSyst. 5, 269–275.

Balhorn, R., 2007. The protamine family of sperm nuclear proteins. Genome Biol. 8, 227.

Banumathy, G., Somaiah, N., Zhang, R., Tang, Y., Hoffmann, J., Andrake, M., Ceulemans, H., Schultz, D., Marmorstein, R., Adams, P.D., 2009. Human UBN1 is an ortholog of yeast Hpc2p and has an essential role in the HIRA/ASF1a chromatin-remodeling pathway in senescent cells. Mol. Cell. Biol. 29, 758–770.

Bao, Y., Konesky, K., Park, Y.J., Rosu, S., Dyer, P.N., Rangasamy, D., Tremethick, D.J., Laybourn, P.J., Luger, K., 2004. Nucleosomes containing the histone variant H2A.Bbd organize only 118 base pairs of DNA. EMBO J. 23, 3314–3324.

Barnhart, M.C., Kuich, P.H., Stellfox, M.E., Ward, J.A., Bassett, E.A., Black, B.E., Foltz, D.R., 2011. HJURP is a CENP-A chromatin assembly factor sufficient to form a functional de novo kinetochore. J. Cell Biol. 194, 229–243.

Barral, S., Morozumi, Y., Tanaka, H., Montellier, E., Govin, J., de Dieuleveult, M., Charbonnier, G., Coute, Y., Puthier, D., Buchou, T., et al., 2017. Histone variant H2A.L.2 guides transition protein-dependent protamine assembly in male germ cells. Mol. Cell 66, 89–101 e108.

Barski, A., Cuddapah, S., Cui, K., Roh, T.Y., Schones, D.E., Wang, Z., Wei, G., Chepelev, I., Zhao, K., 2007. High-resolution profiling of histone methylations in the human genome. Cell 129, 823–837.

Becker, P.B., Wu, C., 1992. Cell-free system for assembly of transcriptionally repressed chromatin from Drosophila embryos. Mol. Cell. Biol. 12, 2241–2249.

Bellelli, R., Belan, O., Pye, V.E., Clement, C., Maslen, S.L., Skehel, J.M., Cherepanov, P., Almouzni, G., Boulton, S.J., 2018. POLE3-POLE4 is a histone H3-h4 chaperone that maintains chromatin integrity during DNA replication. Mol. Cell 72 (112–126), e115.

Belotserkovskaya, R., Oh, S., Bondarenko, V.A., Orphanides, G., Studitsky, V.M., Reinberg, D., 2003. FACT facilitates transcription-dependent nucleosome alteration. Science 301, 1090–1093.

Bhattacharyya, S., Mattiroli, F., Luger, K., 2018. Archaeal DNA on the histone merry-go-round. FEBS J. 285, 3168–3174.

Bina-Stein, M., 1978. Folding of 140-base pair length DNA by a core of arginine-rich histones. J. Biol. Chem. 253, 5213–5219.

Bina-Stein, M., Simpson, R.T., 1977. Specific folding and contraction of DNA by histones H3 and H4. Cell 11, 609–618.

Birnstiel, M.L., 2002. The dawn of gene isolation. Gene 300, 3–11.

Birnstiel, M., Telford, J., Weinberg, E., Stafford, D., 1974. Isolation and some properties of the genes coding for histone proteins. Proc. Natl. Acad. Sci. USA 71, 2900–2904.

Black, B.E., Cleveland, D.W., 2011. Epigenetic centromere propagation and the nature of CENP-a nucleosomes. Cell 144, 471–479.

Bloch, D.P., Godman, G.C., 1955. A microphotometric study of the syntheses of desoxyribonucleic acid and nuclear histone. J. Biophys. Biochem. Cytol. 1, 17–28.

Bönisch, C., Hake, S.B., 2012. Histone H2A variants in nucleosomes and chromatin: more or less stable? Nucleic Acids Res. 40, 10719–10741.

Bonisch, C., Schneider, K., Punzeler, S., Wiedemann, S.M., Bielmeier, C., Bocola, M., Eberl, H.C., Kuegel, W., Neumann, J., Kremmer, E., et al., 2012. H2A.Z.2.2 is an alternatively spliced histone H2A.Z variant that causes severe nucleosome destabilization. Nucleic Acids Res. 40, 5951–5964.

Bonnefoy, E., Orsi, G.A., Couble, P., Loppin, B., 2007. The essential role of Drosophila HIRA for de novo assembly of paternal chromatin at fertilization. PLoS Genet. 3, 1991–2006.

Bonner, W.M., 1975. Protein migration into nuclei. II. Frog oocyte nuclei accumulate a class of microinjected oocyte nuclear proteins and exclude a class of microinjected oocyte cytoplasmic proteins. J. Cell Biol. 64, 431–437.

Bonner, W.M., Wu, R.S., Panusz, H.T., Muneses, C., 1988. Kinetics of accumulation and depletion of soluble newly synthesized histone in the reciprocal regulation of histone and DNA synthesis. Biochemistry 27, 6542–6550.

Bonner, W.M., Redon, C.E., Dickey, J.S., Nakamura, A.J., Sedelnikova, O.A., Solier, S., Pommier, Y., 2008. GammaH2AX and cancer. Nat. Rev. Cancer 8, 957–967.

Boseley, P.G., Bradbury, E.M., Butler-Browne, G.S., Carpenter, B.G., Stephens, R.M., 1976. Physical studies of chromatin. The recombination of histones with DNA. Eur. J. Biochem. 62, 21–31.

Boyer, M., Yutin, N., Pagnier, I., Barrassi, L., Fournous, G., Espinosa, L., Robert, C., Azza, S., Sun, S., Rossmann, M.G., et al., 2009. Giant Marseillevirus highlights the role of amoebae as a melting pot in emergence of chimeric microorganisms. Proc. Natl. Acad. Sci. USA 106, 21848–21853.

Brahma, S., Henikoff, S., 2019. RSC-associated subnucleosomes define MNase-sensitive promoters in yeast. Mol. Cell 73 (238–249), e233.

Brandt, W.F., Strickland, W.N., Von Holt, C., 1974. The primary structure of histone F3 from shark erythrocytes. FEBS Lett. 40, 349–352.

Brewster, N.K., Johnston, G.C., Singer, R.A., 1998. Characterization of the CP complex, an abundant dimer of Cdc68 and Pob3 proteins that regulates yeast transcriptional activation and chromatin repression. J. Biol. Chem. 273, 21972–21979.

Briggs, R., King, T.J., 1952. Transplantation of living nuclei from blastula cells into enucleated frogs' eggs. Proc. Natl. Acad. Sci. USA 38, 455–463.

Brown, D.D., 2004. A tribute to the Xenopus laevis oocyte and egg. J. Biol. Chem. 279, 45291–45299.

Brykczynska, U., Hisano, M., Erkek, S., Ramos, L., Oakeley, E.J., Roloff, T.C., Beisel, C., Schubeler, D., Stadler, M.B., Peters, A.H., 2010. Repressive and active histone methylation mark distinct promoters in human and mouse spermatozoa. Nat. Struct. Mol. Biol. 17, 679–687.

Buchwitz, B.J., Ahmad, K., Moore, L.L., Roth, M.B., Henikoff, S., 1999. A histone-H3-like protein in C. elegans. Nature 401, 547–548.

Bulger, M., Ito, T., Kamakaka, R.T., Kadonaga, J.T., 1995. Assembly of regularly spaced nucleosome arrays by Drosophila chromatin assembly factor 1 and a 56-kDa histone-binding protein. Proc. Natl. Acad. Sci. USA 92, 11726–11730.

Burma, S., Chen, B.P., Murphy, M., Kurimasa, A., Chen, D.J., 2001. ATM phosphorylates histone H2AX in response to DNA double-strand breaks. J. Biol. Chem. 276, 42462–42467.

Buschbeck, M., Hake, S.B., 2017. Variants of core histones and their roles in cell fate decisions, development and cancer. Nat. Rev. Mol. Cell Biol. 18, 299–314.

Buschbeck, M., Uribesalgo, I., Wibowo, I., Rue, P., Martin, D., Gutierrez, A., Morey, L., Guigo, R., Lopez-Schier, H., Di Croce, L., 2009. The histone variant macroH2A is an epigenetic regulator of key developmental genes. Nat. Struct. Mol. Biol. 16, 1074–1079.

Bustin, M., Cole, R.D., 1968. Species and organ specificity in very lysine-rich histones. J. Biol. Chem. 243, 4500–4505.

Camahort, R., Li, B., Florens, L., Swanson, S.K., Washburn, M.P., Gerton, J.L., 2007. Scm3 is essential to recruit the histone h3 variant cse4 to centromeres and to maintain a functional kinetochore. Mol. Cell 26, 853–865.

Camerini-Otero, R.D., Felsenfeld, G., 1977. Supercoiling energy and nucleosome formation: the role of the arginine-rich histone kernel. Nucleic Acids Res. 4, 1159–1181.

Camerini-Otero, R.D., Sollner-Webb, B., Felsenfeld, G., 1976. The organization of histones and DNA in chromatin: evidence for an arginine-rich histone kernel. Cell 8, 333–347.

Campos, E.I., Fillingham, J., Li, G., Zheng, H., Voigt, P., Kuo, W.H., Seepany, H., Gao, Z., Day, L.A., Greenblatt, J.F., et al., 2010. The program for processing newly synthesized histones H3.1 and H4. Nat. Struct. Mol. Biol. 17, 1343–1351.

Cao, S., Bendall, H., Hicks, G.G., Nashabi, A., Sakano, H., Shinkai, Y., Gariglio, M., Oltz, E.M., Ruley, H.E., 2003. The high-mobility-group box protein SSRP1/T160 is essential for cell viability in day 3.5 mouse embryos. Mol. Cell. Biol. 23, 5301–5307.

Carr, A.M., Dorrington, S.M., Hindley, J., Phear, G.A., Aves, S.J., Nurse, P., 1994. Analysis of a histone H2A variant from fission yeast: evidence for a role in chromosome stability. Mol. Gen. Genet. 245, 628–635.

Celeste, A., Petersen, S., Romanienko, P.J., Fernandez-Capetillo, O., Chen, H.T., Sedelnikova, O.A., Reina-San-Martin, B., Coppola, V., Meffre, E., Difilippantonio, M.J., et al., 2002. Genomic instability in mice lacking histone H2AX. Science 296, 922–927.

Cermelli, S., Guo, Y., Gross, S.P., Welte, M.A., 2006. The lipid-droplet proteome reveals that droplets are a protein-storage depot. Curr. Biol. 16, 1783–1795.

Chadwick, B.P., Willard, H.F., 2001. A novel chromatin protein, distantly related to histone H2A, is largely excluded from the inactive X chromosome. J. Cell Biol. 152, 375–384.

Chakravarthy, S., Luger, K., 2006. The histone variant macro-H2A preferentially forms "hybrid nucleosomes". J. Biol. Chem. 281, 25522–25531.

Chakravarthy, S., Bao, Y., Roberts, V.A., Tremethick, D., Luger, K., 2004. Structural characterization of histone H2A variants. Cold Spring Harb. Symp. Quant. Biol. 69, 227–234.

Chakravarthy, S., Gundimella, S.K., Caron, C., Perche, P.Y., Pehrson, J.R., Khochbin, S., Luger, K., 2005. Structural characterization of the histone variant macroH2A. Mol. Cell. Biol. 25, 7616–7624.

Chang, L., Loranger, S.S., Mizzen, C., Ernst, S.G., Allis, C.D., Annunziato, A.T., 1997. Histones in transit: cytosolic histone complexes and diacetylation of H4 during nucleosome assembly in human cells. Biochemistry 36, 469–480.

Chang, E.Y., Ferreira, H., Somers, J., Nusinow, D.A., Owen-Hughes, T., Narlikar, G.J., 2008. MacroH2A allows ATP-dependent chromatin remodeling by SWI/SNF and ACF complexes but specifically reduces recruitment of SWI/SNF. Biochemistry 47, 13726–13732.

Changolkar, L.N., Singh, G., Cui, K., Berletch, J.B., Zhao, K., Disteche, C.M., Pehrson, J.R., 2010. Genome-wide distribution of macroH2A1 histone variants in mouse liver chromatin. Mol. Cell. Biol. 30, 5473–5483.

Chatton, E., 1925. *Pansporella perplexa.* Réflexions sur la biologie et la phylogénie des protozoaites. Ann. Sci. Nat. Zool. 10e serie *VII*, 1–84.

Chatton, E., 1938. Titre et travaux scientifique (1906-1937) de Edouard Chatton. Sette, Sottano, Italy.

Chen, C.C., Dechassa, M.L., Bettini, E., Ledoux, M.B., Belisario, C., Heun, P., Luger, K., Mellone, B.G., 2014. CAL1 is the Drosophila CENP-A assembly factor. J. Cell Biol. 204, 313–329.

Chen, Z., Gabizon, R., Brown, A.I., Lee, A., Song, A., Diaz-Celis, C., Kaplan, C.D., Koslover, E.F., Yao, T., Bustamante, C., 2019. High-resolution and high-accuracy topographic and transcriptional maps of the nucleosome barrier. elife 8, e48281.

Chicoine, L.G., Schulman, I.G., Richman, R., Cook, R.G., Allis, C.D., 1986. Nonrandom utilization of acetylation sites in histones isolated from Tetrahymena. Evidence for functionally distinct H4 acetylation sites. J. Biol. Chem. 261, 1071–1076.

Choe, J., Kolodrubetz, D., Grunstein, M., 1982. The two yeast histone H2A genes encode similar protein subtypes. Proc. Natl. Acad. Sci. USA 79, 1484–1487.

Chong, J.P., Thommes, P., Blow, J.J., 1996. The role of MCM/P1 proteins in the licensing of DNA replication. Trends Biochem. Sci. 21, 102–106.

Clarkson, M.J., Wells, J.R., Gibson, F., Saint, R., Tremethick, D.J., 1999. Regions of variant histone His2AvD required for Drosophila development. Nature 399, 694–697.

Cohen, L.H., Newrock, K.M., Zweidler, A., 1975. Stage-specific switches in histone synthesis during embryogenesis of the sea urchin. Science 190, 994–997.

Cohn, R.H., Lowry, J.C., Kedes, L.H., 1976. Histone genes of the sea urchin (S. purpuratus) cloned in E coli: order, polarity, and strandedness of the five histone-coding and spacer regions. Cell 9, 147–161.

Conde e Silva, N., Black, B.E., Sivolob, A., Filipski, J., Cleveland, D.W., Prunell, A., 2007. CENP-A-containing nucleosomes: easier disassembly versus exclusive centromeric localization. J. Mol. Biol. 370, 555–573.

Coon, J.J., Ueberheide, B., Syka, J.E., Dryhurst, D.D., Ausio, J., Shabanowitz, J., Hunt, D.F., 2005. Protein identification using sequential ion/ion reactions and tandem mass spectrometry. Proc. Natl. Acad. Sci. USA 102, 9463–9468.

Copur, O., Gorchakov, A., Finkl, K., Kuroda, M.I., Muller, J., 2018. Sex-specific phenotypes of histone H4 point mutants establish dosage compensation as the critical function of H4K16 acetylation in Drosophila. Proc. Natl. Acad. Sci. USA 115, 13336–13341.

Core, L., Adelman, K., 2019. Promoter-proximal pausing of RNA polymerase II: a nexus of gene regulation. Genes Dev. 33, 960–982.

Costanzi, C., Pehrson, J.R., 1998. Histone macroH2A1 is concentrated in the inactive X chromosome of female mammals. Nature 393, 599–601.

Cremisi, C., Chestier, A., Yaniv, M., 1978. Assembly of SV40 and polyoma minichromosomes during replication. Cold Spring Harb. Symp. Quant. Biol. 42 (Pt 1), 409–416.

Creyghton, M.P., Markoulaki, S., Levine, S.S., Hanna, J., Lodato, M.A., Sha, K., Young, R.A., Jaenisch, R., Boyer, L.A., 2008. H2AZ is enriched at polycomb complex target genes in ES cells and is necessary for lineage commitment. Cell 135, 649–661.

Cui, B., Liu, Y., Gorovsky, M.A., 2006. Deposition and function of histone H3 variants in Tetrahymena thermophila. Mol. Cell. Biol. 26, 7719–7730.

Daganzo, S.M., Erzberger, J.P., Lam, W.M., Skordalakes, E., Zhang, R., Franco, A.A., Brill, S.J., Adams, P.D., Berger, J.M., Kaufman, P.D., 2003. Structure and function of the conserved core of histone deposition protein Asf1. Curr. Biol. 13, 2148–2158.

Dai, J., Hyland, E.M., Yuan, D.S., Huang, H., Bader, J.S., Boeke, J.D., 2008. Probing nucleosome function: a highly versatile library of synthetic histone H3 and H4 mutants. Cell 134, 1066–1078.

Davie, J.R., Saunders, C.A., Walsh, J.M., Weber, S.C., 1981. Histone modifications in the yeast S. cerevisiae. Nucleic Acids Res 9, 3205–3216.

Day, D.S., Zhang, B., Stevens, S.M., Ferrari, F., Larschan, E.N., Park, P.J., Pu, W.T., 2016. Comprehensive analysis of promoter-proximal RNA polymerase II pausing across mammalian cell types. Genome Biol. 17, 120.

de Groot, C., Houston, J., Davis, B., Gerson-Gurwitz, A., Monen, J., Lara-Gonzalez, P., Oegema, K., Shiau, A.K., Desai, A., 2021. The N-terminal tail of C. elegans CENP-A interacts with KNL-2 and is essential for centromeric chromatin assembly. Mol. Biol. Cell 32, 1193–1201.

De Koning, L., Corpet, A., Haber, J.E., Almouzni, G., 2007. Histone chaperones: an escort network regulating histone traffic. Nat. Struct. Mol. Biol. 14, 997–1007.

De Robertis, E.M., Longthorne, R.F., Gurdon, J.B., 1978. Intracellular migration of nuclear proteins in Xenopus oocytes. Nature 272, 254–256.

Deal, R.B., Henikoff, J.G., Henikoff, S., 2010. Genome-wide kinetics of nucleosome turnover determined by metabolic labeling of histones. Science 328, 1161–1164.

Decanniere, K., Babu, A.M., Sandman, K., Reeve, J.N., Heinemann, U., 2000. Crystal structures of recombinant histones HMfA and HMfB from the hyperthermophilic archaeon Methanothermus fervidus. J. Mol. Biol. 303, 35–47.

Delaney, K., Mailler, J., Wenda, J.M., Gabus, C., Steiner, F.A., 2018. Differential expression of histone H3.3 genes and their role in modulating temperature stress response in Caenorhabditis elegans. Genetics 209, 551–565.

Delange, R.J., 1978. Peptide mapping and amino acid sequencing of histones. Methods Cell Biol. 18, 169–188.

DeLange, R.J., Smith, E.L., 1971. Histones: structure and function. Annu. Rev. Biochem. 40, 279–314.

DeLange, R.J., Williams, L.C., Searcy, D.G., 1981. A histone-like protein (HTa) from Thermoplasma acidophilum. II. Complete amino acid sequence. J. Biol. Chem. 256, 905–911.

Dhillon, N., Kamakaka, R.T., 2000. A histone variant, Htz1p, and a Sir1p-like protein, Esc2p, mediate silencing at HMR. Mol. Cell 6, 769–780.

Dhillon, N., Oki, M., Szyjka, S.J., Aparicio, O.M., Kamakaka, R.T., 2006. H2A.Z functions to regulate progression through the cell cycle. Mol. Cell. Biol. 26, 489–501.

Dillon, S.C., Dorman, C.J., 2010. Bacterial nucleoid-associated proteins, nucleoid structure and gene expression. Nat. Rev. Microbiol. 8, 185–195.

Dilworth, S.M., Black, S.J., Laskey, R.A., 1987. Two complexes that contain histones are required for nucleosome assembly in vitro: role of nucleoplasmin and N1 in Xenopus egg extracts. Cell 51, 1009–1018.

Dimitrov, S., Dasso, M.C., Wolffe, A.P., 1994. Remodeling sperm chromatin in Xenopus laevis egg extracts: the role of core histone phosphorylation and linker histone B4 in chromatin assembly. J. Cell Biol. 126, 591–601.

Dion, M.F., Altschuler, S.J., Wu, L.F., Rando, O.J., 2005. Genomic characterization reveals a simple histone H4 acetylation code. Proc. Natl. Acad. Sci. USA 102, 5501–5506.

Dion, M.F., Kaplan, T., Kim, M., Buratowski, S., Friedman, N., Rando, O.J., 2007. Dynamics of replication-independent histone turnover in budding yeast. Science 315, 1405–1408.

Dorigo, B., Schalch, T., Bystricky, K., Richmond, T.J., 2003. Chromatin fiber folding: requirement for the histone H4 N-terminal tail. J. Mol. Biol. 327, 85–96.

Douet, J., Corujo, D., Malinverni, R., Renauld, J., Sansoni, V., Posavec Marjanovic, M., Cantarino, N., Valero, V., Mongelard, F., Bouvet, P., et al., 2017. MacroH2A histone variants maintain nuclear organization and heterochromatin architecture. J. Cell Sci. 130, 1570–1582.

Downs, J.A., Lowndes, N.F., Jackson, S.P., 2000. A role for Saccharomyces cerevisiae histone H2A in DNA repair. Nature 408, 1001–1004.

Doyen, C.M., Montel, F., Gautier, T., Menoni, H., Claudet, C., Delacour-Larose, M., Angelov, D., Hamiche, A., Bednar, J., Faivre-Moskalenko, C., et al., 2006. Dissection of the unusual structural and functional properties of the variant H2A.Bbd nucleosome. EMBO J. 25, 4234–4244.

Draizen, E.J., Shaytan, A.K., Marino-Ramirez, L., Talbert, P.B., Landsman, D., Panchenko, A.R., 2016. HistoneDB 2.0: a histone database with variants—an integrated resource to explore histones and their variants. Database (Oxford) 2016.

Drane, P., Ouararhni, K., Depaux, A., Shuaib, M., Hamiche, A., 2010. The death-associated protein DAXX is a novel histone chaperone involved in the replication-independent deposition of H3.3. Genes Dev. 24, 1253–1265.

Dunleavy, E.M., Roche, D., Tagami, H., Lacoste, N., Ray-Gallet, D., Nakamura, Y., Daigo, Y., Nakatani, Y., Almouzni-Pettinotti, G., 2009. HJURP is a cell-cycle-dependent maintenance and deposition factor of CENP-A at centromeres. Cell 137, 485–497.

Dunn, C.J., Sarkar, P., Bailey, E.R., Farris, S., Zhao, M., Ward, J.M., Dudek, S.M., Saha, R.N., 2017. Histone hypervariants H2A.Z.1 and H2A.Z.2 play independent and context-specific roles in neuronal activity-induced transcription of Arc/Arg3.1 and other immediate early genes. eNeuro 4. ENEURO.0040-17.2017.

Durrin, L.K., Mann, R.K., Kayne, P.S., Grunstein, M., 1991. Yeast histone H4 N-terminal sequence is required for promoter activation in vivo. Cell 65, 1023–1031.

Earnshaw, W.C., Rothfield, N., 1985. Identification of a family of human centromere proteins using autoimmune sera from patients with scleroderma. Chromosoma 91, 313–321.

Easton, D., Chalkley, R., 1972. High-resolution electrophoretic analysis of the histones from embryos and sperm of Arbacia punctulata. Exp. Cell Res. 72, 502–508.

Eickbush, T.H., Moudrianakis, E.N., 1978. The histone core complex: an octamer assembled by two sets of protein-protein interactions. Biochemistry 17, 4955–4964.

Eickbush, T.H., Godfrey, J.E., Elia, M.C., Moudrianakis, E.N., 1988. H2a-specific proteolysis as a unique probe in the analysis of the histone octamer. J. Biol. Chem. 263, 18972–18978.

Ejlassi-Lassallette, A., Mocquard, E., Arnaud, M.C., Thiriet, C., 2011. H4 replication-dependent diacetylation and Hat1 promote S-phase chromatin assembly in vivo. Mol. Biol. Cell 22, 245–255.

Elsaesser, S.J., Goldberg, A.D., Allis, C.D., 2010. New functions for an old variant: no substitute for histone H3.3. Curr. Opin. Genet. Dev. 20, 110–117.

Elsasser, S.J., Huang, H., Lewis, P.W., Chin, J.W., Allis, C.D., Patel, D.J., 2012. DAXX envelops a histone H3.3-H4 dimer for H3.3-specific recognition. Nature 491, 560–565.

English, C.M., Maluf, N.K., Tripet, B., Churchill, M.E., Tyler, J.K., 2005. ASF1 binds to a heterodimer of histones H3 and H4: a two-step mechanism for the assembly of the H3-H4 heterotetramer on DNA. Biochemistry 44, 13673–13682.

English, C.M., Adkins, M.W., Carson, J.J., Churchill, M.E., Tyler, J.K., 2006. Structural basis for the histone chaperone activity of Asf1. Cell 127, 495–508.

Enomoto, S., McCune-Zierath, P.D., Gerami-Nejad, M., Sanders, M.A., Berman, J., 1997. RLF2, a subunit of yeast chromatin assembly factor-I, is required for telomeric chromatin function in vivo. Genes Dev. 11, 358–370.

Erives, A.J., 2017. Phylogenetic analysis of the core histone doublet and DNA topo II genes of Marseilleviridae: evidence of proto-eukaryotic provenance. Epigenetics Chromatin 10, 55.

Ernst, S.G., Miller, H., Brenner, C.A., Nocente-McGrath, C., Francis, S., McIsaac, R., 1987. Characterization of a cDNA clone coding for a sea urchin histone H2A variant related to the H2A.F/Z histone protein in vertebrates. Nucleic Acids Res. 15, 4629–4644.

Faast, R., Thonglairoam, V., Schulz, T.C., Beall, J., Wells, J.R., Taylor, H., Matthaei, K., Rathjen, P.D., Tremethick, D.J., Lyons, I., 2001. Histone variant H2A.Z is required for early mammalian development. Curr. Biol. 11, 1183–1187.

Fan, J.Y., Gordon, F., Luger, K., Hansen, J.C., Tremethick, D.J., 2002. The essential histone variant H2A.Z regulates the equilibrium between different chromatin conformational states. Nat. Struct. Biol. 9, 172–176.

Fan, J.Y., Rangasamy, D., Luger, K., Tremethick, D.J., 2004. H2A.Z alters the nucleosome surface to promote HP1alpha-mediated chromatin fiber folding. Mol. Cell 16, 655–661.

Fei, J., Torigoe, S.E., Brown, C.R., Khuong, M.T., Kassavetis, G.A., Boeger, H., Kadonaga, J.T., 2015. The prenucleosome, a stable conformational isomer of the nucleosome. Genes Dev. 29, 2563–2575.

Flaus, A., Downs, J.A., Owen-Hughes, T., 2021. Histone isoforms and the oncohistone code. Curr. Opin. Genet. Dev. 67, 61–66.

Foltman, M., Evrin, C., De Piccoli, G., Jones, R.C., Edmondson, R.D., Katou, Y., Nakato, R., Shirahige, K., Labib, K., 2013. Eukaryotic replisome components cooperate to process histones during chromosome replication. Cell Rep. 3, 892–904.

Foltz, D.R., Jansen, L.E., Bailey, A.O., Yates 3rd, J.R., Bassett, E.A., Wood, S., Black, B.E., Cleveland, D.W., 2009. Centromere-specific assembly of CENP-a nucleosomes is mediated by HJURP. Cell 137, 472–484.

Formosa, T., 2008. FACT and the reorganized nucleosome. Mol. BioSyst. 4, 1085–1093.

Fowler, E., Farb, R., El-Saidy, S., 1982. Distribution of the core histones H2A.H2B.H3 and H4 during cell replication. Nucleic Acids Res. 10, 735–748.

Franklin, S.G., Zweidler, A., 1977. Non-allelic variants of histones 2a, 2b and 3 in mammals. Nature 266, 273–275.

Freeman, L., Kurumizaka, H., Wolffe, A.P., 1996. Functional domains for assembly of histones H3 and H4 into the chromatin of Xenopus embryos. Proc. Natl. Acad. Sci. USA 93, 12780–12785.

Fujita, Y., Hayashi, T., Kiyomitsu, T., Toyoda, Y., Kokubu, A., Obuse, C., Yanagida, M., 2007. Priming of centromere for CENP-A recruitment by human hMis18alpha, hMis18beta, and M18BP1. Dev. Cell 12, 17–30.

Gabrielli, F., Baglioni, C., 1975. Maternal messenger RNA and histone synthesis in embryos of the surf clam Spisula solidissima. Dev. Biol. 43, 254–263.

Gabrielli, F., Hancock, R., Faber, A.J., 1981. Characterisation of a chromatin fraction bearing pulse-labelled RNA. 2. Quantification of histones and high-mobility-group proteins. Eur. J. Biochem. 120, 363–369.

Gaillard, P.H.L., Martini, E.M., Kaufman, P.D., Stillman, B., Moustacchi, E., Almouzni, G., 1996. Chromatin assembly coupled to DNA repair: a new role for chromatin assembly factor I. Cell 86, 887–896.

Galvani, A., Courbeyrette, R., Agez, M., Ochsenbein, F., Mann, C., Thuret, J.Y., 2008. In vivo study of the nucleosome assembly functions of ASF1 histone chaperones in human cells. Mol. Cell. Biol. 28, 3672–3685.

Gambus, A., Jones, R.C., Sanchez-Diaz, A., Kanemaki, M., van Deursen, F., Edmondson, R.D., Labib, K., 2006. GINS maintains association of Cdc45 with MCM in replisome progression complexes at eukaryotic DNA replication forks. Nat. Cell Biol. 8, 358–366.

Gatewood, J.M., Cook, G.R., Balhorn, R., Schmid, C.W., Bradbury, E.M., 1990. Isolation of four core histones from human sperm chromatin representing a minor subset of somatic histones. J. Biol. Chem. 265, 20662–20666.

Giaimo, B.D., Ferrante, F., Herchenrother, A., Hake, S.B., Borggrefe, T., 2019. The histone variant H2A.Z in gene regulation. Epigenetics Chromatin 12, 37.
Gilbert, S.F., Barresi, M.J.F., 2016. Developmental Biology. Sinauer Associates.
Goldberg, A.D., Banaszynski, L.A., Noh, K.M., Lewis, P.W., Elsaesser, S.J., Stadler, S., Dewell, S., Law, M., Guo, X., Li, X., et al., 2010. Distinct factors control histone variant H3.3 localization at specific genomic regions. Cell 140, 678–691.
Gordon, F., Luger, K., Hansen, J.C., 2005. The core histone N-terminal tail domains function independently and additively during salt-dependent oligomerization of nucleosomal arrays. J. Biol. Chem. 280, 33701–33706.
Gornik, S.G., Ford, K.L., Mulhern, T.D., Bacic, A., McFadden, G.I., Waller, R.F., 2012. Loss of nucleosomal DNA condensation coincides with appearance of a novel nuclear protein in dinoflagellates. Curr. Biol. 22, 2303–2312.
Goshima, G., Wollman, R., Goodwin, S.S., Zhang, N., Scholey, J.M., Vale, R.D., Stuurman, N., 2007. Genes required for mitotic spindle assembly in Drosophila S2 cells. Science 316, 417–421.
Govin, J., Escoffier, E., Rousseaux, S., Kuhn, L., Ferro, M., Thevenon, J., Catena, R., Davidson, I., Garin, J., Khochbin, S., et al., 2007. Pericentric heterochromatin reprogramming by new histone variants during mouse spermiogenesis. J. Cell Biol. 176, 283–294.
Greaves, I.K., Rangasamy, D., Ridgway, P., Tremethick, D.J., 2007. H2A.Z contributes to the unique 3D structure of the centromere. Proc. Natl. Acad. Sci. USA 104, 525–530.
Green, G.R., Poccia, D.L., 1985. Phosphorylation of sea urchin sperm H1 and H2B histones precedes chromatin decondensation and H1 exchange during pronuclear formation. Dev. Biol. 108, 235–245.
Griffith, J.D., 1976. Visualization of prokaryotic DNA in a regularly condensed chromatin-like fiber. Proc. Natl. Acad. Sci. USA 73, 563–567.
Groth, A., Ray-Gallet, D., Quivy, J.P., Lukas, J., Bartek, J., Almouzni, G., 2005. Human Asf1 regulates the flow of S phase histones during replicational stress. Mol. Cell 17, 301–311.
Groth, A., Corpet, A., Cook, A.J., Roche, D., Bartek, J., Lukas, J., Almouzni, G., 2007. Regulation of replication fork progression through histone supply and demand. Science 318, 1928–1931.
Grover, P., Asa, J.S., Campos, E.I., 2018. H3-H4 histone chaperone pathways. Annu. Rev. Genet. 52, 109–130.
Guillemette, B., Bataille, A.R., Gevry, N., Adam, M., Blanchette, M., Robert, F., Gaudreau, L., 2005. Variant histone H2A.Z is globally localized to the promoters of inactive yeast genes and regulates nucleosome positioning. PLoS Biol. 3, e384.
Gunesdogan, U., Jackle, H., Herzig, A., 2010. A genetic system to assess in vivo the functions of histones and histone modifications in higher eukaryotes. EMBO Rep. 11, 772–776.
Gurard-Levin, Z.A., Quivy, J.P., Almouzni, G., 2014. Histone chaperones: assisting histone traffic and nucleosome dynamics. Annu. Rev. Biochem. 83, 487–517.
Gurdon, J.B., 1962a. Adult frogs derived from the nuclei of single somatic cells. Dev. Biol. 4, 256–273.
Gurdon, J.B., 1962b. The developmental capacity of nuclei taken from intestinal epithelium cells of feeding tadpoles. J. Embryol. Exp. Morpholog. 10, 622–640.
Hake, S.B., Garcia, B.A., Kauer, M., Baker, S.P., Shabanowitz, J., Hunt, D.F., Allis, C.D., 2005. Serine 31 phosphorylation of histone variant H3.3 is specific to regions bordering centromeres in metaphase chromosomes. Proc. Natl. Acad. Sci. USA 102, 6344–6349.
Hammoud, S.S., Nix, D.A., Zhang, H., Purwar, J., Carrell, D.T., Cairns, B.R., 2009. Distinctive chromatin in human sperm packages genes for embryo development. Nature 460, 473–478.
Han, W., Li, X., Fu, X., 2011. The macro domain protein family: structure, functions, and their potential therapeutic implications. Mutat. Res. 727, 86–103.
Harp, J.M., Hanson, B.L., Timm, D.E., Bunick, G.J., 2000. Asymmetries in the nucleosome core particle at 2.5 A resolution. Acta Crystallogr D Biol Crystallogr 56, 1513–1534.
Hartley, P.D., Madhani, H.D., 2009. Mechanisms that specify promoter nucleosome location and identity. Cell 137, 445–458.
Harvey, R.P., Whiting, J.A., Coles, L.S., Krieg, P.A., Wells, J.R., 1983. H2A.F: an extremely variant histone H2A sequence expressed in the chicken embryo. Proc. Natl. Acad. Sci. USA 80, 2819–2823.
Hatch, C.L., Bonner, W.M., 1988. Sequence of cDNAs for mammalian H2A.Z, an evolutionarily diverged but highly conserved basal histone H2A isoprotein species. Nucleic Acids Res. 16, 1113–1124.
Hatch, C.L., Bonner, W.M., Moudrianakis, E.N., 1983. Minor histone 2A variants and ubiquinated forms in the native H2A:H2B dimer. Science 221, 468–470.
Hayashi, T., Fujita, Y., Iwasaki, O., Adachi, Y., Takahashi, K., Yanagida, M., 2004. Mis16 and Mis18 are required for CENP-A loading and histone deacetylation at centromeres. Cell 118, 715–729.

Henikoff, S., Furuyama, T., 2012. The unconventional structure of centromeric nucleosomes. Chromosoma 121, 341–352.

Henikoff, S., Smith, M.M., 2015. Histone variants and epigenetics. Cold Spring Harb. Perspect. Biol. 7, a019364.

Henikoff, S., Ahmad, K., Platero, J.S., van Steensel, B., 2000. Heterochromatic deposition of centromeric histone H3-like proteins. Proc. Natl. Acad. Sci. USA 97, 716–721.

Henneman, B., van Emmerik, C., van Ingen, H., Dame, R.T., 2018. Structure and function of archaeal histones. PLoS Genet. 14, e1007582.

Hereford, L., Fahrner, K., Woolford Jr., J., Rosbash, M., Kaback, D.B., 1979. Isolation of yeast histone genes H2A and H2B. Cell 18, 1261–1271.

Hereford, L., Bromley, S., Osley, M.A., 1982. Periodic transcription of yeast histone genes. Cell 30, 305–310.

Herman, T.M., DePamphilis, M.L., Wassarman, P.M., 1981. Structure of chromatin at deoxyribonucleic acid replication forks: location of the first nucleosomes on newly synthesized simian virus 40 deoxyribonucleic acid. Biochemistry 20, 621–630.

Hernandez-Munoz, I., Lund, A.H., van der Stoop, P., Boutsma, E., Muijrers, I., Verhoeven, E., Nusinow, D.A., Panning, B., Marahrens, Y., van Lohuizen, M., 2005. Stable X chromosome inactivation involves the PRC1 Polycomb complex and requires histone MACROH2A1 and the CULLIN3/SPOP ubiquitin E3 ligase. Proc. Natl. Acad. Sci. USA 102, 7635–7640.

Hildebrand, C.E., Walters, R.A., 1976. Rapid assembly of newly synthesized DNA into chromatin subunits prior to joining to small DNA replication intermediates. Biochem. Biophys. Res. Commun. 73, 157–163.

Hill, R.J., Poccia, D.L., Doty, P., 1971. Towards a total macromolecular analysis of sea urchin embryo chromatin. J. Mol. Biol. 61, 445–462.

Hocher, A., Laursen, S.P., Radford, P., Tyson, J., Lambert, C., Stevens, K.M., Picardeau, M., Sockett, R.E., Luger, K., Warnecke, T., 2023. Histone-organized chromatin in bacteria. BioRxiv.

Hodl, M., Basler, K., 2009. Transcription in the absence of histone H3.3. Curr. Biol. 19, 1221–1226.

Hodl, M., Basler, K., 2012. Transcription in the absence of histone H3.2 and H3K4 methylation. Curr. Biol. 22, 2253–2257.

Hoek, M., Stillman, B., 2003. Chromatin assembly factor 1 is essential and couples chromatin assembly to DNA replication in vivo. Proc. Natl. Acad. Sci. USA 100, 12183–12188.

Hoghoughi, N., Barral, S., Curtet, S., Chuffart, F., Charbonnier, G., Puthier, D., Buchou, T., Rousseaux, S., Khochbin, S., 2020. RNA-guided genomic localization of H2A.L.2 histone variant. Cells 9, 474.

Hong, J., Feng, H., Wang, F., Ranjan, A., Chen, J., Jiang, J., Ghirlando, R., Xiao, T.S., Wu, C., Bai, Y., 2014. The catalytic subunit of the SWR1 remodeler is a histone chaperone for the H2A.Z-H2B dimer. Mol. Cell 53, 498–505.

Horikoshi, N., Arimura, Y., Taguchi, H., Kurumizaka, H., 2016. Crystal structures of heterotypic nucleosomes containing histones H2A.Z and H2A. Open Biol. 6, 160127.

Hsiang, M.W., Cole, R.D., 1978. Determination of the primary structures of histones. Methods Cell Biol. 18, 189–228.

Hsu, H.T., Chen, H.M., Yang, Z., Wang, J., Lee, N.K., Burger, A., Zaret, K., Liu, T., Levine, E., Mango, S.E., 2015. TRANSCRIPTION. Recruitment of RNA polymerase II by the pioneer transcription factor PHA-4. Science 348, 1372–1376.

Huang, H., Sabari, B.R., Garcia, B.A., Allis, C.D., Zhao, Y., 2014. SnapShot: histone modifications. Cell 159 (458–458), e451.

Huang, H., Stromme, C.B., Saredi, G., Hodl, M., Strandsby, A., Gonzalez-Aguilera, C., Chen, S., Groth, A., Patel, D.J., 2015. A unique binding mode enables MCM2 to chaperone histones H3-H4 at replication forks. Nat. Struct. Mol. Biol. 22, 618–626.

Hud, N.V., Milanovich, F.P., Balhorn, R., 1994. Evidence of novel secondary structure in DNA-bound protamine is revealed by Raman spectroscopy. Biochemistry 33, 7528–7535.

Irwin, N.A.T., Martin, B.J.E., Young, B.P., Browne, M.J.G., Flaus, A., Loewen, C.J.R., Keeling, P.J., Howe, L.J., 2018. Viral proteins as a potential driver of histone depletion in dinoflagellates. Nat. Commun. 9, 1535.

Ishibashi, T., Li, A., Eirin-Lopez, J.M., Zhao, M., Missiaen, K., Abbott, D.W., Meistrich, M., Hendzel, M.J., Ausio, J., 2010. H2A.Bbd: an X-chromosome-encoded histone involved in mammalian spermiogenesis. Nucleic Acids Res. 38, 1780–1789.

Ishimi, Y., Kikuchi, A., 1991. Identification and molecular cloning of yeast homolog of nucleosome assembly protein I which facilitates nucleosome assembly in vitro. J. Biol. Chem. 266, 7025–7029.

Ishimi, Y., Yasuda, H., Hirosumi, J., Hanaoka, F., Yamada, M., 1983. A protein which facilitates assembly of nucleosome-like structures in vitro in mammalian cells. J. Biochem. 94, 735–744.

Ishimi, Y., Komamura, Y., You, Z., Kimura, H., 1998. Biochemical function of mouse minichromosome maintenance 2 protein. J. Biol. Chem. 273, 8369–8375.

Ito, T., Bulger, M., Kobayashi, R., Kadonaga, J.T., 1996a. Drosophila NAP-1 is a core histone chaperone that functions in ATP-facilitated assembly of regularly spaced nucleosomal arrays. Mol. Cell. Biol. 16, 3112–3124.

Ito, T., Tyler, J.K., Bulger, M., Kobayashi, R., Kadonaga, J.T., 1996b. ATP-facilitated chromatin assembly with a nucleoplasmin-like protein from Drosophila melanogaster. J. Biol. Chem. 271, 25041–25048.

Jackson, V., 1987a. Deposition of newly synthesized histones: misinterpretations due to cross-linking density-labeled proteins with Lomant's reagent. Biochemistry 26, 2325–2334.

Jackson, V., 1987b. Deposition of newly synthesized histones: new histones H2A and H2B do not deposit in the same nucleosome with new histones H3 and H4. Biochemistry 26, 2315–2325.

Jackson, V., Chalkley, R., 1981. A reevaluation of new histone deposition on replicating chromatin. J. Biol. Chem. 256, 5095–5103.

Jackson, V., Granner, D.K., Chalkley, R., 1975. Deposition of histones onto replicating chromosomes. Proc. Natl. Acad. Sci. USA 72, 4440–4444.

Jackson, V., Shires, A., Tanphaichitr, N., Chalkley, R., 1976. Modifications to histones immediately after synthesis. J. Mol. Biol. 104, 471–483.

Jackson, V., Marshall, S., Chalkley, R., 1981. The sites of deposition of newly synthesized histone. Nucleic Acids Res. 9, 4563–4581.

Jackson, J.D., Falciano, V.T., Gorovsky, M.A., 1996. A likely histone H2A.F/Z variant in Saccharomyces cerevisiae. Trends Biochem. Sci. 21, 466–467.

Jacob, Y., Bergamin, E., Donoghue, M.T., Mongeon, V., LeBlanc, C., Voigt, P., Underwood, C.J., Brunzelle, J.S., Michaels, S.D., Reinberg, D., et al., 2014. Selective methylation of histone H3 variant H3.1 regulates heterochromatin replication. Science 343, 1249–1253.

Jasencakova, Z., Scharf, A.N., Ask, K., Corpet, A., Imhof, A., Almouzni, G., Groth, A., 2010. Replication stress interferes with histone recycling and predeposition marking of new histones. Mol. Cell 37, 736–743.

Jiang, X., Soboleva, T.A., Tremethick, D.J., 2020. Short histone H2A variants: small in stature but not in function. Cells 9, 867.

Jin, C., Felsenfeld, G., 2007. Nucleosome stability mediated by histone variants H3.3 and H2A.Z. Genes Dev. 21, 1519–1529.

Jin, C., Zang, C., Wei, G., Cui, K., Peng, W., Zhao, K., Felsenfeld, G., 2009. H3.3/H2A.Z double variant-containing nucleosomes mark 'nucleosome-free regions' of active promoters and other regulatory regions. Nat. Genet. 41, 941–945.

Jones, C.M., Smith, J.C., 2008. An overview of Xenopus development. Methods Mol. Biol. 461, 385–394.

Jumper, J., Evans, R., Pritzel, A., Green, T., Figurnov, M., Ronneberger, O., Tunyasuvunakool, K., Bates, R., Zidek, A., Potapenko, A., et al., 2021. Highly accurate protein structure prediction with AlphaFold. Nature 596, 583–589.

Kalashnikova, A.A., Porter-Goff, M.E., Muthurajan, U.M., Luger, K., Hansen, J.C., 2013. The role of the nucleosome acidic patch in modulating higher order chromatin structure. J. R. Soc. Interface 10, 20121022.

Kalocsay, M., Hiller, N.J., Jentsch, S., 2009. Chromosome-wide Rad51 spreading and SUMO-H2A.Z-dependent chromosome fixation in response to a persistent DNA double-strand break. Mol. Cell 33, 335–343.

Kamakaka, R.T., Bulger, M., Kadonaga, J.T., 1993. Potentiation of RNA polymerase II transcription by Gal4-VP16 during but not after DNA replication and chromatin assembly. Genes Dev. 7, 1779–1795.

Kamakaka, R.T., Bulger, M., Kaufman, P.D., Stillman, B., Kadonaga, J.T., 1996. Postreplicative chromatin assembly by Drosophila and human chromatin assembly factor 1. Mol. Cell. Biol. 16, 810–817.

Kan, P.Y., Caterino, T.L., Hayes, J.J., 2009. The H4 tail domain participates in intra- and internucleosome interactions with protein and DNA during folding and oligomerization of nucleosome arrays. Mol. Cell. Biol. 29, 538–546.

Karczewski, K.J., Francioli, L.C., Tiao, G., Cummings, B.B., Alfoldi, J., Wang, Q., Collins, R.L., Laricchia, K.M., Ganna, A., Birnbaum, D.P., et al., 2020. The mutational constraint spectrum quantified from variation in 141,456 humans. Nature 581, 434–443.

Kaufman, P.D., Kobayashi, R., Kessler, N., Stillman, B., 1995. The p150 and p60 subunits of chromatin assembly factor I: a molecular link between newly synthesized histones and DNA replication. Cell 81, 1105–1114.

Kaufman, P.D., Kobayashi, R., Stillman, B., 1997. Ultraviolet radiation sensitivity and reduction of telomeric silencing in Saccharomyces cerevisiae cells lacking chromatin assembly factor-I. Genes Dev. 11, 345–357.

Kaufman, P.D., Cohen, J.L., Osley, M.A., 1998. Hir proteins are required for position-dependent gene silencing in Saccharomyces cerevisiae in the absence of chromatin assembly factor I. Mol. Cell. Biol. 18, 4793–4806.

Kaya, H., Shibahara, K.I., Taoka, K.I., Iwabuchi, M., Stillman, B., Araki, T., 2001. FASCIATA genes for chromatin assembly factor-1 in arabidopsis maintain the cellular organization of apical meristems. Cell 104, 131–142.

Kayne, P.S., Kim, U.J., Han, M., Mullen, J.R., Yoshizaki, F., Grunstein, M., 1988. Extremely conserved histone H4 N terminus is dispensable for growth but essential for repressing the silent mating loci in yeast. Cell 55, 27–39.

Kedes, L.H., 1979. Histone genes and histone messengers. Annu. Rev. Biochem. 48, 837–870.

Keichline, L.D., Wassarman, P.M., 1979. Structure of chromatin in sea urchin embryos, sperm, and adult somatic cells. Biochemistry 18, 214–219.

Keller, W., Muller, U., Eicken, I., Wendel, I., Zentgraf, H., 1978. Biochemical and ultrastructural analysis of SV40 chromatin. Cold Spring Harb. Symp. Quant. Biol. 42 (Pt 1), 227–244.

Khuong, M.T., Fei, J., Ishii, H., Kadonaga, J.T., 2015. Prenucleosomes and active chromatin. Cold Spring Harb. Symp. Quant. Biol. 80, 65–72.

King, T.J., Briggs, R., 1955. Changes in the nuclei of differentiating gastrula cells, as demonstrated by nuclear transplantation. Proc. Natl. Acad. Sci. USA 41, 321–325.

Kinkade Jr., J.M., Cole, R.D., 1966. A structural comparison of different lysine-rich histones of calf thymus. J. Biol. Chem. 241, 5798–5805.

Kleinschmidt, J.A., Franke, W.W., 1982. Soluble acidic complexes containing histones H3 and H4 in nuclei of Xenopus laevis oocytes. Cell 29, 799–809.

Kleinschmidt, J.A., Fortkamp, E., Krohne, G., Zentgraf, H., Franke, W.W., 1985. Co-existence of two different types of soluble histone complexes in nuclei of Xenopus laevis oocytes. J. Biol. Chem. 260, 1166–1176.

Kleinschmidt, J.A., Seiter, A., Zentgraf, H., 1990. Nucleosome assembly in vitro: separate histone transfer and synergistic interaction of native histone complexes purified from nuclei of Xenopus laevis oocytes. EMBO J. 9, 1309–1318.

Kloeber, J.A., Lou, Z., 2021. Critical DNA damaging pathways in tumorigenesis. Semin. Cancer Biol. 85, 164–184.

Kobor, M.S., Venkatasubrahmanyam, S., Meneghini, M.D., Gin, J.W., Jennings, J.L., Link, A.J., Madhani, H.D., Rine, J., 2004. A protein complex containing the conserved Swi2/Snf2-related ATPase Swr1p deposits histone variant H2A.Z into euchromatin. PLoS Biol. 2, E131.

Kolodrubetz, D., Rykowski, M.C., Grunstein, M., 1982. Histone H2A subtypes associate interchangeably in vivo with histone H2B subtypes. Proc. Natl. Acad. Sci. USA 79, 7814–7818.

Kossel, A., 1896. Über die basischen Stoffe des Zellkerns. S-B Kgl Preuss Akad Wiss 18, 403–408.

Kossel, A., 1896. Über die basischen Stoffe des Zellkerns. Z. Physiol. Chem. 22, 176–287.

Kota, S.K., Feil, R., 2010. Epigenetic transitions in germ cell development and meiosis. Dev. Cell 19, 675–686.

Koumandou, V.L., Wickstead, B., Ginger, M.L., van der Giezen, M., Dacks, J.B., Field, M.C., 2013. Molecular paleontology and complexity in the last eukaryotic common ancestor. Crit. Rev. Biochem. Mol. Biol. 48, 373–396.

Kozlowski, M., Corujo, D., Hothorn, M., Guberovic, I., Mandemaker, I.K., Blessing, C., Sporn, J., Gutierrez-Triana, A., Smith, R., Portmann, T., et al., 2018. MacroH2A histone variants limit chromatin plasticity through two distinct mechanisms. EMBO Rep. 19, e44445.

Krawitz, D.C., Kama, T., Kaufman, P.D., 2002. Chromatin assembly factor I mutants defective for PCNA binding require Asf1/Hir proteins for silencing. Mol. Cell. Biol. 22, 614–625.

Krogan, N.J., Keogh, M.C., Datta, N., Sawa, C., Ryan, O.W., Ding, H., Haw, R.A., Pootoolal, J., Tong, A., Canadien, V., et al., 2003. A Snf2 family ATPase complex required for recruitment of the histone H2A variant Htz1. Mol. Cell 12, 1565–1576.

Krude, T., 1995. Chromatin assembly factor 1 (CAF-1) colocalizes with replication foci in HeLa cell nuclei. Exp. Cell Res. 220, 304–311.

Kurumizaka, H., Kujirai, T., Takizawa, Y., 2021. Contributions of histone variants in nucleosome structure and function. J. Mol. Biol. 433, 166678.

Kustatscher, G., Hothorn, M., Pugieux, C., Scheffzek, K., Ladurner, A.G., 2005. Splicing regulates NAD metabolite binding to histone macroH2A. Nat. Struct. Mol. Biol. 12, 624–625.

Laine, B., Kmiecik, D., Sautiere, P., Biserte, G., Cohen-Solal, M., 1980. Complete amino-acid sequences of DNA-binding proteins HU-1 and HU-2 from Escherichia coli. Eur. J. Biochem. 103, 447–461.

Lamaa, A., Humbert, J., Aguirrebengoa, M., Cheng, X., Nicolas, E., Cote, J., Trouche, D., 2020. Integrated analysis of H2A.Z isoforms function reveals a complex interplay in gene regulation. elife 9, e53375.

Lamour, V., Lecluse, Y., Desmaze, C., Spector, M., Bodescot, M., Aurias, A., Osley, M.A., Lipinski, M., 1995. A human homolog of the S. cerevisiae HIR1 and HIR2 transcriptional repressors cloned from the DiGeorge syndrome critical region. Hum. Mol. Genet. 4, 791–799.

Landschulz, W.H., Johnson, P.F., McKnight, S.L., 1988. The leucine zipper: a hypothetical structure common to a new class of DNA binding proteins. Science 240, 1759–1764.
Landsman, D., 1996. Histone H1 in Saccharomyces cerevisiae: a double mystery solved? Trends Biochem. Sci. 21, 287–288.
Laskey, R.A., Earnshaw, W.C., 1980. Nucleosome assembly. Nature 286, 763–767.
Laskey, R.A., Mills, A.D., Morris, N.R., 1977. Assembly of SV40 chromatin in a cell-free system from Xenopus eggs. Cell 10, 237–243.
Laskey, R.A., Honda, B.M., Mills, A.D., Finch, J.T., 1978. Nucleosomes are assembled by an acidic protein which binds histones and transfers them to DNA. Nature 275, 416–420.
Laursen, S.P., Bowerman, S., Luger, K., 2021. Archaea: the final frontier of chromatin. J. Mol. Biol. 433, 166791.
Le, S., Davis, C., Konopka, J.B., Sternglanz, R., 1997. Two new S-phase-specific genes from Saccharomyces cerevisiae. Yeast 13, 1029–1042.
Leach, T.J., Mazzeo, M., Chotkowski, H.L., Madigan, J.P., Wotring, M.G., Glaser, R.L., 2000. Histone H2A.Z is widely but nonrandomly distributed in chromosomes of Drosophila melanogaster. J. Biol. Chem. 275, 23267–23272.
Lee, J.T., 2011. Gracefully ageing at 50, X-chromosome inactivation becomes a paradigm for RNA and chromatin control. Nat. Rev. Mol. Cell Biol. 12, 815–826.
Lee, K.P., Baxter, H.J., Guillemette, J.G., Lawford, H.G., Lewis, P.N., 1982. Structural studies on yeast nucleosomes. Can. J. Biochem. 60, 379–388.
Leffak, I.M., 1984. Conservative segregation of nucleosome core histones. Nature 307, 82–85.
Lejeune, E., Bortfeld, M., White, S.A., Pidoux, A.L., Ekwall, K., Allshire, R.C., Ladurner, A.G., 2007. The chromatin-remodeling factor FACT contributes to centromeric heterochromatin independently of RNAi. Curr. Biol. 17, 1219–1224.
Leung, A., Cheema, M., Gonzalez-Romero, R., Eirin-Lopez, J.M., Ausio, J., Nelson, C.J., 2016. Unique yeast histone sequences influence octamer and nucleosome stability. FEBS Lett. 590, 2629–2638.
Lewis, J.D., Song, Y., de Jong, M.E., Bagha, S.M., Ausio, J., 2003. A walk though vertebrate and invertebrate protamines. Chromosoma 111, 473–482.
Lewis, P.W., Elsaesser, S.J., Noh, K.M., Stadler, S.C., Allis, C.D., 2010. Daxx is an H3.3-specific histone chaperone and cooperates with ATRX in replication-independent chromatin assembly at telomeres. Proc. Natl. Acad. Sci. USA 107, 14075–14080.
Li, J.J., Kelly, T.J., 1984. Simian virus 40 DNA replication in vitro. Proc. Natl. Acad. Sci. USA 81, 6973–6977.
Li, W.T., Sandman, K., Pereira, S.L., Reeve, J.N., 2000. MJ1647, an open reading frame in the genome of the hyperthermophile Methanococcus jannaschii, encodes a very thermostable archaeal histone with a C-terminal extension. Extremophiles 4, 43–51.
Li, B., Pattenden, S.G., Lee, D., Gutierrez, J., Chen, J., Seidel, C., Gerton, J., Workman, J.L., 2005. Preferential occupancy of histone variant H2AZ at inactive promoters influences local histone modifications and chromatin remodeling. Proc. Natl. Acad. Sci. USA 102, 18385–18390.
Lifton, R.P., Goldberg, M.L., Karp, R.W., Hogness, D.S., 1978. The organization of the histone genes in Drosophila melanogaster: functional and evolutionary implications. Cold Spring Harb. Symp. Quant. Biol. 42 (Pt 2), 1047–1051.
Liu, X., Li, B., Gorovsky, M.A., 1996. Essential and nonessential histone H2A variants in Tetrahymena thermophila. Mol. Cell. Biol. 16, 4305–4311.
Liu, C.P., Xiong, C., Wang, M., Yu, Z., Yang, N., Chen, P., Zhang, Z., Li, G., Xu, R.M., 2012a. Structure of the variant histone H3.3-H4 heterodimer in complex with its chaperone DAXX. Nat. Struct. Mol. Biol. 19, 1287–1292.
Liu, W.H., Roemer, S.C., Port, A.M., Churchill, M.E., 2012b. CAF-1-induced oligomerization of histones H3/H4 and mutually exclusive interactions with Asf1 guide H3/H4 transitions among histone chaperones and DNA. Nucleic Acids Res. 40, 11229–11239.
Liu, S., Xu, Z., Leng, H., Zheng, P., Yang, J., Chen, K., Feng, J., Li, Q., 2017. RPA binds histone H3-H4 and functions in DNA replication-coupled nucleosome assembly. Science 355, 415–420.
Liu, Y., Bisio, H., Toner, C.M., Jeudy, S., Philippe, N., Zhou, K., Bowerman, S., White, A., Edwards, G., Abergel, C., et al., 2021. Virus-encoded histone doublets are essential and form nucleosome-like structures. Cell 184 (4237–4250), e4219.
Lolas, I.B., Himanen, K., Gronlund, J.T., Lynggaard, C., Houben, A., Melzer, M., Van Lijsebettens, M., Grasser, K.D., 2010. The transcript elongation factor FACT affects Arabidopsis vegetative and reproductive development and genetically interacts with HUB1/2. Plant J. 61, 686–697.

Long, M., Sun, X., Shi, W., Yanru, A., Leung, S.T.C., Ding, D., Cheema, M.S., MacPherson, N., Nelson, C.J., Ausio, J., et al., 2019. A novel histone H4 variant H4G regulates rDNA transcription in breast cancer. Nucleic Acids Res. 47, 8399–8409.

Loppin, B., Bonnefoy, E., Anselme, C., Laurencon, A., Karr, T.L., Couble, P., 2005. The histone H3.3 chaperone HIRA is essential for chromatin assembly in the male pronucleus. Nature 437, 1386–1390.

Lou, Z., Minter-Dykhouse, K., Franco, S., Gostissa, M., Rivera, M.A., Celeste, A., Manis, J.P., van Deursen, J., Nussenzweig, A., Paull, T.T., et al., 2006. MDC1 maintains genomic stability by participating in the amplification of ATM-dependent DNA damage signals. Mol. Cell 21, 187–200.

Louie, A.J., Candido, E.P., Dixon, G.H., 1974. Enzymatic modifications and their possible roles in regulating the binding of basic proteins to DNA and in controlling chromosomal structure. Cold Spring Harb. Symp. Quant. Biol. 38, 803–819.

Louters, L., Chalkley, R., 1984. In vitro exchange of nucleosomal histones H2a and H2b. Biochemistry 23, 547–552.

Luger, K., Mader, A.W., Richmond, R.K., Sargent, D.F., Richmond, T.J., 1997. Crystal structure of the nucleosome core particle at 2.8 A resolution [see comments]. Nature 389, 251–260.

Luk, E., Vu, N.D., Patteson, K., Mizuguchi, G., Wu, W.H., Ranjan, A., Backus, J., Sen, S., Lewis, M., Bai, Y., et al., 2007. Chz1, a nuclear chaperone for histone H2AZ. Mol. Cell 25, 357–368.

Luk, E., Ranjan, A., Fitzgerald, P.C., Mizuguchi, G., Huang, Y., Wei, D., Wu, C., 2010. Stepwise histone replacement by SWR1 requires dual activation with histone H2A.Z and canonical nucleosome. Cell 143, 725–736.

Maayan, I., Cohmer, S., 2012. John Bertrand Gurdon (1933-). In: Embryo Project Encyclopedia. http://embryo.asu.edu/handle/10776/3945.

Mackay, S., Newrock, K.M., 1982. Histone subtypes and switches in synthesis of histone subtypes during Ilyanassa development. Dev. Biol. 93, 430–437.

Madamba, E.V., Berthet, E.B., Francis, N.J., 2017. Inheritance of histones H3 and H4 during DNA replication in vitro. Cell Rep. 21, 1361–1374.

Maeshima, K., Hihara, S., Eltsov, M., 2010. Chromatin structure: does the 30-nm fibre exist in vivo? Curr. Opin. Cell Biol. 22, 291–297.

Malay, A.D., Umehara, T., Matsubara-Malay, K., Padmanabhan, B., Yokoyama, S., 2008. Crystal structures of fission yeast histone chaperone Asf1 complexed with the Hip1 B-domain or the Cac2 C terminus. J. Biol. Chem. 283, 14022–14031.

Malik, H.S., Henikoff, S., 2003. Phylogenomics of the nucleosome. Nat. Struct. Biol. 10, 882–891.

Malone, E.A., Clark, C.D., Chiang, A., Winston, F., 1991. Mutations in SPT16/CDC68 suppress cis- and trans-acting mutations that affect promoter function in Saccharomyces cerevisiae. Mol. Cell. Biol. 11, 5710–5717.

Mandl, B., Brandt, W.F., Superti-Furga, G., Graninger, P.G., Birnstiel, M.L., Busslinger, M., 1997. The five cleavage-stage (CS) histones of the sea urchin are encoded by a maternally expressed family of replacement histone genes: functional equivalence of the CS H1 and frog H1M (B4) proteins. Mol. Cell. Biol. 17, 1189–1200.

Mann, R.K., Grunstein, M., 1992. Histone H3 N-terminal mutations allow hyperactivation of the yeast GAL1 gene in vivo. EMBO J. 11, 3297–3306.

Mannironi, C., Bonner, W.M., Hatch, C.L., 1989. H2A.X. a histone isoprotein with a conserved C-terminal sequence, is encoded by a novel mRNA with both DNA replication type and polyA 3′ processing signals. Nucleic Acids Res. 17, 9113–9126.

Marinov, G.K., Lynch, M., 2015. Diversity and divergence of dinoflagellate histone proteins. G3 (Bethesda) 6, 397–422.

Martire, S., Banaszynski, L.A., 2020. The roles of histone variants in fine-tuning chromatin organization and function. Nat. Rev. Mol. Cell Biol. 21, 522–541.

Martire, S., Gogate, A.A., Whitmill, A., Tafessu, A., Nguyen, J., Teng, Y.C., Tastemel, M., Banaszynski, L.A., 2019. Phosphorylation of histone H3.3 at serine 31 promotes p300 activity and enhancer acetylation. Nat. Genet. 51, 941–946.

Marzluff, W.F., Gongidi, P., Woods, K.R., Jin, J., Maltais, L.J., 2002. The human and mouse replication-dependent histone genes. Genomics 80, 487–498.

Marzluff, W.F., Sakallah, S., Kelkar, H., 2006. The sea urchin histone gene complement. Dev. Biol. 300, 308–320.

Marzluff, W.F., Wagner, E.J., Duronio, R.J., 2008. Metabolism and regulation of canonical histone mRNAs: life without a poly(A) tail. Nat. Rev. Genet. 9, 843–854.

Matsubara, K., Sano, N., Umehara, T., Horikoshi, M., 2007. Global analysis of functional surfaces of core histones with comprehensive point mutants. Genes Cells 12, 13–33.

Matsuda, R., Hori, T., Kitamura, H., Takeuchi, K., Fukagawa, T., Harata, M., 2010. Identification and characterization of the two isoforms of the vertebrate H2A.Z histone variant. Nucleic Acids Res. 38, 4263–4273.

Mattiroli, F., Bhattacharyya, S., Dyer, P.N., White, A.E., Sandman, K., Burkhart, B.W., Byrne, K.R., Lee, T., Ahn, N.G., Santangelo, T.J., et al., 2017a. Structure of histone-based chromatin in Archaea. Science 357, 609–612.

Mattiroli, F., Gu, Y., Yadav, T., Balsbaugh, J.L., Harris, M.R., Findlay, E.S., Liu, Y., Radebaugh, C.A., Stargell, L.A., Ahn, N.G., et al., 2017b. DNA-mediated association of two histone-bound complexes of yeast chromatin assembly factor-1 (CAF-1) drives tetrasome assembly in the wake of DNA replication. elife 6.

Mavrich, T.N., Jiang, C., Ioshikhes, I.P., Li, X., Venters, B.J., Zanton, S.J., Tomsho, L.P., Qi, J., Glaser, R.L., Schuster, S.C., et al., 2008. Nucleosome organization in the Drosophila genome. Nature 453, 358–362.

Maxson Jr., R.E., Egzie, J.C., 1980. Expression of maternal and paternal histone genes during early cleavage stages of the echinoderm hybrid Strongylocentrotus purpuratus x Lytechinus pictus. Dev. Biol. 74, 335–342.

McBurney, K.L., Leung, A., Choi, J.K., Martin, B.J., Irwin, N.A., Bartke, T., Nelson, C.J., Howe, L.J., 2016. Divergent residues within histone H3 dictate a unique chromatin structure in Saccharomyces cerevisiae. Genetics 202, 341–349.

McKay, D.J., Klusza, S., Penke, T.J., Meers, M.P., Curry, K.P., McDaniel, S.L., Malek, P.Y., Cooper, S.W., Tatomer, D.C., Lieb, J.D., et al., 2015. Interrogating the function of metazoan histones using engineered gene clusters. Dev. Cell 32, 373–386.

McKittrick, E., Gafken, P.R., Ahmad, K., Henikoff, S., 2004. Histone H3.3 is enriched in covalent modifications associated with active chromatin. Proc. Natl. Acad. Sci. USA 101, 1525–1530.

McKnight, S.L., Miller Jr., O.L., 1977. Electron microscopic analysis of chromatin replication in the cellular blastoderm Drosophila melanogaster embryo. Cell 12, 795–804.

McKnight, S.L., Bustin, M., Miller Jr., O.L., 1978. Electron microscopic analysis of chromosome metabolism in the Drosophila melanogaster embryo. Cold Spring Harb. Symp. Quant. Biol. 42 (Pt 2), 741–754.

Megee, P.C., Morgan, B.A., Mittman, B.A., Smith, M.M., 1990. Genetic analysis of histone H4: essential role of lysines subject to reversible acetylation. Science 247, 841–845.

Mejlvang, J., Feng, Y., Alabert, C., Neelsen, K.J., Jasencakova, Z., Zhao, X., Lees, M., Sandelin, A., Pasero, P., Lopes, M., et al., 2014. New histone supply regulates replication fork speed and PCNA unloading. J. Cell Biol. 204, 29–43.

Mello, J.A., Sillje, H.H., Roche, D.M., Kirschner, D.B., Nigg, E.A., Almouzni, G., 2002. Human Asf1 and CAF-1 interact and synergize in a repair-coupled nucleosome assembly pathway. EMBO Rep. 3, 329–334.

Meneghini, M.D., Wu, M., Madhani, H.D., 2003. Conserved histone variant H2A.Z protects euchromatin from the ectopic spread of silent heterochromatin. Cell 112, 725–736.

Miescher, F., 1874. Das Protamin, eine neue organische Base aus den Samenfädendes Rheinlachses. Ber. Dtsch. Chem. Ges. 7, 376–379.

Millar, C.B., Xu, F., Zhang, K., Grunstein, M., 2006. Acetylation of H2AZ Lys 14 is associated with genome-wide gene activity in yeast. Genes Dev. 20, 711–722.

Mito, Y., Henikoff, J.G., Henikoff, S., 2005. Genome-scale profiling of histone H3.3 replacement patterns. Nat. Genet. 37, 1090–1097.

Mito, Y., Henikoff, J.G., Henikoff, S., 2007. Histone replacement marks the boundaries of cis-regulatory domains. Science 315, 1408–1411.

Mizuguchi, G., Shen, X., Landry, J., Wu, W.H., Sen, S., Wu, C., 2004. ATP-driven exchange of histone H2AZ variant catalyzed by SWR1 chromatin remodeling complex. Science 303, 343–348.

Mizuguchi, G., Xiao, H., Wisniewski, J., Smith, M.M., Wu, C., 2007. Nonhistone Scm3 and histones CenH3-H4 assemble the core of centromere-specific nucleosomes. Cell 129, 1153–1164.

Moggs, J.G., Grandi, P., Quivy, J.P., Jonsson, Z.O., Hubscher, U., Becker, P.B., Almouzni, G., 2000. A CAF-1-PCNA-mediated chromatin assembly pathway triggered by sensing DNA damage. Mol. Cell. Biol. 20, 1206–1218.

Molaro, A., Young, J.M., Malik, H.S., 2018. Evolutionary origins and diversification of testis-specific short histone H2A variants in mammals. Genome Res. 28, 460–473.

Moosmann, A., Campsteijn, C., Jansen, P.W., Nasrallah, C., Raasholm, M., Stunnenberg, H.G., Thompson, E.M., 2011. Histone variant innovation in a rapidly evolving chordate lineage. BMC Evol. Biol. 11, 208.

Moreno-Andres, D., Yokoyama, H., Scheufen, A., Holzer, G., Lue, H., Schellhaus, A.K., Weberruss, M., Takagi, M., Antonin, W., 2020. VPS72/YL1-mediated H2A.Z deposition is required for nuclear reassembly after mitosis. Cells 9, 1702.

Morgan, B.A., Mittman, B.A., Smith, M.M., 1991. The highly conserved N-terminal domains of histones H3 and H4 are required for normal cell cycle progression. Mol. Cell. Biol. 11, 4111–4120.

Morse, R.H., Cantor, C.R., 1985. Nucleosome core particles suppress the thermal untwisting of core DNA and adjacent linker DNA. Proc. Natl. Acad. Sci. USA 82, 4653–4657.

Morse, R.H., Pederson, D.S., Dean, A., Simpson, R.T., 1987. Yeast nucleosomes allow thermal untwisting of DNA. Nucleic Acids Res. 15, 10311–10330.

Mylonas, C., Lee, C., Auld, A.L., Cisse, I.I., Boyer, L.A., 2021. A dual role for H2A.Z.1 in modulating the dynamics of RNA polymerase II initiation and elongation. Nat. Struct. Mol. Biol. 28, 435–442.

Nakanishi, S., Sanderson, B.W., Delventhal, K.M., Bradford, W.D., Staehling-Hampton, K., Shilatifard, A., 2008. A comprehensive library of histone mutants identifies nucleosomal residues required for H3K4 methylation. Nat. Struct. Mol. Biol. 15, 881–888.

Natsume, R., Eitoku, M., Akai, Y., Sano, N., Horikoshi, M., Senda, T., 2007. Structure and function of the histone chaperone CIA/ASF1 complexed with histones H3 and H4. Nature 446, 338–341.

Nelson, D.A., 1982. Histone acetylation in baker's yeast. Maintenance of the hyperacetylated configuration in log phase protoplasts. J. Biol. Chem. 257, 1565–1568.

Nelson, T., Wiegand, R., Brutlag, D., 1981. Ribonucleic acid and other polyanions facilitate chromatin assembly in vitro. Biochemistry 20, 2594–2601.

Newport, J., Kirschner, M., 1982. A major developmental transition in early Xenopus embryos: I. Characterization and timing of cellular changes at the midblastula stage. Cell 30, 675–686.

Nightingale, K., Dimitrov, S., Reeves, R., Wolffe, A.P., 1996. Evidence for a shared structural role for HMG1 and linker histones B4 and H1 in organizing chromatin. EMBO J. 15, 548–561.

Nili, H.A., Mozdarani, H., Aleyasin, A., 2009. Correlation of sperm DNA damage with protamine deficiency in Iranian subfertile men. Reprod. BioMed. Online 18, 479–485.

Nishida, H., Oshima, T., 2017. Archaeal histone distribution is associated with archaeal genome base composition. J. Gen. Appl. Microbiol. 63, 28–35.

Norris, D., Osley, M.A., 1987. The two gene pairs encoding H2A and H2B play different roles in the Saccharomyces cerevisiae life cycle. Mol. Cell. Biol. 7, 3473–3481.

Nusinow, D.A., Hernandez-Munoz, I., Fazzio, T.G., Shah, G.M., Kraus, W.L., Panning, B., 2007a. Poly(ADP-ribose) polymerase 1 is inhibited by a histone H2A variant, MacroH2A, and contributes to silencing of the inactive X chromosome. J. Biol. Chem. 282, 12851–12859.

Nusinow, D.A., Sharp, J.A., Morris, A., Salas, S., Plath, K., Panning, B., 2007b. The histone domain of macroH2A1 contains several dispersed elements that are each sufficient to direct enrichment on the inactive X chromosome. J. Mol. Biol. 371, 11–18.

Obri, A., Ouararhni, K., Papin, C., Diebold, M.L., Padmanabhan, K., Marek, M., Stoll, I., Roy, L., Reilly, P.T., Mak, T.W., et al., 2014. ANP32E is a histone chaperone that removes H2A.Z from chromatin. Nature 505, 648–653.

Ohlenbusch, H.H., Olivera, B.M., Tuan, D., Davidson, N., 1967. Selective dissociation of histones from calf thymus nucleoprotein. J. Mol. Biol. 25, 299–315.

Ohsumi, K., Katagiri, C., 1991. Occurrence of H1 subtypes specific to pronuclei and cleavage-stage cell nuclei of anuran amphibians. Dev. Biol. 147, 110–120.

O'Rand, M.G., Richardson, R.T., Zimmerman, L.J., Widgren, E.E., 1992. Sequence and localization of human NASP: conservation of a Xenopus histone-binding protein. Dev. Biol. 154, 37–44.

Orphanides, G., LeRoy, G., Chang, C.H., Luse, D.S., Reinberg, D., 1998. FACT, a factor that facilitates transcript elongation through nucleosomes. Cell 92, 105–116.

Orphanides, G., Wu, W.H., Lane, W.S., Hampsey, M., Reinberg, D., 1999. The chromatin-specific transcription elongation factor FACT comprises human SPT16 and SSRP1 proteins. Nature 400, 284–288.

Osley, M.A., Lycan, D., 1987. Trans-acting regulatory mutations that alter transcription of Saccharomyces cerevisiae histone genes. Mol. Cell. Biol. 7, 4204–4210.

Palmer, D.K., Margolis, R.L., 1985. Kinetochore components recognized by human autoantibodies are present on mononucleosomes. Mol. Cell. Biol. 5, 173–186.

Palmer, D.K., O'Day, K., Wener, M.H., Andrews, B.S., Margolis, R.L., 1987. A 17-kD centromere protein (CENP-A) copurifies with nucleosome core particles and with histones. J. Cell Biol. 104, 805–815.

Palmer, D.K., O'Day, K., Trong, H.L., Charbonneau, H., Margolis, R.L., 1991. Purification of the centromere-specific protein CENP-A and demonstration that it is a distinctive histone. Proc. Natl. Acad. Sci. USA 88, 3734–3738.

Pandey, N.B., Chodchoy, N., Liu, T.J., Marzluff, W.F., 1990. Introns in histone genes alter the distribution of 3' ends. Nucleic Acids Res. 18, 3161–3170.

Papoutsopoulou, S., Nikolakaki, E., Chalepakis, G., Kruft, V., Chevaillier, P., Giannakouros, T., 1999. SR protein-specific kinase 1 is highly expressed in testis and phosphorylates protamine 1. Nucleic Acids Res. 27, 2972–2980.

Pardal, A.J., Fernandes-Duarte, F., Bowman, A.J., 2019. The histone chaperoning pathway: from ribosome to nucleosome. Essays Biochem. 63, 29–43.

Park, Y.J., Luger, K., 2006. The structure of nucleosome assembly protein 1. Proc. Natl. Acad. Sci. USA 103, 1248–1253.

Park, E.C., Szostak, J.W., 1990. Point mutations in the yeast histone H4 gene prevent silencing of the silent mating type locus HML. Mol. Cell. Biol. 10, 4932–4934.

Park, Y.J., Dyer, P.N., Tremethick, D.J., Luger, K., 2004. A new fluorescence resonance energy transfer approach demonstrates that the histone variant H2AZ stabilizes the histone octamer within the nucleosome. J. Biol. Chem. 279, 24274–24282.

Pasque, V., Gillich, A., Garrett, N., Gurdon, J.B., 2011. Histone variant macroH2A confers resistance to nuclear reprogramming. EMBO J. 30, 2373–2387.

Patthy, L., Smith, E.L., 1973. Histone 3. V. The amino acid sequence of pea embryo histone 3. J. Biol. Chem. 248, 6834–6840.

Paull, T.T., Rogakou, E.P., Yamazaki, V., Kirchgessner, C.U., Gellert, M., Bonner, W.M., 2000. A critical role for histone H2AX in recruitment of repair factors to nuclear foci after DNA damage. Curr. Biol. 10, 886–895.

Pehrson, J.R., Fried, V.A., 1992. MacroH2A, a core histone containing a large nonhistone region. Science 257, 1398–1400.

Pehrson, J.R., Fuji, R.N., 1998. Evolutionary conservation of histone macroH2A subtypes and domains. Nucleic Acids Res. 26, 2837–2842.

Pehrson, J.R., Changolkar, L.N., Costanzi, C., Leu, N.A., 2014. Mice without macroH2A histone variants. Mol. Cell. Biol. 34, 4523–4533.

Pengelly, A.R., Copur, O., Jackle, H., Herzig, A., Muller, J., 2013. A histone mutant reproduces the phenotype caused by loss of histone-modifying factor Polycomb. Science 339, 698–699.

Pereira, S.L., Grayling, R.A., Lurz, R., Reeve, J.N., 1997. Archaeal nucleosomes. Proc. Natl. Acad. Sci. USA 94, 12633–12637.

Perry, C.A., Annunziato, A.T., 1989. Influence of histone acetylation on the solubility, H1 content and DNase I sensitivity of newly assembled chromatin. Nucleic Acids Res. 17, 4275–4291.

Pina, B., Suau, P., 1987. Changes in histones H2A and H3 variant composition in differentiating and mature rat brain cortical neurons. Dev. Biol. 123, 51–58.

Pineiro, M., Puerta, C., Palacian, E., 1991. Yeast nucleosomal particles: structural and transcriptional properties. Biochemistry 30, 5805–5810.

Poccia, D.L., Green, G.R., 1992. Packaging and unpackaging the sea urchin sperm genome. Trends Biochem. Sci. 17, 223–227.

Poccia, D., Greenough, T., Green, G.R., Nash, E., Erickson, J., Gibbs, M., 1984. Remodeling of sperm chromatin following fertilization: nucleosome repeat length and histone variant transitions in the absence of DNA synthesis. Dev. Biol. 104, 274–286.

Posavec Marjanovic, M., Hurtado-Bages, S., Lassi, M., Valero, V., Malinverni, R., Delage, H., Navarro, M., Corujo, D., Guberovic, I., Douet, J., et al., 2017. MacroH2A1.1 regulates mitochondrial respiration by limiting nuclear NAD(+) consumption. Nat. Struct. Mol. Biol. 24, 902–910.

Prendergast, J.A., Murray, L.E., Rowley, A., Carruthers, D.R., Singer, R.A., Johnston, G.C., 1990. Size selection identifies new genes that regulate Saccharomyces cerevisiae cell proliferation. Genetics 124, 81–90.

Prescott, D.M., 1966. The syntheses of total macronuclear protein, histone, and DNA during the cell cycle in Euplotes eurystomus. J. Cell Biol. 31, 1–9.

Prior, C.P., Cantor, C.R., Johnson, E.M., Allfrey, V.G., 1980. Incorporation of exogenous pyrene-labeled histone into Physarum chromatin: a system for studying changes in nucleosomes assembled in vivo. Cell 20, 597–608.

Probst, A.V., Desvoyes, B., Gutierrez, C., 2020. Similar yet critically different: the distribution, dynamics and function of histone variants. J. Exp. Bot. 71, 5191–5204.

Qiu, Y., Tereshko, V., Kim, Y., Zhang, R., Collart, F., Yousef, M., Kossiakoff, A., Joachimiak, A., 2006. The crystal structure of Aq_328 from the hyperthermophilic bacteria Aquifex aeolicus shows an ancestral histone fold. Proteins 62, 8–16.

Rabinowitz, M., 1941. Studies on the cytology and early embryology of the egg *Drosophila melanogaster*. J. Morphol. 69, 1–49.

Radman-Livaja, M., Verzijlbergen, K.F., Weiner, A., van Welsem, T., Friedman, N., Rando, O.J., van Leeuwen, F., 2011. Patterns and mechanisms of ancestral histone protein inheritance in budding yeast. PLoS Biol. 9, e1001075.

Rai, T.S., Puri, A., McBryan, T., Hoffman, J., Tang, Y., Pchelintsev, N.A., van Tuyn, J., Marmorstein, R., Schultz, D.C., Adams, P.D., 2011. Human CABIN1 is a functional member of the human HIRA/UBN1/ASF1a histone H3.3 chaperone complex. Mol. Cell. Biol. 31, 4107–4118.

Raisner, R.M., Hartley, P.D., Meneghini, M.D., Bao, M.Z., Liu, C.L., Schreiber, S.L., Rando, O.J., Madhani, H.D., 2005. Histone variant H2A.Z marks the 5′ ends of both active and inactive genes in euchromatin. Cell 123, 233–248.

Rangasamy, D., Berven, L., Ridgway, P., Tremethick, D.J., 2003. Pericentric heterochromatin becomes enriched with H2A.Z during early mammalian development. EMBO J. 22, 1599–1607.

Rangasamy, D., Greaves, I., Tremethick, D.J., 2004. RNA interference demonstrates a novel role for H2A.Z in chromosome segregation. Nat. Struct. Mol. Biol. 11, 650–655.

Ranjan, A., Mizuguchi, G., FitzGerald, P.C., Wei, D., Wang, F., Huang, Y., Luk, E., Woodcock, C.L., Wu, C., 2013. Nucleosome-free region dominates histone acetylation in targeting SWR1 to promoters for H2A.Z replacement. Cell 154, 1232–1245.

Rasmussen, T.P., Huang, T., Mastrangelo, M.A., Loring, J., Panning, B., Jaenisch, R., 1999. Messenger RNAs encoding mouse histone macroH2A1 isoforms are expressed at similar levels in male and female cells and result from alternative splicing. Nucleic Acids Res. 27, 3685–3689.

Rasmussen, T.P., Wutz, A.P., Pehrson, J.R., Jaenisch, R.R., 2001. Expression of Xist RNA is sufficient to initiate macrochromatin body formation. Chromosoma 110, 411–420.

Rathke, C., Barckmann, B., Burkhard, S., Jayaramaiah-Raja, S., Roote, J., Renkawitz-Pohl, R., 2010. Distinct functions of Mst77F and protamines in nuclear shaping and chromatin condensation during Drosophila spermiogenesis. Eur. J. Cell Biol. 89, 326–338.

Ray-Gallet, D., Quivy, J.P., Scamps, C., Martini, E.M., Lipinski, M., Almouzni, G., 2002. HIRA is critical for a nucleosome assembly pathway independent of DNA synthesis. Mol. Cell 9, 1091–1100.

Ray-Gallet, D., Woolfe, A., Vassias, I., Pellentz, C., Lacoste, N., Puri, A., Schultz, D.C., Pchelintsev, N.A., Adams, P.D., Jansen, L.E., et al., 2011. Dynamics of histone H3 deposition in vivo reveal a nucleosome gap-filling mechanism for H3.3 to maintain chromatin integrity. Mol. Cell 44, 928–941.

Redon, C., Pilch, D., Rogakou, E., Sedelnikova, O., Newrock, K., Bonner, W., 2002. Histone H2A variants H2AX and H2AZ. Curr. Opin. Genet. Dev. 12, 162–169.

Richet, N., Liu, D., Legrand, P., Velours, C., Corpet, A., Gaubert, A., Bakail, M., Moal-Raisin, G., Guerois, R., Compper, C., et al., 2015. Structural insight into how the human helicase subunit MCM2 may act as a histone chaperone together with ASF1 at the replication fork. Nucleic Acids Res. 43, 1905–1917.

Ricketts, M.D., Frederick, B., Hoff, H., Tang, Y., Schultz, D.C., Singh Rai, T., Grazia Vizioli, M., Adams, P.D., Marmorstein, R., 2015. Ubinuclein-1 confers histone H3.3-specific-binding by the HIRA histone chaperone complex. Nat. Commun. 6, 7711.

Ridgway, P., Brown, K.D., Rangasamy, D., Svensson, U., Tremethick, D.J., 2004. Unique residues on the H2A.Z containing nucleosome surface are important for Xenopus laevis development. J. Biol. Chem. 279, 43815–43820.

Rill, R.L., Livolant, F., Aldrich, H.C., Davidson, M.W., 1989. Electron microscopy of liquid crystalline DNA: direct evidence for cholesteric-like organization of DNA in dinoflagellate chromosomes. Chromosoma 98, 280–286.

Rivera-Casas, C., Gonzalez-Romero, R., Cheema, M.S., Ausio, J., Eirin-Lopez, J.M., 2016. The characterization of macroH2A beyond vertebrates supports an ancestral origin and conserved role for histone variants in chromatin. Epigenetics 11, 415–425.

Rizzo, P.J., Nooden, L.D., 1972. Chromosomal proteins in the dinoflagellate alga Gyrodinium cohnii. Science 176, 796–797.

Robbins, E., Borun, T.W., 1967. The cytoplasmic synthesis of histones in hela cells and its temporal relationship to DNA replication. Proc. Natl. Acad. Sci. USA 57, 409–416.

Rogakou, E.P., Pilch, D.R., Orr, A.H., Ivanova, V.S., Bonner, W.M., 1998. DNA double-stranded breaks induce histone H2AX phosphorylation on serine 139. J. Biol. Chem. 273, 5858–5868.

Rogakou, E.P., Boon, C., Redon, C., Bonner, W.M., 1999. Megabase chromatin domains involved in DNA double-strand breaks in vivo. J. Cell Biol. 146, 905–916.

Roulland, Y., Ouararhni, K., Naidenov, M., Ramos, L., Shuaib, M., Syed, S.H., Lone, I.N., Boopathi, R., Fontaine, E., Papai, G., et al., 2016. The flexible ends of CENP-A nucleosome are required for mitotic fidelity. Mol. Cell 63, 674–685.

Rufiange, A., Jacques, P.E., Bhat, W., Robert, F., Nourani, A., 2007. Genome-wide replication-independent histone H3 exchange occurs predominantly at promoters and implicates H3 K56 acetylation and Asf1. Mol. Cell 27, 393–405.

Ruhl, D.D., Jin, J., Cai, Y., Swanson, S., Florens, L., Washburn, M.P., Conaway, R.C., Conaway, J.W., Chrivia, J.C., 2006. Purification of a human SRCAP complex that remodels chromatin by incorporating the histone variant H2A.Z into nucleosomes. Biochemistry 45, 5671–5677.

Ruiz-Carrillo, A., Wangh, L.J., Allfrey, V.G., 1975. Processing of newly synthesized histone molecules. Science 190, 117–128.

Ruiz-Carrillo, A., Jorcano, J.L., Eder, G., Lurz, R., 1979. In vitro core particle and nucleosome assembly at physiological ionic strength. Proc. Natl. Acad. Sci. USA 76, 3284–3288.

Russev, G., Hancock, R., 1981. Formation of hybrid nucleosomes cantaining new and old histones. Nucleic Acids Res. 9, 4129–4137.

Russev, G., Hancock, R., 1982. Assembly of new histones into nucleosomes and their distribution in replicating chromatin. Proc. Natl. Acad. Sci. USA 79, 3143–3147.

Ryan, D.P., Tremethick, D.J., 2018. The interplay between H2A.Z and H3K9 methylation in regulating HP1alpha binding to linker histone-containing chromatin. Nucleic Acids Res. 46, 9353–9366.

Rykowski, M.C., Wallis, J.W., Choe, J., Grunstein, M., 1981. Histone H2B subtypes are dispensable during the yeast cell cycle. Cell 25, 477–487.

Saavedra, R.A., Huberman, J.A., 1986. Both DNA topoisomerases I and II relax 2 micron plasmid DNA in living yeast cells. Cell 45, 65–70.

Sagan, L., 1967. On the origin of mitosing cells. J. Theor. Biol. 14, 255–274.

Sakai, A., Schwartz, B.E., Goldstein, S., Ahmad, K., 2009. Transcriptional and developmental functions of the H3.3 histone variant in Drosophila. Curr. Biol. 19, 1816–1820.

Sanchez-Pulido, L., Pidoux, A.L., Ponting, C.P., Allshire, R.C., 2009. Common ancestry of the CENP-A chaperones Scm3 and HJURP. Cell 137, 1173–1174.

Sandman, K., Reeve, J.N., 2006. Archaeal histones and the origin of the histone fold. Curr. Opin. Microbiol. 9, 520–525.

Sandman, K., Krzycki, J.A., Dobrinski, B., Lurz, R., Reeve, J.N., 1990. HMf, a DNA-binding protein isolated from the hyperthermophilic archaeon Methanothermus fervidus, is most closely related to histones. Proc. Natl. Acad. Sci. USA 87, 5788–5791.

Sandman, K., Grayling, R.A., Dobrinski, B., Lurz, R., Reeve, J.N., 1994. Growth-phase-dependent synthesis of histones in the archaeon Methanothermus fervidus. Proc. Natl. Acad. Sci. USA 91, 12624–12628.

Sanematsu, F., Takami, Y., Barman, H.K., Fukagawa, T., Ono, T., Shibahara, K.I., Nakayama, T., 2006. Asf1 is required for viability and chromatin assembly during DNA replication in vertebrate cells. J. Biol. Chem. 281, 13817–13827.

Santisteban, M.S., Arents, G., Moudrianakis, E.N., Smith, M.M., 1997. Histone octamer function in vivo: mutations in the dimer-tetramer interfaces disrupt both gene activation and repression. EMBO J. 16, 2493–2506.

Santisteban, M.S., Kalashnikova, T., Smith, M.M., 2000. Histone H2A.Z regulats transcription and is partially redundant with nucleosome remodeling complexes. Cell 103, 411–422.

Sapp, J., 2005. The prokaryote-eukaryote dichotomy: meanings and mythology. Microbiol. Mol. Biol. Rev. 69, 292–305.

Sauer, P.V., Timm, J., Liu, D., Sitbon, D., Boeri-Erba, E., Velours, C., Mucke, N., Langowski, J., Ochsenbein, F., Almouzni, G., et al., 2017. Insights into the molecular architecture and histone H3-H4 deposition mechanism of yeast Chromatin assembly factor 1. elife 6, e23474.

Sauer, P.V., Gu, Y., Liu, W.H., Mattiroli, F., Panne, D., Luger, K., Churchill, M.E., 2018. Mechanistic insights into histone deposition and nucleosome assembly by the chromatin assembly factor-1. Nucleic Acids Res. 46, 9907–9917.

Scacchetti, A., Becker, P.B., 2021. Variation on a theme: Evolutionary strategies for H2A.Z exchange by SWR1-type remodelers. Curr. Opin. Cell Biol. 70, 1–9.

Scacchetti, A., Schauer, T., Reim, A., Apostolou, Z., Campos Sparr, A., Krause, S., Heun, P., Wierer, M., Becker, P.B., 2020. Drosophila SWR1 and NuA4 complexes are defined by DOMINO isoforms. elife 9, 56325.

Schaffner, W., Kunz, G., Daetwyler, H., Telford, J., Smith, H.O., Birnstiel, M.L., 1978. Genes and spacers of cloned sea urchin histone DNA analyzed by sequencing. Cell 14, 655–671.

Schermer, U.J., Korber, P., Horz, W., 2005. Histones are incorporated in trans during reassembly of the yeast PHO5 promoter. Mol. Cell 19, 279–285.

Schneiderman, J.I., Orsi, G.A., Hughes, K.T., Loppin, B., Ahmad, K., 2012. Nucleosome-depleted chromatin gaps recruit assembly factors for the H3.3 histone variant. Proc. Natl. Acad. Sci. USA 109, 19721–19726.

Schones, D.E., Cui, K., Cuddapah, S., Roh, T.Y., Barski, A., Wang, Z., Wei, G., Zhao, K., 2008. Dynamic regulation of nucleosome positioning in the human genome. Cell 132, 887–898.

Schulz, L.L., Tyler, J.K., 2006. The histone chaperone ASF1 localizes to active DNA replication forks to mediate efficient DNA replication. FASEB J. 20, 488–490.

Schuster, T., Han, M., Grunstein, M., 1986. Yeast histone H2A and H2B amino termini have interchangeable functions. Cell 45, 445–451.

Schwartz, B.E., Ahmad, K., 2005. Transcriptional activation triggers deposition and removal of the histone variant H3.3. Genes Dev. 19, 804–814.

Schwartzentruber, J., Korshunov, A., Liu, X.Y., Jones, D.T., Pfaff, E., Jacob, K., Sturm, D., Fontebasso, A.M., Quang, D.A., Tonjes, M., et al., 2012. Driver mutations in histone H3.3 and chromatin remodelling genes in paediatric glioblastoma. Nature 482, 226–231.

Seal, R.L., Denny, P., Bruford, E.A., Gribkova, A.K., Landsman, D., Marzluff, W.F., McAndrews, M., Panchenko, A.R., Shaytan, A.K., Talbert, P.B., 2022. A standardized nomenclature for mammalian histone genes. Epigenetics Chromatin 15, 34.

Seale, R.L., 1975. Assembly of DNA and protein during replication in HeLa cells. Nature 255, 247–249.

Seale, R.L., 1976. Studies on the mode of segregation of histone nu bodies during replication in HeLa cells. Cell 9, 423–429.

Seale, R.L., 1981. In vivo assembly of newly synthesized histones. Biochemistry 20, 6432–6437.

Seale, R.L., Aronson, A.I., 1973. Chromatin-associated proteins of the developing sea urchin embryo. II. Acid-soluble proteins. J Mol Biol 75, 647–658.

Seale, R.L., Simpson, R.T., 1975. Effects of cycloheximide on chromatin biosynthesis. J. Mol. Biol. 94, 479–501.

Seidman, M.M., Garon, C.F., Salzman, N.P., 1978. The relationship of SV40 replicating chromosomes to two forms of the non-replicating SV40. Nucleic Acids Res. 5, 2877–2893.

Senshu, T., Yamada, F., 1980. Involvement of cytoplasmic soluble fraction in the assembly of nucleosome-like materials under near physiological conditions. J. Biochem. 87, 1658–1668.

Sharp, J.A., Fouts, E.T., Krawitz, D.C., Kaufman, P.D., 2001. Yeast histone deposition protein Asf1p requires Hir proteins and PCNA for heterochromatic silencing. Curr. Biol. 11, 463–473.

Shaytan, A.K., Landsman, D., Panchenko, A.R., 2015. Nucleosome adaptability conferred by sequence and structural variations in histone H2A-H2B dimers. Curr. Opin. Struct. Biol. 32, 48–57.

Shibahara, K., Stillman, B., 1999. Replication-dependent marking of DNA by PCNA facilitates CAF-1-coupled inheritance of chromatin. Cell 96, 575–585.

Shukla, M.S., Syed, S.H., Goutte-Gattat, D., Richard, J.L., Montel, F., Hamiche, A., Travers, A., Faivre-Moskalenko, C., Bednar, J., Hayes, J.J., et al., 2011. The docking domain of histone H2A is required for H1 binding and RSC-mediated nucleosome remodeling. Nucleic Acids Res. 39, 2559–2570.

Siegel, T.N., Hekstra, D.R., Kemp, L.E., Figueiredo, L.M., Lowell, J.E., Fenyo, D., Wang, X., Dewell, S., Cross, G.A., 2009. Four histone variants mark the boundaries of polycistronic transcription units in Trypanosoma brucei. Genes Dev. 23, 1063–1076.

Singer, M.S., Kahana, A., Wolf, A.J., Meisinger, L.L., Peterson, S.E., Goggin, C., Mahowald, M., Gottschling, D.E., 1998. Identification of high-copy disruptors of telomeric silencing in Saccharomyces cerevisiae. Genetics 150, 613–632.

Sinha, D., Shogren-Knaak, M.A., 2010. Role of direct interactions between the histone H4 Tail and the H2A core in long range nucleosome contacts. J. Biol. Chem. 285, 16572–16581.

Sitbon, D., Boyarchuk, E., Dingli, F., Loew, D., Almouzni, G., 2020. Histone variant H3.3 residue S31 is essential for Xenopus gastrulation regardless of the deposition pathway. Nat. Commun. 11, 1256.

Smith, M.M., 1984. The organization of the yeast histone genes. In: Stein, G.S., Stein, J.L., Marzluff, W.F. (Eds.), Histone Genes: Structure, Organization, and Regulation. Jon Wiley & Sons, New York, p. 3.33.

Smith, M.M., 2002. Histone variants and nucleosome deposition pathways. Mol. Cell 9, 1158–1160.

Smith, M.M., Andresson, O.S., 1983. DNA sequences of yeast H3 and H4 histone genes from two non-allelic gene sets encode identical H3 and H4 proteins. J. Mol. Biol. 169, 663–690.

Smith, M.M., Murray, K., 1983. Yeast H3 and H4 histone messenger RNAs are transcribed from two non-allelic gene sets. J. Mol. Biol. 169, 641–661.

Smith, S., Stillman, B., 1989. Purification and characterization of CAF-I, a human cell factor required for chromatin assembly during DNA replication in vitro. Cell 58, 15–25.

Smith, R.C., Dworkin-Rastl, E., Dworkin, M.B., 1988. Expression of a histone H1-like protein is restricted to early Xenopus development. Genes Dev. 2, 1284–1295.

Soares, D.J., Sandman, K., Reeve, J.N., 2000. Mutational analysis of archaeal histone-DNA interactions. J. Mol. Biol. 297, 39–47.

Sobel, R.E., Cook, R.G., Perry, C.A., Annunziato, A.T., Allis, C.D., 1995. Conservation of deposition-related acetylation sites in newly synthesized histones H3 and H4. Proc. Natl. Acad. Sci. USA 92, 1237–1241.

Soboleva, T.A., Nekrasov, M., Pahwa, A., Williams, R., Huttley, G.A., Tremethick, D.J., 2011. A unique H2A histone variant occupies the transcriptional start site of active genes. Nat. Struct. Mol. Biol. 19, 25–30.

Soboleva, T.A., Parker, B.J., Nekrasov, M., Hart-Smith, G., Tay, Y.J., Tng, W.Q., Wilkins, M., Ryan, D., Tremethick, D.J., 2017. A new link between transcriptional initiation and pre-mRNA splicing: the RNA binding histone variant H2A.B. PLoS Genet. 13, e1006633.

Sogo, J.M., Stahl, H., Koller, T., Knippers, R., 1986. Structure of replicating simian virus 40 minichromosomes. The replication fork, core histone segregation and terminal structures. J. Mol. Biol. 189, 189–204.

Sollner-Webb, B., Camerini-Otero, R.D., Felsenfeld, G., 1976. Chromatin structure as probed by nucleases and proteases: evidence for the central role of histones H3 and H4. Cell 9, 179–193.

Song, Y., He, F., Xie, G., Guo, X., Xu, Y., Chen, Y., Liang, X., Stagljar, I., Egli, D., Ma, J., et al., 2007. CAF-1 is essential for Drosophila development and involved in the maintenance of epigenetic memory. Dev. Biol. 311, 213–222.

Spadafora, C., Oudet, P., Chambon, P., 1978. The same amount of DNA is organized in in vitro-assembled nucleosomes irrespective of the origin of the histones. Nucleic Acids Res. 5, 3479–3489.

Stanier, R.Y., Van Niel, C.B., 1962. The concept of a bacterium. Arch. Mikrobiol. 42, 17–35.

Stargell, L.A., Bowen, J., Dadd, C.A., Dedon, P.C., Davis, M., Cook, R.G., Allis, C.D., Gorovsky, M.A., 1993. Temporal and spatial association of histone H2A variant hv1 with transcriptionally competent chromatin during nuclear development in Tetrahymena thermophila. Genes Dev. 7, 2641–2651.

Starich, M.R., Sandman, K., Reeve, J.N., Summers, M.F., 1996. NMR structure of HMfB from the hyperthermophile, Methanothermus fervidus, confirms that this archaeal protein is a histone. J. Mol. Biol. 255, 187–203.

Stedman, E., Stedman, E., 1951. The basic proteins of cell nuclei. Phil. Trans. R. Soc. London 235, 565–595.

Stein, A., Whitlock Jr., J.P., Bina, M., 1979. Acidic polypeptides can assemble both histones and chromatin in vitro at physiological ionic strength. Proc. Natl. Acad. Sci. USA 76, 5000–5004.

Stein, G.S., Stein, J.L., Marzluff, W.F., 1984. Histone Genes: Structure, Organization, and Regulation. John Wiley & Sons, New York.

Stillman, B., 1986. Chromatin assembly during SV40 DNA replication in vitro. Cell 45, 555–565.

Stoler, S., Keith, K.C., Curnick, K.E., Fitzgerald-Hayes, M., 1995. A mutation in CSE4, an essential gene encoding a novel chromatin-associated protein in yeast, causes chromosome nondisjunction and cell cycle arrest at mitosis. Genes Dev. 9, 573–586.

Stoler, S., Rogers, K., Weitze, S., Morey, L., Fitzgerald-Hayes, M., Baker, R.E., 2007. Scm3, an essential Saccharomyces cerevisiae centromere protein required for G2/M progression and Cse4 localization. Proc. Natl. Acad. Sci. USA 104, 10571–10576.

Stucki, M., Clapperton, J.A., Mohammad, D., Yaffe, M.B., Smerdon, S.J., Jackson, S.P., 2005. MDC1 directly binds phosphorylated histone H2AX to regulate cellular responses to DNA double-strand breaks. Cell 123, 1213–1226.

Stuwe, T., Hothorn, M., Lejeune, E., Rybin, V., Bortfeld, M., Scheffzek, K., Ladurner, A.G., 2008. The FACT Spt16 "peptidase" domain is a histone H3-H4 binding module. Proc. Natl. Acad. Sci. USA 105, 8884–8889.

Sullivan, K.F., Hechenberger, M., Masri, K., 1994. Human CENP-A contains a histone H3 related histone fold domain that is required for targeting to the centromere. J. Cell Biol. 127, 581–592.

Sun, L., Pierrakeas, L., Li, T., Luk, E., 2020. Thermosensitive nucleosome editing reveals the role of DNA sequence in targeted histone variant deposition. Cell Rep. 30 (257–268), e255.

Sures, I., Lowry, J., Kedes, L.H., 1978. The DNA sequence of sea urchin (S. purpuratus) H2A, H2B and H3 histone coding and spacer regions. Cell 15, 1033–1044.

Suto, R.K., Clarkson, M.J., Tremethick, D.J., Luger, K., 2000. Crystal structure of a nucleosome core particle containing the variant histone H2A.Z. Nat. Struct. Biol. 7, 1121–1124.

Sutton, A., Bucaria, J., Osley, M.A., Sternglanz, R., 2001. Yeast ASF1 protein is required for cell cycle regulation of histone gene transcription. Genetics 158, 587–596.

Swaminathan, J., Baxter, E.M., Corces, V.G., 2005. The role of histone H2Av variant replacement and histone H4 acetylation in the establishment of Drosophila heterochromatin. Genes Dev. 19, 65–76.

Tachiwana, H., Kagawa, W., Shiga, T., Osakabe, A., Miya, Y., Saito, K., Hayashi-Takanaka, Y., Oda, T., Sato, M., Park, S.Y., et al., 2011a. Crystal structure of the human centromeric nucleosome containing CENP-A. Nature 476, 232–235.

Tachiwana, H., Osakabe, A., Shiga, T., Miya, Y., Kimura, H., Kagawa, W., Kurumizaka, H., 2011b. Structures of human nucleosomes containing major histone H3 variants. Acta Crystallogr D Biol Crystallogr 67, 578–583.

Tagami, H., Ray-Gallet, D., Almouzni, G., Nakatani, Y., 2004. Histone H3.1 and H3.3 complexes mediate nucleosome assembly pathways dependent or independent of DNA synthesis. Cell 116, 51–61.

Takahashi, K., Chen, E.S., Yanagida, M., 2000. Requirement of Mis6 centromere connector for localizing a CENP-A-like protein in fission yeast. Science 288, 2215–2219.

Talbert, P.B., Henikoff, S., 2010. Histone variants—ancient wrap artists of the epigenome. Nat. Rev. Mol. Cell Biol. 11, 264–275.

Talbert, P.B., Henikoff, S., 2017. Histone variants on the move: substrates for chromatin dynamics. Nat. Rev. Mol. Cell Biol. 18, 115–126.

Talbert, P.B., Ahmad, K., Almouzni, G., Ausio, J., Berger, F., Bhalla, P.L., Bonner, W.M., Cande, W.Z., Chadwick, B.P., Chan, S.W., et al., 2012. A unified phylogeny-based nomenclature for histone variants. Epigenetics Chromatin 5, 7.

Talbert, P.B., Meers, M.P., Henikoff, S., 2019. Old cogs, new tricks: the evolution of gene expression in a chromatin context. Nat. Rev. Genet. 20, 283–297.

Talbert, P.B., Armache, K.J., Henikoff, S., 2022. Viral histones: pickpocket's prize or primordial progenitor? Epigenetics Chromatin 15, 21.

Tan, B.C., Chien, C.T., Hirose, S., Lee, S.C., 2006. Functional cooperation between FACT and MCM helicase facilitates initiation of chromatin DNA replication. EMBO J. 25, 3975–3985.

Tang, Y., Poustovoitov, M.V., Zhao, K., Garfinkel, M., Canutescu, A., Dunbrack, R., Adams, P.D., Marmorstein, R., 2006. Structure of a human ASF1a-HIRA complex and insights into specificity of histone chaperone complex assembly. Nat. Struct. Mol. Biol. 13, 921–929.

Thakar, A., Gupta, P., Ishibashi, T., Finn, R., Silva-Moreno, B., Uchiyama, S., Fukui, K., Tomschik, M., Ausio, J., Zlatanova, J., 2009. H2A.Z and H3.3 histone variants affect nucleosome structure: biochemical and biophysical studies. Biochemistry 48, 10852–10857.

Thambirajah, A.A., Dryhurst, D., Ishibashi, T., Li, A., Maffey, A.H., Ausio, J., 2006. H2A.Z stabilizes chromatin in a way that is dependent on core histone acetylation. J. Biol. Chem. 281, 20036–20044.

Thatcher, T.H., Gorovsky, M.A., 1994. Phylogenetic analysis of the core histones H2A, H2B, H3, and H4. Nucleic Acids Res. 22, 174–179.

Thatcher, T.H., MacGaffey, J., Bowen, J., Horowitz, S., Shapiro, D.L., Gorovsky, M.A., 1994. Independent evolutionary origin of histone H3.3-like variants of animals and Tetrahymena. Nucleic Acids Res. 22, 180–186.

Timinszky, G., Till, S., Hassa, P.O., Hothorn, M., Kustatscher, G., Nijmeijer, B., Colombelli, J., Altmeyer, M., Stelzer, E.H., Scheffzek, K., et al., 2009. A macrodomain-containing histone rearranges chromatin upon sensing PARP1 activation. Nat. Struct. Mol. Biol. 16, 923–929.

Timmers, H.T.M., Tora, L., 2018. Transcript buffering: a balancing act between mRNA synthesis and mRNA degradation. Mol Cell 72, 10–17.

Tolstorukov, M.Y., Goldman, J.A., Gilbert, C., Ogryzko, V., Kingston, R.E., Park, P.J., 2012. Histone variant H2A.Bbd is associated with active transcription and mRNA processing in human cells. Mol. Cell 47, 596–607.

Torigoe, S.E., Urwin, D.L., Ishii, H., Smith, D.E., Kadonaga, J.T., 2011. Identification of a rapidly formed nonnucleosomal histone-DNA intermediate that is converted into chromatin by ACF. Mol. Cell 43, 638–648.

Torne, J., Ray-Gallet, D., Boyarchuk, E., Garnier, M., Le Baccon, P., Coulon, A., Orsi, G.A., Almouzni, G., 2020. Two HIRA-dependent pathways mediate H3.3 de novo deposition and recycling during transcription. Nat. Struct. Mol. Biol. 27, 1057–1068.

Tramantano, M., Sun, L., Au, C., Labuz, D., Liu, Z., Chou, M., Shen, C., Luk, E., 2016. Constitutive turnover of histone H2A.Z at yeast promoters requires the preinitiation complex. elife 5, e14243.

Truong, D.M., Boeke, J.D., 2017. Resetting the yeast epigenome with human nucleosomes. Cell 171 (1508–1519), e1513.

Tsunaka, Y., Fujiwara, Y., Oyama, T., Hirose, S., Morikawa, K., 2016. Integrated molecular mechanism directing nucleosome reorganization by human FACT. Genes Dev. 30, 673–686.

Tye, B.K., 1994. The MCM2-3-5 proteins: are they replication licensing factors? Trends Cell Biol. 4, 160–166.

Tyler, J.K., Bulger, M., Kamakaka, R.T., Kobayashi, R., Kadonaga, J.T., 1996. The p55 subunit of Drosophila chromatin assembly factor 1 is homologous to a histone deacetylase-associated protein. Mol. Cell. Biol. 16, 6149–6159.

Tyler, J.K., Adams, C.R., Chen, S.R., Kobayashi, R., Kamakaka, R.T., Kadonaga, J.T., 1999. The RCAF complex mediates chromatin assembly during DNA replication and repair. Nature 402, 555–560.

Tyler, J.K., Collins, K.A., Prasad-Sinha, J., Amiott, E., Bulger, M., Harte, P.J., Kobayashi, R., Kadonaga, J.T., 2001. Interaction between the Drosophila CAF-1 and ASF1 chromatin assembly factors. Mol. Cell. Biol. 21, 6574–6584.

Ukogu, O.A., Smith, A.D., Devenica, L.M., Bediako, H., McMillan, R.B., Ma, Y., Balaji, A., Schwab, R.D., Anwar, S., Dasgupta, M., et al., 2020. Protamine loops DNA in multiple steps. Nucleic Acids Res. 48, 6108–6119.

Umehara, T., Chimura, T., Ichikawa, N., Horikoshi, M., 2002. Polyanionic stretch-deleted histone chaperone cia1/Asf1p is functional both in vivo and in vitro. Genes Cells 7, 59–73.

Updike, D.L., Mango, S.E., 2006. Temporal regulation of foregut development by HTZ-1/H2A.Z and PHA-4/FoxA. PLoS Genet. 2, e161.

Ura, K., Nightingale, K., Wolffe, A.P., 1996. Differential association of HMG1 and linker histones B4 and H1 with dinucleosomal DNA: structural transitions and transcriptional repression. EMBO J. 15, 4959–4969.

Urban, M.K., Zweidler, A., 1983. Changes in nucleosomal core histone variants during chicken development and maturation. Dev. Biol. 95, 421–428.

Valencia-Sanchez, M.I., Abini-Agbomson, S., Wang, M., Lee, R., Vasilyev, N., Zhang, J., De Ioannes, P., La Scola, B., Talbert, P., Henikoff, S., et al., 2021. The structure of a virus-encoded nucleosome. Nat. Struct. Mol. Biol. 28, 413–417.

van Daal, A., Elgin, S.C., 1992. A histone variant, H2AvD, is essential in Drosophila melanogaster. Mol. Biol. Cell 3, 593–602.

van Daal, A., White, E.M., Gorovsky, M.A., Elgin, S.C., 1988. Drosophila has a single copy of the gene encoding a highly conserved histone H2A variant of the H2A.F/Z type. Nucleic Acids Res. 16, 7487–7497.

van Daal, A., White, E.M., Elgin, S.C., Gorovsky, M.A., 1990. Conservation of intron position indicates separation of major and variant H2As is an early event in the evolution of eukaryotes. J. Mol. Evol. 30, 449–455.

van Holde, K.E., 1988. Chromatin. Springer-Verlag.

Venters, B.J., Pugh, B.F., 2009. A canonical promoter organization of the transcription machinery and its regulators in the Saccharomyces genome. Genome Res. 19, 360–371.

Verreault, A., Kaufman, P.D., Kobayashi, R., Stillman, B., 1996. Nucleosome assembly by a complex of CAF-1 and acetylated histones H3/H4. Cell 87, 95–104.

Viens, A., Mechold, U., Brouillard, F., Gilbert, C., Leclerc, P., Ogryzko, V., 2006. Analysis of human histone H2AZ deposition in vivo argues against its direct role in epigenetic templating mechanisms. Mol. Cell. Biol. 26, 5325–5335.

Vogler, C., Huber, C., Waldmann, T., Ettig, R., Braun, L., Izzo, A., Daujat, S., Chassignet, I., Lopez-Contreras, A.J., Fernandez-Capetillo, O., et al., 2010. Histone H2A C-terminus regulates chromatin dynamics, remodeling, and histone H1 binding. PLoS Genet. 6, e1001234.

Wallis, J.W., Hereford, L., Grunstein, M., 1980. Histone H2B genes of yeast encode two different proteins. Cell 22, 799–805.

Wallis, J.W., Rykowski, M., Grunstein, M., 1983. Yeast histone H2B containing large amino terminus deletions can function in vivo. Cell 35, 711–719.

Waterborg, J.H., 2012. Evolution of histone H3: emergence of variants and conservation of post-translational modification sites. Biochem. Cell Biol. 90, 79–95.

Weber, C.M., Henikoff, J.G., Henikoff, S., 2010. H2A.Z nucleosomes enriched over active genes are homotypic. Nat. Struct. Mol. Biol. 17, 1500–1507.

Weber, C.M., Ramachandran, S., Henikoff, S., 2014. Nucleosomes are context-specific, H2A.Z-modulated barriers to RNA polymerase. Mol. Cell 53, 819–830.

Weintraub, H., 1976. Cooperative alignment of nu bodies during chromosome replication in the presence of cycloheximide. Cell 9, 419–422.

Wenkert, D., Allis, C.D., 1984. Timing of the appearance of macronuclear-specific histone variant hv1 and gene expression in developing new macronuclei of Tetrahymena thermophila. J. Cell Biol. 98, 2107–2117.

West, M.H., Bonner, W.M., 1980. Histone 2A, a heteromorphous family of eight protein species. Biochemistry 19, 3238–3245.

White, C.L., Suto, R.K., Luger, K., 2001. Structure of the yeast nucleosome core particle reveals fundamental changes in internucleosome interactions. EMBO J. 20, 5207–5218.

Whittle, C.M., McClinic, K.N., Ercan, S., Zhang, X., Green, R.D., Kelly, W.G., Lieb, J.D., 2008. The genomic distribution and function of histone variant HTZ-1 during C. elegans embryogenesis. PLoS Genet. 4, e1000187.

Wilhelm, F.X., Wilhelm, M.L., Erard, M., Duane, M.P., 1978. Reconstitution of chromatin: assembly of the nucleosome. Nucleic Acids Res. 5, 505–521.

Willhoft, O., Ghoneim, M., Lin, C.L., Chua, E.Y.D., Wilkinson, M., Chaban, Y., Ayala, R., McCormack, E.A., Ocloo, L., Rueda, D.S., et al., 2018. Structure and dynamics of the yeast SWR1-nucleosome complex. Science 362, aat7716.

Wittmeyer, J., Formosa, T., 1997. The Saccharomyces cerevisiae DNA polymerase alpha catalytic subunit interacts with Cdc68/Spt16 and with Pob3, a protein similar to an HMG1-like protein. Mol. Cell. Biol. 17, 4178–4190.

Woese, C.R., Fox, G.E., 1977. Phylogenetic structure of the prokaryotic domain: the primary kingdoms. Proc. Natl. Acad. Sci. USA 74, 5088–5090.

Wolffe, A.P., 1995. Centromeric chromatin. Histone deviants. Curr. Biol. 5, 452–454.

Wolffe, A.P., 1998. Chromatin: Structure & Function, third ed. Academic Press, San Diego.

Wolffe, A.P., Pruss, D., 1996. Deviant nucleosomes: the functional specialization of chromatin. Trends Genet. 12, 58–62.

Wong, L.H., McGhie, J.D., Sim, M., Anderson, M.A., Ahn, S., Hannan, R.D., George, A.J., Morgan, K.A., Mann, J.R., Choo, K.H., 2010. ATRX interacts with H3.3 in maintaining telomere structural integrity in pluripotent embryonic stem cells. Genome Res. 20, 351–360.

Worcel, A., Han, S., Wong, M.L., 1978. Assembly of newly replicated chromatin. Cell 15, 969–977.

Wratting, D., Thistlethwaite, A., Harris, M., Zeef, L.A., Millar, C.B., 2012. A conserved function for the H2A.Z C terminus. J. Biol. Chem. 287, 19148–19157.

Wu, R.S., Bonner, W.M., 1981. Separation of basal histone synthesis from S-phase histone synthesis in dividing cells. Cell 27, 321–330.

Wu, R.S., Nishioka, D., Bonner, W.M., 1982. Differential conservation of histone 2A variants between mammals and sea urchins. J. Cell Biol. 93, 426–431.

Wu, J.Y., Ribar, T.J., Cummings, D.E., Burton, K.A., McKnight, G.S., Means, A.R., 2000. Spermiogenesis and exchange of basic nuclear proteins are impaired in male germ cells lacking Camk4. Nat. Genet. 25, 448–452.

Wu, W.H., Alami, S., Luk, E., Wu, C.H., Sen, S., Mizuguchi, G., Wei, D., Wu, C., 2005. Swc2 is a widely conserved H2AZ-binding module essential for ATP-dependent histone exchange. Nat. Struct. Mol. Biol. 12, 1064–1071.

Wu, G., Broniscer, A., McEachron, T.A., Lu, C., Paugh, B.S., Becksfort, J., Qu, C., Ding, L., Huether, R., Parker, M., et al., 2012. Somatic histone H3 alterations in pediatric diffuse intrinsic pontine gliomas and non-brainstem glioblastomas. Nat. Genet. 44, 251–253.

Xu, M., Long, C., Chen, X., Huang, C., Chen, S., Zhu, B., 2010. Partitioning of histone H3-H4 tetramers during DNA replication-dependent chromatin assembly. Science 328, 94–98.

Xu, Y., Ayrapetov, M.K., Xu, C., Gursoy-Yuzugullu, O., Hu, Y., Price, B.D., 2012. Histone H2A.Z controls a critical chromatin remodeling step required for DNA double-strand break repair. Mol. Cell 48, 723–733.

Yamada, S., Kugou, K., Ding, D.Q., Fujita, Y., Hiraoka, Y., Murakami, H., Ohta, K., Yamada, T., 2018. The histone variant H2A.Z promotes initiation of meiotic recombination in fission yeast. Nucleic Acids Res. 46, 609–620.

Ye, X., Franco, A.A., Santos, H., Nelson, D.M., Kaufman, P.D., Adams, P.D., 2003. Defective S phase chromatin assembly causes DNA damage, activation of the S phase checkpoint, and S phase arrest. Mol. Cell 11, 341–351.

Yen, K., Vinayachandran, V., Pugh, B.F., 2013. SWR-C and INO80 chromatin remodelers recognize nucleosome-free regions near +1 nucleosomes. Cell 154, 1246–1256.

Yu, L., Gorovsky, M.A., 1997. Constitutive expression, not a particular primary sequence, is the important feature of the H3 replacement variant hv2 in Tetrahymena thermophila. Mol. Cell. Biol. 17, 6303–6310.

Yung, P.Y., Stuetzer, A., Fischle, W., Martinez, A.M., Cavalli, G., 2015. Histone H3 serine 28 is essential for efficient polycomb-mediated gene repression in Drosophila. Cell Rep. 11, 1437–1445.

Zhang, H., Roberts, D.N., Cairns, B.R., 2005. Genome-wide dynamics of Htz1, a histone H2A variant that poises repressed/basal promoters for activation through histone loss. Cell 123, 219–231.

Zhang, K., Gao, Y., Li, J., Burgess, R., Han, J., Liang, H., Zhang, Z., Liu, Y., 2016. A DNA binding winged helix domain in CAF-1 functions with PCNA to stabilize CAF-1 at replication forks. Nucleic Acids Res. 44, 5083–5094.

Zhang, W., Zhang, X., Xue, Z., Li, Y., Ma, Q., Ren, X., Zhang, J., Yang, S., Yang, L., Wu, M., et al., 2019. Probing the function of metazoan histones with a systematic library of H3 and H4 mutants. Dev. Cell 48 (406–419), e405.

Zhou, J., Fan, J.Y., Rangasamy, D., Tremethick, D.J., 2007. The nucleosome surface regulates chromatin compaction and couples it with transcriptional repression. Nat. Struct. Mol. Biol. 14, 1070–1076.

Zilberman, D., Coleman-Derr, D., Ballinger, T., Henikoff, S., 2008. Histone H2A.Z and DNA methylation are mutually antagonistic chromatin marks. Nature 456, 125–129.

Zlatanova, J., Thakar, A., 2008. H2A.Z: view from the top. Structure 16, 166–179.

Zweidler, A., 1984. Core histone variants of the mouse: primary structure and differential expression. In: Stein, G.S., Stein, J.L., Marzluff, W.F. (Eds.), Histone Genes: Structure, Organization, and Regulation. Jon Wiley & Sons, Inc., New York, pp. 339–371.

PART 2

Chromatin function

CHAPTER

6

Chromatin remodeling

Abbreviations

ACF ATP-utilizing chromatin assembly and remodeling factor
ARP actin-related protein
BAF BRG1-associated factor
BRG1 brahma-related gene 1
BRM brahma gene
CHD chromodomain helicase DNA-binding
ChIP chromatin immunoprecipitation
CHRAC chromatin accessibility complex
HSF heat shock factor
ISWI imitation switch
NDR nucleosome-depleted region
NuRD nucleosome remodeling and deacetylase
NURF nucleosome remodeling factor
RSC Remodels the Structure of Chromatin
SHL superhelix location
sin switch-independent
snf sucrose nonfermenting
spt suppressor of Ty insertion
ssn suppressor of *snf*
UAS upstream activating sequence

Molecular machines that utilize the energy of ATP hydrolysis to alter histone-DNA interactions were discovered in the 1990s. First identified as global regulators of transcription in yeast, chromatin remodelers would eventually be revealed to be ubiquitous among eukaryotes, functioning primarily in DNA transcription and repair. Their discovery was a major revelation and prompted a flood of research on their distribution, mechanisms, and functional roles.

The SWI/SNF complex

In 1990, nucleosomes were understood to be inhibitory to transcription initiation, though less so for elongation (Kornberg and Lorch, 1991). Various means were proposed for how cells might solve this problem, including competition between histones and transcription factors for binding DNA (Felsenfeld, 1992), distortion of chromatin structure by torsional stress

Chromatin. https://doi.org/10.1016/B978-0-12-814809-9.00006-8

(Van Holde, 1988), and intermediary factors acting between transcriptional activators and the basic transcription apparatus (Kornberg and Lorch, 1991). While a focused search for such an intermediary factor did yield the Mediator complex (Kornberg, 2005), the first chromatin remodeler identified, the SWI/SNF complex, was discovered in large part fortuitously through the convergence of two independent genetic screens conducted in budding yeast. (A note on nomenclature: the literature variously references the Swi/Snf and the SWI/SNF complex among other, less commonly used appellations. I have chosen to use the upper case designation for this and other remodeling complexes; lower case designations are employed for single subunit remodelers in yeast.) One of these screens was aimed at identifying *SWI* genes involved in mating type switching and the other sought to identify sucrose nonfermenting, or *SNF*, genes required for derepression of the glucose-repressed *SUC2* gene. A third screen that yielded insight into chromatin remodeling was aimed at discovering mutants that suppressed gene repression resulting from insertion of transposable elements, or parts thereof, upstream of the *HIS4* gene. Characterization of *spt*, or suppressor of Ty insertion, mutants identified in this screen provided additional evidence that *SWI* and *SNF* gene products worked at least in part to overcome repressive effects of chromatin (Winston and Carlson, 1992). Purification and characterization of the multisubunit SWI/SNF complex established its activity as an ATP-dependent chromatin remodeling machine. Homologs to SWI/SNF subunits, especially the ATPase or "motor" subunit Swi2, were found in yeast and in other eukaryotes, and methods developed to monitor chromatin remodeling by SWI/SNF were extended to test those candidates and, along with structural data, to elucidate their mechanisms of action. Eventually genome-wide approaches permitted the action of chromatin remodelers on complete genomes to be ascertained, and mutations in chromatin remodelers were found to be linked to human disease, especially cancer. This is the pathway that has led to our current knowledge of chromatin remodelers; a more detailed recounting now follows.

Yeast genetics points to a chromatin remodeling machine

swi *and sin mutants*

Switching between **a** and **α** mating types in haploid yeast cells requires DNA cleavage by the HO endonuclease at the *MAT* locus, from which either the *MATa* or *MATα* genes are expressed. (If this rings a bell, it's because the topic also came up in Chapter 3 in the discussion of yeast heterochromatin.) The *HO* gene is expressed in haploid but not diploid cells, in mother cells but not daughters (i.e., the smaller, budded cells) following cell division, and only for a short period in late G1. This complex regulation in a unicellular eukaryote intrigued investigators (particularly the labs of Ira Herskowitz and Kim Nasmyth) interested in using it as a model for understanding gene regulation during cellular differentiation in metazoans. In yeast with a wild type *HO* gene, haploid spores derived from **a/α** diploids produce cells that rapidly switch mating type and self-mate, giving rise to colonies in which nearly all cells are **a/α** diploids and therefore are unable to mate. (For this reason, yeast strains used in the lab are ho^- to maintain their stability as haploids.) Genes involved in mating type switching were thus identified by screening for *HO* positive colonies that retained the ability to mate (Haber and Garvik, 1977; Stern et al., 1984). (Some yeast genetic trickery was used to exclude mutants in which *HO* itself was defective.) The resulting mutants were sorted into five complementation groups termed *SWI1-5* (Stern et al., 1984). Rather remarkably, Northern blotting showed that all five mutants were defective in transcription of the *HO* gene, and monitoring expression of an *HO-lacZ* reporter gene confirmed this deficiency. A subsequent screen found five additional *SWI* mutants that were defective in *HO-lacZ* expression (Breeden and Nasmyth, 1987).

The *HO* gene was known to be inactive in **a/α** diploid yeast cells, and genetic evidence suggested it was under negative control. This suggested that the *SWI* genes could encode factors that served to overcome this negative regulation in haploid cells and might be dispensable in the absence of their putative target. Suppressor screens were therefore undertaken to identify such a target or targets by mutagenizing *swi*− yeast with ethyl methanesulfonate (EMS) and finding mutants that expressed *HO-lacZ*, easily identified by the blue color of mutant colonies on X-gal indicator plates (Nasmyth et al., 1987; Sternberg et al., 1987). Suppressor mutants were identified that restored *HO-lacZ* expression in *swi*$^{-}$ yeast to 30%–150% of the level seen in wild type cells, and were termed *sdi* or *sin*, for switch independent, mutants; the latter designation eventually stuck. The *sin* mutants varied in their ability to suppress specific *swi* mutants; in particular, *swi1*, *swi2*, and *swi3* were suppressed only by *sin1* and *sin2* mutations, which did not suppress other *swi* mutants, and were accordingly suggested to have a common function. These three genes along with *SWI10/SNF5* were eventually found to encode subunits of the SWI/SNF complex, while *SWI4*, *SWI5*, and *SWI6* encode transcriptional activators required for *HO* activation, and *SWI7*, *SWI8*, and *SWI9* encode the Ada2, Ngg1/Ada3, and Gcn5 subunits of the histone-modifying SAGA complex, which is discussed in Chapter 7.

snf *and* ssn *mutants*

A second screen, begun by Marian Carlson while working with David Botstein and continued at her own lab at Columbia University, identified mutants defective in expression of the *SUC2* gene, which encodes invertase and is necessary for yeast to utilize sucrose or raffinose as a primary carbon source (Carlson et al., 1981; Neigeborn and Carlson, 1984). Similarly to the *HO* gene, evidence indicated that *SUC2* expression is repressed in media containing glucose by a negative regulatory mechanism. Mutants were identified by exposing yeast cells to EMS, growing thousands of single colonies on glucose-containing plates, and replica plating onto plates containing sucrose or raffinose instead of glucose. Mutants failing to grow on the latter media were confirmed by additional testing, including measurement of invertase activity, and were designated sucrose nonfermenting, or *snf*, mutants. Extragenic suppressor mutants, termed *ssn* mutants, were isolated and, similarly to the *swi* and *sin* mutants, were used to divide *snf* mutants into functional groups (Neigeborn et al., 1986). One group comprised *SNF2*, *SNF5*, and *SNF6*, and cloning of *SNF6* led to the finding that it was important for transcriptional activity of a variety of promoters, therefore suggesting that *SNF2*, *SNF5*, and *SNF6* could contribute to transcription in a global or semiglobal fashion (Estruch and Carlson, 1990). These three genes, like *SWI1-3*, encode SWI/SNF subunits.

spt *mutants*

The search for Suppressor of Ty and δ insertion, or *spt*, mutants was begun by Fred Winston as a postdoctoral fellow in Gerry Fink's lab and continued at Harvard Medical School (Winston et al., 1984). Ty1 is a transposable element that can, at very low frequency, be copied and inserted into new genomic locations via an RNA intermediate (Curcio et al., 2015). Insertion of Ty1 or of Ty1 long terminal repeat (δ element), into a promoter region can affect the regulation of the gene adjacent to that promoter. *Spt* mutants were identified as suppressing the His^{-} phenotype resulting from insertion of a δ element upstream of the

HIS4 promoter (specifically, the *his4-912δ* allele, which is His$^-$ at 23°C); *spt* mutants allow growth on media lacking histidine. Genetic experiments suggested that the δ element interfered with *HIS4* expression by competing or interfering with the *HIS4* promoter (Hirschman et al., 1988), and consistent with this idea, transcription from Ty elements was reduced in *spt* mutants. Two of the genes identified in this screen were found to encode histones H2A and H2B, and additional experiments indicated that altered stoichiometry of H2A-H2B or H3-H4 dimers could alter gene expression (Clark-Adams et al., 1988). This was one of the first indications that alterations in chromatin structure could affect transcription in vivo. Identities between *spt* mutants and suppressors of *swi* and *snf* mutations, described in the next section, forged connections among these three mutant categories and supported the notion that *swi* and *snf* genes served to overcome repressive effects mediated at least in part by chromatin.

Connections between swi, snf, *and* spt *mutants and chromatin*

It is not surprising that none of the *SPT* genes encode SWI/SNF subunits; unlike the *SWI* and *SNF* genes, which are required for gene activity or release from repression (of the *HO* or *SUC2* genes), *SPT* genes enforce repression on *HIS4* resulting from Ty1 or δ element insertion. Consistent with this complementary role, *SIN1* was found to be identical to *SPT2*, and *SPT6* to *SSN20* (Clark-Adams and Winston, 1987; Kruger and Herskowitz, 1991; Neigeborn et al., 1987). Other experiments revealed additional connections that pointed to a multisubunit complex, composed of subunits encoded by *SWI* and *SNF* genes, that functioned as a global regulator of transcription by overcoming chromatin-mediated repression. Cross-comparison of the effects of mutations in *SNF2*, *SNF5*, and *SNF6* and in *SWI1-3* showed similar phenotypes, such as impairment in the ability to grow on nonfermentable carbon sources and to undergo sporulation, alterations in *SUC2* and Ty-driven gene expression, and with regard to suppression by *spt* mutations of phenotypes caused by Ty or δ insertions (Happel et al., 1991; Hirschhorn et al., 1992; Laurent et al., 1990, 1991; Malone et al., 1991). Cloning and sequencing revealed *SWI2* to be identical to *SNF2*, and *SWI10* to *SNF5* (Laurent et al., 1990, 1991; Yoshimoto and Yamashita, 1991). (These cited papers reported on the sequences of *SNF2* and *SNF5*; the sequencing of *SWI2* and *SWI10* was apparently never published in the scientific literature. The matching identities instead were cited as personal communication from Kim Nasmyth (Peterson and Herskowitz, 1992; Winston and Carlson, 1992) and eventually recorded in the *Saccharomyces* Genome Database (Cherry et al., 2012).) Sequencing of *SWI1* revealed it to be identical to *ADR6*, which is required for transcription of *ADH1* and *ADH2* in low glucose medium, and this dependence was seen also for *SNF2/SWI2* and *SWI3* (Peterson and Herskowitz, 1992; Taguchi and Young, 1987). Additional diversely regulated promoters, such as *MFα1*, encoding one of the yeast mating type factors; *INO1*, induced in medium containing low levels of inositol; and the divergently transcribed *GAL1* and *GAL10* genes also exhibited dependence on multiple *SWI* and *SNF* genes, as did expression of reporters driven by the *Drosophila* transcriptional activators fushi tarazu and bicoid, and the rat glucocorticoid receptor, when expressed in yeast (Laurent and Carlson, 1992; Laurent et al., 1990; Peterson and Herskowitz, 1992; Peterson et al., 1991; Yoshinaga et al., 1992). Altogether, these results implicated the products of *SWI1*, *SNF2/SWI2*, *SWI3*, *SNF5*, and *SNF6* in facilitating transcription of a diverse set of genes in yeast.

In spite of the apparently widespread dependence of gene transcription on these *SWI* and *SNF* genes, their products were clearly distinct from universally required transcription factors such as Pol II; some genes, such as *CLN2* and *TUB2*, showed no dependence on *SWI/*

SNF genes for their transcription (Laurent et al., 1990; Peterson and Herskowitz, 1992), whereas, for example, all mRNA synthesis was eliminated in yeast harboring a temperature sensitive mutation in a Pol II subunit when shifted to the nonpermissive temperature (Nonet et al., 1987). Moreover, Snf2/Swi2 and Snf5 proteins behaved like transcriptional activators in "artificial recruitment" experiments. In these experiments, Swi and Snf proteins were fused to the LexA DNA-binding domain and found to activate reporter genes having LexA-binding sites in their promoters (Laurent et al., 1990, 1991). Initial evidence did not indicate direct binding to DNA of the Swi or Snf proteins, suggesting that they most likely functioned as intermediary factors, or "coactivators," that are recruited by primary, DNA-binding activators such as Gal4 to facilitate downstream steps in transcriptional activation (Laurent et al., 1990, 1991; Peterson and Herskowitz, 1992). In support of this possibility, Swi3 could be immunoprecipitated from yeast extracts by a fragment of the transcriptionally activating glucocorticoid receptor, but not if the yeast were *swi1*$^-$ or *swi2*$^-$ (Yoshinaga et al., 1992).

The first clues suggesting a connection between regulation of transcription by Swi/Snf proteins and chromatin structure were tenuous but tantalizing. The discovery that *SPT11* and *SPT12* encoded histones H2A and H2B led to the finding that changes in histone dosage suppressed the interference with *HIS4* transcription caused by insertion of the Ty δ element (Clark-Adams et al., 1988). The finding that specific *spt* mutations were found in genes identified as suppressors of *swi* mutations (*SPT2/SIN1*) (Kruger and Herskowitz, 1991) or *snf* mutations (*SPT6/SSN20*) (Clark-Adams and Winston, 1987; Neigeborn et al., 1987) further suggested that *SWI* and *SNF* genes might be involved in overcoming repressive chromatin structures. Even more tellingly, deletion of *HTA1-HTB1*, one of the two copies of the gene cassettes encoding H2A and H2B, suppressed defective *SUC2* expression and the Gal$^-$ and Ino$^-$ phenotypes of *snf2*, *snf5*, and *snf6* yeast, directly linking chromatin to the *SNF* genes (Clark-Adams et al., 1988; Hirschhorn et al., 1992). Sin1/Spt2 was found to be a nuclear protein able to bind DNA in a nonsequence-specific fashion, with a modest homology to mammalian protein HMG1 (high mobility group 1), a nonhistone chromosomal protein (Kruger and Herskowitz, 1991). As little was known regarding the function of the HMG proteins, aside from their characterization as chromatin-associated proteins that could be extracted by 0.35 M NaCl, this finding provided only modest insight into Sin1 function or, correspondingly, that of the Swi proteins. It did, however, forge another connection between those proteins and chromatin. More exciting was the discovery of the identity between *SIN2* and the histone H3-encoding gene *HHT1*. *SIN2*, similarly to *SIN1*, was identified by transforming *swi1Δ* sin2-1 yeast harboring an *HO-lacZ* reporter with a genomic library and screening for colonies with decreased β-galactosidase activity (Kruger et al., 1995). A single clone was obtained, leading to the identification of the *sin2-1* mutation as *hht1-1*, which expresses histone H3 having histidine substituted for arginine at position 116 (position 115 if the initiating methionine is discounted). The mutation is semidominant to *SIN2*$^+$, as it was uncovered in the presence of a wild type copy of the gene encoding histone H3 (recall that yeast have two copies of each histone gene), and also conferred an Spt$^-$ phenotype. Additional sin mutants were identified in histone H3 and H4 by transforming *swi1Δ* yeast with plasmids harboring *HHT2* or *HHF2* mutagenized with hydroxylamine, and a T118I mutation identified by this route was also identified as a *sin* mutant in an independent study (Prelich and Winston, 1993). The *sin2* mutants that were identified exhibited only partial relief of the phenotypes resulting from deletion of *SWI1*, with *HO-lacZ* activity increasing only to ~5%–10% the level seen in *SWI1+* yeast, but this was increased to 66%–100% when *sin* mutants of histone H4 were the only source of this histone (Kruger et al., 1995; Wechser et al., 1997). Point mutations conferring partial suppression of *swi* phenotypes were clustered in the L1-L2 region of the $(H3\text{-}H4)_2$ tetramer, close to the pseudodyad axis and at sites suggested to affect interactions of H3-H4 with H2A-H2B dimers or with DNA (Fig. 6.1) (see Sidebar: More on *sin* mutants).

FIG. 6.1 Histone *sin* mutants. Four histone H3 residues altered in *sin* mutants (E105, A111, R116, and T118) are shown as red spheres overlaying yellow ribbon depiction of H3, and two histone H4 sites of *sin* mutations (V43 and R45) are shown as pink spheres overlaying blue ribbon depiction of H4. *White ribbon* represents H2A and the *light green ribbon* is H2B. PDB 1AOI structure rendered courtesy of Janice Pata.

Sidebar: More on *sin* mutants

Insight into how *sin* mutations in histones H3 and H4 could alter chromatin structure to bypass the requirement for chromatin remodeling was sought by way of structural and biochemical studies. Support for altered chromatin structure caused by *sin* mutations derived from a report that in yeast cells expressing the R45H *sin* mutant as the only source of histone H4, chromatin exhibited increased accessibility to MNase and to endogenously expressed *E. coli dam* methyltransferase (as described in Chapter 4) (Wechser et al., 1997). In vitro studies revealed that nucleosomes could be efficiently reconstituted using the *sin* mutants R116H and T118I in histone H3, but that sensitivity to MNase was increased and the rotational positioning seen with wild type histones reconstituted on 5S DNA (see Chapter 4) was disrupted in these mutants (Kurumizaka and Wolffe, 1997). Residue T118 forms a hydrogen bond with R45 of histone H4, which is inserted into the minor groove of DNA at superhelix (SHL) ± 0.5, half of a helical turn from the outward-facing minor groove at the pseudodyad, while R116 is also in close proximity to DNA. Nucleosomes reconstituted using the weaker E105K H3 *sin* mutant, in contrast to R116H and T118I, conferred no apparent differences relative to wild type H3 in MNase sensitivity or rotational positioning. X-ray crystal structures of nucleosome core particles containing the H3 R116H, T118I, or H4 R45 *sin* mutants could be superimposed on the wild type structure, indicating no significant structural alterations were induced by these mutations, but *sin* mutations H3 R116H, H4 R45C, and H4 V43I caused histone dissociation to occur at lower salt concentration than from wild type nucleosomes and increased the propensity for the histone octamer to slide relative to the nucleosomal DNA (Muthurajan et al., 2004).

A caveat with regard to the biochemical and structural investigations of histone *sin* mutants is that the mutations were engineered into recombinant *Xenopus* histones rather than histones from yeast. Yeast H3 and H4 are 84% and 92% identical to those from *Xenopus*, and nucleosome core particles reconstituted using recombinant histones corresponding to those from yeast and *Xenopus* yield very similar structures as determined by X-ray crystallography. However, yeast nucleosomes show decreased stability in the face of increased temperature or salt concentration compared to those from chicken or calf thymus, as well as a decreased constraint against thermal untwisting of DNA (Lee et al., 1982; Morse et al., 1987; Pineiro et al., 1991). It therefore remains possible that *sin* mutations engineered into yeast histones might confer biochemical or structural changes not seen in the context of *Xenopus* histones, as the in vivo tests of histone *sin* mutants suggest.

Three additional *sin* mutants were recovered in a screen for histone H3 mutants that caused derepression of the inducible *CHA1* promoter (He et al., 2008). Increased MNase sensitivity was seen at the *SWI/SNF*-dependent *ADH2* and *CHA1* promoters, but not at the nondependent *PHO5* promoter, in the H3 mutants. One of these, an A111G mutation, occurs in close proximity to T118 and is close to the nucleosome dyad. A human testis-specific variant, H3T, is characterized by substitution of valine for Ala111, and nucleosome core particles reconstituted using H3T were unstable at 0.6 M NaCl, in contrast to wild type NCPs that were stable at 0.8 M NaCl (Tachiwana et al., 2010). Val111 is not found in other H3 variants, and thus may represent a naturally occurring *sin* variant of histone H3 that serves a testis-specific function. Examination of genome-wide nucleosome occupancy using high-resolution microarrays did not indicate substantial changes in occupancy in the A111G mutant in yeast, suggesting that this mutation affects stability rather than structure (He et al., 2008). The effects of other histone *sin* mutants on genome-wide nucleosome occupancy seem not to have been examined, nor has their ability to suppress defects in chromatin remodelers other than the Swi/Snf complex in yeast or of SWI/SNF in mammalian cells been investigated to date.

The identity of *SIN2* and *HHT1* was first reported at meetings, such as the 1992 Gordon Research Conference on Nuclear Proteins, Chromatin Structure, and Gene Regulation, and in the literature as "unpublished data." The excitement generated from the reported identity even prior to its publication in 1995 stemmed from its felicitous accommodation of a model in which the *SWI* and *SNF* genes encoded the subunits of a complex that, somehow, could overcome the barrier to transcription conferred by nucleosomes; mutant alleles of *HHT1* must encode variants that altered chromatin structure in a way that allowed transcription without the need for accessory help to overcome that barrier. This model quickly found support from both in vivo and biochemical experiments.

The SWI/SNF complex as a chromatin remodeling machine

By 1990, it was well established that transcriptional activation was accompanied by alteration of chromatin structure at many genes. This was first observed as the induction of sites that were cleaved strongly by DNase I, referred to as DNase hypersensitive sites, and later discovered to correspond, in some cases, to loss or altered positioning of nucleosomes (Almer et al., 1986; Benezra et al., 1986; Elgin, 1988). Uncovering the mechanism by which such changes in chromatin structure occurred and determining whether those changes were a cause or effect of transcriptional activation were major goals of researchers. The proteins encoded by the *SWI* and *SNF* genes that acted as global regulators of transcription provided the basis for a tangible model. A logical target to test whether the *SWI/SNF* genes played a role in this "chromatin remodeling" was the *SUC2* locus, as its dependence on the *SWI/SNF* genes was thoroughly established. Fred Winston and colleagues used MNase digestion followed by indirect end-labeling to monitor the chromatin structure of the *SUC2* promoter and observed sites that were cleaved by MNase between the upstream activating sequence (UAS) and coding region under derepressing conditions (low glucose) and in deproteinized genomic DNA, but that were protected from cleavage in *snf5* and *snf2/swi2* mutants, in which *SUC2* is not induced (Fig. 6.2) (Hirschhorn et al., 1992). Deletion of the *HTA1-HTB1* locus (leaving the second copies of the genes encoding H2A and H2B intact), which suppresses the *swi/snf* phenotypes, partially restored MNase cleavage in *swi*$^-$ yeast at these sites, including two sites close to the TATA element. Moreover, cleavage by MNase was seen at these same sites in yeast in

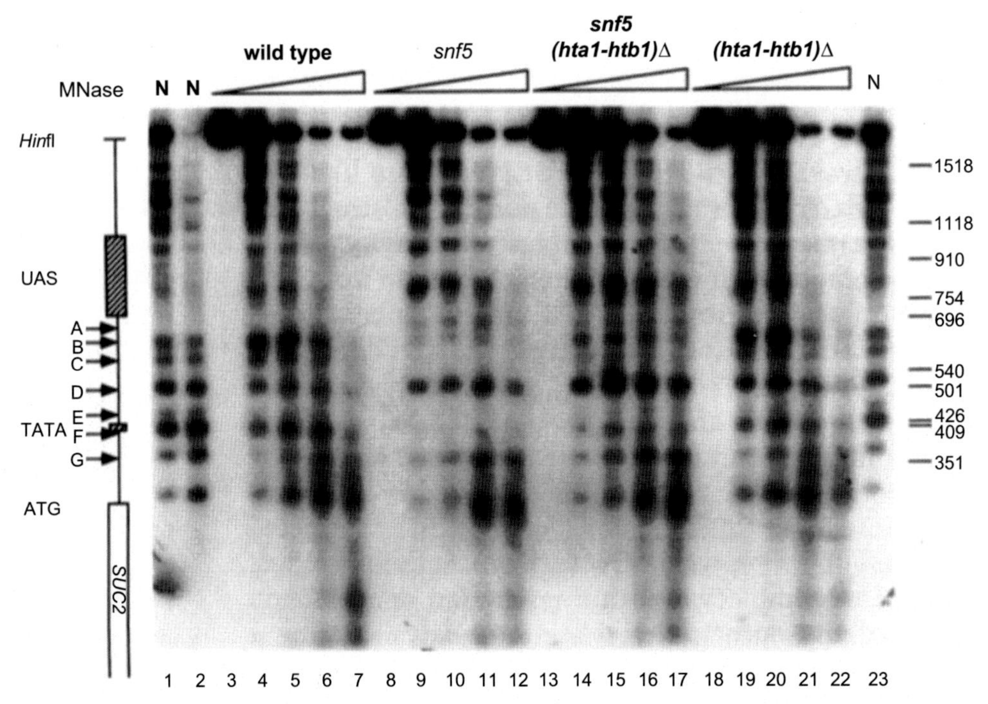

FIG. 6.2 Chromatin remodeling of the induced *SUC2* promoter requires SWI/SNF. MNase digestion of yeast grown in low glucose medium was followed by indirect end-labeling to visualize the *SUC2* promoter region. Lanes 1–2 and 23 are naked DNA; lanes 3–22 are chromatin from wild type, *snf5*, *snf5 (hta1-htb1)*Δ, and *(hta1-htb1)*Δ as indicated, with increasing MNase concentration in each sample from left (no MNase) to right. Location of size standards is shown on the right, and that of the upstream activating sequence, TATA element, and *SUC2* open reading frame on the left. Note especially the cleavage at the TATA element seen in all samples except for *snf5*. *From Hirschhorn, J.N., Brown, S.A., Clark, C.D., Winston, F., 1992. Evidence that SNF2/SWI2 and SNF5 activate transcription in yeast by altering chromatin structure. Genes Dev. 6, 2288–2298.*

which the *SUC2* TATA element was mutated, resulting in transcription being reduced to the same nearly undetectable levels seen in *snf5* mutant yeast. Protection was again observed in the *snf5* mutant, and the sites were cleaved by MNase in *snf5 (hta1-htb1)Δ* yeast. These experiments provided strong evidence that the *SWI/SNF* genes were important for altering repressive chromatin structure at *SUC2* to allow transcription, and further indicated that the observed *SWI/SNF*-dependent alteration in chromatin structure that allowed MNase cleavage preceded recruitment of TBP and thus was independent of transcription. The latter conclusion was consistent with experiments showing that the alterations in promoter chromatin structure accompanying *PHO5* activation were unaffected by mutating the TATA element to disable the transcriptional response (Fascher et al., 1993). Later work corroborated these results by revealing additional examples of genes whose activation resulted in altered chromatin structure that depended on the SWI/SNF complex, such as *PHO8* and *RNR3* (Gregory et al., 1999; Sharma et al., 2003).

While these in vivo experiments provided powerful evidence for chromatin remodeling depending, at least in some cases, on the *SWI/SNF* genes, they could not rigorously rule out indirect effects. For example, it remained possible that *swi/snf* mutations abrogated expression of some other factor that was necessary for induction and altered chromatin structure of *SUC2* and other genes. Moreover, in this pre-ChIP era, investigators lacked the technical means even to test whether SWI/SNF directly associated with its putative targets. To overcome such objections, biochemical experiments demonstrating that the products of the *SWI/SNF* genes could alter nucleosome structure or facilitate nucleosome disruption were necessary.

Several lines of evidence, including the similarity of the phenotypes observed for mutants in individual or multiple *SWI/SNF* genes, along with the observation that transcriptional activation by fusions of Swi or Snf proteins with LexA generally (with exceptions) depended on other of the *SWI/SNF* genes, suggested that the products of five *SWI/SNF* genes could act together as a complex (Laurent et al., 1991; Peterson and Herskowitz, 1992). These were not completely compelling arguments, as some *swi1*, *swi2/snf2*, *swi3*, *snf5*, and *snf6* differed substantially in their effects on activated transcription or their sensitivity to extragenic suppressor mutants, and scenarios not invoking a SWI/SNF complex could also account for observations cited as support for a multiprotein SWI/SNF complex (Laurent et al., 1991; Peterson and Herskowitz, 1992; Yoshinaga et al., 1992). Nevertheless, biochemical fractionation of yeast extracts, using Western blotting to follow the Swi/Snf proteins, revealed that Swi1, Swi2/Snf2, Swi3, Snf5, and Snf6 copurified as a large (~2 MDa) complex that included at least four additional proteins (Cairns et al., 1994; Peterson et al., 1994).

The isolation of the yeast SWI/SNF complex allowed its effect on reconstituted chromatin templates to be tested in vitro. Initial experiments focused on binding of derivatives of the activator Gal4 to a reconstituted nucleosome (Cote et al., 1994). Binding of Gal4-AH, a shortened derivative of Gal4 (more readily produced for in vitro experiments than the 881 aa native Gal4) to two nucleosome-length DNA templates (153 and 154 bp) was monitored by gel shift and DNase I footprinting (Fig. 6.3). The templates differed in the location of the Gal4-binding site, at 32 bp or 43 bp from the fragment's end, and one of the templates exhibited strong rotational positioning of the reconstituted nucleosome. In both cases, binding to the nucleosomal templates was seen only at very high concentrations of Gal4-AH, on

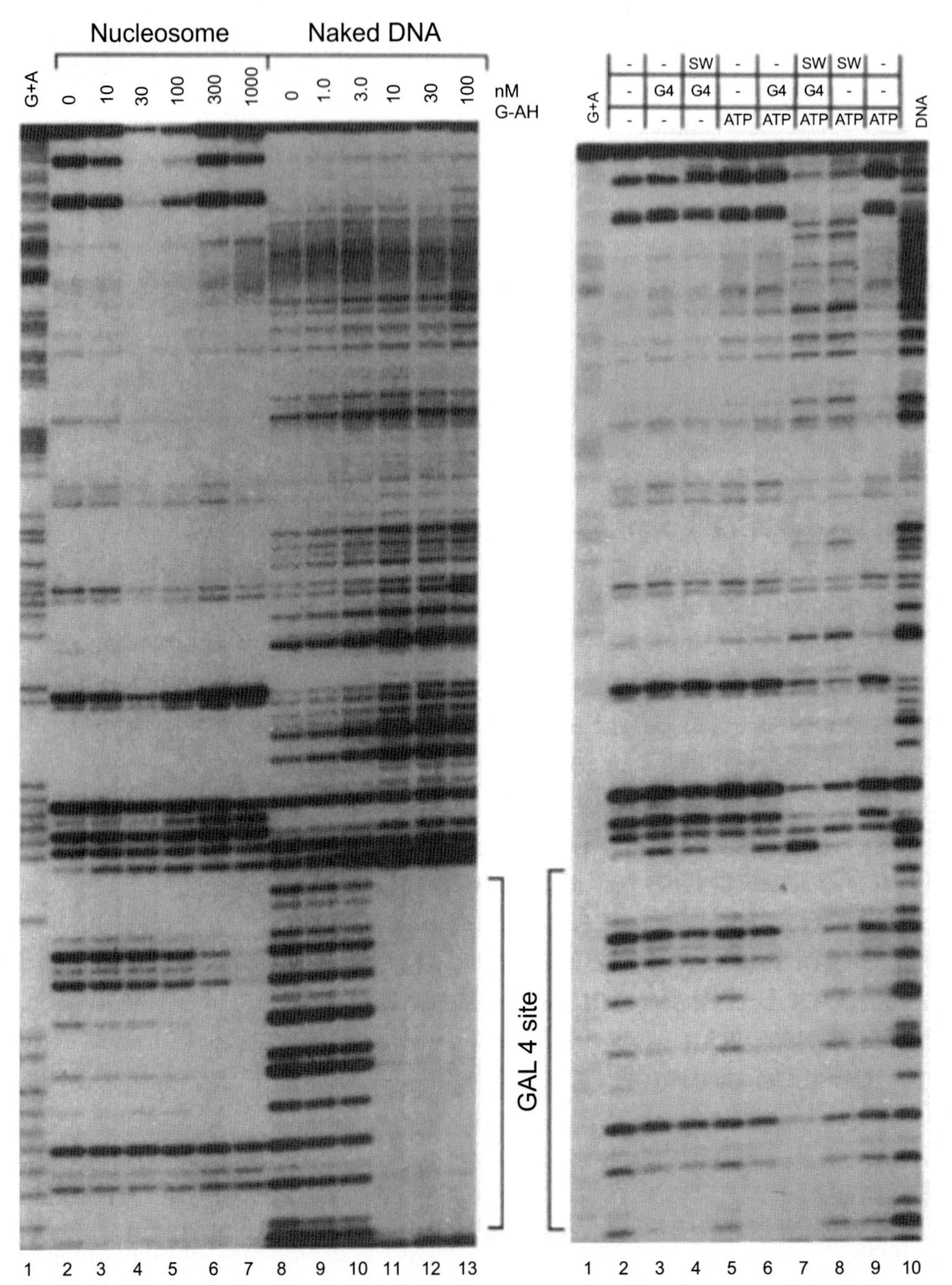

FIG. 6.3 Binding of Gal4-AH to a nucleosomal site enhanced by SWI/SNF in vitro. Left panel: DNase I digestion of end-labeled naked DNA and of a reconstituted nucleosome, as indicated, in the presence of increasing amounts of Gal4-AH from left to right; leftmost lane is G + A reactions. Note the 10 bp periodicity of cleavages in the nucleosome, indicating rotational positioning, and the protection over the Gal4-binding site seen at lower levels of Gal4-AH in naked DNA than in nucleosomal DNA. Right panel: DNase digestion of naked DNA (lane 10) and of the same nucleosomal DNA as in the left panel in the presence or absence of SWI/SNF, Gal4-AH, and ATP, as indicated at the top. Note the strong protection over the Gal4-binding site seen only in the presence of all three components (lane 7), and the altered cleavage (partial loss of protection) seen in the presence of SWI/SNF and ATP (lane 8). *From Cote, J., Quinn, J., Workman, J.L., Peterson, C.L., 1994. Stimulation of GAL4 derivative binding to nucleosomal DNA by the yeast SWI/SNF complex. Science 265, 53–60.*

the order of 100-fold higher than required for binding to the naked DNA templates. Inclusion of SWI/SNF in the binding reaction stimulated Gal4-AH binding 10–30-fold, and this was also seen using a Gal4 derivative lacking an activation domain. Contemporaneously, a similar approach indicated that a partially purified mammalian SWI/SNF complex, later more stringently purified, was also able to facilitate binding of Gal4 derivatives as well as TBP to nucleosomal templates in an ATP-dependent fashion, as described below (see "SWI/SNF in other species" section) (Imbalzano et al., 1994; Kwon et al., 1994; Wang et al., 1996a).

The addition of yeast SWI/SNF without Gal4 induced marked alterations in histone-DNA contacts in the template on which the nucleosome was strongly positioned, as seen by disruption of the characteristic 10bp periodic DNase cleavage pattern. Importantly, this disruption, as well as the facilitated binding of Gal4-AH, depended on ATP hydrolysis. Snf2/Swi2 had been shown to possess motifs resembling those found in DNA-dependent ATPases and DNA helicases, and a mutation in the Walker A motif (a phosphate-binding motif [GXGKT] common to many ATPases (Walker et al., 1982)) that eliminated ATPase activity in vitro and abrogated function in vivo also prevented SWI/SNF from facilitating the binding of Gal4-AH to a nucleosomal template in vitro (Cairns et al., 1994; Cote et al., 1994; Laurent et al., 1993). The resemblance of this region of Snf2/Swi2 to known helicases raised the possibility that it might function by translocating along the DNA in chromatin and displacing histones in the process. However, biochemical assays showed that both bacterially expressed Snf2/Swi2 and the intact SWI/SNF complex lacked helicase activity but possessed potent ATPase activity that was stimulated by double-stranded or nucleosomal DNA (Cote et al., 1994; Laurent et al., 1993). Together with additional evidence, these results established the SWI/SNF complex as a "molecular machine" that could bind to a nucleosome and use the energy of ATP hydrolysis to alter nucleosome structure in a way that facilitated binding of a sequence-specific DNA-binding protein.

More remodeling machines

The SWI/SNF complex turned out to be the founding member (from a historical, not evolutionary, perspective) of a family of ATP-utilizing chromatin remodelers. Additional remodelers were discovered by various routes. At the time of discovery and initial characterization of SWI/SNF, the revolution in molecular biology had reached a phase of early maturity in which the identification, cloning, and sequencing of mutant yeast genes was becoming relatively commonplace. In the early to mid-1980s, cross-hybridization using Southern blots was used to aid cloning of the highly conserved histone genes across species; by 1990 the method was being routinely employed to identify more distantly related homologs of cloned genes by using hybridization conditions with relaxed stringency. These technical advances, coupled with increasing amounts of inferred protein sequence data and expanding availability of computational resources, allowed identification and analysis of protein homologs both across and within species, which contributed greatly to the discovery of relatives of the SWI/SNF complex.

SWI/SNF in other species

Notwithstanding the utility of cross-hybridization to identify homologs of SWI/SNF subunits in other species, the *Drosophila* brahma protein was discovered as a functional counterpart of Snf2/Swi2 by a more circuitous route. The discovery of brahma as a chromatin remodeling protein began with a genetic screen for suppressors of *Polycomb* mutations that cause dysregulation of homeotic genes, leading to developmental defects such as transformations of antenna to leg (the famous *Antennapedia*, or *Antp*, mutation) (Kennison and Tamkun, 1988). Cloning and sequencing of one of these suppressors, named *brahma* or *brm*, revealed four regions of high sequence similarity of the encoded protein (40%–55%) to the yeast Snf2/Swi2 subunit (Tamkun et al., 1992). This sequence similarity was noteworthy inasmuch as there were functional parallels as well; just as the yeast SWI/SNF complex appeared to function to overcome repressive effects due to chromatin, brahma evidently counteracted the repressive action of *Polycomb*, which was also believed to confer repression through a chromatin-mediated mechanism. (*Polycomb*-mediated repression is discussed at length in Chapter 8.) Cross-hybridization experiments indicated that *brm* was the most closely related *Drosophila* gene to *SNF2/SWI2* (Elfring et al., 1994). This functional similarity was underscored by the demonstration that a chimeric protein in which the ATPase domain of brahma was substituted for that of Snf2/Swi2 partially restored SWI/SNF function in *swi2*$^{-}$ yeast.

Searches for human homologs of *brm* resulted in identification of two related genes named *hbrm* (human *brm*) and *BRG1* (*brm*-related gene 1) (Khavari et al., 1993; Muchardt and Yaniv, 1993). The ATPase domain from BRG1, like that from brahma, could functionally substitute for that of Snf2/Swi2 in yeast, and this complementation was abrogated by mutation of the lysine residue corresponding to K798 in the Snf2/Swi2 Walker A motif. BRG1 carrying this same mutation functioned as a dominant negative mutant in mammalian cells, as its expression reduced transcription in a promoter-specific fashion; similarly, *hbrm* expression in cell lines having low endogenous levels of BRM enhanced expression from a GR-dependent promoter but not from the cytomegalovirus promoter. Using an anti-BRG1 antibody, two large, multisubunit complexes were partially purified from HeLa cell nuclear extracts and shown by DNase I footprinting experiments to alter nucleosome structure and facilitate binding of Gal4 and TATA-binding protein to nucleosomal templates in an ATP-dependent fashion (Imbalzano et al., 1994; Kwon et al., 1994). These results were corroborated by a more stringent purification, eliminating the possibility that a contaminating activity such as another remodeler could have produced the observed nucleosome disruption (Wang et al., 1996a). This more stringent purification, relying on conventional biochemical fractionation together with immunoaffinity methods, led to identification of several BRG1-associated factors, or BAFs, from mammalian cells that together with BRG1 or BRM (now also known as SMARCA4 and SMARCA2, respectively) comprise the mammalian versions of SWI/SNF (Wang et al., 1996a,b). These BAFs included homologs to the yeast Swi3 and Snf5 subunits as well as additional subunits unique to mammalian cells. BRG1 and BRM were found to reside in distinct BAF complexes, as they came to be known, with similar subunit composition; purification of complexes from different cell lines revealed distinct subunit compositions. This heterogeneity in complex composition in some cases reflected tissue-specific expression of subunit homologs. For example, while Baf60a (homologous to the yeast SWI/SNF subunit Swp73, also known as Swi12) was widely expressed across tissue types, Baf60b and Baf60c were preferentially found in muscle and pancreatic cells, respectively. These results suggested the existence in mammals of a multiplicity of chromatin remodeling complexes related to the SWI/SNF complex that were adapted to the needs of specific cell types, and this was confirmed by later discoveries of specialized BAF complexes that play essential roles in differentiation into and maintenance of specific cell types (Ho and Crabtree, 2010). Cell-type specific BAF complexes have been found in cardiac cells, neural progenitors, and embryonic stem cells, among others (Ho et al., 2009; Lessard et al., 2007; Lickert et al., 2004).

Continued research allowed categorization of BAF complexes into three types: canonical BAF, or cBAF; polybromo-associated BAF, or PBAF; and noncanonical BAF, or ncBAF, also known as GBAF for its incorporation of GLTSCR1 or GLTSCR1L (Alpsoy and Dykhuizen, 2018; Kaeser et al., 2008; Mashtalir et al., 2018; Nie et al., 2000). These closely related complexes are distinguished by their subunit composition (Fig. 6.4) and differ in their genome-wide localization, their specific function in the cell, and in how their targeting and remodeling activities are affected by the presence of histone modifications and histone variants in the nucleosome (Mashtalir et al., 2021; Michel et al., 2018; Pan et al., 2018; Varga et al., 2021). Most of the various BAF complexes can assemble with either the BRG1 or BRM motor subunits, adding further diversity; the functional implications of BAF complex diversity remain to be fully explored.

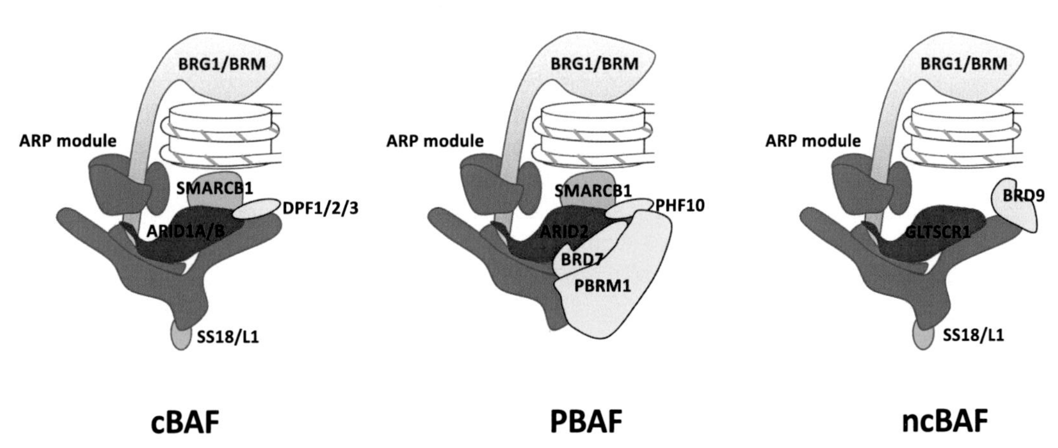

FIG. 6.4 Structures of the three principal BAF subtypes, canonical BAF, polybromo-associated BAF, and noncanonical BAF (also known as GBAF). Shared subunits, not all indicated by name here, are shown in *blue*; subunits unique to one BAF subtype are shown in *yellow*; and subunits common to two of the three subtypes are shown in *green*. The ARID1A/B, ARID2, and GLTSCR1 or GLTSCR1L subunits, colored *red*, are important branch points in BAF assembly, dictating incorporation of subtype-defining subunits.

BAF mutations and cancer

BAF subunits have been discovered to act as tumor suppressors, with the encoding genes found to be mutated in >20% of human tumor samples, implicating aberrant chromatin remodeling as predisposing to or supporting the cancerous phenotype (Centore et al., 2020; Kadoch et al., 2013). The first association of a human SWI/SNF mutation with cancer came from a study of malignant rhabdoid tumors (MRTs), a highly aggressive pediatric cancer that was known to be accompanied by deletion of a part of the long arm of Chr 22 (Versteege et al., 1998). Mapping of the deleted region from 13 patient-derived cell lines led to identification of bi-allelic loss-of-function mutations in the gene encoding hSNF5, also known as INI1/BAF47/SMARCB1. This discovery was corroborated by the finding that 100% of 29 rhabdoid tumors examined carried *ini1* mutations (Biegel et al., 1999). In mice, homozygous loss of *SNF5* was embryonic lethal, while 100% of heterozygotes developed tumors having characteristics of MRT (Roberts et al., 2000). In line with the apparent sufficiency for MRT genesis of loss of hSNF5 function, genome-wide SNP analysis and exome sequencing indicated that MRTs were not characterized by genome rearrangements or, indeed, additional mutations (Hasselblatt et al., 2013; Lee et al., 2012). Genome-wide analysis in cell lines derived from MRTs revealed decreased occupancy at enhancers and promoters by hSWI/SNF, resulting in defective gene expression (Nakayama et al., 2017; Wang et al., 2017). The reason for the remarkable etiological specificity of hSNF5 mutation is clearly a complex question and remains to be explained. Since the discovery of hSNF5 as a tumor suppressor, multiple additional hSWI/SNF subunits have been found to be mutated in various types of cancer (Helming et al., 2014; Kadoch and Crabtree, 2013); there is hope that knowledge of these mutations and the mechanisms by which they contribute to predisposition to or maintenance of the cancerous state may eventually allow novel therapeutic approaches (Centore et al., 2020; Varga et al., 2021).

RSC: A close cousin of the SWI/SNF complex

Even as the SWI/SNF complex was undergoing intensive characterization, several observations suggested that it was not alone in its role as a chromatin remodeler. First, although chromatin remodeling accompanying transcriptional activation of the *SUC2* promoter was shown to depend on SWI/SNF, the *PHO5* promoter, which served as paradigm for chromatin remodeling during gene activation, did not depend on SWI/SNF for its activation nor for accompanying nucleosome loss at its promoter (Gaudreau et al., 1997). Second, contemporaneous research using biochemical approaches to identify activities capable of altering chromatin structure was leading to discovery of chromatin remodeling activities distinct from SWI/SNF, as described below. Third, reports of homologs of SWI/SNF complex subunits pointed to the possibility of related complexes.

One of the first homologs of SWI/SNF subunits reported was *STH1*, which was identified by cross-hybridization to a *SNF2* probe in a Southern blot and, unlike the subunits of SWI/SNF, was shown to be essential for viability (Laurent et al., 1992). Following the sequencing of the complete yeast genome, a BLAST search revealed homologs to four additional SWI/SNF subunits, three of which were determined to be essential (Cairns et al., 1996). Biochemical fractionation of yeast extracts, monitored using antibodies to Sth1 and another SWI/SNF subunit homolog, led to purification of the RSC, or Remodels the Structure of Chromatin, complex. Six subunits of the complex were found to be identical to or have homologs in the SWI/SNF complex, and as its name implies, RSC altered nucleosome structure in an ATP-dependent manner, as shown by changes in the DNase digestion pattern (Cairns et al., 1996, 1998). In contrast to SWI/SNF, RSC is essential (as already mentioned for specific subunits) and is far more abundant than SWI/SNF: whereas SWI/SNF was estimated to be present at ~100–200 copies per haploid yeast cell, RSC was determined to be at least 10-fold more abundant.

Another distinguishing feature of RSC is that it exists as two distinct complexes differing only by which of the paralogs Rsc1 and Rsc2 it contains (Cairns et al., 1999). Yeast cells lacking either Rsc1 or Rsc2 are viable and exhibit phenotypic differences, whereas the double deletion is inviable, showing that the RSC1 and RSC2 complexes perform both overlapping and distinct functions in the cell. Recent work suggests that the two RSC complexes differ in their substrate specificity (Schlichter et al., 2020). RSC1 was better able to facilitate sliding of a nucleosome reconstituted with the 5S rDNA nucleosome positioning sequence (that is, movement of the histone octamer relative to the incorporated DNA sequence) than RSC2, and histone mutations that favor unwrapping of the outermost DNA of the nucleosome were synthetically lethal with *rsc1Δ* yeast mutants, but not with *rsc2Δ*. Reconstituted nucleosomes harboring a histone mutation, H3 R40A, predicted to confer partial unwrapping showed reduced remodeling by RSC2 compared to RSC1. The "fragile" nucleosomes discussed in Chapter 4 have been reported to be partially unwrapped and are associated with RSC (Brahma and Henikoff, 2019). RSC facilitates directional sliding of the −1 and +1 nucleosomes to widen the NDR, as also discussed in Chapter 4. The Rsc1/Rsc2 paralogs, which arose during the whole genome duplication that occurred in *S. cerevisiae* ~150 million years ago, may have evolved distinct mechanisms to remodel nucleosomes prone to partial unwrapping: RSC1 possesses intrinsically superior ability to slide partially unwrapped nucleosomes, while RSC2 appears to work in conjunction with the nucleosome-binding Hmo1 protein, which stabilizes "fragile," partially unwrapped nucleosomes to allow remodeling (Schlichter et al., 2020).

An expanded family of remodelers

Research conducted contemporaneously with the discovery and initial characterization of the SWI/SNF complex led to the identification of additional chromatin remodeling machines. A large body of research has allowed remodelers to be categorized as belonging to one of four large subfamilies: SWI/SNF, ISWI (imitation switch), CHD (chromodomain helicase DNA-binding), and INO80 (Fig. 6.5). These remodeling complexes all have ATPase subunits belonging to the RNA/DNA helicase superfamily 2 (hence the designation of "subfamilies")

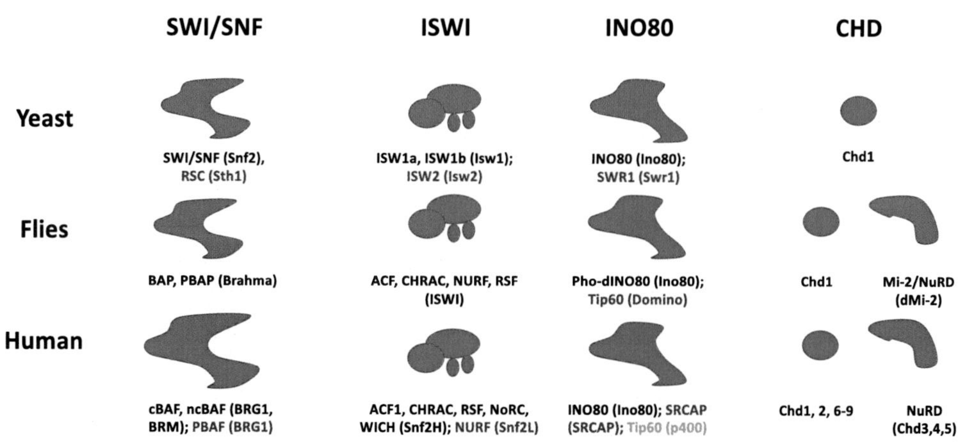

FIG. 6.5 Diversity of chromatin remodeling complexes in budding yeast, *Drosophila*, and humans. The ATPase "motor" subunit in each complex is indicated in parentheses; complexes containing alternative ATPases in a given subfamily are shown in *red font*. SWI/SNF, INO80, and Mi-2/NuRD complexes contain 6–15 subunits, while ISWI complexes all contain 2–4 subunits including the ATPase.

but are distinguished by the organization of specific motifs within the ATPase subunit (Fig. 6.6) (Flaus et al., 2006).

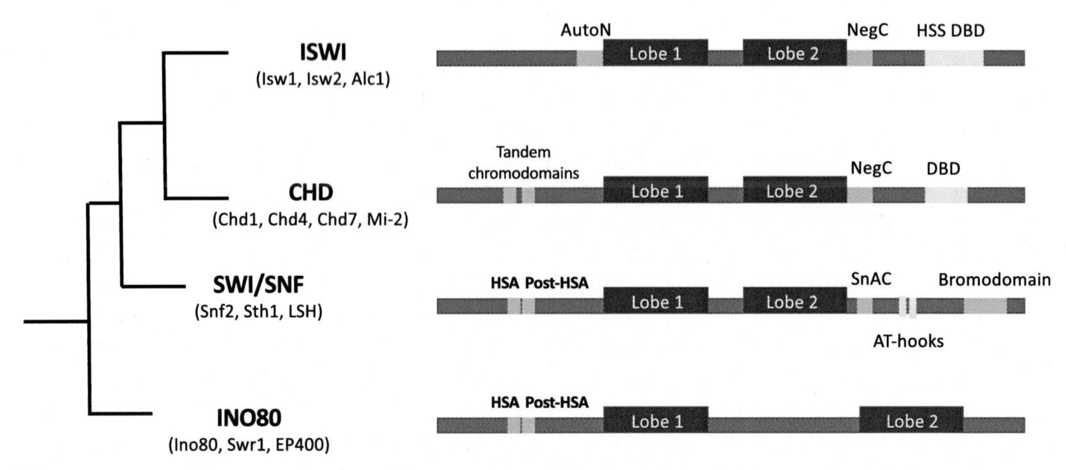

FIG. 6.6 Structure and evolutionary relationship of the ATPase subunits of the principal chromatin remodeling complexes. Left, evolutionary tree of the group from the SF2 superfamily of RNA/DNA helicases giving rise to the main chromatin remodelers (see Flaus et al., 2006). Right, domain organization of the indicated ATPases. The RecA-like lobes 1 and 2 are located in the translocase-ATPase domain, which is sufficient to carry out DNA translocation. The two lobes are separated by an insertion that is extended in the Ino80 family. *Yellow* marks DNA-binding domains (DBD), including the helicase-SANT-slide (HSS) domain. *Green* marks regulatory domains, including the helicase-SANT-associated (HSA) domain, the Snf2 ATP coupling (SnAC) domain, and the AutoN and NegC negative regulatory domains whose repressive effect is alleviated upon binding to the nucleosome. Unique to SWI/SNF members is a bromodomain with affinity for acetylated histones. For additional detail, see text and Clapier et al., 2017.

Within these subfamilies, multiple remodeler subtypes, defined by variability both in the ATPase (e.g., BRG1 vs BRM) and associated subunits, are found in many higher eukaryotes that have specific functions relating to cell type or development. How the first examples of these additional remodelers were discovered, the mechanisms by which they alter nucleosome structure, and the functions they serve in vivo are summarized in the following sections. Some chromatin remodelers in addition to those described below, such as Fun30, ddm1, and Alc1, among others, have been studied less extensively (Ahel et al., 2009; Awad et al., 2010; Gottschalk et al., 2009) and are not discussed further here.

Remodelers have not, to date, been found in archaea, despite the presence of histones and nucleosome-like structures, as described in Chapter 5. Luger and colleagues have carried out cryo-EM structural determinations and molecular dynamics simulations that indicate that the extended archaeal histone-DNA "hypernucleosomes," also referred to as "archaeasomes," undergo sporadic stretching motions akin to a "slinky" toy, or spiral accordion, that allows increased accessibility to DNA without disrupting histone-DNA contacts (Bowerman et al., 2021). The dynamics of archaeal chromatin may thus circumvent the necessity for chromatin remodelers, although it remains possible that they simply have yet to be discovered.

ISWI

In contrast to the genetic screens that led to the identification of the SWI/SNF complex, the first members of the ISWI family of chromatin remodelers were discovered through biochemical experiments. The use of extracts from *Xenopus* eggs and *Drosophila* embryos to assemble chromatin with physiological spacing between nucleosomes and the discovery of histone chaperones in such extracts were described in Chapters 4 and 5. These same extracts also featured in the discovery of ISWI-based complexes as chromatin remodelers. Working independently, three laboratories sought to determine whether *Drosophila* embryo extracts harbored activities that could facilitate changes in chromatin structure and increase accessibility to nucleosomal DNA. The Wu and Becker groups studied chromatin remodeling in the context of the *Drosophila hsp70* and *hsp26* genes, respectively, both of which had been well characterized in terms of chromatin structure alterations accompanying their induction by heat shock in vivo, while the Kadonaga lab employed an artificial template containing binding sites for the widely studied Gal4 activator protein. Using digestion by MNase or DNase I to monitor changes in chromatin structure facilitated by embryo extracts, remodeling of the *hsp26* and *hsp70* templates was induced upon addition of *Drosophila* GAGA factor or heat shock factor (HSF), and upon addition of Gal4 for the artificial template (Pazin et al., 1994; Tsukiyama et al., 1994; Varga-Weisz et al., 1995; Wall et al., 1995). Chromatin remodeling in all cases depended on ATP hydrolysis; moreover, accessibility to restriction enzymes without addition of GAGA factor or HSF was enhanced in the presence of ATP but not by nonhydrolyzable analogs, and altered nucleosome spacing induced by a change in salt concentration also required ATP (Varga-Weisz et al., 1995). Altogether, these results indicated activities present in the extracts that could remodel chromatin in an energy-requiring process; the altered spacing seen with GAGA or HSF addition suggested remodeling could occur at least in some cases in a nontargeted fashion and could reflect nucleosome sliding, in which histones moved along the DNA template to occupy different positions.

The discovery of ATP-dependent chromatin remodeling activities in the *Drosophila* extracts prompted efforts to purify the relevant factors or complexes. This required an assay to follow the activity through the steps of biochemical purification. Two approaches were taken. In one, treatment of the extracts with the detergent Sarkosyl following chromatin assembly was found to prevent GAGA-mediated nucleosome disruption; re-addition of extract after washout of detergent restored GAGA-mediated, ATP-dependent chromatin remodeling (Tsukiyama and Wu, 1995). Following this protocol, biochemical fractionation of *Drosophila* embryo extracts was thus accompanied by assays for chromatin remodeling using either GAGA-mediated nucleosome disruption, monitored by MNase digestion, or increased accessibility to restriction enzymes, resulting in purification of nucleosome remodeling factor (NURF) and chromatin accessibility complex (CHRAC) (Tsukiyama and Wu, 1995; Varga-Weisz et al., 1997). A distinct approach was taken by the Kadonaga lab, which focused on the ATP-dependent chromatin assembly properties of the *Drosophila* extract, using supercoiling and MNase assays described in Chapter 5 to purify a complex they designated as ACF (ATP-utilizing chromatin assembly and remodeling factor) (Ito et al., 1997). As the name implies, this complex possessed chromatin assembly activity, when supplemented with either the CAF-1 or NAP-1 chaperones, and was also able to facilitate binding of Gal4 to a chromatin template, with both processes depending on ATP.

ISWI was first identified as a component of NURF, and subsequently found also in ACF and CHRAC (Ito et al., 1997; Tsukiyama et al., 1995; Varga-Weisz et al., 1997). *ISWI*, for Imitation Switch, had been identified as a weaker homolog to the yeast *SWI2* gene in the search that identified *brm* as the closest *Drosophila* homolog (Elfring et al., 1994). While it was clear that *ISWI* encoded a protein that belonged to the Snf2/Swi2 family of ATPases, its ATPase domain could not confer function to Snf2/Swi2, in contrast to that of Brahma. Moreover, a human homolog, hSNF2L, that was identified in a sequencing effort directed at chromosomal translocations (Okabe et al., 1992), exhibited homology to ISWI along its entire length, whereas the homology to Snf2/Swi2 and Brahma was limited to the ATPase domain (Fig. 6.6). These findings along with identification of additional Snf2/Swi2 relatives in *C. elegans*, *Arabidopsis*, and *S. cerevisiae* suggested the existence of multiple chromatin remodelers, possibly serving a variety of functions.

Despite having similar properties and a common derivation from *Drosophila* embryo extracts, NURF displayed some functional differences compared to ACF and CHRAC, and this was corroborated by their distinctive subunit compositions (Fig. 6.5). While ACF and CHRAC were found to differ from each other only by the presence of two additional subunits in CHRAC, corresponding to proteins that are expressed only in early embryogenesis, NURF contains three subunits in addition to ISWI, none of which are present in ACF/CHRAC (Becker and Workman, 2013). How the different subunits associated with ISWI in NURF compared to ACF/CHRAC confer mechanistic differences in nucleosome remodeling remains to be determined.

The availability of the complete yeast genome sequence enabled a search for ISWI homologs in *S. cerevisiae*, leading to the identification of Isw1 and Isw2 (Tsukiyama et al., 1999). Isw1 is present in two distinct complexes, Isw1a and Isw1b, both of which differ in associated subunits from the Isw2 complex (Vary Jr. et al., 2003). The three yeast ISWI complexes behave differently with respect to chromatin remodeling, nucleosome spacing, and in vivo function, and the activities of both Isw1- and Isw2-containing complexes are lost in mutants analogous to the Snf2/Swi2 K798A mutation that inactivates SWI/SNF ATPase activity and remodeling capability. Subsequent studies revealed an array of ISWI complexes across eukaryotes, with roles in transcriptional initiation and elongation, replication, nucleosome assembly, and centromere maintenance (Clapier and Cairns, 2009; Yadon and Tsukiyama, 2011).

INO80

INO80 was first identified in a screen for *S. cerevisiae* mutants defective in transcriptional activation from a promoter element that binds the Ino2/Ino4 heterodimer, necessary for expression of genes required for utilization of inositol (Ebbert et al., 1999). Cloning revealed Ino80 to have regions of homology to the Snf2/Swi2 ATPase family, and analysis of gene expression in *ino80* yeast revealed it to contribute to expression of multiple genes that did not depend on Ino2/Ino4. Expression of Ino80 with a mutated ATPase motif failed to complement an *ino80* mutant, and the cohort of genes exhibiting Ino80 dependence overlapped with but was distinct from genes that depended on Snf2/Swi2. Working independently, Carl Wu's laboratory identified *INO80* in a search of the *S. cerevisiae* genome for genes similar to the *Drosophila ISWI* gene (Shen et al., 2000). Their results corroborated the earlier study; in addition, a complex of about 12 polypeptides (eventually, 15) containing Ino80 was purified (initially named "Ino80.com," perhaps in a nod to the "dot com" era of the late 1990s) and found to contain actin and three actin-related proteins, or "ARPs" (see Sidebar: Arps and remodelers). In addition, two proteins related to the bacterial helicase RuvB, Rvb1 and Rvb2, were present in the complex, and in contrast to SWI/SNF or the *Drosophila* NURF complex, the INO80 complex was found to display helicase activity, as shown by its ability to displace a single-stranded DNA oligomer from a complementary strand. The ability of the INO80 complex to remodel chromatin structure was tested using an in vitro assay, in which the artificial activator Gal4-VP16 stimulated transcription from a chromatinized template. Both the INO80 complex and NURF stimulated transcription more than 10-fold, although the effect was not measured for a naked DNA template and dependence on ATP was not reported. Both complexes also modestly (~twofold) enhanced cleavage by a restriction enzyme (BamHI) at a site between the TATA element and the Gal4-binding site of the chromatin template.

Sidebar: Arps and remodelers

While actin was long known and much-studied for its structural and mechanical roles in the cytoplasm, evidence for its presence and eventually for its functional roles in the cell nucleus began to accumulate in the 1990s (Rando et al., 2000). Following the sequencing of the yeast genome, homology searches revealed a family of actin-related proteins, or Arps (Poch and Winsor, 1997). While the three Arps most closely related to actin, Arp1-3, as well as the more distantly related Arp10, function in the cytoplasm, Arp4-9 as well actin itself were found to be integral subunits of both chromatin remodeling and histone modification complexes. Two of these, Arp7 and Arp9, were first found to be present both in RSC and SWI/SNF complexes (Cairns et al., 1998; Peterson et al., 1998), while the human SWI/SNF complex was also shown to contain the Arp4 ortholog Baf73 as well as β-actin (also sometimes referred to as nuclear actin, or N-actin) (Zhao et al., 1998). Subsequent studies revealed specific association of Arps and actin with remodelers: SWI/SNF and RSC contain an Arp7/9 heterodimer, SWR1 contains Arp4 and Arp6 as well as actin, while INO80 contains Arp4, 5, and 8 together with actin (Cairns et al., 1998; Mizuguchi et al., 2004; Peterson et al., 1998; Shen et al., 2000). The human TIP60 complex, which possesses both remodeling and histone acetylation activities, also contains Baf73 and β-actin, and the histone acetylation complex NuA4 contains Arp4 and actin (Galarneau et al., 2000; Ikura et al., 2000). Biochemical experiments revealed Arps to interact with the motor proteins of remodelers via their helicase-SANT-associated domains (Fig. 6.6) (Szerlong et al., 2008; Wu et al., 2005; Yang et al., 2007; Zhao et al., 1998), and genetic experiments were generally consistent with Arps functioning in chromatin remodeling (Cairns et al., 1998; Shen et al., 2003; Szerlong et al., 2003). In mammals, the ARP module (Fig. 6.4) is important for cellular differentiation and development, and mutations in Baf73 have been linked to cancer (Centore et al., 2020; Ronan et al., 2013). High-resolution cryo-EM structures allow precise placement of Arps and actin in remodeler complexes and suggest a common mechanistic role, in which the ARP module links the ATPase motor with subunits that contact DNA, thereby coupling ATP hydrolysis with translocation and the building up of torsional strain in the DNA (Jungblut et al., 2020). Consistent with a regulatory role for Arps in coupling the enzymatic activity of remodelers with binding to extranucleosomal DNA, the INO80 ARP module has been demonstrated to be important for nucleosome spacing on a genome-wide basis (Oberbeckmann et al., 2021).

As with SWI/SNF, the INO80 complex is not required for yeast viability, and expression of multiple genes is affected by its loss, as mentioned above. However, *ino80Δ* yeast exhibited increased sensitivity to DNA-damaging agents, whereas *swi/snf* mutants did not, and INO80 complexes were later shown to be important for double-strand break (DSB) repair in plants and mammals (Fritsch et al., 2004; Shen et al., 2000; Wu et al., 2007). The role of the INO80 complex in DNA repair in yeast was supported by experiments showing that deletion of the Arp5 or Arp8 subunits similarly resulted in hypersensitivity to DNA damaging agents; indirect effects such as deficiency in checkpoint response or reduced expression of genes important for DNA damage response were examined and found to be unlikely to account for the increased sensitivity (Morrison et al., 2004; van Attikum et al., 2004). ChIP experiments demonstrated that the INO80 complex associated with engineered DSBs, that this association depended on γ-H2A.X present at DSBs (discussed in Chapter 5), and that the interaction with γ-H2A.X depended on specific INO80 complex subunits, namely Nhp10 and Arp4 (Downs et al., 2004; Morrison et al., 2004; van Attikum et al., 2004). Nucleosome eviction in the vicinity of a DSB accompanies repair, as assayed by MNase cleavage and ChIP assays, and eviction depends on the INO80 complex (Tsukuda et al., 2005; van Attikum et al., 2007). Although these findings suggested a direct role for the INO80 complex in removing nucleosomes at sites of DSB repair, it was recently reported that simultaneous depletion of SWI/SNF and RSC prevented nucleosome eviction at DSBs, indicating redundant roles for these remodelers in repair as well (Peritore et al., 2021). As INO80 has been implicated in multiple steps in DSB repair, it is possible that its requirement for nucleosome eviction occurs prior to engagement of SWI/SNF and RSC. Numerous studies conducted since the early work implicating the INO80 complex in DNA repair have addressed the mechanisms by which remodelers assist in the various avenues by which DNA damage is repaired and the roles of distinct remodelers in yeast and metazoans, and while much progress has been made, many questions remain (Karl et al., 2021).

Closely related to Ino80 is its paralog, Swr1. The yeast SWR1 complex, discussed in Chapter 5, shares four subunits with the INO80 complex but functions to replace canonical H2A in nucleosomes with the H2A.Z variant by H2A-H2B dimer exchange. Corresponding complexes in flies and mammals perform the same exchange function (Ruhl et al., 2006; Scacchetti et al., 2020). The converse exchange reaction, substitution of H2A-H2B dimers for H2A.Z-H2B dimers in nucleosomes, was reported to be catalyzed by the INO80 complex, but this proved to be controversial (Au-Yeung and Horvath, 2018; Jeronimo et al., 2015; Papamichos-Chronakis et al., 2011; Tramantano et al., 2016; Wang et al., 2016; Watanabe and Peterson, 2016). Recently, single particle tracking was used in conjunction with rapid depletion, using the anchor away protocol, of candidate exchange factors to determine the responsible factor(s) for replacement of nucleosomal H2A.Z with H2A in yeast (Ranjan et al., 2020). Depletion of Ino80 together with the SWR1 subunit Swc5 did not result in any increase in DNA-bound H2A.Z, while a moderate increase in bound H2A.Z was seen upon depletion of Swc5 together with Rbp1, the large subunit of RNA polymerase II. These results suggest that Pol II passage through +1 nucleosomes, which are enriched in H2A.Z, may be the major source of H2A.Z replacement with H2A. This is also consistent with earlier work indicating that assembly of the preinitiation complex is important for this exchange (Tramantano et al., 2016). However, given the complexity of this approach and the challenges to earlier findings, it seems premature to decide the matter settled.

Chd1 and the Mi-2/NuRD complexes

Discovery of the CHD family of chromatin remodelers had its genesis in part as an offshoot of the first large-scale genome sequencing efforts. Robert ("Bob") Perry was in the last few years of a distinguished scientific career, spent almost entirely at the Fox Chase Cancer Center in Philadelphia, when his laboratory serendipitously cloned the gene for the murine protein they named CHD-1, for "chromodomain-helicase-DNA-binding protein 1" (Delmas et al., 1993; Schibler et al., 2013). Seeking to clone κY, a DNA-binding protein involved in expression of some immunoglobulin genes, they instead obtained a clone with distinct DNA-binding properties that also, upon sequencing, was found to contain a chromodomain and a domain homologous to the Snf2/Swi2 helicase/nucleotide-binding domain. The chromodomain comprised a 30 amino acid sequence that had been discovered as being highly conserved between the *Drosophila* Polycomb and heterochromatin protein 1 (HP1) proteins (Paro and Hogness, 1991). The simultaneous presence of these domains, together with identification of multiple encoding mRNA species that varied in different cell types and the use of a "zoo blot" to show conservation of Chd1 across mammalian species, suggested a widespread role for Chd1 in transcriptional regulation. Further characterization revealed Chd1 to bind preferentially to AT-rich sequences and also to be present in *Drosophila*, where it showed preferential localization to decondensed interbands and transcriptionally active "puffs" on polytene chromosomes (Stokes and Perry, 1995; Stokes et al., 1996). The laboratory of Francis Collins, who led the sequencing of the human genome at the National Institutes of Health, then apparently stumbled on a clone containing sequence homologous to the murine *CHD1* gene, which led to further identification of homologs in organisms ranging from budding and fission yeast to *C. elegans* to vertebrates, including three human homologs (Woodage et al., 1997).

Uniquely among chromatin remodelers, Chd1 functions without accessory subunits, although it interacts directly with factors involved in transcriptional elongation (Clapier and Cairns, 2009). In flies and vertebrate organisms, additional Chd family members are found as subunits of the Mi-2/NuRD (nucleosome remodeling and deacetylase) complexes. The Mi2 complex was first isolated from *Xenopus* egg extracts, the benefits of which were extolled in Chapter 5, in an effort to identify and characterize complexes possessing histone deacetylase activity (Wade et al., 1998). By monitoring release of radiolabeled acetate from acetylated histones, the authors fractionated extracts and identified an abundant deacetylase activity that copurified with several other proteins as a large, multisubunit complex. Two of the subunits were identified as *Xenopus* homologs of a known histone deacetylase, Rpd3, and of a *Drosophila* protein known to associate with Rpd3 as part of a histone deacetylase complex. Sequencing of a proteolytic fragment of a 240 kDa peptide present in the isolated complex revealed identity to human Chd3 and Chd4 (Woodage et al., 1997), which had originally been identified as a human autoantigen and found to be present in a complex with proteins close in molecular weight to those found in the *Xenopus* complex (Nilasena et al., 1995; Seelig et al., 1995). The Mi2 complex was shown to have ATPase activity, as expected, but unlike remodelers from the SWI/SNF, ISWI, and INO80 families, the activity was stimulated by nucleosomes but not by naked DNA. Subsequently, a similar complex was purified from human cell lines and shown to possess ATP-dependent chromatin remodeling activity as well as histone deacetylase activity (Tong et al., 1998; Xue et al., 1998; Zhang et al., 1998).

Since this early work, CHD proteins have been revealed to comprise a large family functioning in a variety of processes, including transcriptional activation and repression, elongation, chromatin compaction, and DNA repair. Nine CHD proteins are present in humans, with additional variety conferred by alternate splicing and posttranslational modification; these CHD proteins are particularly important in developmental pathways. Correspondingly, mutations in CHD proteins have been implicated in a variety of cancers and developmental disorders. Human CHD proteins have recently been the subject of an exhaustive review; a supplementary table includes information on protein-binding partners, phenotypes of CHD mutants in the mouse, and associated human disorders (Alendar and Berns, 2021). Chd1 has been implicated in organizing genome-wide nucleosome occupancy in yeast, as discussed in Chapter 4 and again further on in this chapter.

Structure and mechanism of remodelers

The large size and complex composition of remodelers rendered them refractory to structural analysis at the time of their identification. Consequently, early work focused on biochemical assays aimed at understanding the mechanism by which they converted the energy derived from ATP hydrolysis to altering histone-DNA contacts. Initial experiments were aimed at distinguishing two likely mechanisms for the altered accessibility to nucleosomal DNA conferred by remodelers: nucleosome disassembly, in which histones are lost entirely from the DNA, or nucleosome mobilization, also referred to as sliding, in which the histone octamer is moved relative to the DNA sequence. Treatment with SWI/SNF of a reconstituted nucleosome array in which a central nucleosome occupied five Gal4-binding sites and was flanked by five positioned nucleosomes resulted in nucleosome remodeling throughout the array, but nucleosome loss after removal of SWI/SNF was only seen at the central nucleosome and only when Gal4 (actually the Gal4-AH derivative mentioned earlier) was present during the remodeling reaction (Owen-Hughes et al., 1996). A subsequent experiment reported that SWI/SNF remodeling of a single nucleosome ligated onto longer naked DNA fragments resulted in movement of the histone octamer away from its original location in *cis* (Whitehouse et al., 1999). Movement of the octamer was blocked by incorporation of a four-way (Holliday) junction into the DNA template, implicating nucleosome sliding catalyzed by SWI/SNF. Similar findings were reported for remodeling by NURF, CHRAC, and their respective Iswi "motors" (Hamiche et al., 1999; Langst et al., 1999). These latter experiments used short (248 and 359 bp) fragments to assemble nucleosomes in vitro; reconstitutions resulted in a small number (two or five) of preferred positions for the resulting nucleosomes. These variously positioned nucleosomes could be separated on polyacrylamide gels and remained stable in the absence of remodelers, as their electrophoretic mobility was preserved after isolation from gels. CHRAC was observed to mobilize nucleosomes from a position near the fragment end, while Iswi in isolation displayed the opposite behavior (Fig. 6.7); thus, CHRAC subunits exerted a major effect on nucleosome movement effected by Iswi (Langst et al., 1999). Similarly, NURF catalyzed movement of nucleosomes occupying different positions on a 359 bp fragment preferentially to a single major site, while Iswi on its own displayed much lower activity and little propensity to move nucleosomes to any preferred site (Hamiche et al., 1999). NURF-dependent mobilization was sequence dependent, as nucleosomes reconstituted onto a 5S rDNA sequence and occupying multiple positions (as discussed in Chapter 4) were unaffected by NURF treatment. Neither CHRAC nor NURF catalyzed octamer transfer to naked DNA, and octamer sliding catalyzed by NURF was supported by observation of fragments protected against digestion by Exo III at very short reaction times corresponding to imputed nucleosome occupancy between the starting and final positions of the sliding reaction.

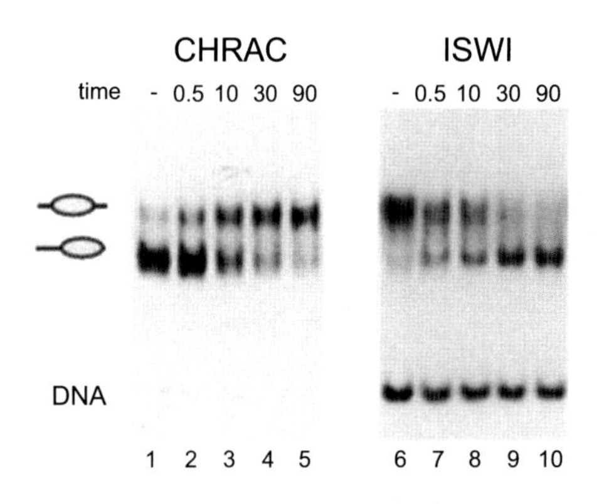

FIG. 6.7 Nucleosome sliding by CHRAC and Iswi. Nucleosomes adopting two distinct positions on a 248bp fragment were purified and treated for the indicated times with CHRAC or recombinant Iswi for the times indicated, and then analyzed by polyacrylamide gel electrophoresis. *From Langst, G., Bonte, E.J., Corona, D.F., Becker, P.B., 1999. Nucleosome movement by CHRAC and ISWI without disruption or trans-displacement of the histone octamer. Cell 97, 843–852.*

In contrast to the behavior of the Iswi-containing NURF and CHRAC complexes, RSC proved capable of catalyzing the transfer of a histone octamer from a nucleosome to naked DNA in *trans* (Lorch et al., 1999). SWI/SNF could also effect octamer transfer, as suggested by the apparent loss of a nucleosome in competition with binding by Gal4-AH, but transfer was found to be 50–100-fold less efficient than sliding (Whitehouse et al., 1999). Such sliding is consistent with the in vivo action of remodelers in determining NDR width, as related in Chapter 4, through relatively modest effects on nucleosome positioning. Nucleosome disassembly at gene promoters during transcriptional activation, facilitated by SWI/SNF, has also been documented in vivo; we discuss this in some detail in Chapter 8.

Biochemical studies of nucleosome remodeling similar to those described above, in combination with single molecule studies and eventually supported by high-resolution structural results, have culminated in a detailed model for the mechanism by which remodelers translocate or evict histone octamers from DNA (Bowman, 2019; Clapier et al., 2017; Jungblut et al., 2020; Markert and Luger, 2021; Yan and Chen, 2020). The essential features of this model are common to all four classes of remodeler, as might reasonably be expected given their highly related ATPase subunits. The ATPase in each case functions as a translocase, holding onto DNA and moving the DNA in single base pair increments (Deindl et al., 2013; Harada et al., 2016); by fastening itself also to the histone octamer, through direct contacts and/or contacts with auxiliary subunits, the ATPase adopts the octamer as a frame of reference, so the DNA is moved relative to the octamer. This creates tension that can be relieved either by additional movement of the DNA helix relative to the octamer, resulting in nucleosome sliding, or by nucleosome disassembly (Fig. 6.8). Eviction may depend on translocation efficiency: when sufficiently rapid, translocation may disrupt multiple histone-DNA contacts, leading to histone eviction (Clapier et al., 2016, 2017). An alternative model proposes that

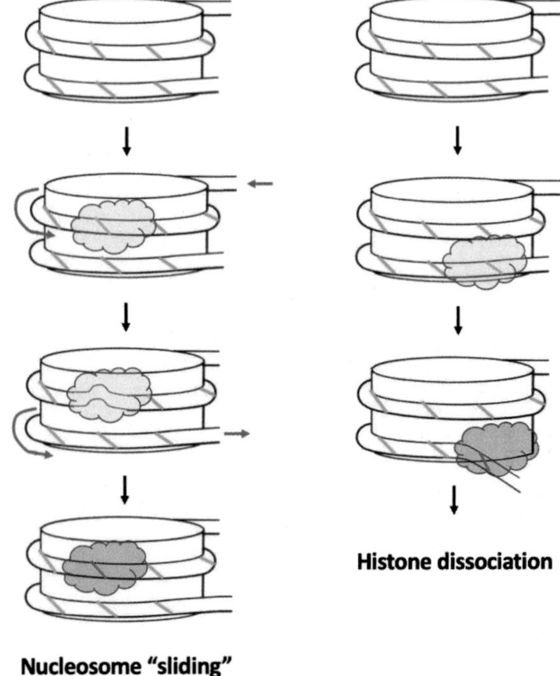

FIG. 6.8 Alternative mechanisms of nucleosome remodeling. Left, association of a remodeler at SHL 2 draws in 1 bp of the DNA helix from the proximal side of the nucleosome core particle; translocation accompanying ATP hydrolysis releases the small "bulge" toward the distal side of the NCP, resulting in a net translocation of 1 bp. Right, association of INO80 at SHL +6 results in ATP-driven release of DNA from the outermost region of the NCP.

nucleosome sliding may result in a collision with the neighboring nucleosome, leading to its unspooling and ultimate disassembly (Clapier et al., 2017; Dechassa et al., 2010; Engeholm et al., 2009). How SWI/SNF and RSC, both of which are capable of catalyzing nucleosome disassembly, achieve this efficiency while other remodelers do not, remains an open question.

Translocase activity was first inferred in experiments showing that both RSC and ISWI could displace one strand of triple helical DNA and exhibited ATPase activity dependent on the length of the DNA substrate (Saha et al., 2002; Whitehouse et al., 2003). The translocase function of remodelers was confirmed and extended to additional chromatin remodelers in numerous subsequent studies, using both biochemical assays and single-molecule approaches (Clapier et al., 2017). Thus, although the ATPase subunits of chromatin remodelers did not behave as helicases, being incapable of duplex separation, they could engage and move along a double helical DNA segment, suggesting a plausible mechanism by which they could move a histone octamer relative to the associated DNA. This translocation is achieved by movement of the two RecA-like lobes common to the ATPases (first determined for the *Sulfolobus solfataricus* Rad54 enzyme, and so named for their resemblance to the ATPase domain of *E. coli* RecA), coupled with cycles of ATP binding and hydrolysis (Durr et al., 2005; Singleton et al., 2007). These two lobes contact the same strand of the double helix, and their movement coupled to ATP hydrolysis creates a sort of inch-worming mechanism that moves the translocase relative to the double helix. The single-stranded association of the translocases makes them vulnerable, with respect to movement along the helix, to single-stranded gaps in sequence, a property that has been exploited in mechanistic studies.

DNase I and hydroxyl radical footprinting, along with cross-linking experiments, indicated binding of NURF (containing ISW1), ISW2, RSC, and SWI/SNF near SHL 2, that is, at two helical turns from the dyad axis of the nucleosome (Dechassa et al., 2008; Kagalwala et al., 2004; Saha et al., 2005; Schwanbeck et al., 2004). This conclusion was supported by experiments showing that single-stranded gaps in this same region, but not at other nucleosomal sites, inhibited translocation of nucleosomal DNA by these remodelers as well as by SWI/SNF, presumably reflecting either inability of the translocating ATPase to navigate the gap or the necessity for torsional strain in the DNA for nucleosome mobility (Kagalwala et al., 2004; Saha et al., 2005; Schwanbeck et al., 2004; Zofall et al., 2006). A recent spate of high-resolution structures of remodelers in complex with nucleosome core particles, driven by the revolution in cryo-EM resolution, has confirmed some of the earlier findings, such as the binding of remodelers near SHL 2, while adding a wealth of new detail (reviewed in (Bowman, 2019; Jungblut et al., 2020; Markert and Luger, 2021; Yan and Chen, 2020)). Structural determinations were aided by the use of ADP-BeF_3, which mimics bound ATP but inhibits catalytic activity (Kagawa et al., 2004). Reported structures include those for CHD1, ISWI, and the isolated Snf2 subunit of SWI/SNF, as well as the larger INO80, RSC, SWR1, and SWI/SNF (both yeast SWI/SNF and mammalian BAF and PBAF) complexes (Table 6.1 and Fig. 6.9)

TABLE 6.1 Structural studies of remodelers engaged with nucleosomes or nucleosome core particles.

Remodeler	State	ATPase-nucleosome contact	Additional contact(s)	Reference
Snf2 (yeast)	ADP, ADP-BeFx; apo	SHL 2 or SHL 6		Li et al. (2019) and Liu et al. (2017)
SWI/SNF	ADP-BeFx, ATPγS	SHL +2	SHL +6	Han et al. (2020)
cBAF	apo	SHL 2		He et al. (2020) and Mashtalir et al. (2020)
PBAF	ADP-BeFx	SHL 2		Yuan et al. (2022)
RSC	apo; ADP-BeFx	SHL +2	SHL −6, linker DNA	Patel et al. (2019), Wagner et al. (2020), and Ye et al. (2019)
Snf2h (ISWI)	ADP-BeFx	SHL 2		Armache et al. (2019)
ISWI	apo; ADP, ADP-BeFx	SHL 2		Chittori et al. (2019) and Yan et al. (2016, 2019)
INO80	ADP-BeF3; apo	SHL +6/7	SHL +2/3, linker DNA	Ayala et al. (2018) and Eustermann et al. (2018)
SWR1	ADP-BeF3	SHL 2		Willhoft et al. (2018)
Chd1	ADP-BeF3; apo	SHL +2	SHL −6/7, linker DNA	Farnung et al. (2017), Sundaramoorthy et al. (2018), and Nodelman et al. (2022)
Chd4	AMP-PNP	SHL +2	SHL −6	Farnung et al. (2020)

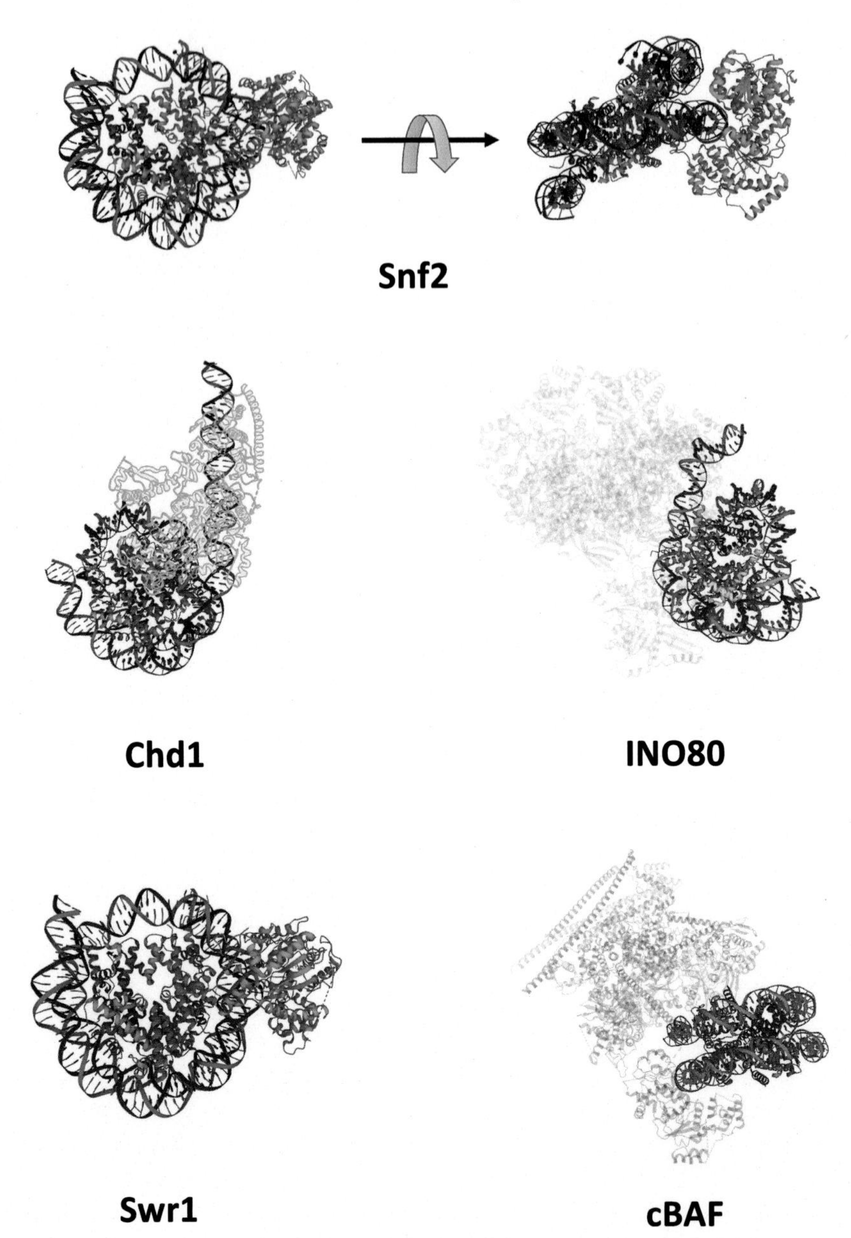

FIG. 6.9 Remodeler-nucleosome structures. Snf2 (green)-nucleosome is shown in face and side view, and the large INO80 and BAF complexes are lightened to allow visualization of the nucleosome core particle. PDB files and relevant references used are: Snf2 (5X0Y; Liu et al., 2017); Chd1 (5O9G; Farnung et al., 2017); INO80 (6FML; Eustermann et al., 2018); Swr1 (6GEJ; Yan et al., 2019); cBAF (6LTJ; He et al., 2020).

(Armache et al., 2019; Ayala et al., 2018; Chittori et al., 2019; Eustermann et al., 2018; Farnung et al., 2017, 2020; Han et al., 2020; He et al., 2020; Li et al., 2019; Liu et al., 2017; Mashtalir et al., 2020; Nodelman et al., 2022; Patel et al., 2019; Sundaramoorthy et al., 2018; Wagner et al., 2020; Willhoft et al., 2018; Yan et al., 2016, 2019; Ye et al., 2019; Yuan et al., 2022). These structures reveal both commonalities and unique features of the various chromatin remodelers.

Numerous contacts are made by remodelers with both the histones and nucleosomal DNA, and in some cases with adjacent linker DNA (Fig. 6.9). With the exception of INO80, the catalytic domain of the ATPase subunit of all remodelers contacts DNA at SHL 2; the catalytic domain of INO80 instead contacts SHL 6, allowing it to peel the terminal 15bp of the core particle away from the histone octamer (Fig. 6.8 and Table 6.1). This peeling away has been suggested to favor destabilization of the nucleosome, eventually facilitating replacement of H2A.Z-H2B dimers with H2A-H2B dimers (Yan and Chen, 2020). However, as noted above, whether INO80 facilitates this exchange is controversial; furthermore, the closely related SWR1 complex (sharing four subunits with INO80) is well established as exchanging H2A.Z-H2B dimers for H2A-H2B dimers, and its motor protein makes primary contact with SHL +2, not SHL +6 (Willhoft et al., 2018). SHL +2 is in close proximity to SHL −6 (and by symmetry, SHL −2 with SHL +6), and several remodelers contact both sites, while Snf2 has been observed to bind in alternative modes either to SHL 2 or SHL 6 (Table 6.1). Additional contacts are made between remodeler subunits and the acidic patch in the case of INO80, SWR1, BAF, and PBAF. In the case of SWI/SNF and PBAF, this interaction occurs through the Snf2 ATP coupling, or SnAC domain, which was discovered as being required both for regulation of ATPase activity of and nucleosome binding by Snf2 (Sen et al., 2013). Contacts are also made between remodelers and the histone H4 tail, which is in close proximity to SHL 2, in the cases of Snf2, Chd1, Chd4, and ISWI. The functional significance of this contact is supported by earlier findings that removal of the H4 tail, but no other histone tail, inhibits activity of the ISWI-containing NURF and CHRAC complexes, while activity of ISW2 and ACF is reduced on nucleosomes lacking the H4 tail or the region thereof (amino acids 16–19) that interact with the acidic patch (Clapier et al., 2001; Dang et al., 2006; Hamiche et al., 2001; Racki et al., 2014).

These multiple contacts between remodelers and the nucleosome serve multiple purposes: first, they situate the catalytic domain on the nucleosome in a manner consistent with mechanistically efficient remodeling; second, they permit both binding and activity of remodelers to be modulated by specific features of the nucleosome, such as histone modifications or the presence of variants; and third, they provide a mechanism for relief of remodeler autoinhibition. The mechanism of remodeling and its modulation by features of the nucleosome are discussed below. With regard to autoinhibition, an indication that the human BRG1-containing SWI/SNF (or BAF) complex could be subject to repressive mechanisms derived from the demonstration that the complex was not constitutively active with respect to its ATPase activity during the cell cycle, with activity repressed in G2/M by phosphorylation of subunits (Sif et al., 1998). The *Drosophila* ISWI protein possesses a region similar to the H4 tail, including an exact match to the RHRK sequence comprising the basic region that makes internucleosomal contact with the H2A acidic patch (Clapier and Cairns, 2012). Mutation of this region of ISWI revealed it to be autoinhibitory, as the mutant was threefold more active than the wild type ISWI, as measured by ATPase activity, and was unaffected by the presence of an H4 tail peptide that stimulated activity of the wild type protein. Binding of the catalytic domain of ISWI to the nucleosome causes the catalytic domain to move away from the autoinhibitory region, designated as the AutoN domain (Fig. 6.6) (Yan et al., 2019). In addition, repression by a second autoinhibitory domain designated as NegC is relieved by interaction between ISWI and the nucleosomal linker DNA (Clapier and Cairns, 2012; Yan et al., 2019). Similar relief of autoinhibition by nucleosomal binding has been seen for Snf2h, Snf2, and Chd1 (Gamarra et al., 2018; Hauk et al., 2010; Xia et al., 2016; Yan et al., 2019).

One model, favored early on, for the mechanism by which remodelers mobilize nucleosomes stipulates that as a consequence of the contacts between remodelers and the nucleosome (including adjacent linker in some cases), DNA translocation driven by a given remodeler causes DNA to move relative to the histone octamer. Translocation occurring at SHL 2 (for all remodelers save INO80 and the alternative mode of SWI/SNF) would then result in one base pair being pulled into this region from the adjacent helical turn; propagation of this movement, which could occur by sequential interruption of DNA-histone contacts, would lead to linker DNA being pulled into the nucleosome on one side and ejected on the other (Fig. 6.8). Recent high-resolution structures of Chd1, ISWI, and Snf2, together with single molecule studies, have led to refinement and some revision of this model (Bowman, 2019; Reyes et al., 2021; Yan and Chen, 2020). The catalytic ATPase subunits of remodelers can adopt a closed or open conformation, in which the RecA-like lobes are respectively close together or more widely separated; the transition between the two, coupled to ATP hydrolysis, results in the inchworm movement mentioned earlier. Remarkably, cross-linking studies show that both closed and open states of Chd1 shift nucleosomal DNA upon binding, even without ATP hydrolysis (Winger et al., 2018). High-resolution structures of ISWI and Snf2 reveal that in the open state, either bound by ADP or unbound, binding pulls one base pair of DNA into the catalytic domain from the side at which DNA enters the nucleosome (that is, proximal rather than distal to the dyad) (Li et al., 2019; Yan et al., 2019). The two strands appear not to be equally distorted; however, the tracking strand of the double helix that is directly contacted by the lobes is pulled away from the histone octamer surface, while the other strand, called the guide strand, shows less distortion. This results in a tilting of base pairs in the helix that propagates out to the edge of the nucleosome. This twist defect is resolved in the transition to the closed structure, observed using the ATP mimic ADP-BeF$_3$. Additional support for this mechanism comes from Förster resonance energy transfer (FRET) experiments showing that only the tracking strand is shifted when Snf2 is bound in the open state, while the guide strand catches up in the transition to the closed state (Li et al., 2019; Sabantsev et al., 2019). This second step would result in the DNA helix being undertwisted by 1 bp on the side of SHL 2 closer to the DNA entry site and overtwisted on the distal side (closer to the pseudodyad). A variation on this mechanism is suggested by a recent structural determination of Chd1 bound to a nucleosome in the open state, which revealed a one base pair bulge of the tracking strand at the catalytic domain and a complementary bulge of the guide strand one helical turn away (Nodelman et al., 2022). In spite of the apparent elegance and detail of such models, however, it should be stated that the alterations in DNA helical structure revealed in the cryo-EM structures are subtle and require further corroboration.

The apparent twist defects caused by remodeler binding to SHL 2 followed by the ATP-catalyzed transition from open to closed state could be resolved by DNA being pulled in from the entry site, and pushed around the octamer on the other side. Such a "twist diffusion" model was proposed as an alternative to a "looping out" mechanism in which a much larger bulge of one helical turn of DNA transits rapidly around the nucleosome (van Holde and Yager, 2003). Cross-linking experiments combined with gel electrophoresis, together with FRET, indicate that DNA enters and exits the nucleosomes in small steps of 1–2 up to ~7 bp, with movement of DNA into the nucleosome decoupled from the exit step (Zhong et al., 2020). However, the mechanism by which a twist defect or small DNA bulge originating at SHL 2 would transit around the nucleosome remains unclear, particularly when the energetics of DNA bending in the nucleosome are considered.

INO80 and SWR1, which in contrast to Chd1, SWI/SNF, and ISWI remodelers do not slide nucleosomes, appear to operate by a variation on this twist diffusion mechanism. As already mentioned, the motor protein of INO80 binds not to SHL 2, but to SHL +6. Binding of INO80, as for other remodelers, induces a bulge at the motor-binding site (Eustermann et al., 2018); however, the resolution of the structure did not allow the size of the bulge, and whether it corresponds to one strand or two drawn into the remodeler, to be ascertained. SWR1, like INO80, binds to both SHL +2 and SHL +6, but differs in that an auxiliary subunit (from the ARP6 module) binds to SHL +6, while it is the motor protein that binds to SHL +2 (Willhoft et al., 2018). Binding of SWR1 in the closed state induces a 1 bp bulge at SHL +2, but in contrast to other remodelers, the bulge involves both DNA strands, so that no twist defect is induced. Whether and how this difference contributes to the different outcome of SWR1-induced remodeling—exchange of H2A.Z-H2B for H2A-H2B—currently remains unresolved.

A remarkable aspect of this highly detailed picture of remodeler action is the special nature of SHL 2 of the nucleosome. By definition, one helical turn connects each of the SHLs in the nucleosome. At most locations, the structure of the helical turn is fixed, as evidenced by nucleosomes containing different sequences having identical structures (Davey et al., 2002; Luger et al., 1997; Vasudevan et al., 2010). DNA in contact with SHL 2, however, can accommodate either 10 or 11 bp, thereby lowering the energetic barrier to introduction of a twist defect upon remodeler binding and activity (Bilokapic et al., 2018; Luger et al., 1997; Markert and Luger, 2021; Ong et al., 2007). Thus, the multiple contacts made between large chromatin remodelers and the nucleosome position the catalytic domain precisely where the resultant alteration in DNA structure can best be accommodated.

The multiple contacts between remodeler and nucleosome also allow remodeler activity to be "tuned" by alterations in nucleosome structure. Variation in nucleosome remodeling can be effected by changes in recruitment or in the activity of the bound remodeler. For example, loss or tetra-acetylation of the H4 tail eliminates or reduces remodeling by Isw2 and Chd1 without affecting their binding to a nucleosome (Dang et al., 2006; Ferreira et al., 2007). The effect of histone modifications, mutations, and incorporation of variant histones, alone or in combination, on nucleosome remodeling has been examined in high-throughput fashion (Dann et al., 2017; Mashtalir et al., 2021). Muir and colleagues developed bar-coded libraries of over 100 variant nucleosomes including variant histones or specific modified or mutant histones; these were made by production of recombinant histones and construction of specifically modified histones using synthesized peptides and sophisticated peptide ligation methods (Nguyen et al., 2014). Each variant nucleosome was associated with a specific DNA sequence tag (the bar-code) appended onto the Widom 601 positioning sequence, or variants thereof. Remodeling activity was assessed by a restriction enzyme accessibility assay in which a restriction site incorporated into the initially assembled nucleosome is exposed upon remodeling, with the relative abundance of bar-codes, measured by high-throughput sequencing, providing a readout of the relative remodeling activity toward variant nucleosomes. Both positive and negative effects of variant nucleosomes were seen on the activities of ISWI, CHD, and SWI/SNF family remodelers, and variable effects were seen for different ISWI remodelers, indicating a modulating influence of associated subunits (Dann et al., 2017). Remodelers of all three types tested required the acidic patch for maximal activity. Similarly, the three mammalian BAF complexes showed both positive and negative modulation in both binding affinity and remodeling activity exerted by variant nucleosomes, and often differed in the effects observed (Mashtalir et al., 2021). Integrating these biochemical observations with in vivo effects on localization and activity of remodelers remains a substantial challenge for the field.

Function and localization of remodelers

Chromatin remodelers participate in all aspects of eukaryotic cell function involving DNA, including transcription, replication, repair, and recombination, in addition to their roles in chromatin assembly and nucleosome spacing, and their participation in restructuring chromatin by facilitating incorporation of histone variants, most notably H2A.Z (Becker and Workman, 2013; Clapier and Cairns, 2009). Which remodelers perform which functions varies according to organism and genomic location and process, reflecting the contingent nature of evolution. Remodelers frequently have been found to function redundantly; several examples of this have already been discussed. Novel functions continue to be uncovered: for example, the Iswi-containing ACF complex in *Neurospora crassa* functions to position nucleosomes that repress transcription in facultative heterochromatin (Wiles et al., 2022). Consistent with the origins of our knowledge of chromatin remodelers, much of the research efforts in the decades following has centered on their role in transcriptional activation, which is the focus of this section.

The experiments that led to the discovery of the SWI/SNF chromatin remodeling complex in yeast implicated its activity as important for the activation of inducible genes such as *SUC2*, *HO*, *ADH1*, and others. In parallel with the genetic evidence for this role, the two ISWI-containing complexes, ACF and NURF, were shown to facilitate in vitro transcription of chromatin templates assembled using *Drosophila* extracts (Ito et al., 1997; Mizuguchi et al., 1997). The templates used included binding sites for Gal4, and disruption of chromatin structure was shown to occur only in the presence of Gal4 derivatives, ATP, and the remodeling complex. Mizuguchi et al. further directly demonstrated ~30-fold enhancement by NURF of transcriptional activation by a Gal4-HSF fusion protein with a chromatin template, but not with naked DNA. Closely following these reports, a mammalian SWI/SNF complex purified from erythroid cells was shown to disrupt chromatin structure and facilitate in vitro transcription of a chromatin template in an ATP-dependent fashion, while exhibiting no such enhancement of transcription of a naked DNA template in a study aimed at identifying factors required for activation of the chromatin-assembled β-globin promoter (Fig. 6.10) (Armstrong et al., 1998). The complex, named E-RC1 for EKLF coactivator-remodeling complex 1 (EKLF, or erythroid Kruppel-like factor, being a transcriptional activator that regulates erythropoiesis), was then found to include BRG1 and several BAF subunits, thus implicating the mammalian SWI/SNF

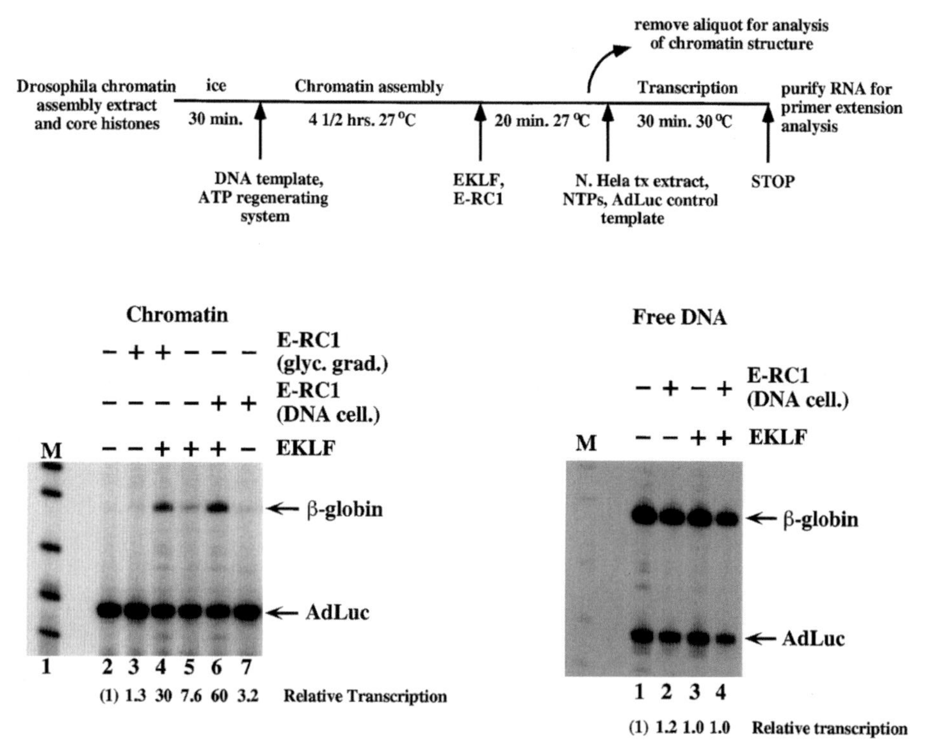

FIG. 6.10 The erythroid BAF chromatin remodeling complex, E-RC1, facilitates transcription of a chromatin template. The protocol at the top was used to assay transcription of the β-globin gene assembled into chromatin (left panel) or as naked DNA (right panel); a naked DNA template in which the luciferase gene is driven by the adenovirus major late promoter was used as a control in both experiments. Transcription of the unassembled β-globin gene did not depend on E-RC1 or the transcription factor EKLF, while the chromatin assembled β-globin gene was transcribed most efficiently when EKLF and E-RC1 were both included. *From Armstrong, J.A., Bieker, J.J., Emerson, B.M., 1998. A SWI/SNF-related chromatin remodeling complex, E-RC1, is required for tissue-specific transcriptional regulation by EKLF in vitro. Cell 95, 93–104.*

complex as functioning in transcriptional activation. An early hint of functional specificity came from the observation that E-RC1 could not facilitate transcriptional activation of a chromatin-assembled template in vitro via a different activator, TFE-3.

The complexity of SWI/SNF function in transcription was also apparent early on: the effect of deleting individual SWI/SNF subunits on expression of *lacZ* reporter fusions in yeast cells varied with the activator, the number of activator-binding sites, the target promoter, and the subunit deleted (Laurent and Carlson, 1992). Microarray studies indicated that expression of only a small fraction of expressed genes in yeast, whether grown in rich or minimal media, was affected by deletion of SWI/SNF subunits, with no identifiable features characterizing those genes showing maximal dependence (Holstege et al., 1998; Sudarsanam et al., 2000). Inducible genes also varied in their dependence, although highly inducible genes appear most dependent on SWI/SNF (Qiu et al., 2016). Even genes depending on the same activator vary in their dependence on SWI/SNF; *PHO5* and *PHO8* both require the activator Pho4 for their induction under conditions of low phosphate, but *PHO8* depends more strongly on SWI/SNF for chromatin remodeling and transcriptional activation (Gregory et al., 1999). This complexity no doubt reflects at least in part the existence of multiple chromatin remodelers with potential redundancies in function. Adding a further layer of complication, genetic approaches also demonstrated redundancy between SWI/SNF and the histone modification complex, SAGA (discussed in Chapter 8) (Gregory et al., 1999; Pollard and Peterson, 1997; Roberts and Winston, 1997; Sudarsanam and Winston, 2000). Despite numerous studies of transcriptional dependence on the SWI/SNF complex in yeast and other eukaryotes, it remains the case that the dependence of a particular gene on SWI/SNF, or any other remodeler, for its transcriptional activation cannot be predicted de novo.

In vitro, SWI/SNF was shown to be capable of remodeling chromatin to facilitate binding of both activators and general transcription factors (in particular, TBP). However, these experiments used concentrations of SWI/SNF far higher than found in yeast cells (Tsukiyama and Wu, 1997), raising the question as to how the limited number of SWI/SNF complexes in a cell could be directed to target sites in chromatin in living cells. The idea that activators functioned by recruiting components of the transcription machinery via protein-protein interactions crystallized in the 1990s (Ptashne and Gann, 1997), and it was not difficult to extend this model to imagine activators recruiting chromatin remodelers that would facilitate eviction of nucleosomes from proximal promoter regions to allow access to transcription factors. Envisioning a mechanism for remodelers facilitating access by transcriptional activators to nucleosomal sites was a little more challenging—how would a remodeler provide specific access to an inaccessible site? That is, how would the remodeler "know" where to go? However, experiments showing that nucleosomal DNA exhibited dynamic behavior, such that occluded DNA sites were occasionally exposed, with the outermost DNA being most accessible (Polach and Widom, 1995; Vettese-Dadey et al., 1994), provided a rationale for such a mechanism: transient exposure of a binding site for an activator could allow initial access, followed by recruitment of SWI/SNF or another accessory factor by the bound activator that could then stabilize binding and nucleosome disruption either through remodeling activity or cooperative binding (Miller and Widom, 2003; Morse, 2003).

Experiments conducted in yeast provided evidence for SWI/SNF altering chromatin structure to facilitate access to both activators and the general transcription machinery. High-resolution mapping of nucleosomes at the repressed *SUC2* promoter revealed positioned nucleosomes both over the TATA element and at upstream activating sites, and both regions exhibited increased accessibility to MNase upon *SUC2* induction in wild type, but not in *swi*$^-$, yeast (Gavin and Simpson, 1997; Wu and Winston, 1997). These results suggested that SWI/SNF aided access to both UAS and proximal promoter regions; however, as the primary activator responsible for *SUC2* induction was (and is) not known, it remained possible that initial activator binding occurred in a linker region, as was shown for Pho4 binding to the *PHO5* promoter during its induction (Svaren et al., 1994), which could allow SWI/SNF-independent activator binding followed by recruitment of SWI/SNF and remodeling of the surrounding chromatin. Direct examination of binding by DMS footprinting of the activator Gal4 to nucleosomal and nonnucleosomal sites in constructs based on the *GAL1-10* promoter, together with monitoring the ability of Gal4 to activate transcription from the same sites, indicated little dependence on SWI/SNF at high affinity sites in either circumstance, nor for low affinity sites in a context where positioned nucleosomes were absent (Burns and Peterson, 1997). Gal4 binding accompanied by nucleosome disruption also occurred independently of SWI/SNF at a nucleosomal site in an assay based on the TRP1ARS1 plasmid, as described in Chapter 4, while transcription of a reporter gene having a nucleosomal TATA element exhibited much greater dependence on SWI/SNF than one having an accessible proximal promoter region (Ryan et al., 1998). The latter result suggested a role for SWI/SNF in transcriptional activation at a step subsequent to activator binding, which was corroborated by experiments showing that binding of Pho4, the primary activator of *PHO8*, occurred in the absence of SWI/SNF activity (as shown by dimethyl sulfate footprinting), while chromatin remodeling of the proximal promoter and transcriptional induction did not (Gregory et al., 1999). Similarly, binding of the requisite transcription factor Swi5 occurred independently of SWI/SNF (in this case demonstrated by ChIP) at the SWI/SNF-dependent *HO* promoter (Cosma et al., 1999), again indicating a role for chromatin remodeling subsequent to activator binding. Numerous additional studies substantiated activator-dependent recruitment of chromatin remodeling complexes in both yeast and higher eukaryotes (Sudarsanam and Winston, 2000); this topic is discussed further in Chapter 8.

Insight into the determinants that govern whether a particular gene is a target for remodeling by SWI/SNF emerged from genome-wide investigations into transcriptional regulation and nucleosome positioning. Pioneering studies from Kevin Struhl had indicated the existence of distinct types of TATA elements conferring constitutive or inducible transcription, as exemplified by the *HIS3* promoter (Struhl, 1986). TBP in yeast was eventually discovered to be delivered to gene promoters by either the TFIID or SAGA complex (the latter discussed at length in the next chapter), with the mechanism by which TBP is delivered generally depending on the nature of the TATA element (Kuras et al., 2000; Lee et al., 2000; Li et al., 2000). Analysis of a collection of genome-wide datasets pertaining to various aspects of gene regulation and chromatin structure revealed that ~90% of yeast genes lacked a match to a consensus TATA element and were predominantly transcribed by a mechanism involving TFIID, while ~10% had a consensus TATA element in their promoters and were SAGA-dominated (Basehoar et al., 2004; Huisinga and Pugh, 2004). (The details and interpretation of this categorization have been revisited in subsequent work, but the regulatory pathways associated with them still apply (Donczew et al., 2020; Rhee and Pugh, 2012).) These two categories of genes differed markedly in their association with chromatin regulators such as histone modification enzymes, H2A.Z, and SWI/SNF, with such regulators being preferentially associated with SAGA-dominated genes. The SAGA-dominated cohort is also enriched for highly inducible genes such as those involved in stress response, while constitutively expressed genes such as those encoding ribosomal proteins are TFIID-dominated. Genome-wide analyses of nucleosome positioning revealed a strong coupling of this categorization with chromatin structure: despite the general pattern of a nucleosome-depleted region flanked by arrays of positioned nucleosomes (Fig. 4.14), about one-fourth of all yeast genes showed a very narrow NDR and no evidence of surrounding nucleosome arrays, and this cohort was highly enriched for genes whose transcriptional regulation varied with environmental conditions, including those involved in stress response (Lee et al., 2007; Tirosh and Barkai, 2008).

These findings suggest that the rough division of yeast genes into those dominated by SAGA or TFIID, with distinct characteristic nucleosome positioning patterns, may also reflect a division of labor among chromatin remodelers. As discussed in Chapter 4, ISW1, INO80,

RSC, and CHD1, together with general regulatory factors such as Reb1, Abf1, and Rap1, cooperate to establish NDRs by adjusting nucleosome position at promoters, while SWI/SNF appears more dedicated to evicting promoter-occupying nucleosomes at inducible genes. These roles may prove to be conserved across evolution, as deletion of SNF2H in murine ES cells resulted in increased nucleosome spacing, while deletion of BRG1 did not (Barisic et al., 2019). Dedication of SWI/SNF to nucleosome disassembly of induced genes is also consistent with the relatively low abundance of SWI/SNF in yeast cells (Cairns et al., 1996), as transcriptional activation induced by a specific environmental perturbation typically involves at most a few hundred genes, and often fewer. In addition to establishing NDRs at promoters, ISWI, CHD1, and INO80 dictate nucleosome spacing in the surrounding arrays, as demonstrated by a combination of in vivo and in vitro approaches (see Chapter 4) (Prajapati et al., 2020). Using in vitro assembly of genomic chromatin with purified remodelers, Korber and colleagues have reported that ISWI, CHD1, and INO80 remodeling complexes each mobilize nucleosomes into arrays having linker lengths specified by the particular remodeler; the spacing characteristic of a given remodeler is likely governed by subunits auxiliary to the ATPase, as spacing set by INO80 could be altered by mutation of such subunits (Oberbeckmann et al., 2021). In contrast to ISWI, CHD1, and INO80, RSC contributes to the width of the NDR, but does not affect spacing of the surrounding arrays (Ganguli et al., 2014). While nucleosome occupancy patterns are unaltered in *snf2Δ* yeast, deletion of *SNF2* does result in increased nucleosome occupancy at the NDRs of highly transcribed genes when Sth1, the motor protein of RSC, is also depleted (Rawal et al., 2018). Cooperativity between RSC and SWI/SNF in NDR formation was seen at both constitutively active genes and those activated by Gcn4 in response to starvation. Altogether, nucleosome positioning, occupancy, and spacing in yeast are governed by a complex interplay among chromatin remodelers and general regulatory factors that includes functional redundancy, cooperativity, and antagonism (Kubik et al., 2019; Prajapati et al., 2020).

Genome-wide location analysis of chromatin remodelers also supports functional roles of chromatin remodelers that are both overlapping and distinct. Determination of remodeler localization by ChIP-seq or related methods poses two principal challenges: first, remodeler association may be transient, with rapid loss from chromatin following remodeling (Kim et al., 2021; Tilly et al., 2021); and second, artifactual ChIP-seq signals are often seen at highly transcribed regions or promoters, potentially confounding interpretation of association of remodelers with coding sequences (Jain et al., 2015; Park et al., 2013; Teytelman et al., 2013). One way of addressing the first problem is to use an inactive mutant, thus "freezing" the association of the remodeler with chromatin; this approach was used to map Isw2 to early meiotic genes, which were known to be sites of Isw2 remodeling activity, as well as to tDNA genes (Gelbart et al., 2005). (A subsequent study reported that association of Isw1, Isw2, and Chd1 with transcription factor-binding sites occurred independently of catalytic activity (Zentner et al., 2013).) To guard against artifactual ChIP-seq signals, control experiments can be conducted using strains or cells in which the target is depleted or the epitope tag employed as a ChIP target is absent. Although this issue is now fairly widely recognized, interpretations made in some research reports published before recognition of the artifact may need to be reconsidered.

A novel approach taken by the Pugh lab to investigate the genome-wide localization of remodelers was to digest cross-linked yeast chromatin with MNase to mononucleosomes, followed by immunoprecipitation to isolate NCPs associated with a particular remodeler and deep sequencing of the associated DNA fragments (Yen et al., 2012). The results indicated a remarkable variation in the distribution of nucleosome-associated remodelers relative to transcription start sites (Fig. 6.11):

- RSC and SWI/SNF subunits were associated with the first three genic nucleosomes (i.e., nucleosomes +1 to +3) and with upstream nucleosomes (preferential association with nucleosomes −1 and −2, respectively).
- Isw1 and Ino80 were broadly distributed across coding sequences, while Isw2 showed preferred occupancy at nucleosome +1.
- Subunits of the two Isw1-containing complexes ISW1a and ISW1b showed distinct occupancy patterns, with Ioc3 (ISW1a) being enriched at nucleosome +1 and Ioc4 (ISW1b) at nucleosomes +2 to +4.
- Isw1 and Ino80 were also associated with the 3′ end of genes, whereas SWI/SNF and RSC were not.

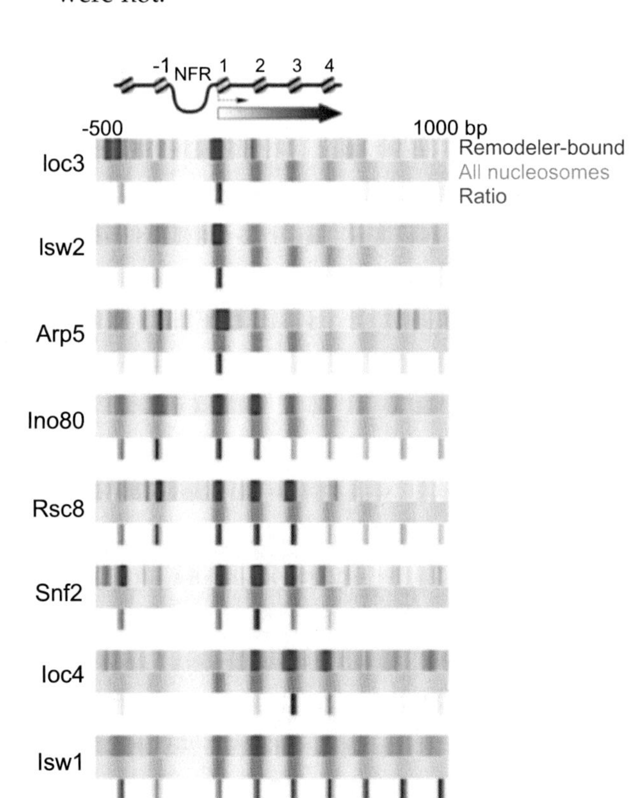

FIG. 6.11 Relative enrichment of nucleosomal association with remodeler subunits, as monitored by immunoprecipitation of cross-linked nucleosomes. Data for remodeler-bound nucleosomes was aligned to the TSS and averaged *(blue)*; data for all nucleosomes is shown in *gray*, and the ratio is indicated by the *red bars*. *From Yen, K., Vinayachandran, V., Batta, K., Koerber, R.T., Pugh, B.F., 2012. Genome-wide nucleosome specificity and directionality of chromatin remodelers. Cell 149, 1461–1473.*

These findings are generally consistent with other reports, with some variability likely due to different methods employed (e.g., ChIP-exo of remodelers associated with mononucleosomes vs using methods not involving MNase digestion), and provide further support for the disparate roles of remodelers in establishing patterns of nucleosome occupancy in yeast (Ng et al., 2002; Ramachandran et al., 2015; Rawal et al., 2018; Tirosh et al., 2010; Whitehouse et al., 2007; Zentner et al., 2013). In mammalian cells, ChIP-seq experiments reveal that the three SWI/SNF complexes are targeted to distinct sites: canonical BAF principally associates with enhancers, PBAF with promoters, and ncBAF with CTCF-binding sites (Mathur et al., 2017; Michel et al., 2018; Nakayama et al., 2017; Pan et al., 2019; Wang et al., 2017). Correct targeting depends on ATPase activity at many, but not all, sites (Pan et al., 2019), and is mediated by specific subunits; for example, loss of ARID1A or SMARCB1 (BAF47) in cancers results in mistargeting of BAF complexes (Mathur et al., 2017; Nakayama et al., 2017). Rapid depletion of ARID1A in murine ESCs by use of an inducible degron resulted in decreased chromatin accessibility at thousands of enhancer sites with coincident loss at many of these sites of the coactivator EP300, supporting a direct role for ARID1A in targeting BAF with consequent remodeling of chromatin to create accessible sites (Blumli et al., 2021). Similarly, acute depletion of SMARCA4 (BRG1) in Hap1 cells resulted in loss of accessibility to BAF-controlled sites in less than one cell cycle (Schick et al., 2021). BAF complexes additionally appear to function at distinct targets than ISWI, as depletion of the motor subunits BRG1 and SNF2H in murine ES cells resulted in different effects on transcription factor binding, in addition to SNF2H but not BRG1 affecting nucleosome spacing (Barisic et al., 2019).

Remodelers are recruited to their targets by varied mechanisms. Yeast Chd1, Isw1, and Isw2 require linker DNA of ~30–85 bp to mobilize nucleosomes in vitro, and it is likely that this affinity is important in directing them to NDRs in vivo (Gangaraju and Bartholomew, 2007; McKnight et al., 2011; Zentner et al., 2013; Zofall et al., 2004). Isw2 has also been reported to be recruited by the repressor protein Ume6 and the activator Swi6 in yeast through interactions between Ume6/Swi6 and the accessory subunit Itc1 (Donovan et al., 2021). SWR1, which catalyzes exchange of H2A.Z-H2B dimers for H2A-H2B dimers and was discussed at length in Chapter 5, is directed to NDRs by virtue of affinity of its Swc2 subunit for nucleosomes associated with long linkers, as well as its affinity for nucleosomes with acetylated tails. INO80 is recruited to double-strand breaks in DNA via interaction of subunits Nhp10 and Arp4 with γ-H2A.X, as discussed earlier.

Consistent with its role in remodeling chromatin during transcriptional induction, early studies indicated that both yeast and human SWI/SNF interact with activation domains of transcriptional activators (Armstrong et al., 1998; Fryer and Archer, 1998; Neely et al., 1999, 2002; Yudkovsky et al., 1999). Two studies reported diminished interaction of activation domains with yeast SWI/SNF lacking all or part of its Snf5 subunit, but differed in whether activation domains also interacted with Swi1 (Prochasson et al., 2003; Sen et al., 2017). Comparison of genome-wide expression in yeast by RNA-seq and SWI/SNF occupancy by ChIP-seq revealed reduced occupancy of Snf2 in deletion mutants of the regulatory module comprising Snf5, Snf6, Snf12, and Swi3; changes in occupancy were sometimes, but not always, accompanied by changes in expression, and vice versa (Dutta et al., 2017). The existence of a considerable fraction of genes whose expression is altered in *snf2Δ* yeast that were not affected in *snf5Δ* yeast implies the existence of multiple recruitment mechanisms in vivo. Finally, it must be recognized that remodelers function in the context of chromatin templates that can be altered by histone modifications and incorporation of variant histones.

References

Ahel, D., Horejsi, Z., Wiechens, N., Polo, S.E., Garcia-Wilson, E., Ahel, I., Flynn, H., Skehel, M., West, S.C., Jackson, S.P., et al., 2009. Poly(ADP-ribose)-dependent regulation of DNA repair by the chromatin remodeling enzyme ALC1. Science 325, 1240–1243.

Alendar, A., Berns, A., 2021. Sentinels of chromatin: chromodomain helicase DNA-binding proteins in development and disease. Genes Dev. 35, 1403–1430.

Almer, A., Rudolph, H., Hinnen, A., Horz, W., 1986. Removal of positioned nucleosomes from the yeast PHO5 promoter upon PHO5 induction releases additional upstream activating DNA elements. EMBO J. 5, 2689–2696.

Alpsoy, A., Dykhuizen, E.C., 2018. Glioma tumor suppressor candidate region gene 1 (GLTSCR1) and its paralog GLTSCR1-like form SWI/SNF chromatin remodeling subcomplexes. J. Biol. Chem. 293, 3892–3903.

Armache, J.P., Gamarra, N., Johnson, S.L., Leonard, J.D., Wu, S., Narlikar, G.J., Cheng, Y., 2019. Cryo-EM structures of remodeler-nucleosome intermediates suggest allosteric control through the nucleosome. elife 8, e46057.

Armstrong, J.A., Bieker, J.J., Emerson, B.M., 1998. A SWI/SNF-related chromatin remodeling complex, E-RC1, is required for tissue-specific transcriptional regulation by EKLF in vitro. Cell 95, 93–104.

Au-Yeung, N., Horvath, C.M., 2018. Histone H2A.Z suppression of interferon-stimulated transcription and antiviral immunity is modulated by GCN5 and BRD2. iScience 6, 68–82.

Awad, S., Ryan, D., Prochasson, P., Owen-Hughes, T., Hassan, A.H., 2010. The Snf2 homolog Fun30 acts as a homodimeric ATP-dependent chromatin-remodeling enzyme. J. Biol. Chem. 285, 9477–9484.

Ayala, R., Willhoft, O., Aramayo, R.J., Wilkinson, M., McCormack, E.A., Ocloo, L., Wigley, D.B., Zhang, X., 2018. Structure and regulation of the human INO80-nucleosome complex. Nature 556, 391–395.

Barisic, D., Stadler, M.B., Iurlaro, M., Schubeler, D., 2019. Mammalian ISWI and SWI/SNF selectively mediate binding of distinct transcription factors. Nature 569, 136–140.

Basehoar, A.D., Zanton, S.J., Pugh, B.F., 2004. Identification and distinct regulation of yeast TATA box-containing genes. Cell 116, 699–709.

Becker, P.B., Workman, J.L., 2013. Nucleosome remodeling and epigenetics. Cold Spring Harb. Perspect. Biol. 5, a017905.

Benezra, R., Cantor, C.R., Axel, R., 1986. Nucleosomes are phased along the mouse beta-major globin gene in erythroid and nonerythroid cells. Cell 44, 697–704.

Biegel, J.A., Zhou, J.Y., Rorke, L.B., Stenstrom, C., Wainwright, L.M., Fogelgren, B., 1999. Germ-line and acquired mutations of INI1 in atypical teratoid and rhabdoid tumors. Cancer Res. 59, 74–79.

Bilokapic, S., Strauss, M., Halic, M., 2018. Structural rearrangements of the histone octamer translocate DNA. Nat. Commun. 9, 1330.

Blumli, S., Wiechens, N., Wu, M.Y., Singh, V., Gierlinski, M., Schweikert, G., Gilbert, N., Naughton, C., Sundaramoorthy, R., Varghese, J., et al., 2021. Acute depletion of the ARID1A subunit of SWI/SNF complexes reveals distinct pathways for activation and repression of transcription. Cell Rep. 37, 109943.

Bowerman, S., Wereszczynski, J., Luger, K., 2021. Archaeal chromatin 'slinkies' are inherently dynamic complexes with deflected DNA wrapping pathways. elife 10, e65587.

Bowman, G.D., 2019. Uncovering a new step in sliding nucleosomes. Trends Biochem. Sci. 44, 643–645.

Brahma, S., Henikoff, S., 2019. RSC-associated subnucleosomes define MNase-sensitive promoters in yeast. Mol. Cell 73, 238–249.e233.

Breeden, L., Nasmyth, K., 1987. Cell cycle control of the yeast HO gene: cis- and trans-acting regulators. Cell 48, 389–397.

Burns, L.G., Peterson, C.L., 1997. The yeast SWI-SNF complex facilitates binding of a transcriptional activator to nucleosomal sites in vivo. Mol. Cell. Biol. 17, 4811–4819.

Cairns, B.R., Kim, Y.J., Sayre, M.H., Laurent, B.C., Kornberg, R.D., 1994. A multisubunit complex containing the SWI1/ADR6, SWI2/SNF2, SWI3, SNF5, and SNF6 gene products isolated from yeast. Proc. Natl. Acad. Sci. U. S. A. 91, 1950–1954.

Cairns, B.R., Lorch, Y., Li, Y., Zhang, M., Lacomis, L., Erdjument-Bromage, H., Tempst, P., Du, J., Laurent, B., Kornberg, R.D., 1996. RSC, an essential, abundant chromatin-remodeling complex. Cell 87, 1249–1260.

Cairns, B.R., Erdjument-Bromage, H., Tempst, P., Winston, F., Kornberg, R.D., 1998. Two actin-related proteins are shared functional components of the chromatin-remodeling complexes RSC and SWI/SNF. Mol. Cell 2, 639–651.

Cairns, B.R., Schlichter, A., Erdjument-Bromage, H., Tempst, P., Kornberg, R.D., Winston, F., 1999. Two functionally distinct forms of the RSC nucleosome-remodeling complex, containing essential AT hook, BAH, and bromodomains. Mol. Cell 4, 715–723.

Carlson, M., Osmond, B.C., Botstein, D., 1981. Mutants of yeast defective in sucrose utilization. Genetics 98, 25–40.

Centore, R.C., Sandoval, G.J., Soares, L.M.M., Kadoch, C., Chan, H.M., 2020. Mammalian SWI/SNF chromatin remodeling complexes: emerging mechanisms and therapeutic strategies. Trends Genet. 36, 936–950.

Cherry, J.M., Hong, E.L., Amundsen, C., Balakrishnan, R., Binkley, G., Chan, E.T., Christie, K.R., Costanzo, M.C., Dwight, S.S., Engel, S.R., et al., 2012. Saccharomyces genome database: the genomics resource of budding yeast. Nucleic Acids Res. 40, D700–D705.

Chittori, S., Hong, J., Bai, Y., Subramaniam, S., 2019. Structure of the primed state of the ATPase domain of chromatin remodeling factor ISWI bound to the nucleosome. Nucleic Acids Res. 47, 9400–9409.

Clapier, C.R., Cairns, B.R., 2009. The biology of chromatin remodeling complexes. Annu. Rev. Biochem. 78, 273–304.

Clapier, C.R., Cairns, B.R., 2012. Regulation of ISWI involves inhibitory modules antagonized by nucleosomal epitopes. Nature 492, 280–284.

Clapier, C.R., Langst, G., Corona, D.F., Becker, P.B., Nightingale, K.P., 2001. Critical role for the histone H4 N terminus in nucleosome remodeling by ISWI. Mol. Cell. Biol. 21, 875–883.

Clapier, C.R., Kasten, M.M., Parnell, T.J., Viswanathan, R., Szerlong, H., Sirinakis, G., Zhang, Y., Cairns, B.R., 2016. Regulation of DNA translocation efficiency within the chromatin remodeler RSC/Sth1 potentiates nucleosome sliding and ejection. Mol. Cell 62, 453–461.

Clapier, C.R., Iwasa, J., Cairns, B.R., Peterson, C.L., 2017. Mechanisms of action and regulation of ATP-dependent chromatin-remodelling complexes. Nat. Rev. Mol. Cell Biol. 18, 407–422.

Clark-Adams, C.D., Winston, F., 1987. The SPT6 gene is essential for growth and is required for delta-mediated transcription in Saccharomyces cerevisiae. Mol. Cell. Biol. 7, 679–686.

Clark-Adams, C.D., Norris, D., Osley, M.A., Fassler, J.S., Winston, F., 1988. Changes in histone gene dosage alter transcription in yeast. Genes Dev. 2, 150–159.

Cosma, M.P., Tanaka, T., Nasmyth, K., 1999. Ordered recruitment of transcription and chromatin remodeling factors to a cell cycle- and developmentally regulated promoter. Cell 97, 299–311.

Cote, J., Quinn, J., Workman, J.L., Peterson, C.L., 1994. Stimulation of GAL4 derivative binding to nucleosomal DNA by the yeast SWI/SNF complex. Science 265, 53–60.

Curcio, M.J., Lutz, S., Lesage, P., 2015. The Ty1 LTR-retrotransposon of budding yeast. Microbiol. Spectr. 3, 1–35.

Dang, W., Kagalwala, M.N., Bartholomew, B., 2006. Regulation of ISW2 by concerted action of histone H4 tail and extranucleosomal DNA. Mol. Cell. Biol. 26, 7388–7396.

Dann, G.P., Liszczak, G.P., Bagert, J.D., Muller, M.M., Nguyen, U.T.T., Wojcik, F., Brown, Z.Z., Bos, J., Panchenko, T., Pihl, R., et al., 2017. ISWI chromatin remodellers sense nucleosome modifications to determine substrate preference. Nature 548, 607–611.

Davey, C.A., Sargent, D.F., Luger, K., Maeder, A.W., Richmond, T.J., 2002. Solvent mediated interactions in the structure of the nucleosome core particle at 1.9 A resolution. J. Mol. Biol. 319, 1097–1113.

Dechassa, M.L., Zhang, B., Horowitz-Scherer, R., Persinger, J., Woodcock, C.L., Peterson, C.L., Bartholomew, B., 2008. Architecture of the SWI/SNF-nucleosome complex. Mol. Cell. Biol. 28, 6010–6021.

Dechassa, M.L., Sabri, A., Pondugula, S., Kassabov, S.R., Chatterjee, N., Kladde, M.P., Bartholomew, B., 2010. SWI/SNF has intrinsic nucleosome disassembly activity that is dependent on adjacent nucleosomes. Mol. Cell 38, 590–602.

Deindl, S., Hwang, W.L., Hota, S.K., Blosser, T.R., Prasad, P., Bartholomew, B., Zhuang, X., 2013. ISWI remodelers slide nucleosomes with coordinated multi-base-pair entry steps and single-base-pair exit steps. Cell 152, 442–452.

Delmas, V., Stokes, D.G., Perry, R.P., 1993. A mammalian DNA-binding protein that contains a chromodomain and an SNF2/SWI2-like helicase domain. Proc. Natl. Acad. Sci. U. S. A. 90, 2414–2418.

Donczew, R., Warfield, L., Pacheco, D., Erijman, A., Hahn, S., 2020. Two roles for the yeast transcription coactivator SAGA and a set of genes redundantly regulated by TFIID and SAGA. elife 9, e50109.

Donovan, D.A., Crandall, J.G., Truong, V.N., Vaaler, A.L., Bailey, T.B., Dinwiddie, D., Banks, O.G., McKnight, L.E., McKnight, J.N., 2021. Basis of specificity for a conserved and promiscuous chromatin remodeling protein. elife 10, e64061.

Downs, J.A., Allard, S., Jobin-Robitaille, O., Javaheri, A., Auger, A., Bouchard, N., Kron, S.J., Jackson, S.P., Cote, J., 2004. Binding of chromatin-modifying activities to phosphorylated histone H2A at DNA damage sites. Mol. Cell 16, 979–990.

Durr, H., Korner, C., Muller, M., Hickmann, V., Hopfner, K.P., 2005. X-ray structures of the *Sulfolobus solfataricus* SWI2/SNF2 ATPase core and its complex with DNA. Cell 121, 363–373.

Dutta, A., Sardiu, M., Gogol, M., Gilmore, J., Zhang, D., Florens, L., Abmayr, S.M., Washburn, M.P., Workman, J.L., 2017. Composition and function of mutant Swi/Snf complexes. Cell Rep. 18, 2124–2134.
Ebbert, R., Birkmann, A., Schuller, H.J., 1999. The product of the SNF2/SWI2 paralogue INO80 of *Saccharomyces cerevisiae* required for efficient expression of various yeast structural genes is part of a high-molecular-weight protein complex. Mol. Microbiol. 32, 741–751.
Elfring, L.K., Deuring, R., McCallum, C.M., Peterson, C.L., Tamkun, J.W., 1994. Identification and characterization of Drosophila relatives of the yeast transcriptional activator SNF2/SWI2. Mol. Cell. Biol. 14, 2225–2234.
Elgin, S.C., 1988. The formation and function of DNase I hypersensitive sites in the process of gene activation. J. Biol. Chem. 263, 19259–19262.
Engeholm, M., de Jager, M., Flaus, A., Brenk, R., van Noort, J., Owen-Hughes, T., 2009. Nucleosomes can invade DNA territories occupied by their neighbors. Nat. Struct. Mol. Biol. 16, 151–158.
Estruch, F., Carlson, M., 1990. SNF6 encodes a nuclear protein that is required for expression of many genes in *Saccharomyces cerevisiae*. Mol. Cell. Biol. 10, 2544–2553.
Eustermann, S., Schall, K., Kostrewa, D., Lakomek, K., Strauss, M., Moldt, M., Hopfner, K.P., 2018. Structural basis for ATP-dependent chromatin remodelling by the INO80 complex. Nature 556, 386–390.
Farnung, L., Vos, S.M., Wigge, C., Cramer, P., 2017. Nucleosome-Chd1 structure and implications for chromatin remodelling. Nature 550, 539–542.
Farnung, L., Ochmann, M., Cramer, P., 2020. Nucleosome-CHD4 chromatin remodeler structure maps human disease mutations. elife 9, e56178.
Fascher, K.D., Schmitz, J., Horz, W., 1993. Structural and functional requirements for the chromatin transition at the PHO5 promoter in *Saccharomyces cerevisiae* upon PHO5 activation. J. Mol. Biol. 231, 658–667.
Felsenfeld, G., 1992. Chromatin as an essential part of the transcriptional mechanism. Nature 355, 219–224.
Ferreira, H., Flaus, A., Owen-Hughes, T., 2007. Histone modifications influence the action of Snf2 family remodelling enzymes by different mechanisms. J. Mol. Biol. 374, 563–579.
Flaus, A., Martin, D.M., Barton, G.J., Owen-Hughes, T., 2006. Identification of multiple distinct Snf2 subfamilies with conserved structural motifs. Nucleic Acids Res. 34, 2887–2905.
Fritsch, O., Benvenuto, G., Bowler, C., Molinier, J., Hohn, B., 2004. The INO80 protein controls homologous recombination in *Arabidopsis thaliana*. Mol. Cell 16, 479–485.
Fryer, C.J., Archer, T.K., 1998. Chromatin remodelling by the glucocorticoid receptor requires the BRG1 complex. Nature 393, 88–91.
Galarneau, L., Nourani, A., Boudreault, A.A., Zhang, Y., Heliot, L., Allard, S., Savard, J., Lane, W.S., Stillman, D.J., Cote, J., 2000. Multiple links between the NuA4 histone acetyltransferase complex and epigenetic control of transcription. Mol. Cell 5, 927–937.
Gamarra, N., Johnson, S.L., Trnka, M.J., Burlingame, A.L., Narlikar, G.J., 2018. The nucleosomal acidic patch relieves auto-inhibition by the ISWI remodeler SNF2h. elife 7, e35322.
Gangaraju, V.K., Bartholomew, B., 2007. Dependency of ISW1a chromatin remodeling on extranucleosomal DNA. Mol. Cell. Biol. 27, 3217–3225.
Ganguli, D., Chereji, R.V., Iben, J.R., Cole, H.A., Clark, D.J., 2014. RSC-dependent constructive and destructive interference between opposing arrays of phased nucleosomes in yeast. Genome Res. 24, 1637–1649.
Gaudreau, L., Schmid, A., Blaschke, D., Ptashne, M., Horz, W., 1997. RNA polymerase II holoenzyme recruitment is sufficient to remodel chromatin at the yeast PHO5 promoter. Cell 89, 55–62.
Gavin, I.M., Simpson, R.T., 1997. Interplay of yeast global transcriptional regulators Ssn6p-Tup1p and Swi-Snf and their effect on chromatin structure. EMBO J. 16, 6263–6271.
Gelbart, M.E., Bachman, N., Delrow, J., Boeke, J.D., Tsukiyama, T., 2005. Genome-wide identification of Isw2 chromatin-remodeling targets by localization of a catalytically inactive mutant. Genes Dev. 19, 942–954.
Gottschalk, A.J., Timinszky, G., Kong, S.E., Jin, J., Cai, Y., Swanson, S.K., Washburn, M.P., Florens, L., Ladurner, A.G., Conaway, J.W., et al., 2009. Poly(ADP-ribosyl)ation directs recruitment and activation of an ATP-dependent chromatin remodeler. Proc. Natl. Acad. Sci. U. S. A. 106, 13770–13774.
Gregory, P.D., Schmid, A., Zavari, M., Munsterkotter, M., Horz, W., 1999. Chromatin remodelling at the PHO8 promoter requires SWI-SNF and SAGA at a step subsequent to activator binding. EMBO J. 18, 6407–6414.
Haber, J.E., Garvik, B., 1977. A new gene affecting the efficiency of mating-type interconversions in homothallic strains of *Saccharomyces cerevisiae*. Genetics 87, 33–50.

Hamiche, A., Sandaltzopoulos, R., Gdula, D.A., Wu, C., 1999. ATP-dependent histone octamer sliding mediated by the chromatin remodeling complex NURF. Cell 97, 833–842.

Hamiche, A., Kang, J.G., Dennis, C., Xiao, H., Wu, C., 2001. Histone tails modulate nucleosome mobility and regulate ATP-dependent nucleosome sliding by NURF. Proc. Natl. Acad. Sci. U. S. A. 98, 14316–14321.

Han, Y., Reyes, A.A., Malik, S., He, Y., 2020. Cryo-EM structure of SWI/SNF complex bound to a nucleosome. Nature 579, 452–455.

Happel, A.M., Swanson, M.S., Winston, F., 1991. The SNF2, SNF5 and SNF6 genes are required for Ty transcription in *Saccharomyces cerevisiae*. Genetics 128, 69–77.

Harada, B.T., Hwang, W.L., Deindl, S., Chatterjee, N., Bartholomew, B., Zhuang, X., 2016. Stepwise nucleosome translocation by RSC remodeling complexes. elife 5, e10051.

Hasselblatt, M., Isken, S., Linge, A., Eikmeier, K., Jeibmann, A., Oyen, F., Nagel, I., Richter, J., Bartelheim, K., Kordes, U., et al., 2013. High-resolution genomic analysis suggests the absence of recurrent genomic alterations other than SMARCB1 aberrations in atypical teratoid/rhabdoid tumors. Genes Chromosom. Cancer 52, 185–190.

Hauk, G., McKnight, J.N., Nodelman, I.M., Bowman, G.D., 2010. The chromodomains of the Chd1 chromatin remodeler regulate DNA access to the ATPase motor. Mol. Cell 39, 711–723.

He, Q., Yu, C., Morse, R.H., 2008. Dispersed mutations in histone H3 that affect transcriptional repression and chromatin structure of the CHA1 promoter in yeast. Eukaryot. Cell 7, 1649–1660.

He, S., Wu, Z., Tian, Y., Yu, Z., Yu, J., Wang, X., Li, J., Liu, B., Xu, Y., 2020. Structure of nucleosome-bound human BAF complex. Science 367, 875–881.

Helming, K.C., Wang, X., Roberts, C.W.M., 2014. Vulnerabilities of mutant SWI/SNF complexes in cancer. Cancer Cell 26, 309–317.

Hirschhorn, J.N., Brown, S.A., Clark, C.D., Winston, F., 1992. Evidence that SNF2/SWI2 and SNF5 activate transcription in yeast by altering chromatin structure. Genes Dev. 6, 2288–2298.

Hirschman, J.E., Durbin, K.J., Winston, F., 1988. Genetic evidence for promoter competition in Saccharomyces cerevisiae. Mol. Cell. Biol. 8, 4608–4615.

Ho, L., Crabtree, G.R., 2010. Chromatin remodelling during development. Nature 463, 474–484.

Ho, L., Ronan, J.L., Wu, J., Staahl, B.T., Chen, L., Kuo, A., Lessard, J., Nesvizhskii, A.I., Ranish, J., Crabtree, G.R., 2009. An embryonic stem cell chromatin remodeling complex, esBAF, is essential for embryonic stem cell self-renewal and pluripotency. Proc. Natl. Acad. Sci. U. S. A. 106, 5181–5186.

Holstege, F.C., Jennings, E.G., Wyrick, J.J., Lee, T.I., Hengartner, C.J., Green, M.R., Golub, T.R., Lander, E.S., Young, R.A., 1998. Dissecting the regulatory circuitry of a eukaryotic genome. Cell 95, 717–728.

Huisinga, K.L., Pugh, B.F., 2004. A genome-wide housekeeping role for TFIID and a highly regulated stress-related role for SAGA in *Saccharomyces cerevisiae*. Mol. Cell 13, 573–585.

Ikura, T., Ogryzko, V.V., Grigoriev, M., Groisman, R., Wang, J., Horikoshi, M., Scully, R., Qin, J., Nakatani, Y., 2000. Involvement of the TIP60 histone acetylase complex in DNA repair and apoptosis. Cell 102, 463–473.

Imbalzano, A.N., Kwon, H., Green, M.R., Kingston, R.E., 1994. Facilitated binding of TATA-binding protein to nucleosomal DNA. Nature 370, 481–485.

Ito, T., Bulger, M., Pazin, M.J., Kobayashi, R., Kadonaga, J.T., 1997. ACF, an ISWI-containing and ATP-utilizing chromatin assembly and remodeling factor. Cell 90, 145–155.

Jain, D., Baldi, S., Zabel, A., Straub, T., Becker, P.B., 2015. Active promoters give rise to false positive 'Phantom Peaks' in ChIP-seq experiments. Nucleic Acids Res. 43, 6959–6968.

Jeronimo, C., Watanabe, S., Kaplan, C.D., Peterson, C.L., Robert, F., 2015. The histone chaperones FACT and Spt6 restrict H2A.Z from intragenic locations. Mol. Cell 58, 1113–1123.

Jungblut, A., Hopfner, K.P., Eustermann, S., 2020. Megadalton chromatin remodelers: common principles for versatile functions. Curr. Opin. Struct. Biol. 64, 134–144.

Kadoch, C., Crabtree, G.R., 2013. Reversible disruption of mSWI/SNF (BAF) complexes by the SS18-SSX oncogenic fusion in synovial sarcoma. Cell 153, 71–85.

Kadoch, C., Hargreaves, D.C., Hodges, C., Elias, L., Ho, L., Ranish, J., Crabtree, G.R., 2013. Proteomic and bioinformatic analysis of mammalian SWI/SNF complexes identifies extensive roles in human malignancy. Nat. Genet. 45, 592–601.

Kaeser, M.D., Aslanian, A., Dong, M.Q., Yates 3rd, J.R., Emerson, B.M., 2008. BRD7, a novel PBAF-specific SWI/SNF subunit, is required for target gene activation and repression in embryonic stem cells. J. Biol. Chem. 283, 32254–32263.

Kagalwala, M.N., Glaus, B.J., Dang, W., Zofall, M., Bartholomew, B., 2004. Topography of the ISW2-nucleosome complex: insights into nucleosome spacing and chromatin remodeling. EMBO J. 23, 2092–2104.

Kagawa, R., Montgomery, M.G., Braig, K., Leslie, A.G., Walker, J.E., 2004. The structure of bovine F1-ATPase inhibited by ADP and beryllium fluoride. EMBO J. 23, 2734–2744.

Karl, L.A., Peritore, M., Galanti, L., Pfander, B., 2021. DNA double strand break repair and its control by nucleosome remodeling. Front. Genet. 12, 821543.

Kennison, J.A., Tamkun, J.W., 1988. Dosage-dependent modifiers of polycomb and antennapedia mutations in Drosophila. Proc. Natl. Acad. Sci. U. S. A. 85, 8136–8140.

Khavari, P.A., Peterson, C.L., Tamkun, J.W., Mendel, D.B., Crabtree, G.R., 1993. BRG1 contains a conserved domain of the SWI2/SNF2 family necessary for normal mitotic growth and transcription. Nature 366, 170–174.

Kim, J.M., Visanpattanasin, P., Jou, V., Liu, S., Tang, X., Zheng, Q., Li, K.Y., Snedeker, J., Lavis, L.D., Lionnet, T., et al., 2021. Single-molecule imaging of chromatin remodelers reveals role of ATPase in promoting fast kinetics of target search and dissociation from chromatin. elife 10, e69387.

Kornberg, R.D., 2005. Mediator and the mechanism of transcriptional activation. Trends Biochem. Sci. 30, 235–239.

Kornberg, R.D., Lorch, Y., 1991. Irresistible force meets immovable object: transcription and the nucleosome. Cell 67, 833–836.

Kruger, W., Herskowitz, I., 1991. A negative regulator of HO transcription, SIN1 (SPT2), is a nonspecific DNA-binding protein related to HMG1. Mol. Cell. Biol. 11, 4135–4146.

Kruger, W., Peterson, C.L., Sil, A., Coburn, C., Arents, G., Moudrianakis, E.N., Herskowitz, I., 1995. Amino acid substitutions in the structured domains of histones H3 and H4 partially relieve the requirement of the yeast SWI/SNF complex for transcription. Genes Dev. 9, 2770–2779.

Kubik, S., Bruzzone, M.J., Challal, D., Dreos, R., Mattarocci, S., Bucher, P., Libri, D., Shore, D., 2019. Opposing chromatin remodelers control transcription initiation frequency and start site selection. Nat. Struct. Mol. Biol. 26, 744–754.

Kuras, L., Kosa, P., Mencia, M., Struhl, K., 2000. TAF-containing and TAF-independent forms of transcriptionally active TBP in vivo. Science 288, 1244–1248.

Kurumizaka, H., Wolffe, A.P., 1997. Sin mutations of histone H3: influence on nucleosome core structure and function. Mol. Cell. Biol. 17, 6953–6969.

Kwon, H., Imbalzano, A.N., Khavari, P.A., Kingston, R.E., Green, M.R., 1994. Nucleosome disruption and enhancement of activator binding by a human SW1/SNF complex. Nature 370, 477–481.

Langst, G., Bonte, E.J., Corona, D.F., Becker, P.B., 1999. Nucleosome movement by CHRAC and ISWI without disruption or trans-displacement of the histone octamer. Cell 97, 843–852.

Laurent, B.C., Carlson, M., 1992. Yeast SNF2/SWI2, SNF5, and SNF6 proteins function coordinately with the gene-specific transcriptional activators GAL4 and Bicoid. Genes Dev. 6, 1707–1715.

Laurent, B.C., Treitel, M.A., Carlson, M., 1990. The SNF5 protein of *Saccharomyces cerevisiae* is a glutamine- and proline-rich transcriptional activator that affects expression of a broad spectrum of genes. Mol. Cell. Biol. 10, 5616–5625.

Laurent, B.C., Treitel, M.A., Carlson, M., 1991. Functional interdependence of the yeast SNF2, SNF5, and SNF6 proteins in transcriptional activation. Proc. Natl. Acad. Sci. U. S. A. 88, 2687–2691.

Laurent, B.C., Yang, X., Carlson, M., 1992. An essential *Saccharomyces cerevisiae* gene homologous to SNF2 encodes a helicase-related protein in a new family. Mol. Cell. Biol. 12, 1893–1902.

Laurent, B.C., Treich, I., Carlson, M., 1993. The yeast SNF2/SWI2 protein has DNA-stimulated ATPase activity required for transcriptional activation. Genes Dev. 7, 583–591.

Lee, K.P., Baxter, H.J., Guillemette, J.G., Lawford, H.G., Lewis, P.N., 1982. Structural studies on yeast nucleosomes. Can. J. Biochem. 60, 379–388.

Lee, T.I., Causton, H.C., Holstege, F.C., Shen, W.C., Hannett, N., Jennings, E.G., Winston, F., Green, M.R., Young, R.A., 2000. Redundant roles for the TFIID and SAGA complexes in global transcription. Nature 405, 701–704.

Lee, W., Tillo, D., Bray, N., Morse, R.H., Davis, R.W., Hughes, T.R., Nislow, C., 2007. A high-resolution atlas of nucleosome occupancy in yeast. Nat. Genet. 39, 1235–1244.

Lee, R.S., Stewart, C., Carter, S.L., Ambrogio, L., Cibulskis, K., Sougnez, C., Lawrence, M.S., Auclair, D., Mora, J., Golub, T.R., et al., 2012. A remarkably simple genome underlies highly malignant pediatric rhabdoid cancers. J. Clin. Invest. 122, 2983–2988.

Lessard, J., Wu, J.I., Ranish, J.A., Wan, M., Winslow, M.M., Staahl, B.T., Wu, H., Aebersold, R., Graef, I.A., Crabtree, G.R., 2007. An essential switch in subunit composition of a chromatin remodeling complex during neural development. Neuron 55, 201–215.

Li, X.Y., Bhaumik, S.R., Green, M.R., 2000. Distinct classes of yeast promoters revealed by differential TAF recruitment. Science 288, 1242–1244.

Li, M., Xia, X., Tian, Y., Jia, Q., Liu, X., Lu, Y., Li, M., Li, X., Chen, Z., 2019. Mechanism of DNA translocation underlying chromatin remodelling by Snf2. Nature 567, 409–413.

Lickert, H., Takeuchi, J.K., Von Both, I., Walls, J.R., McAuliffe, F., Adamson, S.L., Henkelman, R.M., Wrana, J.L., Rossant, J., Bruneau, B.G., 2004. Baf60c is essential for function of BAF chromatin remodelling complexes in heart development. Nature 432, 107–112.

Liu, X., Li, M., Xia, X., Li, X., Chen, Z., 2017. Mechanism of chromatin remodelling revealed by the Snf2-nucleosome structure. Nature 544, 440–445.

Lorch, Y., Zhang, M., Kornberg, R.D., 1999. Histone octamer transfer by a chromatin-remodeling complex. Cell 96, 389–392.

Luger, K., Mader, A.W., Richmond, R.K., Sargent, D.F., Richmond, T.J., 1997. Crystal structure of the nucleosome core particle at 2.8 A resolution [see comments]. Nature 389, 251–260.

Malone, E.A., Clark, C.D., Chiang, A., Winston, F., 1991. Mutations in SPT16/CDC68 suppress cis- and trans-acting mutations that affect promoter function in *Saccharomyces cerevisiae*. Mol. Cell. Biol. 11, 5710–5717.

Markert, J., Luger, K., 2021. Nucleosomes meet their remodeler match. Trends Biochem. Sci. 46, 41–50.

Mashtalir, N., D'Avino, A.R., Michel, B.C., Luo, J., Pan, J., Otto, J.E., Zullow, H.J., McKenzie, Z.M., Kubiak, R.L., St Pierre, R., et al., 2018. Modular organization and assembly of SWI/SNF family chromatin remodeling complexes. Cell 175, 1272–1288.e1220.

Mashtalir, N., Suzuki, H., Farrell, D.P., Sankar, A., Luo, J., Filipovski, M., D'Avino, A.R., St Pierre, R., Valencia, A.M., Onikubo, T., et al., 2020. A structural model of the endogenous human BAF complex informs disease mechanisms. Cell 183, 802–817.e824.

Mashtalir, N., Dao, H.T., Sankar, A., Liu, H., Corin, A.J., Bagert, J.D., Ge, E.J., D'Avino, A.R., Filipovski, M., Michel, B.C., et al., 2021. Chromatin landscape signals differentially dictate the activities of mSWI/SNF family complexes. Science 373, 306–315.

Mathur, R., Alver, B.H., San Roman, A.K., Wilson, B.G., Wang, X., Agoston, A.T., Park, P.J., Shivdasani, R.A., Roberts, C.W., 2017. ARID1A loss impairs enhancer-mediated gene regulation and drives colon cancer in mice. Nat. Genet. 49, 296–302.

McKnight, J.N., Jenkins, K.R., Nodelman, I.M., Escobar, T., Bowman, G.D., 2011. Extranucleosomal DNA binding directs nucleosome sliding by Chd1. Mol. Cell. Biol. 31, 4746–4759.

Michel, B.C., D'Avino, A.R., Cassel, S.H., Mashtalir, N., McKenzie, Z.M., McBride, M.J., Valencia, A.M., Zhou, Q., Bocker, M., Soares, L.M.M., et al., 2018. A non-canonical SWI/SNF complex is a synthetic lethal target in cancers driven by BAF complex perturbation. Nat. Cell Biol. 20, 1410–1420.

Miller, J.A., Widom, J., 2003. Collaborative competition mechanism for gene activation in vivo. Mol. Cell. Biol. 23, 1623–1632.

Mizuguchi, G., Tsukiyama, T., Wisniewski, J., Wu, C., 1997. Role of nucleosome remodeling factor NURF in transcriptional activation of chromatin. Mol. Cell 1, 141–150.

Mizuguchi, G., Shen, X., Landry, J., Wu, W.H., Sen, S., Wu, C., 2004. ATP-driven exchange of histone H2AZ variant catalyzed by SWR1 chromatin remodeling complex. Science 303, 343–348.

Morrison, A.J., Highland, J., Krogan, N.J., Arbel-Eden, A., Greenblatt, J.F., Haber, J.E., Shen, X., 2004. INO80 and gamma-H2AX interaction links ATP-dependent chromatin remodeling to DNA damage repair. Cell 119, 767–775.

Morse, R.H., 2003. Getting into chromatin: how do transcription factors get past the histones? Biochem. Cell Biol. 81, 101–112.

Morse, R.H., Pederson, D.S., Dean, A., Simpson, R.T., 1987. Yeast nucleosomes allow thermal untwisting of DNA. Nucleic Acids Res. 15, 10311–10330.

Muchardt, C., Yaniv, M., 1993. A human homologue of *Saccharomyces cerevisiae* SNF2/SWI2 and Drosophila brm genes potentiates transcriptional activation by the glucocorticoid receptor. EMBO J. 12, 4279–4290.

Muthurajan, U.M., Bao, Y., Forsberg, L.J., Edayathumangalam, R.S., Dyer, P.N., White, C.L., Luger, K., 2004. Crystal structures of histone sin mutant nucleosomes reveal altered protein-DNA interactions. EMBO J. 23, 260–271.

Nakayama, R.T., Pulice, J.L., Valencia, A.M., McBride, M.J., McKenzie, Z.M., Gillespie, M.A., Ku, W.L., Teng, M., Cui, K., Williams, R.T., et al., 2017. SMARCB1 is required for widespread BAF complex-mediated activation of enhancers and bivalent promoters. Nat. Genet. 49, 1613–1623.

Nasmyth, K., Stillman, D., Kipling, D., 1987. Both positive and negative regulators of HO transcription are required for mother-cell-specific mating-type switching in yeast. Cell 48, 579–587.

Neely, K.E., Hassan, A.H., Wallberg, A.E., Steger, D.J., Cairns, B.R., Wright, A.P., Workman, J.L., 1999. Activation domain-mediated targeting of the SWI/SNF complex to promoters stimulates transcription from nucleosome arrays. Mol. Cell 4, 649–655.

Neely, K.E., Hassan, A.H., Brown, C.E., Howe, L., Workman, J.L., 2002. Transcription activator interactions with multiple SWI/SNF subunits. Mol. Cell. Biol. 22, 1615–1625.

Neigeborn, L., Carlson, M., 1984. Genes affecting the regulation of SUC2 gene expression by glucose repression in Saccharomyces cerevisiae. Genetics 108, 845–858.

Neigeborn, L., Rubin, K., Carlson, M., 1986. Suppressors of SNF2 mutations restore invertase derepression and cause temperature-sensitive lethality in yeast. Genetics 112, 741–753.

Neigeborn, L., Celenza, J.L., Carlson, M., 1987. SSN20 is an essential gene with mutant alleles that suppress defects in SUC2 transcription in *Saccharomyces cerevisiae*. Mol. Cell. Biol. 7, 672–678.

Ng, H.H., Robert, F., Young, R.A., Struhl, K., 2002. Genome-wide location and regulated recruitment of the RSC nucleosome-remodeling complex. Genes Dev. 16, 806–819.

Nguyen, U.T., Bittova, L., Muller, M.M., Fierz, B., David, Y., Houck-Loomis, B., Feng, V., Dann, G.P., Muir, T.W., 2014. Accelerated chromatin biochemistry using DNA-barcoded nucleosome libraries. Nat. Methods 11, 834–840.

Nie, Z., Xue, Y., Yang, D., Zhou, S., Deroo, B.J., Archer, T.K., Wang, W., 2000. A specificity and targeting subunit of a human SWI/SNF family-related chromatin-remodeling complex. Mol. Cell. Biol. 20, 8879–8888.

Nilasena, D.S., Trieu, E.P., Targoff, I.N., 1995. Analysis of the Mi-2 autoantigen of dermatomyositis. Arthritis Rheum. 38, 123–128.

Nodelman, I.M., Das, S., Faustino, A.M., Fried, S.D., Bowman, G.D., Armache, J.P., 2022. Nucleosome recognition and DNA distortion by the Chd1 remodeler in a nucleotide-free state. Nat. Struct. Mol. Biol. 29, 121–129.

Nonet, M., Scafe, C., Sexton, J., Young, R., 1987. Eucaryotic RNA polymerase conditional mutant that rapidly ceases mRNA synthesis. Mol. Cell. Biol. 7, 1602–1611.

Oberbeckmann, E., Niebauer, V., Watanabe, S., Farnung, L., Moldt, M., Schmid, A., Cramer, P., Peterson, C.L., Eustermann, S., Hopfner, K.P., et al., 2021. Ruler elements in chromatin remodelers set nucleosome array spacing and phasing. Nat. Commun. 12, 3232.

Okabe, I., Bailey, L.C., Attree, O., Srinivasan, S., Perkel, J.M., Laurent, B.C., Carlson, M., Nelson, D.L., Nussbaum, R.L., 1992. Cloning of human and bovine homologs of SNF2/SWI2: a global activator of transcription in yeast *S. cerevisiae*. Nucleic Acids Res. 20, 4649–4655.

Ong, M.S., Richmond, T.J., Davey, C.A., 2007. DNA stretching and extreme kinking in the nucleosome core. J. Mol. Biol. 368, 1067–1074.

Owen-Hughes, T., Utley, R.T., Cote, J., Peterson, C.L., Workman, J.L., 1996. Persistent site-specific remodeling of a nucleosome array by transient action of the SWI/SNF complex. Science 273, 513–516.

Pan, J., Meyers, R.M., Michel, B.C., Mashtalir, N., Sizemore, A.E., Wells, J.N., Cassel, S.H., Vazquez, F., Weir, B.A., Hahn, W.C., et al., 2018. Interrogation of mammalian protein complex structure, function, and membership using genome-scale fitness screens. Cell Syst. 6, 555–568.e557.

Pan, J., McKenzie, Z.M., D'Avino, A.R., Mashtalir, N., Lareau, C.A., St Pierre, R., Wang, L., Shilatifard, A., Kadoch, C., 2019. The ATPase module of mammalian SWI/SNF family complexes mediates subcomplex identity and catalytic activity-independent genomic targeting. Nat. Genet. 51, 618–626.

Papamichos-Chronakis, M., Watanabe, S., Rando, O.J., Peterson, C.L., 2011. Global regulation of H2A.Z localization by the INO80 chromatin-remodeling enzyme is essential for genome integrity. Cell 144, 200–213.

Park, D., Lee, Y., Bhupindersingh, G., Iyer, V.R., 2013. Widespread Misinterpretable ChIP-seq Bias in yeast. PLoS One 8, e83506.

Paro, R., Hogness, D.S., 1991. The Polycomb protein shares a homologous domain with a heterochromatin-associated protein of Drosophila. Proc. Natl. Acad. Sci. U. S. A. 88, 263–267.

Patel, A.B., Moore, C.M., Greber, B.J., Luo, J., Zukin, S.A., Ranish, J., Nogales, E., 2019. Architecture of the chromatin remodeler RSC and insights into its nucleosome engagement. elife 8, e54449.

Pazin, M.J., Kamakaka, R.T., Kadonaga, J.T., 1994. ATP-dependent nucleosome reconfiguration and transcriptional activation from preassembled chromatin templates. Science 266, 2007–2011.

Peritore, M., Reusswig, K.U., Bantele, S.C.S., Straub, T., Pfander, B., 2021. Strand-specific ChIP-seq at DNA breaks distinguishes ssDNA versus dsDNA binding and refutes single-stranded nucleosomes. Mol. Cell 81, 1841–1853.e1844.

Peterson, C.L., Herskowitz, I., 1992. Characterization of the yeast SWI1, SWI2, and SWI3 genes, which encode a global activator of transcription. Cell 68, 573–583.

Peterson, C.L., Kruger, W., Herskowitz, I., 1991. A functional interaction between the C-terminal domain of RNA polymerase II and the negative regulator SIN1. Cell 64, 1135–1143.

Peterson, C.L., Dingwall, A., Scott, M.P., 1994. Five SWI/SNF gene products are components of a large multisubunit complex required for transcriptional enhancement. Proc. Natl. Acad. Sci. U. S. A. 91, 2905–2908.

Peterson, C.L., Zhao, Y., Chait, B.T., 1998. Subunits of the yeast SWI/SNF complex are members of the actin-related protein (ARP) family. J. Biol. Chem. 273, 23641–23644.

Pineiro, M., Puerta, C., Palacian, E., 1991. Yeast nucleosomal particles: structural and transcriptional properties. Biochemistry 30, 5805–5810.

Poch, O., Winsor, B., 1997. Who's who among the *Saccharomyces cerevisiae* actin-related proteins? A classification and nomenclature proposal for a large family. Yeast 13, 1053–1058.

Polach, K.J., Widom, J., 1995. Mechanism of protein access to specific DNA sequences in chromatin: a dynamic equilibrium model for gene regulation. J. Mol. Biol. 254, 130–149.

Pollard, K.J., Peterson, C.L., 1997. Role for ADA/GCN5 products in antagonizing chromatin-mediated transcriptional repression. Mol. Cell. Biol. 17, 6212–6222.

Prajapati, H.K., Ocampo, J., Clark, D.J., 2020. Interplay among ATP-dependent chromatin remodelers determines chromatin organisation in yeast. Biology (Basel) 9, 190.

Prelich, G., Winston, F., 1993. Mutations that suppress the deletion of an upstream activating sequence in yeast: involvement of a protein kinase and histone H3 in repressing transcription in vivo. Genetics 135, 665–676.

Prochasson, P., Neely, K.E., Hassan, A.H., Li, B., Workman, J.L., 2003. Targeting activity is required for SWI/SNF function in vivo and is accomplished through two partially redundant activator-interaction domains. Mol. Cell 12, 983–990.

Ptashne, M., Gann, A., 1997. Transcriptional activation by recruitment. Nature 386, 569–577.

Qiu, H., Chereji, R.V., Hu, C., Cole, H.A., Rawal, Y., Clark, D.J., Hinnebusch, A.G., 2016. Genome-wide cooperation by HAT Gcn5, remodeler SWI/SNF, and chaperone Ydj1 in promoter nucleosome eviction and transcriptional activation. Genome Res. 26, 211–225.

Racki, L.R., Naber, N., Pate, E., Leonard, J.D., Cooke, R., Narlikar, G.J., 2014. The histone H4 tail regulates the conformation of the ATP-binding pocket in the SNF2h chromatin remodeling enzyme. J. Mol. Biol. 426, 2034–2044.

Ramachandran, S., Zentner, G.E., Henikoff, S., 2015. Asymmetric nucleosomes flank promoters in the budding yeast genome. Genome Res. 25, 381–390.

Rando, O.J., Zhao, K., Crabtree, G.R., 2000. Searching for a function for nuclear actin. Trends Cell Biol. 10, 92–97.

Ranjan, A., Nguyen, V.Q., Liu, S., Wisniewski, J., Kim, J.M., Tang, X., Mizuguchi, G., Elalaoui, E., Nickels, T.J., Jou, V., et al., 2020. Live-cell single particle imaging reveals the role of RNA polymerase II in histone H2A.Z eviction. elife 9, e55667.

Rawal, Y., Chereji, R.V., Qiu, H., Ananthakrishnan, S., Govind, C.K., Clark, D.J., Hinnebusch, A.G., 2018. SWI/SNF and RSC cooperate to reposition and evict promoter nucleosomes at highly expressed genes in yeast. Genes Dev. 32, 695–710.

Reyes, A.A., Marcum, R.D., He, Y., 2021. Structure and function of chromatin remodelers. J. Mol. Biol. 433, 166929.

Rhee, H.S., Pugh, B.F., 2012. Genome-wide structure and organization of eukaryotic pre-initiation complexes. Nature 483, 295–301.

Roberts, S.M., Winston, F., 1997. Essential functional interactions of SAGA, a Saccharomyces cerevisiae complex of Spt, Ada, and Gcn5 proteins, with the Snf/Swi and Srb/mediator complexes. Genetics 147, 451–465.

Roberts, C.W., Galusha, S.A., McMenamin, M.E., Fletcher, C.D., Orkin, S.H., 2000. Haploinsufficiency of Snf5 (integrase interactor 1) predisposes to malignant rhabdoid tumors in mice. Proc. Natl. Acad. Sci. U. S. A. 97, 13796–13800.

Ronan, J.L., Wu, W., Crabtree, G.R., 2013. From neural development to cognition: unexpected roles for chromatin. Nat. Rev. Genet. 14, 347–359.

Ruhl, D.D., Jin, J., Cai, Y., Swanson, S., Florens, L., Washburn, M.P., Conaway, R.C., Conaway, J.W., Chrivia, J.C., 2006. Purification of a human SRCAP complex that remodels chromatin by incorporating the histone variant H2A.Z into nucleosomes. Biochemistry 45, 5671–5677.

Ryan, M.P., Jones, R., Morse, R.H., 1998. SWI-SNF complex participation in transcriptional activation at a step subsequent to activator binding. Mol. Cell. Biol. 18, 1774–1782.

Sabantsev, A., Levendosky, R.F., Zhuang, X., Bowman, G.D., Deindl, S., 2019. Direct observation of coordinated DNA movements on the nucleosome during chromatin remodelling. Nat. Commun. 10, 1720.

Saha, A., Wittmeyer, J., Cairns, B.R., 2002. Chromatin remodeling by RSC involves ATP-dependent DNA translocation. Genes Dev. 16, 2120–2134.

Saha, A., Wittmeyer, J., Cairns, B.R., 2005. Chromatin remodeling through directional DNA translocation from an internal nucleosomal site. Nat. Struct. Mol. Biol. 12, 747–755.

Scacchetti, A., Schauer, T., Reim, A., Apostolou, Z., Campos Sparr, A., Krause, S., Heun, P., Wierer, M., Becker, P.B., 2020. Drosophila SWR1 and NuA4 complexes are defined by DOMINO isoforms. elife 9, e56325.

Schibler, U., Meyehaus, O., Zaret, K.S., 2013. Robert Perry 1931-2013. Cell 154, 953–954.

Schick, S., Grosche, S., Kohl, K.E., Drpic, D., Jaeger, M.G., Marella, N.C., Imrichova, H., Lin, J.G., Hofstatter, G., Schuster, M., et al., 2021. Acute BAF perturbation causes immediate changes in chromatin accessibility. Nat. Genet. 53, 269–278.

Schlichter, A., Kasten, M.M., Parnell, T.J., Cairns, B.R., 2020. Specialization of the chromatin remodeler RSC to mobilize partially-unwrapped nucleosomes. elife 9, e58130.

Schwanbeck, R., Xiao, H., Wu, C., 2004. Spatial contacts and nucleosome step movements induced by the NURF chromatin remodeling complex. J. Biol. Chem. 279, 39933–39941.

Seelig, H.P., Moosbrugger, I., Ehrfeld, H., Fink, T., Renz, M., Genth, E., 1995. The major dermatomyositis-specific Mi-2 autoantigen is a presumed helicase involved in transcriptional activation. Arthritis Rheum. 38, 1389–1399.

Sen, P., Vivas, P., Dechassa, M.L., Mooney, A.M., Poirier, M.G., Bartholomew, B., 2013. The SnAC domain of SWI/SNF is a histone anchor required for remodeling. Mol. Cell. Biol. 33, 360–370.

Sen, P., Luo, J., Hada, A., Hailu, S.G., Dechassa, M.L., Persinger, J., Brahma, S., Paul, S., Ranish, J., Bartholomew, B., 2017. Loss of Snf5 induces formation of an aberrant SWI/SNF complex. Cell Rep. 18, 2135–2147.

Sharma, V.M., Li, B., Reese, J.C., 2003. SWI/SNF-dependent chromatin remodeling of RNR3 requires TAF(II)s and the general transcription machinery. Genes Dev. 17, 502–515.

Shen, X., Mizuguchi, G., Hamiche, A., Wu, C., 2000. A chromatin remodelling complex involved in transcription and DNA processing. Nature 406, 541–544.

Shen, X., Ranallo, R., Choi, E., Wu, C., 2003. Involvement of actin-related proteins in ATP-dependent chromatin remodeling. Mol. Cell 12, 147–155.

Sif, S., Stukenberg, P.T., Kirschner, M.W., Kingston, R.E., 1998. Mitotic inactivation of a human SWI/SNF chromatin remodeling complex. Genes Dev. 12, 2842–2851.

Singleton, M.R., Dillingham, M.S., Wigley, D.B., 2007. Structure and mechanism of helicases and nucleic acid translocases. Annu. Rev. Biochem. 76, 23–50.

Stern, M., Jensen, R., Herskowitz, I., 1984. Five SWI genes are required for expression of the HO gene in yeast. J. Mol. Biol. 178, 853–868.

Sternberg, P.W., Stern, M.J., Clark, I., Herskowitz, I., 1987. Activation of the yeast HO gene by release from multiple negative controls. Cell 48, 567–577.

Stokes, D.G., Perry, R.P., 1995. DNA-binding and chromatin localization properties of CHD1. Mol. Cell. Biol. 15, 2745–2753.

Stokes, D.G., Tartof, K.D., Perry, R.P., 1996. CHD1 is concentrated in interbands and puffed regions of Drosophila polytene chromosomes. Proc. Natl. Acad. Sci. U. S. A. 93, 7137–7142.

Struhl, K., 1986. Constitutive and inducible *Saccharomyces cerevisiae* promoters: evidence for two distinct molecular mechanisms. Mol. Cell. Biol. 6, 3847–3853.

Sudarsanam, P., Winston, F., 2000. The Swi/Snf family nucleosome-remodeling complexes and transcriptional control. Trends Genet. 16, 345–351.

Sudarsanam, P., Iyer, V.R., Brown, P.O., Winston, F., 2000. Whole-genome expression analysis of snf/swi mutants of *Saccharomyces cerevisiae*. Proc. Natl. Acad. Sci. U. S. A. 97, 3364–3369.

Sundaramoorthy, R., Hughes, A.L., El-Mkami, H., Norman, D.G., Ferreira, H., Owen-Hughes, T., 2018. Structure of the chromatin remodelling enzyme Chd1 bound to a ubiquitinylated nucleosome. elife 7, e35720.

Svaren, J., Schmitz, J., Horz, W., 1994. The transactivation domain of Pho4 is required for nucleosome disruption at the PHO5 promoter. EMBO J. 13, 4856–4862.

Szerlong, H., Saha, A., Cairns, B.R., 2003. The nuclear actin-related proteins Arp7 and Arp9: a dimeric module that cooperates with architectural proteins for chromatin remodeling. EMBO J. 22, 3175–3187.

Szerlong, H., Hinata, K., Viswanathan, R., Erdjument-Bromage, H., Tempst, P., Cairns, B.R., 2008. The HSA domain binds nuclear actin-related proteins to regulate chromatin-remodeling ATPases. Nat. Struct. Mol. Biol. 15, 469–476.

Tachiwana, H., Kagawa, W., Osakabe, A., Kawaguchi, K., Shiga, T., Hayashi-Takanaka, Y., Kimura, H., Kurumizaka, H., 2010. Structural basis of instability of the nucleosome containing a testis-specific histone variant, human H3T. Proc. Natl. Acad. Sci. U. S. A. 107, 10454–10459.

Taguchi, A.K., Young, E.T., 1987. The identification and characterization of ADR6, a gene required for sporulation and for expression of the alcohol dehydrogenase II isozyme from Saccharomyces cerevisiae. Genetics 116, 523–530.

Tamkun, J.W., Deuring, R., Scott, M.P., Kissinger, M., Pattatucci, A.M., Kaufman, T.C., Kennison, J.A., 1992. Brahma: a regulator of Drosophila homeotic genes structurally related to the yeast transcriptional activator SNF2/SWI2. Cell 68, 561–572.

Teytelman, L., Thurtle, D.M., Rine, J., van Oudenaarden, A., 2013. Highly expressed loci are vulnerable to misleading ChIP localization of multiple unrelated proteins. Proc. Natl. Acad. Sci. U. S. A. 110, 18602–18607.

Tilly, B.C., Chalkley, G.E., van der Knaap, J.A., Moshkin, Y.M., Kan, T.W., Dekkers, D.H., Demmers, J.A., Verrijzer, C.P., 2021. In vivo analysis reveals that ATP-hydrolysis couples remodeling to SWI/SNF release from chromatin. elife 10, e69424.

Tirosh, I., Barkai, N., 2008. Two strategies for gene regulation by promoter nucleosomes. Genome Res. 18, 1084–1091.

Tirosh, I., Sigal, N., Barkai, N., 2010. Widespread remodeling of mid-coding sequence nucleosomes by Isw1. Genome Biol. 11, R49.

Tong, J.K., Hassig, C.A., Schnitzler, G.R., Kingston, R.E., Schreiber, S.L., 1998. Chromatin deacetylation by an ATP-dependent nucleosome remodelling complex. Nature 395, 917–921.

Tramantano, M., Sun, L., Au, C., Labuz, D., Liu, Z., Chou, M., Shen, C., Luk, E., 2016. Constitutive turnover of histone H2A.Z at yeast promoters requires the preinitiation complex. elife 5, e14243.

Tsukiyama, T., Wu, C., 1995. Purification and properties of an ATP-dependent nucleosome remodeling factor. Cell 83, 1011–1020.

Tsukiyama, T., Wu, C., 1997. Chromatin remodeling and transcription. Curr. Opin. Genet. Dev. 7, 182–191.

Tsukiyama, T., Becker, P.B., Wu, C., 1994. ATP-dependent nucleosome disruption at a heat-shock promoter mediated by binding of GAGA transcription factor. Nature 367, 525–532.

Tsukiyama, T., Daniel, C., Tamkun, J., Wu, C., 1995. ISWI, a member of the SWI2/SNF2 ATPase family, encodes the 140 kDa subunit of the nucleosome remodeling factor. Cell 83, 1021–1026.

Tsukiyama, T., Palmer, J., Landel, C.C., Shiloach, J., Wu, C., 1999. Characterization of the imitation switch subfamily of ATP-dependent chromatin-remodeling factors in *Saccharomyces cerevisiae*. Genes Dev. 13, 686–697.

Tsukuda, T., Fleming, A.B., Nickoloff, J.A., Osley, M.A., 2005. Chromatin remodelling at a DNA double-strand break site in *Saccharomyces cerevisiae*. Nature 438, 379–383.

van Attikum, H., Fritsch, O., Hohn, B., Gasser, S.M., 2004. Recruitment of the INO80 complex by H2A phosphorylation links ATP-dependent chromatin remodeling with DNA double-strand break repair. Cell 119, 777–788.

van Attikum, H., Fritsch, O., Gasser, S.M., 2007. Distinct roles for SWR1 and INO80 chromatin remodeling complexes at chromosomal double-strand breaks. EMBO J. 26, 4113–4125.

Van Holde, K.E., 1988. Chromatin. Springer-Verlag.

van Holde, K., Yager, T., 2003. Models for chromatin remodeling: a critical comparison. Biochem. Cell Biol. 81, 169–172.

Varga, J., Kube, M., Luck, K., Schick, S., 2021. The BAF chromatin remodeling complexes: structure, function, and synthetic lethalities. Biochem. Soc. Trans. 49, 1489–1503.

Varga-Weisz, P.D., Blank, T.A., Becker, P.B., 1995. Energy-dependent chromatin accessibility and nucleosome mobility in a cell-free system. EMBO J. 14, 2209–2216.

Varga-Weisz, P.D., Wilm, M., Bonte, E., Dumas, K., Mann, M., Becker, P.B., 1997. Chromatin-remodelling factor CHRAC contains the ATPases ISWI and topoisomerase II. Nature 388, 598–602.

Vary Jr., J.C., Gangaraju, V.K., Qin, J., Landel, C.C., Kooperberg, C., Bartholomew, B., Tsukiyama, T., 2003. Yeast Isw1p forms two separable complexes in vivo. Mol. Cell. Biol. 23, 80–91.

Vasudevan, D., Chua, E.Y.D., Davey, C.A., 2010. Crystal structures of nucleosome core particles containing the '601' strong positioning sequence. J. Mol. Biol. 403, 1–10.

Versteege, I., Sevenet, N., Lange, J., Rousseau-Merck, M.F., Ambros, P., Handgretinger, R., Aurias, A., Delattre, O., 1998. Truncating mutations of hSNF5/INI1 in aggressive paediatric cancer. Nature 394, 203–206.

Vettese-Dadey, M., Walter, P., Chen, H., Juan, L.J., Workman, J.L., 1994. Role of the histone amino termini in facilitated binding of a transcription factor, GAL4-AH, to nucleosome cores. Mol. Cell. Biol. 14, 970–981.

Wade, P.A., Jones, P.L., Vermaak, D., Wolffe, A.P., 1998. A multiple subunit Mi-2 histone deacetylase from *Xenopus laevis* cofractionates with an associated Snf2 superfamily ATPase. Curr. Biol. 8, 843–846.

Wagner, F.R., Dienemann, C., Wang, H., Stutzer, A., Tegunov, D., Urlaub, H., Cramer, P., 2020. Structure of SWI/SNF chromatin remodeller RSC bound to a nucleosome. Nature 579, 448–451.

Walker, J.E., Saraste, M., Runswick, M.J., Gay, N.J., 1982. Distantly related sequences in the alpha- and beta-subunits of ATP synthase, myosin, kinases and other ATP-requiring enzymes and a common nucleotide binding fold. EMBO J. 1, 945–951.

Wall, G., Varga-Weisz, P.D., Sandaltzopoulos, R., Becker, P.B., 1995. Chromatin remodeling by GAGA factor and heat shock factor at the hypersensitive Drosophila hsp26 promoter in vitro. EMBO J. 14, 1727–1736.

Wang, W., Cote, J., Xue, Y., Zhou, S., Khavari, P.A., Biggar, S.R., Muchardt, C., Kalpana, G.V., Goff, S.P., Yaniv, M., et al., 1996a. Purification and biochemical heterogeneity of the mammalian SWI-SNF complex. EMBO J. 15, 5370–5382.

Wang, W., Xue, Y., Zhou, S., Kuo, A., Cairns, B.R., Crabtree, G.R., 1996b. Diversity and specialization of mammalian SWI/SNF complexes. Genes Dev. 10, 2117–2130.

Wang, F., Ranjan, A., Wei, D., Wu, C., 2016. Comment on "a histone acetylation switch regulates H2A.Z deposition by the SWR-C remodeling enzyme". Science 353, 358.

Wang, X., Lee, R.S., Alver, B.H., Haswell, J.R., Wang, S., Mieczkowski, J., Drier, Y., Gillespie, S.M., Archer, T.C., Wu, J.N., et al., 2017. SMARCB1-mediated SWI/SNF complex function is essential for enhancer regulation. Nat. Genet. 49, 289–295.

Watanabe, S., Peterson, C.L., 2016. Response to comment on "a histone acetylation switch regulates H2A.Z deposition by the SWR-C remodeling enzyme". Science 353, 358.

Wechser, M.A., Kladde, M.P., Alfieri, J.A., Peterson, C.L., 1997. Effects of sin- versions of histone H4 on yeast chromatin structure and function. EMBO J. 16, 2086–2095.

Whitehouse, I., Flaus, A., Cairns, B.R., White, M.F., Workman, J.L., Owen-Hughes, T., 1999. Nucleosome mobilization catalysed by the yeast SWI/SNF complex. Nature 400, 784–787.

Whitehouse, I., Stockdale, C., Flaus, A., Szczelkun, M.D., Owen-Hughes, T., 2003. Evidence for DNA translocation by the ISWI chromatin-remodeling enzyme. Mol. Cell. Biol. 23, 1935–1945.

Whitehouse, I., Rando, O.J., Delrow, J., Tsukiyama, T., 2007. Chromatin remodelling at promoters suppresses antisense transcription. Nature 450, 1031–1035.

Wiles, E.T., Mumford, C.C., McNaught, K.J., Tanizawa, H., Selker, E.U., 2022. The ACF chromatin-remodeling complex is essential for polycomb repression. elife 11, e77595.

Willhoft, O., Ghoneim, M., Lin, C.L., Chua, E.Y.D., Wilkinson, M., Chaban, Y., Ayala, R., McCormack, E.A., Ocloo, L., Rueda, D.S., et al., 2018. Structure and dynamics of the yeast SWR1-nucleosome complex. Science 362, eaat7716.

Winger, J., Nodelman, I.M., Levendosky, R.F., Bowman, G.D., 2018. A twist defect mechanism for ATP-dependent translocation of nucleosomal DNA. elife 7, e34100.

Winston, F., Carlson, M., 1992. Yeast SNF/SWI transcriptional activators and SPT/SIN chromatin connection. Trends Genet. 8, 387–391.

Winston, F., Chaleff, D.T., Valent, B., Fink, G.R., 1984. Mutations affecting Ty-mediated expression of the HIS4 gene of *Saccharomyces cerevisiae*. Genetics 107, 179–197.

Woodage, T., Basrai, M.A., Baxevanis, A.D., Hieter, P., Collins, F.S., 1997. Characterization of the CHD family of proteins. Proc. Natl. Acad. Sci. U. S. A. 94, 11472–11477.

Wu, L., Winston, F., 1997. Evidence that Snf-Swi controls chromatin structure over both the TATA and UAS regions of the SUC2 promoter in *Saccharomyces cerevisiae*. Nucleic Acids Res. 25, 4230–4234.

Wu, W.H., Alami, S., Luk, E., Wu, C.H., Sen, S., Mizuguchi, G., Wei, D., Wu, C., 2005. Swc2 is a widely conserved H2AZ-binding module essential for ATP-dependent histone exchange. Nat. Struct. Mol. Biol. 12, 1064–1071.

Wu, S., Shi, Y., Mulligan, P., Gay, F., Landry, J., Liu, H., Lu, J., Qi, H.H., Wang, W., Nickoloff, J.A., et al., 2007. A YY1-INO80 complex regulates genomic stability through homologous recombination-based repair. Nat. Struct. Mol. Biol. 14, 1165–1172.

Xia, X., Liu, X., Li, T., Fang, X., Chen, Z., 2016. Structure of chromatin remodeler Swi2/Snf2 in the resting state. Nat. Struct. Mol. Biol. 23, 722–729.
Xue, Y., Wong, J., Moreno, G.T., Young, M.K., Cote, J., Wang, W., 1998. NURD, a novel complex with both ATP-dependent chromatin-remodeling and histone deacetylase activities. Mol. Cell 2, 851–861.
Yadon, A.N., Tsukiyama, T., 2011. SnapShot: chromatin remodeling: ISWI. Cell 144, 453.e451.
Yan, L., Chen, Z., 2020. A unifying mechanism of DNA translocation underlying chromatin remodeling. Trends Biochem. Sci. 45, 217–227.
Yan, L., Wang, L., Tian, Y., Xia, X., Chen, Z., 2016. Structure and regulation of the chromatin remodeller ISWI. Nature 540, 466–469.
Yan, L., Wu, H., Li, X., Gao, N., Chen, Z., 2019. Structures of the ISWI-nucleosome complex reveal a conserved mechanism of chromatin remodeling. Nat. Struct. Mol. Biol. 26, 258–266.
Yang, X., Zaurin, R., Beato, M., Peterson, C.L., 2007. Swi3p controls SWI/SNF assembly and ATP-dependent H2A-H2B displacement. Nat. Struct. Mol. Biol. 14, 540–547.
Ye, Y., Wu, H., Chen, K., Clapier, C.R., Verma, N., Zhang, W., Deng, H., Cairns, B.R., Gao, N., Chen, Z., 2019. Structure of the RSC complex bound to the nucleosome. Science 366, 838–843.
Yen, K., Vinayachandran, V., Batta, K., Koerber, R.T., Pugh, B.F., 2012. Genome-wide nucleosome specificity and directionality of chromatin remodelers. Cell 149, 1461–1473.
Yoshimoto, H., Yamashita, I., 1991. The GAM1/SNF2 gene of Saccharomyces cerevisiae encodes a highly charged nuclear protein required for transcription of the STA1 gene. Mol. Gen. Genet. 228, 270–280.
Yoshinaga, S.K., Peterson, C.L., Herskowitz, I., Yamamoto, K.R., 1992. Roles of SWI1, SWI2, and SWI3 proteins for transcriptional enhancement by steroid receptors. Science 258, 1598–1604.
Yuan, J., Chen, K., Zhang, W., Chen, Z., 2022. Structure of human chromatin-remodelling PBAF complex bound to a nucleosome. Nature 605, 166–171.
Yudkovsky, N., Logie, C., Hahn, S., Peterson, C.L., 1999. Recruitment of the SWI/SNF chromatin remodeling complex by transcriptional activators. Genes Dev. 13, 2369–2374.
Zentner, G.E., Tsukiyama, T., Henikoff, S., 2013. ISWI and CHD chromatin remodelers bind promoters but act in gene bodies. PLoS Genet. 9, e1003317.
Zhang, Y., LeRoy, G., Seelig, H.P., Lane, W.S., Reinberg, D., 1998. The dermatomyositis-specific autoantigen Mi2 is a component of a complex containing histone deacetylase and nucleosome remodeling activities. Cell 95, 279–289.
Zhao, K., Wang, W., Rando, O.J., Xue, Y., Swiderek, K., Kuo, A., Crabtree, G.R., 1998. Rapid and phosphoinositol-dependent binding of the SWI/SNF-like BAF complex to chromatin after T lymphocyte receptor signaling. Cell 95, 625–636.
Zhong, Y., Paudel, B.P., Ryan, D.P., Low, J.K.K., Franck, C., Patel, K., Bedward, M.J., Torrado, M., Payne, R.J., van Oijen, A.M., et al., 2020. CHD4 slides nucleosomes by decoupling entry- and exit-side DNA translocation. Nat. Commun. 11, 1519.
Zofall, M., Persinger, J., Bartholomew, B., 2004. Functional role of extranucleosomal DNA and the entry site of the nucleosome in chromatin remodeling by ISW2. Mol. Cell. Biol. 24, 10047–10057.
Zofall, M., Persinger, J., Kassabov, S.R., Bartholomew, B., 2006. Chromatin remodeling by ISW2 and SWI/SNF requires DNA translocation inside the nucleosome. Nat. Struct. Mol. Biol. 13, 339–346.

CHAPTER

7

Histone modifications

Abbreviations

ATAC	Ada2a-containing
CBP	CREB-binding protein
ChIP	chromatin immunoprecipitation
COMPASS	Complex of Proteins Associated with Set1
DUB	Deubiquitinylation
GlcNAc	β-*N*-acetylglucosamine
GNAT	Gcn5-related *N*-acetyltransferase
HAT	histone acetyltransferase
HDAC	histone deacetylase
HDM	histone demethylase
HMT	histone methyltransferase
KAT	lysine acetyltransferase
KMT	lysine methyltransferase
MLL	Mixed Lineage Leukemia
PCAF	p300/CBP-associated factor
PcG	polycomb group
PRMT	protein arginine methyltransferase
RSC	remodels the structure of chromatin
SAGA	Spt-Ada-Gcn5-acetyltransferase
SAM	S-adenosylmethionine
Taf	TBP associated factor
TBP	TATA-binding protein

[F]ull elucidation of the mechanisms and consequences of posttranslational modification may require many years of careful effort. ***Chromatin, van Holde, p. 118.***

Overview

Histone modifications have been, unavoidably, alluded to several times already in earlier chapters. This area of chromatin research has exploded since the discovery of Gcn5, a positive regulator of transcription, as a histone acetyltransferase, and Rpd3, a repressor, as a histone deacetylase in 1996 (Brownell et al., 1996; Taunton et al., 1996). We are now aware of a dizzying array of modifications to which the histones are subject; the heterogeneity of chromatin structures proposed by the Stedmans to arise from a multitude of histones stems instead

Chromatin. **https://doi.org/10.1016/B978-0-12-814809-9.00002-0**

from their decoration by methyl, acetyl, phosphate, ubiquitinyl, glycosyl, crotonyl, ADP-ribosyl, and other moieties at numerous sites in all four core histones as well as linker histones (Fig. 7.1; Cavalieri, 2021; Zhao and Garcia, 2015). Investigations of histone modification have often been followed by the recognition that many nonhistone proteins are also subject to the same modification; for example, more than 2000 acetylation targets have been identified by high throughput application of mass spectrometry (Kori et al., 2017; Sheikh and Akhtar, 2019). Most of these modifications are reversible, with differing dynamics, adding another dimension to variation in chromatin structure.

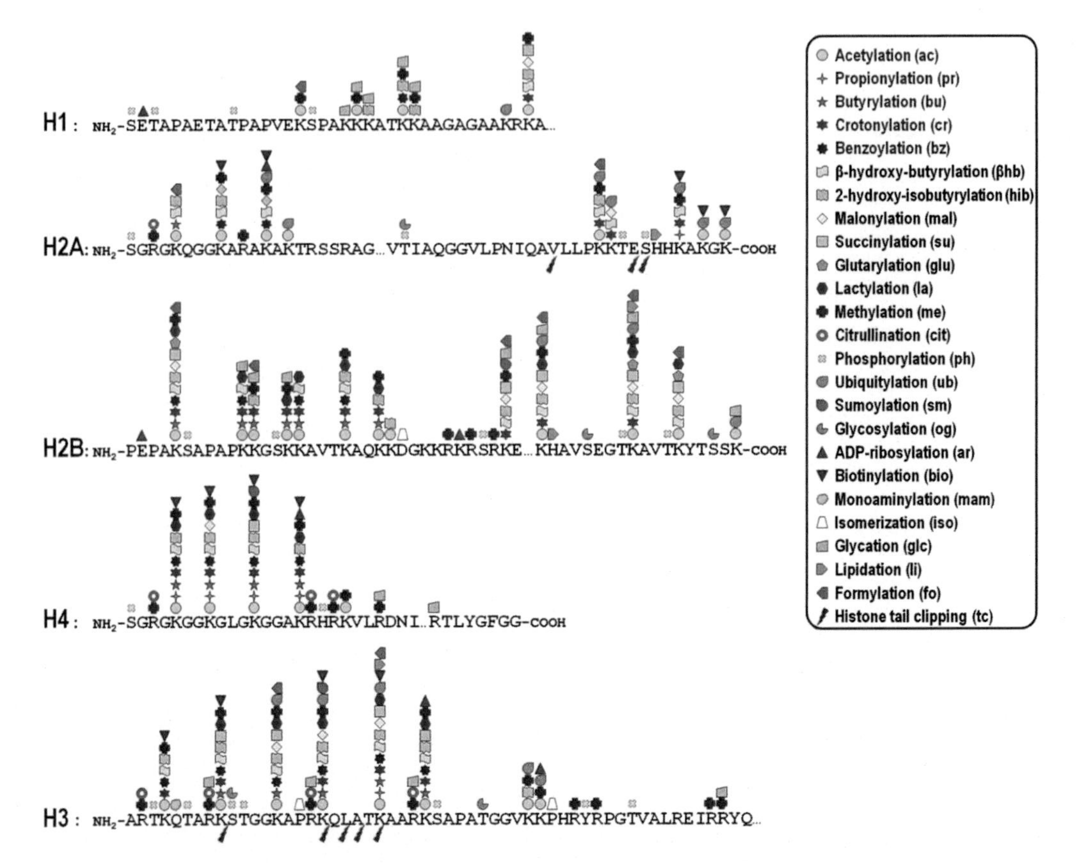

FIG. 7.1 Known sites of histone modification. *Reproduced from Cavalieri, V., 2021. The expanding constellation of histone post-translational modifications in the epigenetic landscape. Genes (Basel) 12, 1596.*

To cover this area comprehensively would be an impossible task. Instead, the intent of this chapter is to provide an overview of the types and functions of the various modifications, including some historical context and discussion of the methods used in their investigation. Regarding nomenclature, histone acetyltransferases were initially abbreviated as HATs; the subsequent realization that they acted on many nonhistone substrates led to their frequently being referred to instead as lysine acetyltransferases, or KATs. In the context of discussions of their action on chromatin, however, the HAT designation is still often used, and I will follow that convention here. In addition, new names were proposed for histone acetyltransferases, methyltransferases, and demethylases in 2007, with the aim of providing a rational naming system that takes into account enzymatic targets and evolutionary relatedness (Allis et al., 2007). The proposed naming system has not been universally adopted by the field, and I will also stubbornly stick with original names, referring to Gcn5 and PCAF rather than KAT2, and to *Drosophila* Su(var)3-9 (see Chapter 3) rather than KMT1. In my view, the old names have value for the reason that they are associated with the stories of their discoveries, and stories help us remember.

Given the scope of this topic, it is easy to get lost in the weeds, and it may therefore be helpful to begin with some general observations and principals:

- How modifications are studied typically follows, more or less, the paradigm established for histone acetylation by Gcn5 and deacetylation by Rpd3. A given modification is determined, for example acetylation of lysine 9 of histone H3 (H3K9ac), either by amino acid sequencing or mass spectrometric analysis. A candidate enzyme (or enzymes) responsible for that modification is identified in some organism—oftentimes yeast—and tested for its ability to confer the modification in vitro, and by determining the effect of its loss or mutation on the specific modification. (Redundancy is sometimes observed, requiring multiple candidate modifiers to be deleted or mutated; alternative techniques are required if the modifier is essential for viability.) Most often the modifying enzyme is found in a protein complex, often very large and comprising many subunits, in living cells, and the impact of this incorporation requires further investigation. Phenotypes caused by loss or mutation (especially to a catalytically inactive state) of the modifier are determined and compared, when possible, to those caused by mutation of the target site, as described in Chapter 5. Genome-wide distribution of the modification is determined, and homologs of the modifier are sought in other organisms (especially metazoans). Structural, biochemical, and genetic approaches are applied to understand mechanism of the modifier and function of the modification. Connections with human disease are explored.
- Modifications are generally considered in terms of the enzymes that confer them, the enzymes that reverse them, and the proteins or complexes whose affinity for nucleosomal surfaces are affected by the modification. These are referred to as *writers*, *erasers*, and *readers* (Ruthenburg et al., 2007).
- Modifications can be divided according to their location in the nucleosome. Early work focused on modifications occurring in the unstructured N-terminal regions, or tails, of the histones (Chapter 2); technological advances in the use of mass spectrometry led to the discovery of modifications occurring in the structured, or globular, region of the nucleosome. These latter modifications can be further divided into those occurring in the lateral surface of the nucleosome, where the histones interact most closely with DNA, and the face of the nucleosome (Mersfelder and Parthun, 2006). Some modifications are conferred in the cytoplasm and are important for nucleosome assembly, but the majority occur in the nucleus at assembled nucleosomes.
- Modifications exert their functions in two ways: by affecting chromatin structure, either at individual nucleosomes or by modulating chromatin compaction, and by affecting interactions of other proteins (readers) with chromatin (Turner, 1993). Histone modifications affect all categories of processes involving DNA: transcription, replication, repair, recombination, and of course its packaging into chromatin. Moreover, as the levels of modification cofactors such as acetyl-CoA and *S*-adenosylmethionine (SAM) are

determined by metabolic pathways, modifications can reflect and in turn influence the metabolic state of the cell (Suganuma and Workman, 2018).

- No simple rules exist regarding functional effects of a given modification. Although early work briefly supported the idea that histone acetylation might positively affect transcription while histone methylation might be repressive, exceptions to such generalizations were soon discovered. Nonetheless, specific modifications at specific sites are often associated with predictable functional effects; for example, methylation of H3K9 and H3K27 is associated with repressive heterochromatin. Even so, function depends on a host of variables, and contextual effects of cell type and genomic features can confound efforts at prediction.
- Additional complexity arises from the intersection of histone modifications with histone variants, the potential combinatorial effects of modifications (the "histone code" (Strahl and Allis, 2000)), and the interplay between specific histone modifications and "writers" affecting other histone marks—referred to as histone cross-talk.
- Histone modification enzymes frequently are found in complex with associated proteins. These complexes, which can be quite large (megadalton or larger), sometimes include more than one modifying enzyme, and sometimes carry out functions that do not depend on the catalytic activity of their histone modifier (Morgan and Shilatifard, 2020). In addition, histone modification enzymes frequently also modify additional, nonhistone targets (Downey, 2021; Herz et al., 2013; Shvedunova and Akhtar, 2022). The latter topic is beyond the scope of this book, but is the subject of many reviews in the literature.
- Although all four core histones, as well as linker histones, are subject to extensive posttranslational modification, investigations into functional effects and mechanisms have been largely focused on the inner histones, H3 and H4. Discussion of specific modifications in this chapter is also heavily weighted toward these histones; modifications of H2A and H2B, and variants thereof, have been reviewed in the literature (Cavalieri, 2021; Wyrick and Parra, 2009; Zhao and Garcia, 2015).
- The study of histone modifications has to a considerable extent been rebranded as "epigenetics." This rebranding has been controversial, with varied definitions of the term and disagreements over the criteria applied to establish heritability. Epigenetics is discussed in Chapter 8.

Early history of protein and histone modification

As with the recognition that the molecular basis of heredity resides in the double helix of DNA, progress in understanding the mechanism, scope, and function of posttranslational modification of proteins was for many years slow and difficult (Keenan et al., 2021). The first posttranslational modification of proteins to be identified was phosphorylation; the absence of phosphorus from known amino acids implied that its presence as a modification to purified protein, apparently first found in caseinogen[a] (Hammarsten, 1877), could be established by simple elemental analysis. Phoebus Levene (the same Phoebus Levene who promoted the tetranucleotide hypothesis discussed in Chapter 1) has been credited as having identified phosphate in vitellin in 1906 (e.g., https://en.wikipedia.org/wiki/Protein_phosphorylation); however, although Levene later cites his 1906 paper as describing

[a]Casein even now is sold by scientific supply houses as "Hammarsten grade."

the isolation from vitellin of a substance containing about 10% phosphorus and about 14% nitrogen, the cited paper includes no mention of phosphorus in any of the analyses presented (Levene and Alsberg, 1906; Lipmann and Levene, 1932). It seems likely that the intended citation was to an earlier paper that is cited by Rimington, along with two other cited studies, as representing early attempts to isolate the phosphorus-containing substance derived from caseinogen (Levene and Alsberg, 1901; Rimington, 1927). Rimington conducted studies aimed at improving on the earlier work, which utilized highly impure peptic digests, by performing chemical analyses on purified caseinogen; he concluded that phosphate was likely to be incorporated into protein via ester linkages to a hydroxyl-containing component and appears to have favored hydroxyglutamic and hydroxybutyric acids as likely partners (Rimington, 1927). Similar studies on vitellin were reported by Posternak, working out of his private lab in Geneva, who favored serine as the phosphate attachment site (Posternak, 1927; Posternak and Posternak, 1927, 1928); Posternak's published work evidently suffered, however, from a lack of experimental detail, making comparison with other studies difficult (Rimington, 1927). In any event, definitive identification of phosphoserine in vitellin was achieved by Lipmann and Levene in 1932 (Lipmann and Levene, 1932). Another two decades would pass before phosphorylation was established as an enzymatically catalyzed posttranslational modification (Burnett and Kennedy, 1954).

Numerous additional posttranslational modifications of proteins are now known; their identification and characterization have been greatly abetted by the use of mass spectrometry and development of antibodies recognizing specific modifications (Keenan et al., 2021). Characterization of histone modifications followed a pace similar to that of proteins in general, while attribution of function can be argued to have lagged behind, as the role of protein phosphorylation in modulating enzyme activity and in signaling pathways was being elucidated while remaining enigmatic for the histones (Pawson and Scott, 2005). Similarly, advances in protein sequencing and analysis allowed identification of additional types of histone modification, such as arginine methylation and lysine acetylation and methylation, before the particle model of the nucleosome was formulated, but the functions of these modifications remained purely a matter for speculation (DeLange and Smith, 1971). Not until 1996 did the field of histone modifications burst into bloom, seeded by the near simultaneous identification of a histone acetyltransferase that functioned in transcriptional activation and, conversely, a histone deacetylase with repressive capability. These discoveries unleashed a flood of functional, biochemical, and structural studies on histone modifications that led to reconsiderations of mechanisms of inheritance, the latter gathered together under the sometimes amorphous term of epigenetics.

Early speculations on the function of histone modifications tended to focus on their likely impact on histone-DNA interactions, with particular emphasis on the idea that neutralizing the positively charged lysine residues in the amino-terminal regions (Fig. 7.2) might drastically affect interaction with the negatively charged DNA. This view has largely been superseded by one that considers these modifications in terms of the enzymes that confer them ("writers"), the complementary enzymes that remove them ("erasers"), and the proteins whose interactions with chromatin are affected by a given modification ("readers"). Vince Allfrey and colleagues in 1964 suggested that altered histone-DNA binding caused by histone acetylation "would allow a means of switching-on or -off RNA synthesis at different times, and at different loci of the chromosomes" (Allfrey et al., 1964). The scope of functional effects of histone modifications has far surpassed this prescient notion and is now

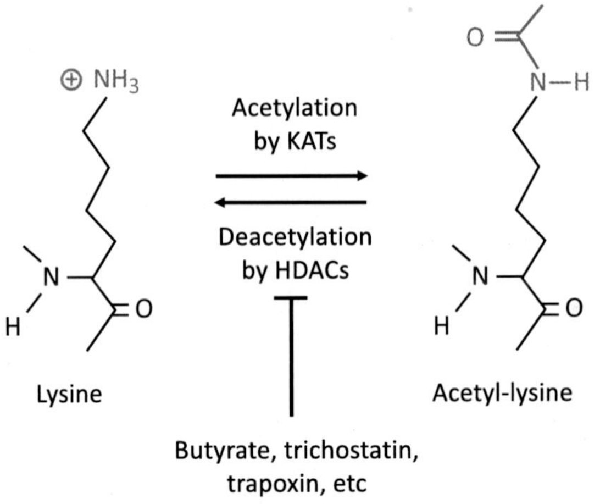

FIG. 7.2 The reversible acetylation of lysine neutralizes its positive charge. Butyrate, trichostatin, trapoxin, and other agents inhibit histone deacetylases.

firmly engrained in our understanding of DNA transactions in eukaryotes. This chapter reviews the research that led to this revolution in ideas regarding the role of histone modifications. Of necessity, only a small fraction of the literature on this topic will be mentioned; the goal is to provide sufficient knowledge to provide context for further investigations of those aspects of special interest to the reader.

Histone acetylation and deacetylation

Histone acetylation before Gcn5

An early method to identify the N-terminal amino acid in a protein chain was by reaction of the α-amino moiety with fluorodinitrobenzene, followed by hydrolysis of the peptide and chromatographic analysis. Phillips noted that this method resulted in a low yield of N-terminal amino acids when applied to histone preparations; when taken together with the presence of acetyl groups, as determined in chemical analyses, this suggested that the acetyl groups could be present as N-terminal substituents of the histones, thus blocking reaction with fluorodinitrobenzene (Phillips, 1963). Eventually it was established that N-terminal acetylation, which had been observed previously in tobacco-mosaic-virus and other proteins (Narita, 1958), occurred at the N-terminal serine residue of H2A and H4 and was implemented in the cytoplasm soon after protein synthesis (Liew et al., 1970; Phillips, 1968). Following the Phillips report, ^{14}C-labeled acetate was shown to be incorporated into histones in isolated nuclei, independent of protein synthesis (Allfrey et al., 1964). The presence of ε-*N*-acetyllysine, which had not previously been identified in proteins, in isolated histone H4 (called f2a1 at the time) was eventually discovered by the labs of Vince Allfrey and James Bonner (Fig. 7.2; DeLange et al., 1969; Gershey et al., 1968). Gershey et al. incubated calf thymus nuclei with ^{14}C-labeled sodium acetate and found that all of the recovered radioactivity associated with isolated f2a1 was present in the form of ε-*N*-acetyllysine; DeLange et al. found ε-*N*-acetyllysine at residue 16 of H4 by biochemical and chromatographic analysis.

Acetylation of H4 in trout testis was reported to occur entirely in the N-terminal region, at lysine residues 5, 8, 12, and 16, by radioisotope incorporation and chromatographic analysis (Candido and Dixon, 1971); similarly, ε-*N*-acetyllysine was detected in the amino terminal regions of H2A and H3 by automated protein sequencing analysis (Candido and Dixon, 1972). Subsequent studies by many research groups revealed the pattern of lysine acetylation in the histone tails, particularly of H3 and H4, to be conserved across species and cell types, from calf thymus to *Tetrahymena* to yeast (van Holde, 1988). Only much later were modifications in the structured region of the nucleosome, including acetylation of H3 K56, discovered and studied in detail (discussed below).

As part of their initial study on histone acetylation, Allfrey and colleagues undertook an ambitious and imaginative experiment aimed at testing whether histone function could be affected by acetylation (Allfrey et al., 1964). Previous studies using isolated nuclei or purified histone-DNA complexes had indicated an inhibitory effect of histones on RNA synthesis; however, when histones were first acetylated by limited treatment with acetic anhydride and assembled into complexes with DNA by salt dialysis, the inhibitory effect on RNA synthesis, effected by a preparation of calf thymus nuclei as the polymerase source, was substantially reduced. While the nature of the inhibitory histone-DNA complex was not understood at this point, the idea that reducing the positive charge of histone lysine residues by acetylation could loosen histone-DNA contacts and allow increased enzyme activity appeared reasonable and potentially offered a means of reversibly allowing DNA transcription in the cell. This idea was pursued in various investigations in the following years. Various fractionation procedures were employed in efforts to separate actively transcribed regions from bulk chromatin, typically involving nuclease digestion and separation of soluble from insoluble chromatin following incubation with Mg^{+2} (Gottesfeld and Butler, 1977; Gottesfeld et al., 1974). Employing these same protocols, the more transcriptionally active chromatin was found to exhibit increased incorporation of labeled acetate and to be enriched for acetylated histones (Davie and Candido, 1978; Levy-Wilson et al., 1977, 1979; Perry and Chalkley, 1981). Similarly, levels of histone acetylation also positively correlated with sensitivity to nuclease digestion, which itself was associated with actively transcribed chromatin (Davie and Candido, 1980; Nelson et al., 1978; Perry and Chalkley, 1981; Sealy and Chalkley, 1978a). Additional correlative evidence for a positive role for histone acetylation in transcription derived from studies of *Tetrahymena*, which found little or no histone acetylation in chromatin from the transcriptionally inactive micronucleus, but considerable acetylation of all four core histones in the transcriptionally active macronucleus (Gorovsky et al., 1973; Vavra et al., 1982).

The correlation between actively transcribed regions of chromatin and histone acetylation was considerably tightened by experiments employing antibodies against acetylated histones (see Sidebar: Antibodies to modified histones). Hebbes and Crane-Robinson raised immunogen against chemically acetylated histone H4 that recognized ε-acetyllysine regardless of the surrounding sequence and could bind to acetylated lysine residues in a nucleosomal context (Hebbes et al., 1988). Chromatin fragments from chick embryo erythrocytes, produced by limited MNase digestion and immunopurified using this antibody, were found to be enriched by 15- to 30-fold for the active β-D-globin gene but not for the inactive ovalbumin gene. This result, together with a follow-up study demonstrating that the active chicken β-globin chromosomal domain exhibited both hyperacetylated histones and DNase sensitivity (Hebbes et al., 1994), ruled out the possibility that these properties might copurify even while residing on distinct loci.

Sidebar: Antibodies to modified histones

Antibodies against specific histone modifications, especially in combination with ChIP, have been instrumental in chromatin biology research. Interest in the extent and specificity of acetylation of histone lysine residues, raised by the correlation with transcription and further elevated by observations of altered acetylation levels during embryogenesis (Allfrey et al., 1964; Chambers and Shaw, 1984; Christensen et al., 1984), predated the identification of Gcn5 and Rpd3 as HAT and histone deacetylase (HDAC), but was highlighted and crystallized by those discoveries. The first efforts at obtaining acetylation-specific antibodies against the histones focused on H4, as available evidence suggested a relatively high fraction of this histone was acetylated in vivo. Chemically acetylated full-length H4 or the purified 37 amino acid N-terminal peptide fragment, purified tri-acetylated H4, and synthetic peptides corresponding to the N-terminal region of *Tetrahymena* H4 that incorporated acetyllysine residues were variously used as immunogens in the earliest studies (Hebbes et al., 1988; Lin et al., 1989; Muller et al., 1987; Pfeffer et al., 1986). The resulting antisera preferentially bound more highly acetylated histones, though in one case the sera was not specific for histones but recognized acetyllysine in nonhistone proteins as well (Hebbes et al., 1988).

Bryan Turner's lab then sought to obtain antibodies specific for individual acetylated sites by raising antisera against a tetra-acetylated peptide corresponding to the N-terminal 20 residues of H4; sera were tested in ELISA competition assays against four singly acetylated peptides (Fig. 7.3A) and were deemed specific to a particular acetylated residue if only that corresponding peptide was inhibitory (Turner and Fellows, 1989). The resulting antisera were used in several of the seminal studies on histone acetylation and HAT isolation, including the demonstration of hyperacetylation of H4K16 on the male X chromosome in *Drosophila*, the hypoacetylation of the inactive X chromosome in mammals, and the characterization of deacetylase activity in the first isolated yeast HDAC complexes (Jeppesen and Turner, 1993; Rundlett et al., 1996; Turner et al., 1992). The adoption of ChIP for the characterization of chromatin heightened interest in obtaining antibodies that could be applied to its use, and other laboratories, most notably those of Michael Grunstein and Dave Allis, were producing dozens of antibodies to specific histone modifications by the late 1990s.

Following their 1996 Gcn5 paper (Brownell et al., 1996), the Allis lab was inundated with requests for antibodies that they had developed, and continued developing, against variously acetylated and nonacetylated histone tails (Braunstein et al., 1996; Lin et al., 1989), to the point that a full-time technician was employed to send out requested antibodies. This led to marketing of antibodies against acetylated histones by Upstate Biotechnology, headquartered in Lake Placid, New York, beginning in 1996, and development of new antibodies, both polyclonal and monoclonal, against a variety of specific histone modifications as well as against HATs and HDACs. Antibodies having exquisite specificity were developed, for example, distinguishing trimethylated from dimethylated H3K4 or asymmetrically from symmetrically dimethylated H4Arg3 (Fig. 7.12; Santos-Rosa et al., 2002). Other companies soon joined as purveyors, and today a huge variety of ChIP-validated antibodies against specific histone modifications are available from multiple sources.

As with most new techniques, technical issues and potential for particular artifacts were recognized over time, resulting in refinements. Regarding antibody specificity, the Grunstein lab reported that the Turner protocol was insufficient for determining antibody specificity; an additional requirement was that an antibody specific for a particular lysine residue should fail to bind to histones from yeast in which the given residue was mutated to arginine (Fig. 7.3B; Suka et al., 2001). A panel of

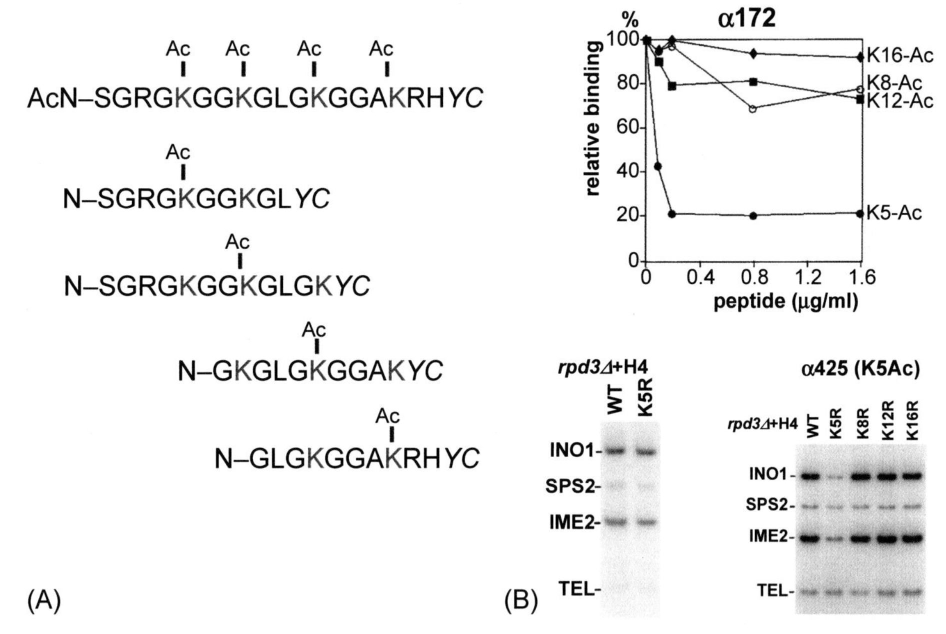

FIG. 7.3 (A) Acetylated peptides corresponding to the 18 N-terminal amino acids of histone H4 used by Turner and Fellows (1989) to generate antibodies. The C-terminal *YC* residues do not correspond to the native H4 sequence. (B) Although the α172 antibody raised against an acetylated H4 peptide appears specific to singly acetylated K5-Ac, based on efficient competition by only the K5-Ac peptide in an ELISA assay (top), no diminution in ChIP signal is seen in yeast harboring an H4K5R mutation relative to wild type (bottom left). The α425 antibody did show specificity toward K5ac (bottom right), as enrichment for the *INO1* and *IME2* promoter sequences in *rpd3Δ* yeast was reduced only in H4K5R yeast. *From Suka, N., Suka, Y., Carmen, A.A., Wu, J., Grunstein, M., 2001. Highly specific antibodies determine histone acetylation site usage in yeast heterochromatin and euchromatin. Mol. Cell 8, 473.*

12 new antibodies against singly acetylated residues in all four core histones was generated and used in conjunction with ChIP to examine the effects of deletion of various HATs and HDACs, in some cases raising doubts about previously published work (see "Specificity and genomic targeting of HDACs" section). Other issues include the potential interference of antibody binding by neighboring modifications—for example, inhibition of anti-H3K9ac binding when H3S10 was phosphorylated—cross-reactivity, and nonspecific immunoprecipitation of highly transcribed regions in yeast and *Drosophila* (Bock et al., 2011; Fuchs et al., 2011; Fuchs and Strahl, 2011; Jain et al., 2015; Park et al., 2013; Teytelman et al., 2013). These concerns have led to the use of additional means for validating specificity of antibodies used for ChIP, such as the use of arrays spotted with peptides corresponding to histone amino-terminal (and other) sequences harboring various modifications, alone and in combination (Bock et al., 2011; Fuchs et al., 2011), and the use of semisynthetic nucleosomes and implementation of a web site repository for antibody validation data (Rothbart et al., 2015; Shah et al., 2018). Unfortunately, the web site appears possibly to be moribund, as no new information has been entered since 2018, and in general the reliability and specificity of the >1000 antibodies available for ChIP must rely on the manufacturer's quality control and, in some cases, additional validation by the investigating researchers.

In this same time period, evidence began to accumulate that acetylation of specific lysine residues, rather than general levels of acetylation, was important for function. Nonrandom acetylation of H3 and H4 was observed in both antibody-based and microsequencing studies, the latter taking advantage of the absence of irreversible N-terminal histone acetylation in *Tetrahymena* to allow sequence determination (Chicoine et al., 1986; Couppez et al., 1987; Turner and Fellows, 1989). Polyclonal sera produced against histone H4 having specifically acetylated individual lysine residues was used in immunostaining experiments allowed interrogation of whole chromosomes for the presence of specifically acetylated lysines (Turner et al., 1992) (see Sidebar: Antibodies to modified histones). These experiments revealed distinct patterns of individual acetylated sites over polytene chromosomes from *Drosophila* larvae; most remarkably, H4K16ac was preferentially associated with the X chromosome in male larvae, but not in females (Fig. 7.4). *Drosophila* are subject to dosage compensation in which the single X chromosome present in males is transcribed at higher levels (about 2X) than the two female X chromosomes; this raised the possibility that H4 acetylation at Lys16 might play a special role in this hyperactivation. Conversely, acetylated H4 was absent from the inactive X chromosomes in mammals, where dosage compensation is effected by inactivation of one of the two female X chromosomes (Jeppesen and Turner, 1993), and the heterochromatic telomeres and silent mating type cassettes of budding yeast (discussed in Chapter 3) also displayed low levels of histone acetylation relative to chromosomal regions containing active genes (Braunstein et al., 1993).

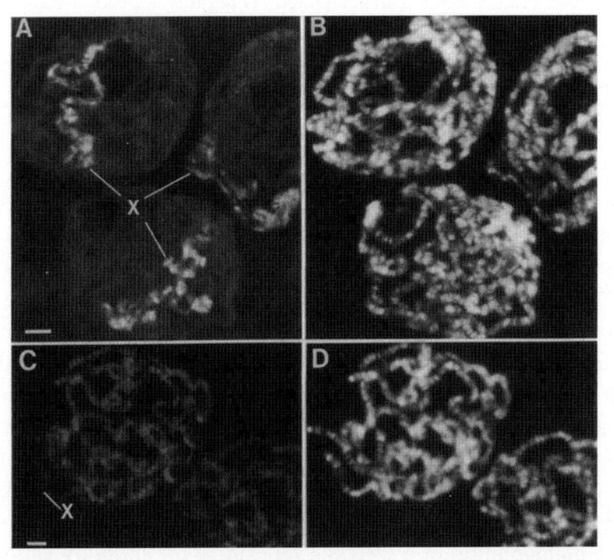

FIG. 7.4 Enrichment of H4K12ac on the *Drosophila* male X chromosome. Polytene nuclei from male (A) and female (C) cells were subjected to immunofluorescent labeling using an antibody to H4K16ac; corresponding Hoechst staining is shown in (B) and (D). Note the intense staining of the X chromosomes in male but not female nuclei, contrasted with uniform staining of all chromosomes with Hoechst dye. *From Turner, B.M., Birley, A.J., Lavender, J., 1992. Histone H4 isoforms acetylated at specific lysine residues define individual chromosomes and chromatin domains in Drosophila polytene nuclei. Cell 69, 375.*

While these studies suggested a provocative connection between histone acetylation and transcription, progress in understanding molecular details and mechanism, and indeed in obtaining compelling evidence for a causal link between transcription and acetylation of histones, was stymied by the murky nature of the chromatin fractions studied and by the absence of any knowledge of the enzymes that facilitated the acetylation reaction. Efforts were made to purify histone acetyltransferases, or HATs, by testing fractionated cellular extracts (using standard biochemical protocols, such as size-exclusion and ion affinity columns) for their ability to incorporate radiolabeled acetate from acetyl-CoA into purified histone substrate (Attisano and Lewis, 1990; Belikoff et al., 1980; Sures and Gallwitz, 1980; Travis et al., 1984). HAT activities purified from calf thymus, yeast, and porcine liver all were found to be present in very low abundance, stymying efforts at their purification and identification. This hurdle was eventually overcome by three different labs using completely different strategies, resulting in the cloning of a cytoplasmic HAT and the *Tetrahymena* homolog of the yeast nuclear HAT, Gcn5.

Paradigm shift part 1: A histone acetyltransferase regulates transcription

The discovery and molecular biology of Gcn5 and the SAGA complex have been described in reviews coauthored by several of the principals involved in these discoveries (Brownell and Allis, 2021; Grant et al., 2021; Soffers and Workman, 2020).

Sternglanz and colleagues adopted an entirely different strategy from the fractionation procedures used previously in efforts to purify HAT enzymes and succeeded in cloning the yeast cytoplasmic Hat1 through a combination of brute force and good luck (Kleff et al., 1995). Rather than taking a purely biochemical approach, they screened extracts from ~250 temperature yeast mutants, seeking mutants defective in acetylation of a peptide corresponding to the first 28 amino acids of histone H4 when assayed at the restrictive temperature. One mutant exhibited a 40% decrease in acetyltransferase activity; cloning of the mutant gene, named *HAT1*, allowed a deletion mutant to be constructed, which showed an identical decrease in HAT activity. Aside from the decrease in HAT activity, no other phenotypic changes relative to the wild type were observed in the deletion mutant. Remarkably, the temperature sensitivity of the starting strain did not segregate with the *hat1-1* mutation; discovery of *HAT1* in the screen was apparently a lucky coincidence. The partially purified Hat1 protein was found predominantly to acetylate Lys12 of histone H4, consistent with its independent identification by more standard biochemical methods as the major cytoplasmic HAT in yeast (see Chapter 5 Sidebar: Modifications of newly synthesized histones) (Parthun et al., 1996). In addition, Hat1 could acetylate free but not nucleosomal H4, and was associated with a second protein, Hat2, that is a close relative of RbAp48, a component of the chromatin assembly factor CAF-1 discussed in Chapter 5. These experiments demonstrated that HAT enzymes could possess great specificity, and that their activity could be modulated by partner proteins and could differ toward free and nucleosomal histones, corroborating earlier results showing differential activity of partially purified HAT activity from rat hepatoma tissue culture cells toward free compared to nucleosomal histones (Garcea and Alberts, 1980).

HAT enzymatic activities were categorized as being type A (nuclear) or type B (cytoplasmic), and while the cloning of the type B Hat1 enzyme paved the way for new investigations into chromatin assembly and HAT structure and mechanism, the identification of the yeast transcriptional regulatory factor Gcn5 as a type A HAT was a seismic event in chromatin biology (Brownell et al., 1996). As recounted in a recent retrospective, a critical element in this discovery was the use of a gel activity assay (Fig. 7.5; Brownell and Allis, 1995, 2021). In this assay, adapted from a method first used to identify protein kinases (Kameshita and Fujisawa, 1989), an extract containing the activity of interest—in this case, HAT activity—is fractionated on SDS-polyacrylamide gels in which histones are added prior to polymerization. Following electrophoresis, a series of washes is used to remove the SDS, denature (using urea) and renature the fractionated proteins in the gel, and infuse the gel with [^{3}H]-acetyl-CoA. The radiolabeled acetyl-CoA that is unattached to protein is then washed out, and if fortune smiles on the investigator, the enzyme sought is successfully renatured and able to transfer the labeled acetate moiety to the gel-embedded histone. This results in a band corresponding to the labeled, acetylated histone at the location of the HAT in the gel. As a control, the experiment is run without adding histones to the gel or with a nonsubstrate such as bovine serum albumin; in some cases (particularly in experiments assaying for kinase activity), automodification may be detected in such control experiments.

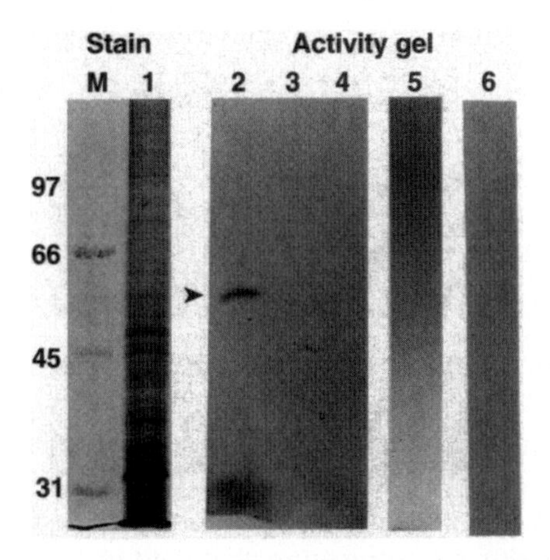

FIG. 7.5 The activity gel assay used to identify Gcn5 as a histone acetyltransferase. A partially purified HAT activity was subjected to gel electrophoresis in gels saturated with histones (lanes 2–4), BSA (lane 5), or no substrate (lanes 1 and 6) and processed as described in the text. Lane M contains marker and lane 1 was silver stained; the partially purified HAT activity was inactivated prior to electrophoresing samples in lanes 3 and 4. Note the band at 55 kDa in lane 2, later identified as the *Tetrahymena* homolog of yeast Gcn5. *From Brownell, J.E., Allis, C.D., 1995. An activity gel assay detects a single, catalytically active histone acetyltransferase subunit in Tetrahymena macronuclei. Proc. Natl. Acad. Sci. USA 92, 6364.*

Application of this gel activity assay to a preparation of highly purified macronuclei from *Tetrahymena*, which were known to contain extensively acetylated histones (Vavra et al., 1982), resulted in labeling of a single band corresponding to a protein of about 55 kDa (Fig. 7.5; Brownell and Allis, 1995). Scaling up the growth and preparation procedures eventually allowed purification of about 5 μg of the 55 kDa band from 150 L of cultured *Tetrahymena*, which was sufficient to conduct peptide microsequencing (Brownell et al., 1996). Cloning and sequencing of the *Tetrahymena* gene encoding the p55 HAT revealed it to be a homolog of Gcn5 from yeast. Identification was corroborated by expression of recombinant yeast Gcn5 in *E. coli*, which exhibited HAT activity in the same gel activity assay used to identify the *Tetrahymena* homolog.

The implications of this discovery were immense and immediately apparent. Gcn5 was already known as a putative transcriptional coactivator; its identification as a histone acetyltransferase directly coupled transcriptional activation to histone modification. To fully appreciate the impact of this connection, it may be helpful to recount the development of the coactivator model and the identification of Gcn5.

The recruitment model for transcriptional activation, which postulated that activators stimulated transcription by bringing the transcription machinery to promoters, was germinating at this time, and the idea of coactivators was a natural corollary to this model (Ptashne and Gann, 1997). The existence of coactivators, also referred to as adaptors in the early literature, was first hypothesized based on two types of experiments (Guarente, 1995). In one approach, overexpression of strong transcriptional activation domains such as those of the yeast activator Gal4 or the herpesvirus protein VP16 was found to suppress stimulation of transcription by unrelated activators both in vitro and in vivo; similarly, cross-inhibition was observed upon overexpression of steroid hormone receptors (Gill and Ptashne, 1988; Kelleher et al., 1990; Meyer et al., 1989; Pugh and Tjian, 1990; Triezenberg et al., 1988). This phenomenon, initially termed *squelching* by Gill and Ptashne, was first suggested to reflect titration of a general transcription factor (Gill and Ptashne, 1988; Meyer et al., 1989), and later refined to explicitly hypothesize the existence of a coactivator distinct from the basic transcription machinery (Berger et al., 1990; Kelleher et al., 1990). This latter view was supported by experiments showing that stimulation of transcription by activators in vitro was observed when partially purified fractions of TFIID from human or *Drosophila* cells were used, but not when purified recombinant TBP was employed; similar experiments indicated a missing component in attempts to reconstitute in vitro activated transcription in a yeast system, and it was inferred that the missing element was an adaptor or mediator that bridged activators with the basic transcription machinery (Kim et al., 1994; Pugh and Tjian, 1990; Thompson et al., 1993).

These approaches eventually led to the discovery of the Mediator complex, an essential coactivator conserved across eukaryotes that bridges activators and the general transcription machinery (Kornberg, 2005). One of the first putative coactivators identified was a B-cell restricted activity, termed OCA-B, capable of enhancing transcription activated by Oct-1 or Oct-2 in a promoter-specific fashion (Luo et al., 1992); coactivators were later discovered that functioned specifically with ligand-activated nuclear receptors (York and O'Malley, 2010). Another major coactivator complex identified through this line of inquiry was the Gcn5-containing SAGA complex (Grant et al., 2021). Berger, Guarente, and colleagues made clever use of squelching in a genetic assay to search for putative coactivators (Berger et al., 1992). In this assay, yeast mutants were sought that suppressed the toxic effect of GAL4-VP16 expression, shown to be closely coupled to its properties as a strong transcriptional activator, the rationale being that mutations that abrogated the interaction of the putative coactivator with the VP16 activation domain would alleviate its toxicity. The screen identified *ada* mutants (for *a*lteration/*d*eficiency in *a*ctivation, perhaps not coincidentally resonating with the idea of an *ada*ptor) that also reduced transcriptional activation by some, but not all, activators in vivo. Biochemical experiments revealed physical interactions in vitro among the cloned Ada proteins, and co-IP and yeast two-hybrid experiments supported their association in vivo (Candau and Berger, 1996; Horiuchi et al., 1995; Marcus et al., 1994). Physical interactions were also observed between Ada proteins and transcriptional activators found to be impaired in *ada*$^{-}$ yeast, and also with TBP (Barlev et al., 1995; Chiang et al., 1996; Melcher and Johnston, 1995; Silverman et al., 1994). Mutations in Ada proteins that abrogated these various physical interactions impaired their function in vivo (Candau and Berger, 1996; Candau et al., 1997). Taken together, these observations lent strong support to the coactivator model, and specifically supported the existence of a multisubunit complex containing the Ada proteins that interacted with activators and TBP to facilitate transcription. Cloning of *ADA4* revealed it to be identical to *GCN5* (Marcus et al., 1994). *GCN5* had first been found in a screen for yeast *gcn*, or *g*eneral *c*ontrol *n*onderepressible, mutants defective in general control of amino acid biosynthesis, in which starvation for a single amino acid activates multiple biosynthetic pathways, and was later shown to be required for wild type levels of transcriptional activation by Gcn4 (Georgakopoulos and Thireos, 1992; Penn et al., 1983). Thus, the identification of Gcn5 as a nuclear HAT immediately suggested the recruitment of a histone modifying enzyme as an integral component of a coactivator complex, supporting the Allfrey hypothesis put forward three decades earlier (Fig. 7.6). The nearly simultaneous discovery of a transcriptional corepressor, Rpd3, as a histone deacetylase (recounted farther on) put an exclamation point on this novel conception; together, these discoveries ushered in a new era in chromatin biology.

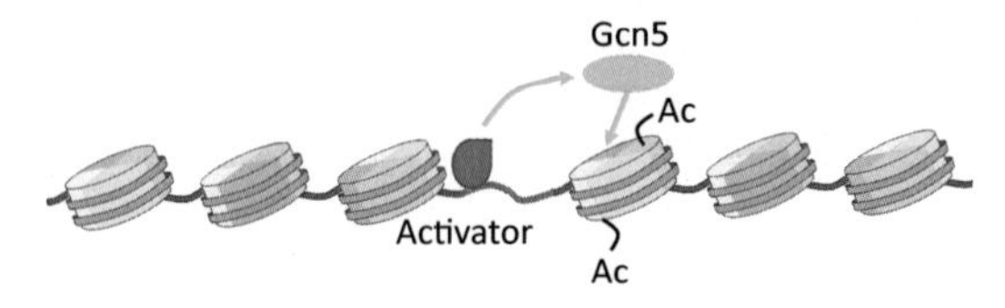

FIG. 7.6 Model for Gcn5 as a transcriptional coactivator.

Gcn5 and the SAGA complex

The discovery of Gcn5 as a histone acetyltransferase immediately raised questions that propelled a rush of research over the next several years, with seminal findings seeming to arrive on a monthly basis. Many of these follow-up studies were collaborative efforts, particularly among the Allis, Berger, Roth (now Dent), Winston, and Workman labs. The most pressing questions included:

- Which histones and which lysine residues did Gcn5 target?
- Did its activity differ for nucleosomal compared to free histones?
- Was its enzymatic activity important for its coactivator function?
- Did Gcn5 target proteins other than the histones for acetylation in vivo?
- What other proteins (e.g., the Ada proteins) were associated with Gcn5, and what were their functions?
- Were there homologs of Gcn5 in metazoans, and if so, how did they compare in their complexation and function to the yeast protein?
- Where and how were Gcn5 and its associated proteins targeted?
- And finally, the big mechanistic question: how did histone acetylation, whether mediated by Gcn5 or other HAT activities, facilitate transcriptional activation?

The gel activity assay used to identify *Tetrahymena* p55 and yeast Gcn5 as HATs used as substrate a mixture of all four core histones from calf thymus and thus provided no information on the specificity of the enzyme toward individual histones or individual residues within the histones (Brownell and Allis, 1995). To address this issue, purified histones were used as substrate for recombinant yeast Gcn5 (rGcn5) and analyzed for acetylation by direct microsequencing (Kuo et al., 1996). (These first experiments, and the isolation of the SAGA and Ada complexes described below, were conducted using the budding yeast system, because of its genetic tractability and the availability of its complete genome sequence, which was completed in 1995.) Histone H3 was highly preferred as a substrate for rGcn5-mediated acetylation over the other core histones; slight acetylation of H4 was observed and virtually no acetylation of H2A or H2B was seen. Acetylation was also site-specific, with H3K14 (lysine 14 of histone H3) by far the preferred site, while Lys8, or Lys8 and Lys16, were acetylated when yeast or *Tetrahymena* H4, respectively, was used as substrate. Combined with earlier findings that H4K5 and K12 were acetylated soon after histone synthesis in the cytoplasm (see Chapter 5 Sidebar: Modifications of newly synthesized histones), these findings demonstrated a marked difference in specificity of Gcn5 compared to the type B cytoplasmic HAT.

Recombinant Gcn5 was initially reported to be unable to acetylate nucleosomal histones, but a later report found that by modifying salt concentrations, both nucleosome core particles and nucleosome arrays could be acetylated by Gcn5 (Kuo et al., 1996; Tse et al., 1998; Yang et al., 1996b). Nonetheless, differences in the acetylation patterns were seen when free or nucleosomal histones were subject to in vitro modification by Gcn5, harkening back to early work showing differing reactivity of an acetyltransferase activity from rat hepatoma cells toward free and nucleosomal histones (Garcea and Alberts, 1980; Tse et al., 1998). Purification of the SAGA and Ada complexes, described below, revealed that Gcn5 was capable of acetylating nucleosomal histones when associated with its in vivo partner proteins (Grant et al., 1997; Pollard and Peterson, 1997; Ruiz-Garcia et al., 1997; Utley et al., 1998), and that its HAT activity was further modulated, in terms of both activity and specificity, by the associated subunits in the complexes, especially Ada2 and Ada3 (Balasubramanian et al., 2002; Grant et al., 1997; Sendra et al., 2000).

A critical question following the discovery of Gcn5 as a HAT was whether its acetyltransferase activity was required for its function as a coactivator. An affirmative answer to this question was first indicated by the demonstration that deletion mutants of Gcn5 that lacked HAT activity were unable to complement growth and transcriptional activation of a reporter gene by Gal4-VP16 in *gcn5Δ* yeast (Candau et al., 1997). The deletion mutants used in this study encompassed fairly large regions of Gcn5; a more targeted approach followed, in which Gcn5 was mutated by substitution of single or tri-amino acids (i.e., "alanine scanning") and tested for HAT activity and transcriptional activation of a reporter gene. The two activities were strongly correlated, revealing a close connection between HAT activity and coactivator function (Fig. 7.7; Kuo et al., 1998; Wang et al., 1998). The connection was further supported by ChIP experiments showing that activation of the Gcn5-dependent *HIS3* gene was accompanied by increased acetylation of nucleosomal H3 in the promoter region (Fig. 7.8; Kuo et al., 1998). This increased acetylation depended on the DNA-binding activator Gcn4 and on Gcn5 HAT activity, but was not abrogated by a TATA element mutation that eliminated transcription, demonstrating that histone acetylation preceded active transcription and did not require TBP association with the promoter (Kuo et al., 1998, 2000). Transcriptional activation abetted by HATs was supported by in vitro experiments using chromatinized templates and was shown to depend on activator-mediated recruitment and HAT acetyltransferase activity (Ikeda et al., 1999; Utley et al., 1998; Wallberg et al., 1999). Later work found decreased levels of several Gcn5-dependent transcripts in yeast mutants having substitutions of H3 Lys residues in the N-terminal region, consistent with acetylation of these regions contributing to transcriptional activation (Yu et al., 2006). However,

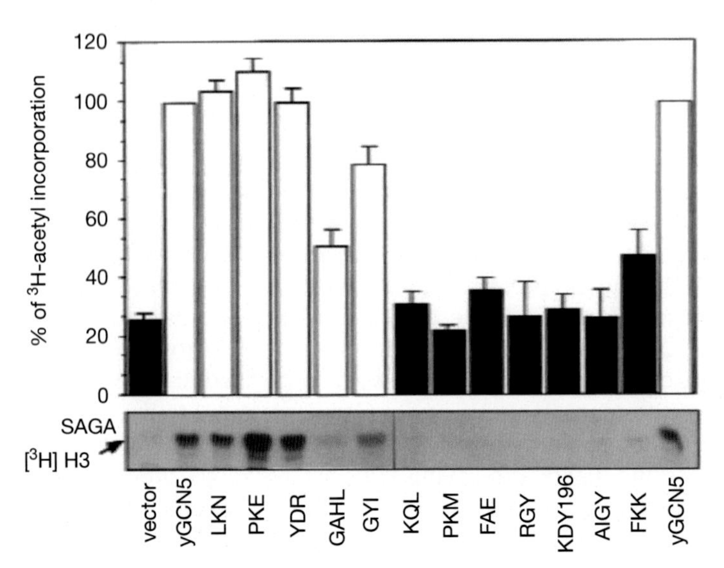

FIG. 7.7 Histone acetyltransferase activity measured for SAGA complexes prepared from *gcn5Δ* yeast expressing wild type or mutant *GCN5* or an empty vector control ("vector"). Activity was monitored by incubating free histones and [^{3}H]-acetylCoA with SAGA complexes and quantifying [^{3}H] incorporation by scintillation counting (graph) or visualizing by gel electrophoresis and fluorography (bottom). Mutants are labeled by the three or four amino acids substituted with alanine residues; mutants exhibiting wild type activity in their ability to support transcription and growth are shown as *white bars*, and defective mutants as *black bars*. *From Wang, L., Liu, L., Berger, S.L., 1998. Critical residues for histone acetylation by Gcn5, functioning in Ada and SAGA complexes, are also required for transcriptional function in vivo. Genes Dev. 12, 640.*

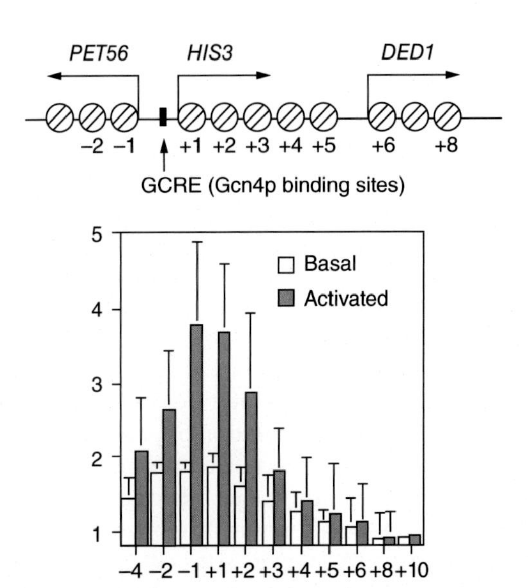

FIG. 7.8 Activation by Gcn4 is accompanied by local histone acetylation. Sonicated chromatin from yeast grown under activating or nonactivating (basal) conditions for Gcn4 activation was used as substrate for ChIP using antibody against diacetylated histone H3 (H3K9ac and H3K14ac). Acetylation levels were quantified by PCR, using primers encompassing positioned nucleosomes over the *HIS3* and flanking genes as shown at the top. Relative acetylation levels were expressed as the ratio of ChIP'd DNA to input DNA for each primer pair, relative to the IP/input ratio for a control region (*ACT1*). *From Kuo, M.-H., Vom Baur, E., Struhl, K., Allis, C.D., 2000. Gcn4 activator targets Gcn5 histone acetyltransferase to specific promoters independently of transcription. Mol. Cell 6, 1309.*

the effect was not localized to any specific residue, in line with genome-wide analyses of Lys substitution mutants in the H4 tail (Dion et al., 2005), and a close analysis indicated that acetylation could function, at least in part, to overcome a repressive effect of the H3 and H4 amino termini (Steinfeld et al., 2007; Yu et al., 2006).

Taken together, the above results were consistent with a model in which a DNA-binding activator recruited Gcn5 to acetylate histone H3 lysine residues in promoter nucleosomes, thereby facilitating transcriptional activation of the target gene. However, whether it was specifically and exclusively acetylation of *histones* that facilitated activation was not conclusively demonstrated in these experiments; acetylation of nonhistone proteins could also have contributed. Epistasis experiments comparing the effects of *GCN5* deletion with Lys substitution mutations in the histone tails alone or together could shed light on this issue. Unfortunately, such studies are rendered difficult by the synthetic lethality of deletion and substitution mutants in the H3 and H4 tails with deletion of *GCN5* (Zhang et al., 1998), although acute depletion protocols could allow such studies (see Chapter 4 Sidebar: Deletions and depletions).

Acetylation of nonhistone proteins by enzymes initially found to acetylate histones was reported soon after identification of Gcn5 as a HAT. The first such example was the transcriptional activator p53, which was shown to be acetylated by the coactivator p300, a relative of Gcn5, both in vitro and in human cells, resulting in enhanced affinity for its sequence-specific binding site (Gu and Roeder, 1997). Biochemical experiments showed that PCAF and p300 were capable of acetylating the general transcription factors TFIIEβ and TFIIF (Imhof et al., 1997), while the SAGA and Ada complexes were reported to acetylate the chromatin-associated protein Sin1 (see Chapter 6) in vitro (Pollard and Peterson, 1997). Since these early results, mass spectrometry has been used in combination with isotopic substitution and genetic manipulations to identify global targets of specific HATs and of acetylation in general—the "acetylome" (Choudhary et al., 2014; Downey et al., 2015; Feller et al., 2015; Fournier et al., 2016; Narita et al., 2019). The list of nonhistone targets of Gcn5 and PCAF in a variety of organisms—predominantly budding yeast and mammals—now includes dozens of proteins involved in transcription, cell cycle, and metabolism, among other processes (Downey, 2021; Narita et al., 2019). Notably, acetylation of subunits of both the SWI/SNF and RSC remodeling complexes by Gcn5 has been demonstrated, and in the case of SWI/SNF was shown to affect the dynamics of its association with active gene promoters (Kim et al., 2010; VanDemark et al., 2007). Clearly, the potential impact of nonhistone acetylation must be borne in mind when interpreting the results of HAT inactivation in vivo.

The SAGA complex

Purification of the putative Gcn5-containing coactivator complex was pursued by several labs, resulting in identification of three distinct complexes having molecular weight of about 200, 800, and 1.8 MDa (Grant et al., 1997; Horiuchi et al., 1997; Pollard and Peterson, 1997; Saleh et al., 1997). In addition to findings that incorporation of Gcn5 into the large complex enhanced acetylation of nucleosomal histones and expanded substrate specificity, analysis of the subunit composition yielded additional surprises. Cloning of *ADA5* and *SPT20*, the latter identified in the screen described in Chapter 6 that also uncovered suppressors of *swi/snf* mutations, revealed them to be identical (Marcus et al., 1996; Roberts and Winston, 1996), and Spt20, along with Spt3, Spt7, and Spt8 were found to be constituents of the 1.8 MDa complex, which was consequently named SAGA for Spt-Ada-Gcn5-acetyltransferase. The smaller, 0.8 MDa complex was named the Ada complex and was eventually determined to comprise only the HAT module found in SAGA (see below) along with two additional subunits, Ahc1 and Ahc2 (Fig. 7.9; Grant et al., 1997). The *spt* mutants had been grouped according to

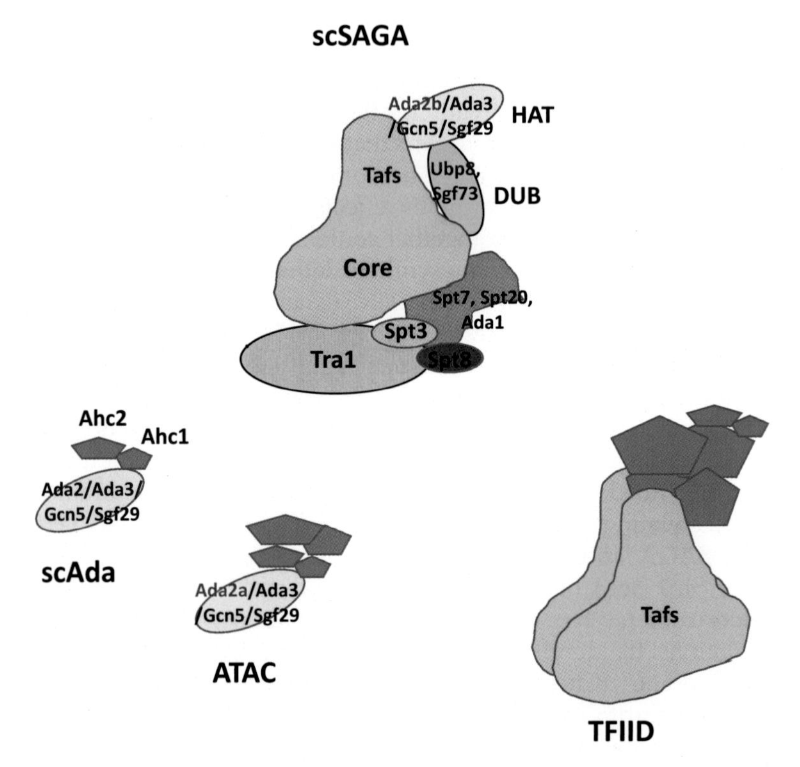

FIG. 7.9 Schematic depiction of the modular structure of SAGA from budding yeast (scSAGA) and related complexes. The size and arrangement of modules are very roughly representative; for a more accurate structure of scSAGA, see Fig. 7.10. The core module is divided in the cartoon between the Taf-containing portion and the remainder containing Spt proteins and Ada1. Spt8 and Spt3 comprise the TBP-binding module in budding yeast, and Spt8 is colored red to highlight its unique presence in scSAGA; it is absent from SAGA in metazoans. Subunits mentioned in the text are indicated; for a complete enumeration, see Helmlinger and Tora (2017). Subunits unique to other complexes (scADA, ATAC, and TFIID) are shown in blue. Tafs present in SAGA in single copy are present in two copies each in TFIID (sometimes as paralogs) and are arranged differently. Ada2a and Ada2b (Ada2 in yeast) are highlighted to indicate their differential presence in metazoan SAGA and ATAC.

phenotype, and *spt3*, *spt7*, *spt8*, and *spt20* mutants exhibited similar phenotypes related to the function of TBP, which itself was identified as *SPT15* (Eisenmann et al., 1989). Moreover, a mutant TBP encoded by *spt15-21* was suppressed by mutations in *SPT3*, suggesting a direct interaction between the two (Eisenmann et al., 1992). This interaction was eventually confirmed by cross-linking and structural studies that revealed in detail the molecular features responsible for TBP association with SAGA and identified Spt3 and Spt8 as a TBP-binding module (Eisenmann et al., 1992, 1994; Papai et al., 2020; Sterner et al., 1999; Wang et al., 2020a). Thus, the presence of this TBP group of Spt proteins in SAGA, together with previously reported indirect interactions between Ada2 and TBP, potentially implicated SAGA in both histone acetylation and TBP recruitment. This link was further emphasized by the discovery of several Tafs (TBP associated factors, present in TFIID) as subunits of the yeast SAGA complex and of the orthologous human PCAF complex (Grant et al., 1998a; Ogryzko et al., 1998). (Gene duplication has led to paralogous Tafs being present in SAGA and TFIID in many organisms; for example, in metazoans the TAF5 and TAF6 paralogs, TAF5L and TAF6L, are specific to SAGA (Helmlinger and Tora, 2017).) Further supporting the role of SAGA as a coactivator, genetic evidence using the *spt15-21* mutant, together with ChIP, supported recruitment of SAGA by Gal4, and SAGA-dependent recruitment of TBP, at genes activated by growth in galactose (Bhaumik and Green, 2001; Larschan and Winston, 2001). High-resolution cryo-EM structures of SAGA, purified from yeast, reveal a substructure that contains Taf6-Taf9, Taf10-Spt7, and Taf12-Ada1 heterodimers and is similar to a structure found in TFIID (Papai et al., 2020; Wang et al., 2020a). Moreover, SAGA and TFIID bind TBP via structurally similar "octamer-like folds," so named for their resemblance to the nucleosomal histone fold structural motif (Chapter 2). These various lines of evidence suggested that the SAGA complex might have functions overlapping with or complementary to TFIID. Genome-wide analyses have since crystallized and refined this notion.

Delivery of TBP to active gene promoters is a critical step in gene activation. The first indication that this might be accomplished via multiple pathways in yeast came from ChIP experiments suggesting the existence of two categories of genes: those that exhibited high Taf occupancy that correlated with TBP occupancy, and those that showed low Taf occupancy in spite of high TBP occupancy (Kuras et al., 2000; Li et al., 2000b). These findings were subsequently expanded in a genome-wide analysis, as discussed in Chapter 6, that indicated that while the majority (85%–90%) of yeast genes depend strongly on TFIID for their transcription (in addition to sharing other properties), the remaining 10%–15% of genes, enriched for stress-inducible genes and showing increased dependence on SWI/SNF, were more highly dependent on SAGA (Donczew et al., 2020; Huisinga and Pugh, 2004; Rhee and Pugh, 2012; Tirosh and Barkai, 2008). This categorization of genes as being TFIID-dominated or SAGA-dominated is not a true dichotomy but rather a continuum, as many genes exhibit properties that are ambiguous with regard to the relevant properties. Nonetheless, considerable evidence supports two distinct pathways in yeast for recruitment of TBP to active gene promoters, one depending more heavily on SAGA and one on TFIID. Whether these two pathways are operative in metazoan cells, potentially divided between TFIID and PCAF, is presently not clear, particularly as Spt8 of the TBP-binding module of SAGA in budding yeast is absent from metazoan SAGA complexes (Helmlinger et al., 2021; Helmlinger and Tora, 2017). Moreover, the extent to which SAGA is responsible for TBP recruitment in budding yeast has yet to be determined on a genome-wide basis.

Analysis of the subunit composition of SAGA also revealed a connection with the SWI/SNF complex: cloning of *SWI7*, which was identified in a screen for mutants defective in mating type switching (Chapter 6), revealed it to be identical to *ADA3*, and *SWI8* and *SWI9* were then found to be the same as *ADA2* and *GCN5*, respectively (Pollard and Peterson, 1997). This connection was underscored by genetic data showing that simultaneous deletion of various SAGA and SWI/SNF components led to synthetic lethality, indicating that SAGA and SWI/SNF overlap in providing a function that is essential to viability (Pollard and Peterson, 1997; Roberts and Winston, 1997).

Metazoan homologs of yeast Gcn5 and Ada2 were first identified by sequence similarity, leading eventually to definition of SAGA complex orthologs (Candau et al., 1996; Yang et al., 1996b). Two paralogs having 75% and 86% sequence identity to yGcn5, and 73% to each other, were found in human cells, one of which was found to interact with two heavily studied coactivators, p300 and CBP (CREB-binding protein, where CREB is the cAMP response element binding protein), and thus dubbed PCAF (initially P/CAF), for p300/CBP associated factor (Yang et al., 1996b). CBP and p300 are closely related in sequence and were known to be targets of adenovirus E1A, expression of which induced transformation of cells to an uncontrolled proliferative—that is, cancerous—state; additionally, they were known to interact with various DNA-binding transcription factors, including nuclear receptors. The presence of a small region of 50 amino acids within p300 and CBP that shared sequence similarity with yeast Ada2 suggested they might be part of a HAT complex related to SAGA; cloning of PCAF as a Gcn5 homolog and the demonstration by pulldown and co-IP experiments that it interacted with p300/CBP both in vitro and in vivo, confirmed this notion (Yang et al., 1996b). Like yeast Gcn5, PCAF and the second identified homolog, hGCN5 (which exists as two isoforms created by alternative splicing, Gcn5S and Gcn5L), both preferentially acetylated H3 and H4 in a mixture of free histones. PCAF showed a strict preference for acetylation of H3 in the context of a nucleosome, while hGCN5 was first reported to have no activity toward nucleosomal histones (Yang et al., 1996b). However, it was later discovered that the expressed form of hGCN5 used in this report represented a splice variant that lacked an N-terminal extension of ~356 amino acids very closely related to an extension found in PCAF but absent from yGcn5; the larger and more prevalent isoform of GCN5 expressed in both human and murine cells, GCN5L, was found to be fully capable of acetylating nucleosomal H3 and H4 (Xu et al., 1998).

Gcn5 relatives were identified in multiple species and found to belong to a superfamily of Gcn5-related *N*-acetyltransferases, or GNATs (Neuwald and Landsman, 1997; Reifsnyder et al., 1996). Evolutionary studies indicate an ancient origin for the GNAT superfamily, with representatives found in protozoans, archaebacteria, and across all eukaryotes (Aravind et al., 2014; Neuwald and Landsman, 1997). This multitude of KATs acetylate a wide variety of substrates, including thousands of nonhistone proteins in eukaryotes (Downey, 2021; Narita et al., 2019).

Identification and cloning of homologs of other SAGA components revealed widespread conservation of the complex coupled with variation tied to novel functional roles in multicellular organisms (Chen and Dent, 2021; Sterner and Berger, 2000). As one example, two paralogous Ada2 proteins, termed Ada2a and Ada2b, were found in *Drosophila* and were shown to be present in distinct complexes (Kusch et al., 2003; Muratoglu et al., 2003). Ada2b is contained in SAGA, while Ada2a is found uniquely in a novel complex, named Ada Two A Containing (ATAC) (Fig. 7.9). ATAC complexes were also found in mammalian cells, and ATAC and SAGA complexes were shown to regulate distinct targets and exert distinct functions in both *Drosophila* and mammals (Guelman et al., 2009; Kusch et al., 2003; Muratoglu et al., 2003; Nagy et al., 2010).

Evidence for expanded specificity of Gcn5 toward histone lysine residues in vivo was observed in experiments using yeast strains in which specific histone residues were mutated (see Chapter 5; Zhang et al., 1998). Consistent with in vitro studies, H3 was a major target for Gcn5-mediated acetylation. Acetylation of H3K9 and H3K18, assayed by Western blotting, was decreased in *gcn5Δ* yeast, in contrast to its nearly absolute preference for Lys14 of H3 (vis a vis other H3 Lys residues) in vitro. In addition, mutation of H3K9 and H4K8 and K16 to Arg, preserving their positive charge but preventing acetylation, was tolerated in otherwise wild type yeast, but was lethal when combined with *gcn5Δ*, implying that there must be additional targets for acetylation by Gcn5. Later work using ChIP-seq showed H3K9ac peaks at essentially all active yeast promoters; these peaks were lost in *gcn5Δ* yeast (Bonnet et al., 2014). In HeLa cells, siRNA knockdown of *ADA3*, resulting in inactivation of both SAGA and ATAC complexes, reduced but did not eliminate H3K9ac peaks present at all active promoters, while H3K9ac and H3K14ac were most strongly depleted upon knockdown of *Gcn5* in *Drosophila* KC cells or in an *Ada2b* null mutant that specifically inactivates SAGA (Bonnet et al., 2014; Feller et al., 2015; Pankotai et al., 2005). Knockdown and mutagenesis experiments targeting subunits specific to ATAC indicated that the *Drosophila* complex preferentially targets H4K5, H4K12, and H4K16, whereas the effect of knockdown of the mammalian ATAC complex was strongest for H3 acetylation (Ciurciu et al., 2006; Guelman et al., 2009; Nagy et al., 2010; Pankotai et al., 2005; Suganuma et al., 2008; Wang et al., 2008a). Overall, then, Gcn5-containing complexes principally target H3 lysine residues, particularly H3K9 and H3K14, for acetylation, but in some contexts (*Drosophila* ATAC) may prefer H4 lysine residues as substrate. Additional studies on these complexes from an array of organisms, together with structural studies, should provide more insight into the determinants of the target specificity of Gcn5 HAT activity.

The large size and complexity of the SAGA complex suggested that its function might extend beyond histone acetylation. The first indications of distinct functions of SAGA subunits emerged from genetic studies in yeast showing that different *ada* and *spt* mutants exhibited different phenotypes: mutations or deletions of HAT module or TBP module subunits exhibited relatively mild effects, while other mutants such as *spt7* or *spt20/ada5* disrupted integrity of the complex and resulted in more severe phenotypes (Marcus et al., 1996; Sterner et al., 1999). An especially telling finding was that mutations that disrupted SAGA integrity were more severe than *gcn5Δ* mutants, directly implying SAGA function in more than histone acetylation. Physical studies also supported multiple roles for SAGA: a modular structure was indicated from an EM structural determination of the human TFTC (TBP-free Taf-containing) complex, which contains homologs of multiple SAGA constituents including Gcn5, although the locations of subunits within the 3.5 Å resolution of the structure were undetermined (Brand et al., 1999a,b). The EM structure of SAGA, affinity purified from yeast using a TAP-tagged Spt7 subunit, revealed it to be similar in size and shape to TFTC; immunolabeling combined with the use of deletion mutants allowed assignment of several subunits to the structure, including Tra1, Gcn5, Spt3, and Spt20 (Fig. 7.10; Wu et al., 2004). Further support for and refinement of this modular structure derived from a computational approach coupled to proteomics analysis using wild type yeast and strains deleted for individual SAGA subunits (Lee et al., 2011). Analysis of SAGA substructures followed, at increasing resolution, culminating in high-resolution structures that confirmed the modular nature of the complex while providing exquisite molecular detail (Fig. 7.10; Diaz-Santin et al., 2017; Han et al., 2014; Helmlinger et al., 2021; Kohler et al., 2010; Papai et al., 2020; Samara et al., 2010; Setiaputra et al., 2015; Sharov et al., 2017; Wang et al., 2020a). As a consequence of these many investigations, the yeast SAGA complex is now understood to consist of five distinct modules (Fig. 7.9; Helmlinger et al., 2021):

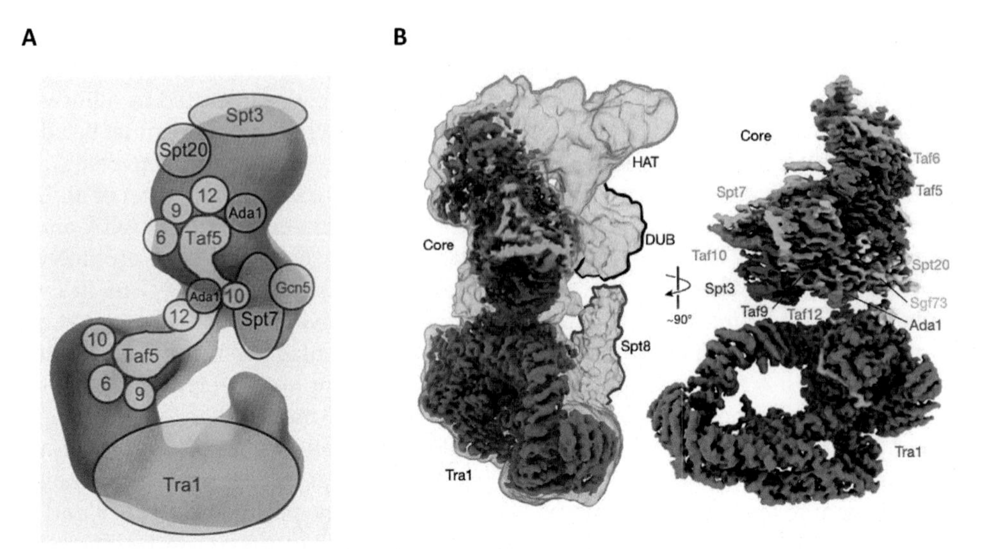

FIG. 7.10 Structure of the *Saccharomyces cerevisiae* SAGA complex (A) based on EM and (B) based on high-resolution cryo-EM. *From (A) Wu, P.Y., Ruhlmann, C., Winston, F., Schultz, P., 2004. Molecular architecture of the S. cerevisiae SAGA complex. Mol. Cell 15, 199; (B) Wang, H., Dienemann, C., Stutzer, A., Urlaub, H., Cheung, A.C.M., Cramer, P., 2020a. Structure of the transcription coactivator SAGA. Nature 577, 717.*

- A HAT module, comprising Gcn5, Ada2, Ada3, and Sfg29.
- A TBP-binding module, comprising Spt3 and Spt8.
- The core structural module, which includes Ada1 and the Taf and Spt subunits arranged in a structure reminiscent of the histone octamer and containing histone folds, as described earlier.
- Tra1, which interacts directly with activators to effect SAGA recruitment.
- A deubiquitinylation, or DUB, module, first identified in a proteomic analysis of Taf-containing complexes, and eventually shown to target mono-ubiquitinated histone H2B (Daniel et al., 2004; Henry et al., 2003; Ingvarsdottir et al., 2005; Sanders et al., 2002).

As already alluded to, this modular structure is broadly conserved across eukaryotes, although it differs in detail in other species (e.g., two splicing factors, SF3B3 and SF3B5, copurify with metazoan SAGA complexes but are absent from SAGA in yeast and plants (Stegeman et al., 2016; Vermeulen et al., 2010).

Tra1 was first identified as a SAGA constituent in yeast through biochemical investigations and recognized as a homolog of the mammalian TRRAP protein, which had been identified as an essential cofactor for activation by the oncogenic TFs c-myc and E1A/E2F (Grant et al., 1998b; McMahon et al., 1998; Saleh et al., 1998). Tra1 and TRRAP were shown to interact physically with activators, based on cross-linking, FRET, and mutational analysis (Bhaumik et al., 2004; Brown et al., 2001; Fishburn et al., 2005; Klein et al., 2003; Reeves and Hahn, 2005). In addition, a *tra1* mutant defective for interaction with Gal4 prevented recruitment of Tra1 and did not support transcriptional activation by Gal4 (Lin et al., 2012). These findings supported a role for Tra1 in activator-mediated recruitment of SAGA to active gene promoters. However, Tra1 is an exceptionally large protein, comprising 3744 amino acids, and was discovered also to be associated with the NuA4 HAT complex, discussed below (Grant et al., 1998b). This shared function rendered genetic analysis difficult in *S. cerevisiae*. However, in *S. pombe* Tra1 has diverged into two paralogs, Tra1 and Tra2, which associate with SAGA and NuA4, respectively (Helmlinger et al., 2011). A *tra1Δ* mutant in *S. pombe* was viable, in contrast to the essentiality of Tra1 in *S. cerevisiae*, and abrogated SAGA recruitment at and affected expression of only a subset of promoters. Evidence also supports Tra1-independent modes of SAGA recruitment in *S. cerevisiae*, as additional SAGA subunits have been found to interact with Gcn4 and the VP16 activation domain (Fishburn et al., 2005; Klein et al., 2003; Natarajan et al., 1998; Reeves and Hahn, 2005; Uesugi et al., 1997). Moreover, at some promoters efficient recruitment of Tra1 itself also depends on additional SAGA subunits (Qiu et al., 2005). Thus, Tra1 is not required universally for SAGA recruitment, but rather supports recruitment by specific activators and likely has additional regulatory roles within SAGA.

Ubp8 and Sgf73, both components of the DUB module, were first found to be associated with SAGA in a proteomic analysis aimed at identifying proteins that interacted with Taf subunits of TFIID (Sanders et al., 2002). Subsequent studies demonstrated the deubiquitinase activity of Ubp8 in the context of SAGA, targeted toward ubiquitinated Lys123 of histone H2B, and showed that this activity influenced transcriptional activation of SAGA-regulated genes (Daniel et al., 2004; Henry et al., 2003; Ingvarsdottir et al., 2005). The catalytic activity of Ubp8 was reported to depend on its incorporation into SAGA (Kohler et al., 2010; Lee et al., 2005; Samara et al., 2010). This dependence may be specific to yeast, as in humans and *Drosophila*, the DUB module is anchored to SAGA by Ataxin-7 and does not depend on this subunit for its catalytic activity (Mohan et al., 2014). Furthermore, Ubp8 appears to be present but not associated with SAGA in the filamentous fungi *Aspergillus nidulans* and *Fusarium fujikoroi*, and evidence exists suggesting that it may function independently of SAGA in plants, budding yeast, and *Drosophila* (Georgakopoulos et al., 2013; Li et al., 2017; Lim et al., 2013; Moraga and Aquea, 2015; Rosler et al., 2016; Srivastava et al., 2015). In yeast, ChIP-seq studies indicate that the DUB module acts to remove ubiquitin from H2B over the transcribed regions of active genes (Bonnet et al., 2014); comparative studies may help to resolve SAGA-dependent and -independent roles of Ubp8 in different organisms.

Structurally, both the DUB and HAT modules appear relatively flexible, potentially enhancing their ability to search regions of chromatin to which SAGA associates for appropriate substrates (Papai et al., 2020; Wang et al., 2020a). It is also interesting to note that Sgf11, a component of the DUB module, targets association of this module with nucleosomes through interaction of a cluster of Arg residues with the H2A-H2B acidic patch—yet another example of this nucleosomal motif as a critical interface for protein-protein interactions (Morgan et al., 2016).

Histone acetylation and transcriptional activation: Mechanisms

By what mechanism could histone acetylation facilitate transcriptional activation? In a cogent analysis published 3 years prior to Brownell and Allis's landmark discovery, Turner divided likely mechanisms into two broad categories: (1) direct alteration of chromatin structure caused by histone acetylation and (2) interactions between nonhistone proteins and chromatin favored by histone acetylation (Turner, 1993). With regard to the first category, histone N-termini were known to be required for higher-order compaction of chromatin (Chapter 3), and hyperacetylated chromatin from HeLa cells treated with butyrate (which inhibits histone deacetylation, as discussed below) was digested more rapidly than control chromatin by

DNase I, suggesting that acetylation might counter repressive compaction of chromatin. The physical properties of nucleosome core particles, however, showed little difference between hyperacetylated and control samples (Ausio and van Holde, 1986; Bauer et al., 1994; Simpson, 1978). The idea that acetylation might affect the interactions of nonhistone proteins with chromatin was indirectly supported by experiments showing that silencing of yeast mating type genes, as well as nucleosome positioning conferred by the α2 repressor protein (Chapter 4), were both disrupted by mutations in a small region of H4, specifically the Lys-Arg-His residues at positions 16–18 (Johnson et al., 1990; Kayne et al., 1988; Roth et al., 1992). Such mutations seemed unlikely to confer substantial structural changes. Moreover, mutations in *SIR3* were identified that suppressed the loss of silencing in point mutations in the H4 tail, suggesting a possible direct interaction that could be modulated by acetylation (Johnson et al., 1990). Subsequent studies confirmed a direct interaction between Sir3 and the H4 tail that was altered by H4 acetylation, as discussed in Chapter 3. More generally, specific protein motifs were discovered to exhibit affinity for particular histone modifications (and in some cases, for those same modifications on nonhistone proteins as well), demonstrating the importance of histone modifications in modulating interactions of nonhistone proteins with chromatin.

One of the first such motifs to be identified was the bromodomain, a module of ~110 amino acids that was initially seen in the *Drosophila* brahma protein (Brm, a component of the SWI/SNF complex; Chapter 6) and subsequently found by homology in multiple other proteins, including some implicated in transcriptional regulation such as Snf2/Swi2 and Spt7 (Haynes et al., 1992; Jeanmougin et al., 1997; Tamkun et al., 1992). A total of nine bromodomain-containing proteins containing a total of 14 bromodomains are present in yeast, all of which have regulatory roles in chromatin structure and/or transcription, while the human genome encodes 42 such proteins containing a total of ~60 bromodomains (Fujisawa and Filippakopoulos, 2017; Sanchez and Zhou, 2009; Zhang et al., 2010).

Discovery of a bromodomain in Gcn5, and subsequently in additional HATs, heightened interest in uncovering the function of this intriguing motif (Brownell et al., 1996; Winston and Allis, 1999). An NMR study reported specific interaction between the bromodomain from PCAF with a polypeptide corresponding to 11 residues from the H4 amino terminal tail seen only when the peptide was acetylated at Lys8; interaction was also seen with an H3 peptide having sequence corresponding to residues 9–20 and acetylated at Lys14 (Dhalluin et al., 1999). Similarities were seen between interactions of Ac-Lys with hydrophobic residues in the bromodomain and hydrophobic residues in Hat1, which otherwise did not resemble the bromodomain in sequence or structure, underscoring the importance of the acetyl group in the interaction (Dhalluin et al., 1999; Dutnall et al., 1998). In a separate study, peptides corresponding to the N-terminal tails of histones H3 and H4 were found to bind to Gcn5 in pulldown assays, with association depending on the Gcn5 HAT domain (Ornaghi et al., 1999). This was perhaps surprising, as the recombinant histone tails utilized were unacetylated, and the measured K_D for the acetylated H4 peptide with the PCAF bromodomain in the NMR experiment was only on the order of 0.3–0.4 mM. However, a follow-up study reported NMR data and a crystal structure that supported preferential binding of the Gcn5 bromodomain to the same acetylated H4 peptide used in the earlier pulldown study, but also found interactions with residues near the acetylated site that could account for a weaker association with the nonacetylated peptide (Owen et al., 2000).

The affinity of bromodomains for acetylated histone tails thus supported the idea that histone acetylation could serve to enhance recruitment of bromodomain-containing components of the transcription machinery to promoters. This idea found support from ChIP experiments showing recruitment of, or remodeling by, SWI/SNF following, and in some cases dependent on, acetylation of promoter nucleosomes in yeast and mammalian cells (Agalioti et al., 2000; Reinke et al., 2001; Reinke and Horz, 2003; Soutoglou and Talianidis, 2002). Sequential action of HAT and remodeling activity was recapitulated in vitro, using mammalian cell-derived assembled chromatin templates (Dilworth et al., 2000). Furthermore, the SWI/SNF complex was retained on arrays of nucleosomes acetylated by SAGA, but retention was abrogated when the bromodomain present in Snf2/Swi2 was deleted (Hassan et al., 2001, 2002). The importance of the bromodomain was corroborated in living cells by experiments demonstrating that its deletion or mutation in Gcn5 impaired transcription from and chromatin remodeling of a hybrid promoter driven by Gcn4 (Syntichaki et al., 2000). In addition, deletion of either the Gcn5 or Snf2/Swi2 bromodomain impaired growth of yeast using raffinose as carbon source, requiring expression of *SUC2*, and the Snf2/Swi2 bromodomain was required for SWI/SNF recruitment to the activated *SUC2* promoter (Hassan et al., 2002). Although this result gives the impression that the Snf2/Swi2 bromodomain is critical for SWI/SNF recruitment to chromatin templates following acetylation by Gcn5, a later study found that cooperative interactions with subunits of both SAGA and the Mediator complex were critical for SWI/SNF recruitment at some target genes, that requirement for specific SAGA subunits for SWI/SNF recruitment varied at different promoters, and that in some cases there was no dependence at all on Gcn5 for SWI/SNF recruitment (Yoon et al., 2003).

In addition to Swi2, bromodomains are found in RSC subunits Sth1, Rsc1, Rsc2, and Rsc4. Bromodomain-containing proteins present in human cells are involved in multiple processes including transcriptional activation, chromatin remodeling, and histone modification, and have attracted recent interest as potential targets for cancer therapeutic applications (Cairns et al., 1996; Fujisawa and Filippakopoulos, 2017; Zaware and Zhou, 2019). Moreover, a second protein motif, the YEATS domain (present in Yaf9, ENL, AF9, Taf14, and Sas5, variously found in chromatin remodeling and histone modification complexes) also associates with Ac-Lys, as well as with the related acylation-based modification of crotonylation (Andrews et al., 2016; Li et al., 2014, 2016). Protein domains with affinity for histone modifications, including methylation as well as acetylation, are summarized in Table 7.1 (Musselman et al., 2012).

TABLE 7.1 Major readers of histone modifications.[a]

Modification	Reader	Examples	References
LysAc	Bromodomain	TAF1; Gcn5; Rsc4	Dhalluin et al. (1999)
	YEATS	AF9, TAF14, ENL	Li et al. (2014)
LysMe	Chromodomain	HP1, Chd1	Bannister et al. (2001) and Lachner et al. (2001)
	Tudor	53BP1, Crb2	Huyen et al. (2004) and Sanders et al. (2004)
	PWWP	NSD2, Dnmt3	Maurer-Stroh et al. (2003)

Continued

TABLE 7.1 Major readers of histone modifications—cont'd

Modification	Reader	Examples	References
	MBT	L3MBTL1	Trojer et al. (2007)
	WD-40 repeat	WDR5 (in MLL1, MLL2, hSet1)	Wysocka et al. (2005)
	Plant homeodomain finger (PHD finger)	BPTF (NURF subunit), TAF3	Li et al. (2006) and Pena et al. (2006)
ArgMe	Tudor[b]	SMN, SPF30	Cheng et al. (2007) and Cote and Richard (2005)
SerP	14-3-3	14-3-3	Macdonald et al. (2005)

[a] *This is not a complete listing; other readers include some that are variations on those in the table (e.g., TTD, or tandem Tudor domains) and others that are more or less closely related to those shown (e.g., Chromo-barrel and zinc finger-CW domains).*
[b] *Distinct classes of Tudor domain-containing proteins exhibit affinity for LysMe or ArgMe (Bedford and Clarke, 2009).*

Histone acetylation facilitates assembly of a variety of factors contributing to chromatin remodeling and transcriptional activation at active promoters. But what of the bromodomain present in Gcn5 itself? While the initial recruitment of SAGA to sites lacking acetylated nucleosomes evidently occurs independently of the Gcn5 bromodomain, in vitro experiments indicated that the bromodomain enhances acetylation of secondary nucleosomal sites, likely by prolonging retention of SAGA at nucleosomal substrates (Cieniewicz et al., 2014; Li and Shogren-Knaak, 2009). The affinity of bromodomains for acetylated lysine residues is not limited to those within histones; in some instances, acetylation of sites within a bromodomain-containing protein leads to intramolecular interactions that result in dissociation of the protein from the original acetylated, nucleosomal target (Kim et al., 2010; VanDemark et al., 2007).

Although considerable evidence supports a positive role for Gcn5-mediated acetylation of histones, especially of H3K9 and H3K14, in transcriptional activation at many genes, its contribution generally appears more analog than digital. It is also highly variable; dependence on Gcn5 and on lysine residues of the histone H3 and H4 for transcriptional activation by Gcn4, the paradigmatic Gcn5-dependent activator, varies for specific promoters, and transcription of some Gcn4-dependent genes shows no change in *gcn5Δ* yeast (Swanson et al., 2003; Yu et al., 2006). Thus, histone acetylation conferred by Gcn5 more resembles a tuning dial than an on-off switch at most genes to which it is targeted.

More HATs

Underscoring their importance, multiple HATs are present in cells, varying in target specificity and function. For example, yeast possess eight nuclear HATs, while mammalian cells have about a dozen (Allis et al., 2007; Yang, 2004). Many excellent reviews have been published regarding HATs and KATs (Ali et al., 2018; Marmorstein and Zhou, 2014; Sheikh and Akhtar, 2019; Shvedunova and Akhtar, 2022). A particularly comprehensive coverage of the early history of HAT discovery is given by Sterner and Berger (2000). Below, distinct families of HATs and examples thereof are briefly discussed.

CBP/p300

Following the discovery of Gcn5 as a histone acetyltransferase, additional HATs were identified at a rapid clip (Table 7.2). As discussed earlier, mammalian PCAF was identified

TABLE 7.2 Major histone acetyltransferases.

HAT	Complex	Family	Major target	References
Hat1[a]			H4K5,12	Feller et al. (2015), Kleff et al. (1995), and Parthun et al. (1996)
Gcn5	SAGA	GNAT	H3K9,14	Bonnet et al. (2014) and Feller et al. (2015)
	ADA (yeast)		H3K9,14	
	ATAC (metazoan)		Various	
PCAF			H3K9	Voss and Thomas (2018)
CBP			Numerous	
P300			Numerous	
TIP60/Esa1	TIP60/NuA4	MYST	Various; mainly H2A and H4	Allard et al. (1999) and Clarke et al. (1999)
MOZ/Sas3	NuA3	MYST	H3K9,14,23	Voss and Thomas (2018)
MORF		MYST	H3K14,23	Voss and Thomas (2018)
HBO1	BRPF	MYST	H3	Lalonde et al. (2013)
HBO1	JADE	MYST	H4	Lalonde et al. (2013)
MOF	MSL (*Drosophila*)	MYST	H4K16	Hilfiker et al. (1997) and Radzisheuskaya et al. (2021)
MOF	NSL (mammals)	MYST	H4K5,8	Radzisheuskaya et al. (2021)

[a] *Mainly cytoplasmic.*

as a Gcn5 homolog that interacts with the highly related coactivator proteins p300 and CBP, which themselves interact with E1A (among many other factors). Pulldown experiments conducted in an effort to identify additional possible HATs recruited by E1A resulted in the finding that p300 and CBP themselves were capable of acetylating histones, whether isolated from HeLa cells or expressed recombinantly (Bannister and Kouzarides, 1996; Ogryzko et al., 1996). The sequences of CBP and p300 did not place them in the GNAT family of HATs mentioned earlier, although several short motifs were identified in common between CBP and PCAF. Target specificity also contrasted to that of other HATs, as p300/CBP acetylated all four histones whether nucleosomal or not, with only modest discrimination as to histone or lysine residue (Bannister and Kouzarides, 1996; Ogryzko et al., 1996; Schiltz et al., 1999). Tethering CBP to a promoter activated a reporter gene in mammalian cells, and activation correlated with HAT activity in a panel of CBP mutants, underscoring the importance of acetyltransferase activity for function (Martinez-Balbas et al., 1998).

MYST family

A second family of HATs, grouped together by sequence homology but distinct from the GNAT family to which Gcn5 belongs, comprises the MYST proteins. This family was named after the founding members MOZ, Ybf2/Sas3, Sas2, and Tip60 (Borrow et al., 1996; Reifsnyder et al., 1996). MOZ is a human protein that was discovered to be fused with CBP at a chromosomal breakpoint associated with acute myeloid leukemia, and is the human homolog of the yeast Sas3 protein. Ybf2/Sas3 and Sas2 ("Something about silencing") were found in a screen as genetically interacting with Sir1 in silencing transcription of the silent mating type loci in *S. cerevisiae*, and Tip60 is a human protein that was discovered via a two-hybrid assay to interact with the Tat activator protein of HIV-1 (Kamine et al., 1996). All four proteins were noted to have atypical zinc finger motifs and sequences indicative of likely acetyltransferase activity, based on their homology to the HAT domain motif of the GNAT family. (The lack of homology to other domains characteristic of GNAT proteins led to MYST family proteins being characterized as distinct.) Additional MYST family members were identified by homology and vary widely in function; these and the founding members are summarized in the following paragraphs.

MOF: The *Drosophila* MOF (males absent on the first) protein functions in dosage compensation in flies, in which expression of genes from the single male X chromosome is about twice that from each of the two female X chromosomes (Kelley and Kuroda, 1995). MOF is a constituent of the male-specific lethal (MSL) complex, together with the *roX1* and *roX2* transcripts and four additional proteins that, like MOF, were discovered based on the male-specific lethality of loss of function alleles (Lucchesi, 1998). As discussed earlier, the *Drosophila* male X chromosome is characterized by acetylation of H4K16; consistent with its homology to known HATs, Mof specifically acetylates nucleosomal H4K16 in vitro, while an MSL complex harboring a mutant MOF protein was unable to do so (Smith et al., 2000a). Although loss-of-function mutants of *mof* or other members of the MSL complex induce male-specific lethality, *mof* male mutant flies survive to the third instar larval or prepupal stage and can be subjected to cytological analysis. Such mutant flies were shown to lack the characteristic H4K16ac modification on their X chromosome (Hilfiker et al., 1997). Thus, MOF mediates a highly specific acetylation event that is critical for dosage compensation in flies.

MORF: MOZ-related factor (MORF) was identified by its homology to MOZ, one of the founding MYST members (Champagne et al., 1999). MOZ and MORF both exhibit HAT activity but differ with regard to specificity: while the recombinantly expressed HAT domain of MORF preferentially acetylates H3 and H4 as free histones and nucleosomal H4, that of MOZ also acetylates free H2A but is unable to acetylate nucleosomal histones (Champagne et al., 1999, 2001). MOZ and MORF have roles in numerous aspects of development and disease through their function as coactivators (Yang, 2015).

HBO1: This protein was discovered via its interaction with the human replication origin binding protein, ORC1, and recognized as homologous to MYST family proteins. It was consequently named HBO1 for histone acetyltransferase binding to ORC (Iizuka and Stillman, 1999). HBO1 is found in two distinct complexes with ING4/5 and EAP6, one with JADE1/2/3 and the other with BRPF1/2/3. The two complexes differ in their specificity, with the JADE complex preferentially acetylating nucleosomal H4 and the BRPF complex targeting nucleosomal H3 (Lalonde et al., 2013). Consistent with its ORC1 interaction, HBO1 has been implicated in DNA replication and has also been found to function as a coactivator (Iizuka et al., 2006; Lan and Wang, 2020; Miotto and Struhl, 2008, 2010, 2011). HBO1 is essential for postgastrulation murine embryonic development, and HBO1-deficient embryos exhibited a > 90% reduction in H3K14 acetylation (Kueh et al., 2011). ChIP-seq experiments revealed HBO1 to be enriched near the transcription start sites of active genes in HeLa cells, but while knockout of *HBO1* reduced acetylation of H3K14 genome-wide, little effect was observed on Pol II association (Xiao et al., 2021).

Esa1/TIP60 and NuA4: Esa1 was identified in budding yeast by its homology to MYST proteins and shown to be an essential HAT required for cell cycle progression (Clarke et al., 1999; Smith et al., 1998). Esa1 turned out to belong to one of the four HAT complexes isolated by Grant and coworkers in their first characterization of SAGA (Grant et al., 1997). While three of the complexes (SAGA, ADA, and NuA3, discussed below) targeted H3 and to much lesser degree, H2B, the fourth preferentially acetylated H2A and H4; purification and characterization of the H4-targeting complex revealed Esa1 as its catalytic subunit and confirmed its specificity for H4 and H2A both in vitro and in vivo (Allard et al., 1999; Clarke et al., 1999; Vogelauer et al., 2000). The complex was named NuA4, for nucleosomal acetyltransferase of histone H4; the cognate human complex is generally referred to as the TIP60 complex after its catalytic subunit, which is one of the founding MYST member proteins. Like SAGA, the NuA4 complex was shown to be conserved across eukaryotes and to function as a coactivator in in vitro assays using chromatinized templates and in an acetylation-dependent fashion (Allard et al., 1999; Doyon et al., 2004; Ikeda et al., 1999; Utley et al., 1998; Wallberg et al., 1999). In vivo experiments have corroborated the role of NuA4 in transcriptional activation, where it participates in a complex interplay with chromatin remodelers and transcription factors to facilitate assembly of the preinitiation complex (e.g., Bruzzone et al., 2018; Durant and Pugh, 2007).

Despite the limited sequence similarity between members of the MYST and GNAT families, the X-ray crystal structure of Esa1 revealed its catalytic core and interaction with acetyl-CoA nearly to superimpose with those of Gcn5 (Hodawadekar and Marmorstein, 2007). The NuA4 complex shares with SAGA a modular structure that includes subunits that are also present in other complexes that regulate chromatin and transcription. Four modules are assembled around the Eaf1 subunit of NuA4. The Piccolo module includes Esa1 together with three additional subunits and also exists as a discrete subcomplex of NuA4; it also bears close similarity to the human HBO1 complex (Avvakumov and Cote, 2007; Boudreault et al., 2003). The SWR1-C module shares its four subunits with the SWR1 complex that remodels chromatin and facilitates exchange of H2A.Z for H2A within assembled nucleosomes (Chapter 5; Auger et al., 2008). The three subunit TINTIN module, like Piccolo, is present as a discrete entity in yeast and facilitates association of Pol II (Rossetto et al., 2014), while Tra1 is shared with SAGA and facilitates activator-dependent recruitment, as discussed above. Two of the SWR1-C subunits are also shared with the INO80 remodeling complex, and a TINTIN subunit is shared with the Rpd3S histone deacetylase complex discussed later. Remarkably, taking into consideration the existence of the Piccolo-NuA4 and TINTIN modules as distinct entities, only Eaf1, which serves as a scaffold for assembly of the other modules, is unique to NuA4.

Cryo-EM structures have been obtained of the NuA4 core complex, comprising the four-subunit Piccolo subcomplex and one additional subunit, and of NuA4 purified from yeast and bound to a nucleosome with a linker of ~25bp (Qu et al., 2022; Xu et al., 2016). The resulting structures reveal interactions between residues within the HAT domain that include the Piccolo module and the linker DNA and nucleosomal acidic patch. The latter interactions were shown to be important by experiments demonstrating reduced HAT activity in the presence of the acidic patch-binding protein, LANA (see Chapter 3). The two major modules of NuA4, one being the HAT module and the other the transcription activator-binding module that includes Tra1, exhibit relative flexibility, likely facilitating the ability of the NuA4 complex to interact simultaneously with multiple partners. The interactions of NuA4 with the face of the nucleosome, through the HAT domain, position the catalytic center near the histone H4 tail, accounting for its targeting specificity. The high degree of conservation between subunits of NuA4 and the mammalian TIP60 complex suggests that the latter is likely to exhibit strong structural similarity.

Sas3 and NuA3: Purification of the NuA3 complex, first identified as one of three HAT complexes principally targeting histone H3 in yeast, revealed its catalytic subunit to be Sas3, a homolog of the human MOZ protein (John et al., 2000). Although NuA3 shared with SAGA and NuA4 the ability to function as a coactivator in in vitro assays using chromatinized templates, it was not found to interact directly with activators (Ikeda et al., 1999; Steger et al., 1998). Instead, NuA3 is targeted to specific regions in chromatin, where it principally acetylates H3K14, by interaction of two of its subunits, Yng1 and Pdp3 respectively, with trimethylated Lys4 of H3 (H3K4me3) and H3K36me3, in addition to contacts made with unmodified histones (Gilbert et al., 2014; Martin et al., 2006a,b; Taverna et al., 2006). Analysis of deletion mutants corresponding to NuA3 subunits indicates that NuA3 plays a relatively minor role in determining mRNA transcript levels, but is important in the dynamics of gene expression (Kim et al., 2020; Lenstra et al., 2011).

Rtt109 and H3K56ac

The fungal-specific HAT Rtt109 was one of the first histone modification enzymes, along with the methyltransferase Dot1, found to act on lysine residues not situated in the unstructured amino terminal regions. It owed its discovery to advances in mass spectrometry that permitted identification of histone modifications located far from the 20–30 N-terminal amino acids susceptible to analysis by Edman degradation (Mersfelder and Parthun, 2006; Zhang et al., 2003a). The development of methods to obtain sensitive and highly accurate molecular weights of peptide fragments from purified histones allows modifications to be identified by comparing expected and observed molecular weights of the corresponding fragments (Lu et al., 2021). By this means, the H3K56ac modification was discovered as an abundant modification in asynchronous cultures of budding yeast, accounting for about a quarter of all H3 molecules (Hyland et al., 2005; Masumoto et al., 2005; Ozdemir et al., 2005; Xu et al., 2005; Zhang et al., 2003a). Two labs found Rtt109, first identified as a Regulator of Ty1 Transposition (Chapter 6; Scholes et al., 2001), to be the responsible HAT for this modification by screening extracts from a library of yeast deletion mutants for the ability to confer the H3K56ac modification (assayed by Western blotting using an antibody specific for H3K56ac) (Han et al., 2007; Schneider et al., 2006). A third group identified Rtt109 by a different route (Driscoll et al., 2007): prompted by the known linkage between Ty1 retrotransposition and DNA damage repair, uncharacterized *RTT* genes were examined for possible roles in damage repair. Yeast deleted for *RTT109* were found to be sensitive to replication stress and exhibited similar phenotypes to yeast lacking the histone chaperone Asf1

(Chapter 5), which had been shown to be deficient in H3K56 acetylation; these findings led to experiments that demonstrated acetylation of H3K56 by recombinant Rtt109. Subsequent work revealed that Rtt109 also targets H3K9 and H3K27; a *gcn5Δ rtt109Δ* mutant completely lacks acetylation of these residues (Burgess et al., 2010; Fillingham et al., 2008).

The K56 residue of H3 lies on the lateral surface of the nucleosome, near the DNA entry/exit point, and is inaccessible in the absence of unwrapping of the DNA from the histone octamer. Correspondingly, its acetylation occurs prior to assembly, and the HAT activity of Rtt109 is enhanced by association with either of the chaperones Asf1 or Vps75 (Driscoll et al., 2007; Tang et al., 2008). (See Chapter 5 for discussion of Asf1 as a histone chaperone and nucleosome assembly factor.) Asf1 and Vps75 tightly associate with $(H3\text{-}H4)_2$ tetramers and H3-H4 dimers, respectively (Bowman et al., 2011; English et al., 2006), and the respective complexes exhibit different specificities: Asf1 association with Rtt109 facilitates H3K56 acetylation, while Vps75 favors H3K9 and H3K27 acetylation (D'Arcy and Luger, 2011).

Consistent with the association of the K56ac modification with newly synthesized H3, Rtt109 and H3K56ac are important for nucleosome assembly following replication or DNA repair (Chen et al., 2008; Li et al., 2008). Biochemical, genetic, and structural studies have provided remarkable detail regarding the mechanism by which this modification functions in the complex nucleosome assembly pathway (Chapter 5). First, H3K56ac increases affinity for the $Rtt101^{Mms1}$-containing ubiquitin ligase complex, resulting in ubiquitination of H3 that decreases its affinity for Asf1, facilitating handoff of H3-H4 to CAF1 (Han et al., 2013); second, H3K56 acetylation increases affinity of H3-H4 for CAF1 and the chaperone Rtt106 (Fazly et al., 2012; Li et al., 2008; Su et al., 2012).

Although Rtt109 is found only in fungi, metazoan cells also possess acetylated H3K56, with the modification carried out by CBP/p300 (Das et al., 2009; Tjeertes et al., 2009). Rtt109 lacks sequence similarity to other HATs, but its catalytic domain is structurally similar to that of CBP/p300 (Tang et al., 2008). While numerous studies have contributed to a detailed, but still incomplete, understanding of the mechanism by which acetylation of H3K56 facilitates nucleosome assembly in yeast, less is known regarding its functions and the mechanisms by which they operate in metazoan cells (D'Arcy and Luger, 2011; Zhang et al., 2018).

Putative HATs

Numerous proteins have been tested for lysine acetyltransferase activity since the discovery of HAT activity in Gcn5. However, many of these putative KATs exhibit only weak activity in vitro, and despite the sequence similarity of some candidates to well-established KATs, the in vivo activity of a considerable fraction of these candidates as bona fide KATs or HATs is suspect. In humans, for example, although 37 proteins have been reported to possess KAT activity, convincing evidence of this has been reported for only 14 (Sheikh and Akhtar, 2019). Of the well documented KATs, the only ones targeting histones—that is, functioning as HATs—belong to or have acetyltransferase domains with homology to those present in the GNAT, p300/CBP, and MYST families discussed above. For historical interest, early examples of candidate HATs that have not been well corroborated in the years since their discoveries are summarized below.

Nuclear receptor coactivators: SRC-1 (steroid receptor coactivator-1), also known as NCoA-1, was identified in a yeast two-hybrid screen as a coactivator able to enhance transcriptional activation by the human progesterone receptor and was found to be effective as a coactivator for multiple nuclear receptors including glucocorticoid receptor, estrogen receptor, and thyroid hormone receptor (Onate et al., 1995). Prompted by the discovery of HAT activity by other mammalian coactivators, SRC-1 was tested for HAT activity and found to acetylate H3 and H4 both as free and nucleosomal histones (Spencer et al., 1997). A similar screen identified a second nuclear receptor coactivator, termed ACTR at the time (also variously named pCIP, A1B1, TRAM-1, and RAC3) and now referred to as SRC-3; ACTR, like SRC-1, also exhibited HAT activity with preference for H3 and H4 (Chen et al., 1997). However, the functional significance of this HAT activity was unclear from early on, as inactivation of the HAT activity of SRC-1 did not have an impact on its ability to enhance transcriptional activation of a chromatinized template in vitro and was shown to be dispensable for its coactivator function in vivo (Liu et al., 2001; Sheppard et al., 2001). While the discovery of HAT activity of SRC-1 and SRC-3 generated considerable early excitement, the targets and significance of this activity remain poorly understood to this day. However, the CLOCK protein, which governs circadian rhythm, is closely related to family of nuclear receptor coactivators to which SRC-1 and SRC-3 belong and possesses HAT activity that resides in a domain homologous to those found in MYST proteins and that is essential for its function (Doi et al., 2006).

Taf1 (TAFII250)/TFIID: Taf1 (then known as TAFII250) was reported to exhibit HAT activity and also was found apparently to be capable of acetylating the general transcription factors TFIIβ and TFIIF (Mizzen et al., 1996). However, Taf1 does not share any resemblance to other known HATs and was not found to contribute significantly to bulk histone acetylation in yeast (Durant and Pugh, 2006). The possibility has been raised that the HAT activity measured using Taf1 expressed in insect cells or immunoprecipitated from HeLa extracts may have arisen from a contaminant (Timmers, 2021).

TFIIC: Two and possibly three subunits of the Pol III transcription factor TFIIIC, which assists in transcription of short noncoding transcripts such as those for tRNAs and 5S rRNA, were reported to exhibit HAT activity as assessed by in-gel assays (Hsieh et al., 1999; Kundu et al., 1999). This, as with the purported HAT activity of Taf1, raised the interesting prospect of intrinsic HAT activity of general transcription factors contributing to their function; however, as with Taf1, convincing evidence to support this notion has remained elusive (Sheikh and Akhtar, 2019). Recently, elevated H3K18ac levels were reported at Alu elements in human T47D cells accompanying TFIIIC association following serum starvation, and this acetylation was reduced upon siRNA knockdown of the C1 subunit (Ferrari et al., 2020). However, this effect might also have resulted from recruitment of CBP, known to interact with TFIIIC (Mertens and Roeder, 2008), and requires further substantiation.

Histone deacetylation before Rpd3

Histone deacetylation was first inferred from pulse-label experiments showing that an initial incorporation of radiolabeled acetate into histones was followed by decreased levels of incorporation in both calf thymus cells and rat liver (Gershey et al., 1968; Pogo et al., 1968; Vidali et al., 1968). An activity was found in calf thymus extract that facilitated removal of acetyl groups from histones and was sensitive to proteolysis, suggestive of an enzymatic activity (Inoue and Fujimoto, 1969). However, despite characterization of putative histone deacetylase (HDAC) activities by multiple labs, including in some cases approximate molecular weights (Hay and Candido, 1983; Inoue and Fujimoto, 1970; Sanchez del Pino et al., 1994; Vidali et al., 1972), purification and molecular identification was not achieved for over 20 years.

Quantitative labeling studies, sometimes exploiting the use of deacetylase inhibitors as described below, revealed the turnover of acetyl groups through competing acetylation and deacetylation reactions to be quite rapid, varying from 3 min to an hour or so depending on the species, cell type, histone, and site (Chestier and Yaniv, 1979; Covault and Chalkley, 1980; Jackson et al., 1975; Waterborg, 2001; Wilhelm and McCarty, 1970b). This rapid turnover raised concern that the measured levels of histone acetylation might change during isolation, leading to inaccurate estimates of the extent and nature of this modification. A partial solution to this problem was achieved by the use of deacetylase inhibitors. The first of these to find widespread application was sodium butyrate, whose ability to inhibit histone deacetylation was discovered in work prompted by earlier studies demonstrating pleiotropic effects of butyrate on cell proliferation and differentiation (Candido et al., 1978; Riggs et al., 1977; Sealy and Chalkley, 1978b; Vidali et al., 1978). Although the use of butyrate was exploited in many studies, such as those linking histone acetylation to nuclease sensitivity mentioned earlier, the same pleiotropic effects that led to its discovery as an inhibitor also potentially confounded interpretations of results based on its use (Yoshida et al., 1995). Nonetheless, inhibition of histone deacetylation by sodium butyrate was useful in studies of the dynamics, extent, and effects of histone acetylation.

Use of sodium butyrate as an HDAC inhibitor was largely superseded by discovery of inhibitory compounds that acted with far greater potency and specificity. HDACs have been found to be misregulated in various cancers and neurodegenerative disorders, and a large effort has been devoted to developing and testing new drugs to target the specific HDACs involved; a number of these new drugs are now in clinical trials (DiBello et al., 2022). Two of the first such inhibitors to be discovered, trapoxin and trichostatin, are microbially derived products first identified for their antitumorigenic properties and shown to inhibit HDAC activity at nanomolar concentrations, about five orders of magnitude lower than the millimolar concentrations required for butyrate to exert its inhibitory effect (Kijima et al., 1993; Yoshida et al., 1990). Both compounds are structurally related to acetyllysine, allowing them to bind to HDAC active sites and block catalytic activity (DiBello et al., 2022; Hassig and Schreiber, 1997). Moreover, trapoxin and trichostatin proved instrumental in the first identification of a histone deacetylase as a homolog to the yeast corepressor, Rpd3 (Taunton et al., 1996).

Paradigm shift part 2: A histone deacetylase is a transcriptional repressor

The presumed high affinity of trapoxin and trichostatin for histone deacetylase was exploited by Schreiber and colleagues in the identification of the mammalian HDAC, HD1 (Taunton et al., 1996). Tritium-labeled trapoxin was used to follow trapoxin-binding activity during fractionation of extracts from bovine thymus (a similar strategy to that used for identification of ligand-binding nuclear receptors), which paralleled HDAC activity and was competed by addition of trichostatin. An affinity matrix based on a specially synthesized trapoxin mimic allowed eventual isolation and sequencing of two proteins. One, RbAp48, had previously been identified as binding to the retinoblastoma gene product, Rb, while the other was ~60% identical to the yeast corepressor Rpd3. The human homolog of this second protein was also cloned and named HD1 (histone deacetylase 1; later referred to as HDAC1). Concurrently with this work, two distinct complexes having histone deacetylase activity were isolated from yeast, one containing Rpd3 and the other containing a novel protein that was named Hda1 (Rundlett et al., 1996). Deacetylase activity of both complexes was inhibited by trichostatin, with the Rpd3-containing complex being less sensitive than the Hda1 complex, and deletion of either *RPD3* or *HDA1* caused global increases in acetylation levels of H3 and H4. In a third study, a protein binding the mammalian repressor YY1 was identified and also found to have sequence homology to Rpd3; this protein, named mRpd3 (mammalian Rpd3; later renamed HDAC2), was found to act as a repressor when artificially recruited as a Gal4 fusion, and its interaction with YY1 was required for YY1-mediated repression (Yang et al., 1996a).

The yeast *RPD3* (*r*educed *p*otassium *d*ependence) gene was first identified in a screen in which growth in media containing low levels of potassium depended on up-regulation of the gene encoding the low-affinity potassium transporter Trk2 (Vidal and Gaber, 1991). Several subsequent screens implicated Rpd3 in transcriptional repression at unrelated genes, suggesting a broad role as a corepressor (Bowdish and Mitchell, 1993; McKenzie et al., 1993; Stillman et al., 1994). This finding, together with the discovery of Gcn5 as a HAT, established a new paradigm in which histone acetyltransferases and deacetylases could be recruited to gene targets to respectively activate or inhibit expression. The impact this had on chromatin research is remarkable: prior to 1996, fewer than 100 papers were published on the topic of histone deacetylation, while from 1996 to 2014, more than 15,000 papers were focused on that subject (Seto and Yoshida, 2014).

Like Gcn5, the yeast Rpd3 protein was found to inhabit a protein complex, as was Hda1 (Rundlett et al., 1996). Three additional genes based on sequence similarity were named *HOS1-3* (Hda One Similar) and eventually demonstrated to encode bona fide HDACs (Carmen et al., 1999; Pijnappel et al., 2001; Robyr et al., 2002). Evidence for recruitment of HDACs by DNA-binding repressors and inhibition of transcription through histone deacetylation was obtained by a similar route to that taken for Gcn5:

- Extracts from yeast expressing Rpd3 harboring mutations in the putative catalytic domain were defective in deacetylase activity, and the corresponding yeast mutants exhibited decreased repressive capability (Kadosh and Struhl, 1998a).
- Repression of reporter genes was seen upon artificial recruitment of Rpd3 via a LexA-Rpd3 fusion and was accompanied by decreased levels of histone acetylation in the region to which it was artificially recruited as well as at endogenous promoters regulated by Rpd3 (Kadosh and Struhl, 1997, 1998b; Rundlett et al., 1998).
- Promoter elements from *INO1*, involved in inositol metabolism, and *IME2*, involved in meiosis, were found to direct Rpd3-dependent repression to a reporter gene, and the known DNA-binding repressor Ume6 was identified as binding to the repressive element (Kadosh and Struhl, 1997; Rundlett et al., 1998).
- Repression conferred by another yeast corepressor complex, the Cyc8-Tup1 complex (also known as Ssn6-Tup1), requires HDAC activity and the histone amino termini and is accompanied by decreased histone acetylation near its sites of recruitment (Bone and Roth, 2001; Davie et al., 2002); interaction of the Cyc8-Tup1 complex with HDACs was also reported (Davie et al., 2003; Wu et al., 2001).

The above results were obtained using budding yeast, but experiments conducted contemporaneously also implicated HDAC activity in transcriptional corepression in mammalian cells. An entry into these latter studies was provided by findings that in yeast, Rpd3 functions through a physical association with Sin3, which was identified in some of the same screens as Rpd3 and shown to operate in the same genetic pathway (Kadosh and Struhl, 1997, 1998b; Stillman et al., 1994; Vidal and Gaber, 1991). In addition, the mammalian homologs of Sin3, mSin3A and mSin3B, act as corepressors (Ayer et al., 1995; Schreiber-Agus et al., 1995). Thus, Sin3 was implicated as a partner for the HDAC-mediated repression conferred by Rpd3 and its mammalian homologs, permitting immediate exploitation of previous work centered on Sin3 to guide experiments aimed at understanding corepression through histone deacetylation. Corepression through HDAC activity in mammalian cells was then corroborated by several labs:

- HDAC1 and HDAC2 (originally referred to as HD1 and mRpd3, respectively) were found associated with mSin3-containing corepressor complexes functioning with various DNA-binding repressors, including the Mad-Max heterodimer, YY1, and nuclear receptors such as the thyroid hormone receptor and retinoic acid receptor, which can act as both activators and repressors (Alland et al., 1997; Ayer et al., 1995; Hassig et al., 1997; Heinzel et al., 1997; Laherty et al., 1997; Nagy et al., 1997; Schreiber-Agus et al., 1995; Yang et al., 1996a).
- A complex containing mSin3A and a close homolog to HDAC1 was shown to be required for repression via the Mad-Max heterodimer, and to possess deacetylase activity that is inhibited by trapoxin; transcription of a reporter gene repressed by mSin3A was relieved by trapoxin treatment (Hassig et al., 1997).
- mSin3A-dependent repression via YY1 or Mad-Max was shown to depend on its association with HDAC1/2 and to be sensitive to trichostatin (Laherty et al., 1997; Yang et al., 1996a).

More HDACs and their complexes

Following these early pioneering studies, additional HDACs were identified by sequence homology searches (Seto and Yoshida, 2014). Histone deacetylases have now been categorized into four major classes based on sequence similarity (Table 7.3). The 11 members of

TABLE 7.3 HDACs.

			Complex	
	S. cerevisiae	**Human**	***S. cerevisiae***	**Human**
Class I	Rpd3	HDAC1,2	Sin3 (Rpd3S, Rpd3L)	Sin3, CoRest, Mi-2/NuRD, MiDAC
		HDAC3		NCoR, SMRT
	Hos1-3	HDAC8	SET3 (Hos2)	
Class II	Hda1	HDAC4, 5, 7, 9 (Iia); HDAC6, 10 (Iib)		
Class IV		HDAC11		
Class III (sirtuins)	Sir2, Hst1-4	SIR1-7	Sir2/3/4; RENT; SET3 (Hst1)	

the HDAC family, divided among class I and II, are Zn^{2+}-dependent deacetylases that share their ancestry with prokaryotic acetylpolyamine amidohydrolases (Leipe and Landsman, 1997; Yang and Seto, 2008), while the class III deacetylases, or sirtuins, belong to the deoxyhypusine synthase-like NAD/FAD-binding superfamily (Seto and Yoshida, 2014). (HDAC11, discovered after designation of the class III family, is the sole member of the class IV deacetylases.) Not all class members are represented throughout eukaryotic species, and not all of these enzymes target histones; for example, HDAC11 is a lysine fatty acid deacetylase, and HDAC6 targets multiple proteins in the cytoplasm, while other HDACs target both histones and nonhistone proteins (Hubbert et al., 2002; Kutil et al., 2018; Milazzo et al., 2020; Moreno-Yruela et al., 2018; Yang and Seto, 2008). Of the class II HDACs, HDAC4,5,7, and 9 have very weak histone deacetylase activity due to replacement of a Tyr residue, conserved among all other HDACs, with a His residue (Lahm et al., 2007). These HDACs are categorized as class IIa HDACs and may principally function by targeting non-histone substrates or by facilitating recruitment of corepressor complexes (Milazzo et al., 2020) In contrast, the class IIb HDAC6 and HDAC10 appear to target nonhistone targets in the cytoplasm (Lee et al., 2022). Thus, the bona fide histone deacetylases are essentially restricted to the class I and class III (or sirtuin) HDACs.

Just as HATs are found in various complexes whose accessory subunits regulate activity and targeting, the class I HDACs are also found in large complexes, some of whose subunits are highly conserved while others exhibit more species specificity (Table 7.3). As with HAT complexes, some HDAC complexes possess additional activities; for example, the NuRD complex, discussed in Chapter 6, functions both as an ATP-dependent chromatin remodeler and a histone deacetylase. HDAC complexes are briefly summarized below; for a more detailed discussion, see (Milazzo et al., 2020; Yang and Seto, 2008).

Class I HDAC complexes

Rpd3-Sin3: Sin3 was the first clearly identified protein associated with Rpd3, and Sin3 complexes are widely distributed, being found in fungi, plants, and animals. Yeast contain two distinct Sin3 complexes, called Rpd3S (Rpd3 small) and Rpd3L (Rpd3 large), that contain both common and distinct subunits and function in transcription elongation and initiation, respectively (Carrozza et al., 2005; Keogh et al., 2005), as discussed in Chapter 8. The mammalian paralogs Sin3A and Sin3B are present in distinct complexes that can contain either HDAC1 or HDAC2 and which function in transcriptional regulation and interact with retinoblastoma protein (Rb) to control the cell cycle (Giacinti and Giordano, 2006).

CoREST: CoREST was identified as a corepressor required for the repressive function of REST (RE1 silencing transcription factor), which represses genes, such as the brain type II sodium channel gene, that are expressed at high levels in neural cells but not elsewhere (Andres et al., 1999). CoREST is found together with HDAC1/HDAC2 in a complex that also contains LSD1, the first identified histone demethylase, and these two activities enhance one another's function at target sites through a positive feedback loop (Hakimi et al., 2002; Humphrey et al., 2001; Lee et al., 2006; Song et al., 2020; You et al., 2001). Consistent with the role of the CoREST complex in regulation of neuronal genes and other lineage-specific genes (Saleque et al., 2007), the complex is not present in fungi or plants.

Mi-2/NuRD complex: The Mi-2/NuRD (nucleosome remodeling and deacetylase) complex, discussed in Chapter 6, was first isolated from *Xenopus* egg extracts and found also to contain a member of the CHD family of chromatin remodelers (Wade et al., 1998). (Mi-2α and Mi-2β, first discovered as disease-associated autoantigens, were later found to be homologs of CHD1 and renamed CHD3 and CHD4.) Related complexes are found in mammals, plants, flies, and worms and possess both HDAC (either HDAC1 or HDAC2) and remodeler activities. The Mi-2/NuRD complex and NuA4/Tip60 complex, discussed earlier, are the only known examples containing both remodeler and histone modification activities. In addition, mammalian NuRD complexes contain either MBD2 or MBD3 (Le Guezennec et al., 2006). While both MBD2 and MBD3 belong to the family of methyl CpG-binding proteins, the two proteins differ markedly in their affinity for methylated DNA (Hendrich and Bird, 1998; Hendrich and Tweedie, 2003), and the mammalian MBD2-NuRD complex exhibits a greater preference for methylated CpG residues than does MBD3-NuRD (Le Guezennec et al., 2006; Leighton and Williams, 2019). In mammalian cells, early work indicated that NuRD associates with regions containing heavily methylated CpG clusters and represses nearby genes (Klose and Bird, 2006; Nan et al., 1998; Ng et al., 1999). ChIP-seq studies have supported this model in part, showing enrichment of MBD2 to be correlated with the density of methylated CpG residues in both murine ESCs and the MCF-7 breast cancer cell line (Baubec et al., 2013; Menafra et al., 2014). However, both MBD2 and MBD3 also bind to an additional set of sites lacking CpG methylation and possessing hallmarks of tissue-specific active regulatory elements (Baubec et al., 2013; Menafra et al., 2014), and the NuRD complex has been implicated in regulation of gene activation in some contexts (Leighton and Williams, 2019). One study reported that the NuRD subunits Mi-2β and HDAC1 were associated only with the unmethylated CpG clusters bound by MBD2 and MBD3 in murine ESCs, but this may reflect the presence in ESCs of an isoform of MBD2 that fails to associate with other NuRD subunits and functions as a dominant negative inhibitor of NuRD-mediated silencing (Baubec et al., 2013; Lu et al., 2014). Overall, the emerging picture is one in which NuRD exists in multiple configurations containing variant isoforms of MBD2 and MBD3 as well as paralogs of other constituents, and participates in both gene activation and repression via association with CpG-rich elements (Leighton and Williams, 2019).

N-CoR/SMRT: In contrast to the Sin3-Rpd3, NuRD, and CoREST complexes, the deacetylase activity of the nuclear receptor corepressors N-CoR (nuclear receptor corepressor) and SMRT (silencing mediator of retinoic and thyroid receptors) is conferred by HDAC3 rather than HDAC1/2. N-CoR and SMRT were first discovered as corepressors through their interactions with the unliganded thyroid hormone and retinoic acid receptors and are closely related in sequence (Chen and Evans, 1995; Horlein et al., 1995). These two corepressors were shown to be associated with histone deacetylase activities that were first reported as HDAC1/2 (Alland et al., 1997; Heinzel et al., 1997; Nagy et al., 1997); later analysis, combining immunoaffinity purification of N-CoR/SMRT and complementarily HDAC3 with mass spectrometry, revealed HDAC3 to be an integral component of the corresponding stable complexes (Guenther et al., 2000; Li et al., 2000a; Wen et al., 2000; Zhang et al., 2002a). The N-CoR/SMRT complexes have been implicated in gene regulation connected with diverse processes, including inflammation, development, and circadian rhythms (Mottis et al., 2013).

MiDAC, MIER, and RERE: The mitotic deacetylase complex, MiDAC, comprises a tetramer of a subcomplex containing HDAC1 and two other subunits, and was discovered in the course of a large-scale study aimed at identifying inhibitors of HDACs within their complexes (Bantscheff et al., 2011). MiDAC associates with cyclin A, suggesting a role in the cell cycle, and like other HDAC complexes functions in developmental gene regulation (Lee et al., 2022; Pagliuca et al., 2011).

MIER1 (Mesoderm induction early response) was identified by its activation during mesoderm induction in *Xenopus* and shown to repress transcription by recruiting HDAC1/2 (Ding et al., 2003; Paterno et al., 1997). The RERE (Arg-Gln, or RE, dipeptide repeat) complex was discovered in the same study as was MiDAC, and the RERE protein shares with MIER1/2/3 an ELM-SANT domain responsible for association with HDAC1/2 (Bantscheff et al., 2011; Lee et al., 2022). Both the MIER and RERE complexes have been found to be involved in developmental processes, but remain relatively understudied.

Class III HDACs: The sirtuins

Transcriptional silencing at the yeast silent mating type loci, *HML* and *HMR*, and at telomeres served as a paradigm for the study of gene regulation, providing key insights into the nature of heterochromatin and its spread, as discussed in Chapter 3. The finding that one of the silent information regulators, Sir2, functioned as a histone deacetylase was a seminal discovery that was critical to understanding the spreading of repressive heterochromatin in yeast (Fig. 3.13); in addition, investigations of the enzymatic properties of Sir2 and its metazoan relatives have led to novel insights into metabolism and aging.

An early connection between Sir2 and histone acetylation stemmed from ChIP experiments examining histone acetylation at the silenced *HML* and *HMR* loci and at telomeres in yeast (Braunstein et al., 1993). The investigation was prompted by the many early studies indicating a correlation between histone acetylation and gene activity, and by the findings that mutations in the amino terminal region of histone H4 resulted in a loss of telomeric and mating type silencing in yeast (Johnson et al., 1990; Kayne et al., 1988; Megee et al., 1990; Park and Szostak, 1990). Using antibodies against an acetylated peptide corresponding to the H4 tail, the authors found that histone H4 was hypoacetylated at these silenced loci in wild type yeast, but was restored to normal levels (i.e., levels seen at transcriptionally active loci) in *sir* mutant yeast. In addition, overexpression of *SIR2*, but not *SIR3*, resulted in decreased global levels of acetylation of all four core histones. Sir2 was also of special interest as it was known to belong to a family of proteins conserved from bacteria to humans and functioning in cell cycle progression and genomic stability (Brachmann et al., 1995). These results raised the possibility that Sir2 could be a histone deacetylase or alternatively could inhibit HAT activity. However, initial efforts to test this notion were frustrated, as no difference in histone deacetylase activity was observed using extracts from yeast overexpressing *SIR2* or control cells (Braunstein et al., 1993).

A critical clue emerged from investigations into the cobalamin biosynthesis pathway in *Salmonella typhimurium* that led to the cloning of a gene, *cobB*, whose sequence placed its product in the Sir2 family of proteins (Tsang and Escalante-Semerena, 1998). As CobB was thought to be a pyridine nucleotide transferase, yeast Sir2 and human SIRT1 were tested for enzymatic activity and reported to transfer radioactivity from [^{32}P]NAD (nicotinamide adenine dinucleotide) to bovine serum albumin (Frye, 1999) and to histones and Sir2 itself (Tanny et al., 1999). Although the latter study suggested that Sir2 could function as an ADP-ribosyltransferase, further testing instead revealed it to deacetylate histones or peptides corresponding to acetylated N-terminal regions of H3 and H4 when NAD$^+$ was included in the reaction (Imai et al., 2000; Landry et al., 2000b). Further supporting the connection between NAD utilization and Sir2-dependent silencing, yeast harboring a mutated *NPT1* gene, important for NAD$^+$ synthesis, had reduced NAD$^+$ levels and exhibited phenotypes similar to *sir2Δ* mutants (Smith et al., 2000b).

These findings, together with the earlier phylogenetic analysis of the Sir2 family, established the existence of a family of HDAC enzymes distinct from the Rpd3-related HDAC family. As discussed in Chapter 3, deacetylation of H4K16ac by Sir2 enhances association of Sir3 with telomeric chromatin to facilitate spreading of telomeric heterochromatin. Sir2 forms a heterotrimer with Sir3 and Sir4 at silenced regions and also functions in the nucleolus as a constituent of the RENT (regulator of nucleolar silencing and telophase exit) complex, where it functions as a histone H3 and H4 deacetylase to regulate transcription of rRNA (Gartenberg and Smith, 2016).

In *Saccharomyces cerevisiae*, the sirtuins comprise Sir2 and its homologs, Hst1-4 (Homologous to Sir Two) (Brachmann et al., 1995). These proteins and their homologs in related fungi principally target histones for deacetylation and function in gene repression, chromosome condensation, and genome stability (particularly in the suppression of recombination among repeated rDNA sequences), and have recently been the subject of a comprehensive review (Zhao and Rusche, 2022). Although their functions are generally preserved among fungal species, differences have evolved in the details; for example, heterochromatin silencing by Sir2 occurs via similar mechanisms in *S. cerevisiae*, *K. lactus*, and *C. glabrata*, but by a completely different mechanism in *S. pombe* that resembles that seen at mammalian centromeric heterochromatin (Zhao and Rusche, 2022). In contrast, in mammalian cells only SIRT1 and SIRT6 target histones, as well as a host of transcription factors and signaling factors, for deacetylation, while the remaining five sirtuins are localized in the cytoplasm, mitochondria, and nucleolus where they function in deacetylation and ADP-ribosylation of a variety of proteins (Houtkooper et al., 2012).

The NAD$^+$ dependence of the sirtuins results from its cleavage into nicotinamide and ADP-ribose, which then serves as acceptor for the acetyl group, generating 2′-*O*-acetyl-ADP ribose (Fig. 7.11; Landry et al., 2000a; Tanner et al., 2000). Because of the central role of NAD$^+$ as a cofactor for many enzymes involved in cellular metabolism and as a redox carrier, sirtuins are positioned to connect the metabolic state of the cell to regulation of gene expression and a host of other processes. The regulation of fat and glucose metabolism by sirtuins in response to changes in cellular metabolic state has been a major investigative topic, and the connection between sirtuin activity, caloric restriction, and aging has prompted research into whether pharmacological perturbation of sirtuin activity might someday be used to treat aging-related diseases or even to extend the "healthspan" of human life (Baur et al., 2012).

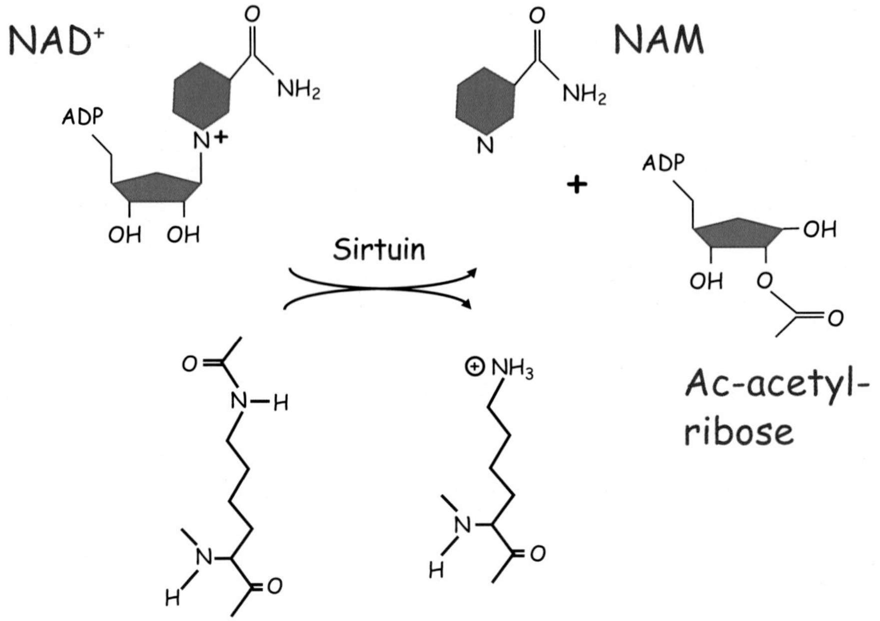

FIG. 7.11 Lysine deacetylation by sirtuins. Protonated nicotinamide adenine dinucleotide (NAD^+) is cleaved to yield nicotinamide (NAM) and ADP-ribose, which serves as acceptor for the acetyl group.

Specificity and genomic targeting of HDACs

The majority of HDACs exhibit low activity in isolation and require incorporation into their native complexes for full deacetylase activity. For example, of the mammalian class I HDACs, only HDAC8 exhibits substantial deacetylase activity in isolation, and in yeast, Hos3 is the exception, retaining deacetylase activity even when expressed in bacteria (Carmen et al., 1999; Guenther et al., 2001; Hu et al., 2000; Lechner et al., 2000; Li et al., 2000a; Wen et al., 2000; Zhang et al., 1999). Thus, as for HAT enzymes, in vitro assays aimed at determining target sites on histones must be conducted with isolated complexes (for example, by immunoaffinity purification (Johnson et al., 2002)) to be meaningful. This presents challenges, as any contaminating activity may affect results; moreover, a given HDAC may be present in multiple complexes, some complexes contain multiple enzymatic activities, and efficiency can also vary depending on the precise nature of the substrate, including the presence of other modifications (see "Cross-talk" section) (Strahl and Briggs, 2021).

Given these caveats, it is not surprising that conflicting results on the preferred targets of various HDACs have been reported (Seto and Yoshida, 2014). Methods used to characterize the action of HDACs at specific residues have advanced, however, allowing more accurate determinations to be carried out. Early work generally involved preparing radiolabeled histones, either individually isolated or nucleosomal, via nonenzymatic acetylation using acetyl-CoA or the use of extracts that contained acetyltransferase activity (Fujimoto and Segawa, 1973; Paik et al., 1970; Sanchez del Pino et al., 1994); deacetylase activity was then monitored by following release of radioactivity over time. Mass spectrometry, applied either to peptides corresponding to acetylated histone amino-termini or to entire acetylated histones, allowed simultaneous determination of the efficiency of deacetylation at multiple sites (Carmen et al., 1999), while the use of antibodies toward specific acetylated residues permitted assessment of the effects of loss or inactivation of HDAC activity both in vitro and in living cells (Johnson et al., 2002; Suka et al., 2001). More recently, construction of "designer" histones and nucleosomes bearing precise and specific modifications has been enabled by the use of novel genetic and chemical strategies (Neumann et al., 2009; Nguyen et al., 2014; Wang et al., 2020b, 2022). An additional improvement in approach is the rapid inhibition or depletion of modifying activities rather than chronic deletion (see Chapter 4 Sidebar: Deletions and depletions); this is particularly salient for HDACs, which in many cases are known to be redundant in function.

Several studies reported deacetylation by class I HDACs of multiple acetylated lysine residues of H3 and H4 as free histones, nucleosomal histones, and peptides corresponding to acetylated histone tails, with efficiency varying depending on the specific HDAC and residue (e.g., Grozinger et al., 1999; Johnson et al., 2002; Vermeulen et al., 2004). A more recent study used a semisynthetic approach to prepare nucleosomes mono-acetylated at H3K9, H3K14, H3K18, H3K23, and H3K27, and compared deacetylation of these nucleosomes, as well as the corresponding free histone H3 variants, by immunoaffinity purified CoREST, NuRD, Sin3B, MiDAC, and SMRT complexes (Wang et al., 2020b). The results indicated similar kinetics of deacetylation by the various complexes for acetylated H3 substrates, but widely varying rates and site selectivity in a nucleosomal context. Similarly, mononucleosomes containing H2B acetylated at four distinct sites were deacetylated with variable kinetics by several HDAC complexes (Wang et al., 2022). Sirtuins appear to exhibit greater selectivity: yeast Sir2 preferentially deacetylates H3K9, H3K14, and H4K16, while mammalian SIRT1 targets H3K9 and H4K16 (Imai et al., 2000; Vaquero et al., 2004).

Targeting of multiple lysine residues by both HAT and HDAC complexes was in accord with mutagenesis studies that generally showed additive effects of mutations of lysine residues on phenotype and gene expression, rather than effects attributable to specific sites (Dion et al., 2005; Kayne et al., 1988; Megee et al., 1990; Park and Szostak, 1990). Evidence that deacetylation of the histone tails, and not some other target, is responsible for Rpd3-mediated repression was provided by microarray analysis of gene expression in yeast mutants lacking the H3 or H4 tails, or harboring Lys to Gln mutations that mimic the uncharged, deacetylated state, alone or together with deletion of *RPD3* (Sabet et al., 2004). This study showed strong overlap between genes derepressed in *rpd3Δ* yeast and in H3 or H4 tail mutants, whether deletions or K→Q mutations, and a greatly reduced effect of the *rpd3Δ* mutation on genome-wide transcription when the histone tails were deleted or mutated, suggesting a common pathway for repression mediated by Rpd3 and the H3 and H4 tails.

The discovery of HDACs as corepressors occurred at the time during which ChIP was first being widely adopted and microarrays were first being implemented for genome-wide interrogation of gene expression and localization of DNA-binding proteins (Chapter 4). Both methods were leveraged to gain further insight into histone deacetylation as a corepression mechanism. Globally increased acetylation was observed at multiple residues of both histones H3 and H4 upon deletion of *RPD3* or *HDA1*, with largest effects in *rpd3Δ* cells at H4K5 and K12, as seen by Western blotting using polyclonal antibodies raised against acetylated peptides (Rundlett et al., 1996). Following the discovery that a Sin3-Rpd3 complex could be recruited by Ume6 to repress gene activation, ChIP was used to examine H3 and H4 acetylation in wild type yeast and in mutants lacking specific HATs or HDACs, both at individual genes and eventually genome-wide. While initial studies suggested preferential deacetylation of H4K5 at genes repressed by Rpd3 (Kadosh and Struhl, 1998b; Rundlett et al., 1998), a later study using antibodies with improved specificity (see Sidebar: Antibodies to modified histones) showed that Rpd3 targeted multiple sites on all four core histones for deacetylation at the repressed *INO1* and *IME2* promoters, with the strongest effect at H4K5 and K12, consistent with in vitro results (Suka et al., 2001). This latter study also reported decreased acetylation of lysine residues in the H3 and H2B tails in *gcn5Δ rpd3Δ* yeast relative to *rpd3Δ* yeast, and similarly found multiple sites targeted by Esa1 (utilizing a temperature sensitive *esa1* mutant), with strongest effect at H4 and H2B. Altered acetylation levels were observed across broad regions (22 kb in total, at three loci), suggesting global and possibly untargeted deacetylation conferred by Rpd3 and acetylation by Gcn5 (Vogelauer et al., 2000); independent evidence for broad regions of histone acetylation by Gcn5 supported widespread activity of HATs (Kuo et al., 2000). These results supported a model in which the steady-state levels of histone acetylation, long known to reflect a balance between acetylase and deacetylase activities (Chestier and Yaniv, 1979; Covault and Chalkley, 1980; Gershey et al., 1968; Vidali et al., 1968; Wilhelm and McCarty, 1970a,b), result from opposing effects of HATs and HDACs at specific loci. A subsequent study demonstrated these opposing activities to be rapid, with inactivation of HAT or HDAC activity resulting in decreased or increased histone acetylation, as assayed by ChIP, within 1.5–8 min (Katan-Khaykovich and Struhl, 2002).

Histone methylation and demethylation

Research on protein methylation, with an emphasis on histone methylation, has been covered in a pair of engaging historical reviews (Murn and Shi, 2017; Paik et al., 2007). Protein methylation was first observed in an analysis of bacterial flagellin, which uncovered an amino acid whose chromatographic behavior did not correspond to that of any known amino acid; further analysis revealed the novel entity to be ε-*N*-methyllysine (Ambler and Rees, 1959). A subsequent report by Ambler and colleagues uncovered a gene not encoding flagellin that was required for the presence of ε-*N*-methyllysine in flagellin, leading to speculation that an enzymatic activity might be responsible for lysine acetylation in fully formed proteins (Stocker et al., 1961). Following the first report of methylated histones (Murray, 1964), Allfrey and colleagues provided further support for posttranslational methylation. In the same publication in which they presented evidence for posttranslational acetylation of histones, they showed that inhibition of histone synthesis by puromycin did not prevent incorporation of radiolabeled methyl from methionine into histones in cell extracts (Allfrey et al., 1964). Several labs then pursued investigations of protein and histone methylation, identifying enzymatic activities capable of transferring a methyl group from *S*-adenosyl-methionine to histones and other proteins albeit without, as for acetyltransferases, obtaining sufficient quantities of purified protein to allow molecular identification (Comb et al., 1966; Kaye and Sheratzky, 1969; Kim and Paik, 1965; Liss and Edelstein, 1967; Liss and Maxam, 1967; Paik and Kim, 1967a, 1968). Arginine as well as lysine was shown to be subject to methylation, with lysine being capable of accepting one, two, or three methyl groups, and arginine one or two methyl groups, with the latter occurring symmetrically or asymmetrically (Figs. 7.12 and 7.13;

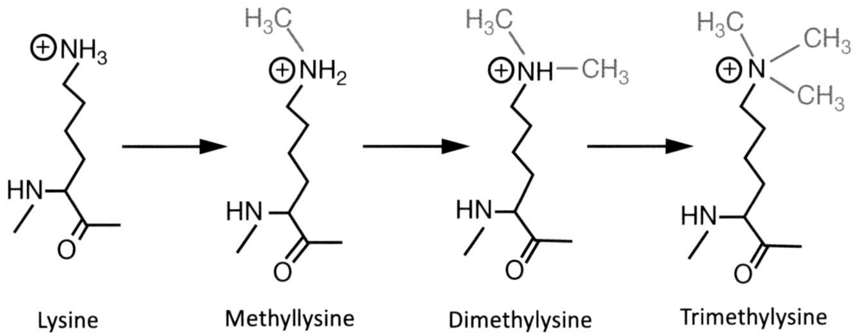

FIG. 7.12 Methylation of lysine.

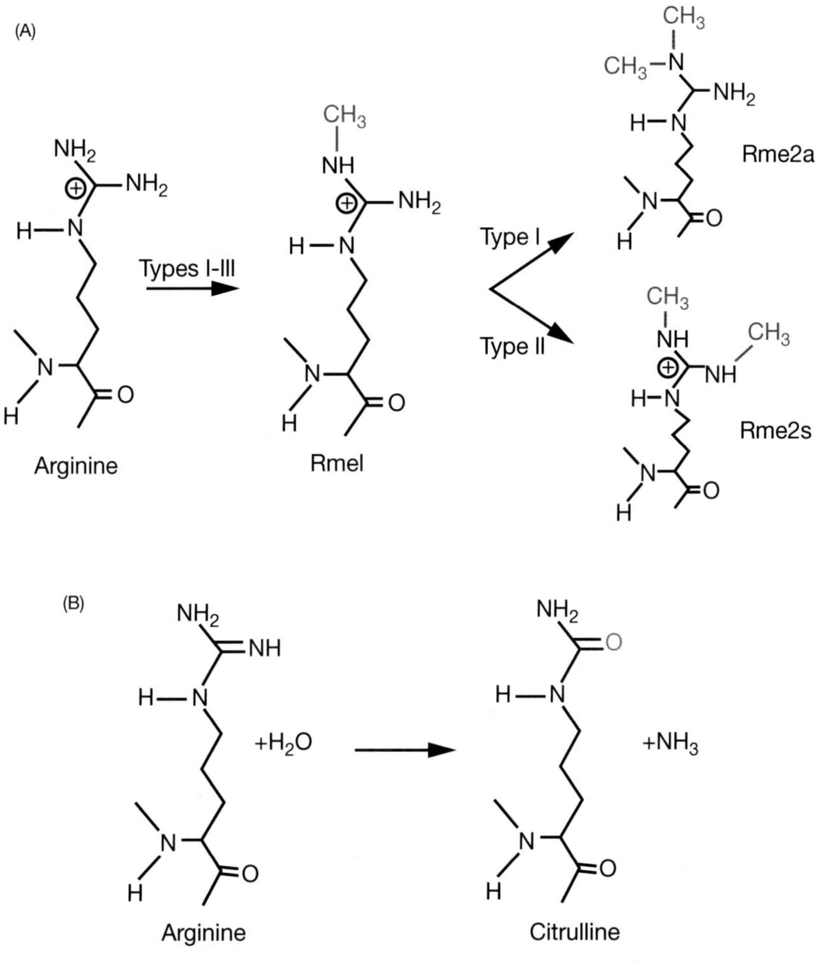

FIG. 7.13 (A) Arginine methylation. (B) Deamidation of arginine to yield citrulline.

Hempel and Lange, 1968; Hempel et al., 1968; Kakimoto and Akazawa, 1970; Paik and Kim, 1967b, 1968). Murray had speculated, based on the relatively low (compared to flagellin) and heterogeneous methylation of the various histone fractions, that histone methylation might occur with high specificity. Supporting this idea, the first sequencing of histones revealed specific methylation at H4K20, H3K9, and H3K27 of calf thymus; dimethylated lysine (H4K20me2) predominated at H4, while all three derivatives were seen for H3K9 and K27 (DeLange et al., 1969, 1973; Ogawa et al., 1969). These modified residues were found by various research groups to be present in high abundance across a spectrum of organisms, with later studies reporting from 40% to 80% of mammalian histones to be dimethylated at H3K9, H3K27, H3K36, and H4K20 (Jung et al., 2010; Schotta et al., 2008; van Holde, 1988). These early findings, together with observations of differing kinetics of methylation of H3 and H4 in vivo suggested that multiple histone methyltransferases might be present in the cell (Thomas et al., 1975).

As with histone acetylation, following the advances in research on histone methylation in the 1960s, a slowdown ensued, with little progress made from the early 1970s until the 1990s (Paik et al., 2007). Two major impediments slowed the field. First, although enzymatic activities present in cell extracts provided an explanation for posttranslational modifications, whether the same activities conferring methylation of histones in vitro were responsible for methylating free histones or intact chromatin in vivo remained unproven. Second, the physiological significance of histone methylation (and of protein methylation in general) was unclear and difficult to test without knowledge of the specific enzymes responsible. Moreover, histone methylation took a back seat to acetylation in terms of general interest, as it was acetylation, not methylation, that was found to correlate with transcriptional activity (Allfrey et al., 1964).

This changed in the 1990s, as novel molecular biological tools and approaches opened new avenues of investigation. Molecular identification of protein arginine methyltransferases was achieved concurrently in yeast and mammalian cells in 1996 (Gary et al., 1996; Henry and Silver, 1996; Lin et al., 1996). Although neither of these enzymes were initially reported to target histones in vivo, both the mammalian enzyme, dubbed PRMT1 for protein arginine methyltransferase 1, and the yeast homolog, Hmt1 (also named Rmt1) were shown to modify Arg residues in protein substrates to all three methylated derivatives, and in a follow-up study, the yeast enzyme was found to facilitate nuclear export of hnRNP proteins, thus providing the first evidence for physiological relevance of protein arginine methylation (Shen et al., 1998).

Histone methyltransferases

It is difficult to overstate the sea change that occurred in the worldview of histone methylation in the chromatin biology research community in the late 1990s. Ken van Holde, reviewing research on histone methylation in his 1988 text on chromatin, tentatively concluded that methylation of H3 and H4 most likely "is not involved in dynamic cell processes" (van Holde, 1988, p. 125). This conclusion was based on metabolic labeling experiments indicating that essentially no ongoing methylation of H3 or H4 occurred in nondividing cells from adult rat brain or from mature avian erythrocytes, in contrast to the clearly dynamic acetylation of histones (Duerre et al., 1977; Sung et al., 1977). It may be that examination of terminally differentiated cells painted a misleading picture; nonetheless, it required innovations in the use of mass spectrometry and antibodies raised against specific histone modifications to effect a major revision in this viewpoint.

The first identification of a bona fide histone methyltransferase, or HMT, occurred in 1999, when a yeast two hybrid screen aimed at identifying proteins interacting with a region of the nuclear receptor coactivator GRIP1/TIF2 important for transcriptional activation revealed a novel protein with extensive homology to known protein arginine methyltransferases, including PRMT1 (Chen et al., 1999). The protein, named CARM1 (coactivator-associated arginine methyltransferase 1; also referred to as PRMT4, being the fourth protein arginine methyltransferase to be identified), methylated histone H3 in vitro using *S*-adenosylmethionine as cofactor, and enhanced transcription by nuclear receptors in transient transfection assays only when the coactivator GRIP1 was also present. A mutant CARM1 defective for H3 methylation failed to support coactivator function, and it was later shown that PRMT1 also functioned as a coactivator, cooperating with CARM1 in a stepwise process involving multiple histone modification events during transcriptional activation (An et al., 2004; Koh et al., 2001; Wang et al., 2001b). CARM1 was initially designated as an arginine methyltransferase on the basis of its homology to known PRMTs together with its demonstrated ability to methylate free histones, and only later shown to methylate R2, R17, and R26 of histone H3, while PRMT1 was found to methylate H4R3 (Schurter et al., 2001; Strahl et al., 2001; Wang et al., 2001b).

The discovery of CARM1 and PRMT1 as coactivators linked histone methylation, like histone acetylation, with a positive effect on transcription. This correlation was further supported by observations of H3K4 methylation in the transcriptionally active macronucleus, but not the transcriptionally inactive micronucleus from *Tetrahymena*; H3K4me was also found in HeLa cells and in yeast, where it was enriched in nucleosomes that contained acetylated histones (Strahl et al., 1999). H3K4 methylation was subsequently found to correlate with histone H3 acetylation, itself associated with active transcription, at the chicken β-globin promoter, and to be associated with transcriptionally active euchromatin in fission yeast (Litt et al., 2001; Noma et al., 2001). These findings extended the correlation between transcriptional activity and histone methylation to lysine, in addition to arginine, residues.

Soon after the discovery of CARM1 and H3K4 methylation, however, things got more complicated. In a study of SU(VAR)3-9, implicated in heterochromatin function through genetic studies (see Chapter 3), Jenuwein, Allis, and colleagues searched for sequences exhibiting homology to the ~130 amino acid SET domain common to SU(VAR)3-9, E(Z) (enhancer of zeste), and Trithorax, all three of which contribute to position effect in *Drosophila* (see "Heterochromatin" section in Chapter 3) (Rea et al., 2000). Finding six plant methyltransferases included among the homologous sequences, the investigators tested human SUV39H1 and murine Suv39h1 (one of two homologs of *Drosophila* SU(VAR)3-9) for HMT activity and found that recombinant SUV39H1 methylated free histones with strong preference for H3K9. Subsequent investigations demonstrated the function of this highly specific modification: histone H3 methylated at K9, but not K4, was strongly bound by the mammalian heterochromatin protein 1 (HP1) and by the *S. pombe* homolog Swi6 in vitro, and elimination of methyltransferase activity of SU(VAR)3-9 or the *S. pombe* homolog Clr4 in vivo resulted in loss of HP1/Swi6 localization and concomitant loss of heterochromatic silencing (Bannister et al., 2001; Lachner et al., 2001; Nakayama et al., 2001). (Budding yeast lack methylated H3K9 and utilize a distinct mechanism, involving the Sir proteins, for their very limited regions of silenced heterochromatin (Chapter 3).) These and related findings led to a model for the spreading and maintenance of constitutive heterochromatin in eukaryotes (Fig. 3.13). Methylation of H3K9 was also found in short order to be involved in repression of euchromatic genes and mammalian X-chromosome inactivation (Boggs et al., 2002; Heard et al., 2001; Mermoud et al., 2002; Nielsen et al., 2001; Peters et al., 2002; Shi et al., 2003; Vandel et al., 2001). Thus, while CARM1-mediated methylation of Arg and H3K4 methylation both correlated with transcriptional activation, the connection of methylation of H3K9 with heterochromatin and downregulated genes in euchromatin explicitly indicated a repressive function.

In the early 2000s the floodgates opened, as a host of HMTs capable, in toto, of methylating all canonical Arg and Lys residues in histones H3 and H4 were reported (Greer and Shi, 2012). As was the case for CARM1, PRMT1, and SUV39H1, most HMTs targeted only one or a small number of potential histone target residues; nonetheless, additional, nonhistone targets have increasingly emerged with passing time. Adding to this complexity, some HMTs were discovered to direct specific methylation states at their targets with distinct functional outcomes (Fig. 7.14). For example, SetDB1 monomethylates H3K9 to provide a substrate for SuvH39H1/H2, which can act on H3K9me1 but not unmethylated H3K9 to produce H3K9me2 and H3K9me3 (Loyola et al., 2009). As with HATs, research on HMTs has exploded since their first identification, including investigations on their involvement in development and disease, with regard to both histone and nonhistone targets (Bedford and Clarke, 2009; Cavalieri, 2021; Hyun et al., 2017; Zhao and Garcia, 2015). The depth and detail of investigations into histone modifications, particularly methylation, can be overwhelming. In an effort not to get entirely lost in this welter of information, I will summarize only a little of what has been uncovered by this research, with the intention of highlighting some generalities that have been uncovered and providing a basic background for the reader interested in pursuing this topic further. Such readers may wish to consult (Greer and Shi, 2012), who provide a list of HMTs and their targets known as of 2012, along with supplementary references.

Histone arginine methylation

Although CARM1 was the first HMT to be identified as such, histone arginine methylation has received somewhat less attention than lysine methylation. The major sites of histone arginine methylation are H3R2, R8, R17, R26, and R42, and H4R3. Additional sites have been identified but without attribution of function, including some in H2A (Zhao and Garcia, 2015). Nine PRMTs have been identified in mammals; PRMT1-8 have been shown to target histones in vitro or in vivo, while PRMT9 is a nonhistone arginine methyltransferase (Fulton et al., 2021; Hadjikyriacou et al., 2015). Only two PRMTs known to target histones are present in budding yeast: the previously mentioned Hmt1, which is homologous to mammalian PRMT1, and Hsl7, which is homologous to and replaceable by mammalian PRMT5 (Lee et al., 2000; Low and Wilkins, 2012). These PRMTs are classified according to the products that their actions result in (Fig. 7.13A). Type I PRMTs catalyze formation of monomethylated and asymmetrically dimethylated arginine (me1 and me2a); Type II yield monomethylated and symmetrically dimethylated arginine (me1 and me2s); and Type III yield only monomethylated products. The symmetrically and asymmetrically dimethylated products are most often distinguished through the use of specific antibodies.

Recombinant PRMTs are capable of methylating free peptides corresponding to histone sequences, intact free histones, and histone octamers, with varying specificities (Fulton et al., 2021). Whether and how PRMTs methylate nucleosomal histones in vivo remains unclear. A recent study reported that recombinant PRMT1, PRMT3, PRMT5, PRMT6, PRMT7, and PRMT8 preferentially methylated histone H3 in the context of histone octamers, while PRMT4/CARM1 preferentially methylated histone H4. None of the recombinant PRMTs methylated nucleosomal histones under the conditions employed (Fulton et al., 2021). This result contrasts with an earlier report in which treatment of oligonucleosomes with purified PRMT1 enhanced their subsequent acetylation by a nuclear extract, indirectly suggesting arginine methylation of nucleosomal histones by PRMT1 (Huang et al., 2005). Enhanced acetylation was observed after treatment of oligonucleosomes with PRMT1 even when PRMT1 was inactivated prior to incubation of the treated oligonucleosomes with the extract, strongly suggesting that they were indeed methylated by PRMT1 in vitro, although this was not directly demonstrated (Huang et al., 2005). It seems likely that most, and perhaps all, PRMTs require association with auxiliary proteins in order to methylate nucleosomal histones, but this remains to be determined.

Functionally, histone arginine methylation has been found in several instances to influence transcription indirectly by affecting other histone modifications. Asymmetric dimethylation of H3R2 by PRMT6 in mammalian cells was shown to inhibit trimethylation of H3K4, thereby acting to repress transcription (Guccione et al., 2007; Hyllus et al., 2007; Iberg et al., 2008). In budding yeast, ChIP-chip experiments showed anticorrelation between H3R2me2a and H3K4me3, with the former being present in heterochromatic regions and inactive euchromatic genes, as well as gene bodies, whereas H3K4me3 was seen at 3′ ends of active genes (Kirmizis et al., 2007). An H3R2A mutation abrogated H3K4 trimethylation by Set1 in vivo and in vitro, and binding of Spp1, which is required for trimethylation but not for mono- or dimethylation of H3K4 (see "Histone lysine methylation" section), to a peptide substrate was inhibited by the H3R2me2a modification. How targeting of H3R2me2a occurs is, however, currently unknown (Zhang et al., 2019); furthermore, as the impact of H3K4me3 on transcriptional activation is murky (see "H3K4 lysine methylation" section), the extent to which H3R2me2a controls genome-wide expression or transcriptional competence is unclear.

Symmetric dimethylation of H3R2, remarkably, has an opposite effect to asymmetric dimethylation: H3R2me2s, installed by PRMT5, inhibits binding of RBB7, a component of three distinct corepressor complexes, and enhances binding of the coactivator component WDR5 (Migliori et al., 2012). The yeast PRMT5 homolog Hsl7 similarly catalyzes H4R3me2s but may cooperate with Rpd3 in transcriptional repression (Ryu et al., 2019).

Histone lysine methylation

Relative to histone acetyltransferases, the enzymes that catalyze methylation of lysine residues tend to be specific in their targeting (Hyun et al., 2017; Shvedunova and Akhtar, 2022; Weinert et al., 2018; Zhao and Garcia, 2015). HATs, we have seen, although exhibiting preference for particular histones, typically target multiple lysine residues. For example, the yeast SAGA complex acetylates multiple sites in histone H3, while the NuA4 complex principally acetylates H4K5 and K12, and the mammalian p300/CBP complexes target multiple lysine residues in both H3 and H4. Correspondingly (with some exceptions), overall levels of histone acetylation in a given region of chromatin are more important for function, primarily transcriptional activation, than the acetylation status of any individual site. (H4K16ac represents something of an exception to this generalization in its role in dosage compensation in *Drosophila* and the requirement for its deacetylation for interaction with Sir3 in the formation of telomeric heterochromatin in budding yeast.) In contrast, histone lysine methyltransferases nearly universally target single sites (Fig. 7.14), and the effects of methylation are site specific. For these reasons, histone lysine methylation is discussed here in the context of specific sites rather than the enzyme families conferring those modifications.

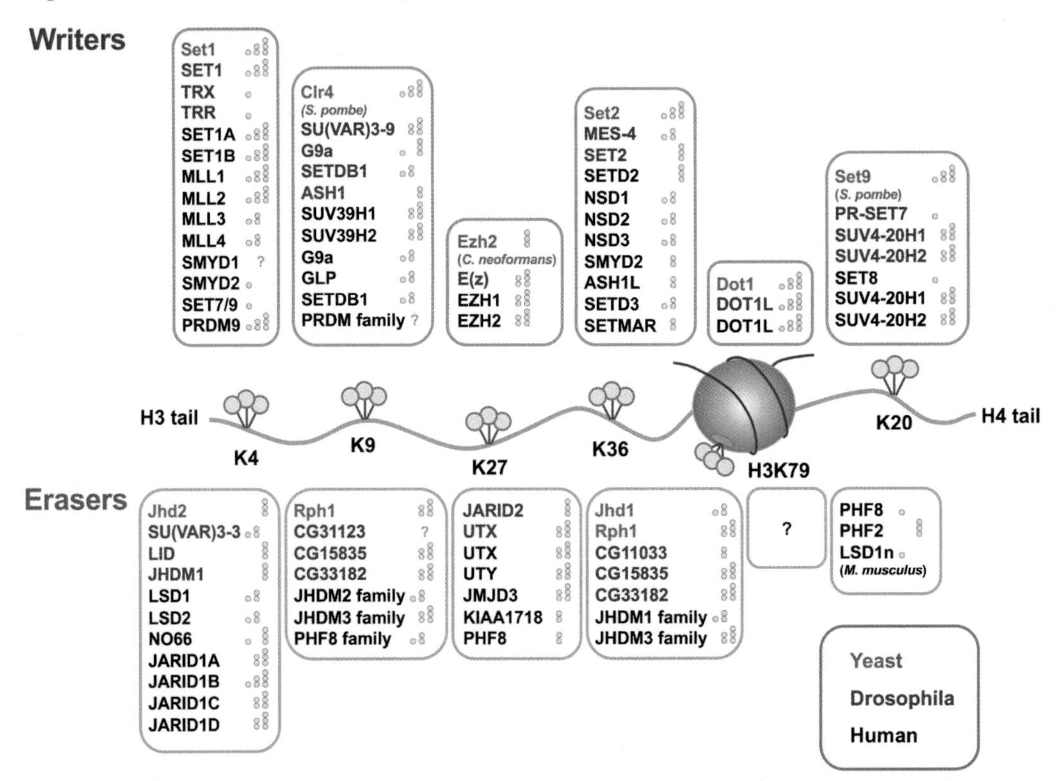

FIG. 7.14 Lysine methyltransferases ("writers") and demethylases ("erasers") for the major methylation sites in histones H3 and H4, as discussed in the text. Methylation specificity is indicated by one, two, or three circles corresponding to mono-, di-, or trimethylation. *From Hyun, K., Jeon, J., Park, K., Kim, J., 2017. Writing, erasing and reading histone lysine methylations. Exp. Mol. Med. Exp Mol Med 49, e324.*

The most intensively studied sites of histone lysine methylation are H3K4, K9, K27, K36, and K79, and H4K20. These sites are, of course, extremely highly conserved (Chapter 5), as are the enzymes that methylate them, although it is worth noting here (again) that *S. cerevisiae* completely lacks methylation of H3K9 and also of H3K27. Functionally, the phenotypes resulting from methylation at these sites are conserved in a broad sense, for example by impacting transcription in a positive or negative fashion, but evolution has coopted these functions in a variety of ways, adapting them to roles in development, cell cycle, rDNA silencing, DNA methylation, meiotic recombination, and programmed DNA elimination, among other processes, not all of which will be discussed here. Methylation of H3K4 and H3K36 is generally associated with sites of active transcription, while H3K9, H3K27, and H4K20 methylation is characteristic of repressed heterochromatin. All of the enzymes conferring methyl marks at these five sites belong to the SET domain family (Herz et al., 2013; Hyun et al., 2017). Only H3K79 methylation is conferred by a non-SET domain HKMT, called Dot1 in yeast and Dot1L in metazoans. H3K79me functions in telomeric silencing in yeast and has also been associated with DNA repair.

The SET domain was shown to be essential for the catalytic activity of the first histone lysine methyltransferase to be characterized, SET1 (Rea et al., 2000). The ubiquity of the SET domain in KMT enzymes (other than Dot1/Dot1L) prompted numerous structural studies focused on this domain, which elucidated the principles of substrate recognition, cofactor (*S*-adenosylmethionine) binding, product specificity (mono-, di-, or trimethylation), and catalysis (reviewed in Chammas et al., 2020; Poulard et al., 2021; Schapira, 2011; Shi et al., 2017). Initial crystal structures of the SET domain enzymes Clr4 from *S. pombe*, Dim-5 from *N. crassa*, and a large fragment from human SET7/9 indicated the presence of a channel that could accommodate the histone peptide target and a pocket likely to be a binding site for *S*-adenosylmethionine (Min et al., 2002; Wilson et al., 2002; Zhang et al., 2002b). Subsequent studies in which SET domain proteins or fragments thereof were cocrystallized with target peptides and cofactor confirmed these assignments and provided structural bases for substrate recognition and product specificity (Schapira, 2011). Target specificity is dictated by interactions between the target histone peptide and variable residues in the lysine-binding channel (Couture et al., 2005; Wu et al., 2010; Xiao et al., 2003a). In addition, a "Phe/Tyr switch" in close proximity to the target lysine governs whether the primary product is the mono-, di-, or trimethylated lysine. This was first demonstrated by replacing a bulky tyrosine residue with phenylanine in the H3K4 monomethylase SET7/9, thereby converting it to a dimethylase; conversely, a F to Y replacement in the corresponding site in the H3K9 HMT DIM-5 from *N. crassa* blocked the trimethylation seen with wild type enzyme (Zhang et al., 2003b). These results were corroborated for DIM-5 and the additional H3K9 MTase G9a both in vitro and in murine ESCs and *N. crassa*, for human SET7/9, and for the H4K20 monomethylase SET8 (originally known as Pr-SET7), firmly establishing the existence of a "Phe/Tyr switch" that dictates product specificity of SET domain-containing HMTases (Collins et al., 2005; Couture et al., 2005, 2008; Wu et al., 2010; Xiao et al., 2003a). Adjacent to the peptide binding channel is a pocket in which the cofactor *S*-adenosylmethionine binds to allow close proximity between the transferred methyl group and the ε-amino group of the acceptor lysine (Del Rizzo et al., 2010; Jiao and Liu, 2015; Zhang et al., 2003b).

In the following section, brief overviews of some of the research on the six major sites of histone methylation are presented. These summaries are by no means comprehensive; readers interested in further detail on any of the specific methylation sites are advised to consult reviews cited in these sections, or to track down more recent reviews by checking citations of relevant publications. Cross-talk between these methylation sites and other modifications is discussed later on, as are their roles in transcriptional regulation (Chapter 8).

H3K4: The discovery that the SET domain protein SUV39H1 methylates H3K9, together with the absence of H3K9 methylation in *S. cerevisiae*, prompted Allis and colleagues to test the ability of the half dozen SET domain-containing proteins present in budding yeast to methylate H3K4 (Briggs et al., 2001). Deletion of *SET1*, but not of any other SET-domain containing protein, abolished H3K4 methylation in vivo. While recombinant Set1 was unable to methylate histones in vitro, a complex isolated from yeast containing Set1 and seven additional proteins methylated H3K4; the complex was dubbed COMPASS, for Complex of Proteins Associated with Set1 (Krogan et al., 2002; Miller et al., 2001; Roguev et al., 2001).

Set1 is conserved across eukaryotes (Roguev et al., 2001). While COMPASS is the only Set1-containing complex present in budding yeast, multiple, nonredundant Set1-containing complexes are found in *Drosophila* and mammals (Shilatifard, 2012). In mammals, these comprise the SET1A and SET1B complexes and four MLL complexes designated MLL1-4. The MLL complexes derive their name from the Mixed Lineage Leukemia (*MLL*) gene, which was identified as being associated with the chromosomal breakpoint in a subset of acute leukemias and eventually shown to function as an H3K4 methyltransferase in the context of its associated partners (Crump and Milne, 2019). This forged one of the earliest connections between chromatin and cancer and generated intense and continuing interest in the function and dysfunction of the MLL complexes.

An additional HMT targeting H3K4 was purified from HeLa nuclear extracts independently by the laboratories of Danny Reinberg and Yi Zhang (a former student of Reinberg's), and named Set7 or Set9 (Nishioka et al., 2002a; Wang et al., 2001a). Now referred to as SetD7, this enzyme is distinct in its lineage from Set1 and the MLL enzymes and catalyzes only monomethylation of H3K4 (Xiao et al., 2003a; Zhang et al., 2003b). SetD7 also targets numerous nonhistone proteins, with p53 being first to be identified, prompting Herz et al. to suggest that histone methylation was not likely to be its primary function (Chuikov et al., 2004; Herz et al., 2013). In support of this notion, depletion of MLL3/4 in mammalian cells or the *Drosophila* homolog Trr in flies resulted in global reduction in H3K4me1 levels, indicating at most a minor role for SetD7 in conferring monomethylation of H3K4me (Herz et al., 2012; Hu et al., 2013).

Less closely related to the SET1A/B and MLL proteins, but still belonging to the SET domain family, are the PRDM (PRDF1 and RIZ1 homology domain containing) proteins (Di Tullio et al., 2022; Hohenauer and Moore, 2012), one of which, PRDM9, catalyzes H3K4 trimethylation. PRDM9, initially named Meisetz, was discovered in a search for transcription factors controlling the initiation and progression of meiosis (Hayashi et al., 2005). Cloning of the gene for Meisetz/PRDM9 revealed it to contain a PR/SET domain in its amino terminus and a C2H2 (named for the two Cys and two His residues that chelate Zn^{2+}) zinc finger motif in its carboxy terminus, prompting Hayashi and colleagues to test for HMTase activity. Recombinant Meisetz/PRDM9 preferentially methylated histone H3 from calf thymus, but displayed no MTase activity toward H3 peptides, suggesting that its activity might be restricted to H3 already bearing modifications. Consistent with this notion, Meisetz/PRDM9 specifically increased H3K4me3 when incubated with *S*-adenosylmethionine and whole cell lysates containing histones. Meisetz/PRDM9 expression is restricted to female germ cells entering meiotic prophase and to postnatal testis, and functions in the generation of hotspots for recombination (Hochwagen and Marais, 2010). This function is evidently widely conserved, as H3K4me3 is enriched at meiotic recombination hotspots in yeast despite the absence of PRDM family proteins, and inactivation of Set1 resulted in decreased formation of double strand breaks during meiosis (Borde et al., 2009).

PRDM family members are characterized by the presence of the PR/SET domain that is essential for MTase activity in SET and MLL proteins, a RIZ1 (now called PRMD2) zinc finger domain, and an array of C2H2 zinc fingers (Di Tullio et al., 2022; Hohenauer and Moore, 2012). The zinc fingers allow sequence-directed binding of PRDMs and thus facilitate recruitment to specific targets. Of the 17 recognized PRDM proteins present in human cells (inclusion of two additional proteins, FOG-1 and FOG-2, that share structural homology with the PRDM proteins has been proposed (Di Tullio et al., 2022)), only six have been reported to directly methylate histones. Many of the remainder have been reported to support histone methylation via recruitment of known HMTs (Di Tullio et al., 2022). In addition to PRDM9, PRDM7, which is highly homologous to PRDM9, and PRDM16 have been reported to methylate H3K4 (Blazer et al., 2016; Zhou et al., 2016a). The PRDM family has evolved more recently than the SET and MLL HMTs, and its members support an array of developmental processes (Di Tullio et al., 2022; Hohenauer and Moore, 2012). Reflecting their more specialized roles relative to the SET and MLL proteins, PRDM family members have been implicated in numerous cancers and developmental disorders.

A critical step to investigating the function of H3K4 methylation was the development of antibodies that distinguished H3K4me2 from H3K4me3 (Santos-Rosa et al., 2002). ChIP experiments employing these antibodies confirmed the essential role of Set1 in conferring H3K4 methylation; more importantly, they showed H3K4me3 to be strictly associated with transcriptionally active genes, while H3K4me2 was present at both active and inactive euchromatic genes. Consistent with the evolutionary conservation of SET1-containing complexes, ChIP-chip and ChIP-seq experiments revealed H3K4me3 enrichment at the 5′ end and promoters of active genes in cells from yeast, humans, and plants, while H3K4me1 and H3K4me2 were observed throughout coding regions (Barski et al., 2007; Bernstein et al., 2002, 2005; Liu et al., 2005; Pokholok et al., 2005; Zhang et al., 2009). H3K4me1 was later shown to be enriched at enhancer sequences (Heintzman et al., 2007). The existence of multiple methylated states of H3K4 immediately raised the problem of how a single methyltransferase could specify different levels of methylation at specific genomic locations. Using extracts prepared from yeast deletion mutants, Shilatifard and colleagues showed that specific COMPASS components are required for di- and trimethylation of H3K4; for example, absence of Sdc1/Cps25 diminished H3K4me2 and abolished H3K4me3, whereas H3K4me3 was reduced with little effect on mono- and dimethylation in the absence of Spp1/Cps40 or Bre2/Cps60 (Schneider et al., 2005). Differential effects on mono-, di-, and trimethylation of H3K4 upon deletion of COMPASS subunits were also reported by others (Dehe et al., 2006; Morillon et al., 2005). Depletion of the human homolog of Spp1, Cfp1, also results in reduced levels of H3K4me3, demonstrating functional conservation (Thomson et al., 2010).

Recruitment of Set1 and installation of H3K4me3 in budding yeast occurs through a complex pathway involving interactions of COMPASS components with Pol II, Pol II elongation factors, and nascent RNA, as well as cross-talk with a distinct modification on histone H2B; this process is discussed in the section on "Cross-talk" in this chapter and in the context of Transcribing through Chromatin in Chapter 8. Recruitment and H3K4me3 formation in mammals follow a different pathway. Many mammalian promoters are associated with sequences enriched for CpG dinucleotides that are unmethylated, in contrast to the high degree of methylation characteristic of CpGs located elsewhere in the genome; these regions are referred to as CpG islands (see Chapter 8, section on DNA Methylation). Interaction between Cfp1 and unmethylated CpG regions recruits Set1 to embedded promoters, where it catalyzes trimethylation of H3K4, with variable effect on transcription (Brown et al., 2017; Clouaire et al., 2014; Thomson et al., 2010).

The close coupling of H3K4me3 with promoters of active genes suggests a possible causative role in transcriptional activation. A plausible mechanism for such a role is provided by studies showing interactions between H3K4me3 and specific motifs (plant homeodomain (PHD) fingers, chromodomains, or Tudor domains) present in TFIID, HAT complexes, and chromatin remodeling complexes (Pray-Grant et al., 2005; Ruthenburg et al., 2007; Sims and Reinberg, 2006; Vermeulen et al., 2007). The importance of H3K4 trimethylation is clear, particularly in metazoans, as loss of Cfp1 results in embryonic lethality in mice and prevents differentiation of ESCs (Carlone et al., 2005; Carlone and Skalnik, 2001). However, several results argue against a general causative role for H3K4me3 in transcriptional activation:

- Cfp1 loss or deficiency decreases H3K4me3 but results in only modest transcriptional changes in murine ESCs (Clouaire et al., 2012, 2014). Both increased and decreased transcription were observed at specific genes.
- Deletion of *SPP1* in yeast resulted in only modest changes in transcript levels or nascent transcription (Howe et al., 2017; Lenstra et al., 2011; Margaritis et al., 2012).
- Several studies have reported recruitment of SET1/MLL complexes and trimethylation of H3K4 *after* recruitment of the preinitiation complex and expression of mRNA (Howe et al., 2017).
- Cfp1 recruitment and H3K4me3 were seen at engineered CpG islands without accompanying elevated recruitment of Pol II (Thomson et al., 2010).

While these results argue against a generally required instructive role for H3K4me3 in transcriptional activation, such a requirement does appear to hold for specific cases. Loss of H3K4me3 affects activation of a subset of genes induced by DNA damage in human HCT116 cells; recruitment of Taf3 by H3K4me3 appears to be a critical step in activation (Lauberth et al., 2013). In murine oocytes, absence of Cfp1 resulted in decreased H3K4me3 levels and a severe decrease in global mRNA expression (Yu et al., 2017). And while only modest dependence on H3K4me3 for transcription is observed in exponentially growing yeast, a more widespread dependence was seen in replicatively aging yeast cells (Cruz et al., 2018). Altogether, these various findings indicate a context-dependent role for H3K4me3 in transcriptional activation, which may in many instances be superseded by independent, redundant mechanisms. Attaining a full understanding of the factors governing the contribution of H3K4 trimethylation in transcriptional activation, as well as in additional processes in which it has been implicated such as DNA repair (Faucher and Wellinger, 2010; Pena et al., 2008), remains a challenge for the field.

H3K9: The identification of SU(VAR)3-9 and its homologs (collectively referred to as Suv39h) as H3K9 methyltransferases and the involvement of methylated H3K9 in heterochromatin formation were introduced earlier in this chapter and in Chapter 3. Soon after the discovery of SU(VAR)3-9 as a histone lysine methyltransferase, two additional HMTs targeting H3K9, G9a and SetDB1, were identified. G9a was identified as a candidate HMT by virtue of its containing a SET domain flanked by cysteine-rich regions, motifs that had been shown to be required for the catalytic activity of SUV39H1 (Rea et al., 2000), and demonstrated in recombinant form to methylate both H3K9 and H3K27 of free histones (Tachibana et al., 2001). SetDB1 was discovered in a yeast two-hybrid assay as interacting with the corepressor KAP-1 (Schultz et al., 2002); the presence of a SET domain prompted the authors to test for HMT activity, which was observed in SetDB1 immunopurified from HEK293 or baculovirus infected S19 cells but not from recombinant SetDB1, suggesting that posttranslational modification or association with ancillary factors might be necessary for catalytic activity. The murine homolog of SetDB1, ESET, was discovered as an interactor with the *ets*-related ERG protein (Yang et al., 2002). Like Suv39h, SetDB1 was shown to be specific for H3K9 (Schultz et al., 2002).

Suv39h and SetDB1 both function in the maintenance of constitutive heterochromatin by directing trimethylation of H3K9, while G9a forms heterodimers with another protein, GLP (G9a-like protein), to mono- and dimethylate H3K9 in euchromatic regions to repress transcription, consistent with the mostly nonoverlapping nuclear localization of Suv39h1 and G9a (Rice et al., 2003; Tachibana et al., 2001, 2002). Similarly, dSETDB1 and SU(VAR)3-9 govern heterochromatin maintenance in *Drosophila* at distinct chromosomal regions and different development stages, while G9a again functions to repress euchromatic genes (Brower-Toland et al., 2009; Yoon et al., 2008).

The mechanisms for heterochromatin formation and maintenance, respectively involving recruitment of Suv39h directed by transposable element sequences or HP1a, have been discussed at length (Chapter 3, Fig. 3.13). Recruitment of G9a and consequent mono- and dimethylation of H3K9 to form facultative heterochromatin in otherwise euchromatic regions is directed by interaction with various DNA-binding transcriptional repressors (Duan et al., 2005; Nishida et al., 2007; Ogawa et al., 2002; Ueda et al., 2006). Domains enriched for H3K9me1/2 in euchromatin can be highly localized or can spread to regions a few Mb in size; the latter domains are referred to as large organized chromatin K9 modifications, or LOCKs (Wen et al., 2009). Although a spreading mechanism depending on interaction of ankyrin repeats in G9a and GLP was proposed, this has not been definitively established; neither is it known what causes such spreading from the site of recruitment to occur in some cases but not in others (Collins and Cheng, 2010).

In addition to G9a, SetDB1 and Suv39h, four of the PRDM proteins (introduced in the preceding section) have been reported to methylate H3K9 (Kim et al., 2003; Powers et al., 2016). PRDM2, originally known as RIZ1 (retinoblastoma protein-interacting zinc finger 1) was first identified through its interaction with the retinoblastoma protein, and acts as a tumor suppressor in various tissues (Donczew et al., 2020; Kim et al., 2003). PRDM9 principally methylates H3K4 to promote meiotic recombination, as detailed above, but recombinant PRDM9 and the isolated PR/SET domain were also able to trimethylate H3K9 in the context of a H3(1-19) peptide, with highest specificity for the dimethylated substrate (Powers et al., 2016). Whether this modification is effected by PRDM9 in vivo is unclear. PRDM16, which functions in the maintenance of adult stem cells in the hematopoietic and neural lineages and in development of brown adipose tissue, was first reported (along with the closely related PRDM3) to methylate H3K9, using a short peptide containing that site (Pinheiro et al., 2012). However, a later report showed H3K4 methylation by PRDM16 in vitro using both free and nucleosomal histones, as well as increased H3K4me2 upon recruitment of PRDM16 in vivo (Zhou et al., 2016a). PRDM8 is involved in neural circuit assembly and neocortical development and was reported to methylate H3K9 in a single study (Eom et al., 2009; Inoue et al., 2015; Ross et al., 2012). PRDM8 is strongly repressed in adult murine neural stem cells by the polycomb group (PcG) complex, which represses transcription of developmental lineage genes through trimethylation of H3K27 as discussed in detail in Chapter 8 (Ganapathi et al., 2018), raising the intriguing possibility that it might function to repress genes required for self-renewal during neural development. Currently, however, the targets and mechanisms by which many of the PRDM members, including PRDM8, exert their functions in development remain largely unexplored.

H3K27: As with H3K9, methylation of H3K27 is a marker of heterochromatin and gene repression in metazoans and plants. Although H3K27 methylation does not occur in *S. cerevisiae* or *S. pombe*, it is present in other unicellular eukaryotes (Shaver et al., 2010). Methylation of H23K27 is carried out by a single enzyme (sometimes expressed as paralogs or isoforms created by alternative splicing), called Enhancer of Zeste, or E(z), in *Drosophila*, or EZH2 in mammals. Enhancer of zeste, it will be recalled, is a founding member of the SET domain family, and is also a member of the PcG complex, which is discussed at length in Chapter 8. Based on genetic experiments over a span of several decades, the PcG complex was known to be important in gene repression during development of *Drosophila*; thus, the discovery of the SET domain protein SU(VAR)3-9 as an H3K9 MTase raised suspicion that EZH2 (I will use this term to include also the *Drosophila* E(z) protein for simplicity) might also methylate histones to create a repressive chromatin environment. Methyltransferase activity by EZH2 toward H3 was demonstrated simultaneously by multiple labs. While recombinant EZH2 exhibited no detectable HMTase activity, such activity was observed using a complex containing EZH2 and as few as two additional PcG components (in some cases purified from cells, in others expressed recombinantly) (Cao et al., 2002; Cao and Zhang, 2004; Czermin et al., 2002; Kuzmichev et al., 2002; Muller et al., 2002). Structural studies revealed that EZH2 contains an autoinhibitory domain that binds to its lysine-binding cleft and impedes catalysis; relief of autoinhibition occurs upon association with the two PRC2 subunits EED and SUZ12, which interact with EZH2 to facilitate stable repositioning of the autoinhibitory subdomain (Antonysamy et al., 2013; Jiao and Liu, 2015; Wu et al., 2013). Two of the original studies reported H3K27 being highly preferred as the site of methylation (Cao et al., 2002; Muller et al., 2002), while two found H3K9 methylation in addition (Czermin et al., 2002; Kuzmichev et al., 2002). Later work established the H3K27 specificity of the EZH2-containing complex, named PRC2; it is possible that the utilization of a peptide rather than nucleosomal histones in one of the latter studies, and of affinity-purified complex possibly contaminated with an H3K9 MTase in the other, resulted in the observations of H3K9 methylation.

The specificity of EZH2 for H3K27 over H3K9 is impressive in light of the closely related sequences surrounding the lysine residue: TARK$_9$ST vs AARK$_{27}$SA, underscoring the specificity conferred by the various salt bridge, van der Waals, and hydrogen bonding interactions between target peptide and nonconserved residues in the peptide binding channel (Wu et al., 2010; Xiao et al., 2003a). A similar discriminatory mechanism underlies the specificities of two "readers" of H3K9me3 and H3K27me3. HP1 and Polycomb recognize H3K9me3 and H3K27me3, respectively, through their chromodomains to facilitate formation of constitutive heterochromatin or to enforce localized gene repression. Domain swaps showed HP1-Polycomb chimeric proteins exhibited nuclear localization dictated by the respective chromodomain, as observed by immunofluorescence in *Drosophila* polytene chromosomes and S2 cells (Fischle et al., 2003; Platero et al., 1995). Structural analyses revealed an aromatic cage around the modified lysine residue, with contacts N-terminal to Lys9 or Lys27 dictating specific binding to HP1 or Pc, respectively (Fischle et al., 2003; Jacobs and Khorasanizadeh, 2002; Min et al., 2003; Nielsen et al., 2002).

H3K36: Set2 was identified as a histone methyltransferase in budding yeast by following the activity of biochemically fractionated extracts; mass spectrometric analysis of fractions exhibiting preferential methylation of nucleosomal histone H3 identified Set2 (Strahl et al., 2002). In vitro experiments using recombinant Set2 and chicken nucleosomes revealed high selectivity for H3K36. An antibody specific for methylated H3K36 was then developed, which allowed demonstration that deletion of *SET2*, and no other SET domain protein-coding gene, eliminated H3K36me in vivo. Although a role in transcriptional repression was initially ascribed to Set2 based on reporter gene assays, subsequent investigations revealed it to function in transcriptional elongation. Set2 was found to associate with Pol II, and specifically with phosphorylated isoforms characteristic of elongating polymerase, and both Set2 and methylated H3K36 were seen in ChIP assays to associate with coding regions of actively transcribed genes (Li et al., 2003; Schaft et al., 2003; Xiao et al., 2003b). The association of methylated H3K36 with transcribed coding regions in yeast was corroborated genome-wide using microarrays (Pokholok et al., 2005) and was seen in metazoan cells as well (Mikkelsen et al., 2007). This association proved a useful guidepost to identify conserved long noncoding RNA species in mammalian cells over a background of nonconserved, noncoding RNA species that likely represent transcriptional "noise" (Guttman et al., 2009).

H3K36 methylation was eventually shown to be implemented over coding regions in yeast via an intricate mechanism involving Pol II and associated elongation factors, including the HDAC Rpd3, and to prevent spurious intragenic transcriptional initiation, as discussed in Chapter 8. H3K36 methylation and the modifying enzymes have been found to function, directly or indirectly, in replication, recombination, DNA damage response, and alternative splicing (Lam et al., 2022; McDaniel and Strahl, 2017; Wagner and Carpenter, 2012). These various functions are achieved in part by suppressing exchange of newly synthesized histones into transcribed chromatin, by antagonizing H3K27 methylation, and by enforcing histone deacetylation over transcribed DNA (Chapter 8) (Bender et al., 2006; Smolle et al., 2012; Venkatesh et al., 2012; Yuan et al., 2011).

In yeast, as already mentioned, Set2 is the only H3K36 methyltransferase. In higher eukaryotes, additional HMTs also target H3K36 in vitro and in vivo (Fig. 7.14) and vary in their specificity for producing mono-, di-, or trimethylated product (Hyun et al., 2017; Lam et al., 2022; Wagner and Carpenter, 2012). NSD1 (nuclear receptor-binding SET domain-containing protein 1) was initially found in a screen seeking coregulators of the retinoic acid receptor, and was subsequently shown to target H3K36 for mono- and dimethylation, as was also found for its close relatives NSD2 and NSD3 (Huang et al., 1998; Li et al., 2009; Lucio-Eterovic et al., 2010; Qiao et al., 2011; Rayasam et al., 2003). Early studies reported inconsistent results with regard to NSD target specificity, but varied in using peptides, free histones, histone octamers, or nucleosomal substrates. In one study, preferential methylation of H3K36 by NSD2 was seen when nucleosomal substrates were used, but a histone octamer substrate was primarily acetylated at H4K44; however, if short single- or double-stranded DNA was included in the reaction with histone octamer, H3K36 was preferentially methylated (Li et al., 2009). Specificity for H3K36 mono- and dimethylation by NSD2 was shown convincingly in a study that used recombinant nucleosomes for in vitro analysis and that also showed that overexpression or knockdown of NSD2 in vivo affected levels of H3K36me2 but not H3K36me3 (Kuo et al., 2011). Structural and biochemical analyses of Set2, NSD2, and NSD3 have subsequently revealed the mechanism underlying specificity dictated by the nucleosome (Bilokapic and Halic, 2019; Li et al., 2009, 2021a; Qiao et al., 2011; Sato et al., 2021; Wang et al., 2015; Yang et al., 2016); a detailed discussion of these structural findings is presented in Lam et al. (2022). First, an autoinhibitory loop was discovered to prevent interaction between H3K36 and bound cofactor in the absence of a nucleosomal substrate; the inhibitory interaction is relieved upon binding a nucleosome. Second, multiple specific contacts with linker DNA and nucleosomal histones were observed to precisely position the enzymes to target K36. Autoinhibitory mechanisms have been noted previously for EZH2 and the chromatin remodeler ISWI (Chapter 6); this commonality suggests that such autoinhibitory mechanisms are likely important to prevent untargeted modification or remodeling of nucleosomes. In addition, the autoinhibitory domain regulates the activity and specificity of Set2 in budding yeast (Wang et al., 2015).

In experiments to test the product specificity of HMTs that target H3K36 in vivo, knockdown of SETD2, the mammalian Set2 homolog, reduced H3K36me3 but not the mono- or dimethyl forms in human and murine cells, whereas knockdown of NSD1 reduced all methylated K36 at two specific targets (*BMP4* and *ZFP36L1*) assayed by ChIP in MCF7 cells (Edmunds et al., 2008; Lucio-Eterovic et al., 2010; Nimura et al., 2009; Yuan et al., 2009). Similarly, the NSD1 homolog in *Drosophila* and *C. elegans*, MES-4, is responsible for H3K36me1/2, while SETD2 is important for H3K36 trimethylation (Andersen and Horvitz, 2007; Bell et al., 2007; Bender et al., 2006; Edmunds et al., 2008; Furuhashi et al., 2010; Rechtsteiner et al., 2010). Although these results suggest that mono- or dimethylation of H3K36 by NSD enzymes is required to provide substrate capable of being trimethylated by SETD2 and its homologs, knockdown of NSD2 in human cell lines reduced H3K36me2 levels without affecting H3K36me3 (Kuo et al., 2011). It seems likely that these enzymes, as well as ASH1L and SMYD2, function redundantly and interdependently to establish the various states of H3K36 methylation in a cell- and organismal-context specific fashion (Lam et al., 2022; Wagner and Carpenter, 2012). Differences in conserved Phe and Tyr residues near the catalytic site between human SETD2 and the homologs in flies and yeast (both fission and budding), potentially related to the Phe/Tyr switch mentioned earlier, have also been suggested to account for organismal differences in product specificity (Lam et al., 2022). SETD2 also apparently requires association with an ancillary protein, HnRNP-L (heterogeneous nuclear ribonucleoprotein L), found to exist in complex with SETD2, for its function in human cells, as knockdown of HnRNP-L reduced H3K36me3 in HeLa and HEK293 cells to a similar extent as knockdown of SETD2 (Yuan et al., 2009).

H3K79: H3K79 is an outlier among histone methylation sites, as it lies on the structured surface of the nucleosomal disk, as part of the loop between the first and second α-helices, rather than in the extended N-termini; and its methylase, Dot1 in yeast and Dot1L (Dot 1-like) in metazoans, lacks a SET domain. Methylation of H3K79 by Dot1 was first reported in three independent lines of investigation. Gottschling and coworkers first identified Dot1 (disruptor of telomeric silencing 1) in a screen for genes whose overexpression disrupted telomeric silencing (Singer et al., 1998). The effect on telomeric silencing was confirmed by a microarray analysis that found increased expression of telomere proximal genes in *dot1Δ* yeast, while a separate study reported an altered distribution of Sir3 by chromosomal staining in the absence of Dot1 (Hughes et al., 2000; San-Segundo and Roeder, 2000). Recognition of homology between Dot1 and SAM-dependent methyltransferases, including the arginine MT PRMT1, along with the discovery by mass spectrometry of a novel site of methylation at K79 in HeLa cells, flies, chicken, and yeast (Feng et al., 2002), prompted testing of Dot1 for HMT activity (Lacoste et al., 2002; Ng et al., 2002a; van Leeuwen et al., 2002). Both recombinant and isolated Dot1 were found exclusively to methylate H3K79 in nucleosomes but not in free histones. Deletion of Dot1 resulted in complete loss of H3K79 methylation in yeast, flies, and mice, demonstrating it to be the sole H3K79 MT (Jones et al., 2008; Shanower et al., 2005; van Leeuwen et al., 2002).

H3K79 methylation by Dot1 was first proposed to prevent widespread Sir3 binding to genomic chromatin in budding yeast and thereby limit Sir3 binding to its proper targets in heterochromatin, consistent with the high degree of H3K79 methylation, comprising ~90% of all H3 in yeast, and with the redistribution of Sir3 in *dot1Δ* yeast (Ng et al., 2002a; van Leeuwen et al., 2002). Supporting this interpretation, binding of Sir3 to an H3K79-including peptide was reduced by H3K79 methylation, and binding of Sir3 to nucleosomes was enhanced in *dot1Δ* yeast and diminished by an H3K79A mutation (Altaf et al., 2007; Onishi et al., 2007). Later work refined this view by showing that the *URA3* reporter gene used to assay telomeric silencing was subject to an indirect effect independent of heterochromatin function, and that changes in Sir3 occupancy were more highly targeted than initially believed; nonetheless, a role for H3K79 methylation by Dot1 in telomeric silencing retained support (Rossmann et al., 2011; Takahashi et al., 2011). Dot1 function in telomeric silencing was also observed in metazoans: murine ES cells deficient in Dot1L exhibited defects in centromeric and telomeric heterochromatin, while mutation of Dot1L in *Drosophila* resulted in defective telomeric silencing but did not affect centric heterochromatin (Jones et al., 2008; Shanower et al., 2005). Although these results indicate broad conservation of function for Dot1L and H3K79 methylation, clearly the mechanisms must differ, as Sir3 is unique to fungi.

H3K79 methylation was also linked to actively transcribed loci in early studies using yeast, *Drosophila*, and mammalian cells (Ng et al., 2003; Okada et al., 2005; Schubeler et al., 2004; Vakoc et al., 2006). Contrary results were reported in two early genome-wide analyses, but these proved to be due to a technical artifact, and subsequent genome-wide analyses demonstrated association of H3K79me2/3 over transcribed sequences, with enrichment correlating with expression level (Lee et al., 2018; Nguyen and Zhang, 2011; Steger et al., 2008). Mammalian Dot1L was shown to interact with the carboxy-terminal domain of Pol II and to associate with elongation factors in complexes isolated from mammalian cells, suggesting a role for Dot1L in transcriptional elongation (Jonkers et al., 2014; Kim et al., 2012; Nguyen and Zhang, 2011; Veloso et al., 2014; Wood et al., 2018). In support of such a role, knockdown of the Dot1L-associated ENL protein resulted in decreased H3K79 methylation and elongation efficiency in HEK293 cells (Mueller et al., 2007). However, a recent study found no substantial defect in transcriptional elongation in *DOT1L* knockout murine ES cells, and additionally found evidence for Dot1L function in elongation (in collaboration with the super elongation complex) and cell-fate determination that was independent of its catalytic activity (Cao et al., 2020). A methylation-independent role for Dot1L in facilitating histone exchange over transcribed regions has also been reported (Lee et al., 2018). Dot1L may also function in the transcription of repeat sequences embedded in heterochromatin (Chapter 3), potentially providing a link between its roles in heterochromatin and transcriptional activity (Malla et al., 2021).

H3K79 methylation also functions in DNA repair and control of checkpoints in the cell cycle (Giannattasio et al., 2005; San-Segundo and Roeder, 2000). These functions depend at least in part on methylation of H3K79 affecting recruitment of the checkpoint adaptor protein Rad9; however, as with the function of H3K79 methylation in heterochromatin structure and transcription, the molecular mechanisms underlying its role in checkpoint control and DNA repair remain incompletely understood (Nguyen and Zhang, 2011; Separovich and Wilkins, 2021).

Finally, it should be noted that H3K79 di- and trimethylation depends on the prior ubiquitination of H2BK120 (in mammals; H3K123 in yeast) (Briggs et al., 2002; McGinty et al., 2008; Ng et al., 2002b; Shahbazian et al., 2005), as discussed later in the "Cross-talk" section.

H4K20: H4K20 was discovered as a methylated residue in the first complete sequence analysis of a histone, but it was not until 33 years later that the responsible methyltransferase was identified (DeLange et al., 1969; Fang et al., 2002; Nishioka et al., 2002b; Ogawa et al., 1969). SET8, initially also named PR-Set7, was purified from HeLa cells by using standard biochemical methods to fractionate nuclear extracts and tracking SAM-dependent methylation of free or nucleosomal histones. Mass spectrometry led to identification of the SET domain-containing protein, which was found to be highly specific for H4K20 methylation: methylation of nucleosomal H4 by recombinant SET8 was almost eliminated when an H4K20A mutant was used.

Initial studies employing immunostaining of *Drosophila* polytene chromosomes indicated that H4K20 methylation was associated with condensed regions of chromatin (Nishioka et al., 2002b). However, although both the polyclonal and monoclonal antibodies developed for these experiments were raised against a dimethylated K20 H4 peptide, their specificity was only compared to the unmethylated peptide, so the methylation status of the stained chromosomal regions remained uncertain. Studies on H3K9 methylation going on at this same time revealed the product specificity of HMTs and the significance of distinct methylation states. First, mice lacking different H3K9 methyltransferases were found to exhibit distinct phenotypes (Peters et al., 2001, 2002; Tachibana et al., 2002). Following these results, Allis, Jenuwein, and colleagues pioneered the use of antibodies against specific methylated isoforms to show differential localization of H3K9me1/2 compared to H3K9me3 as well as distinct product specificities of G9a, Suv39h1, and Suv39h2, as described earlier (Peters et al., 2003; Rice et al., 2003). A similar approach directed at investigation of H4K20 methylation revealed H4K20me3 association with pericentric heterochromatin in murine embryonic fibroblasts (MEFs), while H4K20me1 and H4K20me2 exhibited more diffuse nuclear staining (Schotta et al., 2004). To identify the MTase responsible for generation of H4K20me3, a candidate approach was taken. H4K20 methylation had been observed in metazoan organisms and *S. pombe*, but not in budding yeast; accordingly, 12 SET domain proteins common to the former species but lacking in budding yeast were expressed in MEFs, and two were found to localize to heterochromatin foci. These experiments led to identification of the mammalian H4K20 MTases Suv4-20h1 and Suv4-20h2; the enzyme is present as a single homolog in *Drosophila* and *S. pombe*. H4K20me3 generation was found to require close interplay with H3K9me3; Suv4-20h2 interacts with HP1, and depletion of Suv39h or HP1 in *Drosophila* or mice results in diminished Suv4-20 localization to heterochromatin and globally reduced levels of H4K20me3 (Schotta et al., 2004; Yang et al., 2008). The role of Suv4-20 in conferring H3K20me3 was corroborated by structural and biochemical studies showing SET8 to be a monomethylase, with product specificity governed by the Phe/Tyr switch discussed earlier (Couture et al., 2005; Xiao et al., 2005). Knockdown of Suv4-20 in *Drosophila* resulted in decreased levels of H4K20me2 and H4K20me3 and corresponding increase in H4K20me1, consistent with a model in which SET8 monomethylates H4K20 to provide substrate for Suv4-20 (Yang et al., 2008).

A remarkable aspect of H4K20 methylation is the high proportion of nucleosomes harboring this modification in vivo. Mass spectrometry revealed 80%–90% of all H4K20 residues to be methylated in both *Drosophila* S2 cells and human HeLa cells, with the vast majority being dimethylated and 5% or less being mono- or trimethylated (Evertts et al., 2013; Pesavento et al., 2008; Schotta et al., 2008; Yang et al., 2008). Despite this extremely high background of H4K20me2, H4K20 methylation also functions in DNA repair (Corvalan and Coller, 2021). Identification of the confusingly named Set9 (the *S. pombe* SET8 homolog) as the sole H4K20 MTase in *S. pombe* led to the surprising finding that H4K20 methylation does not play a role in heterochromatin function or gene regulation, but rather is important in the DNA damage response, where it serves to recruit Crb2, a homolog of the mammalian p53 activator 53BP1, which initiates the DNA repair process (Botuyan et al., 2006; Sanders et al., 2004). In mammalian cells, increased H4K20 dimethylation at sites of double-stranded breakage was similarly found to be important for recruitment of 53BP1 (Pei et al., 2011; Sanders et al., 2004). Despite structural data showing interaction of Crb2 and 53BP1 with H4K20me2, it remains unclear how this interaction is restricted to, or enhanced at, sites of double-strand breaks (Corvalan and Coller, 2021). H4K20 methylation also has been implicated in cell cycle regulation and replication (Evertts et al., 2013; Houston et al., 2008; Huen et al., 2008; Pesavento et al., 2008); although some of the underlying mechanisms have been uncovered, a full understanding of the role of this modification in these processes remains to be elucidated (Corvalan and Coller, 2021).

Histone demethylases

The surge in research on histone methylation raised anew the question of the reverse reaction. Early research identified and partially purified an activity from nuclear and mitochondrial extracts from rat kidney capable of demethylating histone lysine residues (Paik and Kim, 1973, 1974). However, although it seemed likely that the responsible activity could correspond to ε-alkyllysinase (EC1.5.3.4), which had been shown to demethylate free ε-alkyllysine in a reaction in which the methyl group was oxidized to formaldehyde (Kim et al., 1964), efforts at rigorously demonstrating this to be the case, or to prove that histones were a substrate in vivo, were unsuccessful (Paik et al., 2007). Evidence both for and against turnover of methylated histones was reported in early work: pulse-labeling experiments indicated slow or no turnover of histone methylation in dividing cells (Borun et al., 1972; Byvoet et al., 1972; Honda et al., 1975), while a study using feline kidney cells, which rarely divide and are enriched for the demethylase activity reported by Paik and Kim, revealed methylation half-lives much shorter than rates of histone biosynthesis, indicating reversal of histone methylation (Hempel et al., 1979).

These varying reports on the reversibility of histone methylation in vivo and the failure to definitively identify a bona fide histone demethylase raised skepticism as to whether histone demethylation occurred in vivo, or whether alternative mechanisms, such as histone replacement or "clipping" of amino-terminal regions containing methylated arginine or lysine might be responsible for removal of modified residues (Bannister et al., 2002). The notion of histone turnover outside of S phase as a mechanism of removing methylated histones was supported by the recent, at the time, demonstration of replacement of canonical H3 by the variant H3.3, discussed in Chapter 5 (Ahmad and Henikoff, 2002), while an older report of a physiologically regulated proteolytic trimming of the six amino terminal residues of H3 in *Tetrahymena* suggested the possibility of a clipping mechanism (Allis et al., 1980). Further tilting bias against dynamic methylation and demethylation of histones akin to acetylation-deacetylation was the association of H3K9 methylation with heterochromatin, whose static nature seemed consistent with an irreversible mark. Nonetheless, early work mentioned above indicated at least a low level of histone demethylation, and it seemed likely that methylation would need to be reversed in cases where it functioned in gene regulation in a euchromatic context (Bannister et al., 2002).

Demethylation won out. Two distinct enzymatic mechanisms for removal of methyl groups from histone arginine and lysine residues were uncovered in 2004, and a third family of histone lysine demethylases using yet another mechanism was reported two years later. Studies on the family of peptidylarginine deiminases (PADs), Ca^{+2}-dependent enzymes that catalyze the conversion of arginine to citrulline (Fig. 7.13B), led to the discovery that human PAD V was localized to the nucleus and that treatment of neutrophils or HL-60 granulocytes with calcium resulted in substantial histone deimination, as seen by Western blotting with an antibody against modified citrulline residues (Hagiwara et al., 2002; Nakashima et al., 2002). Subsequently, this same enzyme, referred to as PAD4 or PADI4, apparently due to its homology to rodent PAD IV and the absence of a designated human PAD IV, was found to convert monomethylated Arg in histone H4 to citrulline with release of methylamine (Cuthbert et al., 2004; Wang et al., 2004b). PAD4 was able to target Arg residues methylated by CARM1 and PRMT1, and ChIP experiments indicated its recruitment coincided with loss of H4R3 methylation and appearance of citrullinated H4 during downregulation following hormone induction of the *pS2* promoter. These experiments provided evidence for PAD4 countering arginine methylation of H3 and H4 to reverse transcriptional activation, but left open some major questions. First, PAD4 could deiminate unmethylated or monomethylated arginine to yield citrulline, which cannot be methylated, but could not act on dimethylated arginine. Second, there was no obvious enzymatic means of reversing PAD4 action to allow a cycle of methylation and demethylation. A member of the jumonji family of histone lysine demethylases discussed below, JMJD6, was reported to function as a histone arginine demethylase, but later work indicated its main function was to hydroxylate an RNA splicing factor (Chang et al., 2007; Webby et al., 2009). Although biochemical studies indicate that the catalytic domains of several JmjC proteins are capable of acting as histone arginine demethylases in vitro, whether and how recurrent cycles of histone arginine methylation and methyl group removal are implemented in vivo remains unclear (Zhang et al., 2019).

Lysine demethylation, in contrast, has emerged as a highly regulated process controlled by a multitude of lysine histone demethylases. Shi and colleagues identified the first bona fide histone demethylase LSD1 (lysine specific demethylase 1; the reader can imagine the clever review titles this nomenclature triggered) and demonstrated it to be a corepressor acting at several neuron specific genes (Shi et al., 2004). Amine oxidases had been proposed as potential HDMs that could remove methyl groups from lysine or arginine via an oxidation reaction (Bannister et al., 2002), and a protein identified as a component of several corepressor complexes was found to share homology with FAD-dependent (flavin adenine dinucleotide-dependent) amine oxidases, marking it as a candidate HDM. Shi et al. showed that this protein was able to demethylate H3K4me2 but not H3K4me3, nor was demethylation observed for several other methylated residues, including H3K9, H3K27, H3K36, or Arg residues on H3 or H4. Knockdown of LSD1 using RNAi resulted in increased expression and an increase in associated H3K4me2 at target genes, supporting a model in which recruitment of LSD1 as a constituent of a corepressor complex contributed to gene repression by demethylation of H3K4.

The inability of LSD1 to demethylate trimethylated lysine residues likely stemmed from a requirement for a protonated nitrogen at its site of action, common to flavin-containing amine monoxidases (Shi et al., 2004). This, together with the marked specificity of LSD1 (also known as KMD1A) and the apparent absence of LSD1 homologs in *S. cerevisiae* in spite of known methylation of H3K4, H3K36, and H3K79, stimulated searches for additional HDMs. Zhang and colleagues developed an assay to screen fractionated HeLa extracts for HDM activity, based on the premise that demethylation of methylated lysine residues was likely to yield formaldehyde in an oxidative reaction requiring Fe(II) and α-ketoglutarate, as observed for demethylation of 1-methyl-adenine (Tsukada et al., 2006). Utilization of this assay led to identification of a protein known previously only through bioinformatic searches; the protein sequence included several well-characterized domains, including a JmjC domain common to a large number of eukaryotic proteins and suggested to be a possible indicator of chromatin regulatory activity. The JmjC domain was the only identified domain found to be essential for enzymatic activity, and it had been previously suggested that some JmjC domain-containing proteins might be able to reverse mono-, di-, and trimethylation of histones via hydroxylation of the methyl groups (Trewick et al., 2005). Accordingly, the protein was named JmjC domain-containing histone demethylase 1, or JHDM1. JHDM1 showed high specificity for dimethylated H3K36; identification of additional JmjC domain-containing HDMs with distinct specificities quickly followed, including enzymes capable of demethylating trimethylated H3K9 and H3K36 (Fig. 7.14; Cloos et al., 2006; Fodor et al., 2006; Klose et al., 2006; Whetstine et al., 2006).

HDMs targeting methylated H3K9 present an interesting example of feedback between modifications. As recounted earlier, methylation of H3K4 and H3K9 was identified early on as being associated respectively with transcriptional activation and with repressed heterochromatin and was found in ChIP experiments to be localized in the genome in mutually exclusive fashion (Barski et al., 2007; Noma et al., 2001). One means to achieve such exclusivity would be mutual inhibition of the relevant HMTs by the opposing methylation mark, and indeed, early work indicated that methylation of H3K9 by Suv39h1 was inhibited by methylated H3K4 (Nishioka et al., 2002a). However, this contention was not supported by subsequent work, and although other examples of such cross-inhibition were later reported, the effects were modest and in some cases were not supported by other studies (Binda et al., 2010; Nishioka et al., 2002a; Wu et al., 2010). Another mechanism to achieve mutually exclusive methylation is by stimulating demethylase activity toward a specific site by methylation at another, and this mechanism has received support for HDMs targeting H3K9. The demethylases PHF8 and JHDM3C (also known as KDM4C) both target methylated H3K9 via their jumonji domains, and bind H3K4me3 via distinct PHD and tandem Tudor domains, respectively (Horton et al., 2010; Pack et al., 2016). Correspondingly, demethylation of H3K9 on peptides, and in the case of JHDM3C on nucleosomal substrates, by these HDMs was shown to be enhanced by H3K4me3 (Horton et al., 2010; Pack et al., 2016). Although these examples help to explain the segregation of methylated H3K4 and H3K9, a fuller mechanistic understanding will require a better definition of the recruitment and dynamics of the involved HMTs and HDMs.

Other modifications

The preceding sections on histone acetylation and methylation illuminate some of the advances and setbacks that have led to our current understanding of these modifications, as well as the widespread and complex functions of specific modifications. Histones are subject to numerous additional modifications that are no less important to cell function, identity, and development. These additional modifications are discussed more briefly in the following sections.

Phosphorylation

Phosphorylated serine was identified in H1 and H3 in 1966, during the period when chromatin research was mainly focused on the chemical characterization of chromatin components (Kleinsmith et al., 1966; Ord and Stocken, 1966). Many early studies reported phosphorylation of serine residues, and in some cases of threonine residues, in histones, but these investigations were limited by analytical approaches that allowed the possibility that contamination of purified histones with other phosphorylated proteins could skew results, or by reliance on in vitro phosphorylation reactions that might not reflect the in vivo status of histones (DeLange and Smith, 1971; van Holde, 1988). Protein sequencing, and eventually mass spectrometry approaches and the application of antibodies recognizing specific histone modifications, allowed unambiguous characterization of phosphorylation of core and linker histones at serine, threonine, and tyrosine residues (Fig. 7.1; Zhao and Garcia, 2015). Phosphorylation of histidine and lysine residues via P-N linkages was also reported in early studies (Chen et al., 1974; Smtih et al., 1973; van Holde, 1988). However, although these modifications are established in both prokaryotes and eukaryotes, they have proved difficult to study due to the lability of the P-N linkage, and phosphorylation of residues outside of Ser, Thr, and Tyr were not noted in a comprehensive survey of histone modifications identified by mass spectrometry (Zhao and Garcia, 2015). Novel methods for producing synthetic peptides containing these modifications may allow production of pertinent antibodies to allow more definitive testing of their in vivo prevalence (Bertran-Vicente et al., 2014; Marmelstein et al., 2017).

Alterations in phosphorylation of histone H1 and H3 were observed in early work and have been intensively investigated. Cell cycle studies using synchronized CHO cells or *Physarum polycephalum*, which can be grown in synchronized culture, found increasing phosphorylation of H1 and H3 during interphase, with peak levels occurring in mitotic cells before decreasing again afterward (Bradbury et al., 1974; Gurley et al., 1978). Direct sequence analysis demonstrated that H3 was specifically phosphorylated at Ser10 in mammalian cells, while multiple residues in H1 were found to be phosphorylated, but the specific sites were not identified (Ajiro et al., 1981; Ajiro and Nishimoto, 1985; Hohmann, 1979; Paulson and Taylor, 1982). H3S10 phosphorylation was later found also to accompany mitosis in *Tetrahymena*, indicating evolutionary conservation of this modification, and an antibody developed against H3S10p allowed rigorous identification of mitotic cells in diverse eukaryotes (Hendzel et al., 1997; Wei et al., 1998).

Phosphorylation of linker histone and of H3S10 is effected by distinct kinases and appears to regulate chromatin condensation during mitosis by distinct mechanisms. Human histone

H1 exists in multiple subtypes in mammalian cells; these are phosphorylated at serine residues during interphase and additionally phosphorylated at threonine residues during mitosis by multiple kinases (Liao and Mizzen, 2016; Sarg et al., 2006; Zheng et al., 2010). Phosphorylation occurs at (S/T)PX(K/R) sites in the C-terminal domains of H1. In contrast to histone acetylation and methylation, which exert their effects primarily by raising or lowering the affinity of effector proteins, phosphorylation at these sites reduces the association of linker histone with DNA by affecting electrostatic interactions involving a small domain of the protein in what has been dubbed a "charge patch" effect (Dou and Gorovsky, 2000; Dou et al., 1999). One might reasonably expect that such release of linker histone from DNA would result in *decondensation* of chromatin rather than the observed mitotic condensation, given the normal role of linker histones in chromatin compaction (Chapter 3). In fact, dephosphorylation of linker histone is indeed associated with chromatin compaction in conjugating *Tetrahymena* macronuclei (which are normally transcriptionally active, but shut down during conjugation), and in terminally differentiated chicken erythrocytes (Aubert et al., 1991; Lin et al., 1991). In another example, dephosphorylation of a sperm-specific histone H1 accompanies condensation of chromatin in sea urchin sperm, and phosphorylation occurs again following fertilization in parallel with chromatin decondensation (Green and Poccia, 1985). Perhaps even more surprisingly, depletion of linker histone from *Xenopus* egg extracts did not affect the ability of the extracts to assemble and condense chromatin to produce metaphase chromosomes (Dasso et al., 1994; Ohsumi et al., 1993). Knockout of micronuclear (MicLH) or macronuclear (H1) linker histone in *Tetrahymena* did not impact viability, but resulted in less condensed mitotic chromosomes in the respective organelle (Shen et al., 1995). These varied findings suggest that linker histone phosphorylation likely contributes to mitotic chromosome condensation by loosening association to allow access by nonhistone proteins, such as the SMC proteins mentioned in Chapter 3, that effect chromatin compaction. Evidently, the importance of linker histones and their phosphorylation to such compaction varies among different organisms.

The mechanism by which H3S10 phosphorylation contributes to mitotic chromosome condensation is completely distinct from that used by the linker histones and was uncovered during the flurry of discovery regarding histone modifications in the late 1990s and early 2000s. First, a strain of *Tetrahymena* expressing only an S10A mutant of histone H3 was shown to exhibit defective chromosome segregation and nuclear division; the defect was limited to micronuclei and not seen in the amitotically dividing macronuclei, which also do not exhibit H3 phosphorylation during replication in wild type *Tetrahymena* (Allis and Gorovsky, 1981; Wei et al., 1999). Mutation of H3S10 was later shown also to result in defective chromosome segregation in *S. pombe* (Mellone et al., 2003). The kinase responsible for H3S10 phosphorylation during mitosis in budding yeast was then identified as Ipl1, a member of the aurora kinase family, based on its expression pattern during the cell cycle and the abnormal chromosome segregation observed in *ipl1* mutants, and demonstrated using in vitro assays and a *ts* mutant in vivo (Hsu et al., 2000). H3S10 phosphorylation was found to be counteracted by the Glc7 phosphatase in yeast; in *C. elegans*, these roles are performed by the aurora kinase AIR-2 and two Glc7 homologs, and in both organisms, the balance of kinase and phosphatase activities is essential for proper mitosis. NIMA, a kinase known to be required for entry into mitosis of cells from the filamentous fungus *Aspergillus nidulans*, was found to be required for H3S10 phosphorylation in that organism, again underscoring the broad conservation of its contribution to the mitotic program (De Souza et al., 2000). An anomalous finding in these studies was the lack of effect of an H3S10A mutant on cell division or growth in *S. cerevisiae*; it was suggested that the presence in the H2B tail, which is also phosphorylated during mitosis, of serine residues embedded in sequences similar to those surrounding H3S10, might act redundantly with H3S10 phosphorylation (Hsu et al., 2000). In support of this contention, it was noted that these serine residues were absent from H2B in *Tetrahymena*, where S10A proved essential to mitotic chromosome condensation. To my knowledge, the telling experiment of mutating or deleting the H2B tail in conjunction with the H3S10A modification has not been performed.

The mechanism by which H3S10p contributes to mitotic chromosome condensation emerged from investigation of HP1 binding to heterochromatin in mammalian tissue culture cells (Fischle et al., 2005). HP1 was known to dissociate from chromatin during mitosis, but Western blotting and mass spectrometric analysis revealed no change in the H3K9me3 modification responsible for its association with chromatin. Instead, association of HP1 with mitotic chromatin was discovered to be inhibited by a dual H3K9me3S10p modification, and depletion of the aurora B kinase that phosphorylates H3S10p caused HP1 to be retained during mitosis (Fischle et al., 2005). While the precise function of HP1 eviction remains unclear, Fischle et al. reasonably speculated that, similarly to explanations proffered for the role of linker histone removal in mitosis, removal of HP1 might be necessary for association of factors needed for normal chromosome condensation and segregation. The role of mitotic H3S10 phosphorylation in evicting HP1 also provided a potential alternative explanation for the lack of phenotype of the H3S10A mutation in budding yeast, as *S. cerevisiae* lacks both HP1 and H3K9 methylation. However, this still begs the question as to why mitotic phosphorylation of H3S10 has been preserved in budding yeast; perhaps it is simply an evolutionary relic, although this seems somewhat unlikely.

Phosphorylation of H3 was also observed in response to growth factors and other inducers of early response genes (Mahadevan et al., 1991). This was shown to occur at H3S10, and the kinase was found to be Rsk-2 by using the same gel activity assays as had been used in the discovery of Gcn5 as a HAT (Sassone-Corsi et al., 1999). *RSK-2* mutations were known to be linked to Coffin-Lowry syndrome, a rare genetic disorder characterized by craniofacial and skeletal abnormalities, and cells from a patient failed to phosphorylate H3 in response to growth factor stimulation, although H3S10p was detected during mitosis. Multiple kinases targeting H3S10 at induced genes in interphase cells in metazoans have since been identified, where H3S10p appears to facilitate transcriptional elongation (Hartzog and Tamkun, 2007). Numerous other core histone serine and threonine residues undergo phosphorylation (Fig. 7.1; Millan-Zambrano et al., 2022; Zhao and Garcia, 2015); a thorough survey and discussion is presented in Banerjee and Chakravarti (2011).

Tyrosine phosphorylation, although long appreciated as integral to many signal transduction pathways, plays a more limited role than serine/threonine phosphorylation in the context of histone modifications. Allis and colleagues, investigating the regulation of γ-H2A.X in the DNA damage response (see Chapter 5), discovered that Tyr142 at the extreme C-terminus of H2A.X is phosphorylated in murine and *Xenopus* cells, and that phosphorylation decreases in response to DNA damage (Xiao et al., 2009). Construction of a murine cell line expressing FLAG-tagged H2A.X allowed purification of associated proteins, leading to identification of the responsible kinase as Williams-Beuren transcription factor (WSTF), which associates with the chromatin remodeler SNF2H, a member of the ISWI family (Chapter 6). A subsequent study identified the mammalian EYA protein, a homolog of the *eyes absent* protein first identified as being required at an early stage of development of the *Drosophila* compound eye, as a tyrosine phosphatase required for dephosphorylation of H2A.X Y142 during the DNA damage response (Cook et al., 2009). Exposure of cells to high-dose ionizing radiation normally provokes an apoptotic response; remarkably, this response was reduced about 6× in MEF cells expressing only an H2A.X Y142F mutant. Thus, the balance of phosphorylation and dephosphorylation at H2A.X Y142 appears to govern the decision of cells to follow pathways leading to repair and survival or to cell death. This same decision pathway also evidently operates in development, as retinal progenitor cells in *Drosophila eya* mutants are diverted from the normal differentiation pathway, instead undergoing programmed cell death (Bonini et al., 1993). More recently, H2A.X Y142 dephosphorylation has been implicated in silencing transcription during the DNA damage response (Ji et al., 2019).

Among core histones, phosphorylation of H3Y41 was found to impact binding of HP1α in human hematopoietic cells (Dawson et al., 2009). Additional sites of tyrosine phosphorylation in the structured (nontail) region of H3 and H4 have been reported, but these have received less attention (Cavalieri, 2021; Millan-Zambrano et al., 2022; Singh et al., 2009). Histone tyrosine phosphorylation, as well as the role of histone tyrosines in early structural and biophysical studies of the nucleosome, have been reviewed by Singh and Gunjan (2011).

Histone ubiquitination and sumoylation

Ubiquitin (Fig. 7.15), a protein 76 amino acids in length so named for its presence and extremely high sequence conservation across eukaryotes (Goldstein et al., 1975), is best known

(A)

MQIFVKTLTGKTITLEVEPSDTIENVKAKIQDKEGIPPDQQRLIFAGKQLEDGRTLSDYNIQKESTLHLVLRLRGG

(B)

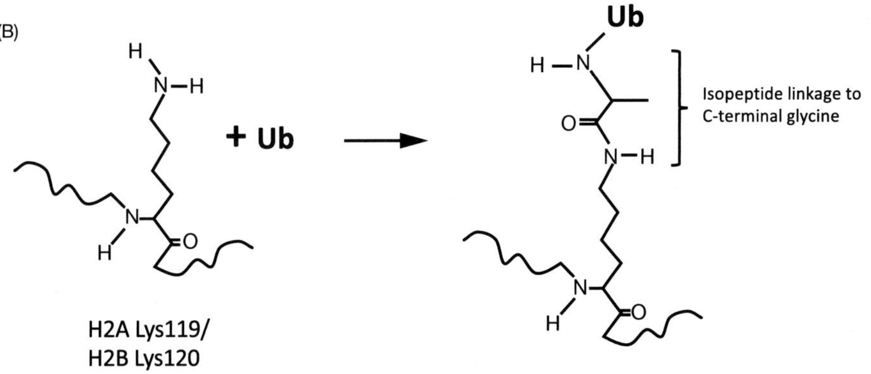

+ Ub

H2A Lys119/
H2B Lys120

Ub

Isopeptide linkage to
C-terminal glycine

FIG. 7.15 Histone ubiquitination. (A) Amino acid sequence of human ubiquitin. Lysine residues that can act as sites of isopeptide linkages to allow polyubiquitin chain formation are highlighted. (B) Mono-ubiquitination of H2A and H2B occurs via an isopeptide linkage between the C-terminal glycine of ubiquitin and Lys119 or Lys120 of mammalian H2A and H2B, respectively (or H2BK123 in yeast).

for its role in signaling protein degradation. Polyubiquitination, in which covalent linkage between the C-terminal glycine of ubiquitin and internal lysine residues forms an extended chain, targets proteins for destruction by the proteasome; in contrast, histones are subject to mono-ubiquitination as described below. Ubiquitination (also referred to as ubiquitylation) requires the action of three enzymes. First, an E1 ubiquitin-activating enzyme forms a thioester linkage with the C-terminus of ubiquitin via an active site cysteine; second, the activated ubiquitin is transferred to an E2 ubiquitin conjugating enzyme; and third, an E3 ubiquitin ligase facilitates transfer of the activated ubiquitin to a substrate, with specificity dictated by interaction between the E3 enzyme and substrate. Thus, ubiquitination of specific histone lysine residues is determined by the identity of the E3 enzyme and the specific E2 enzyme with which it interacts.

As a side note, ubiquitin is only one of a family of ubiquitin-like proteins that are conjugated to a vast array of proteins to impact a wide variety of processes in eukaryotes (Kerscher et al., 2006). This family includes the SUMO proteins, discussed below; conjugation of ubiquitin-like proteins occurs via enzyme cascades similar to that used for ubiquitin attachment but employing distinct, evolutionarily related enzymes (Hochstrasser, 2009). Based on weak similarities between enzymes involved in ubiquitin conjugation and prokaryotic sulfurtransfer enzymes, and parallels in their mechanisms of action, it has been proposed that conjugation of ubiquitin-like proteins evolved from prokaryotic sulfurtransferases or related enzymes, and that ubiquitin conjugation was widespread in the last common eukaryotic ancestor (Hochstrasser, 2009).

In spite of the fundamental difference between mono-ubiquitination of histones and the polyubiquitination that results in targeting proteins to the proteasome, there is an interesting historical connection between protein degradation mediated by polyubiquitination and chromatin. The 2004 Nobel Prize in Chemistry was awarded to Irwin Rose, Avram Hershko, and Aaron Ciechanover for their elucidation of ubiquitin-mediated proteolysis, but many felt that were it not for the stricture that no more than three scientists may share a Nobel Prize, Alex Varshavsky would have joined these honorees. Varshavsky and colleagues' seminal discoveries related to ubiquitin-mediated proteolysis include the first demonstration (in collaboration with Ciechanover) that ubiquitin conjugation was required for degradation of short-lived proteins in vivo and was essential for viability and cell cycle progression; and discovery of the N-end rule, which dictates specificity in ubiquitin-mediated proteolysis based on the identity of the exposed N-terminal amino acid in a given protein, leading to protein half-lives varying from less than 3 min to over 20 h (Bachmair et al., 1986; Ciechanover et al., 1984; Finley et al., 1984). This work had its origins in chromatin studies, inasmuch as Varshavsky's pursuit of ubiquitination stemmed from an initial interest in ubiquitinated H2A (e.g., Levinger and Varshavsky, 1982).

Histone ubiquitination was discovered through identification of a strange protein first named "A-24," which analysis revealed to contain two distinct N-termini (Goldknopf et al., 1975; Orrick et al., 1973). Mapping of tryptic peptides, followed by sequencing from the two N-termini, revealed A-24 to contain all of H2A together with ubiquitin (Hunt and Dayhoff, 1977; Olson et al., 1976). Further analysis showed that ubiquitin and H2A were linked via an isopeptide bond formed between the C-terminal glycine of ubiquitin and the ε-amino group of H2AK119, and A-24 was thus redesignated as ubiquitinated H2A, or uH2A (Fig. 7.15; Goldknopf and Busch, 1977). uH2A was demonstrated to be incorporated into nucleosomes, which were indistinguishable from canonical nucleosomes with respect to enzymatic digestion (Goldknopf et al., 1977; Kleinschmidt and Martinson, 1981).

The protein complex responsible for ubiquitin ligation to H2A in human cells was reported in 2004 to include several proteins from the PcG, a family of proteins known to play a critical role in cell differentiation and development throughout metazoans (Wang et al., 2004a). Cell type-specific repression during development is maintained by PcG proteins, while activation is conversely maintained by the Trithorax group; both processes are intimately connected to histone modifications, including H2AK119ub, and are discussed at length in Chapter 8.

H2B was reported to be ubiquitinated near its C-terminus, at about 10% the level measured for H2A, in 1980 (West and Bonner, 1980), and was discovered over 20 years later to be required for methylation of H3K4 in the first example of *trans*-histone cross-talk. Interplay between histone modifications, such as the cross-inhibition between H3K9me and H3S10p, prompted investigation as to whether other modifications would affect H3K4 methylation in budding yeast (Sun and Allis, 2002). While mutations that affected histone acetylation levels did not affect H3K4 methylation, deletion of several genes known to affect silencing at telomeres and rDNA, which is also affected by deletion of the H3K4 methyltransferase Set1, led to the finding that *rad6Δ* yeast completely lacked H3K4 methylation as detected by Western blotting using an antibody against H3K4me2/3. Rad6 had previously been shown to be responsible for H2B123ub in yeast (Robzyk et al., 2000), and an H3K123R mutation also abrogated H3K4 methylation in vivo (Sun and Allis, 2002). An independent investigation, using an exhaustive screen of deletion mutants, also found Rad6 as necessary for H3K4 methylation and showed H3K4 methylation to be lost in an H2BK123R mutant (Dover et al., 2002). Ubiquitination of H2BK123 was also shown to be required for H3K79 methylation, but not for that of H3K36 (Briggs et al., 2002; Ng et al., 2002b). Subsequent investigations showed that while H2BK123ub was necessary for both H3K4 and H3K79 di- and trimethylation, it was not needed for monomethylation at these sites (Dehe et al., 2005; Shahbazian et al., 2005). Similar *trans*-histone modification dependence appears to occur in human cells, although the difficulty in creating histone mutant strains or of completely abolishing H2BK123 ubiquitination has prevented determination as to whether the dependence is absolute (Kim et al., 2005, 2009; Zhu et al., 2005).

While *trans*-regulation of histone methylation by H2B ubiquitination is clearly a critical function of the latter modification, it seems unlikely that its initial appearance in evolution could be based on that role. Examination of susceptibility of chromatin from yeast having or lacking H2BK123ub to digestion by MNase and DNase indicated that ubiquitination stabilized nucleosome structure, with deubiquitination functioning to facilitate release of H2A-H2B dimers during transcription (Chandrasekharan et al., 2009). In contrast, in vitro experiments indicated that ubiquitination of H2BK120 resulted in a modest *destabilization* of the nucleosome, with enhancement of H2A-H2B eviction (Krajewski et al., 2018). The substantial differences in approach between these two investigations—in vitro vs in yeast, and the inherently lower stability of the yeast nucleosome relative to one constructed with bacterially produced *Xenopus* histones (Chapter 5)—may account for the disparate results. H2B ubiquitination has been implicated in transcriptional regulation and transcriptional elongation, particularly in connection with FACT and histone eviction accompanying passage of Pol II, as well as in DNA repair (Chandrasekharan et al., 2010; Chen et al., 2022). Additional sites of ubiquitination have been identified on H2A and H2B, as well as on H3; these are present at lower abundance and have received less attention, but have been implicated in facilitating DNA methylation and heterochromatin formation (Chen et al., 2022).

SUMO (small ubiquitin-like modifier) refers to a family of proteins that are conjugated to lysine residues in target proteins via an isopeptide bond in a process analogous to that for ubiquitin, although the pathways and protein machinery involved are distinct. SUMO family proteins are comparable in size to the histones themselves; the lone SUMO protein present in yeast (mammalian cells have five), Smt3, is 11.6 kDa, even larger than ubiquitin (8.6 kDa). Nevertheless, despite both SUMO and ubiquitin adducts dwarfing the size of other histone modifications, they differ in function, likely reflecting specific interactions with partner proteins. A compelling example of this specificity is provided by an experiment in which sumoylation engineered to occur at H2BK123 did not substitute for H2BK123ub in facilitating methylation of H3K4 or H3K79 (Chandrasekharan et al., 2009).

Sumoylation was discovered about 20 years later than ubiquitination, owing to the lability of the linkage during protein isolation, and the first report of histone sumoylation appeared in 2003 (Shiio and Eisenman, 2003). Immunopurification of either FLAG-tagged histone H4 expressed in human 293T cells, or of native acetylated H4, followed by Western blotting, resulting in detection of a slower migrating species corresponding to sumoylated H4. Weaker indication of sumoylation of the other core histones was also observed, and subsequent work confirmed sumoylation of all four core histones as well as variants H2A.X, H2A.Z, cenH3, and H1 (Ryu and Hochstrasser, 2021). While research on histone sumoylation has intensified in just the past decade, it is clear that there exists a complex interplay between sumoylation and other histone modifications, with impacts on transcription, DNA repair, and chromosome segregation.

Both sumoylation and ubiquitination are reversible modifications, with removal of ubiquitin and sumo moieties carried out by diverse families of proteases capable of cleaving the isopeptide bond at specific modified sites (Mevissen and Komander, 2017; Wilkinson and Henley, 2010). The most notable example related to chromatin is the deubiquitinase module of SAGA noted earlier in this chapter. This module includes the ubiquitin protease Ubp8 in budding yeast and USP22 in mammalian cells, both of which catalyze removal of ubiquitin from H2BK120/123 (Atanassov et al., 2016; Henry et al., 2003; Sanders et al., 2002). Global effects on H2Bub levels of deletion, depletion, or catalytic inactivation of SAGA-associated deubiquitinases, as well as effects of disconnecting the DUB module from SAGA, have been reported, with variable results (Soffers and Workman, 2020). Transcription appears largely unaffected by manipulations that alter H2Bub levels at transcribed genes; an alternative potential role for H2B ubiquitination is to act as a reservoir for ubiquitin, thereby maintaining the cellular balance of free and conjugated ubiquitin (El-Saafin et al., 2022).

ADP-ribosylation, glycosylation, and more

ADP-ribosylation of proteins occurs by cleavage and transfer of ADP-ribose from nicotinamide adenine dinucleotide by ADP-ribosyltransferases and can be reversed by specific hydrolases. Subsequent conjugation of additional ADP-ribose moieties, catalyzed by diphtheria toxin-like ADP-ribosyl transferases (ARTDs) (also known as poly(ADP-ribose) polymerases, or PARPs), can result in branched or linear poly(ADP-ribose) chains (Fig. 7.16).

ADP-ribosylation of histones was first shown definitively by demonstration that injection of ^{14}C-ribose into rats resulted in radiolabel incorporation into histones (Ueda et al., 1975).

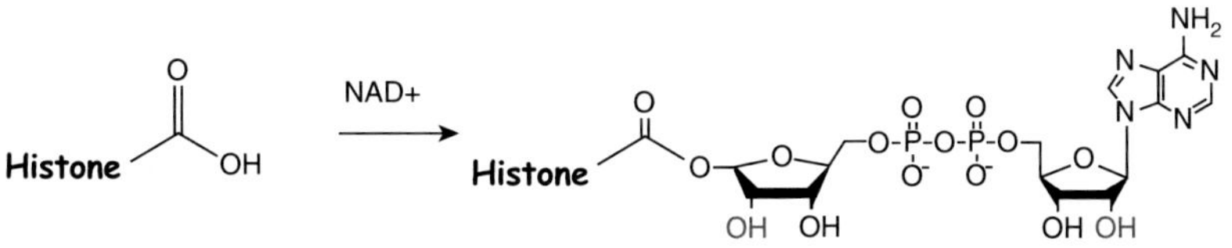

FIG. 7.16 ADP-ribosylation can occur via linkages to the carboxyl groups of Asp or Glu, as well as via linkages to Ser hydroxyl or Lys primary amine groups. Further polymerization to yield poly-ADP-ribosyl modification can occur via the indicated OH groups of the attached ADP-ribosyl moiety.

ADP-ribosylation occurs predominantly at linker histone H1 and has also been observed at Asp, Glu, Lys, and Ser residues in all four core histones (Adamietz et al., 1978; Adamietz and Rudolph, 1984; Boulikas, 1988; Leidecker et al., 2016; Levy-Wilson, 1983; Messner et al., 2010; Ogata et al., 1980a,b). The modification is present at very low levels in normally growing cells but increases markedly during the DNA damage response, and has been studied principally in that context (Adamietz and Rudolph, 1984; Karch et al., 2017; Kreimeyer et al., 1984). A mass spectrometric analysis showed that nearly all accessible nucleosomal Glu/Asp residues undergo ADP-ribosylation in response to DNA damage, making it unlikely that any individual site carries out a specific function in the damage response (Karch et al., 2017). A more likely interpretation is that the overall extent of modification serves to create a more open chromatin structure amenable to access by the repair machinery. Histone ADP-ribosylation has been reported to increase accessibility of DNA in chromatin to DNase I and restriction endonucleases and thereby facilitate transcriptional activation during the inflammatory response (Martinez-Zamudio and Ha, 2012). Thus, the effect of histone ADP-ribosylation may be similar to acetylation of lysine residues, where overall levels appear more important than acetylation of any one specific site (Dion et al., 2005), as opposed to the site-specific effects of histone methylation discussed earlier.

Another modification, GlcNAcylation, occurs upon addition of a single β-*N*-acetylglucosamine, or GlcNAc, monosaccharide unit to the hydroxy group of select serine or threonine residues. Evidence for histone glycosylation was first obtained from studies in which *Tetrahymena* was shown to incorporate radioactivity into histones when fed radiolabeled fucose, but the nature of the modification was not ascertained (Levy-Wilson, 1983). All four core histones were reported to be subject to O-glycosylation, or more accurately O-GlcNAcylation, using biochemical and analytical approaches, including Western blotting, in 2010–11, by which time numerous other proteins were known to be O-GlcNAcylated (Sakabe et al., 2010; Zhang et al., 2011). (Mass spectrometry is inefficient at detecting O-GlcNAcylated residues, due to the lability of the glycosidic bond.) Modification was seen in the unstructured tails (at H2AT101 and H3S10), at the lateral surface of the nucleosome, and at S47 of H4, which is situated in loop 1 of the histone fold domain. These findings indicate some specificity in targeting of O-GlcNAcylation in the histones, similar to most histone modifications other than ADP-ribosylation, despite the apparent existence in mammalian cells of only a single *O*-GlcNAc transferase (OGT) and *O*-GlcNAcase (OGA)

(Gao et al., 2001; Kreppel et al., 1997). Evidently, targeting is dictated by associated subunits, and levels of *O*-GlcNAc must also depend on the cycling of the action of OGT and OGA at specific sites (Hart et al., 2007).

Both of the papers that first reported histone O-GlcNAcylation also noted its variation during the cell cycle, suggesting a role in that process (Sakabe et al., 2010; Zhang et al., 2011). O-GlcNAcylation uses as substrate UDP-GlcNAc, which among high energy, small compounds in the cell is second in abundance only to ATP (Hart et al., 2007). Correspondingly, O-GlcNAcylation of histones and other proteins varies in response to changes in cellular metabolism, leading to altered gene expression, thereby potentially impacting critical processes such as cell proliferation and differentiation (Hardiville and Hart, 2016; Parween et al., 2022; Sheikh et al., 2021). Parsing the precise role of histone O-GlcNAcylation in such processes in the presence of a multitude of transcription factors, histone modifiers, and other proteins that share the modification remains a daunting task.

The list of known histone modifications has proliferated greatly in the past decade and now includes such exotica as crotonylation, biotinylation, lipidation, formylation, lactylation, dopaminylation, and others, ensuring no shortage for chromatin researchers of subject matter or complex interactions to study for years to come (Cavalieri, 2021; Chan and Maze, 2020).

Cross-talk

"Cross-talk" in the context of histone modifications refers to a particular modification having a direct influence, positive or negative, on the implementation of a different modification. Cross-talk in this sense implies some degree of causality, in which the mechanism by which a specific modification is conferred is affected by another modification, as distinct from, for example, two modifications that are independently conferred but co-occur in transcriptionally active chromatin.

Several examples of cross-talk have already been discussed, and some specific instances are given in Table 7.4. A very early example derives from an experiment in which exposure of HeLa cells to butyrate, leading to hyperacetylation of histones, resulted in increased Ca^{2+}-dependent phosphorylation of H3 (Whitlock et al., 1983). Kinase extracted from HeLa cells phosphorylated purified H3 equally well whether or not it was acetylated. Phosphorylated H3 was enriched in chromatin accessible to DNase I, and the authors suggested histone acetylation, which was known to increase nuclease accessibility, could similarly increase access to the relevant kinase. More sophisticated determination of the mechanistic underpinnings of cross-talk had to wait for identification of the enzymes involved, in order to allow detailed biochemical and structural analyses, along with genetic approaches to test involvement of specific sites of modification.

The simplest mechanism underlying cross-talk is steric interference, in which recognition of a potential modification site by the "writer" enzyme or enzyme complex is blocked by the presence of another modification at the same or a nearby site. For example, methylation of Lys9 in an H3 N-terminal peptide by Suv39h1 in vitro is prevented by Lys9 acetylation or

TABLE 7.4 Histone cross-talk.

			References
Positive effect			
Acetylation		H3 phosphorylation	Whitlock et al. (1983)
H4K16ac		H3K79me2/3	Valencia-Sanchez et al. (2021)
H2BK123ub		H3K79me2/3	Briggs et al. (2002) and Ng et al. (2002b)
H2BK123ub		H3K4me	Dover et al. (2002) and Sun and Allis (2002)
H3K9me3	*Enhances*	H4K20me3	Schotta et al. (2004)
H2BS112og		H2B120ub	Fujiki et al. (2011)
H3S10p		H3K14,18,23ac	Cheung et al. (2000) and Lo et al. (2001)
H3T11p		H3K9ac	Shimada et al. (2008)
H3K18,23ac		H3R17me	Daujat et al. (2002)
Inhibitory effect			
H3R2me	*Inhibits*	H3K4me	Guccione et al. (2007) and Hyllus et al. (2007)
H3K9me		H3S10ph	Rea et al. (2000)
H3S10ph		H3K9me	Rea et al. (2000)

Ser10 phosphorylation (Rea et al., 2000). Conversely, increased affinity of a modifying enzyme for a modified substrate can enhance its activity. For example, kinetic assays showed activity of Gcn5 to be increased about 10-fold on an H3 peptide containing phosphorylated Ser10 compared to the unmodified peptide, and H3S10 phosphorylation stimulated H3K14 acetylation in both mammalian and yeast cells (Cheung et al., 2000; Lo et al., 2000).

More complicated mechanisms are at play in cross-talk operating in *trans*, as in the requirement for H2B ubiquitination for H3K4 and H3K79 di- and trimethylation. As discussed earlier, this requirement was stringently established in yeast through demonstration that the E2 ubiquitin conjugating enzyme Rad6 responsible for ubiquitination of H2BK123 in yeast was also required for di- and trimethylation of H3K4 and H3K79, and by the elimination of these latter modifications in H3K123R mutant yeast. Insight into the mechanistic basis for these examples of *trans*-histone cross-talk has derived from biochemical and high-resolution cryo-EM studies, aided by chemical methods for construction of specifically ubiquitinated nucleosomes. With regard to H3K4 methylation, binding of the Set1-containing COMPASS complex to an unmodified nucleosome results in an autoinhibitory configuration of Set1 and consequent inability to methylate H3K4; this autoinhibition is relieved in the presence of H2BK123ub through contacts between the ubiquitin moiety and another COMPASS subunit, while the affinity of COMPASS is unaffected (Hsu et al., 2019). Dot1L binds in a completely distinct manner to the nucleosome and also makes specific contacts with ubiquitin in

H2BK120ub-containing nucleosomes, consistent with previous biochemical studies (Anderson et al., 2019; Holt et al., 2015; Jang et al., 2019; McGinty et al., 2008; Valencia-Sanchez et al., 2019; Worden et al., 2019; Yao et al., 2019; Zhou et al., 2016b). While the contacts between Dot1L and ubiquitin do not evidently increase Dot1L affinity for the nucleosome (although one study reported otherwise (Yao et al., 2019)), they appear to result in allosteric changes in the conformation of both Dot1L and the configuration of H3K79, reorienting K79 to a more accessible state and increasing Dot1L methylase activity by facilitating a more focused interaction with the target site (Anderson et al., 2019; Valencia-Sanchez et al., 2019; Worden et al., 2019).

An additional layer of *trans*-histone cross-talk pertains to Dot1L. Deletions and point mutations in the H4 tail, specifically in residues 16–19 (KRHR), reduced H3K79 methylation by Dot1 in yeast and in vitro methylation activity of both yeast and human enzymes (Altaf et al., 2007; Fingerman et al., 2007; Lee et al., 2018; McGinty et al., 2009). Recent cryo-EM structures confirm interaction between Dot1L and the H4 tail, and direct interaction between H4K16ac and Dot1 was shown to increase the H3K79 MTase activity of yeast Dot1 (Lee et al., 2018; Valencia-Sanchez et al., 2021; Worden et al., 2019). A picture emerges in regulation of telomeric silencing in yeast (see Fig. 3.13) in which specific readers of histone modifications, together with modification cross-talk, lead to formation of either telomeric heterochromatin or to euchromatin. In the former case, Sir2 and its partner Sir proteins bind and deacetylate H4K16, lowering affinity and activity of Dot1 and facilitating spreading of Sir-mediated heterochromatin, while in the remaining ~90% of the genome, modification yielding H4K16ac and H2B123ub facilitates Dot1 methylation of H3K79, which together with H4K16ac inhibits association of Sir2.

The discovery of cross-talk between histone modifications has subsequently served as a paradigm for similar mechanisms occurring with nonhistone proteins. An early and elegant example was the discovery that methylation of the yeast kinetochore subunit, Dam1, by Set1 inhibits phosphorylation of adjacent serine residues (Zhang et al., 2005). Proper chromosome segregation during mitosis depends on a balance between phosphorylation and dephosphorylation of Dam1; thus, cross-talk between methylation and phosphorylation of Dam1 is important for the correct execution of cell division. Moreover, the Set1-mediated methylation of Dam1 also requires ubiquitination of H2BK123 (Latham et al., 2011). Whether the mechanism underlying this requirement involves relief of Set1 autoinhibition, as is the case for H3K4 methylation, is not currently known.

Cross-talk also occurs between histone methylation and DNA methylation, the latter discussed at greater length in Chapter 8. The first examples of this cross-talk were discovered in studies of DNA methylation in the filamentous fungus *Neurospora crassa* and the model flowering plant *Arabidopsis thaliana*. Cloning of *dim-5*, discovered in a screen for mutants defective in DNA methylation in *N. crassa*, revealed the wild type gene to encode a SET domain that harbored a nonsense mutation in the mutant (Tamaru and Selker, 2001). Recombinant DIM-5 was able to methylate H3K9 when provided with *S*-adenosylmethionine and free histones as substrate, and expression of H3K9L or H3K9R mutants in *N. crassa* acted dominantly to phenocopy the *dim-5* mutation. Similarly, a screen for mutants defective for DNA methylation in *A. thaliana* uncovered *KRYPTONITE*, an H3K9 methyltransferase, so named because the mutation suppressed DNA methylation-dependent silencing at the *SUPERMAN* locus

(Jackson et al., 2002). (For readers who do not follow the DC comic universe, kryptonite (specifically green kryptonite) is fatal to Superman.) Loss-of-function *kryptonite* mutants exhibited loss of cytosine methylation similar to that seen in mutants in the DNA methyltransferase gene *CMT3*. CMT3 was shown to interact with an *Arabidopsis* homolog of HP1, which was in turn shown to interact with an H3K9 methylated peptide, indicating a pathway in which DNA methylation by CMT3 recruited to heterochromatin could reinforce heterochromatin repression. Interactions have since been reported between domains of mammalian DNA methyltransferases and their associated proteins and specific histone modifications, including H3K9me3 (Li et al., 2021b). Correspondingly, genome-wide studies have indicated a positive correlation between methylation of H3K9 and of DNA (Fu et al., 2020; Meissner et al., 2008).

The histone code

Even before the full scope and complexity of histone modifications became apparent, the variety of modifications and modifiers emerging from early studies prompted proposal of a histone, or epigenetic, code, a sometimes controversial but extremely popular formulation positing that the combinatorial possibilities of histone modifications could instruct chromatin in a nearly inexhaustible range of outcomes (Strahl and Allis, 2000; Turner, 2000). In their now famous paper on "The language of histone covalent modifications," published in *Nature* in 2000, Strahl and Allis recognize predecessors to their formulation:

> ... growing awareness of the remarkable diversity and biological specificity associated with distinct patterns of covalent histone marks has caused us and others to favour the view that a histone "language" may be encoded on these tail domains that is read by other proteins or protein modules. We refer to this language as the "histone code" ...

Indeed, a central tenet of the histone code paradigm is that modifications act not simply by altering chromatin structure, but by facilitating or inhibiting interactions with regulatory proteins involved in transcription and other processes involving DNA; the idea of modified histones acting as receptors in this way was proposed by Turner (as discussed earlier in this chapter) and others cited by Strahl and Allis (Loidl, 1994; Lopez-Rodas et al., 1993; Tordera et al., 1993; Turner, 1993).

Later in their article, Strahl and Allis explicitly posed the histone code hypothesis as stating

> ... that multiple histone modifications, acting in a combinatorial or sequential fashion on one or multiple histone tails, specify unique downstream functions ...

Thus, focusing on the H3 tail, they propose that the unmodified tail could encode a repressive function through binding of Sir3/Sir4 or Tup1, while H3K14ac, written by Gcn5, could recruit bromodomain-containing proteins to facilitate transcriptional activation. Combinatorial modifications could allow a single mark to contribute to disparate functions: H3S10 could lead to chromosome condensation or to transcriptional activation, depending on whether it occurred in conjunction with H3S28p or H3K14ac.

While the idea of a histone code has by now been cited literally thousands of times, it has also met with some resistance (Kurdistani and Grunstein, 2003). Among the objections raised are that the histone code is not a "code" in the same sense as used in ciphers or the genetic code (Henikoff, 2005), and that in some cases histone modification *follows* initiation of the process to which it was postulated to contribute, underscoring the maxim that correlation does not imply causation (Henikoff and Shilatifard, 2011). Moreover, early genome-wide determinations of the location of histone modifications argued against the potential complexity of combinatorial modifications that was a principal feature of the histone code hypothesis. First, microarray analysis of the effects on transcription of all 15 possible Lys to Arg mutations of H4K5, 8, 12, and 16 in yeast showed only H4K16 to produce changes in expression that were relatively uncorrelated with changes seen in the other mutants. In contrast, mutation of H4K5, 8, and 12 showed highly correlated changes that were additive in double and triple mutants, arguing for nonspecific, cumulative effects likely due to alterations in chromatin structure ascribable to charge neutralization by acetylation (Fig. 7.2; Dion et al., 2005; Henikoff, 2005). Two other reports examined histone modifications genome-wide in cells from *Drosophila*, human, and mouse cells and showed modifications mostly to follow a binary distribution, being present either in transcribed or nontranscribed chromatin, with high correlations among those characteristic of one or the other class (Bernstein et al., 2005; Schubeler et al., 2004). Additional evidence for histone modifications functioning in a relatively simple fashion to facilitate or establish a cellular memory of *on* or *off* states of transcription has been summarized in a recent review (Talbert and Henikoff, 2021).

None of these objections has had much impact on the use of the term "histone code," which likely has permanently entered the lexicon of chromatin biology. One might argue that these and other objections are largely semantic and might have been avoided if the histone code had been presented as a paradigm or organizing principal rather than as a hypothesis. In fact, the original formulation lacked the precision usually associated with a strong hypothesis, and no indication was given as to how it might be falsifiable. Nonetheless, the concept has proven its utility as a framework for studies of histone modifications and has helped to crystallize the distinction between modifications acting as receptors leading to specific outcomes vs causing structural alterations (especially destabilization) in the nucleosome that do not depend on the particular residue modified.

The histone code concept also contributed to the notion that histone modifications could serve as landmarks for functional elements in chromatin. This idea, which harkens back to the use of DNase sensitive sites as indicators of active promoters (Gross and Garrard, 1988), does not depend on the cause and effect relationship between histone modifications and the associated genomic functions, while still relying on their close association. Early efforts to utilize histone modifications in this way focused on identifying enhancer elements in metazoans. These elements, which contain binding sites for the activators that initiate transcription of associated promoters, can operate at distances of hundreds of kilobases and therefore can be difficult to identify. Examination of 30 Mb of noncontiguous human genomic sequence by ChIP-chip of histone modifications, Pol II, Taf1, and the histone acetyltransferase CBP/p300 at high resolution indicated enrichment of H3K4me3 at active promoters (marked by the presence of Pol II and Taf1) and of H3K4me1 at enhancers (marked by the presence of CBP/p300) (Heintzman et al., 2007). Subsequent studies extended and expanded on this work, discovering H3K27ac as characteristic of active enhancers and revealing the extent of modifications associated with active genes to reflect the level of activity (Creyghton et al., 2010; Heintzman et al., 2009; Rajagopal et al., 2014; Wang et al., 2008b). Specific modification patterns were also found associated with alternative splicing, and the histone variants H3.3 and H2A.Z also were seen to exhibit preferential association with particular functional elements, as discussed in Chapter 5 (Agirre et al., 2021; Rajagopal et al., 2014). In addition to extensive use in identifying functional enhancers, "chromatin signatures" have helped in identification of long noncoding RNAs, microRNA promoters, and direct interactions between transcription factors and DNA (Gordan et al., 2009; Guttman et al., 2009; Ozsolak et al., 2008).

In a related approach, chromatin states were categorized according to the histone modifications and associated proteins present (Filion et al., 2010; Kharchenko et al., 2011). Van Steensel and colleagues used microarrays probing the *Drosophila* genome at ~300 bp resolution in conjunction with DamID analysis of 40 chromatin-related proteins and ChIP of four histone methylation states, along with additional published data sets, to identify distinct chromatin types in *Drosophila* Kc167 tissue culture cells (Filion et al., 2010). (In the DamID approach, a protein of interest is fused to a prokaryotic Dam methyltransferase, which methylates adenine in GATC sites (van Steensel et al., 2001); the resulting stable methylation allows detection of even transient interactions.) The proteins used for the categorization included modifying enzymes, heterochromatin proteins, chromatin remodeler subunits, and transcription factors, among others. Using statistical methods (a Hidden Markov model and principal components analysis), analysis of the binding data for all factors resulted in classification of the genome into five chromatin types. Two types of euchromatin were identified, each enriched in transcriptionally active genes and a number of factors that were depleted from the remaining three classes, while differing in the presence of several proteins belonging to chromatin modifying or remodeling complexes and in the relative enrichment for H3K36me3 over transcribed regions. The three remaining chromatin types comprised two known varieties of heterochromatin, one constitutive (characterized by H3K9me2 and HP1) and the other facultative (characterized by PcG proteins and H3K27me3), and a third type covering near half of the genome and containing the majority of unexpressed genes but lacking the characteristic features of constitutive or facultative heterochromatin. This latter chromatin type appears not to be a passive container of silent genes, but rather a distinct environment unconducive to gene activity, as monitoring of the expression of a reporter gene integrated into a large number of genomic sites revealed enhanced silencing when integrated into this latter type relative to the genome-wide average, and lower even than integrants into either of the two classical heterochromatin types. A second study, incorporating additional histone modification localization data, further subdivided the *Drosophila* genome into chromatin types, including five distinct euchromatin classes and comprising nine types in total (Kharchenko et al., 2011).

It should be emphasized that the classifications resulting from the approach described above are not discrete; there is overlap among the factors binding to different chromatin types, and the "boundaries" distinguishing one class from another are fuzzy at best. Nonetheless, the approach has utility in providing models for considering how chromatin modifications and the proteins associated with them contribute to transcriptional activity and other DNA transactions on a genome-wide level. For example, similar approaches have been used to investigate the relationship between histone modifications and the 3D organization of chromosomes (Chapter 3; Boettiger et al., 2016; Di Pierro et al., 2017). The continued application of data on genome-wide association and three-dimensional chromosome organization with modeling approaches, combined with novel CRISPR-based methods for installing specific chromatin modifications at defined genomic locations (Nakamura et al., 2021; Thakore et al., 2016), should add greatly to our understanding of the impact of histone modifications to genomic function in coming years.

References

Adamietz, P., Rudolph, A., 1984. ADP-ribosylation of nuclear proteins in vivo. Identification of histone H2B as a major acceptor for mono- and poly(ADP-ribose) in dimethyl sulfate-treated hepatoma AH 7974 cells. J. Biol. Chem. 259, 6841–6846.

Adamietz, P., Bredehorst, R., Hilz, H., 1978. ADP-ribosylated histone H1 from HeLa cultures. Fundamental differences to (ADP-ribose)n-histone H1 conjugates formed into vitro. Eur. J. Biochem. 91, 317–326.

Agalioti, T., Lomvardas, S., Parekh, B., Yie, J., Maniatis, T., Thanos, D., 2000. Ordered recruitment of chromatin modifying and general transcription factors to the IFN-beta promoter. Cell 103, 667–678 (in process citation).

Agirre, E., Oldfield, A.J., Bellora, N., Segelle, A., Luco, R.F., 2021. Splicing-associated chromatin signatures: a combinatorial and position-dependent role for histone marks in splicing definition. Nat. Commun. 12, 682.

Ahmad, K., Henikoff, S., 2002. The histone variant H3.3 marks active chromatin by replication-independent nucleosome assembly. Mol. Cell 9, 1191–1200.

Ajiro, K., Nishimoto, T., 1985. Specific site of histone H3 phosphorylation related to the maintenance of premature chromosome condensation. Evidence for catalytically induced interchange of the subunits. J. Biol. Chem. 260, 15379–15381.

Ajiro, K., Borun, T.W., Shulman, S.D., McFadden, G.M., Cohen, L.H., 1981. Comparison of the structures of human histone 1A and 1B and their intramolecular phosphorylation sites during the HeLa S-3 cell cycle. Biochemistry 20, 1454–1464.

Ali, I., Conrad, R.J., Verdin, E., Ott, M., 2018. Lysine acetylation goes global: from epigenetics to metabolism and therapeutics. Chem. Rev. 118, 1216–1252.

Alland, L., Muhle, R., Hou Jr., H., Potes, J., Chin, L., Schreiber-Agus, N., DePinho, R.A., 1997. Role for N-CoR and histone deacetylase in Sin3-mediated transcriptional repression. Nature 387, 49–55.

Allard, S., Utley, R.T., Savard, J., Clarke, A., Grant, P., Brandl, C.J., Pillus, L., Workman, J.L., Cote, J., 1999. NuA4, an essential transcription adaptor/histone H4 acetyltransferase complex containing Esa1p and the ATM-related cofactor Tra1p. EMBO J. 18, 5108–5119.

Allfrey, V.G., Faulkner, R., Mirsky, A.E., 1964. Acetylation and methylation of histones and their possible role in the regulation of RNA synthesis. Proc. Natl. Acad. Sci. USA 51, 786–794.

Allis, C.D., Gorovsky, M.A., 1981. Histone phosphorylation in macro- and micronuclei of *Tetrahymena thermophila*. Biochemistry 20, 3828–3833.

Allis, C.D., Bowen, J.K., Abraham, G.N., Glover, C.V., Gorovsky, M.A., 1980. Proteolytic processing of histone H3 in chromatin: a physiologically regulated event in Tetrahymena micronuclei. Cell 20, 55–64.

Allis, C.D., Berger, S.L., Cote, J., Dent, S., Jenuwien, T., Kouzarides, T., Pillus, L., Reinberg, D., Shi, Y., Shiekhattar, R., et al., 2007. New nomenclature for chromatin-modifying enzymes. Cell 131, 633–636.

Altaf, M., Utley, R.T., Lacoste, N., Tan, S., Briggs, S.D., Cote, J., 2007. Interplay of chromatin modifiers on a short basic patch of histone H4 tail defines the boundary of telomeric heterochromatin. Mol. Cell 28, 1002–1014.

Ambler, R.P., Rees, M.W., 1959. Epsilon-N-methyl-lysine in bacterial flagellar protein. Nature 184, 56–57.

An, W., Kim, J., Roeder, R.G., 2004. Ordered cooperative functions of PRMT1, p300, and CARM1 in transcriptional activation by p53. Cell 117, 735–748.

Andersen, E.C., Horvitz, H.R., 2007. Two *C. elegans* histone methyltransferases repress lin-3 EGF transcription to inhibit vulval development. Development 134, 2991–2999.

Anderson, C.J., Baird, M.R., Hsu, A., Barbour, E.H., Koyama, Y., Borgnia, M.J., McGinty, R.K., 2019. Structural basis for recognition of ubiquitylated nucleosome by Dot1L methyltransferase. Cell Rep. 26, 1681–1690.e1685.

Andres, M.E., Burger, C., Peral-Rubio, M.J., Battaglioli, E., Anderson, M.E., Grimes, J., Dallman, J., Ballas, N., Mandel, G., 1999. CoREST: a functional corepressor required for regulation of neural-specific gene expression. Proc. Natl. Acad. Sci. USA 96, 9873–9878.

Andrews, F.H., Shinsky, S.A., Shanle, E.K., Bridgers, J.B., Gest, A., Tsun, I.K., Krajewski, K., Shi, X., Strahl, B.D., Kutateladze, T.G., 2016. The Taf14 YEATS domain is a reader of histone crotonylation. Nat. Chem. Biol. 12, 396–398.

Antonysamy, S., Condon, B., Druzina, Z., Bonanno, J.B., Gheyi, T., Zhang, F., MacEwan, I., Zhang, A., Ashok, S., Rodgers, L., et al., 2013. Structural context of disease-associated mutations and putative mechanism of autoinhibition revealed by X-ray crystallographic analysis of the EZH2-SET domain. PLoS One 8, e84147.

Aravind, L., Burroughs, A.M., Zhang, D., Iyer, L.M., 2014. Protein and DNA modifications: evolutionary imprints of bacterial biochemical diversification and geochemistry on the provenance of eukaryotic epigenetics. Cold Spring Harb. Perspect. Biol. 6, a016063.

Atanassov, B.S., Mohan, R.D., Lan, X., Kuang, X., Lu, Y., Lin, K., McIvor, E., Li, W., Zhang, Y., Florens, L., et al., 2016. ATXN7L3 and ENY2 coordinate activity of multiple H2B deubiquitinases important for cellular proliferation and tumor growth. Mol. Cell 62, 558–571.

Attisano, L., Lewis, P.N., 1990. Purification and characterization of two porcine liver nuclear histone acetyltransferases. J. Biol. Chem. 265, 3949–3955.

Aubert, D., Garcia, M., Benchaibi, M., Poncet, D., Chebloune, Y., Verdier, G., Nigon, V., Samarut, J., Mura, C.V., 1991. Inhibition of proliferation of primary avian fibroblasts through expression of histone H5 depends on the degree of phosphorylation of the protein. J. Cell Biol. 113, 497–506.

Auger, A., Galarneau, L., Altaf, M., Nourani, A., Doyon, Y., Utley, R.T., Cronier, D., Allard, S., Cote, J., 2008. Eaf1 is the platform for NuA4 molecular assembly that evolutionarily links chromatin acetylation to ATP-dependent exchange of histone H2A variants. Mol. Cell. Biol. 28, 2257–2270.

Ausio, J., van Holde, K.E., 1986. Histone hyperacetylation: its effects on nucleosome conformation and stability. Biochemistry 25, 1421–1428.

Avvakumov, N., Cote, J., 2007. The MYST family of histone acetyltransferases and their intimate links to cancer. Oncogene 26, 5395–5407.

Ayer, D.E., Lawrence, Q.A., Eisenman, R.N., 1995. Mad-Max transcriptional repression is mediated by ternary complex formation with mammalian homologs of yeast repressor Sin3. Cell 80, 767–776.

Bachmair, A., Finley, D., Varshavsky, A., 1986. In vivo half-life of a protein is a function of its amino-terminal residue. Science 234, 179–186.

Balasubramanian, R., Pray-Grant, M.G., Selleck, W., Grant, P.A., Tan, S., 2002. Role of the Ada2 and Ada3 transcriptional coactivators in histone acetylation. J. Biol. Chem. 277, 7989–7995.

Banerjee, T., Chakravarti, D., 2011. A peek into the complex realm of histone phosphorylation. Mol. Cell. Biol. 31, 4858–4873.

Bannister, A.J., Kouzarides, T., 1996. The CBP co-activator is a histone acetyltransferase. Nature 384, 641–643.

Bannister, A.J., Zegerman, P., Partridge, J.F., Miska, E.A., Thomas, J.O., Allshire, R.C., Kouzarides, T., 2001. Selective recognition of methylated lysine 9 on histone H3 by the HP1 chromo domain. Nature 410, 120–124.

Bannister, A.J., Schneider, R., Kouzarides, T., 2002. Histone methylation: dynamic or static? Cell 109, 801–806.

Bantscheff, M., Hopf, C., Savitski, M.M., Dittmann, A., Grandi, P., Michon, A.M., Schlegl, J., Abraham, Y., Becher, I., Bergamini, G., et al., 2011. Chemoproteomics profiling of HDAC inhibitors reveals selective targeting of HDAC complexes. Nat. Biotechnol. 29, 255–265.

Barlev, N.A., Candau, R., Wang, L., Darpino, P., Silverman, N., Berger, S.L., 1995. Characterization of physical interactions of the putative transcriptional adaptor, ADA2, with acidic activation domains and TATA-binding protein. J. Biol. Chem. 270, 19337–19344.

Barski, A., Cuddapah, S., Cui, K., Roh, T.Y., Schones, D.E., Wang, Z., Wei, G., Chepelev, I., Zhao, K., 2007. High-resolution profiling of histone methylations in the human genome. Cell 129, 823–837.

Baubec, T., Ivanek, R., Lienert, F., Schubeler, D., 2013. Methylation-dependent and -independent genomic targeting principles of the MBD protein family. Cell 153, 480–492.

Bauer, W.R., Hayes, J.J., White, J.H., Wolffe, A.P., 1994. Nucleosome structural changes due to acetylation. J. Mol. Biol. 236, 685–690.

Baur, J.A., Ungvari, Z., Minor, R.K., Le Couteur, D.G., de Cabo, R., 2012. Are sirtuins viable targets for improving healthspan and lifespan? Nat. Rev. Drug Discov. 11, 443–461.

Bedford, M.T., Clarke, S.G., 2009. Protein arginine methylation in mammals: who, what, and why. Mol. Cell 33, 1–13.
Belikoff, E., Wong, L.J., Alberts, B.M., 1980. Extensive purification of histone acetylase A, the major histone N-acetyl transferase activity detected in mammalian cell nuclei. J. Biol. Chem. 255, 11448–11453.
Bell, O., Wirbelauer, C., Hild, M., Scharf, A.N., Schwaiger, M., MacAlpine, D.M., Zilbermann, F., van Leeuwen, F., Bell, S.P., Imhof, A., et al., 2007. Localized H3K36 methylation states define histone H4K16 acetylation during transcriptional elongation in Drosophila. EMBO J. 26, 4974–4984.
Bender, L.B., Suh, J., Carroll, C.R., Fong, Y., Fingerman, I.M., Briggs, S.D., Cao, R., Zhang, Y., Reinke, V., Strome, S., 2006. MES-4: an autosome-associated histone methyltransferase that participates in silencing the X chromosomes in the C. elegans germ line. Development 133, 3907–3917.
Berger, S.L., Cress, W.D., Cress, A., Triezenberg, S.J., Guarente, L., 1990. Selective inhibition of activated but not basal transcription by the acidic activation domain of VP16: evidence for transcriptional adaptors. Cell 61, 1199.
Berger, S.L., Pina, B., Silverman, N., Marcus, G.A., Agapite, J., Regier, J.L., Triezenberg, S.J., Guarente, L., 1992. Genetic isolation of ADA2: a potential transcriptional adaptor required for function of certain acidic activation domains. Cell 70, 251–265.
Bernstein, B.E., Humphrey, E.L., Erlich, R.L., Schneider, R., Bouman, P., Liu, J.S., Kouzarides, T., Schreiber, S.L., 2002. Methylation of histone H3 Lys 4 in coding regions of active genes. Proc. Natl. Acad. Sci. USA 99, 8695–8700.
Bernstein, B.E., Kamal, M., Lindblad-Toh, K., Bekiranov, S., Bailey, D.K., Huebert, D.J., McMahon, S., Karlsson, E.K., Kulbokas 3rd, E.J., Gingeras, T.R., et al., 2005. Genomic maps and comparative analysis of histone modifications in human and mouse. Cell 120, 169–181.
Bertran-Vicente, J., Serwa, R.A., Schumann, M., Schmieder, P., Krause, E., Hackenberger, C.P., 2014. Site-specifically phosphorylated lysine peptides. J. Am. Chem. Soc. 136, 13622–13628.
Bhaumik, S.R., Green, M.R., 2001. SAGA is an essential in vivo target of the yeast acidic activator Gal4p. Genes Dev. 15, 1935–1945.
Bhaumik, S.R., Raha, T., Aiello, D.P., Green, M.R., 2004. In vivo target of a transcriptional activator revealed by fluorescence resonance energy transfer. Genes Dev. 18, 333–343.
Bilokapic, S., Halic, M., 2019. Nucleosome and ubiquitin position Set2 to methylate H3K36. Nat. Commun. 10, 3795.
Binda, O., LeRoy, G., Bua, D.J., Garcia, B.A., Gozani, O., Richard, S., 2010. Trimethylation of histone H3 lysine 4 impairs methylation of histone H3 lysine 9: regulation of lysine methyltransferases by physical interaction with their substrates. Epigenetics 5, 767–775.
Blazer, L.L., Lima-Fernandes, E., Gibson, E., Eram, M.S., Loppnau, P., Arrowsmith, C.H., Schapira, M., Vedadi, M., 2016. PR domain-containing protein 7 (PRDM7) is a histone 3 lysine 4 trimethyltransferase. J. Biol. Chem. 291, 13509–13519.
Bock, I., Dhayalan, A., Kudithipudi, S., Brandt, O., Rathert, P., Jeltsch, A., 2011. Detailed specificity analysis of antibodies binding to modified histone tails with peptide arrays. Epigenetics 6, 256–263.
Boettiger, A.N., Bintu, B., Moffitt, J.R., Wang, S., Beliveau, B.J., Fudenberg, G., Imakaev, M., Mirny, L.A., Wu, C.T., Zhuang, X., 2016. Super-resolution imaging reveals distinct chromatin folding for different epigenetic states. Nature 529, 418–422.
Boggs, B.A., Cheung, P., Heard, E., Spector, D.L., Chinault, A.C., Allis, C.D., 2002. Differentially methylated forms of histone H3 show unique association patterns with inactive human X chromosomes. Nat. Genet. 30, 73–76.
Bone, J.R., Roth, S.Y., 2001. Recruitment of the yeast Tup1p-Ssn6p repressor is associated with localized decreases in histone acetylation. J. Biol. Chem. 276, 1808–1813.
Bonini, N.M., Leiserson, W.M., Benzer, S., 1993. The eyes absent gene: genetic control of cell survival and differentiation in the developing Drosophila eye. Cell 72, 379–395.
Bonnet, J., Wang, C.Y., Baptista, T., Vincent, S.D., Hsiao, W.C., Stierle, M., Kao, C.F., Tora, L., Devys, D., 2014. The SAGA coactivator complex acts on the whole transcribed genome and is required for RNA polymerase II transcription. Genes Dev. 28, 1999–2012.
Borde, V., Robine, N., Lin, W., Bonfils, S., Geli, V., Nicolas, A., 2009. Histone H3 lysine 4 trimethylation marks meiotic recombination initiation sites. EMBO J. 28, 99–111.
Borrow, J., Stanton Jr., V.P., Andresen, J.M., Becher, R., Behm, F.G., Chaganti, R.S., Civin, C.I., Disteche, C., Dube, I., Frischauf, A.M., et al., 1996. The translocation t(8;16)(p11;p13) of acute myeloid leukaemia fuses a putative acetyltransferase to the CREB-binding protein. Nat. Genet. 14, 33–41.
Borun, T.W., Pearson, D., Paik, W.K., 1972. Studies of histone methylation during the HeLa S-3 cell cycle. J. Biol. Chem. 247, 4288–4298.

Botuyan, M.V., Lee, J., Ward, I.M., Kim, J.E., Thompson, J.R., Chen, J., Mer, G., 2006. Structural basis for the methylation state-specific recognition of histone H4-K20 by 53BP1 and Crb2 in DNA repair. Cell 127, 1361–1373.

Boudreault, A.A., Cronier, D., Selleck, W., Lacoste, N., Utley, R.T., Allard, S., Savard, J., Lane, W.S., Tan, S., Cote, J., 2003. Yeast enhancer of polycomb defines global Esa1-dependent acetylation of chromatin. Genes Dev. 17, 1415–1428.

Boulikas, T., 1988. At least 60 ADP-ribosylated variant histones are present in nuclei from dimethylsulfate-treated and untreated cells. EMBO J. 7, 57–67.

Bowdish, K.S., Mitchell, A.P., 1993. Bipartite structure of an early meiotic upstream activation sequence from *Saccharomyces cerevisiae*. Mol. Cell. Biol. 13, 2172–2181.

Bowman, A., Ward, R., Wiechens, N., Singh, V., El-Mkami, H., Norman, D.G., Owen-Hughes, T., 2011. The histone chaperones Nap1 and Vps75 bind histones H3 and H4 in a tetrameric conformation. Mol. Cell 41, 398–408.

Brachmann, C.B., Sherman, J.M., Devine, S.E., Cameron, E.E., Pillus, L., Boeke, J.D., 1995. The SIR2 gene family, conserved from bacteria to humans, functions in silencing, cell cycle progression, and chromosome stability. Genes Dev. 9, 2888–2902.

Bradbury, E.M., Inglis, R.J., Matthews, H.R., Langan, T.A., 1974. Molecular basis of control of mitotic cell division in eukaryotes. Nature 249, 553–556.

Brand, M., Leurent, C., Mallouh, V., Tora, L., Schultz, P., 1999a. Three-dimensional structures of the TAFII-containing complexes TFIID and TFTC. Science 286, 2151–2153.

Brand, M., Yamamoto, K., Staub, A., Tora, L., 1999b. Identification of TATA-binding protein-free TAFII-containing complex subunits suggests a role in nucleosome acetylation and signal transduction. J. Biol. Chem. 274, 18285–18289.

Braunstein, M., Rose, A.B., Holmes, S.G., Allis, C.D., Broach, J.R., 1993. Transcriptional silencing in yeast is associated with reduced nucleosome acetylation. Genes Dev. 7, 592–604.

Braunstein, M., Sobel, R.E., Allis, C.D., Turner, B.M., Broach, J.R., 1996. Efficient transcriptional silencing in *Saccharomyces cerevisiae* requires a heterochromatin histone acetylation pattern. Mol. Cell. Biol. 16, 4349–4356.

Briggs, S.D., Bryk, M., Strahl, B.D., Cheung, W.L., Davie, J.K., Dent, S.Y., Winston, F., Allis, C.D., 2001. Histone H3 lysine 4 methylation is mediated by Set1 and required for cell growth and rDNA silencing in *Saccharomyces cerevisiae*. Genes Dev. 15, 3286–3295.

Briggs, S.D., Xiao, T., Sun, Z.W., Caldwell, J.A., Shabanowitz, J., Hunt, D.F., Allis, C.D., Strahl, B.D., 2002. Gene silencing: trans-histone regulatory pathway in chromatin. Nature 418, 498.

Brower-Toland, B., Riddle, N.C., Jiang, H., Huisinga, K.L., Elgin, S.C., 2009. Multiple SET methyltransferases are required to maintain normal heterochromatin domains in the genome of *Drosophila melanogaster*. Genetics 181, 1303–1319.

Brown, C.E., Howe, L., Sousa, K., Alley, S.C., Carrozza, M.J., Tan, S., Workman, J.L., 2001. Recruitment of HAT complexes by direct activator interactions with the ATM-related Tra1 subunit. Science 292, 2333–2337.

Brown, D.A., Di Cerbo, V., Feldmann, A., Ahn, J., Ito, S., Blackledge, N.P., Nakayama, M., McClellan, M., Dimitrova, E., Turberfield, A.H., et al., 2017. The SET1 complex selects actively transcribed target genes via multivalent interaction with CpG Island chromatin. Cell Rep. 20, 2313–2327.

Brownell, J.E., Allis, C.D., 1995. An activity gel assay detects a single, catalytically active histone acetyltransferase subunit in Tetrahymena macronuclei. Proc. Natl. Acad. Sci. USA 92, 6364–6368.

Brownell, J.E., Allis, C.D., 2021. HAT discovery: heading toward an elusive goal with a key biological assist. Biochim. Biophys. Acta Gene Regul. Mech. 1864, 194605.

Brownell, J.E., Zhou, J., Ranalli, T., Kobayashi, R., Edmondson, D.G., Roth, S.Y., Allis, C.D., 1996. Tetrahymena histone acetyltransferase A: a homolog to yeast Gcn5p linking histone acetylation to gene activation. Cell 84, 843–851.

Bruzzone, M.J., Grunberg, S., Kubik, S., Zentner, G.E., Shore, D., 2018. Distinct patterns of histone acetyltransferase and mediator deployment at yeast protein-coding genes. Genes Dev. 32, 1252–1265.

Burgess, R.J., Zhou, H., Han, J., Zhang, Z., 2010. A role for Gcn5 in replication-coupled nucleosome assembly. Mol. Cell 37, 469–480.

Burnett, G., Kennedy, E.P., 1954. The enzymatic phosphorylation of proteins. J. Biol. Chem. 211, 969–980.

Byvoet, P., Shepherd, G.R., Hardin, J.M., Noland, B.J., 1972. The distribution and turnover of labeled methyl groups in histone fractions of cultured mammalian cells. Arch. Biochem. Biophys. 148, 558–567.

Cairns, B.R., Lorch, Y., Li, Y., Zhang, M., Lacomis, L., Erdjument-Bromage, H., Tempst, P., Du, J., Laurent, B., Kornberg, R.D., 1996. RSC, an essential, abundant chromatin-remodeling complex. Cell 87, 1249–1260.

Candau, R., Berger, S.L., 1996. Structural and functional analysis of yeast putative adaptors. Evidence for an adaptor complex in vivo. J. Biol. Chem. 271, 5237–5245.

Candau, R., Moore, P.A., Wang, L., Barlev, N., Ying, C.Y., Rosen, C.A., Berger, S.L., 1996. Identification of human proteins functionally conserved with the yeast putative adaptors ADA2 and GCN5. Mol. Cell. Biol. 16, 593–602.

Candau, R., Zhou, J.X., Allis, C.D., Berger, S.L., 1997. Histone acetyltransferase activity and interaction with ADA2 are critical for GCN5 function in vivo. EMBO J. 16, 555–565.

Candido, E.P., Dixon, G.H., 1971. Sites of in vivo acetylation in trout testis histone IV. J. Biol. Chem. 246, 3182–3188.

Candido, E.P., Dixon, G.H., 1972. Amino-terminal sequences and sites of in vivo acetylation of trout-testis histones 3 and IIb 2. Proc. Natl. Acad. Sci. USA 69, 2015–2019.

Candido, E.P., Reeves, R., Davie, J.R., 1978. Sodium butyrate inhibits histone deacetylation in cultured cells. Cell 14, 105–113.

Cao, R., Zhang, Y., 2004. SUZ12 is required for both the histone methyltransferase activity and the silencing function of the EED-EZH2 complex. Mol. Cell 15, 57–67.

Cao, R., Wang, L., Wang, H., Xia, L., Erdjument-Bromage, H., Tempst, P., Jones, R.S., Zhang, Y., 2002. Role of histone H3 lysine 27 methylation in polycomb-group silencing. Science 298, 1039–1043.

Cao, K., Ugarenko, M., Ozark, P.A., Wang, J., Marshall, S.A., Rendleman, E.J., Liang, K., Wang, L., Zou, L., Smith, E.R., et al., 2020. DOT1L-controlled cell-fate determination and transcription elongation are independent of H3K79 methylation. Proc. Natl. Acad. Sci. USA 117, 27365–27373.

Carlone, D.L., Skalnik, D.G., 2001. CpG binding protein is crucial for early embryonic development. Mol. Cell. Biol. 21, 7601–7606.

Carlone, D.L., Lee, J.H., Young, S.R., Dobrota, E., Butler, J.S., Ruiz, J., Skalnik, D.G., 2005. Reduced genomic cytosine methylation and defective cellular differentiation in embryonic stem cells lacking CpG binding protein. Mol. Cell. Biol. 25, 4881–4891.

Carmen, A.A., Griffin, P.R., Calaycay, J.R., Rundlett, S.E., Suka, Y., Grunstein, M., 1999. Yeast HOS3 forms a novel trichostatin A-insensitive homodimer with intrinsic histone deacetylase activity. Proc. Natl. Acad. Sci. USA 96, 12356–12361.

Carrozza, M.J., Li, B., Florens, L., Suganuma, T., Swanson, S.K., Lee, K.K., Shia, W.J., Anderson, S., Yates, J., Washburn, M.P., et al., 2005. Histone H3 methylation by Set2 directs deacetylation of coding regions by Rpd3S to suppress spurious intragenic transcription. Cell 123, 581–592.

Cavalieri, V., 2021. The expanding constellation of histone post-translational modifications in the epigenetic landscape. Genes (Basel) 12, 1596.

Chambers, S.A., Shaw, B.R., 1984. Levels of histone H4 diacetylation decrease dramatically during sea urchin embryonic development and correlate with cell doubling rate. J. Biol. Chem. 259, 13458–13463.

Chammas, P., Mocavini, I., Di Croce, L., 2020. Engaging chromatin: PRC2 structure meets function. Br. J. Cancer 122, 315–328.

Champagne, N., Bertos, N.R., Pelletier, N., Wang, A.H., Vezmar, M., Yang, Y., Heng, H.H., Yang, X.J., 1999. Identification of a human histone acetyltransferase related to monocytic leukemia zinc finger protein. J. Biol. Chem. 274, 28528–28536.

Champagne, N., Pelletier, N., Yang, X.J., 2001. The monocytic leukemia zinc finger protein MOZ is a histone acetyltransferase. Oncogene 20, 404–409.

Chan, J.C., Maze, I., 2020. Nothing is yet set in (hi)stone: novel post-translational modifications regulating chromatin function. Trends Biochem. Sci. 45, 829–844.

Chandrasekharan, M.B., Huang, F., Sun, Z.W., 2009. Ubiquitination of histone H2B regulates chromatin dynamics by enhancing nucleosome stability. Proc. Natl. Acad. Sci. USA 106, 16686–16691.

Chandrasekharan, M.B., Huang, F., Sun, Z.W., 2010. Histone H2B ubiquitination and beyond: regulation of nucleosome stability, chromatin dynamics and the trans-histone H3 methylation. Epigenetics 5, 460–468.

Chang, B., Chen, Y., Zhao, Y., Bruick, R.K., 2007. JMJD6 is a histone arginine demethylase. Science 318, 444–447.

Chen, Y.C., Dent, S.Y.R., 2021. Conservation and diversity of the eukaryotic SAGA coactivator complex across kingdoms. Epigenetics Chromatin 14, 26.

Chen, J.D., Evans, R.M., 1995. A transcriptional co-repressor that interacts with nuclear hormone receptors. Nature 377, 454–457.

Chen, C.C., Smith, D.L., Bruegger, B.B., Halpern, R.M., Smith, R.A., 1974. Occurrence and distribution of acid-labile histone phosphates in regenerating rat liver. Biochemistry 13, 3785–3789.

Chen, H., Lin, R.J., Schiltz, R.L., Chakravarti, D., Nash, A., Nagy, L., Privalsky, M.L., Nakatani, Y., Evans, R.M., 1997. Nuclear receptor coactivator ACTR is a novel histone acetyltransferase and forms a multimeric activation complex with P/CAF and CBP/p300. Cell 90, 569–580.

Chen, D., Ma, H., Hong, H., Koh, S.S., Huang, S.M., Schurter, B.T., Aswad, D.W., Stallcup, M.R., 1999. Regulation of transcription by a protein methyltransferase. Science 284, 2174–2177.

Chen, C.C., Carson, J.J., Feser, J., Tamburini, B., Zabaronick, S., Linger, J., Tyler, J.K., 2008. Acetylated lysine 56 on histone H3 drives chromatin assembly after repair and signals for the completion of repair. Cell 134, 231–243.

Chen, J.J., Stermer, D., Tanny, J.C., 2022. Decoding histone ubiquitylation. Front. Cell Dev. Biol. 10, 968398.

Cheng, D., Cote, J., Shaaban, S., Bedford, M.T., 2007. The arginine methyltransferase CARM1 regulates the coupling of transcription and mRNA processing. Mol. Cell 25, 71–83.

Chestier, A., Yaniv, M., 1979. Rapid turnover of acetyl groups in the four core histones of simian virus 40 minichromosomes. Proc. Natl. Acad. Sci. USA 76, 46–50.

Cheung, P., Tanner, K.G., Cheung, W.L., Sassone-Corsi, P., Denu, J.M., Allis, C.D., 2000. Synergistic coupling of histone H3 phosphorylation and acetylation in response to epidermal growth factor stimulation. Mol. Cell 5, 905–915.

Chiang, Y.C., Komarnitsky, P., Chase, D., Denis, C.L., 1996. ADR1 activation domains contact the histone acetyltransferase GCN5 and the core transcriptional factor TFIIB. J. Biol. Chem. 271, 32359–32365.

Chicoine, L.G., Schulman, I.G., Richman, R., Cook, R.G., Allis, C.D., 1986. Nonrandom utilization of acetylation sites in histones isolated from Tetrahymena. Evidence for functionally distinct H4 acetylation sites. J. Biol. Chem. 261, 1071–1076.

Choudhary, C., Weinert, B.T., Nishida, Y., Verdin, E., Mann, M., 2014. The growing landscape of lysine acetylation links metabolism and cell signalling. Nat. Rev. Mol. Cell Biol. 15, 536–550.

Christensen, M.E., Rattner, J.B., Dixon, G.H., 1984. Hyperacetylation of histone H4 promotes chromatin decondensation prior to histone replacement by protamines during spermatogenesis in rainbow trout. Nucleic Acids Res. 12, 4575–4592.

Chuikov, S., Kurash, J.K., Wilson, J.R., Xiao, B., Justin, N., Ivanov, G.S., McKinney, K., Tempst, P., Prives, C., Gamblin, S.J., et al., 2004. Regulation of p53 activity through lysine methylation. Nature 432, 353–360.

Ciechanover, A., Finley, D., Varshavsky, A., 1984. Ubiquitin dependence of selective protein degradation demonstrated in the mammalian cell cycle mutant ts85. Cell 37, 57–66.

Cieniewicz, A.M., Moreland, L., Ringel, A.E., Mackintosh, S.G., Raman, A., Gilbert, T.M., Wolberger, C., Tackett, A.J., Taverna, S.D., 2014. The bromodomain of Gcn5 regulates site specificity of lysine acetylation on histone H3. Mol. Cell. Proteomics 13, 2896–2910.

Ciurciu, A., Komonyi, O., Pankotai, T., Boros, I.M., 2006. The Drosophila histone acetyltransferase Gcn5 and transcriptional adaptor Ada2a are involved in nucleosomal histone H4 acetylation. Mol. Cell. Biol. 26, 9413–9423.

Clarke, A.S., Lowell, J.E., Jacobson, S.J., Pillus, L., 1999. Esa1p is an essential histone acetyltransferase required for cell cycle progression. Mol. Cell. Biol. 19, 2515–2526.

Cloos, P.A., Christensen, J., Agger, K., Maiolica, A., Rappsilber, J., Antal, T., Hansen, K.H., Helin, K., 2006. The putative oncogene GASC1 demethylates tri- and dimethylated lysine 9 on histone H3. Nature 442, 307–311.

Clouaire, T., Webb, S., Skene, P., Illingworth, R., Kerr, A., Andrews, R., Lee, J.H., Skalnik, D., Bird, A., 2012. Cfp1 integrates both CpG content and gene activity for accurate H3K4me3 deposition in embryonic stem cells. Genes Dev. 26, 1714–1728.

Clouaire, T., Webb, S., Bird, A., 2014. Cfp1 is required for gene expression-dependent H3K4 trimethylation and H3K9 acetylation in embryonic stem cells. Genome Biol. 15, 451.

Collins, R., Cheng, X., 2010. A case study in cross-talk: the histone lysine methyltransferases G9a and GLP. Nucleic Acids Res. 38, 3503–3511.

Collins, R.E., Tachibana, M., Tamaru, H., Smith, K.M., Jia, D., Zhang, X., Selker, E.U., Shinkai, Y., Cheng, X., 2005. In vitro and in vivo analyses of a Phe/Tyr switch controlling product specificity of histone lysine methyltransferases. J. Biol. Chem. 280, 5563–5570.

Comb, D.G., Sarkar, N., Pinzino, C.J., 1966. The methylation of lysine residues in protein. J. Biol. Chem. 241, 1857–1862.

Cook, P.J., Ju, B.G., Telese, F., Wang, X., Glass, C.K., Rosenfeld, M.G., 2009. Tyrosine dephosphorylation of H2AX modulates apoptosis and survival decisions. Nature 458, 591–596.

Corvalan, A.Z., Coller, H.A., 2021. Methylation of histone 4's lysine 20: a critical analysis of the state of the field. Physiol. Genomics 53, 22–32.

Cote, J., Richard, S., 2005. Tudor domains bind symmetrical dimethylated arginines. J. Biol. Chem. 280, 28476–28483.
Couppez, M., Martin-Ponthieu, A., Sautiere, P., 1987. Histone H4 from cuttlefish testis is sequentially acetylated. Comparison with acetylation of calf thymus histone H4. J. Biol. Chem. 262, 2854–2860.
Couture, J.F., Collazo, E., Brunzelle, J.S., Trievel, R.C., 2005. Structural and functional analysis of SET8, a histone H4 Lys-20 methyltransferase. Genes Dev. 19, 1455–1465.
Couture, J.F., Dirk, L.M., Brunzelle, J.S., Houtz, R.L., Trievel, R.C., 2008. Structural origins for the product specificity of SET domain protein methyltransferases. Proc. Natl. Acad. Sci. USA 105, 20659–20664.
Covault, J., Chalkley, R., 1980. The identification of distinct populations of acetylated histone. J. Biol. Chem. 255, 9110–9116.
Creyghton, M.P., Cheng, A.W., Welstead, G.G., Kooistra, T., Carey, B.W., Steine, E.J., Hanna, J., Lodato, M.A., Frampton, G.M., Sharp, P.A., et al., 2010. Histone H3K27ac separates active from poised enhancers and predicts developmental state. Proc. Natl. Acad. Sci. USA 107, 21931–21936.
Crump, N.T., Milne, T.A., 2019. Why are so many MLL lysine methyltransferases required for normal mammalian development? Cell. Mol. Life Sci. 76, 2885–2898.
Cruz, C., Della Rosa, M., Krueger, C., Gao, Q., Horkai, D., King, M., Field, L., Houseley, J., 2018. Tri-methylation of histone H3 lysine 4 facilitates gene expression in ageing cells. elife 7, e34081.
Cuthbert, G.L., Daujat, S., Snowden, A.W., Erdjument-Bromage, H., Hagiwara, T., Yamada, M., Schneider, R., Gregory, P.D., Tempst, P., Bannister, A.J., et al., 2004. Histone deimination antagonizes arginine methylation. Cell 118, 545–553.
Czermin, B., Melfi, R., McCabe, D., Seitz, V., Imhof, A., Pirrotta, V., 2002. Drosophila enhancer of Zeste/ESC complexes have a histone H3 methyltransferase activity that marks chromosomal polycomb sites. Cell 111, 185–196.
Daniel, J.A., Torok, M.S., Sun, Z.W., Schieltz, D., Allis, C.D., Yates 3rd, J.R., Grant, P.A., 2004. Deubiquitination of histone H2B by a yeast acetyltransferase complex regulates transcription. J. Biol. Chem. 279, 1867–1871.
D'Arcy, S., Luger, K., 2011. Understanding histone acetyltransferase Rtt109 structure and function: how many chaperones does it take? Curr. Opin. Struct. Biol. 21, 728–734.
Das, C., Lucia, M.S., Hansen, K.C., Tyler, J.K., 2009. CBP/p300-mediated acetylation of histone H3 on lysine 56. Nature 459, 113–117.
Dasso, M., Dimitrov, S., Wolffe, A.P., 1994. Nuclear assembly is independent of linker histones. Proc. Natl. Acad. Sci. USA 91, 12477–12481.
Daujat, S., Bauer, U.M., Shah, V., Turner, B., Berger, S., Kouzarides, T., 2002. Crosstalk between CARM1 methylation and CBP acetylation on histone H3. Curr. Biol. 12, 2090–2097.
Davie, J.R., Candido, E.P., 1978. Acetylated histone H4 is preferentially associated with template-active chromatin. Proc. Natl. Acad. Sci. USA 75, 3574–3577.
Davie, J.R., Candido, E.P., 1980. DNase I sensitive chromatin is enriched in the acetylated species of histone H4. FEBS Lett. 110, 164–168.
Davie, J.K., Trumbly, R.J., Dent, S.Y., 2002. Histone-dependent association of Tup1-Ssn6 with repressed genes in vivo. Mol. Cell. Biol. 22, 693–703.
Davie, J.K., Edmondson, D.G., Coco, C.B., Dent, S.Y., 2003. Tup1-Ssn6 interacts with multiple class I histone deacetylases in vivo. J. Biol. Chem. 278, 50158–50162.
Dawson, M.A., Bannister, A.J., Gottgens, B., Foster, S.D., Bartke, T., Green, A.R., Kouzarides, T., 2009. JAK2 phosphorylates histone H3Y41 and excludes HP1alpha from chromatin. Nature 461, 819–822.
De Souza, C.P., Osmani, A.H., Wu, L.P., Spotts, J.L., Osmani, S.A., 2000. Mitotic histone H3 phosphorylation by the NIMA kinase in Aspergillus nidulans. Cell 102, 293–302.
Dehe, P.M., Pamblanco, M., Luciano, P., Lebrun, R., Moinier, D., Sendra, R., Verreault, A., Tordera, V., Geli, V., 2005. Histone H3 lysine 4 mono-methylation does not require ubiquitination of histone H2B. J. Mol. Biol. 353, 477–484.
Dehe, P.M., Dichtl, B., Schaft, D., Roguev, A., Pamblanco, M., Lebrun, R., Rodriguez-Gil, A., Mkandawire, M., Landsberg, K., Shevchenko, A., et al., 2006. Protein interactions within the Set1 complex and their roles in the regulation of histone 3 lysine 4 methylation. J. Biol. Chem. 281, 35404–35412.
Del Rizzo, P.A., Couture, J.F., Dirk, L.M., Strunk, B.S., Roiko, M.S., Brunzelle, J.S., Houtz, R.L., Trievel, R.C., 2010. SET7/9 catalytic mutants reveal the role of active site water molecules in lysine multiple methylation. J. Biol. Chem. 285, 31849–31858.
DeLange, R.J., Smith, E.L., 1971. Histones: structure and function. Annu. Rev. Biochem. 40, 279–314.
DeLange, R.J., Fambrough, D.M., Smith, E.L., Bonner, J., 1969. Calf and pea histone IV. II. The complete amino acid sequence of calf thymus histone IV; presence of epsilon-N-acetyllysine. J. Biol. Chem. 244, 319–334.

DeLange, R.J., Hooper, J.A., Smith, E.L., 1973. Histone 3. 3. Sequence studies on the cyanogen bromide peptides; complete amino acid sequence of calf thymus histone 3. J. Biol. Chem. 248, 3261–3274.

Dhalluin, C., Carlson, J.E., Zeng, L., He, C., Aggarwal, A.K., Zhou, M.M., 1999. Structure and ligand of a histone acetyltransferase bromodomain. Nature 399, 491–496.

Di Pierro, M., Cheng, R.R., Lieberman Aiden, E., Wolynes, P.G., Onuchic, J.N., 2017. De novo prediction of human chromosome structures: epigenetic marking patterns encode genome architecture. Proc. Natl. Acad. Sci. USA 114, 12126–12131.

Di Tullio, F., Schwarz, M., Zorgati, H., Mzoughi, S., Guccione, E., 2022. The duality of PRDM proteins: epigenetic and structural perspectives. FEBS J. 289, 1256–1275.

Diaz-Santin, L.M., Lukoyanova, N., Aciyan, E., Cheung, A.C., 2017. Cryo-EM structure of the SAGA and NuA4 coactivator subunit Tra1 at 3.7 angstrom resolution. elife 6, e28384.

DiBello, E., Noce, B., Fioravanti, R., Mai, A., 2022. Current HDAC inhibitors in clinical trials. Chimia 76, 448–453.

Dilworth, F.J., Fromental-Ramain, C., Yamamoto, K., Chambon, P., 2000. ATP-driven chromatin remodeling activity and histone acetyltransferases act sequentially during transactivation by RAR/RXR in vitro. Mol. Cell 6, 1049–1058.

Ding, Z., Gillespie, L.L., Paterno, G.D., 2003. Human MI-ER1 alpha and beta function as transcriptional repressors by recruitment of histone deacetylase 1 to their conserved ELM2 domain. Mol. Cell. Biol. 23, 250–258.

Dion, M.F., Altschuler, S.J., Wu, L.F., Rando, O.J., 2005. Genomic characterization reveals a simple histone H4 acetylation code. Proc. Natl. Acad. Sci. USA 102, 5501–5506.

Doi, M., Hirayama, J., Sassone-Corsi, P., 2006. Circadian regulator CLOCK is a histone acetyltransferase. Cell 125, 497–508.

Donczew, R., Warfield, L., Pacheco, D., Erijman, A., Hahn, S., 2020. Two roles for the yeast transcription coactivator SAGA and a set of genes redundantly regulated by TFIID and SAGA. elife 9.

Dou, Y., Gorovsky, M.A., 2000. Phosphorylation of linker histone H1 regulates gene expression in vivo by creating a charge patch. Mol. Cell 6, 225–231.

Dou, Y., Mizzen, C.A., Abrams, M., Allis, C.D., Gorovsky, M.A., 1999. Phosphorylation of linker histone H1 regulates gene expression in vivo by mimicking H1 removal. Mol. Cell 4, 641–647.

Dover, J., Schneider, J., Tawiah-Boateng, M.A., Wood, A., Dean, K., Johnston, M., Shilatifard, A., 2002. Methylation of histone H3 by COMPASS requires ubiquitination of histone H2B by Rad6. J. Biol. Chem. 277, 28368–28371.

Downey, M., 2021. Non-histone protein acetylation by the evolutionarily conserved GCN5 and PCAF acetyltransferases. Biochim. Biophys. Acta Gene Regul. Mech. 1864, 194608.

Downey, M., Johnson, J.R., Davey, N.E., Newton, B.W., Johnson, T.L., Galaang, S., Seller, C.A., Krogan, N., Toczyski, D.P., 2015. Acetylome profiling reveals overlap in the regulation of diverse processes by sirtuins, gcn5, and esa1. Mol. Cell. Proteomics 14, 162–176.

Doyon, Y., Selleck, W., Lane, W.S., Tan, S., Cote, J., 2004. Structural and functional conservation of the NuA4 histone acetyltransferase complex from yeast to humans. Mol. Cell. Biol. 24, 1884–1896.

Driscoll, R., Hudson, A., Jackson, S.P., 2007. Yeast Rtt109 promotes genome stability by acetylating histone H3 on lysine 56. Science 315, 649–652.

Duan, Z., Zarebski, A., Montoya-Durango, D., Grimes, H.L., Horwitz, M., 2005. Gfi1 coordinates epigenetic repression of p21Cip/WAF1 by recruitment of histone lysine methyltransferase G9a and histone deacetylase 1. Mol. Cell. Biol. 25, 10338–10351.

Duerre, J.A., Wallwork, J.C., Quick, D.P., Ford, K.M., 1977. In vitro studies on the methylation of histones in rat brain nuclei. J. Biol. Chem. 252, 6981–6985.

Durant, M., Pugh, B.F., 2006. Genome-wide relationships between TAF1 and histone acetyltransferases in *Saccharomyces cerevisiae*. Mol. Cell. Biol. 26, 2791–2802.

Durant, M., Pugh, B.F., 2007. NuA4-directed chromatin transactions throughout the *Saccharomyces cerevisiae* genome. Mol. Cell. Biol. 27, 5327–5335.

Dutnall, R.N., Tafrov, S.T., Sternglanz, R., Ramakrishnan, V., 1998. Structure of the histone acetyltransferase Hat1: a paradigm for the GCN5-related N-acetyltransferase superfamily. Cell 94, 427–438.

Edmunds, J.W., Mahadevan, L.C., Clayton, A.L., 2008. Dynamic histone H3 methylation during gene induction: HYPB/Setd2 mediates all H3K36 trimethylation. EMBO J. 27, 406–420.

Eisenmann, D.M., Dollard, C., Winston, F., 1989. SPT15, the gene encoding the yeast TATA binding factor TFIID, is required for normal transcription initiation in vivo. Cell 58, 1183–1191.

Eisenmann, D.M., Arndt, K.M., Ricupero, S.L., Rooney, J.W., Winston, F., 1992. SPT3 interacts with TFIID to allow normal transcription in *Saccharomyces cerevisiae*. Genes Dev. 6, 1319–1331.

Eisenmann, D.M., Chapon, C., Roberts, S.M., Dollard, C., Winston, F., 1994. The *Saccharomyces cerevisiae* SPT8 gene encodes a very acidic protein that is functionally related to SPT3 and TATA-binding protein. Genetics 137, 647–657.

El-Saafin, F., Devys, D., Johnsen, S.A., Vincent, S.D., Tora, L., 2022. SAGA-dependent histone H2Bub1 deubiquitination is essential for cellular ubiquitin balance during embryonic development. Int. J. Mol. Sci. 23, 7459.

English, C.M., Adkins, M.W., Carson, J.J., Churchill, M.E., Tyler, J.K., 2006. Structural basis for the histone chaperone activity of Asf1. Cell 127, 495–508.

Eom, G.H., Kim, K., Kim, S.M., Kee, H.J., Kim, J.Y., Jin, H.M., Kim, J.R., Kim, J.H., Choe, N., Kim, K.B., et al., 2009. Histone methyltransferase PRDM8 regulates mouse testis steroidogenesis. Biochem. Biophys. Res. Commun. 388, 131–136.

Evertts, A.G., Manning, A.L., Wang, X., Dyson, N.J., Garcia, B.A., Coller, H.A., 2013. H4K20 methylation regulates quiescence and chromatin compaction. Mol. Biol. Cell 24, 3025–3037.

Fang, J., Feng, Q., Ketel, C.S., Wang, H., Cao, R., Xia, L., Erdjument-Bromage, H., Tempst, P., Simon, J.A., Zhang, Y., 2002. Purification and functional characterization of SET8, a nucleosomal histone H4-lysine 20-specific methyltransferase. Curr. Biol. 12, 1086–1099.

Faucher, D., Wellinger, R.J., 2010. Methylated H3K4, a transcription-associated histone modification, is involved in the DNA damage response pathway. PLoS Genet. 6, e1001082.

Fazly, A., Li, Q., Hu, Q., Mer, G., Horazdovsky, B., Zhang, Z., 2012. Histone chaperone Rtt106 promotes nucleosome formation using (H3-H4)2 tetramers. J. Biol. Chem. 287, 10753–10760.

Feller, C., Forne, I., Imhof, A., Becker, P.B., 2015. Global and specific responses of the histone acetylome to systematic perturbation. Mol. Cell 57, 559–571.

Feng, Q., Wang, H., Ng, H.H., Erdjument-Bromage, H., Tempst, P., Struhl, K., Zhang, Y., 2002. Methylation of H3-lysine 79 is mediated by a new family of HMTases without a SET domain. Curr. Biol. 12, 1052–1058.

Ferrari, R., de Llobet Cucalon, L.I., Di Vona, C., Le Dilly, F., Vidal, E., Lioutas, A., Oliete, J.Q., Jochem, L., Cutts, E., Dieci, G., et al., 2020. TFIIIC binding to Alu elements controls gene expression via chromatin looping and histone acetylation. Mol. Cell 77, 475–487.e411.

Filion, G.J., van Bemmel, J.G., Braunschweig, U., Talhout, W., Kind, J., Ward, L.D., Brugman, W., de Castro, I.J., Kerkhoven, R.M., Bussemaker, H.J., et al., 2010. Systematic protein location mapping reveals five principal chromatin types in Drosophila cells. Cell 143, 212–224.

Fillingham, J., Recht, J., Silva, A.C., Suter, B., Emili, A., Stagljar, I., Krogan, N.J., Allis, C.D., Keogh, M.C., Greenblatt, J.-F., 2008. Chaperone control of the activity and specificity of the histone H3 acetyltransferase Rtt109. Mol. Cell. Biol. 28, 4342–4353.

Fingerman, I.M., Li, H.C., Briggs, S.D., 2007. A charge-based interaction between histone H4 and Dot1 is required for H3K79 methylation and telomere silencing: identification of a new trans-histone pathway. Genes Dev. 21, 2018–2029.

Finley, D., Ciechanover, A., Varshavsky, A., 1984. Thermolability of ubiquitin-activating enzyme from the mammalian cell cycle mutant ts85. Cell 37, 43–55.

Fischle, W., Wang, Y., Jacobs, S.A., Kim, Y., Allis, C.D., Khorasanizadeh, S., 2003. Molecular basis for the discrimination of repressive methyl-lysine marks in histone H3 by polycomb and HP1 chromodomains. Genes Dev. 17, 1870–1881.

Fischle, W., Tseng, B.S., Dormann, H.L., Ueberheide, B.M., Garcia, B.A., Shabanowitz, J., Hunt, D.F., Funabiki, H., Allis, C.D., 2005. Regulation of HP1-chromatin binding by histone H3 methylation and phosphorylation. Nature 438, 1116–1122.

Fishburn, J., Mohibullah, N., Hahn, S., 2005. Function of a eukaryotic transcription activator during the transcription cycle. Mol. Cell 18, 369–378.

Fodor, B.D., Kubicek, S., Yonezawa, M., O'Sullivan, R.J., Sengupta, R., Perez-Burgos, L., Opravil, S., Mechtler, K., Schotta, G., Jenuwein, T., 2006. Jmjd2b antagonizes H3K9 trimethylation at pericentric heterochromatin in mammalian cells. Genes Dev. 20, 1557–1562.

Fournier, M., Orpinell, M., Grauffel, C., Scheer, E., Garnier, J.M., Ye, T., Chavant, V., Joint, M., Esashi, F., Dejaegere, A., et al., 2016. KAT2A/KAT2B-targeted acetylome reveals a role for PLK4 acetylation in preventing centrosome amplification. Nat. Commun. 7, 13227.

Frye, R.A., 1999. Characterization of five human cDNAs with homology to the yeast SIR2 gene: Sir2-like proteins (sirtuins) metabolize NAD and may have protein ADP-ribosyltransferase activity. Biochem. Biophys. Res. Commun. 260, 273–279.

Fu, K., Bonora, G., Pellegrini, M., 2020. Interactions between core histone marks and DNA methyltransferases predict DNA methylation patterns observed in human cells and tissues. Epigenetics 15, 272–282.

Fuchs, S.M., Strahl, B.D., 2011. Antibody recognition of histone post-translational modifications: emerging issues and future prospects. Epigenomics 3, 247–249.

Fuchs, S.M., Krajewski, K., Baker, R.W., Miller, V.L., Strahl, B.D., 2011. Influence of combinatorial histone modifications on antibody and effector protein recognition. Curr. Biol. 21, 53–58.

Fujiki, R., Hashiba, W., Sekine, H., Yokoyama, A., Chikanishi, T., Ito, S., Imai, Y., Kim, J., He, H.H., Igarashi, K., et al., 2011. GlcNAcylation of histone H2B facilitates its monoubiquitination. Nature 480, 557–560.

Fujimoto, D., Segawa, K., 1973. Enzymatic deacetylation of f2a2 histone. FEBS Lett. 32, 59–61.

Fujisawa, T., Filippakopoulos, P., 2017. Functions of bromodomain-containing proteins and their roles in homeostasis and cancer. Nat. Rev. Mol. Cell Biol. 18, 246–262.

Fulton, M.D., Cao, M., Ho, M.C., Zhao, X., Zheng, Y.G., 2021. The macromolecular complexes of histones affect protein arginine methyltransferase activities. J. Biol. Chem. 297, 101123.

Furuhashi, H., Takasaki, T., Rechtsteiner, A., Li, T., Kimura, H., Checchi, P.M., Strome, S., Kelly, W.G., 2010. Trans-generational epigenetic regulation of *C. elegans* primordial germ cells. Epigenetics Chromatin 3, 15.

Ganapathi, M., Boles, N.C., Charniga, C., Lotz, S., Campbell, M., Temple, S., Morse, R.H., 2018. Effect of Bmi1 over-expression on gene expression in adult and embryonic murine neural stem cells. Sci. Rep. 8, 7464.

Gao, Y., Wells, L., Comer, F.I., Parker, G.J., Hart, G.W., 2001. Dynamic O-glycosylation of nuclear and cytosolic proteins: cloning and characterization of a neutral, cytosolic beta-N-acetylglucosaminidase from human brain. J. Biol. Chem. 276, 9838–9845.

Garcea, R.L., Alberts, B.M., 1980. Comparative studies of histone acetylation in nucleosomes, nuclei, and intact cells. Evidence for special factors which modify acetylase action. J. Biol. Chem. 255, 11454–11463.

Gartenberg, M.R., Smith, J.S., 2016. The nuts and bolts of transcriptionally silent chromatin in *Saccharomyces cerevisiae*. Genetics 203, 1563–1599.

Gary, J.D., Lin, W.J., Yang, M.C., Herschman, H.R., Clarke, S., 1996. The predominant protein-arginine methyltransferase from *Saccharomyces cerevisiae*. J. Biol. Chem. 271, 12585–12594.

Georgakopoulos, T., Thireos, G., 1992. Two distinct yeast transcriptional activators require the function of the GCN5 protein to promote normal levels of transcription. EMBO J. 11, 4145–4152.

Georgakopoulos, P., Lockington, R.A., Kelly, J.M., 2013. The Spt-Ada-Gcn5 acetyltransferase (SAGA) complex in Aspergillus nidulans. PLoS One 8, e65221.

Gershey, E.L., Vidali, G., Allfrey, V.G., 1968. Chemical studies of histone acetylation. The occurrence of epsilon-N-acetyllysine in the f2a1 histone. J. Biol. Chem. 243, 5018–5022.

Giacinti, C., Giordano, A., 2006. RB and cell cycle progression. Oncogene 25, 5220–5227.

Giannattasio, M., Lazzaro, F., Plevani, P., Muzi-Falconi, M., 2005. The DNA damage checkpoint response requires histone H2B ubiquitination by Rad6-Bre1 and H3 methylation by Dot1. J. Biol. Chem. 280, 9879–9886.

Gilbert, T.M., McDaniel, S.L., Byrum, S.D., Cades, J.A., Dancy, B.C., Wade, H., Tackett, A.J., Strahl, B.D., Taverna, S.D., 2014. A PWWP domain-containing protein targets the NuA3 acetyltransferase complex via histone H3 lysine 36 trimethylation to coordinate transcriptional elongation at coding regions. Mol. Cell. Proteomics 13, 2883–2895.

Gill, G., Ptashne, M., 1988. Negative effect of the transcriptional activator GAL4. Nature 334, 721–724.

Goldknopf, I.L., Busch, H., 1977. Isopeptide linkage between nonhistone and histone 2A polypeptides of chromosomal conjugate-protein A24. Proc. Natl. Acad. Sci. USA 74, 864–868.

Goldknopf, I.L., Taylor, C.W., Baum, R.M., Yeoman, L.C., Olson, M.O., Prestayko, A.W., Busch, H., 1975. Isolation and characterization of protein A24, a "histone-like" non-histone chromosomal protein. J. Biol. Chem. 250, 7182–7187.

Goldknopf, I.L., French, M.F., Musso, R., Busch, H., 1977. Presence of protein A24 in rat liver nucleosomes. Proc. Natl. Acad. Sci. USA 74, 5492–5495.

Goldstein, G., Scheid, M., Hammerling, U., Schlesinger, D.H., Niall, H.D., Boyse, E.A., 1975. Isolation of a polypeptide that has lymphocyte-differentiating properties and is probably represented universally in living cells. Proc. Natl. Acad. Sci. USA 72, 11–15.

Gordan, R., Hartemink, A.J., Bulyk, M.L., 2009. Distinguishing direct versus indirect transcription factor-DNA interactions. Genome Res. 19, 2090–2100.

Gorovsky, M.A., Pleger, G.L., Keevert, J.B., Johmann, C.A., 1973. Studies on histone fraction F2A1 in macro- and micronuclei of *Tetrahymena pyriformis*. J. Cell Biol. 57, 773–781.

Gottesfeld, J.M., Butler, P.J., 1977. Structure of transcriptionally-active chromatin subunits. Nucleic Acids Res. 4, 3155–3173.

Gottesfeld, J.M., Garrard, W.T., Bagi, G., Wilson, R.F., Bonner, J., 1974. Partial purification of the template-active fraction of chromatin: a preliminary report. Proc. Natl. Acad. Sci. USA 71, 2193–2197.

Grant, P.A., Duggan, L., Cote, J., Roberts, S.M., Brownell, J.E., Candau, R., Ohba, R., Owen-Hughes, T., Allis, C.D., Winston, F., et al., 1997. Yeast Gcn5 functions in two multisubunit complexes to acetylate nucleosomal histones: characterization of an Ada complex and the SAGA (Spt/Ada) complex. Genes Dev. 11, 1640–1650.

Grant, P.A., Schieltz, D., Pray-Grant, M.G., Steger, D.J., Reese, J.C., Yates 3rd, J.R., Workman, J.L., 1998a. A subset of TAF(II)s are integral components of the SAGA complex required for nucleosome acetylation and transcriptional stimulation. Cell 94, 45–53.

Grant, P.A., Schieltz, D., Pray-Grant, M.G., Yates 3rd, J.R., Workman, J.L., 1998b. The ATM-related cofactor Tra1 is a component of the purified SAGA complex. Mol. Cell 2, 863–867.

Grant, P.A., Winston, F., Berger, S.L., 2021. The biochemical and genetic discovery of the SAGA complex. Biochim. Biophys. Acta Gene Regul. Mech. 1864, 194669.

Green, G.R., Poccia, D.L., 1985. Phosphorylation of sea urchin sperm H1 and H2B histones precedes chromatin decondensation and H1 exchange during pronuclear formation. Dev. Biol. 108, 235–245.

Greer, E.L., Shi, Y., 2012. Histone methylation: a dynamic mark in health, disease and inheritance. Nat. Rev. Genet. 13, 343–357.

Gross, D.S., Garrard, W.T., 1988. Nuclease hypersensitive site in chromatin. Annu. Rev. Biochem. 57, 159–197.

Grozinger, C.M., Hassig, C.A., Schreiber, S.L., 1999. Three proteins define a class of human histone deacetylases related to yeast Hda1p. Proc. Natl. Acad. Sci. USA 96, 4868–4873.

Gu, W., Roeder, R.G., 1997. Activation of p53 sequence-specific DNA binding by acetylation of the p53 C-terminal domain. Cell 90, 595–606.

Guarente, L., 1995. Transcriptional coactivators in yeast and beyond. Trends Biochem. Sci. 20, 517–521.

Guccione, E., Bassi, C., Casadio, F., Martinato, F., Cesaroni, M., Schuchlautz, H., Luscher, B., Amati, B., 2007. Methylation of histone H3R2 by PRMT6 and H3K4 by an MLL complex are mutually exclusive. Nature 449, 933–937.

Guelman, S., Kozuka, K., Mao, Y., Pham, V., Solloway, M.J., Wang, J., Wu, J., Lill, J.R., Zha, J., 2009. The double-histone-acetyltransferase complex ATAC is essential for mammalian development. Mol. Cell. Biol. 29, 1176–1188.

Guenther, M.G., Lane, W.S., Fischle, W., Verdin, E., Lazar, M.A., Shiekhattar, R., 2000. A core SMRT corepressor complex containing HDAC3 and TBL1, a WD40-repeat protein linked to deafness. Genes Dev. 14, 1048–1057.

Guenther, M.G., Barak, O., Lazar, M.A., 2001. The SMRT and N-CoR corepressors are activating cofactors for histone deacetylase 3. Mol. Cell. Biol. 21, 6091–6101.

Gurley, L.R., D'Anna, J.A., Barham, S.S., Deaven, L.L., Tobey, R.A., 1978. Histone phosphorylation and chromatin structure during mitosis in Chinese hamster cells. Eur. J. Biochem. 84, 1–15.

Guttman, M., Amit, I., Garber, M., French, C., Lin, M.F., Feldser, D., Huarte, M., Zuk, O., Carey, B.W., Cassady, J.P., et al., 2009. Chromatin signature reveals over a thousand highly conserved large non-coding RNAs in mammals. Nature 458, 223–227.

Hadjikyriacou, A., Yang, Y., Espejo, A., Bedford, M.T., Clarke, S.G., 2015. Unique features of human protein arginine methyltransferase 9 (PRMT9) and its substrate RNA splicing factor SF3B2. J. Biol. Chem. 290, 16723–16743.

Hagiwara, T., Nakashima, K., Hirano, H., Senshu, T., Yamada, M., 2002. Deimination of arginine residues in nucleophosmin/B23 and histones in HL-60 granulocytes. Biochem. Biophys. Res. Commun. 290, 979–983.

Hakimi, M.A., Bochar, D.A., Chenoweth, J., Lane, W.S., Mandel, G., Shiekhattar, R., 2002. A core-BRAF35 complex containing histone deacetylase mediates repression of neuronal-specific genes. Proc. Natl. Acad. Sci. USA 99, 7420–7425.

Hammarsten, O., 1877. Zur Kenntn. d. Caseins, etc. (Uppsala).

Han, J., Zhou, H., Horazdovsky, B., Zhang, K., Xu, R.M., Zhang, Z., 2007. Rtt109 acetylates histone H3 lysine 56 and functions in DNA replication. Science 315, 653–655.

Han, J., Zhang, H., Zhang, H., Wang, Z., Zhou, H., Zhang, Z., 2013. A Cul4 E3 ubiquitin ligase regulates histone hand-off during nucleosome assembly. Cell 155, 817–829.

Han, Y., Luo, J., Ranish, J., Hahn, S., 2014. Architecture of the *Saccharomyces cerevisiae* SAGA transcription coactivator complex. EMBO J. 33, 2534–2546.

Hardiville, S., Hart, G.W., 2016. Nutrient regulation of gene expression by O-GlcNAcylation of chromatin. Curr. Opin. Chem. Biol. 33, 88–94.

Hart, G.W., Housley, M.P., Slawson, C., 2007. Cycling of O-linked beta-N-acetylglucosamine on nucleocytoplasmic proteins. Nature 446, 1017–1022.

Hartzog, G.A., Tamkun, J.W., 2007. A new role for histone tail modifications in transcription elongation. Genes Dev. 21, 3209–3213.

Hassan, A.H., Neely, K.E., Workman, J.L., 2001. Histone acetyltransferase complexes stabilize swi/snf binding to promoter nucleosomes. Cell 104, 817–827.

Hassan, A.H., Prochasson, P., Neely, K.E., Galasinski, S.C., Chandy, M., Carrozza, M.J., Workman, J.L., 2002. Function and selectivity of bromodomains in anchoring chromatin-modifying complexes to promoter nucleosomes. Cell 111, 369–379.

Hassig, C.A., Schreiber, S.L., 1997. Nuclear histone acetylases and deacetylases and transcriptional regulation: HATs off to HDACs. Curr. Opin. Chem. Biol. 1, 300–308.

Hassig, C.A., Fleischer, T.C., Billin, A.N., Schreiber, S.L., Ayer, D.E., 1997. Histone deacetylase activity is required for full transcriptional repression by mSin3A. Cell 89, 341–347.

Hay, C.W., Candido, E.P., 1983. Histone deacetylase. Association with a nuclease resistant, high molecular weight fraction of HeLa cell chromatin. J. Biol. Chem. 258, 3726–3734.

Hayashi, K., Yoshida, K., Matsui, Y., 2005. A histone H3 methyltransferase controls epigenetic events required for meiotic prophase. Nature 438, 374–378.

Haynes, S.R., Dollard, C., Winston, F., Beck, S., Trowsdale, J., Dawid, I.B., 1992. The bromodomain: a conserved sequence found in human, Drosophila and yeast proteins. Nucleic Acids Res. 20, 2603.

Heard, E., Rougeulle, C., Arnaud, D., Avner, P., Allis, C.D., Spector, D.L., 2001. Methylation of histone H3 at Lys-9 is an early mark on the X chromosome during X inactivation. Cell 107, 727–738.

Hebbes, T.R., Thorne, A.W., Crane-Robinson, C., 1988. A direct link between core histone acetylation and transcriptionally active chromatin. EMBO J. 7, 1395–1402.

Hebbes, T.R., Clayton, A.L., Thorne, A.W., Crane-Robinson, C., 1994. Core histone hyperacetylation co-maps with generalized DNase I sensitivity in the chicken beta-globin chromosomal domain. EMBO J. 13, 1823–1830.

Heintzman, N.D., Stuart, R.K., Hon, G., Fu, Y., Ching, C.W., Hawkins, R.D., Barrera, L.O., Van Calcar, S., Qu, C., Ching, K.A., et al., 2007. Distinct and predictive chromatin signatures of transcriptional promoters and enhancers in the human genome. Nat. Genet. 39, 311–318.

Heintzman, N.D., Hon, G.C., Hawkins, R.D., Kheradpour, P., Stark, A., Harp, L.F., Ye, Z., Lee, L.K., Stuart, R.K., Ching, C.W., et al., 2009. Histone modifications at human enhancers reflect global cell-type-specific gene expression. Nature 459, 108–112.

Heinzel, T., Lavinsky, R.M., Mullen, T.M., Soderstrom, M., Laherty, C.D., Torchia, J., Yang, W.M., Brard, G., Ngo, S.D., Davie, J.R., et al., 1997. A complex containing N-CoR, mSin3 and histone deacetylase mediates transcriptional repression. Nature 387, 43–48.

Helmlinger, D., Tora, L., 2017. Sharing the SAGA. Trends Biochem. Sci. 42, 850–861.

Helmlinger, D., Marguerat, S., Villen, J., Swaney, D.L., Gygi, S.P., Bahler, J., Winston, F., 2011. Tra1 has specific regulatory roles, rather than global functions, within the SAGA co-activator complex. EMBO J. 30, 2843–2852.

Helmlinger, D., Papai, G., Devys, D., Tora, L., 2021. What do the structures of GCN5-containing complexes teach us about their function? Biochim. Biophys. Acta Gene Regul. Mech. 1864, 194614.

Hempel, K., Lange, H.W., 1968. Nepsilon-methylated lysine in histones from chicken erythrocytes. Hoppe Seylers Z. Physiol. Chem. 349, 603–607.

Hempel, K., Lange, H.W., Birkofer, L., 1968. Epsilon-N-trimethyllysine, a new amino acid in histones. Naturwissenschaften 55, 37.

Hempel, K., Thomas, G., Roos, G., Stocker, W., Lange, H.W., 1979. N epsilon-methyl groups on the lysine residues in histones turn over independently of the polypeptide backbone. Hoppe Seylers Z. Physiol. Chem. 360, 869–876.

Hendrich, B., Bird, A., 1998. Identification and characterization of a family of mammalian methyl-CpG binding proteins. Mol. Cell. Biol. 18, 6538–6547.

Hendrich, B., Tweedie, S., 2003. The methyl-CpG binding domain and the evolving role of DNA methylation in animals. Trends Genet. 19, 269–277.

Hendzel, M.J., Wei, Y., Mancini, M.A., Van Hooser, A., Ranalli, T., Brinkley, B.R., Bazett-Jones, D.P., Allis, C.D., 1997. Mitosis-specific phosphorylation of histone H3 initiates primarily within pericentromeric heterochromatin during G2 and spreads in an ordered fashion coincident with mitotic chromosome condensation. Chromosoma 106, 348–360.

Henikoff, S., 2005. Histone modifications: combinatorial complexity or cumulative simplicity? Proc. Natl. Acad. Sci. USA 102, 5308–5309.

Henikoff, S., Shilatifard, A., 2011. Histone modification: cause or cog? Trends Genet. 27, 389–396.
Henry, M.F., Silver, P.A., 1996. A novel methyltransferase (Hmt1p) modifies poly(A)+-RNA-binding proteins. Mol. Cell. Biol. 16, 3668–3678.
Henry, K.W., Wyce, A., Lo, W.S., Duggan, L.J., Emre, N.C., Kao, C.F., Pillus, L., Shilatifard, A., Osley, M.A., Berger, S.-L., 2003. Transcriptional activation via sequential histone H2B ubiquitylation and deubiquitylation, mediated by SAGA-associated Ubp8. Genes Dev. 17, 2648–2663.
Herz, H.M., Mohan, M., Garruss, A.S., Liang, K., Takahashi, Y.H., Mickey, K., Voets, O., Verrijzer, C.P., Shilatifard, A., 2012. Enhancer-associated H3K4 monomethylation by Trithorax-related, the Drosophila homolog of mammalian Mll3/Mll4. Genes Dev. 26, 2604–2620.
Herz, H.M., Garruss, A., Shilatifard, A., 2013. SET for life: biochemical activities and biological functions of SET domain-containing proteins. Trends Biochem. Sci. 38, 621–639.
Hilfiker, A., Hilfiker-Kleiner, D., Pannuti, A., Lucchesi, J.C., 1997. mof, a putative acetyl transferase gene related to the Tip60 and MOZ human genes and to the SAS genes of yeast, is required for dosage compensation in Drosophila. EMBO J. 16, 2054–2060.
Hochstrasser, M., 2009. Origin and function of ubiquitin-like proteins. Nature 458, 422–429.
Hochwagen, A., Marais, G.A., 2010. Meiosis: a PRDM9 guide to the hotspots of recombination. Curr. Biol. 20, R271–R274.
Hodawadekar, S.C., Marmorstein, R., 2007. Chemistry of acetyl transfer by histone modifying enzymes: structure, mechanism and implications for effector design. Oncogene 26, 5528–5540.
Hohenauer, T., Moore, A.W., 2012. The Prdm family: expanding roles in stem cells and development. Development 139, 2267–2282.
Hohmann, P., 1979. Species-specific variations in H1 histone phosphopeptides. J. Biol. Chem. 254, 9022–9029.
Holt, M.T., David, Y., Pollock, S., Tang, Z., Jeon, J., Kim, J., Roeder, R.G., Muir, T.W., 2015. Identification of a functional hotspot on ubiquitin required for stimulation of methyltransferase activity on chromatin. Proc. Natl. Acad. Sci. USA 112, 10365–10370.
Honda, B.M., Candido, P.M., Dixon, G.H., 1975. Histone methylation. Its occurrence in different cell types and relation to histone H4 metabolism in developing trout testis. J. Biol. Chem. 250, 8686–8689.
Horiuchi, J., Silverman, N., Marcus, G.A., Guarente, L., 1995. ADA3, a putative transcriptional adaptor, consists of two separable domains and interacts with ADA2 and GCN5 in a trimeric complex. Mol. Cell. Biol. 15, 1203–1209.
Horiuchi, J., Silverman, N., Pina, B., Marcus, G.A., Guarente, L., 1997. ADA1, a novel component of the ADA/GCN5 complex, has broader effects than GCN5, ADA2, or ADA3. Mol. Cell. Biol. 17, 3220–3228.
Horlein, A.J., Naar, A.M., Heinzel, T., Torchia, J., Gloss, B., Kurokawa, R., Ryan, A., Kamei, Y., Soderstrom, M., Glass, C.K., et al., 1995. Ligand-independent repression by the thyroid hormone receptor mediated by a nuclear receptor co-repressor. Nature 377, 397–404.
Horton, J.R., Upadhyay, A.K., Qi, H.H., Zhang, X., Shi, Y., Cheng, X., 2010. Enzymatic and structural insights for substrate specificity of a family of jumonji histone lysine demethylases. Nat. Struct. Mol. Biol. 17, 38–43.
Houston, S.I., McManus, K.J., Adams, M.M., Sims, J.K., Carpenter, P.B., Hendzel, M.J., Rice, J.C., 2008. Catalytic function of the PR-Set7 histone H4 lysine 20 monomethyltransferase is essential for mitotic entry and genomic stability. J. Biol. Chem. 283, 19478–19488.
Houtkooper, R.H., Pirinen, E., Auwerx, J., 2012. Sirtuins as regulators of metabolism and healthspan. Nat. Rev. Mol. Cell Biol. 13, 225–238.
Howe, F.S., Fischl, H., Murray, S.C., Mellor, J., 2017. Is H3K4me3 instructive for transcription activation? BioEssays 39, 1–12.
Hsieh, Y.J., Kundu, T.K., Wang, Z., Kovelman, R., Roeder, R.G., 1999. The TFIIIC90 subunit of TFIIIC interacts with multiple components of the RNA polymerase III machinery and contains a histone-specific acetyltransferase activity. Mol. Cell. Biol. 19, 7697–7704.
Hsu, J.Y., Sun, Z.W., Li, X., Reuben, M., Tatchell, K., Bishop, D.K., Grushcow, J.M., Brame, C.J., Caldwell, J.A., Hunt, D.F., et al., 2000. Mitotic phosphorylation of histone H3 is governed by Ipl1/aurora kinase and Glc7/PP1 phosphatase in budding yeast and nematodes. Cell 102, 279–291.
Hsu, P.L., Shi, H., Leonen, C., Kang, J., Chatterjee, C., Zheng, N., 2019. Structural basis of H2B ubiquitination-dependent H3K4 methylation by COMPASS. Mol. Cell 76, 712–723.e714.
Hu, E., Chen, Z., Fredrickson, T., Zhu, Y., Kirkpatrick, R., Zhang, G.F., Johanson, K., Sung, C.M., Liu, R., Winkler, J., 2000. Cloning and characterization of a novel human class I histone deacetylase that functions as a transcription repressor. J. Biol. Chem. 275, 15254–15264.

Hu, D., Gao, X., Morgan, M.A., Herz, H.M., Smith, E.R., Shilatifard, A., 2013. The MLL3/MLL4 branches of the COMPASS family function as major histone H3K4 monomethylases at enhancers. Mol. Cell. Biol. 33, 4745–4754.

Huang, N., vom Baur, E., Garnier, J.M., Lerouge, T., Vonesch, J.L., Lutz, Y., Chambon, P., Losson, R., 1998. Two distinct nuclear receptor interaction domains in NSD1, a novel SET protein that exhibits characteristics of both corepressors and coactivators. EMBO J. 17, 3398–3412.

Huang, S., Litt, M., Felsenfeld, G., 2005. Methylation of histone H4 by arginine methyltransferase PRMT1 is essential in vivo for many subsequent histone modifications. Genes Dev. 19, 1885–1893.

Hubbert, C., Guardiola, A., Shao, R., Kawaguchi, Y., Ito, A., Nixon, A., Yoshida, M., Wang, X.F., Yao, T.P., 2002. HDAC6 is a microtubule-associated deacetylase. Nature 417, 455–458.

Huen, M.S., Sy, S.M., van Deursen, J.M., Chen, J., 2008. Direct interaction between SET8 and proliferating cell nuclear antigen couples H4-K20 methylation with DNA replication. J. Biol. Chem. 283, 11073–11077.

Hughes, T.R., Marton, M.J., Jones, A.R., Roberts, C.J., Stoughton, R., Armour, C.D., Bennett, H.A., Coffey, E., Dai, H., He, Y.D., et al., 2000. Functional discovery via a compendium of expression profiles. Cell 102, 109–126.

Huisinga, K.L., Pugh, B.F., 2004. A genome-wide housekeeping role for TFIID and a highly regulated stress-related role for SAGA in *Saccharomyces cerevisiae*. Mol. Cell 13, 573–585.

Humphrey, G.W., Wang, Y., Russanova, V.R., Hirai, T., Qin, J., Nakatani, Y., Howard, B.H., 2001. Stable histone deacetylase complexes distinguished by the presence of SANT domain proteins CoREST/kiaa0071 and Mta-L1. J. Biol. Chem. 276, 6817–6824.

Hunt, L.T., Dayhoff, M.O., 1977. Amino-terminal sequence identity of ubiquitin and the nonhistone component of nuclear protein A24. Biochem. Biophys. Res. Commun. 74, 650–655.

Huyen, Y., Zgheib, O., Ditullio Jr., R.A., Gorgoulis, V.G., Zacharatos, P., Petty, T.J., Sheston, E.A., Mellert, H.S., Stavridi, E.S., Halazonetis, T.D., 2004. Methylated lysine 79 of histone H3 targets 53BP1 to DNA double-strand breaks. Nature 432, 406–411.

Hyland, E.M., Cosgrove, M.S., Molina, H., Wang, D., Pandey, A., Cottee, R.J., Boeke, J.D., 2005. Insights into the role of histone H3 and histone H4 core modifiable residues in *Saccharomyces cerevisiae*. Mol. Cell. Biol. 25, 10060–10070.

Hyllus, D., Stein, C., Schnabel, K., Schiltz, E., Imhof, A., Dou, Y., Hsieh, J., Bauer, U.M., 2007. PRMT6-mediated methylation of R2 in histone H3 antagonizes H3 K4 trimethylation. Genes Dev. 21, 3369–3380.

Hyun, K., Jeon, J., Park, K., Kim, J., 2017. Writing, erasing and reading histone lysine methylations. Exp. Mol. Med. 49, e324.

Iberg, A.N., Espejo, A., Cheng, D., Kim, D., Michaud-Levesque, J., Richard, S., Bedford, M.T., 2008. Arginine methylation of the histone H3 tail impedes effector binding. J. Biol. Chem. 283, 3006–3010.

Iizuka, M., Stillman, B., 1999. Histone acetyltransferase HBO1 interacts with the ORC1 subunit of the human initiator protein. J. Biol. Chem. 274, 23027–23034.

Iizuka, M., Matsui, T., Takisawa, H., Smith, M.M., 2006. Regulation of replication licensing by acetyltransferase Hbo1. Mol. Cell. Biol. 26, 1098–1108.

Ikeda, K., Steger, D.J., Eberharter, A., Workman, J.L., 1999. Activation domain-specific and general transcription stimulation by native histone acetyltransferase complexes. Mol. Cell. Biol. 19, 855–863.

Imai, S., Armstrong, C.M., Kaeberlein, M., Guarente, L., 2000. Transcriptional silencing and longevity protein Sir2 is an NAD-dependent histone deacetylase. Nature 403, 795–800.

Imhof, A., Yang, X.J., Ogryzko, V.V., Nakatani, Y., Wolffe, A.P., Ge, H., 1997. Acetylation of general transcription factors by histone acetyltransferases. Curr. Biol. 7, 689–692.

Ingvarsdottir, K., Krogan, N.J., Emre, N.C., Wyce, A., Thompson, N.J., Emili, A., Hughes, T.R., Greenblatt, J.F., Berger, S.L., 2005. H2B ubiquitin protease Ubp8 and Sgf11 constitute a discrete functional module within the *Saccharomyces cerevisiae* SAGA complex. Mol. Cell. Biol. 25, 1162–1172.

Inoue, A., Fujimoto, D., 1969. Enzymatic deacetylation of histone. Biochem. Biophys. Res. Commun. 36, 146–150.

Inoue, A., Fujimoto, D., 1970. Histone deacetylase from calf thymus. Biochim. Biophys. Acta 220, 307–316.

Inoue, M., Iwai, R., Yamanishi, E., Yamagata, K., Komabayashi-Suzuki, M., Honda, A., Komai, T., Miyachi, H., Kitano, S., Watanabe, C., et al., 2015. Deletion of Prdm8 impairs development of upper-layer neocortical neurons. Genes Cells 20, 758–770.

Jackson, V., Shires, A., Chalkley, R., Granner, D.K., 1975. Studies on highly metabolically active acetylation and phosphorylation of histones. J. Biol. Chem. 250, 4856–4863.

Jackson, J.P., Lindroth, A.M., Cao, X., Jacobsen, S.E., 2002. Control of CpNpG DNA methylation by the KRYPTONITE histone H3 methyltransferase. Nature 416, 556–560.

Jacobs, S.A., Khorasanizadeh, S., 2002. Structure of HP1 chromodomain bound to a lysine 9-methylated histone H3 tail. Science 295, 2080–2083.

Jain, D., Baldi, S., Zabel, A., Straub, T., Becker, P.B., 2015. Active promoters give rise to false positive 'Phantom Peaks' in ChIP-seq experiments. Nucleic Acids Res. 43, 6959–6968.

Jang, S., Kang, C., Yang, H.S., Jung, T., Hebert, H., Chung, K.Y., Kim, S.J., Hohng, S., Song, J.J., 2019. Structural basis of recognition and destabilization of the histone H2B ubiquitinated nucleosome by the DOT1L histone H3 Lys79 methyltransferase. Genes Dev. 33, 620–625.

Jeanmougin, F., Wurtz, J.M., Le Douarin, B., Chambon, P., Losson, R., 1997. The bromodomain revisited. Trends Biochem. Sci. 22, 151–153.

Jeppesen, P., Turner, B.M., 1993. The inactive X chromosome in female mammals is distinguished by a lack of histone H4 acetylation, a cytogenetic marker for gene expression. Cell 74, 281–289.

Ji, J.H., Min, S., Chae, S., Ha, G.H., Kim, Y., Park, Y.J., Lee, C.W., Cho, H., 2019. De novo phosphorylation of H2AX by WSTF regulates transcription-coupled homologous recombination repair. Nucleic Acids Res. 47, 6299–6314.

Jiao, L., Liu, X., 2015. Structural basis of histone H3K27 trimethylation by an active polycomb repressive complex 2. Science 350, aac4383.

John, S., Howe, L., Tafrov, S.T., Grant, P.A., Sternglanz, R., Workman, J.L., 2000. The something about silencing protein, Sas3, is the catalytic subunit of NuA3, a yTAF(II)30-containing HAT complex that interacts with the Spt16 subunit of the yeast CP (Cdc68/Pob3)-FACT complex. Genes Dev. 14, 1196–1208.

Johnson, L.M., Kayne, P.S., Kahn, E.S., Grunstein, M., 1990. Genetic evidence for an interaction between SIR3 and histone H4 in the repression of the silent mating loci in *Saccharomyces cerevisiae*. Proc. Natl. Acad. Sci. USA 87, 6286–6290.

Johnson, C.A., White, D.A., Lavender, J.S., O'Neill, L.P., Turner, B.M., 2002. Human class I histone deacetylase complexes show enhanced catalytic activity in the presence of ATP and co-immunoprecipitate with the ATP-dependent chaperone protein Hsp70. J. Biol. Chem. 277, 9590–9597.

Jones, B., Su, H., Bhat, A., Lei, H., Bajko, J., Hevi, S., Baltus, G.A., Kadam, S., Zhai, H., Valdez, R., et al., 2008. The histone H3K79 methyltransferase Dot1L is essential for mammalian development and heterochromatin structure. PLoS Genet. 4, e1000190.

Jonkers, I., Kwak, H., Lis, J.T., 2014. Genome-wide dynamics of Pol II elongation and its interplay with promoter proximal pausing, chromatin, and exons. elife 3, e02407.

Jung, H.R., Pasini, D., Helin, K., Jensen, O.N., 2010. Quantitative mass spectrometry of histones H3.2 and H3.3 in Suz12-deficient mouse embryonic stem cells reveals distinct, dynamic post-translational modifications at Lys-27 and Lys-36. Mol. Cell. Proteomics 9, 838–850.

Kadosh, D., Struhl, K., 1997. Repression by Ume6 involves recruitment of a complex containing Sin3 corepressor and Rpd3 histone deacetylase to target promoters. Cell 89, 365–371.

Kadosh, D., Struhl, K., 1998a. Histone deacetylase activity of Rpd3 is important for transcriptional repression in vivo. Genes Dev. 12, 797–805.

Kadosh, D., Struhl, K., 1998b. Targeted recruitment of the Sin3-Rpd3 histone deacetylase complex generates a highly localized domain of repressed chromatin in vivo. Mol. Cell. Biol. 18, 5121–5127.

Kakimoto, Y., Akazawa, S., 1970. Isolation and identification of N-G,N-G- and N-G,N'-G-dimethyl-arginine, N-epsilon-mono-, di-, and trimethyllysine, and glucosylgalactosyl- and galactosyl-delta-hydroxylysine from human urine. J. Biol. Chem. 245, 5751–5758.

Kameshita, I., Fujisawa, H., 1989. A sensitive method for detection of calmodulin-dependent protein kinase II activity in sodium dodecyl sulfate-polyacrylamide gel. Anal. Biochem. 183, 139–143.

Kamine, J., Elangovan, B., Subramanian, T., Coleman, D., Chinnadurai, G., 1996. Identification of a cellular protein that specifically interacts with the essential cysteine region of the HIV-1 Tat transactivator. Virology 216, 357–366.

Karch, K.R., Langelier, M.F., Pascal, J.M., Garcia, B.A., 2017. The nucleosomal surface is the main target of histone ADP-ribosylation in response to DNA damage. Mol. BioSyst. 13, 2660–2671.

Katan-Khaykovich, Y., Struhl, K., 2002. Dynamics of global histone acetylation and deacetylation in vivo: rapid restoration of normal histone acetylation status upon removal of activators and repressors. Genes Dev. 16, 743–752.

Kaye, A.M., Sheratzky, D., 1969. Methylation of protein (histone) in vitro: enzymic activity from the soluble fraction of rat organs. Biochim. Biophys. Acta 190, 527–538.

Kayne, P.S., Kim, U.J., Han, M., Mullen, J.R., Yoshizaki, F., Grunstein, M., 1988. Extremely conserved histone H4 N terminus is dispensable for growth but essential for repressing the silent mating loci in yeast. Cell 55, 27–39.

Keenan, E.K., Zachman, D.K., Hirschey, M.D., 2021. Discovering the landscape of protein modifications. Mol. Cell 81, 1868–1878.

Kelleher 3rd, R.J., Flanagan, P.M., Kornberg, R.D., 1990. A novel mediator between activator proteins and the RNA polymerase II transcription apparatus. Cell 61, 1209–1215.

Kelley, R.L., Kuroda, M.I., 1995. Equality for X chromosomes. Science 270, 1607–1610.

Keogh, M.C., Kurdistani, S.K., Morris, S.A., Ahn, S.H., Podolny, V., Collins, S.R., Schuldiner, M., Chin, K., Punna, T., Thompson, N.J., et al., 2005. Cotranscriptional set2 methylation of histone H3 lysine 36 recruits a repressive Rpd3 complex. Cell 123, 593–605.

Kerscher, O., Felberbaum, R., Hochstrasser, M., 2006. Modification of proteins by ubiquitin and ubiquitin-like proteins. Annu. Rev. Cell Dev. Biol. 22, 159–180.

Kharchenko, P.V., Alekseyenko, A.A., Schwartz, Y.B., Minoda, A., Riddle, N.C., Ernst, J., Sabo, P.J., Larschan, E., Gorchakov, A.A., Gu, T., et al., 2011. Comprehensive analysis of the chromatin landscape in *Drosophila melanogaster*. Nature 471, 480–485.

Kijima, M., Yoshida, M., Sugita, K., Horinouchi, S., Beppu, T., 1993. Trapoxin, an antitumor cyclic tetrapeptide, is an irreversible inhibitor of mammalian histone deacetylase. J. Biol. Chem. 268, 22429–22435.

Kim, S., Paik, W.K., 1965. Studies on the origin of epsilon-N-methyl-L-lysine in protein. J. Biol. Chem. 240, 4629–4634.

Kim, S., Benoiton, L., Paik, W.K., 1964. Epsilon-alkyllysinase. Purification and properties of the enzyme. J. Biol. Chem. 239, 3790–3796.

Kim, Y.J., Bjorklund, S., Li, Y., Sayre, M.H., Kornberg, R.D., 1994. A multiprotein mediator of transcriptional activation and its interaction with the C-terminal repeat domain of RNA polymerase II. Cell 77, 599–608.

Kim, K.C., Geng, L., Huang, S., 2003. Inactivation of a histone methyltransferase by mutations in human cancers. Cancer Res. 63, 7619–7623.

Kim, J., Hake, S.B., Roeder, R.G., 2005. The human homolog of yeast BRE1 functions as a transcriptional coactivator through direct activator interactions. Mol. Cell 20, 759–770.

Kim, J., Guermah, M., McGinty, R.K., Lee, J.S., Tang, Z., Milne, T.A., Shilatifard, A., Muir, T.W., Roeder, R.G., 2009. RAD6-mediated transcription-coupled H2B ubiquitylation directly stimulates H3K4 methylation in human cells. Cell 137, 459–471.

Kim, J.H., Saraf, A., Florens, L., Washburn, M., Workman, J.L., 2010. Gcn5 regulates the dissociation of SWI/SNF from chromatin by acetylation of Swi2/Snf2. Genes Dev. 24, 2766–2771.

Kim, S.K., Jung, I., Lee, H., Kang, K., Kim, M., Jeong, K., Kwon, C.S., Han, Y.M., Kim, Y.S., Kim, D., et al., 2012. Human histone H3K79 methyltransferase DOT1L protein [corrected] binds actively transcribing RNA polymerase II to regulate gene expression. J. Biol. Chem. 287, 39698–39709.

Kim, J.H., Yoon, C.Y., Jun, Y., Lee, B.B., Lee, J.E., Ha, S.D., Woo, H., Choi, A., Lee, S., Jeong, W., et al., 2020. NuA3 HAT antagonizes the Rpd3S and Rpd3L HDACs to optimize mRNA and lncRNA expression dynamics. Nucleic Acids Res. 48, 10753–10767.

Kirmizis, A., Santos-Rosa, H., Penkett, C.J., Singer, M.A., Vermeulen, M., Mann, M., Bahler, J., Green, R.D., Kouzarides, T., 2007. Arginine methylation at histone H3R2 controls deposition of H3K4 trimethylation. Nature 449, 928–932.

Kleff, S., Andrulis, E.D., Anderson, C.W., Sternglanz, R., 1995. Identification of a gene encoding a yeast histone H4 acetyltransferase. J. Biol. Chem. 270, 24674–24677.

Klein, J., Nolden, M., Sanders, S.L., Kirchner, J., Weil, P.A., Melcher, K., 2003. Use of a genetically introduced cross-linker to identify interaction sites of acidic activators within native transcription factor IID and SAGA. J. Biol. Chem. 278, 6779–6786.

Kleinschmidt, A.M., Martinson, H.G., 1981. Structure of nucleosome core particles containing uH2A (A24). Nucleic Acids Res. 9, 2423–2431.

Kleinsmith, L.J., Allfrey, V.G., Mirsky, A.E., 1966. Phosphoprotein metabolism in isolated lymphocyte nuclei. Proc. Natl. Acad. Sci. USA 55, 1182–1189.

Klose, R.J., Bird, A.P., 2006. Genomic DNA methylation: the mark and its mediators. Trends Biochem. Sci. 31, 89–97.

Klose, R.J., Yamane, K., Bae, Y., Zhang, D., Erdjument-Bromage, H., Tempst, P., Wong, J., Zhang, Y., 2006. The transcriptional repressor JHDM3A demethylates trimethyl histone H3 lysine 9 and lysine 36. Nature 442, 312–316.

Koh, S.S., Chen, D., Lee, Y.H., Stallcup, M.R., 2001. Synergistic enhancement of nuclear receptor function by p160 coactivators and two coactivators with protein methyltransferase activities. J. Biol. Chem. 276, 1089–1098.

Kohler, A., Zimmerman, E., Schneider, M., Hurt, E., Zheng, N., 2010. Structural basis for assembly and activation of the heterotetrameric SAGA histone H2B deubiquitinase module. Cell 141, 606–617.

Kori, Y., Sidoli, S., Yuan, Z.F., Lund, P.J., Zhao, X., Garcia, B.A., 2017. Proteome-wide acetylation dynamics in human cells. Sci. Rep. 7, 10296.

Kornberg, R.D., 2005. Mediator and the mechanism of transcriptional activation. Trends Biochem. Sci. 30, 235–239.

Krajewski, W.A., Li, J., Dou, Y., 2018. Effects of histone H2B ubiquitylation on the nucleosome structure and dynamics. Nucleic Acids Res. 46, 7631–7642.

Kreimeyer, A., Wielckens, K., Adamietz, P., Hilz, H., 1984. DNA repair-associated ADP-ribosylation in vivo. Modification of histone H1 differs from that of the principal acceptor proteins. J. Biol. Chem. 259, 890–896.

Kreppel, L.K., Blomberg, M.A., Hart, G.W., 1997. Dynamic glycosylation of nuclear and cytosolic proteins. Cloning and characterization of a unique O-GlcNAc transferase with multiple tetratricopeptide repeats. J. Biol. Chem. 272, 9308–9315.

Krogan, N.J., Dover, J., Khorrami, S., Greenblatt, J.F., Schneider, J., Johnston, M., Shilatifard, A., 2002. COMPASS, a histone H3 (lysine 4) methyltransferase required for telomeric silencing of gene expression. J. Biol. Chem. 277, 10753–10755.

Kueh, A.J., Dixon, M.P., Voss, A.K., Thomas, T., 2011. HBO1 is required for H3K14 acetylation and normal transcriptional activity during embryonic development. Mol. Cell. Biol. 31, 845–860.

Kundu, T.K., Wang, Z., Roeder, R.G., 1999. Human TFIIIC relieves chromatin-mediated repression of RNA polymerase III transcription and contains an intrinsic histone acetyltransferase activity. Mol. Cell. Biol. 19, 1605–1615.

Kuo, M.H., Brownell, J.E., Sobel, R.E., Ranalli, T.A., Cook, R.G., Edmondson, D.G., Roth, S.Y., Allis, C.D., 1996. Transcription-linked acetylation by Gcn5p of histones H3 and H4 at specific lysines. Nature 383, 269–272.

Kuo, M.H., Zhou, J., Jambeck, P., Churchill, M.E., Allis, C.D., 1998. Histone acetyltransferase activity of yeast Gcn5p is required for the activation of target genes in vivo. Genes Dev. 12, 627–639.

Kuo, M.-H., Vom Baur, E., Struhl, K., Allis, C.D., 2000. Gcn4 activator targets Gcn5 histone acetyltransferase to specific promoters independently of transcription. Mol. Cell 6, 1309–1320.

Kuo, A.J., Cheung, P., Chen, K., Zee, B.M., Kioi, M., Lauring, J., Xi, Y., Park, B.H., Shi, X., Garcia, B.A., et al., 2011. NSD2 links dimethylation of histone H3 at lysine 36 to oncogenic programming. Mol. Cell 44, 609–620.

Kuras, L., Kosa, P., Mencia, M., Struhl, K., 2000. TAF-containing and TAF-independent forms of transcriptionally active TBP in vivo. Science 288, 1244–1248.

Kurdistani, S.K., Grunstein, M., 2003. Histone acetylation and deacetylation in yeast. Nat. Rev. Mol. Cell Biol. 4, 276–284.

Kusch, T., Guelman, S., Abmayr, S.M., Workman, J.L., 2003. Two Drosophila Ada2 homologues function in different multiprotein complexes. Mol. Cell. Biol. 23, 3305–3319.

Kutil, Z., Novakova, Z., Meleshin, M., Mikesova, J., Schutkowski, M., Barinka, C., 2018. Histone deacetylase 11 is a fatty-acid deacylase. ACS Chem. Biol. 13, 685–693.

Kuzmichev, A., Nishioka, K., Erdjument-Bromage, H., Tempst, P., Reinberg, D., 2002. Histone methyltransferase activity associated with a human multiprotein complex containing the enhancer of Zeste protein. Genes Dev. 16, 2893–2905.

Lachner, M., O'Carroll, D., Rea, S., Mechtler, K., Jenuwein, T., 2001. Methylation of histone H3 lysine 9 creates a binding site for HP1 proteins. Nature 410, 116–120.

Lacoste, N., Utley, R.T., Hunter, J.M., Poirier, G.G., Cote, J., 2002. Disruptor of telomeric silencing-1 is a chromatin-specific histone H3 methyltransferase. J. Biol. Chem. 277, 30421–30424.

Laherty, C.D., Yang, W.M., Sun, J.M., Davie, J.R., Seto, E., Eisenman, R.N., 1997. Histone deacetylases associated with the mSin3 corepressor mediate mad transcriptional repression. Cell 89, 349–356.

Lahm, A., Paolini, C., Pallaoro, M., Nardi, M.C., Jones, P., Neddermann, P., Sambucini, S., Bottomley, M.J., Lo Surdo, P., Carfi, A., et al., 2007. Unraveling the hidden catalytic activity of vertebrate class IIa histone deacetylases. Proc. Natl. Acad. Sci. USA 104, 17335–17340.

Lalonde, M.E., Avvakumov, N., Glass, K.C., Joncas, F.H., Saksouk, N., Holliday, M., Paquet, E., Yan, K., Tong, Q., Klein, B.J., et al., 2013. Exchange of associated factors directs a switch in HBO1 acetyltransferase histone tail specificity. Genes Dev. 27, 2009–2024.

Lam, U.T.F., Tan, B.K.Y., Poh, J.J.X., Chen, E.S., 2022. Structural and functional specificity of H3K36 methylation. Epigenetics Chromatin 15, 17.

Lan, R., Wang, Q., 2020. Deciphering structure, function and mechanism of lysine acetyltransferase HBO1 in protein acetylation, transcription regulation, DNA replication and its oncogenic properties in cancer. Cell. Mol. Life Sci. 77, 637–649.

Landry, J., Slama, J.T., Sternglanz, R., 2000a. Role of NAD(+) in the deacetylase activity of the SIR2-like proteins. Biochem. Biophys. Res. Commun. 278, 685–690.

Landry, J., Sutton, A., Tafrov, S.T., Heller, R.C., Stebbins, J., Pillus, L., Sternglanz, R., 2000b. The silencing protein SIR2 and its homologs are NAD-dependent protein deacetylases. Proc. Natl. Acad. Sci. USA 97, 5807–5811.

Larschan, E., Winston, F., 2001. The S. cerevisiae SAGA complex functions in vivo as a coactivator for transcriptional activation by Gal4. Genes Dev. 15, 1946–1956.

Latham, J.A., Chosed, R.J., Wang, S., Dent, S.Y., 2011. Chromatin signaling to kinetochores: transregulation of Dam1 methylation by histone H2B ubiquitination. Cell 146, 709–719.

Lauberth, S.M., Nakayama, T., Wu, X., Ferris, A.L., Tang, Z., Hughes, S.H., Roeder, R.G., 2013. H3K4me3 interactions with TAF3 regulate preinitiation complex assembly and selective gene activation. Cell 152, 1021–1036.

Le Guezennec, X., Vermeulen, M., Brinkman, A.B., Hoeijmakers, W.A., Cohen, A., Lasonder, E., Stunnenberg, H.G., 2006. MBD2/NuRD and MBD3/NuRD, two distinct complexes with different biochemical and functional properties. Mol. Cell. Biol. 26, 843–851.

Lechner, T., Carrozza, M.J., Yu, Y., Grant, P.A., Eberharter, A., Vannier, D., Brosch, G., Stillman, D.J., Shore, D., Workman, J.L., 2000. Sds3 (suppressor of defective silencing 3) is an integral component of the yeast Sin3[middle dot]Rpd3 histone deacetylase complex and is required for histone deacetylase activity. J. Biol. Chem. 275, 40961–40966.

Lee, J.H., Cook, J.R., Pollack, B.P., Kinzy, T.G., Norris, D., Pestka, S., 2000. Hsl7p, the yeast homologue of human JBP1, is a protein methyltransferase. Biochem. Biophys. Res. Commun. 274, 105–111.

Lee, K.K., Florens, L., Swanson, S.K., Washburn, M.P., Workman, J.L., 2005. The deubiquitylation activity of Ubp8 is dependent upon Sgf11 and its association with the SAGA complex. Mol. Cell. Biol. 25, 1173–1182.

Lee, M.G., Wynder, C., Bochar, D.A., Hakimi, M.A., Cooch, N., Shiekhattar, R., 2006. Functional interplay between histone demethylase and deacetylase enzymes. Mol. Cell. Biol. 26, 6395–6402.

Lee, K.K., Sardiu, M.E., Swanson, S.K., Gilmore, J.M., Torok, M., Grant, P.A., Florens, L., Workman, J.L., Washburn, M.P., 2011. Combinatorial depletion analysis to assemble the network architecture of the SAGA and ADA chromatin remodeling complexes. Mol. Syst. Biol. 7, 503.

Lee, S., Oh, S., Jeong, K., Jo, H., Choi, Y., Seo, H.D., Kim, M., Choe, J., Kwon, C.S., Lee, D., 2018. Dot1 regulates nucleosome dynamics by its inherent histone chaperone activity in yeast. Nat. Commun. 9, 240.

Lee, K., Whedon, S.D., Wang, Z.A., Cole, P.A., 2022. Distinct biochemical properties of the class I histone deacetylase complexes. Curr. Opin. Chem. Biol. 70, 102179.

Leidecker, O., Bonfiglio, J.J., Colby, T., Zhang, Q., Atanassov, I., Zaja, R., Palazzo, L., Stockum, A., Ahel, I., Matic, I., 2016. Serine is a new target residue for endogenous ADP-ribosylation on histones. Nat. Chem. Biol. 12, 998–1000.

Leighton, G., Williams Jr., D.C., 2019. The methyl-CpG-binding domain 2 and 3 proteins and formation of the nucleosome remodeling and deacetylase complex. J. Mol. Biol. 432, 1624–1639.

Leipe, D.D., Landsman, D., 1997. Histone deacetylases, acetoin utilization proteins and acetylpolyamine amidohydrolases are members of an ancient protein superfamily. Nucleic Acids Res. 25, 3693–3697.

Lenstra, T.L., Benschop, J.J., Kim, T., Schulze, J.M., Brabers, N.A., Margaritis, T., van de Pasch, L.A., van Heesch, S.A., Brok, M.O., Groot Koerkamp, M.J., et al., 2011. The specificity and topology of chromatin interaction pathways in yeast. Mol. Cell 42, 536–549.

Levene, P.A., Alsberg, T., 1901. Zur chemie der paranucleinsäure. Z. Physiol. Chem. 31, 543–555.

Levene, P.A., Alsberg, T., 1906. The cleavage products of vitellin. J. Biol. Chem. 2, 127–133.

Levinger, L., Varshavsky, A., 1982. Selective arrangement of ubiquitinated and D1 protein-containing nucleosomes within the Drosophila genome. Cell 28, 375–385.

Levy-Wilson, B., 1983. Glycosylation, ADP-ribosylation, and methylation of tetrahymena histones. Biochemistry 22, 484–489.

Levy-Wilson, B., Gjerset, R.A., McCarthy, B.J., 1977. Acetylation and phosphorylation of Drosophila histones. Distribution of acetate and phosphate groups in fractionated chromatin. Biochim. Biophys. Acta 475, 168–175.

Levy-Wilson, B., Watson, D.C., Dixon, G.H., 1979. Multiacetylated forms of H4 are found in a putative transcriptionally competent chromatin fraction from trout testis. Nucleic Acids Res. 6, 259–274.

Li, S., Shogren-Knaak, M.A., 2009. The Gcn5 bromodomain of the SAGA complex facilitates cooperative and cross-tail acetylation of nucleosomes. J. Biol. Chem. 284, 9411–9417.

Li, J., Wang, J., Wang, J., Nawaz, Z., Liu, J.M., Qin, J., Wong, J., 2000a. Both corepressor proteins SMRT and N-CoR exist in large protein complexes containing HDAC3. EMBO J. 19, 4342–4350.

Li, X.Y., Bhaumik, S.R., Green, M.R., 2000b. Distinct classes of yeast promoters revealed by differential TAF recruitment. Science 288, 1242–1244.

Li, B., Howe, L., Anderson, S., Yates 3rd, J.R., Workman, J.L., 2003. The Set2 histone methyltransferase functions through the phosphorylated carboxyl-terminal domain of RNA polymerase II. J. Biol. Chem. 278, 8897–8903.

Li, H., Ilin, S., Wang, W., Duncan, E.M., Wysocka, J., Allis, C.D., Patel, D.J., 2006. Molecular basis for site-specific readout of histone H3K4me3 by the BPTF PHD finger of NURF. Nature 442, 91–95.

Li, Q., Zhou, H., Wurtele, H., Davies, B., Horazdovsky, B., Verreault, A., Zhang, Z., 2008. Acetylation of histone H3 lysine 56 regulates replication-coupled nucleosome assembly. Cell 134, 244–255.

Li, Y., Trojer, P., Xu, C.F., Cheung, P., Kuo, A., Drury 3rd, W.J., Qiao, Q., Neubert, T.A., Xu, R.M., Gozani, O., et al., 2009. The target of the NSD family of histone lysine methyltransferases depends on the nature of the substrate. J. Biol. Chem. 284, 34283–34295.

Li, Y., Wen, H., Xi, Y., Tanaka, K., Wang, H., Peng, D., Ren, Y., Jin, Q., Dent, S.Y., Li, W., et al., 2014. AF9 YEATS domain links histone acetylation to DOT1L-mediated H3K79 methylation. Cell 159, 558–571.

Li, Y., Sabari, B.R., Panchenko, T., Wen, H., Zhao, D., Guan, H., Wan, L., Huang, H., Tang, Z., Zhao, Y., et al., 2016. Molecular coupling of histone crotonylation and active transcription by AF9 YEATS domain. Mol. Cell 62, 181–193.

Li, X., Seidel, C.W., Szerszen, L.T., Lange, J.J., Workman, J.L., Abmayr, S.M., 2017. Enzymatic modules of the SAGA chromatin-modifying complex play distinct roles in Drosophila gene expression and development. Genes Dev. 31, 1588–1600.

Li, W., Tian, W., Yuan, G., Deng, P., Sengupta, D., Cheng, Z., Cao, Y., Ren, J., Qin, Y., Zhou, Y., et al., 2021a. Molecular basis of nucleosomal H3K36 methylation by NSD methyltransferases. Nature 590, 498–503.

Li, Y.L., Chen, X., Lu, C., 2021b. The interplay between DNA and histone methylation: molecular mechanisms and disease implications. EMBO Rep. 22, e51803.

Liao, R., Mizzen, C.A., 2016. Interphase H1 phosphorylation: regulation and functions in chromatin. Biochim. Biophys. Acta 1859, 476–485.

Liew, C.C., Haslett, G.W., Allfrey, V.G., 1970. N-acetyl-seryl-tRNA and polypeptide chain initiation during histone biosynthesis. Nature 226, 414–417.

Lim, S., Kwak, J., Kim, M., Lee, D., 2013. Separation of a functional deubiquitylating module from the SAGA complex by the proteasome regulatory particle. Nat. Commun. 4, 2641.

Lin, R., Leone, J.W., Cook, R.G., Allis, C.D., 1989. Antibodies specific to acetylated histones document the existence of deposition- and transcription-related histone acetylation in Tetrahymena. J. Cell Biol. 108, 1577–1588.

Lin, R., Cook, R.G., Allis, C.D., 1991. Proteolytic removal of core histone amino termini and dephosphorylation of histone H1 correlate with the formation of condensed chromatin and transcriptional silencing during Tetrahymena macronuclear development. Genes Dev. 5, 1601–1610.

Lin, W.J., Gary, J.D., Yang, M.C., Clarke, S., Herschman, H.R., 1996. The mammalian immediate-early TIS21 protein and the leukemia-associated BTG1 protein interact with a protein-arginine N-methyltransferase. J. Biol. Chem. 271, 15034–15044.

Lin, L., Chamberlain, L., Zhu, L.J., Green, M.R., 2012. Analysis of Gal4-directed transcription activation using Tra1 mutants selectively defective for interaction with Gal4. Proc. Natl. Acad. Sci. USA 109, 1997–2002.

Lipmann, F.A., Levene, P.A., 1932. Serinephosphoric acid obtained on hydrolysis of vitellinic acid. J. Biol. Chem. 98, 109–114.

Liss, M., Edelstein, L.M., 1967. Evidence for the enzymatic methylation of crystalline ovalbumin preparations. Biochem. Biophys. Res. Commun. 26, 497–504.

Liss, M., Maxam, A.M., 1967. Methylation of ovalbumin and human serum albumin by a purified enzyme from calf spleen. Biochim. Biophys. Acta 140, 555–557.

Litt, M.D., Simpson, M., Gaszner, M., Allis, C.D., Felsenfeld, G., 2001. Correlation between histone lysine methylation and developmental changes at the chicken beta-globin locus. Science 293, 2453–2455.

Liu, Z., Wong, J., Tsai, S.Y., Tsai, M.J., O'Malley, B.W., 2001. Sequential recruitment of steroid receptor coactivator-1 (SRC-1) and p300 enhances progesterone receptor-dependent initiation and reinitiation of transcription from chromatin. Proc. Natl. Acad. Sci. USA 98, 12426–12431.

Liu, C.L., Kaplan, T., Kim, M., Buratowski, S., Schreiber, S.L., Friedman, N., Rando, O.J., 2005. Single-nucleosome mapping of histone modifications in *S. cerevisiae*. PLoS Biol. 3, e328.

Lo, W.S., Trievel, R.C., Rojas, J.R., Duggan, L., Hsu, J.Y., Allis, C.D., Marmorstein, R., Berger, S.L., 2000. Phosphorylation of serine 10 in histone H3 is functionally linked in vitro and in vivo to Gcn5-mediated acetylation at lysine 14. Mol. Cell 5, 917–926.

Lo, W.S., Duggan, L., Emre, N.C., Belotserkovskya, R., Lane, W.S., Shiekhattar, R., Berger, S.L., 2001. Snf1—a histone kinase that works in concert with the histone acetyltransferase Gcn5 to regulate transcription. Science 293, 1142–1146.

Loidl, P., 1994. Histone acetylation: facts and questions. Chromosoma 103, 441–449.

Lopez-Rodas, G., Brosch, G., Georgieva, E.I., Sendra, R., Franco, L., Loidl, P., 1993. Histone deacetylase. A key enzyme for the binding of regulatory proteins to chromatin. FEBS Lett. 317, 175–180.

Low, J.K., Wilkins, M.R., 2012. Protein arginine methylation in *Saccharomyces cerevisiae*. FEBS J. 279, 4423–4443.

Loyola, A., Tagami, H., Bonaldi, T., Roche, D., Quivy, J.P., Imhof, A., Nakatani, Y., Dent, S.Y., Almouzni, G., 2009. The HP1alpha-CAF1-SetDB1-containing complex provides H3K9me1 for Suv39-mediated K9me3 in pericentric heterochromatin. EMBO Rep. 10, 769–775.

Lu, Y., Loh, Y.H., Li, H., Cesana, M., Ficarro, S.B., Parikh, J.R., Salomonis, N., Toh, C.X., Andreadis, S.T., Luckey, C.J., et al., 2014. Alternative splicing of MBD2 supports self-renewal in human pluripotent stem cells. Cell Stem Cell 15, 92–101.

Lu, C., Coradin, M., Porter, E.G., Garcia, B.A., 2021. Accelerating the Field of epigenetic histone modification through mass spectrometry-based approaches. Mol. Cell. Proteomics 20, 100006.

Lucchesi, J.C., 1998. Dosage compensation in flies and worms: the ups and downs of X-chromosome regulation. Curr. Opin. Genet. Dev. 8, 179–184.

Lucio-Eterovic, A.K., Singh, M.M., Gardner, J.E., Veerappan, C.S., Rice, J.C., Carpenter, P.B., 2010. Role for the nuclear receptor-binding SET domain protein 1 (NSD1) methyltransferase in coordinating lysine 36 methylation at histone 3 with RNA polymerase II function. Proc. Natl. Acad. Sci. USA 107, 16952–16957.

Luo, Y., Fujii, H., Gerster, T., Roeder, R.G., 1992. A novel B cell-derived coactivator potentiates the activation of immunoglobulin promoters by octamer-binding transcription factors. Cell 71, 231–241.

Macdonald, N., Welburn, J.P., Noble, M.E., Nguyen, A., Yaffe, M.B., Clynes, D., Moggs, J.G., Orphanides, G., Thomson, S., Edmunds, J.W., et al., 2005. Molecular basis for the recognition of phosphorylated and phosphoacetylated histone h3 by 14-3-3. Mol. Cell 20, 199–211.

Mahadevan, L.C., Willis, A.C., Barratt, M.J., 1991. Rapid histone H3 phosphorylation in response to growth factors, phorbol esters, okadaic acid, and protein synthesis inhibitors. Cell 65, 775–783.

Malla, A.B., Yu, H., Kadimi, S., Lam, T., Cox, A.L., Smith, Z.D., Lesch, B.J., 2021. DOT1L bridges transcription and heterochromatin formation at pericentromeres. bioRxiv.

Marcus, G.A., Silverman, N., Berger, S.L., Horiuchi, J., Guarente, L., 1994. Functional similarity and physical association between GCN5 and ADA2: putative transcriptional adaptors. EMBO J. 13, 4807–4815.

Marcus, G.A., Horiuchi, J., Silverman, N., Guarente, L., 1996. ADA5/SPT20 links the ADA and SPT genes, which are involved in yeast transcription. Mol. Cell. Biol. 16, 3197–3205.

Margaritis, T., Oreal, V., Brabers, N., Maestroni, L., Vitaliano-Prunier, A., Benschop, J.J., van Hooff, S., van Leenen, D., Dargemont, C., Geli, V., et al., 2012. Two distinct repressive mechanisms for histone 3 lysine 4 methylation through promoting 3′-end antisense transcription. PLoS Genet. 8, e1002952.

Marmelstein, A.M., Moreno, J., Fiedler, D., 2017. Chemical approaches to studying labile amino acid phosphorylation. Top. Curr. Chem. 375, 22.

Marmorstein, R., Zhou, M.M., 2014. Writers and readers of histone acetylation: structure, mechanism, and inhibition. Cold Spring Harb. Perspect. Biol. 6, a018762.

Martin, D.G., Baetz, K., Shi, X., Walter, K.L., MacDonald, V.E., Wlodarski, M.J., Gozani, O., Hieter, P., Howe, L., 2006a. The Yng1p plant homeodomain finger is a methyl-histone binding module that recognizes lysine 4-methylated histone H3. Mol. Cell. Biol. 26, 7871–7879.

Martin, D.G., Grimes, D.E., Baetz, K., Howe, L., 2006b. Methylation of histone H3 mediates the association of the NuA3 histone acetyltransferase with chromatin. Mol. Cell. Biol. 26, 3018–3028.

Martinez-Balbas, M.A., Bannister, A.J., Martin, K., Haus-Seuffert, P., Meisterernst, M., Kouzarides, T., 1998. The acetyltransferase activity of CBP stimulates transcription. EMBO J. 17, 2886–2893.

Martinez-Zamudio, R., Ha, H.C., 2012. Histone ADP-ribosylation facilitates gene transcription by directly remodeling nucleosomes. Mol. Cell. Biol. 32, 2490–2502.

Masumoto, H., Hawke, D., Kobayashi, R., Verreault, A., 2005. A role for cell-cycle-regulated histone H3 lysine 56 acetylation in the DNA damage response. Nature 436, 294–298.

Maurer-Stroh, S., Dickens, N.J., Hughes-Davies, L., Kouzarides, T., Eisenhaber, F., Ponting, C.P., 2003. The Tudor domain 'Royal Family': Tudor, plant agenet, chromo, PWWP and MBT domains. Trends Biochem. Sci. 28, 69–74.

McDaniel, S.L., Strahl, B.D., 2017. Shaping the cellular landscape with Set2/SETD2 methylation. Cell. Mol. Life Sci. 74, 3317–3334.

McGinty, R.K., Kim, J., Chatterjee, C., Roeder, R.G., Muir, T.W., 2008. Chemically ubiquitylated histone H2B stimulates hDot1L-mediated intranucleosomal methylation. Nature 453, 812–816.

McGinty, R.K., Kohn, M., Chatterjee, C., Chiang, K.P., Pratt, M.R., Muir, T.W., 2009. Structure-activity analysis of semisynthetic nucleosomes: mechanistic insights into the stimulation of Dot1L by ubiquitylated histone H2B. ACS Chem. Biol. 4, 958–968.

McKenzie, E.A., Kent, N.A., Dowell, S.J., Moreno, F., Bird, L.E., Mellor, J., 1993. The centromere and promoter factor, 1, CPF1, of *Saccharomyces cerevisiae* modulates gene activity through a family of factors including SPT21, RPD1 (SIN3), RPD3 and CCR4. Mol. Gen. Genet. 240, 374–386.

McMahon, S.B., Van Buskirk, H.A., Dugan, K.A., Copeland, T.D., Cole, M.D., 1998. The novel ATM-related protein TRRAP is an essential cofactor for the c-Myc and E2F oncoproteins. Cell 94, 363–374.

Megee, P.C., Morgan, B.A., Mittman, B.A., Smith, M.M., 1990. Genetic analysis of histone H4: essential role of lysines subject to reversible acetylation. Science 247, 841–845.

Meissner, A., Mikkelsen, T.S., Gu, H., Wernig, M., Hanna, J., Sivachenko, A., Zhang, X., Bernstein, B.E., Nusbaum, C., Jaffe, D.B., et al., 2008. Genome-scale DNA methylation maps of pluripotent and differentiated cells. Nature 454, 766–770.

Melcher, K., Johnston, S.A., 1995. GAL4 interacts with TATA-binding protein and coactivators. Mol. Cell. Biol. 15, 2839–2848.

Mellone, B.G., Ball, L., Suka, N., Grunstein, M.R., Partridge, J.F., Allshire, R.C., 2003. Centromere silencing and function in fission yeast is governed by the amino terminus of histone H3. Curr. Biol. 13, 1748–1757.

Menafra, R., Brinkman, A.B., Matarese, F., Franci, G., Bartels, S.J., Nguyen, L., Shimbo, T., Wade, P.A., Hubner, N.C., Stunnenberg, H.G., 2014. Genome-wide binding of MBD2 reveals strong preference for highly methylated loci. PLoS One 9, e99603.

Mermoud, J.E., Popova, B., Peters, A.H., Jenuwein, T., Brockdorff, N., 2002. Histone H3 lysine 9 methylation occurs rapidly at the onset of random X chromosome inactivation. Curr. Biol. 12, 247–251.

Mersfelder, E.L., Parthun, M.R., 2006. The tale beyond the tail: histone core domain modifications and the regulation of chromatin structure. Nucleic Acids Res. 34, 2653–2662.

Mertens, C., Roeder, R.G., 2008. Different functional modes of p300 in activation of RNA polymerase III transcription from chromatin templates. Mol. Cell. Biol. 28, 5764–5776.

Messner, S., Altmeyer, M., Zhao, H., Pozivil, A., Roschitzki, B., Gehrig, P., Rutishauser, D., Huang, D., Caflisch, A., Hottiger, M.O., 2010. PARP1 ADP-ribosylates lysine residues of the core histone tails. Nucleic Acids Res. 38, 6350–6362.

Mevissen, T.E.T., Komander, D., 2017. Mechanisms of deubiquitinase specificity and regulation. Annu. Rev. Biochem. 86, 159–192.

Meyer, M.E., Gronemeyer, H., Turcotte, B., Bocquel, M.T., Tasset, D., Chambon, P., 1989. Steroid hormone receptors compete for factors that mediate their enhancer function. Cell 57, 433–442.

Migliori, V., Muller, J., Phalke, S., Low, D., Bezzi, M., Mok, W.C., Sahu, S.K., Gunaratne, J., Capasso, P., Bassi, C., et al., 2012. Symmetric dimethylation of H3R2 is a newly identified histone mark that supports euchromatin maintenance. Nat. Struct. Mol. Biol. 19, 136–144.

Mikkelsen, T.S., Ku, M., Jaffe, D.B., Issac, B., Lieberman, E., Giannoukos, G., Alvarez, P., Brockman, W., Kim, T.K., Koche, R.P., et al., 2007. Genome-wide maps of chromatin state in pluripotent and lineage-committed cells. Nature 448, 553–560.

Milazzo, G., Mercatelli, D., Di Muzio, G., Triboli, L., De Rosa, P., Perini, G., Giorgi, F.M., 2020. Histone deacetylases (HDACs): evolution, specificity, role in transcriptional complexes, and pharmacological actionability. Genes (Basel) 11, 556.

Millan-Zambrano, G., Burton, A., Bannister, A.J., Schneider, R., 2022. Histone post-translational modifications—cause and consequence of genome function. Nat. Rev. Genet. 23, 563–580.

Miller, T., Krogan, N.J., Dover, J., Erdjument-Bromage, H., Tempst, P., Johnston, M., Greenblatt, J.F., Shilatifard, A., 2001. COMPASS: a complex of proteins associated with a trithorax-related SET domain protein. Proc. Natl. Acad. Sci. USA 98, 12902–12907.

Min, J., Zhang, X., Cheng, X., Grewal, S.I., Xu, R.M., 2002. Structure of the SET domain histone lysine methyltransferase Clr4. Nat. Struct. Biol. 9, 828–832.

Min, J., Zhang, Y., Xu, R.M., 2003. Structural basis for specific binding of polycomb chromodomain to histone H3 methylated at Lys 27. Genes Dev. 17, 1823–1828.

Miotto, B., Struhl, K., 2008. HBO1 histone acetylase is a coactivator of the replication licensing factor Cdt1. Genes Dev. 22, 2633–2638.

Miotto, B., Struhl, K., 2010. HBO1 histone acetylase activity is essential for DNA replication licensing and inhibited by Geminin. Mol. Cell 37, 57–66.

Miotto, B., Struhl, K., 2011. JNK1 phosphorylation of Cdt1 inhibits recruitment of HBO1 histone acetylase and blocks replication licensing in response to stress. Mol. Cell 44, 62–71.

Mizzen, C.A., Yang, X.J., Kokubo, T., Brownell, J.E., Bannister, A.J., Owen-Hughes, T., Workman, J., Wang, L., Berger, S.L., Kouzarides, T., et al., 1996. The TAF(II)250 subunit of TFIID has histone acetyltransferase activity. Cell 87, 1261–1270.

Mohan, R.D., Dialynas, G., Weake, V.M., Liu, J., Martin-Brown, S., Florens, L., Washburn, M.P., Workman, J.L., Abmayr, S.M., 2014. Loss of Drosophila Ataxin-7, a SAGA subunit, reduces H2B ubiquitination and leads to neural and retinal degeneration. Genes Dev. 28, 259–272.

Moraga, F., Aquea, F., 2015. Composition of the SAGA complex in plants and its role in controlling gene expression in response to abiotic stresses. Front. Plant Sci. 6, 865.

Moreno-Yruela, C., Galleano, I., Madsen, A.S., Olsen, C.A., 2018. Histone deacetylase 11 is an epsilon-N-myristoyllysine hydrolase. Cell Chem. Biol. 25, 849–856.e848.

Morgan, M.A.J., Shilatifard, A., 2020. Reevaluating the roles of histone-modifying enzymes and their associated chromatin modifications in transcriptional regulation. Nat. Genet. 52, 1271–1281.

Morgan, M.T., Haj-Yahya, M., Ringel, A.E., Bandi, P., Brik, A., Wolberger, C., 2016. Structural basis for histone H2B deubiquitination by the SAGA DUB module. Science 351, 725–728.

Morillon, A., Karabetsou, N., Nair, A., Mellor, J., 2005. Dynamic lysine methylation on histone H3 defines the regulatory phase of gene transcription. Mol. Cell 18, 723–734.

Mottis, A., Mouchiroud, L., Auwerx, J., 2013. Emerging roles of the corepressors NCoR1 and SMRT in homeostasis. Genes Dev. 27, 819–835.

Mueller, D., Bach, C., Zeisig, D., Garcia-Cuellar, M.P., Monroe, S., Sreekumar, A., Zhou, R., Nesvizhskii, A., Chinnaiyan, A., Hess, J.L., et al., 2007. A role for the MLL fusion partner ENL in transcriptional elongation and chromatin modification. Blood 110, 4445–4454.

Muller, S., Isabey, A., Couppez, M., Plaue, S., Sommermeyer, G., Van Regenmortel, M.H., 1987. Specificity of antibodies raised against triacetylated histone H4. Mol. Immunol. 24, 779–789.

Muller, J., Hart, C.M., Francis, N.J., Vargas, M.L., Sengupta, A., Wild, B., Miller, E.L., O'Connor, M.B., Kingston, R.E., Simon, J.A., 2002. Histone methyltransferase activity of a drosophila polycomb group repressor complex. Cell 111, 197–208.

Muratoglu, S., Georgieva, S., Papai, G., Scheer, E., Enunlu, I., Komonyi, O., Cserpan, I., Lebedeva, L., Nabirochkina, E., Udvardy, A., et al., 2003. Two different Drosophila ADA2 homologues are present in distinct GCN5 histone acetyltransferase-containing complexes. Mol. Cell. Biol. 23, 306–321.

Murn, J., Shi, Y., 2017. The winding path of protein methylation research: milestones and new frontiers. Nat. Rev. Mol. Cell Biol. 18, 517–527.

Murray, K., 1964. The occurrence of epsilon-N-methyl lysine in histones. Biochemistry 3, 10–15.

Musselman, C.A., Lalonde, M.E., Cote, J., Kutateladze, T.G., 2012. Perceiving the epigenetic landscape through histone readers. Nat. Struct. Mol. Biol. 19, 1218–1227.

Nagy, L., Kao, H.Y., Chakravarti, D., Lin, R.J., Hassig, C.A., Ayer, D.E., Schreiber, S.L., Evans, R.M., 1997. Nuclear receptor repression mediated by a complex containing SMRT, mSin3A, and histone deacetylase. Cell 89, 373–380.

Nagy, Z., Riss, A., Fujiyama, S., Krebs, A., Orpinell, M., Jansen, P., Cohen, A., Stunnenberg, H.G., Kato, S., Tora, L., 2010. The metazoan ATAC and SAGA coactivator HAT complexes regulate different sets of inducible target genes. Cell. Mol. Life Sci. 67, 611–628.

Nakamura, M., Gao, Y., Dominguez, A.A., Qi, L.S., 2021. CRISPR technologies for precise epigenome editing. Nat. Cell Biol. 23, 11–22.

Nakashima, K., Hagiwara, T., Yamada, M., 2002. Nuclear localization of peptidylarginine deiminase V and histone deimination in granulocytes. J. Biol. Chem. 277, 49562–49568.

Nakayama, J., Rice, J.C., Strahl, B.D., Allis, C.D., Grewal, S.I., 2001. Role of histone H3 lysine 9 methylation in epigenetic control of heterochromatin assembly. Science 292, 110–113.

Nan, X., Ng, H.H., Johnson, C.A., Laherty, C.D., Turner, B.M., Eisenman, R.N., Bird, A., 1998. Transcriptional repression by the methyl-CpG-binding protein MeCP2 involves a histone deacetylase complex. Nature 393, 386–389.

Narita, K., 1958. Isolation of acetylseryltyrosine from the chymotryptic digests of proteins of five strains of tobacco mosaic virus. Biochim. Biophys. Acta 30, 352–359.

Narita, T., Weinert, B.T., Choudhary, C., 2019. Functions and mechanisms of non-histone protein acetylation. Nat. Rev. Mol. Cell Biol. 20, 156–174.

Natarajan, K., Jackson, B.M., Rhee, E., Hinnebusch, A.G., 1998. yTAFII61 has a general role in RNA polymerase II transcription and is required by Gcn4p to recruit the SAGA coactivator complex. Mol. Cell 2, 683–692.

Nelson, D.A., Perry, M., Sealy, L., Chalkley, R., 1978. DNAse I preferentially digests chromatin containing hyperacetylated histones. Biochem. Biophys. Res. Commun. 82, 1346–1353.

Neumann, H., Hancock, S.M., Buning, R., Routh, A., Chapman, L., Somers, J., Owen-Hughes, T., van Noort, J., Rhodes, D., Chin, J.W., 2009. A method for genetically installing site-specific acetylation in recombinant histones defines the effects of H3 K56 acetylation. Mol. Cell 36, 153–163.

Neuwald, A.F., Landsman, D., 1997. GCN5-related histone N-acetyltransferases belong to a diverse superfamily that includes the yeast SPT10 protein. Trends Biochem. Sci. 22, 154–155.

Ng, H.H., Zhang, Y., Hendrich, B., Johnson, C.A., Turner, B.M., Erdjument-Bromage, H., Tempst, P., Reinberg, D., Bird, A., 1999. MBD2 is a transcriptional repressor belonging to the MeCP1 histone deacetylase complex. Nat. Genet. 23, 58–61.

Ng, H.H., Feng, Q., Wang, H., Erdjument-Bromage, H., Tempst, P., Zhang, Y., Struhl, K., 2002a. Lysine methylation within the globular domain of histone H3 by Dot1 is important for telomeric silencing and sir protein association. Genes Dev. 16, 1518–1527.

Ng, H.H., Xu, R.M., Zhang, Y., Struhl, K., 2002b. Ubiquitination of histone H2B by Rad6 is required for efficient Dot1-mediated methylation of histone H3 lysine 79. J. Biol. Chem. 277, 34655–34657.

Ng, H.H., Ciccone, D.N., Morshead, K.B., Oettinger, M.A., Struhl, K., 2003. Lysine-79 of histone H3 is hypomethylated at silenced loci in yeast and mammalian cells: a potential mechanism for position-effect variegation. Proc. Natl. Acad. Sci. USA 100, 1820–1825.

Nguyen, A.T., Zhang, Y., 2011. The diverse functions of Dot1 and H3K79 methylation. Genes Dev. 25, 1345–1358.

Nguyen, U.T., Bittova, L., Muller, M.M., Fierz, B., David, Y., Houck-Loomis, B., Feng, V., Dann, G.P., Muir, T.W., 2014. Accelerated chromatin biochemistry using DNA-barcoded nucleosome libraries. Nat. Methods 11, 834–840.

Nielsen, S.J., Schneider, R., Bauer, U.M., Bannister, A.J., Morrison, A., O'Carroll, D., Firestein, R., Cleary, M., Jenuwein, T., Herrera, R.E., et al., 2001. Rb targets histone H3 methylation and HP1 to promoters. Nature 412, 561–565.

Nielsen, P.R., Nietlispach, D., Mott, H.R., Callaghan, J., Bannister, A., Kouzarides, T., Murzin, A.G., Murzina, N.V., Laue, E.D., 2002. Structure of the HP1 chromodomain bound to histone H3 methylated at lysine 9. Nature 416, 103–107.

Nimura, K., Ura, K., Shiratori, H., Ikawa, M., Okabe, M., Schwartz, R.J., Kaneda, Y., 2009. A histone H3 lysine 36 trimethyltransferase links Nkx2-5 to Wolf-Hirschhorn syndrome. Nature 460, 287–291.

Nishida, M., Kato, M., Kato, Y., Sasai, N., Ueda, J., Tachibana, M., Shinkai, Y., Yamaguchi, M., 2007. Identification of ZNF200 as a novel binding partner of histone H3 methyltransferase G9a. Genes Cells 12, 877–888.

Nishioka, K., Chuikov, S., Sarma, K., Erdjument-Bromage, H., Allis, C.D., Tempst, P., Reinberg, D., 2002a. Set9, a novel histone H3 methyltransferase that facilitates transcription by precluding histone tail modifications required for heterochromatin formation. Genes Dev. 16, 479–489.

Nishioka, K., Rice, J.C., Sarma, K., Erdjument-Bromage, H., Werner, J., Wang, Y., Chuikov, S., Valenzuela, P., Tempst, P., Steward, R., et al., 2002b. PR-Set7 is a nucleosome-specific methyltransferase that modifies lysine 20 of histone H4 and is associated with silent chromatin. Mol. Cell 9, 1201–1213.

Noma, K., Allis, C.D., Grewal, S.I., 2001. Transitions in distinct histone H3 methylation patterns at the heterochromatin domain boundaries. Science 293, 1150–1155.

Ogata, N., Ueda, K., Hayaishi, O., 1980a. ADP-ribosylation of histone H2B. Identification of glutamic acid residue 2 as the modification site. J. Biol. Chem. 255, 7610–7615.

Ogata, N., Ueda, K., Kagamiyama, H., Hayaishi, O., 1980b. ADP-ribosylation of histone H1. Identification of glutamic acid residues 2, 14, and the COOH-terminal lysine residue as modification sites. J. Biol. Chem. 255, 7616–7620.

Ogawa, Y., Quagliarotti, G., Jordan, J., Taylor, C.W., Starbuck, W.C., Busch, H., 1969. Structural analysis of the glycine-rich, arginine-rich histone. 3. Sequence of the amino-terminal half of the molecule containing the modified lysine residues and the total sequence. J. Biol. Chem. 244, 4387–4392.

Ogawa, H., Ishiguro, K., Gaubatz, S., Livingston, D.M., Nakatani, Y., 2002. A complex with chromatin modifiers that occupies E2F- and Myc-responsive genes in G0 cells. Science 296, 1132–1136.

Ogryzko, V.V., Schiltz, R.L., Russanova, V., Howard, B.H., Nakatani, Y., 1996. The transcriptional coactivators p300 and CBP are histone acetyltransferases. Cell 87, 953–959.

Ogryzko, V.V., Kotani, T., Zhang, X., Schiltz, R.L., Howard, T., Yang, X.J., Howard, B.H., Qin, J., Nakatani, Y., 1998. Histone-like TAFs within the PCAF histone acetylase complex. Cell 94, 35–44.
Ohsumi, K., Katagiri, C., Kishimoto, T., 1993. Chromosome condensation in Xenopus mitotic extracts without histone H1. Science 262, 2033–2035.
Okada, Y., Feng, Q., Lin, Y., Jiang, Q., Li, Y., Coffield, V.M., Su, L., Xu, G., Zhang, Y., 2005. hDOT1L links histone methylation to leukemogenesis. Cell 121, 167–178.
Olson, M.O., Goldknopf, I.L., Guetzow, K.A., James, G.T., Hawkins, T.C., Mays-Rothberg, C.J., Busch, H., 1976. The NH2- and COOH-terminal amino acid sequence of nuclear protein A24. J. Biol. Chem. 251, 5901–5903.
Onate, S.A., Tsai, S.Y., Tsai, M.J., O'Malley, B.W., 1995. Sequence and characterization of a coactivator for the steroid hormone receptor superfamily. Science 270, 1354–1357.
Onishi, M., Liou, G.G., Buchberger, J.R., Walz, T., Moazed, D., 2007. Role of the conserved Sir3-BAH domain in nucleosome binding and silent chromatin assembly. Mol. Cell 28, 1015–1028.
Ord, M.G., Stocken, L.A., 1966. Metabolic properties of histones from rat liver and thymus gland. Biochem. J. 98, 888–897.
Ornaghi, P., Ballario, P., Lena, A.M., Gonzalez, A., Filetici, P., 1999. The bromodomain of Gcn5p interacts in vitro with specific residues in the N terminus of histone H4. J. Mol. Biol. 287, 1–7.
Orrick, L.R., Olson, M.O., Busch, H., 1973. Comparison of nucleolar proteins of normal rat liver and Novikoff hepatoma ascites cells by two-dimensional polyacrylamide gel electrophoresis. Proc. Natl. Acad. Sci. USA 70, 1316–1320.
Owen, D.J., Ornaghi, P., Yang, J.C., Lowe, N., Evans, P.R., Ballario, P., Neuhaus, D., Filetici, P., Travers, A.A., 2000. The structural basis for the recognition of acetylated histone H4 by the bromodomain of histone acetyltransferase gcn5p. EMBO J. 19, 6141–6149.
Ozdemir, A., Spicuglia, S., Lasonder, E., Vermeulen, M., Campsteijn, C., Stunnenberg, H.G., Logie, C., 2005. Characterization of lysine 56 of histone H3 as an acetylation site in *Saccharomyces cerevisiae*. J. Biol. Chem. 280, 25949–25952.
Ozsolak, F., Poling, L.L., Wang, Z., Liu, H., Liu, X.S., Roeder, R.G., Zhang, X., Song, J.S., Fisher, D.E., 2008. Chromatin structure analyses identify miRNA promoters. Genes Dev. 22, 3172–3183.
Pack, L.R., Yamamoto, K.R., Fujimori, D.G., 2016. Opposing chromatin signals direct and regulate the activity of lysine demethylase 4C (KDM4C). J. Biol. Chem. 291, 6060–6070.
Pagliuca, F.W., Collins, M.O., Lichawska, A., Zegerman, P., Choudhary, J.S., Pines, J., 2011. Quantitative proteomics reveals the basis for the biochemical specificity of the cell-cycle machinery. Mol. Cell 43, 406–417.
Paik, W.K., Kim, S., 1967a. Enzymatic methylation of protein fractions from calf thymus nuclei. Biochem. Biophys. Res. Commun. 29, 14–20.
Paik, W.K., Kim, S., 1967b. Epsilon-N-dimethyllysine in histones. Biochem. Biophys. Res. Commun. 27, 479–483.
Paik, W.K., Kim, S., 1968. Protein methylase I. Purification and properties of the enzyme. J. Biol. Chem. 243, 2108–2114.
Paik, W.K., Kim, S., 1973. Enzymatic demethylation of calf thymus histones. Biochem. Biophys. Res. Commun. 51, 781–788.
Paik, W.K., Kim, S., 1974. Epsilon-alkyllysinase. New assay method, purification, and biological significance. Arch. Biochem. Biophys. 165, 369–378.
Paik, W.K., Pearson, D., Lee, H.W., Kim, S., 1970. Nonenzymatic acetylation of histones with acetyl-CoA. Biochim. Biophys. Acta 213, 513–522.
Paik, W.K., Paik, D.C., Kim, S., 2007. Historical review: the field of protein methylation. Trends Biochem. Sci. 32, 146–152.
Pankotai, T., Komonyi, O., Bodai, L., Ujfaludi, Z., Muratoglu, S., Ciurciu, A., Tora, L., Szabad, J., Boros, I., 2005. The homologous Drosophila transcriptional adaptors ADA2a and ADA2b are both required for normal development but have different functions. Mol. Cell. Biol. 25, 8215–8227.
Papai, G., Frechard, A., Kolesnikova, O., Crucifix, C., Schultz, P., Ben-Shem, A., 2020. Structure of SAGA and mechanism of TBP deposition on gene promoters. Nature 577, 711–716.
Park, E.C., Szostak, J.W., 1990. Point mutations in the yeast histone H4 gene prevent silencing of the silent mating type locus HML. Mol. Cell. Biol. 10, 4932–4934.
Park, D., Lee, Y., Bhupindersingh, G., Iyer, V.R., 2013. Widespread misinterpretable ChIP-seq bias in yeast. PLoS One 8, e83506.
Parthun, M.R., Widom, J., Gottschling, D.E., 1996. The major cytoplasmic histone acetyltransferase in yeast: links to chromatin replication and histone metabolism. Cell 87, 85–94.

Parween, S., Alawathugoda, T.T., Prabakaran, A.D., Dheen, S.T., Morse, R.H., Emerald, B.S., Ansari, S.A., 2022. Nutrient sensitive protein O-GlcNAcylation modulates the transcriptome through epigenetic mechanisms during embryonic neurogenesis. Life Sci. Alliance 5, e202201385.

Paterno, G.D., Li, Y., Luchman, H.A., Ryan, P.J., Gillespie, L.L., 1997. cDNA cloning of a novel, developmentally regulated immediate early gene activated by fibroblast growth factor and encoding a nuclear protein. J. Biol. Chem. 272, 25591–25595.

Paulson, J.R., Taylor, S.S., 1982. Phosphorylation of histones 1 and 3 and nonhistone high mobility group 14 by an endogenous kinase in HeLa metaphase chromosomes. J. Biol. Chem. 257, 6064–6072.

Pawson, T., Scott, J.D., 2005. Protein phosphorylation in signaling—50 years and counting. Trends Biochem. Sci. 30, 286–290.

Pei, H., Zhang, L., Luo, K., Qin, Y., Chesi, M., Fei, F., Bergsagel, P.L., Wang, L., You, Z., Lou, Z., 2011. MMSET regulates histone H4K20 methylation and 53BP1 accumulation at DNA damage sites. Nature 470, 124–128.

Pena, P.V., Davrazou, F., Shi, X., Walter, K.L., Verkhusha, V.V., Gozani, O., Zhao, R., Kutateladze, T.G., 2006. Molecular mechanism of histone H3K4me3 recognition by plant homeodomain of ING2. Nature 442, 100–103.

Pena, P.V., Hom, R.A., Hung, T., Lin, H., Kuo, A.J., Wong, R.P., Subach, O.M., Champagne, K.S., Zhao, R., Verkhusha, V.V., et al., 2008. Histone H3K4me3 binding is required for the DNA repair and apoptotic activities of ING1 tumor suppressor. J. Mol. Biol. 380, 303–312.

Penn, M.D., Galgoci, B., Greer, H., 1983. Identification of AAS genes and their regulatory role in general control of amino acid biosynthesis in yeast. Proc. Natl. Acad. Sci. USA 80, 2704–2708.

Perry, M., Chalkley, R., 1981. The effect of histone hyperacetylation on the nuclease sensitivity and the solubility of chromatin. J. Biol. Chem. 256, 3313–3318.

Pesavento, J.J., Yang, H., Kelleher, N.L., Mizzen, C.A., 2008. Certain and progressive methylation of histone H4 at lysine 20 during the cell cycle. Mol. Cell. Biol. 28, 468–486.

Peters, A.H., O'Carroll, D., Scherthan, H., Mechtler, K., Sauer, S., Schofer, C., Weipoltshammer, K., Pagani, M., Lachner, M., Kohlmaier, A., et al., 2001. Loss of the Suv39h histone methyltransferases impairs mammalian heterochromatin and genome stability. Cell 107, 323–337.

Peters, A.H., Mermoud, J.E., O'Carroll, D., Pagani, M., Schweizer, D., Brockdorff, N., Jenuwein, T., 2002. Histone H3 lysine 9 methylation is an epigenetic imprint of facultative heterochromatin. Nat. Genet. 30, 77–80.

Peters, A.H., Kubicek, S., Mechtler, K., O'Sullivan, R.J., Derijck, A.A., Perez-Burgos, L., Kohlmaier, A., Opravil, S., Tachibana, M., Shinkai, Y., et al., 2003. Partitioning and plasticity of repressive histone methylation states in mammalian chromatin. Mol. Cell 12, 1577–1589.

Pfeffer, U., Ferrari, N., Vidali, G., 1986. Availability of hyperacetylated H4 histone in intact nucleosomes to specific antibodies. J. Biol. Chem. 261, 2496–2498.

Phillips, D.M., 1963. The presence of acetyl groups of histones. Biochem. J. 87, 258–263.

Phillips, D.M., 1968. N-terminal acetyl-peptides from two calf thymus histones. Biochem. J. 107, 135–138.

Pijnappel, W.W., Schaft, D., Roguev, A., Shevchenko, A., Tekotte, H., Wilm, M., Rigaut, G., Seraphin, B., Aasland, R., Stewart, A.F., 2001. The S. cerevisiae SET3 complex includes two histone deacetylases, Hos2 and Hst1, and is a meiotic-specific repressor of the sporulation gene program. Genes Dev. 15, 2991–3004.

Pinheiro, I., Margueron, R., Shukeir, N., Eisold, M., Fritzsch, C., Richter, F.M., Mittler, G., Genoud, C., Goyama, S., Kurokawa, M., et al., 2012. Prdm3 and Prdm16 are H3K9me1 methyltransferases required for mammalian heterochromatin integrity. Cell 150, 948–960.

Platero, J.S., Hartnett, T., Eissenberg, J.C., 1995. Functional analysis of the chromo domain of HP1. EMBO J. 14, 3977–3986.

Pogo, B.G., Pogo, A.O., Allfrey, V.G., Mirsky, A.E., 1968. Changing patterns of histone acetylation and RNA synthesis in regeneration of the liver. Proc. Natl. Acad. Sci. USA 59, 1337–1344.

Pokholok, D.K., Harbison, C.T., Levine, S., Cole, M., Hannett, N.M., Lee, T.I., Bell, G.W., Walker, K., Rolfe, P.A., Herbolsheimer, E., et al., 2005. Genome-wide map of nucleosome acetylation and methylation in yeast. Cell 122, 517–527.

Pollard, K.J., Peterson, C.L., 1997. Role for ADA/GCN5 products in antagonizing chromatin-mediated transcriptional repression. Mol. Cell. Biol. 17, 6212–6222.

Posternak, S., 1927. The phosphorus nucleus of caseinogen. Biochem. J. 21, 289.

Posternak, S., Posternak, T., 1927. Préparation des polypeptides contenant les noyaux phosphoré et ferrique de l'ovovitelline. Compt. Rend. Acad. 184, 909.

Posternak, S., Posternak, T., 1928. Sur la labilité des chaînes d'acides sérine phosphoriques et sur une réaction générales des tyrines. Compt. Rend. Acad. 187, 313.

Poulard, C., Noureddine, L.M., Pruvost, L., Le Romancer, M., 2021. Structure, activity, and function of the protein lysine methyltransferase G9a. Life (Basel) 11, 1082.

Powers, N.R., Parvanov, E.D., Baker, C.L., Walker, M., Petkov, P.M., Paigen, K., 2016. The meiotic recombination activator PRDM9 trimethylates both H3K36 and H3K4 at recombination hotspots in vivo. PLoS Genet. 12, e1006146.

Pray-Grant, M.G., Daniel, J.A., Schieltz, D., Yates 3rd, J.R., Grant, P.A., 2005. Chd1 chromodomain links histone H3 methylation with SAGA- and SLIK-dependent acetylation. Nature 433, 434–438.

Ptashne, M., Gann, A., 1997. Transcriptional activation by recruitment. Nature 386, 569–577.

Pugh, B.F., Tjian, R., 1990. Mechanism of transcriptional activation by Sp1: evidence for coactivators. Cell 61, 1187–1197.

Qiao, Q., Li, Y., Chen, Z., Wang, M., Reinberg, D., Xu, R.M., 2011. The structure of NSD1 reveals an autoregulatory mechanism underlying histone H3K36 methylation. J. Biol. Chem. 286, 8361–8368.

Qiu, H., Hu, C., Zhang, F., Hwang, G.J., Swanson, M.J., Boonchird, C., Hinnebusch, A.G., 2005. Interdependent recruitment of SAGA and Srb mediator by transcriptional activator Gcn4p. Mol. Cell. Biol. 25, 3461–3474.

Qu, K., Chen, K., Wang, H., Li, X., Chen, Z., 2022. Structure of the NuA4 acetyltransferase complex bound to the nucleosome. Nature 610, 569–574.

Radzisheuskaya, A., Shliaha, P.V., Grinev, V.V., Shlyueva, D., Damhofer, H., Koche, R., Gorshkov, V., Kovalchuk, S., Zhan, Y., Rodriguez, K.L., et al., 2021. Complex-dependent histone acetyltransferase activity of KAT8 determines its role in transcription and cellular homeostasis. Mol. Cell 81, 1749–1765.e1748.

Rajagopal, N., Ernst, J., Ray, P., Wu, J., Zhang, M., Kellis, M., Ren, B., 2014. Distinct and predictive histone lysine acetylation patterns at promoters, enhancers, and gene bodies. G3 (Bethesda) 4, 2051–2063.

Rayasam, G.V., Wendling, O., Angrand, P.O., Mark, M., Niederreither, K., Song, L., Lerouge, T., Hager, G.L., Chambon, P., Losson, R., 2003. NSD1 is essential for early post-implantation development and has a catalytically active SET domain. EMBO J. 22, 3153–3163.

Rea, S., Eisenhaber, F., O'Carroll, D., Strahl, B.D., Sun, Z.W., Schmid, M., Opravil, S., Mechtler, K., Ponting, C.P., Allis, C.D., et al., 2000. Regulation of chromatin structure by site-specific histone H3 methyltransferases. Nature 406, 593–599.

Rechtsteiner, A., Ercan, S., Takasaki, T., Phippen, T.M., Egelhofer, T.A., Wang, W., Kimura, H., Lieb, J.D., Strome, S., 2010. The histone H3K36 methyltransferase MES-4 acts epigenetically to transmit the memory of germline gene expression to progeny. PLoS Genet. 6, e1001091.

Reeves, W.M., Hahn, S., 2005. Targets of the Gal4 transcription activator in functional transcription complexes. Mol. Cell. Biol. 25, 9092–9102.

Reifsnyder, C., Lowell, J., Clarke, A., Pillus, L., 1996. Yeast SAS silencing genes and human genes associated with AML and HIV-1 Tat interactions are homologous with acetyltransferases. Nat. Genet. 14, 42–49.

Reinke, H., Horz, W., 2003. Histones are first hyperacetylated and then lose contact with the activated PHO5 promoter. Mol. Cell 11, 1599–1607.

Reinke, H., Gregory, P.D., Horz, W., 2001. A transient histone hyperacetylation signal marks nucleosomes for remodeling at the PHO8 promoter in vivo. Mol. Cell 7, 529–538.

Rhee, H.S., Pugh, B.F., 2012. Genome-wide structure and organization of eukaryotic pre-initiation complexes. Nature 483, 295–301.

Rice, J.C., Briggs, S.D., Ueberheide, B., Barber, C.M., Shabanowitz, J., Hunt, D.F., Shinkai, Y., Allis, C.D., 2003. Histone methyltransferases direct different degrees of methylation to define distinct chromatin domains. Mol. Cell 12, 1591–1598.

Riggs, M.G., Whittaker, R.G., Neumann, J.R., Ingram, V.M., 1977. N-butyrate causes histone modification in HeLa and friend erythroleukaemia cells. Nature 268, 462–464.

Rimington, C., 1927. The phosphorus of caseinogen: constitution of phosphopeptone. Biochem. J. 21, 1187–1193.

Roberts, S.M., Winston, F., 1996. SPT20/ADA5 encodes a novel protein functionally related to the TATA-binding protein and important for transcription in *Saccharomyces cerevisiae*. Mol. Cell. Biol. 16, 3206–3213.

Roberts, S.M., Winston, F., 1997. Essential functional interactions of SAGA, a *Saccharomyces cerevisiae* complex of Spt, Ada, and Gcn5 proteins, with the Snf/Swi and Srb/mediator complexes. Genetics 147, 451–465.

Robyr, D., Suka, Y., Xenarios, I., Kurdistani, S.K., Wang, A., Suka, N., Grunstein, M., 2002. Microarray deacetylation maps determine genome-wide functions for yeast histone deacetylases. Cell 109, 437–446.

Robzyk, K., Recht, J., Osley, M.A., 2000. Rad6-dependent ubiquitination of histone H2B in yeast. Science 287, 501–504.

Roguev, A., Schaft, D., Shevchenko, A., Pijnappel, W.W., Wilm, M., Aasland, R., Stewart, A.F., 2001. The *Saccharomyces cerevisiae* Set1 complex includes an Ash2 homologue and methylates histone 3 lysine 4. EMBO J. 20, 7137–7148.

Rosler, S.M., Kramer, K., Finkemeier, I., Humpf, H.U., Tudzynski, B., 2016. The SAGA complex in the rice pathogen fusarium fujikuroi: structure and functional characterization. Mol. Microbiol. 102, 951–974.

Ross, S.E., McCord, A.E., Jung, C., Atan, D., Mok, S.I., Hemberg, M., Kim, T.K., Salogiannis, J., Hu, L., Cohen, S., et al., 2012. Bhlhb5 and Prdm8 form a repressor complex involved in neuronal circuit assembly. Neuron 73, 292–303.

Rossetto, D., Cramet, M., Wang, A.Y., Steunou, A.L., Lacoste, N., Schulze, J.M., Cote, V., Monnet-Saksouk, J., Piquet, S., Nourani, A., et al., 2014. Eaf5/7/3 form a functionally independent NuA4 submodule linked to RNA polymerase II-coupled nucleosome recycling. EMBO J. 33, 1397–1415.

Rossmann, M.P., Luo, W., Tsaponina, O., Chabes, A., Stillman, B., 2011. A common telomeric gene silencing assay is affected by nucleotide metabolism. Mol. Cell 42, 127–136.

Roth, S.Y., Shimizu, M., Johnson, L., Grunstein, M., Simpson, R.T., 1992. Stable nucleosome positioning and complete repression by the yeast alpha 2 repressor are disrupted by amino-terminal mutations in histone H4. Genes Dev. 6, 411–425.

Rothbart, S.B., Dickson, B.M., Raab, J.R., Grzybowski, A.T., Krajewski, K., Guo, A.H., Shanle, E.K., Josefowicz, S.Z., Fuchs, S.M., Allis, C.D., et al., 2015. An interactive database for the assessment of histone antibody specificity. Mol. Cell 59, 502–511.

Ruiz-Garcia, A.B., Sendra, R., Pamblanco, M., Tordera, V., 1997. Gcn5p is involved in the acetylation of histone H3 in nucleosomes. FEBS Lett. 403, 186–190.

Rundlett, S.E., Carmen, A.A., Kobayashi, R., Bavykin, S., Turner, B.M., Grunstein, M., 1996. HDA1 and RPD3 are members of distinct yeast histone deacetylase complexes that regulate silencing and transcription. Proc. Natl. Acad. Sci. USA 93, 14503–14508.

Rundlett, S.E., Carmen, A.A., Suka, N., Turner, B.M., Grunstein, M., 1998. Transcriptional repression by UME6 involves deacetylation of lysine 5 of histone H4 by RPD3. Nature 392, 831–835.

Ruthenburg, A.J., Allis, C.D., Wysocka, J., 2007. Methylation of lysine 4 on histone H3: intricacy of writing and reading a single epigenetic mark. Mol. Cell 25, 15–30.

Ryu, H.Y., Hochstrasser, M., 2021. Histone sumoylation and chromatin dynamics. Nucleic Acids Res. 49, 6043–6052.

Ryu, H.Y., Duan, R., Ahn, S.H., 2019. Yeast symmetric arginine methyltransferase Hsl7 has a repressive role in transcription. Res. Microbiol. 170, 222–229.

Sabet, N., Volo, S., Yu, C., Madigan, J.P., Morse, R.H., 2004. Genome-wide analysis of the relationship between transcriptional regulation by Rpd3p and the histone H3 and H4 amino termini in budding yeast. Mol. Cell. Biol. 24, 8823–8833.

Sakabe, K., Wang, Z., Hart, G.W., 2010. Beta-N-acetylglucosamine (O-GlcNAc) is part of the histone code. Proc. Natl. Acad. Sci. USA 107, 19915–19920.

Saleh, A., Lang, V., Cook, R., Brandl, C.J., 1997. Identification of native complexes containing the yeast coactivator/repressor proteins NGG1/ADA3 and ADA2. J. Biol. Chem. 272, 5571–5578.

Saleh, A., Schieltz, D., Ting, N., McMahon, S.B., Litchfield, D.W., Yates 3rd, J.R., Lees-Miller, S.P., Cole, M.D., Brandl, C.J., 1998. Tra1p is a component of the yeast Ada.Spt transcriptional regulatory complexes. J. Biol. Chem. 273, 26559–26565.

Saleque, S., Kim, J., Rooke, H.M., Orkin, S.H., 2007. Epigenetic regulation of hematopoietic differentiation by Gfi-1 and Gfi-1b is mediated by the cofactors CoREST and LSD1. Mol. Cell 27, 562–572.

Samara, N.L., Datta, A.B., Berndsen, C.E., Zhang, X., Yao, T., Cohen, R.E., Wolberger, C., 2010. Structural insights into the assembly and function of the SAGA deubiquitinating module. Science 328, 1025–1029.

Sanchez del Pino, M.M., Lopez-Rodas, G., Sendra, R., Tordera, V., 1994. Properties of the yeast nuclear histone deacetylase. Biochem. J. 303 (Pt 3), 723–729.

Sanchez, R., Zhou, M.M., 2009. The role of human bromodomains in chromatin biology and gene transcription. Curr. Opin. Drug Discov. Devel. 12, 659–665.

Sanders, S.L., Jennings, J., Canutescu, A., Link, A.J., Weil, P.A., 2002. Proteomics of the eukaryotic transcription machinery: identification of proteins associated with components of yeast TFIID by multidimensional mass spectrometry. Mol. Cell. Biol. 22, 4723–4738.

Sanders, S.L., Portoso, M., Mata, J., Bahler, J., Allshire, R.C., Kouzarides, T., 2004. Methylation of histone H4 lysine 20 controls recruitment of Crb2 to sites of DNA damage. Cell 119, 603–614.

San-Segundo, P.A., Roeder, G.S., 2000. Role for the silencing protein Dot1 in meiotic checkpoint control. Mol. Biol. Cell 11, 3601–3615.

Santos-Rosa, H., Schneider, R., Bannister, A.J., Sherriff, J., Bernstein, B.E., Emre, N.C., Schreiber, S.L., Mellor, J., Kouzarides, T., 2002. Active genes are tri-methylated at K4 of histone H3. Nature 419, 407–411.

Sarg, B., Helliger, W., Talasz, H., Forg, B., Lindner, H.H., 2006. Histone H1 phosphorylation occurs site-specifically during interphase and mitosis: identification of a novel phosphorylation site on histone H1. J. Biol. Chem. 281, 6573–6580.

Sassone-Corsi, P., Mizzen, C.A., Cheung, P., Crosio, C., Monaco, L., Jacquot, S., Hanauer, A., Allis, C.D., 1999. Requirement of Rsk-2 for epidermal growth factor-activated phosphorylation of histone H3. Science 285, 886–891.

Sato, K., Kumar, A., Hamada, K., Okada, C., Oguni, A., Machiyama, A., Sakuraba, S., Nishizawa, T., Nureki, O., Kono, H., et al., 2021. Structural basis of the regulation of the normal and oncogenic methylation of nucleosomal histone H3 Lys36 by NSD2. Nat. Commun. 12, 6605.

Schaft, D., Roguev, A., Kotovic, K.M., Shevchenko, A., Sarov, M., Shevchenko, A., Neugebauer, K.M., Stewart, A.F., 2003. The histone 3 lysine 36 methyltransferase, SET2, is involved in transcriptional elongation. Nucleic Acids Res. 31, 2475–2482.

Schapira, M., 2011. Structural chemistry of human SET domain protein methyltransferases. Curr. Chem. Genomics 5, 85–94.

Schiltz, R.L., Mizzen, C.A., Vassilev, A., Cook, R.G., Allis, C.D., Nakatani, Y., 1999. Overlapping but distinct patterns of histone acetylation by the human coactivators p300 and PCAF within nucleosomal substrates. J. Biol. Chem. 274, 1189–1192.

Schneider, J., Wood, A., Lee, J.S., Schuster, R., Dueker, J., Maguire, C., Swanson, S.K., Florens, L., Washburn, M.P., Shilatifard, A., 2005. Molecular regulation of histone H3 trimethylation by COMPASS and the regulation of gene expression. Mol. Cell 19, 849–856.

Schneider, J., Bajwa, P., Johnson, F.C., Bhaumik, S.R., Shilatifard, A., 2006. Rtt109 is required for proper H3K56 acetylation: a chromatin mark associated with the elongating RNA polymerase II. J. Biol. Chem. 281, 37270–37274.

Scholes, D.T., Banerjee, M., Bowen, B., Curcio, M.J., 2001. Multiple regulators of Ty1 transposition in *Saccharomyces cerevisiae* have conserved roles in genome maintenance. Genetics 159, 1449–1465.

Schotta, G., Lachner, M., Sarma, K., Ebert, A., Sengupta, R., Reuter, G., Reinberg, D., Jenuwein, T., 2004. A silencing pathway to induce H3-K9 and H4-K20 trimethylation at constitutive heterochromatin. Genes Dev. 18, 1251–1262.

Schotta, G., Sengupta, R., Kubicek, S., Malin, S., Kauer, M., Callen, E., Celeste, A., Pagani, M., Opravil, S., De La Rosa-Velazquez, I.A., et al., 2008. A chromatin-wide transition to H4K20 monomethylation impairs genome integrity and programmed DNA rearrangements in the mouse. Genes Dev. 22, 2048–2061.

Schreiber-Agus, N., Chin, L., Chen, K., Torres, R., Rao, G., Guida, P., Skoultchi, A.I., DePinho, R.A., 1995. An amino-terminal domain of Mxi1 mediates anti-Myc oncogenic activity and interacts with a homolog of the yeast transcriptional repressor SIN3. Cell 80, 777–786.

Schubeler, D., MacAlpine, D.M., Scalzo, D., Wirbelauer, C., Kooperberg, C., van Leeuwen, F., Gottschling, D.E., O'Neill, L.P., Turner, B.M., Delrow, J., et al., 2004. The histone modification pattern of active genes revealed through genome-wide chromatin analysis of a higher eukaryote. Genes Dev. 18, 1263–1271.

Schultz, D.C., Ayyanathan, K., Negorev, D., Maul, G.G., Rauscher 3rd, F.J., 2002. SETDB1: a novel KAP-1-associated histone H3, lysine 9-specific methyltransferase that contributes to HP1-mediated silencing of euchromatic genes by KRAB zinc-finger proteins. Genes Dev. 16, 919–932.

Schurter, B.T., Koh, S.S., Chen, D., Bunick, G.J., Harp, J.M., Hanson, B.L., Henschen-Edman, A., Mackay, D.R., Stallcup, M.R., Aswad, D.W., 2001. Methylation of histone H3 by coactivator-associated arginine methyltransferase 1. Biochemistry 40, 5747–5756.

Sealy, L., Chalkley, R., 1978a. DNA associated with hyperacetylated histone is preferentially digested by DNase I. Nucleic Acids Res. 5, 1863–1876.

Sealy, L., Chalkley, R., 1978b. The effect of sodium butyrate on histone modification. Cell 14, 115–121.

Sendra, R., Tse, C., Hansen, J.C., 2000. The yeast histone acetyltransferase A2 complex, but not free Gcn5p, binds stably to nucleosomal arrays. J. Biol. Chem. 275, 24928–24934.

Separovich, R.J., Wilkins, M.R., 2021. Ready, SET, go: post-translational regulation of the histone lysine methylation network in budding yeast. J. Biol. Chem. 297, 100939.

Setiaputra, D., Ross, J.D., Lu, S., Cheng, D.T., Dong, M.Q., Yip, C.K., 2015. Conformational flexibility and subunit arrangement of the modular yeast Spt-Ada-Gcn5 acetyltransferase complex. J. Biol. Chem. 290, 10057–10070.

Seto, E., Yoshida, M., 2014. Erasers of histone acetylation: the histone deacetylase enzymes. Cold Spring Harb. Perspect. Biol. 6, a018713.

Shah, R.N., Grzybowski, A.T., Cornett, E.M., Johnstone, A.L., Dickson, B.M., Boone, B.A., Cheek, M.A., Cowles, M.W., Maryanski, D., Meiners, M.J., et al., 2018. Examining the roles of H3K4 methylation states with systematically characterized antibodies. Mol. Cell 72, 162–177.e167.

Shahbazian, M.D., Zhang, K., Grunstein, M., 2005. Histone H2B ubiquitylation controls processive methylation but not monomethylation by Dot1 and Set1. Mol. Cell 19, 271–277.

Shanower, G.A., Muller, M., Blanton, J.L., Honti, V., Gyurkovics, H., Schedl, P., 2005. Characterization of the grappa gene, the Drosophila histone H3 lysine 79 methyltransferase. Genetics 169, 173–184.

Sharov, G., Voltz, K., Durand, A., Kolesnikova, O., Papai, G., Myasnikov, A.G., Dejaegere, A., Ben Shem, A., Schultz, P., 2017. Structure of the transcription activator target Tra1 within the chromatin modifying complex SAGA. Nat. Commun. 8, 1556.

Shaver, S., Casas-Mollano, J.A., Cerny, R.L., Cerutti, H., 2010. Origin of the polycomb repressive complex 2 and gene silencing by an E(z) homolog in the unicellular alga Chlamydomonas. Epigenetics 5, 301–312.

Sheikh, B.N., Akhtar, A., 2019. The many lives of KATs—detectors, integrators and modulators of the cellular environment. Nat. Rev. Genet. 20, 7–23.

Sheikh, M.A., Emerald, B.S., Ansari, S.A., 2021. Stem cell fate determination through protein O-GlcNAcylation. J. Biol. Chem. 296, 100035.

Shen, X., Yu, L., Weir, J.W., Gorovsky, M.A., 1995. Linker histones are not essential and affect chromatin condensation in vivo. Cell 82, 47–56.

Shen, E.C., Henry, M.F., Weiss, V.H., Valentini, S.R., Silver, P.A., Lee, M.S., 1998. Arginine methylation facilitates the nuclear export of hnRNP proteins. Genes Dev. 12, 679–691.

Sheppard, H.M., Harries, J.C., Hussain, S., Bevan, C., Heery, D.M., 2001. Analysis of the steroid receptor coactivator 1 (SRC1)-CREB binding protein interaction interface and its importance for the function of SRC1. Mol. Cell. Biol. 21, 39–50.

Shi, Y., Sawada, J., Sui, G., Affar el, B., Whetstine, J.R., Lan, F., Ogawa, H., Luke, M.P., Nakatani, Y., Shi, Y., 2003. Coordinated histone modifications mediated by a CtBP co-repressor complex. Nature 422, 735–738.

Shi, Y., Lan, F., Matson, C., Mulligan, P., Whetstine, J.R., Cole, P.A., Casero, R.A., Shi, Y., 2004. Histone demethylation mediated by the nuclear amine oxidase homolog LSD1. Cell 119, 941–953.

Shi, Y., Wang, X.X., Zhuang, Y.W., Jiang, Y., Melcher, K., Xu, H.E., 2017. Structure of the PRC2 complex and application to drug discovery. Acta Pharmacol. Sin. 38, 963–976.

Shiio, Y., Eisenman, R.N., 2003. Histone sumoylation is associated with transcriptional repression. Proc. Natl. Acad. Sci. USA 100, 13225–13230.

Shilatifard, A., 2012. The COMPASS family of histone H3K4 methylases: mechanisms of regulation in development and disease pathogenesis. Annu. Rev. Biochem. 81, 65–95.

Shimada, M., Niida, H., Zineldeen, D.H., Tagami, H., Tanaka, M., Saito, H., Nakanishi, M., 2008. Chk1 is a histone H3 threonine 11 kinase that regulates DNA damage-induced transcriptional repression. Cell 132, 221–232.

Shvedunova, M., Akhtar, A., 2022. Modulation of cellular processes by histone and non-histone protein acetylation. Nat. Rev. Mol. Cell Biol. 23, 329–349.

Silverman, N., Agapite, J., Guarente, L., 1994. Yeast ADA2 protein binds to the VP16 protein activation domain and activates transcription. Proc. Natl. Acad. Sci. USA 91, 11665–11668.

Simpson, R.T., 1978. Structure of chromatin containing extensively acetylated H3 and H4. Cell 13, 691–699.

Sims 3rd, R.J., Reinberg, D., 2006. Histone H3 Lys 4 methylation: caught in a bind? Genes Dev. 20, 2779–2786.

Singer, M.S., Kahana, A., Wolf, A.J., Meisinger, L.L., Peterson, S.E., Goggin, C., Mahowald, M., Gottschling, D.E., 1998. Identification of high-copy disruptors of telomeric silencing in *Saccharomyces cerevisiae*. Genetics 150, 613–632.

Singh, R.K., Gunjan, A., 2011. Histone tyrosine phosphorylation comes of age. Epigenetics 6, 153–160.

Singh, R.K., Kabbaj, M.H., Paik, J., Gunjan, A., 2009. Histone levels are regulated by phosphorylation and ubiquitylation-dependent proteolysis. Nat. Cell Biol. 11, 925–933.

Smith, E.R., Eisen, A., Gu, W., Sattah, M., Pannuti, A., Zhou, J., Cook, R.G., Lucchesi, J.C., Allis, C.D., 1998. ESA1 is a histone acetyltransferase that is essential for growth in yeast. Proc. Natl. Acad. Sci. USA 95, 3561–3565.

Smith, E.R., Pannuti, A., Gu, W., Steurnagel, A., Cook, R.G., Allis, C.D., Lucchesi, J.C., 2000a. The drosophila MSL complex acetylates histone H4 at lysine 16, a chromatin modification linked to dosage compensation. Mol. Cell. Biol. 20, 312–318.

Smith, J.S., Brachmann, C.B., Celic, I., Kenna, M.A., Muhammad, S., Starai, V.J., Avalos, J.L., Escalante-Semerena, J.C., Grubmeyer, C., Wolberger, C., et al., 2000b. A phylogenetically conserved NAD+-dependent protein deacetylase activity in the Sir2 protein family. Proc. Natl. Acad. Sci. USA 97, 6658–6663.

Smolle, M., Venkatesh, S., Gogol, M.M., Li, H., Zhang, Y., Florens, L., Washburn, M.P., Workman, J.L., 2012. Chromatin remodelers Isw1 and Chd1 maintain chromatin structure during transcription by preventing histone exchange. Nat. Struct. Mol. Biol. 19, 884–892.

Smtih, D.L., Bruegger, B.B., Halpern, R.M., Smith, R.A., 1973. New histone kinases in nuclei of rat tissues. Nature 246, 103–104.

Soffers, J.H.M., Workman, J.L., 2020. The SAGA chromatin-modifying complex: the sum of its parts is greater than the whole. Genes Dev. 34, 1287–1303.

Song, Y., Dagil, L., Fairall, L., Robertson, N., Wu, M., Ragan, T.J., Savva, C.G., Saleh, A., Morone, N., Kunze, M.B.A., et al., 2020. Mechanism of crosstalk between the LSD1 demethylase and HDAC1 deacetylase in the CoREST complex. Cell Rep. 30, 2699–2711.e2698.

Soutoglou, E., Talianidis, I., 2002. Coordination of PIC assembly and chromatin remodeling during differentiation-induced gene activation. Science 295, 1901–1904.

Spencer, T.E., Jenster, G., Burcin, M.M., Allis, C.D., Zhou, J., Mizzen, C.A., McKenna, N.J., Onate, S.A., Tsai, S.Y., Tsai, M.J., et al., 1997. Steroid receptor coactivator-1 is a histone acetyltransferase. Nature 389, 194–198.

Srivastava, R., Rai, K.M., Pandey, B., Singh, S.P., Sawant, S.V., 2015. Spt-Ada-Gcn5-acetyltransferase (SAGA) complex in plants: genome wide identification, evolutionary conservation and functional determination. PLoS One 10, e0134709.

Stegeman, R., Spreacker, P.J., Swanson, S.K., Stephenson, R., Florens, L., Washburn, M.P., Weake, V.M., 2016. The spliceosomal protein SF3B5 is a novel component of Drosophila SAGA that functions in gene expression independent of splicing. J. Mol. Biol. 428, 3632–3649.

Steger, D.J., Eberharter, A., John, S., Grant, P.A., Workman, J.L., 1998. Purified histone acetyltransferase complexes stimulate HIV-1 transcription from preassembled nucleosomal arrays. Proc. Natl. Acad. Sci. USA 95, 12924–12929.

Steger, D.J., Lefterova, M.I., Ying, L., Stonestrom, A.J., Schupp, M., Zhuo, D., Vakoc, A.L., Kim, J.E., Chen, J., Lazar, M.-A., et al., 2008. DOT1L/KMT4 recruitment and H3K79 methylation are ubiquitously coupled with gene transcription in mammalian cells. Mol. Cell. Biol. 28, 2825–2839.

Steinfeld, I., Shamir, R., Kupiec, M., 2007. A genome-wide analysis in *Saccharomyces cerevisiae* demonstrates the influence of chromatin modifiers on transcription. Nat. Genet. 39, 303–309.

Sterner, D.E., Berger, S.L., 2000. Acetylation of histones and transcription-related factors. Microbiol. Mol. Biol. Rev. 64, 435–459.

Sterner, D.E., Grant, P.A., Roberts, S.M., Duggan, L.J., Belotserkovskaya, R., Pacella, L.A., Winston, F., Workman, J.L., Berger, S.L., 1999. Functional organization of the yeast SAGA complex: distinct components involved in structural integrity, nucleosome acetylation, and TATA-binding protein interaction. Mol. Cell. Biol. 19, 86–98.

Stillman, D.J., Dorland, S., Yu, Y., 1994. Epistasis analysis of suppressor mutations that allow HO expression in the absence of the yeast SW15 transcriptional activator. Genetics 136, 781–788.

Stocker, B.A.D., McDonough, M.W., Ambler, R.P., 1961. A gene determining presence or absence of epsilon-N-methyl-lysine in *Salmonella* flagellar protein. Nature 189, 556–558.

Strahl, B.D., Allis, C.D., 2000. The language of covalent histone modifications. Nature 403, 41–45.

Strahl, B.D., Briggs, S.D., 2021. The SAGA continues: the rise of cis- and trans-histone crosstalk pathways. Biochim. Biophys. Acta Gene Regul. Mech. 1864, 194600.

Strahl, B.D., Ohba, R., Cook, R.G., Allis, C.D., 1999. Methylation of histone H3 at lysine 4 is highly conserved and correlates with transcriptionally active nuclei in Tetrahymena. Proc. Natl. Acad. Sci. USA 96, 14967–14972.

Strahl, B.D., Briggs, S.D., Brame, C.J., Caldwell, J.A., Koh, S.S., Ma, H., Cook, R.G., Shabanowitz, J., Hunt, D.F., Stallcup, M.R., et al., 2001. Methylation of histone H4 at arginine 3 occurs in vivo and is mediated by the nuclear receptor coactivator PRMT1. Curr. Biol. 11, 996–1000.

Strahl, B.D., Grant, P.A., Briggs, S.D., Sun, Z.W., Bone, J.R., Caldwell, J.A., Mollah, S., Cook, R.G., Shabanowitz, J., Hunt, D.F., et al., 2002. Set2 is a nucleosomal histone H3-selective methyltransferase that mediates transcriptional repression. Mol. Cell. Biol. 22, 1298–1306.

Su, D., Hu, Q., Li, Q., Thompson, J.R., Cui, G., Fazly, A., Davies, B.A., Botuyan, M.V., Zhang, Z., Mer, G., 2012. Structural basis for recognition of H3K56-acetylated histone H3-H4 by the chaperone Rtt106. Nature 483, 104–107.

Suganuma, T., Workman, J.L., 2018. Chromatin and metabolism. Annu. Rev. Biochem. 87, 27–49.

Suganuma, T., Gutierrez, J.L., Li, B., Florens, L., Swanson, S.K., Washburn, M.P., Abmayr, S.M., Workman, J.L., 2008. ATAC is a double histone acetyltransferase complex that stimulates nucleosome sliding. Nat. Struct. Mol. Biol. 15, 364–372.

Suka, N., Suka, Y., Carmen, A.A., Wu, J., Grunstein, M., 2001. Highly specific antibodies determine histone acetylation site usage in yeast heterochromatin and euchromatin. Mol. Cell 8, 473–479.
Sun, Z.W., Allis, C.D., 2002. Ubiquitination of histone H2B regulates H3 methylation and gene silencing in yeast. Nature 418, 104–108.
Sung, M.T., Harford, J., Bundman, M., Vidalakas, G., 1977. Metabolism of histones in avian erythroid cells. Biochemistry 16, 279–285.
Sures, I., Gallwitz, D., 1980. Histone-specific acetyltransferases from calf thymus. Isolation, properties, and substrate specificity of three different enzymes. Biochemistry 19, 943–951.
Swanson, M.J., Qiu, H., Sumibcay, L., Krueger, A., Kim, S.J., Natarajan, K., Yoon, S., Hinnebusch, A.G., 2003. A multiplicity of coactivators is required by Gcn4p at individual promoters in vivo. Mol. Cell. Biol. 23, 2800–2820.
Syntichaki, P., Topalidou, I., Thireos, G., 2000. The Gcn5 bromodomain co-ordinates nucleosome remodelling. Nature 404, 414–417.
Tachibana, M., Sugimoto, K., Fukushima, T., Shinkai, Y., 2001. Set domain-containing protein, G9a, is a novel lysine-preferring mammalian histone methyltransferase with hyperactivity and specific selectivity to lysines 9 and 27 of histone H3. J. Biol. Chem. 276, 25309–25317.
Tachibana, M., Sugimoto, K., Nozaki, M., Ueda, J., Ohta, T., Ohki, M., Fukuda, M., Takeda, N., Niida, H., Kato, H., et al., 2002. G9a histone methyltransferase plays a dominant role in euchromatic histone H3 lysine 9 methylation and is essential for early embryogenesis. Genes Dev. 16, 1779–1791.
Takahashi, Y.H., Schulze, J.M., Jackson, J., Hentrich, T., Seidel, C., Jaspersen, S.L., Kobor, M.S., Shilatifard, A., 2011. Dot1 and histone H3K79 methylation in natural telomeric and HM silencing. Mol. Cell 42, 118–126.
Talbert, P.B., Henikoff, S., 2021. The Yin and Yang of histone marks in transcription. Annu. Rev. Genomics Hum. Genet. 22, 147–170.
Tamaru, H., Selker, E.U., 2001. A histone H3 methyltransferase controls DNA methylation in Neurospora crassa. Nature 414, 277–283.
Tamkun, J.W., Deuring, R., Scott, M.P., Kissinger, M., Pattatucci, A.M., Kaufman, T.C., Kennison, J.A., 1992. Brahma: a regulator of Drosophila homeotic genes structurally related to the yeast transcriptional activator SNF2/SWI2. Cell 68, 561–572.
Tang, Y., Holbert, M.A., Wurtele, H., Meeth, K., Rocha, W., Gharib, M., Jiang, E., Thibault, P., Verreault, A., Cole, P.A., et al., 2008. Fungal Rtt109 histone acetyltransferase is an unexpected structural homolog of metazoan p300/CBP. Nat. Struct. Mol. Biol. 15, 738–745.
Tanner, K.G., Landry, J., Sternglanz, R., Denu, J.M., 2000. Silent information regulator 2 family of NAD- dependent histone/protein deacetylases generates a unique product, 1-O-acetyl-ADP-ribose. Proc. Natl. Acad. Sci. USA 97, 14178–14182.
Tanny, J.C., Dowd, G.J., Huang, J., Hilz, H., Moazed, D., 1999. An enzymatic activity in the yeast Sir2 protein that is essential for gene silencing. Cell 99, 735–745.
Taunton, J., Hassig, C.A., Schreiber, S.L., 1996. A mammalian histone deacetylase related to the yeast transcriptional regulator Rpd3p. Science 272, 408–411.
Taverna, S.D., Ilin, S., Rogers, R.S., Tanny, J.C., Lavender, H., Li, H., Baker, L., Boyle, J., Blair, L.P., Chait, B.T., et al., 2006. Yng1 PHD finger binding to H3 trimethylated at K4 promotes NuA3 HAT activity at K14 of H3 and transcription at a subset of targeted ORFs. Mol. Cell 24, 785–796.
Teytelman, L., Thurtle, D.M., Rine, J., van Oudenaarden, A., 2013. Highly expressed loci are vulnerable to misleading ChIP localization of multiple unrelated proteins. Proc. Natl. Acad. Sci. USA 110, 18602–18607.
Thakore, P.I., Black, J.B., Hilton, I.B., Gersbach, C.A., 2016. Editing the epigenome: technologies for programmable transcription and epigenetic modulation. Nat. Methods 13, 127–137.
Thomas, G., Lange, H.W., Hempel, K., 1975. Kinetics of histone methylation in vivo and its relation to the cell cycle in Ehrlich ascites tumor cells. Eur. J. Biochem. 51, 609–615.
Thompson, C.M., Koleske, A.J., Chao, D.M., Young, R.A., 1993. A multisubunit complex associated with the RNA polymerase II CTD and TATA-binding protein in yeast. Cell 73, 1361–1375.
Thomson, J.P., Skene, P.J., Selfridge, J., Clouaire, T., Guy, J., Webb, S., Kerr, A.R., Deaton, A., Andrews, R., James, K.D., et al., 2010. CpG islands influence chromatin structure via the CpG-binding protein Cfp1. Nature 464, 1082–1086.
Timmers, H.T.M., 2021. SAGA and TFIID: friends of TBP drifting apart. Biochim. Biophys. Acta Gene Regul. Mech. 1864, 194604.
Tirosh, I., Barkai, N., 2008. Two strategies for gene regulation by promoter nucleosomes. Genome Res. 18, 1084–1091.
Tjeertes, J.V., Miller, K.M., Jackson, S.P., 2009. Screen for DNA-damage-responsive histone modifications identifies H3K9Ac and H3K56Ac in human cells. EMBO J. 28, 1878–1889.

Tordera, V., Sendra, R., Perez-Ortin, J.E., 1993. The role of histones and their modifications in the informative content of chromatin. Experientia 49, 780–788.

Travis, G.H., Colavito-Shepanski, M., Grunstein, M., 1984. Extensive purification and characterization of chromatin-bound histone acetyltransferase from *Saccharomyces cerevisiae*. J. Biol. Chem. 259, 14406–14412.

Trewick, S.C., McLaughlin, P.J., Allshire, R.C., 2005. Methylation: lost in hydroxylation? EMBO Rep. 6, 315–320.

Triezenberg, S.J., Kingsbury, R.C., McKnight, S.L., 1988. Functional dissection of VP16, the trans-activator of herpes simplex virus immediate early gene expression. Genes Dev. 2, 718–729.

Trojer, P., Li, G., Sims 3rd, R.J., Vaquero, A., Kalakonda, N., Boccuni, P., Lee, D., Erdjument-Bromage, H., Tempst, P., Nimer, S.D., et al., 2007. L3MBTL1, a histone-methylation-dependent chromatin lock. Cell 129, 915–928.

Tsang, A.W., Escalante-Semerena, J.C., 1998. CobB, a new member of the SIR2 family of eucaryotic regulatory proteins, is required to compensate for the lack of nicotinate mononucleotide: 5,6-dimethylbenzimidazole phosphoribosyltransferase activity in cobT mutants during cobalamin biosynthesis in *Salmonella typhimurium* LT2. J. Biol. Chem. 273, 31788–31794.

Tse, C., Georgieva, E.I., Ruiz-Garcia, A.B., Sendra, R., Hansen, J.C., 1998. Gcn5p, a transcription-related histone acetyltransferase, acetylates nucleosomes and folded nucleosomal arrays in the absence of other protein subunits. J. Biol. Chem. 273, 32388–32392.

Tsukada, Y., Fang, J., Erdjument-Bromage, H., Warren, M.E., Borchers, C.H., Tempst, P., Zhang, Y., 2006. Histone demethylation by a family of JmjC domain-containing proteins. Nature 439, 811–816.

Turner, B.M., 1993. Decoding the nucleosome. Cell 75, 5–8.

Turner, B.M., 2000. Histone acetylation and an epigenetic code. BioEssays 22, 836–845.

Turner, B.M., Fellows, G., 1989. Specific antibodies reveal ordered and cell-cycle-related use of histone-H4 acetylation sites in mammalian cells. Eur. J. Biochem. 179, 131–139.

Turner, B.M., Birley, A.J., Lavender, J., 1992. Histone H4 isoforms acetylated at specific lysine residues define individual chromosomes and chromatin domains in Drosophila polytene nuclei. Cell 69, 375–384.

Ueda, K., Omachi, A., Kawaichi, M., Hayaishi, O., 1975. Natural occurrence of poly(ADP-ribosyl) histones in rat liver. Proc. Natl. Acad. Sci. USA 72, 205–209.

Ueda, J., Tachibana, M., Ikura, T., Shinkai, Y., 2006. Zinc finger protein Wiz links G9a/GLP histone methyltransferases to the co-repressor molecule CtBP. J. Biol. Chem. 281, 20120–20128.

Uesugi, M., Nyanguile, O., Lu, H., Levine, A.J., Verdine, G.L., 1997. Induced alpha helix in the VP16 activation domain upon binding to a human TAF. Science 277, 1310–1313.

Utley, R.T., Ikeda, K., Grant, P.A., Cote, J., Steger, D.J., Eberharter, A., John, S., Workman, J.L., 1998. Transcriptional activators direct histone acetyltransferase complexes to nucleosomes. Nature 394, 498–502.

Vakoc, C.R., Sachdeva, M.M., Wang, H., Blobel, G.A., 2006. Profile of histone lysine methylation across transcribed mammalian chromatin. Mol. Cell. Biol. 26, 9185–9195.

Valencia-Sanchez, M.I., De Ioannes, P., Wang, M., Vasilyev, N., Chen, R., Nudler, E., Armache, J.P., Armache, K.J., 2019. Structural basis of Dot1L stimulation by histone H2B lysine 120 ubiquitination. Mol. Cell 74, 1010–1019. e1016.

Valencia-Sanchez, M.I., De Ioannes, P., Wang, M., Truong, D.M., Lee, R., Armache, J.P., Boeke, J.D., Armache, K.J., 2021. Regulation of the Dot1 histone H3K79 methyltransferase by histone H4K16 acetylation. Science 371, eabc6663.

van Holde, K.E., 1988. Chromatin. Springer-Verlag.

van Leeuwen, F., Gafken, P.R., Gottschling, D.E., 2002. Dot1p modulates silencing in yeast by methylation of the nucleosome core. Cell 109, 745–756.

van Steensel, B., Delrow, J., Henikoff, S., 2001. Chromatin profiling using targeted DNA adenine methyltransferase. Nat. Genet. 27, 304–308.

Vandel, L., Nicolas, E., Vaute, O., Ferreira, R., Ait-Si-Ali, S., Trouche, D., 2001. Transcriptional repression by the retinoblastoma protein through the recruitment of a histone methyltransferase. Mol. Cell. Biol. 21, 6484–6494.

VanDemark, A.P., Kasten, M.M., Ferris, E., Heroux, A., Hill, C.P., Cairns, B.R., 2007. Autoregulation of the rsc4 tandem bromodomain by gcn5 acetylation. Mol. Cell 27, 817–828.

Vaquero, A., Scher, M., Lee, D., Erdjument-Bromage, H., Tempst, P., Reinberg, D., 2004. Human SirT1 interacts with histone H1 and promotes formation of facultative heterochromatin. Mol. Cell 16, 93–105.

Vavra, K.J., Allis, C.D., Gorovsky, M.A., 1982. Regulation of histone acetylation in Tetrahymena macro- and micronuclei. J. Biol. Chem. 257, 2591–2598.

Veloso, A., Kirkconnell, K.S., Magnuson, B., Biewen, B., Paulsen, M.T., Wilson, T.E., Ljungman, M., 2014. Rate of elongation by RNA polymerase II is associated with specific gene features and epigenetic modifications. Genome Res. 24, 896–905.

Venkatesh, S., Smolle, M., Li, H., Gogol, M.M., Saint, M., Kumar, S., Natarajan, K., Workman, J.L., 2012. Set2 methylation of histone H3 lysine 36 suppresses histone exchange on transcribed genes. Nature 489, 452–455.

Vermeulen, M., Carrozza, M.J., Lasonder, E., Workman, J.L., Logie, C., Stunnenberg, H.G., 2004. In vitro targeting reveals intrinsic histone tail specificity of the Sin3/histone deacetylase and N-CoR/SMRT corepressor complexes. Mol. Cell. Biol. 24, 2364–2372.

Vermeulen, M., Mulder, K.W., Denissov, S., Pijnappel, W.W., van Schaik, F.M., Varier, R.A., Baltissen, M.P., Stunnenberg, H.G., Mann, M., Timmers, H.T., 2007. Selective anchoring of TFIID to nucleosomes by trimethylation of histone H3 lysine 4. Cell 131, 58–69.

Vermeulen, M., Eberl, H.C., Matarese, F., Marks, H., Denissov, S., Butter, F., Lee, K.K., Olsen, J.V., Hyman, A.A., Stunnenberg, H.G., et al., 2010. Quantitative interaction proteomics and genome-wide profiling of epigenetic histone marks and their readers. Cell 142, 967–980.

Vidal, M., Gaber, R.F., 1991. RPD3 encodes a second factor required to achieve maximum positive and negative transcriptional states in *Saccharomyces cerevisiae*. Mol. Cell. Biol. 11, 6317–6327.

Vidali, G., Gershey, E.L., Allfrey, V.G., 1968. Chemical studies of histone acetylation. The distribution of epsilon-N-acetyllysine in calf thymus histones. J. Biol. Chem. 243, 6361–6366.

Vidali, G., Boffa, L.C., Allfrey, V.G., 1972. Properties of an acidic histone-binding protein fraction from cell nuclei. Selective precipitation and deacetylation of histones F2A1 and F3. J. Biol. Chem. 247, 7365–7373.

Vidali, G., Boffa, L.C., Bradbury, E.M., Allfrey, V.G., 1978. Butyrate suppression of histone deacetylation leads to accumulation of multiacetylated forms of histones H3 and H4 and increased DNase I sensitivity of the associated DNA sequences. Proc. Natl. Acad. Sci. USA 75, 2239–2243.

Vogelauer, M., Wu, J., Suka, N., Grunstein, M., 2000. Global histone acetylation and deacetylation in yeast. Nature 408, 495–498.

Voss, A.K., Thomas, T., 2018. Histone lysine and genomic targets of histone acetyltransferases in mammals. BioEssays 40, e1800078.

Wade, P.A., Jones, P.L., Vermaak, D., Wolffe, A.P., 1998. A multiple subunit Mi-2 histone deacetylase from Xenopus laevis cofractionates with an associated Snf2 superfamily ATPase. Curr. Biol. 8, 843–846.

Wagner, E.J., Carpenter, P.B., 2012. Understanding the language of Lys36 methylation at histone H3. Nat. Rev. Mol. Cell Biol. 13, 115–126.

Wallberg, A.E., Neely, K.E., Gustafsson, J.A., Workman, J.L., Wright, A.P., Grant, P.A., 1999. Histone acetyltransferase complexes can mediate transcriptional activation by the major glucocorticoid receptor activation domain. Mol. Cell. Biol. 19, 5952–5959.

Wang, L., Liu, L., Berger, S.L., 1998. Critical residues for histone acetylation by Gcn5, functioning in Ada and SAGA complexes, are also required for transcriptional function in vivo. Genes Dev. 12, 640–653.

Wang, H., Cao, R., Xia, L., Erdjument-Bromage, H., Borchers, C., Tempst, P., Zhang, Y., 2001a. Purification and functional characterization of a histone H3-lysine 4-specific methyltransferase. Mol. Cell 8, 1207–1217.

Wang, H., Huang, Z.Q., Xia, L., Feng, Q., Erdjument-Bromage, H., Strahl, B.D., Briggs, S.D., Allis, C.D., Wong, J., Tempst, P., et al., 2001b. Methylation of histone H4 at arginine 3 facilitating transcriptional activation by nuclear hormone receptor. Science 293, 853–857.

Wang, H., Wang, L., Erdjument-Bromage, H., Vidal, M., Tempst, P., Jones, R.S., Zhang, Y., 2004a. Role of histone H2A ubiquitination in polycomb silencing. Nature 431, 873–878.

Wang, Y., Wysocka, J., Sayegh, J., Lee, Y.H., Perlin, J.R., Leonelli, L., Sonbuchner, L.S., McDonald, C.H., Cook, R.G., Dou, Y., et al., 2004b. Human PAD4 regulates histone arginine methylation levels via demethylimination. Science 306, 279–283.

Wang, Y.L., Faiola, F., Xu, M., Pan, S., Martinez, E., 2008a. Human ATAC is a GCN5/PCAF-containing acetylase complex with a novel NC2-like histone fold module that interacts with the TATA-binding protein. J. Biol. Chem. 283, 33808–33815.

Wang, Z., Zang, C., Rosenfeld, J.A., Schones, D.E., Barski, A., Cuddapah, S., Cui, K., Roh, T.Y., Peng, W., Zhang, M.Q., et al., 2008b. Combinatorial patterns of histone acetylations and methylations in the human genome. Nat. Genet. 40, 897–903.

Wang, Y., Niu, Y., Li, B., 2015. Balancing acts of SRI and an auto-inhibitory domain specify Set2 function at transcribed chromatin. Nucleic Acids Res. 43, 4881–4892.

Wang, H., Dienemann, C., Stutzer, A., Urlaub, H., Cheung, A.C.M., Cramer, P., 2020a. Structure of the transcription coactivator SAGA. Nature 577, 717–720.

Wang, Z.A., Millard, C.J., Lin, C.L., Gurnett, J.E., Wu, M., Lee, K., Fairall, L., Schwabe, J.W., Cole, P.A., 2020b. Diverse nucleosome site-selectivity among histone deacetylase complexes. elife 9, e57663.

Wang, Z.A., Whedon, S.D., Wu, M., Wang, S., Brown, E.A., Anmangandla, A., Regan, L., Lee, K., Du, J., Hong, J.Y., et al., 2022. Histone H2B deacylation selectivity: exploring chromatin's dark matter with an engineered sortase. J. Am. Chem. Soc. 144, 3360–3364.

Waterborg, J.H., 2001. Dynamics of histone acetylation in *Saccharomyces cerevisiae*. Biochemistry 40, 2599–2605.

Webby, C.J., Wolf, A., Gromak, N., Dreger, M., Kramer, H., Kessler, B., Nielsen, M.L., Schmitz, C., Butler, D.S., Yates 3rd, J.R., et al., 2009. Jmjd6 catalyses lysyl-hydroxylation of U2AF65, a protein associated with RNA splicing. Science 325, 90–93.

Wei, Y., Mizzen, C.A., Cook, R.G., Gorovsky, M.A., Allis, C.D., 1998. Phosphorylation of histone H3 at serine 10 is correlated with chromosome condensation during mitosis and meiosis in Tetrahymena. Proc. Natl. Acad. Sci. USA 95, 7480–7484.

Wei, Y., Yu, L., Bowen, J., Gorovsky, M.A., Allis, C.D., 1999. Phosphorylation of histone H3 is required for proper chromosome condensation and segregation. Cell 97, 99–109.

Weinert, B.T., Narita, T., Satpathy, S., Srinivasan, B., Hansen, B.K., Scholz, C., Hamilton, W.B., Zucconi, B.E., Wang, W.W., Liu, W.R., et al., 2018. Time-resolved analysis reveals rapid dynamics and broad scope of the CBP/p300 acetylome. Cell 174, 231–244.e212.

Wen, Y.D., Perissi, V., Staszewski, L.M., Yang, W.M., Krones, A., Glass, C.K., Rosenfeld, M.G., Seto, E., 2000. The histone deacetylase-3 complex contains nuclear receptor corepressors. Proc. Natl. Acad. Sci. USA 97, 7202–7207.

Wen, B., Wu, H., Shinkai, Y., Irizarry, R.A., Feinberg, A.P., 2009. Large histone H3 lysine 9 dimethylated chromatin blocks distinguish differentiated from embryonic stem cells. Nat. Genet. 41, 246–250.

West, M.H., Bonner, W.M., 1980. Histone 2B can be modified by the attachment of ubiquitin. Nucleic Acids Res. 8, 4671–4680.

Whetstine, J.R., Nottke, A., Lan, F., Huarte, M., Smolikov, S., Chen, Z., Spooner, E., Li, E., Zhang, G., Colaiacovo, M., et al., 2006. Reversal of histone lysine trimethylation by the JMJD2 family of histone demethylases. Cell 125, 467–481.

Whitlock Jr., J.P., Galeazzi, D., Schulman, H., 1983. Acetylation and calcium-dependent phosphorylation of histone H3 in nuclei from butyrate-treated HeLa cells. J. Biol. Chem. 258, 1299–1304.

Wilhelm, J.A., McCarty, K.S., 1970a. Partial characterization of the histones and histone acetylation in cell cultures. Cancer Res. 30, 409–417.

Wilhelm, J.A., McCarty, K.S., 1970b. The uptake and turnover of acetate in HeLa cell histone fractions. Cancer Res. 30, 418–425.

Wilkinson, K.A., Henley, J.M., 2010. Mechanisms, regulation and consequences of protein SUMOylation. Biochem. J. 428, 133–145.

Wilson, J.R., Jing, C., Walker, P.A., Martin, S.R., Howell, S.A., Blackburn, G.M., Gamblin, S.J., Xiao, B., 2002. Crystal structure and functional analysis of the histone methyltransferase SET7/9. Cell 111, 105–115.

Winston, F., Allis, C.D., 1999. The bromodomain: a chromatin-targeting module? Nat. Struct. Biol. 6, 601–604.

Wood, K., Tellier, M., Murphy, S., 2018. DOT1L and H3K79 methylation in transcription and genomic stability. Biomol. Ther. 8, 11.

Worden, E.J., Hoffmann, N.A., Hicks, C.W., Wolberger, C., 2019. Mechanism of cross-talk between H2B ubiquitination and H3 methylation by Dot1L. Cell 176, 1490–1501.e1412.

Wu, J., Suka, N., Carlson, M., Grunstein, M., 2001. TUP1 utilizes histone H3/H2B-specific HDA1 deacetylase to repress gene activity in yeast. Mol. Cell 7, 117–126.

Wu, P.Y., Ruhlmann, C., Winston, F., Schultz, P., 2004. Molecular architecture of the *S. cerevisiae* SAGA complex. Mol. Cell 15, 199–208.

Wu, H., Min, J., Lunin, V.V., Antoshenko, T., Dombrovski, L., Zeng, H., Allali-Hassani, A., Campagna-Slater, V., Vedadi, M., Arrowsmith, C.H., et al., 2010. Structural biology of human H3K9 methyltransferases. PLoS One 5, e8570.

Wu, H., Zeng, H., Dong, A., Li, F., He, H., Senisterra, G., Seitova, A., Duan, S., Brown, P.J., Vedadi, M., et al., 2013. Structure of the catalytic domain of EZH2 reveals conformational plasticity in cofactor and substrate binding sites and explains oncogenic mutations. PLoS One 8, e83737.

Wyrick, J.J., Parra, M.A., 2009. The role of histone H2A and H2B post-translational modifications in transcription: a genomic perspective. Biochim. Biophys. Acta 1789, 37–44.

Wysocka, J., Swigut, T., Milne, T.A., Dou, Y., Zhang, X., Burlingame, A.L., Roeder, R.G., Brivanlou, A.H., Allis, C.D., 2005. WDR5 associates with histone H3 methylated at K4 and is essential for H3 K4 methylation and vertebrate development. Cell 121, 859–872.

Xiao, B., Jing, C., Wilson, J.R., Walker, P.A., Vasisht, N., Kelly, G., Howell, S., Taylor, I.A., Blackburn, G.M., Gamblin, S.J., 2003a. Structure and catalytic mechanism of the human histone methyltransferase SET7/9. Nature 421, 652–656.

Xiao, T., Hall, H., Kizer, K.O., Shibata, Y., Hall, M.C., Borchers, C.H., Strahl, B.D., 2003b. Phosphorylation of RNA polymerase II CTD regulates H3 methylation in yeast. Genes Dev. 17, 654–663.

Xiao, B., Jing, C., Kelly, G., Walker, P.A., Muskett, F.W., Frenkiel, T.A., Martin, S.R., Sarma, K., Reinberg, D., Gamblin, S.J., et al., 2005. Specificity and mechanism of the histone methyltransferase Pr-Set7. Genes Dev. 19, 1444–1454.

Xiao, A., Li, H., Shechter, D., Ahn, S.H., Fabrizio, L.A., Erdjument-Bromage, H., Ishibe-Murakami, S., Wang, B., Tempst, P., Hofmann, K., et al., 2009. WSTF regulates the H2A.X DNA damage response via a novel tyrosine kinase activity. Nature 457, 57–62.

Xiao, Y., Li, W., Yang, H., Pan, L., Zhang, L., Lu, L., Chen, J., Wei, W., Ye, J., Li, J., et al., 2021. HBO1 is a versatile histone acyltransferase critical for promoter histone acylations. Nucleic Acids Res. 49, 8037–8059.

Xu, W., Edmondson, D.G., Roth, S.Y., 1998. Mammalian GCN5 and P/CAF acetyltransferases have homologous amino-terminal domains important for recognition of nucleosomal substrates. Mol. Cell. Biol. 18, 5659–5669.

Xu, F., Zhang, K., Grunstein, M., 2005. Acetylation in histone H3 globular domain regulates gene expression in yeast. Cell 121, 375–385.

Xu, P., Li, C., Chen, Z., Jiang, S., Fan, S., Wang, J., Dai, J., Zhu, P., Chen, Z., 2016. The NuA4 core complex acetylates nucleosomal histone H4 through a double recognition mechanism. Mol. Cell 63, 965–975.

Yang, X.J., 2004. The diverse superfamily of lysine acetyltransferases and their roles in leukemia and other diseases. Nucleic Acids Res. 32, 959–976.

Yang, X.J., 2015. MOZ and MORF acetyltransferases: molecular interaction, animal development and human disease. Biochim. Biophys. Acta 1853, 1818–1826.

Yang, X.J., Seto, E., 2008. The Rpd3/Hda1 family of lysine deacetylases: from bacteria and yeast to mice and men. Nat. Rev. Mol. Cell Biol. 9, 206–218.

Yang, W.M., Inouye, C., Zeng, Y., Bearss, D., Seto, E., 1996a. Transcriptional repression by YY1 is mediated by interaction with a mammalian homolog of the yeast global regulator RPD3. Proc. Natl. Acad. Sci. USA 93, 12845–12850.

Yang, X.J., Ogryzko, V.V., Nishikawa, J., Howard, B.H., Nakatani, Y., 1996b. A p300/CBP-associated factor that competes with the adenoviral oncoprotein E1A. Nature 382, 319–324.

Yang, L., Xia, L., Wu, D.Y., Wang, H., Chansky, H.A., Schubach, W.H., Hickstein, D.D., Zhang, Y., 2002. Molecular cloning of ESET, a novel histone H3-specific methyltransferase that interacts with ERG transcription factor. Oncogene 21, 148–152.

Yang, H., Pesavento, J.J., Starnes, T.W., Cryderman, D.E., Wallrath, L.L., Kelleher, N.L., Mizzen, C.A., 2008. Preferential dimethylation of histone H4 lysine 20 by Suv4-20. J. Biol. Chem. 283, 12085–12092.

Yang, S., Zheng, X., Lu, C., Li, G.M., Allis, C.D., Li, H., 2016. Molecular basis for oncohistone H3 recognition by SETD2 methyltransferase. Genes Dev. 30, 1611–1616.

Yao, T., Jing, W., Hu, Z., Tan, M., Cao, M., Wang, Q., Li, Y., Yuan, G., Lei, M., Huang, J., 2019. Structural basis of the crosstalk between histone H2B monoubiquitination and H3 lysine 79 methylation on nucleosome. Cell Res. 29, 330–333.

Yoon, S., Qiu, H., Swanson, M.J., Hinnebusch, A.G., 2003. Recruitment of SWI/SNF by Gcn4p does not require Snf2p or Gcn5p but depends strongly on SWI/SNF integrity, SRB mediator, and SAGA. Mol. Cell. Biol. 23, 8829–8845.

Yoon, J., Lee, K.S., Park, J.S., Yu, K., Paik, S.G., Kang, Y.K., 2008. dSETDB1 and SU(VAR)3-9 sequentially function during germline-stem cell differentiation in *Drosophila melanogaster*. PLoS One 3, e2234.

York, B., O'Malley, B.W., 2010. Steroid receptor coactivator (SRC) family: masters of systems biology. J. Biol. Chem. 285, 38743–38750.

Yoshida, M., Kijima, M., Akita, M., Beppu, T., 1990. Potent and specific inhibition of mammalian histone deacetylase both in vivo and in vitro by trichostatin A. J. Biol. Chem. 265, 17174–17179.

Yoshida, M., Horinouchi, S., Beppu, T., 1995. Trichostatin A and trapoxin: novel chemical probes for the role of histone acetylation in chromatin structure and function. BioEssays 17, 423–430.

You, A., Tong, J.K., Grozinger, C.M., Schreiber, S.L., 2001. CoREST is an integral component of the CoREST-human histone deacetylase complex. Proc. Natl. Acad. Sci. USA 98, 1454–1458.

Yu, C., Palumbo, M.J., Lawrence, C.E., Morse, R.H., 2006. Contribution of the histone H3 and H4 amino termini to Gcn4p- and Gcn5p-mediated transcription in yeast. J. Biol. Chem. 281, 9755–9764.

Yu, C., Fan, X., Sha, Q.Q., Wang, H.H., Li, B.T., Dai, X.X., Shen, L., Liu, J., Wang, L., Liu, K., et al., 2017. CFP1 regulates histone H3K4 trimethylation and developmental potential in mouse oocytes. Cell Rep. 20, 1161–1172.

Yuan, W., Xie, J., Long, C., Erdjument-Bromage, H., Ding, X., Zheng, Y., Tempst, P., Chen, S., Zhu, B., Reinberg, D., 2009. Heterogeneous nuclear ribonucleoprotein L is a subunit of human KMT3a/Set2 complex required for H3 Lys-36 trimethylation activity in vivo. J. Biol. Chem. 284, 15701–15707.

Yuan, W., Xu, M., Huang, C., Liu, N., Chen, S., Zhu, B., 2011. H3K36 methylation antagonizes PRC2-mediated H3K27 methylation. J. Biol. Chem. 286, 7983–7989.

Zaware, N., Zhou, M.M., 2019. Bromodomain biology and drug discovery. Nat. Struct. Mol. Biol. 26, 870–879.

Zhang, W., Bone, J.R., Edmondson, D.G., Turner, B.M., Roth, S.Y., 1998. Essential and redundant functions of histone acetylation revealed by mutation of target lysines and loss of the Gcn5p acetyltransferase. EMBO J. 17, 3155–3167.

Zhang, Y., Ng, H.H., Erdjument-Bromage, H., Tempst, P., Bird, A., Reinberg, D., 1999. Analysis of the NuRD subunits reveals a histone deacetylase core complex and a connection with DNA methylation. Genes Dev. 13, 1924–1935.

Zhang, J., Kalkum, M., Chait, B.T., Roeder, R.G., 2002a. The N-CoR-HDAC3 nuclear receptor corepressor complex inhibits the JNK pathway through the integral subunit GPS2. Mol. Cell 9, 611–623.

Zhang, X., Tamaru, H., Khan, S.I., Horton, J.R., Keefe, L.J., Selker, E.U., Cheng, X., 2002b. Structure of the Neurospora SET domain protein DIM-5, a histone H3 lysine methyltransferase. Cell 111, 117–127.

Zhang, L., Eugeni, E.E., Parthun, M.R., Freitas, M.A., 2003a. Identification of novel histone post-translational modifications by peptide mass fingerprinting. Chromosoma 112, 77–86.

Zhang, X., Yang, Z., Khan, S.I., Horton, J.R., Tamaru, H., Selker, E.U., Cheng, X., 2003b. Structural basis for the product specificity of histone lysine methyltransferases. Mol. Cell 12, 177–185.

Zhang, K., Lin, W., Latham, J.A., Riefler, G.M., Schumacher, J.M., Chan, C., Tatchell, K., Hawke, D.H., Kobayashi, R., Dent, S.Y., 2005. The Set1 methyltransferase opposes Ipl1 aurora kinase functions in chromosome segregation. Cell 122, 723–734.

Zhang, X., Bernatavichute, Y.V., Cokus, S., Pellegrini, M., Jacobsen, S.E., 2009. Genome-wide analysis of mono-, di- and trimethylation of histone H3 lysine 4 in Arabidopsis thaliana. Genome Biol. 10, R62.

Zhang, Q., Chakravarty, S., Ghersi, D., Zeng, L., Plotnikov, A.N., Sanchez, R., Zhou, M.M., 2010. Biochemical profiling of histone binding selectivity of the yeast bromodomain family. PLoS One 5, e8903.

Zhang, S., Roche, K., Nasheuer, H.P., Lowndes, N.F., 2011. Modification of histones by sugar beta-N-acetylglucosamine (GlcNAc) occurs on multiple residues, including histone H3 serine 10, and is cell cycle-regulated. J. Biol. Chem. 286, 37483–37495.

Zhang, L., Serra-Cardona, A., Zhou, H., Wang, M., Yang, N., Zhang, Z., Xu, R.M., 2018. Multisite substrate recognition in Asf1-dependent acetylation of histone H3 K56 by Rtt109. Cell 174, 818–830.e811.

Zhang, J., Jing, L., Li, M., He, L., Guo, Z., 2019. Regulation of histone arginine methylation/demethylation by methylase and demethylase (review). Mol. Med. Rep. 19, 3963–3971.

Zhao, Y., Garcia, B.A., 2015. Comprehensive catalog of currently documented histone modifications. Cold Spring Harb. Perspect. Biol. 7, a025064.

Zhao, G., Rusche, L.N., 2022. Sirtuins in epigenetic silencing and control of gene expression in model and pathogenic fungi. Annu. Rev. Microbiol. 76, 157–178.

Zheng, Y., John, S., Pesavento, J.J., Schultz-Norton, J.R., Schiltz, R.L., Baek, S., Nardulli, A.M., Hager, G.L., Kelleher, N.L., Mizzen, C.A., 2010. Histone H1 phosphorylation is associated with transcription by RNA polymerases I and II. J. Cell Biol. 189, 407–415.

Zhou, B., Wang, J., Lee, S.Y., Xiong, J., Bhanu, N., Guo, Q., Ma, P., Sun, Y., Rao, R.C., Garcia, B.A., et al., 2016a. PRDM16 suppresses MLL1r leukemia via intrinsic histone methyltransferase activity. Mol. Cell 62, 222–236.

Zhou, L., Holt, M.T., Ohashi, N., Zhao, A., Muller, M.M., Wang, B., Muir, T.W., 2016b. Evidence that ubiquitylated H2B corrals hDot1L on the nucleosomal surface to induce H3K79 methylation. Nat. Commun. 7, 10589.

Zhu, B., Zheng, Y., Pham, A.D., Mandal, S.S., Erdjument-Bromage, H., Tempst, P., Reinberg, D., 2005. Monoubiquitination of human histone H2B: the factors involved and their roles in HOX gene regulation. Mol. Cell 20, 601–611.

CHAPTER

8

Chromatin and transcription

Abbreviations

5mC	5-methylcytosine
Ad MLP	adenovirus major late promoter
ANT-C	antennapedia complex
BX-C	bithorax complex
CGI	CpG island
ChIP	chromatin immunoprecipitation
DPN	depleted promoter nucleosome
EM	electron microscopy
ESC	embryonic stem cell
GR	glucocorticoid receptor
MMTV	mouse mammary tumor virus
MNase	micrococcal nuclease
NDF	nucleosome-displacing factor
NDR	nucleosome-depleted region
NF1	nuclear factor 1
NTP	nucleotide triphosphate
OPN	occupied promoter nucleosome
PcG	polycomb group
PIC	preinitiation complex
PRC	polycomb repressor complex
PRE	polycomb group responsive element
RSC	remodels the structure of chromatin
SHL	superhelix location
SV40	simian virus 40
TBP	TATA-binding protein
TF	transcription factor
TrxG	trithorax group
UAS	upstream activator sequence

Chromatin. **https://doi.org/10.1016/B978-0-12-814809-9.00009-3**

The preceding chapters have laid out our knowledge of the structures of nucleosomes and chromatin fibers, and how those structures are affected by histone variants, histone modifications, and remodelers. Much of the research on these topics was conducted in the service of understanding the impact of chromatin on processes involving DNA, most particularly transcription. Some of the ways in which chromatin and proteins interacting with chromatin have evolved to allow and regulate transcription have been touched on in earlier chapters; in this chapter, a transcription-centric perspective is adopted and some of the manifold aspects by which chromatin affects this process are explored.

Transcription: A brief review

Many of the scientists who have studied transcriptional regulation were motivated to do so by a desire to understand the mechanisms underlying the differentiation of a single fertilized cell to yield a multitude of cell types, each requiring the establishment and maintenance of its own distinct transcriptional program. To understand how genes could be turned on or off, early investigators focused on prokaryotic phenomena that they hoped might provide insight relevant to the presumably more complex processes in metazoans. Studies on carbon source utilization in *E. coli* and on the decision by bacteriophage lambda to replicate and lyse its host cells or to remain in a dormant, lysogenic state until induced by an external signal such as UV light, led to major insights into transcriptional regulation (Jacob and Monod, 1961; Ptashne, 2004). Key principles that emerged from these and related studies, including the control of gene expression by proteins binding to specific DNA sequences, the importance of cooperative interactions, and the requirement for distinct programs to establish and maintain a particular transcriptional state, did indeed turn out to apply to eukaryotic organisms. These early studies on bacterial transcriptional regulation thus provided a solid foundation for investigations of gene regulation in eukaryotes over the following decades.

While gene regulation in prokaryotes and eukaryotes share common principles, transcription in eukaryotes was gradually revealed to be far more complex with respect to both the necessary components and the milieu in which it takes place (Taatjes et al., 2004). Prokaryotic genomes are more compact than those of eukaryotes, are not contained within nuclei, and are not packaged into chromatin. Transcription in *E. coli* is performed by a single RNA polymerase that is directed to specific promoter sequences by any of a small number of sigma factors with which it associates to form a holoenzyme (Mejia-Almonte et al., 2020). In contrast, eukaryotes possess three RNA polymerases that, respectively, transcribe the large ribosomal RNA genes (RNA polymerase I, or Pol I), mRNA-encoding genes (Pol II), or small noncoding RNA genes, including tRNA genes (Pol III). Furthermore, transcription of mRNA-encoding genes by Pol II requires the assemblage of a large preinitiation complex (PIC) that includes a host of general transcription factors, including the multicomponent transcription factors TFIIA, TFIIB, TFIID, TFIIE, and TFIIH (Fig. 8.1).

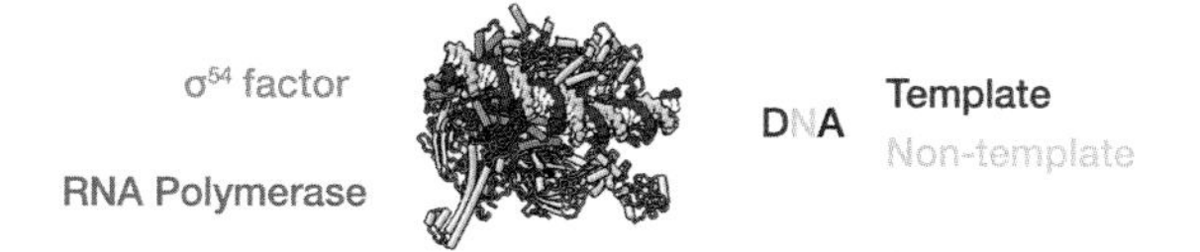

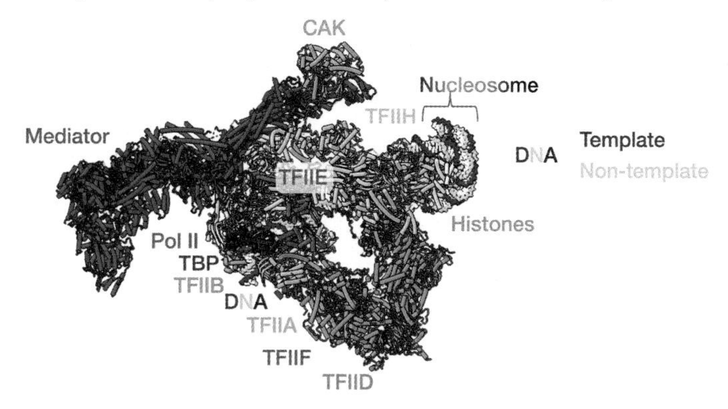

FIG. 8.1 Comparison of prokaryotic and eukaryotic transcription preinitiation complexes. Accessory transcription factors are colored as indicated. DNA in the Pol II initiation complex can be partially seen from *left*, where it engages with TBP, to *right*, where it wraps around the histones in the +1 nucleosome. "CAK" is the CDK (cyclin-dependent kinase) activating kinase module of TFIIH. *PDB files and relevant references used are 6GH6, Glyde et al. 2018. Structures of bacterial RNA polymerase complexes reveal the mechanism of DNA loading and transcription initiation. Mol. Cell 70, 1111–1120; 8GXS, Chen et al. 2022. Structures of +1 nucleosome-bound PIC-Mediator complex. Science 378, 62–68; 8BVW, Abril-Garrido et al. 2023. Structural basis of transcription reduction by a promoter-proximal +1 nucleosome. Mol. Cell 83, 1790–1809. Courtesy of Julio Abril-Garrido.*

Ancillary factors that regulate transcription are also much more complex in eukaryotes than in prokaryotes. In prokaryotes, transcription is often facilitated by proteins that bind to specific upstream promoter sequences and aid polymerase binding at otherwise weak binding sites through direct cooperative interactions, while repressors most often function by direct occlusion of polymerase binding sites (Mejia-Almonte et al., 2020; Ptashne, 2004). In eukaryotes, activation of many genes—especially those whose expression varies with cell type or upon environmental perturbation—depends on activator proteins that bind to specific sequences, termed enhancers, that reside from hundreds to thousands of base pairs distant from the promoters they control. Many hundreds of such activators (~1600 in human cells (Lambert et al., 2018)) act at thousands of enhancers in metazoan cells and function not only by directly stimulating association of the transcription machinery but also by recruiting co-activators that do not directly bind DNA but play critical roles in transcriptional activation. One such co-activator is the Mediator complex, which is recruited by activators and facilitates PIC formation (Fig. 8.1) (Kornberg, 2005; Malik and Roeder, 2010). The Mediator complex, although specific to eukaryotes, functions through mechanisms not related to chromatin, and so is not considered further here. Other co-activators include chromatin remodelers and modifiers, as discussed in Chapters 6 and 7. Similarly to activators, repressor proteins can function in eukaryotic nuclei not only by directly occluding the binding of factors needed for transcription but also by recruiting co-repressors that function in various ways to inhibit transcription, including by modifying or remodeling chromatin. These co-activators and co-repressors are unique to eukaryotes.

Early investigations

Edward and Ellen Stedman close their classic 1951 paper by hypothesizing that "the basic proteins of cell nuclei are gene inhibitors, each histone or protamin being capable of suppressing the activities of specific groups of genes" (Stedman and Stedman, 1951). Thus, the notion of gene-specific regulation by chromatin preceded knowledge of the structure of both DNA and nucleosomes. The simple addition of histones to isolated thymus nuclei was shown to inhibit RNA synthesis (Allfrey and Mirsky, 1962), but this effect was nonspecific, as other enzymatic processes were also suppressed and transcriptional inhibition was also observed upon the addition of polycations such as polylysine (Allfrey et al., 1963; Barr and Butler, 1963). A more convincing demonstration of a specific effect due to chromatin was reported by Huang and Bonner, who showed that synthesis of RNA from chromatin by DNA-dependent RNA polymerase, both prepared from pea plant embryos, yielded only about 20% of the product obtained using deproteinized DNA (Huang and Bonner, 1962). Furthermore, reconstitution of chromatin from isolated histones and DNA by salt dialysis resulted in a soluble nucleohistone complex having a T_m increased from that of naked DNA (79°C vs. 70°C) and a greatly decreased ability to support RNA synthesis. Naked DNA added to the reconstituted template supported RNA synthesis, ruling out a direct repressive effect of the histones on the polymerase. These experiments, highly sophisticated for their time, provided strong evidence that chromatin templates were inherently refractory to transcription and that histones might accordingly have a role in gene regulation.

The discovery of the nucleosome as the fundamental unit of chromatin, and the ability to reconstitute chromatin from purified components and to assess its structure using enzymatic and biophysical methods, prompted new investigations into the effect of chromatin on transcription. Assembly of SV40 or bacteriophage T7 DNA into chromatin inhibited transcription by both prokaryotic and eukaryotic RNA polymerases, yielding shorter transcripts and reducing the rate of chain elongation relative to that seen using naked DNA (Meneguzzi et al., 1979; Wasylyk and Chambon, 1979; Wasylyk et al., 1979; Williamson and Felsenfeld, 1978). Some disruption or displacement of nucleosomes appeared to accompany transcriptional elongation, as seen by increased susceptibility of a chromatinized SV40 template to digestion at a unique *Eco*RI site during active transcription (Wasylyk and Chambon, 1980). Inhibition of transcriptional elongation was relieved at higher salt concentrations (but not sufficiently high to cause histone dissociation), suggesting that weakening histone-DNA interactions could facilitate polymerase passage (Brooks and Green, 1977; Wasylyk et al., 1979; Williamson and Felsenfeld, 1978). At the same time, hybridization experiments showed that mRNA-encoding sequences were not depleted relative to total DNA in nucleosomal DNA from rat liver nuclei, while transcribed globin sequences from chick erythrocytes were similarly shown to be incorporated into nucleosomes (Lacy and Axel, 1975; Weintraub and Groudine, 1976). These experiments indicated that DNA that was transcribed in living cells was associated with nucleosomes. This conclusion was further supported by electron microscopy (EM) experiments showing transcribing Pol II in close association with nucleosomal DNA (Foe et al., 1976). Together these and other findings suggested that cells had evolved mechanisms to cope with the inhibitory effects of chromatin (Morse, 1992).

Prokaryotic polymerases and RNA polymerase III

Mechanistic interpretation of the transcriptional impediment posed by chromatin in vitro was limited by the heterogeneous (and largely uncharacterized at the time) nature of the chromatin templates, the lack of specific initiation sites, and the relatively small fraction of templates that were actually transcribed. Two advances greatly benefited studies of chromatin and transcription at this time. One was the development of systems for assembling chromatin templates with spacing and density more closely resembling the physiological state than could be obtained by salt dialysis methods (Chapter 4) (Becker and Wu, 1992; Glikin et al., 1984; Kamakaka et al., 1993; Laskey et al., 1977); the second advance was in methods for preparing cellular extracts that allowed efficient and precise initiation in vitro by eukaryotic RNA polymerase II and III (Birkenmeier et al., 1978; Conaway et al., 1987; Dignam et al., 1983; Manley et al., 1980; Weil et al., 1979). In addition, phage RNA polymerases that function as single-subunit enzymes were cloned and expressed using nascent recombinant technology, allowing studies of their ability to transcribe chromatin templates in vitro.

Using these newly available reagents and protocols, it was shown that a highly processive polymerase from the bacteriophage SP6, together with either short, linear templates or a closed circular plasmid subjected to restriction enzyme digestion to exclude nonnucleosomal templates (Fig. 8.2), could not initiate transcription on a nucleosomal template but was capable of transcribing through a nucleosome (Lorch et al., 1987; Losa and Brown, 1987; Morse, 1989). SP6 RNA polymerase had previously been shown to transcribe 5S rDNA sequences occupied by the Pol III transcription complex from *Xenopus* (Wolffe et al., 1986); this was notable as the complex occluded ~50 bp of sequence from DNase I digestion, indicating extensive contacts with DNA, and yet was not lost from the template following transcription by SP6. It therefore seemed possible that nucleosomal histones, while preventing engagement by polymerases during transcriptional initiation just as they prevented digestion by endonucleases, might allow passage by transiting polymerases, just as they did not completely prevent exonucleolytic digestion (Chapter 4). However, prokaryotic RNA polymerases—SP6 in particular—appeared to be more highly processive than the much larger, multisubunit eukaryotic polymerases, and had not evolved to function with chromatin templates. Might eukaryotic polymerases have evolved innate mechanisms allowing them to initiate transcription directly on nucleosomal DNA?

The 5S rRNA gene, transcribed by Pol III, provided an attractive system for the investigation of transcription of a chromatin template by a eukaryotic RNA polymerase

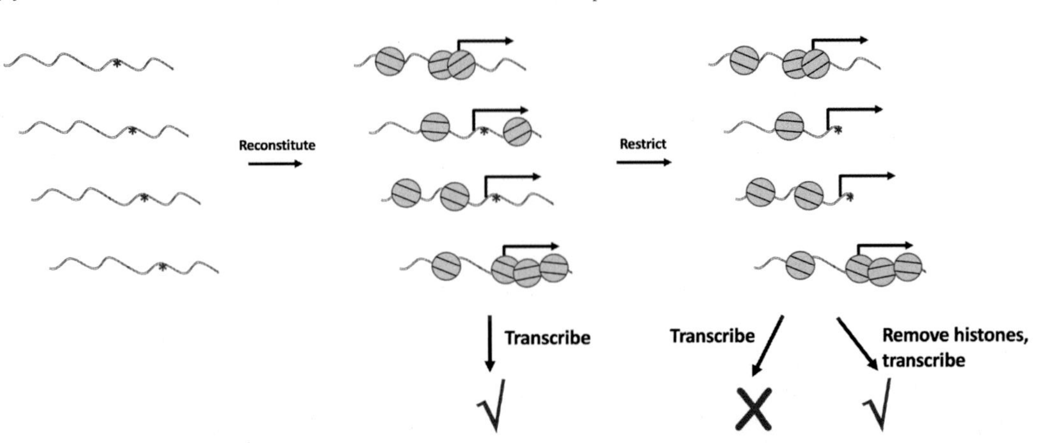

FIG. 8.2 Scheme for testing the effect on transcription of incorporation of the proximal promoter region of a gene into a nucleosome. Following assembly into chromatin, templates are subjected to restriction enzyme digestion (site designated with a *red asterisk*); templates having accessible sites are cleaved close to their initiation sites and cannot produce full-length transcripts, while nucleosomal sites remain uncut. Transcription by SP6 RNA polymerase, Pol III, and Pol II was eliminated following restriction digestion, indicating that incorporation of the initiation site into a nucleosome inhibited transcription. Removal of histones after restriction digestion allowed transcription, demonstrating that full-length templates capable of being transcribed were still present. Transcripts were also produced using the reconstituted templates prior to restriction digestion; these evidently arose from those templates lacking a nucleosome over the initiation site. A similar protocol, using a restriction site in the transcribed region downstream from the initiation site, allows testing for transcription through a nucleosome. *See Morse, R.H. 1989. Nucleosomes inhibit both transcriptional initiation and elongation by RNA polymerase III in vitro. EMBO J. 8, 2343–2351; Laybourn, P.J., Kadonaga, J.T. 1992. Threshold phenomena and long-distance activation of transcription by RNA polymerase II. Science 257, 1682–1685 for details.*

(see Sidebar: Transcription by Pol III). The abundant stores of TFIIIA and Pol III in *Xenopus* oocytes permitted facile preparation of extracts capable of programming the 5S rRNA gene for transcription, with accurate initiation occurring on the majority of templates under appropriate reaction conditions (Birkenmeier et al., 1978). Moreover, the 5S rDNA sequences were known to favor a specifically positioned nucleosome (Chapter 4) (Hayes et al., 1990; Simpson and Stafford, 1983). A generally inhibitory effect on transcription was seen upon reconstitution of plasmids harboring the 5S rRNA gene into chromatin in vitro, and with isolated SV40 minichromosome derivatives containing a 5S rRNA gene (Bogenhagen et al., 1982; Gottesfeld and Bloomer, 1982; Lassar et al., 1985). A more direct determination of the effect of nucleosomes on initiation and elongation by Pol III was achieved using a protocol in which nonnucleosomal templates were eliminated by restriction enzyme digestion (Fig. 8.2) (Morse, 1989). This and subsequent studies demonstrated that nucleosomes incorporating sequences critical for initiation prevented transcription by

Pol III, while elongation was substantially inhibited only by longer arrays of nucleosomes or by compaction of the chromatin template (Clark and Wolffe, 1991; Felts et al., 1990; Hansen and Wolffe, 1992; Morse, 1989).

Sidebar: Transcription by Pol III

Eukaryotes possess three RNA polymerases, designated I, II, and III, first identified by Robert Roeder as chromatographically distinct entities that differ in their DNA template preferences and optimal metal ion concentrations (Roeder, 2019; Roeder and Rutter, 1969). Exploiting the differential sensitivity of the three polymerases to the mushroom toxin α-amanitin, Roeder and colleagues demonstrated that Pol I was responsible for the synthesis of the large ribosomal RNA transcripts, while Pol II transcribed genes encoding mRNAs and Pol III synthesized the much smaller 5S rRNA and tRNAs (Weinmann et al., 1974; Weinmann and Roeder, 1974).

Like the large ribosomal RNA genes, 5S DNA was first isolated not by modern cloning methods but by CsCl gradient centrifugation, based on its high GC content and high copy number (Brown, 1994; Brown et al., 1971), and the isolated gene was in fact the first eukaryotic gene to be completely sequenced (Fedoroff and Brown, 1978; Miller et al., 1978). A transatlantic collaboration by mail between Brown and John Gurdon led to the discovery that 5S DNA was accurately transcribed when injected into oocyte nuclei, and an extract from *Xenopus* germinal vesicles was subsequently found to reproduce this faithful transcription (Birkenmeier et al., 1978; Brown and Gurdon, 1977). These experiments set the stage for analysis of the DNA sequence elements required for accurate transcription of the 5S rRNA gene by Pol III. The cloned gene was inserted into a plasmid, and derivatives were prepared in which sequences at the 5′ or 3′ end were progressively deleted (Bogenhagen et al., 1980; Sakonju et al., 1980). Transcription of these templates in the germinal vesicle extract yielded results that were surprising and bizarre: a correctly initiated and terminated transcript was produced even as deletions from the 5′ end encroached more than 30 bp into the coding sequence, and while 3′ deletions resulted in altered termination, correct initiation was seen until deletions invaded more than 20 bp into the gene. The clear implication of these findings was that the DNA sequence extending from ~+45–95 bp in the gene, dubbed the internal control region, was critical for 5S rDNA transcription. This finding was soon validated by experiments showing that TFIIIA, found to bind tightly to 5S rDNA and required for its accurate transcription, footprinted precisely to this region (Engelke et al., 1980; Sakonju et al., 1981). Subsequent investigations by several labs eventually revealed that binding of TFIIIA permits rapid binding of TFIIIC to form a stable complex; subsequent binding of TFIIIB forms a complex that is recognized by Pol III to allow transcription (Fig. 8.3) (Bieker et al., 1985; Geiduschek and Tocchini-Valentini, 1988; Kassavetis et al., 1990; Lassar et al., 1983; Talyzina et al., 2023). The tRNA genes do not require TFIIIA for their transcription by Pol III; rather, binding of TFIIIC to specific sequence elements, again situated internal to the tRNA genes, is followed by association of TFIIIB and recognition by Pol III.

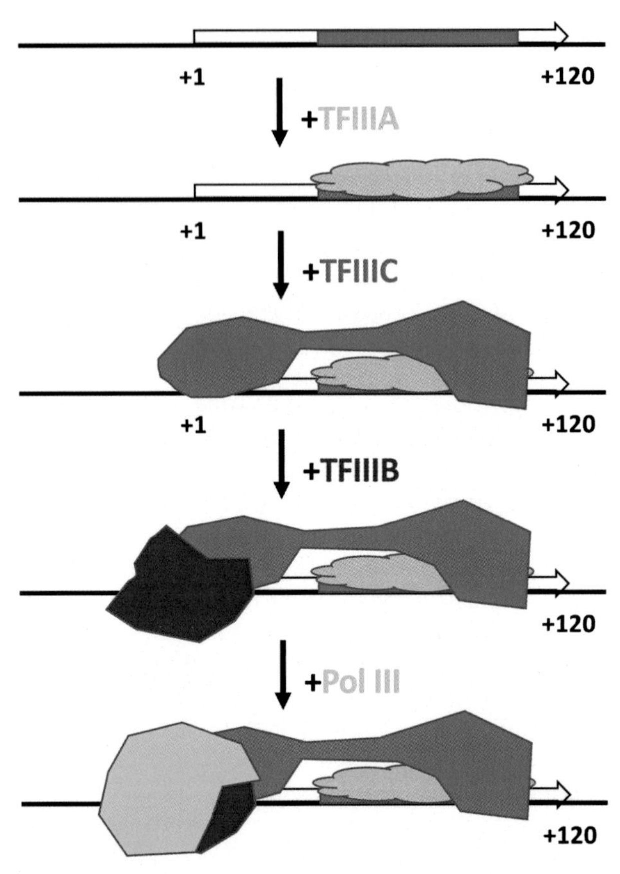

FIG. 8.3 Mechanism of transcription of the 5S rRNA gene by Pol III. See text for details.

Although research on Pol III transcription has fallen out of favor (having largely been mined out!) in the decades since these early studies, investigations of 5S rDNA transcription occupy a special place in the history of molecular biology, providing a bridge of sorts between fundamental discoveries made in bacterial systems, such as the *E. coli lac* repressor and phage lambda, and the increased complexity of transcription by RNA polymerase II.

The inhibition of transcription initiation seen upon incorporation of the 5S rRNA gene into a nucleosome was consistent with prior and concurrent studies pointing to a competition between the Pol III transcription complex and histones for the same sequences. Transcription of the 5S rRNA gene is initiated by binding of TFIIIA to an internal control region contained within the transcribed gene (Fig. 8.3) (see Sidebar: Transcription by Pol III). The TFIIIA:5S complex is recognized by TFIIIC, binding of which results in a stable complex that is resistant to 0.6 M NaCl and is unaffected by the presence of competitor DNA (Bogenhagen et al., 1982; Wolffe and Brown, 1988). Binding of TFIIIB then creates a platform for Pol III association and ensuing transcription. The preassembled complex is refractory to inhibition by subsequent assembly into nucleosomes in vitro; conversely, nucleosome assembly prior to the addition of the Pol III transcription factors inhibits transcription (Bogenhagen et al., 1982; Felts et al., 1990; Shimamura et al., 1988; Tremethick et al., 1990). An early study reported that TFIIIA was sufficient to prevent nucleosome-mediated inhibition, but later work found the complete transcription complex to be required and suggested that small amounts of TFIIIB and TFIIIC present in the TFIIIA preparation used previously may have accounted for the discrepancy (Felts et al., 1990; Gottesfeld and Bloomer, 1982; Tremethick et al., 1990).

In vivo experiments pointed to mechanisms that create a permissive environment for the transcription of tRNA and 5S rRNA genes. As recounted in Chapter 4, a tRNA reporter gene engineered to be in competition with a nucleosome positioning sequence was transcribed when introduced into yeast, and the predicted positioned nucleosome was not formed; a 2bp mutation in the TFIIIC binding site that rendered the gene inactive resulted in the formation of a positioned nucleosome that incorporated the tRNA start site (Morse et al., 1992). Genome-wide determinations of nucleosome occupancy revealed gene bodies of 5S rRNA and tRNA genes to be depleted of nucleosomes in yeast and metazoan cells (Helbo et al., 2017; Kumar and Bhargava, 2013; Mavrich et al., 2008; Parnell et al., 2008). Furthermore, 5S rRNA maxigenes of sufficient length to accommodate several nucleosomes were transcribed by Pol III in yeast, despite the inhibitory effect of multiple nucleosomes on transcriptional elongation in vitro (Felts et al., 1990). Increased nucleosome occupancy at tRNA genes was found in both *isw1* and *rsc* mutants, indicating that these chromatin remodelers are important in establishing permissive chromatin structures in yeast (Kumar and Bhargava, 2013; Parnell et al., 2008; Schlichter et al., 2020). Whether chromatin remodelers similarly facilitate the formation of nucleosome-depleted regions (NDRs) at genes transcribed by Pol III in metazoans is not currently known.

A series of elegant experiments, principally undertaken in Don Brown's lab at the Carnegie Institute, revealed a role for chromatin in the developmental regulation of a family of 5S rRNA genes in *Xenopus*, with linker histones playing a central part (Wolffe and Brown, 1988). Three types of 5S rRNA genes are expressed in oocytes, which accumulate large stores of ribosomes in preparation for fertilization and rapid embryonic growth (Almouzni and Wolffe, 1993a). The three types are encoded as simple tandem repeats: the major oocyte class is present in ~20,000 copies, the trace oocyte class in 1300 copies, and the somatic 5S rRNA gene in 400 copies. In somatic cells, over 95% of 5S rRNA is synthesized from the somatic genes, despite the 50-fold excess of the oocyte version. The selective repression of oocyte 5S rRNA genes is a complex process due in part to the differential stability of the transcription complex and limiting transcription factor availability during development but is reinforced by a specific chromatin structure (Wolffe and Brown, 1988). The first indication of 5S rRNA regulation by chromatin was derived from the analysis of in vitro transcription of 5S rRNA genes from *X. laevis* cultured cells (Bogenhagen et al., 1982; Schlissel and Brown, 1984). Chromatin prepared from oocyte germinal vesicles was permissive for transcription of both oocyte and somatic 5S rRNA genes using extracts from either oocytes or somatic cells, whereas transcription was observed only for somatic 5S rRNA genes when chromatin from erythrocytes or cultured cells was used (Bogenhagen et al., 1982). Washing chromatin from tissue culture cells with high salt allowed transcription of the oocyte 5S rRNA genes, as did depletion of the linker histone H1 by exposure to 0.6M NaCl; adding H1 back to the depleted chromatin restored inhibition (Schlissel and Brown, 1984). Oocytes do not express a canonical linker histone, but instead contain a linker histone variant originally called B4 and now designated H1.8 that binds nucleosomes with lower affinity than canonical H1 and is less effective at inducing formation of chromatin structure that is repressive to transcription (see also Chapter 5) (Nightingale et al., 1996; Smith et al., 1988; Ura et al., 1996). Correspondingly, addition of histone H1 to chromatin isolated from the nuclei of oocytes selectively repressed oocyte 5S rRNA genes (Wolffe, 1989), and experiments with reconstituted chromatin further supported these findings, showing that a 5S rRNA gene on a plasmid was inhibited by nucleosomes assembled at low density (one nucleosome per 215bp) only when a linker histone was added (Shimamura et al., 1989). In sum, the 5S rRNA gene system not only provided early evidence for a specific repressive effect of nucleosomes on transcription through inhibition of the binding of transcription factors but also indicated the potential of chromatin to confer precise and specific gene regulation in a developmental capacity.

Pol II

Advances in understanding the mechanistic basis of Pol II transcription made in the 1980s permitted questions regarding chromatin and transcription to be framed in much more sophisticated terms. While the early experiments described above indicated inhibitory effects of nucleosomes on transcription by Pol II, this was typically monitored by measuring the incorporation of nucleotides or generation of nonspecifically initiated transcripts, and therefore did not yield information on specific initiation events. The factors and events leading to precise initiation by Pol II began to be established in the 1980s, and the discovery of activation domains in transcription factors led to a focus on their mode of action, including in the context of chromatin. In addition, there was something of a schism between the communities of researchers engaged in studies of transcriptional regulation and those focused on chromatin, with some of the former contingent dismissing the relevance of studying transcription of chromatin templates (Kadonaga, 2019). This outlook was fostered in part by experiments showing increased transcription in vitro in the presence of sequence-specific transcriptional activators on naked DNA templates, suggesting that regulation might occur solely through interactions involving the transcription machinery and DNA. Moreover, the observations cited above that sequences that were transcribed in vivo were nonetheless associated with nucleosomes raised the possibility that in living cells chromatin was somehow generally transparent to the transcription machinery. Reasonable objections were also raised to interpreting both in vivo and in vitro experiments as indicating a regulatory role for chromatin (see the discussion at the end of Paranjape et al. (1994)).

In *The Sun Also Rises*, one of Hemingway's characters is asked how he went bankrupt. "Two ways," is the famous response. "Gradually, then suddenly." The same may be said of the path by which the notion was dispelled that chromatin was "a mere nuisance," in which "[t]rans-acting factors need only get to their sites first and hold on, and the histones can then be ignored" (Felsenfeld, 1992). (Felsenfeld was not advocating for that point of view!) The gradual stage was marked by increasing evidence from in vitro studies that nucleosomes impacted transcription by Pol II in ways that reflected complex interactions between transcription factors and chromatin, by experiments showing that histones similarly affected transcription in vivo in a fashion suggesting regulatory roles beyond being a simple and uniform impediment, and by the discovery of chromatin remodeling enzymes that functioned as global regulators of transcription. The sudden stage came with the revelation that Gcn5 and Rpd3, previously identified as a transcriptional co-activator and a co-repressor, were enzymes that acetylated and deacetylated histones (Chapter 7).

Nucleosomal inhibition of transcription can be overcome by activators: In vitro studies

One of the first eukaryotic promoters shown to allow specific transcriptional initiation by Pol II in vitro was the major late promoter of adenovirus (Ad MLP) (Manley et al., 1980; Weil et al., 1979). Specific initiation from the Ad MLP on a plasmid could be monitored using an assay in which hybridization of transcripts to a labeled, single-stranded DNA probe conferred protection against nuclease digestion, providing a novel tool for in vitro studies of transcription not just with "naked" DNA templates but also with chromatin. The first experiments testing the effect of chromatin on transcription from the Ad MLP showed that plasmid templates assembled into nucleosomes using *Xenopus* oocyte extracts were refractory to transcription at high nucleosome density but not at low density; similar results were obtained using plasmid templates in which the SV40 early promoter or the rabbit β-globin promoter was placed under control of the SV40 enhancer (Knezetic and Luse, 1986; Sergeant et al., 1984). Using a highly defined system in which a positioned nucleosome incorporated the Ad MLP on a short linear DNA fragment (the same system used to demonstrate nucleosomal inhibition of transcription initiation by SP6 polymerase), Lorch et al. observed essentially complete inhibition of transcription initiation by Pol II (using a rat liver extract) (Lorch et al., 1987). Later work showed similar inhibition when a nucleosome occupied the start site of the *Drosophila Kruppel* gene, using the protocol of Fig. 8.2 (Laybourn and Kadonaga, 1991), and when a nucleosome incorporated the TATA element or start site of the yeast *CYC1* promoter, using a yeast extract to support transcription (Lorch et al., 1992).

Studies aimed at dissecting the mechanism by which nucleosomes inhibited transcription by Pol II revealed a dependence on the specific molecular characteristics of the transcription unit. Similarly to results for Pol III, incubation of templates with extracts known to assemble active transcription complexes prior to chromatin assembly prevented inhibition of transcription from the Ad MLP (Knezetic et al., 1988; Matsui, 1987; Workman and Roeder, 1987). Depletion of TFIID from the HeLa nuclear extract used to assemble the transcription complex prevented suppression of nucleosomal inhibition, while the addition of a fractionated extract enriched for TFIID or of purified yeast TFIID overcame the inhibition, suggesting a direct competition between nucleosomes and the TFIID-dependent assembly of the PIC (Meisterernst et al., 1990; Workman and Roeder, 1987). In contrast, nucleosomal inhibition of transcription from the peptide IX promoter from adenovirus was not suppressed by preincubation with HeLa cell nuclear extract, while inhibition was relieved for transcription of the Ad MLP on the same plasmid template (Matsui, 1987). Although the Ad MLP is upregulated at late stages of adenovirus infection, it is also expressed early, while peptide IX expression is restricted to the middle stages, and is therefore apparently controlled by products of early gene expression. Matsui thus suggested as one possibility that some promoter-specific component that was missing from the HeLa cell extract used to assemble transcription complexes, possibly an early gene product, was needed to assemble a stable complex resistant to nucleosome assembly on the peptide IX promoter. A conceptually similar idea had been raised earlier in a study in which chromatin assembly was conducted on a template containing the chicken β^A promoter, which drives globin expression in adult chicken red blood cells (Emerson and Felsenfeld, 1984). Prior incubation of the template in an extract from 9-day-old chicken erythrocytes resulted in the generation of a DNase hypersensitive site in the assembled chromatin in the same region as seen in cells actively expressing the β^A gene, while incubation with extracts from nonexpressing cells failed to generate the hypersensitive site. These results hinted at a potential regulatory capacity for chromatin beyond one of simple and uniform repression.

Foundational studies on eukaryotic gene activation conducted in the early 1980s set the stage for investigating transcription mechanisms in the context of chromatin. Analysis of promoter deletion mutants identified regulatory elements in yeast and viral gene promoters (McKnight and Kingsbury, 1982; Struhl, 1981), leading to the discovery of enhancer elements capable of inducing high levels of transcription from distant sites (Banerji et al., 1981; Moreau et al., 1981). These elements were shown to function through the binding of specific *trans*-acting factors termed transcriptional activators (reviewed in (Kadonaga, 2004)). Most remarkably, activators were shown often to have a modular structure, with distinct and independent domains responsible for binding to DNA and for activating transcription. This was first demonstrated for the yeast Gal4 and Gcn4 activators in a series of papers from the Ptashne (Gal4) and Struhl (Gcn4) labs identifying small regions of these proteins able to bind to their recognition sites, and separate domains able to confer transcriptional activation when fused to their cognate DNA-binding domains or to heterologous (in fact, prokaryotic) DNA-binding domains (Brent and Ptashne, 1985; Hope and Struhl, 1986; Keegan et al., 1986; Ma and Ptashne, 1987). In vitro transcription experiments employing naked DNA templates suggested that *trans*-acting factors might function by facilitating the assembly of activation transcription complexes (Abmayr et al., 1988). Testing the effect of such activators in the context of chromatin was clearly of great interest.

Early in vitro experiments pointed to a role for activators in overcoming the repressive effects of chromatin to allow transcriptional activation. While prior incubation of a reporter template based on the Ad MLP with TFIID prevented subsequent nucleosome assembly from inhibiting transcription, simultaneous incubation did not (Workman and Roeder, 1987). The inhibition seen in the latter case could be relieved by including regulatory factors (rabies virus immediate-early protein or upstream stimulatory factor (USF)) during the assembly reactions (Workman et al., 1988, 1990). This suggested that such factors might function to tip the balance between nucleosome assembly and transcription complex formation toward the latter. Activation domains seemed likely to be critical players in this role. The chimeric protein Gal4-VP16, in which the DNA binding domain from Gal4 was fused to the activation domain of the herpesvirus regulatory protein VP16, was a popular choice for examining the effect of activator proteins on transcription in vitro (in conjunction with reporter genes driven by promoters containing Gal4 binding sites), due to its small size and potent activity (similarly, Gal4-VP16 was used in early in vitro experiments with chromatin remodelers (Chapter 6)) (Sadowski et al., 1988). While the inclusion of a nonactivating Gal4 derivative during nucleosome assembly did not prevent inhibition, the inclusion of Gal4-VP16 or another chimeric activator, Gal4-AH, partially relieved inhibition (Workman et al., 1991). Similar findings were reported using short linear fragments as templates and a transcriptional extract derived from yeast: a nucleosome occluding the TATA element or transcription start site inhibited transcription, and inhibition could be relieved by the addition of Gal4-VP16 either concurrently with or after chromatin assembly (Lorch et al., 1992). However, when transcription was performed using purified components, Gal4-VP16 failed to relieve nucleosomal inhibition, indicating the presence in the extract of factors in addition to those required for transcription on naked DNA templates that were required for the effect of Gal4-VP16. In an entirely distinct system, transcriptional repression was seen when chromatin assembly was coupled with DNA replication in *Xenopus* oocytes but was relieved by the addition of Gal4-VP16 before or after chromatin assembly (Almouzni and Wolffe, 1993b).

Other studies also supported a role for activators in counteracting nucleosome-mediated repression, while also indicating that effects could vary depending on the particular system employed. For example, while simultaneous incubation with TFIID and a chromatin assembly extract resulted in repression of transcription from the Ad MLP, a similar protocol using a *Drosophila* nuclear assembly extract and the *hsp26* promoter resulted in a "potentiated" template that was transcribed upon the addition of heat shock factor (Becker et al., 1991). Differing conclusions were also reached regarding the role of linker histone, with some studies reporting greater activator-mediated relief for H1-containing chromatin than for chromatin comprising only nucleosome cores, while others observed no difference (Kamakaka et al., 1993; Laybourn and Kadonaga, 1991; Sandaltzopoulos et al., 1994). Such differences likely stemmed from the many experimental variables, including the nature of the extract, the template utilized, and the time at which the template was exposed to the various components.

In spite of these differences, these experiments generally supported an important conclusion: the effect exerted by activators in vivo, where activated transcription could be increased a thousand-fold relative to the nonactivated state, was more closely mimicked in vitro on chromatin templates than on naked DNA, as was the ability of activators to stimulate transcription from sites at distances hundreds of bp from promoters (Kamakaka et al., 1993; Laybourn and Kadonaga, 1992). Stimulation of transcription in vitro by activators was typically two- to tenfold when naked templates were used, but was increased by close to an order of magnitude for chromatin templates (Fig. 8.4) (Laybourn and Kadonaga, 1991; Workman et al., 1990, 1991). This increased stimulation was not due to higher levels of transcription on chromatin templates, but rather to a lower "basal" level of transcription from chromatin compared to naked DNA in the absence of an activator. Thus, it appeared that activators functioned in part to relieve nucleosomal repression, while also sometimes additionally functioning in "true" activation by increasing the rate or efficiency of transcription (Felsenfeld, 1992; Kadonaga, 2019). In contrast, transcription in prokaryotes takes place on an accessible DNA template, reflecting a different "ground state" that requires only recognition of promoters by the polymerase holoenzyme, and prokaryotic activators generally function by facilitating engagement of the polymerase holoenzyme via direct cooperative interactions (Dove et al., 1997; Struhl, 1999).

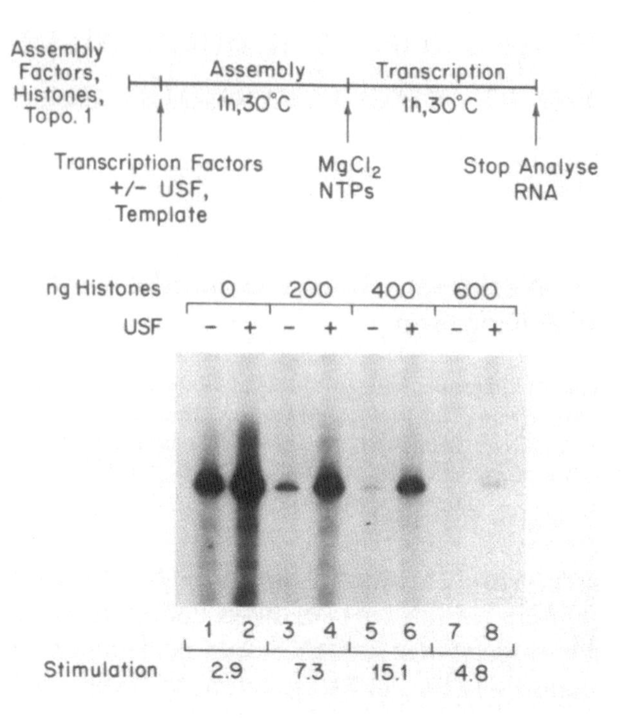

FIG. 8.4 Enhanced effect of the transcriptional activator USF on a chromatin template compared to naked DNA. *Top*, experimental protocol; the template contained the adenovirus major late promoter upstream of a G-less cassette that produces a 390nt transcript when CTP is omitted from the NTPs added to allow transcription. *Bottom*, visualization of labeled transcripts by gel electrophoresis and autoradiography. The fold stimulation of transcription by USF is increased by packaging of the template into chromatin. *From Workman, J.L., Roeder, R.G., and Kingston, R.E. 1990. An upstream transcription factor, USF (MLTF), facilitates the formation of preinitiation complexes during in vitro chromatin assembly. EMBO J. 9, 1299–1308.*

These ideas were the beginnings of a new paradigm for eukaryotic transcription. The discoveries of chromatin remodelers in the early 1990s, followed by the identification of histone-modifying enzymes as coactivators and corepressors, crystallized the notion that activators have two jobs in eukaryotes: to overcome repressive effects of chromatin and to facilitate the assembly of the transcription complex. Later work uncovered additional complexities, such as the presence of paused polymerase in metazoan cells and the need for mechanisms to ensure productive elongation.

Not surprisingly, however, the initial conception of nucleosomes constituting barriers to transcription by occupying TATA elements or transcription start sites (this could be termed "the single nucleosome blockade model"), which activators and coactivators functioned to overcome, proved to be overly simplistic. The early studies on which this conception was predicated were biased toward highly inducible promoters such as those for the globin genes and in viral genomes, for good reason: promoters in this class provided excellent models for understanding how genes were turned on or off during development or following a viral infection, which was of central interest to researchers. In addition, studies favored genes that were expressed at very high levels when induced, for the simple reason that this made experiments easier. However, this bias skewed models for transcriptional regulation toward a single class of genes, leaving genes expressed constitutively or at low levels relatively understudied. As one example of this bias, literature reviews and textbooks generally asserted that eukaryotic gene promoters contained TATA elements, whereas we now know that in both yeast and metazoans, only about 20% of gene promoters contain a consensus TATA sequence, and those that do are indeed members of the class of highly regulated genes (Basehoar et al., 2004; Haberle and Stark, 2018; Huisinga and Pugh, 2004; Lenhard et al., 2012; Ohler et al., 2002; Rach et al., 2011).

Also complicating the simple idea that nucleosomes could block initiation and that this block could be relieved by activators was the discovery of repressive mechanisms that coopted chromatin, such as the recruitment of Rpd3 as corepressor and the interaction of Sir proteins with the histone tails to repress transcription over telomeric regions in budding yeast. Another such mechanism that has been the subject of extensive study is repression by the polycomb group (PcG) complex, discussed later in this chapter. The existence of such mechanisms, including the formation of repressive heterochromatin, implies that occlusion by a single well-placed nucleosome is not always sufficient for maximal gene repression.

Despite these complications, in vitro transcription studies using chromatinized templates, in conjunction with in vivo studies discussed in the next section, helped to establish the idea that chromatin was more than a uniform and inert packaging material. Moreover, it seemed likely that if nucleosomes were not transparent to the transcriptional machinery, cells must have evolved mechanisms to overcome their repressive potential, and these mechanisms were likely to have been coopted as novel mechanisms for transcriptional regulation. These early in vitro studies provided a foundation for later models that incorporated chromatin remodelers, histone modifying enzymes, and histone variants in a complex process that we now know involves many dozens of proteins.

In vivo studies

Concurrently with in vitro studies aimed at elucidating the mechanisms by which chromatin is transcribed by Pol II, in vivo studies increasingly supported a critical regulatory role for chromatin in living cells. As is evident from reviews published during this accelerated period of discovery, in vivo and in vitro studies were mutually informative (Felsenfeld, 1992; Grunstein, 1992). Two avenues of in vivo investigation were especially influential: focused studies on chromatin organization and transcriptional regulation of individual genes, and studies examining the effects of perturbations of histone gene expression on gene regulation in the budding yeast *Saccharomyces cerevisiae*. Research on the effects of histone mutations in *S. cerevisiae* was discussed at length in Chapter 5; here, early experiments using this approach that are relevant to chromatin and gene regulation are recounted. With regard to studies on individual gene promoters, the yeast *PHO5* gene and the mouse mammary tumor virus (MMTV) promoter were especially informative and are discussed below.

Perturbation of histone gene expression in budding yeast

The first experiments to investigate the impact of chromatin on transcription by examining the effect of histone mutants took advantage of two important features of budding yeast: they are relatively efficient at homologous recombination, enabling gene editing, and they contain only two copies of each of the histone genes (Hereford et al., 1979; Hinnen et al., 1978; Scherer and Davis, 1979; Smith and Andresson, 1983). Experiments described in Chapter 5 demonstrated that while H2B is essential, yeast tolerate the loss of the amino-terminal region of either H2A or H2B but not their simultaneous deletion (Schuster et al., 1986; Wallis et al., 1983). The effect of loss of H2B on transcription was examined using a glucose shut-off protocol in which its expression was placed under the control of the *GAL1-10* promoter (Han et al., 1987). The resulting strain expressed H2B and grew relatively normally when cultured in a galactose-containing medium, but when the medium was changed to provide glucose as a carbon source, H2B expression was repressed and the cells arrested in mitosis, failing to undergo chromosome segregation, although chromosome replication was observed. Transcription was assessed by measuring incorporation of [^{3}H]-adenine into RNA following H2B shutoff; the interpretation was complicated by the shift in the growth medium, which affects transcription, but it was clear that RNA synthesis was not prevented. Induction of the *CUP1* gene by the addition of $CuSO_4$ was unaffected 30 min after H2B shutoff; however, the status of genes that would normally be repressed was not examined in these initial experiments.

A more sophisticated analysis of transcriptional effects was subsequently conducted using yeast engineered to allow shutoff of H4 synthesis. This led to similar phenotypes as the shutoff of H2B, with cells replicating their DNA but arresting soon after and exhibiting a disrupted chromatin structure characterized by increased nuclease accessibility and an altered plasmid topology indicative of decreased nucleosome density (Kim et al., 1988). As with H2B, the shutoff of H4 synthesis did not result in global changes in transcription, and induction of *CUP1* was essentially normal. However, examination of the inducible *PHO5* gene showed it to be expressed even in the growth medium containing high phosphate levels, where it was normally repressed. Moreover, the *PHO5* promoter and those of two additional inducible genes, *GAL1* and *CYC1*, all drove the expression of a *lacZ* reporter gene under normally repressive conditions upon H4 depletion and accompanying nucleosome loss (Han and Grunstein, 1988; Han et al., 1988). Derepression occurred even when promoter mutants lacking upstream activating sequence (UAS) elements were used, supporting the conclusion that it was directly caused by the altered chromatin structure rather than adventitious functioning of activators, and nucleosome positioning over the *PHO5* promoter was disrupted upon H4 shutoff. These findings, together with those from Fred Winston's lab showing that inactivation of the *HIS4* promoter by a Ty1 insertion was suppressed by altered histone stoichiometry (discussed in Chapter 6) (Clark-Adams et al., 1988), provided the first evidence that nucleosomes could act as specific repressors of transcription in vivo. The observations that in vivo manipulations of histone stoichiometry and chromatin structure resulted in the loss of repression at specific promoters directly implicated chromatin structure as a controlling element in transcriptional regulation. As with results from in vitro studies of transcription of nucleosomal templates, these results further implied that some mechanism must exist in living cells to overcome nucleosomal repression, most likely depending on transcriptional activators (Grunstein, 1992; Han and Grunstein, 1988).

Results from deletions and mutations of the histone H3 and H4 amino termini further strengthened the argument for histones acting as regulators of transcription (see also Chapter 5). The discovery that yeast lacking the highly conserved H4 amino terminus (specifically, amino acids 4–28) were viable was startling; the derepression seen at the silent mating type loci, and later at genes located near the telomeres, was also unexpected (Aparicio et al., 1991; Kayne et al., 1988). Subsequent studies identified specific amino acids in the H4 tail required for heterochromatin-mediated silencing in yeast and uncovered the mechanism by which the H4 amino terminus collaborates with the Sir proteins to facilitate heterochromatin formation and gene silencing, as discussed in Chapter 3 (Gartenberg and Smith, 2016; Johnson et al., 1990; Megee et al., 1990; Park and Szostak, 1990). Conversely, the H4 amino terminus was also implicated in specific gene activation: transcript levels of genes induced by galactose were reduced up to 20-fold by loss of the H4 tail, but were essentially unaffected by loss of other histone tails; *PHO5* induction was also diminished in yeast lacking the H4 tail (Durrin et al., 1991). Underscoring the specificity of histone function in gene regulation, deletions and mutations in the H3 tail caused hyperactivation of the *GAL* genes while leaving *PHO5* induction unaffected (Mann and Grunstein, 1992).

Later work expanded on these early studies by examining the genome-wide effect on transcript levels of histone tail deletions and mutations, and of decreased overall nucleosome occupancy induced by altered stoichiometry (Dion et al., 2005; Martin et al., 2004; Parra et al., 2006; Sabet et al., 2003; Wyrick et al., 1999). Depleting histone H4 altered transcript levels for about a quarter of all yeast genes, with ~15% being upregulated and ~10% downregulated (Wyrick et al., 1999); amino-terminal deletions of individual histones similarly affected transcript levels of a significant minority of genes (Martin et al., 2004; Sabet et al., 2003). It would perhaps be interesting to test the effects of such mutations and altered nucleosome occupancy on the association of TBP and Pol II genome-wide, rather than monitoring transcript levels, which can be subject to feedback mechanisms such that altered *transcription* is not always reflected by alterations in *transcript levels* (Timmers and Tora, 2018). However, such an approach would still not resolve the difficulty of distinguishing indirect effects of chronic absence of histone tails, or of achieving rapid removal of histones from chromatin in vivo. Perhaps novel methods for rapid protein degradation or in vivo selective proteolysis will eventually be adapted to investigate more rigorously histone function in transcriptional regulation in vivo, in metazoan as well as protozoan cells.

Model promoters: PHO5 *and MMTV*

A major step forward in determining the relationship between transcriptional activation and chromatin structure came from the development of methods using nucleases to monitor DNA accessibility in chromatin. The introduction of the indirect end-label technique for identifying sites accessible to nucleases (Chapter 4) led to the discovery of hypersensitive sites in active gene promoters, such as induced heat shock genes in *Drosophila*. These sites were quickly recognized as probable binding sites for regulatory proteins (Elgin, 1988; Wu et al., 1979). Recognizing the advantages inherent in utilizing budding yeast as an experimental system, Wolfram Hörz and colleagues began a systematic exploration of the changes in chromatin structure, and the mechanisms responsible for those changes, during transcriptional activation of the *PHO5* gene of *S. cerevisiae* (Korber and Barbaric, 2014). The *PHO* genes, identified by the Oshima group (Oshima, 1982), comprise about 20 genes that are regulated by intracellular phosphate concentrations. Studies by the Hörz lab initially focused on *PHO5* and later also on *PHO8*, both of which are repressed by high phosphate and induced upon phosphate starvation.

Indirect end-label experiments using DNase I, MNase, and restriction enzymes revealed the repressed *PHO5* promoter to be packaged into positioned nucleosomes that are disrupted upon induction in low phosphate medium. Nuclease cleavage sites in the repressed *PHO5* promoter were separated by 160–180 bp regions that were accessible to nucleases in DNA but protected in repressed chromatin, consistent with an array of positioned nucleosomes—one of the first examples of nucleosomes occupying specific sequences in vivo, with immediate implications for a potential regulatory role (Fig. 8.5; see also Fig. 4.15) (Almer and Horz, 1986). Comparison of the inferred positions of nucleosomes over the inactive *PHO5* promoter with DNA sequences identified as critical for activation revealed one of the regulatory elements, UASp1, to coincide with a DNase hypersensitive site located between positioned nucleosomes −2 and −3 (Almer et al., 1986; Rudolph and Hinnen, 1987). In contrast, a second activation site, UASp2, was found to be incorporated into nucleosome −2 in the repressed promoter. Subsequently, UASp1 and UASp2 were each shown by DNase footprinting to bind the activator proteins Pho2 and Pho4 in vitro (Vogel et al., 1989), and by dimethylsulfate (DMS) footprinting to bind Pho4 at both sites under activating but not repressive conditions (Venter et al., 1994). In addition, the essential TATA element was mapped near the center of the putative −1 nucleosome, and later revised to a location closer to the nucleosome edge (Small et al., 2014). Induction of *PHO5* was accompanied by increased accessibility to nucleases throughout the promoter region, suggesting that activation resulted not just in randomization of nucleosome positioning, but in removal of nucleosomes from the activated promoter (Almer et al., 1986). An alternative explanation that the increased nuclease accessibility resulted from some structural change in promoter nucleosomes was eventually ruled out by experiments demonstrating loss of histone association in ChIP experiments and a change in the topology of circular minichromosomes containing the *PHO5* promoter upon induction that was consistent with nucleosome loss (Boeger et al., 2003, 2004; Jessen et al., 2006; Korber et al., 2004; Reinke and Horz, 2003).

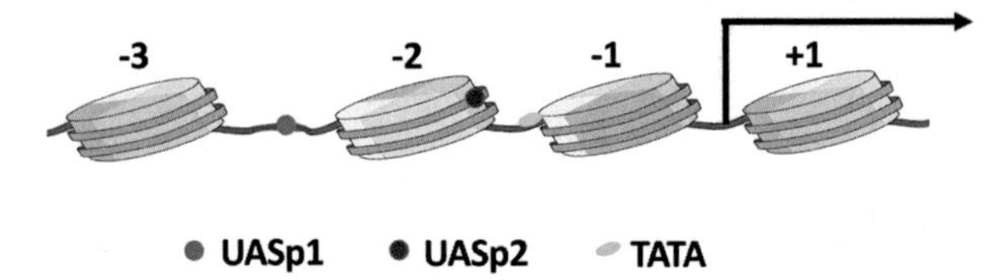

FIG. 8.5 Schematic of the repressed yeast *PHO5* promoter. Transcription initiates as indicated by the arrow. See text for details.

A key question that arose from these studies was one of cause and effect: were changes in chromatin structure accompanying transcriptional activation required for transcription, or were they a consequence, perhaps resulting from the action of the transcription machinery on the chromatin template? This question was addressed by several labs by monitoring chromatin structure at promoters rendered transcriptionally inactive by mutation of critical *cis*-acting elements such as the TATA element. In several cases, including the *PHO5* promoter and the SWI/SNF-dependent *SUC2* promoter discussed in Chapter 6, remodeling of promoter chromatin structure was observed in such mutants, indicating that chromatin perturbation was independent of transcriptional activation and thus likely to be a prerequisite (Axelrod et al., 1993; Fascher et al., 1993; Hirschhorn et al., 1992; Lee and Garrard, 1992; Morgan and Whitlock Jr., 1992; Verdin et al., 1993). As discussed in Chapter 6, chromatin remodeling was eventually discovered to be carried out in many cases by dedicated ATP-dependent complexes recruited by transcriptional activators.

The paradigm of inactive promoters being packaged by positioned nucleosomes that are disrupted upon gene activation was reinforced by additional examples, such as the yeast heat shock gene *HSP82*, the mouse β-globin promoter, and the MMTV promoter (discussed below) (Benezra et al., 1986; Richard-Foy and Hager, 1987; Szent-Gyorgyi et al., 1987). However, this paradigm did not apply universally—for example, regulatory sites in the *Drosophila hsp26* promoter were shown to be accessible regardless of its activation, and activation did not disrupt a positioned nucleosome in the promoter (Elgin, 1988). Nonetheless, the transitions in chromatin structure seen in multiple systems, together with the altered gene expression resulting from in vivo manipulations of chromatin, made the prospect of understanding the underlying mechanism irresistible, and the *PHO5* gene proved to be an excellent vehicle for this undertaking.

The contribution of the two key regulatory elements in the *PHO5* promoter, UASp1 and UASp2, to nucleosome disruption of the activated *PHO5* promoter was examined by using DMS footprinting and nuclease digestion in yeast harboring *pho5* promoter mutations or expressing mutant versions of Pho4. Of the two cooperatively binding activators that bind to these elements, Pho2 and Pho4 (Hirst et al., 1994), Pho4 was shown to be critical for activation and chromatin disruption, as its overexpression could compensate for the absence of Pho2, while the converse was not the case (Fascher et al., 1990). Thus, Pho4 was identified as the primary actor in the chromatin remodeling and activation of the *PHO5* promoter. The accessible UASp1 element proved to be critical, as nucleosome disruption and *PHO5* activation were abrogated under normally activating conditions in a mutant lacking this element (Fascher et al., 1993; Venter et al., 1994). Moreover, binding of Pho4 to UASp2, detected by DMS footprinting in wild-type cells at the activated promoter, was lost in the UASp1Δ mutant, indicating that its binding was occluded by nucleosome −2 (Venter et al., 1994). This occlusion was not absolute, as overexpression of Pho4 resulted in binding and concomitant nucleosome disruption even in mutants lacking UASp1 (Venter et al., 1994). Later work revealed that nucleosome occupancy at the *PHO5* promoter is heterogeneous both under repressed and activating conditions (see Fig. 4.15) (Boeger et al., 2008; Jessen et al., 2006; Small et al., 2014); thus, overexpression of Pho4 may facilitate its binding to UASp2 in the absence of UASp1 via increased binding to transiently accessible sites (Korber and Barbaric, 2014; Mao et al., 2011).

Another major finding was that Pho4 lacking its activation domain could bind UASp1, but chromatin disruption and binding to UASp2 were no longer observed (Svaren et al., 1994). Replacement of the Pho4 activation domain by that from VP16 restored chromatin disruption and *PHO5* activation, implying that activation domains might generally be capable of remodeling chromatin. A previous study had similarly reported a role for the Gal4 activation domain in nucleosome displacement upon induction of the *GAL1* promoter (Axelrod et al., 1993). This latter study determined nucleosome occupancy by monitoring UV-induced photoproducts of DNA, which are affected by DNA conformation and thus exhibit altered patterns in the presence of bound proteins (Selleck and Majors, 1987). The results on *PHO5*, using more accessible and direct methods to monitor nucleosome occupancy, provided convincing evidence in a second, independent system for the importance of activation domains in overcoming the repressive effects of nucleosomes in vivo.

Altogether, investigations using the *PHO5* promoter pointed to a scenario in which initial binding of Pho4 (and Pho2) to the accessible UASp1 resulted in disruption of nucleosome −2, thus allowing Pho4 binding to UASp2 and consequent transcriptional activation. Later work showed that while UASp1 is critical, it is not sufficient for full chromatin remodeling and activation of *PHO5*: mutation of UASp2 resulted in an intermediate level of remodeling, with a corresponding reduction in *PHO5* activation (Ertel et al., 2010; Mao et al., 2011). Furthermore, using ChEC-seq, in which MNase is fused to a protein of interest to assess occupancy (Chapter 4 Sidebar: ChIP-seq variations), TBP binding to the TATA element was observed under activating but not repressed conditions, and was lost in a UASp1Δ mutant (Mao et al., 2011). Single-molecule determination of nucleosome occupancy assessed by protection against DNA methylation, as discussed in Chapter 4, revealed stronger disruption during *PHO5* activation of nucleosomes −2 and −3 flanking UASp1 than of more distal nucleosomes, supporting a proximate effect of the Pho4 activation domain (Jessen et al., 2006). The importance of the nucleosome configuration of the *PHO5* promoter for its proper regulation was further supported by experiments in which the DNA sequence incorporated into nucleosome −2 was replaced by sequences that increased or decreased nucleosome stability; in the former case, nucleosome −2 persisted under activating conditions and *PHO5* induction was severely curtailed, while in the latter case, *PHO5* was constitutively activated, albeit at low levels that were increased under activating conditions (Straka and Horz, 1991).

Finally, it was demonstrated that *PHO5* promoter activation and nucleosome disruption did not require DNA replication (Schmid et al., 1992). Although disruption of nucleosomes by the passage of DNA polymerase during replication was once considered a likely mechanism for allowing access of the transcription machinery to chromatin (Svaren and Chalkley, 1990), the discovery of chromatin remodeling complexes and dynamic turnover of nucleosomes throughout the cell cycle has undercut the rationale for such a mechanism. Overexpression of the yeast TF Rfx1 was reported to decrease nucleosome occupancy at an engineered site in cycling yeast but not in α-factor arrested yeast, consistent with a requirement for DNA replication for binding to a site occluded by a nucleosome (Yan et al., 2018). However, to my knowledge, no examples of replication being required for chromatin remodeling involved in transcriptional activation in a physiological context have been reported. With regard to histone turnover in vivo, the study of the *PHO5* promoter again played an important historical role: as described in Chapter 5 (section on H3.3), a dual-label protocol was used to demonstrate that nucleosome reassembly after repression of the activated *PHO5* promoter is achieved using histones from the soluble cellular pool, rather than those that packaged the promoter prior to its activation (Schermer et al., 2005). This same protocol was subsequently used to demonstrate the turnover of histones in chromatin outside of DNA replication (Dion et al., 2007; Jamai et al., 2007; Rufiange et al., 2007).

The scenario described above raises an obvious question: how does Pho4 binding to UASp1 disrupt the surrounding chromatin in an activation domain-dependent manner? A first approach to addressing this question was to test a panel of *pho4* mutants for their ability to remodel chromatin and to activate transcription, with the aim of separating the two functions (McAndrew et al., 1998). This effort did not succeed, however, as the effects of mutations on nucleosome disruption and activation of *PHO5* were highly correlated, and no mutants capable of remodeling chromatin but not of activating transcription were found. This was not entirely surprising, as direct recruitment of the SWI/SNF complex via fusions with the Gal4 or LexA DNA-binding domains was known to be sufficient for transcriptional activation of reporter genes in yeast (e.g., Laurent et al., 1991; Chapter 6). Moreover, activation domains had been shown to be capable of recruiting chromatin remodeling complexes (Chapter 6). It thus seemed possible that activation domain-dependent recruitment of a chromatin remodeler by Pho4 could be sufficient for at least some degree of *PHO5* induction.

A focused search for an essential factor among known chromatin remodelers for activation of *PHO5* proved, however, to be frustrating. Individual deletion mutants of potential remodelers showed little effect on *PHO5* activation or remodeling, although some effects were seen in the kinetics of induction or under suboptimal conditions (Barbaric et al., 2007; Gaudreau et al., 1997; Gregory et al., 1998; Huang and O'Shea, 2005). Multiple remodelers, including SWI/SNF, ISWI, RSC, and Chd1 were found to affect *PHO5* remodeling under specific conditions, pointing to a complex mechanism with substantial redundancy, while chromatin modifiers and histone chaperones also were found to have impacts (Korber and Barbaric, 2014). Redundancy was firmly established by demonstration that remodeling under normal inducing conditions was abrogated by simultaneous loss of RSC and Isw1/Chd1 or RSC and Snf2 (Musladin et al., 2014). How multiple remodelers are recruited during activation of *PHO5*, and whether all are recruited under normal induction conditions, remains unknown. Studies of additional inducible genes in yeast have revealed varied requirements for associated remodeling: some genes, such as *SUC2*, depend strongly on SWI/SNF (Carlson et al., 1981; Hirschhorn et al., 1992); some depend on SWI/SNF and additional chromatin remodelers and modifiers (for example, *INO1* depends on SWI/SNF and Ino80) (Ebbert et al., 1999; Ford et al., 2007; Peterson and Herskowitz, 1992; Shen et al., 2000); and some, like *PHO5* and *CHA1*, depend on SWI/SNF under specific conditions or for proper kinetics, but exhibit normal remodeling under inducing conditions even in its absence (Ansari et al., 2014; He et al., 2008; Moreira and Holmberg, 1998). Most remarkably, *PHO8*, which like *PHO5* depends on Pho4 and undergoes disruption of promoter nucleosomes upon activation, exhibits a strong dependence on SWI/SNF for its activation and chromatin remodeling (Gregory et al., 1999). As emphasized in Chapter 6, current knowledge still does not enable the prediction of which chromatin remodeling or modifying complexes participate in the activation of a given gene, nor is the mechanism by which multiple such coactivators are employed generally understood. However, ChIP combined with the judicious use of mutants has allowed the formulation of models by which the ordered recruitment of such coactivators facilitates transcriptional activation, as discussed further on.

Studies of MMTV, a single-stranded RNA retrovirus that integrates into the mouse genome as double-stranded DNA, were initially undertaken because of its connection with naturally occurring mammary adenocarcinomas (Bittner, 1936, 1942). The discovery that treatment of cells harboring integrated retroviral DNA with glucocorticoids caused a rapid increase in the production of viral RNA raised interest in studying MMTV as a model for hormone-stimulated transcriptional induction, which was known to be important in multiple physiological contexts (Hager et al., 1984; Parks et al., 1974; Ringold et al., 1977). Evidence that MMTV induction involved chromatin structural changes was first suggested by DNase digestion experiments, which localized a hypersensitive site seen upon hormone activation to a region that bound the glucocorticoid receptor (GR) in vitro and that acted as a hormone-dependent enhancer element in vivo (Zaret and Yamamoto, 1984). Similar findings were reported for the estrogen-responsive chicken vitellogenin gene (Burch and Weintraub, 1983).

At this juncture, the mechanism by which enhancer elements stimulated transcription from distant locations was mysterious, and it was speculated that the hypersensitivity induced by hormone might specify a change in chromatin structure that could facilitate entry by RNA polymerase, either directly or through propagation of chromatin structural alterations to more remote sites close to the transcription start (Zaret and Yamamoto, 1984). A subsequent investigation sought to identify proteins that bound to the MMTV promoter upon induction by using an exonuclease protection assay, in which bound proteins block processive 3′ to 5′ digestion of DNA by *E. coli* Exo III (see Chapter 4) (Cordingley et al., 1987). Using a chimeric MMTV LTR-*ras*^{H} fusion, two major hormone-dependent sites of protection were identified, one at about 80 bp upstream of the *ras*^{H} ORF and another slightly downstream of the cap site. A sequence at the upstream site was a close match for the binding site for NF1 (nuclear factor 1), which had first been identified as important for the replication of adenovirus and was known to bind to this region of the MMTV promoter (Fig. 8.6). Protection of this region by NF1 binding was further supported by experiments showing that NF1 bound to the MMTV promoter in vitro yielded an identical exonuclease "stop" (Cordingley et al., 1987). The downstream protection was suggested to correspond to a transcription factor or factors such as TFIID binding to the initiation site in the active promoter. Somewhat surprisingly, protection from exonuclease digestion was not seen at upstream (−190 bp) GR binding sites, although protection was detectable in vitro; this was attributed to the lability of GR binding to chromatin in vivo. The results of these experiments led Hager and colleagues to propose a model in which hormone activation of GR led to recruitment of NF1 and additional transcription factors at the promoter, rather than activation of a preformed transcription complex (Cordingley et al., 1987). Similar models were envisioned by other researchers in the transcriptional regulation community, based on observations of cooperativity between binding of transcription factors to DNA and consideration of protein-protein interactions (e.g., Ptashne, 1986; Sawadogo and Roeder, 1985); nonetheless, this formulation was remarkably prescient.

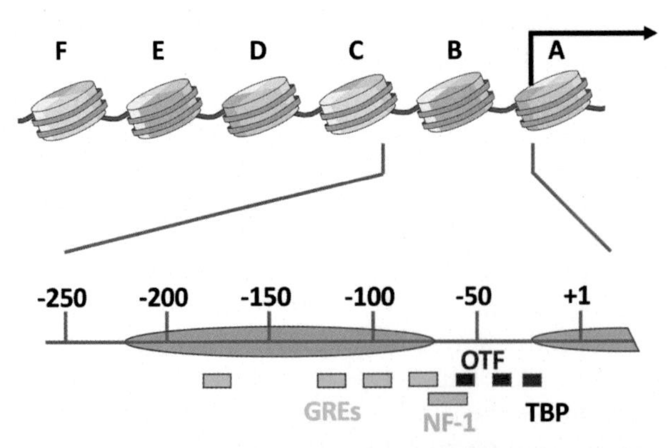

FIG. 8.6 Nucleosome positioning over the MMTV promoter. The region around Nucleosome B is expanded at the bottom; shown are binding sites for GR, NF-1, the transcription factor OTF, and TBP. In contrast to the *PHO5* promoter in yeast, which lacks canonical linker histone, the MMTV promoter nucleosomes contain histone H1. Moreover, despite the location of OTF in a linker region and NF-1 at the edge of nucleosome B, access to these TFs is blocked in the absence of GR binding.

Hager and colleagues additionally proposed that "binding of the receptor at the GRE [could result] in modification of the nucleoprotein template, exposing the binding site for NF1" (Cordingley et al., 1987). Determination of nucleosome positions using MNase and the less sequence-specific chemical reagent, methidiumpropyl-EDTA-Fe(II) (MPE-Fe(II)) (see Chapter 4), revealed an array of six strongly positioned nucleosomes over the uninduced MMTV promoter (Fig. 8.6) (Richard-Foy and Hager, 1987). Mapping nucleosome position in metazoan cells is much more challenging than doing so for budding yeast, as the 100-fold increase in genome size renders identification of specific cleavage sites in indirect end-label experiments far more difficult. In this case, this problem was circumvented by incorporating the MMTV promoter into stably replicating episomes derived from bovine papillomavirus; these episomes are maintained at ~200 copies/cell, resulting in considerable improvement in signal to noise in autoradiographic images. Nucleosome positioning over the MMTV promoter was later confirmed by applying a primer extension assay to nucleosomal DNA obtained from MNase digestion (Bresnick et al., 1992b) and was shown to be consistent in various cell lines, even when determined at a single integrated copy (Truss et al., 1995). As mentioned in Chapter 4, nucleosome positioning over the MMTV promoter was eventually found to arise not from nucleosomes having precise translational positions, but rather from overlapping positions that are rotationally equivalent but differ by a small number of helical turns in their translational positions (see Chapter 4) (Fragoso et al., 1995).

Two major findings emerged from the experiments mapping TF binding and nucleosome positioning: first, the primary binding sites identified for GR as well as NF1 were fully or partially contained in nucleosome B, implying that initial binding of activated GR occurred on a nucleosomal template (Fig. 8.6). Second, nucleosome B exhibited increased accessibility to MPE-Fe(II) as well as to DNase upon hormone induction, while the other five positioned nucleosomes were essentially unchanged (Richard-Foy and Hager, 1987; Zaret and Yamamoto, 1984). Similar findings were made in investigations of the rat tyrosine aminotransferase (TAT) gene, also inducible by the hormone-induced GR: an array of positioned nucleosomes was found encompassing the region of the GR-binding enhancer, and activation was accompanied by perturbation of only two of these nucleosomes (Carr and Richard-Foy, 1990; Reik et al., 1991). These results raised the possibility that binding of activated GR to nucleosome B in the MMTV promoter (and similarly to a positioned nucleosome in the rat TAT enhancer) could lead to nucleosome disruption, thereby allowing access to NF1 and subsequent assembly of the transcription complex.

To investigate the mechanism by which GR binds to the MMTV promoter upon hormone induction, the ability of GR to bind to sites within nucleosomes reconstituted in vitro using fragments from the MMTV promoter was assessed (Perlmann and Wrange, 1988; Pina et al., 1990). DNase footprinting revealed strong rotational positioning of a nucleosome reconstituted using a DNA fragment including most of the B nucleosome sequence, with translational positioning essentially indistinguishable from that observed in vivo. Remarkably, the addition of GR resulted in alterations in the DNase footprint consistent with GR binding, while leaving the periodic cleavage pattern characteristic of rotational nucleosomal positioning undisturbed; the clear conclusion was that GR could form a ternary complex by binding to its site in a nucleosomal context (Fig. 8.7). GR binds its site as a dimer, and its ability to bind to nucleosomal DNA made sense from a structural standpoint as the contact sites of the two monomer receptors are a helical turn apart and thus face in the same direction.

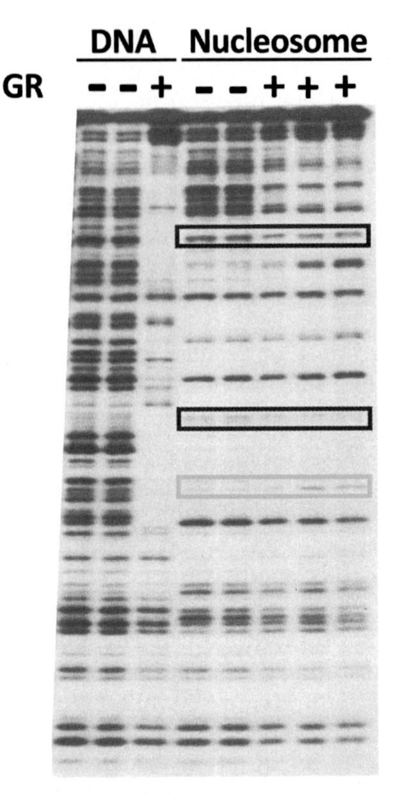

FIG. 8.7 DNase footprinting of GR bound to naked or nucleosomal DNA. A 199bp DNA fragment including MMTV sequences from −198 to −58 relative to the starting ATG was end-labeled and GR added or not, as indicated, to the naked or reconstituted nucleosomal template, prior to DNase I digestion and gel electrophoresis. Note the clear pattern of protection due to multiple GR binding sites (Fig. 8.6) in the naked DNA sample when GR is present, and the periodic cutting pattern for the nucleosomal template. The *red boxes* indicate sites where GR binding to the nucleosomal template confers protection against cleavage by DNase I, while the *green box* indicates a site where enhanced DNase I cleavage is seen. *From Perlmann, T., Wrange, O. 1988. Specific glucocorticoid receptor binding to DNA reconstituted in a nucleosome. EMBO J. 7, 3073–3079.*

Correspondingly, in vitro binding of GR to nucleosomal DNA was seen to depend on both the rotational and translational positions; binding to outward-facing nucleosomal sites showed only a two- to fivefold reduction in affinity relative to naked DNA, whereas an inward-facing site at the dyad was completely refractory to GR binding (Li and Wrange, 1995; Pina et al., 1990). DNase footprinting of the MMTV promoter in vivo revealed rotational positioning favorable for GR binding, consistent with that obtained upon in vitro reconstitution (Truss et al., 1995). Much more recently, ChIP-seq combined with MNase-seq has provided evidence for widespread hormone-activated GR binding to nucleosomal sites in vivo (Johnson et al., 2018). Other sites that bind GR reside in NDRs, while sites that exhibit cell-type specific GR responsiveness are almost all nucleosomal but are refractory to GR binding in unresponsive cells. Whether the failure of GR to bind to these latter sites is due to the rotational orientation of the nucleosome is not currently known.

In contrast to GR, NF1 does not bind to a nucleosomal site in vitro, and binding of NF1 in vivo is not observed in the absence of hormone stimulation (Archer et al., 1991; Cordingley et al., 1987; Eisfeld et al., 1997; Pina et al., 1990; Truss et al., 1995). GR does not cooperate with NF1 in binding assays in vitro on naked templates, nor do the two factors synergize in in vitro transcription assays, ruling out cooperative binding via protein-protein interactions (Bruggemeier et al., 1990; Kalff et al., 1990). The binding of NF1 is essential for normal levels of MMTV transcription, as mutation of the NF1 binding site impairs MMTV transcription by activated GR and results in diminished chromatin remodeling and decreased occupancy of TBP (Hebbar and Archer, 2003; Miksicek et al., 1987). Altogether, these results pointed to a mechanism of hormone-induced transcriptional activation in which chromatin plays a central part.

Additional support for chromatin-mediated regulation of the MMTV promoter was provided by two distinct lines of experimentation. The first derived from observations that in contrast to chromosomally integrated MMTV, or MMTV carried on stably replicating plasmid episomes, transiently transfected MMTV is packaged into nucleosomes that are randomly positioned (Archer et al., 1992; Hebbar and Archer, 2008). This difference allowed the comparison of an ordered to a disordered chromatin structure on transcription factor binding and transcriptional output in vivo. A transiently transfected MMTV promoter driving a luciferase gene reporter and a stably replicating MMTV promoter driving the chloramphenicol transferase reporter in the same cells were both induced by hormone-activated GR, and both exhibited protection against Exo III consistent with NF1 binding after hormone treatment (Archer et al., 1992). However, only the transiently transfected MMTV promoter showed protection against exonuclease digestion at the NF1 binding site in the absence of hormone. These experiments, together with the in vitro experiments cited above, indicated that the chromatin structure at stably replicating MMTV promoters—specifically, nucleosome B—prevented NF1 access, and that binding of activated GR was necessary to facilitate its binding. The observation that transcription was nevertheless also hormone-dependent for the transiently transfected, NF1-bound MMTV promoter indicated that GR was also critical for the assembly of a productive transcription complex. Consistent with this idea, a weak protection against Exo III digestion at the TATA element was observed in the transiently transfected template only in the presence of hormone (Archer et al., 1992; Lee and Archer, 1994). A later study showed that the progesterone receptor, acting via the same promoter element as GR, was able to activate transcription from a transiently transfected MMTV promoter but not from a stably replicating copy, and was not able to effect the change in accessibility on the stably replicating template that was seen with GR upon hormone activation (Smith et al., 1997). Differences in regulation between transiently transfected templates having disorganized chromatin structure and their stably replicating counterparts having well-ordered chromatin were reported in other systems as well, underscoring the regulatory impact of chromatin in vivo (Smith and Hager, 1997).

The second line of support for the importance of chromatin structure in MMTV regulation came from experiments in which an MMTV-driven reporter was injected into *Xenopus* oocytes, where it was assembled into chromatin and could be activated by the addition of hormone (Perlmann and Wrange, 1991). The addition of nonspecific DNA that competed for the available histone pool resulted in an altered chromatin structure on the MMTV template, including disruption of the rotational positioning seen in the absence of a competitor. This alteration in chromatin structure had functional consequences, as transcription from the MMTV promoter in the absence of hormone was greatly increased. Thus, as in the experiments in which histone type or expression was manipulated in budding yeast, altered chromatin structure had a substantial impact on transcriptional regulation.

Additional studies addressed the mechanism by which GR binding led to the disruption of nucleosome B and the nature of the remodeled chromatin. The increased accessibility to DNA in nucleosome B in the presence of hormone was suggested as arising either from histone displacement or a structural alteration of the nucleosome (Cordingley et al., 1987). UV cross-linking followed by an immunoadsorption assay (closely related to later ChIP protocols) indicated that H2B was bound at the nucleosome B region even after GR induction, while H1 content was reduced, arguing against histone eviction (Bresnick et al., 1992a). High-resolution analysis of nuclease protection revealed continued protection against MNase digestion of nucleosome B after hormone treatment, but increased accessibility to restriction enzymes and DNase at a small region near the nucleosome B dyad, also indicating an altered nucleosome configuration rather than histone eviction (Truss et al., 1995).

Transcriptional activation by GR expressed in yeast was shown to depend on SWI/SNF soon after the initial characterization of SWI/SNF, as mentioned in Chapter 6, and this dependence was soon thereafter demonstrated in mammalian cells (Muchardt and Yaniv, 1993; Yoshinaga et al., 1992). A partially purified SWI/SNF complex was then shown to facilitate nucleosome disruption and concomitant NF1 binding to a nucleosomal template in vitro, further strengthening the case for SWI/SNF facilitating hormone-dependent nucleosome remodeling by GR (Ostlund Farrants et al., 1997). Co-immunoprecipitation demonstrated in vivo interaction between GR and Brg1, one of the two mammalian homologs of Snf2, the ATPase subunit of the yeast SWI/SNF complex, and interruption of that interaction by competition with the progesterone receptor reduced activation of MMTV on a stably replicating episome but not with a transiently transfected template, implying that Brg1 was required for the remodeling step necessary for activation (Fryer and Archer, 1998). Hormone-dependent Brg1 recruitment to an array of MMTV promoters was later visualized by combining RNA FISH (fluorescence in situ hybridization, to identify the promoter array) and indirect immunofluorescence; recruitment of the Brg1 homolog, Brm (Chapter 6), was also observed (Johnson et al., 2008). Manipulating cellular levels of Brg1 altered restriction enzyme accessibility to nucleosome B of the MMTV promoter, consistent with Brg1-dependent remodeling. More recently, genome-wide analysis of GR binding and nucleosome occupancy showed that Brg1 was recruited to all hormone-responsive sites, with an accompanying shift in nucleosome position away from GR binding sites (Johnson et al., 2018).

The mechanistic parallel between induction of the MMTV and *PHO5* promoters is clear, in spite of differences. At *PHO5*, binding of Pho4 to an accessible site leads to disruption of an adjacent positioned nucleosome, thereby enabling binding of Pho4 to the occluded UASp2 and allowing assembly of a transcription complex and productive transcription. At the MMTV promoter, binding of the activated GR to a site that is nucleosomal but nevertheless accessible leads to perturbation of that same nucleosome, thus allowing binding of NF1 and recruitment of the transcription machinery. More recent work has revealed additional nuance to the initial binding events accompanying transcriptional activation by hormone induction of steroid receptors; this is discussed in the section on "Transcription factor access" below.

Ordered recruitment

We have encountered the concept of ordered recruitment in transcriptional activation, without naming it as such, in previous discussions of chromatin remodeling and histone modifications. The concept refers simply to a required temporal order of recruitment for productive assembly of the transcription machinery. This idea was explicitly laid out in a classic paper in which sequential binding of TFIID, TFIIA, TFIIB, RNA polymerase II, and TFIIE was shown to be required for transcription in an in vitro system: binding of each component was a prerequisite for the one following, and all five needed to be in place to allow transcription (Buratowski et al., 1989). The discovery in the 1990s of complexes in the cell dedicated to histone modifications and ATP-dependent chromatin remodeling, together with in vitro and in vivo results on the repressive effects of nucleosomes, led to the acceptance of a new paradigm in which the action of such complexes was required for transcriptional activation. An obvious implication was that transcriptional activation was likely to occur in at least two stages, as overcoming the repressive effects of nucleosomes must precede assembly of the transcription machinery. The discovery that interactions of proteins with nucleosomes could be modulated by posttranslational modification of the histones crystallized this notion further, pointing to mechanisms in which the action of histone modifiers could be a prerequisite to the recruitment of chromatin remodelers or other components of the transcription machinery (e.g., see Table 7.4).

The advent of ChIP and the availability of a multitude of antibodies against modified histones and components of the transcription machinery allowed the molecular underpinnings of sequential recruitment during transcriptional activation to be explored in detail. The first paper to use the term "ordered recruitment" relative to transcription in its title was a study of the successive recruitment of transcription factors, SWI/SNF, and SAGA during transcriptional activation of the *HO* endonuclease promoter in budding yeast (Cosma et al., 1999). The endonuclease encoded by *HO* is responsible for mating-type switching in haploid yeast (as outlined in Chapter 6), and its expression is subject to an exceptionally complex regulation. Following cell division, *HO* is expressed transiently in mother cells but not in daughters. SWI/SNF and SAGA were identified as being required for *HO* activation in mutant screens for yeast defective in mating-type switching (Chapter 6), together with sequence-specific activators encoded by *SWI4*, *SWI5*, and *SWI6* (Fig. 8.8). These activators and co-activators provided a starting point for the determination of the interdependencies and order of events during *HO* activation, which could be followed temporally following release from metaphase arrest.

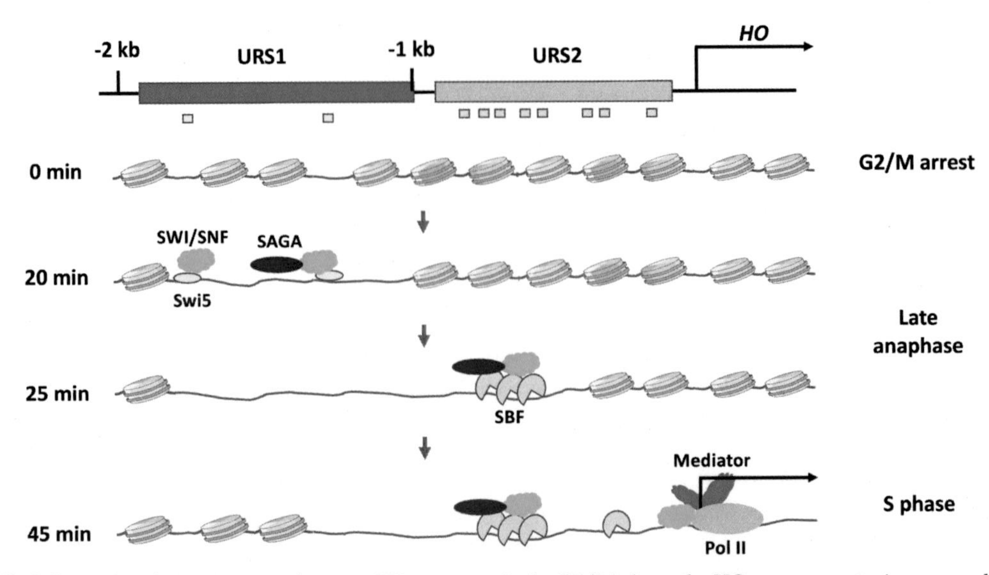

FIG. 8.8 Ordered recruitment at the yeast *HO* promoter. In the G2/M phase, the *HO* gene promoter is repressed and packaged into nucleosomes. Twenty minutes after release from G2/M arrest, in late anaphase, Swi5 binds to sites in URS1 *(yellow rectangles, top)* and recruits SWI/SNF and SAGA with concomitant loss of nucleosomes over URS1. Swi5 binding is followed by the SWI/SNF- and SAGA-dependent removal of nucleosomes and binding of SBF (the Swi4-Swi6 heterodimer) to sites *(orange boxes, top)* in URS2. SBF recruits SWI/SNF and SAGA, eventually leading to productive transcription. The histone chaperones FACT and Asf1 are also involved in nucleosome eviction from URS2 and the 5′ region of the coding sequence but are not shown here. *For additional detail, see Stillman, D.J. 2013. Dancing the cell cycle two-step: regulation of yeast G1-cell-cycle genes by chromatin structure. Trends Biochem. Sci. 38, 467–475.*

The *HO* gene promoter is unusual in yeast, as it contains two upstream regulatory sequence regions, URS1 and URS2, with the more distal URS1 encompassing elements from 1000 to 1900bp upstream of the transcription start site (Fig. 8.8) (Stillman, 2013). Swi4 and Swi6 were known to bind as a heterodimer, termed SBF (Swi4-Swi6 box factor), and to act as a transcriptional activator; ChIP showed that recruitment of Swi4 to URS2 depended not only on Swi6 but also on Swi5, Gcn5, and SWI/SNF (Fig. 8.8) (Cosma et al., 1999). ChIP experiments over a time course after the release of cells from metaphase showed Swi5 binding at URS1 occurring transiently at late anaphase. Consistent with the binding of Swi5 being the initiating event in *HO* activation, association of SWI/SNF, SAGA (Gcn5 and Ada2), and Swi4-Swi6 was abrogated in a *swi5Δ* mutant. Recruitment of SWI/SNF followed shortly after Swi5 binding, and association of the SAGA subunit Ada2 was seen shortly thereafter (Cosma et al., 1999); later work showed SWI/SNF and SAGA association at URS1 to be interdependent (Stillman, 2013). Recruitment of SAGA and SWI/SNF to URS1 is followed by binding of SBF to URS2 and recruitment of SWI/SNF and SAGA. Finally, Mediator, TBP, and Pol II are recruited to allow transcription of *HO* (Cosma et al., 1999). Remarkably, Swi5 associates with URS2 only very briefly—on the order of 5min—before being degraded, while SAGA and SWI/SNF remain bound considerably longer despite the absence of their initial recruiter. Swi5 also binds to URS2 in daughter cells, but binding to URS1 by Ash1, restricted to daughter cells, prevents recruitment of SWI/SNF and SAGA so that *HO* is not expressed. Subsequent studies resulted in some minor revisions to this scenario, mainly due to effects of the *ash1* background used in the original study (Mitra et al., 2006; Stillman, 2013; Takahata et al., 2009). The molecular details of *HO* activation are in any case of secondary importance to the general reader; rather, *HO* activation serves as a case study to illustrate the complexity of the interactions and events that accompany transcriptional activation, and the methods used to elucidate those interactions and events.

With regard to the role of chromatin in *HO* activation, later work demonstrated histone eviction at sequential steps and indicated that the chromatin structure of the *HO* promoter was important in determining the ordered recruitment required for its activation. To test this latter idea, promoter sequences were swapped between *HO* and *CLN2*, which like *HO* is activated by binding of the Swi4-Swi6 heterodimer but unlike *HO* harbors its Swi4-Swi6 binding sites in an NDR (Bai et al., 2010; Yarrington et al., 2016). Replacement of this NDR with sequence from URS2 of *HO* containing seven Swi4-Swi6 sites within three positioned nucleosomes (Fig. 8.8) reduced expression such that only 8% of cells expressed a *CLN2*-driven GFP reporter in a given cell cycle. Conversely, chimeric promoters made by replacing URS2 segments from the *HO* promoter with sequences from the *CLN2* NDR exhibited lower nucleosome occupancy and no longer required the URS1 sequence at which Swi5 binding normally initiates activation.

Numerous examples of ordered recruitment have been documented since the now-classic study of *HO*; a few early examples are enumerated in Table 8.1 (although it should be noted that not all of the details of recruitment for the mammalian examples have been independently corroborated) (Biddick and Young, 2009; Cosma, 2002). In both yeast and metazoans, promoter-dependent variability is seen in the dependence on co-activators, including chromatin remodelers and modifiers, in the interdependence of transcription factors, and in

TABLE 8.1 Examples of gene activation by ordered recruitment.

Gene	Organism	Induction stimulus	Nucleosome remodeling?	Critical players	Reference(s)
PHO8	*Saccharomyces cerevisiae*	Low phosphate	Yes; eviction	Pho4, SAGA, SWI/SNF	Gregory et al. (1999), Reinke et al. (2001)
GAL1	*S. cerevisiae*	Galactose	Yes, but not conclusively characterized	Gal4, SAGA; TBP recruited by SAGA	Bhaumik and Green (2001), Bryant and Ptashne (2003), Larschan and Winston (2001)
α-1 antitrypsin	Mammals	Differentiation	Yes	HNF1-α, HNF-4α, CBP/PCAF, SWI/SNF	Soutoglou and Talianidis (2002)
IFN-β	Mammals	Viral infection	Yes; sliding	TFs and HMGI(Y) in enhanceosome; Gcn5; PolII/CBP; SWI/SNF	Agalioti et al. (2000), Lomvardas and Thanos (2001)

the sequence of events leading to productive transcription. Variable dependence on co-activators has been particularly well documented for promoters activated by Gcn4 in yeast, as exhaustively analyzed by the Hinnebusch lab (Govind et al., 2005; Qiu et al., 2016, 2004; Swanson et al., 2003). Oftentimes dependence on a given remodeler or modifier is not absolute, but is reflected in altered dynamics of activation when the remodeler or modifier is absent. Different promoters driven by the same activator can vary drastically in co-activator requirement and the sequence of events leading to activation. A classic example of this is the contrast between the *PHO5* and *PHO8* promoters, both driven by Pho4: *PHO8* depends strongly on SWI/SNF for its remodeling and activation, while *PHO5* does not, as mentioned earlier (Gregory et al., 1999); moreover, co-activator dependence at *PHO5* depends on the extent of phosphate depletion (Dhasarathy and Kladde, 2005). Dependence on SWI/SNF for promoter activation by Gal4 varies with the strength of the Gal4 binding site (Burns and Peterson, 1997), and altered chromatin structure at *HO* affects transcriptional activation, as just discussed. In metazoans, changing the location of the TATA element and core promoter relative to a positioned nucleosome has been reported to alter the requirements for and mechanism of transcriptional activation (Lomvardas and Thanos, 2002; Martinez-Campa et al., 2004). While the utility of this bewildering complexity is not always readily apparent, one might speculate that it has afforded grist for the evolution of the widely variable responses to endogenous and exogenous stimuli that are characteristic of eukaryotic cells.

Transcription factor access in a chromatin milieu

Before the discovery of eukaryotic transcriptional activators and the uncovering of their mechanistic roles in transcription in the 1980s and 1990s, the problem posed by the packaging of DNA into nucleosomes was viewed mainly as one of a potential block to polymerase entry. Hence, the earliest experiments addressing this issue employed purified bacteriophage RNA polymerases or partially purified eukaryotic RNA polymerase II, as we have discussed. With the change in paradigm wrought by the discovery of enhancers (UASs in yeast) and their cognate activators, and the identification of general transcription factors such as TFIID, the problem was reframed: how do activator proteins gain access to binding sites potentially incorporated into nucleosomes to facilitate assembly of a PIC at the proximal promoter, which might also be occluded by chromatin?

Early studies suggested three possible mechanisms by which sequence-specific transcription factors, as well as the general transcription machinery (in particular, TFIID/TBP and Pol II), could gain access sites in chromatin (Fig. 8.9). First, a binding site may be constitutively accessible, either residing in a linker region, as with Pho4 at the yeast *PHO5* promoter, or in the larger accessible region provided by an NDR. Second, a transcription factor may, like GR, be capable of binding its site in the context of a nucleosome, with or without consequent nucleosome disruption. Third, a factor may be inhibited from binding to a particular nucleosomal site but have its binding enabled by nucleosome perturbation or disruption caused by another factor.

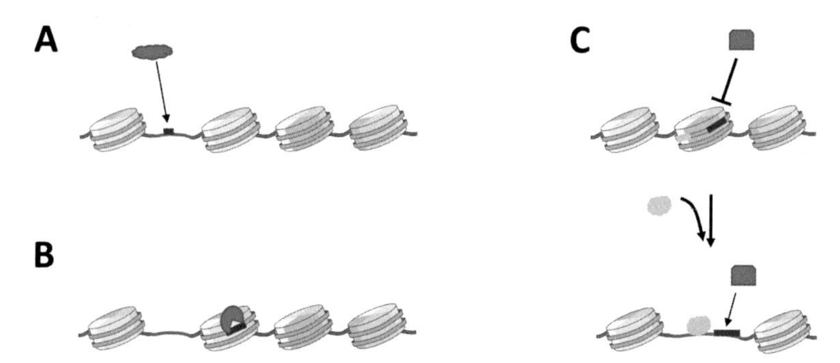

FIG. 8.9 Possible mechanisms for TF access to chromatin. (A) A TF may have its binding site in a nonnucleosomal location. (B) A TF may be capable of recognizing its site on the surface of the nucleosome. (C) A TF may be unable to bind its site in a nucleosome, but a second factor *(orange)* may displace the nucleosome or prevent its formation and thus allow binding.

Nucleosome exclusion

It was generally assumed, and in a few cases observed, that the promoters of constitutively active genes such as those encoding ribosomal proteins must be permanently accessible to the transcription machinery, a mechanism termed "persistent exclusion" (Workman and Buchman, 1993). Constitutive accessibility was observed for a number of inducible gene promoters even under noninducing conditions. For example, the inducible *GAL1-10* and *CHA1* promoters in yeast were found to have activator binding sites (UASs) that were accessible even when the activators (Gal4 and Cha4, respectively) were absent, and heat shock promoters in *Drosophila* were similarly shown to possess constitutive DNase hypersensitivity at sites bound by activators only after heat shock (Elgin, 1988; Fedor et al., 1988; Lohr and Hopper, 1985; Moreira and Holmberg, 1998; Thomas and Elgin, 1988; Wu, 1984). Later studies revealed that the formation of such accessible sites is often facilitated by accessory factors (GAGA factor in the case of the *Drosophila* heat shock promoters; general regulatory factors and chromatin remodelers at yeast promoters) and comprise either histone-free regions or altered nucleosome structures facilitated by remodelers such as RSC (Chapter 4) (Brahma and Henikoff, 2019, 2020; Fedor et al., 1988; Floer et al., 2010; Granok et al., 1995; Tsukiyama et al., 1994).

An early potential explanation proffered for the formation of DNase hypersensitive sites was the presence of unusual DNA structures unfavorable for incorporation into chromatin, such as cruciforms, Z-DNA, or highly supercoiled domains (Eissenberg et al., 1985; Weintraub, 1985). Although this notion was not borne out experimentally, the effect of homopolymeric DNA tracts, especially poly(dA-dT), on transcription factor access deserves mention. These tracts resist deformation more than mixed sequence DNA and are relatively unfavorable for nucleosome formation, as discussed at length in Chapter 4. Their effect on accessibility and transcription in vivo has also been the subject of considerable research. A seminal study showed that a naturally occurring run of T residues just upstream of the binding site for Gcn4 in the *HIS3* promoter of budding yeast increased accessibility and facilitated activation in vivo (although Gcn4 binding was not assessed directly) (Iyer and Struhl, 1995):

- Longer poly(dA-dT) tracts increased *HIS3* transcription, while shorter tracts caused a decrease.
- The orientation of poly(dA-dT) did not matter, and poly(dG-dC) also stimulated transcription.
- The stimulatory effect of poly(dA-dT) decreased with increasing intracellular concentration of Gcn4.
- Accessibility of DNA close to the Gcn4 binding site to *Hinf*I or *E. coli dam* methylase expressed in vivo was increased by poly(dA-dT).
- Maintaining a mixed population of yeast cells with varied lengths of poly(dA-dT) under selection for *HIS3* expression resulted in enrichment for cells with longer poly(dA-dT) tracts.

Subsequent investigations showed nucleosome occupancy to be lower genome-wide at poly(dA-dT) tracts in both budding yeast and mammalian cells, although not in *S. pombe* (Lantermann et al., 2010; Lee et al., 2007; Moyle-Heyrman et al., 2013; Ozsolak et al., 2007; Whitehouse et al., 2007; Yuan et al., 2005). These studies, together with prior work on the effects of homopolymeric tracts on nucleosome stability and the strength of yeast promoters, provided compelling evidence that these elements act to facilitate access to sites in chromatin and thereby to enhance transcriptional activation. However, while the structural features of poly(dA-dT) tracts do influence nucleosome occupancy, much of the effect seen in budding yeast can be attributed to their stimulatory and directional effect on nucleosome sliding catalyzed by RSC, as discussed in Chapter 4 (Barnes and Korber, 2021).

These same genome-wide studies revealed that persistent exclusion—or more accurately, persistent low nucleosome occupancy—is in fact widespread in promoters of budding yeast, such that most TF binding sites located in promoters, whether active or not, are located in regions of low nucleosome occupancy (Bernstein et al., 2004; Lee et al., 2004, 2007; Yuan et al., 2005). As discussed in Chapter 4, these NDRs generally are the products of collaboration between general regulatory factors, chromatin remodelers, and structural effects of specific sequences. Numerous studies have provided evidence for the significance of NDRs for promoter function (and for TF binding sites located in linker regions, as for *PHO5*). As one example, the large-scale study by Segal and colleagues described in Chapter 4 (Fig. 4.18) demonstrated that binding sites for several TFs placed into a closed chromatin context resulted in lower factor occupancy and decreased ability to drive transcription than when placed into an NDR (Levo et al., 2017).

Low nucleosome occupancy may not only affect activator binding; a few reports indicate that it may in some cases render activators superfluous. Both the constitutively expressed yeast *AKY2* and *PFY1* promoters were reported not to require classical activator proteins for their expression, and both were found to possess nucleosome-free gaps at their core promoters (Angermayr et al., 2003, 2002). In another example, the DNA damage-inducible yeast *RNR3* gene was converted to being constitutively active by insertion of poly(dA-dT) tracts flanking a TATA element that is incorporated into a nucleosome in the native uninduced promoter, resulting in loss of the TATA-incorporating nucleosome (Zhang and Reese, 2007). A recent study suggests that a large number of genes in yeast may be characterized by accessible core promoters and do not require classical activators to promote transcription (Rossi et al., 2021). High-resolution characterization of the binding of a large number of activator and coactivator proteins and components of the transcription machinery in budding yeast concluded that ~2500 genes, most of which are transcribed at low levels, are associated with PICs but are unbound by classical activators, with their activity depending on the proximal promoter being nucleosome-free. This proposition is further supported by the observation that the Mediator coactivator complex remains closely associated with the core promoter of this gene set under conditions in which it is found farther upstream at genes having well-characterized UASs, where it is recruited by activators (Morse, 2022). Nonetheless, despite the preceding evidence, definitive demonstration of mRNA transcription occurring in vivo without the need for an activator remains to be shown; further studies will be needed to establish the extent to which this phenomenon occurs and the circumstances that allow it.

The situation with regard to nucleosome exclusion is somewhat different in metazoans. The majority of yeast gene promoters, whether active or inactive, possess NDRs. In contrast, mammalian gene promoters that are expressed or occupied by a PIC frequently possess NDRs, but inactive genes typically do not (Ozsolak et al., 2007; Valouev et al., 2011). Paused RNA polymerase contributes to low nucleosome occupancy at many genes, as recounted in Chapter 4 (section on Barrier elements, nucleosome spacing, and chromatin remodelers), through a mechanism that is not well understood. Although less is known about how transcription factor access is established at constitutively active genes in metazoans, recent work suggests that a factor that binds to CpG islands may contribute. CpG islands are sequences containing clusters of unmethylated CpG residues, first identified in the mouse genome by their susceptibility to cleavage by the methylation-sensitive restriction endonuclease *Hpa* II (see section on DNA methylation farther on) (Bird et al., 1985). Subsequent studies revealed that the majority of Pol II transcripts in mammalian genomes initiate at promoters containing CpG islands (Deaton and Bird, 2011). A search for factors binding to unmethylated CGCG elements, which are enriched in the promoters of genes highly expressed across human tissues, identified BANP (Btg3-associated nuclear protein) as a methylation-sensitive TF that binds to a 9bp palindromic element with a central CGCG (Grand et al., 2021). ChIP-seq of BANP revealed its binding to be enriched at CpG islands linked to genes whose products participate in essential cellular functions such as metabolism and cell cycle. Binding sites were characterized by high accessibility, as assessed by ATAC-seq and MNase-seq, and appear to function at least weakly as barrier elements, with nucleosome occupancy exhibiting periodicity extending outwards from binding sites. Accessibility was diminished upon inducible degradation of BANP in mouse ES cells, as was expression of associated genes. Thus, BANP may function to maintain open chromatin in the promoters of constitutively active genes in metazoans, similar to the role of Rap1 at ribosomal protein genes in budding yeast (Chapter 4) (Lieb et al., 2001; Morse, 2000). The mechanism by which it does so and whether other such factors exist (as not all essential and constitutively active metazoan genes are associated with BANP) remain to be seen.

Nucleosome binders and pioneer factors

A special class of transcription factors that are able to engage binding sites in transcriptionally inactive chromatin to initiate gene activation programs during cell differentiation has been designated as pioneer factors (Bulyk et al., 2023; Mayran and Drouin, 2018; Zaret, 2020; Zaret and Carroll, 2011). The concept of pioneer factors may be considered to have emerged from that of master gene regulators, TFs that first bind to developmentally expressed genes to initiate a series of events that culminate in the stable, tissue-specific expression of those genes (e.g., Herskowitz, 1989; Ohno, 1979; Tapscott et al., 1988). Investigations detailed below led to the realization that proteins that first associate with enhancer sequences of developmentally regulated genes often reorganize chromatin structure to promote access to additional sequence-specific transcription factors in a cooperative process (Fig. 8.9C), culminating in lineage-specific gene expression. This latter property of accessing binding sites in closed chromatin and facilitating binding of additional TFs in a hierarchical process is now often considered a defining characteristic of pioneer factors (Bulyk et al., 2023). Definitions of exactly what defines a pioneer factor have varied somewhat in the literature; in its most extreme form, the term implies a qualitative distinction between TFs able to gain access to nucleosomal sites and those that cannot. We shall see that the pioneering qualities of TFs are better viewed as representing a continuum than a discrete property. In addition, although gene expression in yeast is not subject to developmental programming in the way that is seen in metazoan cells, it is useful to consider yeast TFs that open chromatin to allow access to other TFs together with classic metazoan pioneer factors (and yeast TFs capable of opening chromatin have indeed been explicitly referred to as pioneer factors (Bulyk et al., 2023; Zaret, 2020)).

The first indication of the special relationship to chromatin of TFs that initiate developmentally regulated gene expression came from a study in which nucleosome positioning and TF occupancy in the liver-specific enhancer of the *alb1* gene were analyzed in mouse embryonic liver cells (McPherson et al., 1993). The *alb1* gene was chosen on the basis of its highly preferential expression in the liver, and the enhancer was singled out for study so that the effect of factor binding on nucleosome organization could be examined independently of interactions of the transcription machinery occurring at the distant promoter sequences. Previous work had identified critical elements in the liver-specific *alb1* enhancer, including binding sites for HN3Fα and HNF3β, which are expressed at the onset of liver development (Sasaki and Hogan, 1993). Surprisingly, high-resolution analysis of nucleosome positioning and TF occupancy, enabled by the technical innovation of ligation-mediated PCR (Mueller and Wold, 1989), revealed that in mouse embryonic liver cells the enhancer was packaged in three positioned nucleosomes, two of which also exhibited protection of sequences essential for *alb1* activation, similar to the interaction of GR with nucleosomal DNA described earlier (McPherson et al., 1993). In contrast, nucleosome positioning was not seen in nuclei from tissues not expressing *alb1* and not exhibiting protection against DNase digestion of TF binding sites.

Two important implications followed from these results: first, that transcription factors were bound to the surface of a nucleosome without consequent disruption in vivo; and second, that binding of these same TFs imposed an ordered chromatin structure on the *alb1* enhancer. The latter conclusion was supported by in vitro nucleosome assembly experiments, using the *Drosophila* extract system described in Chapter 5 (Becker and Wu, 1992): nucleosome reconstitution onto a plasmid harboring the 830bp *alb1* enhancer sequence resulted in positioning of the TF-binding nucleosomes essentially identical to that seen in mouse embryonic liver cells, and positioning was not seen when the HNF3 (now known as FoxA) binding site was mutated (McPherson et al., 1993). Subsequent investigation revealed that while multiple TFs bound the *alb1* liver-specific enhancer at early stages of differentiation in which the gene is active, only binding sites for FoxA/HNF3 and GATA TFs were occupied in undifferentiated gut endoderm, when *alb1* is not active but is primed for activation (Gualdi et al., 1996). These findings pointed to FoxA/HNF3 as the responsible entity for initiating binding to and dictating nucleosome positioning over the *alb1* liver-specific enhancer (with a presumptive *Drosophila* HNF3-related factor performing the same function in the assembly experiments). Structural studies provided a rationale for this capability by showing that the DNA-binding domain of FoxA/HNF3 resembles a DNA-binding motif present in linker histone H5 dubbed a "winged helix" (Clark et al., 1993; Ramakrishnan et al., 1993). Follow-up studies showed directly that FoxA1/HNF3α bound a nucleosomal template with higher affinity than naked DNA and that its binding in vitro on a fragment corresponding to the *alb1* liver-specific enhancer resulted in nucleosome positioning similar to that seen in liver cell nuclei (Cirillo et al., 1998; Cirillo and Zaret, 1999; Shim et al., 1998). These results, together with its early expression during hepatic development, cemented the identification of FoxA1/HNF3α as a pioneer factor. The functional impact of FoxA1/HNF3α binding was highlighted by in vitro studies showing that FoxA1/HNF3α was capable of binding to its site in reconstituted chromatin containing linker histones, resulting in decompaction of the surrounding region, and in vivo experiments showing that FoxA/HNF3 binding displaces linker histones genome-wide (Cirillo et al., 2002; Iwafuchi-Doi et al., 2016).

Subsequent studies have revealed pioneer factors functioning in both plant and animal development as well as at steroid hormone inducible genes and genes involved in circadian control (Balsalobre and Drouin, 2022; He et al., 2010; Menet et al., 2014; Michael and Thoma, 2021; Vernimmen and Bickmore, 2015; Zaret, 2020). Examples include:

- GATA3 and GATA4, which were identified along with FoxA/HNF3 as pioneer factors acting at the liver-specific *alb1* enhancer, albeit more weakly (Cirillo et al., 2002).
- Oct4 and Sox2, which are critical for maintaining pluripotency during early mammalian development and which function as pioneer factors during reprogramming of fibroblasts to pluripotency (Soufi et al., 2012).
- The tumor suppressors p53 and p63, which in this context play a role in specification of epithelial enhancers in mammalian cells (Lin-Shiao et al., 2019; Sammons et al., 2015).
- Zelda from *Drosophila* and β-catenin in frogs. Zelda plays a critical role in promoting enhancer accessibility during zygotic gene activation in *Drosophila* (Brennan et al., 2023; Schulz et al., 2015; Sun et al., 2015), while β-catenin functions in *Xenopus* to establish transcriptional competence to developmentally regulated enhancers prior to the mid-blastula transition, at which point active transcription begins in the developing zygote (Blythe et al., 2010). The binding of β-catenin to enhancers in *Xenopus* results in the recruitment of the histone arginine methyltransferase PRMT2; this recruitment appears necessary and sufficient for later activation of the target genes.
- LEAFY, a master regulator in plants that promotes floral fate (Jin et al., 2021).

The first step in pioneer factor action is its binding to a site in closed chromatin, either at an individual nucleosome or in heterochromatin. While some TFs are clearly able to bind to a nucleosomal surface (e.g., GR) or to outcompete histones for binding to DNA (Rap1 in yeast) and others are nearly entirely impeded (NF1 in mammals and Gcn4 in yeast), the categorization of TFs with regard to nucleosome binding is blurrier in some cases than this would suggest. This issue has been addressed in two recent high throughput studies of TF binding that have provided new insights into structural features important for TF binding to nucleosomal templates (reviewed in Luzete-Monteiro and Zaret, 2022; Michael and Thoma, 2021). Cramer and colleagues compared the binding of over 200 human TFs to nucleosomal and naked DNA templates, with the results corroborating the generally inhibitory effect of nucleosomes on TF binding but at the same time revealing that many TFs can in fact bind nucleosomal DNA, albeit at reduced affinity and with particular structural constraints (Zhu et al., 2018). In this work, a collection of human TFs was subjected to a SELEX (systematic evolution of ligands by exponential enrichment) protocol using nucleosomal templates (Fig. 8.10). DNA fragments of

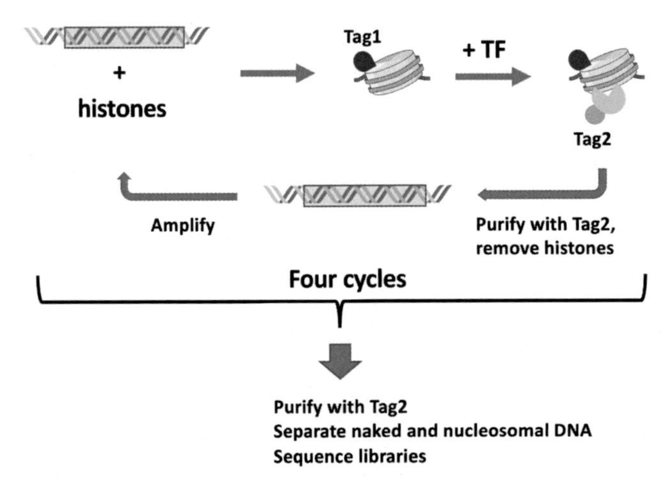

FIG. 8.10 Scheme for the enrichment of DNA sequences permissive for TF binding to nucleosomes, dubbed NCAP-SELEX (nucleosome consecutive affinity purification-systematic enrichment of ligands by exponential enrichment). A library of "randomized" DNA sequences (boxed region, upper left) is assembled into nucleosomes that include H2A bearing an affinity tag; the nucleosome "library" is then purified using the affinity tag and incubated with a TF bearing a second affinity tag, allowing purification of nucleosomes that bind the TF. DNA is prepared from the TF-bound nucleosomes and amplified and the cycle is repeated a total of four times; a final cycle is followed by the separation of naked and nucleosomal DNA, the former resulting from nucleosomes from which histones are dissociated upon TF binding. *Zhu, F., Farnung, L., Kaasinen, E., Sahu, B., Yin, Y., Wei, B., Dodonova, S.O., Nitta, K.R., Morgunova, E., Taipale, M., et al. 2018. The interaction landscape between transcription factors and the nucleosome. Nature 562, 76–81.*

147 bp or 200 bp, containing 101 bp or 154 bp regions of randomized sequence, were packaged into nucleosomes and allowed to interact with TFs. Sequences bound by TFs were recovered, the DNA was purified and amplified, and the process was repeated. After four cycles of amplification, the TF-bound sequences were separated into nucleosomal and naked DNA, reflecting sequences that allowed simultaneous occupancy by histones and a particular TF or for which TF binding resulted in histone eviction. Although the number of distinct sequences employed was far from covering the entire sequence space ($\sim 10^{66}$ for 100 bp), it

was large enough to yield information on the relative binding to naked and nucleosomal DNA and the structural features allowing binding of individual TFs. Moreover, the use of the 200 bp fragment allowed comparison of binding to nucleosomal and naked DNA on the same template, while the 147 bp fragment, accommodating only nucleosomal DNA (or nearly so), allowed determination of the site-specific preferences of TFs in a nucleosomal context. Of the >400 TFs tested, 233 yielded information on binding; most of those that failed did so due to poor expression. The results from these 233 TFs yielded several major insights:

- Not surprisingly, nucleosomes generally inhibited TF binding. This was especially apparent from the observation that TF binding to the 200 bp nucleosomal fragment was lowest in the center and highest at the ends.
- TFs varied considerably in their binding to nucleosomes, as judged by their binding to the center vs. the edges of the 200 bp nucleosomal fragment. Measuring nucleosome occupancy around binding sites for 20 TFs in K562 cells by MNase-seq indicated a significant correlation with the in vitro data, indicating nucleosomal inhibition of TF binding in vivo.
- Among the 209 TFs exhibiting significant binding to nucleosomal (147 bp) templates, four major modes of binding were seen:
    - End preference was observed among TFs whose binding modes require interaction with DNA over more than 180° of the DNA helical turn, and would therefore be occluded by histones. Most of these TFs, including basic helix-loop-helix (bHLH), basic region/leucine zipper motif (bZIP), and many of the numerous zinc finger factors found in humans, likely can be considered as strongly disfavoring binding to nucleosomal DNA. Rather, the transient release of DNA-histone contacts at the nucleosome periphery, sometimes referred to as "breathing," could allow interaction (Li et al., 2005; Polach and Widom, 1995), and for TFs with sufficient affinity, the resultant binding would prevent reassociation of the DNA ends with the histones. This category was by far the most populous of the four, consistent with the generally inhibitory effect of nucleosomes on TF binding. A few of these TFs are also classified as periodic binders (see below) and therefore are likely to be genuine nucleosome-binding proteins able to contact solvent-accessible recognition sites that are in complex with the histones.
    - Periodic preference was seen for several TFs that, like GR, have recognition sites consisting of short motifs separated by an integral number of helical turns of DNA and thus are able to access binding sites facing away from the histone core. Two interesting examples of this are factors called EOMES and PITX; these TFs mainly contact DNA via the minor or major groove, respectively, and correspondingly exhibit periodic binding preferences offset by ~5 bp, or a half helical turn.
    - Some TFs showed preferred binding at the nucleosome dyad, where the DNA is not occluded by a neighboring DNA helix. Evidence from MNase-seq and MNase-ChIP indicated that one such TF, RFX5, also preferred dyad sites in vivo.
    - A few TFs belonging to the T-box family were found to contact sites separated by 80 bp and therefore situated adjacently on neighboring gyres of the nucleosomal DNA; these were referred to as gyre-spanning TFs. However, analysis of available ChIP-seq data did not yield evidence for such binding occurring in vivo, so its relevance remains an open question.

- TF binding typically facilitated nucleosome dissociation; dissociation varied according to the mode of TF binding and the position of the TF binding site in the nucleosome.

In another study, the quantitative determination of binding constants for a small group of pioneer factors revealed only a modest reduction in affinity for nucleosomal templates relative to naked DNA, from 0.1–1 nM to 2–6 nM (Fernandez Garcia et al., 2019). A similar assessment of an additional 24 TFs, selected on the basis of binding experiments using a protein microarray, identified TFs that bound well (but with a range of affinities) to both nucleosomal and naked DNA or only to naked DNA. No TFs were found to bind only nucleosomal templates, although some did show higher affinity to nucleosomal than naked DNA (Fernandez Garcia et al., 2019). TFs that bound well to nucleosomal templates, and thus qualified as pioneer factors, had in common binding to short or partial DNA sequence motifs by short α-helices, while nonbinders utilized unstructured regions and/or β-sheets. Structural studies on individual pioneer factors have similarly found binding to short, outward-facing nucleosomal sequences, near the DNA entry–exit site of the nucleosome, or to the dyad (Fig. 8.11) (Dodonova et al., 2020; Echigoya et al., 2020; Michael et al., 2020; Soufi et al., 2015; Tanaka et al., 2020; Yu and Buck, 2019).

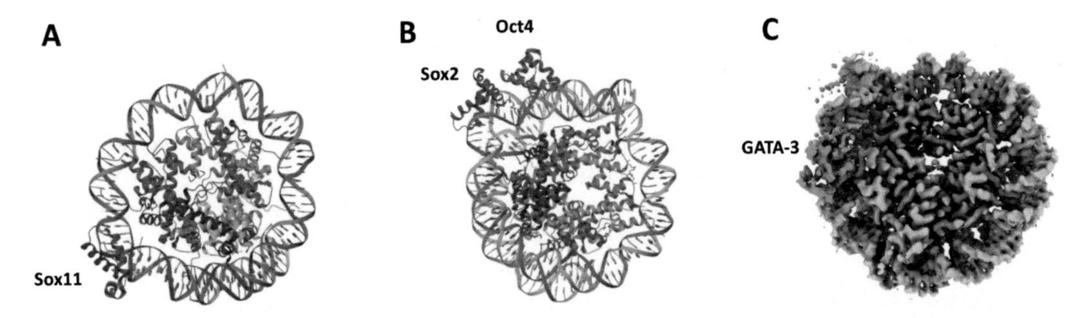

FIG. 8.11 Structures of TFs bound to nucleosome core particles. (A) Sox11 bound at SHL +2 (PDB ID 6T7A); (B) Oct4 and Sox2 bound at SHL +6 (PDB ID 6YOV); (C) GATA-3 bound at SHL +5.5 (EMD ID 0783). *(A) Dodonova, S.O., Zhu, F., Dienemann, C., Taipale, J., Cramer, P. 2020. Nucleosome-bound SOX2 and SOX11 structures elucidate pioneer factor function. Nature 580, 669-672. (B) Michael, A.K., Grand, R.S., Isbel, L., Cavadini, S., Kozicka, Z., Kempf, G., Bunker, R.D., Schenk, A.D., Graff-Meyer, A., Pathare, G.R., et al. 2020. Mechanisms of OCT4-SOX2 motif readout on nucleosomes. Science 368, 1460–1465. (C) Tanaka, H., Takizawa, Y., Takaku, M., Kato, D., Kumagawa, Y., Grimm, S.A., Wade, P.A., and Kurumizaka, H. (2020). Interaction of the pioneer transcription factor GATA3 with nucleosomes. Nat. Commun. 11, 4136.*

Two studies have reported direct interactions between pioneer factors and histones that are important for nucleosome binding. In the first, chromatin opening by FoxA family members in an in vitro assay depended on the presence of a short α-helical region that directly interacted with histones as shown by cross-linking experiments (Iwafuchi et al., 2020). In the second, single-molecule methods were used to compare binding at the DNA entry-exit site of the nucleosome by two yeast TFs, Cbf1 and Pho4 (Donovan et al., 2023a). Despite both of these TFs having basic helix-loop-helix (bHLH) DNA-binding domains with nearly identical sequence specificity, Cbf1 has characteristics of a pioneer factor while Pho4, as we have seen, does not (Yan et al., 2018). Consistent with this difference, in vitro occupancy of a nucleosomal site compared to naked DNA was greatly inhibited for Pho4 but not for Cbf1.

Cryo-EM of Cbf1 bound to a nucleosome revealed interactions of the bHLH domain with the histone octamer, and swapping the relevant domains of Cbf1 and Pho4 demonstrated that this region of Cbf1 was necessary and sufficient for stable binding to nucleosomal sites both in vitro and in vivo. In addition, point mutants of the critical region of Cbf1 abrogated its ability to bind to nucleosomes; similarly, mutations of the FoxA histone-interacting short α-helical region prevented chromatin opening in vitro (Donovan et al., 2023a; Iwafuchi et al., 2020).

While in vitro experiments have provided important insights into the structural features that govern TF binding to nucleosomes, these studies are necessarily performed in the absence of the numerous factors potentially interacting with pioneer factors in the cellular milieu. Indeed, various studies demonstrate the importance of additional factors in chromatin opening by pioneer factors in vivo. In our earlier discussion of the MMTV promoter, we saw that the GR can remodel chromatin in conjunction with Brg1, and several other metazoan pioneer factors, including GATA3, Oct4, and LEAFY, also recruit Swi/Snf homologs (Jin et al., 2021; King and Klose, 2017; Takaku et al., 2016). Collaboration between pioneer factors and mammalian SWI/SNF has been recapitulated in an in vitro study in which accessibility to DNA in an H1-compacted nucleosome array conferred by PU.1 was enhanced by recruitment of cBAF (Frederick et al., 2023). The cooperation of pioneer factors with auxiliary factors is also supported by a delayed increase in chromatin accessibility following pioneer factor binding in vivo (Li et al., 2018; Mayran et al., 2018; Sardina et al., 2018). These and other results suggest that stable association and chromatin opening by metazoan pioneer factors may depend on the recruitment of chromatin remodelers; alternatively, remodelers may transiently sample genomic sites, with prolonged accessibility only occurring at sites of TF binding (Ahmad et al., 2024).

In yeast, several TFs function similarly to metazoan pioneer factors in that they are capable of displacing histones from their binding sites to create regions of open chromatin, as discussed in Chapter 4. These factors have been referred to as nucleosome displacing factors, or NDFs, to distinguish them from their metazoan counterparts (Yan et al., 2018). Yeast NDFs appear able to outcompete histones for binding, thereby creating local regions of open chromatin, without directly recruiting chromatin remodelers. The NDFs Abf1 and Rap1 are capable of displacing a positioned nucleosome in vivo independently of their activation domains, and a high-throughput investigation found that depletion of chromatin remodelers did not affect the ability of NDFs to create NDRs surrounding their binding sites, but did influence the size of the NDR in many cases (Chen et al., 2022; Yarragudi et al., 2004; Yu et al., 2001). Instead, NDFs in some cases collaborate with RSC to create NDRs, with RSC association governed by direct binding of its Rsc3 and Rsc30 subunits to GC-rich motifs (Badis et al., 2008; Kubik et al., 2018; Mivelaz et al., 2020).

Displacement of histones by NDFs is likely to depend on the location of the binding site in the nucleosome. Biophysical studies have shown that Reb1 and Rap1 can bind nucleosomes at the DNA entry-exit sites without displacement of histones (Donovan et al., 2019; Mivelaz et al., 2020). A caveat of these studies is that nucleosomes were reconstituted with recombinant human histones (Donovan et al., 2019; Mivelaz et al., 2020). Nucleosomes containing yeast histones differ from mammalian nucleosomes, with biophysical properties indicating a looser structure (Chapter 5); it is thus possible that binding of GRFs to yeast nucleosomes, even in the absence of remodelers, might be more disruptive than seen with nucleosomes derived from human histones. Nonetheless, these in vitro findings are consistent with a report that Rap1 co-purified, after cross-linking, with >200 promoter nucleosomes from yeast where its binding localized to the outermost two helical turns of nucleosomal DNA, with the recognition site facing away from the histones (Koerber et al., 2009). Reb1-bound nucleosomes were also detected, and Reb1 binding was enriched near the entry-exit sites and was associated with a 12 bp decrease in nucleosomal protection against MNase digestion, not seen for Rap1. However, this latter study also reported ~70% of Rap1-bound loci being at non-nucleosomal sites, and a Rap1 site placed in the center of a nucleosome-positioning sequence rendered the surrounding region accessible to MNase (Koerber et al., 2009; Yu and Morse, 1999). In addition, in vivo residence times for Rap1, measured by ChIP-seq after induction of epitope-tagged Rap1, revealed long residence times at many sites consistent with binding to free DNA (Lickwar et al., 2012). Taken together, these results suggest that binding of Reb1 and Rap1 near the periphery of the nucleosome is compatible with retention of histones, but more centrally located binding is not, resulting in a competition between binding of histones and the GRF at such sites.

A key insight into the mechanism by which some TFs stably bind to nucleosomal DNA emerged from kinetic studies of the binding of the yeast NDFs Reb1 and Cbf1 to nucleosomal DNA close to the DNA entry-exit site (Donovan et al., 2019). This study found that while binding of these factors to nucleosomal sites is reduced relative to naked DNA, the dissociation rate is reduced even more so, resulting in stable binding. The slow dissociation of Cbf1 depends on the histone-interacting bHLH region, thus providing a molecular basis for the slow dissociation of NDFs relative to other TFs such as Pho4 (Donovan et al., 2023a). FRAP (fluoroscence recovery after photobleaching) experiments showed that Reb1 exchanges more slowly at its sites in yeast nuclei than does the RSC component Sth1 or the high mobility group (HMG) protein Nhp6a; however, comparison with TFs that are not NDFs was not performed (Donovan et al., 2019). Nonetheless, exchange times were similar to those reported for FoxA in mammalian cell nuclei, and exchange of the pioneer factors FoxA and Oct4 was observed to be slower than that of other TFs (Plachta et al., 2011; Sekiya et al., 2009). These studies open new avenues for addressing the mechanisms by which pioneer factors bind to and potentially disrupt nucleosomes, and raise additional questions: does the "dissociation rate compensation mechanism" seen for yeast nucleosome-binding TFs (Donovan et al., 2019) pertain to mammalian pioneer factors? How is this mechanism affected by the precise location of the TF binding site in the nucleosome? Are additional mechanisms at play? Doubtless, these and other questions prompted by these studies will be addressed in the near future.

Based on our discussion so far, TFs may access their sites in a chromatin environment by binding at linker or NDRs or by binding in a fashion compatible with nucleosome structure, particularly at surface-exposed sites, at the nucleosome dyad, or at the nucleosome periphery, facilitated by the transient release of histone-DNA contacts. Binding of TFs at nucleosome-compatible sites, or to peripheral sites exposed by transient histone release even when binding is not compatible with histone association, may result in the recruitment of chromatin remodelers that shift nucleosome positions or evict histones. TF binding may also be facilitated by altered nucleosome structure, as in nucleosomes containing H3.3 and H2A.Z variants (Chapter 5) (Jin and Felsenfeld, 2007; Jin et al., 2009) or by species-specific histones that decrease nucleosome stability, as in yeast (though any effect of yeast histones on TF binding in chromatin remains to be demonstrated). An unusual example of a specialized nucleosome structure facilitating TF binding is for Amt1 in the pathogenic yeast *C. glabrata* (Zhu and Thiele, 1996). This TF is important for rapid auto-induction of the *AMT1* gene in response to elevated levels of copper, which Amt1 sequesters and thereby prevents toxicity. The binding site for Amt1 is adjacent to a poly(dA-dT) element that is important for the rapid kinetics of the transcriptional response to elevated copper levels, and which imparts a modest, but significant distortion to the nucleosome that enhances Amt1 binding (Bao et al., 2006; White and Luger, 2004).

While transient DNA unpeeling from the histones can explain TF binding near the DNA entry-exit point of the nucleosome, how pioneer factors access more internal sites is a more difficult question. One possible mechanism, mentioned earlier, is that DNA replication provides a window for TF access. The binding of Pho4 and Gal4 to nucleosomal sites in yeast has been shown to occur in the absence of replication (Balasubramanian and Morse, 1999; Schmid et al., 1992). More sophisticated investigations into the role of replication in TF access are challenging, as the rapid movement of the replication fork necessitates temporal resolution of newly replicated chromatin on the order of minutes to analyze events within a few kb of the replication fork (Blumenthal et al., 1974; Rabinowitz, 1941). Two relatively recent studies have examined this issue in budding yeast and *Drosophila*. In the first, pulse-labeling of newly replicated DNA in *Drosophila* S2 cells with EdU (ethynyl deoxyuridine) was followed by cross-linking and attachment of a biotin tag to the EdU-labeled DNA; MNase treatment followed by pull-down with streptavidin beads allowed purification of nucleosomes containing DNA within 10–20 kb of the replication fork (Ramachandran and Henikoff, 2016). This study revealed a relatively "flat" nucleosome landscape that was reorganized over time into the stereotypical patterns surrounding transcription start sites characteristic of mature chromatin, consistent with contemporaneous reports of transient occupancy of promoters by nucleosomes following replication in budding yeast (Fennessy and Owen-Hughes, 2016; Vasseur et al., 2016). In contrast to mature chromatin, newly replicated chromatin exhibited nucleosomes occupying promoter NDRs and a lack of protection characteristic of TF and PIC occupancy. The authors thus suggest that TFs, in collaboration with chromatin remodelers, act to re-establish nucleosome positioning after DNA replication. In the second study, budding yeast were synchronized in late G1 phase and sampled for analysis of TF binding over 3-min intervals following release (Bar-Ziv et al., 2020). ChEC-seq experiments then showed that the NDFs Abf1, Rap1, and Reb1 were displaced transiently during replication with binding re-established in 10–20 min. Since this binding is not faster than the reassembly of nucleosomes, which re-form close to the replication fork (Chapter 5), the results imply that these GRFs somehow compete successfully with histones for occupancy of their binding sites following replication. Whether newly replicated chromatin is more susceptible than mature chromatin to TF binding, and the mechanism by which TFs such as the yeast NDFs stably occupy their binding sites following replication, remains to be determined.

As mentioned at the beginning of this section, the characteristics that define a pioneer factor are best considered to fall in a continuum rather than reflecting a qualitative distinction between TFs able to gain access to nucleosomal sites and those that cannot. In both yeast and mammalian cells, TFs span a range in their ability to bind to a nucleosomal site, which depends not only on the factor itself but also on the affinity to its site and its abundance in the cell (Levo et al., 2017; Yan et al., 2018; Yu and Morse, 1999; Zhu et al., 2018), and multiple modes of nucleosomal binding have been observed for metazoan TFs (Luzete-Monteiro and Zaret, 2022; Zhu et al., 2018). In metazoans, the "classical" pioneer TF FoxA1 may be assisted in binding to chromatin by steroid receptors in some circumstances (Paakinaho et al., 2019; Swinstead et al., 2016) (but see Glont et al., 2019; Zaret et al., 2016). The strong pioneer factor hypothesis has also been challenged in a study in which expression of either FoxA1 or the nonpioneer factor HNF4A was sufficient to activate endoderm-specific expression in K562 lymphoblast cells, in contrast to the expected dependence on FoxA1 for HNF4A binding (Hansen et al., 2022).

Taking these various exceptions to a strict definition of "pioneer factor" into account, it is probably best not to take the designation in extremis, but rather as a recognition that TFs vary in their ability to contend with nucleosomes, with multiple parameters influencing the effect of nucleosomal occlusion on TF binding in individual cases. These factors include the nature of the DNA-binding domain, the location of the binding site in the nucleosome, the abundance and affinity of the factor for the specific site, the identity of the histone or histone variants, and the presence of modifications in the nucleosome. The problem of transcription factor access to sites in chromatin remains a highly active area of research (Isbel et al., 2022).

Facilitated accessibility

Clustering of TF binding sites at metazoan enhancers and promoters, recognized since the early days of "promoter-bashing" of globin genes and genes controlling pattern formation in *Drosophila*, among others, was eventually shown to be prevalent at human gene promoters on a genomic scale (The ENCODE Project Consortium, 2012; Li et al., 2011). While TF clustering at regulatory sites occurs at a somewhat lower frequency in invertebrates and yeast, it is still the case that many enhancers and promoters bind multiple TFs (Gerstein et al., 2010; Hughes and de Boer, 2013; Roy et al., 2010). Such clustering can facilitate transcriptional activation through protein–protein interactions between TFs that lead to cooperative binding or by differential interactions between individual TFs and downstream effectors (coactivators or the transcription machinery) that have synergistic effects on transcriptional output (Carey, 1998). In addition, DNA-binding architectural proteins may alter the configuration of DNA to promote transcriptional activation. Prototypical examples of this occur at the *Xenopus* ribosomal gene promoter, where binding of the transcription factor xUBF creates a looped structure that facilitates transcription, and at the IFN-β (interferon beta) promoter, where HMG proteins bend a 55bp stretch of DNA to facilitate cooperative assembly of three distinct TFs, creating a structure dubbed an enhanceosome and activating IFN-β transcription in response to viral induction (Bazett-Jones et al., 1994; Falvo et al., 1995; Thanos and Maniatis, 1995).

Nucleosomes themselves may sometimes similarly serve an architectural function in gene activation. Two early examples of this were reported at the *Drosophila hsp26* and *Xenopus vitellogenin B1* promoters, in which distal binding sites for TFs are brought into proximity by an intervening positioned nucleosome in what has been described as a "static loop" (Elgin, 1988; Schild et al., 1993; Thomas and Elgin, 1988). In the case of *hsp26*, deletion, inversion, or replacement of the 168bp nucleosomal sequence by a sequence from a bacteriophage did not alter *hsp26* induction when introduced into flies, while altering the sequence to produce a shift in nucleosome positioning reduced inducibility more than fivefold (Lu et al., 1995).

These results indicated that a proper chromatin structure is indeed important for full induction of *hsp26*, and that juxtaposing the distal elements directly or via an intervening nucleosome was equally effective in allowing normal induction. However, few cases in which such static loops contribute to transcriptional activation in vivo have been reported. Another potential mechanism for enhancement of TF function on nucleosomal templates is the proximal binding of TFs on the curved surface of the nucleosome that would be inhibited by steric clash on naked DNA. Enhanced binding by GR and NF1 at the MMTV promoter when incorporated into a nucleosome was inferred from biochemical experiments showing that the two TFs competed for binding on naked DNA in vitro, together with experiments showing that simultaneous binding to the MMTV promoter by GR and NF1 expressed in yeast, and NF1-dependent hormone induction, was compromised by histone depletion (Chavez and Beato, 1997). However, to my knowledge, this mechanism has yet to be supported by structural data.

A better-documented mechanism by which TFs can synergize is by one TF creating an accessible site in chromatin to facilitate the binding of another TF that acts as a primary activator (Fig. 8.9C). This mechanism defines pioneer factors, as we have just discussed; the TFs whose binding follows and depends on pioneer factors have been referred to as "settlers." Specific examples are FoxA1 facilitating binding of GATA-4 and GR facilitating binding of NF1, the latter mediated by recruitment of Brg1, the catalytic subunit of the mammalian SWI/SNF complex (Archer et al., 1992; Cirillo and Zaret, 1999; Fryer and Archer, 1998).

Early in vitro reports supported the concept of facilitated binding of TFs to chromatin that had been gleaned from in vivo work, demonstrating that it stemmed from the inherent structural dynamics of nucleosomes. Experiments in which the ability of TFs to bind to sites in reconstituted nucleosome core particles in vitro was assessed by gel shift assays and DNase footprinting revealed varied abilities of TFs to bind to nucleosomal sites, as discussed in Chapter 4. The incorporation of binding sites for multiple TFs (USF, NF-κB, and two Gal4 derivatives) into a nucleosomal template resulted in cooperative binding between pairs of factors; no such cooperative effect was seen with naked DNA templates (Adams and Workman, 1995). In these experiments, one of the TF binding sites was situated near the periphery of the core particle; previous work had shown binding to more interior sites was greatly inhibited, a finding recently supported quantitatively by single-molecule fluorescence measurements showing that accessibility of Gal4 to nucleosomal DNA in vitro is over 100-fold lower for the central 80 bp than for the outermost 30 bp (Donovan et al., 2023b; Vettese-Dadey et al., 1994).

A series of papers from the Widom lab provided additional mechanistic insight and a quantitative basis for facilitated binding of TFs to chromatin. By assessing the ability of restriction enzymes to cleave sites located at various locations within reconstituted nucleosomes, it was demonstrated that DNA spontaneously disengages from the histones, with transient exposure being highest nearest the nucleosome periphery and decreasing toward the dyad (Polach and Widom, 1995). Spontaneous unwrapping was later confirmed and quantified by stopped-flow fluorescence resonance energy transfer (FRET), in which energy transfer between a fluorescence donor at the DNA end and an acceptor attached to the histone core was used to monitor the extent of disengagement of peripheral DNA from the histones (Li et al., 2005; Li and Widom, 2004). These experiments were performed using a DNA template having a binding site for the bacterial repressor LexA near the periphery, and in the presence of high concentrations of LexA; in this protocol, binding of LexA captures the unwrapped state, and at very high concentrations, it is unwrapping rather than LexA concentration that is rate-limiting. The results showed the dynamics of unwrapping to be very rapid, occurring at a rate of about $4\,s^{-1}$ (or on a time scale of ~250 ms), with rewrapping occurring within ~10–50 ms. The rate of unwrapping can be increased by tenfold or more by H3K56 acetylation and effects of DNA sequence in the entry-exit region of the nucleosome, allowing modulation of accessibility in vivo (North et al., 2012).

The transient unwrapping of DNA from the histones at the nucleosome periphery suggested a mechanism for cooperative binding of TFs independent of direct interactions, as well as for stabilization of individual TF binding (Miller and Widom, 2003; Morse, 2003; Polach and Widom, 1996). The binding of a TF upon transient site exposure could be followed by the recruitment of a chromatin remodeler, leading to nucleosome sliding or histone eviction and thus resulting in stable TF binding. Similarly, facilitated binding of TFs to nucleosomal sites was envisioned as occurring through initial TF binding to the outermost region of the nucleosome, enabled by transient DNA-histone dissociation, which in turn would lower the free energy cost of DNA unwrapping to allow binding of a second TF (Polach and Widom, 1996). Such a mechanism would not depend on interaction between TFs or on recruitment of chromatin remodelers or modifiers. An early example proposed to utilize this mechanism, later dubbed a "collaborative competition mechanism" (Miller and Widom, 2003), was activation of the *HIS4* gene in budding yeast in response to amino acid starvation. The *HIS4* promoter contains binding sites for Rap1, Gcn4, and the heterodimeric activator Bas1/Bas2, and while Rap1 was found to bind the *HIS4* promoter in the absence of Gcn4 and Bas1/Bas2, it was not sufficient for *HIS4* activation (Devlin et al., 1991). Conversely, activation by either Gcn4 or Bas1/Bas2 required the Rap1 binding site, and sensitivity to cleavage by MNase at the Gcn4 and Bas1/Bas2 binding sites was greatly diminished when the Rap1 binding site was mutated. Together, these findings suggested that Rap1 functioned to open chromatin and allow binding and activation by Gcn4 and Bas1/Bas2. As recounted in Chapter 4, Rap1 was then shown to be capable of outcompeting histones for occupancy of its binding site in vivo, and this capability, as well as that for the NDF Abf1, was demonstrated to be independent of its activation domain (Yarragudi et al., 2004; Yu et al., 2001). In addition, both Rap1 and Abf1 were found to facilitate activation by distinct TFs at several loci, independent of the Rap1 and Abf1 activation domains, and *HIS4* promoter mutations in which the separation between Rap1 and Gcn4 binding sites was increased by 5 bp or 10 bp did not impair growth on medium lacking histidine, indicating that direct interaction between Rap1 and Gcn4 was not responsible for their cooperative effect on transcription (Yarragudi et al., 2004; Yu and Morse, 1999). Thus, both of these GRFs exhibited properties consistent with their behaving as pioneer factors in facilitating binding of TF "settlers" without recruitment of ancillary factors such as chromatin remodelers. In further support of the collaborative competition mechanism for cooperative TF binding to nucleosomal sites, transcriptional activation of the yeast *HIS3* gene by Gcn4 was stimulated 1.4- to 1.8-fold by expression of LexA when a LexA binding site was placed in close proximity to the Gcn4 binding site (Miller and Widom, 2003). ChIP experiments demonstrated increased Gcn4 occupancy in the presence of LexA; since LexA, a bacterial DNA-binding repressor, did not stimulate transcription on its own and never encounters Gcn4 in nature, it was inferred that it stimulated transcription by enhancing Gcn4 binding by increasing accessibility to chromatin. More recently, the high-throughput study by Levo et al. (2017) described in Chapter 4 (see Fig. 4.18) demonstrated enhanced gene expression from templates in which binding sites for nucleosome-depleting factors such as Abf1 or Rap1, which do not exhibit strong transcriptional activation capability on their own, were combined with TFs such as Gcn4 that have poor ability to activate transcription from nucleosomal binding sites. Another high-throughput study used a methylation protection assay (see Chapter 4) to simultaneously map the occupancy of TFs and nucleosomes in murine cells and reported a high frequency for co-occupancy of TFs in regions of high nucleosome occupancy (Sonmezer et al., 2021). Altogether, these studies support a role for facilitated binding of TFs by a mechanism involving nucleosome disruption by pioneer factors (NDFs in yeast). Future studies are likely to probe the role of chromatin remodelers, histone modifications, and histone variants in this process.

The permissive chromatin structure for TF binding created by pioneer factors appears generally to require their continued binding. Transcript levels of a reporter gene driven by the *HIS4* promoter decreased to near background levels upon loss of Rap1 binding, using a temperature-sensitive mutant of *rap1* (Yarragudi et al., 2004). More recently, optogenetic inactivation of the *Drosophila* pioneer factor Zelda resulted in downregulation of hundreds of genes during the zygotic-to-maternal transition, demonstrating a requirement for continued Zelda activity (McDaniel et al., 2019). A requirement for continued activity of pioneer factors for facilitated binding has also been shown in mouse ES cells, where depletion of Oct4 resulted in decreased accessibility and cofactor occupancy at target sites (Friman et al., 2019; King and Klose, 2017). In this latter instance, opening of chromatin by Oct4 requires recruitment of the remodeler Brg1; the requirement for continuous action by Oct4 is therefore consistent with other studies demonstrating an ongoing requirement for chromatin remodeling activity to maintain accessibility, both in yeast and in mammalian cells (Biggar and Crabtree, 1999; Iurlaro et al., 2021; Schick et al., 2021; Sudarsanam et al., 1999). Indeed, depletion of chromatin remodelers that act to establish NDRs in yeast results in rapid alterations in promoter chromatin structure on a genome-wide level (Kubik et al., 2019), although, as discussed above, occupancy by NDFs in yeast appears little affected upon depletion of individual remodelers (Chen et al., 2022). These various results, demonstrating transient site exposure of nucleosomal DNA and the necessity for continued action of pioneer factors, whether via recruitment of remodelers or direct action on histone-DNA contacts, emphasize the need to consider chromatin as a dynamic entity in its effects on transcriptional activation.

Promoter chromatin structure reflects function

By the early 2000s, there was no question as to the importance of chromatin structure in transcriptional regulation. The advent of computational and experimental methods for analyzing chromatin structure and nucleosome occupancy on a genome-wide basis (Chapter 4) provided means to investigate connections between chromatin structure and gene regulation at a level far beyond what could be achieved by examining individual genes. Two early studies employing computational approaches categorized yeast promoters according to predicted nucleosome occupancy or DNA bendability, in both cases noting correlation of these properties with the presence or the absence of consensus TATA elements (Ioshikhes et al., 2006; Tirosh et al., 2007). (For discussion on promoters lacking or containing consensus TATA elements, and their differential regulation by TFIID and SAGA complexes, see Chapters 6 and 7.)

Similar analysis of experimentally determined nucleosome occupancies across the budding yeast genome revealed four major clusters of genes exhibiting distinct patterns of nucleosome occupancy and enriched for different functional gene categories (Fig. 8.12A and B) (see Sidebar: K-means clustering) (Lee et al., 2007). A subsequent analysis based on nucleosome occupancy data of Lee et al. (2007) identified two major classes of promoters (Tirosh and Barkai, 2008). One class was characterized by an NDR proximal to the transcription start site, with high nucleosome occupancy farther upstream, while the second class exhibited high nucleosome occupancy close to the transcription start site (TSS) and lower occupancy, and less well-positioned (fuzzier) nucleosomes upstream (Fig. 8.12C). Although the genes comprising the two classes did not differ significantly in expression level, they differed in other properties: genes having high nucleosome occupancy (termed "occupied promoter nucleosome," or OPN genes) near the TSS exhibited greater variability in expression under different conditions (i.e., greater transcriptional plasticity), and also in response to defects in chromatin regulators, than "depleted promoter nucleosome" (DPN) genes. OPN genes also exhibited higher histone turnover and were enriched for TATA-containing genes, diverged more rapidly in expression across related species, and exhibited greater cell-to-cell expression, or "noise." In addition, TATA elements were underrepresented among DPN genes, whose other characteristics also generally contrasted with those of OPN genes. An independent analysis based on MNase-seq data arrived at similar conclusions (Field et al., 2008), while another report linked the connection between TATA-containing promoters and transcriptional plasticity to preferential control by chromatin regulators (Choi and Kim, 2008).

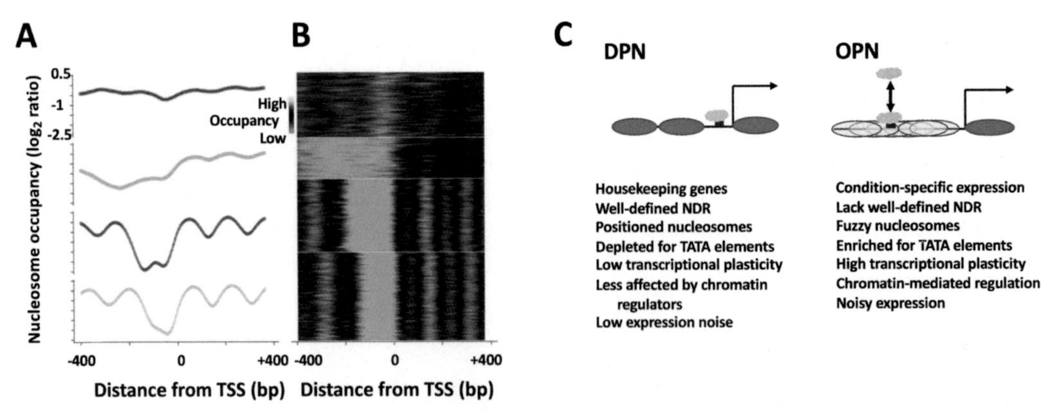

FIG. 8.12 K-means clustering of nucleosome occupancy surrounding the TSS in budding yeast grown in a rich medium. (A) Average nucleosome occupancy, defined as the $\log_2$ ratio of the signal for nucleosomal DNA over the signal for genomic DNA, for the four identified clusters. The same range was used for all four plots; for simplicity, only the top graph is labeled. (B) Heat map of nucleosome occupancy for the four clusters, respectively, containing 1211, 766, 1374, and 1663 genes; each horizontal band represents one gene. (C) Two promoter classes differ in chromatin structure. Depleted nucleosome occupancy (DPN) promoters are characterized by an NDR often containing TF binding sites *(red rectangle)* that are constitutively accessible to TFs *(orange oval)*, while occupied promoter nucleosome (OPN) promoters are occupied by fuzzier nucleosomes and lack a well-defined NDR. TF binding at OPN promoters may vary conditionally with a corresponding effect on nucleosome occupancy. See text for additional detail. *(A) and (B) From Lee et al., 2007. A high-resolution atlas of nucleosome occupancy in yeast. Nat. Gen. 39, 1235.*

Sidebar: K-means clustering

Clustering algorithms used in constructing phylogenetic trees were soon applied to the analysis of large-scale gene expression and related genome-wide data sets (Eisen et al., 1998; Ioshikhes et al., 2006; Lee et al., 2007). Cluster analysis can be applied to any data set that can be represented as a set of vectors, such as expression levels over time for individual genes (each gene represented by a vector containing values for each time) or nucleosome occupancy values across a small region of the genome (e.g., within a few hundred bp of transcription start sites). Clustering can be performed on a hierarchical basis, in which a tree is created having branches with length reflecting the similarity between vectors, as in phylogenetic analysis, or by K-means clustering, which seeks to identify a specific number (designated as "K") of clusters containing similar "objects"—that is, vectors representing genes or genomic regions, as in Fig. 8.12. When performing K-means clustering, typically K is varied until increasing it further does not result in the generation of novel clusters. Results are generally presented in the form of heat maps, as seen in Fig. 8.12, in which individual genes or regions are depicted as horizontal lines and the colors are scaled according to the values across the relevant region (or time range, etc.).

These studies suggested that while promoter chromatin structure does not dictate expression levels, it might be important in influencing regulatory properties. Influencing is not the same as governing; some TATA-containing genes exhibiting high plasticity belong to the DPN class and some OPN genes lack TATA elements and/or exhibit low plasticity. In fact, the full-scale analysis revealed that transcriptional plasticity, expression divergence, sensitivity to chromatin regulators, and expression noise are impacted both by promoter chromatin structure and by the presence or absence of a TATA element, evidently by distinct mechanisms. Thus, the DPN/OPN spectrum, like the TATA-containing and TATA-lacking spectrum, is not a strict dichotomy, with some overlap existing among gene properties in spite of the strong associated enrichments (Tirosh and Barkai, 2008).

Comparative studies of promoter chromatin structure and gene expression across related species support a conserved role for chromatin structure in gene regulation, while also showing that effects of chromatin structure can be superseded by other factors, such as the presence or absence of TF binding sites. Comparison of the closely related (diverged ~5 million years ago) yeast species *S. cerevisiae* and *S. paradoxus*, and interspecific hybrids between the two (see Fig. 4.16), indicated that alterations in nucleosome occupancy at promoters were not generally associated with gene expression changes, consistent with the comparable expression levels found among DPN and OPN genes (Tirosh et al., 2009, 2010). However, examples were also seen in which decreased nucleosome occupancy at TF binding sites was accompanied by mutation of the TF binding site, suggesting compensatory mutations, and interspecies alterations of nucleosome occupancy were disfavored at regulatory sites (Tirosh et al., 2010). Moreover, modest correlations were reported between changes in nucleosome occupancy and gene expression in budding yeast under disparate growth conditions (Shivaswamy et al., 2008; Tirosh et al., 2010; Zawadzki et al., 2009).

Two additional studies also supported a regulatory role for promoter chromatin structure based on evolutionary comparisons. Determination of genome-wide nucleosome occupancy across 12 species of *Hemiascomycota* yeasts encompassing >200 million years of evolutionary change revealed a general conservation of the dichotomy between DPN and OPN genes and their respective linkages to genes exhibiting low or high transcriptional plasticity (grouped as growth genes and stress genes) (Tsankov et al., 2010). Most tellingly, large-scale phenotypic changes were accompanied by changes in chromatin organization of the relevant gene sets. The most notable example relates to a change in respiratory behavior that occurred at the evolutionary break marked by a whole-genome duplication (WGD). Species that diverged prior to the WGD grow by respiration in a high glucose medium, and activate genes required for such growth; conversely, species diverging after the WGD grow mainly by fermentation and repress those same genes. Remarkably, the promoter chromatin structure of such respiration genes was found to have undergone a large-scale transition from an open chromatin structure (i.e., DPN) to a more closed structure (OPN) at the WGD (Field et al., 2009; Tsankov et al., 2010). A subsequent study compared nucleosome occupancy at orthologous promoters of three *Saccharomyces* species and *Candida glabrata*, which diverged less recently than the other three species ($\sim$50 million years versus $\sim$10–20 million years) while still belonging to the clade that emerged subsequent to the WGD (Tsui et al., 2011). In agreement with Tsankov et al., this study found nucleosome occupancy patterns at orthologous promoters to be well conserved among the *Saccharomyces* species. However, considerable divergence was seen for orthologs in *C. glabrata*, even while gene expression levels and response to different stresses were conserved. Altogether, these studies indicate a significant but nonexclusive role for nucleosome occupancy patterns in governing gene expression patterns; changes in promoter sequence, coupled with changes in promoter chromatin structure, evidently allow multiple pathways to the same gene expression properties. Similarly, changes in promoter sequences across evolution, including gain or loss of TF binding sites, have occurred in the face of preserved patterns of gene expression (Fisher et al., 2006; Ludwig et al., 1998; Romano and Wray, 2003; Takahashi et al., 1999; Tirosh et al., 2008).

The linkage between promoter chromatin structure and gene regulation is not restricted to yeast, as analysis of metazoan promoters also reveals correlations among promoter sequence, chromatin structure, and functional properties (Fig. 8.13). Analysis of 5′ capped

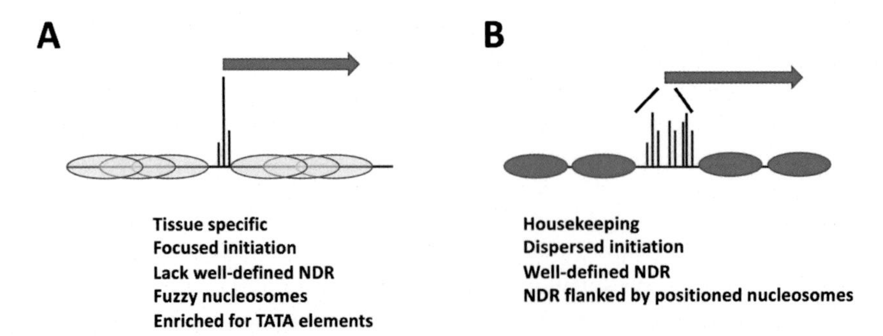

FIG. 8.13 Two major promoter types in metazoans. (A) Genes having focused initiation sites, enriched for genes expressed in specific tissues or cell types. (B) Genes with dispersed initiation sites, enriched for housekeeping genes. Genes expressing developmentally expressed TFs show a similar structure to that shown in (B) but exhibit focused initiation in *Drosophila*.

sequences of mRNAs from mammals and *Drosophila* revealed two major types of promoters, one giving rise to focused transcripts arising from transcription start sites occurring in a narrow window and the other producing transcripts originating over a more dispersed region. (These two promoter categories were originally divided according to CpG methylation status in vertebrate cells (Carninci et al., 2006); a subsequent study that also included data from *Drosophila* genome-wide analyses reported that a sharper functional distinction was seen based on narrow vs. dispersed 5′ mRNA ends (Rach et al., 2011)). These two types of promoters were, respectively, enriched for genes exhibiting developmentally regulated expression (analogous to genes exhibiting transcriptional plasticity in yeast) and for housekeeping genes; focused promoters were also enriched for TATA elements (Carninci et al., 2006; Lenhard et al., 2012; Yamashita et al., 2005). Analysis of nucleosome occupancy patterns revealed that the promoters associated with dispersed TSSs and housekeeping genes are characterized by an NDR flanked by well-positioned nucleosomes, whereas focused promoters generally lack a well-defined NDR and are associated with fuzzier nucleosomes (Rach et al., 2011). Later studies uncovered another class of promoters, enriched for genes encoding TFs important in patterning and morphogenesis, that shares with housekeeping genes an NDR flanked by well-positioned nucleosomes, but differs in giving rise to dispersed initiation sites in mammals while frequently exhibiting focused transcript start sites in flies (Haberle and Stark, 2018).

Additional evidence for the functional importance of promoter chromatin structure in gene regulation derives from a study of the action of chromatin remodelers in *Drosophila* (Hendy et al., 2022). In this study, essential subunits of chromatin remodelers were rapidly depleted in *D. melanogaster* S2 cells (see Chapter 4 Sidebar: Deletions and depletions), and the effects on nascent transcription were monitored by using precision run-on sequencing (Core et al., 2008). Each of the four major remodelers was examined (Chapter 6). Depletion of Chd1 did not significantly affect nascent transcription, while depletion of SWI/SNF complexes (both BAF and pBAF, via the common subunit Snr1) resulted in downregulation of ~1400 genes, many of which were severely impacted with two- to 10-fold reduction in nascent transcription. The effects of depletion of Iswi or Ino80 differed from SWI/SNF, with approximately equal numbers of genes being upregulated and downregulated (Iswi) or genes being preferentially upregulated (Ino80), with most showing less than twofold change in expression. SWI/SNF depletion primarily affected developmentally regulated genes; in fact, most such genes were downregulated by SWI/SNF depletion. Housekeeping genes were enriched among those downregulated upon Iswi depletion, although the effects were weaker and less widespread than those seen upon SWI/SNF depletion. SWI/SNF was preferentially localized to the enhancers, but not promoters, of developmentally regulated genes, and its depletion correspondingly reduced accessibility at those enhancers while not affecting nucleosome occupancy at proximal promoters. In contrast, Iswi and Ino80 did not exhibit any preferential localization; however, both affected nucleosome occupancy patterns at housekeeping promoters. Overall, these results indicate segregated activity of remodelers at these two classes of genes. How the action of SWI/SNF at enhancers, which often are located kilobases distant from the promoters at which they act, is connected to the nucleosome occupancy pattern at proximal promoters remains a topic for future investigation, as does the mechanism by which remodeler activity is partitioned between the two classes.

Chromatin-mediated gene regulation in development: The polycomb group and trithorax group complexes

Repression mediated by heterochromatin was discussed at length in Chapter 3. Here, we discuss an additional example of chromatin-mediated gene regulation that emerged from studies of pattern formation and developmental gene regulation during *Drosophila melanogaster* embyrogenesis, culminating in the discovery of the conserved PcG and trithorax group (TrxG) complexes that regulate cell differentiation through their action on chromatin (Chetverina et al., 2020; Grimaud et al., 2006; Kassis et al., 2017; Schuettengruber et al., 2017). This work had its origins in the early 20th century in the lab of Thomas Hunt Morgan, where studies on the inheritance of a mutation affecting eye color provided powerful evidence for chromosomes as the carriers of hereditary traits (Morgan, 1910; Mukherjee, 2016; Sturtevant, 2001). The popularization of *Drosophila* as a model organism for genetic studies that began in Morgan's lab, together with its short generation time and ease of characterization of abnormal anatomy, led to studies by Ed Lewis in which the genetic loci responsible for the specification of the identities of the body segments covering the posterior two-thirds of the fruit fly were determined by combining classical genetics and cytology, and eventually by molecular techniques (Lewis, 1978).

Lewis's studies, performed over a span of 50 years, did not originally have a goal of elucidating developmental processes, but rather were undertaken as a route to investigating the origins of new genes (Crow and Bender, 2004; Duncan and Montgomery, 2002; Lewis, 1951). This research sought to study closely linked genes that exhibited similar effects when mutated, and over considerable time landed on a group of genes controlling segmental identity. The *Drosophila* embryo develops according to a well-defined body plan, in which a cascade of regulatory events begins with the definition of posterior and anterior ends of an ellipsoid, syncitial embryo and ultimately leads to the formation of distinct thoracic and abdominal segments having characteristic features such as antenna, wings, and legs (Fig. 8.14). Lewis's experiments focused on mutants that caused homeotic transformations, in which the characteristics of individual segments were altered to resemble those of more posterior or anterior segments, and eventually led to the identification of a small group of tightly linked genes that controlled segmental identity in the *Drosophila* thorax and abdomen. Lewis named this group of genes the bithorax complex (BX-C) (Lewis, 1978). He also recognized that the Polycomb (*Pc*) gene encoded a repressor of BX-C genes that acted in *trans*. *Pc* was first identified by Pam Lewis (Ed's wife) via a mutant that produced ectopic sex combs on the second and third legs of adult male flies instead of their normally restricted presence on the first pair of thoracic legs (Lewis, 1947), and subsequent experiments showed that homozygous *Pc* mutants caused the transformation of thoracic segments and abdominal segments 1–7 to resemble the most posterior abdominal segment 8. This led Ed Lewis to suggest that the expression of gene products specifying the identity of Abd8 must be repressed in Abd1-7, and that *Pc* was required for this repression (Lewis, 1978). Previous hypotheses to explain mutations leading to extra-sex-combs phenotypes involved interruption of a putative

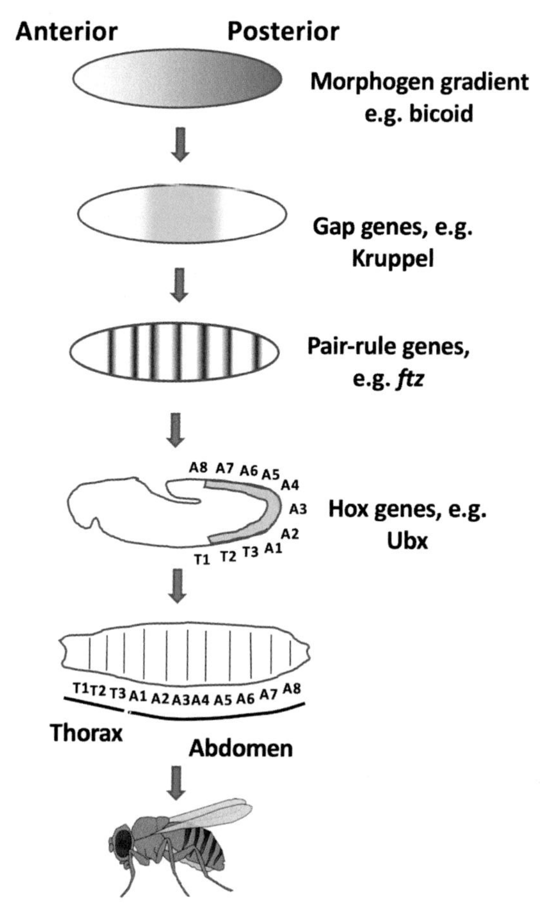

FIG. 8.14 Simplified schematic of *Drosophila* development. The body plan of the adult fly is initiated in the embryo by the expression of genes that establish the anterior-posterior (A-P) axis and segmental identities, including morphogen gradient genes, gap genes, pair-rule genes, and the Hox genes. The combined effects of these patterning genes lead to the specification of the three thoracic segments, T1–T3, and the eight abdominal segments, A1–A8, as well as their A-P orientation. *For a full description of the sequence of events determining the pattern of the adult fly, see Gilbert S.F., Barresi, M.J.F., 2016. Developmental Biology, Sinauer Associates.*

anterior-posterior pattern gradient in the thorax or an imbalance in the proliferation of specific cell types leading to altered morphology (Kassis et al., 2017). The idea that development fates were controlled by a global repressor, epitomized by *Pc*, was a radical suggestion at the time and led to a paradigm shift in thinking about the implementation of embryonic form on a molecular level.

A second complex of linked genes governing the *Drosophila* body plan, the Antennapedia complex, or ANT-C, is also regulated by PcG genes. The BX-C and ANT-C genes together comprise the *Drosophila* Hox genes. Hox is short for homeobox, a motif defining a set of transcription factors; these transcription factors in turn control the expression of genes that define segmental identity. For simplicity, we restrict discussion here to BX-C. Additional mutants that behaved similarly to *Pc* mutants were identified in genetic screens, along with mutants that enhanced or suppressed *Pc* mutations, leading to the suggestion that they be collectively referred to as PcG products (Jürgens, 1985). Mutations in any of the ten tested PcG genes were shown to cause ectopic expression of both *abdA* and *AbdB* genes, consistent with the action of a common repressive pathway dependent on the known PcG genes (Simon et al., 1992). PcG genes now number about 15, depending on the precise criteria used for their inclusion, somewhat fewer than initially estimated (Jürgens, 1985; Kassis et al., 2017; Landecker et al., 1994).

The proposed role of *Pc* as a repressor of BX-C genes suggested the possible existence of an opposing global activator. A mutant exhibiting the phenotypes characteristic of a defect in such an activator, resembling BX-C loss-of-function mutants, was first identified serendipitously in an unrelated mutant screen and was named *trithorax* (*trx*) (Ingham, 1981, 1983). (This spontaneously arising mutant was nearly lost when Ingham, a graduate student at the time, preserved his original transformed flies in ethanol under the assumption that the observed mutant phenotype reflected the mutations used in his cross. Fortunately, he was quickly disabused of this misconception by his advisor and was able to recover additional mutants from the few pupae still being produced by his original cultures (Ingham, 1998).) Subsequent screens identified mutants with phenotypes similar to *trx* mutants, and these positive regulators of Hox gene expression came to be known as TrxG genes; additional TrxG members were identified as suppressors of PcG gene mutations or through candidate gene screens (Kassis et al., 2017; Kennison and Tamkun, 1988; Kingston and Tamkun, 2014). TrxG mutants, like PcG genes, also regulate ANT-C gene expression, but in addition affect expression of genes outside of the BX-C and ANT-C loci, reflecting a more general role in activation.

The molecular genetics revolution instigated an explosion in the field of *Drosophila* developmental genetics in the 1980s. A great number of experiments at this time examined interactions among genes identified as regulating body plan and segmental identity; to the outsider (this author, for instance), this research at times seemed arcane and the results difficult to integrate into a coherent model. Eventually, however, determination of the spatial and temporal expression and interdependence of transcription factors acting in the earliest stages of development, such as *fushi tarazu* (*ftz*) and *bicoid*, led to a picture in which sequential expression of regulators specified embryonic patterning (Fig. 8.14). Most intriguingly, PcG and TrxG genes were discovered not to be required for establishing activation or repression of genes specifying cell identity but were instead required later in development for maintenance of the patterns initiated by the earliest acting TFs, after those TFs were no longer present (Fig. 8.15) (Akam, 1987). For example, the PcG genes *esc* and *E(z)* are not required to initiate spatially restricted expression of Hox genes, whereas mutants exhibit ectopic expression later in development (Jones and Gelbart, 1990; Struhl and Akam, 1985). Conversely, mutations in TrxG genes resulted in homeotic transformations consistent with failure to maintain expression of BX-C genes activated early in embryogenesis (Kassis et al., 2017; Kennison, 1993).

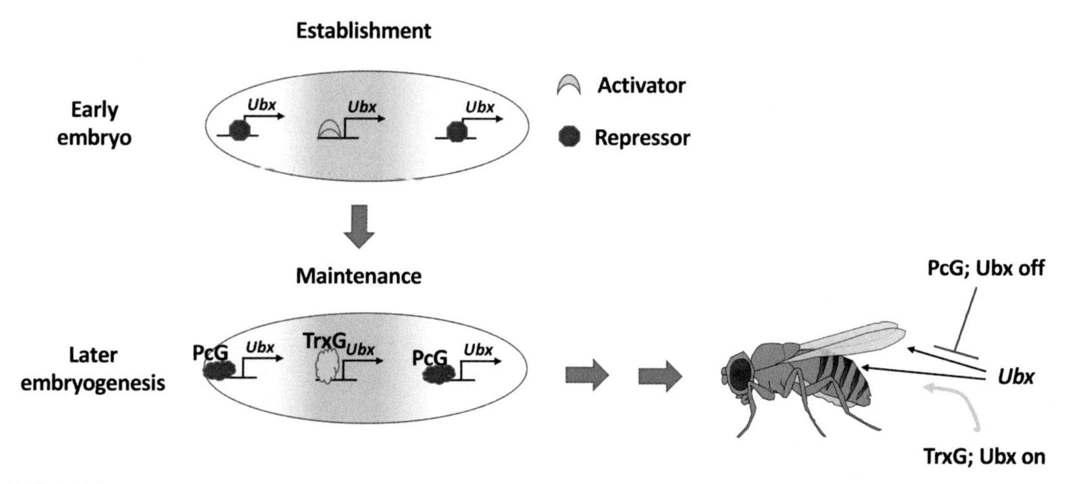

FIG. 8.15 Hox gene expression in the early embryo is first controlled by transcriptional activators and repressors that bind to specific promoter elements but is later governed by TrxG or PcG complexes.

The maintenance of cellular memory of *on* and *off* states of gene expression by the PcG and TrxG genes captured the attention of researchers in the chromatin field (Paro, 1990; Simon, 1995). Polycomb was found to share a homologous domain of 48 aa with the heterochromatin protein HP1, leading to the suggestion that PcG proteins might cooperate to package DNA into repressive chromatin structures (Paro, 1990; Paro and Hogness, 1991). HP1 was discovered to be encoded by *Su(var)2–5*, a suppressor of position effect variegation (PEV), in which the expression of the *white* gene and corresponding eye color phenotype in *D. melanogaster* vary according to its proximity to condensed heterochromatin (discussed in Chapter 3) (Elgin and Reuter, 2013). Several additional genes identified through such studies were also found to have homeotic effects and thus classified as PcG or TrxG members (Wu et al., 1989). The connection between heterochromatin and regulation by PcG and TrxG was reinforced by the discovery that the suppressor of PEV, Su(var)3-9, shared a homologous domain with the PcG member E(z) and Trithorax (Jones and Gelbart, 1993; Tschiersch et al., 1994); this domain was named the SET domain after these three founding members and was later revealed to possess histone methyltransferase activity (Chapter 7).

Cloning and sequencing of PcG and TrxG members yielded accumulating evidence for their functions involving chromatin-mediated mechanisms. As detailed in Chapter 6, the TrxG member brahma (brm) was found to share several regions of homology with Snf2/Swi2, the ATPase subunit of the yeast SWI/SNF complex, just as the *SWI/SNF* genes were being discovered to encode subunits of a multisubunit complex that acted as a global regulator of transcription through its action on chromatin (Tamkun et al., 1992). Two years later came the unexpected discovery that the *Drosophila Trithorax-like*, or *trl*, gene encoded GAGA factor (Farkas et al., 1994). At about this same time, GAGA factor was found to disrupt nucleosome organization in an ATP-dependent fashion, as discussed in Chapter 6 (Tsukiyama et al., 1994); these findings implicated TrxG members as facilitating activation of genes, including BX-C members, by a mechanism that depended at least in part on chromatin remodeling.

By the end of the 1990s and early 2000s, it was clear that PcG and TrxG proteins functioned as multiprotein complexes containing not only chromatin remodelers but also histone modifiers (Tables 8.2 and 8.3). Early genetic and biochemical studies had provided clues that PcG proteins might function together in a large complex. For example, PcG proteins were observed in proximity on polytene chromosomes, and co-immunoprecipitation experiments also pointed to interactions among PcG proteins (DeCamillis et al., 1992; Franke et al., 1992). Purification of complexes containing either of two Flag-tagged PcG proteins (polyhomeotic and Posterior sex combs) revealed an association with several other PcG proteins, confirming the notion that PcG proteins function in a common pathway (Shao et al., 1999). The complex was named Polycomb Repressor Complex 1 (PRC1) and was capable of inducing compaction of chromatin templates and inhibiting chromatin remodeling by SWI/SNF. Several lines of evidence suggested that not all PcG proteins acted through PRC1, and a second PcG protein-containing complex was subsequently purified and named PRC2 (Ng et al., 2000).

TABLE 8.2 PcG components.[a]

	Drosophila	Mouse	Molecular feature(s)	Molecular function
PRC2 core components	E(z) (enhancer of zeste)	EZH1/EZH2	SET domain	H3K27 methylation
	Esc (extra sex combs)	EED		H3K27me3 reader
	Suz(12) (suppressor of zeste 12)	SUZ12	Zinc finger	PRC2 stability
	p55/Nurf55/Caf1	RBBP4/RBBP7	Histone-binding domain	Nucleosome binding
PRC1 core components	Pc (polycomb)	CBX2,4,6,7,8	Chromodomain	H3K27me3 binding
	Ph (polyhomeotic)	PHC1-3	Zinc finger	Chromatin compaction
	Sce (sex combs extra)/dRing (really interesting new gene)	Ring1A/B	RING zinc finger	E3 ubiquitin ligase (H2AK119ub)
	Psc (posterior sex combs)	Bmi1/PCGF4	Zinc finger	Enhance catalytic activity
		Mel18/PCGF2	HTH domain	PRC1 stability
	Scm (sex comb on midleg)[b]	SCMH1/SCMH2		May link PRC1 and PRC2
PcG "adaptors"	Pho (pleiohomeotic), PHOL	YY1	Zinc finger	PRC1 and PRC2 recruitment
	dSFMBT	SFMBT	Heterodimerizes with PHO	PRC1 and PRC2 recruitment
	PSQ (pipsqueak)		BTB-POZ domain	PRC1 and PRC2 recruitment

[a] *Not included in this table are components of noncanonical mammalian PRC1 and PRC2 complexes. For an exhaustive tabulation, see Piunti and Shilatifard (2021).*

[b] *SCM is not, strictly speaking, a core subunit of PRC1 but co-purifies with* Drosophila *and human PRC1; it also is found associated with chromatin independently of PRC1.*

TABLE 8.3 TrxG components.[a]

Drosophila	Human	Molecular function
TRX/TRR	MLL1-4	H3K4 methylation
ASH1 (absent, small, or homeotic)	MILL4, hSET1, hASH1	
BRM	BRG1/HBRM	Chromatin remodeling
OSA	BAF250	
MOR	BAF155, 170	
SNR1	hSNF5/INI1	
KIS (Kismet)	CHD7	
VTD (Verthandi)	RAD21	Cohesin subunit
TRL (Trithorax-like)	BTBD14B	Transcription factor
ASH2	hASH2L	

[a] *Not all proteins classified as TrxG members are included. See Kingston and Tamkun (2014) for additional detail.*

Several histone modification enzymes and proteins with affinity for specific modified histone residues—that is, "writers" and "readers"—were identified among the constituents of PRC1 and PRC2 (Tables 8.2 and 8.3). E(z) is both a founding member of the SET domain family of proteins and a component of PRC2; the discovery of the histone methyltransferase activity of SU(VAR)3-9 led to the eventual demonstration of specific trimethylation of H3K27 by E(z) and its mammalian homologs, as described in Chapter 7 (section on Histone lysine methylation). The H3K27 methyltransferase activity of PRC2 is critical to its function, as a catalytic *e(z)* mutant fails to repress Hox genes (Cao et al., 2002; Czermin et al., 2002; Muller et al., 2002). Moreover, *Drosophila* larvae in which histone H3 was replaced with an H3K27R mutant phenocopied PRC2 mutants, and murine embryonic stem cells (ESCs) in which all H3 genes were converted to H3K27R-expressing mutants by CRISPR base editing showed close phenotypic and transcriptional correspondence to PRC2 null mutants (Pengelly et al., 2013; Sankar et al., 2022). Conversely, Trx, another founding member of the SET domain family, is homologous to Set1 in yeast and MLL1-4 in mammals, which confer the activating H3K4me3 mark on chromatin (Chapter 7, section on Histone lysine methylation). (However, Trx may exert its function through monomethylation rather than trimethylation of H3K4: a *trx* catalytic mutant, while failing to complement *trx* mutant flies, exhibited decreased H3K4me1 levels but unchanged H3K4me3 levels in vivo (Smith et al., 2004; Tie et al., 2014).) In addition, PRC1 was found to catalyze the ubiquitination of H2AK119 (H2AK118 in *Drosophila*), while TrxG complexes, purified in a similar manner to PcG complexes, were shown to include three SWI/SNF subunits, including brm (Tables 8.2 and 8.3) (Cao et al., 2005; Kingston and Tamkun, 2014; Wang et al., 2004a). The presence of histone modifiers associated with gene repression and their readers in PcG complexes, and of an H3K4 methyltransferase and a chromatin remodeler as TrxG complex components, solidified a model in which maintenance of cellular memory of active or repressed states of the Hox genes is governed by opposing activities of the PcG and TrxG proteins acting through chromatin (Fig. 8.16).

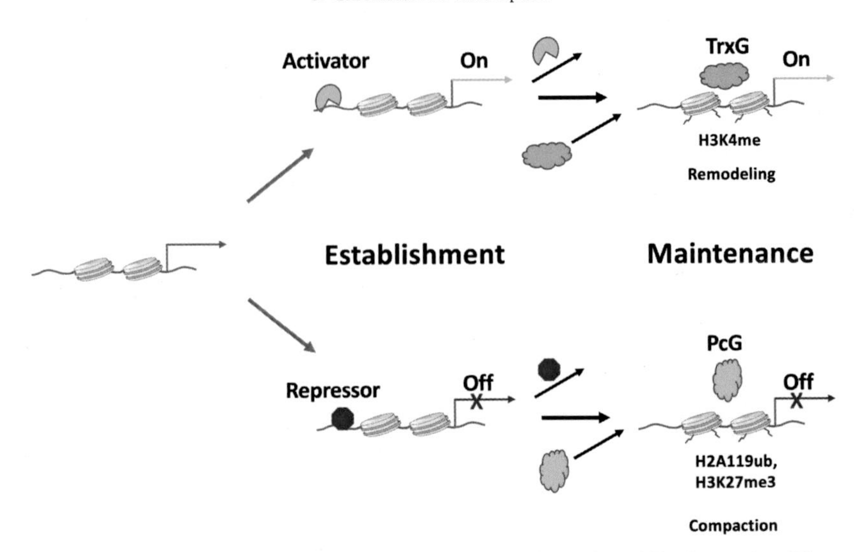

FIG. 8.16 Simplified model for maintenance of gene expression by PcG and TrxG proteins. The *on* pathway is maintained by TrxG proteins that include chromatin remodelers and an H3K4 methyltransferase, while the *off* pathway is enforced by PcG-mediated H2A119ubiquitination, H3K27 trimethylation, and chromatin compaction.

Conservation of the *Drosophila* BX-C and ANT-C Hox gene clusters in mammals was recognized in 1989 (Duboule and Dolle, 1989; Graham et al., 1989); correspondingly, ortholog-containing PRC1 and PRC2 complexes are present in metazoan species and plants, although considerable divergence is seen for individual components (Whitcomb et al., 2007). These have been studied most intensively in mammals. As in flies, studies in mice showed that PcG genes are crucial regulators of Hox gene expression and implementation of the embryonic body plan (Alkema et al., 1995; Schumacher et al., 1996; van der Lugt et al., 1994; Yu et al., 1995). Regulation by PcG is, however, considerably more complex in mammals than in *Drosophila*, as multiple variants of both PRC1 and PRC2 complexes are present, differing in their patterns of expression and targeting mechanisms (Blackledge and Klose, 2021; Gao et al., 2012). This variation is touched on below in the context of the mechanism of PcG repression.

Interest in regulation by PcG members was further heightened by the discovery of their central role in regulating somatic stem cell differentiation. Multipotent stem cell populations present in mature mammals can divide to give rise to a limited repertoire of cell types such as in the hematopoietic or neural lineages while also self-renewing to maintain the stem cell lineage (Fig. 8.17). Bmi-1, a member of the PRC1 complex, was shown to be required for self-renewal of hematopoietic and neural stem cells, while the PRC2 component EZH2 (the mammalian homolog of E(z)) is involved in differentiation of muscle and blood cell precursors (Lessard and Sauvageau, 2003; Molofsky et al., 2003; Park et al., 2003). These findings, together with the development of methods for the isolation and cultivation of totipotent ESCs, which are capable of developing into all tissue types of the adult organism (Keller, 2005), prompted examination of the role of PcG members in gene expression in human and murine ESCs. Genome-wide localization experiments revealed that genes associated with PcG

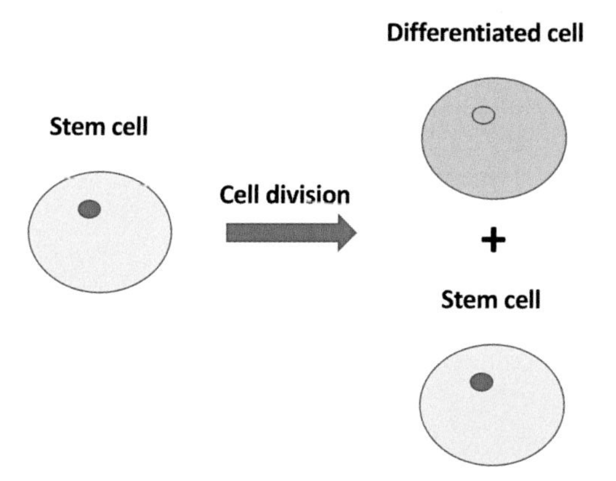

FIG. 8.17 Stem cell division yields a differentiated cell (which may further differentiate as it divides) and another stem cell again capable of producing differentiated progeny.

components in ESCs were associated with chromatin enriched for the H3K27me3 modification, were derepressed upon knockdown of PcG members, and were activated during cell differentiation (Boyer et al., 2006; Bracken et al., 2006; Lee et al., 2006; Tolhuis et al., 2006). Subsequent work showed that PcG members govern the differentiation of somatic stem cells in a similar fashion (Sauvageau and Sauvageau, 2010). Multiple variants of PRC1 and PRC2 complexes differing in specific subunits contribute to the regulation of different types of somatic stem cells (Piunti and Shilatifard, 2021). The literature concerning regulation of embryonic and adult stem cell differentiation by PcG members can fairly be described as vast; readers interested in exploring this topic further are referred to several excellent reviews (Aloia et al., 2013; Jaenisch and Young, 2008; Ming and Song, 2011; Pietersen and van Lohuizen, 2008; Rajasekhar and Begemann, 2007; Spivakov and Fisher, 2007; Surface et al., 2010).

In addition to revealing the overlap in ESCs between association of PcG proteins and H3K27me3, as well as H2AK119ub, investigations of the localization of histone modifications in ESCs uncovered a surprise: large regions of H3K27me3-enriched chromatin overlapped with regions harboring the H3K4me3 modification normally associated with actively transcribed genes (Azuara et al., 2006; Bernstein et al., 2006; Pan et al., 2007; Zhao et al., 2007). Such "bivalent domains" were also discovered in somatic stem cells and fibroblasts (Mikkelsen et al., 2007; Oguro et al., 2010). These bivalent marks occur at genes that are silent or expressed at low levels and are resolved upon differentiation to either H3K27me3 or H3K4me3 in lineages in which the associated genes are, respectively, repressed or activated (Mikkelsen et al., 2007).

With regard to the mechanism of PcG-mediated repression, recruitment of PcG complex proteins in *Drosophila* occurs at PcG response elements, or PREs, first identified by their ability to confer PcG-dependent repression on reporter genes (Muller and Bienz, 1991; Simon et al., 1993; Zink et al., 1991). Detailed studies of specific PREs revealed core PRE sequences of a few hundred base pairs, characterized by closely spaced binding sites for PRC components GAGA, Zeste, and Pho (Ringrose and Paro, 2004). Subsequent investigations identified several hundred PREs in the *Drosophila* genome, often extending into "PC domains" up to

150 kb in size, based on PcG member association (Schuettengruber et al., 2009; Schwartz et al., 2006; Tolhuis et al., 2006). An early study posited a model in which PcG-mediated repression was initiated by recruitment of PRC2 via interaction between Pho, the only PcG member initially found to exhibit sequence-specific binding to DNA, and E(z), followed by recruitment of PRC1 via interaction between the chromodomain of Polycomb and the H3K27me3 modification installed by E(z) (Wang et al., 2004b). However, later studies revealed a more complicated picture in which multiple sequence-specific DNA-binding proteins can recruit PRC2 and PRC1, with recruitment depending on cooperative interactions among both PcG members and on additional factors (GAGA factor and Pipsqueak/GAGA), and is influenced by the context of the binding site (Kahn et al., 2016; Kassis et al., 2017; Schuettengruber et al., 2014). Moreover, PRC1 recruitment to some PREs has been reported to occur in the absence of PRC2, and PRC2 association has been reported in the absence of PRC1 in ESCs (Kahn et al., 2016; Ku et al., 2008).

Recruitment of PcG complexes differs in mammals. In contrast to the sharp peaks of PcG proteins observed at PREs in flies in ChIP-seq experiments, occupancy is instead seen over broad regions enriched for H3K27me3 (e.g., Schorderet et al., 2013), and identification of DNA sequences conferring PRE-like properties in mammalian cells did not reveal a common recruitment mechanism (Kuroda et al., 2020; Schorderet et al., 2013; Sing et al., 2009; Woo et al., 2010, 2013). Instead, mammalian PRC1 and PRC2 can be recruited by distinct mechanisms that depend on the specific composition of the PRC, reflecting disparate roles played by variant PRC1 and PRC2 complexes in distinct cell types and developmental stages (Blackledge and Klose, 2021). These mechanisms include sequence-specific binding by PRC components, such as PCGF6 in the vPRC1 complex, or through transient interaction with TFs; RNA-mediated recruitment, as by Xist at the inactive X chromosome; and recruitment to CpG islands (see section on DNA methylation later in this chapter) (Blackledge and Klose, 2021). Once initiated, PcG complex association then spreads via binding of both PRC1 and PRC2 to H3K27me3, deposited by PRC2, or to H2AK119ub, deposited by PRC1, in a positive feedback mechanism reminiscent of that responsible for spreading of heterochromatin (Fig. 3.13) (Loh and Veenstra, 2022; Margueron et al., 2009).

While the detailed mechanism of repression implemented by the PcG complex association remains incompletely understood, it seems clear that a central feature is chromatin compaction (Simon and Kingston, 2013; Verrijzer, 2022). PRC1 association with chromatin results in compaction as visualized by EM both in vitro and in vivo, and the resulting chromatin is refractory to remodeling by SWI/SNF (Eskeland et al., 2010; Francis et al., 2004, 2001). Mutations that result in loss of chromatin compaction, such as deletion of the Ring1A/B subunits, are accompanied by derepression of PcG target genes in murine ESCs and during embryonic development (Eskeland et al., 2010; Lau et al., 2017). H2AK119 ubiquitylation appears critical to chromatin compaction and PcG-mediated repression in mammalian ESCs, but is not required for chromatin compaction or PcG-mediated repression in flies (Blackledge et al., 2020; Boyle et al., 2020; Eskeland et al., 2010; Pengelly et al., 2015; Tamburri et al., 2020). H2AK119 ubiquitylation appears to affect the dynamics of PcG-mediated repression, with the balance between ubiquitinated and unmodified H2A affecting chromatin compaction and likely being important for developmental stage-specific repression by PRC1 (Bonnet et al., 2022; Tsuboi et al., 2018; Verrijzer, 2022).

Given their central role in governing cell differentiation, it is perhaps not surprising that PcG members have been implicated in various cancers (Piunti and Shilatifard, 2021). For example, Bmi-1, a PRC1 subunit, is a critical regulator of the cell cycle inhibitors p16 and p19, and was first discovered in a screen for genes that cooperate with Myc in B cell transformation (Molofsky et al., 2003; Park et al., 2003; van Lohuizen et al., 1991). Another remarkable example, mentioned at the conclusion of Chapter 5, is the finding that pediatric diffuse intrinsic pontine glioma (DIPG) patients were found to harbor at high frequency a mutation converting Lys27 of H3.3 to Met (Schwartzentruber et al., 2012; Wu et al., 2012). A likely link to PcG-mediated regulation was immediately suspected, but some aspects were puzzling: Why was the K27M mutation in *H3F3A*, encoding H3.3, prevalent, while other K27 mutations were not observed? How were H3.3G34R and H3.3G34V, also observed, connected? What confined this pathway to uncontrolled cell proliferation to the tissues involved in glioma?

Partial answers to some of these questions have been uncovered. H3.3K27M acts as a dominant mutation by inhibiting PRC2 function, resulting in globally decreased levels of H3K27me2 and H3K27me3 in mouse and human cells (Chan et al., 2013; Lewis et al., 2013). Ectopic expression of all possible H3K27 mutants revealed the effect to be observed for K27M and to a lesser extent K27I, but for no other mutants (Lewis et al., 2013); a subsequent study found a single example of an H3.3K27I patient mutation associated with DIPG (Castel et al., 2015). Peptide binding and structural studies indicated a strong affinity of H3K27M for the catalytic site of the PRC2 component EZH2, suggesting a model in which the mutant H3.3, despite its relatively low abundance, sequesters PRC2 and inhibits its genome-wide activity (Bender et al., 2013; Jiao and Liu, 2015; Justin et al., 2016; Lewis et al., 2013).

Despite the overall decrease in H3K27me2/3, levels were increased at some loci, as was association of EZH2, resulting in altered expression of numerous genes involved in cancer pathways. It appears that the H3.3K27M mutant may mimic a recently discovered, naturally occurring inhibitor of PRC2 function named CATACOMB (Hubner et al., 2019; Jain et al., 2019; Piunti et al., 2019). H3.3G34V/R mutations, also found in the original cohorts in which oncogenic H3.3K27M was identified, differ in the tumor types in which they are found and appear to exert their effects by a distinct mechanism, with H3.3G34R acting by impeding recruitment of the co-repressor ZMYND11 to highly expressed genes in forebrain cells, resulting in uncontrolled proliferation (Bressan et al., 2021; Mackay et al., 2017; Sturm et al., 2012). Similarly, although the majority of K27M oncogenic H3 mutants have been identified in the H3.3 variant, H3.1K27M oncogenic mutations have also been observed but in tumors that differ clinically and molecularly from those harboring H3.3K27M (Mohammad and Helin, 2017). Clinical applications of these findings are being explored: an H3K27M peptide vaccine showed a specific immune response resulting in control of H3K27M+ tumors in mice, and has been tested in a Phase I clinical trial in patients with H3K27M+ diffuse midline gliomas with encouraging results (Grassl et al., 2023; Ochs et al., 2017). The study of oncohistones has accelerated rapidly since their original discovery and will doubtless yield important insights into the role of chromatin in developmental processes, and one may hope lead to clinical applications, in years to come.

Transcribing through chromatin

The recognition of the nucleosome core particle as the fundamental unit of chromatin reframed the problem of how Pol II could navigate a chromatin template. As discussed earlier in this chapter, the first investigations of Pol II transcription of chromatin templates revealed an inhibitory effect of nucleosomes, but shed no light on the mechanism of inhibition (Wasylyk and Chambon, 1979; Williamson and Felsenfeld, 1978). Experiments aimed at separating the effects of nucleosomes on transcriptional initiation and elongation, also discussed earlier, revealed that while transcription initiation by both bacteriophage SP6 RNA polymerase and eukaryotic RNA polymerase III was completely inhibited by nucleosomes, elongation could proceed, albeit at reduced efficiency relative to naked DNA (Felts et al., 1990; Hansen and Wolffe, 1992; Knezetic and Luse, 1986; Lorch et al., 1987; Losa and Brown, 1987; Morse, 1989).

A major advance in examining the effect of nucleosomes on transcriptional elongation by Pol II was made by Izban and Luse (1991), who generated stable elongation complexes by assembling Pol II initiation complexes on the adenovirus major late promoter on plasmid templates and allowing elongation to proceed in the presence of three of the four rNTPs. This produced complexes stalled at residue +15, just before the first C residue in the transcript. Remarkably, these stalled complexes are stable to the detergent sarkosyl, while other proteins are removed in its presence. The resulting complexes were washed with sarkosyl, assembled into chromatin using a *Xenopus* egg extract, purified, and incubated with the complete set of rNTPs to allow elongation to proceed. Production of full-length radiolabeled transcripts, visualized by gel electrophoresis, was strongly inhibited on chromatinized templates and was only modestly improved by the addition of Pol II elongation factors (Izban and Luse, 1991, 1992). Short transcripts arising from transcribed chromatin indicated that polymerase pausing occurred at specific sites; this pausing was attributed to the underlying DNA sequence, as pause sites did not depend on nucleosome positioning (Izban and Luse, 1991). (However, as we shall see, later work revealed that Pol II elongation is impeded at specific locations in the nucleosome.) Thus, in contrast to the ability of the small, extremely processive polymerase from bacteriophages to transcribe nucleosomal DNA, elongation by the much larger and more complex purified Pol II was severely inhibited in vitro.

With regard to the mechanism by which bacteriophage RNA polymerase could transcribe through a nucleosome, two early studies differed as to whether traversal of a nucleosome by SP6 RNA polymerase caused histone dissociation (Lorch et al., 1987; Losa and Brown, 1987). This difference was suggested to be due to differing stability of the nucleosome on the templates used, with the more stable nucleosomal template retaining histones following polymerase passage (Lorch et al., 1988); however, interpretation may have been complicated by the formation of an RNA-DNA hybrid on the in vitro template (Clark, 1995). This issue was eventually settled by experiments recounted below that showed retention of histones following transcription of nucleosomes by bacteriophage RNA polymerase in vitro.

Key to determining the fate of the histones following polymerase passage through nucleosomal templates was a series of papers from the Felsenfeld lab. These studies examined the mechanism in detail by using a strategy in which short linear fragments were reconstituted into nucleosomes having defined positions, such that the initiation site was accessible, and then either transcribed directly or after insertion into plasmids. Transcription by SP6 polymerase through a single positioned nucleosome resulted in an upstream shift of nucleosome position, as assayed by the protection of specific restriction sites before and after polymerase transit (Clark and Felsenfeld, 1992). Subsequent studies demonstrated that the upstream transfer by 40–95bp of nucleosomes transcribed by SP6 RNA polymerase occurred in *cis*, and a spooling mechanism was suggested in which nucleosomal DNA loops out during transcription and reengages with the histones following polymerase passage; thus, nucleosomes were proposed not to dissociate during transcription (O'Donohue et al., 1994; Studitsky et al., 1994). Pause sites seen using nucleosomal templates generally coincided with sites seen with naked DNA for both SP6 and T7 RNA polymerases, as previously reported for Pol II (Izban and Luse, 1991; O'Neill et al., 1992; Studitsky et al., 1995). A kinetic analysis, using paused elongation complexes on nucleosomal templates, revealed little pausing over the first 25 bp of nucleosomal DNA or after the transcribing polymerase reached the nucleosome dyad, and it was suggested that pausing was due to difficulty in transcribing the looped-out segment of DNA over the region from +25 to about +72 of the nucleosomal DNA (Studitsky et al., 1995). Further evidence for looping out of nucleosomal DNA during transcription derived from cryo-EM observation of paused complexes in which DNA both ahead of and behind the transcribing polymerase was seen in contact with the same histone octamer (Bednar et al., 1999).

While these early studies provided initial clues as to possible mechanisms by which RNA polymerase could transit chromatinized templates, it was clearly essential to investigate eukaryotic RNA polymerase directly. The first such test was conducted with RNA polymerase III using a nucleosomal template similar to the one used to examine transcription by SP6 RNA polymerase (Studitsky et al., 1997). As with SP6 polymerase, Pol III was able to transcribe through a single nucleosome, with the histone octamer transferred directly to the DNA behind the transcribing polymerase. Because Pol III is principally responsible for transcription of tRNAs and 5S rRNA, both of which are short transcripts whose coding sequences are partly occupied by Pol III transcription factors, it seemed likely that Pol III would not have to contend with nucleosomes in vivo, as was later borne out by genome-wide determination of nucleosome occupancy (e.g., Mavrich et al., 2008). Nonetheless, the ability of both prokaryotic RNA polymerases and Pol III to navigate nucleosomal templates demonstrated that nucleosomes did not constitute an absolute barrier to invasion by a processive enzyme—consistent also with the ability of *E. coli* Exonuclease III to digest nucleosomal DNA.

Two early ideas regarding the mechanism by which polymerase transits nucleosomes and the fate of transcribed nucleosomes involved the "splitting" of nucleosomes into half-nucleosomes, and the potential contribution of torsional stress induced by the transcribing polymerase on nucleosome stability (reviewed in Morse, 1992; Thoma, 1991). The idea that nucleosomes might split into half-nucleosomes upon being transcribed was based on nuclease accessibility, microscopic imaging, and exposure of a normally buried Cys residue in H3 to exogenous reagents (Bazett-Jones et al., 1996; Chen and Allfrey, 1987; Chen et al., 1990; Lee and Garrard, 1991; Prior et al., 1980). Some of these observations may have resulted from the loss of H2A–H2B dimers or from conformational changes less drastic than the putative splitting or unfolding of nucleosomes. In any event, subsequent biochemical experiments and recent cryo-EM structural results on complexes of nucleosomes with transcribing Pol II have not supported the splitting of nucleosomes.

A possible contribution of torsional stress to transcription through nucleosomes was based on findings, both theoretical and experimental in nature, that transcription generated positive supercoils ahead of and negative supercoils behind the transcribing polymerase (Liu and Wang, 1987; Pruss and Drlica, 1989). Positive supercoiling is unfavorable for nucleosome formation, while negative supercoiling is favorable; thus, the torsional stress resulting from polymerase movement could result in nucleosome loss in advance and reformation behind (Clark and Felsenfeld, 1991). However, experiments demonstrating that canonical nucleosomes could be formed on positively supercoiled templates, and that transcription through nucleosomes took place on short linear templates on which induced torsional stress was likely to be minimal, cast doubt on this mechanism (Clark and Felsenfeld, 1991; Felsenfeld et al., 2000). Single-molecule experiments examining the effect of torsional stress on nucleosomal templates formed on the Widom 601 nucleosome positioning sequence (Chapter 4) showed only modest effects on unwrapping of the DNA from the histones, but did indicate that positive supercoiling resulted in the loss of H2A–H2B dimers (Sheinin et al., 2013). Consistent with this result, inhibition of topoisomerase activity in *Drosophila* nuclei increased Pol II pausing just downstream of the +1 nucleosome, altered elongation kinetics, and caused a modest increase in depletion of H2A–H2B dimers from the +1 nucleosome of transcribed genes (Ramachandran et al., 2017; Teves and Henikoff, 2014). Thus, torsional stress generated by transcribing Pol II may contribute to removing nucleosomal roadblocks, but the magnitude of this contribution is not currently known.

Transcription by Pol II of short DNA fragments assembled into a single nucleosome downstream of arrested elongation complexes revealed both similarities and differences compared to bacteriophage polymerases and Pol III (Kireeva et al., 2002). At physiological salt concentration (150mM KCl) or lower, transcriptional elongation was almost completely blocked; experiments examining effects on elongation were therefore carried out at 300mM KCl, at which concentration about 30% of the arrested elongation complexes produced full-length transcripts after addition of NTPs. Consistent with earlier work (Izban and Luse, 1991), pause sites on nucleosomal templates were observed that coincided with those seen on naked DNA, but additional pause sites were also seen. Biochemical and single-molecule studies reported major pause sites on nucleosomal templates transcribed by Pol II at 15 and 45bp, with the strength of the pause site varying depending on the DNA sequence (Bintu et al., 2012; Bondarenko et al., 2006; Hodges et al., 2009; Kireeva et al., 2005). These pause sites correspond to the leading edge of the large Pol II complex encountering the two strongest histone-DNA contacts in the nucleosome, at the nucleosome dyad and 40bp upstream of the dyad, as determined in single-molecule experiments measuring the force required to "unzip" nucleosomal DNA (Hall et al., 2009; Kulaeva et al., 2013; Teves et al., 2014). Similar pausing was inferred to occur in vivo by comparing genome-wide, high-resolution localization of Pol II with nucleosome occupancy data (Churchman and Weissman, 2011; Kwak et al., 2013; Weber et al., 2014). An additional barrier ~13bp into the nucleosome was reported specifically for the +1 nucleosome in *Drosophila*, a pause site not observed in vitro, in budding yeast, or for nucleosomes at gene bodies in *Drosophila* (Weber et al., 2014). The nature of this highly specific barrier at the +1 nucleosome, which gave rise to a pause about three times greater in magnitude than seen at downstream nucleosomes, remains unclear.

Similarly to bacteriophage RNA polymerase and Pol III, histones were retained following the passage of Pol II, through a mechanism in which the transcribing polymerase is engaged with a short intranucleosomal DNA loop (Kireeva et al., 2002; Kulaeva et al., 2009). Rapid nucleosome reassembly had previously been supported by electron microscopic observations of nucleosomes in close proximity to stalled polymerase molecules and nascent RNA chains in SV40 minichromosomes and in the highly transcribed Balbiani ring genes from *Chironomus tentans* salivary glands (Björkroth et al., 1988; De Bernardin et al., 1986), and also by topological measurements showing no change in topology of closed circular minichromosomes harboring the highly transcribed *HSP26* gene following its induction in yeast (Pederson and Morse, 1990).

A novel feature of Pol II transcription through a nucleosome was the loss of an H2A–H2B dimer, resulting in the production of a hexasome containing a single H2A–H2B dimer together with the $(H3–H4)_2$ tetramer (Kireeva et al., 2002). This product was first observed as a novel nucleoprotein complex migrating faster than intact nucleosomes by native polyacrylamide gel electrophoresis and was shown to be a hexasome by demonstrating that it migrated equivalently to a hexasome reconstituted on the same fragment and migrated at the position of an intact nucleosome after addition of H2A–H2B dimers. Earlier studies had also suggested that transcription could result in H2A–H2B-deficient nucleosomes:

- About 15% of nucleosomes isolated from mouse myeloma cells were able to form a complex with Pol II in vitro, and these were enriched in transcribed sequences and depleted of H2A–H2B (Baer and Rhodes, 1983).
- H2A–H2B-deficient nucleosomes were transcribed more efficiently than intact particles by both *E. coli* RNA polymerase and Pol II (Gonzalez and Palacian, 1989).
- Depletion of H2A–H2B was seen in transcriptionally active genes in *Drosophila* in cross-linking experiments (Nacheva et al., 1989).
- Exchange rates for nucleosomal H2A–H2B were increased by transcription, whereas little effect was observed on the dynamics of H3–H4 association (Jackson, 1990).

Subsequent observations using atomic force microscopy of particles whose size matched that of hexasomes following in vitro transcription of nucleosomal templates by Pol II provided further corroboration for H2A–H2B depletion (Bintu et al., 2011). In vivo results also support the loss of H2A–H2B at transcribed regions. Increased lability of H2A–H2B dimers compared to H3–H4 is of course consistent with the biochemical and structural properties of the nucleosome (Chapter 2), and exchange rates for H2B were much greater than for H3 genome-wide in yeast (Jamai et al., 2007). More direct evidence for the loss of H2A–H2B dimers during transcription was provided by experiments monitoring the size of DNA fragments resulting from MNase digestion in budding yeast (Cole et al., 2014). These experiments yielded evidence for the presence of subnucleosomal particles together with a decrease in mononucleosomes over transcribed regions of genes strongly induced under histidine starvation conditions; this was corroborated by ChIP results showing H2B to be depleted compared to H4 at these same regions. Similarly, an examination of the size and localization of DNA fragments produced by MNase digestion in *Drosophila* cells identified subnucleosomal particles at the +1 nucleosome of transcribed genes (Ramachandran et al., 2017). These particles were shown to lack an H2A–H2B dimer by performing sequential ChIP-seq using cells expressing two versions of tagged H2A; recovery of subnucleosomal DNA (length 90, 103, and 112 bp) after the second IP was about half that seen for full length (147 bp) nucleosomal DNA corresponding to +1 nucleosomes, supporting the depletion of H2A–H2B dimers by elongating Pol II. (The ~50% recovery seen in sequential ChIP for shorter fragments presumably represents partially unwound nucleosomes containing the full complement of histones.)

Recent cryo-EM experiments demonstrate directly the loss of an H2A–H2B dimer at the proximal side of the nucleosome by elongating Pol II in the presence of elongation factors (discussed below), and a partial unwrapping of hexasomal DNA leaving ~90 bp in contact with the histones (Farnung et al., 2022). In addition, cryo-EM structures, together with biochemical experiments, confirm strong pause sites at SHL−5, when ~20 bp of nucleosomal DNA loses contact with the histones, and at SHL−1, near the nucleosome dyad, and a lack of pause sites downstream of the dyad (Fig. 8.18) (Kujirai et al., 2018; Weber et al., 2014).

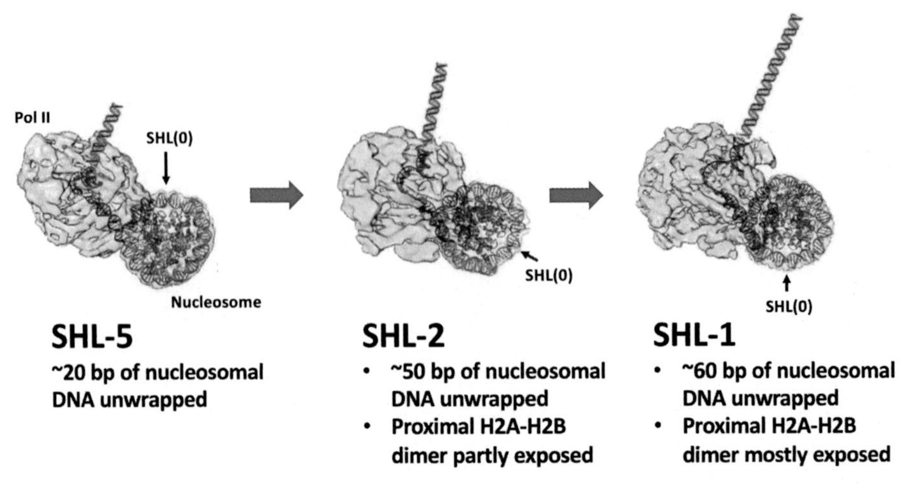

FIG. 8.18 Cryo-EM images of Pol II elongation through a nucleosomal template. The sequence shown is derived from structures in which Pol II is paused at the major pause sites at SHL−5 and SHL−1, and an additional minor pause at SHL−2. SHL(O) denotes the nucleosome pseudodyad. The straight red double helix represents the already transcribed, trailing DNA and was computationally added to the structures. Note that no ancillary factors were present in these structures. *Adapted from Kujirai, T., Ehara, H., Fujino, Y., Shirouzu, M., Sekine, S.I., Kurumizaka, H. 2018. Structural basis of the nucleosome transition during RNA polymerase II passage. Science 362, 595–598.*

Nucleosomes in complex with stalled elongation complexes were constructed by using a Widom 601 nucleosome-positioning sequence modified to lack thymidine residues (i.e. a "T-less cassette") up to the natural pause site at SHL−1, one helical turn upstream of the nucleosome dyad (Ehara et al., 2019; Kujirai et al., 2018), thereby preventing elongation past SHL−1 in the absence of ATP. Alternatively, a natural block to elongation occurring at this juncture was utilized to produce stalled elongation complexes (Farnung et al., 2018). Transcription was initiated using either a DNA-RNA hybrid "tail" upstream of the nucleosome or a "bubble" created by ligating a sequence having an unpaired region to the nucleosome. Because cryo-EM analysis entails the computational separation of distinct classes of image, structures of complexes paused at different sites could be elucidated. Strong pausing was seen at SHL−1 and SHL−5, consistent with in vitro biochemical experiments and in vivo observations discussed above. Kujirai et al. additionally obtained structures of complexes paused at SHL−2 and SHL−6, the latter just prior to polymerase entry to the nucleosome (Kujirai et al., 2018). The observed paused complexes correspond to barriers to elongation at sites of strong histone–DNA contacts, either with H2A–H2B (SHL−5) or H3–H4 (SHL−6, SHL−2, and SHL−1). Consistent with this notion, nucleosomes assembled using histone *sin* mutants (Chapter 6) having weakened histone-DNA interactions exhibit decreased inhibition to Pol II transit (Hsieh et al., 2010; Luse et al., 2011). Follow-up cryo-EM studies have reported structures of elongation complexes comprising Pol II together with various elongation factors at distinct stages of traversing a nucleosomal template (Table 8.4). These studies provide insight into the molecular details of the mechanism by which the histones are peeled away from DNA and reassembled behind Pol II in its transit of a nucleosomal template, and have been the topic of several recent reviews (Farnung, 2023; Francette et al., 2021; Kujirai and Kurumizaka, 2020; Mohamed et al., 2022; Robert and Jeronimo, 2023).

TABLE 8.4 Cryo-EM studies of Pol II elongation complexes.

Accessory factors[a]	Source[b]	SHL locations (Pol II front edge)[c]	Reference
Spt4–Spt5, Elf1, TFIIS	*K. pastoris*	NA	Ehara et al. (2017)
DSIF, NELF	Human (factors) and *Sus scrofa* (Pol II)	NA	Vos et al. (2018)
DSIF, PAF1c, SPT6	Human (factors) and *S. scrofa* (Pol II)	NA	Vos et al. (2020)
DSIF, PAF1c, RTF1, SPT6	Human (factors) and *S. scrofa* (Pol II)	NA	Vos et al. (2020)
None	*Saccharomyces cerevisiae*	−6	Farnung et al. (2018)
None	*K. pastoris*	−6, −5, −2, −1	Kujirai et al. (2018)
Elf1, Spt4–Spt5, TFIIS[d]	*K. pastoris*	−1, −5	Ehara et al. (2019)
Spt4–Spt5, Chd1, FACT[e]	*S. cerevisiae*	−4[g]	Farnung et al. (2021)
SPT6, DSIF, PAF1c, RTF1, TFIIS	Human (factors) and *S. scrofa* (Pol II)	−0.5	Filipovski et al. (2022)
SPT6, DSIF, PAF1c, RTF1, TFIIS[d,f]	Human (factors) and *S. scrofa* (Pol II)	−1	Farnung et al. (2022)
Spt6, Spn1, Elf1, Spt4–Spt5, Paf1C, FACT, TFIIS[d]	*K. pastoris*	−1, 0, +1, +6[h]	Ehara et al. (2022)

[a] *DSIF is the mammalian homolog of yeast Spt4–Spt5. Rtf1 is an integral component of yeast (including* K. pastoris*) Paf1C but is more loosely associated with mammalian PAF1c and was included as indicated.*
[b] *Human histones were used in the studies employing other components from* K. pastoris*, and recombinant histones from* Xenopus laevis *were used in the other studies.*
[c] *The superhelical location of the leading edge of Pol II, where SHL 0 is where the minor groove of DNA faces inward at the nucleosome dyad. The active site of Pol II is 15 bp behind the front edge. The label "NA" designates studies that did not use a nucleosome template.*
[d] *TFIIS included in elongation reaction but not present in the structure.*
[e] *Complexes contained either Chd1 or FACT.*
[f] *Structure with TFIIS obtained by adding TFIIS to the elongation complex.*
[g] *Hexasome containing backtracked Pol II.*
[h] *Two distinct structures were obtained for Pol II stalled at SHL 0 of the original nucleosome position, one with the original position preserved and one in which the nucleosome has moved 17 bp downstream. Two structures were also obtained at SHL +1, one containing the intact histone octamer and the other a hexasome.*

Elongation factors

Our discussion thus far has mentioned only in passing a critical aspect of elongation by Pol II in vivo: the contribution of accessory factors that participate in the transcription of chromatin in vivo. Findings that sequences transcribed in vivo were packaged into nucleosomes despite the impediment posed by nucleosomes to transcriptional elongation by Pol II in vitro suggested that some factor or factors in addition to Pol II must be necessary to facilitate transcriptional elongation in the cell. Subsequent investigations uncovered several distinct entities that function in transcriptional elongation through chromatin, as follows:

TFIIS: Originally purified from Ehrlich ascites tumor cells by virtue of its stimulatory activity toward transcription by Pol II, TFIIS (first known as "SII") was found to aid elongation by overcoming myriad barriers that result in pausing of Pol II, such as specific DNA sequences or lesions in the transcribed strand (Fish and Kane, 2002; Sekimizu et al., 1976). Upon encountering such barriers, Pol II backtracks, resulting in the extrusion of the 3′ end of the nascent RNA chain; recovery requires cleavage of the extruded RNA, which is stimulated by TFIIS interaction with the active site of Pol II (Cheung and Cramer, 2011). Backtracking by Pol II was shown to occur at nucleosomal pause sites, and nucleosomal inhibition of Pol II elongation in vitro was substantially suppressed by TFIIS but also required salt concentrations sufficient to partially destabilize histone-DNA contacts (Kireeva et al., 2005); only very modest stimulation was observed with TFIIS alone (Chang and Luse, 1997; Ehara et al., 2019; Izban and Luse, 1992; Kireeva et al., 2005). TFIIS was also required for a cooperative effect between two other factors, Elf1 and Spt4–Spt5 (discussed below) on transcriptional elongation of a nucleosomal template (Ehara et al., 2019). Pol II backtracking was observed just prior to reaching the nucleosome dyad and close to the stalling site induced by the use of a T-less cassette in a cryo-EM study; the addition of TFIIS resulted in cleavage of the extruded RNA, consistent with its facilitation of elongation through a nucleosomal template (Farnung et al., 2022).

RNA polymerase pausing and backtracking also occur in prokaryotes; resolution is achieved with help from the Gre transcription factors that, like TFIIS, facilitate cleavage of the extruded RNA (Abdelkareem et al., 2019). This functional homology (which is not reflected in structural homology) underscores the fact that although TFIIS assists Pol II traversal of chromatin, it does not directly alter nucleosome structure or influence Pol II interaction with the nucleosome.

Spt4–Spt5/DSIF: DSIF, or DRB sensitivity-inducing factor, was first identified in HeLa cells through its ability to confer sensitivity to an inhibitor of transcriptional elongation by Pol II (Wada et al., 1998). DRB (5,6-dichloro-1-β-D-ribofuranosylbenzimidazole), which was initially found as an inhibitor of heterogeneous nuclear RNA synthesis in metazoan cells, was later discovered to inhibit transcriptional elongation in vivo and in vitro using a HeLa cell extract, but did not affect transcription using a system reconstituted from purified components (Chodosh et al., 1989). Handa and colleagues leveraged these findings to assay for a factor conferring DRB sensitivity in vitro and identified a complex from HeLa cell extracts composed of two proteins of MW 14 and 160 kD that proved to be homologs of Spt4 and Spt5 from yeast (Wada et al., 1998). Spt4 and Spt5 had previously been discovered in the screen for suppressors of Ty insertion (*spt* mutants; Chapter 6) and were hypothesized to function in the establishment or maintenance of chromatin structure supporting active transcription (Swanson and Winston, 1992). Sequence similarity between Spt5 and the prokaryotic elongation factor NusG, together with phenotypic tests and physical association between Spt5 and Pol II, suggested that Spt4–Spt5/DSIF likely facilitated transcriptional elongation (Hartzog et al., 1998).

Spt5 and Pol II were found to colocalize on *Drosophila* polytene chromosomes by immunofluorescence, and high-resolution ChIP-chip and ChIP-seq studies showed Spt4–Spt5 association with transcribed gene bodies in budding yeast and mammalian cells, supporting a conserved role for Spt4–Spt5 in elongation (Andrulis et al., 2000; Kaplan et al., 2000; Mayer et al., 2010; Rahl et al., 2010; Shetty et al., 2017; Uzun et al., 2021). Structural and biochemical studies have confirmed this role, showing that Spt4–Spt5 associates with the Pol II elongation complex (Bernecky et al., 2017; Ehara et al., 2017; Farnung et al., 2018; Grohmann et al., 2011).

Analysis of Pol II occupancy, which increases at the 5′ end of active genes, in *spt4Δ* yeast, or after depletion of Spt5, together with observations of a downstream shift in nucleosome positioning over gene bodies in *spt4Δ* yeast, suggested that Spt4–Spt5 functions to aid transit of the initially elongating Pol II complex through the +2 nucleosome (Shetty et al., 2017; Uzun et al., 2021). Consistent with a conserved role in elongation, rapid depletion of Spt5 in mammalian cells resulted in decreased production of nascent transcripts and an accumulation of Pol II in the early gene body region (Henriques et al., 2018; Hu et al., 2021). In addition to its role in facilitating elongation, DSIF regulates the pausing of Pol II near the TSS in metazoans (discussed in Chapter 4), and thus has distinct functions in metazoans and yeast (Decker, 2021; Song and Chen, 2022). Recent in vitro studies also demonstrate that Spt4–Spt5 relieves the pausing of Pol II on nucleosomal templates and synergizes with Elf1 to facilitate the transcription of nucleosomal DNA (see below) (Crickard et al., 2017; Ehara et al., 2019). Altogether, then, a role for DSIF in facilitating Pol II elongation through nucleosomes near the TSS has been firmly established; mechanistic details as to how this function is achieved are beginning to emerge from high-resolution cryo-EM studies (Ehara et al., 2019; Farnung et al., 2022; Filipovski et al., 2022).

Spt6: Spt4, Spt5, and Spt6 were initially hypothesized to function as a complex regulating chromatin structure during transcription, based on their similar mutant phenotypes and on co-IP experiments showing interaction between Spt5 and Spt6 (Swanson and Winston, 1992). Subsequent work not only demonstrated genetic interactions of all three genes with Pol II mutants affecting elongation but also showed that the complex of Spt4 and Spt5 did not include Spt6 (Hartzog et al., 1998). This suggested closely related but nonidentical roles for these proteins in transcriptional elongation through chromatin, and this has been borne out in structural studies showing that Spt6 clamps DSIF to Pol II in the Pol II elongation complex (Ehara et al., 2017; Vos et al., 2018a,b). Additional support for a positive role for Spt6 in transcriptional elongation derived from experiments showing its localization on transcribed gene bodies in *Drosophila* and yeast (Andrulis et al., 2000; Ivanovska et al., 2011; Kaplan et al., 2000; Kim et al., 2004; Krogan et al., 2002; Mayer et al., 2010). Moreover, inactivation or depletion of Spt6 impairs Pol II processivity in vivo in yeast, *Drosophila*, and human cells (Ardehali et al., 2009; Narain et al., 2021; Pathak et al., 2018; Zumer et al., 2021).

A specific role for Spt6 in the transcription of chromatin was initially suggested by the similar phenotypes of *spt6* and histone gene mutants (the latter discussed in Chapter 5) and gained additional support with findings that Spt6 physically interacts with histones H3 and H4 and is capable of facilitating nucleosome assembly in vitro (Bortvin and Winston, 1996; Winston and Carlson, 1992). Additional in vivo evidence for regulation of chromatin structure by Spt6 was provided by observations of increased sensitivity to MNase and altered nucleosome positioning over transcribed regions in *spt6* mutants in both *S. cerevisiae* and *S. pombe* (DeGennaro et al., 2013; Doris et al., 2018; Ivanovska et al., 2011; Jeronimo et al., 2019; Kaplan et al., 2003; van Bakel et al., 2013). In human cells, rapid depletion of Spt6 resulted in Pol II accumulation at the barrier to elongation created by the +1 nucleosome (Narain et al., 2021; Zumer et al., 2021).

The ability of Spt6 to facilitate nucleosome assembly in vitro suggested that Spt6 might function in nucleosome disassembly and/or reassembly during transcription. This notion found unexpected support in the observation of cryptic initiation from within transcribed sequences in *spt6* mutants (Kaplan et al., 2003); this phenomenon is discussed in more detail below.

Structural, biochemical, and genetic experiments revealed Spt6 to cooperate with FACT (FAcilitates Chromatin Transcription), DSIF, and Iws1 (also known as Spn1; FACT and Iws1/Spn1 are discussed below) to facilitate retention of histones during traversal of nucleosomal DNA by Pol II (Ehara et al., 2022; Farnung et al., 2022; Filipovski et al., 2022; Gopalakrishnan and Winston, 2021; Miller et al., 2023). In addition to being important to preventing initiation from so-called cryptic promoters within transcribed gene bodies (discussed below), proper reassembly of nucleosomes following transit of Pol II is also important for preserving localization of histone variants and histone modifications. In *S. pombe*, methylated H3K4 is lost from transcribed loci in *spt6* mutants and replaced by H3K56ac, characteristic of newly incorporated H3 (Kato et al., 2013), while in budding yeast, mutations disrupting activity of Spt6 or FACT result in both mislocalization of H2A.Z and scrambling of histone modifications (Jeronimo et al., 2019, 2015). Finally, it should be mentioned that Spt6 functions in both yeast and mammalian cells in DNA replication and guarding genome integrity, as well as playing additional roles in transcription including initiation and co-transcriptional mRNA processing (Miller et al., 2023).

Elf1 and Iws1 (Spn1 in S. cerevisiae): Elf1 was identified in a screen for yeast mutants exhibiting synthetic lethality with combined mutations in *SPT6* and *DST1* (encoding TFIIS) (Prather et al., 2005). Mutations in several other elongation factors were also found to be synthetically lethal with *elf1Δ* mutations, and high-resolution ChIP-chip experiments showed Elf1 as well as Spn1 to localize over transcribed gene bodies in budding yeast (Mayer et al., 2010). The mammalian homolog of Elf1, ELOF1, also facilitates transcriptional elongation in vivo, and is important for transcription-coupled DNA repair (Geijer et al., 2021; van der Weegen et al., 2021). Spn1/Iws1 was identified first in yeast in two independent studies, one using a targeted screen for mutants suppressing a postinitiation defect in transcription and the other a large-scale proteomics screen for proteins associating with various elongation factors (Fischbeck et al., 2002; Krogan et al., 2002). The latter report designated the new protein as Iws1, for Interacts with Spt6, and this is the name that has been used for its metazoan homologs (Miller et al., 2023). Elf1 and Iws1/Spn1 each show high evolutionary conservation across eukaryotes, and both have been situated in the context of an elongation complex that includes FACT, Spt4–Spt5, Spt6, and the Paf1 complex (discussed below) (Ehara et al., 2022). Elf1 was observed to cooperate with Spt4–Spt5 in traversal of a nucleosomal template by Pol II in vitro, as when both factors were present, stalling at SHL−2 and SHL−6 was alleviated, while facilitation of elongation on a nucleosomal template by Elf1 or Spt4–Spt5 alone was much less pronounced (Ehara et al., 2019). (TFIIS is also required for alleviation of nucleosomal inhibition by these factors, as mentioned earlier.) Structural data suggest that Elf1 and Spt4–Spt5/DSIF together prevent the trapping of nucleosomal DNA by elongating Pol II to alleviate nucleosome-induced pausing, while Iws1 may facilitate Pol II entry to the nucleosome by binding H3–H4 and disrupting histone–DNA contacts in advance of the approaching Pol II (Kujirai and Kurumizaka, 2020; Miller et al., 2023).

FACT: FACT was discussed in Chapter 5 in its role as an essential and evolutionarily conserved histone chaperone, with a preference for binding H2A–H2B, in the context of nucleosome assembly and disassembly. FACT was isolated from HeLa cells through its ability to facilitate transcript elongation in vitro on a chromatin template assembled using a *Drosophila* assembly extract, in a reaction using purified Pol II and the general transcription factors

(Orphanides et al., 1998). FACT comprises a heterodimer of SSRP1 (Pob3 in yeast) and Spt16 (Orphanides et al., 1999); thus, like several other of the elongation factors discussed here, FACT was implicated in regulating transcription in vivo through the effect of one of its components in suppressing the inhibitory effect of a Ty element insertion on downstream transcription (Malone et al., 1991). FACT co-localizes with Pol II over transcribed gene bodies in yeast and *Drosophila* (Kim et al., 2004; Mason and Struhl, 2003; Mayer et al., 2010; Pathak et al., 2018; Saunders et al., 2003); however, FACT-deficient yeast did not show elongation defects monitored at specific loci, perhaps due to redundancy with other elongation factors (Biswas et al., 2006; Fleming et al., 2008; LeRoy et al., 2019; Mason and Struhl, 2003).

FACT primarily engages H2A–H2B in its role in replication (Chapter 5), and the same is true for its transcription elongation function. Removal of an H2A–H2B dimer from a nucleosomal template by FACT during polymerase passage was found to facilitate elongation in one in vitro study, which also demonstrated histone chaperone activity of FACT, while another investigation reported alleviation of polymerase pausing by FACT at sites within intact nucleosomes but not in $(H3–H4)_2$ tetramer particles (Belotserkovskaya et al., 2003; Hsieh et al., 2013). The latter study also demonstrated reduced disruption of nucleosomes (monitored by production of free DNA) after polymerase passage when FACT was present and concluded that interactions of FACT with the H2A–H2B dimer at the proximal side of the nucleosome facilitated both passage of Pol II and efficient reassembly of the nucleosome behind the elongating polymerase (Hsieh et al., 2013). A role for FACT in facilitating nucleosome retention during transcription was supported by genetic interactions with nucleosome assembly factors, by its requirement to prevent cryptic initiation from within transcribed gene bodies (discussed below), and by the mislocalization of H2A.Z and scrambling of histone modifications over transcribed regions in *spt16* mutants (Formosa et al., 2002; Jeronimo et al., 2019, 2015; Kaplan et al., 2003). Structural and biochemical studies have revealed molecular details of the interaction of FACT with H2A–H2B (Kemble et al., 2015), and a cryo-EM study of a FACT-nucleosome complex revealed a structure resembling a unicycle in which the human FACT heterodimer sits atop the DNA-wrapped $(H3–H4)_2$ tetrasome and one or two H2A–H2B dimers (in a hexasome or nucleosome) form flanking pedals, with both SSRP and Spt16 subunits contacting nucleosomal DNA and histones (Liu et al., 2020). Cryo-EM of FACT in conjunction with a Pol II elongation complex on a nucleosomal template revealed FACT binding to an H2A–H2B dimer that was exposed by partial unwrapping of nucleosomal DNA; this binding stabilized the association of H2A–H2B with the transcribed nucleosome (Farnung et al., 2021). A recent *tour de force* investigation determined high-resolution cryo-EM structures of elongating Pol II, in association with Spt6, Spn1, Elf1, Spt4–Spt5, the Paf1 complex (discussed below), and FACT (all from the yeast *K. pastoris*), stalled at multiple positions in a transcribed nucleosome (Ehara et al., 2022). Analysis of elongation complexes paused at 42, 49, 58, and 115 bp from the nucleosome entry site (corresponding to the leading edge of polymerase being located near the dyad, at SHL−1, SHL0, and SHL+1, or farther downstream at SHL+6) revealed initial stabilization of the H2A–H2B dimer by interaction with Spt16; transcription further into the nucleosome was accompanied by transfer of the histone octamer or hexamer to DNA upstream of Pol II (Fig. 8.19).

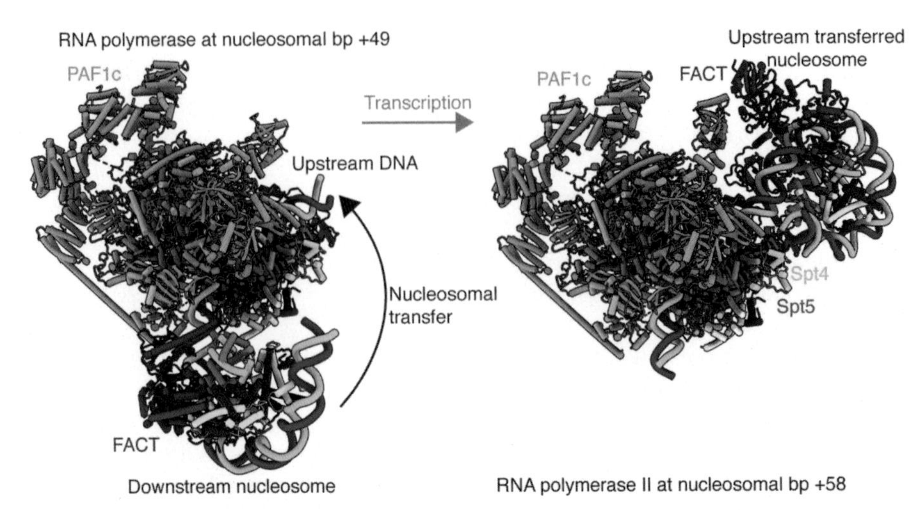

FIG. 8.19 Nucleosome transfer during Pol II passage visualized by cryo-EM structures. Pol II, in complex with Spt4-Spt5-Spt6-PaflC-Elfl-Spnl-FACT, paused at SHL(0) *(left)* and SHL+1, at which point the nucleosome has been transferred to the trailing DNA upstream of Pol II. *From Farnung, L. 2023. Nucleosomes unwrapped: Structural perspectives on transcription through chromatin. Curr. Opin. Struct. Biol. 82, 102690.*

Consistent with the early studies showing in vitro transcription of nucleosomal DNA by Pol II in the absence of accessory factors (albeit at elevated salt concentrations) and reassembly of nucleosomes behind the elongating polymerase, structural and biochemical studies have revealed multiple pathways for transcription through nucleosomes and nucleosome reassembly. The chromatin remodeler Chd1 can collaborate with a Pol II elongation complex to facilitate transcription through a nucleosome, and a structure in which DNA both upstream and downstream of a Pol II elongation complex contacts the histone octamer core was observed in the absence of FACT (Farnung et al., 2021; Filipovski et al., 2022). This latter structure suggests a mechanism for nucleosome retention independent of FACT. The likely existence of multiple pathways for transcribing chromatin and preserving nucleosome structure provides a possible explanation for the relatively minor effects of FACT impairment on elongation in vivo mentioned earlier. Moreover, it may also be relevant in cells in which FACT is absent or present at low levels (LeRoy et al., 2019).

In addition to its roles in replication of chromatin and in facilitating polymerase passage through chromatin and reassembly of nucleosomes during transcriptional elongation, FACT has been implicated in regulating transcriptional initiation and polymerase pausing; all of the various functions of FACT are likely related to its ability to facilitate nucleosome disruption and reassembly through its association with the H2A–H2B dimer (Formosa and Winston, 2020).

Paf1C: The founding members of the yeast Paf1 complex, Paf1 and Cdc73, were implicated in gene expression through their physical association with immobilized Pol II, along with other known components of the transcription machinery present in yeast cell extracts (Wade et al., 1996). Inspired by reports of Pol II forming a holoenzyme in association with the Mediator complex and by altered expression of a subset of genes in *paf1* and *cdc73* mutants, the Jaehning lab initially theorized that the Paf1 complex, now designated Paf1C in yeast and PAF1C or PAF1c in metazoans, might similarly function as a transcriptional coactivator (Shi et al., 1997, 1996). Later investigations, however, identified additional proteins belonging to Paf1C and revealed genetic and physical interactions between Paf1C constituents and known elongation factors, pointing to a role for Paf1C in transcriptional elongation together with Spt4–Spt5 and FACT (Costa and Arndt, 2000; Krogan et al., 2002; Mueller and Jaehning, 2002; Squazzo et al., 2002). ChIP experiments confirmed the association of Paf1C components with coding regions at select genes and genome-wide, and in vitro transcription assays using yeast extracts showed elongation on a naked DNA template was reduced by ~6fold using extracts from *paf1Δ* or *cdc73Δ* cells (Krogan et al., 2002; Pokholok et al., 2002; Rondon et al., 2004). Human PAF1c was also observed to stimulate transcriptional elongation in an in vitro assay on a chromatin template, and a synergistic effect was observed between PAF1c and TFIIS, although the stimulatory effect seen with chromatin was not compared with a naked DNA template in this study (Kim et al., 2010). In vivo, genome-wide monitoring of nascent transcription in mouse myoblast cells showed that depletion of Paf1 resulted in decreased Pol II elongation rates (Hou et al., 2019). Although these results and structural studies (see below) indicate a conserved role for the Paf1 complex in transcriptional elongation, metazoan PAF1c differs from the yeast complex in some respects: the subunit Rtf1 is an integral component of yeast Paf1C but is more loosely associated with the mammalian complex, the two complexes differ in their distribution over transcribed regions, and the mammalian complex has also been implicated in regulating polymerase pausing, which is not observed in yeast (Francette et al., 2021; Kim et al., 2010; Mayer et al., 2010; Yang et al., 2016; Zhu et al., 2005).

Histone methyltransferases were identified at about the same time that the role of Paf1C in transcriptional elongation emerged, and a screen in which extracts from ~4800 nonessential yeast gene deletion mutants were tested by Western blotting revealed Paf1 and Rtf1 to be essential for methylation of H3K4 and H3K79 (Krogan et al., 2003a). An independent study obtained evidence connecting recruitment of the H3K4 methyltransferase to transcriptional elongation, prompting experiments showing loss of H3K4me3 in *paf1Δ* and *rtf1Δ* yeast (Ng et al., 2003b). The discovery that H2B ubiquitination was required for methylation of H3K4 and H3K79 then led to the demonstration that Paf1C was essential for conferring the H2Bub modification (see Chapter 7, section on Cross-talk) (Ng et al., 2003a; Wood et al., 2003). The mechanism by which Paf1C facilitates H2B ubiquitination in yeast involves recruitment of the ubiquitin-conjugating enzyme Rad6 and the ubiquitin ligase Bre1, and primarily resides in Rtf1; the protein domains and interactions responsible have been dissected in detail (Francette et al., 2021; Van Oss et al., 2016).

In addition to (or contributing to) its roles in transcriptional elongation and histone modifications, the Paf1 complex has also been implicated in phosphorylation of Ser2 of the Pol II C-terminal domain (CTD), a critical event in transcriptional elongation, to splicing, RNA cleavage, and polyadenylation, and to recruitment of the chromatin remodelers Chd1 and Ino80 (Francette et al., 2021). The complexity of the web of interactions responsible for the multifunctional nature of the Paf1 complex is underscored and to some extent illuminated by high-resolution cryo-EM studies of variants of the transcriptional elongation complex, assembled with naked DNA or nucleosomal templates, supplemented by information regarding protein-protein interactions derived from cross-linking studies (Farnung, 2023;

Francette et al., 2021; Kujirai and Kurumizaka, 2020; Mohamed et al., 2022). An elongation complex comprising Pol II together with Spt6, Spn1, Spt4–Spt5, Elf1, FACT, and Paf1C (from *K. pastoris*) showed the Pol II surface to be almost entirely covered by these accessory factors, consistent with the cooperative interactions among them mentioned in previous discussion (Ehara et al., 2022). With regard to the role of the Paf1 complex in the transcription of chromatin templates, the Ctr9 subunit from the human complex was closely apposed to DNA at SHL−2 in the partially unraveled hexasome structure and was found to contact DNA in a structure in which the transcribed nucleosome is rewrapped behind Pol II (Fig. 8.19); this together with contacts made by Pol II stabilizes the rewrapped nucleosome in *cis* (Farnung et al., 2022; Filipovski et al., 2022).

An additional factor deserving mention in the context of the present discussion is the Elongator complex. Elongator was initially identified in yeast as a three-subunit complex associating with the hyperphosphorylated form of Pol II characteristic of the elongating enzyme, and its subunits were named Elp1, Elp2, and Elp3 (Otero et al., 1999). However, despite phenotypic evidence for a role in transcriptional regulation and for HAT activity toward H3 and H4, Elongator was found to be principally localized to the cytoplasm and was not detected over transcribed open reading frames, whereas other elongation factors were, raising serious doubts as to whether "Elongator" actually functioned in transcriptional elongation (Pokholok et al., 2002). Additional roles for Elongator emerged from subsequent studies, and it now appears that its main, and possibly exclusive, function in yeast as well as in metazoans is as a tRNA-modifying enzyme (Karlsborn et al., 2014; Lin et al., 2019; Svejstrup, 2007).

Cryptic transcriptional initiation

Elongating Pol II associates with an array of factors that not only facilitate its traversal of nucleosomal DNA, but are critical for reassembly of nucleosomes in its wake. This reassembly is important in preventing aberrant transcriptional initiation from within genic coding sequences, and a complex and conserved machinery has evolved to ensure the suppression of such "cryptic" intragenic initiation.

Cryptic transcription arising from defects in intragenic chromatin structure should be distinguished from pervasive transcription occurring throughout eukaryotes, which was discovered in large-scale sequencing efforts soon after the initial discovery of cryptic intragenic transcription initiation in yeast (Jensen et al., 2013; Smolle and Workman, 2013). Pervasive transcripts in yeast have been categorized according to whether they are found in nonmutant cells (stable unannotated transcripts, or SUTs) or under conditions of impaired RNA degradation pathways (cryptic unstable transcripts, or CUTs, and Xrn1-sensitive unstable transcripts, or XUTs), although there is some overlap in these categories. Pervasive transcripts generally do not encode proteins, and like the cryptic intragenic transcripts discussed below, nearly always arise from initiation sites depleted of nucleosomes.

Cryptic transcripts were discovered in an investigation in which the effect of an *spt6 ts* mutant on transcript levels was examined using microarray technology (Kaplan et al., 2003). Northern analysis of several transcripts conducted to corroborate microarray results revealed a surprise: in many cases novel transcripts smaller than the parent transcript being assessed were detected, suggesting aberrant initiation or termination events. Further analysis of one such short transcript revealed it to arise from within the *FLO8* coding sequence, originating from a cluster of initiation sites 1.7kb downstream of the *FLO8* initiating ATG codon. The same novel short transcript was also observed in several additional yeast mutants that had previously been found to allow transcription of the *SUC2* promoter lacking its UAS, specifically in mutants affecting the level of H2A and H2B, and in an *spt16* mutant, thus also implicating FACT in suppressing intragenic transcription. A likely TATA element was noted 54bp upstream of the intragenic *FLO8* ATG and was shown to be required for generation of the short *FLO8* transcript. These findings suggested that the aberrant transcript might result from the unmasking of an internal promoter that was normally occluded by nucleosomes. Support for this idea came from experiments showing increased MNase sensitivity in the *spt6 ts* mutant compared to wild-type cells that was discernible for total cellular chromatin, but was more pronounced for the *FLO8* coding region and less so for the nontranscribed *GAL1* locus. The increased MNase sensitivity was suppressed by the *rpb1-1 ts* mutation that inactivates Pol II at the restrictive temperature, showing that it depended on active transcription. A subsequent exhaustive screen identified mutations in 50 genes that allow cryptic transcription, including the core histones and genes regulating their expression, and genes encoding proteins involved in chromatin remodeling, nucleosome assembly, and transcriptional elongation (Cheung et al., 2008). Cryptic transcripts were widespread in these mutants, with over 1000 genes exhibiting aberrantly initiated transcripts having intragenic origin. High-resolution determination of transcription start sites in *spt6* yeast later revised this number upwards, revealing ~6000 upregulated intragenic initiation sites on transcript sense strands in addition to upregulated initiation sites at sense and antisense promoter sites and antisense intragenic sites (Doris et al., 2018). Nucleosome occupancy and positioning were perturbed on a genome-wide scale in *spt6* mutant yeast; moreover, delocalization of TFIIB with spreading into transcribed regions was observed, and intragenic start sites were associated with sequences enriched for features characteristic of canonical TSSs, consistent with promiscuous assembly of PICs leading to cryptic initiation (Doris et al., 2018; Pattenden et al., 2010). An important consequence of the loss of Spt6 function was downregulation of transcription from the majority of genic promoters, with accompanying decreased association of TFIIB (Doris et al., 2018). This finding led the authors to suggest that in *spt6* mutant yeast, mislocalization of limiting components of the PIC results in decreased PIC assembly at genic promoters and concomitantly decreased transcription. Kornberg and Lorch have taken this idea to its logical conclusion, arguing that the primary role of the nucleosome is not in packaging the genome to fit the nuclear volume, but in preventing mislocalization of the transcription apparatus, whether to repressed promoter sites or intragenic locations, where maintenance of chromatin structure that is refractory to PIC assembly is enforced by Spt6 and associated factors (Kornberg and Lorch, 2020).

Following the discovery of cryptic intragenic initiation in yeast, several labs reported experiments revealing its suppression to depend on an elaborate network of interactions centered on restoring a repressive chromatin structure following the passage of Pol II (Fig. 8.20). In this mechanism, suppression of cryptic intragenic transcription depends on methylation of H3K36 by Set2 at nucleosomes present in transcribed chromatin; Set2 is recruited to

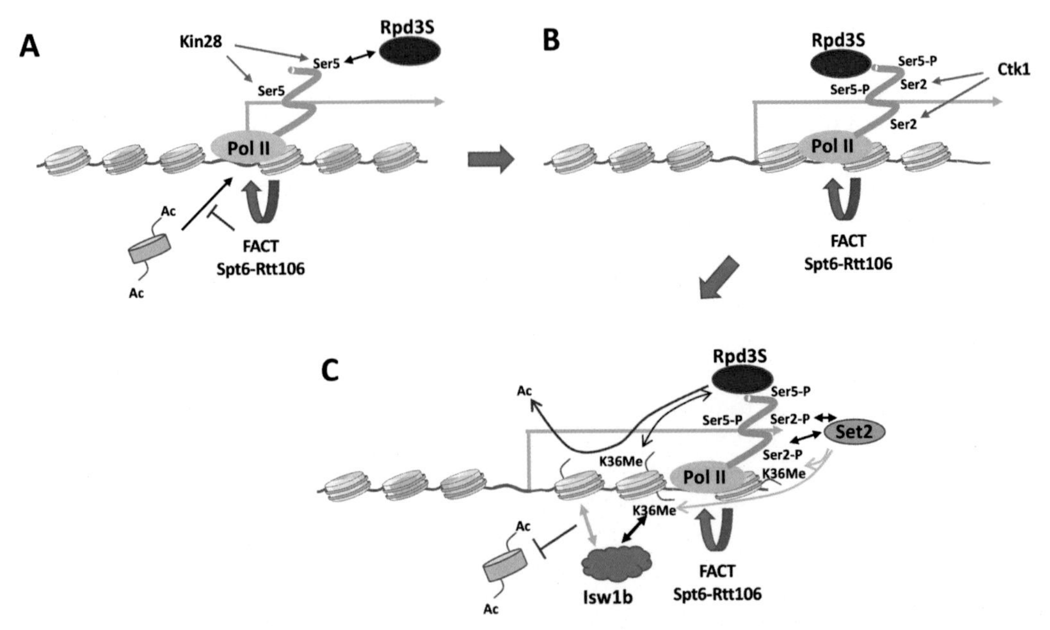

FIG. 8.20 Inhibition of cryptic intragenic transcription by the Set2-Rpd3S pathway. The green arrow represents the transcribed region, with Pol II proceeding from the initiation site in (A) into the coding region in (B and C). (A) Nucleosome reassembly behind transcribing Pol II facilitated by FACT and Spt6 *(curved blue arrow)* blocks incorporation of free, acetylated histones, while Ser5 phosphorylation of the Pol II CTD by Kin28 (CDK7 in mammals) facilitates association of Rpd3S. (B) As Pol II elongation begins, Ser2 of the CTD is phosphorylated in yeast by Ctk1 (and by Bur1 near the initiation site); in metazoans, multiple CDKs may contribute to Ser2 phosphorylation. (C) Set2 is recruited by association with Ser2-P and catalyzes di- and trimethylation of H3K36 *(orange arrows)*; this modification leads to recruitment of Isw1b, which promotes retention of nucleosomes, and stimulates the HDAC activity of Rpd3S toward nearby histones. Together, these events act to ensure a "closed" chromatin configuration that is refractory to cryptic initiation. *(B) Jeronimo, C., Collin, P., Robert, F. 2016. The RNA polymerase CTD: the increasing complexity of a low-complexity protein domain. J. Mol. Biol. 428, 2607–2622.*

transcribed chromatin by Pol II that is phosphorylated at Ser2 of the CTD of the large subunit of elongating Pol II (Krogan et al., 2003b; Li et al., 2003; Schaft et al., 2003; Xiao et al., 2003). The small Rpd3-containing complex, Rpd3S, is recruited to transcribed chromatin by elongating Pol II via its characteristic Ser-5 phosphorylated CTD and is activated by di- and tri-methylated H3K36-containing nucleosomes, whereupon it deacetylates nucleosomal histones that are acetylated during the course of Pol II transit (Drouin et al., 2010; Govind et al., 2010). This deacetylation counters the more open chromatin structure facilitated by histone acetylation and thus suppresses access by the transcription machinery that results in aberrant transcription initiation. In addition, H3K36me3 suppresses histone exchange, consistent with turnover of histones H3 and H4 at all but the most highly transcribed coding regions being lower than that observed at promoters (Dion et al., 2007; Rufiange et al., 2007). This suppression of histone exchange is effected in part by the reduced affinity of histone chaperones that could remove histones and increased affinity of the remodeler Isw1b, which promotes nucleosome retention, for H3K36me3-containing chromatin (Venkatesh et al., 2012; Venkatesh and Workman, 2015).

A variety of genetic and biochemical experiments led to the formulation of this scenario. The histone deacetylase Rpd3, discussed at length in Chapter 7, was initially reported to be present in a large protein complex, and later found to belong to two distinct complexes of ~0.6 and 1.2 MDa (Kasten et al., 1997; Lechner et al., 2000; Rundlett et al., 1996). While several proteins interacting with Rpd3 were known, the composition of the two complexes was not. Two independent investigations succeeded in determining the composition of both the large (Rpd3L) and small (Rpd3S) complexes, and found two subunits, Eaf3 and Rco1, that were present in the Rpd3S but not the Rpd3L complex (Carrozza et al., 2005; Keogh et al., 2005). One of the proteins unique to Rpd3S, Eaf3, is also a component of the NuA4 HAT, and *eaf3Δ* yeast correspondingly exhibit decreased acetylation of histones in intergenic regions (Reid et al., 2004). However, *eaf3Δ* yeast also show increased acetylation in coding regions, and Eaf3 possesses a chromodomain, suggesting that it might be responsible for the recruitment of Rpd3S to these regions by interaction with methylated H3K36. This idea was supported by a variety of evidence:

- A comparison of changes in gene expression of 382 yeast deletion strains with *eaf3Δ* and *rco1Δ* yeast revealed the most closely correlated mutant to be *set2Δ* (Keogh et al., 2005).
- Increased acetylation of histones H3 and H4 was seen by ChIP over coding regions of several transcribed genes, but not at promoters, in *eaf3Δ* and *set2Δ* yeast, as well as in H3K36A mutant yeast and yeast expressing Eaf3 lacking its chromodomain (Carrozza et al., 2005; Joshi and Struhl, 2005; Keogh et al., 2005; Reid et al., 2004).
- Mononucleosomes from wild-type but not *set2Δ* yeast were able to precipitate Eaf3 from yeast cell extracts, but failed to do so when the Eaf3 chromodomain was mutated (Keogh et al., 2005).
- Set2 recruitment to gene bodies did not depend on Eaf3, and H3K36me2 over the transcribed *STE11* coding region was diminished in *set2Δ* yeast but not in *eaf3Δ* yeast, showing that H3K36 methylation by Set2 occurred upstream of Rpd3S recruitment (Carrozza et al., 2005; Keogh et al., 2005).
- Short transcripts characteristic of cryptic intragenic initiation at *FLO8* and *STE11* genes were observed in mutants affecting Rpd3S, including an *eaf1Δchd* mutant lacking the Eaf1 chromodomain, and in *set2Δ* and H3K36A mutant yeast (Carrozza et al., 2005).
- A number of control experiments showed that increased acetylation, Rpd3S recruitment, and cryptic initiation were specific to mutations affecting the Set2-Rpd3S pathway, and were not observed in mutants affecting Rpd3L or the histone methyltransferases Dot1 and Set1 (Carrozza et al., 2005; Joshi and Struhl, 2005; Keogh et al., 2005).

These results firmly established the Set2-Rpd3S pathway as a suppressor of cryptic intragenic transcription in yeast. Subsequent studies, as often is the case, resulted in some revisions to this model (Molenaar and van Leeuwen, 2022). It was initially reported that in addition to increasing cryptic transcription and histone acetylation over gene bodies, *set2Δ* and H3K36A mutations also resulted in reduced Eaf3 occupancy at the *FLO8* coding sequence, leading to the conclusion that Rpd3S recruitment required H36 methylation by Set2 (Carrozza et al., 2005). However, a later study reported little or no effect on the occupancy of the Rpd3S-specific subunit Rco1 at the *STE11* and *FLO8* genes in *set2Δ* or *eaf3Δ* yeast (Govind et al., 2010). A contemporaneous genome-wide analysis found Rco1 associated with a subset of transcribed genes that also bound Rpd3L at their 5′ ends, with only about a third of that subset exhibiting a reduction, but not complete loss, of Rco1 association in *set2Δ*, *eaf3-chd-Δ*, and H3K36A mutants (Drouin et al., 2010). Instead, both studies found that recruitment depended on the Ser5-phosphorylated form of Pol II, with Drouin et al. also reporting a strong dependence on Spt4. These results pointed to a mechanism in which Rpd3S is recruited to the gene bodies principally by the elongating, phosphorylated Pol II, while its deacetylase activity requires interaction with methylated histones (Fig. 8.20). In addition, a distinct pathway that involves the Set3 histone deacetylase complex and its interaction with H3K4me2 was also shown to suppress cryptic transcription initiation from within transcribed gene bodies, but at sites distinct from those suppressed by the Set2-Rpd3S pathway (Kim et al., 2012).

Cryptic intragenic initiation has also been found in metazoans and plants, although aspects of the mechanism differ from yeast (Fig. 8.20). Knockdown of SETD2 in HeLa cells resulted in decreased H3K36me2/3 over downstream regions of transcribed genes and concomitant cryptic initiation in at least 11% of active genes, despite histone acetylation being unchanged either genome-wide or at gene regions exhibiting reduced H3K36me3 levels (Carvalho et al., 2013). (In these experiments, cryptic initiation was inferred by observation of Pol II occupancy increasing at downstream exons while remaining unchanged at the first exon and promoter region, rather than by Northern analysis; this permitted analysis based on ChIP-seq data.) Knockdown of SETD2 also resulted in reduced recruitment of FACT subunits at transcribed gene bodies and altered histone dynamics, consistent with FACT facilitating reassembly of nucleosomes following Pol II passage and thereby inhibiting cryptic initiation. Another study demonstrated that in mouse ESCs, the H3K4 demethylase KDM5B is recruited to transcribed regions in part through its association with the Eaf3 ortholog MRG15 to suppress cryptic intragenic initiation (Xie et al., 2011); in addition, recruitment of the DNA methyltransferase Dnmt3b to gene bodies through interaction with H3K36me3 was also shown to suppress cryptic intragenic transcription (Neri et al., 2017). FACT-dependent suppression of intragenic transcription has also been observed in plants, but again is likely to differ mechanistically from yeast inasmuch as transcribed regions are associated with H3K36me2 rather than H3K36me3, which instead is localized to the 5′ ends of genes in *Arabidopsis* (Mahrez et al., 2016; Nielsen et al., 2019). In contrast to the preceding examples, pervasive transcriptional initiation from gene bodies was observed in wild-type *Drosophila* and was unaffected in flies expressing only H3K36R (Meers et al., 2017). Clearly, gaps remain in our understanding of the regulation of transcription from intragenic sites (Molenaar and van Leeuwen, 2022); moreover, there are as yet no reports of an in vitro system that recapitulates the phenomenon, whether based on yeast or metazoan components.

Epigenetics

The term "epigenetics" was coined by Conrad Waddington in 1942, and was originally conceived as a way to denote the study of the mechanisms by which genotype leads to phenotype, especially in the developing embryo (Waddington, 1942) (reprinted as Waddington, 2012). This usage explicitly emphasizes the relationship to epigenesis, the theory that all of the structures found in a fully formed organism arise de novo following the union of sperm and egg, in contrast to the preformation theory in which a tiny, fully formed organism is present from the start of embryogenesis. Over the years, the term has diverged considerably from its original meaning, as discussed by Haig (2012). Through the 1980s, "epigenetics" was a rarely employed term used principally to designate the means by which cells having the same genotype could exhibit distinct phenotypes, whether through environmental perturbations or intrinsic processes (as in development); it was not necessarily tied to heritability. With the discovery of gene regulation by DNA methylation, "epigenetics" was adopted to refer to "Nuclear inheritance which is not based on differences in DNA sequence," or similar formulations, with particular emphasis on the heritability of CpG methylation as described below (Holliday, 1987, 1994). Notably, heritability was variably used to indicate inheritance following cell division (mitotic heritability), or generational inheritance necessitating transmission through the germline (meiotic heritability). Coincident with the discovery of the role of histone modifications in gene regulation, "epigenetics" gained greatly in popularity to designate any modification to DNA or chromatin that affected gene expression (Fig. 8.21), and can fairly be considered as having been adopted as a marketing tool "in the competition for grants, citations, and tenure" (Haig, 2012). The widespread employment of "epigenetics" led, not surprisingly, to sometimes sloppy usage, which provoked occasional caustic, though accurate, commentary (Ptashne, 2013a,b).

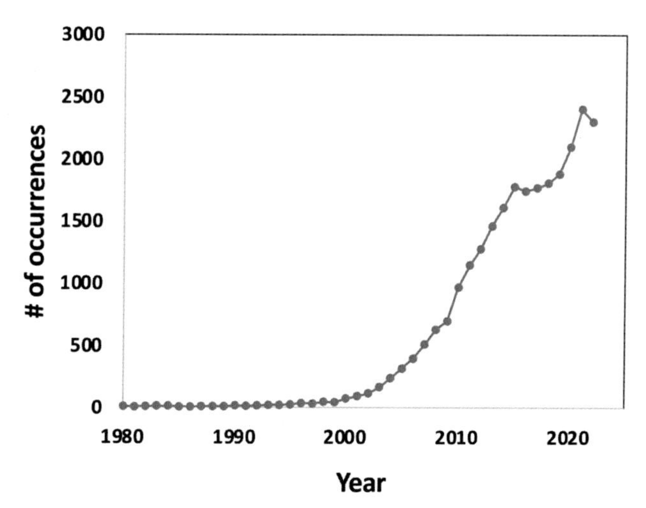

FIG. 8.21 Number of occurrences of the term "epigenetic" in article titles in Pubmed from 1980 through 2022. A related graph was published by Haig (2012).

"Epigenetics" continues to be a sometimes slippery label; for example, epigenetics has been defined as:

- "the study of changes in gene function that are mitotically and/or meiotically heritable and that do not entail a change in DNA sequence" (Dupont et al., 2009)
- "the inheritance of alternative chromatin states in the absence of changes in the DNA sequence" (Cavalli and Heard, 2019)
- "the study of molecules and mechanisms that can perpetuate alternative gene activity states in the context of the same DNA sequence" (Fitz-James and Cavalli, 2022)
- "the study of how your behaviors and environment can cause changes that affect the way your genes work" (https://www.cdc.gov/genomics/disease/epigenetics.htm)

Rather than obsessing over the precise meaning of epigenetics (and whether cytosine methylation leaves DNA sequence truly unchanged), as entertaining as that might be, it seems more useful to focus on mechanisms that affect the translation of genotype to phenotype—gene regulation, in modern parlance (and not intending to disregard posttranscriptional events)—and whether and how specific patterns of gene expression, and their resulting phenotypes, can be transmitted mitotically and through the germline. In regard to the latter, "transgenerational epigenetic inheritance" has been the subject of more stringent consideration, following early claims of epigenetic inheritance which in some cases have not stood up to close scrutiny (Fitz-James and Cavalli, 2022; Horsthemke, 2018). Transgenerational inheritance is distinguished from intergenerational inheritance: the latter refers to inheritance by the immediate offspring and potentially the F2 generation, if the F1 generation and its germline are exposed to the precipitating environment in utero, while true transgenerational inheritance requires continued inheritance through subsequent generations (Fitz-James and Cavalli, 2022; Miska and Ferguson-Smith, 2016). This is because an environmental perturbation that results in an epigenetic modification—say, a change in DNA methylation—exposes the first generation or the germline that produces it, and potentially the germline of the F1 generation, but may or may not be carried forward beyond that point.

A thorough review of epigenetic inheritance is beyond the scope of the present discussion. Instead, we consider two mechanisms of such inheritance that are germane to chromatin structure and function: DNA methylation and histone modifications. Noncoding RNA species have also been implicated in germline transmission of phenotypes (Boskovic and Rando, 2018; Fitz-James and Cavalli, 2022), but are not discussed here.

DNA methylation

A comprehensive historical perspective on DNA methylation is provided in a recent review (Mattei et al., 2022). Similarly to histone modifications, the modified base 5-methylcytosine (5mC) (Fig. 8.22A) was discovered, first in *Mycobacterium tuberculosis* and later in calf thymus, long before its functional significance yielded to inquiry (Hotchkiss, 1948; Johnson and Coghill, 1925; Wyatt, 1950). (It should be mentioned that 5mC is not the only known DNA base modification; for example, 6-methyladenine has also been observed in both prokaryotes and eukaryotes (Boulias and Greer, 2022)). The elucidation of the bacterial restriction-modification system in the 1960s, in which DNA methylation functions as part of a protective mechanism against bacteriophage invasion, stimulated interest in cytosine methylation as a possible gene regulatory mechanism in metazoans, leading to several such mechanisms being proposed in the 1970s (Mattei et al., 2022). Early biochemical results indicated that nearly all cytosine methylation occurred in the context of CpG dinucleotides in mammalian cells, although 5mC was also observed in a non-CpG context in plants (Grippo et al., 1968; Guseinov and Vanyushin, 1975; Sinsheimer, 1954). Not all eukaryotes harbor 5mC; while up to ~10% of all cytosines are methylated in mammalian cells, the modication is absent in yeast and the nematode *C. elegans*, and is present in at most trace levels in insects, as determined using a sensitive mass spectrometry assay (Capuano et al., 2014; Varma et al., 2022). A consequence of CpG methylation in mammals is a relative depletion of CpG dinucleotides in the genome, due to deamination of 5mC to yield thymine (Bird, 1980).

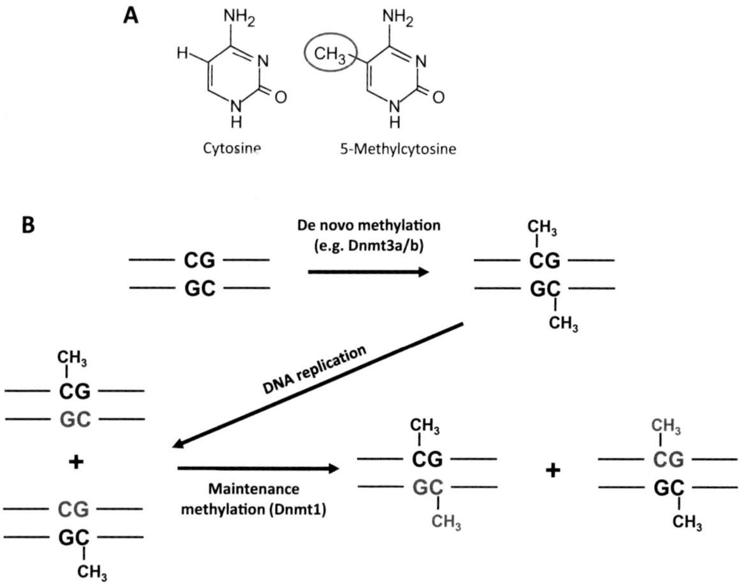

FIG. 8.22 (A) Structures of cytosine and 5-methylcytosme. (B) Select cytosines are methylated by de novo methylases (Dnmt3a/b in mammals). Newly replicated hemimethylated DNA is recognized by the maintenance methylase Dnmt1, preserving the fully methylated state.

Progress in DNA methylation research accelerated with the discovery of restriction enzymes that differed in their ability to cleave DNA sequences harboring methylated CpG sites, allowing determination of the methylation status at specific genomic regions. For example, the isoschizomers *Hpa*II and *Msp*I both cleave CCGG sites, but HpaII only cleaves unmethylated sites, whereas MspI cleaves whether or not the central CpG is methylated. Examination of CpG methylation at differentially regulated genes revealed 5mC to be correlated with gene repression, and inhibition of cytosine methylation by treatment with 5-azacytidine was shown to induce cell differentiation, supporting a gene regulatory role for 5mC (Mattei et al., 2022; McGhee and Ginder, 1979; Taylor and Jones, 1979). Subsequent experiments in which methylated and unmethylated plasmids harboring reporter genes were introduced into cultured cells lent firm support to CpG methylation as a repressive signal (e.g., Busslinger et al., 1983; Kruczek and Doerfler, 1983; McGeady et al., 1983). DNA methylation was eventually shown to contribute to repressive mechanisms in X-chromosome inactivation, imprinting (in which paternal and maternal alleles are differentially expressed), and developmental gene regulation.

A key discovery made in the 1980s was the existence of CpG islands, or CGIs. These comprise genomic regions enriched in unmethylated CpG dinucleotides, and were discovered by use of the methylation-sensitive restriction enzyme HpaII (Bird et al., 1985). Digestion of murine genomic DNA from various tissues with HpaII followed by end-labeling revealed that the large majority of genomic DNA was uncut by HpaII, and therefore methylated, but ~1% comprised regions of clustered, unmethylated CpG sites that were first termed HpaII tiny fragments. These ~30,000 CGIs contain *Hpa* II sites at about 15 times the genomic frequency and do not exhibit the CpG deficiency seen at other regions of the genome, consistent with a general lack of methylation. CGIs were found in proximity to genes, in spite of the low density of genes in the mammalian genome, but were not associated with genes expressed in a tissue-specific fashion, suggesting that they might correspond to regions continually occupied by transcription factors at constitutively active genes (Bird, 1986). Subsequent genome-wide elucidation of CGIs revealed a division between promoters in the human genome characterized by high or low CpG content, the former comprising CGI-associated genes and being enriched for broadly expressed genes (Carninci et al., 2006; Saxonov et al., 2006). Categorization of mammalian promoters according to chromatin structure and focused vs. broad initiation sites (Fig. 8.13) showed a modest enrichment for CGIs in promoters characterized by dispersed initiation sites, and enriched in housekeeping genes, over those having focused initiation and enriched in developmentally regulated genes (~75% vs. ~50%–60%, depending on the definition used for CGIs) (Rach et al., 2011). Later work on the genome-wide distribution of 5mC took advantage of bisulfite sequencing, in which cytosine, but not 5mC, is converted to uracil and then replaced by thymine during PCR amplification; sequencing of the resulting material allows distinction of C from 5mC (Frommer et al., 1992; Suzuki and Bird, 2008). The method has been widely applied to mammalian and plant genomes; its use in conjunction with DNA methyltransferases for determining nucleosome occupancy was discussed in Chapter 4.

With regard to the maintenance of DNA methylation following mitosis, a critical feature of the CpG sequence is its symmetry in the DNA double helix (Fig. 8.22B). This symmetry prompted the proposal of a mechanism for the inheritance of methylation patterns, in which the hemimethylated CpG doublets resulting from DNA replication would be recognized by a "maintenance" methyltransferase (Fig. 8.22B) (Holliday and Pugh, 1975; Riggs, 1975). Clonal inheritance of CpG methylation was subsequently demonstrated, and the maintenance methyltransferase responsible, Dnmt1, was purified and cloned (Bestor et al., 1988; Gruenbaum et al., 1982; Stein et al., 1982). Knockout of *DNMT1* in murine cells eliminated nearly all cytosine methylation, but a small amount of residual methylation pointed to the existence of methyltransferase enzymes capable of de novo DNA methylation (Li et al., 1992). Homology searches led to the identification of the de novo DNA methyltransferases Dnmt3a and Dnmt3b, which were found to be most active early in development (Okano et al., 1998).

The efficiency of maintenance methylation is underscored by a study of DNA methylation in the yeast *Cryptococcus neoformans* (Catania et al., 2020). This organism possesses only one known DNA methyltransferase, Dnmt5, having lost the de novo MTase DnmtX present in ancestral species on the order of 100 million years ago. Dnmt5 functions strictly as a maintenance methyltransferase: conditional shutoff in *C. neoformans* results in loss of CG methylation, which is not restored upon re-expression of Dnmt5. From these experiments, the authors deduce a methylation half-life of ~7500 generations and suggest that this low rate of loss combined with selective pressure explains the retention of conserved methylation patterns in *C. neoformans* despite the ancient loss of any de novo DNA methyltransferase.

Investigations into the mechanism by which DNA methylation could influence gene expression led to the identification of proteins that recognize sequences containing 5mC. The first of these was dubbed MeCP1 (Meehan et al., 1989); a second methylation "reader," MeCP2, was discovered to be a component of the Mi-2/NuRD complex, which contains both histone deacetylase and chromatin remodeling activities, as discussed in Chapters 6 and 7 (Nan et al., 1998; Wade et al., 1998). Additional 5mC binding proteins, dubbed methyl-binding domain proteins (MBDs), were later identified; MBD2 and MBD3 are also found associated with the Mi-2/NuRD complex and other remodelers and histone modifiers (Chapter 7) (Fitz-James and Cavalli, 2022). These findings implicated histone deacetylation as one mechanism by which DNA methylation could be coupled to gene repression. Cross-talk between DNA methylation and histone modifications leading to the formation of repressive heterochromatin in *N. crassa* and *A. thaliana* was discussed in Chapter 7. Another important mechanism by which DNA methylation can influence transcription is by decreasing the affinity of transcription factors having CG doublets within their recognition sequences (Yin et al., 2017).

In contrast to mitotic inheritance of CpG methylation, the issue of germline inheritance was not so clear. Early work, corroborated over time, revealed erasure of most CpG methylation following fertilization, with de novo methylation subsequently occurring in the time frame from embryonic implantation through gastrulation in mice; erasure of 5mC marks was also shown to occur during gametogenesis (Monk, 1987; Morgan et al., 2005; Tang et al., 2016). Nonetheless, other studies identified examples of germline inheritance of DNA methylation; for example, differential methylation of homologous chromosomes observed in somatic tissues was seen to be heritable in humans (Silva and White, 1988). However, what was really sought was a bit more: researchers were interested in learning whether there existed heritable, distinct states of DNA methylation that affected phenotype. Two renowned examples are the *Agouti variable yellow* (A^{vy}) and *Axin fused* ($Axin^{Fu}$) in the mouse (Bertozzi and Ferguson-Smith, 2020). Variable and heritable expression of these "epialleles," identical in DNA sequence but differing in methylation status, lead, respectively, to altered coat color and a "tail-kink" phenotype. Both A^{vy} and $Axin^{Fu}$ alleles are characterized by repetitive element (intracisternal A particle, or IAP) insertions that include elements that can drive transcription; the variable expression of A^{vy} and $Axin^{Fu}$ alleles is caused by variable methylation of these elements, which survives passage through the germline (Morgan et al., 1999; Rakyan et al., 2003). Phenotypic variation attributable to DNA methylation has also been observed to be transmitted through several generations in plants, affecting traits such as flowering time, floral symmetry, and pathogen resistance; however, as with the methylation state of IAPs in mice, genetic background effects can interact strongly with epialleles in determining phenotype (Fitz-James and Cavalli, 2022). Transgenerational inheritance of methylation state in plants appears more straightforward mechanistically than in metazoans, as the extensive reprogramming seen in metazoans is not observed in plants (Gehring, 2019; Morgan et al., 2005; Quadrana and Colot, 2016).

A putative example of transgenerational inheritance of an altered phenotype based on environmental exposure that garnered widespread popular attention is connected to the Dutch Hunger Winter at the conclusion of WW II. Individuals exposed to famine in utero, as well as the offspring of prenatally undernourished fathers, exhibited phenotypes related to altered fat metabolism; effects were also seen in grandchildren of undernourished mothers (Painter et al., 2008; Veenendaal et al., 2013). Moreover, individuals exposed to famine early in pregnancy exhibited altered levels of DNA methylation 60 years later, indicating that changes in the prenatal environment can result in altered DNA methylation that persists throughout the life of the individual (Heijmans et al., 2008). However, evidence for effects in the F3 generation—that is, for grandmaternal exposure to famine in utero—was not observed (Heard and Martienssen, 2014). In related studies, a famine from 1867 to 1869 in the northern Swedish town of Överkalix was seen to result in effects on disease and longevity in the grandchildren of men and women exposed to famine in their prepubertal years (Bygren et al., 2001; Kaati et al., 2007). However, no effects were seen after the F2 generation in this and other examples, and molecular data is not available in these cases (Gonzalez-Rodriguez et al., 2023).

A recent innovation promises to open new avenues for the study of transgenerational inheritance depending on DNA methylation (Takahashi et al., 2017). In this approach, insertion of a DNA fragment bearing a drug resistance marker and lacking CpG sequences is inserted into a CpG island in murine ESCs. The insertion results in methylation of previously unmethylated CpG sites within the now interrupted CGI, possibly caused by the failure of zinc finger proteins that normally provide protection to bind to these sequences (Fig. 8.23) (Clouaire et al., 2012; Wu et al., 2011). Removal of the CpG-lacking cassette by Cre-loxP-mediated recombination did not eliminate the novel methylation; cytosine methylation induced by the insertion was maintained for 20 cell passages. Even more remarkably, a follow-up study showed that in transgenic mice created by injection into early embryos of ESCs harboring the edited and now methylated CGIs, the methylation status (using two promoters, for the *Ankrd26* and *Ldlr* genes) was preserved in the resulting adult mice and transmitted to offspring through at least four generations (Fig. 8.23) (Takahashi et al., 2023). Methylation was associated with gene silencing; silencing of *Ankrd26* resulted in obesity and this was heritable along with CGI methylation. The inheritance of CGI methylation patterns was not, surprisingly, due to a failure to erase cytosine methylation during gametogenic and early embryonic reprogramming; rather, it appears that some unknown factor must be responsible for preserving a molecular memory of the methylation pattern. It is possible that interactions with retrotransposons and/or CTCF binding sites present in the regions that were engineered may contribute to this phenomenon (McGraw and Kimmins, 2023). It will be important to corroborate this apparent example of 5mC-mediated transgenerational inheritance and to determine whether and to what extent promoter context and the altered phenotype itself contribute to the potential for germline transmission.

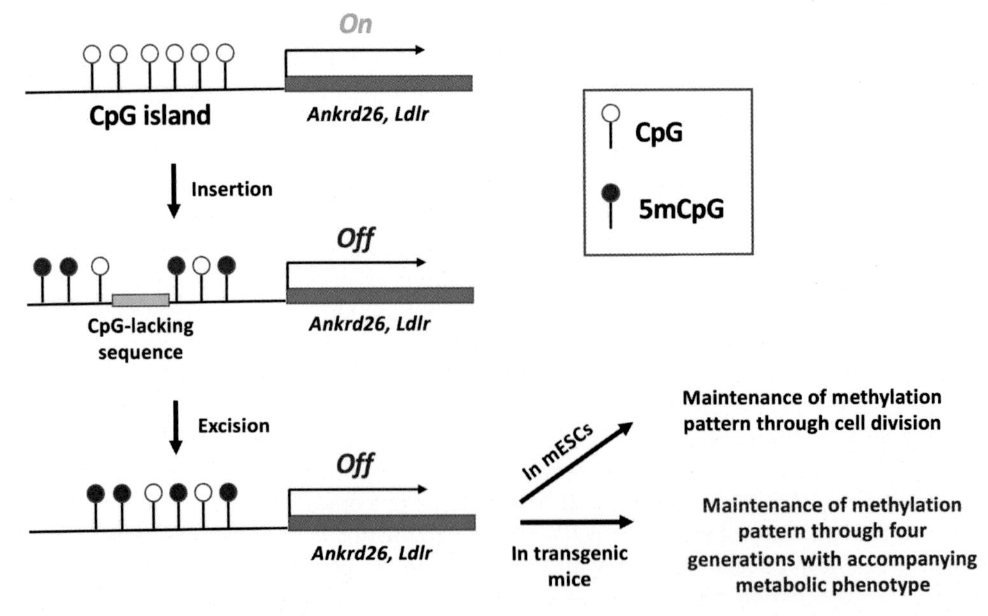

FIG. 8.23 Transgenerational inheritance of an induced methylation state. See text for details.

Finally, it should be mentioned that as with histone methylation, DNA can be actively demethylated. Compelling evidence for active DNA demethylation came from experiments showing that the paternal genome of murine zygotes is demethylated independently of DNA replication (Mayer et al., 2000). This finding led to an intensive search for a DNA demethylase, eventually culminating in the discovery of the conversion of 5mC to 5-hydroxymethylcytosine (5hmC) by the Ten Eleven Translocation dioxygenase, TET1 (Ooi and Bestor, 2008; Tahiliani et al., 2009). Conversion of 5hmC to cytosine then occurs through replicative dilution or DNA repair mechanisms (Wu and Zhang, 2017). The mechanism by which TET1, as well as the de novo methylases Dnmt3a and Dnmt3b, acts at specific genomic sites is an active area of investigation; targeting of the latter two enzymes is influenced by the presence of nucleosomes and their modifications (Baubec et al., 2015; Weinberg et al., 2019).

Chromatin-mediated inheritance

The surge of interest in chromatin in the early 1970s led researchers to consider whether chromatin might play a role in inheritance. Even before the structure of the nucleosome was elucidated, Tsanev and Sendov theorized that "specific histone arrangements which block and at the same time specify the initiation sites of different genetic units" might explain "stable changes in the protein pattern leading to different cellular types" (Tsanev and Sendov, 1971). Weintraub and colleagues reported the stable propagation of DNase hypersensitive sites during cell division in the absence of the initiating inducing stimulus and speculated on mechanisms involving chromatin structure that might contribute to such mitotic transmission (Groudine and Weintraub, 1982; Weintraub, 1985).

Experimental examples of putatively chromatin-mediated epigenetic inheritance were also reported before the inherited chromatin structures themselves had been elucidated. One example involved repression of the silent mating type loci in budding yeast, which requires the Sir1–4 proteins, as discussed in Chapter 3. While silencing is lost in *sir2*, *sir3*, and *sir4* mutants, *sir1* mutants exhibit a leaky phenotype in which partial silencing is retained. Close analysis of single cells mutant in *sir1* revealed a heritable, metastable phenotype in which silencing or the lack thereof was stably transmitted to progeny over multiple cell divisions, with only occasional switching (Pillus and Rine, 1989). Altered repression of *HML* and *HMR* had also been observed in mutants having changes in histone dosage (Clark-Adams et al., 1988; Kayne et al., 1988), leading to the suggestion that "the repressed and derepressed states [might be considered as] reflections of different protein complexes that contribute to the structure of chromatin at these sites" (Pillus and Rine, 1989). Epigenetic inheritance of gene silencing through a mechanism likely based on chromatin structure was later also reported in the context of telomeric silencing (Chapter 3) (Gottschling et al., 1990) and in fission yeast (Grewal and Klar, 1996); the latter inheritance is now well understood mechanistically, as discussed below.

A criterion employed by many researchers for true epigenetic inheritance is the continued transmission of a molecular or macroscopic phenotype in the absence of the initiating signal. Application of this stricture would rule out the stable transmission of states that rely on simple feedback loops, for example, the expression of a master transcription factor that stimulates its own production while also specifying a particular cellular state. One of the first examples of such transmission was reported by Cavalli and Paro, who demonstrated that the *on* and *off* states implemented by TrxG and PcG complexes in *Drosophila* could be transmitted to subsequent generations. To this end, transgenic flies were constructed in which a dual *lacZ* and mini-*white* reporter were placed downstream of a PRE, *Fab-7*, derived from the bithorax complex (Cavalli and Paro, 1998). In addition, a binding site for the yeast transcriptional activator Gal4 (known to function as an activator when expressed in *Drosophila*) was placed between *Fab-7* and the *lacZ* reporter (Fig. 8.24). The introduction of this construct into flies that also harbored the *GAL4* gene under the control of a heat shock element allowed transient activation of the *lacZ* and mini-*white* reporters by a pulse of heat shock. Control embryos allowed to develop in the absence of heat shock yielded adults having dark yellow/orange, slightly variegated eyes, due to repression of mini-*white*; induction of a pulse of Gal4 during

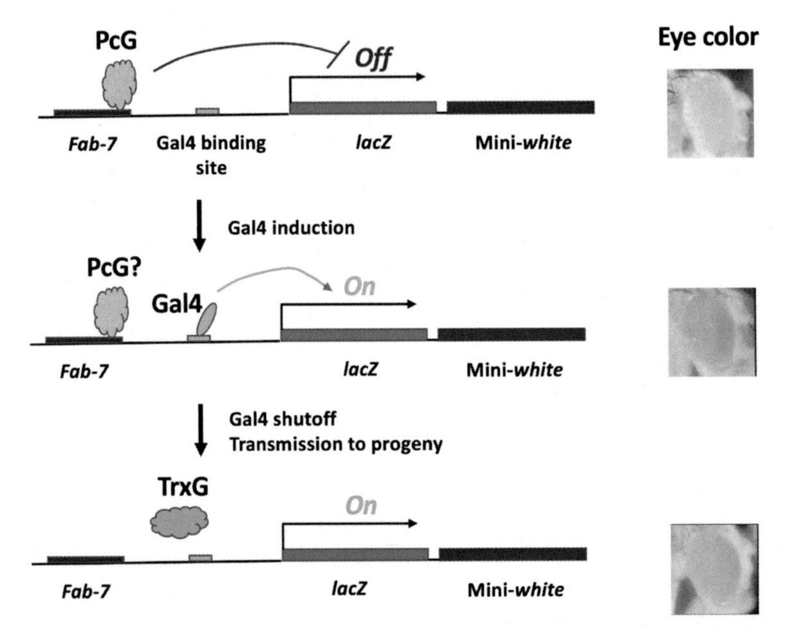

FIG. 8.24 Scheme employed by (Cavalli and Paro 1998) to test epigenetic inheritance in *Drosophila.* The mini-*white* gene was placed under the control of the *Fab-7* element, resulting in PcG-mediated repression and a white eye phenotype (*white* refers to the mutant, or off, phenotype). Expression of Gal4 resulted in the expression of mini-*white* and corresponding red eye phenotype. (The status of PcG association after Gal4 induction was not determined.) This phenotype was maintained through embryogenesis after the shutoff of Gal4 and was transmitted to progeny when red-eyed flies were crossed, with 27% of progeny in the F3 generation exhibiting the red eye phenotype, which depended on TrxG. *Images of fly eyes taken from Fig. 7 of Cavalli, G., Paro, R. 1998. The Drosophila Fab-7 chromosomal element conveys epigenetic inheritance during mitosis and meiosis. Cell 93, 505-518.*

embryogenesis gave rise to flies having red eyes, indicating that the memory of activation of mini-*white* by Gal4 was retained through development. Mitotic transmission of the activated state depended on *trx* but not on *Pc* (Cavalli and Paro, 1999). Crosses of red-eyed flies obtained using this protocol revealed variable transmission of the phenotype, with about 25% of progeny having red eyes. Meiotic transmission was seen even in flies in which the *GAL4* gene had been eliminated by crossing, demonstrating that the primary signal was dispensable for the maintenance of the induced state (Cavalli and Paro, 1999). Although inheritance of the red-eyed phenotype was eventually lost after successive meioses, this was the first demonstration of transgenerational inheritance (albeit not 100% faithful) in *Drosophila*.

Galvanized by the findings that histone modifiers could act as coactivators and corepressors (Chapter 7), chromatin researchers speculated that histone modifications could act as carriers of heritable epigenetic information (Jeppesen, 1997; Turner, 2000). Although acetylation was posited as a potential carrier of epigenetic information, the rapid turnover of this modification in fungi, plant, and animal cells posed difficulties for such a mechanism (Waterborg, 2002). In contrast, histone methylation was observed to be a far more stable modification (Chapter 7) (Borun et al., 1972; Byvoet et al., 1972; Thomas et al., 1972); moreover, a plausible mechanism for cellular inheritance of histone methylation was suggested by the affinity of the machinery responsible for methylation of H3K9, the characteristic mark of constitutive heterochromatin (Chapters 3 and 7), for that same modification (Jenuwein and Allis, 2001). Similarly, PcG components were reported to be "readers" of both the H2A119ub and H3K27me2/3 mark conferred by the PcG complex, suggesting a self-reinforcing feedback cycle that could allow propagation of PcG-dependent domains of facultative heterochromatin (Fig. 8.25) (Hansen et al., 2008;

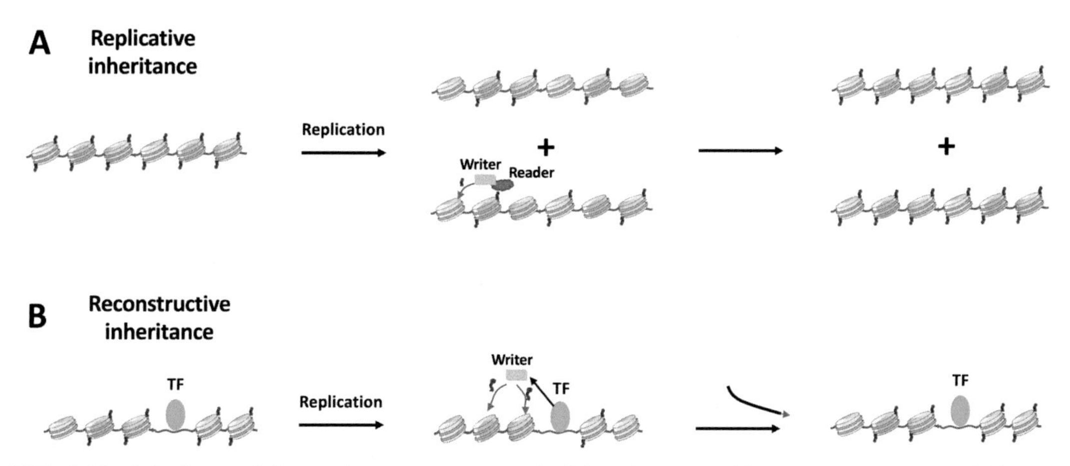

FIG. 8.25 Inheritance of chromatin states may occur via (A) replicative or (B) reconstructive means. Following DNA synthesis, parental H3–H4 tetramers with their modifications are distributed in nearly equal proportions between leading and lagging strands with retention of positional information. A modification may provide a platform for a reader protein that in turn associates with a writer that confers the same modification, resulting in replicative inheritance of the modification pattern. In reconstructive inheritance, the modification is partially or completely lost following DNA synthesis and is restored via a secondary mechanism, such as a transcription factor as shown here or a small RNA molecule.

Margueron et al., 2009). This proposed mechanism for inheritance of a specific chromatin state is conceptually similar to the maintenance of DNA methylation described above, and has been termed "replicative inheritance" (Miska and Ferguson-Smith, 2016). Replicative inheritance is facilitated by the dispersive segregation of parental histones to both leading and lagging strands following DNA replication (Chapter 5) (Petryk et al., 2018; Xu et al.,

2010; Yu et al., 2018), and by parental histones with their modifications being re-deposited at or close to their original location (Escobar et al., 2019; Reveron-Gomez et al., 2018; Schlissel and Rine, 2019); reviewed in Stewart-Morgan et al. (2020). Nearly symmetric segregation of parental histones has been observed in both budding yeast and mouse ESCs, although the mechanisms by which this occurs differ for these examples (Petryk et al., 2018; Yu et al., 2018). However, not all modifications are necessarily faithfully re-deposited close to their location in the parental cell; while domains containing H3K4me3, H3K9me3, H3K27me3, and H3K36me3 were retained following replication in HeLa cells and mouse ESCs, histones in active euchromatin were dispersed following replication (Escobar et al., 2019; Reveron-Gomez et al., 2018). Furthermore, the importance of replicative inheritance appears to vary depending on circumstance: in budding yeast, shortening of heterochromatic regions, which would be predicted to increase the likelihood of loss of heterochromatic silencing through replicative inheritance, had no such effect, and only minor effects were seen in mutants causing disruption of H3–H4 tetramer inheritance (Saxton and Rine, 2019). In contrast, a mutation that disrupted symmetric segregation of histones in mouse ESCs affected mitotic inheritance of histone modifications, leading to altered expression of genes governed by H3K9me3 or H3K27me3 modifications (Wen et al., 2023; Wenger et al., 2023). While more studies are clearly needed, results so far suggest that replicative inheritance is most likely to apply to domains marked by histone methylation, and perhaps especially to regions of repressive heterochromatin in metazoan cells.

In an alternative mechanism dubbed "reconstructive inheritance," the epigenetic mark is erased during mitosis or meiosis and is re-established via a "memory" implemented by ancillary molecular signals (Fig. 8.25) (Fitz-James and Cavalli, 2022; Miska and Ferguson-Smith, 2016). A possible, and surprising, example of the latter may be the retention of methylation of CpG islands following the insertion and removal of a fragment lacking CpG doublets, as described above, inasmuch as it appears that the resulting methylation is lost during embryogenesis and restored later on (Takahashi et al., 2023). While replicative inheritance satisfies the generally accepted definition of epigenetic inheritance, reconstructive inheritance is more problematic in this regard, as re-establishment may require sequence-specific signals encoded in transcription factors or small RNA molecules.

Replicative inheritance of chromatin states has been most thoroughly examined for facultative heterochromatin mediated by PcG and H3K27 methylation in flies and mammals, and by Swi6/HP1 and H3K9 methylation in fission yeast. Two early studies reported that PcG-mediated silencing was lost upon excision of PREs, indicating a continued requirement for these specific sequences (Busturia et al., 1997; Sengupta et al., 2004). These findings were corroborated by later studies in *Drosophila.* In these experiments, a transgene in which repression of a reporter gene was conferred by a PRE from the *Ubx* locus was engineered to allow excision of the PRE; loss of the PRE resulted in loss of silencing over subsequent cell divisions (Coleman and Struhl, 2017; Laprell et al., 2017). However, loss of silencing followed different kinetics in these two studies, in one case being estimated at about 12% per cell division and in the other about 50%. (Note that these results do not contradict the findings of Cavalli and Paro described earlier: in that study, the transient expression of the Gal4 activator resulted in a heritable *On* state that did not require PcG components, while the sequence requirements for transmission of the default, PcG-repressed state were not examined.)

Studies examining the inheritance of PcG-mediated repression in mammalian cells produced stronger evidence for mitotic transmission of the silenced state, but again showed some variability, with two studies reporting a low level of erosion of silencing during cell division. In the first study, recruitment of PRC2, via a fusion protein comprising the Gal4 DNA-binding domain and the PRC2 component EED (Table 8.2), resulted in repression of a reporter gene and H3K27 methylation in human diploid fibroblasts that was maintained over at least four cell divisions after depletion of the initiating fusion protein (Hansen et al., 2008). A subsequent study established a PcG-dependent repressive domain in Chinese hamster ovary cells using a fusion of EED to the reverse Tet repressor, which binds to Tet operator sequences only in the presence of doxycycline (Bintu et al., 2016). Monitoring of a fluorescent reporter under control of the artificially recruited PcG complex in single cells revealed reactivation upon removal of doxycycline occurring in a stochastic fashion, contrasting with the stable repression when the de novo methyltransferase, Dnmt3b, was similarly transiently recruited. Quantitation by flow cytometry, however, uncovered the existence of a fraction of cells in which PcG-mediated repression was stably maintained, and a three-state model was proposed in which PcG recruitment results first in the conversion of the active state to a reversibly silenced state, with the latter being converted to an irreversibly silenced state by an unknown mechanism. A third, more recent study reported the silencing of a reporter gene in murine ESCs via recruitment of either canonical or variant PRC1 complexes, but stable repression after removal of the initiating signal (using the TET^{OFF} system) was seen only for canonical PRC1 (Moussa et al., 2019). Again, a fraction of cells (23% after ~12 cell divisions) exhibited loss of repression after removal of the initiating signal. In sum, then, experiments in *Drosophila* and mammalian cell lines indicate that PcG-mediated repression can be stably inherited through mitosis, but with lower fidelity than seen for silencing mediated by DNA methylation, and with possible contributions of DNA sequence and chromosomal context.

Replicative inheritance of heterochromatin has also been intensively studied in the fission yeast *S. pombe* (Grewal, 2023; Martienssen and Moazed, 2015). Formation of repressive heterochromatin at the silent *mat* region in *S. pombe* is initiated by recruitment of the H3K9 methyltransferase Clr4, the *S. pombe* homolog of *Drosophila* SU(VAR)3-9 and mammalian Suv39h, as part of a complex that is targeted to the *mat* region by small RNA transcripts originating from the *cenH* element (Grewal, 2023). Mutants lacking Clr4 do not form repressive heterochromatin, and domains of silent heterochromatin can be established by artificial recruitment of the SET domain of Clr4 (Kagansky et al., 2009). Spreading of heterochromatin involves interactions of both Clr4 and Swi6/HP1 with H3K9me2 as well as recruitment of histone deacetylases (Grewal, 2023). Similarly to the clonally heritable silencing and repression seen in *S. cerevisiae sir1* mutants discussed earlier, under certain conditions populations of *S. pombe* cells can be generated that differ only in the presence or absence of repressive heterochromatin at the *mat* region (Hall et al., 2002; Nakayama et al., 2000). These distinct states are not only stably transmitted through cell division in haploid cells, but are maintained in diploid cells arising from the mating of the two cell types and are faithfully transmitted to haploid progeny following sporulation (Grewal and Klar, 1996; Hall et al., 2002). These findings represent some of the strongest evidence extant for epigenetic inheritance independent of DNA methylation (which is absent in *S. pombe*).

To examine the heritability of heterochromatin in *S. pombe* in the absence of the initiating signal, two independent investigations recruited a TetR-Clr4 fusion protein upstream of an *ade6*$^{+}$ reporter (Fig. 8.26) (Audergon et al., 2015; Ragunathan et al., 2015). Recruitment of Clr4 created a domain of heterochromatin enriched for H3K9me2 and repressed *ade6*$^{+}$, as seen by the presence of pink colonies, whereas cells expressing *ade6*$^{+}$ produce white colonies. Silencing induced by TetR-Clr4 binding was lost following the addition of tetracycline, but was maintained if the H3K9 demethylase Epe1 was absent (Fig. 8.26) (Audergon et al., 2015; Ragunathan et al., 2015). Thus, while inheritance of repressive heterochromatin can occur in fission yeast in the absence of the initiating signal, this is normally prevented by the presence of a histone demethylase. Heritability of H3K9me3-enriched heterochromatic domains has also been reported in murine cells through multiple cell divisions following removal of the initiating signal; in this case, heterochromatin formation was initiated by transient recruitment of HP1α (Hathaway et al., 2012).

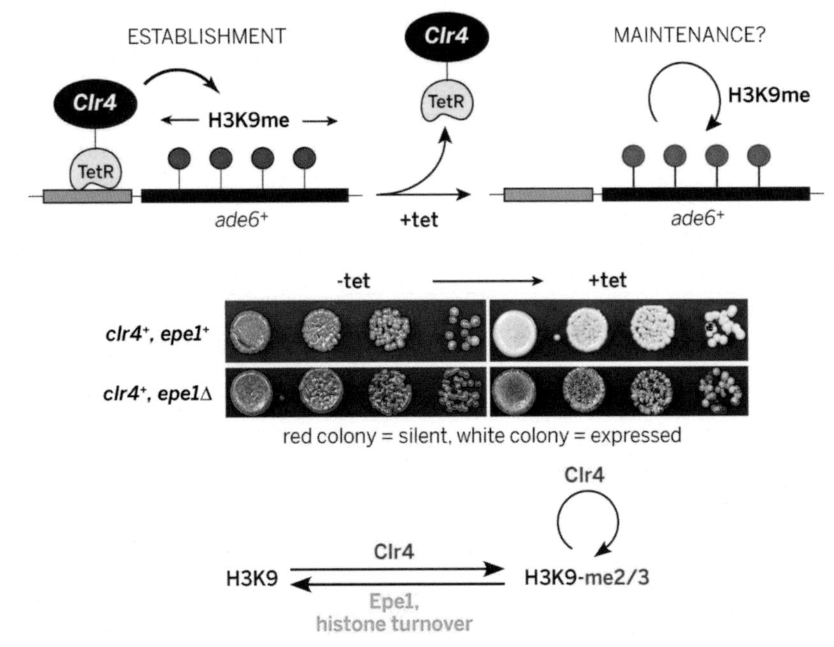

FIG. 8.26 Testing for epigenetic inheritance in *S. pombe*. See text for details. *From Ragunathan, K., Jih, G., Moazed, D. 2015. Epigenetics. Epigenetic inheritance uncoupled from sequence-specific recruitment. Science 348, 1258699.*

Transgenerational replicative inheritance of methylated DNA sequences in metazoans entails the additional complication, as compared to mitotic transmission, of requiring protection against reprogramming during gametogenesis (e.g., Li et al., 2008). Transgenerational replicative inheritance of histone modifications faces similar challenges, as reprogramming of modifications occurs during oogenesis and in the fertilized zygote, and the majority of nucleosomes are lost when histones are replaced by protamines during spermatogenesis

(Chapter 5, Sidebar on Protamines) (Ben Maamar et al., 2021b; Fraser and Lin, 2016; Gatewood et al., 1990; Gu et al., 2010; Stewart et al., 2015). As with DNA methylation, however, some regions can escape such reprogramming: a minor portion of the genome remains packaged into nucleosomes in sperm, and the modification pattern of those nucleosomes is important for development following fertilization (Hammoud et al., 2009; Siklenka et al., 2015; Teperek et al., 2016). Recent studies have implicated altered histone retention and DNA methylation in sperm associated with transgenerational inheritance of pathologies in mice exposed to pesticides (Beck et al., 2021; Ben Maamar et al., 2021a). It will be important to corroborate these results and to determine whether heritable epigenetic changes caused by environmental stimuli can also occur in the female germline.

References

Abdelkareem, M., Saint-Andre, C., Takacs, M., Papai, G., Crucifix, C., Guo, X., Ortiz, J., Weixlbaumer, A., 2019. Structural basis of transcription: RNA polymerase backtracking and its reactivation. Mol. Cell 75 (298–309), e294.

Abmayr, S.M., Workman, J.L., Roeder, R.G., 1988. The pseudorabies immediate early protein stimulates in vitro transcription by facilitating TFIID: promoter interactions. Genes Dev. 2, 542–553.

Adams, C.C., Workman, J.L., 1995. Binding of disparate transcriptional activators to nucleosomal DNA is inherently cooperative. Mol. Cell. Biol. 15, 1405–1421.

Agalioti, T., Lomvardas, S., Parekh, B., Yie, J., Maniatis, T., Thanos, D., 2000. Ordered recruitment of chromatin modifying and general transcription factors to the IFN-beta promoter. Cell 103, 667–678.

Ahmad, K., Brahma, S., Henikoff, S., 2024. Epigenetic pioneering by SWI/SNF family remodelers. Mol. Cell 84, 194–201.

Akam, M., 1987. The molecular basis for metameric pattern in the Drosophila embryo. Development 101, 1–22.

Alkema, M.J., van der Lugt, N.M., Bobeldijk, R.C., Berns, A., van Lohuizen, M., 1995. Transformation of axial skeleton due to overexpression of bmi-1 in transgenic mice. Nature 374, 724–727.

Allfrey, V.G., Mirsky, A.E., 1962. Evidence for the complete DNA-dependence of RNA synthesis in isolated thymus nuclei. Proc. Natl. Acad. Sci. USA 48, 1590–1596.

Allfrey, V.G., Littau, V.C., Mirsky, A.E., 1963. On the role of histones in regulation ribonucleic acid synthesis in the cell nucleus. Proc. Natl. Acad. Sci. USA 49, 414–421.

Almer, A., Horz, W., 1986. Nuclease hypersensitive regions with adjacent positioned nucleosomes mark the gene boundaries of the PHO_5/PHO_3 locus in yeast. EMBO J. 5, 2681–2687.

Almer, A., Rudolph, H., Hinnen, A., Horz, W., 1986. Removal of positioned nucleosomes from the yeast PHO_5 promoter upon PHO_5 induction releases additional upstream activating DNA elements. EMBO J. 5, 2689–2696.

Almouzni, G., Wolffe, A.P., 1993a. Nuclear assembly, structure, and function: the use of Xenopus in vitro systems. Exp. Cell Res. 205, 1–15.

Almouzni, G., Wolffe, A.P., 1993b. Replication-coupled chromatin assembly is required for the repression of basal transcription in vivo. Genes Dev. 7, 2033–2047.

Aloia, L., Di Stefano, B., Di Croce, L., 2013. Polycomb complexes in stem cells and embryonic development. Development 140, 2525–2534.

Andrulis, E.D., Guzman, E., Doring, P., Werner, J., Lis, J.T., 2000. High-resolution localization of Drosophila Spt5 and Spt6 at heat shock genes in vivo: roles in promoter proximal pausing and transcription elongation. Genes Dev. 14, 2635–2649.

Angermayr, M., Oechsner, U., Gregor, K., Schroth, G.P., Bandlow, W., 2002. Transcription initiation in vivo without classical transactivators: DNA kinks flanking the core promoter of the housekeeping yeast adenylate kinase gene, AKY2, position nucleosomes and constitutively activate transcription. Nucleic Acids Res. 30, 4199–4207.

Angermayr, M., Oechsner, U., Bandlow, W., 2003. Reb1p-dependent DNA bending effects nucleosome positioning and constitutive transcription at the yeast profilin promoter. J. Biol. Chem. 278, 17918–17926.

Ansari, S.A., Paul, E., Sommer, S., Lieleg, C., He, Q., Daly, A.Z., Rode, K.A., Barber, W.T., Ellis, L.C., LaPorta, E., et al., 2014. Mediator, TATA-binding protein, and RNA polymerase II contribute to low histone occupancy at active gene promoters in yeast. J. Biol. Chem. 289, 14981–14995.

Aparicio, O.M., Billington, B.L., Gottschling, D.E., 1991. Modifiers of position effect are shared between telomeric and silent mating-type loci in *S. cerevisiae*. Cell 66, 1279–1287.

Archer, T.K., Cordingley, M.G., Wolford, R.G., Hager, G.L., 1991. Transcription factor access is mediated by accurately positioned nucleosomes on the mouse mammary tumor virus promoter. Mol. Cell. Biol. 11, 688–698.

Archer, T.K., Lefebvre, P., Wolford, R.G., Hager, G.L., 1992. Transcription factor loading on the MMTV promoter: a bimodal mechanism for promoter activation. Science 255, 1573–1576.

Ardehali, M.B., Yao, J., Adelman, K., Fuda, N.J., Petesch, S.J., Webb, W.W., Lis, J.T., 2009. Spt6 enhances the elongation rate of RNA polymerase II in vivo. EMBO J. 28, 1067–1077.

Audergon, P.N., Catania, S., Kagansky, A., Tong, P., Shukla, M., Pidoux, A.L., Allshire, R.C., 2015. Epigenetics. Restricted epigenetic inheritance of H3K9 methylation. Science 348, 132–135.

Axelrod, J.D., Reagan, M.S., Majors, J., 1993. GAL4 disrupts a repressing nucleosome during activation of GAL1 transcription in vivo. Genes Dev. 7, 857–869.

Azuara, V., Perry, P., Sauer, S., Spivakov, M., Jorgensen, H.F., John, R.M., Gouti, M., Casanova, M., Warnes, G., Merkenschlager, M., et al., 2006. Chromatin signatures of pluripotent cell lines. Nat. Cell Biol. 8, 532–538.

Badis, G., Chan, E.T., van Bakel, H., Pena-Castillo, L., Tillo, D., Tsui, K., Carlson, C.D., Gossett, A.J., Hasinoff, M.J., Warren, C.L., et al., 2008. A library of yeast transcription factor motifs reveals a widespread function for Rsc3 in targeting nucleosome exclusion at promoters. Mol. Cell 32, 878–887.

Baer, B.W., Rhodes, D., 1983. Eukaryotic RNA polymerase II binds to nucleosome cores from transcribed genes. Nature 301, 482–488.

Bai, L., Charvin, G., Siggia, E.D., Cross, F.R., 2010. Nucleosome-depleted regions in cell-cycle-regulated promoters ensure reliable gene expression in every cell cycle. Dev. Cell 18, 544–555.

Balasubramanian, B., Morse, R.H., 1999. Binding of Gal4p and bicoid to nucleosomal sites in yeast in the absence of replication. Mol. Cell. Biol. 19, 2977–2985.

Balsalobre, A., Drouin, J., 2022. Pioneer factors as master regulators of the epigenome and cell fate. Nat. Rev. Mol. Cell Biol. 23, 449–464.

Banerji, J., Rusconi, S., Schaffner, W., 1981. Expression of a beta-globin gene is enhanced by remote SV40 DNA sequences. Cell 27, 299–308.

Bao, Y., White, C.L., Luger, K., 2006. Nucleosome core particles containing a poly(dA.dT) sequence element exhibit a locally distorted DNA structure. J. Mol. Biol. 361, 617–624.

Barbaric, S., Luckenbach, T., Schmid, A., Blaschke, D., Horz, W., Korber, P., 2007. Redundancy of chromatin remodeling pathways for the induction of the yeast PHO_5 promoter in vivo. J. Biol. Chem. 282, 27610–27621.

Barnes, T., Korber, P., 2021. The active mechanism of nucleosome depletion by poly(dA:dT) tracts in vivo. Int. J. Mol. Sci. 22.

Barr, G.C., Butler, J.A., 1963. Histones and gene function. Nature 199, 1170–1172.

Bar-Ziv, R., Brodsky, S., Chapal, M., Barkai, N., 2020. Transcription factor binding to replicated DNA. Cell Rep. 30 (3989–3995), e3984.

Basehoar, A.D., Zanton, S.J., Pugh, B.F., 2004. Identification and distinct regulation of yeast TATA box-containing genes. Cell 116, 699–709.

Baubec, T., Colombo, D.F., Wirbelauer, C., Schmidt, J., Burger, L., Krebs, A.R., Akalin, A., Schubeler, D., 2015. Genomic profiling of DNA methyltransferases reveals a role for DNMT3B in genic methylation. Nature 520, 243–247.

Bazett-Jones, D.P., Leblanc, B., Herfort, M., Moss, T., 1994. Short-range DNA looping by the Xenopus HMG-box transcription factor, xUBF. Science 264, 1134–1137.

Bazett-Jones, D.P., Mendez, E., Czarnota, G.J., Ottensmeyer, F.P., Allfrey, V.G., 1996. Visualization and analysis of unfolded nucleosomes associated with transcribing chromatin. Nucleic Acids Res. 24, 321–329.

Beck, D., Ben Maamar, M., Skinner, M.K., 2021. Integration of sperm ncRNA-directed DNA methylation and DNA methylation-directed histone retention in epigenetic transgenerational inheritance. Epigenetics Chromatin 14, 6.

Becker, P.B., Wu, C., 1992. Cell-free system for assembly of transcriptionally repressed chromatin from Drosophila embryos. Mol. Cell. Biol. 12, 2241–2249.

Becker, P.B., Rabindran, S.K., Wu, C., 1991. Heat shock-regulated transcription in vitro from a reconstituted chromatin template. Proc. Natl. Acad. Sci. USA 88, 4109–4113.

Bednar, J., Studitsky, V.M., Grigoryev, S.A., Felsenfeld, G., Woodcock, C.L., 1999. The nature of the nucleosomal barrier to transcription: direct observation of paused intermediates by electron cryomicroscopy. Mol. Cell 4, 377–386.

Belotserkovskaya, R., Oh, S., Bondarenko, V.A., Orphanides, G., Studitsky, V.M., Reinberg, D., 2003. FACT facilitates transcription-dependent nucleosome alteration. Science 301, 1090–1093.

Ben Maamar, M., Beck, D., Nilsson, E.E., Kubsad, D., Skinner, M.K., 2021a. Epigenome-wide association study for glyphosate induced transgenerational sperm DNA methylation and histone retention epigenetic biomarkers for disease. Epigenetics 16, 1150–1167.

Ben Maamar, M., Nilsson, E.E., Skinner, M.K., 2021b. Epigenetic transgenerational inheritance, gametogenesis and germline development dagger. Biol. Reprod. 105, 570–592.

Bender, S., Tang, Y., Lindroth, A.M., Hovestadt, V., Jones, D.T., Kool, M., Zapatka, M., Northcott, P.A., Sturm, D., Wang, W., et al., 2013. Reduced H3K27me3 and DNA hypomethylation are major drivers of gene expression in K27M mutant pediatric high-grade gliomas. Cancer Cell 24, 660–672.

Benezra, R., Cantor, C.R., Axel, R., 1986. Nucleosomes are phased along the mouse beta-major globin gene in erythroid and nonerythroid cells. Cell 44, 697–704.

Bernecky, C., Plitzko, J.M., Cramer, P., 2017. Structure of a transcribing RNA polymerase II-DSIF complex reveals a multidentate DNA-RNA clamp. Nat. Struct. Mol. Biol. 24, 809–815.

Bernstein, B.E., Liu, C.L., Humphrey, E.L., Perlstein, E.O., Schreiber, S.L., 2004. Global nucleosome occupancy in yeast. Genome Biol. 5, R62.

Bernstein, B.E., Mikkelsen, T.S., Xie, X., Kamal, M., Huebert, D.J., Cuff, J., Fry, B., Meissner, A., Wernig, M., Plath, K., et al., 2006. A bivalent chromatin structure marks key developmental genes in embryonic stem cells. Cell 125, 315–326.

Bertozzi, T.M., Ferguson-Smith, A.C., 2020. Metastable epialleles and their contribution to epigenetic inheritance in mammals. Semin. Cell Dev. Biol. 97, 93–105.

Bestor, T., Laudano, A., Mattaliano, R., Ingram, V., 1988. Cloning and sequencing of a cDNA encoding DNA methyltransferase of mouse cells. The carboxyl-terminal domain of the mammalian enzymes is related to bacterial restriction methyltransferases. J. Mol. Biol. 203, 971–983.

Bhaumik, S.R., Green, M.R., 2001. SAGA is an essential in vivo target of the yeast acidic activator Gal4p. Genes Dev. 15, 1935–1945.

Biddick, R., Young, E.T., 2009. The disorderly study of ordered recruitment. Yeast 26, 205–220.

Bieker, J.J., Martin, P.L., Roeder, R.G., 1985. Formation of a rate-limiting intermediate in 5S RNA gene transcription. Cell 40, 119–127.

Biggar, S.R., Crabtree, G.R., 1999. Continuous and widespread roles for the Swi-Snf complex in transcription. EMBO J. 18, 2254–2264.

Bintu, L., Kopaczynska, M., Hodges, C., Lubkowska, L., Kashlev, M., Bustamante, C., 2011. The elongation rate of RNA polymerase determines the fate of transcribed nucleosomes. Nat. Struct. Mol. Biol. 18, 1394–1399.

Bintu, L., Ishibashi, T., Dangkulwanich, M., Wu, Y.Y., Lubkowska, L., Kashlev, M., Bustamante, C., 2012. Nucleosomal elements that control the topography of the barrier to transcription. Cell 151, 738–749.

Bintu, L., Yong, J., Antebi, Y.E., McCue, K., Kazuki, Y., Uno, N., Oshimura, M., Elowitz, M.B., 2016. Dynamics of epigenetic regulation at the single-cell level. Science 351, 720–724.

Bird, A.P., 1980. DNA methylation and the frequency of CpG in animal DNA. Nucleic Acids Res. 8, 1499–1504.

Bird, A.P., 1986. CpG-rich islands and the function of DNA methylation. Nature 321, 209–213.

Bird, A., Taggart, M., Frommer, M., Miller, O.J., Macleod, D., 1985. A fraction of the mouse genome that is derived from islands of nonmethylated, CpG-rich DNA. Cell 40, 91–99.

Birkenmeier, E.H., Brown, D.D., Jordan, E., 1978. A nuclear extract of Xenopus laevis oocytes that accurately transcribes 5S RNA genes. Cell 15, 1077–1086.

Biswas, D., Dutta-Biswas, R., Mitra, D., Shibata, Y., Strahl, B.D., Formosa, T., Stillman, D.J., 2006. Opposing roles for Set2 and yFACT in regulating TBP binding at promoters. EMBO J. 25, 4479–4489.

Bittner, J.J., 1936. Some possible effects of nursing on the mammary gland tumor incidence in mice. Science 84, 162.

Bittner, J.J., 1942. The milk-influence of breast tumors in mice. Science 95, 462–463.

Björkroth, B., Ericsson, C., Lamb, M.M., Daneholt, B., 1988. Structure of the chromatin axis during transcription. Chromosoma 96, 333–340.

Blackledge, N.P., Klose, R.J., 2021. The molecular principles of gene regulation by polycomb repressive complexes. Nat. Rev. Mol. Cell Biol. 22, 815–833.

Blackledge, N.P., Fursova, N.A., Kelley, J.R., Huseyin, M.K., Feldmann, A., Klose, R.J., 2020. PRC1 catalytic activity is central to polycomb system function. Mol. Cell 77 (857–874), e859.

Blumenthal, A.B., Kriegstein, H.J., Hogness, D.S., 1974. The units of DNA replication in *Drosophila melanogaster* chromosomes. Cold Spring Harb. Symp. Quant. Biol. 38, 205–223.
Blythe, S.A., Cha, S.W., Tadjuidje, E., Heasman, J., Klein, P.S., 2010. beta-Catenin primes organizer gene expression by recruiting a histone H3 arginine 8 methyltransferase, Prmt2. Dev. Cell 19, 220–231.
Boeger, H., Griesenbeck, J., Strattan, J.S., Kornberg, R.D., 2003. Nucleosomes unfold completely at a transcriptionally active promoter. Mol. Cell 11, 1587–1598.
Boeger, H., Griesenbeck, J., Strattan, J.S., Kornberg, R.D., 2004. Removal of promoter nucleosomes by disassembly rather than sliding in vivo. Mol. Cell 14, 667–673.
Boeger, H., Griesenbeck, J., Kornberg, R.D., 2008. Nucleosome retention and the stochastic nature of promoter chromatin remodeling for transcription. Cell 133, 716–726.
Bogenhagen, D.F., Sakonju, S., Brown, D.D., 1980. A control region in the center of the 5S RNA gene directs specific initiation of transcription: II. The 3′ border of the region. Cell 19, 27–35.
Bogenhagen, D.F., Wormington, W.M., Brown, D.D., 1982. Stable transcription complexes of Xenopus 5S RNA genes: a means to maintain the differentiated state. Cell 28, 413–421.
Bondarenko, V.A., Steele, L.M., Ujvari, A., Gaykalova, D.A., Kulaeva, O.I., Polikanov, Y.S., Luse, D.S., Studitsky, V.M., 2006. Nucleosomes can form a polar barrier to transcript elongation by RNA polymerase II. Mol. Cell 24, 469–479.
Bonnet, J., Boichenko, I., Kalb, R., Le Jeune, M., Maltseva, S., Pieropan, M., Finkl, K., Fierz, B., Muller, J., 2022. PR-DUB preserves polycomb repression by preventing excessive accumulation of H2Aub1, an antagonist of chromatin compaction. Genes Dev. 36, 1046–1061.
Bortvin, A., Winston, F., 1996. Evidence that Spt6p controls chromatin structure by a direct interaction with histones. Science 272, 1473–1476.
Borun, T.W., Pearson, D., Paik, W.K., 1972. Studies of histone methylation during the HeLa S-3 cell cycle. J. Biol. Chem. 247, 4288–4298.
Boskovic, A., Rando, O.J., 2018. Transgenerational epigenetic inheritance. Annu. Rev. Genet. 52, 21–41.
Boulias, K., Greer, E.L., 2022. Means, mechanisms and consequences of adenine methylation in DNA. Nat. Rev. Genet. 23, 411–428.
Boyer, L.A., Plath, K., Zeitlinger, J., Brambrink, T., Medeiros, L.A., Lee, T.I., Levine, S.S., Wernig, M., Tajonar, A., Ray, M.K., et al., 2006. Polycomb complexes repress developmental regulators in murine embryonic stem cells. Nature 441, 349–353.
Boyle, S., Flyamer, I.M., Williamson, I., Sengupta, D., Bickmore, W.A., Illingworth, R.S., 2020. A central role for canonical PRC1 in shaping the 3D nuclear landscape. Genes Dev. 34, 931–949.
Bracken, A.P., Dietrich, N., Pasini, D., Hansen, K.H., Helin, K., 2006. Genome-wide mapping of polycomb target genes unravels their roles in cell fate transitions. Genes Dev. 20, 1123–1136.
Brahma, S., Henikoff, S., 2019. RSC-associated subnucleosomes define MNase-sensitive promoters in yeast. Mol. Cell 73 (238–249), e233.
Brahma, S., Henikoff, S., 2020. Epigenome regulation by dynamic nucleosome unwrapping. Trends Biochem. Sci. 45, 13–26.
Brennan, K.J., Weilert, M., Krueger, S., Pampari, A., Liu, H.Y., Yang, A.W.H., Morrison, J.A., Hughes, T.R., Rushlow, C.A., Kundaje, A., et al., 2023. Chromatin accessibility in the *Drosophila* embryo is determined by transcription factor pioneering and enhancer activation. Dev. Cell 58 (1898–1916), e1899.
Brent, R., Ptashne, M., 1985. A eukaryotic transcriptional activator bearing the DNA specificity of a prokaryotic repressor. Cell 43, 729–736.
Bresnick, E.H., Bustin, M., Marsaud, V., Richard-Foy, H., Hager, G.L., 1992a. The transcriptionally-active MMTV promoter is depleted of histone H1. Nucleic Acids Res. 20, 273–278.
Bresnick, E.H., Rories, C., Hager, G.L., 1992b. Evidence that nucleosomes on the mouse mammary tumor virus promoter adopt specific translational positions. Nucleic Acids Res. 20, 865–870.
Bressan, R.B., Southgate, B., Ferguson, K.M., Blin, C., Grant, V., Alfazema, N., Wills, J.C., Marques-Torrejon, M.A., Morrison, G.M., Ashmore, J., et al., 2021. Regional identity of human neural stem cells determines oncogenic responses to histone H3.3 mutants. Cell Stem Cell 28, 877–893 e879.
Brooks, T.L., Green, M.H., 1977. The sv40 transcription complex. I. Effect of viral chromatin proteins on endogenous RNA polymerase activity. Nucleic Acids Res. 4, 4261–4277.
Brown, D.D., 1994. Some genes were isolated and their structure studied before the recombinant DNA era. BioEssays 16, 139–143.

Brown, D.D., Gurdon, J.B., 1977. High-fidelity transcription of 5S DNA injected into Xenopus oocytes. Proc. Natl. Acad. Sci. USA 74, 2064–2068.

Brown, D.D., Wensink, P.C., Jordan, E., 1971. Purification and some characteristics of 5S DNA from Xenopus laevis. Proc. Natl. Acad. Sci. USA 68, 3175–3179.

Bruggemeier, U., Rogge, L., Winnacker, E.L., Beato, M., 1990. Nuclear factor I acts as a transcription factor on the MMTV promoter but competes with steroid hormone receptors for DNA binding. EMBO J. 9, 2233–2239.

Bryant, G.O., Ptashne, M., 2003. Independent recruitment in vivo by gal4 of two complexes required for transcription. Mol. Cell 11, 1301–1309.

Bulyk, M.L., Drouin, J., Harrison, M.M., Taipale, J., Zaret, K.S., 2023. Pioneer factors – key regulators of chromatin and gene expression. Nat. Rev. Genet. 24, 809–815.

Buratowski, S., Hahn, S., Guarente, L., Sharp, P.A., 1989. Five intermediate complexes in transcription initiation by RNA polymerase II. Cell 56, 549–561.

Burch, J.B., Weintraub, H., 1983. Temporal order of chromatin structural changes associated with activation of the major chicken vitellogenin gene. Cell 33, 65–76.

Burns, L.G., Peterson, C.L., 1997. The yeast SWI-SNF complex facilitates binding of a transcriptional activator to nucleosomal sites in vivo. Mol. Cell. Biol. 17, 4811–4819.

Busslinger, M., Hurst, J., Flavell, R.A., 1983. DNA methylation and the regulation of globin gene expression. Cell 34, 197–206.

Busturia, A., Wightman, C.D., Sakonju, S., 1997. A silencer is required for maintenance of transcriptional repression throughout *Drosophila* development. Development 124, 4343–4350.

Bygren, L.O., Kaati, G., Edvinsson, S., 2001. Longevity determined by paternal ancestors' nutrition during their slow growth period. Acta Biotheor. 49, 53–59.

Byvoet, P., Shepherd, G.R., Hardin, J.M., Noland, B.J., 1972. The distribution and turnover of labeled methyl groups in histone fractions of cultured mammalian cells. Arch. Biochem. Biophys. 148, 558–567.

Cao, R., Wang, L., Wang, H., Xia, L., Erdjument-Bromage, H., Tempst, P., Jones, R.S., Zhang, Y., 2002. Role of histone H3 lysine 27 methylation in polycomb-group silencing. Science 298, 1039–1043.

Cao, R., Tsukada, Y., Zhang, Y., 2005. Role of Bmi-1 and Ring1A in H2A ubiquitylation and Hox gene silencing. Mol. Cell 20, 845–854.

Capuano, F., Mulleder, M., Kok, R., Blom, H.J., Ralser, M., 2014. Cytosine DNA methylation is found in *Drosophila melanogaster* but absent in Saccharomyces cerevisiae, *Schizosaccharomyces pombe*, and other yeast species. Anal. Chem. 86, 3697–3702.

Carey, M., 1998. The enhanceosome and transcriptional synergy. Cell 92, 5–8.

Carlson, M., Osmond, B.C., Botstein, D., 1981. Mutants of yeast defective in sucrose utilization. Genetics 98, 25–40.

Carninci, P., Sandelin, A., Lenhard, B., Katayama, S., Shimokawa, K., Ponjavic, J., Semple, C.A., Taylor, M.S., Engstrom, P.G., Frith, M.C., et al., 2006. Genome-wide analysis of mammalian promoter architecture and evolution. Nat. Genet. 38, 626–635.

Carr, K.D., Richard-Foy, H., 1990. Glucocorticoids locally disrupt an array of positioned nucleosomes on the rat tyrosine aminotransferase promoter in hepatoma cells. Proc. Natl. Acad. Sci. USA 87, 9300–9304.

Carrozza, M.J., Li, B., Florens, L., Suganuma, T., Swanson, S.K., Lee, K.K., Shia, W.J., Anderson, S., Yates, J., Washburn, M.P., et al., 2005. Histone H3 methylation by Set2 directs deacetylation of coding regions by Rpd3S to suppress spurious intragenic transcription. Cell 123, 581–592.

Carvalho, S., Raposo, A.C., Martins, F.B., Grosso, A.R., Sridhara, S.C., Rino, J., Carmo-Fonseca, M., de Almeida, S.F., 2013. Histone methyltransferase SETD2 coordinates FACT recruitment with nucleosome dynamics during transcription. Nucleic Acids Res. 41, 2881–2893.

Castel, D., Philippe, C., Calmon, R., Le Dret, L., Truffaux, N., Boddaert, N., Pages, M., Taylor, K.R., Saulnier, P., Lacroix, L., et al., 2015. Histone H3F3A and HIST1H3B K27M mutations define two subgroups of diffuse intrinsic pontine gliomas with different prognosis and phenotypes. Acta Neuropathol. 130, 815–827.

Catania, S., Dumesic, P.A., Pimentel, H., Nasif, A., Stoddard, C.I., Burke, J.E., Diedrich, J.K., Cook, S., Shea, T., Geinger, E., et al., 2020. Evolutionary persistence of DNA methylation for millions of years after ancient loss of a de novo methyltransferase. Cell 180 (263–277), e220.

Cavalli, G., Heard, E., 2019. Advances in epigenetics link genetics to the environment and disease. Nature 571, 489–499.

Cavalli, G., Paro, R., 1998. The *Drosophila* Fab-7 chromosomal element conveys epigenetic inheritance during mitosis and meiosis. Cell 93, 505–518.

Cavalli, G., Paro, R., 1999. Epigenetic inheritance of active chromatin after removal of the main transactivator. Science 286, 955–958.

Chan, K.M., Fang, D., Gan, H., Hashizume, R., Yu, C., Schroeder, M., Gupta, N., Mueller, S., James, C.D., Jenkins, R., et al., 2013. The histone H3.3K27M mutation in pediatric glioma reprograms H3K27 methylation and gene expression. Genes Dev. 27, 985–990.

Chang, C.H., Luse, D.S., 1997. The H3/H4 tetramer blocks transcript elongation by RNA polymerase II in vitro. J. Biol. Chem. 272, 23427–23434.

Chavez, S., Beato, M., 1997. Nucleosome-mediated synergism between transcription factors on the mouse mammary tumor virus promoter. Proc. Natl. Acad. Sci. USA 94, 2885–2890.

Chen, T.A., Allfrey, V.G., 1987. Rapid and reversible changes in nucleosome structure accompany the activation, repression, and superinduction of murine fibroblast protooncogenes c-fos and c-myc. Proc. Natl. Acad. Sci. USA 84, 5252–5256.

Chen, T.A., Sterner, R., Cozzolino, A., Allfrey, V.G., 1990. Reversible and irreversible changes in nucleosome structure along the c-fos and c-myc oncogenes following inhibition of transcription. J. Mol. Biol. 212, 481–493.

Chen, H., Kharerin, H., Dhasarathy, A., Kladde, M., Bai, L., 2022. Partitioned usage of chromatin remodelers by nucleosome-displacing factors. Cell Rep. 40, 111250.

Chetverina, D.A., Lomaev, D.V., Erokhin, M.M., 2020. Polycomb and trithorax group proteins: the long road from mutations in drosophila to use in medicine. Acta Nat. 12, 66–85.

Cheung, A.C., Cramer, P., 2011. Structural basis of RNA polymerase II backtracking, arrest and reactivation. Nature 471, 249–253.

Cheung, V., Chua, G., Batada, N.N., Landry, C.R., Michnick, S.W., Hughes, T.R., Winston, F., 2008. Chromatin- and transcription-related factors repress transcription from within coding regions throughout the *Saccharomyces cerevisiae* genome. PLoS Biol. 6, e277.

Chodosh, L.A., Fire, A., Samuels, M., Sharp, P.A., 1989. 5,6-Dichloro-1-beta-D-ribofuranosylbenzimidazole inhibits transcription elongation by RNA polymerase II in vitro. J. Biol. Chem. 264, 2250–2257.

Choi, J.K., Kim, Y.J., 2008. Epigenetic regulation and the variability of gene expression. Nat. Genet. 40, 141–147.

Churchman, L.S., Weissman, J.S., 2011. Nascent transcript sequencing visualizes transcription at nucleotide resolution. Nature 469, 368–373.

Cirillo, L.A., Zaret, K.S., 1999. An early developmental transcription factor complex that is more stable on nucleosome core particles than on free DNA. Mol. Cell 4, 961–969.

Cirillo, L.A., McPherson, C.E., Bossard, P., Stevens, K., Cherian, S., Shim, E.Y., Clark, K.L., Burley, S.K., Zaret, K.S., 1998. Binding of the winged-helix transcription factor HNF3 to a linker histone site on the nucleosome. EMBO J. 17, 244–254.

Cirillo, L.A., Lin, F.R., Cuesta, I., Friedman, D., Jarnik, M., Zaret, K.S., 2002. Opening of compacted chromatin by early developmental transcription factors HNF3 (FoxA) and GATA-4. Mol. Cell 9, 279–289.

Clark, D.J., 1995. Transcription through the nucleosome. In: Wolffe, A. (Ed.), The Nucleus. JAI Press, Greenwich, CT, pp. 207–239.

Clark, D.J., Felsenfeld, G., 1991. Formation of nucleosomes on positively supercoiled DNA. EMBO J. 10, 387–395.

Clark, D.J., Felsenfeld, G., 1992. A nucleosome core is transferred out of the path of a transcribing polymerase. Cell 71, 11–22.

Clark, D.J., Wolffe, A.P., 1991. Superhelical stress and nucleosome-mediated repression of 5S RNA gene transcription in vitro. EMBO J. 10, 3419–3428.

Clark, K.L., Halay, E.D., Lai, E., Burley, S.K., 1993. Co-crystal structure of the HNF-3/fork head DNA-recognition motif resembles histone H5. Nature 364, 412–420.

Clark-Adams, C.D., Norris, D., Osley, M.A., Fassler, J.S., Winston, F., 1988. Changes in histone gene dosage alter transcription in yeast. Genes Dev. 2, 150–159.

Clouaire, T., Webb, S., Skene, P., Illingworth, R., Kerr, A., Andrews, R., Lee, J.H., Skalnik, D., Bird, A., 2012. Cfp1 integrates both CpG content and gene activity for accurate H3K4me3 deposition in embryonic stem cells. Genes Dev. 26, 1714–1728.

Cole, H.A., Ocampo, J., Iben, J.R., Chereji, R.V., Clark, D.J., 2014. Heavy transcription of yeast genes correlates with differential loss of histone H2B relative to H4 and queued RNA polymerases. Nucleic Acids Res. 42, 12512–12522.

Coleman, R.T., Struhl, G., 2017. Causal role for inheritance of H3K27me3 in maintaining the OFF state of a *Drosophila* HOX gene. Science 356.
Conaway, J.W., Bond, M.W., Conaway, R.C., 1987. An RNA polymerase II transcription system from rat liver. Purification of an essential component. J. Biol. Chem. 262, 8293–8297.
Cordingley, M.G., Riegel, A.T., Hager, G.L., 1987. Steroid-dependent interaction of transcription factors with the inducible promoter of mouse mammary tumor virus in vivo. Cell 48, 261–270.
Core, L.J., Waterfall, J.J., Lis, J.T., 2008. Nascent RNA sequencing reveals widespread pausing and divergent initiation at human promoters. Science 322, 1845–1848.
Cosma, M.P., 2002. Ordered recruitment: gene-specific mechanism of transcription activation. Mol. Cell 10, 227–236.
Cosma, M.P., Tanaka, T., Nasmyth, K., 1999. Ordered recruitment of transcription and chromatin remodeling factors to a cell cycle- and developmentally regulated promoter. Cell 97, 299–311.
Costa, P.J., Arndt, K.M., 2000. Synthetic lethal interactions suggest a role for the *Saccharomyces cerevisiae* Rtf1 protein in transcription elongation. Genetics 156, 535–547.
Crickard, J.B., Lee, J., Lee, T.H., Reese, J.C., 2017. The elongation factor Spt4/5 regulates RNA polymerase II transcription through the nucleosome. Nucleic Acids Res. 45, 6362–6374.
Crow, J.F., Bender, W., 2004. Edward B. Lewis, 1918–2004. Genetics 168, 1773–1783.
Czermin, B., Melfi, R., McCabe, D., Seitz, V., Imhof, A., Pirrotta, V., 2002. Drosophila enhancer of Zeste/ESC complexes have a histone H3 methyltransferase activity that marks chromosomal polycomb sites. Cell 111, 185–196.
De Bernardin, W., Koller, T., Sogo, J.M., 1986. Structure of in-vivo transcribing chromatin as studied in simian virus 40 minichromosomes. J. Mol. Biol. 191, 469–482.
Deaton, A.M., Bird, A., 2011. CpG islands and the regulation of transcription. Genes Dev. 25, 1010–1022.
DeCamillis, M., Cheng, N.S., Pierre, D., Brock, H.W., 1992. The polyhomeotic gene of *Drosophila* encodes a chromatin protein that shares polytene chromosome-binding sites with polycomb. Genes Dev. 6, 223–232.
Decker, T.M., 2021. Mechanisms of Transcription Elongation Factor DSIF (Spt4-Spt5). J. Mol. Biol. 433, 166657.
DeGennaro, C.M., Alver, B.H., Marguerat, S., Stepanova, E., Davis, C.P., Bahler, J., Park, P.J., Winston, F., 2013. Spt6 regulates intragenic and antisense transcription, nucleosome positioning, and histone modifications genome-wide in fission yeast. Mol. Cell. Biol. 33, 4779–4792.
Devlin, C., Tice-Baldwin, K., Shore, D., Arndt, K.T., 1991. RAP1 is required for BAS1/BAS2- and GCN4-dependent transcription of the yeast HIS4 gene. Mol. Cell. Biol. 11, 3642–3651.
Dhasarathy, A., Kladde, M.P., 2005. Promoter occupancy is a major determinant of chromatin remodeling enzyme requirements. Mol. Cell. Biol. 25, 2698–2707.
Dignam, J.D., Lebovitz, R.M., Roeder, R.G., 1983. Accurate transcription initiation by RNA polymerase II in a soluble extract from isolated mammalian nuclei. Nucleic Acids Res. 11, 1475–1489.
Dion, M.F., Altschuler, S.J., Wu, L.F., Rando, O.J., 2005. Genomic characterization reveals a simple histone H4 acetylation code. Proc. Natl. Acad. Sci. USA 102, 5501–5506.
Dion, M.F., Kaplan, T., Kim, M., Buratowski, S., Friedman, N., Rando, O.J., 2007. Dynamics of replication-independent histone turnover in budding yeast. Science 315, 1405–1408.
Dodonova, S.O., Zhu, F., Dienemann, C., Taipale, J., Cramer, P., 2020. Nucleosome-bound SOX2 and SOX11 structures elucidate pioneer factor function. Nature 580, 669–672.
Donovan, B.T., Chen, H., Jipa, C., Bai, L., Poirier, M.G., 2019. Dissociation rate compensation mechanism for budding yeast pioneer transcription factors. elife 8.
Donovan, B.T., Chen, H., Eek, P., Meng, Z., Jipa, C., Tan, S., Bai, L., Poirier, M.G., 2023a. Basic helix-loop-helix pioneer factors interact with the histone octamer to invade nucleosomes and generate nucleosome-depleted regions. Mol. Cell 83 (1251–1263), e1256.
Donovan, B.T., Luo, Y., Meng, Z., Poirier, M.G., 2023b. The nucleosome unwrapping free energy landscape defines distinct regions of transcription factor accessibility and kinetics. Nucleic Acids Res. 51, 1139–1153.
Doris, S.M., Chuang, J., Viktorovskaya, O., Murawska, M., Spatt, D., Churchman, L.S., Winston, F., 2018. Spt6 is required for the fidelity of promoter selection. Mol. Cell 72 (687–699), e686.
Dove, S.L., Joung, J.K., Hochschild, A., 1997. Activation of prokaryotic transcription through arbitrary protein-protein contacts. Nature 386, 627–630.
Drouin, S., Laramee, L., Jacques, P.E., Forest, A., Bergeron, M., Robert, F., 2010. DSIF and RNA polymerase II CTD phosphorylation coordinate the recruitment of Rpd3S to actively transcribed genes. PLoS Genet. 6, e1001173.

Duboule, D., Dolle, P., 1989. The structural and functional organization of the murine HOX gene family resembles that of *Drosophila* homeotic genes. EMBO J. 8, 1497–1505.

Duncan, I., Montgomery, G., 2002. E. B. Lewis and the bithorax complex: part I. Genetics 160, 1265–1272.

Dupont, C., Armant, D.R., Brenner, C.A., 2009. Epigenetics: definition, mechanisms and clinical perspective. Semin. Reprod. Med. 27, 351–357.

Durrin, L.K., Mann, R.K., Kayne, P.S., Grunstein, M., 1991. Yeast histone H4 N-terminal sequence is required for promoter activation in vivo. Cell 65, 1023–1031.

Ebbert, R., Birkmann, A., Schuller, H.J., 1999. The product of the SNF2/SWI2 paralogue INO80 of Saccharomyces cerevisiae required for efficient expression of various yeast structural genes is part of a high-molecular-weight protein complex. Mol. Microbiol. 32, 741–751.

Echigoya, K., Koyama, M., Negishi, L., Takizawa, Y., Mizukami, Y., Shimabayashi, H., Kuroda, A., Kurumizaka, H., 2020. Nucleosome binding by the pioneer transcription factor OCT4. Sci. Rep. 10, 11832.

Ehara, H., Yokoyama, T., Shigematsu, H., Yokoyama, S., Shirouzu, M., Sekine, S.I., 2017. Structure of the complete elongation complex of RNA polymerase II with basal factors. Science 357, 921–924.

Ehara, H., Kujirai, T., Fujino, Y., Shirouzu, M., Kurumizaka, H., Sekine, S.I., 2019. Structural insight into nucleosome transcription by RNA polymerase II with elongation factors. Science 363, 744–747.

Ehara, H., Kujirai, T., Shirouzu, M., Kurumizaka, H., Sekine, S.I., 2022. Structural basis of nucleosome disassembly and reassembly by RNAPII elongation complex with FACT. Science 377, eabp9466.

Eisen, M.B., Spellman, P.T., Brown, P.O., Botstein, D., 1998. Cluster analysis and display of genome-wide expression patterns. Proc. Natl. Acad. Sci. USA 95, 14863–14868.

Eisfeld, K., Candau, R., Truss, M., Beato, M., 1997. Binding of NF1 to the MMTV promoter in nucleosomes: influence of rotational phasing, translational positioning and histone H1. Nucleic Acids Res. 25, 3733–3742.

Eissenberg, J.C., Cartwright, I.L., Thomas, G.H., Elgin, S.C., 1985. Selected topics in chromatin structure. Annu. Rev. Genet. 19, 485–536.

Elgin, S.C., 1988. The formation and function of DNase I hypersensitive sites in the process of gene activation. J. Biol. Chem. 263, 19259–19262.

Elgin, S.C., Reuter, G., 2013. Position-effect variegation, heterochromatin formation, and gene silencing in *Drosophila*. Cold Spring Harb. Perspect. Biol. 5, a017780.

Emerson, B.M., Felsenfeld, G., 1984. Specific factor conferring nuclease hypersensitivity at the 5′ end of the chicken adult beta-globin gene. Proc. Natl. Acad. Sci. USA 81, 95–99.

Engelke, D.R., Ng, S.Y., Shastry, B.S., Roeder, R.G., 1980. Specific interaction of a purified transcription factor with an internal control region of 5S RNA genes. Cell 19, 717–728.

Ertel, F., Dirac-Svejstrup, A.B., Hertel, C.B., Blaschke, D., Svejstrup, J.Q., Korber, P., 2010. In vitro reconstitution of PHO5 promoter chromatin remodeling points to a role for activator-nucleosome competition in vivo. Mol. Cell. Biol. 30, 4060–4076.

Escobar, T.M., Oksuz, O., Saldana-Meyer, R., Descostes, N., Bonasio, R., Reinberg, D., 2019. Active and repressed chromatin domains exhibit distinct nucleosome segregation during DNA replication. Cell 179 (953–963), e911.

Eskeland, R., Leeb, M., Grimes, G.R., Kress, C., Boyle, S., Sproul, D., Gilbert, N., Fan, Y., Skoultchi, A.I., Wutz, A., et al., 2010. Ring1B compacts chromatin structure and represses gene expression independent of histone ubiquitination. Mol. Cell 38, 452–464.

Falvo, J.V., Thanos, D., Maniatis, T., 1995. Reversal of intrinsic DNA bends in the IFN beta gene enhancer by transcription factors and the architectural protein HMG I(Y). Cell 83, 1101–1111.

Farkas, G., Gausz, J., Galloni, M., Reuter, G., Gyurkovics, H., Karch, F., 1994. The Trithorax-like gene encodes the *Drosophila* GAGA factor. Nature 371, 806–808.

Farnung, L., 2023. Nucleosomes unwrapped: Structural perspectives on transcription through chromatin. Curr. Opin. Struct. Biol. 82, 102690.

Farnung, L., Vos, S.M., Cramer, P., 2018. Structure of transcribing RNA polymerase II-nucleosome complex. Nat. Commun. 9, 5432.

Farnung, L., Ochmann, M., Engeholm, M., Cramer, P., 2021. Structural basis of nucleosome transcription mediated by Chd1 and FACT. Nat. Struct. Mol. Biol. 28, 382–387.

Farnung, L., Ochmann, M., Garg, G., Vos, S.M., Cramer, P., 2022. Structure of a backtracked hexasomal intermediate of nucleosome transcription. Mol. Cell 82 (3126–3134), e3127.

Fascher, K.D., Schmitz, J., Horz, W., 1990. Role of trans-activating proteins in the generation of active chromatin at the PHO5 promoter in *S. cerevisiae*. EMBO J. 9, 2523–2528.

Fascher, K.D., Schmitz, J., Horz, W., 1993. Structural and functional requirements for the chromatin transition at the PHO_5 promoter in *Saccharomyces cerevisiae* upon PHO_5 activation. J. Mol. Biol. 231, 658–667.
Fedor, M.J., Lue, N.F., Kornberg, R.D., 1988. Statistical positioning of nucleosomes by specific protein-binding to an upstream activating sequence in yeast. J. Mol. Biol. 204, 109–127.
Fedoroff, N.V., Brown, D.D., 1978. The nucleotide sequence of oocyte 5S DNA in Xenopus laevis. I. The AT-rich spacer. Cell 13, 701–716.
Felsenfeld, G., 1992. Chromatin as an essential part of the transcriptional mechanism. Nature 355, 219–224.
Felsenfeld, G., Clark, D., Studitsky, V., 2000. Transcription through nucleosomes. Biophys. Chem. 86, 231–237.
Felts, S.J., Weil, P.A., Chalkley, R., 1990. Transcription factor requirements for in vitro formation of transcriptionally competent 5S rRNA gene chromatin. Mol. Cell. Biol. 10, 2390–2401.
Fennessy, R.T., Owen-Hughes, T., 2016. Establishment of a promoter-based chromatin architecture on recently replicated DNA can accommodate variable inter-nucleosome spacing. Nucleic Acids Res. 44, 7189–7203.
Fernandez Garcia, M., Moore, C.D., Schulz, K.N., Alberto, O., Donague, G., Harrison, M.M., Zhu, H., Zaret, K.S., 2019. Structural features of transcription factors associating with nucleosome binding. Mol. Cell 75 (921–932), e926.
Field, Y., Kaplan, N., Fondufe-Mittendorf, Y., Moore, I.K., Sharon, E., Lubling, Y., Widom, J., Segal, E., 2008. Distinct modes of regulation by chromatin encoded through nucleosome positioning signals. PLoS Comput. Biol. 4, e1000216.
Field, Y., Fondufe-Mittendorf, Y., Moore, I.K., Mieczkowski, P., Kaplan, N., Lubling, Y., Lieb, J.D., Widom, J., Segal, E., 2009. Gene expression divergence in yeast is coupled to evolution of DNA-encoded nucleosome organization. Nat. Genet. 41, 438–445.
Filipovski, M., Soffers, J.H.M., Vos, S.M., Farnung, L., 2022. Structural basis of nucleosome retention during transcription elongation. Science 376, 1313–1316.
Fischbeck, J.A., Kraemer, S.M., Stargell, L.A., 2002. SPN1, a conserved gene identified by suppression of a postrecruitment-defective yeast TATA-binding protein mutant. Genetics 162, 1605–1616.
Fish, R.N., Kane, C.M., 2002. Promoting elongation with transcript cleavage stimulatory factors. Biochim. Biophys. Acta 1577, 287–307.
Fisher, S., Grice, E.A., Vinton, R.M., Bessling, S.L., McCallion, A.S., 2006. Conservation of RET regulatory function from human to zebrafish without sequence similarity. Science 312, 276–279.
Fitz-James, M.H., Cavalli, G., 2022. Molecular mechanisms of transgenerational epigenetic inheritance. Nat. Rev. Genet. 23, 325–341.
Fleming, A.B., Kao, C.F., Hillyer, C., Pikaart, M., Osley, M.A., 2008. H2B ubiquitylation plays a role in nucleosome dynamics during transcription elongation. Mol. Cell 31, 57–66.
Floer, M., Wang, X., Prabhu, V., Berrozpe, G., Narayan, S., Spagna, D., Alvarez, D., Kendall, J., Krasnitz, A., Stepansky, A., et al., 2010. A RSC/nucleosome complex determines chromatin architecture and facilitates activator binding. Cell 141, 407–418.
Foe, V.E., Wilkinson, L.E., Laird, C.D., 1976. Comparative organization of active transcription units in *Oncopeltus fasciatus*. Cell 9, 131–146.
Ford, J., Odeyale, O., Eskandar, A., Kouba, N., Shen, C.H., 2007. A SWI/SNF- and INO80-dependent nucleosome movement at the INO1 promoter. Biochem. Biophys. Res. Commun. 361, 974–979.
Formosa, T., Winston, F., 2020. The role of FACT in managing chromatin: disruption, assembly, or repair? Nucleic Acids Res. 48, 11929–11941.
Formosa, T., Ruone, S., Adams, M.D., Olsen, A.E., Eriksson, P., Yu, Y., Rhoades, A.R., Kaufman, P.D., Stillman, D.J., 2002. Defects in SPT16 or POB3 (yFACT) in *Saccharomyces cerevisiae* cause dependence on the Hir/Hpc pathway: polymerase passage may degrade chromatin structure. Genetics 162, 1557–1571.
Fragoso, G., John, S., Roberts, M.S., Hager, G.L., 1995. Nucleosome positioning on the MMTV LTR results from the frequency-biased occupancy of multiple frames. Genes Dev. 9, 1933–1947.
Francette, A.M., Tripplehorn, S.A., Arndt, K.M., 2021. The Paf1 complex: a keystone of nuclear regulation operating at the interface of transcription and chromatin. J. Mol. Biol. 433, 166979.
Francis, N.J., Saurin, A.J., Shao, Z., Kingston, R.E., 2001. Reconstitution of a functional core polycomb repressive complex. Mol. Cell 8, 545–556.
Francis, N.J., Kingston, R.E., Woodcock, C.L., 2004. Chromatin compaction by a polycomb group protein complex. Science 306, 1574–1577.
Franke, A., DeCamillis, M., Zink, D., Cheng, N., Brock, H.W., Paro, R., 1992. Polycomb and polyhomeotic are constituents of a multimeric protein complex in chromatin of *Drosophila melanogaster*. EMBO J. 11, 2941–2950.

Fraser, R., Lin, C.J., 2016. Epigenetic reprogramming of the zygote in mice and men: on your marks, get set, go! Reproduction 152, R211–R222.

Frederick, M.A., Williamson, K.E., Fernandez Garcia, M., Ferretti, M.B., McCarthy, R.L., Donahue, G., Luzete Monteiro, E., Takenaka, N., Reynaga, J., Kadoch, C., et al., 2023. A pioneer factor locally opens compacted chromatin to enable targeted ATP-dependent nucleosome remodeling. Nat. Struct. Mol. Biol. 30, 31–37.

Friman, E.T., Deluz, C., Meireles-Filho, A.C., Govindan, S., Gardeux, V., Deplancke, B., Suter, D.M., 2019. Dynamic regulation of chromatin accessibility by pluripotency transcription factors across the cell cycle. elife 8.

Frommer, M., McDonald, L.E., Millar, D.S., Collis, C.M., Watt, F., Grigg, G.W., Molloy, P.L., Paul, C.L., 1992. A genomic sequencing protocol that yields a positive display of 5-methylcytosine residues in individual DNA strands. Proc. Natl. Acad. Sci. USA 89, 1827–1831.

Fryer, C.J., Archer, T.K., 1998. Chromatin remodelling by the glucocorticoid receptor requires the BRG1 complex. Nature 393, 88–91.

Gao, Z., Zhang, J., Bonasio, R., Strino, F., Sawai, A., Parisi, F., Kluger, Y., Reinberg, D., 2012. PCGF homologs, CBX proteins, and RYBP define functionally distinct PRC1 family complexes. Mol. Cell 45, 344–356.

Gartenberg, M.R., Smith, J.S., 2016. The nuts and bolts of transcriptionally silent chromatin in *Saccharomyces cerevisiae*. Genetics 203, 1563–1599.

Gatewood, J.M., Cook, G.R., Balhorn, R., Schmid, C.W., Bradbury, E.M., 1990. Isolation of four core histones from human sperm chromatin representing a minor subset of somatic histones. J. Biol. Chem. 265, 20662–20666.

Gaudreau, L., Schmid, A., Blaschke, D., Ptashne, M., Horz, W., 1997. RNA polymerase II holoenzyme recruitment is sufficient to remodel chromatin at the yeast PHO_5 promoter. Cell 89, 55–62.

Gehring, M., 2019. Epigenetic dynamics during flowering plant reproduction: evidence for reprogramming? New Phytol. 224, 91–96.

Geiduschek, E.P., Tocchini-Valentini, G.P., 1988. Transcription by RNA polymerase III. Annu. Rev. Biochem. 57, 873–914.

Geijer, M.E., Zhou, D., Selvam, K., Steurer, B., Mukherjee, C., Evers, B., Cugusi, S., van Toorn, M., van der Woude, M., Janssens, R.C., et al., 2021. Elongation factor ELOF1 drives transcription-coupled repair and prevents genome instability. Nat. Cell Biol. 23, 608–619.

Gerstein, M.B., Lu, Z.J., Van Nostrand, E.L., Cheng, C., Arshinoff, B.I., Liu, T., Yip, K.Y., Robilotto, R., Rechtsteiner, A., Ikegami, K., et al., 2010. Integrative analysis of the *Caenorhabditis elegans* genome by the modENCODE project. Science 330, 1775–1787.

Glikin, G.C., Ruberti, I., Worcel, A., 1984. Chromatin assembly in Xenopus oocytes: in vitro studies. Cell 37, 33–41.

Glont, S.E., Chernukhin, I., Carroll, J.S., 2019. Comprehensive genomic analysis reveals that the pioneering function of FOXA1 is independent of hormonal signaling. Cell Rep. 26 (2558–2565), e2553.

Gonzalez, P.J., Palacian, E., 1989. Interaction of RNA polymerase II with structurally altered nucleosomal particles. Transcription is facilitated by loss of one H2A.H2B dimer. J. Biol. Chem. 264, 18457–18462.

Gonzalez-Rodriguez, P., Fullgrabe, J., Joseph, B., 2023. The hunger strikes back: an epigenetic memory for autophagy. Cell Death Differ. 30, 1404–1415.

Gopalakrishnan, R., Winston, F., 2021. The histone chaperone Spt6 is required for normal recruitment of the capping enzyme Abd1 to transcribed regions. J. Biol. Chem. 297, 101205.

Gottesfeld, J., Bloomer, L.S., 1982. Assembly of transcriptionally active 5S RNA gene chromatin in vitro. Cell 28, 781–791.

Gottschling, D.E., Aparicio, O.M., Billington, B.L., Zakian, V.A., 1990. Position effect at *S. cerevisiae* telomeres: reversible repression of Pol II transcription. Cell 63, 751–762.

Govind, C.K., Yoon, S., Qiu, H., Govind, S., Hinnebusch, A.G., 2005. Simultaneous recruitment of coactivators by Gcn4p stimulates multiple steps of transcription in vivo. Mol. Cell. Biol. 25, 5626–5638.

Govind, C.K., Qiu, H., Ginsburg, D.S., Ruan, C., Hofmeyer, K., Hu, C., Swaminathan, V., Workman, J.L., Li, B., Hinnebusch, A.G., 2010. Phosphorylated Pol II CTD recruits multiple HDACs, including Rpd3C(S), for methylation-dependent deacetylation of ORF nucleosomes. Mol. Cell 39, 234–246.

Graham, A., Papalopulu, N., Krumlauf, R., 1989. The murine and *Drosophila* homeobox gene complexes have common features of organization and expression. Cell 57, 367–378.

Grand, R.S., Burger, L., Grawe, C., Michael, A.K., Isbel, L., Hess, D., Hoerner, L., Iesmantavicius, V., Durdu, S., Pregnolato, M., et al., 2021. BANP opens chromatin and activates CpG-island-regulated genes. Nature 596, 133–137.

Granok, H., Leibovitch, B.A., Shaffer, C.D., Elgin, S.C., 1995. Chromatin. Ga-ga over GAGA factor. Curr. Biol. 5, 238–241.

Grassl, N., Poschke, I., Lindner, K., Bunse, L., Mildenberger, I., Boschert, T., Jahne, K., Green, E.W., Hulsmeyer, I., Junger, S., et al., 2023. A H3K27M-targeted vaccine in adults with diffuse midline glioma. Nat. Med.
Gregory, P.D., Schmid, A., Zavari, M., Lui, L., Berger, S.L., Horz, W., 1998. Absence of Gcn5 HAT activity defines a novel state in the opening of chromatin at the PHO_5 promoter in yeast. Mol. Cell 1, 495–505.
Gregory, P.D., Schmid, A., Zavari, M., Munsterkotter, M., Horz, W., 1999. Chromatin remodelling at the PHO8 promoter requires SWI-SNF and SAGA at a step subsequent to activator binding. EMBO J. 18, 6407–6414.
Grewal, S.I.S., 2023. The molecular basis of heterochromatin assembly and epigenetic inheritance. Mol. Cell 83, 1767–1785.
Grewal, S.I., Klar, A.J., 1996. Chromosomal inheritance of epigenetic states in fission yeast during mitosis and meiosis. Cell 86, 95–101.
Grimaud, C., Negre, N., Cavalli, G., 2006. From genetics to epigenetics: the tale of polycomb group and trithorax group genes. Chromosom. Res. 14, 363–375.
Grippo, P., Iaccarino, M., Parisi, E., Scarano, E., 1968. Methylation of DNA in developing sea urchin embryos. J. Mol. Biol. 36, 195–208.
Grohmann, D., Nagy, J., Chakraborty, A., Klose, D., Fielden, D., Ebright, R.H., Michaelis, J., Werner, F., 2011. The initiation factor TFE and the elongation factor Spt4/5 compete for the RNAP clamp during transcription initiation and elongation. Mol. Cell 43, 263–274.
Groudine, M., Weintraub, H., 1982. Propagation of globin DNAase I-hypersensitive sites in absence of factors required for induction: a possible mechanism for determination. Cell 30, 131–139.
Gruenbaum, Y., Cedar, H., Razin, A., 1982. Substrate and sequence specificity of a eukaryotic DNA methylase. Nature 295, 620–622.
Grunstein, M., 1992. Histones as regulators of genes. Sci. Am. 267, 68–74B.
Gu, L., Wang, Q., Sun, Q.Y., 2010. Histone modifications during mammalian oocyte maturation: dynamics, regulation and functions. Cell Cycle 9, 1942–1950.
Gualdi, R., Bossard, P., Zheng, M., Hamada, Y., Coleman, J.R., Zaret, K.S., 1996. Hepatic specification of the gut endoderm in vitro: cell signaling and transcriptional control. Genes Dev. 10, 1670–1682.
Guseinov, V.A., Vanyushin, B.F., 1975. Content and localisation of 5-methylcytosine in DNA of healthy and wilt-infected cotton plants. Biochim. Biophys. Acta 395, 229–238.
Haberle, V., Stark, A., 2018. Eukaryotic core promoters and the functional basis of transcription initiation. Nat. Rev. Mol. Cell Biol. 19, 621–637.
Hager, G.L., Richard-Foy, H., Kessel, M., Wheeler, D., Lichtler, A.C., Ostrowski, M.C., 1984. The mouse mammary tumor virus model in studies of glucocorticoid regulation. Recent Prog. Horm. Res. 40, 121–142.
Haig, D., 2012. Commentary: the epidemiology of epigenetics. Int. J. Epidemiol. 41, 13–16.
Hall, I.M., Shankaranarayana, G.D., Noma, K., Ayoub, N., Cohen, A., Grewal, S.I., 2002. Establishment and maintenance of a heterochromatin domain. Science 297, 2232–2237.
Hall, M.A., Shundrovsky, A., Bai, L., Fulbright, R.M., Lis, J.T., Wang, M.D., 2009. High-resolution dynamic mapping of histone-DNA interactions in a nucleosome. Nat. Struct. Mol. Biol. 16, 124–129.
Hammoud, S.S., Nix, D.A., Zhang, H., Purwar, J., Carrell, D.T., Cairns, B.R., 2009. Distinctive chromatin in human sperm packages genes for embryo development. Nature 460, 473–478.
Han, M., Grunstein, M., 1988. Nucleosome loss activates yeast downstream promoters in vivo. Cell 55, 1137–1145.
Han, M., Chang, M., Kim, U.J., Grunstein, M., 1987. Histone H2B repression causes cell-cycle-specific arrest in yeast: effects on chromosomal segregation, replication, and transcription. Cell 48, 589–597.
Han, M., Kim, U.J., Kayne, P., Grunstein, M., 1988. Depletion of histone H4 and nucleosomes activates the PHO_5 gene in *Saccharomyces cerevisiae*. EMBO J. 7, 2221–2228.
Hansen, J.C., Wolffe, A.P., 1992. Influence of chromatin folding on transcription initiation and elongation by RNA polymerase III. Biochemistry 31, 7977–7988.
Hansen, K.H., Bracken, A.P., Pasini, D., Dietrich, N., Gehani, S.S., Monrad, A., Rappsilber, J., Lerdrup, M., Helin, K., 2008. A model for transmission of the H3K27me3 epigenetic mark. Nat. Cell Biol. 10, 1291–1300.
Hansen, J.L., Loell, K.J., Cohen, B.A., 2022. A test of the pioneer factor hypothesis using ectopic liver gene activation. elife 11.
Hartzog, G.A., Wada, T., Handa, H., Winston, F., 1998. Evidence that Spt4, Spt5, and Spt6 control transcription elongation by RNA polymerase II in *Saccharomyces cerevisiae*. Genes Dev. 12, 357–369.
Hathaway, N.A., Bell, O., Hodges, C., Miller, E.L., Neel, D.S., Crabtree, G.R., 2012. Dynamics and memory of heterochromatin in living cells. Cell 149, 1447–1460.

Hayes, J.J., Tullius, T.D., Wolffe, A.P., 1990. The structure of DNA in a nucleosome. Proc. Natl. Acad. Sci. USA 87, 7405–7409.

He, Q., Battistella, L., Morse, R.H., 2008. Mediator requirement downstream of chromatin remodeling during transcriptional activation of CHA1 in yeast. J. Biol. Chem. 283, 5276–5286.

He, H.H., Meyer, C.A., Shin, H., Bailey, S.T., Wei, G., Wang, Q., Zhang, Y., Xu, K., Ni, M., Lupien, M., et al., 2010. Nucleosome dynamics define transcriptional enhancers. Nat. Genet. 42, 343–347.

Heard, E., Martienssen, R.A., 2014. Transgenerational epigenetic inheritance: myths and mechanisms. Cell 157, 95–109.

Hebbar, P.B., Archer, T.K., 2003. Nuclear factor 1 is required for both hormone-dependent chromatin remodeling and transcriptional activation of the mouse mammary tumor virus promoter. Mol. Cell. Biol. 23, 887–898.

Hebbar, P.B., Archer, T.K., 2008. Altered histone H1 stoichiometry and an absence of nucleosome positioning on transfected DNA. J. Biol. Chem. 283, 4595–4601.

Heijmans, B.T., Tobi, E.W., Stein, A.D., Putter, H., Blauw, G.J., Susser, E.S., Slagboom, P.E., Lumey, L.H., 2008. Persistent epigenetic differences associated with prenatal exposure to famine in humans. Proc. Natl. Acad. Sci. USA 105, 17046–17049.

Helbo, A.S., Lay, F.D., Jones, P.A., Liang, G., Gronbaek, K., 2017. Nucleosome positioning and NDR structure at RNA polymerase III promoters. Sci. Rep. 7, 41947.

Hendy, O., Serebreni, L., Bergauer, K., Muerdter, F., Huber, L., Nemcko, F., Stark, A., 2022. Developmental and housekeeping transcriptional programs in *Drosophila* require distinct chromatin remodelers. Mol. Cell 82 (3598–3612), e3597.

Henriques, T., Scruggs, B.S., Inouye, M.O., Muse, G.W., Williams, L.H., Burkholder, A.B., Lavender, C.A., Fargo, D.C., Adelman, K., 2018. Widespread transcriptional pausing and elongation control at enhancers. Genes Dev. 32, 26–41.

Hereford, L., Fahrner, K., Woolford Jr., J., Rosbash, M., Kaback, D.B., 1979. Isolation of yeast histone genes H2A and H2B. Cell 18, 1261–1271.

Herskowitz, I., 1989. A regulatory hierarchy for cell specialization in yeast. Nature 342, 749–757.

Hinnen, A., Hicks, J.B., Fink, G.R., 1978. Transformation of yeast. Proc. Natl. Acad. Sci. USA 75, 1929–1933.

Hirschhorn, J.N., Brown, S.A., Clark, C.D., Winston, F., 1992. Evidence that SNF2/SWI2 and SNF5 activate transcription in yeast by altering chromatin structure. Genes Dev. 6, 2288–2298.

Hirst, K., Fisher, F., McAndrew, P.C., Goding, C.R., 1994. The transcription factor, the Cdk, its cyclin and their regulator: directing the transcriptional response to a nutritional signal. EMBO J. 13, 5410–5420.

Hodges, C., Bintu, L., Lubkowska, L., Kashlev, M., Bustamante, C., 2009. Nucleosomal fluctuations govern the transcription dynamics of RNA polymerase II. Science 325, 626–628.

Holliday, R., 1987. The inheritance of epigenetic defects. Science 238, 163–170.

Holliday, R., 1994. Epigenetics: an overview. Dev. Genet. 15, 453–457.

Holliday, R., Pugh, J.E., 1975. DNA modification mechanisms and gene activity during development. Science 187, 226–232.

Hope, I.A., Struhl, K., 1986. Functional dissection of a eukaryotic transcriptional activator protein, GCN4 of yeast. Cell 46, 885–894.

Horsthemke, B., 2018. A critical view on transgenerational epigenetic inheritance in humans. Nat. Commun. 9, 2973.

Hotchkiss, R.D., 1948. The quantitative separation of purines, pyrimidines, and nucleosides by paper chromatography. J. Biol. Chem. 175, 315–332.

Hou, L., Wang, Y., Liu, Y., Zhang, N., Shamovsky, I., Nudler, E., Tian, B., Dynlacht, B.D., 2019. Paf1C regulates RNA polymerase II progression by modulating elongation rate. Proc. Natl. Acad. Sci. USA 116, 14583–14592.

Hsieh, F.K., Fisher, M., Ujvari, A., Studitsky, V.M., Luse, D.S., 2010. Histone Sin mutations promote nucleosome traversal and histone displacement by RNA polymerase II. EMBO Rep. 11, 705–710.

Hsieh, F.K., Kulaeva, O.I., Patel, S.S., Dyer, P.N., Luger, K., Reinberg, D., Studitsky, V.M., 2013. Histone chaperone FACT action during transcription through chromatin by RNA polymerase II. Proc. Natl. Acad. Sci. USA 110, 7654–7659.

Hu, S., Peng, L., Xu, C., Wang, Z., Song, A., Chen, F.X., 2021. SPT5 stabilizes RNA polymerase II, orchestrates transcription cycles, and maintains the enhancer landscape. Mol. Cell 81 (4425–4439), e4426.

Huang, R.C., Bonner, J., 1962. Histone, a suppressor of chromosomal RNA synthesis. Proc. Natl. Acad. Sci. USA 48, 1216–1222.

Huang, S., O'Shea, E.K., 2005. A systematic high-throughput screen of a yeast deletion collection for mutants defective in PHO_5 regulation. Genetics 169, 1859–1871.

Hubner, J.M., Muller, T., Papageorgiou, D.N., Mauermann, M., Krijgsveld, J., Russell, R.B., Ellison, D.W., Pfister, S.M., Pajtler, K.W., Kool, M., 2019. EZHIP/CXorf67 mimics K27M mutated oncohistones and functions as an intrinsic inhibitor of PRC2 function in aggressive posterior fossa ependymoma. Neuro-Oncology 21, 878–889.

Hughes, T.R., de Boer, C.G., 2013. Mapping yeast transcriptional networks. Genetics 195, 9–36.

Huisinga, K.L., Pugh, B.F., 2004. A genome-wide housekeeping role for TFIID and a highly regulated stress-related role for SAGA in Saccharomyces cerevisiae. Mol. Cell 13, 573–585.

Ingham, P.W., 1981. Trithorax: a new homoeotic mutation of *Drosophila melanogaster*: II. The role oftrx (+) after embryogenesis. Wilehm Roux Arch. Dev. Biol. 190, 365–369.

Ingham, P.W., 1983. Differential expression of bithorax complex genes in the absence of the extra sex combs and trithorax genes. Nature 306, 591–593.

Ingham, P.W., 1998. trithorax and the regulation of homeotic gene expression in *Drosophila*: a historical perspective. Int. J. Dev. Biol. 42, 423–429.

Ioshikhes, I.P., Albert, I., Zanton, S.J., Pugh, B.F., 2006. Nucleosome positions predicted through comparative genomics. Nat. Genet. 38, 1210–1215.

Isbel, L., Grand, R.S., Schubeler, D., 2022. Generating specificity in genome regulation through transcription factor sensitivity to chromatin. Nat. Rev. Genet. 23, 728–740.

Iurlaro, M., Stadler, M.B., Masoni, F., Jagani, Z., Galli, G.G., Schubeler, D., 2021. Mammalian SWI/SNF continuously restores local accessibility to chromatin. Nat. Genet. 53, 279–287.

Ivanovska, I., Jacques, P.E., Rando, O.J., Robert, F., Winston, F., 2011. Control of chromatin structure by spt6: different consequences in coding and regulatory regions. Mol. Cell. Biol. 31, 531–541.

Iwafuchi, M., Cuesta, I., Donahue, G., Takenaka, N., Osipovich, A.B., Magnuson, M.A., Roder, H., Seeholzer, S.H., Santisteban, P., Zaret, K.S., 2020. Gene network transitions in embryos depend upon interactions between a pioneer transcription factor and core histones. Nat. Genet. 52, 418–427.

Iwafuchi-Doi, M., Donahue, G., Kakumanu, A., Watts, J.A., Mahony, S., Pugh, B.F., Lee, D., Kaestner, K.H., Zaret, K.S., 2016. The pioneer transcription factor FoxA maintains an accessible nucleosome configuration at enhancers for tissue-specific gene activation. Mol. Cell 62, 79–91.

Iyer, V., Struhl, K., 1995. Poly(dA:dT), a ubiquitous promoter element that stimulates transcription via its intrinsic DNA structure. EMBO J. 14, 2570–2579.

Izban, M.G., Luse, D.S., 1991. Transcription on nucleosomal templates by RNA polymerase II in vitro: inhibition of elongation with enhancement of sequence-specific pausing. Genes Dev. 5, 683–696.

Izban, M.G., Luse, D.S., 1992. Factor-stimulated RNA polymerase II transcribes at physiological elongation rates on naked DNA but very poorly on chromatin templates. J. Biol. Chem. 267, 13647–13655.

Jackson, V., 1990. In vivo studies on the dynamics of histone-DNA interaction: evidence for nucleosome dissolution during replication and transcription and a low level of dissolution independent of both. Biochemistry 29, 719–731.

Jacob, F., Monod, J., 1961. Genetic regulatory mechanisms in the synthesis of proteins. J. Mol. Biol. 3, 318–356.

Jaenisch, R., Young, R., 2008. Stem cells, the molecular circuitry of pluripotency and nuclear reprogramming. Cell 132, 567–582.

Jain, S.U., Do, T.J., Lund, P.J., Rashoff, A.Q., Diehl, K.L., Cieslik, M., Bajic, A., Juretic, N., Deshmukh, S., Venneti, S., et al., 2019. PFA ependymoma-associated protein EZHIP inhibits PRC2 activity through a H3 K27M-like mechanism. Nat. Commun. 10, 2146.

Jamai, A., Imoberdorf, R.M., Strubin, M., 2007. Continuous histone H2B and transcription-dependent histone H3 exchange in yeast cells outside of replication. Mol. Cell 25, 345–355.

Jensen, T.H., Jacquier, A., Libri, D., 2013. Dealing with pervasive transcription. Mol. Cell 52, 473–484.

Jenuwein, T., Allis, C.D., 2001. Translating the histone code. Science 293, 1074–1080.

Jeppesen, P., 1997. Histone acetylation: a possible mechanism for the inheritance of cell memory at mitosis. BioEssays 19, 67–74.

Jeronimo, C., Watanabe, S., Kaplan, C.D., Peterson, C.L., Robert, F., 2015. The histone chaperones FACT and Spt6 restrict H2A.Z from intragenic locations. Mol. Cell 58, 1113–1123.

Jeronimo, C., Poitras, C., Robert, F., 2019. Histone recycling by FACT and Spt6 during transcription prevents the scrambling of histone modifications. Cell Rep. 28 (1206–1218), e1208.

Jessen, W.J., Hoose, S.A., Kilgore, J.A., Kladde, M.P., 2006. Active PHO5 chromatin encompasses variable numbers of nucleosomes at individual promoters. Nat. Struct. Mol. Biol. 13, 256–263.

Jiao, L., Liu, X., 2015. Structural basis of histone H3K27 trimethylation by an active polycomb repressive complex 2. Science 350, aac4383.

Jin, C., Felsenfeld, G., 2007. Nucleosome stability mediated by histone variants H3.3 and H2A.Z. Genes Dev. 21, 1519–1529.

Jin, C., Zang, C., Wei, G., Cui, K., Peng, W., Zhao, K., Felsenfeld, G., 2009. H3.3/H2A.Z double variant-containing nucleosomes mark 'nucleosome-free regions' of active promoters and other regulatory regions. Nat. Genet. 41, 941–945.

Jin, R., Klasfeld, S., Zhu, Y., Fernandez Garcia, M., Xiao, J., Han, S.K., Konkol, A., Wagner, D., 2021. LEAFY is a pioneer transcription factor and licenses cell reprogramming to floral fate. Nat. Commun. 12, 626.

Johnson, T.B., Coghill, R.D., 1925. Researches on pyrimidines. C111. The discovery of 5-methyl-cytosine in tuberculinic acid, the nucleic acid of the tubercle bacillus. J. Am. Chem. Soc. 47, 2838–2844.

Johnson, L.M., Kayne, P.S., Kahn, E.S., Grunstein, M., 1990. Genetic evidence for an interaction between SIR3 and histone H4 in the repression of the silent mating loci in Saccharomyces cerevisiae. Proc. Natl. Acad. Sci. USA 87, 6286–6290.

Johnson, T.A., Elbi, C., Parekh, B.S., Hager, G.L., John, S., 2008. Chromatin remodeling complexes interact dynamically with a glucocorticoid receptor-regulated promoter. Mol. Biol. Cell 19, 3308–3322.

Johnson, T.A., Chereji, R.V., Stavreva, D.A., Morris, S.A., Hager, G.L., Clark, D.J., 2018. Conventional and pioneer modes of glucocorticoid receptor interaction with enhancer chromatin in vivo. Nucleic Acids Res. 46, 203–214.

Jones, R.S., Gelbart, W.M., 1990. Genetic analysis of the enhancer of zeste locus and its role in gene regulation in Drosophila melanogaster. Genetics 126, 185–199.

Jones, R.S., Gelbart, W.M., 1993. The *Drosophila* polycomb-group gene enhancer of zeste contains a region with sequence similarity to trithorax. Mol. Cell. Biol. 13, 6357–6366.

Joshi, A.A., Struhl, K., 2005. Eaf3 chromodomain interaction with methylated H3-K36 links histone deacetylation to Pol II elongation. Mol. Cell 20, 971–978.

Jürgens, G., 1985. A group of genes controlling the spatial expression of the bithorax complex in *Drosophila*. Nature 316, 153–155.

Justin, N., Zhang, Y., Tarricone, C., Martin, S.R., Chen, S., Underwood, E., De Marco, V., Haire, L.F., Walker, P.A., Reinberg, D., et al., 2016. Structural basis of oncogenic histone H3K27M inhibition of human polycomb repressive complex 2. Nat. Commun. 7, 11316.

Kaati, G., Bygren, L.O., Pembrey, M., Sjostrom, M., 2007. Transgenerational response to nutrition, early life circumstances and longevity. Eur. J. Hum. Genet. 15, 784–790.

Kadonaga, J.T., 2004. Regulation of RNA polymerase II transcription by sequence-specific DNA binding factors. Cell 116, 247–257.

Kadonaga, J.T., 2019. The transformation of the DNA template in RNA polymerase II transcription: a historical perspective. Nat. Struct. Mol. Biol. 26, 766–770.

Kagansky, A., Folco, H.D., Almeida, R., Pidoux, A.L., Boukaba, A., Simmer, F., Urano, T., Hamilton, G.L., Allshire, R.C., 2009. Synthetic heterochromatin bypasses RNAi and centromeric repeats to establish functional centromeres. Science 324, 1716–1719.

Kahn, T.G., Dorafshan, E., Schultheis, D., Zare, A., Stenberg, P., Reim, I., Pirrotta, V., Schwartz, Y.B., 2016. Interdependence of PRC1 and PRC2 for recruitment to polycomb response elements. Nucleic Acids Res. 44, 10132–10149.

Kalff, M., Gross, B., Beato, M., 1990. Progesterone receptor stimulates transcription of mouse mammary tumour virus in a cell-free system. Nature 344, 360–362.

Kamakaka, R.T., Bulger, M., Kadonaga, J.T., 1993. Potentiation of RNA polymerase II transcription by Gal4-VP16 during but not after DNA replication and chromatin assembly. Genes Dev. 7, 1779–1795.

Kaplan, C.D., Morris, J.R., Wu, C., Winston, F., 2000. Spt5 and spt6 are associated with active transcription and have characteristics of general elongation factors in *D. melanogaster*. Genes Dev. 14, 2623–2634.

Kaplan, C.D., Laprade, L., Winston, F., 2003. Transcription elongation factors repress transcription initiation from cryptic sites. Science 301, 1096–1099.

Karlsborn, T., Tukenmez, H., Mahmud, A.K., Xu, F., Xu, H., Bystrom, A.S., 2014. Elongator, a conserved complex required for wobble uridine modifications in eukaryotes. RNA Biol. 11, 1519–1528.

Kassavetis, G.A., Braun, B.R., Nguyen, L.H., Geiduschek, E.P., 1990. S. cerevisiae TFIIIB is the transcription initiation factor proper of RNA polymerase III, while TFIIIA and TFIIIC are assembly factors. Cell 60, 235–245.

Kassis, J.A., Kennison, J.A., Tamkun, J.W., 2017. Polycomb and trithorax group genes in *Drosophila*. Genetics 206, 1699–1725.

Kasten, M.M., Dorland, S., Stillman, D.J., 1997. A large protein complex containing the yeast Sin3p and Rpd3p transcriptional regulators. Mol. Cell. Biol. 17, 4852–4858.

Kato, H., Okazaki, K., Iida, T., Nakayama, J., Murakami, Y., Urano, T., 2013. Spt6 prevents transcription-coupled loss of posttranslationally modified histone H3. Sci. Rep. 3, 2186.

Kayne, P.S., Kim, U.J., Han, M., Mullen, J.R., Yoshizaki, F., Grunstein, M., 1988. Extremely conserved histone H4 N terminus is dispensable for growth but essential for repressing the silent mating loci in yeast. Cell 55, 27–39.

Keegan, L., Gill, G., Ptashne, M., 1986. Separation of DNA binding from the transcription-activating function of a eukaryotic regulatory protein. Science 231, 699–704.

Keller, G., 2005. Embryonic stem cell differentiation: emergence of a new era in biology and medicine. Genes Dev. 19, 1129–1155.

Kemble, D.J., McCullough, L.L., Whitby, F.G., Formosa, T., Hill, C.P., 2015. FACT disrupts nucleosome structure by binding H2A-H2B with conserved peptide motifs. Mol. Cell 60, 294–306.

Kennison, J.A., 1993. Transcriptional activation of Drosophila homeotic genes from distant regulatory elements. Trends Genet. 9, 75–79.

Kennison, J.A., Tamkun, J.W., 1988. Dosage-dependent modifiers of polycomb and antennapedia mutations in *Drosophila*. Proc. Natl. Acad. Sci. USA 85, 8136–8140.

Keogh, M.C., Kurdistani, S.K., Morris, S.A., Ahn, S.H., Podolny, V., Collins, S.R., Schuldiner, M., Chin, K., Punna, T., Thompson, N.J., et al., 2005. Cotranscriptional set2 methylation of histone H3 lysine 36 recruits a repressive Rpd3 complex. Cell 123, 593–605.

Kim, U.-J., Han, M., Kayne, P., Grunstein, M., 1988. Effects of histone H4 depletion on the cell cycle and transcription of *Saccharomyces cerevisiae*. EMBO J. 7, 2211–2219.

Kim, M., Ahn, S.H., Krogan, N.J., Greenblatt, J.F., Buratowski, S., 2004. Transitions in RNA polymerase II elongation complexes at the 3′ ends of genes. EMBO J. 23, 354–364.

Kim, J., Guermah, M., Roeder, R.G., 2010. The human PAF1 complex acts in chromatin transcription elongation both independently and cooperatively with SII/TFIIS. Cell 140, 491–503.

Kim, T., Xu, Z., Clauder-Munster, S., Steinmetz, L.M., Buratowski, S., 2012. Set3 HDAC mediates effects of overlapping noncoding transcription on gene induction kinetics. Cell 150, 1158–1169.

King, H.W., Klose, R.J., 2017. The pioneer factor OCT4 requires the chromatin remodeller BRG1 to support gene regulatory element function in mouse embryonic stem cells. elife 6.

Kingston, R.E., Tamkun, J.W., 2014. Transcriptional regulation by trithorax-group proteins. Cold Spring Harb. Perspect. Biol. 6, a019349.

Kireeva, M.L., Walter, W., Tchernajenko, V., Bondarenko, V., Kashlev, M., Studitsky, V.M., 2002. Nucleosome remodeling induced by RNA polymerase II: loss of the H2A/H2B dimer during transcription. Mol. Cell 9, 541–552.

Kireeva, M.L., Hancock, B., Cremona, G.H., Walter, W., Studitsky, V.M., Kashlev, M., 2005. Nature of the nucleosomal barrier to RNA polymerase II. Mol. Cell 18, 97–108.

Knezetic, J.A., Luse, D.S., 1986. The presence of nucleosomes on a DNA template prevents initiation by RNA polymerase II in vitro. Cell 45, 95–104.

Knezetic, J.A., Jacob, G.A., Luse, D.S., 1988. Assembly of RNA polymerase II preinitiation complexes before assembly of nucleosomes allows efficient initiation of transcription on nucleosomal templates. Mol. Cell. Biol. 8, 3114–3121.

Koerber, R.T., Rhee, H.S., Jiang, C., Pugh, B.F., 2009. Interaction of transcriptional regulators with specific nucleosomes across the *Saccharomyces genome*. Mol. Cell 35, 889–902.

Korber, P., Barbaric, S., 2014. The yeast PHO5 promoter: from single locus to systems biology of a paradigm for gene regulation through chromatin. Nucleic Acids Res. 42, 10888–10902.

Korber, P., Luckenbach, T., Blaschke, D., Horz, W., 2004. Evidence for histone eviction in trans upon induction of the yeast PHO5 promoter. Mol. Cell. Biol. 24, 10965–10974.

Kornberg, R.D., 2005. Mediator and the mechanism of transcriptional activation. Trends Biochem. Sci. 30, 235–239.

Kornberg, R.D., Lorch, Y., 2020. Primary role of the nucleosome. Mol. Cell 79, 371–375.

Krogan, N.J., Kim, M., Ahn, S.H., Zhong, G., Kobor, M.S., Cagney, G., Emili, A., Shilatifard, A., Buratowski, S., Greenblatt, J.F., 2002. RNA polymerase II elongation factors of Saccharomyces cerevisiae: a targeted proteomics approach. Mol. Cell. Biol. 22, 6979–6992.

Krogan, N.J., Dover, J., Wood, A., Schneider, J., Heidt, J., Boateng, M.A., Dean, K., Ryan, O.W., Golshani, A., Johnston, M., et al., 2003a. The Paf1 complex is required for histone H3 methylation by COMPASS and Dot1p: linking transcriptional elongation to histone methylation. Mol. Cell 11, 721–729.

Krogan, N.J., Kim, M., Tong, A., Golshani, A., Cagney, G., Canadien, V., Richards, D.P., Beattie, B.K., Emili, A., Boone, C., et al., 2003b. Methylation of histone H3 by Set2 in *Saccharomyces cerevisiae* is linked to transcriptional elongation by RNA polymerase II. Mol. Cell. Biol. 23, 4207–4218.

Kruczek, I., Doerfler, W., 1983. Expression of the chloramphenicol acetyltransferase gene in mammalian cells under the control of adenovirus type 12 promoters: effect of promoter methylation on gene expression. Proc. Natl. Acad. Sci. USA 80, 7586–7590.

Ku, M., Koche, R.P., Rheinbay, E., Mendenhall, E.M., Endoh, M., Mikkelsen, T.S., Presser, A., Nusbaum, C., Xie, X., Chi, A.S., et al., 2008. Genomewide analysis of PRC1 and PRC2 occupancy identifies two classes of bivalent domains. PLoS Genet. 4, e1000242.

Kubik, S., O'Duibhir, E., de Jonge, W.J., Mattarocci, S., Albert, B., Falcone, J.L., Bruzzone, M.J., Holstege, F.C.P., Shore, D., 2018. Sequence-directed action of RSC remodeler and general regulatory factors modulates+1 nucleosome position to facilitate transcription. Mol. Cell 71, 89-+.

Kubik, S., Bruzzone, M.J., Challal, D., Dreos, R., Mattarocci, S., Bucher, P., Libri, D., Shore, D., 2019. Opposing chromatin remodelers control transcription initiation frequency and start site selection. Nat. Struct. Mol. Biol. 26, 744–754.

Kujirai, T., Kurumizaka, H., 2020. Transcription through the nucleosome. Curr. Opin. Struct. Biol. 61, 42–49.

Kujirai, T., Ehara, H., Fujino, Y., Shirouzu, M., Sekine, S.I., Kurumizaka, H., 2018. Structural basis of the nucleosome transition during RNA polymerase II passage. Science 362, 595–598.

Kulaeva, O.I., Gaykalova, D.A., Pestov, N.A., Golovastov, V.V., Vassylyev, D.G., Artsimovitch, I., Studitsky, V.M., 2009. Mechanism of chromatin remodeling and recovery during passage of RNA polymerase II. Nat. Struct. Mol. Biol. 16, 1272–1278.

Kulaeva, O.I., Hsieh, F.K., Chang, H.W., Luse, D.S., Studitsky, V.M., 2013. Mechanism of transcription through a nucleosome by RNA polymerase II. Biochim. Biophys. Acta 1829, 76–83.

Kumar, Y., Bhargava, P., 2013. A unique nucleosome arrangement, maintained actively by chromatin remodelers facilitates transcription of yeast tRNA genes. BMC Genomics 14, 402.

Kuroda, M.I., Kang, H., De, S., Kassis, J.A., 2020. Dynamic competition of polycomb and trithorax in transcriptional programming. Annu. Rev. Biochem. 89, 235–253.

Kwak, H., Fuda, N.J., Core, L.J., Lis, J.T., 2013. Precise maps of RNA polymerase reveal how promoters direct initiation and pausing. Science 339, 950–953.

Lacy, E., Axel, R., 1975. Analysis of DNA of isolated chromatin subunits. Proc. Natl. Acad. Sci. USA 72, 3978–3982.

Lambert, S.A., Jolma, A., Campitelli, L.F., Das, P.K., Yin, Y., Albu, M., Chen, X., Taipale, J., Hughes, T.R., Weirauch, M.T., 2018. The human transcription factors. Cell 172, 650–665.

Landecker, H.L., Sinclair, D.A., Brock, H.W., 1994. Screen for enhancers of polycomb and polycomblike in *Drosophila melanogaster*. Dev. Genet. 15, 425–434.

Lantermann, A.B., Straub, T., Stralfors, A., Yuan, G.C., Ekwall, K., Korber, P., 2010. *Schizosaccharomyces pombe* genome-wide nucleosome mapping reveals positioning mechanisms distinct from those of Saccharomyces cerevisiae. Nat. Struct. Mol. Biol. 17, 251–257.

Laprell, F., Finkl, K., Muller, J., 2017. Propagation of polycomb-repressed chromatin requires sequence-specific recruitment to DNA. Science 356, 85–88.

Larschan, E., Winston, F., 2001. The *S. cerevisiae* SAGA complex functions in vivo as a coactivator for transcriptional activation by Gal4. Genes Dev. 15, 1946–1956.

Laskey, R.A., Mills, A.D., Morris, N.R., 1977. Assembly of SV40 chromatin in a cell-free system from Xenopus eggs. Cell 10, 237–243.

Lassar, A.B., Martin, P.L., Roeder, R.G., 1983. Transcription of class III genes: formation of preinitiation complexes. Science 222, 740–748.

Lassar, A.B., Hamer, D.H., Roeder, R.G., 1985. Stable transcription complex on a class III gene in a minichromosome. Mol. Cell. Biol. 5, 40–45.

Lau, M.S., Schwartz, M.G., Kundu, S., Savol, A.J., Wang, P.I., Marr, S.K., Grau, D.J., Schorderet, P., Sadreyev, R.I., Tabin, C.J., et al., 2017. Mutation of a nucleosome compaction region disrupts polycomb-mediated axial patterning. Science 355, 1081–1084.

Laurent, B.C., Treitel, M.A., Carlson, M., 1991. Functional interdependence of the yeast SNF2, SNF5, and SNF6 proteins in transcriptional activation. Proc. Natl. Acad. Sci. USA 88, 2687–2691.

Laybourn, P.J., Kadonaga, J.T., 1991. Role of nucleosomal cores and histone H1 in regulation of transcription by RNA polymerase II. Science 254, 238–245.

Laybourn, P.J., Kadonaga, J.T., 1992. Threshold phenomena and long-distance activation of transcription by RNA polymerase II. Science 257, 1682–1685.

Lechner, T., Carrozza, M.J., Yu, Y., Grant, P.A., Eberharter, A., Vannier, D., Brosch, G., Stillman, D.J., Shore, D., Workman, J.L., 2000. Sds3 (suppressor of defective silencing 3) is an integral component of the yeast Sin3[middle dot]Rpd3 histone deacetylase complex and is required for histone deacetylase activity. J. Biol. Chem. 275, 40961–40966.

Lee, H.L., Archer, T.K., 1994. Nucleosome-mediated disruption of transcription factor-chromatin initiation complexes at the mouse mammary tumor virus long terminal repeat in vivo. Mol. Cell. Biol. 14, 32–41.

Lee, M.S., Garrard, W.T., 1991. Transcription-induced nucleosome 'splitting': an underlying structure for DNase I sensitive chromatin. EMBO J. 10, 607–615.

Lee, M.S., Garrard, W.T., 1992. Uncoupling gene activity from chromatin structure: promoter mutations can inactivate transcription of the yeast HSP82 gene without eliminating nucleosome-free regions. Proc. Natl. Acad. Sci. USA 89, 9166–9170.

Lee, C.K., Shibata, Y., Rao, B., Strahl, B.D., Lieb, J.D., 2004. Evidence for nucleosome depletion at active regulatory regions genome-wide. Nat. Genet. 36, 900–905.

Lee, T.I., Jenner, R.G., Boyer, L.A., Guenther, M.G., Levine, S.S., Kumar, R.M., Chevalier, B., Johnstone, S.E., Cole, M.F., Isono, K., et al., 2006. Control of developmental regulators by polycomb in human embryonic stem cells. Cell 125, 301–313.

Lee, W., Tillo, D., Bray, N., Morse, R.H., Davis, R.W., Hughes, T.R., Nislow, C., 2007. A high-resolution atlas of nucleosome occupancy in yeast. Nat. Genet. 39, 1235–1244.

Lenhard, B., Sandelin, A., Carninci, P., 2012. Metazoan promoters: emerging characteristics and insights into transcriptional regulation. Nat. Rev. Genet. 13, 233–245.

LeRoy, G., Oksuz, O., Descostes, N., Aoi, Y., Ganai, R.A., Kara, H.O., Yu, J.R., Lee, C.H., Stafford, J., Shilatifard, A., et al., 2019. LEDGF and HDGF2 relieve the nucleosome-induced barrier to transcription in differentiated cells. Sci. Adv. 5, eaay3068.

Lessard, J., Sauvageau, G., 2003. Bmi-1 determines the proliferative capacity of normal and leukaemic stem cells. Nature 423, 255–260.

Levo, M., Avnit-Sagi, T., Lotan-Pompan, M., Kalma, Y., Weinberger, A., Yakhini, Z., Segal, E., 2017. Systematic investigation of transcription factor activity in the context of chromatin using massively parallel binding and expression assays. Mol. Cell 65 (604–617), e606.

Lewis, P.H., 1947. New mutants. Drosoph. Inf. Serv. 21, 69.

Lewis, E.B., 1951. Pseudoallelism and gene evolution. Cold Spring Harb. Symp. Quant. Biol. 16, 159–174.

Lewis, E.B., 1978. A gene complex controlling segmentation in *Drosophila*. Nature 276, 565–570.

Lewis, P.W., Muller, M.M., Koletsky, M.S., Cordero, F., Lin, S., Banaszynski, L.A., Garcia, B.A., Muir, T.W., Becher, O.J., Allis, C.D., 2013. Inhibition of PRC2 activity by a gain-of-function H3 mutation found in pediatric glioblastoma. Science 340, 857–861.

Li, G., Widom, J., 2004. Nucleosomes facilitate their own invasion. Nat. Struct. Mol. Biol. 11, 763–769.

Li, Q., Wrange, O., 1995. Accessibility of a glucocorticoid response element in a nucleosome depends on its rotational positioning. Mol. Cell. Biol. 15, 4375–4384.

Li, E., Bestor, T.H., Jaenisch, R., 1992. Targeted mutation of the DNA methyltransferase gene results in embryonic lethality. Cell 69, 915–926.

Li, B., Howe, L., Anderson, S., Yates 3rd, J.R., Workman, J.L., 2003. The Set2 histone methyltransferase functions through the phosphorylated carboxyl-terminal domain of RNA polymerase II. J. Biol. Chem. 278, 8897–8903.

Li, G., Levitus, M., Bustamante, C., Widom, J., 2005. Rapid spontaneous accessibility of nucleosomal DNA. Nat. Struct. Mol. Biol. 12, 46–53.

Li, X., Ito, M., Zhou, F., Youngson, N., Zuo, X., Leder, P., Ferguson-Smith, A.C., 2008. A maternal-zygotic effect gene, Zfp57, maintains both maternal and paternal imprints. Dev. Cell 15, 547–557.

Li, X.Y., Thomas, S., Sabo, P.J., Eisen, M.B., Stamatoyannopoulos, J.A., Biggin, M.D., 2011. The role of chromatin accessibility in directing the widespread, overlapping patterns of *Drosophila* transcription factor binding. Genome Biol. 12, R34.

Li, R., Cauchy, P., Ramamoorthy, S., Boller, S., Chavez, L., Grosschedl, R., 2018. Dynamic EBF1 occupancy directs sequential epigenetic and transcriptional events in B-cell programming. Genes Dev. 32, 96–111.

Lickwar, C.R., Mueller, F., Hanlon, S.E., McNally, J.G., Lieb, J.D., 2012. Genome-wide protein-DNA binding dynamics suggest a molecular clutch for transcription factor function. Nature 484, 251–255.

Lieb, J.D., Liu, X., Botstein, D., Brown, P.O., 2001. Promoter-specific binding of Rap1 revealed by genome-wide maps of protein-DNA association. Nat. Genet. 28, 327–334.

Lin, T.Y., Abbassi, N.E.H., Zakrzewski, K., Chramiec-Glabik, A., Jemiola-Rzeminska, M., Rozycki, J., Glatt, S., 2019. The elongator subunit Elp3 is a non-canonical tRNA acetyltransferase. Nat. Commun. 10, 625.

Lin-Shiao, E., Lan, Y., Welzenbach, J., Alexander, K.A., Zhang, Z., Knapp, M., Mangold, E., Sammons, M., Ludwig, K.U., Berger, S.L., 2019. p63 establishes epithelial enhancers at critical craniofacial development genes. Sci. Adv. 5, eaaw0946.

Liu, L.F., Wang, J.C., 1987. Supercoiling of the DNA template during transcription. Proc. Natl. Acad. Sci. USA 84, 7024–7027.

Liu, Y., Zhou, K., Zhang, N., Wei, H., Tan, Y.Z., Zhang, Z., Carragher, B., Potter, C.S., D'Arcy, S., Luger, K., 2020. FACT caught in the act of manipulating the nucleosome. Nature 577, 426–431.

Loh, C.H., Veenstra, G.J.C., 2022. The role of polycomb proteins in cell lineage commitment and embryonic development. Epigenomes 6.

Lohr, D., Hopper, J.E., 1985. The relationship of regulatory proteins and DNase I hypersensitive sites in the yeast GAL1-10 genes. Nucleic Acids Res. 13, 8409–8423.

Lomvardas, S., Thanos, D., 2001. Nucleosome sliding via TBP DNA binding in vivo. Cell 106, 685–696.

Lomvardas, S., Thanos, D., 2002. Modifying gene expression programs by altering core promoter chromatin architecture. Cell 110, 261–271.

Lorch, Y., LaPointe, J.W., Kornberg, R.D., 1987. Nucleosomes inhibit the initiation of transcription but allow chain elongation with the displacement of histones. Cell 49, 203–210.

Lorch, Y., LaPointe, J.W., Kornberg, R.D., 1988. On the displacement of histones from DNA by transcription. Cell 55, 743–744.

Lorch, Y., LaPointe, J.W., Kornberg, R.D., 1992. Initiation on chromatin templates in a yeast RNA polymerase II transcription system. Genes Dev. 6, 2282–2287.

Losa, R., Brown, D.D., 1987. A bacteriophage RNA polymerase transcribes in vitro through a nucleosome core without displacing it. Cell 50, 801–808.

Lu, Q., Wallrath, L.L., Elgin, S.C., 1995. The role of a positioned nucleosome at the *Drosophila melanogaster* hsp26 promoter. EMBO J. 14, 4738–4746.

Ludwig, M.Z., Patel, N.H., Kreitman, M., 1998. Functional analysis of eve stripe 2 enhancer evolution in *Drosophila*: rules governing conservation and change. Development 125, 949–958.

Luse, D.S., Spangler, L.C., Ujvari, A., 2011. Efficient and rapid nucleosome traversal by RNA polymerase II depends on a combination of transcript elongation factors. J. Biol. Chem. 286, 6040–6048.

Luzete-Monteiro, E., Zaret, K.S., 2022. Structures and consequences of pioneer factor binding to nucleosomes. Curr. Opin. Struct. Biol. 75, 102425.

Ma, J., Ptashne, M., 1987. Deletion analysis of GAL4 defines two transcriptional activating segments. Cell 48, 847–853.

Mackay, A., Burford, A., Carvalho, D., Izquierdo, E., Fazal-Salom, J., Taylor, K.R., Bjerke, L., Clarke, M., Vinci, M., Nandhabalan, M., et al., 2017. Integrated molecular meta-analysis of 1,000 pediatric high-grade and diffuse intrinsic pontine glioma. Cancer Cell 32 (520–537), e525.

Mahrez, W., Arellano, M.S., Moreno-Romero, J., Nakamura, M., Shu, H., Nanni, P., Kohler, C., Gruissem, W., Hennig, L., 2016. H3K36ac is an evolutionary conserved plant histone modification that marks active genes. Plant Physiol. 170, 1566–1577.

Malik, S., Roeder, R.G., 2010. The metazoan mediator co-activator complex as an integrative hub for transcriptional regulation. Nat. Rev. Genet. 11, 761–772.

Malone, E.A., Clark, C.D., Chiang, A., Winston, F., 1991. Mutations in SPT16/CDC68 suppress cis- and trans-acting mutations that affect promoter function in *Saccharomyces cerevisiae*. Mol. Cell. Biol. 11, 5710–5717.

Manley, J.L., Fire, A., Cano, A., Sharp, P.A., Gefter, M.L., 1980. DNA-dependent transcription of adenovirus genes in a soluble whole-cell extract. Proc. Natl. Acad. Sci. USA 77, 3855–3859.

Mann, R.K., Grunstein, M., 1992. Histone H3 N-terminal mutations allow hyperactivation of the yeast GAL1 gene in vivo. EMBO J. 11, 3297–3306.
Mao, C., Brown, C.R., Griesenbeck, J., Boeger, H., 2011. Occlusion of regulatory sequences by promoter nucleosomes in vivo. PLoS One 6, e17521.
Margueron, R., Justin, N., Ohno, K., Sharpe, M.L., Son, J., Drury 3rd, W.J., Voigt, P., Martin, S.R., Taylor, W.R., De Marco, V., et al., 2009. Role of the polycomb protein EED in the propagation of repressive histone marks. Nature 461, 762–767.
Martienssen, R., Moazed, D., 2015. RNAi and heterochromatin assembly. Cold Spring Harb. Perspect. Biol. 7, a019323.
Martin, A.M., Pouchnik, D.J., Walker, J.L., Wyrick, J.J., 2004. Redundant roles for histone H3 N-terminal lysine residues in subtelomeric gene repression in *Saccharomyces cerevisiae*. Genetics 167, 1123–1132.
Martinez-Campa, C., Politis, P., Moreau, J.L., Kent, N., Goodall, J., Mellor, J., Goding, C.R., 2004. Precise nucleosome positioning and the TATA box dictate requirements for the histone H4 tail and the bromodomain factor Bdf1. Mol. Cell 15, 69–81.
Mason, P.B., Struhl, K., 2003. The FACT complex travels with elongating RNA polymerase II and is important for the fidelity of transcriptional initiation in vivo. Mol. Cell. Biol. 23, 8323–8333.
Matsui, T., 1987. Transcription of adenovirus 2 major late and peptide IX genes under conditions of in vitro nucleosome assembly. Mol. Cell. Biol. 7, 1401–1408.
Mattei, A.L., Bailly, N., Meissner, A., 2022. DNA methylation: a historical perspective. Trends Genet. 38, 676–707.
Mavrich, T.N., Jiang, C., Ioshikhes, I.P., Li, X., Venters, B.J., Zanton, S.J., Tomsho, L.P., Qi, J., Glaser, R.L., Schuster, S.-C., et al., 2008. Nucleosome organization in the *Drosophila* genome. Nature 453, 358–362.
Mayer, W., Niveleau, A., Walter, J., Fundele, R., Haaf, T., 2000. Demethylation of the zygotic paternal genome. Nature 403, 501–502.
Mayer, A., Lidschreiber, M., Siebert, M., Leike, K., Soding, J., Cramer, P., 2010. Uniform transitions of the general RNA polymerase II transcription complex. Nat. Struct. Mol. Biol. 17, 1272–1278.
Mayran, A., Drouin, J., 2018. Pioneer transcription factors shape the epigenetic landscape. J. Biol. Chem. 293, 13795–13804.
Mayran, A., Khetchoumian, K., Hariri, F., Pastinen, T., Gauthier, Y., Balsalobre, A., Drouin, J., 2018. Pioneer factor Pax7 deploys a stable enhancer repertoire for specification of cell fate. Nat. Genet. 50, 259–269.
McAndrew, P.C., Svaren, J., Martin, S.R., Horz, W., Goding, C.R., 1998. Requirements for chromatin modulation and transcription activation by the Pho4 acidic activation domain. Mol. Cell. Biol. 18, 5818–5827.
McDaniel, S.L., Gibson, T.J., Schulz, K.N., Fernandez Garcia, M., Nevil, M., Jain, S.U., Lewis, P.W., Zaret, K.S., Harrison, M.M., 2019. Continued activity of the pioneer factor zelda is required to drive zygotic genome activation. Mol. Cell 74 (185–195), e184.
McGeady, M.L., Jhappan, C., Ascione, R., Vande Woude, G.F., 1983. In vitro methylation of specific regions of the cloned Moloney sarcoma virus genome inhibits its transforming activity. Mol. Cell. Biol. 3, 305–314.
McGhee, J.D., Ginder, G.D., 1979. Specific DNA methylation sites in the vicinity of the chicken beta-globin genes. Nature 280, 419–420.
McGraw, S., Kimmins, S., 2023. Inheritance of epigenetic DNA marks studied in new mouse model. Nature 615, 800–802.
McKnight, S.L., Kingsbury, R., 1982. Transcriptional control signals of a eukaryotic protein-coding gene. Science 217, 316–324.
McPherson, C.E., Shim, E.Y., Friedman, D.S., Zaret, K.S., 1993. An active tissue-specific enhancer and bound transcription factors existing in a precisely positioned nucleosomal array. Cell 75, 387–398.
Meehan, R.R., Lewis, J.D., McKay, S., Kleiner, E.L., Bird, A.P., 1989. Identification of a mammalian protein that binds specifically to DNA containing methylated CpGs. Cell 58, 499–507.
Meers, M.P., Henriques, T., Lavender, C.A., McKay, D.J., Strahl, B.D., Duronio, R.J., Adelman, K., Matera, A.G., 2017. Histone gene replacement reveals a post-transcriptional role for H3K36 in maintaining metazoan transcriptome fidelity. elife 6.
Megee, P.C., Morgan, B.A., Mittman, B.A., Smith, M.M., 1990. Genetic analysis of histone H4: essential role of lysines subject to reversible acetylation. Science 247, 841–845.
Meisterernst, M., Horikoshi, M., Roeder, R.G., 1990. Recombinant yeast TFIID, a general transcription factor, mediates activation by the gene-specific factor USF in a chromatin assembly assay. Proc. Natl. Acad. Sci. USA 87, 9153–9157.
Mejia-Almonte, C., Busby, S.J.W., Wade, J.T., van Helden, J., Arkin, A.P., Stormo, G.D., Eilbeck, K., Palsson, B.O., Galagan, J.E., Collado-Vides, J., 2020. Redefining fundamental concepts of transcription initiation in bacteria. Nat. Rev. Genet. 21, 699–714.

Meneguzzi, G., Chenciner, N., Milanesi, G., 1979. Transcription of nucleosomal DNA in SV40 minichromosomes by eukaryotic and prokaryotic RNA polymerases. Nucleic Acids Res. 6, 2947–2960.
Menet, J.S., Pescatore, S., Rosbash, M., 2014. CLOCK:BMAL1 is a pioneer-like transcription factor. Genes Dev. 28, 8–13.
Michael, A.K., Thoma, N.H., 2021. Reading the chromatinized genome. Cell 184, 3599–3611.
Michael, A.K., Grand, R.S., Isbel, L., Cavadini, S., Kozicka, Z., Kempf, G., Bunker, R.D., Schenk, A.D., Graff-Meyer, A., Pathare, G.R., et al., 2020. Mechanisms of OCT4-SOX2 motif readout on nucleosomes. Science 368, 1460–1465.
Mikkelsen, T.S., Ku, M., Jaffe, D.B., Issac, B., Lieberman, E., Giannoukos, G., Alvarez, P., Brockman, W., Kim, T.K., Koche, R.P., et al., 2007. Genome-wide maps of chromatin state in pluripotent and lineage-committed cells. Nature 448, 553–560.
Miksicek, R., Borgmeyer, U., Nowock, J., 1987. Interaction of the TGGCA-binding protein with upstream sequences is required for efficient transcription of mouse mammary tumor virus. EMBO J. 6, 1355–1360.
Miller, J.A., Widom, J., 2003. Collaborative competition mechanism for gene activation in vivo. Mol. Cell. Biol. 23, 1623–1632.
Miller, J.R., Cartwright, E.M., Brownlee, G.G., Fedoroff, N.V., Brown, D.D., 1978. The nucleotide sequence of oocyte 5S DNA in Xenopus laevis. II. The GC-rich region. Cell 13, 717–725.
Miller, C.L.W., Warner, J.L., Winston, F., 2023. Insights into Spt6: a histone chaperone that functions in transcription, DNA replication, and genome stability. Trends Genet.
Ming, G.L., Song, H., 2011. Adult neurogenesis in the mammalian brain: significant answers and significant questions. Neuron 70, 687–702.
Miska, E.A., Ferguson-Smith, A.C., 2016. Transgenerational inheritance: models and mechanisms of non-DNA sequence-based inheritance. Science 354, 59–63.
Mitra, D., Parnell, E.J., Landon, J.W., Yu, Y., Stillman, D.J., 2006. SWI/SNF binding to the HO promoter requires histone acetylation and stimulates TATA-binding protein recruitment. Mol. Cell. Biol. 26, 4095–4110.
Mivelaz, M., Cao, A.M., Kubik, S., Zencir, S., Hovius, R., Boichenko, I., Stachowicz, A.M., Kurat, C.F., Shore, D., Fierz, B., 2020. Chromatin fiber invasion and nucleosome displacement by the Rap1 transcription factor. Mol. Cell 77 (488–500), e489.
Mohamed, A.A., Vazquez Nunez, R., Vos, S.M., 2022. Structural advances in transcription elongation. Curr. Opin. Struct. Biol. 75, 102422.
Mohammad, F., Helin, K., 2017. Oncohistones: drivers of pediatric cancers. Genes Dev. 31, 2313–2324.
Molenaar, T.M., van Leeuwen, F., 2022. SETD2: from chromatin modifier to multipronged regulator of the genome and beyond. Cell. Mol. Life Sci. 79, 346.
Molofsky, A.V., Pardal, R., Iwashita, T., Park, I.K., Clarke, M.F., Morrison, S.J., 2003. Bmi-1 dependence distinguishes neural stem cell self-renewal from progenitor proliferation. Nature 425, 962–967.
Monk, M., 1987. Genomic imprinting. Memories of mother and father. Nature 328, 203–204.
Moreau, P., Hen, R., Wasylyk, B., Everett, R., Gaub, M.P., Chambon, P., 1981. The SV40 72 base repair repeat has a striking effect on gene expression both in SV40 and other chimeric recombinants. Nucleic Acids Res. 9, 6047–6068.
Moreira, J.M., Holmberg, S., 1998. Nucleosome structure of the yeast CHA1 promoter: analysis of activation-dependent chromatin remodeling of an RNA-polymerase-II-transcribed gene in TBP and RNA pol II mutants defective in vivo in response to acidic activators. EMBO J. 17, 6028–6038.
Morgan, T.H., 1910. Sex limited inheritance in *Drosophila*. Science 32, 120–122.
Morgan, J.E., Whitlock Jr., J.P., 1992. Transcription-dependent and transcription-independent nucleosome disruption induced by dioxin. Proc. Natl. Acad. Sci. USA 89, 11622–11626.
Morgan, H.D., Sutherland, H.G., Martin, D.I., Whitelaw, E., 1999. Epigenetic inheritance at the agouti locus in the mouse. Nat. Genet. 23, 314–318.
Morgan, H.D., Santos, F., Green, K., Dean, W., Reik, W., 2005. Epigenetic reprogramming in mammals. Hum. Mol. Genet. 14 Spec No 1, R47–R58.
Morse, R.H., 1989. Nucleosomes inhibit both transcriptional initiation and elongation by RNA polymerase III in vitro. EMBO J. 8, 2343–2351.
Morse, R.H., 1992. Transcribed chromatin. Trends Biochem. Sci. 17, 23–26.
Morse, R.H., 2000. RAP, RAP, open up! new wrinkles for RAP1 in yeast. Trends Genet. 16, 51–53.
Morse, R.H., 2003. Getting into chromatin: how do transcription factors get past the histones? Biochem. Cell Biol. 81, 101–112.

Morse, R.H., 2022. Function and dynamics of the Mediator complex: novel insights and new frontiers. Transcription 13, 39–52.

Morse, R.H., Roth, S.Y., Simpson, R.T., 1992. A transcriptionally active tRNA gene interferes with nucleosome positioning in vivo. Mol. Cell. Biol. 12, 4015–4025.

Moussa, H.F., Bsteh, D., Yelagandula, R., Pribitzer, C., Stecher, K., Bartalska, K., Michetti, L., Wang, J., Zepeda-Martinez, J.A., Elling, U., et al., 2019. Canonical PRC1 controls sequence-independent propagation of polycomb-mediated gene silencing. Nat. Commun. 10, 1931.

Moyle-Heyrman, G., Zaichuk, T., Xi, L., Zhang, Q., Uhlenbeck, O.C., Holmgren, R., Widom, J., Wang, J.P., 2013. Chemical map of *Schizosaccharomyces pombe* reveals species-specific features in nucleosome positioning. Proc. Natl. Acad. Sci. USA 110, 20158–20163.

Muchardt, C., Yaniv, M., 1993. A human homologue of *Saccharomyces cerevisiae* SNF2/SWI2 and Drosophila brm genes potentiates transcriptional activation by the glucocorticoid receptor. EMBO J. 12, 4279–4290.

Mueller, C.L., Jaehning, J.A., 2002. Ctr9, Rtf1, and Leo1 are components of the Paf1/RNA polymerase II complex. Mol. Cell. Biol. 22, 1971–1980.

Mueller, P.R., Wold, B., 1989. In vivo footprinting of a muscle specific enhancer by ligation mediated PCR. Science 246, 780–786.

Mukherjee, S., 2016. The Gene: An Intimate History (Scribner).

Muller, J., Bienz, M., 1991. Long range repression conferring boundaries of Ultrabithorax expression in the *Drosophila* embryo. EMBO J. 10, 3147–3155.

Muller, J., Hart, C.M., Francis, N.J., Vargas, M.L., Sengupta, A., Wild, B., Miller, E.L., O'Connor, M.B., Kingston, R.E., Simon, J.A., 2002. Histone methyltransferase activity of a *Drosophila* polycomb group repressor complex. Cell 111, 197–208.

Musladin, S., Krietenstein, N., Korber, P., Barbaric, S., 2014. The RSC chromatin remodeling complex has a crucial role in the complete remodeler set for yeast PHO5 promoter opening. Nucleic Acids Res. 42, 4270–4282.

Nacheva, G.A., Guschin, D.Y., Preobrazhenskaya, O.V., Karpov, V.L., Ebralidse, K.K., Mirzabekov, A.D., 1989. Change in the pattern of histone binding to DNA upon transcriptional activation. Cell 58, 27–36.

Nakayama, J., Klar, A.J., Grewal, S.I., 2000. A chromodomain protein, Swi6, performs imprinting functions in fission yeast during mitosis and meiosis. Cell 101, 307–317.

Nan, X., Ng, H.H., Johnson, C.A., Laherty, C.D., Turner, B.M., Eisenman, R.N., Bird, A., 1998. Transcriptional repression by the methyl-CpG-binding protein MeCP2 involves a histone deacetylase complex. Nature 393, 386–389.

Narain, A., Bhandare, P., Adhikari, B., Backes, S., Eilers, M., Dolken, L., Schlosser, A., Erhard, F., Baluapuri, A., Wolf, E., 2021. Targeted protein degradation reveals a direct role of SPT6 in RNAPII elongation and termination. Mol. Cell 81 (3110–3127), e3114.

Neri, F., Rapelli, S., Krepelova, A., Incarnato, D., Parlato, C., Basile, G., Maldotti, M., Anselmi, F., Oliviero, S., 2017. Intragenic DNA methylation prevents spurious transcription initiation. Nature 543, 72–77.

Ng, J., Hart, C.M., Morgan, K., Simon, J.A., 2000. A *Drosophila* ESC-E(Z) protein complex is distinct from other polycomb group complexes and contains covalently modified ESC. Mol. Cell. Biol. 20, 3069–3078.

Ng, H.H., Dole, S., Struhl, K., 2003a. The Rtf1 component of the Paf1 transcriptional elongation complex is required for ubiquitination of histone H2B. J. Biol. Chem. 278, 33625–33628.

Ng, H.H., Robert, F., Young, R.A., Struhl, K., 2003b. Targeted recruitment of Set1 histone methylase by elongating Pol II provides a localized mark and memory of recent transcriptional activity. Mol. Cell 11, 709–719.

Nielsen, M., Ard, R., Leng, X., Ivanov, M., Kindgren, P., Pelechano, V., Marquardt, S., 2019. Transcription-driven chromatin repression of Intragenic transcription start sites. PLoS Genet. 15, e1007969.

Nightingale, K., Dimitrov, S., Reeves, R., Wolffe, A.P., 1996. Evidence for a shared structural role for HMG1 and linker histones B4 and H1 in organizing chromatin. EMBO J. 15, 548–561.

North, J.A., Shimko, J.C., Javaid, S., Mooney, A.M., Shoffner, M.A., Rose, S.D., Bundschuh, R., Fishel, R., Ottesen, J.J., Poirier, M.G., 2012. Regulation of the nucleosome unwrapping rate controls DNA accessibility. Nucleic Acids Res. 40, 10215–10227.

Ochs, K., Ott, M., Bunse, T., Sahm, F., Bunse, L., Deumelandt, K., Sonner, J.K., Keil, M., von Deimling, A., Wick, W., et al., 2017. K27M-mutant histone-3 as a novel target for glioma immunotherapy. OncoTargets Ther. 6, e1328340.

O'Donohue, M.F., Duband-Goulet, I., Hamiche, A., Prunell, A., 1994. Octamer displacement and redistribution in transcription of single nucleosomes. Nucleic Acids Res. 22, 937–945.

Oguro, H., Yuan, J., Ichikawa, H., Ikawa, T., Yamazaki, S., Kawamoto, H., Nakauchi, H., Iwama, A., 2010. Poised lineage specification in multipotential hematopoietic stem and progenitor cells by the polycomb protein Bmi1. Cell Stem Cell 6, 279–286.

Ohler, U., Liao, G.C., Niemann, H., Rubin, G.M., 2002. Computational analysis of core promoters in the *Drosophila* genome. Genome Biol. 3, RESEARCH0087.

Ohno, S., 1979. Major Sex-Determining Genes. Springer-Verlag, Berlin.

Okano, M., Xie, S., Li, E., 1998. Cloning and characterization of a family of novel mammalian DNA (cytosine-5) methyltransferases. Nat. Genet. 19, 219–220.

O'Neill, T.E., Roberge, M., Bradbury, E.M., 1992. Nucleosome arrays inhibit both initiation and elongation of transcripts by bacteriophage T7 RNA polymerase. J. Mol. Biol. 223, 67–78.

Ooi, S.K., Bestor, T.H., 2008. The colorful history of active DNA demethylation. Cell 133, 1145–1148.

Orphanides, G., LeRoy, G., Chang, C.H., Luse, D.S., Reinberg, D., 1998. FACT, a factor that facilitates transcript elongation through nucleosomes. Cell 92, 105–116.

Orphanides, G., Wu, W.H., Lane, W.S., Hampsey, M., Reinberg, D., 1999. The chromatin-specific transcription elongation factor FACT comprises human SPT16 and SSRP1 proteins. Nature 400, 284–288.

Oshima, Y., 1982. Regulatory circuits for gene expression: the metabolism of galactose and phosphate. In: Strathern, J.N., Jones, E.W., Broach, J.R. (Eds.), The Molecular Biology of the Yeast *Saccharomyces cerevisiae*: Metabolism and Gene Expression. Cold Spring Harbor Laboratory), (Cold Spring Harbor, NY, pp. 159–180.

Ostlund Farrants, A.K., Blomquist, P., Kwon, H., Wrange, O., 1997. Glucocorticoid receptor-glucocorticoid response element binding stimulates nucleosome disruption by the SWI/SNF complex. Mol. Cell. Biol. 17, 895–905.

Otero, G., Fellows, J., Li, Y., de Bizemont, T., Dirac, A.M., Gustafsson, C.M., Erdjument-Bromage, H., Tempst, P., Svejstrup, J.Q., 1999. Elongator, a multisubunit component of a novel RNA polymerase II holoenzyme for transcriptional elongation. Mol. Cell 3, 109–118.

Ozsolak, F., Song, J.S., Liu, X.S., Fisher, D.E., 2007. High-throughput mapping of the chromatin structure of human promoters. Nat. Biotechnol. 25, 244–248.

Paakinaho, V., Swinstead, E.E., Presman, D.M., Grontved, L., Hager, G.L., 2019. Meta-analysis of chromatin programming by steroid receptors. Cell Rep. 28 (3523–3534), e3522.

Painter, R.C., Osmond, C., Gluckman, P., Hanson, M., Phillips, D.I., Roseboom, T.J., 2008. Transgenerational effects of prenatal exposure to the Dutch famine on neonatal adiposity and health in later life. BJOG 115, 1243–1249.

Pan, G., Tian, S., Nie, J., Yang, C., Ruotti, V., Wei, H., Jonsdottir, G.A., Stewart, R., Thomson, J.A., 2007. Whole-genome analysis of histone H3 lysine 4 and lysine 27 methylation in human embryonic stem cells. Cell Stem Cell 1, 299–312.

Paranjape, S.M., Kamakaka, R.T., Kadonaga, J.T., 1994. Role of chromatin structure in the regulation of transcription by RNA polymerase II. Annu. Rev. Biochem. 63, 265–297.

Park, E.C., Szostak, J.W., 1990. Point mutations in the yeast histone H4 gene prevent silencing of the silent mating type locus HML. Mol. Cell. Biol. 10, 4932–4934.

Park, I.K., Qian, D., Kiel, M., Becker, M.W., Pihalja, M., Weissman, I.L., Morrison, S.J., Clarke, M.F., 2003. Bmi-1 is required for maintenance of adult self-renewing haematopoietic stem cells. Nature 423, 302–305.

Parks, W.P., Scolnick, E.M., Kozikowski, E.H., 1974. Dexamethasone stimulation of murine mammary tumor virus expression: a tissue culture source of virus. Science 184, 158–160.

Parnell, T.J., Huff, J.T., Cairns, B.R., 2008. RSC regulates nucleosome positioning at Pol II genes and density at Pol III genes. EMBO J. 27, 100–110.

Paro, R., 1990. Imprinting a determined state into the chromatin of *Drosophila*. Trends Genet. 6, 416–421.

Paro, R., Hogness, D.S., 1991. The polycomb protein shares a homologous domain with a heterochromatin-associated protein of *Drosophila*. Proc. Natl. Acad. Sci. USA 88, 263–267.

Parra, M.A., Kerr, D., Fahy, D., Pouchnik, D.J., Wyrick, J.J., 2006. Deciphering the roles of the histone H2B N-terminal domain in genome-wide transcription. Mol. Cell. Biol. 26, 3842–3852.

Pathak, R., Singh, P., Ananthakrishnan, S., Adamczyk, S., Schimmel, O., Govind, C.K., 2018. Acetylation-dependent recruitment of the FACT complex and its role in regulating Pol II occupancy genome-wide in *Saccharomyces cerevisiae*. Genetics 209, 743–756.

Pattenden, S.G., Gogol, M.M., Workman, J.L., 2010. Features of cryptic promoters and their varied reliance on bromodomain-containing factors. PLoS One 5, e12927.

Pederson, D.S., Morse, R.H., 1990. Effect of transcription of yeast chromatin on DNA topology in vivo. EMBO J. 9, 1873–1881.

Pengelly, A.R., Copur, O., Jackle, H., Herzig, A., Muller, J., 2013. A histone mutant reproduces the phenotype caused by loss of histone-modifying factor polycomb. Science 339, 698–699.
Pengelly, A.R., Kalb, R., Finkl, K., Muller, J., 2015. Transcriptional repression by PRC1 in the absence of H2A monoubiquitylation. Genes Dev. 29, 1487–1492.
Perlmann, T., Wrange, O., 1988. Specific glucocorticoid receptor binding to DNA reconstituted in a nucleosome. EMBO J. 7, 3073–3079.
Perlmann, T., Wrange, O., 1991. Inhibition of chromatin assembly in Xenopus oocytes correlates with derepression of the mouse mammary tumor virus promoter. Mol. Cell. Biol. 11, 5259–5265.
Peterson, C.L., Herskowitz, I., 1992. Characterization of the yeast SWI1, SWI2, and SWI3 genes, which encode a global activator of transcription. Cell 68, 573–583.
Petryk, N., Dalby, M., Wenger, A., Stromme, C.B., Strandsby, A., Andersson, R., Groth, A., 2018. MCM2 promotes symmetric inheritance of modified histones during DNA replication. Science 361, 1389–1392.
Pietersen, A.M., van Lohuizen, M., 2008. Stem cell regulation by polycomb repressors: postponing commitment. Curr. Opin. Cell Biol. 20, 201–207.
Pillus, L., Rine, J., 1989. Epigenetic inheritance of transcriptional states in *S. cerevisiae*. Cell 59, 637–647.
Pina, B., Bruggemeier, U., Beato, M., 1990. Nucleosome positioning modulates accessibility of regulatory proteins to the mouse mammary tumor virus promoter. Cell 60, 719–731.
Piunti, A., Shilatifard, A., 2021. The roles of polycomb repressive complexes in mammalian development and cancer. Nat. Rev. Mol. Cell Biol. 22, 326–345.
Piunti, A., Smith, E.R., Morgan, M.A.J., Ugarenko, M., Khaltyan, N., Helmin, K.A., Ryan, C.A., Murray, D.C., Rickels, R.A., Yilmaz, B.D., et al., 2019. CATACOMB: an endogenous inducible gene that antagonizes H3K27 methylation activity of polycomb repressive complex 2 via an H3K27M-like mechanism. Sci. Adv. 5, eaax2887.
Plachta, N., Bollenbach, T., Pease, S., Fraser, S.E., Pantazis, P., 2011. Oct4 kinetics predict cell lineage patterning in the early mammalian embryo. Nat. Cell Biol. 13, 117–123.
Pokholok, D.K., Hannett, N.M., Young, R.A., 2002. Exchange of RNA polymerase II initiation and elongation factors during gene expression in vivo. Mol. Cell 9, 799–809.
Polach, K.J., Widom, J., 1995. Mechanism of protein access to specific DNA sequences in chromatin: a dynamic equilibrium model for gene regulation. J. Mol. Biol. 254, 130–149.
Polach, K.J., Widom, J., 1996. A model for the cooperative binding of eukaryotic regulatory proteins to nucleosomal target sites. J. Mol. Biol. 258, 800–812.
Prather, D., Krogan, N.J., Emili, A., Greenblatt, J.F., Winston, F., 2005. Identification and characterization of Elf1, a conserved transcription elongation factor in *Saccharomyces cerevisiae*. Mol. Cell. Biol. 25, 10122–10135.
Prior, C.P., Cantor, C.R., Johnson, E.M., Allfrey, V.G., 1980. Incorporation of exogenous pyrene-labeled histone into Physarum chromatin: a system for studying changes in nucleosomes assembled in vivo. Cell 20, 597–608.
Pruss, G.J., Drlica, K., 1989. DNA supercoiling and prokaryotic transcription. Cell 56, 521–523.
Ptashne, M., 1986. Gene regulation by proteins acting nearby and at a distance. Nature 322, 697–701.
Ptashne, M., 2004. A Genetic Switch, Third ed. Cold Spring Harbor Press.
Ptashne, M., 2013a. Epigenetics: core misconcept. Proc. Natl. Acad. Sci. USA 110, 7101–7103.
Ptashne, M., 2013b. Faddish stuff: epigenetics and the inheritance of acquired characteristics. FASEB J. 27, 1–2.
Qiu, H., Hu, C., Yoon, S., Natarajan, K., Swanson, M.J., Hinnebusch, A.G., 2004. An array of coactivators is required for optimal recruitment of TATA binding protein and RNA polymerase II by promoter-bound Gcn4p. Mol. Cell. Biol. 24, 4104–4117.
Qiu, H., Chereji, R.V., Hu, C., Cole, H.A., Rawal, Y., Clark, D.J., Hinnebusch, A.G., 2016. Genome-wide cooperation by HAT Gcn5, remodeler SWI/SNF, and chaperone Ydj1 in promoter nucleosome eviction and transcriptional activation. Genome Res. 26, 211–225.
Quadrana, L., Colot, V., 2016. Plant transgenerational epigenetics. Annu. Rev. Genet. 50, 467–491.
Rabinowitz, M., 1941. Studies on the cytology and early embryology of the egg *Drosophila melanogaster*. J. Morphol. 69, 1–49.
Rach, E.A., Winter, D.R., Benjamin, A.M., Corcoran, D.L., Ni, T., Zhu, J., Ohler, U., 2011. Transcription initiation patterns indicate divergent strategies for gene regulation at the chromatin level. PLoS Genet. 7, e1001274.
Ragunathan, K., Jih, G., Moazed, D., 2015. Epigenetics. Epigenetic inheritance uncoupled from sequence-specific recruitment. Science 348, 1258699.
Rahl, P.B., Lin, C.Y., Seila, A.C., Flynn, R.A., McCuine, S., Burge, C.B., Sharp, P.A., Young, R.A., 2010. c-Myc regulates transcriptional pause release. Cell 141, 432–445.

Rajasekhar, V.K., Begemann, M., 2007. Concise review: roles of polycomb group proteins in development and disease: a stem cell perspective. Stem Cells 25, 2498–2510.

Rakyan, V.K., Chong, S., Champ, M.E., Cuthbert, P.C., Morgan, H.D., Luu, K.V., Whitelaw, E., 2003. Transgenerational inheritance of epigenetic states at the murine Axin(Fu) allele occurs after maternal and paternal transmission. Proc. Natl. Acad. Sci. USA 100, 2538–2543.

Ramachandran, S., Henikoff, S., 2016. Transcriptional regulators compete with nucleosomes post-replication. Cell 165, 580–592.

Ramachandran, S., Ahmad, K., Henikoff, S., 2017. Transcription and remodeling produce asymmetrically unwrapped nucleosomal intermediates. Mol. Cell 68 (1038–1053), e1034.

Ramakrishnan, V., Finch, J.T., Graziano, V., Lee, P.L., Sweet, R.M., 1993. Crystal structure of globular domain of histone H5 and its implications for nucleosome binding. Nature 362, 219–223.

Reid, J.L., Moqtaderi, Z., Struhl, K., 2004. Eaf3 regulates the global pattern of histone acetylation in *Saccharomyces cerevisiae*. Mol. Cell. Biol. 24, 757–764.

Reik, A., Schutz, G., Stewart, A.F., 1991. Glucocorticoids are required for establishment and maintenance of an alteration in chromatin structure: induction leads to a reversible disruption of nucleosomes over an enhancer. EMBO J. 10, 2569–2576.

Reinke, H., Horz, W., 2003. Histones are first hyperacetylated and then lose contact with the activated PHO5 promoter. Mol. Cell 11, 1599–1607.

Reinke, H., Gregory, P.D., Horz, W., 2001. A transient histone hyperacetylation signal marks nucleosomes for remodeling at the PHO8 promoter in vivo. Mol. Cell 7, 529–538.

Reveron-Gomez, N., Gonzalez-Aguilera, C., Stewart-Morgan, K.R., Petryk, N., Flury, V., Graziano, S., Johansen, J.V., Jakobsen, J.S., Alabert, C., Groth, A., 2018. Accurate recycling of parental histones reproduces the histone modification landscape during DNA replication. Mol. Cell 72 (239-249), e235.

Richard-Foy, H., Hager, G.L., 1987. Sequence-specific positioning of nucleosomes over the steroid-inducible MMTV promoter. EMBO J. 6, 2321–2328.

Riggs, A.D., 1975. X inactivation, differentiation, and DNA methylation. Cytogenet. Cell Genet. 14, 9–25.

Ringold, G.M., Yamamoto, K.R., Bishop, J.M., Varmus, H.E., 1977. Glucocorticoid-stimulated accumulation of mouse mammary tumor virus RNA: increased rate of synthesis of viral RNA. Proc. Natl. Acad. Sci. USA 74, 2879–2883.

Ringrose, L., Paro, R., 2004. Epigenetic regulation of cellular memory by the polycomb and trithorax group proteins. Annu. Rev. Genet. 38, 413–443.

Robert, F., Jeronimo, C., 2023. Transcription-coupled nucleosome assembly. Trends Biochem. Sci. 48, 978–992.

Roeder, R.G., 2019. 50+ years of eukaryotic transcription: an expanding universe of factors and mechanisms. Nat. Struct. Mol. Biol. 26, 783–791.

Roeder, R.G., Rutter, W.J., 1969. Multiple forms of DNA-dependent RNA polymerase in eukaryotic organisms. Nature 224, 234–237.

Romano, L.A., Wray, G.A., 2003. Conservation of Endo16 expression in sea urchins despite evolutionary divergence in both cis and trans-acting components of transcriptional regulation. Development 130, 4187–4199.

Rondon, A.G., Gallardo, M., Garcia-Rubio, M., Aguilera, A., 2004. Molecular evidence indicating that the yeast PAF complex is required for transcription elongation. EMBO Rep. 5, 47–53.

Rossi, M.J., Kuntala, P.K., Lai, W.K.M., Yamada, N., Badjatia, N., Mittal, C., Kuzu, G., Bocklund, K., Farrell, N.P., Blanda, T.R., et al., 2021. A high-resolution protein architecture of the budding yeast genome. Nature 592, 309–314.

Roy, S., Ernst, J., Kharchenko, P.V., Kheradpour, P., Negre, N., Eaton, M.L., Landolin, J.M., Bristow, C.A., Ma, L., Lin, M.F., et al., 2010. Identification of functional elements and regulatory circuits by *Drosophila* modENCODE. Science 330, 1787–1797.

Rudolph, H., Hinnen, A., 1987. The yeast PHO5 promoter: phosphate-control elements and sequences mediating mRNA start-site selection. Proc. Natl. Acad. Sci. USA 84, 1340–1344.

Rufiange, A., Jacques, P.E., Bhat, W., Robert, F., Nourani, A., 2007. Genome-wide replication-independent histone H3 exchange occurs predominantly at promoters and implicates H3 K56 acetylation and Asf1. Mol. Cell 27, 393–405.

Rundlett, S.E., Carmen, A.A., Kobayashi, R., Bavykin, S., Turner, B.M., Grunstein, M., 1996. HDA1 and RPD3 are members of distinct yeast histone deacetylase complexes that regulate silencing and transcription. Proc. Natl. Acad. Sci. USA 93, 14503–14508.

Sabet, N., Tong, F., Madigan, J.P., Volo, S., Smith, M.M., Morse, R.H., 2003. Global and specific transcriptional repression by the histone H3 amino terminus in yeast. Proc. Natl. Acad. Sci. USA 100, 4084–4089.

Sadowski, I., Ma, J., Triezenberg, S., Ptashne, M., 1988. GAL4-VP16 is an unusually potent transcriptional activator. Nature 335, 563–564.

Sakonju, S., Bogenhagen, D.F., Brown, D.D., 1980. A control region in the center of the 5S RNA gene directs specific initiation of transcription: I. The 5′ border of the region. Cell 19, 13–25.

Sakonju, S., Brown, D.D., Engelke, D., Ng, S.Y., Shastry, B.S., Roeder, R.G., 1981. The binding of a transcription factor to deletion mutants of a 5S ribosomal RNA gene. Cell 23, 665–669.

Sammons, M.A., Zhu, J., Drake, A.M., Berger, S.L., 2015. TP53 engagement with the genome occurs in distinct local chromatin environments via pioneer factor activity. Genome Res. 25, 179–188.

Sandaltzopoulos, R., Blank, T., Becker, P.B., 1994. Transcriptional repression by nucleosomes but not H1 in reconstituted preblastoderm *Drosophila* chromatin. EMBO J. 13, 373–379.

Sankar, A., Mohammad, F., Sundaramurthy, A.K., Wang, H., Lerdrup, M., Tatar, T., Helin, K., 2022. Histone editing elucidates the functional roles of H3K27 methylation and acetylation in mammals. Nat. Genet. 54, 754–760.

Sardina, J.L., Collombet, S., Tian, T.V., Gomez, A., Di Stefano, B., Berenguer, C., Brumbaugh, J., Stadhouders, R., Segura-Morales, C., Gut, M., et al., 2018. Transcription factors drive Tet2-mediated enhancer demethylation to reprogram cell fate. Cell Stem Cell 23 (727–741), e729.

Sasaki, H., Hogan, B.L., 1993. Differential expression of multiple fork head related genes during gastrulation and axial pattern formation in the mouse embryo. Development 118, 47–59.

Saunders, A., Werner, J., Andrulis, E.D., Nakayama, T., Hirose, S., Reinberg, D., Lis, J.T., 2003. Tracking FACT and the RNA polymerase II elongation complex through chromatin in vivo. Science 301, 1094–1096.

Sauvageau, M., Sauvageau, G., 2010. Polycomb group proteins: multi-faceted regulators of somatic stem cells and cancer. Cell Stem Cell 7, 299–313.

Sawadogo, M., Roeder, R.G., 1985. Interaction of a gene-specific transcription factor with the adenovirus major late promoter upstream of the TATA box region. Cell 43, 165–175.

Saxonov, S., Berg, P., Brutlag, D.L., 2006. A genome-wide analysis of CpG dinucleotides in the human genome distinguishes two distinct classes of promoters. Proc. Natl. Acad. Sci. USA 103, 1412–1417.

Saxton, D.S., Rine, J., 2019. Epigenetic memory independent of symmetric histone inheritance. elife 8.

Schaft, D., Roguev, A., Kotovic, K.M., Shevchenko, A., Sarov, M., Shevchenko, A., Neugebauer, K.M., Stewart, A.F., 2003. The histone 3 lysine 36 methyltransferase, SET2, is involved in transcriptional elongation. Nucleic Acids Res. 31, 2475–2482.

Scherer, S., Davis, R.W., 1979. Replacement of chromosome segments with altered DNA sequences constructed in vitro. Proc. Natl. Acad. Sci. USA 76, 4951–4955.

Schermer, U.J., Korber, P., Horz, W., 2005. Histones are incorporated in trans during reassembly of the yeast PHO5 promoter. Mol. Cell 19, 279–285.

Schick, S., Grosche, S., Kohl, K.E., Drpic, D., Jaeger, M.G., Marella, N.C., Imrichova, H., Lin, J.G., Hofstatter, G., Schuster, M., et al., 2021. Acute BAF perturbation causes immediate changes in chromatin accessibility. Nat. Genet. 53, 269–278.

Schild, C., Claret, F.X., Wahli, W., Wolffe, A.P., 1993. A nucleosome-dependent static loop potentiates estrogen-regulated transcription from the Xenopus vitellogenin B1 promoter in vitro. EMBO J. 12, 423–433.

Schlichter, A., Kasten, M.M., Parnell, T.J., Cairns, B.R., 2020. Specialization of the chromatin remodeler RSC to mobilize partially-unwrapped nucleosomes. elife 9.

Schlissel, M.S., Brown, D.D., 1984. The transcriptional regulation of Xenopus 5s RNA genes in chromatin: the roles of active stable transcription complexes and histone H1. Cell 37, 903–913.

Schlissel, G., Rine, J., 2019. The nucleosome core particle remembers its position through DNA replication and RNA transcription. Proc. Natl. Acad. Sci. USA 116, 20605–20611.

Schmid, A., Fascher, K.D., Horz, W., 1992. Nucleosome disruption at the yeast PHO5 promoter upon PHO5 induction occurs in the absence of DNA replication. Cell 71, 853–864.

Schorderet, P., Lonfat, N., Darbellay, F., Tschopp, P., Gitto, S., Soshnikova, N., Duboule, D., 2013. A genetic approach to the recruitment of PRC2 at the HoxD locus. PLoS Genet. 9, e1003951.

Schuettengruber, B., Ganapathi, M., Leblanc, B., Portoso, M., Jaschek, R., Tolhuis, B., van Lohuizen, M., Tanay, A., Cavalli, G., 2009. Functional anatomy of polycomb and trithorax chromatin landscapes in *Drosophila* embryos. PLoS Biol. 7, e13.

Schuettengruber, B., Oded Elkayam, N., Sexton, T., Entrevan, M., Stern, S., Thomas, A., Yaffe, E., Parrinello, H., Tanay, A., Cavalli, G., 2014. Cooperativity, specificity, and evolutionary stability of polycomb targeting in *Drosophila*. Cell Rep. 9, 219–233.

Schuettengruber, B., Bourbon, H.M., Di Croce, L., Cavalli, G., 2017. Genome regulation by polycomb and trithorax: 70 years and counting. Cell 171, 34–57.

Schulz, K.N., Bondra, E.R., Moshe, A., Villalta, J.E., Lieb, J.D., Kaplan, T., McKay, D.J., Harrison, M.M., 2015. Zelda is differentially required for chromatin accessibility, transcription factor binding, and gene expression in the early Drosophila embryo. Genome Res. 25, 1715–1726.

Schumacher, A., Faust, C., Magnuson, T., 1996. Positional cloning of a global regulator of anterior-posterior patterning in mice. Nature 384, 648.

Schuster, T., Han, M., Grunstein, M., 1986. Yeast histone H2A and H2B amino termini have interchangeable functions. Cell 45, 445–451.

Schwartz, Y.B., Kahn, T.G., Nix, D.A., Li, X.Y., Bourgon, R., Biggin, M., Pirrotta, V., 2006. Genome-wide analysis of polycomb targets in *Drosophila melanogaster*. Nat. Genet. 38, 700–705.

Schwartzentruber, J., Korshunov, A., Liu, X.Y., Jones, D.T., Pfaff, E., Jacob, K., Sturm, D., Fontebasso, A.M., Quang, D.A., Tonjes, M., et al., 2012. Driver mutations in histone H3.3 and chromatin remodelling genes in paediatric glioblastoma. Nature 482, 226–231.

Sekimizu, K., Kobayashi, N., Mizuno, D., Natori, S., 1976. Purification of a factor from Ehrlich ascites tumor cells specifically stimulating RNA polymerase II. Biochemistry 15, 5064–5070.

Sekiya, T., Muthurajan, U.M., Luger, K., Tulin, A.V., Zaret, K.S., 2009. Nucleosome-binding affinity as a primary determinant of the nuclear mobility of the pioneer transcription factor FoxA. Genes Dev. 23, 804–809.

Selleck, S.B., Majors, J., 1987. Photofootprinting in vivo detects transcription-dependent changes in yeast TATA boxes. Nature 325, 173–177.

Sengupta, A.K., Kuhrs, A., Muller, J., 2004. General transcriptional silencing by a polycomb response element in *Drosophila*. Development 131, 1959–1965.

Sergeant, A., Bohmann, D., Zentgraf, H., Weiher, H., Keller, W., 1984. A transcription enhancer acts in vitro over distances of hundreds of base-pairs on both circular and linear templates but not on chromatin-reconstituted DNA. J. Mol. Biol. 180, 577–600.

Shao, Z., Raible, F., Mollaaghababa, R., Guyon, J.R., Wu, C.T., Bender, W., Kingston, R.E., 1999. Stabilization of chromatin structure by PRC1, a polycomb complex. Cell 98, 37–46.

Sheinin, M.Y., Li, M., Soltani, M., Luger, K., Wang, M.D., 2013. Torque modulates nucleosome stability and facilitates H2A/H2B dimer loss. Nat. Commun. 4, 2579.

Shen, X., Mizuguchi, G., Hamiche, A., Wu, C., 2000. A chromatin remodelling complex involved in transcription and DNA processing. Nature 406, 541–544.

Shetty, A., Kallgren, S.P., Demel, C., Maier, K.C., Spatt, D., Alver, B.H., Cramer, P., Park, P.J., Winston, F., 2017. Spt5 plays vital roles in the control of sense and antisense transcription elongation. Mol. Cell 66 (77–88), e75.

Shi, X., Finkelstein, A., Wolf, A.J., Wade, P.A., Burton, Z.F., Jaehning, J.A., 1996. Paf1p, an RNA polymerase II-associated factor in *Saccharomyces cerevisiae*, may have both positive and negative roles in transcription. Mol. Cell. Biol. 16, 669–676.

Shi, X., Chang, M., Wolf, A.J., Chang, C.H., Frazer-Abel, A.A., Wade, P.A., Burton, Z.F., Jaehning, J.A., 1997. Cdc73p and Paf1p are found in a novel RNA polymerase II-containing complex distinct from the Srbp-containing holoenzyme. Mol. Cell. Biol. 17, 1160–1169.

Shim, E.Y., Woodcock, C., Zaret, K.S., 1998. Nucleosome positioning by the winged helix transcription factor HNF3. Genes Dev. 12, 5–10.

Shimamura, A., Tremethick, D., Worcel, A., 1988. Characterization of the repressed 5S DNA minichromosomes assembled in vitro with a high-speed supernatant of Xenopus laevis oocytes. Mol. Cell. Biol. 8, 4257–4269.

Shimamura, A., Sapp, M., Rodriguez-Campos, A., Worcel, A., 1989. Histone H1 represses transcription from minichromosomes assembled in vitro. Mol. Cell. Biol. 9, 5573–5584.

Shivaswamy, S., Bhinge, A., Zhao, Y., Jones, S., Hirst, M., Iyer, V.R., 2008. Dynamic remodeling of individual nucleosomes across a eukaryotic genome in response to transcriptional perturbation. PLoS Biol. 6, e65.

Siklenka, K., Erkek, S., Godmann, M., Lambrot, R., McGraw, S., Lafleur, C., Cohen, T., Xia, J., Suderman, M., Hallett, M., et al., 2015. Disruption of histone methylation in developing sperm impairs offspring health transgenerationally. Science 350, aab2006.

Silva, A.J., White, R., 1988. Inheritance of allelic blueprints for methylation patterns. Cell 54, 145–152.

Simon, J., 1995. Locking in stable states of gene expression: transcriptional control during *Drosophila* development. Curr. Opin. Cell Biol. 7, 376–385.

Simon, J.A., Kingston, R.E., 2013. Occupying chromatin: polycomb mechanisms for getting to genomic targets, stopping transcriptional traffic, and staying put. Mol. Cell 49, 808–824.

Simon, J., Chiang, A., Bender, W., 1992. Ten different polycomb group genes are required for spatial control of the abdA and AbdB homeotic products. Development 114, 493–505.
Simon, J., Chiang, A., Bender, W., Shimell, M.J., O'Connor, M., 1993. Elements of the *Drosophila* bithorax complex that mediate repression by polycomb group products. Dev. Biol. 158, 131–144.
Simpson, R.T., Stafford, D.W., 1983. Structural features of a phased nucleosome core particle. Proc. Natl. Acad. Sci. USA 80, 51–55.
Sing, A., Pannell, D., Karaiskakis, A., Sturgeon, K., Djabali, M., Ellis, J., Lipshitz, H.D., Cordes, S.P., 2009. A vertebrate polycomb response element governs segmentation of the posterior hindbrain. Cell 138, 885–897.
Sinsheimer, R.L., 1954. The action of pancreatic desoxyribonuclease. I. Isolation of mono- and dinucleotides. J. Biol. Chem. 208, 445–459.
Small, E.C., Xi, L., Wang, J.P., Widom, J., Licht, J.D., 2014. Single-cell nucleosome mapping reveals the molecular basis of gene expression heterogeneity. Proc. Natl. Acad. Sci. USA 111, E2462–E2471.
Smith, M.M., Andresson, O.S., 1983. DNA sequences of yeast H3 and H4 histone genes from two non-allelic gene sets encode identical H3 and H4 proteins. J. Mol. Biol. 169, 663–690.
Smith, C.L., Hager, G.L., 1997. Transcriptional regulation of mammalian genes in vivo. A tale of two templates. J. Biol. Chem. 272, 27493–27496.
Smith, R.C., Dworkin-Rastl, E., Dworkin, M.B., 1988. Expression of a histone H1-like protein is restricted to early Xenopus development. Genes Dev. 2, 1284–1295.
Smith, C.L., Htun, H., Wolford, R.G., Hager, G.L., 1997. Differential activity of progesterone and glucocorticoid receptors on mouse mammary tumor virus templates differing in chromatin structure. J. Biol. Chem. 272, 14227–14235.
Smith, S.T., Petruk, S., Sedkov, Y., Cho, E., Tillib, S., Canaani, E., Mazo, A., 2004. Modulation of heat shock gene expression by the TAC1 chromatin-modifying complex. Nat. Cell Biol. 6, 162–167.
Smolle, M., Workman, J.L., 2013. Transcription-associated histone modifications and cryptic transcription. Biochim. Biophys. Acta 1829, 84–97.
Song, A., Chen, F.X., 2022. The pleiotropic roles of SPT5 in transcription. Transcription 13, 53–69.
Sonmezer, C., Kleinendorst, R., Imanci, D., Barzaghi, G., Villacorta, L., Schubeler, D., Benes, V., Molina, N., Krebs, A.-R., 2021. Molecular co-occupancy identifies transcription factor binding cooperativity in vivo. Mol. Cell 81 (255–267), e256.
Soufi, A., Donahue, G., Zaret, K.S., 2012. Facilitators and impediments of the pluripotency reprogramming factors' initial engagement with the genome. Cell 151, 994–1004.
Soufi, A., Garcia, M.F., Jaroszewicz, A., Osman, N., Pellegrini, M., Zaret, K.S., 2015. Pioneer transcription factors target partial DNA motifs on nucleosomes to initiate reprogramming. Cell 161, 555–568.
Soutoglou, E., Talianidis, I., 2002. Coordination of PIC assembly and chromatin remodeling during differentiation-induced gene activation. Science 295, 1901–1904.
Spivakov, M., Fisher, A.G., 2007. Epigenetic signatures of stem-cell identity. Nat. Rev. Genet. 8, 263–271.
Squazzo, S.L., Costa, P.J., Lindstrom, D.L., Kumer, K.E., Simic, R., Jennings, J.L., Link, A.J., Arndt, K.M., Hartzog, G.A., 2002. The Paf1 complex physically and functionally associates with transcription elongation factors in vivo. EMBO J. 21, 1764–1774.
Stedman, E., Stedman, E., 1951. The basic proteins of cell nuclei. Phil. Trans. R. Soc. London 235, 565–595.
Stein, R., Gruenbaum, Y., Pollack, Y., Razin, A., Cedar, H., 1982. Clonal inheritance of the pattern of DNA methylation in mouse cells. Proc. Natl. Acad. Sci. USA 79, 61–65.
Stewart, K.R., Veselovska, L., Kim, J., Huang, J., Saadeh, H., Tomizawa, S., Smallwood, S.A., Chen, T., Kelsey, G., 2015. Dynamic changes in histone modifications precede de novo DNA methylation in oocytes. Genes Dev. 29, 2449–2462.
Stewart-Morgan, K.R., Petryk, N., Groth, A., 2020. Chromatin replication and epigenetic cell memory. Nat. Cell Biol. 22, 361–371.
Stillman, D.J., 2013. Dancing the cell cycle two-step: regulation of yeast G1-cell-cycle genes by chromatin structure. Trends Biochem. Sci. 38, 467–475.
Straka, C., Horz, W., 1991. A functional role for nucleosomes in the repression of a yeast promoter. EMBO J. 10, 361–368.
Struhl, K., 1981. Deletion mapping a eukaryotic promoter. Proc. Natl. Acad. Sci. USA 78, 4461–4465.
Struhl, K., 1999. Fundamentally different logic of gene regulation in eukaryotes and prokaryotes. Cell 98, 1–4.
Struhl, G., Akam, M., 1985. Altered distributions of Ultrabithorax transcripts in extra sex combs mutant embryos of *Drosophila*. EMBO J. 4, 3259–3264.

Studitsky, V.M., Clark, D.J., Felsenfeld, G., 1994. A histone octamer can step around a transcribing polymerase without leaving the template. Cell 76, 371–382.
Studitsky, V.M., Clark, D.J., Felsenfeld, G., 1995. Overcoming a nucleosomal barrier to transcription. Cell 83, 19–27.
Studitsky, V.M., Kassavetis, G.A., Geiduschek, E.P., Felsenfeld, G., 1997. Mechanism of transcription through the nucleosome by eukaryotic RNA polymerase. Science 278, 1960–1963.
Sturm, D., Witt, H., Hovestadt, V., Khuong-Quang, D.A., Jones, D.T., Konermann, C., Pfaff, E., Tonjes, M., Sill, M., Bender, S., et al., 2012. Hotspot mutations in H3F3A and IDH1 define distinct epigenetic and biological subgroups of glioblastoma. Cancer Cell 22, 425–437.
Sturtevant, A.H., 2001. A History of Genetics. Cold Spring Harbor Laboratory Press, Cold Spring Harbor, NY.
Sudarsanam, P., Cao, Y., Wu, L., Laurent, B.C., Winston, F., 1999. The nucleosome remodeling complex, Snf/Swi, is required for the maintenance of transcription in vivo and is partially redundant with the histone acetyltransferase, Gcn5. EMBO J. 18, 3101–3106.
Sun, Y., Nien, C.Y., Chen, K., Liu, H.Y., Johnston, J., Zeitlinger, J., Rushlow, C., 2015. Zelda overcomes the high intrinsic nucleosome barrier at enhancers during *Drosophila* zygotic genome activation. Genome Res. 25, 1703–1714.
Surface, L.E., Thornton, S.R., Boyer, L.A., 2010. Polycomb group proteins set the stage for early lineage commitment. Cell Stem Cell 7, 288–298.
Suzuki, M.M., Bird, A., 2008. DNA methylation landscapes: provocative insights from epigenomics. Nat. Rev. Genet. 9, 465–476.
Svaren, J., Chalkley, R., 1990. The structure and assembly of active chromatin. Trends Genet. 6, 52–56.
Svaren, J., Schmitz, J., Horz, W., 1994. The transactivation domain of Pho4 is required for nucleosome disruption at the PHO5 promoter. EMBO J. 13, 4856–4862.
Svejstrup, J.Q., 2007. Elongator complex: how many roles does it play? Curr. Opin. Cell Biol. 19, 331–336.
Swanson, M.S., Winston, F., 1992. SPT4, SPT5 and SPT6 interactions: effects on transcription and viability in *Saccharomyces cerevisiae*. Genetics 132, 325–336.
Swanson, M.J., Qiu, H., Sumibcay, L., Krueger, A., Kim, S.J., Natarajan, K., Yoon, S., Hinnebusch, A.G., 2003. A multiplicity of coactivators is required by Gcn4p at individual promoters in vivo. Mol. Cell. Biol. 23, 2800–2820.
Swinstead, E.E., Miranda, T.B., Paakinaho, V., Baek, S., Goldstein, I., Hawkins, M., Karpova, T.S., Ball, D., Mazza, D., Lavis, L.D., et al., 2016. Steroid receptors reprogram FoxA1 occupancy through dynamic chromatin transitions. Cell 165, 593–605.
Szent-Gyorgyi, C., Finkelstein, D.B., Garrard, W.T., 1987. Sharp boundaries demarcate the chromatin structure of a yeast heat-shock gene. J. Mol. Biol. 193, 71–80.
Taatjes, D.J., Marr, M.T., Tjian, R., 2004. Regulatory diversity among metazoan co-activator complexes. Nat. Rev. Mol. Cell Biol. 5, 403–410.
Tahiliani, M., Koh, K.P., Shen, Y., Pastor, W.A., Bandukwala, H., Brudno, Y., Agarwal, S., Iyer, L.M., Liu, D.R., Aravind, L., et al., 2009. Conversion of 5-methylcytosine to 5-hydroxymethylcytosine in mammalian DNA by MLL partner TET1. Science 324, 930–935.
Takahashi, H., Mitani, Y., Satoh, G., Satoh, N., 1999. Evolutionary alterations of the minimal promoter for notochord-specific Brachyury expression in ascidian embryos. Development 126, 3725–3734.
Takahashi, Y., Wu, J., Suzuki, K., Martinez-Redondo, P., Li, M., Liao, H.K., Wu, M.Z., Hernandez-Benitez, R., Hishida, T., Shokhirev, M.N., et al., 2017. Integration of CpG-free DNA induces de novo methylation of CpG islands in pluripotent stem cells. Science 356, 503–508.
Takahashi, Y., Morales Valencia, M., Yu, Y., Ouchi, Y., Takahashi, K., Shokhirev, M.N., Lande, K., Williams, A.E., Fresia, C., Kurita, M., et al., 2023. Transgenerational inheritance of acquired epigenetic signatures at CpG islands in mice. Cell 186 (715–731), e719.
Takahata, S., Yu, Y., Stillman, D.J., 2009. FACT and Asf1 regulate nucleosome dynamics and coactivator binding at the HO promoter. Mol. Cell 34, 405–415.
Takaku, M., Grimm, S.A., Shimbo, T., Perera, L., Menafra, R., Stunnenberg, H.G., Archer, T.K., Machida, S., Kurumizaka, H., Wade, P.A., 2016. GATA3-dependent cellular reprogramming requires activation-domain dependent recruitment of a chromatin remodeler. Genome Biol. 17, 36.
Talyzina, A., Han, Y., Banerjee, C., Fishbain, S., Reyes, A., Vafabakhsh, R., He, Y., 2023. Structural basis of TFIIIC-dependent RNA polymerase III transcription initiation. Mol. Cell 83 (2641–2652), e2647.
Tamburri, S., Lavarone, E., Fernandez-Perez, D., Conway, E., Zanotti, M., Manganaro, D., Pasini, D., 2020. Histone H2AK119 mono-ubiquitination is essential for polycomb-mediated transcriptional repression. Mol. Cell 77 (840–856), e845.

Tamkun, J.W., Deuring, R., Scott, M.P., Kissinger, M., Pattatucci, A.M., Kaufman, T.C., Kennison, J.A., 1992. brahma: a regulator of *Drosophila* homeotic genes structurally related to the yeast transcriptional activator SNF2/SWI2. Cell 68, 561–572.

Tanaka, H., Takizawa, Y., Takaku, M., Kato, D., Kumagawa, Y., Grimm, S.A., Wade, P.A., Kurumizaka, H., 2020. Interaction of the pioneer transcription factor GATA3 with nucleosomes. Nat. Commun. 11, 4136.

Tang, W.W., Kobayashi, T., Irie, N., Dietmann, S., Surani, M.A., 2016. Specification and epigenetic programming of the human germ line. Nat. Rev. Genet. 17, 585–600.

Tapscott, S.J., Davis, R.L., Thayer, M.J., Cheng, P.F., Weintraub, H., Lassar, A.B., 1988. MyoD1: a nuclear phosphoprotein requiring a Myc homology region to convert fibroblasts to myoblasts. Science 242, 405–411.

Taylor, S.M., Jones, P.A., 1979. Multiple new phenotypes induced in 10T1/2 and 3T3 cells treated with 5-azacytidine. Cell 17, 771–779.

Teperek, M., Simeone, A., Gaggioli, V., Miyamoto, K., Allen, G.E., Erkek, S., Kwon, T., Marcotte, E.M., Zegerman, P., Bradshaw, C.R., et al., 2016. Sperm is epigenetically programmed to regulate gene transcription in embryos. Genome Res. 26, 1034–1046.

Teves, S.S., Henikoff, S., 2014. Transcription-generated torsional stress destabilizes nucleosomes. Nat. Struct. Mol. Biol. 21, 88–94.

Teves, S.S., Weber, C.M., Henikoff, S., 2014. Transcribing through the nucleosome. Trends Biochem. Sci. 39, 577–586.

Thanos, D., Maniatis, T., 1995. Virus induction of human IFN beta gene expression requires the assembly of an enhanceosome. Cell 83, 1091–1100.

The ENCODE Project Consortium, 2012. An integrated encyclopedia of DNA elements in the human genome. Nature 489, 57–74.

Thoma, F., 1991. Structural changes in nucleosomes during transcription: strip, split or flip? Trends Genet. 7, 175–177.

Thomas, G.H., Elgin, S.C., 1988. Protein/DNA architecture of the DNase I hypersensitive region of the *Drosophila* hsp26 promoter. EMBO J. 7, 2191–2201.

Thomas, G., Lange, H.W., Hempel, K., 1972. Relative stability of lysine-bound methyl groups in arginie-rich histones and their subfrations in Ehrlich ascites tumor cells in vitro. Hoppe Seylers Z Physiol. Chem. 353, 1423–1428.

Tie, F., Banerjee, R., Saiakhova, A.R., Howard, B., Monteith, K.E., Scacheri, P.C., Cosgrove, M.S., Harte, P.J., 2014. Trithorax monomethylates histone H3K4 and interacts directly with CBP to promote H3K27 acetylation and antagonize polycomb silencing. Development 141, 1129–1139.

Timmers, H.T.M., Tora, L., 2018. Transcript buffering: a balancing act between mRNA synthesis and mRNA degradation. Mol. Cell 72, 10–17.

Tirosh, I., Barkai, N., 2008. Two strategies for gene regulation by promoter nucleosomes. Genome Res. 18, 1084–1091.

Tirosh, I., Berman, J., Barkai, N., 2007. The pattern and evolution of yeast promoter bendability. Trends Genet. 23, 318–321.

Tirosh, I., Weinberger, A., Bezalel, D., Kaganovich, M., Barkai, N., 2008. On the relation between promoter divergence and gene expression evolution. Mol. Syst. Biol. 4, 159.

Tirosh, I., Reikhav, S., Levy, A.A., Barkai, N., 2009. A yeast hybrid provides insight into the evolution of gene expression regulation. Science 324, 659–662.

Tirosh, I., Sigal, N., Barkai, N., 2010. Divergence of nucleosome positioning between two closely related yeast species: genetic basis and functional consequences. Mol. Syst. Biol. 6, 365.

Tolhuis, B., de Wit, E., Muijrers, I., Teunissen, H., Talhout, W., van Steensel, B., van Lohuizen, M., 2006. Genome-wide profiling of PRC1 and PRC2 polycomb chromatin binding in *Drosophila melanogaster*. Nat. Genet. 38, 694–699.

Tremethick, D., Zucker, K., Worcel, A., 1990. The transcription complex of the 5 S RNA gene, but not transcription factor IIIA alone, prevents nucleosomal repression of transcription. J. Biol. Chem. 265, 5014–5023.

Truss, M., Bartsch, J., Schelbert, A., Hache, R.J., Beato, M., 1995. Hormone induces binding of receptors and transcription factors to a rearranged nucleosome on the MMTV promoter in vivo. EMBO J. 14, 1737–1751.

Tsanev, R., Sendov, B., 1971. Possible molecular mechanism for cell differentiation in multicellular organisms. J. Theor. Biol. 30, 337–393.

Tsankov, A.M., Thompson, D.A., Socha, A., Regev, A., Rando, O.J., 2010. The role of nucleosome positioning in the evolution of gene regulation. PLoS Biol. 8, e1000414.

Tschiersch, B., Hofmann, A., Krauss, V., Dorn, R., Korge, G., Reuter, G., 1994. The protein encoded by the *Drosophila* position-effect variegation suppressor gene Su(var)3-9 combines domains of antagonistic regulators of homeotic gene complexes. EMBO J. 13, 3822–3831.

Tsuboi, M., Kishi, Y., Yokozeki, W., Koseki, H., Hirabayashi, Y., Gotoh, Y., 2018. Ubiquitination-independent repression of PRC1 targets during neuronal fate restriction in the developing mouse neocortex. Dev. Cell 47 (758–772), e755.

Tsui, K., Dubuis, S., Gebbia, M., Morse, R.H., Barkai, N., Tirosh, I., Nislow, C., 2011. Evolution of nucleosome occupancy: conservation of global properties and divergence of gene-specific patterns. Mol. Cell. Biol. 31, 4348–4355.

Tsukiyama, T., Becker, P.B., Wu, C., 1994. ATP-dependent nucleosome disruption at a heat-shock promoter mediated by binding of GAGA transcription factor. Nature 367, 525–532.

Turner, B.M., 2000. Histone acetylation and an epigenetic code. BioEssays 22, 836–845.

Ura, K., Nightingale, K., Wolffe, A.P., 1996. Differential association of HMG1 and linker histones B4 and H1 with dinucleosomal DNA: structural transitions and transcriptional repression. EMBO J. 15, 4959–4969.

Uzun, U., Brown, T., Fischl, H., Angel, A., Mellor, J., 2021. Spt4 facilitates the movement of RNA polymerase II through the +2 nucleosomal barrier. Cell Rep. 36, 109755.

Valouev, A., Johnson, S.M., Boyd, S.D., Smith, C.L., Fire, A.Z., Sidow, A., 2011. Determinants of nucleosome organization in primary human cells. Nature 474, 516–520.

van Bakel, H., Tsui, K., Gebbia, M., Mnaimneh, S., Hughes, T.R., Nislow, C., 2013. A compendium of nucleosome and transcript profiles reveals determinants of chromatin architecture and transcription. PLoS Genet. 9.

van der Lugt, N.M., Domen, J., Linders, K., van Roon, M., Robanus-Maandag, E., te Riele, H., van der Valk, M., Deschamps, J., Sofroniew, M., van Lohuizen, M., et al., 1994. Posterior transformation, neurological abnormalities, and severe hematopoietic defects in mice with a targeted deletion of the bmi-1 proto-oncogene. Genes Dev. 8, 757–769.

van der Weegen, Y., de Lint, K., van den Heuvel, D., Nakazawa, Y., Mevissen, T.E.T., van Schie, J.J.M., San Martin Alonso, M., Boer, D.E.C., Gonzalez-Prieto, R., Narayanan, I.V., et al., 2021. ELOF1 is a transcription-coupled DNA repair factor that directs RNA polymerase II ubiquitylation. Nat. Cell Biol. 23, 595–607.

van Lohuizen, M., Verbeek, S., Scheijen, B., Wientjens, E., van der Gulden, H., Berns, A., 1991. Identification of cooperating oncogenes in E mu-myc transgenic mice by provirus tagging. Cell 65, 737–752.

Van Oss, S.B., Shirra, M.K., Bataille, A.R., Wier, A.D., Yen, K., Vinayachandran, V., Byeon, I.L., Cucinotta, C.E., Heroux, A., Jeon, J., et al., 2016. The histone modification domain of Paf1 complex subunit Rtf1 directly stimulates H2B ubiquitylation through an interaction with Rad6. Mol. Cell 64, 815–825.

Varma, S.J., Calvani, E., Gruning, N.M., Messner, C.B., Grayson, N., Capuano, F., Mulleder, M., Ralser, M., 2022. Global analysis of cytosine and adenine DNA modifications across the tree of life. elife 11.

Vasseur, P., Tonazzini, S., Ziane, R., Camasses, A., Rando, O.J., Radman-Livaja, M., 2016. Dynamics of nucleosome positioning maturation following genomic replication. Cell Rep. 16, 2651–2665.

Veenendaal, M.V., Painter, R.C., de Rooij, S.R., Bossuyt, P.M., van der Post, J.A., Gluckman, P.D., Hanson, M.A., Roseboom, T.J., 2013. Transgenerational effects of prenatal exposure to the 1944-45 Dutch famine. BJOG 120, 548–553.

Venkatesh, S., Workman, J.L., 2015. Histone exchange, chromatin structure and the regulation of transcription. Nat. Rev. Mol. Cell Biol. 16, 178–189.

Venkatesh, S., Smolle, M., Li, H., Gogol, M.M., Saint, M., Kumar, S., Natarajan, K., Workman, J.L., 2012. Set2 methylation of histone H3 lysine 36 suppresses histone exchange on transcribed genes. Nature 489, 452–455.

Venter, U., Svaren, J., Schmitz, J., Schmid, A., Horz, W., 1994. A nucleosome precludes binding of the transcription factor Pho4 in vivo to a critical target site in the PHO5 promoter. EMBO J. 13, 4848–4855.

Verdin, E., Paras Jr., P., Van Lint, C., 1993. Chromatin disruption in the promoter of human immunodeficiency virus type 1 during transcriptional activation. EMBO J. 12, 3249–3259.

Vernimmen, D., Bickmore, W.A., 2015. The hierarchy of transcriptional activation: from enhancer to promoter. Trends Genet. 31, 696–708.

Verrijzer, C.P., 2022. Goldilocks meets polycomb. Genes Dev. 36, 1043–1045.

Vettese-Dadey, M., Walter, P., Chen, H., Juan, L.J., Workman, J.L., 1994. Role of the histone amino termini in facilitated binding of a transcription factor, GAL4-AH, to nucleosome cores. Mol. Cell. Biol. 14, 970–981.

Vogel, K., Horz, W., Hinnen, A., 1989. The two positively acting regulatory proteins PHO_2 and PHO_4 physically interact with PHO_5 upstream activation regions. Mol. Cell. Biol. 9, 2050–2057.

Vos, S.M., Farnung, L., Urlaub, H., Cramer, P., 2018. Structure of paused transcription complex Pol II-DSIF-NELF. Nature 560, 601–606.

Vos, S.M., Farnung, L., Boehning, M., Wigge, C., Linden, A., Urlaub, H., Cramer, P., 2018a. Structure of activated transcription complex Pol II-DSIF-PAF-SPT6. Nature 560, 607–612.

Vos, S.M., Farnung, L., Urlaub, H., Cramer, P., 2018b. Structure of paused transcription complex Pol II-DSIF-NELF. Nature 560, 601–606.

Vos, S.M., Farnung, L., Linden, A., Urlaub, H., Cramer, P., 2020. Structure of complete Pol II-DSIF-PAF-SPT6 transcription complex reveals RTF1 allosteric activation. Nat. Struct. Mol. Biol. 27, 668–677.

Wada, T., Takagi, T., Yamaguchi, Y., Ferdous, A., Imai, T., Hirose, S., Sugimoto, S., Yano, K., Hartzog, G.A., Winston, F., et al., 1998. DSIF, a novel transcription elongation factor that regulates RNA polymerase II processivity, is composed of human Spt4 and Spt5 homologs. Genes Dev. 12, 343–356.

Waddington, C.H., 1942. The epigenotype. Endeavor 1, 18–20.

Waddington, C.H., 2012. The epigenotype. 1942. Int. J. Epidemiol. 41, 10–13.

Wade, P.A., Werel, W., Fentzke, R.C., Thompson, N.E., Leykam, J.F., Burgess, R.R., Jaehning, J.A., Burton, Z.F., 1996. A novel collection of accessory factors associated with yeast RNA polymerase II. Protein Expr. Purif. 8, 85–90.

Wade, P.A., Jones, P.L., Vermaak, D., Wolffe, A.P., 1998. A multiple subunit Mi-2 histone deacetylase from *Xenopus laevis* cofractionates with an associated Snf2 superfamily ATPase. Curr. Biol. 8, 843–846.

Wallis, J.W., Rykowski, M., Grunstein, M., 1983. Yeast histone H2B containing large amino terminus deletions can function in vivo. Cell 35, 711–719.

Wang, H., Wang, L., Erdjument-Bromage, H., Vidal, M., Tempst, P., Jones, R.S., Zhang, Y., 2004a. Role of histone H2A ubiquitination in polycomb silencing. Nature 431, 873–878.

Wang, L., Brown, J.L., Cao, R., Zhang, Y., Kassis, J.A., Jones, R.S., 2004b. Hierarchical recruitment of polycomb group silencing complexes. Mol. Cell 14, 637–646.

Wasylyk, B., Chambon, P., 1979. Transcription by eukaryotic RNA polymerases A and B of chromatin assembled in vitro. Eur. J. Biochem. 98, 317–327.

Wasylyk, B., Chambon, P., 1980. Studies on the mechanism of transcription of nucleosomal complexes. Eur. J. Biochem. 103, 219–226.

Wasylyk, B., Thevenin, G., Oudet, P., Chambon, P., 1979. Transcription of in vitro assembled chromatin by *Escherichia coli* RNA polymerase. J. Mol. Biol. 128, 411–440.

Waterborg, J.H., 2002. Dynamics of histone acetylation in vivo. A function for acetylation turnover? Biochem. Cell Biol. 80, 363–378.

Weber, C.M., Ramachandran, S., Henikoff, S., 2014. Nucleosomes are context-specific, H2A.Z-modulated barriers to RNA polymerase. Mol. Cell 53, 819–830.

Weil, P.A., Luse, D.S., Segall, J., Roeder, R.G., 1979. Selective and accurate initiation of transcription at the Ad2 major late promotor in a soluble system dependent on purified RNA polymerase II and DNA. Cell 18, 469–484.

Weinberg, D.N., Papillon-Cavanagh, S., Chen, H., Yue, Y., Chen, X., Rajagopalan, K.N., Horth, C., McGuire, J.T., Xu, X., Nikbakht, H., et al., 2019. The histone mark H3K36me2 recruits DNMT3A and shapes the intergenic DNA methylation landscape. Nature 573, 281–286.

Weinmann, R., Roeder, R.G., 1974. Role of DNA-dependent RNA polymerase 3 in the transcription of the tRNA and 5S RNA genes. Proc. Natl. Acad. Sci. USA 71, 1790–1794.

Weinmann, R., Raskas, H.J., Roeder, R.G., 1974. Role of DNA-dependent RNA polymerases II and III in transcription of the adenovirus genome late in productive infection. Proc. Natl. Acad. Sci. USA 71, 3426–3439.

Weintraub, H., 1985. Assembly and propagation of repressed and depressed chromosomal states. Cell 42, 705–711.

Weintraub, H., Groudine, M., 1976. Chromosomal subunits in active genes have an altered conformation. Science 193, 848–856.

Wen, Q., Zhou, J., Tian, C., Li, X., Song, G., Gao, Y., Sun, Y., Ma, C., Yao, S., Liang, X., et al., 2023. Symmetric inheritance of parental histones contributes to safeguarding the fate of mouse embryonic stem cells during differentiation. Nat. Genet. 55, 1555–1566.

Wenger, A., Biran, A., Alcaraz, N., Redo-Riveiro, A., Sell, A.C., Krautz, R., Flury, V., Reveron-Gomez, N., Solis-Mezarino, V., Volker-Albert, M., et al., 2023. Symmetric inheritance of parental histones governs epigenome maintenance and embryonic stem cell identity. Nat. Genet. 55, 1567–1578.

Whitcomb, S.J., Basu, A., Allis, C.D., Bernstein, E., 2007. Polycomb group proteins: an evolutionary perspective. Trends Genet. 23, 494–502.

White, C.L., Luger, K., 2004. Defined structural changes occur in a nucleosome upon Amt1 transcription factor binding. J. Mol. Biol. 342, 1391–1402.

Whitehouse, I., Rando, O.J., Delrow, J., Tsukiyama, T., 2007. Chromatin remodelling at promoters suppresses antisense transcription. Nature 450, 1031–1035.

Williamson, P., Felsenfeld, G., 1978. Transcription of histone-covered T7 DNA by Escherichia coli RNA polymerase. Biochemistry 17, 5695–5705.

Winston, F., Carlson, M., 1992. Yeast SNF/SWI transcriptional activators and SPT/SIN chromatin connection. Trends Genet. 8, 387–391.
Wolffe, A.P., 1989. Dominant and specific repression of Xenopus oocyte 5S RNA genes and satellite I DNA by histone H1. EMBO J. 8, 527–537.
Wolffe, A.P., Brown, D.D., 1988. Developmental regulation of two 5S ribosomal RNA genes. Science 241, 1626–1632.
Wolffe, A.P., Jordan, E., Brown, D.D., 1986. A bacteriophage RNA polymerase transcribes through a Xenopus 5S RNA gene transcription complex without disrupting it. Cell 44, 381–389.
Woo, C.J., Kharchenko, P.V., Daheron, L., Park, P.J., Kingston, R.E., 2010. A region of the human HOXD cluster that confers polycomb-group responsiveness. Cell 140, 99–110.
Woo, C.J., Kharchenko, P.V., Daheron, L., Park, P.J., Kingston, R.E., 2013. Variable requirements for DNA-binding proteins at polycomb-dependent repressive regions in the human HOX clusters. Mol. Cell. Biol.
Wood, A., Schneider, J., Dover, J., Johnston, M., Shilatifard, A., 2003. The Paf1 complex is essential for histone monoubiquitination by the Rad6-Bre1 complex, which signals for histone methylation by COMPASS and Dot1p. J. Biol. Chem. 278, 34739–34742.
Workman, J.L., Buchman, A.R., 1993. Multiple functions of nucleosomes and regulatory factors in transcription. Trends Biochem. Sci. 18, 90–95.
Workman, J.L., Roeder, R.G., 1987. Binding of transcription factor TFIID to the major late promoter during in vitro nucleosome assembly potentiates subsequent initiation by RNA polymerase II. Cell 51, 613–622.
Workman, J.L., Abmayr, S.M., Cromlish, W.A., Roeder, R.G., 1988. Transcriptional regulation by the immediate early protein of pseudorabies virus during in vitro nucleosome assembly. Cell 55, 211–219.
Workman, J.L., Roeder, R.G., Kingston, R.E., 1990. An upstream transcription factor, USF (MLTF), facilitates the formation of preinitiation complexes during in vitro chromatin assembly. EMBO J. 9, 1299–1308.
Workman, J.L., Taylor, I.C., Kingston, R.E., 1991. Activation domains of stably bound GAL4 derivatives alleviate repression of promoters by nucleosomes. Cell 64, 533–544.
Wu, C., 1984. Activating protein factor binds in vitro to upstream control sequences in heat shock gene chromatin. Nature 311, 81–84.
Wu, X., Zhang, Y., 2017. TET-mediated active DNA demethylation: mechanism, function and beyond. Nat. Rev. Genet. 18, 517–534.
Wu, C., Wong, Y.C., Elgin, S.C., 1979. The chromatin structure of specific genes: II. Disruption of chromatin structure during gene activity. Cell 16, 807–814.
Wu, C.T., Jones, R.S., Lasko, P.F., Gelbart, W.M., 1989. Homeosis and the interaction of zeste and white in *Drosophila*. Mol. Gen. Genet. 218, 559–564.
Wu, H., D'Alessio, A.C., Ito, S., Xia, K., Wang, Z., Cui, K., Zhao, K., Sun, Y.E., Zhang, Y., 2011. Dual functions of Tet1 in transcriptional regulation in mouse embryonic stem cells. Nature 473, 389–393.
Wu, G., Broniscer, A., McEachron, T.A., Lu, C., Paugh, B.S., Becksfort, J., Qu, C., Ding, L., Huether, R., Parker, M., et al., 2012. Somatic histone H3 alterations in pediatric diffuse intrinsic pontine gliomas and non-brainstem glioblastomas. Nat. Genet. 44, 251–253.
Wyatt, G.R., 1950. Occurrence of 5-methylcytosine in nucleic acids. Nature 166, 237–238.
Wyrick, J.J., Holstege, F.C., Jennings, E.G., Causton, H.C., Shore, D., Grunstein, M., Lander, E.S., Young, R.A., 1999. Chromosomal landscape of nucleosome-dependent gene expression and silencing in yeast. Nature 402, 418–421.
Xiao, T., Hall, H., Kizer, K.O., Shibata, Y., Hall, M.C., Borchers, C.H., Strahl, B.D., 2003. Phosphorylation of RNA polymerase II CTD regulates H3 methylation in yeast. Genes Dev. 17, 654–663.
Xie, L., Pelz, C., Wang, W., Bashar, A., Varlamova, O., Shadle, S., Impey, S., 2011. KDM5B regulates embryonic stem cell self-renewal and represses cryptic intragenic transcription. EMBO J. 30, 1473–1484.
Xu, M., Long, C., Chen, X., Huang, C., Chen, S., Zhu, B., 2010. Partitioning of histone H3-H4 tetramers during DNA replication-dependent chromatin assembly. Science 328, 94–98.
Yamashita, R., Suzuki, Y., Sugano, S., Nakai, K., 2005. Genome-wide analysis reveals strong correlation between CpG islands with nearby transcription start sites of genes and their tissue specificity. Gene 350, 129–136.
Yan, C., Chen, H., Bai, L., 2018. Systematic study of nucleosome-displacing factors in budding yeast. Mol. Cell 71 (294–305), e294.
Yang, Y., Li, W., Hoque, M., Hou, L., Shen, S., Tian, B., Dynlacht, B.D., 2016. PAF complex plays novel subunit-specific roles in alternative cleavage and polyadenylation. PLoS Genet. 12, e1005794.

Yarragudi, A., Miyake, T., Li, R., Morse, R.H., 2004. Comparison of ABF1 and RAP1 in chromatin opening and transactivator potentiation in the budding yeast Saccharomyces cerevisiae. Mol. Cell. Biol. 24, 9152–9164.

Yarrington, R.M., Goodrum, J.M., Stillman, D.J., 2016. Nucleosomes are essential for proper regulation of a multigated promoter in *Saccharomyces cerevisiae*. Genetics 202, 551–563.

Yin, Y., Morgunova, E., Jolma, A., Kaasinen, E., Sahu, B., Khund-Sayeed, S., Das, P.K., Kivioja, T., Dave, K., Zhong, F., et al., 2017. Impact of cytosine methylation on DNA binding specificities of human transcription factors. Science 356.

Yoshinaga, S.K., Peterson, C.L., Herskowitz, I., Yamamoto, K.R., 1992. Roles of SWI1, SWI2, and SWI3 proteins for transcriptional enhancement by steroid receptors. Science 258, 1598–1604.

Yu, X., Buck, M.J., 2019. Defining TP53 pioneering capabilities with competitive nucleosome binding assays. Genome Res. 29, 107–115.

Yu, L., Morse, R.H., 1999. Chromatin opening and transactivator potentiation by RAP1 in *Saccharomyces cerevisiae*. Mol. Cell. Biol. 19, 5279–5288.

Yu, B.D., Hess, J.L., Horning, S.E., Brown, G.A., Korsmeyer, S.J., 1995. Altered Hox expression and segmental identity in Mll-mutant mice. Nature 378, 505–508.

Yu, L., Sabet, N., Chambers, A., Morse, R.H., 2001. The N-terminal and C-terminal domains of RAP1 are dispensable for chromatin opening and GCN4-mediated HIS4 activation in budding yeast. J. Biol. Chem. 276, 33257–33264.

Yu, C., Gan, H., Serra-Cardona, A., Zhang, L., Gan, S., Sharma, S., Johansson, E., Chabes, A., Xu, R.M., Zhang, Z., 2018. A mechanism for preventing asymmetric histone segregation onto replicating DNA strands. Science 361, 1386–1389.

Yuan, G.C., Liu, Y.J., Dion, M.F., Slack, M.D., Wu, L.F., Altschuler, S.J., Rando, O.J., 2005. Genome-scale identification of nucleosome positions in *S. cerevisiae*. Science 309, 626–630.

Zaret, K.S., 2020. Pioneer Transcription Factors Initiating Gene Network Changes. Annu. Rev. Genet.

Zaret, K.S., Carroll, J.S., 2011. Pioneer transcription factors: establishing competence for gene expression. Genes Dev. 25, 2227–2241.

Zaret, K.S., Yamamoto, K.R., 1984. Reversible and persistent changes in chromatin structure accompany activation of a glucocorticoid-dependent enhancer element. Cell 38, 29–38.

Zaret, K.S., Lerner, J., Iwafuchi-Doi, M., 2016. Chromatin scanning by dynamic binding of pioneer factors. Mol. Cell 62, 665–667.

Zawadzki, K.A., Morozov, A.V., Broach, J.R., 2009. Chromatin-dependent transcription factor accessibility rather than nucleosome remodeling predominates during global transcriptional restructuring in *Saccharomyces cerevisiae*. Mol. Biol. Cell 20, 3503–3513.

Zhang, H., Reese, J.C., 2007. Exposing the core promoter is sufficient to activate transcription and alter coactivator requirement at RNR3. Proc. Natl. Acad. Sci. USA 104, 8833–8838.

Zhao, X.D., Han, X., Chew, J.L., Liu, J., Chiu, K.P., Choo, A., Orlov, Y.L., Sung, W.K., Shahab, A., Kuznetsov, V.A., et al., 2007. Whole-genome mapping of histone H3 Lys4 and 27 trimethylations reveals distinct genomic compartments in human embryonic stem cells. Cell Stem Cell 1, 286–298.

Zhu, Z., Thiele, D.J., 1996. A specialized nucleosome modulates transcription factor access to a *C. glabrata* metal responsive promoter. Cell 87, 459–470.

Zhu, B., Mandal, S.S., Pham, A.D., Zheng, Y., Erdjument-Bromage, H., Batra, S.K., Tempst, P., Reinberg, D., 2005. The human PAF complex coordinates transcription with events downstream of RNA synthesis. Genes Dev. 19, 1668–1673.

Zhu, F., Farnung, L., Kaasinen, E., Sahu, B., Yin, Y., Wei, B., Dodonova, S.O., Nitta, K.R., Morgunova, E., Taipale, M., et al., 2018. The interaction landscape between transcription factors and the nucleosome. Nature 562, 76–81.

Zink, B., Engstrom, Y., Gehring, W.J., Paro, R., 1991. Direct interaction of the polycomb protein with Antennapedia regulatory sequences in polytene chromosomes of *Drosophila melanogaster*. EMBO J. 10, 153–162.

Zumer, K., Maier, K.C., Farnung, L., Jaeger, M.G., Rus, P., Winter, G., Cramer, P., 2021. Two distinct mechanisms of RNA polymerase II elongation stimulation in vivo. Mol. Cell 81 (3096–3109), e3098.

Epilogue

A last word: looking back over the topics covered in this book, the reader will recognize that the knowledge gained and discoveries made in chromatin biology described herein, from the first isolation of *nuclein* to the elucidation of the structure of the nucleosome core particle to understanding the mechanisms used by chromatin remodelers, were nearly uniformly the product of curiosity-driven, basic research. These discoveries utilized a veritable menagerie of models, including *Tetrahymena*, budding and fission yeasts, African clawed frogs, trout, fruit flies, and mice, and both employed and stimulated the development of numerous experimental and theoretical methods.

The accumulated knowledge of chromatin structure and function is now bearing fruit in terms of our understanding of, and potentially identifying treatments for, numerous cancers and other diseases in which proteins comprising or interacting with chromatin are involved. In the early 1980s, a *New York Times* science writer crafting an article on chromatin called Gary Felsenfeld at the NIH and asked how histones impacted human health. His reply was that you couldn't live without them. That was all that was known at the time. Remodelers and histone modifiers are now implicated in multiple types of cancer, and "oncohistones" have been implicated in pediatric gliomas. We have entered a new age in which the application of our knowledge in chromatin biology holds great promise in public health. At the same time, it is impossible to predict what new discoveries may yield to curious minds—as long as the means and the motivation to support research into fundamental questions, without regard for their immediate application, remain.

Index

Note: Page numbers followed by *f* indicate figures, *t* indicate tables, and *b* indicate boxes.

D

I

J

K

L

M

Printed in the United States
by Baker & Taylor Publisher Services